NONDESTRUCTIVE TESTING HANDBOOK

Second Edition

VOLUME 3
RADIOGRAPHY AND RADIATION TESTING

Lawrence E. Bryant
Technical Editor

Paul McIntire
Editor

AMERICAN SOCIETY FOR
NONDESTRUCTIVE TESTING

Library of Congress Cataloging in Publication Data

Radiography and radiation testing.

(Nondestructive testing handbook; v. 3)
Includes bibliographies and index.
1. Radiography, Industrial—Handbooks, manuals, etc.
I. Bryant, Lawrence E. II. McIntire, Paul. III. American Society for Nondestructive Testing. IV. Series: Nondestructive testing handbook (2nd ed.); v. 3.
TA417.25.R327 1985 620.1'1272 84-24438
ISBN 0-931403-00-6

Published by the American Society for Nondestructive Testing

PRINTED IN THE UNITED STATES OF AMERICA

FOREWORD

In preparing this volume of the Nondestructive Testing Handbook, the Penetrating Radiation Committee of ASNT had in mind the needs of the student, the technician and the scientist. Therefore, the book was expanded from its original 14 sections to 20 sections.

More than 10 years ago, under the leadership of John Aman and Bruce Meyer, past chairmen of the Penetrating Radiation Committee, work on this volume was begun. In 1977, Lawrence E. Bryant, Jr. of the Los Alamos National Laboratory and Dana Elliott, also of LANL, accepted positions as Handbook Coordinators for Volume 3.

With their guidance, work on Volume 3 began to show results. In 1980, the committee decided to publish individual Handbook sections in softbound as they became availablc. Authors were notified that sections would be accepted on a first come, first served basis for production editing and publication. Today, four years later, *Radiography and Radiation Testing* is completed and available as a hardbound book.

The authors of this book have taken special care in crediting individual contributors to their sections. We also would like to extend the appreciation of the Penetrating Radiation Committee to all contibutors and authors. Their time and efforts made this Handbook possible.

Two special people deserve our sincerest thanks and heartfelt appreciation: Lawrence E. Bryant, Jr. for his role as Technical Editor of Volume 3, and Dana E. Elliott, who assisted Bryant with the early work on this book.

K. Dieter Markert
Chairman, ASNT Penetrating Radiation Committee

PREFACE

This is the first *Nondestructive Testing Handbook* published solely by ASNT. The final edits and all the printing production were done by society staff, with the intent of reducing costs and controlling deadlines.

In less than two years, the twenty chapters of this volume were printed in an experimental, softbound format; then ASNT contracted the artists, indexers, typesetters and publishers needed for the finished book.

The result, which you now hold, is the source of some pride at ASNT. Bringing Handbook production in-house was a gamble that paid off, providing ASNT members with a high-quality text, efficiently produced at the least possible cost.

It is, though, ASNT's *job* to publish the Handbook. The real praise belongs to the people who volunteered their own time to review the text, produce illustrations and, most importantly, to write the chapters.

None of these are simple tasks. Reviewing manuscript is time-consuming and sometimes tedious. Illustrations must be made to show exactly what they intend, while maintaining standards of design and quality. As difficult as these things are, writing the chapters was the most demanding assignment of all.

To break a topic into its components and to discuss those components in a way that is clear and educational requires a deep and complete understanding of the material. And it is not just brainpower that is needed; because all the authors of this book were volunteers, writing Volume 3 was a task of the heart as well as the mind.

Special thanks and congratulations are due this volume's technical editor, Lawrence E. Bryant of Los Alamos National Laboratory. *Radiography and Radiation Testing* was for many years his project alone. Bryant outlined the entire volume; assigned most of the authors; reviewed, critiqued and wrote manuscripts; generated enthusiasm and carried the whole load, when necessary. He has literally spent thousands of hours checking millions of words for clarity, accuracy and content. Readers of this volume will truly benefit from Bryant's good work and dedication.

* * *

A project of this size naturally generates a number of problems and debates. Foremost among these was the use of SI metric units.

If the authors of this Handbook are a representative sample, then the American radiography community is not yet using SI units in its work; no single manuscript was submitted using sieverts, becquerels, grays, coulombs per kilogram and pascals. Units of length were often submitted in English with metric equivalents, but the *radiographic* SI units were uniformly absent.

Jan van den Andel of Westinghouse Canada served as the SI advisor for this volume, as he did for the second edition's Volumes 1 and 2. His extensive efforts were performed under the pressure of tight deadlines, and his suggestions and conversions were gratefully given the most serious attention.

Throughout this volume, SI units (followed by English equivalents) are used wherever space permits, otherwise the conversion factors for SI units are given.

Although every effort was made to present data in a contemporary way, the primary goal of ASNT was to produce an effective radiography handbook. Our hope is that *Radiography and Radiation Testing* achieves the same high standards set by Dr. Robert McMaster in the other Handbook volumes.

Paul McIntire
Handbook Editor

ACKNOWLEDGMENTS

Radiography and Radiation Testing, Volume 3 of the Nondestructive Testing Handbook's second edition, is an effort by many dedicated volunteers to present state-of-the-art (as well as fundamental) information on the techniques of nondestructive radiographic and radiation testing. Although advances, improvements and changes in emphasis occurred rapidly as we worked, the volume is presented to the reader in the belief that it achieves its stated objective.

Contributors to this book comprise a broad range of NDT authorities. Some of this book's authors were instrumental in writing the first edition of the NDT Handbook. Some have the bulk of their careers ahead of them but are, nonetheless, emerging as leaders in the newest developments of their professions. These volunteer authors are from all over the United States and in several instances from other countries. They represent industry, universities, government, and themselves as individual consultants.

As technological changes accelerate, it is our hope that readers of this book will have gained enough from it to become authors for the third edition.

Contributions from Prior Handbook

There are a number of Sections in this volume dealing with topics that were not in existence (or only in their infancy) in 1959 when the first edition was published. On the other hand, many of the Section titles for this volume were taken directly from the first NDT Handbook. Of these, many have been entirely rewritten. Three new Sections make use of significant amounts of first edition material, but with careful updating. These include: "Radiation and Particle Physics," "Isotope Radiation Sources" and "Film and Paper Radiography."

Contributions by ASNT

This volume was developed under the auspices of ASNT's Penetrating Radiation Committee and the Handbook Development Committee. James Schmitt originally served as director of the latter committee; the former committee was chaired by John Aman, Bruce Meyer and Dieter Markert during the course of this lengthy endeavor. Several other individuals also warrant recognition for their contributions to Volume 3. During the early years of this effort, Dana Elliott, Los Alamos National Laboratory, served as Handbook co-coordinator. Jan van den Andel of Westinghouse Canada, assisted by A.B. Nieberg, served as SI advisor for conversions and units in Volume 3.

Paul McIntire, ASNT's Handbook editor, deserves enormous credit for his contributions to this volume. These include his professionalism, dedication, and many days of long hours without sacrificing his good humor or high standards. The technical editor worked most closely with this individual and expresses his appreciation for these fine qualities and the pleasure of their working relationship. Much the same can be said for Lynne Schmitt who, before June 1982, filled the role of production editor on three of the volume's early chapters. Finally, recognition goes to Rhonda Andrews, ASNT's word processing specialist, who keyboarded most of this volume's manuscript for final editorial review.

Contributions by Technical Societies

Thanks are due to many technical societies for use of their material by a number of our authors and their contributors. Foremost among these societies are the American Society for Nondestructive Testing (ASNT) and the American Society for Testing and Materials (ASTM). The American Nuclear Society (ANS), Society of Motion Picture and Television Engineers (SMPTE), Society of Photo-optical Instrumentation Engineers (SPIE), American Institute of Physics, Society of Photographic Science and Engineering (SPSE), American Society of Radiologic Technologists, American Society for Mechanical Engineers (ASME), American National Standards Institute (ANSI), The British Institute of Nondestructive Testing, American Welding Society (AWS), American Petroleum Institute (API), Canadian Standards Association, Institute of Radio Engineers, Royal Society (Edinburgh), Royal Society (London), Institute of Electrical and Electronic Engineers (IEEE) and the American Society for Metals (ASM) also provided valuable information for this Volume.

Contributions by Industry, Government and Universities

These significant contributors range from large industrial organizations and smaller inspection laboratories to individuals working as consultants. Special thanks are due Varian Associates for portions of the Section titled "High-Energy Radiography" and to Eastman Kodak Company for their contributions to "Film and Paper Radiography." The Risø National Laboratory, Denmark, is also recognized for contributing Part 8 of "Film and Paper Radiography."

Both the United States Department of Energy and its contractor, the University of California's Los Alamos National Laboratory, are gratefully acknowledged for their support of our technical editing of this volume. Similar recognition is due all the organizations listed with the individuals in the following lists of contributors and reviewers.

Lawrence E. Bryant
Volume 3 Technical Editor

Volume 3 Reviewers

Charles Arney, consultant
John P. Barton, N-Ray Engineering, Inc.
Len Baxter
Harold Berger, Industrial Quality, Inc.
Richard Bossi, Sigma Research
Roy Braley, Lawrence Livermore National Laboratory
B.J. Brunty, US Naval Weapons Station, Concord
Lawrence E. Bryant, Los Alamos National Laboratory
Robert Buchanan, Lockheed Missiles & Space Company
William D. Burnett, Sandia National Laboratories
Herb Chapman, Canadian Welding Bureau
Francis Charbonnier, Hewlett-Packard McMinnville Division
Rosalie M. Chiffoleau, US Naval Weapons Station, Concord
Darrell C. Cutforth, Argonne National Laboratory
J.C. Domanus, Risø National Laboratory, Denmark
Dana Elliott, Los Alamos National Laboratory
C.R. Emigh, University of New Mexico
Thomas H. Feiertag, Los Alamos National Laboratory
Donald A. Garrett, National Bureau of Standards
James D. Geis, Ridge, Inc.
Solomon Goldspiel, professional engineer
James W. Guthrie, Sandia National Laboratories
Roger Hadland, Hadland Photonics, UK
Donald J. Hagemaier, McDonnell Douglas Corporation
Jerry Haskins, Lawrence Livermore National Laboratory
Bill Havercraft
Charles Hellier, Brand Examination Services and Testing Co.
Frank A. Iddings, Nuclear Science Center, Louisiana State University
Marv Jacoby, Lockheed Missiles & Space Company
Donald H. Janney, Los Alamos National Laboratory
Ron Jenkins, Phillips Electronic Instruments
John F. Landolt, Rockwell International
Claude LaPerle, General Electric
Gregory A. McDaniel, Presbyterian Hospital of Dallas
W.E.J. McKinney, E.I. DuPont de Nemours & Company
Paul Mengers, Quantex Corp.
Roger A. Morris, Los Alamos National Laboratory
Richard L. Newacheck, Aerotest Operations, Inc.
A.B. Nieberg, Westinghouse Canada
Charles Oien, Sandia National Laboratories
Richard H. Olsher, Los Alamos National Laboratory
Frank Patricelli, Science Applications, Inc.
Richard Quinn, Eastman Kodak Company
William B. Rivkin, Health Physics Associates, Ltd.
Edward H. Ruescher, Southwest Research Institute
James K. Schmitt, Chrysler Corp.
Lynn Schmitt, Wordschmitts, Inc.
Larry Schwalbe, Los Alamos National Laboratory
Sam Snow, Martin-Marietta Co., Oak Ridge
Ronald D. Strong, Los Alamos National Laboratory
Jan van den Andel, Westinghouse Canada
Gerald Wicks, Health Physics Associates, Ltd.
James D. Willenberg, Industrial Testing Consultants, Inc.

Contributors

Winston Boone, Varian Associates
Wayne Carter, Stasuk Testing and Inspection, Ltd.
Dr. J.C. Courtney
Charles H. Goldie, High Voltage Engineering Corp.
Howard Heffan, Naval Weapons Station, Concord
Vaughn Jensen, Hercules, Inc.
Robert Koenke, X-Ray Products Corporation
Mike Newsome
Lester Northeast, DuPont Canada
Mark L. Rock
Russell Schonberg, Schonberg Radiation, Inc.
Sandra C. Shipp
David Walshe, ITT Grinnell Corp.
Pat Walsh, Ontario Hydro

Major contributors to individual chapters are listed alphabetically on the Section title page under the name of the primary author.

SYSTEME INTERNATIONALE (SI) UNITS IN RADIOGRAPHY

The original discoveries of radioactivity helped establish units of measurement based on observation rather than precise physical phenomena. Later scientists who worked with radioactive substances (or who managed to manufacture radioactive beams) again made circumstantial observations which were then used for measurement purposes. This was acceptable at the time, but with our broader understanding of physics and the present tendency to use one unit for one concept, many of the original units have been modified (see Tables 1-4).

Single-Unit Comparisons

The original *curie* was simply the radiation of one gram of radium. Eventually all equivalent radiation from any source was measured with this same unit. The original *roentgen* was the quantity of radiation which would ionize one cubic centimeter of air to one electrostatic unit of electricity of either sign.

It is now known that a curie is equivalent to 37×10^9 disintegrations per second and a roentgen is equivalent to 258 microcoulombs per kilogram of air. This corresponds to 1.61×10^{15} ion pairs per kilogram of air which has then absorbed 8.8 millijoules (mJ) or 0.88 rads.

The roentgen was an intensity unit but was not representative of the dose absorbed by material in the radiation field. The radiation absorbed dose, *rad*, was first created to measure this value and was based on ergs, an energy unit from the old centimeter-gram-second (cgs) system.

In the SI system, radiation units have been given established physical foundations and new names where necessary.

The unit for radioactivity (formerly curie) is the *becquerel* (Bq) which is one disintegration per second. Because billions of disintegrations are required in a useful source, the multiplier prefix *giga* (10^9) is nearly always used and the unit is normally seen as *gigabecquerel* (GBq).

The unit for radiation dose (formerly the rad) is the *gray* (Gy) in the SI system. The gray is useful because it applies to doses which are absorbed by matter at a particular location. It is expressed in energy units per mass of matter or joules per kilogram (J/kg). The mass is normally that of the absorbing body.

The SI system's unit for the dose absorbed by the human body (formerly *rem* for roentgen equivalent man) is similar to the gray but includes quality factors that are dependent on the type of radiation. This absorbed dose has been given the name *sievert* (Sv) but its dimensions are the same as the gray (J/kg).

Combination Units

Roentgens could be measured with an ionization chamber which, when placed one meter from the radiation source, provided a good deal of necessary information (roentgen per hour at one meter per curie, for example). The numbers, though, had limited physical meaning and could not be used for different applications such as high-voltage X-ray machines.

The roentgen per hour (R/h) was used to designate the exposure to an ionizing radiation of the stated value. The SI unit used for this is the sievert (Sv), which is 100 times as large as the unit it replaces. Because the received radiation from 1 R/h was considered about equal to 1 rem, the relationship is now approximated as 1 R/h = 0.01 Gy/h. This is better expressed as 1 R/h = 10 mGy/h with higher levels of radiation in the range of one gray.

A previously popular unit, roentgen per hour at one meter per curie, is expressed in SI units as millisievert per hour at one meter per gigabecquerel, such that:

$$1 \text{ mSv/h at } 1 \text{ m/GBq} = 4 \text{ R/h at } 1 \text{ m/Ci}$$

In this relationship, milliroentgens converts to millisieverts on a 1 to 10 basis.

Exposure charts were often made using curie-minutes at a source-to-film distance in inches squared. This was written Ci-min/in.^{2}. Exposure charts made in SI use gigabecquerel-minutes for a source-to-film distance in centimeters squared, where 1 Ci-min/in.^{2} = 50 GBq-min/cm^2. Table 5 lists some of these new combination units.

Origin and Use of the SI System

The SI system was designed so that all branches of science could use a single set of interrelated measurement units; these established units are modified in specified ways to make them adaptable to the needs of individual disciplines. Without the SI system, this *NDT Handbook* volume could have contained a confusing mix of English units, old cgs metric units and the units preferred by certain localities or scientific specialities.

Aside from the consistency provided by the SI system, there is an additional mathematical advantage to its use. Complex equations always balance in the SI system; base units and their powers, on both sides of the equals sign, are indeed equal. This provides a double-check of accuracy: an equation error will reveal itself not only through an imbalance but through the different units created by the imbalance.

Every effort has been made to include all necessary SI and conversion specifications in the text of this book and in the tables that follow. If questions remain, the reader is referred to the abundance of information available through national standards organizations and to the specialized information compiled by technical societies (see ASTM E-380, *Standard Metric Practice Guide*, for example).

Jan van den Andel
Westinghouse Canada

General SI Units

TABLE 1. Base SI Units

Quantity	Unit Name	Unit Symbol
length	meter	m
mass	kilogram	kg
time	second	s
electric current	ampere	A
thermodynamic temperature	kelvin	K
amount of substance	mole	mol
luminous intensity	candela	cd
plane angle	radian	rad
solid angle	steradian	sr

TABLE 2. Derived SI Units

Quantity	Name	Symbol	Relation to Other SI Units
frequency	hertz	Hz	1/s
force	newton	N	$kg \times m/s^2$
pressure (stress)	pascal	Pa	N/m
energy (work)	joule	J	$N \times m^2$
power	watt	W	J/s
electric charge	coulomb	C	$A \times s$
electric potential	volt	V	W/A
capacitance	farad	F	C/V
electric resistance	ohm	Ω	V/A
conductance	siemens	S	A/V
magnetic flux	weber	Wb	$V \times s$
magnetic flux density	tesla	T	Wb/m^2
inductance	henry	H	Wb/A
temperature	degrees Celsius	°C	K − 273.15
luminous flux	lumen	lm	$cd \times sr$
illuminance	lux	lx	lm/m^2
radioactivity	becquerel	Bq	1/s
radiation absorbed dose	gray	Gy	J/kg
radiation dose equivalent	sievert	Sv	J/kg
time	hour	h (or hr)	3600 s

SI Units For Radiography

TABLE 3. Conversion of Traditional Radiographic Units to SI Units

Traditional Unit	Traditional Abbreviation	Multiplier of Traditional Unit for Conversion to SI Unit	SI Unit	SI Symbol
curie	Ci	37×10^{9}	becquerel	Bq
		37×10^{6}	kilobecquerel	kBq
		37×10^{3}	megabecquerel	MBq
		37	gigabecquerel	GBq
		0.037	terabecquerel	TBq
rem	rem	10^{-2}	sievert	Sv
		10	millisievert	mSv
millirem	mrem	10^{-2}	millisievert	mSv
		10	microsievert	μSv
rad	rad	10^{-2}	gray	Gy
		10	milligray	mGy
millirad	mrad	10^{-2}	milligray	mGy
		10	microgray	μGy
roentgen	R	258×10^{-6}	coulomb/kg	C/kg
milliroentgen	mR	258×10^{-9}	coulomb/kg	C/kg
electron volt	eV	1.6×10^{-19}	joule	J

TABLE 4. Conversion of Radiographic SI Units to Traditional Units

SI Unit	SI Symbol	Multiplier of SI Unit for Conversion to Traditional Unit	Traditional Unit	Traditional Abbreviation
becquerel	Bq	27×10^{-12}	curie	Ci
gigabecquerel	GBq	27×10^{-3}	curie	Ci
terabecquerel	TBq	27	curie	Ci
sievert	Sv	10^{2}	rem	rem
millisievert	mSv	10^{-1}	rem	rem
microsievert	μSv	10^{-1}	millirem	mrem
gray	Gy	10^{2}	rad	rad
milligray	mGy	10^{2}	millirad	millirad
microgray	μGy	10	millirad	millirad
coulomb/kg	C/kg	3.88×10^{3}	roentgen	R

TABLE 5. Combination SI Units

Previous Units	Multiplier of Previous Units For Conversion to SI Units	New Units
R/h at 1 m/Ci	0.25	mSv/h at 1 m/GBq*
Ci-min/in.^{2}	50	GBq-min/cm^2
rad	0.01	gray (GY)
rem	0.01	sievert (Sv)
R/min	0.01	Sv/min or Gy/min

*For effect on humans use mGy/h at 1 m/Bq.

CONTENTS

SECTION 14: REAL-TIME RADIOGRAPHY

SECTION 15: IMAGE DATA ANALYSIS

INTRODUCTION TO RADIOGRAPHY AND RADIATION TESTING

Sam Wenk, Southwest Research Institute, San Antonio, TX

INTRODUCTION

In the years since the publication of the first edition *Nondestructive Testing Handbook*, there has been a vast change in NDT technology, brought about by the increasing complexity of the products to be tested, and by simultaneous changes in NDT's supporting technologies: electronics and engineering materials. The fundamentals, however, remain unchanged.

Radiography, and all other nondestructive tests, have five essential features:[1,2]

1. supplying a suitable form and distribution of energy from an external source to the test object;
2. modifying the energy distribution within the test object as a result of its discontinuities or variations in material properties which correlate to serviceability;
3. detecting the change in energy and intensity by a sensitive detector;
4. indicating or recording the energy intensity measurement from the detector in a form useful for interpretation; and
5. interpreting the indication, and judging the corresponding serviceability of the test object.

Radiographic testing problems are specific and demand specific solutions; as with all nondestructive testing, radiographic tests must be designed for validity and reliability in each individual application. There is no such thing as a general nondestructive test applicable to every kind of material, part or structure, nor to all their functions or operating conditions. Instead, each test design must be based upon a thorough understanding of the nature and function of the part being tested and of the conditions of its service.

For this reason, specific radiographic procedures must be developed, written and adhered to in both the production and the interpretation of the radiographic image. These procedures must be based on applicable specifications, codes and standards, and the interpreter must be thoroughly familiar with their requirements in order to properly assess the image and product quality.

The intent of this NDT Handbook volume is to describe the proper use of radiography in its many industrial applications. The text delineates the important factors and concepts for analysis of critical radiographic problems. Solutions to these problems are best achieved with a thorough understanding of the basic physics of radiography, as presented in this volume. It is also important that the selection of the best nondestructive test be based on a complete familiarity with the comparative capabilities, advantages and limitations of all nondestructive testing methods, particularly those that are complementary to radiography.

As an aid to the proper selection of nondestructive test methods, the following material has been excerpted from the milestone NASA publication SP3079, which contains a total of seventy (70) inspection techniques.

It is perhaps of historical interest that of the eleven members of the National Materials Advisory Board Committee, which established the following classification system, eight were contributors to the *Nondestructive Testing Handbook*, first edition.

PART 1

THE MANAGEMENT OF NDT AND ITS INFLUENCE ON PRODUCT QUALITY AND ECONOMICS

The nondestructive testing profession in the United States is over sixty years old. In 1922, this country's first industrial radiographic laboratory was installed at the Watertown Arsenal by Dr. H.H. Lester, to aid in the development of steel castings for Army ordnance components. In 1927, the Arsenal used radiography to develop and improve fusion welding techniques. From this beginning until the start of World War II, NDT was used primarily for process control and secondarily for quality control. With the advent of World War II, the tremendous escalation of war material production, and the limited resources of NDT personnel and equipment, nondestructive testing was relegated to the role of an inspection tool.

Final inspection of a finished product is an insurance policy which will usually satisfy the customer. However, the important question that management must ask is: Where can NDT give us the competitive edge in quality and cost? There are two cost factors which must be assessed: product manufacturing costs and life-cycle costs. These cost factors are highly dependent on the initial quality and inspectability of critical components and assemblies.

The starting point in developing a cost effective and technically meaningful testing program is material and component purchase specifications. This has been studied by various groups including National Materials Advisory Board (NMAB) Committees on NDT, NMAB 252, and NMAB 337. The following text paraphrases the latter committee's report on *Economic and Management Aspects of Nondestructive Testing, Evaluation and Inspecting in Aerospace Manufacturing*.

There has been a proliferation of military, government, technical society and industrial specifications in recent years and this has caused confusion in interpretation, application and maintenance control of documentation. Properly prepared and properly applied specifications may increase initial NDT costs, but they can result in lower end-product rejection rates and lower life-cycle costs.

As a rule, management should break out those NDT costs directly associated with manufacture. Generally, the NDT function falls under the quality and/or inspection department and the tendency is to include all costs associated with quality control and quality assurance (including vendor surveillance) under one cost code.

To place NDT costs in persepctive with other elements of the total quality assurance program, data developed by the Aerospace Industries Association (AIA) were combined with an industrial sampling to illustrate the relationship of NDT costs to the quality assurance function and to selling costs. NDT costs for three elements (forgings, engines and airframes) were considered and the comparisons are shown in Fig. 1. In no instance did the actual NDT costs approach 1.5 percent of the selling cost of the item. Thus, NDT costs represent a small percentage of the final selling price and a relatively small percentage of the total quality function. A typical industrial quality assurance organization is shown in Fig. 2 and the elements that involve NDT technology are highlighted. When viewed in this context, the cost of the NDT function does not appear excessive.

From the Committee's findings, there appears to be little rationale for reducing the dollars expended on NDT when compared to costs of other quality functions. If life-cycle performance is considered, there is justification for *increasing* the amount and level of NDT in order to improve overall quality, because such an increase can lead to higher reliability, greater serviceability and lower life-cycle costs. In addition, modeling techniques such as engineering and economic analysis may provide major life-cycle cost savings by identifying the effectiveness of nondestructive tests performed on critical components of an aerospace system.

Efforts aimed specifically and exclusively at reducing the costs of NDT associated with the manufacturing sequence will have little overall impact on reducing the total cost of a system. To improve the cost effectiveness of the NDT function, the Committee has made recommendations aimed at the development of a more effective NDT program through the reduction of ineffective or unnecessary tests and the enhancement of technology. In addition, it suggests that a practical and comprehensive NDT program covering all aspects of manufacture, service inspection and maintenance can greatly reduce total life-cycle costs.

Effective NDT Application

Our business climate has changed drastically in recent years. Among other factors, product liability and product recalls mandated by regulatory agencies have strongly influenced liability insurance costs and contingency cash reserves. This in turn should encourage careful management scrutiny of the potential of well-planned NDT and its ability to ensure both product quality and lower operational costs through timely examinations.

To effectively apply NDT, its capabilities must be considered early in the product's design phase. Design engineers and structural analysts are beginning to seriously consider process engineering and

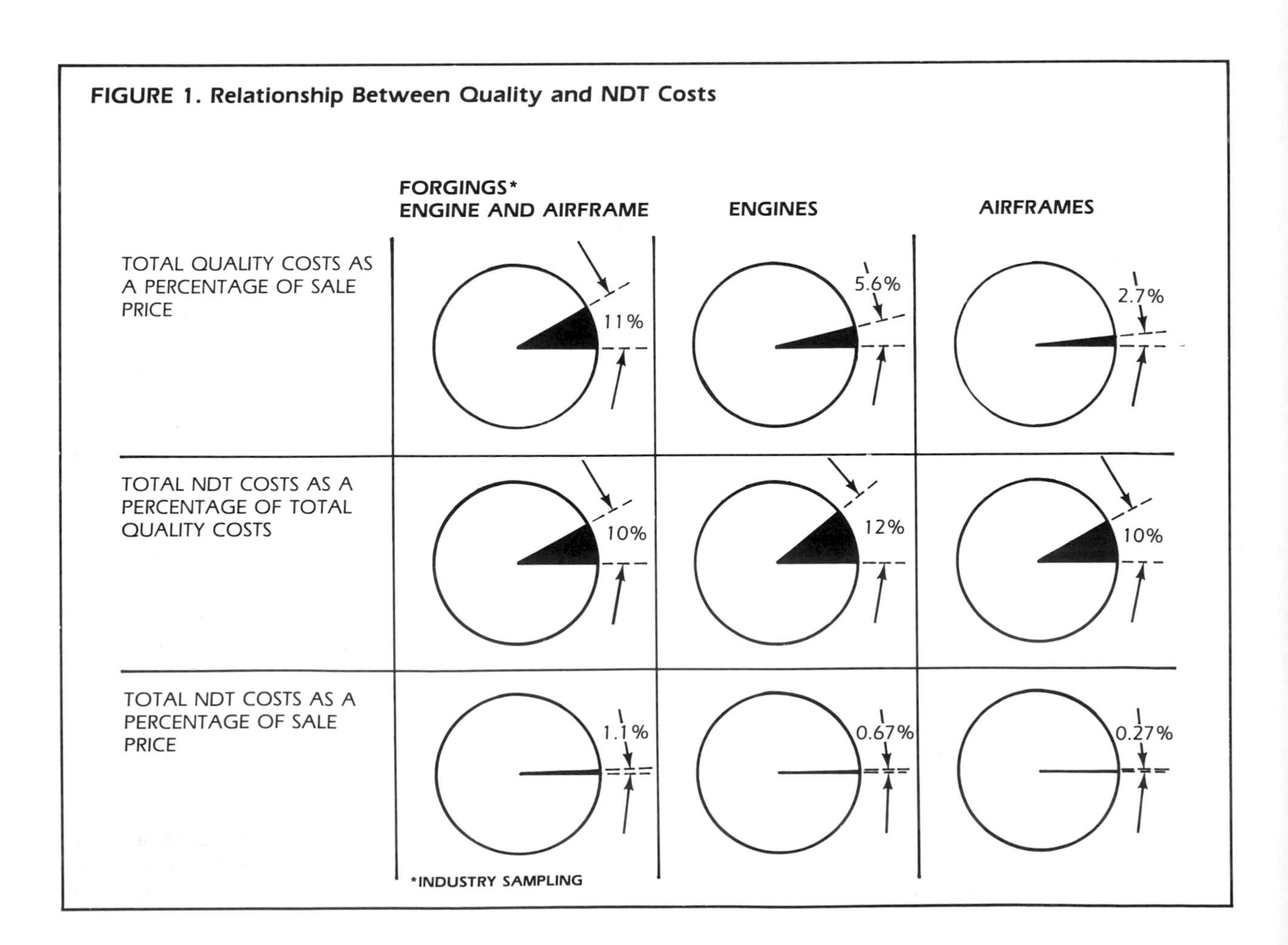

FIGURE 1. Relationship Between Quality and NDT Costs

NDT applications during product design, rather than specifying vague, after-the-fact inspection requirements. Some designers and analysts, though, have little regard for NDT, perhaps because they do not understand its limitations or its capabilities. Designers may consider NDT to have an adverse effect on their designs; design modifications sometimes need to be made because the NDT process cannot reliably detect critical defects located at inaccessible orientations. NDT is a desirable process control tool, but if its application is to be expanded, its positive aspects must be reinforced and the ignorance and prejudice acting as a deterrent to its use must be eliminated.

As a first step in designing for process control by NDT, realistic accept/reject criteria must be established by engineering. What constitutes realistic criteria is not always obvious. Analytical determination of the effects of porosity or nonmetallic inclusions in structural metallic components is difficult, and an empirical determination of these effects might prove to be prohibitively expensive. What is the distribution of voids in nonstructural composite honeycomb that can be tolerated for satisfactory service life? What quality of surface finish must be achieved to make a product acceptable? What level of material anaomalies can be reliably detected by NDT? How must the design be changed to accommodate the NDT procedures? If the correct process controls are to be established, these and similar questions must be appraised realistically and answers agreed upon as early as possible.

One of the most complex problems involves determining when, during the overall fabrication and assembly process, the NDT controls will be most effective and least expensive. It is obvious that if a product is to perform its intended function, the basic raw materials from which it is made must meet the standards presumed by the design. However, it is not always obvious just what process controls should be used to ensure economical application of these criteria.

NDT requirements must be considered at every stage of a product's design. If design involves four phases (conceptual, preliminary design, layout and detail) the influence of NDT begins during the conceptual and preliminary design stages. For example, in the conceptual phase, it must be determined whether the design concept is compatible with NDT requirements. If it is not, then the concept should

FIGURE 2. Typical Manufacturing Quality Inspection Organization

► = SPECIFIC NDT RESPONSIBILITY
> = RELATED NDT RESPONSIBILITY

PRODUCT INSPECTION

1. Tooling inspection
► 2. Detail parts inspection and NDT
► 3. Assembly inspection and NDT
4. Checkout inspection
5. Material review

PROCUREMENT QUALITY CONTROL

1. Purchase order review
> 2. Source inspection
> 3. Receiving inspection
4. Subcontractor control
5. Supplier quality history
6. Shipping inspection

QUALITY ADMINISTRATION AND AUDIT

1. Budgets
2. Audits
3. Correction action
> 4. Manuals and procedures
> 5. Quality planning
6. Administrative activities
7. Complaint investigation

QUALITY ENGINEERING

1. Technical services
2. Materials control lab
> 3. Process control
► 4. NDT research and development
► 5. NDT procedures and certification
> 6. Failure analysis
7. System equipment acceptance test procedures

METROLOGY

1. Maintain company measurement standards
2. Calibrate inspection instruments
3. Calibrate test equipment

be revised. In the preliminary design phase, performance criteria and material selection should be made compatible with NDT. During the layout phase, the inspectability of the product must be determined. It is most important that the design consider the following:

1. fracture mechanics/NDT relationships;
2. safe life/fail safe criteria;
3. tailoring requirements for production needs rather than desires;
4. in-shop and field NDT capabilities;
5. accessibility for inspection;
6. cost/trade-off studies;
7. testing and verifying procedures;
8. characteristic process anomalies.

It is important that the efforts of qualified materials engineering (including processing), stress engineering (including fracture and fatigue), planning (manufacturing), and quality control (NDT) personnel be closely coordinated. Producibility and quality should receive the greatest attention in the detail design phase, but all disciplines must be considered. Areas of complex structure not inspectable because of geometrical constraints must be redesigned or must be designed with full knowledge of uninspectability. NDT is obviously a cost element at this point; but, when properly applied, it could substantially reduce the total life-cycle cost.

The NDT specialist must participate in the design process by assisting the designer in understanding the function of NDT. This can best be accomplished by:

1. providing qualified NDT specialist support during design;
2. revising design handbook data to appropriately cover the NDT function;
3. establishing an NDT guide.

The inaccessibility and, hence, the uninspectability of critical or high stress zones of structural members on a completed assembly can increase costs significantly.

PART 2

QUALITY CONTROL AND QUALITY ASSURANCE

Quality assurance is the establishment of a *program* to guarantee the desired quality level of a product from raw materials through fabrication, final assembly and delivery to the customer. This is accomplished with the judicious preparation of specifications and procedures for material selection, design review and vendor selection. Surveillance through audits, in-house process control and quality checkpoints are also specified by the QA program.

Quality control is the *physical and administrative actions* required to ensure compliance with the quality assurance program. These functions include physical and chemical tests, where appropriate, as well as NDT at appropriate points in the manufacturing cycle. Also included in the quality control function are those administrative actions of documentation needed to establish a record of all quality procedures and their disposition. This is a vital element of protection against product liability actions, and could serve as a basis for lower insurance premiums.

Completeness of final documentation packages is a function of quality assurance, hence document control is also. This is an activity which requires constant attention. In major construction projects, there is frequently a considerable time lag between vendor component NDT and customer record review. In one embarrassing case, the vendor radiographs of a valve casting which had been rejected and scrapped were forwarded to the customer along with documentation showing that the casting was acceptable. Fortunately, the vendor's document control was able to verify that the original casting had been scrapped and also was able to produce the radiographs of the acceptable replacement. This example illustrates another vital function of QA, the followup and disposition of corrective action.

Quality Assurance Program

A good quality assurance program consists of five basic elements.

1. *Prevention.* Here, a formalized plan is required for designing, for inspectability and for cost-effectiveness. This must be a continuing effort.
2. *Control.* This is where documented workmanship standards and compatible procedures are vital for the training of production and quality personnel.
3. *Assurance.* By establishing quality assurance check points, and a rapid information feed-back system, incipient problems can be deterred.
4. *Corrective action.* To effectively implement the feed-back system, a team of specialists must be organized to rapidly assess and implement the necessary corrective action.
5. *Auditing.* The quality assurance program must contain provisions for unbiased third party audits of all aspects of the program, including vendor supplied materials and/or components. Audits can be made on either a scheduled or random basis. This function keeps quality management in touch with reality, consequently it must have management support to audit anything, any time and anywhere in the manufacturing cycle, and to initiate timely corrective action.

Increased interest in this kind of basic program reflects America's growing concern for product quality. The roots of this concern are part of our national heritage. In 1785, Benjamin Franklin wrote the following quality philosophy in *Poor Richard's Almanac:*

> "For want of a nail, the shoe was lost
> For want of a shoe, the horse was lost
> For want of the horse, the rider was lost
> For want of the rider, the battle was lost
> For want of the battle, the kingdom was lost
> And all for the want of a horseshoe nail."

In a competitive marketplace such as ours, the quality of a product directly affects its success and may carry additional far-reaching consequences. Benjamin Franklin realized this, and we must realize it today. You never get a second chance to make a first impression; quality must be built in from the beginning. Nondestructive testing in general, and radiographic testing in particular, have now become essential elements in the vital quality control of our manufactured goods.

PART 3

RELATION OF INDUSTRIAL AND MEDICAL RADIOGRAPHY

In the area of equipment, industrial radiography historically progressed as medical applications progressed, mainly because the size of the medical market was a stimulus for development. While the use of industrial and medical X-ray equipment is somewhat similar, there are different controlling factors. This was pointed out by Roy S. Sharpe in a lecture at a joint meeting of medical and engineering specialists in ultrasonics and radiography at the Royal Society of London:

"Although the problems and technical challenges are somewhat similar in the two fields and many of the diagnostic techniques have independently developed along parallel lines, there are two very important differences which have undoubtedly affected the rate and direction of progress in the two fields. In the medical case, it might be said that there are only two basic products to be inspected, male and female, differing only slightly in constructional detail and surface profile, and with an extremely long evolutionary time-scale. In the engineering case, there is an infinity of different products with designs changing, in many cases, quite dramatically on an extremely short time-scale, compared with reasonable nondestructive testing development times; this considerably complicates the [testing] requirements and provides a serious limitation to technique standardization. Secondly, in the medical case, changes in product design and modification in the manufacturing process are largely outside the terms of reference, or indeed the capability, of the profession; primary research efforts can therefore be directed to the diagnosis and cure of defects. In the case of industrial products, design and production have always been considered the preeminent engineering challenges, with inspection, testing and defect diagnosis relegated to subsidiary roles."

PART 4

APPLICATIONS OF NONDESTRUCTIVE TESTING

Excerpted from the "Nondestructive Evaluation Technique Guidebook" by Alex Vary, Lewis Research Center

Summary

In this Section, a total of 32 individual nondestructive testing (NDT) techniques are described, each in a standardized single-page format for quick reference. Information is presented in a manner that permits easy comparison of the merits and limitations of each technique with respect to various NDT problems.

An NDT technique classification system is also presented and is based on the system adopted by the National Materials Advisory Board (NMAB). The classification system presented here follows the NMAB system closely, with the exception of categories that have been added to cover more advanced techniques presently in use.

This format provides a concise description of each technique, the physical principles involved, objectives of the testing, example applications, limitations of each technique, and key reference material. Some cross-index tabulations are also provided so that appropriate techniques can be applied to particular NDT problems.

Purpose

The intent of this text is to serve as a guide for the application of nondestructive testing (NDT) techniques. The objective is to provide a quick survey of the broad range of available NDT techniques.

Nondestructive testing is a branch of the materials sciences that is concerned with all aspects of the uniformity, quality and serviceability of materials and structures. The science of NDT incorporates all the technology for detection and measurement of significant properties, including discontinuities, in items ranging from research specimens to finished hardware and products. By definition, nondestructive techniques are the means by which materials and structures may be inspected without disruption or impairment of their serviceability. Using NDT, internal properties of hidden flaws are revealed or inferred by appropriate techniques.

NDT is becoming an increasingly vital factor in the effective conduct of research, development, design and manufacturing programs. Only with the appropriate use of NDT techniques can the benefits of advanced materials science be fully realized. However, the information required for appreciating the broad scope of NDT is rather widely scattered in a multitude of publications and reports. This report presents a concise compilation of NDT technique information in a format that briefly reviews the essential data.

The term *technique* as used here refers to the body of specialized procedures, methods and instruments associated with each NDT approach. There are usually many methods or procedures associated with each technique. The following text identifies, classifies and describes these techniques without giving details on application or procedures, thus providing a resumé of each method in a single place, for quick reference.

Mode of Presentation

Classification of Techniques

In its report, the National Materials Advisory Board (NMAB) Ad Hoc Committee on Nondestructive Evaluation (NDE) adopted a classification system that divided techniques into six major categories: visual, penetrating radiation, magnetic-electrical, mechanical vibration, thermal and chemical-electrochemical.[1] A modified version of the

NMAB classification system is presented in this report. Additional categories have been included in order to cover new techniques. The resulting classification system is described in Table 1. The first six categories involve basic physical processes that require the transfer of matter and/or energy with respect to the object being inspected. Two auxiliary categories describe processes that provide for the transfer and accumulation of information, and the evaluation of the raw signals and images common to nondestructive testing methods.

Technique Description Format

In order to describe each technique, a concise method of exposition was adopted. Each page takes the form of a tabulation consisting of six headings *(Method, Principles, Objectives, Applications, Limitations* and *References)* and an illustration. There are standard subheadings in each block to further organize the presentation of information. The following paragraphs explain the mode of presentation and terminology used in the technique description format.

Technique

The technique name appears at the top of each table and is the one commonly used in the literature. However, there are cases where different names are used for the same technique. For these cases, the alternative names are given in the *References* block. Alternative technique names are cross referenced in the *Technique Name Index.*

Method

To further identify each technique, the *Method* block describes the key process and its basic result. This is given in terms of the principal energy, matter or information transfer process involved. The method description usually includes two brief sentences that describe the technique.

Principles

Each technique can be completely characterized in terms of five principal factors:

1. energy source or medium used to probe the object (such as X rays, ultrasonic waves or thermal radiation);
2. nature of the signals, image and/or signature resulting from interaction with the object (attenuation of X rays or reflection of ultrasound, for example);
3. method of detecting or sensing the resultant signals (photo-emulsion, piezoelectric crystal or inductance coil);
4. method of indicating and/or recording the signals (meter deflection, oscilloscope trace or radiograph); and
5. basis for interpreting the results (direct or indirect indication, qualitative or quantitative, and pertinent dependencies).

One or two lines of descriptive terminology corresponding to each of these five factors is given in the *Principles* block.

Objectives

The attributes for which test objects are scrutinized are listed as subheadings in the *Objectives* block. The attributes are:

1. discontinuities and separations (cracks, voids, inclusions, delaminations, etc.);
2. structure (crystalline structure, grain size, segregation, misalignment, etc.);
3. dimensions and metrology (thickness, diameter, gap size, flaw size, etc.);
4. physical and mechanical properties (reflectivity, conductivity, elastic modulus, sonic velocity, etc.);
5. composition and chemical analysis (alloy identification, impurities, elemental distributions, etc.);
6. stress and dynamic response (residual stress, crack growth, wear, vibration, etc.); and
7. signature analysis (image content, frequency spectrum, field configuration, etc.).

Terms used in this block are further defined in Table 2 with respect to specific objectives and specific attributes to be measured, detected and defined.

Applications

The *Applications* block lists practical uses of the technique. Information in this block is divided into three groups and each is covered by two subheadings. The first group includes the materials and the particular forms and features of these materials to which the technique applies. The second group includes on-line process-control and quality-control uses. The second also lists uses for monitoring and/or examining equipment during operation and maintenance. The third group lists representative components, structures, assemblies and systems to which the technique has been applied.

Limitations

The *Limitations* block gives conditions required by the technique, including:

1. conditions to be met for technique application (access, physical contact, preparation, etc.);
2. requirements to adapt the probe or probe medium to the object examined.

This block also identifies factors that:

1. limit the detection and/or characterization of flaws, properties and other attributes;
2. limit the interpretation of signals and/or images generated.

References

The *References* block indicates where additional information may be found. In some cases, adequate standard reference material is unavailable; in those cases, supplemental sources are cited. If available, standards, specifications and bibliographical sources are also listed. Usually, however, the primary reference for each technique will contain a good bibliography. Related or synonymous terms for the technique are also listed in this block, along with closely related techniques. All sources referenced in this block are abbreviated but are fully designated in the *References* listing. A general, supplemental *Bibliography* follows the *Reference* listing at the end of the Introduction.

PART 5

NONDESTRUCTIVE TESTING TECHNIQUE CATALOG

The technique description catalog that follows is organized in accordance with the eight categories given in Table 1.

The catalog is comprehensive, but is not exhaustive. Many instances exist where a number of techniques are so similar that they are combined under one representative heading to avoid unnecessary repetition. There are also a number of techniques that are similar and yet must be listed separately because they are conventionally recognized as separate techniques with unique methodologies (X and gamma radiography, for instance). Moreover, there is necessarily a considerable overlap of some auxiliary and basic techniques. The auxiliary techniques do constitute specialized branches that are distinct from the basic techniques upon which they are based (such as fluoroscopy and X-radiography). The prime criterion for the selection of these techniques was presenting a comprehensive account of the many separate technologies that currently constitute the field of NDT.

Technique Name Index

TABLE 1. Nondestructive Testing Technique Categories

Categories	Objectives
	Basic Categories
Mechanical-optical	Color; crack; dimensions; film thickness; gaging; reflectivity; strain distribution and magnitude; surface finish; surface flaws; through-cracks
Penetrating radiation	Bond separation; cracks; density and chemistry variations; elemental distribution; foreign objects; inclusions; microporosity; misalignment; missing parts; segregation; service degradation; shrinkage; thickness; voids
Electromagnetic-electronic	Alloy content; anisotropy; cavities; cold work, local strain, hardness; composition; contamination; corrosion; cracks; crack depth; crystal structure; electrical and thermal conductivity; flakes; heat treat; hot tears; inclusions; ion concentrations; laps; lattice strain; layer thickness; moisture content; polarization; seams; segregation; shrinkage; state of cure; tensile strength; thickness; unbond
Sonic-ultrasonic	Crack initiation and propagation; cracks, voids; damping factor; degree of cure; degree of impregnation; degree of sintering; delaminations; density; dimensions; elastic moduli; grain size; inclusions; mechanical degradation; misalignment; porosity; radiation degradation; structure of composites; surface stress; tensile, shear and compressive strength; unbonds; wear
Thermal	Bonding; composition; emissivity; heat contours; plating thickness; porosity; reflectivity; stress; thermal conductivity; thickness; voids
Chemical-analytical	Alloy identification; composition; cracks; elemental analysis and distribution; grain size; inclusions; macrostructure; porosity; segregation; surface flaws
	Auxiliary Categories
Image generation	Dimensional variations; dynamic performance; flaw characterization and definition; flaw distribution; flaw propagation; magnetic field configurations
Signal-image analysis	Data selection, processing and presentation; flaw mapping, correlation and identification; image enhancement; separation of multiple variables; signature analysis

TABLE 2. Specific Objectives of Nondestructive Testing Techniques

Main Objectives	Specific Objectives	Specific Attributes Measured or Detected
Discontinuities and separations	Surface flaws	Roughness; scratches; gouges; crazing; pitting; inclusions and imbedded foreign material
	Surface-connected flaws	Cracks; porosity; pinholes; laps; seams; folds; and inclusions
	Internal flaws	Cracks; separations; hot tears; cold shuts; shrinkage; voids; lack of fusion; pores; cavities; delaminations; debonds; poor bonds; inclusions and segregations
Structure	Microstructure	Molecular structure; crystalline structure and/or strain: lattice structure; strain; dislocation; vacancy and/or deformation
	Matrix structure	Grain structure, size, orientation and phases; sinter and/or porosity; impregnation; filler and or reinforcement distribution; anisotropy; inhomogeneity; and segregation
	Small structural flaws	Leaks (lack of seal/throughholes); poor fit; poor contact; loose parts; loose particles; and foreign objects
	Gross structural flaws	Assembly errors; misalignment; poor spacing or ordering; deformation; malformation; and missing parts
Dimensions and metrology	Displacement and/or position	Linear measurement; separation; gap size; flaw size, depth, location and orientation
	Dimensional variations	Uneveness; nonuniformity; eccentricity; shape and contour; size and mass variations (of entire object or part)
	Thickness or density	Film, coating, layer, plating, wall and sheet thickness; density or thickness variations
Physical and mechanical properties	Electrical properties	Resistivity; conductivity; dielectric constant and dissipation factor
	Magnetic properties	Polarization; permeability; ferromagnetism; and cohesive force
	Thermal properties	Conductivity; thermal time constant and thermoelectric potential
	Mechanical properties	Compressive, shear and tensile strength (and moduli); Poisson's ratio; sonic velocity; hardness; temper and embrittlement
	Surface properties	Color; reflectivity; refraction index; and emissivity
Chemical composition and analysis	Elemental analysis	Detection; identification, distribution and/or profile
	Impurity concentrations	Contamination; depletion; doping and diffusants
	Metallurgical content	Variation; alloy identification, verification and sorting
	Physicochemical state	Moisture content; degree of cure; ion concentrations and corrosion; and reaction products

TABLE 2. continued

Stress and dynamic response	Stress; strain and/or fatigue	Heat-treatment, annealing and cold-work effects; residual stress and strain; fatigue damage and life (residual)
	Mechanical damage	Wear; spalling; erosion; and friction effects
	Chemical damage	Corrosion; stress corrosion; and phase transformation
	Other damage	Radiation damage and high-frequency voltage breakdown
	Dynamic performance	Crack initiation and propagation; plastic deformation; creep; excessive motion; vibration; damping; timing of events; and any anomalous behavior
Signature analysis	Electromagnetic field	Potential; strength; field distribution and pattern
	Thermal field	Isotherms; heat contours; temperatures; heat flow; temperature distribution; heat leaks; and hot spots
	Acoustic signature	Noise; vibration characteristics; frequency amplitude; harmonic analysis and or spectrum; sonic and/or ultrasonic emissions
	Radioactive signature	Distribution and diffusion of isotopes and tracers
	Signal or image analysis	Image enhancement and quantization; pattern recognition; densitometry; signal classification, separation and correlation; flaw identification, definition (size; shape) and distribution analysis; flaw mapping and display

TABLE 3. Attributes Measured or Detected with Mechanical-Optical Nondestructive Testing Techniques

Objectives of Nondestructive Testing		Nondestructive Testing Techniques*									
Main Objectives	**Specific Objectives**	**Visual-Optical**	**Holointer-ferometry**	**Photo-elastic Coating**	**Brittle Coating**	**Strain Gage**	**Micro-hardness**	**Liquid Penetrant**	**Volatile Liquid**	**Filtered Particle**	**Leak Detection**
Discontinuities and separations	Surface flaws	A	C	C	-	-	C	-	-	-	-
	Surface-connected flaws	A	B	B	-	-	-	A	B	A	C
	Internal flaws	-	A	C	-	-	-	-	-	-	-
Structure	Microstructure	-	-	-	-	-	-	-	-	-	-
	Matrix structure	-	-	-	-	-	-	-	-	-	-
	Small structural flaws	A	C	-	-	-	-	C	-	-	A
	Gross structural flaws	A	C	-	-	-	-	-	-	-	-
Dimensions and metrology	Displacement and/or position	C	-	-	-	-	-	-	-	-	-
	Dimensional variations	C	-	-	-	-	-	-	-	-	-
	Thickness or density	-	B	-	-	-	-	-	-	-	-
Physical and mechanical properties	Electrical properties	-	-	-	-	-	-	-	-	-	-
	Magnetic properties	-	-	-	-	-	-	-	-	-	-
	Thermal properties	-	-	-	-	-	-	-	-	-	-
	Mechanical properties	-	-	-	-	-	B	-	-	-	-
	Surface properties	A	-	-	-	-	-	-	-	-	-
Chemical composition and analysis	Elemental analysis	-	-	-	-	-	-	-	-	-	-
	Impurity concentrations	-	-	-	-	-	-	-	-	-	-
	Metallurgical content	-	-	-	-	-	-	-	-	-	-
	Physicochemical state	-	-	-	-	-	-	-	-	-	-
Stress and dynamic responses	Stress, strain and/or fatigue	-	A	A	A	A	-	-	-	-	-
	Mechanical damage	A	-	-	-	-	-	-	-	-	-
	Chemical damage	B	-	-	-	-	-	-	-	-	-
	Other damage	-	-	-	-	-	-	-	-	-	-
	Dynamic performance	C**	B	B	-	B	-	-	-	-	-
Signature analysis	Electromagnetic field	-	-	-	-	-	-	-	-	-	-
	Thermal field	-	-	-	-	-	-	-	-	-	-
	Acoustic signature	-	-	-	-	-	-	-	-	-	-
	Radioactive signature	-	-	-	-	-	-	-	-	-	-
	Signal or image analysis	-	-	-	-	-	-	-	-	-	-

***A: very satisfactory technique; B: satisfactory technique; C: restricted usage; D: potential usage; E: experimental.**

****For the visual-optical techniques of high-speed video and high-speed photography, the dynamic performance is very satisfactory.**

TABLE 4. Attributes Measured or Detected with Penetrating Radiation Nondestructive Testing Techniques

Objectives of Nondestructive Testing		Nondestructive Testing Techniques*								
Main Objectives	Specific Objectives	X-radiography	Gamma Radiography	Neutron Radiography	Penetrating Radiometry	Backscatter Radiometry	Autoradiography	Radioactive Penetrant	Positron Annihilation	Tomography
Discontinuities and separations	Surface flaws	-	-	-	-	-	-	-	-	-
	Surface-connected flaws	-	-	-	-	-	-	A	-	B
	Internal flaws	B	C	C	C	-	-	-	-	A
Structure	Microstructure	A**	-	-	-	-	-	-	E	-
	Matrix structure	A**	-	-	-	-	-	-	-	-
	Small structural flaws	B	C	C	-	-	-	-	-	A
	Gross structural flaws	A	A	A	-	-	-	-	-	A
Dimensions and metrology	Displacement and/or position	A	A	A	-	-	-	-	-	B
	Dimensional variations	A	A	A	-	-	-	-	-	B
	Thickness or density	C	C	C	A	A	A	-	-	A
Physical and mechanical properties	Electrical properties	-	-	-	-	-	-	-	-	-
	Magnetic properties	-	-	-	-	-	-	-	-	-
	Thermal properties	-	-	-	-	-	-	-	-	-
	Mechanical properties	-	-	-	-	-	-	-	E	-
	Surface properties	-	-	-	-	-	-	-	-	-
Chemical composition and analysis	Elemental analysis	-	-	-	-	C	B	-	-	E
	Impurity concentrations	-	-	-	-	-	-	-	-	D
	Metallurgical content	-	-	-	-	-	-	-	-	-
	Physicochemical state	-	-	C	-	-	-	-	-	-
Stress and dynamic responses	Stress, strain and/or fatigue	-	-	-	-	-	-	-	B	-
	Mechanical damage	C	D	C	B	-	A	B	-	-
	Chemical damage	-	-	-	D	D	B	C	-	-
	Other damage	-	-	D	-	-	-	-	-	-
	Dynamic performance	A†	-	A††	-	-	-	-	-	-
Signature analysis	Electromagnetic field	-	-	-	-	-	-	-	-	-
	Thermal field	-	-	-	-	-	-	-	-	-
	Acoustic signature	-	-	-	-	-	-	-	-	-
	Radioactive signature	-	-	-	B	-	A	C	-	-
	Signal or image analysis	-	-	-	-	-	-	-	-	A

* A: very satisfactory technique; B: satisfactory technique; C: restricted usage; D: potential usage; E: experimental.
** Microradiography and microfocus X-ray.
† Flash X-ray, high-speed radiography and real-time radiography.
†† Real-time radiography.

TABLE 5. Attributes Measured or Detected with Electromagnetic-Electronic Nondestructive Testing Techniques

Objectives of Nondestructive Testing		Nondestructive Testing Techniques*										
Main Objectives	**Specific Objectives**	**Static Magnetic Field**	**Magnetic Particle**	**Nuclear Magnetic Resonance**	**Barkhausen Effect**	**Eddy Current**	**Electric Current**	**Electrified Particle**	**Corona Discharge**	**Dielectric**	**Exoelectron Emission**	**Microwave Radiation**
Discontinuities and separations	Surface flaws	-	-	-	-	-	-	A	-	-	A	-
	Surface-connected flaws	A	A	D	B	A	A	A	A	C	-	C
	Internal flaws	C	C	-	-	B	B	-	-	-	-	C
Structure	Microstructure	-	-	A	-	-	-	-	-	-	-	-
	Matrix structure	C	-	-	A	C	-	-	-	C	-	-
	Small structural flaws	-	-	-	-	-	-	-	-	-	-	-
	Gross structural flaws	-	-	-	-	-	-	-	-	-	-	-
Dimensions and metrology	Displacement and/or position	-	-	-	-	-	-	-	-	-	-	C
	Dimensional variations	-	-	-	-	-	-	-	-	-	-	-
	Thickness or density	A	-	-	-	A	B	-	-	B	-	A
Physical and mechanical properties	Electrical properties	-	-	-	-	A	A	-	-	A	-	B
	Magnetic properties	A	C	-	-	A	-	-	-	-	-	-
	Thermal properties	-	-	-	-	-	-	-	-	-	-	-
	Mechanical properties	C	-	-	-	-	-	-	-	-	-	-
	Surface properties	-	-	-	-	-	-	-	-	-	A	-
Chemical composition and analysis	Elemental analysis	-	-	C	-	-	-	-	-	-	-	-
	Impurity concentrations	-	-	C	-	-	-	-	C	-	C	C
	Metallurgical content	B	-	-	-	B	-	-	-	-	-	-
	Physicochemical state	-	-	-	-	-	-	-	-	A	-	B
Stress and dynamic responses	Stress, strain and/or fatigue	-	-	C	B	B	C	-	-	-	A	-
	Mechanical damage	-	-	-	-	-	B	-	-	-	-	-
	Chemical damage	-	-	-	-	-	B	-	-	-	-	-
	Other damage	-	-	-	-	-	-	-	-	B	-	-
	Dynamic performance	-	-	-	-	-	B	C	-	-	-	C
Signature analysis	Electromagnetic field	D	C	-	-	-	-	-	-	-	-	-
	Thermal field	-	-	-	-	-	-	-	-	-	-	-
	Acoustic signature	-	-	-	-	-	-	-	-	-	-	-
	Radioactive signature	-	-	-	-	-	-	-	-	-	-	-
	Signal or image analysis	-	-	-	-	-	-	-	-	-	-	-

***A: very satisfactory technique; B: satisfactory technique; C: restricted usage; D: potential usage; E: experimental.**

TABLE 6. Attributes Measured or Detected with Sonic-Ultrasonic Nondestructive Testing Techniques

Objectives of Nondestructive Testing		Nondestructive Testing Techniques*								
Main Objectives	**Specific Objectives**	**Acoustic Impact**	**Sonic Vibration**	**Eddy Sonic Vibration**	**Acoustic Emission**	**Pulse-echo Ultrasonics**	**Transmission Ultrasonics**	**Resonance Ultrasonics**	**Surface-wave Ultrasonics**	**Critical-angle Ultrasonics**
Discontinuities and separations	Surface flaws	-	-	-	-	-	-	-	A	B
	Surface-connected flaws	D	-	-	-	A	C	-	A	-
	Internal flaws	C	C	B	-	A	B	C	-	-
Structure	Microstructure	-	-	-	-	-	-	-	-	-
	Matrix structure	-	-	-	-	B	-	C	-	B
	Small structural flaws	B	B	B	A	-	-	-	-	-
	Gross structural flaws	C	-	-	-	-	-	-	-	-
Dimensions and metrology	Displacement and/or position	-	-	-	-	A	C	-	C	-
	Dimensional variations	B	B	-	-	-	-	-	C	-
	Thickness or density	C	C	-	-	C	C	A	B	-
Physical and mechanical properties	Electrical properties	-	-	-	-	-	-	-	-	-
	Magnetic properties	-	-	-	-	-	-	-	-	-
	Thermal properties	-	-	-	-	-	-	-	-	-
	Mechanical properties	-	A	D	-	B	B	A	-	B
	Surface properties	-	-	-	-	-	-	-	-	-
Chemical composition and analysis	Elemental analysis	-	-	-	-	-	-	-	-	-
	Impurity concentrations	-	-	-	-	-	-	-	-	-
	Metallurgical content	-	-	-	-	-	-	-	-	-
	Physicochemical state	-	-	-	-	-	-	-	-	-
Stress and dynamic responses	Stress, strain and/or fatigue	-	-	D	B	-	-	-	B	A
	Mechanical damage	-	-	D	-	C	-	C	-	-
	Chemical damage	-	-	D	-	-	-	-	-	-
	Other damage	-	-	-	-	-	-	-	-	-
	Dynamic performance	C	C	C	-	A	A	-	-	D
Signature analysis	Electromagnetic field	-	-	-	-	-	-	-	-	-
	Thermal field	-	-	-	-	-	-	-	-	-
	Acoustic signature	D	B	C	A	-	-	-	-	-
	Radioactive signature	-	-	-	-	-	-	-	-	-
	Signal or image analysis	-	-	-	-	-	-	-	-	-

***A: very satisfactory technique; B: satisfactory technique; C: restricted usage; D: potential usage; E: experimental.**

TABLE 7. Attributes Measured or Detected with Thermal Nondestructive Testing Techniques

Objectives of Nondestructive Testing		Nondestructive Testing Techniques*				
Main Objectives	**Specific Objectives**	**Contact Thermometry**	**Thermoelectric Probe**	**Infrared Radiometry**	**Thermochronic**	**Electrothermal**
Discontinuities and separations	Surface flaws	C	-	-	-	-
	Surface-connected flaws	-	B	B	B	-
	Internal flaws	-	-	A	C	B
Structure	Microstructure	-	-	-	-	-
	Matrix structure	-	C	-	-	-
	Small structural flaws	-	-	-	-	-
	Gross structural flaws	-	-	-	-	-
Dimensions and metrology	Displacement and/or position	-	-	-	-	-
	Dimensional variations	-	-	-	-	-
	Thickness or density	C	C	C	C	C
Physical and mechanical properties	Electrical properties	-	-	-	-	C
	Magnetic properties	-	-	-	-	-
	Thermal properties	B	C	A	C	C
	Mechanical properties	D	-	-	-	-
	Surface properties	-	C	A	-	-
Chemical composition and analysis	Elemental analysis	-	-	-	-	-
	Impurity concentrations	-	C	-	-	-
	Metallurgical content	-	-	-	-	-
	Physicochemical state	-	-	-	-	-
Stress and dynamic responses	Stress, strain and/or fatigue	-	-	-	-	-
	Mechanical damage	-	-	-	-	-
	Chemical damage	-	-	-	-	-
	Other damage	-	-	-	-	-
	Dynamic performance	-	-	-	-	C
Signature analysis	Electromagnetic field	-	-	-	-	-
	Thermal field	C	-	A	B	-
	Acoustic signature	-	-	-	-	-
	Radioactive signature	-	-	-	-	-
	Signal or image analysis	-	-	-	-	-

***A: very satisfactory technique; B: satisfactory technique; C: restricted usage; D: potential usage; E: experimental.**

TABLE 8. Attributes Measured or Detected with Chemical-Analytical Nondestructive Testing Techniques

Objectives of Nondestructive Testing		Nondestructive Testing Techniques*										
Main Objectives	Specific Objectives	Chemical Spot Test	Electrolytic Probe	Laser Probe	Ion Scatter	Ion Probe	Auger Analysis	X-ray Fluorescence	X-ray Diffraction	Neutron Activation	Charged-particle Activation	Mossbauer Analysis
Discontinuities and separations	Surface flaws	-	C	-	-	-	-	-	-	-	-	E
	Surface-connected flaws	-	-	-	-	-	-	-	-	-	-	-
	Internal flaws	-	-	-	-	-	-	-	-	-	-	-
Structure	Microstructure	-	-	B	-	B	-	-	B	-	-	A
	Matrix structure	-	-	-	-	-	B	-	B	-	-	-
	Small structural flaws	-	-	-	-	-	-	-	-	-	-	-
	Gross structural flaws	-	-	-	-	-	-	-	-	-	-	-
Dimensions and metrology	Displacement and/or position	-	-	-	-	-	-	-	-	-	-	-
	Dimensional variations	-	-	-	-	-	-	-	-	-	-	-
	Thickness or density	-	-	-	-	-	-	C	-	-	E	D
Physical and mechanical properties	Electrical properties	-	-	-	-	-	-	-	-	-	-	B
	Magnetic properties	-	-	-	-	-	-	-	-	-	-	-
	Thermal properties	-	-	-	-	-	-	-	-	-	-	-
	Mechanical properties	-	-	-	-	-	-	-	-	-	-	-
	Surface properties	-	-	-	-	-	-	-	-	-	-	-
Chemical composition and analysis	Elemental analysis	B	D	C	B	B	C	B	-	B	B	C
	Impurity concentrations	-	D	B	B	B	C	B	C	-	B	-
	Metallurgical content	B	-	C	D	D	-	-	-	-	C	-
	Physicochemical state	-	-	-	-	-	-	-	-	-	-	-
Stress and dynamic responses	Stress, strain and/or fatigue	-	E	-	-	-	E	-	A	-	-	-
	Mechanical damage	-	-	-	-	-	-	-	-	-	-	-
	Chemical damage	-	-	B	B	B	D	B	C	C	-	C
	Other damage	-	-	-	-	-	-	-	-	-	-	-
	Dynamic performance	-	-	-	-	-	-	-	-	-	-	-
Signature analysis	Electromagnetic field	-	-	-	-	-	-	-	-	-	-	-
	Thermal field	-	-	-	-	-	-	-	-	-	-	-
	Acoustic signature	-	-	-	-	-	-	-	-	-	-	-
	Radioactive signature	-	-	-	-	-	-	-	-	-	D	-
	Signal or image analysis	-	-	-	-	-	-	-	-	-	-	-

***A: very satisfactory technique; B: satisfactory technique; C: restricted usage; D: potential usage; E: experimental.**

TABLE 9. Attributes Measured or Detected with Image Generation Nondestructive Testing Techniques

Objectives of Nondestructive Testing		Nondestructive Testing Techniques*									
Main Objectives	Specific Objectives	Photo Imaging	Film Radiography	Xeroradiography	Track-etch Radiography	Fluoroscopy	Video Radiography	Immersion Ultrasonics	Ultrasonic Videography	Ultrasonic Holography	Video Thermography
Discontinuities and separations	Surface flaws	A	-	-	-	-	-	-	-	-	-
	Surface-connected flaws	C	C	-	-	-	-	A	C	B	B
	Internal flaws	-	A	B	C	C	B	A	B	B	B
Structure	Microstructure	-	-	-	E	-	-	-	E	-	-
	Matrix structure	-	-	-	-	-	-	C	D	-	-
	Small structural flaws	-	B	B	-	B	B	C	C	C	-
	Gross structural flaws	-	A	A	-	A	B	-	-	-	-
Dimensions and metrology	Displacement and/or position	-	A	A	-	C	C	C	C	-	-
	Dimensional variations	-	A	B	-	C	B	-	-	-	-
	Thickness or density	-	C	C	-	D	C	C	D	-	C
Physical and mechanical properties	Electrical properties	-	-	-	-	-	-	-	-	-	-
	Magnetic properties	-	-	-	-	-	-	-	-	-	-
	Thermal properties	-	-	-	-	-	-	-	-	-	C
	Mechanical properties	-	-	-	-	-	-	D	-	-	-
	Surface properties	A	-	-	-	-	-	-	-	-	C
Chemical composition and analysis	Elemental analysis	-	-	-	-	-	-	-	-	-	-
	Impurity concentrations	-	-	-	-	-	-	-	-	-	-
	Metallurgical content	-	-	-	-	-	-	-	-	-	-
	Physicochemical state	-	-	-	-	-	-	-	-	-	-
Stress and dynamic responses	Stress, strain and/or fatigue	-	-	-	-	-	-	-	E	-	-
	Mechanical damage	-	D	D	-	D	D	-	-	-	-
	Chemical damage	-	-	-	-	-	-	-	-	-	-
	Other damage	-	-	-	-	-	-	-	-	-	-
	Dynamic performance	-	-	-	-	C	B	-	D	D	-
Signature analysis	Electromagnetic field	-	-	-	-	-	-	-	-	-	-
	Thermal field	C	-	-	-	-	-	-	-	-	A
	Acoustic signature	-	-	-	-	-	-	-	-	-	-
	Radioactive signature	D	A	D	-	D	A	-	-	-	-
	Signal or image analysis	-	-	-	B	-	B	D	B	B	B

***A: very satisfactory technique; B: satisfactory technique; C: restricted usage; D: potential usage; E: experimental.**

Mechanical-Optical Techniques

TABLE 10. Visual-Optical

METHOD	
Key process and basic result	Direct visual and optically aided inspection is applied to object surfaces for indications of flaws and anomalies independently and in combination with other NDT techniques
PRINCIPLES	
Probe medium and/or energy source	Visible light (long wavelength ultraviolet light with fluorescent materials)
Nature of signal and/or signature	Reflected or transmitted photons
Detection and/or sensing method	Eyes, optical aids, magnifiers, borescopes, video and film cameras
Indication and/or recording method	Visual image
Interpretation basis	Direct; used with other techniques for direct interpretation (liquid penetrants, filtered particle, magnetic particle)
OBJECTIVES	
Discontinuities and separations	Cracks, voids, pores and inclusions
Structure	Roughness, grain and film
Dimensions and metrology	Mechanically aided measurements
Physical and mechanical properties	—
Composition and chemical analyses	—
Stress and dynamic responses	Visible responses to stress
Signature analysis	—
APPLICATIONS	
Applicable materials	Indefinite range of materials
Applicable features and forms	Surfaces, layers, films, coatings, entire objects
Process control applications	On-line and off-line monitoring and control
In situ and diagnostic applications	All forms of nondestructive inspection and testing
Example structures and components	Machined parts, internal surfaces, indefinite range of test objects, components assemblies and systems
LIMITATIONS	
Access, contact and/or preparation	Visual access
Probe and object limits	Specialized optical aids usually required
Sensitivity and/or resolution	Various degrees of magnification
Interpretation limits	Requires supplementation with other NDT techniques for flaw discrimination, detection and measurement
Other conditions and limits	
REFERENCES	
Primary source material	Reference 2 (sections 10, 11 and 12)
Bibliographic material	—
Standards and specifications	—
Related terms	—
Related techniques	Borescopy, refractometry, diffractometry, interferometry, reflectometry, microscopy, telescopy, light radiometry, phase-contrast and Schlieren techniques

TABLE 11. Photoelastic Coating

METHOD	
Key process and basic result	Transparent plastic coating on specimen becomes birefringent when substrate is stressed and thereby reveals loading and stress patterns in substructure
PRINCIPLES	
Probe medium and/or energy source	Application of stress and/or loading
Nature of signal and/or signature	Birefringence and fringe patterns
Detection and/or sensing method	Polarized light and reflection polariscope
Indication and/or recording method	Direct visual observation and optical aids
Interpretation basis	Comparative or differential; becomes quantitative with fringe count calibration
OBJECTIVES	
Discontinuities and separations	Cracks, voids, porosity, inclusions and debonds
Structure	—
Dimensions and metrology	Grain- and crystal-scale strain measurements
Physical and mechanical properties	—
Composition and chemical analyses	—
Stress and dynamic responses	Propagation of plastic deformation and cracks
Signature analysis	Magnitude and direction of principal strains and stress concentrations and residual stress
APPLICATIONS	
Applicable materials	Metals, nonmetals and composites
Applicable features and forms	Surfaces of specimens and test objects
Process control applications	—
In situ and diagnostic applications	—
Example structures and components	Experimental design and hardware analysis, rivet holes, welds, airframes, spot-welded and/or bonded structures and bridges
LIMITATIONS	
Access, contact and/or preparation	Surface must be smooth and a good reflector; coating remains on test surface
Probe and object limits	Surface must be accessible to coating and observation
Sensitivity and/or resolution	Strain differences to 0.1 millimeter per meter
Interpretation limits	Errors introduced by differential thermal expansion, temperature variations and/or coating thickness relative to part thickness
Other conditions and limits	—
REFERENCES	
Primary source material	Reference 2 (section 53)
Bibliographic material	—
Standards and specifications	—
Related terms	Birefringent coating
Related techniques	Holographic interferometry, brittle coating and strain gage

TABLE 12. Strain Gage

METHOD	
Key process and basic result	Electric resistance strain gage is bonded to surface of test object to indicate strain; gage consists of fine wire or thin-foil grid layer sandwiched between layers of carrier material
PRINCIPLES	
Probe medium and/or energy source	Electric current
Nature of signal and/or signature	Variation of gage resistance
Detection and/or sensing method	Wheatstone bridge circuit
Indication and/or recording method	Meter indication; potentiometer reading
Interpretation basis	Direct indication which is quantitative and depends on uniformity, standardization and calibration of gages
OBJECTIVES	
Discontinuities and separations	—
Structure	—
Dimensions and metrology	—
Physical and mechanical properties	Microdisplacements, forces, torque, pressure, acceleration and magnetostriction
Composition and chemical analyses	—
Stress and dynamic responses	Stress-strain response and/or properties, creep and crack growth
Signature analysis	—
APPLICATIONS	
Applicable materials	Metals, nonmetals and/or composites
Applicable features and forms	Surfaces
Process control applications	—
In situ and diagnostic applications	Tensile testing, stress analysis and strain monitoring
Example structures and components	Operating turbines, engines, airframes, ship hulls, cranes, earth-moving equipment and pressure vessels
LIMITATIONS	
Access, contact and/or preparation	Cleaned and prepared surface with access to critical area
Probe and object limits	Application and orientation of gages critical
Sensitivity and/or resolution	Measures strain to 1 micrometer per meter
Interpretation limits	Affected by temperature, humidity, moisture and slippage
Other conditions and limits	Gage becomes permanently bonded to equipment
REFERENCES	
Primary source material	Reference 2 (section 54)
Bibliographic material	—
Standards and specifications	—
Related terms	Resistance gage
Related techniques	Photoelastic coating, brittle coating, holographic interferometry and mechanical strain gage

TABLE 13. Liquid Penetrant

METHOD	
Key process and basic result	Test surface is covered with penetrating liquid that seeks surface-connected cracks; liquid in cracks bleeds out to stain powder-coating applied to surface after removal of excess liquid film from surface of test object
PRINCIPLES	
Probe medium and/or energy source	Liquid medium containing dye or fluorescent substance
Nature of signal and/or signature	Capillary bleedout of liquid trapped in flaws
Detection and/or sensing method	Localized staining of applied developer powder
Indication and/or recording method	Direct visual observation for dye; black light for fluorescence
Interpretation basis	Direct indication (dependent on proper methods for application of penetrant and developer)
OBJECTIVES	
Discontinuities and separations	Cracks, pinholes, laps, seams, cold shuts and leaks
Structure	—
Dimensions and metrology	—
Physical and mechanical properties	—
Composition and chemical analyses	—
Stress and dynamic responses	Fatigue cracking and grinding cracks
Signature analysis	—
APPLICATIONS	
Applicable materials	All nonporous, nonabsorbing materials
Applicable features and forms	Surfaces, entire objects and complex shapes
Process control applications	Control step in metal processing and/or joining
In situ and diagnostic applications	Cracks formed during testing and equipment operation
Example structures and components	Weldments, joints, tubing, castings and billets; fuel and liquid-oxygen tanks and vessels, aluminum parts, gas turbine disks and blades, engine mounts and gears
LIMITATIONS	
Access, contact and/or preparation	Access required for surface decontamination and cleaning
Probe and object limits	Discontinuity must be surface-connected and open
Sensitivity and/or resolution	Microcracks to 1 micrometer width
Interpretation limits	False indications from shallow scratches and/or smearing
Other conditions and limits	Porosity of surface may mask important indications; discontinuity depth is not indicated; vapor hazard
REFERENCES	
Primary source material	Reference 2 (sections 5 to 8)
Bibliographic material	—
Standards and specifications	Reference 4, ASTM E-165-65, E-270-68
Related terms	Dye penetrants and fluorescent penetrants
Related techniques	Filtered particle, electrified particle, magnetic particle and radioactive gas penetrant

TABLE 14. Leak Detection

METHOD	
Key process and basic result	Exit of gas from or ingress of gas to sealed enclosure is induced; enclosed or external search gas is used to locate and sense leaks and to estimate leak rate
PRINCIPLES	
Probe medium and/or energy source	Search gas: helium, hydrogen and krypton-85
Nature of signal and/or signature	Leakage
Detection and/or sensing method	Spectrometer, counter and vacuum or pressure gage
Indication and/or recording method	Meter indication; audible signal
Interpretation basis	Direct (standard reference leak required for quantitative indication)
OBJECTIVES	
Discontinuities and separations	Through-holes and cracks
Structure	Porosity and lack of seal
Dimensions and metrology	—
Physical and mechanical properties	—
Composition and chemical analyses	—
Stress and dynamic responses	—
Signature analysis	—
APPLICATIONS	
Applicable materials	Metals and mixed, nonporous materials
Applicable features and forms	Enclosures and seals
Process control applications	Quality control of envelopes and seals
In situ and diagnostic applications	Vacuum leak check of experimental and operating equipment
Example structures and components	Weld, braze and adhesive bonds; glass envelopes; vacuum chambers; elastomer and metal gasket seals; reactor fuel pins; liquid-metal containers and components
LIMITATIONS	
Access, contact and/or preparation	Direct access required to at least one side, indirect to the other
Probe and object limits	Special probe or sniffer; object enclosure usual
Sensitivity and/or resolution	Sensitivity to order of 10^{-7} liter-nanobar per second
Interpretation limits	—
Other conditions and limits	Location and size of leak are usually difficult to detect; smeared metal or contaminants may plug leak passage; radiation and other residual gas hazards are possible
REFERENCES	
Primary source material	Reference 6 (chapter 15)
Bibliographic material	Reference 7
Standards and specifications	Reference 4; ASTM E-425-71, E-427-71, E-432-71
Related terms	Helium leak check and pressure check
Related techniques	Hydrostatic tests

Penetrating Radiation Techniques

TABLE 15. X-radiography

METHOD	
Key process and basic result	Penetrating radiation emitted by X-ray generator is imposed on test object; radiation transmitted or attenuated by test object is used to image or detect internal structure and/or flaws
PRINCIPLES	
Probe medium and/or energy source	X-rays in 10^{-13} to 10^{-9} meter wavelength range
Nature of signal and/or signature	Transmission or attenuation by object variables
Detection and/or sensing method	Photo- or penetrating radiation-sensitive emulsion, fluoroscope and/or radiometer
Indication and/or recording method	Radiographic image; densitometry
Interpretation basis	Direct interpretation (standard penetrameters for image quality indication); control of contrast, resolution and density critical
OBJECTIVES	
Discontinuities and separations	Cracks, porosity, voids and inclusions
Structure	Internal malstructure, misassembly or misalignment
Dimensions and metrology	Thickness, diameter, gap and position
Physical and mechanical properties	Density variations
Composition and chemical analyses	—
Stress and dynamic responses	Wear, corrosion, dynamic behavior
Signature analysis	—
APPLICATIONS	
Applicable materials	Metals, nonmetals, composites and mixed materials
Applicable features and forms	Entire objects or structures; wide range of shapes and sizes
Process control applications	Casting; assembly joining and welding quality control
In situ and diagnostic applications	Service degradation; loose or stray parts and breakage in mechanisms
Example structures and components	Castings; welds; braze distribution; electronic assemblies; rocket propellant encasements; and aircraft, space and automotive components
LIMITATIONS	
Access, contact and/or preparation	Access to opposite sides of test object required
Probe and object limits	Voltage, exposure time and focal spot size critical
Sensitivity and/or resolution	Density and thickness variations to 2 percent or better
Interpretation limits	Sensitivity decreases with increasing thickness
Other conditions and limits	Cracks must be oriented parallel to beam; radiation hazard
REFERENCES	
Primary source material	Reference 2 (sections 17, 21, 24 and 25)
Bibliographic material	Reference 8
Standards and specifications	Reference 4; ASTM E-142-68, E-94-68
Related terms	Film radiography, real-time radiography, flash X-ray, cineradiography, high-speed radiography
Related techniques	Gamma radiography, neutron radiography, autoradiography and microradiography

TABLE 16. Gamma Radiography

METHOD	
Key process and basic result	Penetrating radiation emitted by isotope source is imposed on test object; radiation transmitted or attenuated by test object is used to image or detect internal structure and/or flaws in thick cross sections of dense materials
PRINCIPLES	
Probe medium and/or energy source	Gamma radiation in 10^{-13} to 10^{-9} meter wavelength range
Nature of signal and/or signature	Transmission or attenuation by object variables
Detection and/or sensing method	Photoemulsion and/or radiometer
Indication and/or recording method	Radiographic image; densitometry
Interpretation basis	Direct interpretation (standard penetrameters for image quality indication); control of contrast, resolution and density critical
OBJECTIVES	
Discontinuities and separations	Cracks, porosity, voids and inclusions
Structure	Internal malstructure or malformation
Dimensions and metrology	Dimensional variations and anomalies, thickness and gaps
Physical and mechanical properties	Density variations
Composition and chemical analyses	—
Stress and dynamic responses	—
Signature analysis	Panoramic imaging
APPLICATIONS	
Applicable materials	Usually applied to dense or thick metallic materials
Applicable features and forms	Entire objects or structures; range of shapes and sizes
Process control applications	Shipyard, power plant, pipeline, structures fabrication, and construction quality control
In situ and diagnostic applications	Shipyard, power plant and structures maintenance where thickness or access limit X-ray generators
Example structures and components	Castings, thick or large components and especially configurations not accessible to X-ray generators
LIMITATIONS	
Access, contact and/or preparation	Often requires fitting source inside complex parts
Probe and object limits	Special mechanism for remote extension of source to part
Sensitivity and/or resolution	Density or thickness variations to 2 percent
Interpretation limits	Sensitivity usually less than with X rays
Other conditions and limits	Cracks must be oriented parallel to rays; source energy level uncontrollable and decays with time; radiation hazard
REFERENCES	
Primary source material	Reference 2 (section 15)
Bibliographic material	Reference 8
Standards and specifications	Reference 4; ASTM E-280-68, E-94-68
Related terms	—
Related techniques	X-radiography and neutron radiography

TABLE 17. Neutron Radiography

METHOD	
Key process and basic result	Neutron beam from reactor, accelerator, or isotope source is imposed on test object; neutrons transmitted or attenuated by test object are used to image or detect internal structure and/or flaws that are poorly revealed by X and gamma radiation
PRINCIPLES	
Probe medium and/or energy source	Neutron beam, various energies
Nature of signal and/or signature	Transmission or attenuation by object variables
Detection and/or sensing method	Radiographic film or electronic methods (such as neutron-sensitive image intensifier or fluor observed by video)
Indication and/or recording method	Radiographic image; densitometry
Interpretation basis	Direct interpretation (standard penetrameters for image quality indication); control of contrast resolution and density critical
OBJECTIVES	
Discontinuities and separations	Voids, porosity, inclusions and cracks
Structure	Internal malstructure, anomalies and/or misalignment
Dimensions and metrology	Thickness, diameters and gaps
Physical and mechanical properties	—
Composition and chemical analyses	Contamination; element and/or isotope distribution and/or identification
Stress and dynamic responses	Dynamic behavior (with real-time neutron radiography)
Signature analysis	—
APPLICATIONS	
Applicable materials	Metals, nonmetals, composites and mixed materials
Applicable features and forms	Entire objects or structures; range of shapes and sizes
Process control applications	—
In situ and diagnostic applications	—
Example structures and components	Pyrotechnic devices, resins, plastics, organic substances, honeycomb structures, integrated microcircuits, nuclear fuel elements, radioactive materials, high-density metals, materials containing hydrogen
LIMITATIONS	
Access, contact and/or preparation	Access for interposing object between source and detectors
Probe and object limits	Collimation, filtering and moderation of neutron beam
Sensitivity and/or resolution	Sensitive to different elements or isotopes
Interpretation limits	Sensitivity decreases with increasing thickness
Other conditions and limits	Cracks must be oriented parallel to beam; image quality varies with neutron source; radiation hazard
REFERENCES	
Primary source material	Reference 9
Bibliographic material	Reference 8
Standards and specifications	—
Related terms	Neutrography
Related techniques	X-radiography and gamma radiography

Electromagnetic-Electronic Techniques

TABLE 18. Static Magnetic Field

METHOD	
Key process and basic result	Magnetic field is imposed on test object; field permeates and magnetizes test zone and probe scans and detects field perturbations that are characteristic of surface and/or subsurface flaws and anomalies
PRINCIPLES	
Probe medium and/or energy source	Static magnetic field induced in object
Nature of signal and/or signature	Field gradient and normal or tangential perturbations
Detection and/or sensing method	Rotating or oscillating coil, probes
Indication and/or recording method	Meter deflection; coordinate plot and field map
Interpretation basis	Comparative or differential; requires standard defects for comparison
OBJECTIVES	
Discontinuities and separations	Cracks, inclusions, gouges, scratches, holes and pores
Structure	Magnetic anisotropy
Dimensions and metrology	Thickness
Physical and mechanical properties	Coercive force, magnetic permeability, hardness
Composition and chemical analyses	Compositional variations
Stress and dynamic responses	—
Signature analysis	Magnetic field strength, signature and characteristics
APPLICATIONS	
Applicable materials	Ferromagnetic and permeable metals
Applicable features and forms	Surfaces and substrates; uniform and regular shapes
Process control applications	Feedback sorting
In situ and diagnostic applications	—
Example structures and components	Nonmagnetic coating thickness, depth of case hardening, analysis or carbon content in steels, bearing raceways, gear teeth
LIMITATIONS	
Access, contact and/or preparation	Very near proximity of probe to surface
Probe and object limits	Specialized probes usual for various measurements
Sensitivity and/or resolution	Cracks to 0.03 millimeters
Interpretation limits	Ambiguities arise because of edge and/or lift-off effects
Other conditions and limits	Access to both sides for some thickness measuring; does not discriminate among discontinuity types
REFERENCES	
Primary source material	Reference 2 (sections 33 and 34)
Bibliographic material	References 16 and 17
Standards and specifications	—
Related terms	Magnetic field perturbation and magnetic field test
Related techniques	Magnetic particle and eddy current

TABLE 19. Magnetic Particle

METHOD	
Key process and basic result	Test object or part is magnetized; magnetic powder applied to surface accumulates over regions where magnetic field erupts or emerges as a result of surface or subsurface flaws
PRINCIPLES	
Probe medium and/or energy source	Magnetizing current or field imposed on part
Nature of signal and/or signature	Field distortion or leakage at surface
Detection and/or sensing method	Accumulation and pattern of magnetic powder clusters; pattern in magnetic tape*
Indication and/or recording method	Visual, photography and magnetic tape and/or rubber
Interpretation basis	Direct indication, depends on direction and strength of magnetic field and on powder and vehicle control
OBJECTIVES	
Discontinuities and separations	Cracks, seams, pores and inclusions
Structure	—
Dimensions and metrology	—
Physical and mechanical properties	Permeability variations
Composition and chemical analyses	—
Stress and dynamic responses	—
Signature analysis	—
APPLICATIONS	
Applicable materials	Ferromagnetic materials
Applicable features and forms	Surface and substrate; regular and uniform shapes
Process control applications	Quality control following heat treat, machining, grinding or forming operations
In situ and diagnostic applications	Monitoring service flaws
Example structures and components	Bars, forgings, weldments, extrusions, fasteners, engines components, shafts and gears
LIMITATIONS	
Access, contact and/or preparation	Requires clean and relatively smooth surface
Probe and object limits	Fixturing required to hold and magnetize part
Sensitivity and/or resolution	Cracks to 0.5 millimeter major dimension
Interpretation limits	Field alignment and strength critical
Other conditions and limits	Followup metal removal may be required; part demagnetization may be problematic; removal of powder and vehicle required
REFERENCES	
Primary source material	Reference 2 (sections 30 to 32)
Bibliographic material	Reference 18
Standards and specifications	Reference 4; ASTM E-109-63(71), E-138-63(71), E-269-68
Related terms	Magnetic tape and magnetic rubber
Related techniques	Magnetic field perturbation and eddy current

*Also field set into room-temperature, vulcanizing, magnetic, silicone rubber replication of magnetized part.

TABLE 20. Eddy Current

METHOD	
Key process and basic result	Localized alternating current loop (eddy) is induced in test object; inductive reactance of probe to magnetic field of induced current indicates subsurface flaws
PRINCIPLES	
Probe medium and/or energy source	Localized induced current of 10 Hz to 6 MHz range
Nature of signal and/or signature	Perturbation of induced current and hence induced magnetic field
Detection and/or sensing method	Inductance coil; Hall probe
Indication and/or recording method	Meter deflection; oscilloscope trace
Interpretation basis	Differential or comparative; reference standard required for each type of flaw
OBJECTIVES	
Discontinuities and separations	Cracks, seams, pits and inclusions
Structure	—
Dimensions and metrology	Wall thickness and coating thickness
Physical and mechanical properties	Conductivity and permeability
Composition and chemical analyses	Composition variations
Stress and dynamic responses	Heat-treatment effects
Signature analysis	—
APPLICATIONS	
Applicable materials	Metals, alloys and electroconductors
Applicable features and forms	Subsurfaces, and substrates; regular and uniform shapes
Process control applications	Feedback control for sorting materials and parts
In situ and diagnostic applications	Detection of service degradation
Example structures and components	Tube, wire, ball bearings, nonmetal coatings, train rails and wheels, airframe components, turbine blades, turbine disks and automotive transmission shafts
LIMITATIONS	
Access, contact and/or preparation	No contact but close proximity of probe to surface
Probe and object limits	Probe usually tailored to part geometry
Sensitivity and/or resolution	Cracks to 0.2 millimeter length
Interpretation limits	False indications possible because of mixed variables, edge effects and lift-off (clearance) effects
Other conditions and limits	Low penetration (limited to thin walls or near-surface flaws)
REFERENCES	
Primary source material	Reference 2 (sections 35 to 42)
Bibliographic material	Reference 7
Standards and specifications	Reference 4; ASTM E-309-71, E-243-67T, E-246-71, E-215-67, E-268-68
Related terms	Magnetic reaction analysis and phase-sensitive eddy current
Related techniques	Static magnetic field, magnetic particle and electric current

TABLE 21. Electric Current

METHOD	
Key process and basic result	Current flows through part or zone under test; current strength or density between electrode pair touching surface is affected by inhomogeneities and discontinuities
PRINCIPLES	
Probe medium and/or energy source	Current between pair of surface contacts or probes
Nature of signal and/or signature	Voltage drop and external magnetic field perturbation
Detection and/or sensing method	Potential probe, Hall probe or induction coil
Indication and/or recording method	Potentiometer indication; oscillograph
Interpretation basis	Comparative; requires standard flaws and calibration curves
OBJECTIVES	
Discontinuities and separations	Cracks and inclusions
Structure	Segregations and grain orientation
Dimensions and metrology	Thickness variations
Physical and mechanical properties	Resistivity, conductivity and stress or cold-work
Composition and chemical analyses	—
Stress and dynamic responses	Corrosion, erosion, wear effects, fatigue damage and crack propagation
Signature analysis	—
APPLICATIONS	
Applicable materials	Metallic materials and electroconductors
Applicable features and forms	Surfaces and substrates; uniform, regular areas and shapes
Process control applications	—
In situ and diagnostic applications	Railroad track scanning
Example structures and components	Bars, plate, rails, fastenings and joints; pressure vessels, tanks and hulls
LIMITATIONS	
Access, contact and/or preparation	Good surface contact required
Probe and object limits	Electrode or probe spacing and contact critical
Sensitivity and/or resolution	Can indicate (relative) depth of cracks
Interpretation limits	Dependent upon shape and orientation of discontinuity
Other conditions and limits	Edge-effects and contamination of surface limit utility; may mar surface
REFERENCES	
Primary source material	Reference 2 (section 35)
Bibliographic material	—
Standards and specifications	—
Related terms	Electric current injection
Related techniques	Static magnetic field and eddy current

TABLE 22. Microwave Radiation

METHOD	
Key process and basic result	Continuous or modulated microwave radiation directed at object propagates according to internal state or structure of part
PRINCIPLES	
Probe medium and/or energy source	Electromagnetic radiation of 3 to 0.03 centimeter wavelength
Nature of signal and/or signature	Scatter, reflection and/or attenuation of radiation
Detection and/or sensing method	Microwave guide and crystal detector
Indication and/or recording method	Meter deflection; coordinate plot
Interpretation basis	Comparative, phase-amplitude differentiation; reference standard required
OBJECTIVES	
Discontinuities and separations	Cracks, porosity, holes and debonds
Structure	Inhomogeneity
Dimensions and metrology	Thickness and position
Physical and mechanical properties	Dielectric properties
Composition and chemical analyses	Compositional variations, moisture content and cure
Stress and dynamic responses	Vibrational characteristics
Signature analysis	—
APPLICATIONS	
Applicable materials	Plastic, cellulose, ceramic, liquid and elastomer
Applicable features and forms	Surface, bulk material and coatings
Process control applications	Feedback control of thickness and/or position
In situ and diagnostic applications	—
Example structures and components	Reinforced plastic structures, polyurethane foam, solid (rocket) propellant and motor cases
LIMITATIONS	
Access, contact and/or preparation	No surface contact is required, but positioning of part may be critical
Probe and object limits	Alignment and coupling of waveguide and detector is critical; complex waveguide arrangement usual
Sensitivity and/or resolution	Thickness variations to order of 25 micrometers
Interpretation limits	Spatial resolution for flaws depends on probe (horn) size
Other conditions and limits	Requires metal backing for thickness and position gaging of nonmetals; no microwave hazard at power levels usually used
REFERENCES	
Primary source material	Reference 25 (chapter 13)
Bibliographic material	—
Standards and specifications	—
Related terms	—
Related techniques	Dielectric and corona discharge

Sonic-Ultrasonic Techniques

TABLE 23. Acoustic-Impact

METHOD	
Key process and basic result	Tapping or ringing of object is accomplished by striking it; mechanically applied pulse causes response vibrations indicative of anomalies and/or flaws
PRINCIPLES	
Probe medium and/or energy source	Impact energy from hammer or pulser
Nature of signal and/or signature	Vibrational response and acoustic damping
Detection and/or sensing method	Ear, microphone and/or accelerometer
Indication and/or recording method	Audible sound, meter deflection and oscilloscope trace
Interpretation basis	Comparative; based on sonic signature, vibrational and/or damping response identification
OBJECTIVES	
Discontinuities and separations	Cracks, debonds and delaminations
Structure	Macrostructure variations and anomalies
Dimensions and metrology	Variations of physical dimensions
Physical and mechanical properties	Density, mass and elastic properties
Composition and chemical analyses	—
Stress and dynamic responses	Differential dynamic response and damping capacity
Signature analysis	—
APPLICATIONS	
Applicable materials	Metals, nonmetals and composites
Applicable features and forms	Entire objects, including complex shapes
Process control applications	Off-line component testing
In situ and diagnostic applications	Integrity of fasteners, bonds and cores
Example structures and components	Honeycomb, laminated, brazed and adhesive-bonded structures; bolted or riveted assemblies and automotive components
LIMITATIONS	
Access, contact and/or preparation	Contact, fixturing and support of object required
Probe and object limits	Pulser design and impact point critical
Sensitivity and/or resolution	Low spatial resolution to centimeters
Interpretation limits	Sensitive to ambient and extraneous noise and signals
Other conditions and limits	Mass and/or complexity and impact point influence results; all physical and geometric properties but the one tested must be constant
REFERENCES	
Primary source material	References 2 (section 51) and 26
Bibliographic material	—
Standards and specifications	—
Related terms	Impact, shock, tapping and ringing tests
Related techniques	—

TABLE 24. Sonic Vibration

METHOD	
Key process and basic result	Continuous sonic vibrations are imparted to object; induced natural-frequency vibrations of test object reveal flaws and physical property variations
PRINCIPLES	
Probe medium and/or energy source	Force applied by exciter or transducer
Nature of signal and/or signature	Resonance and/or harmonic response in 0 to 20 kHz range
Detection and/or sensing method	Microphone and accelerometer of holointerferometry
Indication and/or recording method	Meter deflection; oscilloscope or holograph
Interpretation basis	Comparative; frequency-spectrum, Lissajous-pattern or holographic fringe-pattern identification
OBJECTIVES	
Discontinuities and separations	Debonds, delaminations and cracks
Structure	Metallurgical variations
Dimensions and metrology	Variations in physical dimensions
Physical and mechanical properties	Density, elasticity and shear moduli, and Poisson's ratio
Composition and chemical analyses	—
Stress and dynamic responses	Differential dynamic response and/or damping capacity
Signature analysis	—
APPLICATIONS	
Applicable materials	Metals, nonmetals and composites
Applicable features and forms	Entire objects; uniform and regular shapes
Process control applications	On-line bond integrity of entire honeycomb structures
In situ and diagnostic applications	—
Example structures and components	Solid bars, rods and disks; abrasive wheels and rods; turbine blade and disks
LIMITATIONS	
Access, contact and/or preparation	Contact, isolation and support of object required
Probe and object limits	Pulser and probe design to accommodate part
Sensitivity and/or resolution	Spatial resolution to 1 millimeter
Interpretation limits	Flaw shape and location generally not revealed
Other conditions and limits	Object should have definitive vibration modes; influenced by mass and geometry of object
REFERENCES	
Primary source material	References 2 (section 51) and 27
Bibliographic material	—
Standards and specifications	—
Related terms	Vibration, natural-frequency and resonance tests
Related techniques	—

TABLE 25. Acoustic Emission

METHOD	
Key process and basic result	Phonon signals arise from plastic deformation, fracture or phase changes; ultrasonic emission rate and intensity reveals crack initiation, propagation and deformations activated by stressing
PRINCIPLES	
Probe medium and/or energy source	Energy released at deformation or fracture sites
Nature of signal and/or signature	Stress or ultrasonic waves propagating in material
Detection and/or sensing method	Piezoelectric transducer
Indication and/or recording method	Digital counter, meter indication, coordinate plot or waveform
Interpretation basis	Comparative or differential; analysis of emission count rate, amplitude-frequency spectrum, differential signal arrival time, waveform or total emission count
OBJECTIVES	
Discontinuities and separations	—
Structure	—
Dimensions and metrology	—
Physical and mechanical properties	Tensile and fatigue properties, weld properties, metallurgical properties
Composition and chemical analyses	—
Stress and dynamic responses	Crack initiation and propagation; strain rate; friction, wear, spalling and erosion effects; martensitic phase transformation; stress corrosion; and fatigue
Signature analysis	—
APPLICATIONS	
Applicable materials	Metals, nonmetals and composites
Applicable features and forms	Welds, coatings, bonds, substrates and entire objects
Process control applications	Welding and die-forming; pressure (proof) testing
In situ and diagnostic applications	Incipient failure detection in stressed structures; dynamic monitoring
Example structures and components	Fracture specimens; nuclear, cryogenic and pressure vessels; airframe and engine components; and fluid systems
LIMITATIONS	
Access, contact and/or preparation	Contact, acoustic coupling and stressing required
Probe and object limits	Probe coupling, waveguides and arrangement critical
Sensitivity and/or resolution	Sensitivity to crack precursors and microcracking under investigation
Interpretation limits	Multiple probe and computer for flaw location by triangulation required
Other conditions and limits	Poor acoustic channels, noise, temperature effects may hamper effective signal extraction; high ductile materials yield poor signals; requires creation of signature catalog for signal interpretation
REFERENCES	
Primary source material	References 30 and 31
Bibliographic material	—
Standards and specifications	—
Related terms	Stress wave emission
Related techniques	Acoustic impact and sonic vibration

TABLE 26. Pulse-Echo Ultrasonics

METHOD	
Key process and basic result	Ultrasonic pulses are directed into test object; ultrasonic echos and reflections indicate absence, presence and location of flaws, interfaces and/or discontinuities
PRINCIPLES	
Probe medium and/or energy source	Beam of pulsed ultrasound, 20 kHz to 50 MHz in range
Nature of signal and/or signature	Reflection or transmission of pulses or echos
Detection and/or sensing method	Piezoelectric transducer(s)
Indication and/or recording method	Oscilloscope trace and pulse-echo gating
Interpretation basis	Quantitative for flaw and interface location; reference standards required for calibration and flaw characterization
OBJECTIVES	
Discontinuities and separations	Cracks, voids, laminations, inclusions and debonds
Structure or malstructure	Porosity; metallurgical structure and graininess
Dimensions and metrology	Thickness
Physical and mechanical properties	Density and sonic velocity
Composition and chemical analyses	—
Stress and dynamic responses	Crack growth
Signature analysis	—
APPLICATIONS	
Applicable materials	Metals, nonmetals and composites
Applicable features and forms	Substrates, joints and bonds, structure components
Process control applications	Heat treatment, rolling mill, forming and joining quality control
In situ and diagnostic applications	Service degradation monitoring of power plants, pipelines, static structures and aerospace vehicles
Example structures and components	Sheet, plate, bar and tube stock; castings; forgings; welds; airframe and engine components; pressure vessels; and nuclear reactor components
LIMITATIONS	
Access, contact and/or preparation	Access to one side and liquid coupling applied to object
Probe and object limits	Special probes, coupling and alignment fixtures usual
Sensitivity and/or resolution	Flaws to 0.01 millimeter in size
Interpretation limits	Ambiguous signals may arise as a result of scatter effects, multiple reflections and geometric complexity
Other conditions and limits	Small or thin parts are difficult to inspect
REFERENCES	
Primary source material	References 2 (sections 43, 48 and 49) and 32
Bibliographic material	References 33 and 34
Standards and specifications	Reference 4; ASTM E-164-65, E-317-68, E-127-64
Related terms	—
Related techniques	Transmission, resonance, surface-wave, critical-angle, delta, contact and immersion ultrasonics

TABLE 27. Transmission Ultrasonics

METHOD	
Key process and basic result	Continuous, pulsed, or modulated ultrasound transmitted through part is attenuated by flaws and/or interfaces
PRINCIPLES	
Probe medium and/or energy source	Beam of ultrasound (usually 20 kHz to 50 MHz range)
Nature of signal and/or signature	Attenuation or obstruction of ultrasound
Detection and/or sensing method	Piezoelectric transducer
Indication and/or recording method	Meter deflection; oscilloscope trace
Interpretation basis	Comparative or differential; reference standards required for quantitative indications
OBJECTIVES	
Discontinuities and separations	Cracks, voids, laminations and inclusions
Structure	Homogeneity of composite matrix after cure
Dimensions and metrology	Thickness
Physical and mechanical properties	Density, impedance and sonic velocity
Composition and chemical analyses	—
Stress and dynamic responses	—
Signature analysis	—
APPLICATIONS	
Applicable materials	Metals, nonmetals and composites
Applicable features and forms	Uniform, regular and/or flat parts
Process control applications	Quality control of bonds
In situ and diagnostic applications	Monitoring bonds in aircraft trailing-edge surfaces
Example structures and components	Sheet, plate and bar stock; laminated structures
LIMITATIONS	
Access, contact and/or preparation	Coupling and access to two sides
Probe and object limits	Selection, alignment and fixturing of probes
Sensitivity and/or resolution	Flaws to 0.2 millimeter
Interpretation limits	Spurious signals may arise as a result of reflections, dispersion, or resonance
Other conditions and limits	Poor results unless surfaces are uniform and parallel and only one type of defect is present
REFERENCES	
Primary source material	Reference 2 (sections 43 and 49) and 32
Bibliographic material	References 33 and 34
Standards and specifications	—
Related terms	—
Related techniques	Pulse-echo, surface-wave, critical-angle, contact, immersion and delta ultrasonics

TABLE 28. Resonance Ultrasonics

METHOD	
Key process and basic result	Frequency is varied until probe introduces continuous and compressional ultrasonic waves into part at resonant frequencies for thickness gaging
PRINCIPLES	
Probe medium and/or energy source	Continuous-wave ultrasound (20 kHz to 25 MHz range)
Nature of signal and/or signature	Generation of standing waves and resonance
Detection and/or sensing method	Piezoelectric transducer
Indication and/or recording method	Meter indication; oscilloscope trace
Interpretation basis	Quantitative; requires catalog of sonic velocities and/or dimensions for thickness and/or velocity measurements
OBJECTIVES	
Discontinuities and separations	Delaminations and debonds
Structure or malstructure	Matrix structure variations
Dimensions and metrology	Thickness
Physical and mechanical properties	Velocity of sound
Composition and chemical analyses	—
Stress and dynamic responses	—
Signature analysis	—
APPLICATIONS	
Applicable materials	Homogeneous metals and nonmetals
Applicable features and forms	Platelike forms, tubelike forms and uniform parts
Process control applications	—
In situ and diagnostic applications	Blistering, thinning and corrosion of pipelines
Example structures and components	Extruded, drawn, or bored tubes; rolled or milled sheet metal; pressure vessels; ship hulls; boiler tubes; glass, ceramic and rigid plastic parts
LIMITATIONS	
Access, contact and/or preparation	Smooth surface and close coupling preferred
Probe and object limits	Frequency, type and mounting of transducer critical
Sensitivity and/or resolution	Thickness variation to 0.1 percent
Interpretation limits	Spurious indications may arise as a result of extraneous vibrational modes, harmonics, end effects and/or reflections
Other conditions and limits	Taper or irregularities reduce signal value
REFERENCES	
Primary source material	Reference 2 (sections 43 and 50) and 32
Bibliographic material	References 33 and 34
Standards and specifications	Reference 4; ASTM E-133-67
Related terms	—
Related techniques	Pulse-echo, transmission, surface-wave, critical-angle, delta, contact and immersion ultrasonics

Thermal Techniques

TABLE 29. Thermoelectric Probe

METHOD	
Key process and basic result	Voltage produced at probe-part contact point due to thermal gradient indicates variations in surface and/or substrate composition
PRINCIPLES	
Probe medium and/or energy source	Thermal gradient produced by heated probe
Nature of signal and/or signature	Characteristic bimetallic Seebeck voltage
Detection and/or sensing method	Potentiometric circuit containing part and substrate
Indication and/or recording method	Potentiometer
Interpretation basis	Comparative or differential; reference or standard surface required for calibration
OBJECTIVES	
Discontinuities and separations	Porosity
Structure	Segregations and depletions
Dimensions and metrology	Thickness variations
Physical and mechanical properties	Thermoelectric properties
Composition and chemical analyses	Surface chemistry and composition
Stress and dynamic responses	—
Signature analysis	—
APPLICATIONS	
Applicable materials	Metals and electroconductive substrates
Applicable features and forms	Subsurfaces and substrates; coatings
Process control applications	Metal sorting and coating thickness
In situ and diagnostic applications	—
Example structures and components	Diffusion coatings and layers, ceramic-coated metals, P-N junctions in semiconductors and graphite parts
LIMITATIONS	
Access, contact and/or preparation	Uncontaminated surface and tight contact required
Probe and object limits	Probe tip material, hardness and oxidation critical
Sensitivity and/or resolution	Voltages in individual metal grains sensed
Interpretation limits	Contact pressure variations, probe radius variation and dulling limit utility
Other conditions and limits	Primarily experimental technique
REFERENCES	
Primary source material	References 41 and 42
Bibliographic material	Reference 38
Standards and specifications	—
Related terms	—
Related techniques	Thermochromic, infrared radiometry, electrolytic probe and eddy current

Chemical-Analytical Techniques

TABLE 30. Chemical Spot Test

METHOD	
Key process and basic result	Small sample of material is removed from test object to determine its composition; chemical identifications are made by combining specimen particles with series of reagents
PRINCIPLES	
Probe medium and/or energy source	Chemical reaction
Nature of signal and/or signature	Color and/or phase boundary changes; precipitation
Detection and/or sensing method	Visual
Indication and/or recording method	Direct observation
Interpretation basis	Qualitative analysis; standardized procedure for reagent preparation and application sequence
OBJECTIVES	
Discontinuities and separations	—
Structure	—
Dimensions and metrology	—
Physical and mechanical properties	—
Composition and chemical analyses	Elemental composition and alloy identification
Stress and dynamic responses	—
Signature analysis	—
APPLICATIONS	
Applicable materials	Metals and alloys
Applicable features and forms	Surfaces and subsurfaces
Process control applications	Applied prior to joining and fabrication operations
In situ and diagnostic applications	Metallurgical specimens
Example structures and components	Stock material identification; engine components
LIMITATIONS	
Access, contact and/or preparation	Requires semi-microscopic specimen particles from part
Probe and object limits	Special chemical kit of prepared reagents
Sensitivity and/or resolution	Approximately 150 metals and alloys identified
Interpretation limits	Assumes particles taken are representative of entire part
Other conditions and limits	Difficult to establish quantitative values of constituents detected; minute amount of material removed
REFERENCES	
Primary source material	Reference 47
Bibliographic material	—
Standards and specifications	—
Related terms	—
Related techniques	Spark test and spark oxidation

TABLE 31. Laser Probe

METHOD	
Key process and basic result	Laser beam is microfocused on test object to determine composition and/or microstructure; minute quantities of vaporized material are spectroscopically analyzed
PRINCIPLES	
Probe medium and/or energy source	Pulsed laser beam and electroexcitation
Nature of signal and/or signature	Ionized vapor plume sample
Detection and/or sensing method	Spectrometer
Indication and/or recording method	Spectrograph
Interpretation basis	Differential and/or quantitative analysis
OBJECTIVES	
Discontinuities and separations	—
Structure	Grain and/or microstructure; inclusion, grain, grain-boundary analysis
Dimensions and metrology	—
Physical and mechanical properties	—
Composition and chemical analyses	Analysis and distribution of elements and impurities
Stress and dynamic responses	—
Signature analysis	—
APPLICATIONS	
Applicable materials	Metals, nonmetals and composites
Applicable features and forms	Surfaces, subsurfaces, layers, deposits and coatings
Process control applications	—
In situ and diagnostic applications	Metallurgical specimens and mineral analysis
Example structures and components	Aircraft engine components, jet rotor hubs and experimental specimens
LIMITATIONS	
Access, contact and/or preparation	Optical view and access to surface required
Probe and object limits	Focusing and microminiaturization of beam diameter
Sensitivity and/or resolution	Analytical accuracy to 5 percent
Interpretation limits	Depends on control and reproducibility of laser beam diameter and energy
Other conditions and limits	Minute amount of material removed
REFERENCES	
Primary source material	References 49 to 51
Bibliographic material	—
Standards and specifications	—
Related terms	Laser microprobe
Related techniques	Ion probe, ion scatter and chemical spot test

TABLE 32. X-ray Fluorescence

METHOD	
Key process and basic result	X-ray irradiation of specimen surface produces fluorescence and spectrographic scanning of emissions identifies elemental composition
PRINCIPLES	
Probe medium and/or energy source	X-radiation to 100 kV level
Nature of signal and/or signature	Secondary radiation or fluorescence
Detection and/or sensing method	Crystal analyzer, scintillation and ionization counter
Indication and/or recording method	Coordinate plotter
Interpretation basis	Quantitative spectrographic analysis; based on empirical calibration curves and/or standard specimens
OBJECTIVES	
Discontinuities and separations	—
Structure	—
Dimensions and metrology	Thickness
Physical and mechanical properties	—
Composition and chemical analyses	Elemental analysis and impurities
Stress and dynamic responses	Corrosion and carburization effects
Signature analysis	—
APPLICATIONS	
Applicable materials	Solids (metals) and liquids
Applicable features and forms	Surfaces, subsurfaces, coatings, films and layers
Process control applications	—
In situ and diagnostic applications	—
Example structures and components	Metallurgical specimens; raw materials; fuels; solutions; turbine casings, diffusers and flanges; and aircraft components
LIMITATIONS	
Access, contact and/or preparation	Sample or specimen surface preparation necessary for high accuracy
Probe and object limits	Only small specimens and areas can be accommodated
Sensitivity and/or resolution	Sensitive to trace elements to 0.1 percent
Interpretation limits	Lower atomic numbers and high sensitivity require vacuum enclosure of specimens
Other conditions and limits	Radiation hazard
REFERENCES	
Primary source material	References 2 (section 17) and 57
Bibliographic material	—
Standards and specifications	—
Related terms	—
Related techniques	Auger analysis, ion probe, ion scatter and charged-particle activation

TABLE 33. X-ray Diffraction

METHOD	
Key process and basic result	Sample of test object is exposed to X-radiation; scatter radiation intensity varies with diffraction angles characteristic of crystalline species present
PRINCIPLES	
Probe medium and/or energy source	Monochromatic X-radiation
Nature of signal and/or signature	Diffraction pattern
Detection and/or sensing method	Scintillation counter and photoemulsion
Indication and/or recording method	Diffractometric photograph and coordinate plot
Interpretation basis	Analytical; differential; dependent on file of reference patterns
OBJECTIVES	
Discontinuities and separations	—
Structure	Crystal size, orientation, structure and strain; lattice deformation; amorphous structure; phase changes
Dimensions and metrology	—
Physical and mechanical properties	—
Composition and chemical analyses	Chemical reaction results
Stress and dynamic responses	Residual stress
Signature analysis	—
APPLICATIONS	
Applicable materials	Crystalline materials
Applicable features and forms	Surface specimens (deposits and layers)
Process control applications	Manufacture of magnetic and ceramic materials
In situ and diagnostic applications	—
Example structures and components	Electrodeposited materials, drawn wire, rolled sheet, mineral analysis and aircraft components
LIMITATIONS	
Access, contact and/or preparation	Powder samples preferred but also applicable to solid parts
Probe and object limits	Special staging or containment of sample
Sensitivity and/or resolution	Peak-to-background ratio should be greater than 1:1 to detect low-concentration constituents
Interpretation limits	Amorphous constituents may not be detected
Other conditions and limits	Finite crystal sizes in specimen introduce statistical errors in scintillation counting; radiation hazard
REFERENCES	
Primary source material	Reference 2 (section 17)
Bibliographic material	—
Standards and specifications	—
Related terms	—
Related techniques	—

Image Generation Techniques

TABLE 34. X-ray Tomography

METHOD	
Key process and basic result	Multiple X-ray attenuation measurement yields spatial distribution of linear attenuation coefficient in the plane of the radiation
PRINCIPLES	
Probe medium and/or energy source	X-radiation
Nature of signal and/or signature	Attenuated radiation level
Detection and/or sensing method	Any X-ray detector
Indication and/or recording method	CRT with gray-scale intensity
Interpretation basis	Quantitative, based on standards
OBJECTIVES	
Discontinuities and separations	Gaps and voids to resolution limit
Structure or malstructure	—
Dimensions and metrology	Measures dimensions to the limit of the resolution
Physical and mechanical properties	Result proportional to density
Composition and chemical analyses	Result proportional to effective linear attenuation coefficient
Stress and dynamic responses	—
Signature analysis	Image processing beneficial
APPLICATIONS	
Applicable materials	All
Applicable features and forms	Not applicable to large, thin plates
Process control applications	No
In situ and diagnostic applications	Yes
Example structures and components	—
LIMITATIONS	
Access, contact and/or preparation	Need access to all sides of the object
Probe and object limits	—
Sensitivity and/or resolution	0.1% density sensitivity; 0.1 mm resolution
Interpretation limits	Process-dependent artifacts sometimes occur
Other conditions and limits	Radiation hazard
REFERENCES	
Primary source material	—
Bibliographic material	—
Standards and specifications	—
Related terms	—
Related techniques	—

TABLE 35. Film Radiography*

METHOD	
Key process and basic result	A photographic image is produced by passage of X rays, gamma rays and/or electrons from or through test object onto a film; changes produced in the film emulsion are developed to yield a radiographic transparency
PRINCIPLES	
Probe medium and/or energy source	X rays, gamma rays, neutron activation, electrons and photons
Nature of signal and/or signature	Attenuation, transmission or emission of radiation
Detection and/or sensing method	Photosensitive or penetrating radiation-sensitive emulsion
Indication and/or recording method	Radiographic image or transparency
Interpretation basis	Direct indication, comparative or differential; based on library of standard or reference images
OBJECTIVES	
Discontinuities and separations	Cracks, inclusions, porosity, voids and lack of bond
Structure	Misalignment and/or malstructure
Dimensions and metrology	Thickness, diameter, position and spacing
Physical and mechanical properties	Density
Composition and chemical analyses	—
Stress and dynamic responses	Dynamic behavior (with flash X-ray)
Signature analysis	—
APPLICATIONS	
Applicable materials	Metals, nonmetals and composites
Applicable features and forms	Range of objects and features
Process control applications	—
In situ and diagnostic applications	—
Example structures and components	Objects of X, gamma, neutron and autoradiography; used to obtain neutron radiographs from activated transfer foil
LIMITATIONS	
Access, contact and/or preparation	One-side access if autoradiography; two if external source
Probe and object limits	Special filters, screens and/or scintillators needed for image quality
Sensitivity and/or resolution	Resolution ranges to 2000 line pairs per centimeter
Interpretation limits	Image quality impaired by scatter radiation and finite source or focal-spot size gamma fogging; requires control of chemicals and photoprocessing conditions for reproducible results
Other conditions and limits	—
REFERENCES	
Primary source material	Reference 2 (section 20)
Bibliographic material	—
Standards and specifications	—
Related terms	X, gamma and neutron radiography; autoradiography; xeroradiography
Related techniques	Fluoroscopy and video radiography

*See Table 15.

TABLE 36. Xeroradiography

METHOD	
Key process and basic result	An electrostatic image is produced by passage of X or gamma rays through a test object onto a charged layer; the charge-image is transferred xerographically to form an opaque radiograph
PRINCIPLES	
Probe medium and/or energy source	X and gamma rays
Nature of signal and/or signature	Attenuation, transmission or emission of radiation
Detection and/or sensing method	Electrically charged layer
Indication and/or recording method	Radiographic image
Interpretation basis	Direct indication, comparative or differential; based on library of standard or reference images
OBJECTIVES	
Discontinuities and separations	Cracks, inclusions, porosity, voids and lack of bond
Structure	Misalignment and/or malstructure
Dimensions and metrology	Thickness, diameter, position and spacing
Physical and mechanical properties	Density variations
Composition and chemical analyses	—
Stress and dynamic responses	—
Signature analysis	Edge enhancement of low-contrast images
APPLICATIONS	
Applicable materials	Metals, nonmetals and composites
Applicable features and forms	Range of objects and features
Process control applications	—
In situ and diagnostic applications	—
Example structures and components	Objects of X and gamma radiography
LIMITATIONS	
Access, contact and/or preparation	Access to two sides of test object required
Probe and object limits	Practical voltage limited to less than 100 kV
Sensitivity and/or resolution	Thickness sensitivity to 2 percent
Interpretation limits	Xeroradiographic plates are easily damaged
Other conditions and limits	Powder- and/or layer-deficient dots and other artifacts hamper image interpretation
REFERENCES	
Primary source material	Reference 2 (section 12)
Bibliographic material	—
Standards and specifications	—
Related terms	—
Related techniques	Film radiography; fluoroscopy; X and gamma radiography

TABLE 37. Fluoroscopy

METHOD	
Key process and basic result	A fluorescent image is produced by X rays passing through a test object onto a fluorescent layer; an immediate and real-time image showing radiographic details appears on a screen
PRINCIPLES	
Probe medium and/or energy source	X-radiation
Nature of signal and/or signature	Attenuation or transmission of radiation
Detection and/or sensing method	Fluorescent chemical layer
Indication and/or recording method	Fluorescent image
Interpretation basis	Direct indication; comparative; based on visual impressions
OBJECTIVES	
Discontinuities and separations	Cracks, porosity, voids and inclusions
Structure	Macrostructure anomalies and misalignment
Dimensions and metrology	Thickness, diameter, spacing and position
Physical and mechanical properties	Density variations
Composition and chemical analyses	—
Stress and dynamic responses	Effects of distorting forces; dynamic phenomena
Signature analysis	—
APPLICATIONS	
Applicable materials	Metals, nonmetals and composites
Applicable features and forms	Range of objects and features
Process control applications	—
In situ and diagnostic applications	—
Example structures and components	Objects of X-radiography
LIMITATIONS	
Access, contact and/or preparation	Access to two sides of test object required
Probe and object limits	Fluoroscopic enclosure limits object size
Sensitivity and/or resolution	Considerably lower resolution than film radiography
Interpretation limits	Requires low ambient lighting and eye accommodation
Other conditions and limits	Image quality hampered by screen unsharpness, attenuation by windows, mirrors and fluorescence fluctuations
REFERENCES	
Primary source material	Reference 2 (section 19)
Bibliographic material	—
Standards and specifications	—
Related terms	—
Related techniques	X-radiography and gamma radiography, real-time radiography

TABLE 38. Video Radiography

METHOD	
Key process and basic result	X, gamma, or neutron-sensitive vidicon receives radiation transmitted through test object; television monitor displays radiographic image
PRINCIPLES	
Probe medium and/or energy source	Neutron rays, gamma rays or X rays
Nature of signal and/or signature	Transmission or attenuation by object variables
Detection and/or sensing method	Neutron, gamma, or X-ray sensitive vidicon
Indication and/or recording method	Television display
Interpretation basis	Direct observation; in-motion; real-time viewing aids interpretation of image content
OBJECTIVES	
Discontinuities and separations	Voids, porosity, inclusions and cracks
Structure	Structure anomalies and/or misalignment
Dimensions and metrology	Dimensional variations
Physical and mechanical properties	—
Composition and chemical analyses	—
Stress and dynamic responses	Direct observation of internal motions of structure; in-motion viewing of flaws
Signature analysis	—
APPLICATIONS	
Applicable materials	Indefinite range of materials
Applicable features and forms	Bulk materials and entire objects
Process control applications	Real-time viewing of processing and production
In situ and diagnostic applications	In-motion operation of hidden and internal components; airport baggage inspection
Example structures and components	Nuclear fuel pins, casting operations, metal rolling and forming operations, liquid-metal cavitation flow patterns and submerged welding
LIMITATIONS	
Access, contact and/or preparation	Access for interposing object between source and detector
Probe and object limits	Useful apertures limited to several centimeters
Sensitivity and/or resolution	Thickness variations to 4 percent; 1 percent with digital video processing
Interpretation limits	Usually limited to coarse indication of flaws
Other conditions and limits	Inferior to film radiography for fine cracks; may improve with manipulating table
REFERENCES	
Primary source material	References 25 (chapter 8) and 64
Bibliographic material	—
Standards and specifications	—
Related terms	—
Related techniques	X, gamma and neutron radiography; fluoroscopy

TABLE 39. Immersion Ultrasonics

METHOD	
Key process and basic result	Test object is ultrasonically scanned while immersed in liquid; ultrasound interactions with object produce signals that are used to map or image internal flaws
PRINCIPLES	
Probe medium and/or energy source	Beam of pulsed ultrasound (20 kHz to 50 MHz range)
Nature of signal and/or signature	Reflection of transmitted ultrasound
Detection and/or sensing method	Piezoelectric transducer (or transducers)
Indication and/or recording method	Oscillogram; coordinate plots and maps
Interpretation basis	Comparative or differential; quantitative with reference or calibration standards
OBJECTIVES	
Discontinuities and separations	Cracks, voids, laminations, debonds and inclusions
Structure	Porosity and segregations
Dimensions and metrology	Thickness
Physical and mechanical properties	Density, velocity of sound, elastic moduli
Composition and chemical analyses	—
Stress and dynamic responses	—
Signature analysis	—
APPLICATIONS	
Applicable materials	Metals, nonmetals and composites
Applicable features and forms	Subsurface, bulk and internal features
Process control applications	Rolling, casting and forging monitoring
In situ and diagnostic applications	—
Example structures and components	Sheet, plate, bar and tube items; billets and slabs; engine components; and power transmission shafts
LIMITATIONS	
Access, contact and/or preparation	Liquid immersion and access to at least one surface required
Probe and object limits	Small, thin, rough-surface parts are difficult to evaluate
Sensitivity and/or resolution	Discontinuities to 0.01 millimeter, depending on material and frequency used
Interpretation limits	Ambiguous response from scatter and geometric complexity
Other conditions and limits	Geometrically complex and/or nonregular objects require intricate scanning arrangements
REFERENCES	
Primary source material	Reference 2 (sections 44 to 47)
Bibliographic material	—
Standards and specifications	Reference 4; ASTM E-214-68
Related terms	B-scan and C-scan ultrasonics
Related techniques	Ultrasonic videography and ultrasonic holography

TABLE 40. Ultrasonic Videography

METHOD	
Key process and basic result	Object is illuminated with ultrasound; ultrasound over an extensive area of object is detected to form an image similar to X-radiograph
PRINCIPLES	
Probe medium and/or energy source	Continuous or pulsed ultrasound at 1 to 10 MHz
Nature of signal and/or signature	Transmission or attenuation by object variables
Detection and/or sensing method	Piezoelectric plate or crystal
Indication and/or recording method	Direct visual, television monitor and/or cinematography
Interpretation basis	Direct
OBJECTIVES	
Discontinuities and separations	Debonds, lack of bond and delaminations
Structure or malstructure	Microporosity; grain and crystalline structure
Dimensions and metrology	—
Physical and mechanical properties	Stress patterns
Composition and chemical analyses	—
Stress and dynamic responses	In-motion observation of flaws
Signature analysis	—
APPLICATIONS	
Applicable materials	Metals, nonmetals and liquids
Applicable features and forms	Bulk and internals
Process control applications	—
In situ and diagnostic applications	—
Example structures and components	Metal claddings and coatings, welds, spot welds, nuclear fuel plates and electron population in semiconductors
LIMITATIONS	
Access, contact and/or preparation	Immersion of test object required
Probe and object limits	Crystal (diameter) limits area tested to few centimeters
Sensitivity and/or resolution	Typical sensitivity to 0.1 centimeter
Interpretation limits	Ambiguous response from interference fringes due to Fresnel-Fraunhofer effects
Other conditions and limits	—
REFERENCES	
Primary source material	Reference 25 (chapter 3)
Bibliographic material	—
Standards and specifications	—
Related terms	—
Related techniques	Immersion ultrasonics and ultrasonic holography

Image Analysis Techniques

TABLE 41. Video Enhancement

METHOD	
Key process and basic result	Image is examined with video camera that converts image density values into signals that are enhanced electronically; derivative image is displayed on television monitor
PRINCIPLES	
Probe medium and/or energy source	Visible light
Nature of signal and/or signature	Density variations in original image
Detection and/or sensing method	Vidicon camera
Indication and/or recording method	Television display
Interpretation basis	Interpretative aid; depends on selection of enhancement mode
OBJECTIVES	
Discontinuities and separations	Cracks, voids, inclusions, porosity, etc.
Structure	Structure anomalies and/or misalignment
Dimensions and metrology	Dimensional measurements
Physical and mechanical properties	Density
Composition and chemical analyses	—
Stress and dynamic responses	—
Signature analysis	—
APPLICATIONS	
Applicable materials	Radiographs, photographs and metallographs
Applicable features and forms	Transparencies and opaque films
Process control applications	—
In situ and diagnostic applications	Combined with video radiography
Example structures and components	—
LIMITATIONS	
Access, contact and/or preparation	Presentation of image to camera required
Probe and object limits	Special lens systems required for various uses
Sensitivity and/or resolution	Density increments to 0.05 photodensity units
Interpretation limits	Usually senses 32 density levels
Other conditions and limits	—
REFERENCES	
Primary source material	References 73 and 74
Bibliographic material	—
Standards and specifications	—
Related terms	—
Related techniques	Photographic extraction, spot scanner-digitizer and laser filter

REFERENCES

1. McMaster, R.C. and S.A. Wenk. *A Basic Guide for Management's Choice of Nondestructive Tests*. Special Technical Publication, No. 112. Philadelphia, PA: American Society for Testing and Materials (1951).
2. *Nondestructive Testing Handbook*, 1st edition. McMaster, R.C., ed. Volume 1 (1959).
3. McClung, R.W. "1974 ASNT Lester Honor Lecture." *Materials Evaluation*. Volume 33, No. 2 (February 1975): pp 16-19.
4. *Nondestructive Evaluation*. Rep. NMAB-252. National Academy of Sciences, AD-692491 (June 1969).

References for *Nondestructive Evaluation Technique Guide*

1. *Nondestructive Evaluation*. Rep. NMAB-252. National Academy of Sciences, AD-692491 (June 1969).
2. McMaster, R.C., ed. *Nondestructive Testing Handbook*. Vols. 1 and 2. New York: Ronald Press Company (1959).
3. Collier, R.J., C.B. Burichardt and L.H. Lin. *Optical Holography*. Academic Press (1971).
4. *Annual Book of ASTM Standards*. Part 31. Philadelphia, PA: American Society for Testing and Materials (1972).
5. Oaks, A.E. *Nondestructive Testing Techniques of Scout Rocket Motors*. NASA CR-2013 (1972).
6. Egerton, H.B., ed. *Nondestructive Testing — Views, Reviews, Previews*. Oxford University Press (1969).
7. *Nondestructive Testing: Methods, Techniques and Their Applications*. Rep. DDC-TAS-71-58-1. Defense Documentation Center, AD-733850 (December 1971).
8. *Nondestructive Testing: Radiography*. Rep. DDC-TAS-71-54-1. Defense Documentation Center, AD-733860 (November 1971).
9. Berger, H. *Neutron Radiography*. Elsevier Publishing Company (1965).
10. Berk, S. "Radiation Backscattering and Radiation-Induced X-rays for Measuring Surface Composition and Structure." *Materials Evaluation*. Vol. 24, No. 6 (June 1966): pp 309-312.
11. Hendron, J.A., K.K. Groble and W. Wangard Jr. "The Determination of the Resin-to-Glass-Epoxy Structures by Beta Ray Backscattering." *Materials Evaluation*. Vol. 22, No. 5 (May 1964): pp 213-216.
12. Delacy, T.J. *The Use of Radioactive Labeling*. Convair Research Summary CRS-5. General Dynamics Corporation (1970): pp 12-16.
13. Cucchiara, O. and P. Goodman. "Kryptonates: A New Technique for the Detection of Wear." *Materials Evaluation*. Vol. 25, No. 5 (May 1967): pp 109-117.
14. Packer, L.L., W.A. Burton and R. Hecht. *Investigation of Radioactive Gas Radiography for Inspection of Brazed Insert Joints*. Rep. F-930511-1. United Aircraft Research Laboratory (February 1967).
15. Grosskreutz, J.C. and W.E. Millett. "The Effect of Cyclic Deformation on Positron Lifetimes in Copper and Aluminum." *Physics Letters*. Vol. 28a, No. 9 (February 1969): pp 621-622.
16. Gardner, C.G. and J.R. Barton. "Recent Advances in Magnetic Field Methods of Nondestructive Evaluation for Aerospace Applications." *Advanced Technology for Production of Aerospace Engines*. AGARD-CP-64-70 (September 1970): pp 18.1-18.9.
17. Barton, J.R. and J. Lankford. *Magnetic Perturbation Inspection of Inner Bearing Races*. NASA CR-2-55 (1972).
18. Rodgerd, E.H. and C.P. Merhib. *A Report Guide To Magnetic Particle Testing Literature*. Rep. AMRA-MS-65-04. Army Materials Research Agency, AD-617758 (June 1965).
19. Abragam, A. *The Principles of Nuclear Magnetism*. Oxford University Press (1961).

20. Hewitt, R.P. and B. Mazelsky. *Nondestructive Inspection of Reinforced Plastic Structures by the Nuclear Quadrupole Resonance Method.* Rep. ARA-77, AD-645562 (November 1966).
21. Dakin, T.W. and J. Lim. "Corona Measurement and Interpretation." *AIEE Transcript on Power Apparatus and Systems.* Vol. 76 (December 1957): pp 1059-1065.
22. Hendron, J.A., K.K. Groble, R.W. Gruetzmacher, G.O. McClurg and M.W. Retsky. "Corona and Microwave Methods for Detection of Voids in Glass-Epoxy Structures." *Materials Evaluation.* Vol. 22, No. 7 (July 1968): pp 311-314.
23. Zurbick, J.R. "The Mystery of Reinforced Plastics Variability. Nondestructive Testing Holds the Key." *Material Research Standards.* Vol. 8, No. 7 (July 1968): pp 25-36.
24. Hoenig, S.A., W.A. Ott, M.T. Ali and T.A. Russell. *New Techniques in Nondestructive Testing (Exo-Electron Emission Phase).* Part 1. Arizona University, AFML-TR-71-140, AD-748272 (December 1971).
25. Sharpe, R.S., ed. *Research Techniques in Nondestructive Testing.* Academic Press (1970).
26. Schroeer, R., R. Rowand and H. Kamm. "The Acoustic Impact Technique — A Versatile Tool for Nondestructive Evaluation of Aerospace Structures and Components." *Materials Evaluation.* Vol. 28, No. 11 (November 1970): pp 237-243.
27. Botsco, R.J. "Nondestructive Testing of Composite Structures with the Sonic Resonator." *Materials Evaluation.* Vol. 24, No. 11 (November 1966): pp 617-623.
28. Libby, H.L. *Introduction to Electromagnetic Nondestructive Test Methods.* Wiley-Interscience (1971).
29. Botsco, R.J. "The Eddy-Sonic Test Method." *Materials Evaluation.* Vol. 26, No. 2 (February 1968): pp 21-26.
30. *Acoustic Emission.* Special Technical Publication No. 505. Philadelphia, PA: American Society for Testing and Materials (1972).
31. Liptai, R.G., D.O. Harris, R.B. Engle and C.A. Tatro. "Acoustic Emission Techniques in Materials Research." *International Journal of Nondestructive Testing.* Vol. 3, No. 3 (December 1971): pp 215-275.
32. Krautkrammer, J. *Ultrasonic Testing of Materials.* Springer-Verlag (1969).
33. *Nondestructive Testing: Ultrasonics.* Rep. DDC-TAS-71-55-1. Defense Documentation Center, AD-733700 (November 1971).
34. Stenton, F.G. and C.P. Merhib. *A Report Guide to Ultrasonic Testing Literature.* Vol. 6. Rep. AMMRC-MS-69-03. Army Materials and Mechanics Research Center, AD-689455 (April 1969).
35. Firestone, F.A. and J.R. Frederick. "Refinements in Supersonic Reflectoscopy." *Journal of the Acoustics Society of America.* Vol. 18, No. 1 (July 1946): pp 200-211.
36. Hunter, D.O. *Ultrasonic Velocities and Critical-Angle-Method Changes in Irradiated A302-B and A542-B Steels.* Rep. BNWL-988. Battelle-Northwest (April 1969).
37. Rollins, F.R., Jr. "Critical Ultrasonic Reflectivity — A Neglected Tool For Material Evaluation." *Materials Evaluation.* Vol. 24, No. 12 (December 1966): pp 683-689.
38. Merhib, C.P. and E.H. Rodgers. *A Report Guide to Thermal Testing Literature.* Rep. AMRA-MS-64-14. Army Materials Research Agency, AD-612043 (August 1964).
39. Powell, R.W. "The Contribution of Heat Conduction to Nondestructive Testing." *Physics and Nondestructive Testing*, W.J. McGonnagle, ed. Gordon and Breach Science Publishers (1967): chapter 17.
40. Green, D.R. "High Speed Thermal Image Transducer for Practical NDT Applications." *Materials Evaluation.* Vol. 28, No. 5 (May 1970): pp 97-102, 110.
41. Stinebring, R.C. and R. Cannon. *Development of Nondestructive Methods for Evaluating Diffusion-Formed Coatings on Metallic Substrates.* Rep. AVSSD-0267-67-CR. Avco Corporation, AFML-TR-67-178, AD-823889 (October 1967).
42. Stinebring, R.C. *Development and Application of Nondestructive Methods for Evaluation of Diffusion-Formed Coating on Metallic Substrates.* Rep. AVSSD-0116-68-RR. Avco Corporation, AFML-TR-67-178, AD-853493 (April 1969): part 2.
43. Apple, W.R. "Infrared Nondestructive Inspection — A status Report." *Materials Research Standards.* Vol. 9, No. 5 (May 1969): pp 10-13.

44. Brown, S.P. "Cholesteric Crystals for Nondestructive Testing." *Materials Evaluation.* Vol. 26, No. 8 (August 1968): pp 163-166.
45. Woodmansee, W.E. "Cholesteric Liquid Crystals and Their Application to Thermal Nondestructive Testing." *Materials Evaluation.* Vol. 24, No. 10 (October 1966): pp 564-572.
46. McCullough, L.D. and D.R. Green. "Electrothermal Nondestructive Testing on Metal Structures." *Materials Evaluation.* Vol. 30, No. 4 (April 1972): pp 87-91.
47. *Applications of Aerospace Technology in Industry — Nondestructive Testing.* NASA Technology Utilization Office (April 1972): pp 83-85. Also see NASA Tech Brief 70-10520.
48. *Applications of Aerospace Technology in Industry — Nondestructive Testing.* NASA Technology Utilization Office (April 1972): pp 87-88. Also see NASA Tech Brief 68-10378.
49. Snetsinger, K.G. and K. Keil. "Microspectrochemical Analysis of Minerals with the Laser Microprobe." *American Mineralogist.* Vol. 52, Nos. 11 and 12 (November and December 1967): pp 1842-1854.
50. McCormack, J.T. "The Laser Microprobe — A New Metallurgical Tool." *Metals Rev.* (March 1965): pp 6-7.
51. Karyakin, A.V. and V.A. Kaigorodov. "Effect of the Base in Spectrographic Analysis with a Laser." *Journal of Analytic Chemistry (USSR).* Vol. 22, No. 4 (April 1967): pp 444-447.
52. Smith, D.P. "Analysis of Surface Composition with Low-Energy Backscattered Ions." *Surface Science.* Vol. 25 (1971): pp 171-191.
53. Bayard, M. "The Ion Microprobe." *American Laboratory* (April 1971): pp 15-22.
54. "Ion Microprobe Mass Spectrometry." *Progress in Nuclear Energy.* Pergamon Press (1969).
55. Stein, D.F., A. Joshi and R.P. LaForce. "Studies Utilizing Auger Electron Emission Spectroscopy on Temper Embrittlement in Low Alloy Steels." *ASM Transcript.* Vol. 62, No. 3. Metals Park, OH: American Society for Metals (September 1969): pp 776-783.
56. Connell, G.L. and Y.P. Gupta. "Auger Electron Spectroscopy." *Materials Research Standards.* Vol. 11, No. 1 (January 1971): pp 8-13, 38.
57. Simon, H. "X-Ray Fluorescence Analysis as an Aid to Production and Repair of Aircraft Engines." *Advanced Technology for Production of Aerospace Engines.* AGARD-CP-64-70 (April 1970): pp 16.1-12.
58. Kock, R.C. *Activation Analysis Handbook.* Academic Press (1960).
59. Birks, L.S., R.E. Seebola, A.P. Bott and J.S. Grosso. "Excitation of Characteristic X-Rays by Protons, Electrons and Primary X-Rays." *Journal of Applied Physics.* Vol. 35, No. 9 (September 1964): pp 2578-2581.
60. Johansson, T.B., R. Akselsson and S.A.E. Johansson. "X-Ray Analysis: Elemental Trace Analysis at the 10^{-12}g Level." *Nuclear Instrument Methods.* Vol. 84 (1970): pp 141-143.
61. Wertheim, G.K. "Mossbauer Effect." *Principles and Applications.* Academic Press (1964).
62. Furman, S.C., R.W. Darmitzel, C.R. Porter and D.W. Wilson. "Track Etching — Some Novel Applications and Uses." *ANS Transscript.* Vol. 9, No. 2 (November 1966): pp 598-599.
63. Morley, J. *Two New Methods to Increase the Contrast of Track-Etch Neutron Radiographs.* NASA Report TM X-67947 (1971).
64. McMaster, R.C., M.L. Rhoten and J.P. Mitchell. "The X-Ray Vidicon Television Image System." *Materials Evaluation.* Vol. 25, No. 3 (March 1967): pp 46-52.
65. Metherell, A.F., H.M.A. El-Sum and L. Larmore, eds. "Acoustical Holography." *Proceedings of the First International Symposia on Acoustical Holography.* Vol. 1, Plenum Publishing Corporation (1969).
66. Metherell, A.F. and L. Larmore, eds. "Acoustical Holography." *Proceedings of the Second International Symposia on Acoustical Holography.* Vol. 2. Plenum Publishing Corporation (1970).
67. Mehterell, A.F., ed. "Acoustical Holography." *Proceedings of the Third International Symposia on Acoustical Holography.* Vol. 3. Plenum Publishing Corporation (1971).
68. Lohse, K.H. *Application of Photographic Extraction Techniques to Nondestructive Testing of Graphites and Other Materials.* Philco-Ford Corporation, AFML-TR-70-162, AD-879494 (November 1970).

69. Stratton, R.H. and J.J. Sheppard Jr. *A Photographic Technique for Image Enhancement: Pseudocolor Three-Separation Process.* Rep. R-596-PR. Rand Corporation, AD-717143 (October 1970).

70. O'Neill, E.L. "Spatial Filtering in Optics." *IRE Trans. on Information Theory.* Vol. IT-2 (June 1956): pp 56-65.

71. Janney, D.H., B.R. Hunt and R.K. Zeigler. "Concepts of Radiographic Image Enhancement." *Materials Evaluation.* Vol. 30, No. 9 (September 1972): pp 195-199, 203.

72. Selzer, R.H. *Improving Biomedical Image Quality with Computers.* Rep. TR-32-1336. Jet Propulsion Laboratory, California Institute of Technology. NASA CR-97899 (October 1968).

73. Anderson, R.T. and T.J. Delacy. *Nondestructive Testing Applications for Advanced Aerospace Materials and Components.* Rep. GDC-ERR-1324. General Dynamics/Convair (December 1968): pp 5.1-5.7.

74. Vary, A. "Investigation of an Electronic Image Enhancer for Radiographs." *Materials Evaluation.* Vol. 30, No. 12 (December 1972): pp 259-267.

BIBLIOGRAPHY

1. *Introduction to Nondestructive Testing.* Convair Division, General Dynamics Corporation. NASA CR-61204 (1965).

2. *Quality Assurance: Guide to Specifying NDT in Material Life-Cycle Applications.* AMCP 702-11. Army Material Command (1970).

3. *Quality Assurance: Guidance to Nondestructive Testing Techniques:* AMCP 702-10. Army Material Command, AD-728162 (1970).

4. Der Boghosian, S. *A Report Guide to Radiographic Testing Literature.* Vol. 3. AMMRC-MS-68-08. Army Materials and Mechanics Research Center, AD-676835 (1968).

5. Grubinskas, R.C., C.P. Merhib. *A Report Guide to Literature in the Field of Electromagnetic Testing.* AMRA-MS-65-03. Materials Engineering Division, Army Materials Research Agency, AD-615346 (1965).

6. Hulteen, K. and J. Cook, eds. "30-Year Index: Materials Evaluation Vol. 1 through Vol. 30, 1942-1972." *Materials Evaluation.* Vol. 31, No. 1 (January 1973).

7. Meister, R.P., M.D. Randall, D.K. Mitchell, L.P. Williams and H.E. Pattee. *Summary of Nondestructive Testing Theory and Practice.* NASA CR-2120 (1972).

8. Merhib, C.P. *A Report Guide to Fatigue Testing Literature.* AMRA-MS-67-05. Army Materials Research Agency, AD-652881 (1967).

9. Merhib, C.P. and S. Der Boghosian. *A Report Guide to Literature in the Fields of Fluoroscopy and Remote Viewing Techniques.* AMRA-MS-64-13. Materials Technology Division, Army Materials Research Agency, AD-612045 (1964).

10. Morgan, R. *Selected Bibliographic Guide to Conference Papers on Nondestructive Testing, 1955-1967 (with abstracts).* AERE-BIB-164. Harwell, England: Atomic Energy Research Establishment (1968).

11. Neuschaefer, R.W. and J.B. Beal. *Assessment of an Standardization for Quantitative Nondestructive Testing.* NASA TM X-64706 (1972).

12. Rodgers, E.H. *A Report Guide to Ultrasonic Attenuation Literature.* AMRA-MS-65-09. Materials Technology Division, Army Materials Research Agency, AD-627565 (1965).

13. Rodgers, E.H. *A Report Guide to Autoradiographic and Microradiographic Literature.* AMRA-MS-64-10. Materials Technology Division, Army Materials Research Agency, AD-612047 (1964).

14. Rodgers, E.H. and S. Der Boghosian. *A Report Guide to Gamma Radiographic Literature.* AMRA-MS-64-11. Materials Technology Division, Army Materials Research Agency, AD-612042 (1964).

15. Rodgers, E.H. and C.P. Merhib. *A Report Guide to Liquid Penetrant Literature.* AMRA-MS-64-12. Materials Technology Division, Army Materials Research Agency, AD-612044 (1964).

16. Rodgers, E.H. and C.P. Merhib. *A Report Guide to Magnetic Particle Testing Literature.* AMRA-MS-65-04. Materials Technology Division, Army Materials Research Agency, AD-617758 (1965).

17. Urbach, J.C., R. Aprahamian, and B. Brenden, eds. "Imaging Techniques for Testing and Inspection; Proceedings of the Seminar-in-Depth." *SPIE Seminar Proceedings.* Volume 29. Los Angeles, CA: Society of Photo-Instrumentation Engineers (February 1972): pp 14-15.

SECTION 1

RADIATION AND PARTICLE PHYSICS

C. Robert Emigh, Los Alamos National Laboratory, Los Alamos, NM

INTRODUCTION

Lawrence E. Bryant, Jr.
University of California
Los Alamos National Laboratory
Los Alamos, New Mexico, USA 87545

Within the nondestructive testing field, whether the reader is a physicist or technician, a production worker or a researcher, it is quite likely that the individual will be involved with or affected by the basics of radiation and particle physics. Those of us active in the field, particularly in radiography or other radiation methods, are dependent on the principles discussed in this work for every technique developed and every test result obtained.

The basic particles—the electron, proton, and neutron—as well as those not so familiar to most—the positron, meson, and neutrino—are described. Their places in nuclear and atomic structure are developed and with this understanding it is possible to move on to generation of X and gamma rays. The scene then shifts to how electromagnetic waves, neutrons,and electrons our fundamental probes in nondestructive examinations—interact with matter by absorption and scattering.

As advances in the field occur, whether in real-time radiography, industrial tomography, dual energy DXT, or any of the other evolving techniques, these basic principles are always the starting point. Even those of us who work to maintain quality and production in the established techniques of our profession will do well to review the "roots" of our inspection methods—radiation and particle physics.

PART 1

INTRODUCTION TO ELEMENTARY PARTICLES

The Electron

Probably the first indication that electrons existed came from the early experimenters who discovered electrification by rubbing various materials against each other. It was left to Benjamin Franklin in 1750, however, to suggest that the flow of electricity was not a continuous process but rather a flow of many discrete charges. Evidence for Franklin's hypothesis was a long time in arriving, for it was nearly a century later, in 1833, that Michael Faraday announced his laws of electrolysis. Faraday found that it took exactly twice as much electricity to transport 1 gram-mole of a divalent material from one electrode to another, as it did to transport a univalent material. This was certainly evidence that electricity existed in multiples of some unit of charge, but it was not conclusive proof (references 1-5 provide background detail for the topics discussed in this section.)

In 1874, G. Johnstone Stoney calculated from Faraday's work the average charge carried by an ion in solution, assuming all the univalent ions carried the same charge. He called this charge the electron. Although his calculations were far from being correct, he was on the right trail. A few years earlier, Sir William Crookes discovered cathode rays in a partially evacuated discharge tube. Later, Sir Arthur Schuster and Sir J. J. Thomson, in separate experiments, found that these cathode or beta rays were negatively charged electrons traveling at high velocities. From the deflection of the cathode rays by magnetic and electrostatic means, Thomson was able to make a rough measurement of the ratio of the charge to the mass, e/m, for the electron (or beta particle).

Charge on Electron

The first direct attempt to determine the charge on the electron was undertaken by Townsend in 1896 and by Thomson in 1898. Their methods were similar, consisting of condensing water on negative ions and then calculating the number of water droplets and the total charge carried by the condensed cloud. From these data the average charge on each droplet could be calculated.

Many similar measurements were made, but in 1909 Robert A. Millikan undertook an investigation that resulted in one of the outstanding achievements of physics. This experiment, known as the Millikan oil drop experiment, provided conclusive proof of the atomic character of electricity and won the Nobel Prize for the investigator. By using a telescope of high magnification and observing the free fall of charged oil droplets and their rate of rise under the added influence of an electrical field, Millikan calculated the charge on individual droplets with the aid of Stokes's law. He found that the calculated charges on all droplets were exact multiples of a smallest value, 1.5196×10^{-19} coulomb. What more direct evidence of the atomic character of electricity could be desired? Since Millikan's experiment, many other workers have measured the charge of the electron, using more refined equipment and data. Despite this, the experimental values for the charge of an electron have remained substantially the same.

Another series of experiments, starting in 1928, determined the charge of the electron far more precisely than the values obtained by the oil-drop experiments. By X-ray diffraction methods, using ruled gratings or crystal lattice spacings, accurate determinations of Avogadro's number were obtained. Combining Avogadro's number with the Faraday, the charge magnitude of which was known precisely, resulted in an accurate determination of the electronic charge. As time passed, many experiments were performed to determine other characteristics of the electron. The most recent data on electron characteristics are given in Table 1.

TABLE 1. Electron Characteristics

Charge	1.602×10^{-19} C
Rest mass	9.109×10^{-31} kg
Classical radius	2.818×10^{-15} m
Magnetic moment	-9.271×10^{-24} J/T
Compton wavelength	2.426×10^{-12} m

C = coulombs
J/T = joules per tesla

The Proton

The Greek philosophers were apparently the first to theorize on the problem of matter. As early as 530 B.C., the Pythagoreans speculated on the possibility that all matter was made up of a single "element." This element was considered at different times to be water, air, or fire. Later, they taught that all matter was made up of four elements: earth, water, air, and fire. All matter consisted of these basic elements in different proportions under the influence of two contrasting divine powers, one attractive and the other repulsive. Greek philosophers also had the idea that the atoms were in incessant motion, but this conception was based only on speculative philosophy.

It was not until nearly the nineteenth century, when Joule measured the average speed of gaseous molecules in motion, that experimental evidence of the atomic nature of matter was presented. Such motions were demonstrated by Brown in 1827, when particles of inanimate matter suspended in liquids could be seen with a high-power microscope to have irregular wiggling motions. This phenomenon, known as the Brownian movement, gives an approximate image of the movements of gas molecules as postulated by the kinetic theory of gases.

In 1886, Goldstein observed streams of positive rays passing through holes in the cathode of a Crookes's discharge tube and forming luminous rays. These were called Kanalstrahlen or canal rays. They differed from cathode rays in that their direction of travel indicated that they were positive ions. In 1898, Wein produced deflections of these positive rays by electrical and magnetic fields. From these deflections, Wein calculated the ratio of the charge to the mass, e/m, and found the ratio to be much smaller than the corresponding ratio for the electron. In fact, the ratio was dependent on the type of gas used in the discharge tube and was at a maximum for hydrogen.

Thomson was one of the pioneer workers in separating positive rays of different masses by using deflecting fields, which demonstrated the atomic nature of the elements. These and the later, more exacting, measurements by Aston, Bainbridge, and Dempster, using mass spectrographs, revealed the entire isotopic picture of atomic structure. In all cases, the hydrogen ion seemed to be a basic building block and was given the name proton from the Greek work "protos," meaning "first." Its properties are given in Table 2.

TABLE 2. Proton Characteristics

Charge	1.602×10^{-19} C
Rest mass	1.673×10^{-27} kg
Classical radius	1.534×10^{-18} m
Magnetic moment	$+1.409 \times 10^{-26}$ J/T
Compton wavelength	1.321×10^{-15} m

The Neutron

In 1930, Bothe and Becker observed that a very penetrating radiation was emitted when light elements such as beryllium and lithium were bombarded with alpha particles (helium ions). If this emission had been interpreted as being gamma radiation, the half-value layer for lead would have been several centimeters, indicating a gamma energy of approximately 3 MeV. However, Curie and Joliot showed that such an assumption was inconsistent with their experiments, since very high energy recoil protons were obtained from interactions with the penetrating radiation.

In 1932, Chadwick solved this paradox by assuming that the radiation consisted of neutral particles having the same weight as the proton. This assumption made such a convincing picture that it was accepted almost immediately. Chadwick called this particle the neutron. Further experimentation by many physicists and chemists strengthened Chadwick's hypothesis. Neutron characteristics are given in Table 3.

TABLE 3. Neutron Characteristics

Charge	neutral
Rest mass	1.675×10^{-27} kg
Classical radius	1.532×10^{-18} m
Magnetic moment	-9.646×10^{-27} J/T
Compton wavelength	1.319×10^{-15} m

The Positron

The positron was discovered by Anderson in 1932. In cloud-chamber studies, Anderson noted particles whose tracks were identical to those of electrons except that they were deflected in an opposite sense in a magnetic field. Dirac had advanced a theory several years earlier that predicted this particle. Experiments by Blackett and Occhialini in 1933 showed that cosmic rays interacted with matter and gave rise to showers of electrons and positrons in about equal numbers.

Further experimentation gave evidence that the positron is identical with an electron in atomic weight, rest mass, and rest energy, and that its positive charge is equal in numerical value to that of the negative charge of the electron. It has the same angular spin momentum and its magnetic moment is equal but opposite in sign to that of the electron. However, the positrons are extremely short-lived particles (10^{-7} s), capable of existence only while in motion; as soon as they come into the vicinity of electrons, they combine with the latter to form photons (annihilation radiation).

The Neutrino

The neutrino was postulated by Fermi in 1934 to explain an apparent contradiction of the law of conservation of energy in beta emission; that is, the continuous spectra distribution in place of discrete lines as predicted by quantum theory. Fermi's theory has the neutrino, of zero or near-zero mass and no charge, sharing with the beta particle the energy released in the beta decay process. The existence of the neutrino is generally accepted, and evidence has been supplied by Cowan and Reines of the Los Alamos National Laboratory that substantiates the existence of the neutrino.

Mesons

The meson was first postulated in Yukawa in 1934 to explain the nuclear forces of attraction that bind the various particles of the nucleus. In 1936, Anderson noticed these particles in his cloud-chamber studies of cosmic radiation. The particles were described as highly unstable and very short lived, with a mass less than that of protons but greater than that of electrons, and with either a positive or negative electronic charge. Further experimentation revealed several types of mesons, now generally classified as μ mesons, π mesons, and τ mesons. All mesons have an extremely short lifetime, after which they spontaneously disintegrate into an electron or a positron, and a neutrino, There is evidence for the existence of neutral as well as charged mesons. The mesons are listed in Table 4.

TABLE 4. Mesons

Symbol	Mass (kg)	Mass*	Charge**	Mean Life (seconds)
$\mu^{\pm}$	1.8838×10^{-28}	206.84	±1	2.22×10^{-6}
$\pi^{\pm}$	2.4886×10^{-28}	273.23	±1	2.54×10^{-8}
π°	2.4082×10^{-28}	264.4	0	$\sim 10^{-15}$
k^{+}	8.6038×10^{-28}	966.6	+1	$\sim 10^{-8}$
k°	8.8749×10^{-28}	974.4	0	$\sim 10^{-10}$ to 10^{-7}

*In units of electron mass
**In units of electron charge

Other Particles

Many other particles of various masses have been observed and are generally classified as hyperons, but they are very rare and have very short lives. These are listed in Table 5.

TABLE 5. Hyperons

Symbol	Mass (kg)	Mass*	Charge**	Mean Life (seconds)
Λ°	1.9868×10^{-27}	2181.4	0	2.8×10^{-10}
Σ^{+}	2.1201×10^{-27}	2327.7	+1	$\sim 5 \times 10^{-11}$
Σ^{-}	2.1342×10^{-27}	2343.2	−1	$\sim 10^{-10}$
Ξ^{-}	2.3535×10^{-27}	2584	−1	$\sim 10^{-10}$
Ω^{-}	1.5356×10^{-27}	1686	−1	$\sim 10^{-9}$

*In units of electron mass
**In units of electron charge

Antiparticles

In 1955, evidence for the existence of the antiproton was indicated by the experiments of Segre and his collaborators. These particles account for some of the previous inconsistencies of the quantum theory. All particles are believed to have antiparticle analogues and are usually characterized by opposite angular momentum and/or charge.

PART 2

NUCLEAR STRUCTURE

Basic Elements

An element is defined as a substance that cannot be broken chemically into simpler substances. There are just over 100 chemically basic elements, not counting recently announced elements that have not been officially confirmed. All elements have the same general structure: a heavy, positively charged core surrounded by negatively charged electrons. The electrons are generally thought of as existing in orbits around the core in a manner not unlike the planets rotating about the sun. However, the analogy of electrons to planets should not be taken too seriously because the electron orbits are severely restricted by the rules of quantum mechanics, as discussed in the following section.

The positively charged core of each element is considered to include various numbers of protons and neutrons. The simplest core is a single proton with a positive unit of electronic charge (hydrogen). As all atoms tend to be neutral, the positive core will attract and hold negatively charged electrons in orbits. As the number of protons in the core increases, more electrons can be held in stable orbits about the nucleus. It is the number of orbital electrons, especially those in the outermost orbits, that determines the chemical properties of the individual elements.

In the nucleus, the proton and neutron have an unusual attraction for each other and tend to pair together. For this reason, elements have roughly the same number of protons as neutrons, except for some of the lighter elements. The coulomb repulsion of protons in heavy elements, or those with many protons, forces the protons to the outer regions of the core while neutrons tend to fill in the center. Hence, there will be more neutrons than protons in the heavier elements.

Isotopes-Isobars

Isotopes are varieties of the same chemical element, having the same number of protons in the nucleus but different atomic weights or mass number, i.e., a different number of neutrons. Since the number of orbital electrons is the same for various isotopes of the same element, the chemical characteristics are practically identical. Among the 100 or so known elements, there are some 300 different isotopes that are stable and over 500 that are radioactive or unstable. This instability arises because some isotopes have either too many or too few neutrons in the core. These unstable isotopes attempt to reach a stable ratio of neutrons to protons and, in so doing, emit either alpha particles (helium ions) or beta particles of either sign, or capture orbital electrons (i.e., K capture).

Figure 1 shows a trilinear scheme for plotting the various elements and their isotopes. It has the advantage that it constitutes simultaneous A vs. Z, N vs. A, and N vs. Z plots. Z denotes the number of protons in the nucleus, N the number of neutrons, and A the number of nucleons (protons plus neutrons). In Fig. 1 the isotopes of a given element are plotted on one set of diagonals. The isotones, $N =$ constant, are plotted on the other set of diagonals. Isobars, A = constant, are plotted on vertical lines. Elements of the same isotopic number, $N - Z$, are plotted horizontally. If the isotopic number of a radioactive isotope is greater than that of any of its stable isobars, the isotope usually decays by beta emission. If the isotopic number is less than that of any of its stable isobars, the isotope usually decays by K capture or positron emission. If the isotope lies between two stable isobars, it decays by either process. In addition, isotopes decay by alpha emission. These latter isotopes usually have an isotopic number smaller than any of their stable isobars or occupy the region of the chart where there are no stable isotopes.

FIGURE 1. A Trilinear Plot of the Elements and Their Isotopes. The stable isotopes are designated by the filled-in circles; the naturally radioactive isotopes by filled-in squares; the artificially radioactive isotopes by crosses; and the radioactive daughter isotopes of the naturally radioactive isotopes, or the isotopes continually being formed by nuclear reactions, by filled-in triangles. The line connecting the stable isotopes indicates the position of greatest neutron-proton stability.

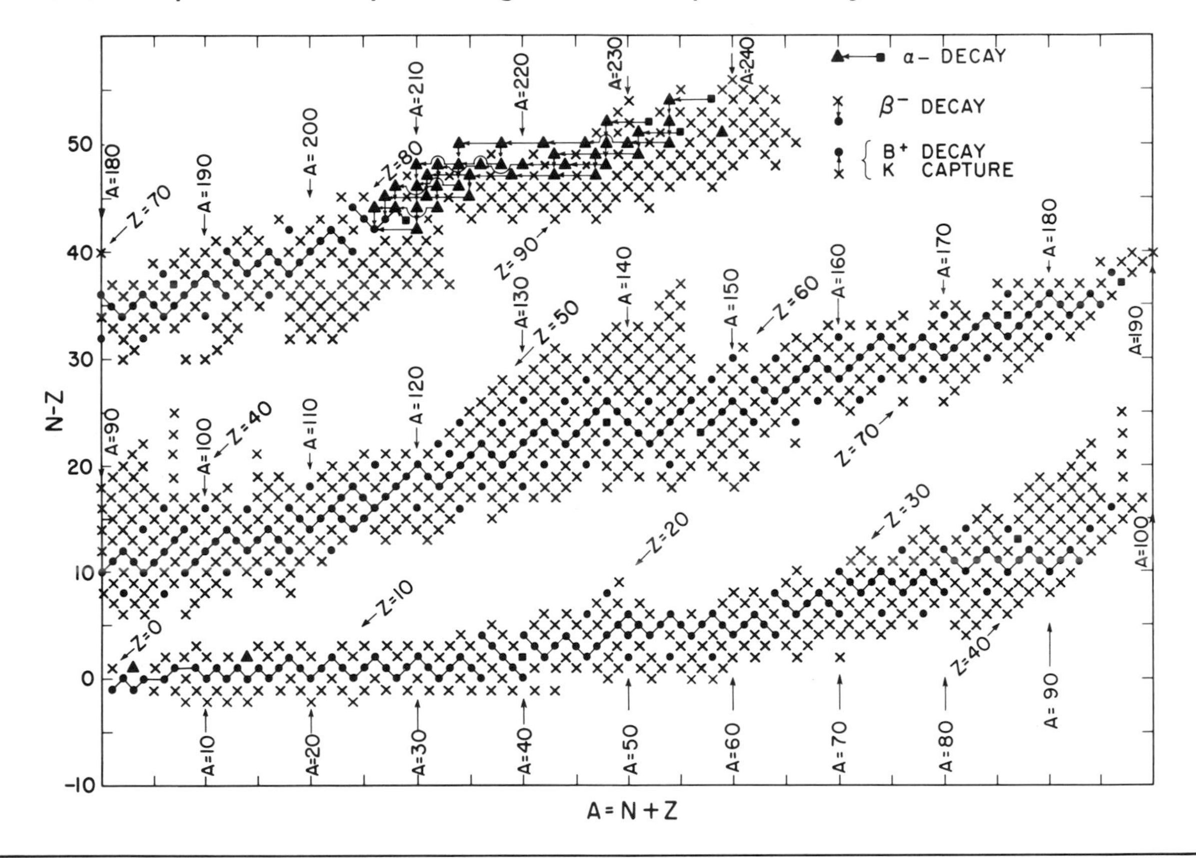

PART 3

ATOMIC STRUCTURE

Basic Concepts

The electrons of an atom are attracted to the nucleus (positive core) and as a result circulate in orbits about the centralized core. Quantum mechanics points out that the electrons may not exist in any random orbit but are restricted to orbits with discrete energy levels. These so-called shells are designated by the letters, K, L, M, N, O, P, and Q, in this order. The K shell, being closest to the core, will contain electrons of highest energy, and consequently is most tightly bound to the atom. All orbital electrons are characterized by certain quantum conditions that restrict their energy, angular momentum, spin, and the number of electrons that may be contained in any one shell. For uranium, the K shell has 2 electrons; L shell, 8; M shell, 18; N shell, 32; O shell, 18; P shell, 12; and Q shell, 2.

Rare Earths and Actinides

In general, as the number of protons in the nucleus increases, the shells are filled in order from K to Q. However, in two pronounced cases this does not occur. In the case of the rare earths, the P shell has 2 electrons, the O shell has either 8 or 9 electrons, and the N shell is being filled. Since the chemical properties of an atom depend only on the number of outermost electrons, the rare earths have the same chemical behavior and therefore occupy the same place in the periodic table. A similar set of circumstances accounts for the position of the actinides; in this case, the P and Q shells are filled but the O shell is not.

Periodic Table

The table of chemical elements is arranged according to a regular periodic recurrence of chemical and physical properties. In this arrangement, the number of the group, 0 to 8, always indicates the number of electrons in the outermost electron subshell. Group 0 contains those elements that have completed shell structures and therefore are very inactive chemically. These are the rare or inert gases. Group 1 contains the alkali metals, which have completed shells plus 1 electron. Group 2 contains the alkaline earth metals, which have completed shells plus 2 electrons. Group 3 contains the aluminum family. Included in this group are the rare earths and actinides; even though they do not have completed inner shells, they have 3 electrons in their outermost shells and thus have the same general chemical properties. The only other group to have been given a name is group 7, usually called the halogens, which have completed shells plus 7 electrons.

Figure 2 shows the periodic table of elements arranged in the order of increasing atomic number, which is indicated in the upper portion of the block. The atomic weight is listed in the lower portion of the block. The letters A and B indicate subgroups having similar chemical and physical properties within the general groups.

Physical Constants

The more important physical constants used in nondestructive testing with penetrating radiation and high-energy particles, which can be used to determine additional useful constants, are given in Table 6.

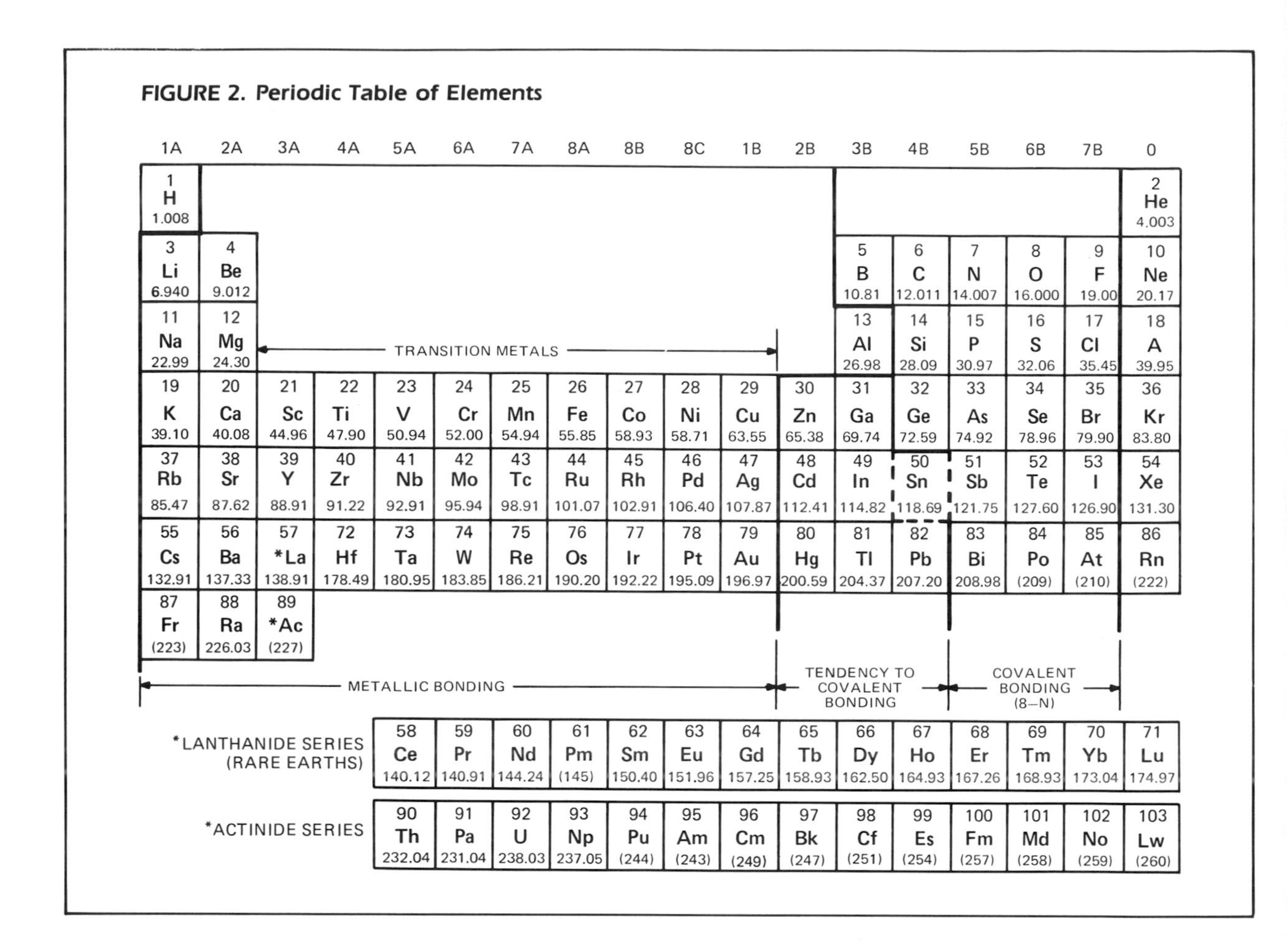

FIGURE 2. Periodic Table of Elements

TABLE 6. Physical Constants in SI Units

Symbol	Name	Value	Units
N_A	Avogadro constant	6.022045×10^{23}	mol^{-1}
c	Speed of light in vacuum	2.997925×10^{8}	$m \cdot s^{-1}$
e	Elementary charge	1.602189×10^{-19}	C
m_e	Electron rest mass	9.109534×10^{-31}	kg
h	Planck constant	6.626176×10^{-34}	$J \cdot Hz^{-1}$
λ_{ce}	Electron Compton wavelength	2.426309×10^{-12}	m
a_o	Bohr radius	$0.5291771 \times 10^{-10}$	m
r_o	Classical electron radius	2.817938×10^{-15}	m
H	Atomic weight of hydrogen	1.008128	u
U	Atomic mass unit	$1.6605655 \times 10^{-27}$	kg
R	Ratio of proton mass to electron mass	1836.15152	—
E_e	Energy of electron mass	0.511004	MeV
E_p	Energy of proton mass	938.2592	MeV
λ_{cp}	Proton Compton wavelength	1.32141×10^{-15}	m
λ_o	Wavelength associated with 1 eV	1.239854×10^{-6}	m
	Energy associated with 1 eV	1.6021917	J

PART 4

ELECTROMAGNETIC RADIATION

The Photon

The German physicist Roentgen discovered X rays in 1895 while working with a high-voltage gaseous discharge tube. Electrons were emitted from a cathode and accelerated toward a target, which they struck with a high velocity. He found that a very penetrating radiation was emitted from this bombarded target. Almost immediately this radiation was used to penetrate the human body and solid materials, and radiology and radiography were well on their way to becoming important parts of their respective fields, medicine and nondestructive testing.

Further research over the many years since Roentgen's discovery indicated that the radiation photon has a dual character, acting sometimes like a particle and at other times like a wave. This duality was hinted at by the quantum theory as put forth by Planck at the turn of the century. Planck postulated that the photon energy was contained in an electromagnetic packet of radiation, or quantum, and was proportional to its frequency. His equation $E=h\nu$ (where E is the energy associated with the quantum, ν is its frequency, and h is Planck's constant) has been used successfully to explain many physical phenomena.

All radiant energy has like characteristics and varies only in frequency. Photons all have equal velocity, have no electric charge, and have no magnetic moment. Photon characteristics are listed in Table 7. A chart of the radiation spectrum is shown in Fig. 3. Only a very small region of the chart is occupied by the visible spectrum.

TABLE 7. Photon Characteristics

Velocity	c
Frequency	$\nu = c/\lambda$
Wavelength	$\lambda = c/\nu$
Mass	$h\nu/c^2$
Momentum	$h\nu/c$
Energy	$h\nu$

X Rays and Gamma Rays

X rays are a form of electromagnetic radiation having wavelengths in the region of 10^{-13} to 10^{-9} m. They are usually produced by allowing a stream of high-energy electrons to impinge on a metallic target, thus producing photons by deceleration of the electrons. (See the discussion under Bremsstrahlung.) Gamma rays are electromagnetic radiation of nuclear origin and have very short wavelengths. Whereas X rays originate in the extranuclear structure of the atom, gamma rays are emitted by atomic nuclei in the state of excitation. The emission of gamma rays usually occurs in close association with the emission of alpha and beta particles.

FIGURE 3. The Electromagnetic Spectrum

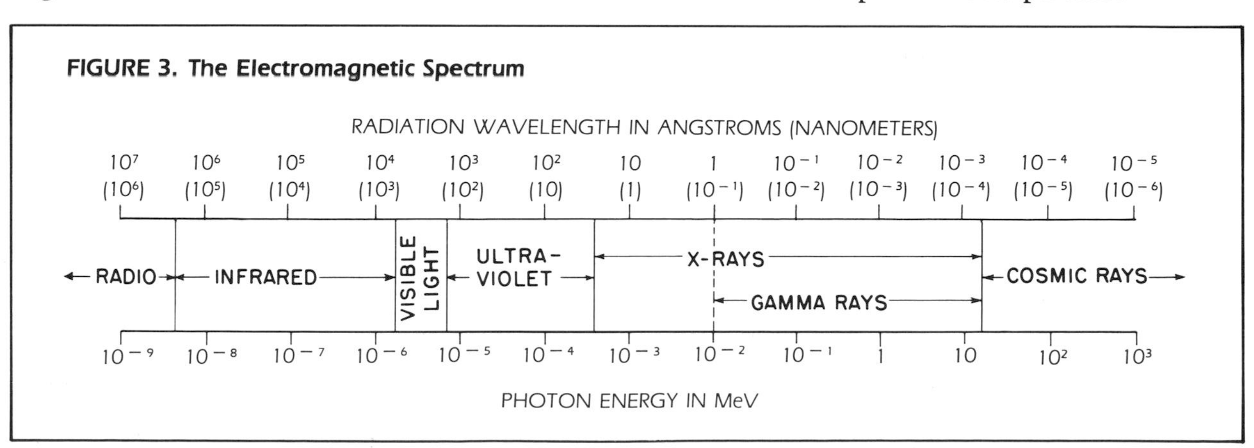

Generation of X rays

X rays are emitted whenever matter is bombarded by a stream of electrons. If it is assumed that an electron starts with zero velocity at the surface of the cathode, its kinetic energy upon arrival at the target of an electrostatic X-ray tube (in joules) is:

$$W = \tfrac{1}{2}\, m_e v^2 = eV \qquad \text{(Eq. 1)}$$

where

W = electron energy, in joules
m_e = mass of the electron
v = electron velocity (small compared with the velocity of light)
e = charge of the electron, in coulombs
V = applied potential difference between the cathode and the target, in volts.

Transformation of Electron Energy into X Rays

When an electron with kinetic energy eV strikes the target of an X-ray tube, the energy may be transformed in several ways. The simplest transformation occurs when the electron interacts directly with the nucleus of a target atom. The electron is stopped by the nucleus, which, due to its heavy mass, is not appreciably disturbed and so gains no energy. Hence, all the kinetic energy of the electron is transformed into a quantum of radiation whose minimum wavelength (in angstroms) is:

$$\lambda_{min} = \frac{hc}{eV} = \frac{12{,}395}{V} \qquad \text{(Eq. 2)}$$

where

h = Planck's constant
c = the velocity of light

Production of Continuous X Radiation

Most of the impinging electrons interact with electrons associated with the target atoms. Only a part of the energy of a high-speed electron is required to remove an electron from an atom. When an impinging electron has lost some of its energy in this way and then is suddenly stopped by an atomic nucleus, the energy that is transformed into an X-ray photon is less than the original kinetic energy, eV, of the electron. The quantum of radiation produced in this manner has a wavelength greater than λ_{min}. In general, X rays of many wavelengths are emitted. The X-ray spectrum is continuous.

When an electron is removed from an atom of the target, the atom is left in an unstable state with greater than normal energy. If an electron replaces the ejected electron, the atom returns to its normal state. It emits one or more photons with an energy hc/λ corresponding to a wavelength λ characteristic of the element. These narrow bands of wavelengths are called characteristic spectral lines. They are of higher intensities than the continuous spectrum. Both phenomena occur at the same time in the operation of an X-ray tube. The complete X-ray spectrum is shown in Fig. 4.

FIGURE 4. Complete X-ray Spectrum

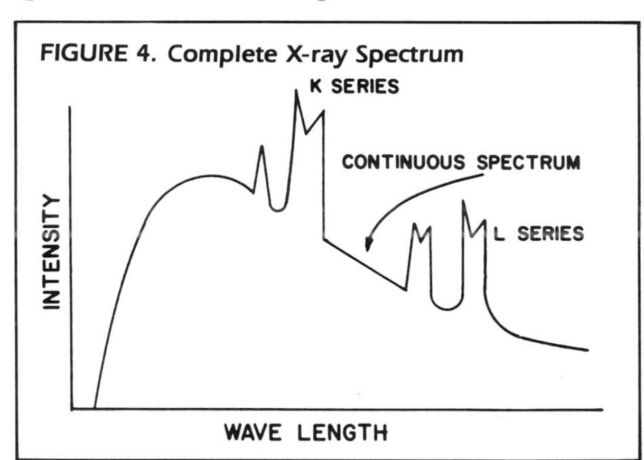

Production of Monochromatic X Radiation

Homogeneous, monochromatic X rays may be obtained from such spectra by using either a double spectrometer or filters.

Double Spectrometer

At a given angular setting θ of a crystal grating with constant spacing d, only rays with a certain wavelength λ can be reflected at the definite angle 2θ from the primary undeviated beam, according to Bragg's law (where n is an integer):

$$n\lambda = 2d \sin\theta. \qquad \text{(Eq. 3)}$$

For a simple cubic lattice the diffracted intensity is

approximately

$$I_n(h,k,l) \sim \frac{1}{n(h^2 + k^2 + l^2)}, \qquad \text{(Eq. 4)}$$

where n is the order of diffraction, and h, k, l are the Miller indices. Consequently, a second spectrometer or other apparatus can be adjusted to receive the purely monochromatic beam. Table 8 shows the Bragg angles for various crystals and $K\alpha$ radiations.

Filters

Some metals absorb more of certain wavelengths than others and show characteristic absorption edges. If a molybdenum-target tube is excited at 30 kV, the spectrum shows the K-series lines superimposed on the continuous spectrum (Fig. 3). If the first part of this band and the $K\beta$ lines could be suppressed, the $K\alpha$ line would be left in essentially undiminished intensity. Zirconium has a K critical absorption wavelength of 0.6888 Å, lying between $K\beta$ and $K\alpha$ wavelengths of molybdenum (i.e., between 0.631 and 0.708 Å). Table 9 shows critical absorption wavelengths for the K, L, and M series. A thin zirconium screen will absorb practically all radiation with wavelengths shorter than 0.6888 Å, but will be transparent to the most intense $K\alpha$ line. In practice, a zirconium filter is used with a molybdenum target, nickel with a copper target, and manganese with an iron target. Table 10 lists filters for obtaining nearly monochromatic X rays.

Generation of Gamma Rays

Large numbers of isotopes have been identified that do not retain their identity for indefinite periods of time. The nuclei of these isotopes disintegrate by the emission of particles or by K electron capture (radioactive disintegration), either of which changes the atomic number of the isotope and may change the atomic weight. The following general types of spontaneous decay of nuclei have been observed: alpha-particle emission, beta- and positron-particle emission, and K capture. The earlier discussion of nuclear particles provides a background for the understanding of nuclear-particle emission. K capture, however, is not quite so straightforward. It occurs when a nucleus captures, or absorbs, one of the electrons in the K shell of the atom.

The emission of an alpha particle decreases the atomic weight by 4 units and the atomic number by 2 units. The emission of either a beta particle, a positron, or K capture leaves the atomic mass essentially unchanged; but the first increases the atomic number by 1 unit, whereas the other two decrease the atomic number by 1 unit. The emission of a proton decreases the atomic mass and number by 1 unit, whereas the emission of a neutron reduces only the atomic mass by 1 unit.

Secondary Effects

Two secondary effects that sometimes play an important role in the radiation observed from a given isotope are annihilation radiation and internal conversion. In the spectra of all positron emitters, gamma radiation with an energy of 0.51 MeV is observed;

TABLE 8. Bragg Angles for Various Crystals and $K\alpha$ Radiations

Crystals	Reflection	Radiation $K\alpha$ (degrees)						
		Mo	Cu	Ni	Co	Fe	Mn	Cr
α-alumina	002	1.8	3.9	4.2	4.6	4.9	5.4	5.8
Gypsum	020	2.7	5.8	6.3	6.8	7.3	8.0	8.7
β-alumina	004	3.6	7.9	8.5	9.1	9.9	10.8	11.8
Pentaerythritol	002	4.7	10.1	10.9	11.8	12.8	13.9	15.2
Quartz	100	6.1	13.3	14.4	15.6	16.9	18.4	20.1
Fluorite	111	6.5	14.1	15.2	16.5	17.9	19.5	21.3
Urea nitrate	002	6.5	14.2	15.3	16.6	18.0	19.6	21.4
Calcite	200	6.7	14.7	15.8	17.1	18.6	20.3	22.2
Rock salt	200	7.2	15.9	17.1	18.5	20.1	21.9	24.0
Diamond	111	9.9	22.0	23.8	25.8	28.1	30.8	39.9

TABLE 9. Critical Absorption Wavelengths in Angstroms

K Series			
Element	Wavelength	Element	Wavelength
12 Mg	9.5304	35 Br	0.9101
13 Al	7.9470	40 Zr	0.6888
17 Cl	4.4027	42 Mo	0.6198
22 Ti	2.4973	45 Rh	0.5338
23 V	2.2690	46 Pd	0.5092
24 Cr	2.0701	47 Ag	0.4858
25 Mn	1.8964	53 I	0.3746
26 Fe	1.7433	56 Ba	0.3315
27 Co	1.6081	74 W	0.17837
28 Ni	1.4880	78 Pt	0.15849
29 Cu	1.3804	79 Au	0.15375
30 Zn	1.2833	82 Pb	0.1413
31 Ga	1.1957		
32 Ge	1.1165		

L Series			
Element	L_I (L_{11})	L_{II} (L_{21})	L_{III} (L_{22})
47 Ag	3.2540	3.5138	3.6983
53 I	2.3887	2.5526	2.7194
56 Ba	2.0662	2.2037	2.3616
74 W	1.0226	1.0735	1.2140
78 Pt	0.8939	0.9340	1.0731
82 Pb	0.7822	0.8152	0.9519
92 U	0.5698	0.5932	0.7231

M Series					
Element	M_I	M_{II}	M_{III}	M_{IV}	M_V
74 W	4.39	4.84	5.46	6.63	6.86
83 Bi	3.106	3.349	3.897	4.583	4.773
90 Th	2.393	2.576	3.064	3.559	3.729
92 U	2.233	2.390	2.879	3.333	3.498

For conversion of angstroms to picometers, multiply values by 10^2.

this radiation is the result of the annihilation of an electron-positron pair. Conservation of momentum leads to the emission of two photons with an energy of 0.51 MeV rather than a single 1.02 MeV photon. During disintegration, the reaction is often accompanied by the release of gamma rays, required to provide the proper binding energies for the residual nucleus. Internal conversion becomes important when a nucleus emits gamma radiation having an energy such that absorption in the electronic structure of the same atom is highly probable. In such a case, the gamma radiation may be strongly absorbed and an electron emitted. When the electronic structure returns to the ground state, characteristic radiation will be emitted, the energy of which will depend on the particular levels involved.

Natural Radioactive Substances

Most of the radioactive isotopes known today are artificially produced by various reactions. Historically, however, radioactive materials were known long before atomic piles and particle accelerators were built.

Radium

Until recently, the most important naturally occurring radioactive isotope was radium. Investigations, following the discovery by Becquerel in 1896 that uranium salts emit some sort of penetrating radiation, led to the isolation of radium and the study of its spectrum. This turned out to be quite complex, with alpha, beta, and gamma radiation of various energies present. Further investigations disclosed that if radium is chemically separated and then chemically analyzed again after a few days, other elements will be present. At the present time it is known that radium is a daughter product of uranium-238 and is a member

TABLE 10. Filters for Obtaining Nearly Monochromatic X-Rays

Target	Lowest Approx. Voltage for K Series (kV)	λ for Kα Doublet (Å)	Filter	Thickness (mm)	Mass (mg/cm²)
Chromium	6	2.287	Vanadium	0.0084	4.8
Iron	7	1.935	Manganese	0.0075	5.5
Copper	9	1.539	Nickel	0.0085	7.6
Molybdenum	20	0.710	Zirconium	0.037	24
Silver	25	0.560	Palladium	0.03	36

of a chain of radioactive elements. Figure 5 is a diagram of this chain. Two other series exist having similar characteristics: one starting with the element uranium-235 and the other starting with the element thulium-232. The uranium-238 series illustrates the essential features of these three series and is presented alone for a clearer diagram.

If radium is separated chemically, only those elements below it in the series can be present from that time on. All the half-lives below radium are very short compared with the half-life of radium; therefore, the daughter products can build up to equilibrium levels. One daughter element, radon, is a gas, so that the radium must be packaged to prevent the escape of this gas and to permit equilibrium to be reached for a given source. Packaged radium has been very widely used in the medical field and for industrial radiography. The presence of the many daughter products leads to a high specific activity for radium, making it a good material for a radiographic source. Radium has a half-life of about 1600 years, which partly offsets its high cost.

Radon

A second way in which radium is used is as a source of radon-222 gas. Radon is allowed to build up to equilibrium with radium and is then removed and packaged. This process can be carried out an indefinite number of times and the resulting source has a very high specific activity. The short half-life of 3.83 days limits the applications for this material.

Artificially Produced Radioactive Isotopes

In recent years a large number of radioactive isotopes has been artificially produced in various types of accelerators and in atomic piles. The atomic pile at Oak Ridge, TN, is the source of a multitude of isotopes possessing various characteristics.

Artificially produced radioactive isotopes are widely used as sources of radiation for radiography, gaging, and as tracers for a variety of measurements not easily made by other methods. The listed references may be consulted for detailed information.

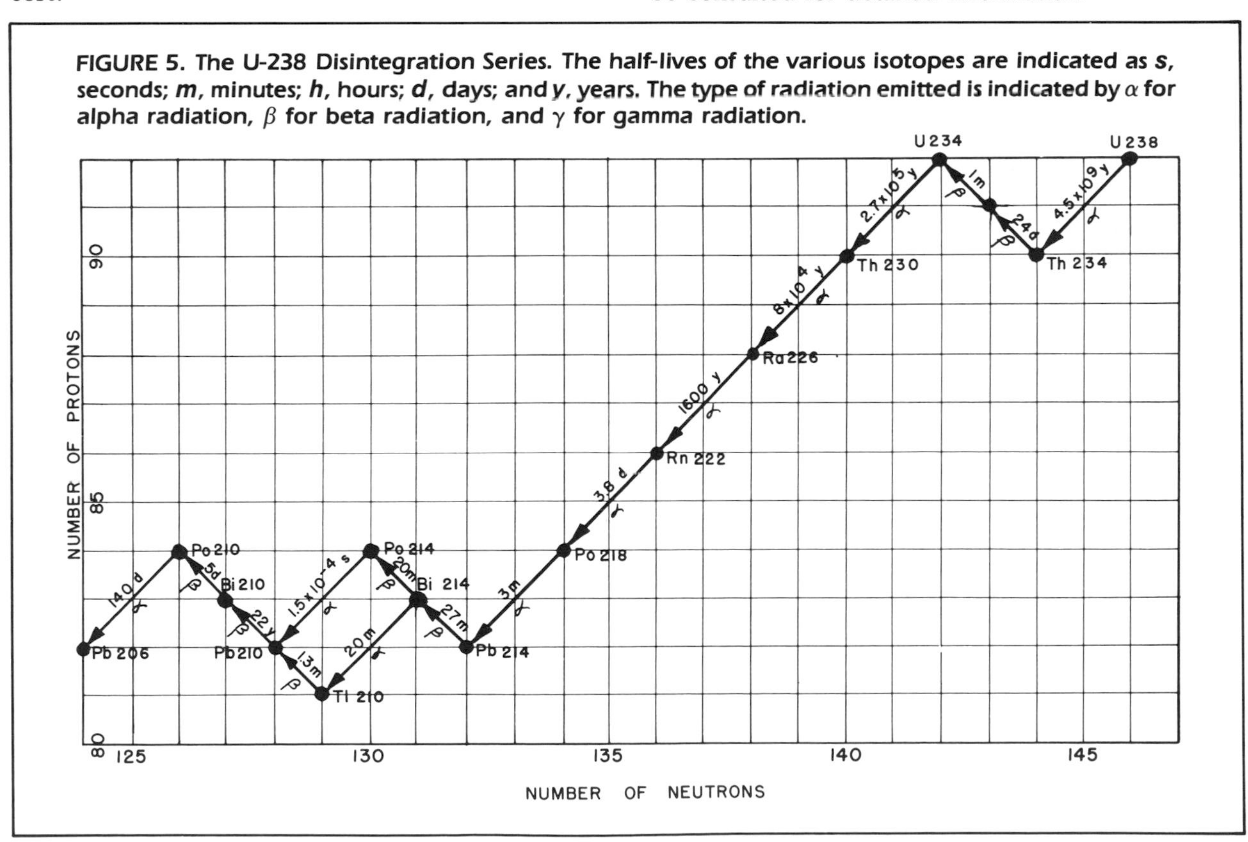

FIGURE 5. The U-238 Disintegration Series. The half-lives of the various isotopes are indicated as *s*, seconds; *m*, minutes; *h*, hours; *d*, days; and *y*, years. The type of radiation emitted is indicated by α for alpha radiation, β for beta radiation, and γ for gamma radiation.

PART 5

RADIATION ABSORPTION

Categories of Absorption

Radiation absorption can be separated into three major categories: the absorption of photons, the absorption of charged particles, and the absorption of neutrons.

The absorption of photons has played the most important role in the field of nondestructive testing. Almost from the very day that Roentgen announced his experimental results, applications have come forth in the fields of radiology, radiography, medicine, experimental physics, and thickness gages, to name a few. Therefore, the major emphasis here is on the physical characteristics of the absorption of X rays.

Absorption of Photons

A beam of X or gamma rays exhibits a characteristic exponential absorption in its passage through matter. This is a consequence of the fact that usually a photon is removed from a beam by a single event. Such an event is a result of the photon's interaction with a nucleus or an orbital electron of the absorbing element and can be classified as one of three predominant types: the photoelectric effect, scattering, and pair production. A fourth but minor contributor to absorption is photodisintegration.

The following analysis of photon absorption assumes a narrow-beam geometry; that is, any photon that is deflected, however small the angle, is considered completely absorbed. Since the number of photons removed from a monoenergetic beam at a thickness x of an absorber is proportional to the intensity at that thickness $I(x)$, the number of atoms per cubic centimeter n, and the incremental thickness of material traversed dx, the change in the beam intensity in dx may be expressed as

$$dI\,(x) = -I\,(x)\; n\sigma\, dx. \qquad \text{(Eq. 5)}$$

Here, σ is a proportionality constant and is interpreted as the total probability (more often referred to as the cross-section because it has the dimensions of an area) per atom for scattering or absorption of a photon of the original energy. Integration of eq. 5 gives

$$I\,(x) = I_o e^{-n\sigma x}, \qquad \text{(Eq. 6)}$$

where I_o is the intensity of the incident beam. The number of atoms per cubic centimeter n can also be expressed as $N\rho/A$, where N is Avogadro's number, A is the atomic weight of the absorption material, and ρ is the density of the material. The exponential factor in eq. 6 can be expressed as

$$-n\sigma x = -\left(\frac{N\,\sigma}{A}\right)\rho x = -\frac{N\,\sigma\,\rho}{A}\,x\,. \qquad \text{(Eq. 7)}$$

where σ is the atomic attenuation coefficient; $N\sigma/A$ (or μ/ρ) is the mass attenuation coefficient; and $N\sigma\rho/A$ (or μ) is the linear attenuation coefficient (see Section 20).

The total attenuation coefficient consists of the sum of the attenuation coefficients for each of the various processes mentioned above. That is, the total atomic attenuation coefficient is given by

$$\sigma = \sigma_{ph} + \sigma_s + \sigma_{pr} + \sigma_{pd}. \qquad \text{(Eq. 8)}$$

Here, σ_{ph} is the attenuation coefficient due to the photoelectric effect, σ_s that due to scattering, σ_{pr} that due to pair production, and σ_{pd} that due to photodisintegration.

Photoelectric Effect

The photoelectric effect is defined as that process in which a photon of energy E_o transfers its total energy to an electron in some shell of an atom (Fig. 6). This energy may be only sufficient to move the electron from one shell to another, or it may be sufficient to remove the electron completely from (i.e., to ionize) the atom. In the latter case, the kinetic energy of the ejected electron is just the difference between the photon's energy and the binding energy of that particular electron in the atom.

FIGURE 6. Photoelectric Interaction of an Incident Photon with an Orbital Electron

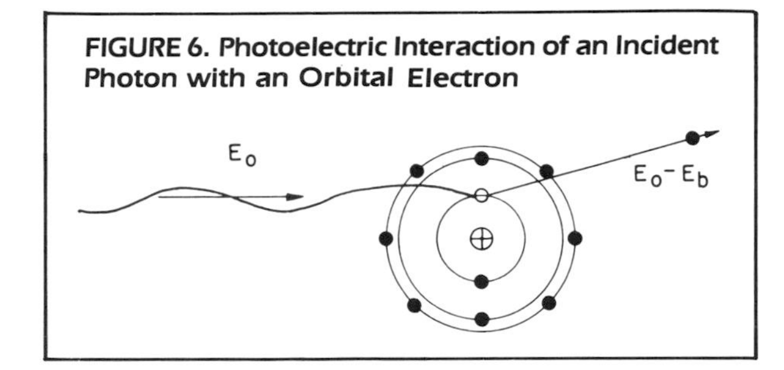

As the photon energy is increased from zero, the photons are absorbed by electrons in deeper-lying shells of the atom. When the photon energy reaches the binding energy of a particular shell of electrons, there is an abrupt increase in the absorption. The energy at which this sharp change occurs for *K* electrons, called the *K* absorption edge, represents a situation where the kinetic energy of the ejected *K* electron is zero. Further increase of the photon energy causes the absorption to decrease approximately inversely with the cube of the energy. The photoelectric attenuation coefficient for an atom may be expressed as the sum of the coefficients representing the contributions of the *K*, *L*, etc., electron shells. The energy dependence for the photoelectric absorption in uranium is shown in Fig. 7. The solid curve represents the total attenuation coefficient and the dotted curves show the components. The total scattering curve includes the Compton effect, coherent scattering, and a correction for the average binding of the electrons in their shell structure. The absorption increases so rapidly with the atomic number Z (between Z^4 and Z^5) that for heavy elements it remains appreciable up to energies of the order of a few million electron volts.

FIGURE 7. Curves for Uranium, Showing the Various Components of the Total Attenuation Coefficient as a Function of Energy

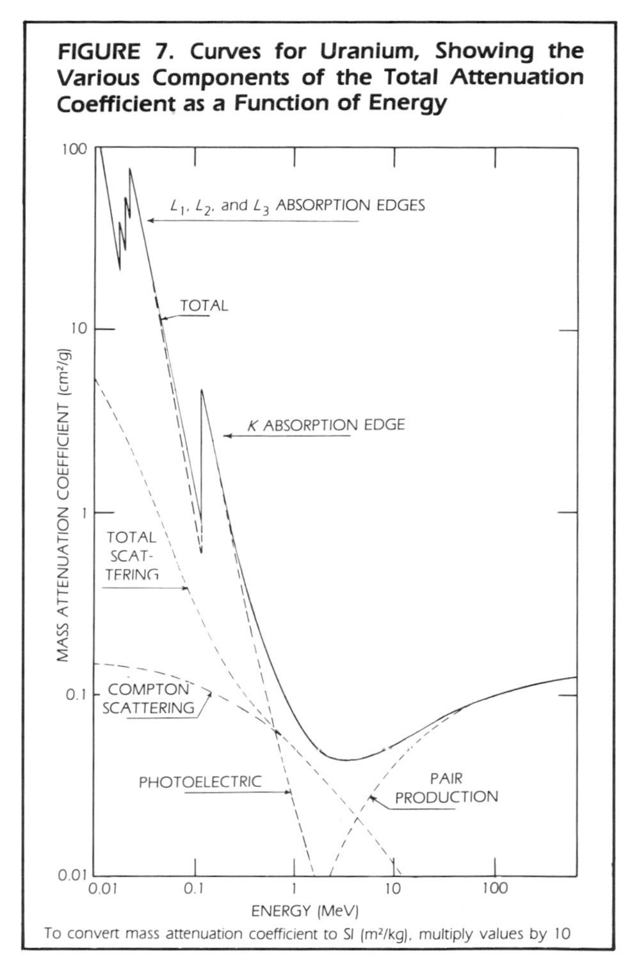

To convert mass attenuation coefficient to SI (m²/kg), multiply values by 10

Scattering of Photons

Upon increasing the photon energy past the *K* edge, the main process contributing to absorption changes from the photoelectric effect to the Compton effect. This is really not true absorption, since part of the photon's energy is not absorbed but merely redirected (Fig. 8). In the Compton effect a photon collides with an electron. Instead of giving up all its energy to the electron as in the photoelectric process, however, the photon only shares its energy with the struck electron (see Table 11). The binding energy of the electron is usually considered negligible compared with the photon's energy.

Compton Scattering

Analysis of the Compton process shows that the energy of a scattered photon is always less than that of the primary photon; the effect is, therefore, one of incoherent scattering. The famous analysis made by Compton was based on the "particle" properties of the photon and the conservation of energy and momentum during the collision. The energy shift predicted depends only on the angle of scattering and not on the nature of the scattering medium, the larger energy shifts being due to the larger angles through which the incident photon is scattered. A complete analysis of the Compton energy-angle relationship and the Klein-Nishina formula for calculating the associated scattering cross sections may be found in the literature.

Figure 7 illustrates the decrease in the attenuation coefficient, attributable to the Compton effect, as the photon energy increases. This decreasing of the number of scattered photons is equivalent, of course, to an increase in the penetrating power of the photons with increasing energy.

FIGURE 8. Compton Scattering in Which an Incident Photon Ejects an Electron and a Lower-Energy Scattered Photon

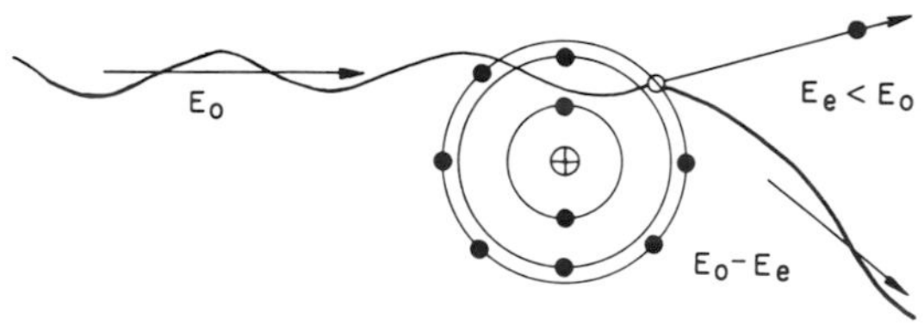

Coherent Scattering

If a photon does not experience an energy shift upon being scattered by an atom, the process is spoken of as being coherent. This phenomenon is often referred to as the Rayleigh process. It can only occur for soft radiation for which the binding energy of the electrons in their atomic shells is important. Classically speaking, the atomic electrons are set into oscillation by the absorption of the incident photon. Then, acting as a common source, they emit a photon of the same frequency as the

TABLE 11. Free Electron Compton Scattering in Uranium. Calculated from the Klein-Nishina formula and the Compton relationship between before-and-after-collision energies of the photon

Energy (MeV)	Cross Sections (Barns: 10^{-28} m^2)			Fraction of Energy Retained by	
	Absorption	Scattering	Total	Photon	Electron
0.01	0.0077	0.629	0.637	0.987	0.013
0.015	0.0138	0.613	0.627	0.978	0.022
0.02	0.0196	0.596	0.616	0.968	0.032
0.03	0.0295	0.566	0.596	0.950	0.050
0.04	0.0380	0.540	0.578	0.934	0.066
0.05	0.0451	0.516	0.561	0.920	0.080
0.06	0.0509	0.495	0.546	0.907	0.093
0.08	0.0610	0.456	0.517	0.882	0.118
0.10	0.0685	0.424	0.493	0.860	0.140
0.15	0.0812	0.363	0.444	0.818	0.182
0.20	0.0886	0.318	0.407	0.781	0.219
0.30	0.0958	0.258	0.354	0.729	0.271
0.40	0.0982	0.219	0.317	0.691	0.309
0.50	0.0986	0.190	0.289	0.657	0.343
0.60	0.0984	0.170	0.268	0.634	0.366
0.80	0.0959	0.139	0.235	0.591	0.409
1.0	0.0929	0.118	0.211	0.559	0.441
1.5	0.0849	0.0867	0.172	0.504	0.496
2.0	0.0777	0.0687	0.146	0.471	0.529
3.0	0.0664	0.0487	0.115	0.423	0.577
4.0	0.0582	0.0378	0.0960	0.394	0.606
5.0	0.0519	0.0309	0.0828	0.373	0.627
6.0	0.0471	0.0261	0.0732	0.357	0.643
8.0	0.0399	0.0200	0.0599	0.334	0.666
10.0	0.0349	0.0161	0.0510	0.316	0.684
15.0	0.0268	0.0109	0.0377	0.289	0.711
20.0	0.0220	0.00823	0.0302	0.273	0.727
30.0	0.0164	0.00556	0.0220	0.253	0.747

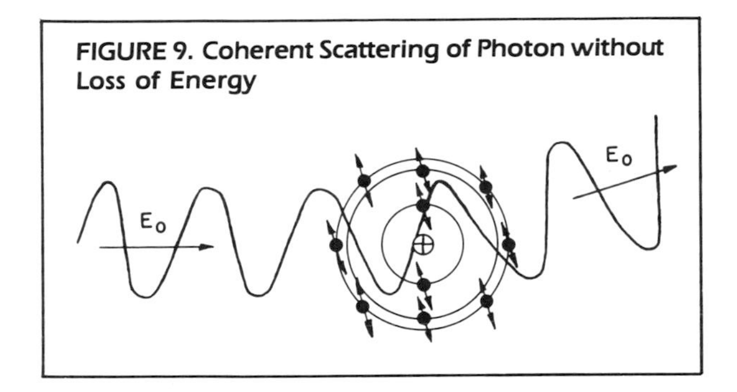

FIGURE 9. Coherent Scattering of Photon without Loss of Energy

incident photon (Fig. 9). The contribution of this type of coherent scattering to the total attenuation coefficient is never greater than about 20 percent.

Coherent scattering can also occur from the atoms in a crystalline structure. However, because of the many ways an atomic system may be ordered in nature and because this type of scattering is primarily contained in very small angles with respect to the incident photon, the effect has little importance in increasing the attenuation coefficient, and therefore it will not be considered further here.

Pair Production

Very high-energy photons are absorbed in matter by a process in which a photon is converted in the electrical field of a nucleus into an electron and a positron (Fig. 10). This is pair production. Since both members of the pair have a nonzero mass, there is a minimum energy corresponding to creation of the rest masses of the two particles, below which pair production cannot occur. The value of this threshold is 1.02 MeV. Any energy in excess of that needed for creating an electron-positron pair appears as kinetic energy of the pair particles. The probability of occurrence of the process increases approximately logarithmically with energy above the threshold value and then levels off for extremely high-energy photons. This process may also occur in the field of an orbital electron and in this case is sometimes referred to as triplet production. However, the third electron, which is often seen because of its large recoil energy, is not created in the process but is the orbital electron. The combined effect of these two pair-production processes is plotted in Fig. 7.

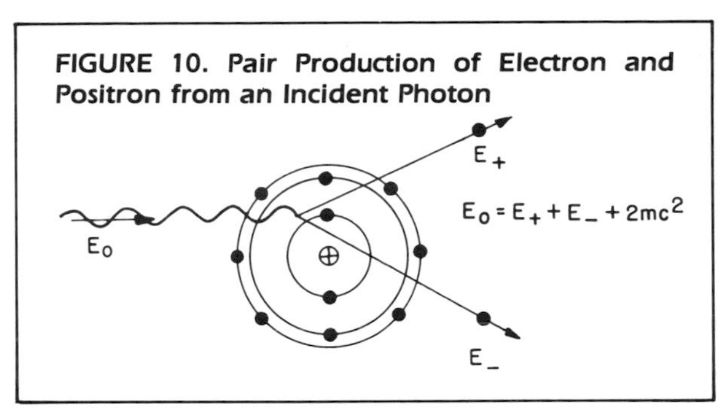

FIGURE 10. Pair Production of Electron and Positron from an Incident Photon

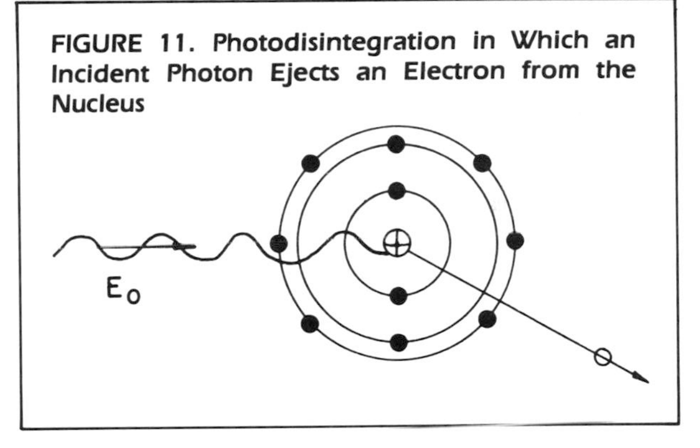

FIGURE 11. Photodisintegration in Which an Incident Photon Ejects an Electron from the Nucleus

Photodisintegration

In addition to the three dominant photon absorption processes already discussed, the effects of photodisintegration of nuclei are of interest. Here, a photon is captured by a nucleus, which then loses one or more of its constituent particles (Fig. 11). This may be thought of as the nuclear analog of the photoelectric effect. This effect, however, is very small, and its first maximum, usually between 10 and 20 MeV, is only a few percent of the attentuation coefficient.

Z Dependence and Energy Dependence

It is apparent that the attenuation coefficient depends very strongly upon the atomic number Z, varying approximately as $Z(Z + 1)$ for pair production, directly with Z for the Compton effect, and between Z^4 and Z^5 for the photoelectric effect. The attenuation coefficient decreases with increasing energy for the photoelectric effect as $E^{-3.5}$, decreases for the Compton effect in the region of interest roughly as E^{-1}, and increases approximately logarithmically with energy for pair production.

Attenuation Coefficients of the Elements

The section dealing with attenuation coefficient tables gives the radiation attenuation coefficients of most of the elements of importance in nondestructive testing. For elements not listed, the attenuation coefficients can be calculated by direct interpolation, using the proper Z dependence and energy dependence for the various components of the total attenuation coefficients. Some care should be exercised in using these tables, since they are based on narrow-beam absorption.

Attenuation Coefficients of Compounds and Mixtures

The mass attenuation coefficient μ/ρ of a compound or a mixture is the average of the mass attenuation coefficients of the constituent elements $\mu/\rho(a)$, $\mu/\rho(b)$, etc., weighted in proportion to their relative abundance by weight, $R(a)$, $R(b)$, etc., i.e.,

$$\mu/\rho = \mu/\rho(a)R(a) + \mu/\rho(b)R(b) + \ldots \qquad \text{(Eq. 9)}$$

Examples of a Compound

Consider the mass attenuation coefficient at 1 MeV for water, H_2O. Since the atomic weight for the hydrogen atom is approximately 1 and that for oxygen is 16, the mass attenuation coefficient for water is

$$\begin{aligned} H_2O &= 0.126\ (2/18) + 0.0636\ (16/18) \\ &= 0.0705\ \text{gram/cm}^2, \end{aligned}$$

where 0.126 and 0.0636 are the mass attenuation coefficients of hydrogen and oxygen, respectively, at 1 MeV.

Examples of a Mixture

Consider the mass attenuation coefficient of 0.02 MeV for air, which in percentages by weight consists primarily of 75.6% N_2, 23.1%, O_2, and 1.3% A. The mass attenuation coefficients are: nitrogen, 0.598 gram/cm^2; oxygen, 0.840 gram/cm^2; and argon, 8.87 gram/cm^2. Therefore, the total mass attenuation coefficient for air at 0.02 MeV is

$$\begin{aligned} \mu/\rho(\text{air}) &= (0.598 \times 0.756) + (0.840 \times 0.231) + (8.87 \times 0.013) \\ &= 0.761\ \text{gram/cm}^2. \end{aligned}$$

Linear Attenuation Coefficient

The linear attenuation coefficient μ of a compound or mixture is the mass attenuation coefficient μ/ρ multiplied by the density ρ of the compound or mixture. Using the aforementioned examples, the linear attenuation coefficients for water and air are, respectively, $0.0705 \times 1 = 0.0705\ \text{cm}^{-1}$, and $0.761 \times 0.0012 = 0.91 \times 10^{-3}\ \text{cm}^{-1}$, where 1 and 0.0012 are the respective densities in grams per cubic centimeter at standard temperature and pressure.

Multiply gram/cm^2 values by 10^{-1} to obtain kg/m^2.

PART 6

ABSORPTION OF CHARGED PARTICLES

Mechanisms of Absorption

As they pass through matter, all charged particles lose energy by ionizing or exciting the atoms and molecules. In addition, radiation loss by high-energy particles, mainly electrons and positrons, is important. This latter effect is often referred to as Bremsstrahlung (continuous-spectrum X-ray emission). These two processes, ionization and Bremsstrahlung, account for practically all the energy loss of charged particles traversing matter.

Ionization

For the most part, charged particles lose very little energy for each ionization or excitation event, compared with their own kinetic energy. Thus, the incident particle suffers little or no deflection during the process (Fig. 12). This type of absorption is basically different from that of X rays, where the photons are considered absorbed in single events.

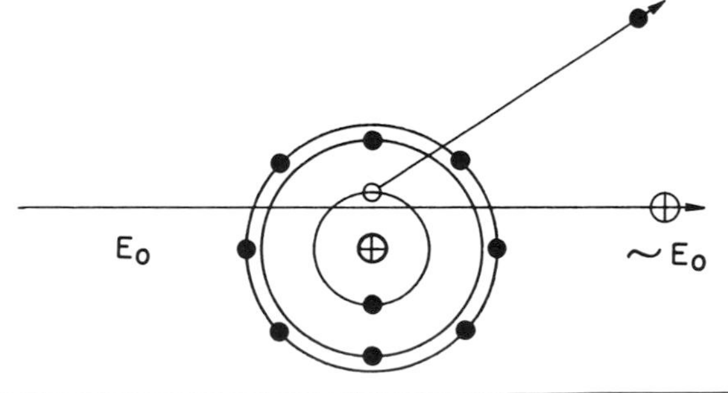

FIGURE 12. Ionization by a Charged Particle That Ejects an Orbital Electron from Atom. Definitions: *Specific ionization* is the number of ion pairs generated by the particle per unit path; *Total ionization* designates the number of ion pairs produced by the particle along its entire path; *Stopping power, F (E)*, of a substance is defined as the energy lost by the particle per unit path, *–dE/dx; Mass stopping power* is defined as the stopping power divided by the density of the substance, *F(E)*/ρ.

Therefore, a somewhat different approach is taken for analyzing the absorption of particles. Range of a particle can be defined from the stopping power by direct integration as

$$R = \int_0^{E_0} \frac{dE}{F(E)} \quad ,$$

where E_0 is the initial energy of the particle.

Ionization by Heavy Particles

In nondestructive testing the need for detailed information regarding absorption of heavy particles is small. Only in the fields of autoradiography and radiation protection is much attention paid to absorption of heavy particles, and then primarily in the energy range between 1 and 10 MeV. Inasmuch as there is no adequate theory for the stopping power for heavy particles in this general energy range, one must rely on experimentally determined values for both the stopping power and the mean range. Allison and Warshaw (Fig. 13) show the mass stopping powers of metals for protons for a representative set of elements. (For a complete discus-

FIGURE 13. The Mass Stopping Power of Several Metals for Protons of Energy Less than 2 MeV

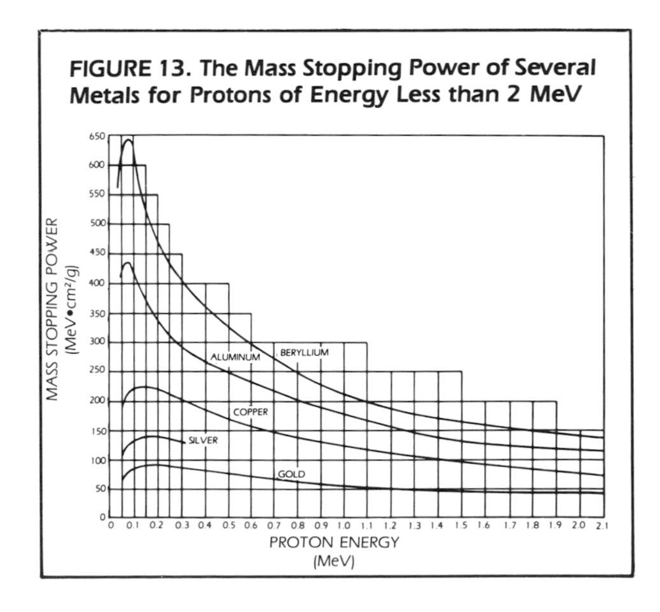

sion of the passage of heavy particles through matter, see references 1 and 2.) The mean ranges for protons in these elements can be determined from the above definition for range and from these experimental curves. For values of mass stopping power in the energy range greater than 2 MeV, but nonrelativistic, the equation

$$-\frac{dE}{dx}\frac{1}{\rho} = \frac{4\pi e^4 z^2 N Z}{Amv^2} \ln\left(\frac{2mv^2}{I_o Z}\right) \quad \text{(Eq. 10)}$$

can be used to calculate the approximate mass stopping power. In this equation, z is the atomic number of the incident particle, A is the atomic weight of the stopping material, N is Avogadro's number, Z is the atomic number of the stopping material, m is the mass of electron, and v is the velocity of the incident particle. I_o has a value ranging from 15 for beryllium to 8.8 for the heavy elements. For extremely high energies this equation must be corrected for the effects of relativity.

Ionization by Electrons

Electrons lose energy in this passage through matter due to excitation and ionization of the bound electrons in the stopping material, just as for the heavy particles. In fact, the energy loss per unit path traveled for a proton and that for an electron of the same velocity, for velocities that are nonrelativistic, are practically the same. Indeed, by omitting the factor 2 in the argument of the logarithm, eq. 10 can be used to calculate the mass stopping power for an electron or positron ($z = 1$) for energies greater than a few kiloelectron volts. For more accurate calculations, this equation must be corrected for the ultimate indistinguishability of the two electrons emerging from the ionizing collision and for the effects of relativity. The first correction is made by multiplying the argument of the logarithm of eq. 10 by $(e/16)^{1/2}$ instead of omitting the factor 2, where e is the base of the natural logarithms. This correction is not made for positrons. The second important correction is that made for the effects of relativity. The reader is referred to the literature for a complete account of this effect. Marshall and Ward show, for example, the range of electrons in aluminum (Fig. 14).

FIGURE 14. Extrapolated Range of Electrons in Aluminum vs. Energy

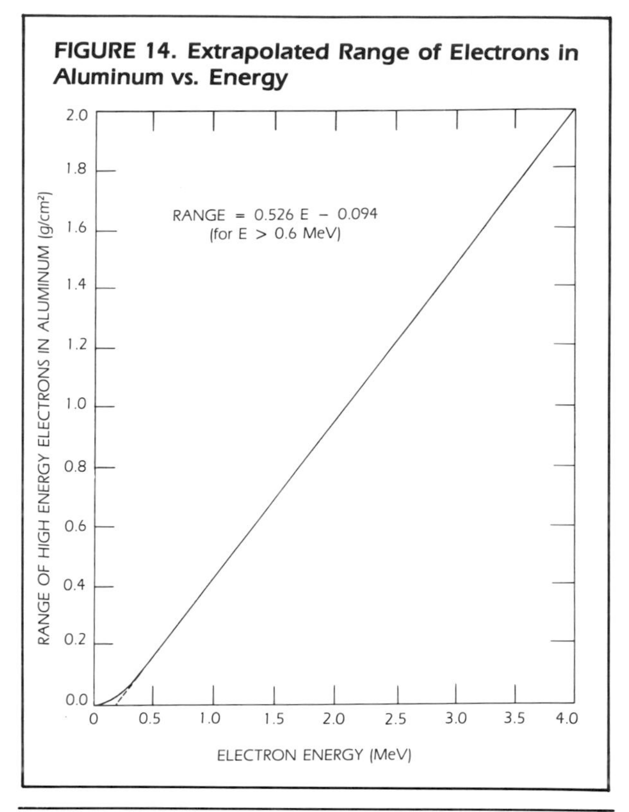

Bremsstrahlung

The energy loss of electrons at high energies is due to an entirely different mechanism; namely, the emission of electromagnetic radiation (Fig. 15). This effect is relatively unimportant for the heavy particles. Because of its small mass, an electron can experience a large deceleration in the electrical field of a nucleus. Consequently, radiation will be emitted. The resulting radiation, or Bremsstrahlung, is the dominant influence in the energy loss of fast electrons. Again, the reader is referred to the literature for a complete account of this effect. As in the case of pair production, which is almost the inverse effect, the cross section for Bremsstrahlung varies approximately with $Z(Z + 1)$, is constant for energies below approximately 1 MeV, increases logarithmically for energies greater than approximately 2 MeV, and then levels off for very high energies to values dependent on the particular stopping material.

Figure 16 shows the energy distribution of the radiation emitted by an electron due to the Brems-

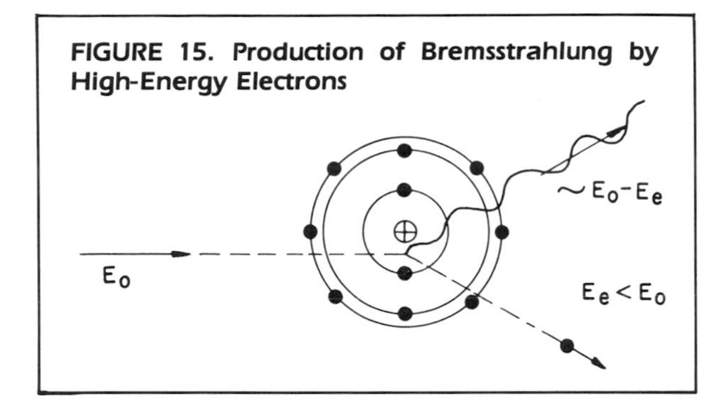

FIGURE 15. Production of Bremsstrahlung by High-Energy Electrons

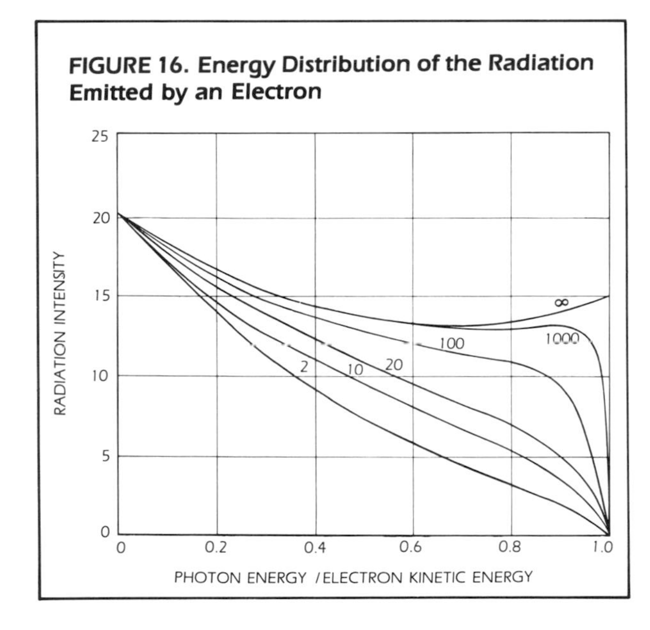

FIGURE 16. Energy Distribution of the Radiation Emitted by an Electron

strahlung in lead. The curves are for lead and include the effect of screening. The ordinate gives the number of photons times the energy of the photons for a unit frequency interval in units of mc^2. The abscissa gives the fractional photon energy in terms of the energy of the primary electron. The numbers on the curves give the primary electron energy in mc^2 units. (Data for curves obtained from Heitler.[9]) Figure 17 shows the cross section for the energy loss per centimeter path length due to Bremsstrahlung as a function of the primary energy of the electron. The unscreened curve is also plotted to show the effect of screening. (Data for curves obtained from Heitler.[9])

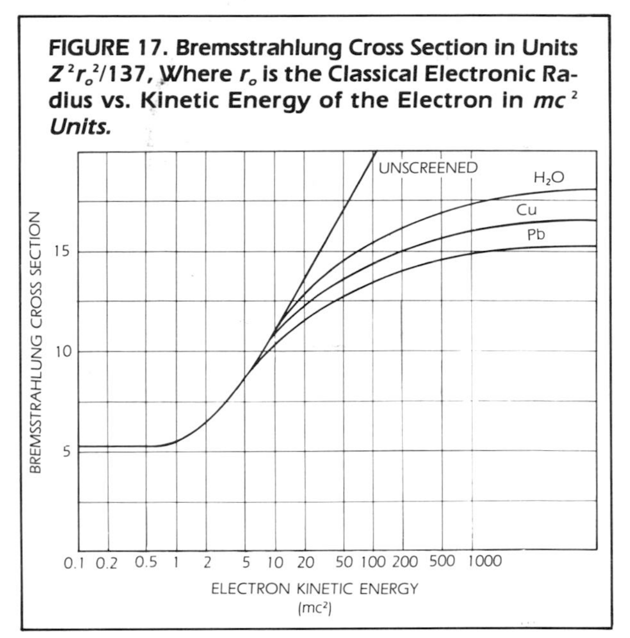

FIGURE 17. Bremsstrahlung Cross Section in Units $Z^2 r_o^2/137$, Where r_o is the Classical Electronic Radius vs. Kinetic Energy of the Electron in mc^2 Units.

Scattering of Charged Particles

The scattering of charged particles results from near collisions of incident particles with the nuclei as well as with the orbital electrons of the stopping material. The resulting average loss of energy from collisions with orbital electrons has been described in the previous paragraph on ionization effects, where scattering angles were considered negligible. For heavy incident particles, the single scattering events due to atomic electrons are never greater than a small fraction of a degree. Occasionally, however, cloud-chamber experiments have shown large-angle, single, scattering events, often with an accompanying track indicating that a collision had taken place with the nucleus. For energies that are not too high, the frequency of such collisions is governed by the well-known Rutherford scattering formula, i.e.,

$$d\phi = \left[\frac{2\,\pi\, e^4 Z^2 z^2}{16\, E^2\, \sin^4(\theta/2)}\right] \sin\theta\, d\theta, \qquad \text{(Eq. 11)}$$

where $d\phi$ is the differential cross section for scattering the incident particle into the solid angle $2\pi \sin\theta d\theta$, z is the charge of the incident particle, Z is the charge of the nucleus, and E is the kinetic energy of the incident particle.

Collisions with Electrons

The case of the incident electron is complicated by the indistinguishability between the incident electron and the orbital electron in the electron-electron scattering events. For this reason, the Rutherford formula is not adequate. By using quantum mechanics, Mott included this effect and derived a similar but somewhat more complicated formula.[15]

Collisions with Nuclei

More important than the electron-electron scattering collisions are collisions of the incident electrons with the nuclei of the stopping material, in which these electrons are scattered with no loss of energy. Since the electrons have such a small mass compared with the nuclei, the electrical field of the atoms can easily deflect the particles through large angles. The scattering will take place in the combined field of the nucleus and the orbital electrons; therefore, the electron distribution will have an appreciable effect on the differential scattering cross sections. In addition to this screening effect by orbital electrons, relativity effects further complicate the theory. Molière has given a theory for calculating such cross sections.[13]

PART 7

SECONDARY RADIATION

Sources of Secondary Radiation

Radiation arising directly from the target of an X-ray tube or an accelerator, or from a radioactive source, is usually referred to as primary radiation. On the other hand, secondary radiation is regarded as those particles or photons that are either produced or degraded in energy or momentum by the interaction with matter from a radiation regarded as primary. Examples of secondary radiation are photoelectrons, Compton recoil electrons, and scattered photons. If a stream of electrons were considered as the primary radiation, then Bremsstrahlung also could be considered as secondary radiation.

Thick Target Bremsstrahlung

In producing Bremsstrahlung radiation for use in the field of radiography or radiology, the spectrum of radiation and the angular dependence are often altered according to the thickness of the target. As the problem is severely complicated by multiple electron scattering, electron-energy degradation in the target, Bremsstrahlung angular distribution, variation of Bremsstrahlung distribution with a decreasing electron energy in the target, and target self-absorption, very few theories have accounted for all of these effects. However, one theory[4] does present an adequate practical approach and is especially useful for the heavier target materials.

Radiographic Effects

This secondary radiation, which is the result of multiple scattering and Bremsstrahlung in a specimen under examination by X rays, presents an important problem to the radiographer. The secondary radiation can very often build up to magnitudes greater than the transmitted primary radiation. This results in a deleterious effect on both the sensitivity and resolution obtainable by radiography. Because of the many scatterings that may occur before an X ray emerges from a specimen under observation and because the primary radiation is heterogeneous, a practical theory that could accurately predict the effects of scatter is close to impossible to formulate. However, some approximate methods of taking scatter into account will be presented.

Broad-Beam Attenuation

Under the conditions that photons which suffered any deflection, however small, were considered absorbed and secondary radiation was considered negligible, the term narrow-beam attenuation was defined earlier in this section. In most cases it is impossible to perform an experiment or measurement under these conditions. Therefore, broad-beam attenuation, which takes scattering into account, must be considered. A lower attenuation is found under broad-beam than under narrow-beam conditions. The loss of beam intensity by scattering may not be so large as first thought because the photons initially aimed at the detector and scattered out of the beam are compensated in part by the deflection toward the detector of photons initially directed otherwise. For this reason the problem becomes one of geometry as well as physics.

Build-up Factor

Because the build up of the secondary radiation plays such an important role, it is convenient to have some measure of its effects with respect to the effects of the primary beam. A build-up factor has been defined for any specific situation as the ratio of the actual response of a detector to the response under ideal narrow-beam conditions. These ratios are often dependent upon the type and size of the source, type and size of the detector, and the usual spacing parameters. A thorough account of the effects of various parameters on the build-up factors can be obtained from reference 7.

Half-Value Layer

One of the most useful and practical means of expressing the quality of a beam of X rays consists of giving the half-value layer; that is, the thickness of some substance necessary to reduce the response of the detector by 50 percent. In case film is used as the detector, for example, the radiographic half-value layer is defined as the thickness of interposed material through which the attenuated radiation will produce the same density on a film as that produced by the unattenuated beam with half the exposure. Under conditions of narrow-beam geometry and a monoenergetic beam of photons, the half-value layer can be calculated from the simple exponential law

$$t_{1/2} = \frac{\ln 2}{\grave{\mu}} = \frac{0.693}{\grave{\mu}} , \qquad \text{(Eq. 12)}$$

where $\grave{\mu}$ is an absorption coefficient and t is the thickness. If $\grave{\mu}$ is the linear absorption coefficient expressed in reciprocal centimeters, then t will be given in centimeters (see refs. 5, 6, and 17).

PART 8

NEUTRON IRRADIATION

Production of Neutrons

The most frequently used source of neutrons is the nuclear reactor. Here the neutrons are formed by the fission process, where an unstable heavy element isotope (uranium or plutonium) formed by neutron capture disintegrates into several elements with the release of several neutrons. This chain reaction reactor produces a continuous Maxwellian energy spectrum with a long energy tail up to about 10 MeV, but peaking in the thermal energy range or somewhat higher depending on the type of reactor. Typical intensities available range from 10^{14} to 10^{18} $ncm^{-2}s^{-1}$.

Artificially produced radioactive neutron sources are of two general types, those formed by the interaction of alpha (α) particles with the light elements, e.g., beryllium, boron, or lithium; or by interaction of gamma (γ) photons with neutron producing targets. Neutrons from these radioactive sources usually have energies in the 1-10 MeV energy range.

Spontaneous fission is a popular source of neutrons. These sources produce a fission spectrum, including unfortunately many alpha particles and gamma radiation for each neutron produced. The use of these neutron sources requires great care and shielding.

Particle accelerators capable of producing beams of protons, deuterons, and alpha particles can produce neutrons by target interactions. An accelerator capable of imparting energies up to 10 MeV can produce monoenergetic neutrons at energies up to 27 MeV. Very high energy accelerators (>100 MeV) can produce large numbers of neutrons from heavy materials due to multiple nuclear interactions within the target material, notably photo-neutron and photo-fission caused by secondary reactions from the primary particle beam. When large particles or elements are formed, the process is usually referred to as spallation. Fusion reaction neutron sources use energetic deuterium ion beams impinging on tritiated targets to produce 14 MeV neutrons. Special, small vacuum-sealed machines are built specifically for this purpose.

Neutron Absorption Process

The absorption of neutrons relies on completely different mechanisms than that of charged particles. The neutron, carrying no charge, does not have any Coulomb interactions, and is free to travel through material until it has a direct collision with a target nucleus or an orbital electron. In the latter case, the electron is so small as to contribute only negligibly to the total absorption. In general, neutrons interact with matter in two ways, the neutron is either scattered by the nucleus or is absorbed into the nucleus.

The Elastic Scattering Process

Here, the neutron collides with the nucleus and bounces off leaving the nucleus unchanged. This type of collision can be treated in a straightforward manner as a mechanical "billiard ball" type of collision. In the collision the energy of the neutron is shared by the nucleus, thus each collision reduces the energy of the neutron. After a number of collisions with the nuclei, the energy is reduced to the same average kinetic energy as that of the absorbing medium. This energy is often referred to as the thermal energy, since it depends primarily upon the temperature. These neutrons are thermal neutrons. As the transfer of energy per collision is greater for light nuclei, low Z materials are used as moderators—that is, they reduce the speed of neutrons in a small number of elastic collisions without absorbing them to any great extent. The thermal energy of a particle is given by $3/2\ kT$, where k is called the Boltzman constant, and at a room temperature of 293K, the thermal energy is about 0.04 eV. The thermal neutrons have a Maxwellian distribution about this mean energy.

The Inelastic Scattering Process

Here the neutron collides with the nucleus leaving the nucleus in an excited state. In this process, the neutron may either stay in the excited state (n,n) as a metastable isomer or will immediately emit gamma radiation $(n,\gamma n)$ and return to the ground or original state.

Neutron Absorption or Radiative Capture

As the neutron has no charge, it can approach the nucleus until the close range attractive forces of the nucleus begin to operate. In this process, the neutron is captured, forming a compound nucleus. Because there is no charge barrier, even the slowest neutron can be readily captured. As the binding energy of a neutron into a compound nucleus is nearly 8 MeV, even the capture of thermal neutrons can result in a highly excited state. This excited nucleus can attain relative stability by ejecting a proton, an alpha particle, or by emitting the excess energy as gamma radiation. When a particle is ejected, the nucleus becomes a new element, then the process is also known as nuclear transmutation. The discovery of transmutation by slow neutrons led to the realization of nuclear fission.

The simplest capture reaction is that of capture of slow neutrons with emission of gamma rays (n,γ). An example is the absorption of a neutron by hydrogen to form deuterium, with an emission of gamma radiation of about 2 MeV. In heavier nuclei, the capture of a slow neutron, followed by the emission of gamma radiation, increases the neutron-to-proton ratio—usually making the nucleus radioactive with decay by electron emission likely.

As the energy of the impinging neutron is made larger, a charged particle can be ejected. However, a charged particle, because of the short-range attractive forces of the nuclei, is hindered from leaving the nucleus and processes such as (n,p), (n,a), (n,d) can only take place when the incident neutron supplies sufficient energy to overcome the binding energies of the particles in the nucleus. For heavy nuclei these forces are appreciable and the requisite neutron energy becomes greater. Thus, for example, a particle ejection is possible only if the neutron has sufficient energy to overcome the binding energy of the α-particle; that is, the neutron must be a fast neutron.

In the (n,α) reaction, the product nucleus contains one neutron and two protons less than the original nucleus. The neutron-to-proton ratio is increased and the transmutation usually produces a radioactive nucleus that decays by the emission of an electron (beta disintegration).

As the energy of the incident neutron approaches 30 MeV, the compound nucleus should be able to eject three neutrons $(n,3n)$ or two neutrons and a proton $(n,2np)$. As the energy becomes even greater, larger particles may be ejected (spallation).

Finally, nuclear fission (n,f), where the nucleus breaks up with the release of several larger particles and several neutrons, can be induced in certain nuclides, such as uranium-235, by neutrons of almost any energy, whereas in other nuclides, fast or energetic neutrons are required.

Nuclear Cross Sections

Because of many reactions possible for absorbing neutrons and their complicated energy and mass dependencies, there is no simple way to present the total absorption effect. However, the probability of any interaction between neutrons and matter can be made qualitative by means of the concept of cross sections. The cross section σ is the effective target area of the nucleus as seen by the impinging neutron of a given energy. The number of interactions per unit time will be $nvN\sigma$, where n is the number of neutrons per unit volume moving with velocity v towards the target of N nuclei. The quantity nv is the neutron flux density. The cross section σ is usually expressed in barns (10^{-24} cm^2).

In discussing the variation of nuclear cross section with energy of the incident neutrons, we can make certain generalizations of a broad character. In general, there are three regions that can be distinguished. First is the low-energy region, which includes the thermal range, where the cross section decreases steadily with increasing neutron energy. The total cross section is the sum of two terms, one due to neutron scattering is quite small and almost constant, the other representing absorption by the nucleus is inversely proportional to the velocity. The energy range is termed the $1/v$ region, where the time spent by the neutron near the nucleus is proportional to $1/v$. Second, following the somewhat indefinite $1/v$ region, many elements exhibit peaks called resonance peaks, where the neutron cross sections rise sharply to high values for certain energies, then fall to lower values again. Depending on the element, the number of such peaks may number three or more. These peaks may be found mostly in the energy range 0.1 to 1 eV, although in a few elements like uranium-238, they may be found up to energies of 10 eV. These reactions are of the (η,γ) type. And third, with neutrons of high energy in the MeV range, the cross sections are very low, less than 10 barns, compared to the very high cross sections of several thousand barns at low energies.

A simple example of the total absorption cross section is that of cadmium, shown in Fig. 18. The $1/v$ region is shown up to about 0.03 eV, the resonance at 0.176 eV, and the low cross section region for energies greater than about 2 MeV.

The dramatic increase in cross sections at the resonance have been worked out by the theory of G. Breit and E. P. Wigner.[3] In its simplicity, if the energy of the neutron is such that a compound nucleus can be formed at or near one of its energy levels, then the probability of capture of these neutrons will be exceptionally high.

All elements do not show the resonant absorption effect; for example, boron has no measurable resonance and the cross section follows the $1/v$ law from 0.01 eV to over 1000 eV. However, its cross section for (n,α) is so large for neutrons of low energy that this reaction is often used for neutron detectors. Table 12 shows the dramatic variation of cross section for absorbing thermal neutrons of some of the better neutron absorbers.

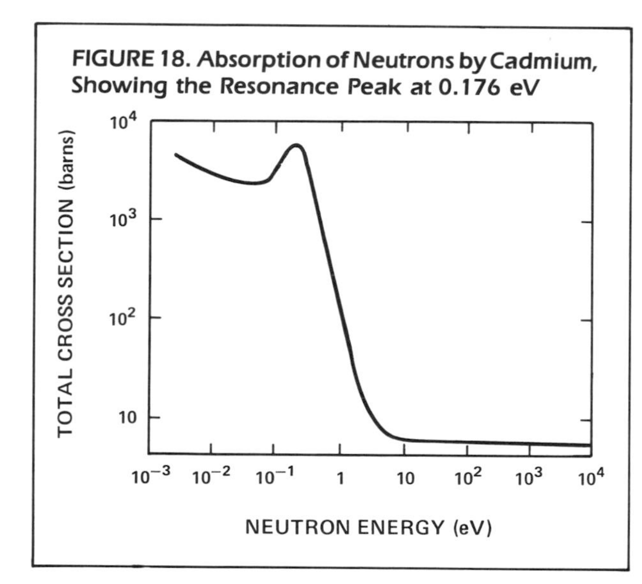

FIGURE 18. Absorption of Neutrons by Cadmium, Showing the Resonance Peak at 0.176 eV

TABLE 12. Capture Cross-Section of Strongly Absorbing Elements (for Neutrons in Approximate Thermal Equilibrium)

Element		$\sigma \cdot 10^{24}$ 300°K	Element		$\sigma \cdot 10^{24}$ 300°K
3	Li	65	62	Sm	4,260
5	B	540	63	Eu	3,400
17	Cl	40	64	Gd	22,200
27	Co	35	66	Dy	1,200
45	Rh	125	67	Ho	340
47	Ag	55	75	Re	90
48	Cd	3,000	77	Ir	285
49	In	300	79	Au	90
			80	Hg	450

REFERENCES

1. Allison, S.K., and S.D. Warshaw. "Passage of Heavy Particles through Matter." *Rev. Mod. Phys.*, 25 (1953): 779.
2. Bethe, H.A., and J. Ashkin. "The Passage of Heavy Particles through Matter." In E. Sergè, ed. *Experimental Nuclear Physics*, 1, pt. 2. New York: Wiley, 1953.
3. Breit, G., and E.P. Wigner. *Phys. Rev.*, 49, 519 (1936); see also Bohr, *Nature*, 137, 344 (1936).
4. Emigh, C.R. "Thick Target Bremsstrahlung Theory." Los Alamos Scientific Laboratory report LA-4097-MS (1970).
5. Fano, U. "Gamma-Ray Attenuation. Part I. Basic Processes." *Nucleonics*, 11, No. 8 (1953): 8.
6. ----. "Gamma Ray Attenuation. Part II. Analysis of Penetration." *Nucleonics*, 11, No. 9 (1953): 55.
7. ----. "Penetration of X- and Gamma-Rays to Extremely Great Depths." *J. Research Natl. Bur. Standards*, 51 (1953): 95.
8. Glasstone, S. *Source Book of Atomic Energy*. New York: D. Van Nostrand, 1950.
9. Heitler, W. *The Quantum Theory of Radiation*. 2d ed. London: Oxford University Press, 1950.
10. Hogerton, J.F., and R.C. Grass, eds. *The Reactor Handbook. I: Physics*. U.S. Atomic Energy Comm., AECD 3645, 1955.
11. Lapp, R.E., and H.L. Andrews. *Nuclear Radiation Physics*. 2d ed. Englewood Cliffs, N.J.: Prentice-Hall, 1954.
12. Marshall, J.S., and A.G. Ward. "Absorption Curves and Ranges for Homogeneous β-Ray." *Can. J. Research*, 15A (1937): 39.
13. Molière, G. "Theorie der Streuung schneller geladener Teilchen. I. Einzelsteuung am abgeschirmten Coulomb-Feld (Theory of the Scattering of Fast Charged Particles. I. Single Scattering on the Shielding Coulomb Field)." *Z. Naturforsch.*, 2a (1947): 133.
14. Morgan, R.H., and K.E. Corrigan. *Handbook of Radiology*. Chicago: Yearbook Publishers, 1955.
15. Mott, N.F., "The Collision between Two Electrons." *Proc. Roy. Soc. (London)*, A126 (1930): 259.
16. Nelms, Ann T. "Graphs of the Compton Energy-Angle Relationship and the Klein-Nishina Formula from 10 KeV to 500 MeV." Natl. Bur. Standards (U.S.) Circ., No. 542, 1953.
17. "Permissible Dose from External Sources of Ionizing Radiation." Natl. Bur. Standards (U.S.), Handbook, No. 59 (1954).
18. Reines, F. and Cowan, C.L. "*The Neutrino*," *Nature*, 178 (1956): 446-49.
19. Stephenson, R. *Introduction to Nuclear Engineering*. New York: McGraw-Hill, 1954.

SECTION 2

CONVENTIONAL ELECTRONIC RADIATION SOURCES

James D. Willenberg, Industrial Testing Consultants, Warren, MI

INTRODUCTION

Technological advances in image processing, flash radiography, real-time radiography, microfocus and metal-ceramic tubes have stimulated increased interest in the electronic production of ionizing radiation. The versatility of an electronic source is difficult to equal with an isotopic source even for specific applications. With the advent of solid state components, many of the objections related to size, weight and durability of electronic X-ray sources have been overcome.

In this chapter, the basic operation, construction and application of currently manufactured equipment will be discussed. The main emphasis will be on industrial X-ray equipment used in radiography, with a lesser emphasis on analytical and high-energy units, which are covered in detail in other chapters of Volume 3. References to older or obsolete units have been kept to a minimum. The reader is referred to the *Nondestructive Testing Handbook*, first edition, for a more detailed treatment of these older units.

PART 1
PHYSICAL PRINCIPLES

Radiation of an unknown type and origin was discovered in the year 1895 by Wilhelm Roentgen. Called X rays by their discoverer, they were soon found to be a form of electromagnetic radiation with extremely short wavelengths.

Conservation of Energy

Electromagnetic theory had long predicted that a charged particle undergoing acceleration would emit radiation. This theory can be used to qualitatively explain the continuous portion of a typical X-ray spectrum (see Fig. 1). Although a complete study of the continuous spectrum requires the use of many disciplines within modern physics, an understanding of the spectrum's basic principles may be gained by considering just one of the fundamental laws of physics, the conservation of energy. As its name implies, the law states that energy can neither be created nor destroyed, although it is possible to change it from one form to another.

In the case of X rays, a stream of quickly moving charged particles, usually electrons, strikes a target material and is brought rapidly to a halt. Much of the electrons' kinetic energy is transformed into heat energy as the stream strikes the target. In fact, except for the case of very high energy generators, almost all of the electrons' kinetic energy (more than 97%) is converted into heat; disposal of this thermal energy is an important design consideration.

Bremsstrahlung

A small portion of the energy will also be given off as packets of electromagnetic radiation called *photons*. The X-ray photons can have energies ranging from zero to a maximum which is determined by: (1) the original kinetic energy of the electron; and (2) by how rapidly the electron is decelerated. This process produces the continuous portion of the X-ray spectrum and is known by the German term *Bremsstrahlung* for braking radiation.

Energies of the electrons (and the X rays) are frequently given in terms of keV (kiloelectron volts) or MeV (million electron volts). The meaning of *electron volt* becomes clear if we consider the charged electron. Under the influence of a voltage difference (technically called a potential difference), charged particles will experience a force which causes them to accelerate. A negatively charged particle, such as an electron, will move from a place of low voltage (−) to a place of high (+) or even zero voltage, and increase its kinetic energy as it does so. Thus the unit, keV, corresponds to the amount of kinetic energy that an electron would gain while moving between two points that differ in voltage by one thousand volts (1 kilovolt or kV). Similarly, an electron would gain 1 MeV of kinetic energy while moving between two points that differ by one million volts. The two points of differing voltage are called the *cathode* (−) and the *anode* (+) and will be discussed later in this Section.

Characteristic X rays

In addition to the Bremsstrahlung there are several *characteristic* peaks in a typical X-ray spectrum. These intensity spikes are caused by interaction between the impinging stream of high speed electrons and the electrons that are bound tightly to the atomic nuclei of the target material. If we picture the simplified model of an atom as a planetary system, with the nucleus of protons and neutrons at the center of the system (see Fig. 2) and the electrons moving in orbits around the nucleus, we can again apply the law of energy conservation to describe the origin of characteristic radiation.

Modern physics predicts that these orbital electrons near the nucleus will have very well-defined energies, with electrons in different orbits having different energy levels. If a vacancy in a particular level were created by knocking an electron from its orbit, an electron from a higher energy level would, after a time, drop down to fill the void. In order

FIGURE 1. Typical X-ray Spectrum

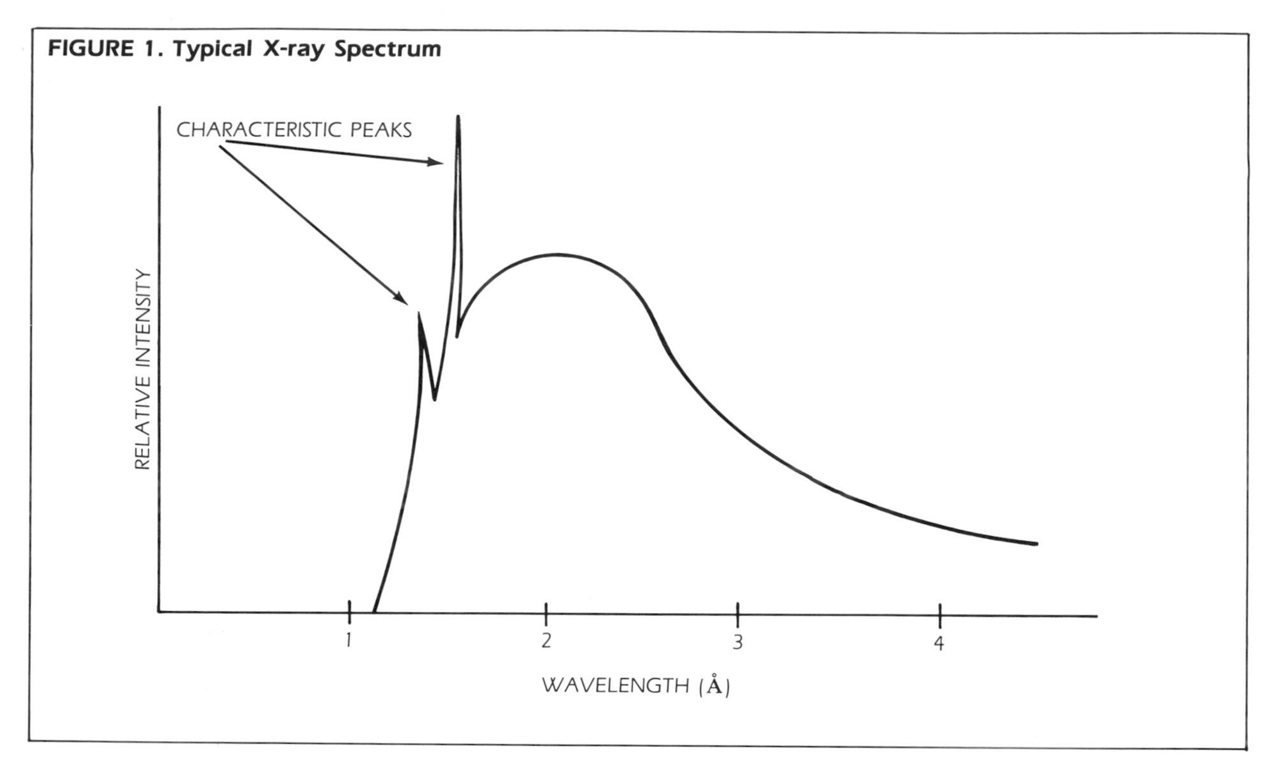

to do this, it would have to lose (emit) energy. Because the energy levels of an atom are well specified, the exact amount of energy lost by an electron making such a transition would also be specified.

Using the conservation of energy law, we recognize that the energy is not really lost but given off as electromagnetic radiation in the form of an X-ray photon. Because each atomic element has its own distinct set of energy levels, the line spectrum produced in such a manner is characteristic of the particular target material. Interest in these characteristic X rays lies in their application to X-ray diffraction and other analytical applications. By a combination of filtration and diffraction, characteristic X rays can be used to produce a monoenergetic beam of X rays.

Thermionic Emission

Two final concepts are of interest in this brief treatment of the basic physics of X-ray production. The first process, thermionic emission, provides electrons which are accelerated to high speeds by the voltage difference between the cathode and anode.

FIGURE 2. Planetary Model of Atomic Structure

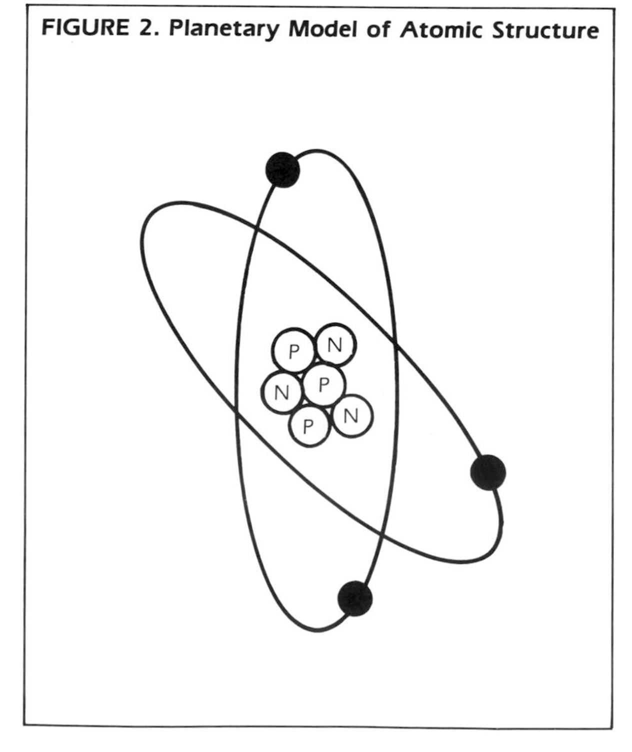

When metal (in this case the X-ray tube filament) is heated to incandescence, a small portion of free electrons is actually able to escape the surface of the material. Without an accelerating voltage, the electrons form a cloud-like *space charge* around the filament. Under the influence of a potential difference, however, the electrons are quickly accelerated toward the anode of the tube.

Absorption

As X rays pass into or through a material, they are absorbed in a manner that depends on: (1) the energy of the radiation; (2) the density of the material; and (3) the atomic number of the material. In equation form, the intensity, I, as a function of thickness, has a standard exponential form given by:

$$I = I_o e^{-\mu x} \qquad \text{(Eq. 1)}$$

where I is the intensity after passing through the material; x is the thickness of the absorbing material; I_o is the initial intensity; and μ is the linear absorption coefficient (characteristic of the material for a particular X-ray energy range). The importance of absorption in selection of materials for tube construction, beam filtration and shielding are discussed later in this Section.

PART 2

BASIC GENERATOR CONSTRUCTION

A conventional X-ray generator consists of three main components: (1) X-ray tube; (2) high voltage source; and (3) control unit.

These components are common to all conventional units and are examined in detail below.

X-ray Tubes

Early X-ray tubes used gas-filled tubes and a cold cathode from which electrons were freed by positive ion bombardment. Modern tubes used in radiography are of the high vacuum variety, allowing for reduction in size, extended tube life and more stable operation.

Electrons are supplied by thermionic emission from the filament. The accelerating potential and the tube current can then be independently varied, with the exception that, at low accelerating voltages, tube current is affected by the space charge which accumulates around the cathode.

Envelope

Envelopes for X-ray tubes are usually of the glass or metal-ceramic type (see Figs. 3 and 4). Glass envelope tubes, although still in common use, are susceptible to thermal and mechanical shock and are being replaced in many industrial applications with the more durable metal-ceramic tubes.

The vacuum envelope of the metal-ceramic tubes consists of a metal cylinder capped on both ends with ceramic disks, usually made of aluminum oxide. These ceramic insulators are designed to allow for more effective use of the insulation characteristics of both the ceramic and the high tension grease used in sealing connections between the high voltage source and the tube. This design permits a reduction in the size of the tube housing that is especially important for higher energy units.

Cathode

The cathode includes the tungsten filament which provides the thermal electrons for acceleration. The filament is usually powered by alternating current (50-60 Hz) from a separately controlled transformer, although in some units the filament current is fixed or automatically controlled to maintain a constant tube current. Normally, filament currents range from one to ten amperes (A). The *tube current*, passing between the cathode and anode by means of the high-speed electrons, ranges from several hundred microamperes (μA) for microfocus units up to 20+ milliamperes (mA) for conventional industrial radiographic units.

Beam Focusing

At times, the filament is located in a recess in the cathode called a *focusing cup*. This surrounds the emerging beam of electrons with an electric field which repels the beam away from the cup wall and into a more localized form. The importance of a well-defined beam of electrons arises from the fact that the sharpness or unsharpness of an image is dependent upon the focal spot size (see Fig. 5). The relationship for geometric unsharpness, U_g, is:

$$U_g = \frac{F}{D/t} \qquad \text{(Eq. 2)}$$

where U_g is a measure of the penumbra effect of the focal spot; F is the focal spot size; D is the distance from the target (focal spot to the object); and t is the thickness of the object.

FIGURE 3. Glass X-ray Tube

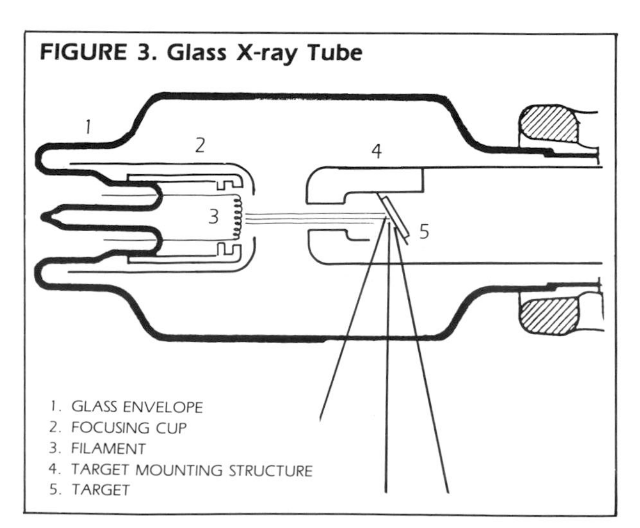

FIGURE 4. Metal-ceramic X-ray Tube

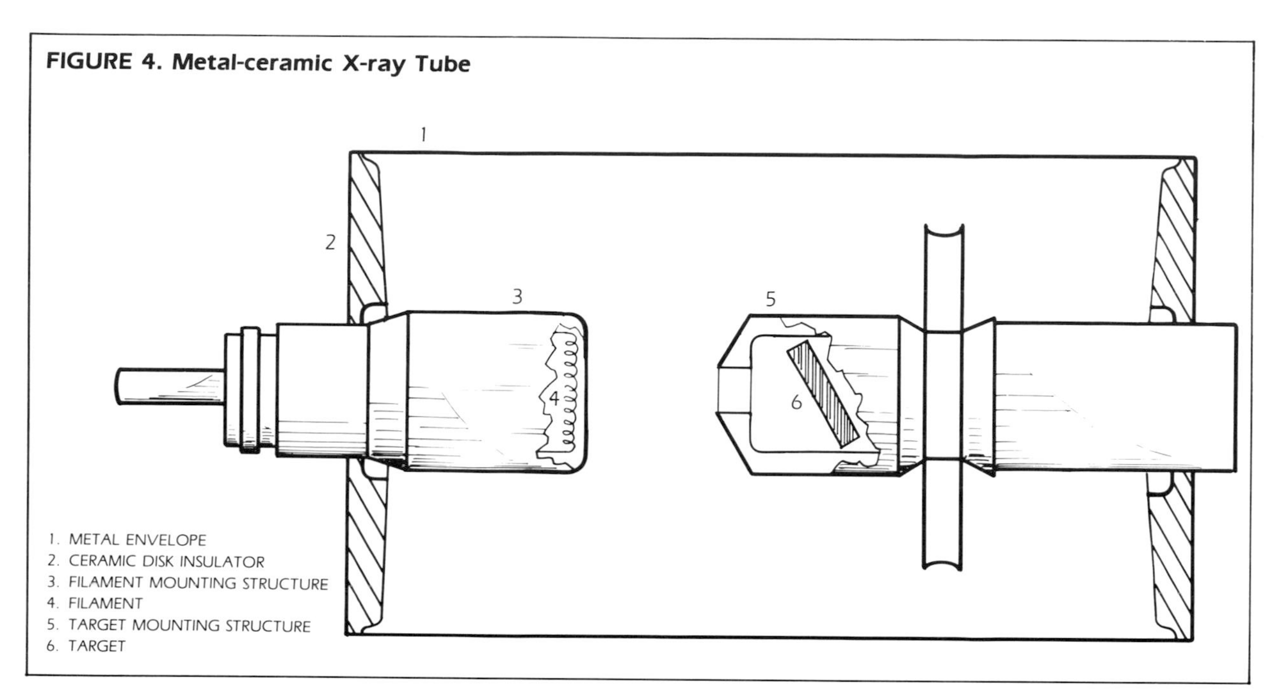

From the equation, we see that U_g increases directly as the focal spot size increases. Because the beam originates at the filament, reduction of the filament size might seem to be a solution to geometrical unsharpness, but this approach is limited by the durability of the filament.

One alternative, called *line focusing*, is to project the approximately rectangular beam produced by the filament onto a target which is angled with respect to the beam (approximately 21 degrees). This projects an X-ray beam that appears to emanate from a focal spot with approximately equal lateral dimensions (see Fig. 6). In practice, this method allows production of units with focal sizes in the range of 1.0 to 3.0 mm (0.04 to 0.12 in.).

By use of a deep focusing cup, advantage can also be taken of the *screen effect* (see Fig. 7). This refers to the removal of the lower energy electrons which are produced during that portion of the AC cycle where the potential difference between cathode and anode is significantly less than maximum. In practice, this improvement is not without cost to the output of the unit. A loss of approximately 25% is experienced in units with high screen effect. This can be compensated for, in part, with higher filament current though this adversely affects the lifetime of the filament. An alternate method of removing low energy components of the electron beam is found in the discussion on constant potential (CP) units.

If still further focus of the beam is desired, as in microfocus radiography and some analytical applications, additional methods may be used, including: (1) conversion of the conventional diode arrangement of cathode and anode into a triode arrangement with a focusing electrode or grid; and (2) electrostatic or magnetic deflection systems.

For the triode arrangement, which is used widely in the microfocus industry, a negative bias up to − 150 V is applied to the third element of the tube to further focus the beam and remove lower energy components. This configuration allows a reduction of beam size, producing focal spots smaller than 50 micrometers, and a subsequent drop in tube current.

In the case of electrostatic deflection, even more elements are included within the envelope; a magnetic deflection system is external to the tube. These types of deflection systems have an additional advantage in the fact that the beam may also be deflected to various areas of the target for added service life. Units which incorporate their own vacuum systems usually allow for replacement of both filament and target components. These types, though formerly limited to analytical and research units, are now available from several manufacturers as standard microfocus and analytical units.

FIGURE 5. Illustration of Geometric Unsharpness

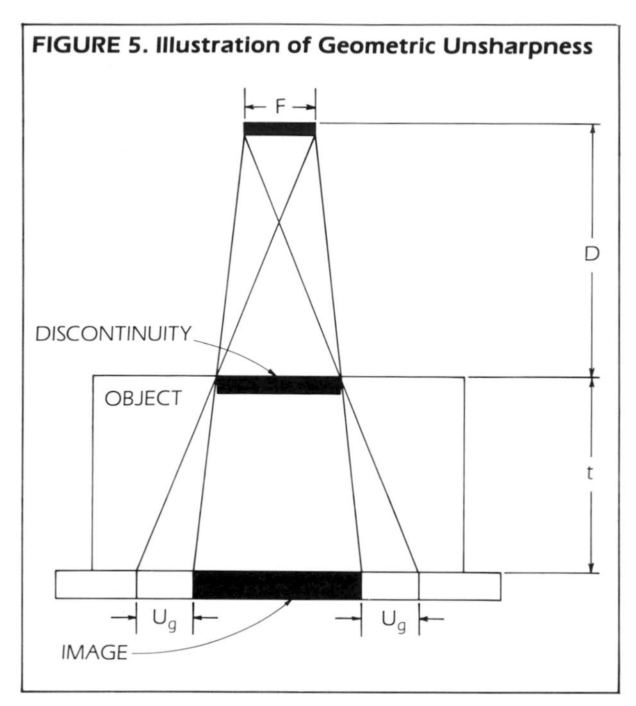

FIGURE 6. Diagram of Line Focusing Set-up

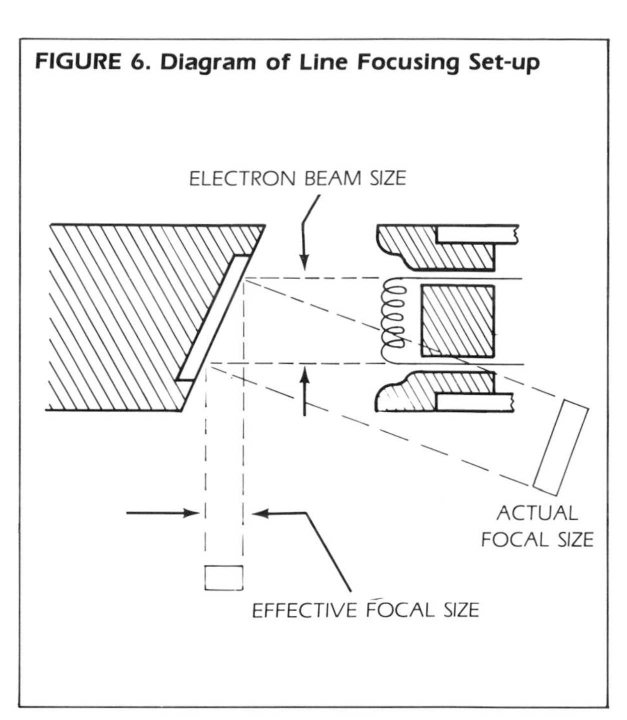

FIGURE 7. Graph of Screen Effect

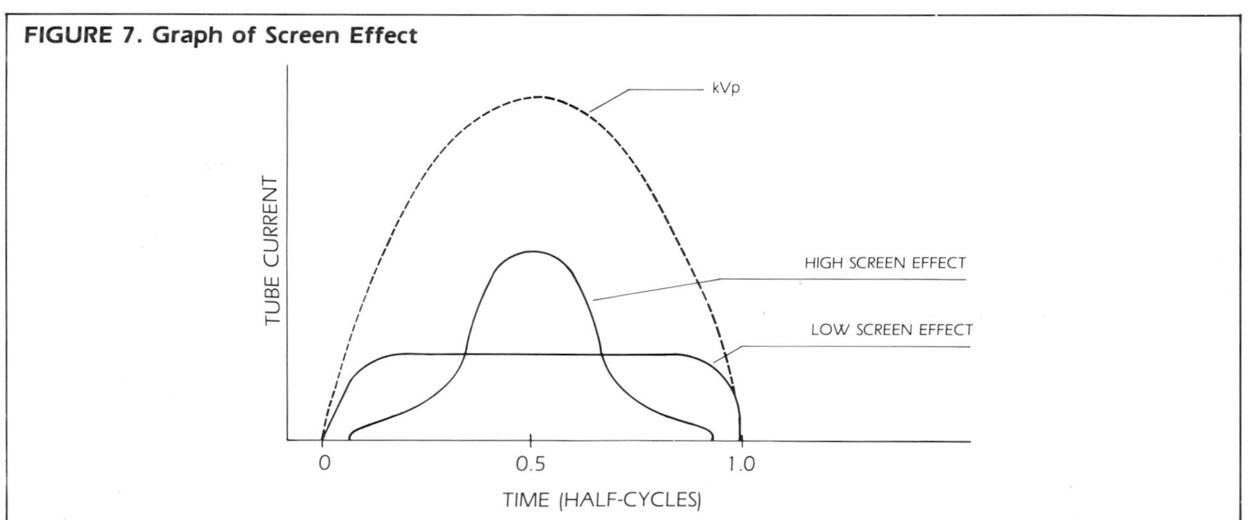

Anode

As mentioned previously, heat is the major form of energy produced as the electrons strike the target. Uncontrolled, this heat would quickly cause the surface of the target to erode, which in turn reduces the definition of the focal spot. In addition, the vaporized target material reduces the high vacuum of the tube and leads to premature failure due to conduction within the tube. To avoid overheating of the target, the anode to which it is attached is composed of a material with high thermal conductivity, such as copper. If the cooling demands are relatively low, as for a low energy unit or intermittent use, cooling is often accomplished by means of a conductor which passes through the tube end for connection to the high voltage source; this allows for radiation of heat into an oil or gas reservoir surrounding the tube. Although not the most efficient, the weight of such a tube is minimal due to the absence of pumps or heat exchangers.

For higher energy units in continuous use, it is usually necessary to cool the anode by injecting coolant directly into it. This is accomplished by hollow construction of the anode conductor.

Another way of alleviating the problem of localized heating of the target is by use of a rotating anode in which the target, a tungsten disk, is driven as shown in Fig. 8. This allows the tube current to be increased by as much as ten times the value for a stationary target. The focal spot on such units can be reduced to less than 1 mm (0.04 in.) for short exposure times; this is of value in medical as well as some specialized industrial applications, including flash radiography.

Target

In radiographic applications, the target is usually tungsten and is bonded to the copper anode. However, analytical units make use of several other target materials to take advantage of the characteristic X rays produced. Some of these materials include copper, iron and cobalt.

The orientation of the target with respect to the electron beam strongly influences the size and shape of the focal spot. Orientations from 0 degrees to 30 degrees are used for various applications. For example, zero is an angle used for panoramic units. An angle of 20 degrees is commonly selected for directional units because, in this case, the distribution of X rays is predominantly in a direction perpendicular to the tube axis. This is shown graphically in Fig. 9. The actual maximum intensity occurs at +12 degrees. For radiography of objects whose lateral

FIGURE 8. Rotating Anode

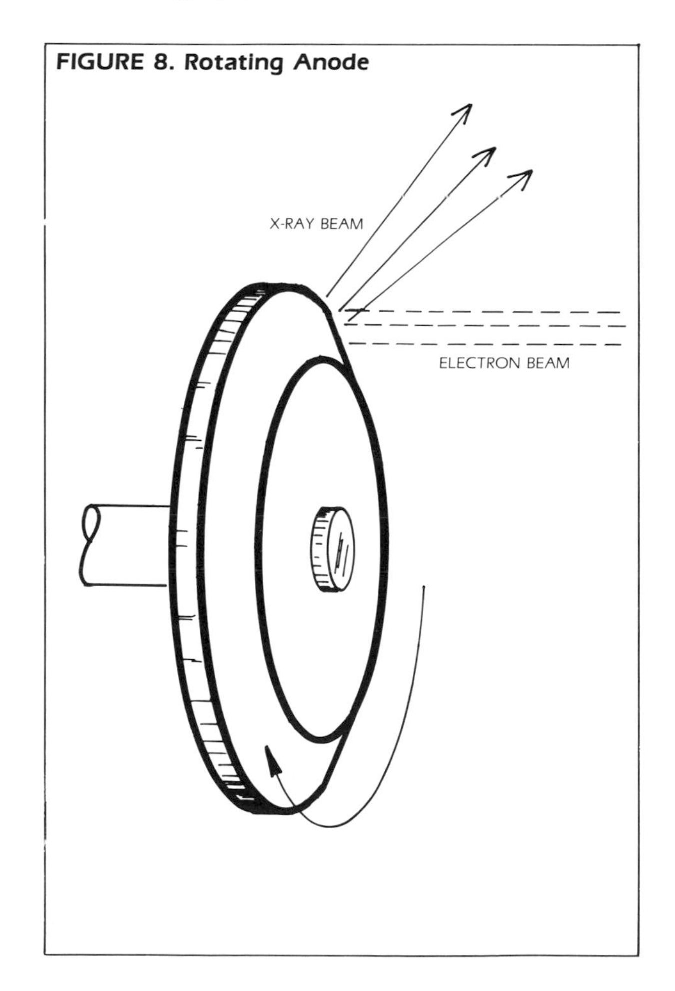

FIGURE 9. X-ray Distribution Graph

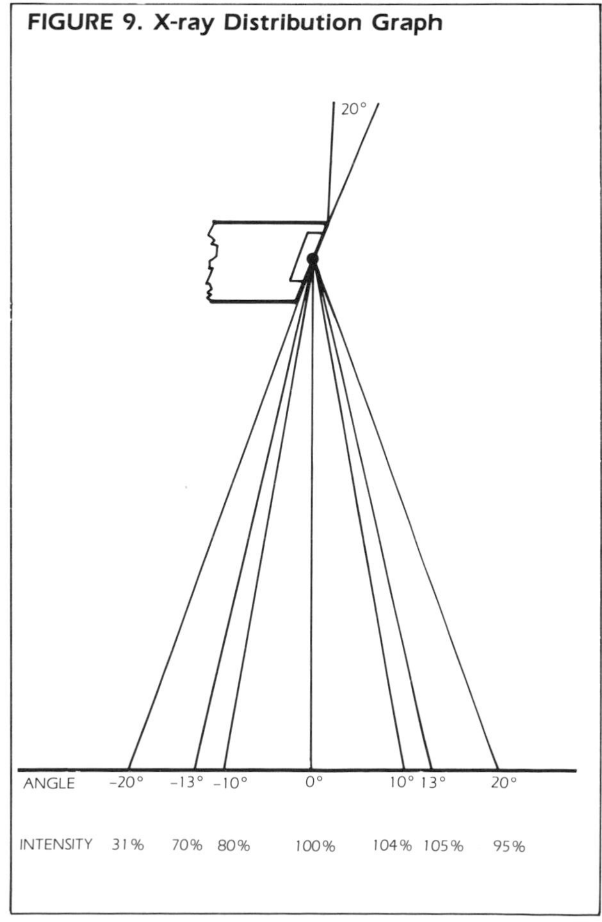

dimensions are less than one-half the focus-to-film distance (objects which subtend an angle of less than 30 degrees), the variation is seldom consequential.

Another cause for intensity variation is the electron beam itself. A cross section of the electron beam from the filament would resemble Fig. 10a. Figure 10b shows a similar representation for a microfocus beam. The beam distribution in Fig. 10b is said to be Gaussian (bell-shaped) because of the shape of the intensity curve. Such a beam profile is required when it is necessary to define very closely spaced objects, such as microcircuitry components.

Hood

Addition of a hood to the anode provides the twofold function of (1) eliminating a portion of the X-ray beam outside the central cone of radiation and (2) electrically shielding the insulating portions of the envelope (glass or ceramic) from charge build-up due to electrons scattered from the tungsten target or released by the photoelectric effect (see Fig. 11).

Removing the unused radiation directly at the anode reduces the amount of radiation shielding that must be provided externally or incorporated into the tube housing. The hood, normally constructed of copper, may have materials with high atomic numbers, such as tungsten, incorporated to increase absorption. The electrical shielding function of the hood may be improved by the addition of a beryllium window over the X-ray port. A window several millimeters in thickness will stop electrons with negligible effect on the overall X-ray beam.

FIGURE 10. Electron Beam Distributions: (a) Conventional Beam; (b) Microfocus Beam

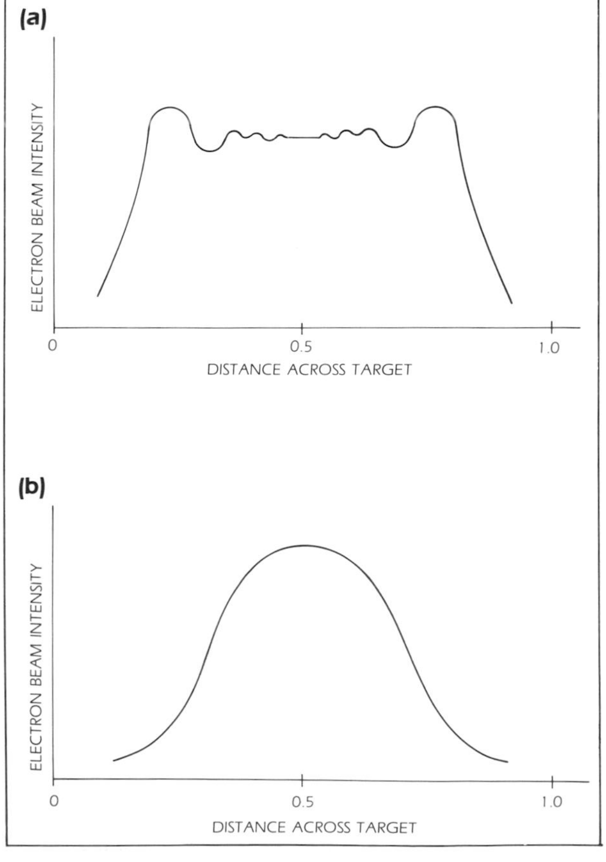

(a)

ELECTRON BEAM INTENSITY

0 0.5 1.0

DISTANCE ACROSS TARGET

(b)

ELECTRON BEAM INTENSITY

0 0.5 1.0

DISTANCE ACROSS TARGET

Rod Anode

The rod anode (sometimes referred to as an *ox-tail*) is another adaptation of the anode. This type of tube arrangement requires special circuit considerations which allow the anode to be grounded. This tube, developed for use through small openings, has been partially replaced by the metal-ceramic tube which can have a diameter of less than 5 cm (2 in.) and tube head diameter as small as 7.6 cm (3 in.). The target of such an *end-grounded* tube can be cooled by circulating water in direct contact with the anode. Beam focusing is often required for longer tubes.

FIGURE 11. Hooded Anode Tube

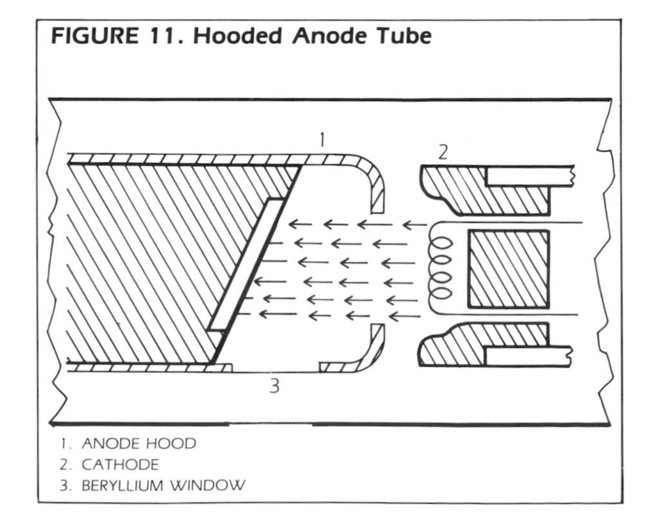

1. ANODE HOOD
2. CATHODE
3. BERYLLIUM WINDOW

Coolant

With the exception of the end-grounded configurations and units designed for low energy output (less than 50 kV), the tube insert is surrounded by an insulating coolant and encased in a housing called the *tube head*.

The coolant may be highly dielectric gas or oil. If oil is used, simple convection may be sufficient for lower output units. For larger units, an oil circulating pump combined with a heat exchanger, either internal or external to the tube head, may be used.

For units making use of a fixed amount of oil in the tube and a circulating pump to circulate it within the tube head, an oil resistant bellows is incorporated to allow for expansion and contraction of the oil. This is not required for gas-filled heads because of the compressibility of insulating gases, but a pressure gage is normally included to monitor possible loss of coolant insulation.

Tank-type Head

The housing itself protects the tube, contains the coolant and forms the structural support for the X-ray tube, electrical connections, fittings, pumps, thermal and high voltage overload sensors and radiation shielding contained in the head. For the tank type unit, the tube head also houses the high voltage and filament transformers. If the unit has separate components, the tube head will also provide for connection to the high voltage source.

FIGURE 12. Standard High Voltage Circuit Designs for Portable Tank-type Units: (a) Cathode Grounded; (b) Center Grounded; (c) Anode Grounded

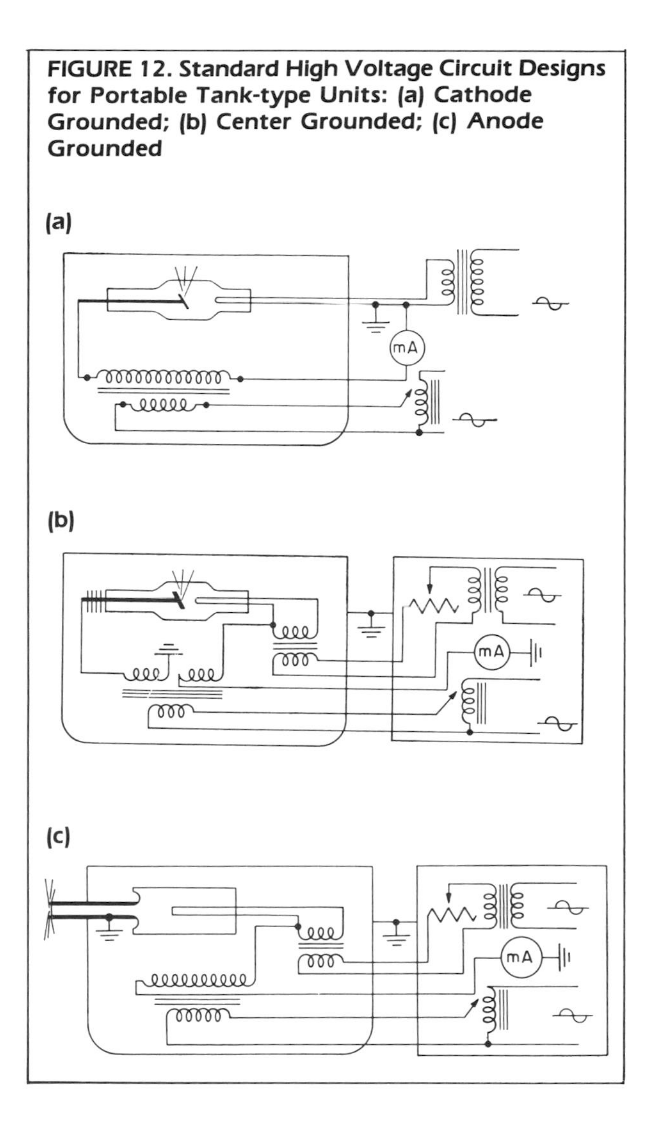

High-Energy Sources

From line voltages in the range of 100 to 250 volts, the high tension circuitry supplies potential differences to the tube from 5 kV to as much as 420 kV for the larger industrial radiographic units. Several standard circuit designs are used for various applications. The portable tank type units generally employ one of the designs shown in Fig. 12.

These circuits are all self-rectified; the X-ray tube itself limits the flow of electrons to one direction in the circuit. While the anode is at negative potential with respect to the cathode, no tube current flows.

One drawback of the self-rectified system is the possibility of tube backfire. If the target or anode overheats, reverse conduction can occur during the negative half-cycle. This type of unit is normally used for tubes producing X rays in the range of 50 to 200 kilovolts peak (kVp) and tube currents from 2 to 8 mA.

Cathode-grounded Circuit

The main advantage of the cathode-grounded configuration is that it allows the filament transformer to be external to the tank, since the cathode is at ground potential and does not require isolation. The tube head can be reduced in size and weight and is often gas-filled to further decrease weight.

FIGURE 13. Typical Anode-grounded Circuits and their Waveforms: (a) Villard Circuit; (b) Graetz Circuit; (c) Greinacker Circuit

(a)

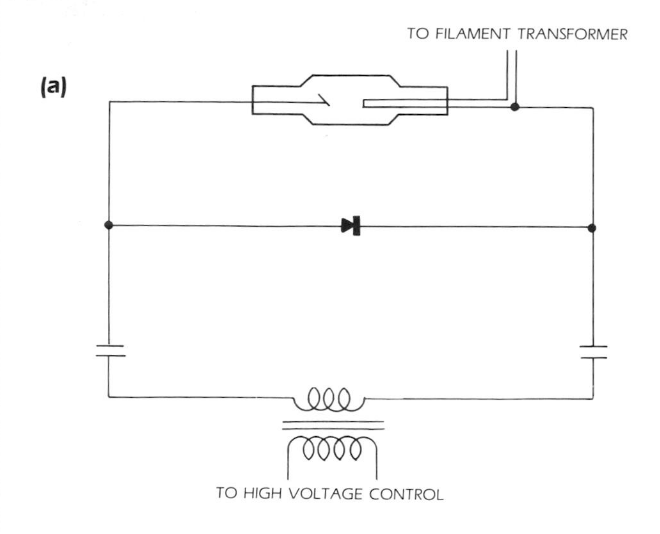

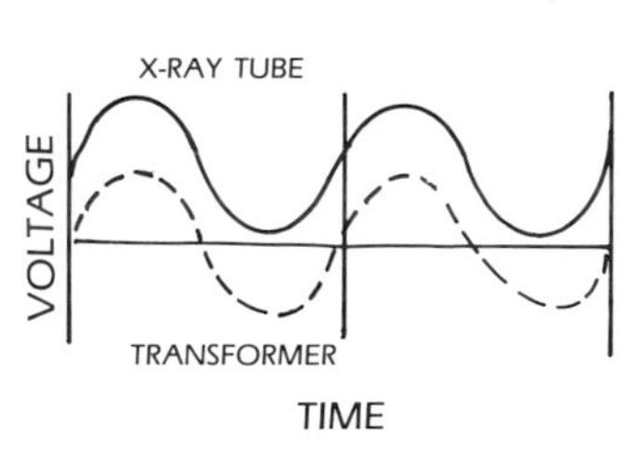

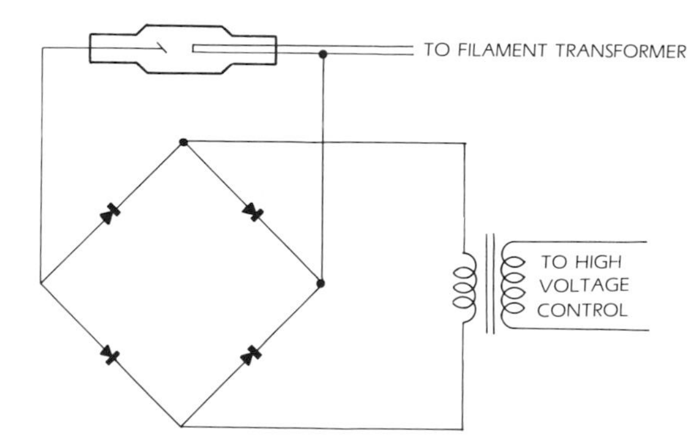

GRAETZ CIRCUIT WAVEFORM

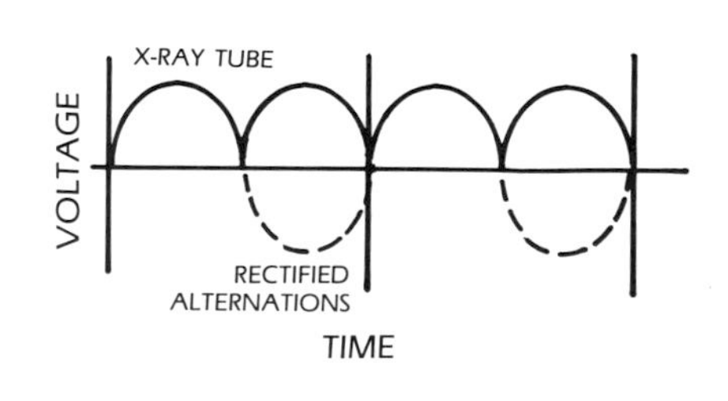

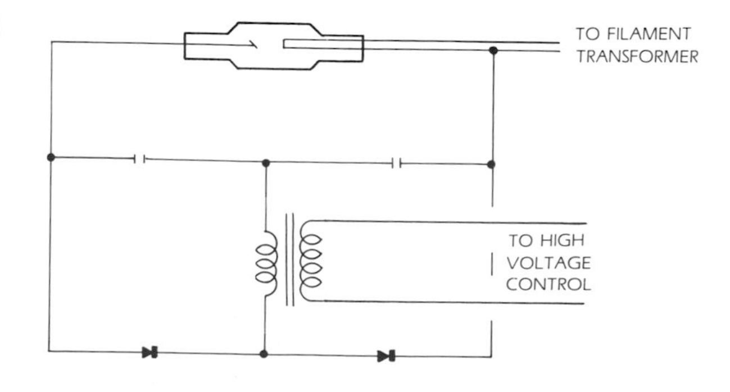

GREINACKER CIRCUIT WAVEFORM

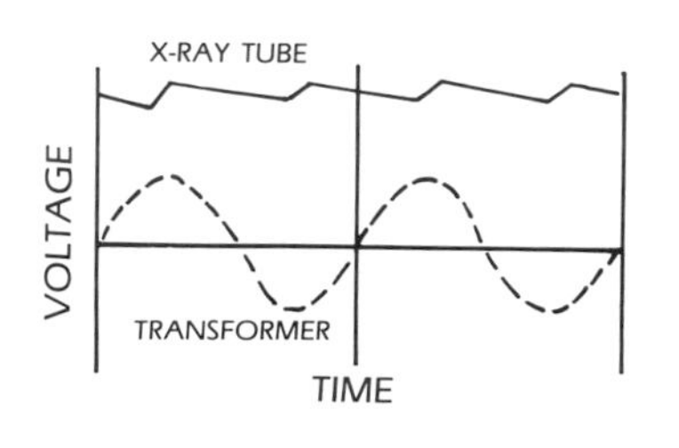

Center-grounded Circuit

Both center-grounded and anode-grounded units require isolated filament transformers, which must be adequately insulated. For the center-grounded unit, this is justified by the reduction of the high tension (HT) transformer insulation. The transformer needs to supply only one half of the potential difference to each electrode, rather than having either the cathode or anode held at ground potential and supplying the entire accelerating voltage to the other electrode.

In the range of 200 to 300 kVp with beam currents up to 15 mA, center-grounded systems can be made smaller than comparable end-grounded units.

Anode-grounded Circuit

For the anode- or end-grounded system, the advantage lies in the specialized use of the rod anode tube or metal-ceramic tube for access through small openings. As mentioned earlier, cooling of the target is also simplified.

With the addition of capacitors and diode rectifiers, the transformer is normally placed in a tank separate from the head. The additional elements allow the current to be rectified by means of valve tubes or solid state diodes and to be filtered and smoothed to provide a more nearly constant accelerating voltage. Several popular circuits and their waveforms are shown in Fig. 13.

Villard Circuit

An extension of the half-wave system, the Villard circuit allows production of accelerating potentials of twice the transformer peak voltage. Capacitors are charged during one half of the cycle and discharged when current passes through the tube, augmenting the voltage produced by the transformer (see Fig. 13a).

Graetz Circuit

The full-wave or Graetz circuit allows use of both halves of the AC cycle with a substantial increase in tube output per unit time. This system is widely used in medical applications but is used less than constant potential (CP) units in industrial applications (see Fig. 13b).

Greinacker Circuit

As can be seen from the output waveform, the Greinacker circuit is of the CP type (see Fig. 13c). Basically a variation of the Villard circuit, in which capacitors are charged during both halves of the cycle, the voltage is not only doubled, but remains near maximum value throughout the cycle. This gives enhanced high-energy output and eliminates the electrical stress placed on the tube and insulation. Enhanced tube life, as well as approximately 30% reduction in exposure times, are the results.

A common misconception is that CP units provide a beam of constant energy X rays. Although the electron beam is nearly monoenergetic, the X rays are produced during random deceleration processes which give the statistical energy distribution shown in Fig. 1. The absence of low energy electrons reduces the number of low energy X rays but does not eliminate them.

Alternative Circuit Designs

A method for improving tube output, which can be used in conjunction with any of the above circuits (including those in tank units), is the use of a higher frequency waveform to power the high tension (HT) transformer. Although this requires additional electronic circuitry or a motor generator, the core of the HT transformer can be reduced in size because of increased reactance at higher frequencies. This can be used to advantage in portable or mobile units. Also, if filtering is to be done, the variation or ripple of the output voltage can be reduced even further.

A variation of this technique is the use of three-phase input power with the HT transformer. Commonly used in medical X-ray generators, this method is now in use by several industrial manufacturers as well.

Another approach is to use an output waveform other than the standard sine wave. Approximate square wave outputs, in conjunction with both phase inversion circuitry and a high frequency transformer, can provide accelerating potentials with extremely low ripple characteristics. Such units are currently available for industrial applications.

HT Connections

One remaining topic in our discussion of conventional high voltage sources is the connection of the

HT transformer to the tube. For tank units, this is not a major consideration because the transformer can be connected directly to the tube electrodes. However, for units with separate components, insulation and connection of leads (which may carry voltages in excess of 200 kV) is an important consideration. The HT cables themselves are shielded to provide protection against electrical shock. Cables used at lower energies are relatively flexible but as the amount of insulation is increased, the flexibility decreases and sharp bends during installation should be avoided.

The cables are inserted into terminations usually made of phenolic (thermosetting plastic) or ceramics and are sealed against air by use of insulating epoxy materials called *potting compound.* The phenolic termination used primarily with glass and lower energy metal-ceramic tubes is of the form shown in Fig. 14a.

FIGURE 14. High Voltage Transformer Terminations: (a) Phenolic Connection; (b) Ceramic Connection

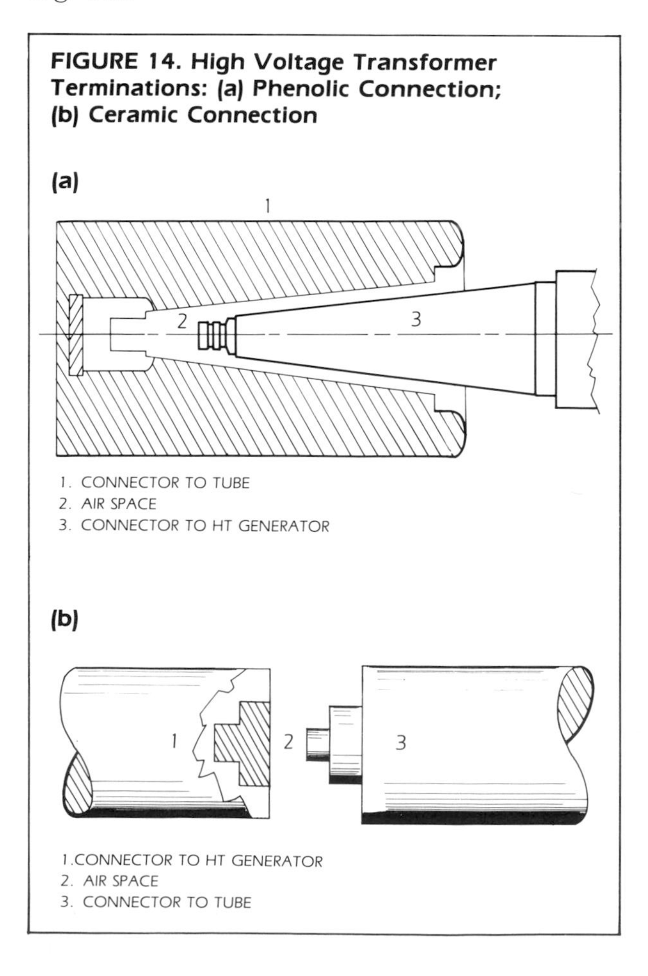

These connectors have rather large dimensions in comparison to the newer style ceramic terminations used with higher energy metal-ceramic tubes (Fig. 14b). Both styles use highly dielectric grease to seal out air at joining surfaces. The long male-female connections tend to trap air and substantially reduce the insulation capabilities of the grease and these joints are normally rated at approximately one-tenth their theoretical values or about 10 kV per centimeter. The linear dimensions of such a termination must be correspondingly increased. At the transformer tank, this size increase is not so important, but at the tube head the increase in size and weight can make the unit very cumbersome and bulky. For a 400 kV tube head, 200 kV is applied to each electrode. This requires a termination of 18 to 20 cm (7 or 8 in.) in length at each end of the tube for proper insulation.

The ceramic insulator pictured in Fig. 14b makes full use of the dielectric strength of the insulating grease by providing rigid, flat mating surfaces which exclude air from the joint. This allows for a substantial decrease in the length of the joint. This design has been incorporated into tubes used by several equipment manufacturers for units up to 420 kV.

As stated above, a transformer is used to provide the potential difference for conventional X-ray units. As the accelerating voltages are increased toward 1 MeV, standard transformer and insulation technologies become inadequate. Although the techniques used in these higher energy X-ray generators will be discussed at length in Section 6, *High-Energy Radiography*, the following text lists the major types and briefly discusses their operating principles.

Resonance Transformer

Resonance transformers are used in conjunction with multisection tubes to produce X rays in the range of 1 to 2 MeV. By resonating the transformer circuitry at a multiple of the input frequency, the ferromagnetic core of the transformer can be eliminated. The weight and bulk of the transformer are reduced not only by the removal of the core, but also by the removal of the insulation necessary to isolate the core from the windings. In the absence of a core, the tube can be placed on the axis of the HT windings. Proper spacing of the active tube segments allows the acceleration of electrons to take place in several intervals instead of the single active region of the conventional tube. Connections to sequential portions of the tube and windings are facilitated by the concentric arrangement, and the tube is also electrically shielded due to its central location. Although bulky in comparison to more modern generators, resonance transformers have proven to be very durable. Many of the original units are still in use some three decades after manufacture.

Van de Graaff Generator

This generator (see Fig. 15) is unique in that the potential difference is produced by mechanically transporting a charge to the high voltage terminal via an insulating belt. The terminal is surrounded by a case which is held at ground potential. Electrons are emitted from an electron gun and pass between the high voltage terminal and the outer case. The potential difference between the terminal and case causes the electrons to accelerate to very high speeds.

Upon striking a target, X rays are produced in large quantities due to the increased efficiency (10% and higher) of all high-energy generators. The accelerating potential is essentially constant and X-ray output of several grays (several hundred rads) per minute at one meter is attainable. Van de Graaff units are commonly used to provide energies of up to 8 MeV; but because of the size of the larger machines, radiographic units seldom produce more than 2.5 MeV. The larger units find applications in research and development and may be used to accelerate particles other than electrons.

Betatron

The betatron, synchrotron and cyclotron, all share a common heritage in the use of magnetic fields. The oldest member of the family, the cyclotron (see Fig. 16) makes use of magnetic deflection to bend a stream of heavy, charged particles, such as completely ionized deuterium, into a spiral path. During each cycle, the particles are accelerated as they pass between oppositely charged hollow electrodes, shaped like the letter D (*dees*). Using this type of arrangement, it is possible to produce deuterons with energies of more than 15 MeV, in dees that are just over 1 m (3 ft) in diameter.

The spiral path of the charged particles is produced by a fixed magnetic field. By increasing the field in synchronization with the acceleration of the particles, it is possible to maintain a circular path. This allows replacement of the polar type magnet with a toroidal (doughnut-shaped) magnet. The resulting synchronized cyclotron or *synchrotron* can produce particles having energies of more than 500 billion electron volts. Although such units are used mainly for elementary particle research and military applications, smaller versions used to accelerate electrons do have radiographic applications.

The betatron is such an adaptation (see Fig. 17). Here, acceleration of the electrons is produced as a direct effect of the increasing magnetic field. As the field of the polar magnet is increased, an electric field is created by the process of induction. The electric field acts much the same as a potential difference

FIGURE 15. Van de Graaff Generator

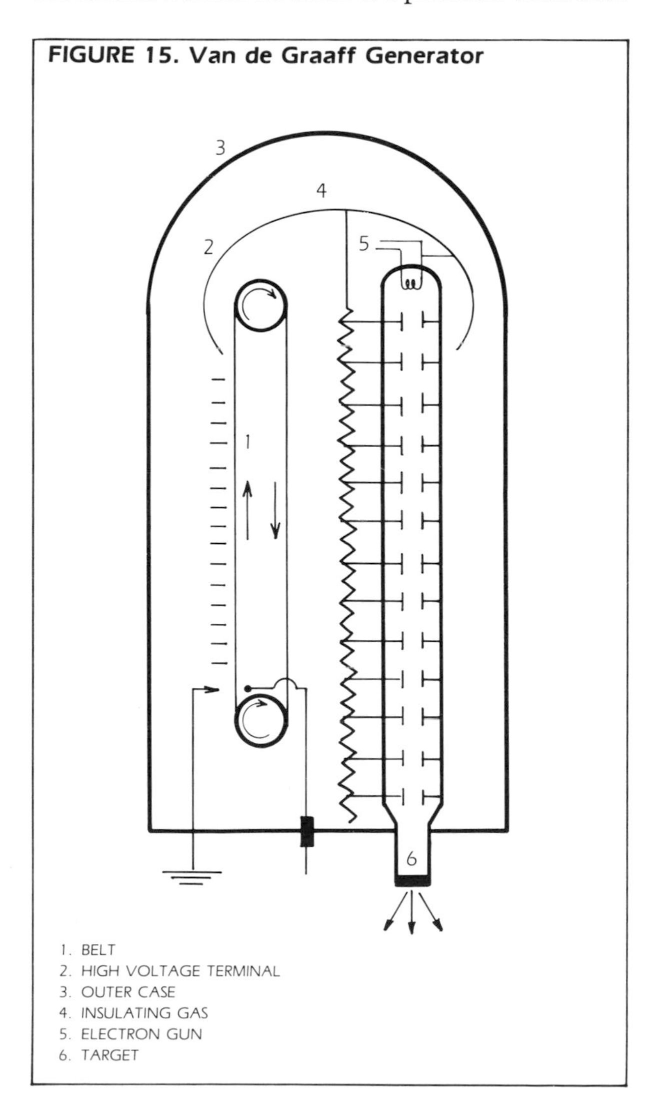

would in accelerating the electrons to high speeds. During each circuit around the betatron, the electrons gain several hundred electron volts of energy. Following more than one hundred thousand cycles, the beam of electrons can have energies of 20 to 50 MeV. The X rays produced by this beam are capable of penetrating 40 cm (15 in.) of steel in relatively short exposure times.

Linear Accelerator

Electrons introduced into the cavity of a linear waveguide (carrying a standing or traveling radiofrequency [rf] wave) will experience an electric field acting axially along the guide. In this field, electrons will increase their kinetic energy at the rate of approximately 1 MeV per 30 cm (12 in.) of guide. This allows for construction of very compact units capable of producing X-ray energies from 2 to 10 MeV with outputs as high as several grays (several hundred rads) per minute at one meter. These units can be used to accelerate particles other than electrons and have been used in materials research and weapons technology as well as in radiography.

FIGURE 16. Cyclotron Dees

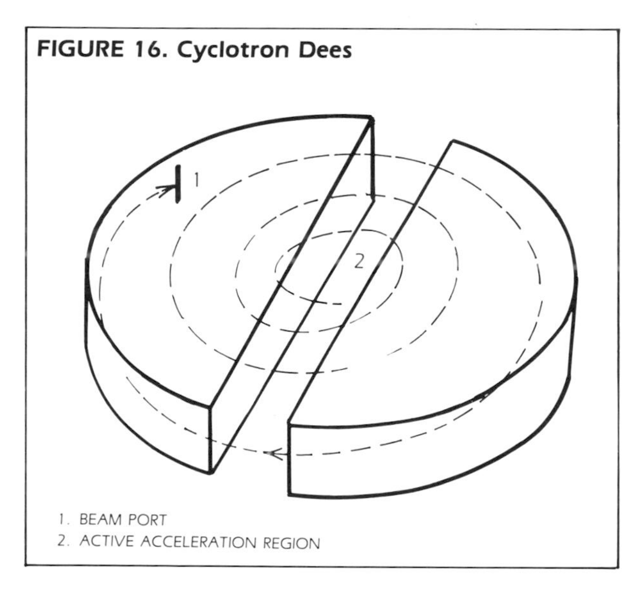

1. BEAM PORT
2. ACTIVE ACCELERATION REGION

FIGURE 17. Betatron

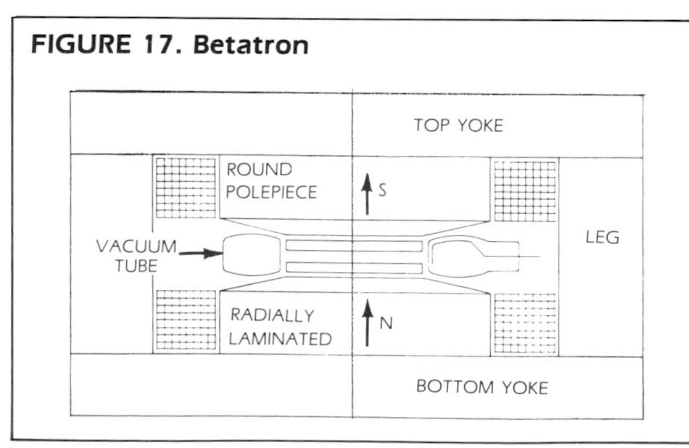

Control Unit

Line voltage is introduced to the unit through the *control* which is often capable, especially for portable units, of accepting 110, 220 or 440 V. Single phase 50 or 60 Hz AC is generally used for portable applications, while fixed units may be single or three-phase in design. Units may also have some adjustment to compensate for line voltage drops such as those caused by use of long extension cords.

Aside from allowing the unit to be turned on and off, controls usually allow for the adjustment and monitoring of the three radiographic variables: exposure time, energy and tube output. The actual controls and monitors may vary in appearance, or may even be absent from a particular unit, but these functions will be performed either manually or automatically in all units.

The control will contain the fuses and circuit breakers; warning and interlock circuitry; various initiation, termination and security switches; and on some later models, computerized memories with digital controls. The particular type of control depends on the range of functions required of the unit, and just as importantly, the skill and training of the operator. For a production facility with standard techniques established, a programmable control unit might be desirable; for an independent test lab, a control which allows easy adjustment and which is rugged and portable enough for field use may be required. Such requirements need to be considered prior to the acquisition of any unit.

Kilovoltage Adjustment

X-ray energy is controlled by adjustment of the voltage supplied to the primary of the HT transformer. This may be done electronically or manually. For manually adjusted units, a *Variac* is commonly used. The Variac is a transformer with a toroidal winding that provides an adjustable output voltage from zero to about 17 percent above line voltage. This allows for continuous adjustment of the voltage between minimum and maximum values. It is usually necessary to reduce the kV setting to its minimum value before an exposure can be initiated. The Variac setting is then increased gradually to the desired value. This avoids electrical stress that would otherwise be experienced by the tube and insulation.

For units which are not continuously variable (for units that are variable in steps), adjustment during an exposure should not be attempted. This introduces transients into the high tension circuitry which can exceed design limits of the tube; arc-over between the cathode and anode can occur, leading to tube failure. Such units normally have provisions for minor adjustment of kilovoltage, as well as an automatic initiation cycle to gradually apply voltage to the tube. Digitally controlled units will also have an initiation cycle as a part of their circuitry.

Programmable systems range in complexity from card to keyboard and will become a more requested item on most control units.

Milliamperage Adjustment

Tube output is controlled by regulating tube current which is, in turn, strongly influenced by filament current. As previously discussed, cathode-grounded units need not have an isolated filament transformer. The filament supply is often contained directly in the control. Monitoring of tube current is done by measuring the current passing between ground and the secondary of the HT transformer. On some units, monitoring and adjustment of the current is done automatically within the control.

For units that use center or anode-grounded systems, the filament transformer must be isolated and the control will allow adjustment of the input voltage to the primary of the filament transformer.

A fact to remember is that kilovoltage and milliamperage settings are not completely independent. At high voltages the electrons are quickly accelerated away from the cathode, but at lower voltages, the X-ray tube operates much less efficiently due to space charge buildup. In addition to space charge, the screen effect also influences tube output. For units rated to operate at 5 mA, the kV meter is calibrated at 5 mA and will not necessarily be accurate at other values. Likewise, if the unit is adjusted to produce 5 mA at 100 kV, the tube current may be above limits when the kV is raised to higher values. Operators should also be aware that some older units did not monitor tube current at all, but instead measured filament current.

PART 3

GENERAL OPERATING RECOMMENDATIONS

Baseline Data

Once a unit is acquired, actual operating characteristics should be checked against the quoted values and recorded for future reference. This *baseline* information is useful not only for technique determination, but also for trouble-shooting purposes.

Equipment Literature

Information available from the manufacturer might include: weight and size, energy, tube current, inherent filtration, focal spot size, normal gas pressure and pressure rise (for gas filled heads), coolant requirements, line current, operating temperature, etc. Service manuals also supply useful information and should be acquired for all units. This information should not simply be kept on file; it should be utilized by personnel operating the unit. A separate specification for the X-ray tube is often available from the manufacturer or from the tube supplier.

Exposure Charts

For any unit that will be utilized in a variety of applications, a set of exposure charts at different output energies is very useful and should be produced for each unit. This should be done under conditions that are representative of procedures at the particular facility. It is necessary to do this for each unit (or, at least, for each type of unit) due to the differences in generator characteristics, tube efficiency and inherent filtration.

To produce a chart for a particular energy range, several exposures of a step wedge are made. All variables including kV, mA, focus-to-film distance, film type, screens, chemistry and processing times are kept constant for each chart; only the exposure time is varied. At least three, preferably five, exposure times are then selected which adequately span the range of possible exposure times. The exposures are made and processed (fresh chemistry is recommended). Density measurements are then made for each thickness represented on the radiograph. Usually densities of 1.5, 2.0 and 2.5 are chosen and the corresponding material thickness and exposure times are plotted on semi-log paper. This is done to convert the exponentially increasing graph into a straight line representation. Production of such a graph not only allows for accurate determination of exposure techniques but also allows the unit performance to be checked at any later date by comparison with established chart values.

Focal Spot Size

As previously discussed, focal spot size and beam intensity are important factors affecting radiographic quality. In addition, many codes require calculation of geometric unsharpness for technique approval. Because the focal size is subject to change due to target wear and variation in the electronic focusing circuitry, periodic measurement is necessary to ensure the accuracy of calculations and, more importantly, the quality of the radiographic image.

For conventional X-ray units up to 300 kV, measurement is done using a standard pinhole aperture. Using a fine grained film without screens, an exposure is made of the aperture. The film is processed and the image is measured, using a 5× to 10× calibrated reticle. To facilitate measurement of smaller focal spots, the aperture is usually placed so that a magnification of 2× is obtained for focal sizes between 1.2 and 2.5 mm (0.05 to 0.1 in.) and a magnification of 3× for focal sizes between 0.3 and 1.2 mm (0.01 to 0.05 in.). Smaller focal spots require an alternative measurement technique in which a standard grid pattern is used in place of the pinhole camera.

Additional Data

Other useful information obtainable through measurement might include:

1. sieverts (roentgens) output at various kV settings;
2. filament current versus tube current;
3. line current drawn at fixed output.

An observant operator will notice other characteristics, such as sounds or control positions. Any change in these characteristics may be just as important as those already mentioned.

Selecting a Unit

In addition to size limitations and mobility, the process of matching the unit to the job is in many cases determined by equipment availability. However, in those instances where a choice exists, including equipment purchase, there are some basic considerations which should be taken into account.

Application

The application or range of applications (including material and thickness, configuration and accessibility, and production rates) forms the first consideration in selection of a unit. Specific applications dictate the techniques to be used and the rate at which the exposures are made.

Applicable Specifications

The government or commercial specification, to which the radiograph is being produced, may also influence equipment selection. As an example, some specifications require various energy ranges for different materials and thicknesses. Quality and sensitivity requirements are also in this category.

Energy Range

Operating a unit continually at maximum output will shorten the time between maintenance. A rule followed by many experienced radiographers is never to exceed 90% of the maximum kV rating of the tube. This provides a measurable increase in tube, transformer, connector and control-unit life.

Likewise, operating a tube at much below half of its maximum output voltage, although not taxing on the electrical components of the unit, is not recommended as the standard mode. Again, availability may require such usage, but a unit used in this manner is usually operating at a much reduced efficiency and may not provide optimum results.

As an example, consider a 200 kV unit. If operated at 70 kV, the inherent filtration of the tube and tube head is such that a large fraction of the low energy photons are absorbed before leaving the housing. Consequently, the tube current and/or the exposure time must be increased to compensate for the loss.

Duty Cycle

In addition to maximum and minimum output energies, duty cycle must also be considered in selection of a unit. Although overheating protection is provided by all major manufacturers, a competent technician will seldom allow the unit to reach cut-off temperature before allowing it to cool. The possibility of localized overheating of target, envelope and other components is enhanced any time the rated duty cycles are exceeded.

In conjunction with production requirements, the above items form the basic considerations in selection of a unit.

A unit should generally be used in the range of 50-90% of its maximum output voltage with exposure times and production rates consistent with the duty cycle and quality.

Tube Warm-up

To avoid thermal shock, arc-over, backfire due to outgassing of the target, or other detrimental effects, it is advisable (and generally recommended by manufacturers) to follow a warm-up procedure when placing a unit into service. The longer the period between uses, the longer the recommended warm-up. Starting at approximately 50% of the maximum and proceeding in 10% increments to maximum output will generally suffice for overnight or weekend periods. The amount of time at each stage should not be much longer than two minutes and should not exceed values consistent with the rated duty cycle. If fluctuations in output are noticed, such as jumps in tube current (mA), the kV setting should be reduced until a stable value is obtained. Warm-up should be continued from this point with a decrease in the size of the increments used.

One definite improvement made by the introduction of programmable controls is that this entire warm-up procedure can be carried out automatically. This leaves the technician free for other duties, and ensures that the warm-up is not rushed. In more sophisticated systems, adjustment will also be made for the amount of time that the tube has been dormant.

Maintenance

The extent of the maintenance performed by the user is, of course, determined by the capabilities of the personnel. Some items of maintenance that can and should be performed on a routine basis, are listed below.

1. *Unit cleanliness.* This includes removal of dirt and oil from the tube head, connectors and control. This should not be considered a cosmetic function by any means. Minutes spent here may avoid hours of downtime for repair. Few controls are totally sealed against dirt or moisture and oil can cause insulation to break down prematurely.
2. *Visual inspection.* Wiping down the unit and power cables also affords an opportunity to inspect the components for damage, such as a broken wire or loose connections. Loose connections and partially broken wires should not be overlooked, because they can introduce transients into the high voltage circuitry. Burnt or loose pins on connectors are equally important. Oil seepage or coolant loss are items which also can be detected at this time.
3. *Fuse replacement.* Use of the correct size is recommended. If the unit repeatedly blows fuses, repair is indicated. Use of fuses larger than specified may overload other critical components.

For advanced maintenance and trouble-shooting efforts, refer to the manufacturer's instructions.

Electrical Safety

X-ray equipment manufactured today must conform to many national and international design requirements. In order for safety features of the equipment to remain operative and effective, proper maintenance and operation are required. All personnel involved with the operation of X-ray equipment should be familiar with the manufacturer's operating instructions and the specific safety features of the machine. General electrical safety considerations are listed below.

Power Sources

Many X-ray units are adaptable to different line voltages. Use the appropriate power cable and connector. The voltage selection dial on the control panel must be set for the power in use. Required wiring changes should be performed in accordance with the manufacturer's instructions.

Grounding

All X-ray machine power cables must be grounded. Do not use an ungrounded power circuit. Replace connectors which have defective or missing ground plugs. Failure to do so is a shock hazard and may be a fire hazard as well.

Fuses

Replacement fuses should be rated at the amperage required by manufacturer's specifications. Damage to components or an electrical hazard to the operator may result from the use of inappropriate fuses.

Control Circuits

Do not bypass or override the overload and overheating circuit breakers or the tube head limit switches. Electrical overload and excessive heat can damage electrical components. Tube head limit switches are required for safe and legal operation of the equipment.

Radiation Safety

The radiation hazards associated with the use of X-ray equipment can be minimized by adherence to state regulations and the manufacturer's operating and maintenance instructions. Written instructions for the safe and reliable use of equipment should be supplied to all operators. Operators should be aware of the physiological hazards of penetrating radiation. In addition, careful attention must be given to personnel monitoring, facility design and radiation survey techniques.

Personnel Monitoring

State regulations usually specify the extent of this monitoring. Film badges or thermoluminescent dosimeters (TLDs) are required for recording accumulated doses of radiation exposure. Use of pocket dosimeters or chambers is often required and is highly recommended. These devices allow for the timely detection of radiation exposure. Dosimeter construction should be appropriate to the energy level of the X rays being utilized. Dosimeter housings made of aluminum or composite materials record radiation exposure to low energy X rays (100 kV or less) more reliably than steel-cased dosimeters. Recordkeeping of the results of personnel monitoring systems must also comply with state regulations.

Facility Design

Facility design comprises two categories, fixed and portable. The design of a fixed facility should be reviewed by a qualified expert. Consideration must be given to the occupancy of adjacent areas, the adequacy of shielding, warning signals, interlocks, tube head position restrictions and radiation monitoring procedures. Conformance to state law is essential and subject to inspection or audit by regulatory agencies.

Fixed shielding is impractical for portable operations. The protection of personnel and the public depends almost exclusively on strict adherence to approved, safe operating procedures. Areas of radiation in excess of 1 mSv per hour (100 mR per hour) must have the perimeter posted with signs displaying the radiation symbol and the words *Danger: High Radiation Area*. The perimeter of areas with radiation levels in excess of 50 μSv per hour (5 mR per hour) must be posted with a sign displaying the radiation symbol and the words *Caution: Radiation Area*. Access to these areas must be controlled by the radiographer.

Surveys

Radiation surveys are an intergral part of the safe operation of X-ray machines. These surveys are conducted to determine the extent of radiation hazard in any given area. The survey meter is a rate instrument which indicates the exposure received per unit time. The most commonly used instruments are the Geiger-Mueller (G-M) and ionization chamber meters. Energy response of the meter should be consistent with the energy of the radiation. A meter suitable for field isotope use may be relatively insensitive to low energy X rays. Likewise, the size of the G-M tube or chamber must also be considered. A large detector will give erroneous values when applied to small radiation leaks.

Quarterly calibration of the survey instrument is usually required by law. Use of a calibration source similar in energy to the measured X rays is preferable. The conscientious use of a calibrated survey meter is the most reliable way to ensure the safe use of X-ray equipment. Failure to use the survey meter is a factor in the majority of all occupational overexposures. For facilities which are not required to perform routine surveys, this function may be performed by an outside service at specified intervals (annually or semi-annually).

BIBLIOGRAPHY

1. Beiser, A. *Concepts of Modern Physics*. New York: McGraw Hill Book Company (1963).
2. Halliday and Resnick. *Fundamentals of Physics*, 2nd edition. New York: John Wiley & Sons, Inc. (1981).
3. *Industrial Radiography Manual*. US Office of Education, Division of Vocational and Technical Education, and US Atomic Energy Commission, Division of Nuclear Education and Training (1968).
4. Andrews and Lapp. *Nuclear Radiation Physics*, 3rd edition. New Jersey: Prentice-Hall, Inc. (1963).
5. *The Nondestructive Testing Handbook*, 1st edition. New York: Ronald Press Company (1959).

SECTION 3

ISOTOPE RADIATION SOURCES

Frank A. Iddings, Nuclear Science Center, Louisiana State University, Baton Rouge, LA

ACKNOWLEDGMENTS

Much of this chapter was re-written based on text contained in the first edition *Nondestructive Testing Handbook.* Special credit is given to the original contributors of that material. The author also acknowledges Mark L. Rock, Sandra C. Shipp and Mike Newsom for their help in obtaining illustrations for this chapter, and Dr. J.C. Courtney for his help with Part 6. The encouragement, patience and assistance of many others is greatly appreciated.

PART 1
RADIOACTIVITY

Historical Background

Until the year 1896 it was generally believed that the *elements* represented the most stable form of matter occurring in nature; nothing was known that could change their characteristics. Chemical compounds could be built up from the elements and by various means transformed into other compounds, but all attempts to reduce the elements to simpler forms were unsuccessful. After the 1896 discovery of X rays by Roentgen in Germany, Becquerel discovered that certain uranium ore gave off penetrating radiation. Under his direction, the Curies isolated and identified radium and polonium which were less stable than the uranium. The unstable elements were described as being *radioactive.*

Rutherford, in England, and Villiard, in France, identified the radiations emitted by radioactive elements. Rutherford also developed the concepts of the elements' isotopes and the existence of the neutron. Around 1930, Cockcroft and Walton, in Rutherford's laboratory, discovered that if atoms of hydrogen or helium were accelerated in a high-voltage generator to several hundred kilovolts energy, their impact upon various target elements would cause some of the targets to become radioactive. The Curies discovered that alpha particles interact with beryllium to produce neutrons. Fermi, in 1934, discovered that such neutron sources could produce radioactivity. His study of neutron-rich radioisotopes of the elements made a whole new realm of unstable isotopes available.

A decade later, with the advent of nuclear reactors that could generate tremendous neutron intensities, *artificial isotopes* became available in very large quantities. The nuclear reactor also produced radioactive isotopes by a previously unknown process: *nuclear fission.* Here an atom of the reactor fuel, uranium-235, captures a neutron and thereupon splits into two fragments that are highly unstable. These immediately eject one or more neutrons which are used to split further uranium nuclei. The remaining portions of the fragments are generally unstable and therefore radioactive; they can be chemically separated from the reactor fuel and used as sources of radiation. These *fission products* are produced in great abundance by present-day reactors. Some of the fission products are useful, valuable radioisotopes for industry, medicine and research while others are radioactive waste.

Natural Radioactivity

All elements higher in atomic number than bismuth are radioactive and result from the decay of either uranium-238, uranium-235 or thorium-232. Radioactive decay of these elements and their decay products involves the nuclear release of either a helium nucleus (alpha particle) or an electron (beta minus). Each emission transforms the parent (original) isotope into a daughter isotope having a different mass and nuclear charge. This process continues until the daughter isotope is stable. The radioactive series always terminates in one of several isotopes of lead. The original parents of the three series have a very low probability of decay or a very long half-life and therefore still exist in nature.

Half-Life

Half-life (T) is the time required for one-half the original number of atoms to decay or change to the daughter atoms. Half-lives of each of the radioisotopes in the uranium-238 series[1] are given in Fig. 1. Half-life describes the probability of decay for large numbers of atoms and is much more commonly used than the actual probability, λ (decay constant), of an atom disintegrating per unit time. The number of atoms disintegrating per unit time can be expressed as λ times the total number of parent atoms (N):

$$\frac{\text{disintegrations}}{\text{time}} = \lambda N$$

The *curie* (Ci) is the unit used to describe 3.7×10^{10} disintegrations per second. The SI system uses the *becquerel* (Bq), which is one disintegration per second. The half-life (T) is related to the decay constant by eq. 1:

$$T = \frac{0.693}{\lambda} \qquad \text{(Eq. 1)}$$

where 0.693 is the natural logarithm of 2. The number of radioactive atoms or the number of atoms decaying per unit time changes exponentially with time.

The rate of decay (or intensity) of radioactivity (I), and the number of atoms at any time, can be expressed in terms of time as numbers of elapsed half-lives:

$$\frac{I}{I_o} = (1/2)^n \qquad \text{(Eq. 2)}$$

or

$$\frac{N}{N_o} = (1/2)^n$$

FIGURE 1. Radioactive Series Decay, Showing Relationship of Radium and Its Decay Products to Its Ultimate Parent, Uranium-238

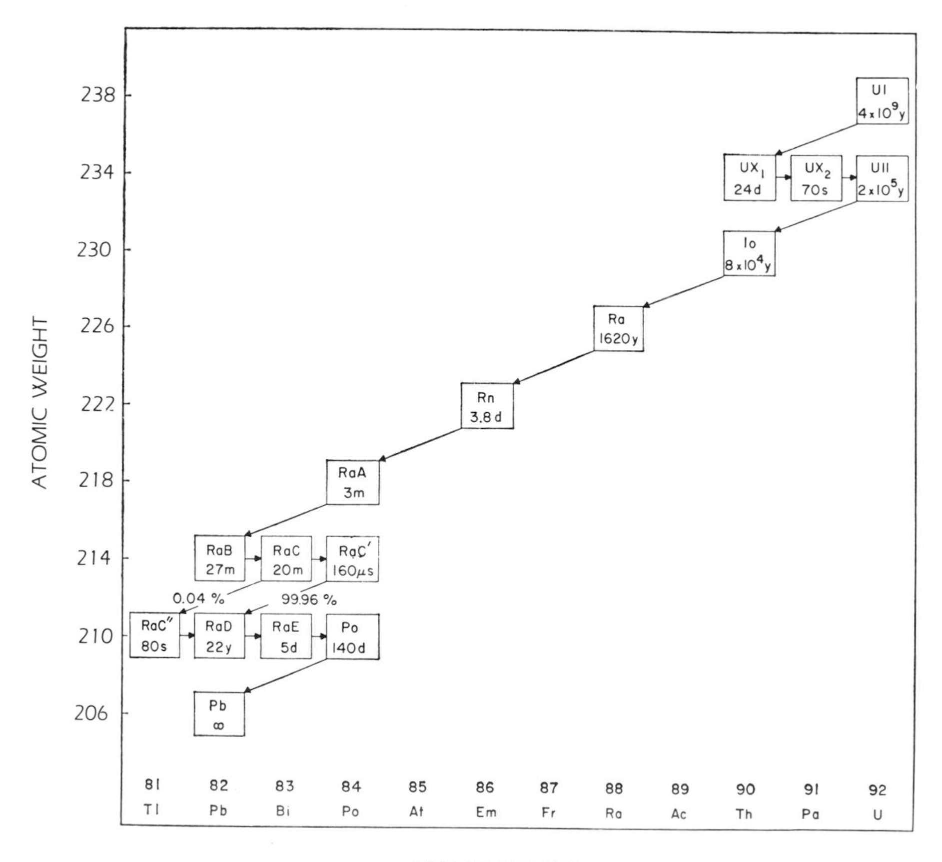

s = seconds d = days m = months y = years

A more convenient expression in terms of time elapsed (t) would be:

$$\frac{I}{I_o} = e^{-\lambda t} = e^{-0.693t/T} \quad \text{(Eq. 3)}$$

or the logarithmic form:

$$\ln\left(\frac{I_o}{I}\right) = 0.693t/T \quad \text{(Eq. 4)}$$

Thus, if intensity of radiation from a quantity of radioactivity is plotted as a function of time on semi-logarithmic coordinates, a straight line results, as shown in Fig. 2.

FIGURE 2. Half-life Plot

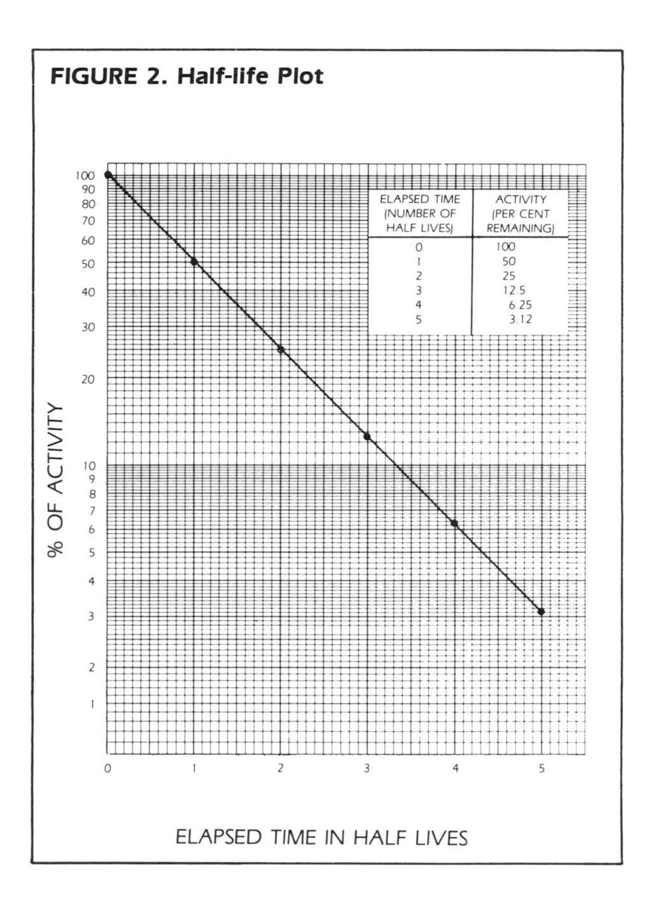

Artificially Produced Radioisotopes

Artificially produced radioisotopes have largely replaced natural radioisotopes, such as radium, for industrial radiography. Presently, artificial production falls into three major categories: (1) neutron activation; (2) fission product separation; and (3) charged particle bombardment.

While other nuclear reactions produce a variety of radioisotopes, neutron bombardment in nuclear reactors constitutes the major method for obtaining industrially important radiographic source materials. Two such nuclear reactions are:

$$^{59}Co + n \rightarrow {}^{60}Co + \gamma$$

and

$$^{191}Ir + n \rightarrow {}^{192}Ir + \gamma$$

Cobalt-59 and iridium-191 exist in nature and are stable. Exposure to the large thermal neutron flux (neutrons with energies less than 0.4 eV) allows stable isotopes to capture a thermal or slow neutron and thus become one mass unit heavier. The energy binding the neutron is lost as a *prompt* or *capture gamma ray*, immediately after the neutron is absorbed by the nucleus. The capture gamma is not the gamma ray utilized in radiography. When, as is the case for both reactions given above, the new isotope is radioactive, the radioactive decay produces the useful gamma rays.

Production of the radioactivity can be predicted by eq. 5:

$$A = Nf\sigma\,(1 - e^{-0.693t_i/T}) \quad \text{(Eq. 5)}$$

where: A is the radioactivity in disintegrations per second; N is the number of target or stable atoms being bombarded; f is the neutron flux in terms of neutrons passing through each square centimeter per second ($ncm^{-2}s^{-1}$); σ is the neutron cross section or the probability of neutron absorption by the target atom nucleus (units are cm^2, an effective area); t_i is the time irradiated in neutron flux; and T is the half-life of the radioactive product.

The above equation is correct only for thin samples of the bombarded material. Absorption of neutrons in the outer layers of the sample (metal pellet) reduces the number of neutrons incident on the interior atoms. This self-shielding of neutrons coupled with self-absorption of gammas released by radioactive atoms inside the sample gives gamma output considerably lower than calculated (see *Effective Curies* below).

Cobalt exists in nature as 100% cobalt-59, so only the cobalt-60 isotope is artificially produced. (An excited state of cobalt-60 is also produced but has a half life of only 10.5 minutes.) Iridium exists as iridium-191 (37.4%) and iridium-193 (62.6%) so that some iridium-194 is also made. Iridium-192 (T = 74.3 days) is the major product after the short lived iridium-194 has decayed away.

The cobalt and iridium are normally irradiated as small metal pellets (1 × 1 mm to 3 × 3 mm [0.04 × 0.04 in. to 0.12 × 0.12 in.]) which can be put together in stainless steel containers to form encapsulated radiographic sources. Typical activities are 5 Ci (200 GBq) for a 1 × 1 mm cobalt pellet and 50 Ci (2,000 GBq) for a 1 × 3 mm iridium pellet, but these values vary widely with irradiation times and reactor flux used.

A less commonly used radiographic source, cesium-137, is produced by a completely different process: fission. The fission of uranium-235 in a nuclear reactor fuel (to produce energy for electricity production or neutrons for radioisotope production) also produces fission fragments or fission products. The reaction can be represented as:

$$^{235}U + n \rightarrow 2 \text{ fission fragments} + 2\tfrac{1}{2}n + 200 \text{ MeV}$$

The two fission fragments are the small atoms into which the uranium atom has been split. Depending upon how the split occurs, two or three neutrons and about 200 MeV in energy are released. The majority of the fragments are close to either mass 95 or 140. The cesium-137 is recovered from used uranium fuel through chemical treatment or reprocessing. The recovered cesium is available in the chemical form CsCl. Conversion to a ceramic or glass form before encapsulation as a radiographic source is now common. The CsCl, similar to NaCl, can cause stress corrosion cracking of the stainless steel capsules and can also easily form an airborne dust.

Some radioisotopes are produced by bombarding stable isotopes with charged particles. High-energy machines such as Van de Graaff and linear accelerators and cyclotrons are being used commercially to produce radioisotopes. Using the deuteron (the nucleus of hydrogen-2) as the bombarding particle can produce products similar to neutron capture in a nuclear reactor by a *(d,p) reaction* (the first particle inside the parentheses is the entering particle and the second particle is the exiting particle).

If a deuteron enters a nucleus and a proton exits, the net result is an isotope that is one neutron heavier than the target isotope; i.e., the same as neutron capture. By bombarding with other particles at selected energies, good yields of very valuable radioisotopes can be produced; these are not readily available by thermal neutron bombardment or recovery from fission fragments. Also, the radioisotope may be obtained carrier free. *Carrier free* means that no stable atoms of that element are present, only the radioactive form is available. This is done by converting the target element into the radioactive form of another element. For example, iron-56 can be converted into cobalt-57 by a (d,n) reaction. After the deuteron bombardment, the cobalt is chemically separated from the iron giving a carrier free source of cobalt-57.

Disintegration Mechanisms

Regardless of how they are produced, radioactive atoms disintegrate by one of five primary modes:

1. emission of an alpha particle (helium nucleus);
2. emission of a beta particle;
3. electron capture or positron emission;
4. emission of a gamma ray (photon); or
5. spontaneous fission.

Emission of a gamma ray photon may follow some of the first three disintegration modes and only rarely occurs alone.

Emission of an Alpha Particle

The relatively heavy nuclei of helium generally carry with them considerable *kinetic energy*, between 2 and 6 MeV. This is the energy that would be acquired if a unit-charged particle were to move across a gap of the stated voltage. Alpha particles are easily stopped by small amounts of matter, such as a sheet of paper.

Emission of a Beta Particle

Beta particles are identical to high-speed electrons, though they may be either positively or negatively charged. They are emitted with energies continuously distributed up to a maximum value characteristic of the particular isotope. The distribution of the number of beta particles, as a function of particle energy, is known as a *Fermi distribution*. It typically rises to a peak at energies of one-third to one-half the maximum energy. Beta particles are usually stopped by thin layers of metal.

Electron Capture

An unstable high atomic number nucleus can capture one of the atomic orbital electrons circling about it. Because the innermost or K-shell electrons are usually caught, the process is often called *K capture.* Though no particle need be emitted from the nucleus after electron capture takes place, the process creates a vacancy in the atomic electron structure which, in being refilled, results in emission of *characteristic X rays.* These X rays form the tangible evidence that the process takes place.

Gamma Ray Emission

A very few radioactive isotopes decay by simple emission of gamma rays, which are electromagnetic radiation similar to monochromatic X rays but of nuclear origin. This process is known as *isomeric transition.* Usually gamma emission follows promptly after disintegration by one of the three processes listed above, rather than being the sole mode of decay. Because gamma rays carry with them neither charge nor rest mass, the isotope remains unchanged in atomic number or mass after emission of gamma radiation. Only the available energy is reduced, leaving the product nucleus in a more stable state.

Gamma rays range in *energy* from a few thousand electron volts (keV) to several million electron volts (MeV), the particular energy being characteristic of the isotope in question. Gamma rays are the chief radiation used in *isotope radiography.*

Spontaneous Fission

Nuclear species with masses greater than 200 can decay by a process of spontaneous fission, where the large nucleus of the atom splits into two smaller nuclei. The process is usually associated with the emission of two to four neutrons. Uranium-238 undergoes spontaneous fission as well as alpha particle decay. The ratio of alpha decay to spontaneous fission for uranium-238 is 600 million alphas to each spontaneous fission. As the mass (and atomic number) of the atom increases, spontaneous fission becomes more predominant. With transuranium isotopes (such as californium-252 and larger), spontaneous fission becomes the most common mode of decay. Such radioisotopes are used as neutron sources. Each microgram of californium-252 yields over two million neutrons per second.

Secondary Processes

During disintegration of a nucleus by one or more of these mechanisms, secondary processes may take place, often resulting in emission of further electromagnetic radiation.

Internal Conversion

When a radioactive nucleus has to eliminate excess energy, an alternate (and competing) mechanism to gamma-ray emission is *internal conversion.* Here the energy which would otherwise appear as a gamma radiation is transferred to an atomic orbital electron that is then ejected with kinetic energy equal to the expected gamma ray minus the electron's binding energy. The vacancy in the atomic structure is filled by an external electron, resulting in production of *characteristic X rays*, just as in the case of electron capture.

Internal conversion becomes more probable at lower energies and in heavy elements. A typical example is the disintegration of thulium-170, where the nucleus, having first lost a beta minus, disposes of its excess energy either by emitting an 84 keV gamma ray or by knocking out an orbital electron. If this electron was in the K shell, the atom emits 52 keV X radiation. In this example, K-shell internal conversion is 60 percent more probable than gamma-ray emission, so that one observes more 52 keV radiation than 84 keV radiation from this isotope.

Generation of Continuous X rays

Another secondary process is the generation of continuous X rays or bremsstrahlung resulting from the deceleration of high-speed electrons in the matter they penetrate. This radiation is identical with that generated in an X-ray tube, where electrons accelerated by the high-voltage electrical field are rapidly decelerated in the target. As its name implies, this radiation consists of all energies from 0 to some maximum value. The intensity is maximum at zero energy, decreasing to zero intensity at an energy equal to that of the electrons generating it.

In practice, the lower-energy radiation is always largely absorbed by the matter that produces it, resulting in an emergent spectrum that shows a maximum around 40 percent of the electron energy. Because the production of continuous X rays is a rather inefficient process, radiation generated by beta rays is usually negligible in comparison with nuclear gamma radiation. Beta and gamma rays are generally produced in approximately equal numbers, while continuous X rays usually represent one percent of the beta-ray intensity. A convenient approximate formula for beta intensity in milliroentgens per hour at one meter per curie (or in microsieverts per hour at one meter per gigabecquerel) is:

$$I = \frac{ZE_o^2}{10}$$

where Z is the atomic number of the absorber and E_o is the maximum energy of the beta rays, in MeV.

In special circumstances, the continuous X radiation becomes important. One example of this is the *pure beta-ray source*, where no gamma radiation is emitted; here the only emergent radiation is continuous X rays. The use of such sources for radiography has been described[2] for sources containing strontium-90. Maximum achievable efficiency appears to be about 10 percent of that from a source emitting equivalent-energy gamma rays.

PART 2

MEASUREMENTS OF ISOTOPES

Definitions

Characteristics of radioactive isotopes are specified by a number of parameters, which are defined in detail below.

The Curie

The *curie* (Ci) is a measure of the disintegration rate of a source. Originally equal to the disintegrations per second in one gram of radium, it is now defined as 3.7×10^{10} disintegrations per second (dps) or 2.22×10^{12} dpm. Clearly, the curie is independent of the mode of decay and of the energy of the disintegration products, so it is not a good measure of the intensity of radioactive materials. However, it is useful in radiographic work for comparing the strengths of various sources of the same isotope. In the SI system, disintegration is measured with the becquerel (Bq), which is one disintegration per second. Because this is a small unit, it is often used with large multipliers; gigabecquerels are common.

The Half-life

The ***half-life***, *T*, of a given radioisotope is the length of time required for the activity to decay to one-half its initial strength. Because any radioactive isotope decays exponentially, it will require a definite time to fall to any given fraction of its initial strength. The strength of a source will decrease one percent in 1/70 of a half-life; the time required to change by a factor of 10 is 10/3 of a half-life. Thus iridium-192, whose half-life is 74 days, loses one percent of its strength per day and will be weaker by a factor of 10 in about eight months.

Gamma-ray Energy

The energy of gamma rays is expressed in kilo electron volts (keV) or in millions of electron volts (MeV) and corresponds with that of the maximum energy or hardest X rays generated by an electronic X-ray tube energized at the stated potential. Because most of the X rays emitted by a tube are at about 40 percent of the maximum energy, gamma rays show characteristics similar to X rays of about twice the stated energy. Thus the penetrating ability of cobalt-60, which averages 1.2 MeV energy, is similar to that of a 3 MeV X-ray generator. Energy should not be confused with intensity; using a rifle-bullet analogy, energy corresponds to the bullet speed while intensity corresponds to the rate of fire.

Specific Activity

The specific activity of an isotopic source, measured in *curies per gram* or *becquerels per gram* of source material, is of importance to radiographers in two ways. A high specific activity indicates that a source of given strength will be of smaller physical size and thus will tend to yield sharper radiographs. Less self-absorption of radiation will take place in this small source, because there is less matter to stop the radiation.

Radiation Intensity

The *roentgen per hour* (R/hr) is a measure of the intensity of radiation and represents the effect of the radiation on a small volume of air at a given point. Strictly, it measures the rate of energy absorption of radiation passing through one cubic centimeter of standard air and so is not a true unit of intensity (which would require the determination of the total energy falling on a unit area). The roentgen per hour is thus the product of the intensity and the absorption coefficient for air. It is proportional to the intensity (within 10 percent) in the energy region of 70 keV to 2 MeV. Isotope radiography lies almost entirely within these limits, so that this very widely used unit can be applied as a measure of intensity.

Exposure to Radiation

Exposure to radiation is measured by *sieverts* (Sv) or *roentgens* (R) and is obtained by multiplying the intensity by the exposure time. Doses to personnel or to integrating detectors such as photographic film can be calculated the same way.

Characteristic Intensity

The *roentgen per hour at one meter per curie* (R/hr-m per curie) is a useful unit for relating the strength of a source (in curies) to its intensity. When the mode of disintegration of an isotope is known, including the energies and abundances of the emergent gamma rays, its R/hr-m per curie can be calculated using Fig. 3. If the necessary information is not known, the quantity is generally determined experimentally. Then, by combining the curie strength of a source with R/hr-m per curie, the source intensity in R/hr-m is obtained.

It should be noted that this procedure is valid only for sources in which self-absorption of the gamma radiation is negligible. However, it is becoming common in the radiographic industry to specify source strength not by actual curies but by effective curies which would give the same radiation output in the absence of self-absorption.

FIGURE 3. Roentgens per Hour at One Meter per Curie as a Function of Gamma-ray Energy, Assuming One Gamma Ray Per Disintegration

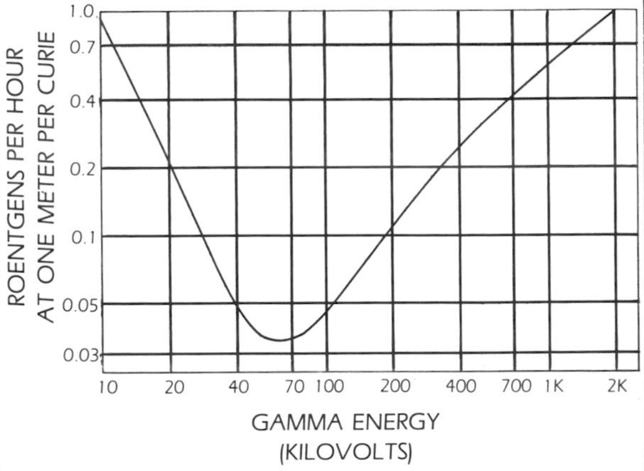

ROENTGENS PER HOUR AT ONE METER PER CURIE

GAMMA ENERGY (KILOVOLTS)

ONE ROENTGEN PER HOUR AT ONE METER PER CURIE EQUALS 250 MICROSIEVERTS PER HOUR AT ONE METER PER GIGABECQUEREL.

Inverse Square Law

The effect of distance on the intensity of radiation is calculable by the inverse square law, represented by eq. 6:

$$I_1(D_1)^2 = I_2(D_2)^2 \qquad \text{(Eq. 6)}$$

The equation shows that intensity varies inversely as the square of the distance from the source. This relationship holds strictly only for point sources in the absence of air or solid materials that scatter or absorb radiation. For gamma radiography, the law can be applied when source and detector are outdoors and separated by less than about 30 m (100 ft) of air. The scattering effects of walls and other indoor objects are not negligible (see Fig. 4), often increasing the observed intensity of a gamma emitter by as much as 30 percent at 3 m (10 ft) source-to-detector distance in a typical room.

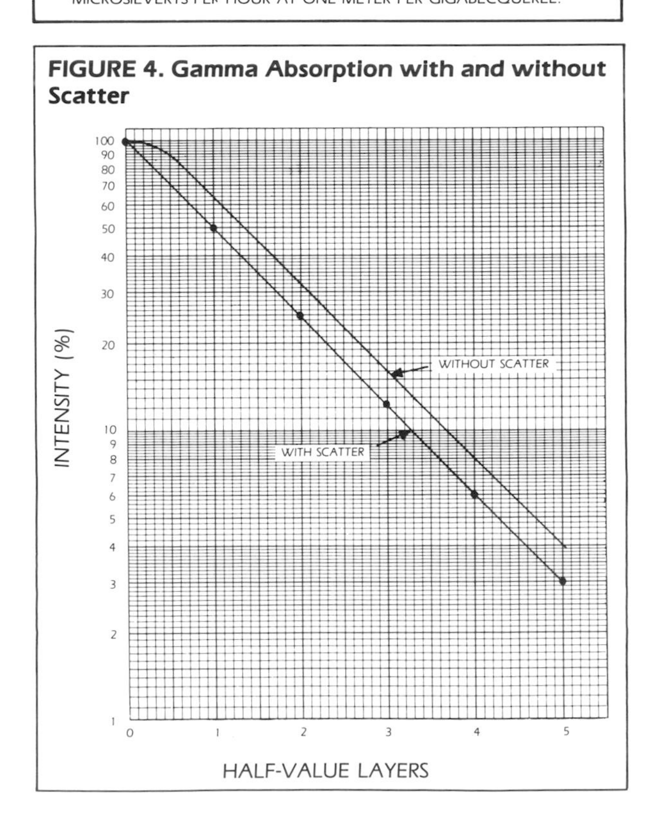

FIGURE 4. Gamma Absorption with and without Scatter

Half-value Layer

Because radiation is absorbed by matter, the intensity at any point will be attenuated if material is interposed between detector and source. This absorption varies with thickness in an approximately exponential manner, giving an equation in a familiar form:

$$I = I_o e^{-\mu t} \qquad \text{(Eq. 7)}$$

or

$$\ln\left(\frac{I_o}{I}\right) = \mu t$$

where: I is the intensity with an absorber of t thickness; I_o is the intensity with no absorber; and μ is the absorption coefficient or fraction absorbed per unit thickness (constant for each energy and absorbing material).

Unit change in thickness reduces the intensity by the same factor, almost independently of the total thickness. Because of this, absorbers are often characterized by their half-value layers (HVL) or the thickness required to halve the initial intensity. This concept is analogous to that of the half-life in radioactive decay. For centimeters, use:

$$\text{HVL} = \frac{1.76}{\mu} \qquad \text{(Eq. 8)}$$

and

$$\text{HVL} = \frac{1.76}{\mu_m}$$

or for inches, use:

$$\text{HVL} = \frac{0.693}{\mu}$$

These equations are correct only for a single gamma energy and a single absorbing material in a collimated (scatter free) measurement system. They can also be rewritten to include density of the absorber (ρ):

$$\mu = \mu_m \rho \qquad \text{(Eq. 9)}$$

$$I = I_o e^{-\mu_m \rho t} \qquad \text{(Eq. 10)}$$

or

$$\ln\left(\frac{I_o}{I}\right) = \mu_m \rho t$$

When scatter of primary gamma rays is considered, the equation becomes more complex. The scatter causes a higher radiation field than predicted by the above equation. This build up of radiation can be approximated by eq. 11:

$$I = I_o B e^{-\mu_m \rho t} \qquad \text{(Eq. 11)}$$

where $B \cong 1 + \mu t$.

PART 3

SPECIFIC ISOTOPES FOR RADIOGRAPHY

Selection of Radiographic Sources

Of the several hundred radioactive isotopes known to exist, a handful have become widely used for radiography; the remainder are unsuitable for a variety of reasons, including short half-life, low available intensity and high cost.

The following discussion gives details on the production and radiation characteristics of four popular radiographic sources (summarized in Fig. 5 and Table 1). Numerous other isotopes are useful radiographically (e.g., iron-59 and tantalum-182) and have no obvious drawbacks; these are not considered here because one or another of the four listed sources can surpass them in all respects. Radium-226 is no longer used for radiography because of the hazards presented by its alpha decay, its gaseous radioactive daughter and because it is a bone-seeking element.

Cobalt-60

Cobalt is a hard, gray, magnetic metal, having a melting point of 1,480 °C (2,700 °F) and a density of 8.9 g/cm³. It is somewhat similar in physical properties to iron. It occurs as a single isotope, cobalt-59, which is transformed into radioactive cobalt-60 after capturing a neutron. This isotope decays with a half-life of 5.27 years by emission of a soft beta ray and two hard gamma rays (see Fig. 6) resulting in the 1.2 MeV average gamma-ray energy. The relatively high slow-neutron capture cross section of 24 barns (one barn equals 10^{-28} m²) makes cobalt-60 one of the most readily available and generally useful isotopes.

FIGURE 5. Disintegration Schemes of Cobalt-60, Thulium-170 and Cesium-137; Diagonal Arrows Represent Beta Rays; Vertical Arrows Represent Gamma Rays; Energies Given in MeV

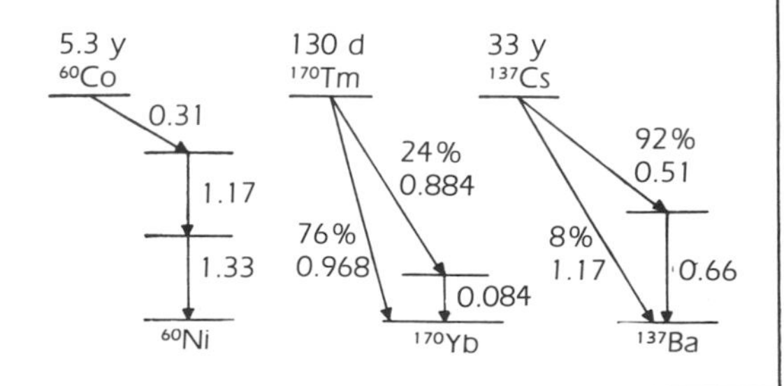

FIGURE 6. Shielding Requirements for Shipment of Cobalt-60 in Interstate Commerce;* Curves Refer to Lead Spheres in Various Size Boxes; Because the Curves Assume Negligible Source Size, Source Diameter Must Be Added to Pig Diameter to Obtain Required Outside Diameter

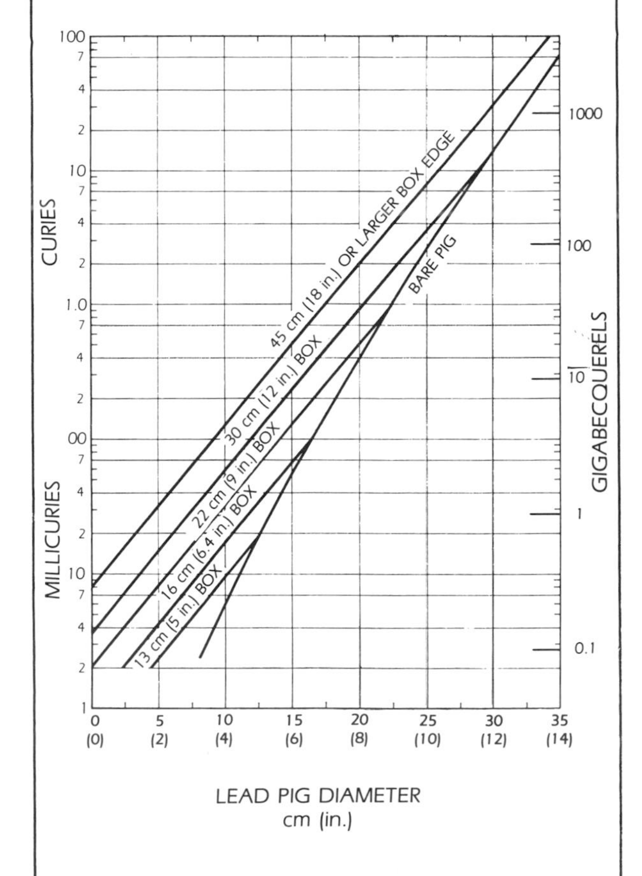

*Lead shielding must be encased in steel to prevent shielding loss in a fire. See Part 6 of this section for more information.

TABLE 1. Characteristics of Four Widely Used Radiographic Isotope Sources

	Element			
Characteristic	**Cobalt**	**Cesium**	**Iridium**	**Thulium**
Isotope	60	137	192	170
Half-life	5.27 years	30.1 years	74.3 days	129 days
Chemical form	Co	CsCl	Ir	Tm_2O_3
Density (g/cm^3)	8.9	3.5	22.4	4
Gamma rays (MeV)	1.33-1.17	0.66	0.31-0.47-0.60	0.084-0.052
Abundance of gamma rays (gamma rays per disintegration)	1.0-1.0	0.92	1.47-0.67-0.27	0.03-0.05
Beta rays (MeV)	0.31	0.5	0.6	1.0
R/hr-m per curie (mSv/h-m per gigabecquerel)	1.35 (310)	0.34 (80)	0.55 (125)	0.0030 (0.7)
Linear self-absorption coefficients (cm^{-1})				
Neutrons	3.0	—	33	1.5
Gammas	0.22	0.10	5.1-2.1-1.4	22.0-17.6
Ultimate specific activity in Ci/g (GBq/g)	1,200 (44,000)	25 (925)	10,000 (370,000)	6,300 (230,000)
Practical specific activity in Ci/g (GBq/g)	50 (1,850)	25 (925)	350 (13,000)	1,000 (37,000)
Practical curies (GBq) per cubic centimeter	450 (17,000)	90 (3,300)	8,000 (300,000)	4,000 (150,000)
Practical R/hr-m (mGy/h-m) per cubic centimeter	600 (6,000)	33 (330)	4,400 (44,000)	·10 (100)
For 50% self-absorption (curies)	200,000	500,000	3,000	2
For 25% self-absorption (curies max)	10,000	25,000	150	0.1
Practical radiographic sources:				
Curies (gigabecquerels)	20 (740)	75 (2,800)	100 (3,700)	50 (1,800)
R/hr-m (mSv/h-m)	27 (270)	30 (300)	60 (600)	0.1 (1)
Approximate diameter [mm (in.)]	3 (0.1)	10 (0.4)	3 (0.1)	3 (0.1)
Uranium shield diameter [cm (in.)]	33 (13)	20 (8)	15 (6)	5 (2)
Typical uranium shield weight [kg (lb)]	225 (500)	54 (120)	20 (45)	1 (2)

Specific Activity

Using eq. 5, calculation of the activity to be expected in 1.6 × 1.6 mm (0.06 × 0.06 in.) pellets of cobalt after irradiation at a flux of 10^{14} $ncm^{-2}s^{-1}$ for one 16-day cycle results in a value of 60 mCi (2 GBq) per pellet, or about 2 Ci (70 GBq) per gram specific activity. A radiographic source usually achieves a higher specific activity (by at least one order of magnitude) by leaving the cobalt in the reactor for many cycles of irradiation. A year (17 cycles) at 10^{14} $ncm^{-2}s^{-1}$ flux will result in the above pellets having about 1 Ci (40 GBq) activity each. Such small pellets can be used either singly or in groups to measure the total activity. In either event, they must be encapsulated to facilitate handling and to prevent small, intensely radioactive particles from being lost.

Applications

Radiographers employ cobalt-60 chiefly for inspection of iron, brass, copper, and other medium-weight metals whose thicknesses are greater than 2.5 cm (1 in.). It can also be used to radiograph thinner sections of heavy materials, such as tantalum or uranium. The isotope is particularly useful to those who normally work with 250 kVp (kilovolts-peak) X-rays and who have occasional demands to investigate sections thicker than 5 cm (2 in.). Cobalt-60 is radiographically equivalent to a 3 MeV X-ray generator, though it is not so intense a source. It can be used to make good radiographs through at least 20 cm (8 in.) of steel.

Shielding

Because of its penetrating radiation, cobalt-60 is a difficult material to shield; the average half-value layer of lead is 13 mm (0.5 in.). Figure 6 shows the shielding necessary to meet the 200 mR/hr (2,000 μSv/h) at the surface, required for shipment on common carriers. The various curves refer to the box size, within which is located the spherical lead shield of stated diameter. Thus an 8 mCi (300 MBq) source will be sufficiently shielded with the distance provided by a 46 cm (18 in.) box, while an 8 Ci (300 GBq) source needs either a 25 cm (10 in.) sphere in the 46 cm (18 in.) box, or a 29 cm (11.5 in.) diameter bare sphere (which will weigh at least 50 percent more).

Additional requirements for packaging and shipment of radioactive isotopes are detailed in the United States *Code of Federal Regulations*, Title 49 (Part 173), Title 10 (Part 71), and the International Atomic Energy Agency *Regulations for Safe Transport of Radioactive Materials*, Safety Series No. 6 (1979) and Safety Series No. 37 (1982). A summary of these materials is found in Part 6 of this Section.

Iridium-192

The 74.3 day half-life isotope iridium-192 is produced by neutron irradiation of the element, a very hard, brittle, white metal of the platinum family that melts at 2,350 °C (4,260 °F) and has a density of 22.4 g/cm³. Natural iridium occurs as two isotopes, 38 percent iridium-191 and 62 percent iridium-193. The lighter isotope yields the desired radioactivity. Iridium-191 possesses a slow-neutron capture cross section of 1,000 barns. This very high value means not only that iridium-192 is generated in large quantities but also that the parent metal is

FIGURE 7. Disintegration Scheme of Iridium-192; Energy Levels in Kilovolts; Numbers in the Arrows are Numbers of Gamma Rays per 100 Disintegrations

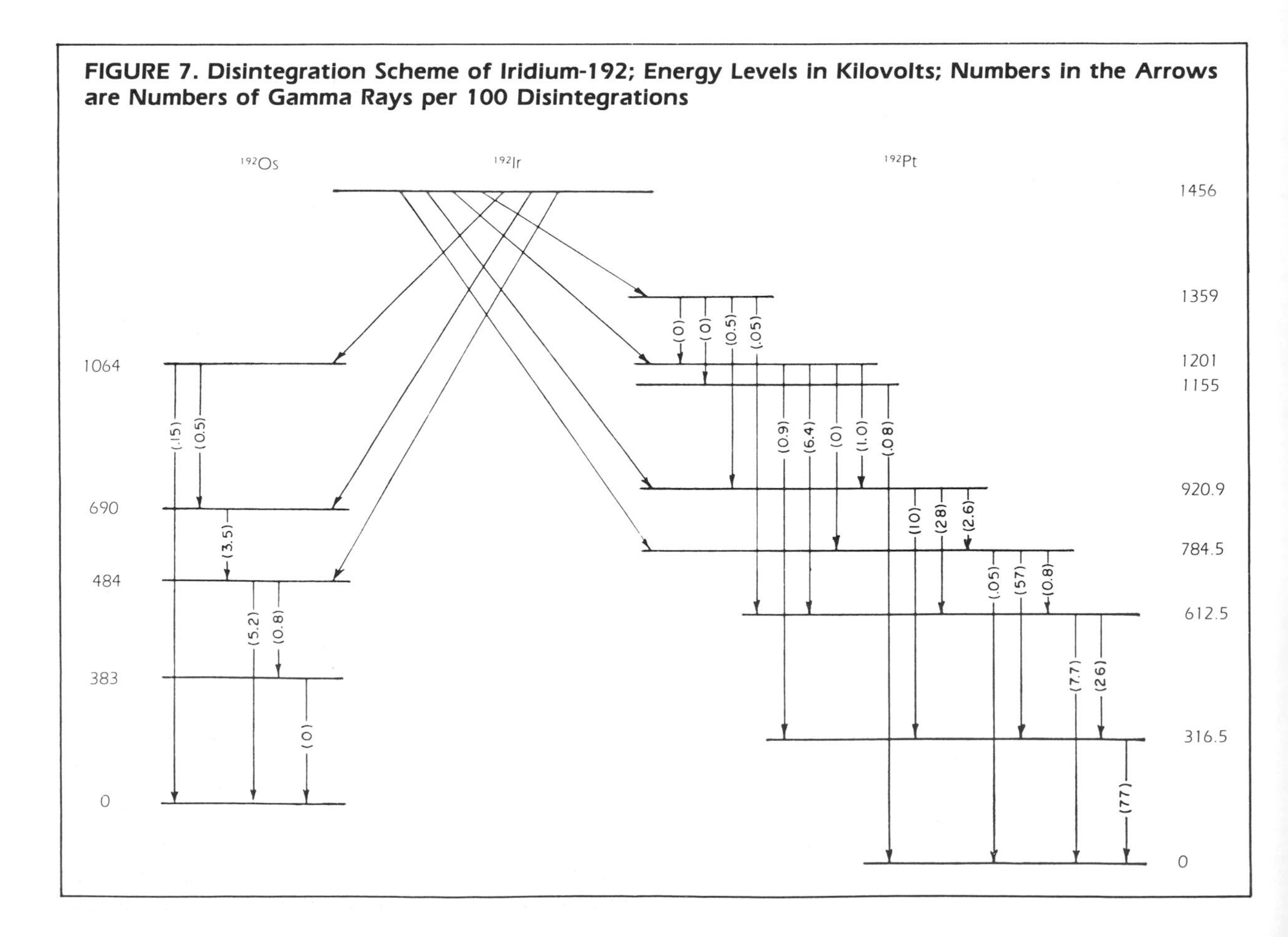

an efficient neutron absorber, the natural element having an absorption cross section of 470 barns. For this reason, the amount of material placed in the reactor must be limited to minimize self-absorption. (Most neutrons are absorbed in the outside layers of large pellets giving little iridium-192 production inside the pellet.)

Decay Processes

Decay of iridium-192 proceeds chiefly by beta-ray emission to platinum-192 but also by electron capture to osmium-192, both of which are stable. At least 24 gamma rays are known; the currently accepted decay scheme[5] is shown in Fig. 7. Computation of the radiation from these gamma rays, together with their abundances, results in agreement with the accepted value of 0.55 R/hr-m per curie (or 125 μSv/h-m per GBq). For radiographic purposes, iridium-192 radiations may be approximated by three gamma rays, shown in Table 2.

Specific Activity

Production of iridium-192 is shown in Fig. 8, which presents the results of computations for 3.2 × 3.2 mm (0.12 × 0.12 in.) metal pellets inserted into the reactor for varying numbers of three-week cycles. These curves reach maxima both because the isotope decays during irradiation and because the target material is being gradually used up. The curies shown are *effective curies* (the values obtained if the gamma-ray output of the pellet in R/hr-m is divided by 0.550). Figure 8 may be used for 1.6 × 1.6 mm (0.06 × 0.06 in) pellets by taking one-fourth of the indicated curies. Thus a maximum specific activity of about 500 Ci per gram of iridium is the most that can be generated in these smaller pellets.

Applications and Shielding

Iridium-192 is used principally for the radiography of steel in sections 3 to 75 mm (0.12 to 3.0 in.) thick, where it produces results similar to those from a 1 MeV X-ray generator. Its relatively low energy permits the use of lead shields weighing well under 50 kg for source strengths of 2,000 to 4,000 GBq (100 lb for source strengths of 50 to 100 Ci), making the isotope ideal for field work where portability and small size are at a premium. Figure 9 gives data on container sizes and lead sphere diameters necessary to shield iridium-192 in interstate commerce (see summary of shipping requirements in Part 6).

TABLE 2. Iridium-192 Radiations

Energy (kV)	Gamma Rays per Disintegration	Percent Gamma	Percent Roentgens
310	1.47	61	47
470	0.67	28	35
600	0.27	11	18

FIGURE 8. Production of Iridium-192 in 3.2 × 3.2 mm (0.12 × 0.12 in.) Metal Pellets for Various Numbers of Three-week Irradiation Cycles; Curies (Gigabecquerels) Are as Measured by Gamma-ray Output; One-fourth the Activity Will Result from Irradiation of 1.6 × 1.6 mm (0.06 × 0.06 in.) Pellets

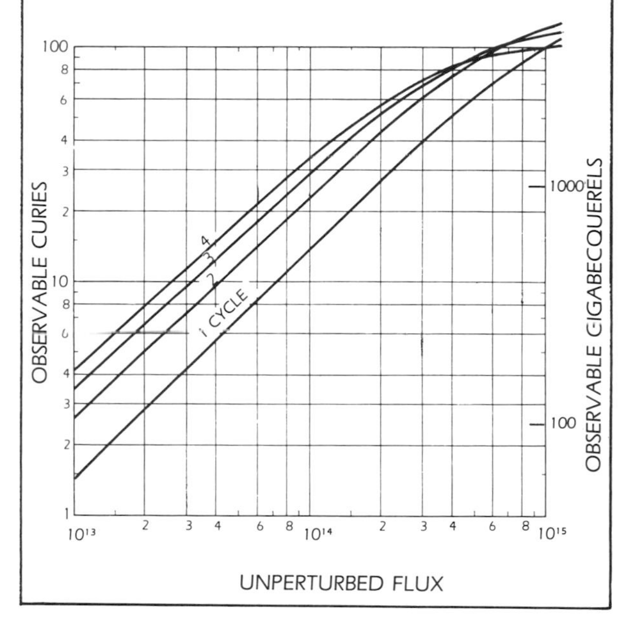

Thulium-170

The element thulium is chemically one of the rare earths, is physically a silvery metal with a density of approximately 9 g/cm³, and consists of the single isotope thulium-169. Because the metal is extremely difficult to produce, the material is generally handled as the oxide Tm_2O_3, either as an encapsulated powder of density approximately 4 g/cm³ or sintered into pellets of density 7 g/cm³. The isotope thulium-169 has a cross section of 120 barns for thermal-neutron capture, thereby producing the 129-day isotope thulium-170. The neutron-absorption cross section is equal to the activation cross section, because no other comparable interactions take place.

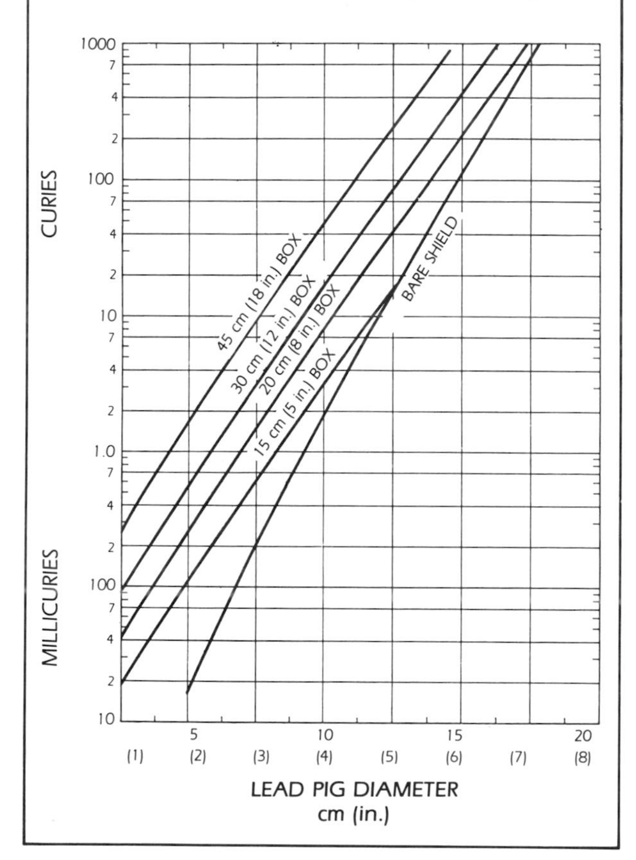

FIGURE 9. Shielding Requirements for Shipment of Iridium-192 in Interstate Commerce; Curves Refer to Lead Spheres in Various Size Boxes and Assume Negligible Source Size (See Part 6 of this Section for Shipping Requirements)

FIGURE 10. Radiation Emergent from 50 mg Tm_2O_3 Source (Compressed 2 × 2 mm [0.08 × 0.08 in.] Pellet) in a 2 cm³ Solution

Decay Processes

Thulium-170 decays with a 129-day half-life by emission of 1 MeV beta particles. In 24 percent of the disintegrations, the nucleus is left in an excited state which is stabilized either (1) by emission of an 84 keV gamma ray or (2) by ejection of an orbital electron (internal conversion). It has been shown[6] that 3.1 percent of the disintegrations result in 84 keV gamma ray emission, 4.9 percent in ejection of a K-shell orbital electron, and 16 percent in ejection of an L- or an M-shell electron.

In refilling those shells, the atom emits X rays characteristic of ytterbium: 52 keV radiation for K-shell, 7 keV for L-shell, and 1 keV for M-shell. The lower-energy quanta are too weak to emerge from the source, and therefore for radiographic purposes it may be assumed that thulium-170 emits 84 keV and 52 keV radiation in 3 percent and 5 percent of the disintegrations, respectively.

In addition to these two soft quanta, a proportion of continuous X radiation is also generated by deceleration of the 1 MeV beta rays in the body of the source. The relative intensity (product of quanta number and energy) of this radiation depends on the source density. The spectra produced by a concentrated and a diluted source of Tm_2O_3 is shown in Fig. 10. The radiation from a metallic thulium source would be expected to show smaller peaks on an even greater continuous X-ray background.

Source Strengths

The strengths of radiographic sources are generally calculated from irradiation conditions by use of data such as shown in Fig. 11; here the strength in curies of thulium-170 is given as a function of source size and reactor exposure. Combining the strength in curies with the radiation intensity as measured by ionization chambers, various experimenters report characteristic intensities varying from 2.5 to 13 mR/h-m per Ci (0.5 to 3 μGy/h-m per GBq) for sources ranging from 50 to 200 mg of Tm_2O_3, and 20 mR/h-m per Ci (5 μGy/h-m per GBq)

FIGURE 11. Production of Thulium-170 in Two Standard Targets for Various Numbers of Three-week Irradiation Cycles; Solid Curves Represent 50 mg (2 × 2 mm [0.08 × 0.08 in.]) Pellets; Dashed Curves Represent 150 mg (3 × 3 mm [0.12 × 0.12 in.])

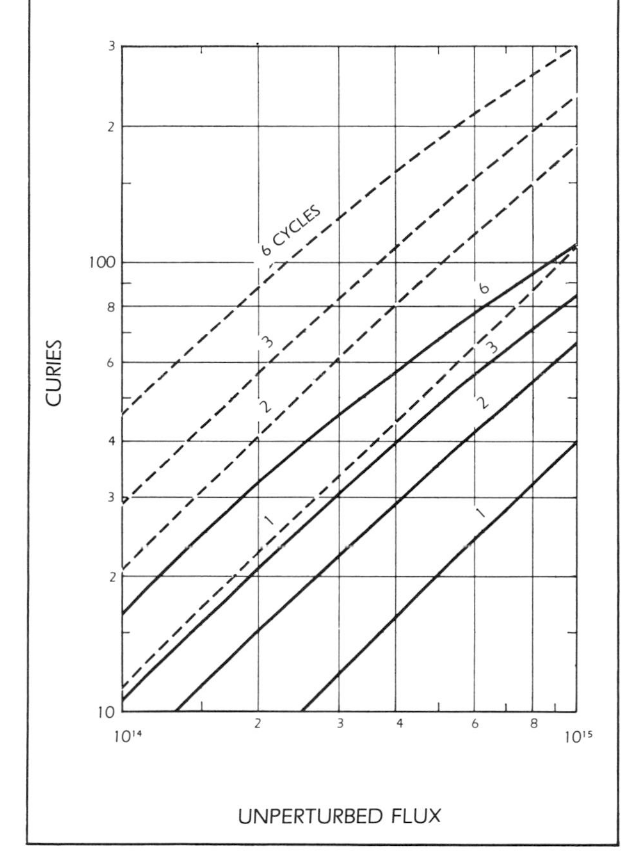

from one 400 mg metallic source.[6] These values may be compared with the calculated 3.0 mR/hr-m per Ci (0.7 μGy/h-m per GBq) for the 84 and 52 keV radiation, assuming the disintegration data shown in ref. 7.

Effect of Source Size

In general, it is found that the larger and more dense the source, the higher the gamma-ray intensity per curie becomes. From the few spectral measurements reported on these sources, the higher intensities per curie in the large sources are attributable to (1) an enhanced continuous X-ray background, and (2) the fact that these X-rays approximate 600 keV in quality.

It appears, therefore, that little is gained by enlarging the physical size of the source if low-energy radiations are desired; not only is the hard background increased, but the soft radiations are more absorbed. For radiography, a compromise must be made between small source size, which produces sharp radiographs, and low density source material. The latter not only permits the desired radiation to escape but also generates less continuous X radiation by lowering the average atomic number.

Applications and Shielding

Because thulium-170 sources can be produced only in limited intensity (150 mR/hr-m of soft radiation is about the maximum, corresponding to 100 mCi of cobalt-60), the isotope has not yet found widespread application in industrial radiography. Its main virtues are small size and portability; a 2.5 cm (1 in.) thick lead shield is sufficient to reduce the radiations from a 50 Ci (2,000 GBq) low-density source to tolerance levels. It can be used for radiography of thicknesses as low as 0.8 mm (0.03 in.) of steel or 13 mm (0.5 in.) of aluminum with 2 percent radiographic sensitivity, and is useful for inspection of internal assemblies such as aerospace components and composite materials.

Cesium-137

Cesium-137 (30.1 year half-life), generally in the chloride form, has been used as a radiographic source. Better materials (less corrosion) for encapsulation presently include ceramics and glass. These materials also reduce loss of the cesium from a capsule with a defective seal. Ninety-two percent of the nuclear disintegrations are by beta particle emission to an excited state of barium-137 which returns to stability by emission of a 0.66 MeV gamma photon. The radiation intensity from this isotope is 0.34 R/hr-m per Ci (80 μGy/h-m per GBq).

Specific Activity

Cesium-137 is one of the most probable products of nuclear fission, resulting from about 6 percent of the fissions. It is therefore generated in great abundance in all nuclear reactors. The difficulty lies in separating the element from the uranium fuel and the other fission products. Once this is done, sources of almost any strength can be produced. Its specific activity is limited by the presence of other isotopes of cesium, principally cesium-133 and cesium-135, resulting in a maximum of 25 Ci (1,000 GBq) per gram of CsCl. Because the density of the compressed salt is approximately 3.5 g/cm^3, sources can be made to 90 Ci (3,500 GBq) per cubic centimeter. Self-absorption is low, with only 30 percent of the intensity of a 50 Ci (2,000 GBq) source being absorbed.

Leakage Hazards

Because CsCl is a soluble powder, special precautions must be taken to keep it from leaking out of its capsule. The technique recommended by the United States Nuclear Regulatory Commission (USNRC) calls for double encapsulation with inner and outer containers made of silver-brazed or heliarc welded Type 316 stainless steel.[8,9]

PART 4

SOURCE HANDLING EQUIPMENT

Requirements

Radiographic sources must be handled so that the radiographic exposure may be made without appreciable exposure to the radiographer. As the required distance between radiographer and exposure device increases, the cost of the equipment increases rapidly. Only low intensity sources are handled with the simplest devices. High intensity sources (such as multicurie iridium-192 and cobalt-60 sources) are best handled with elaborate equipment that permits the radiographer to remain several meters from the exposure device when the source is manipulated.

Classification

Various schemes have been devised to provide remote handling of radiographic sources. Most of the remote exposure devices presently in use fit into one of the following categories: (1) those which move the source from the center to the surface of the shielding container; and (2) those which move the source out of the shielding container to an exposure site some distance away from the shield. The first type generally yields a beam of radiation somewhat restricted in size and direction, while the second type approximates a free source of radiation

FIGURE 12. Diagram of Type 1 Source Handling Device: (a) Source Stored; (b) Source Exposed

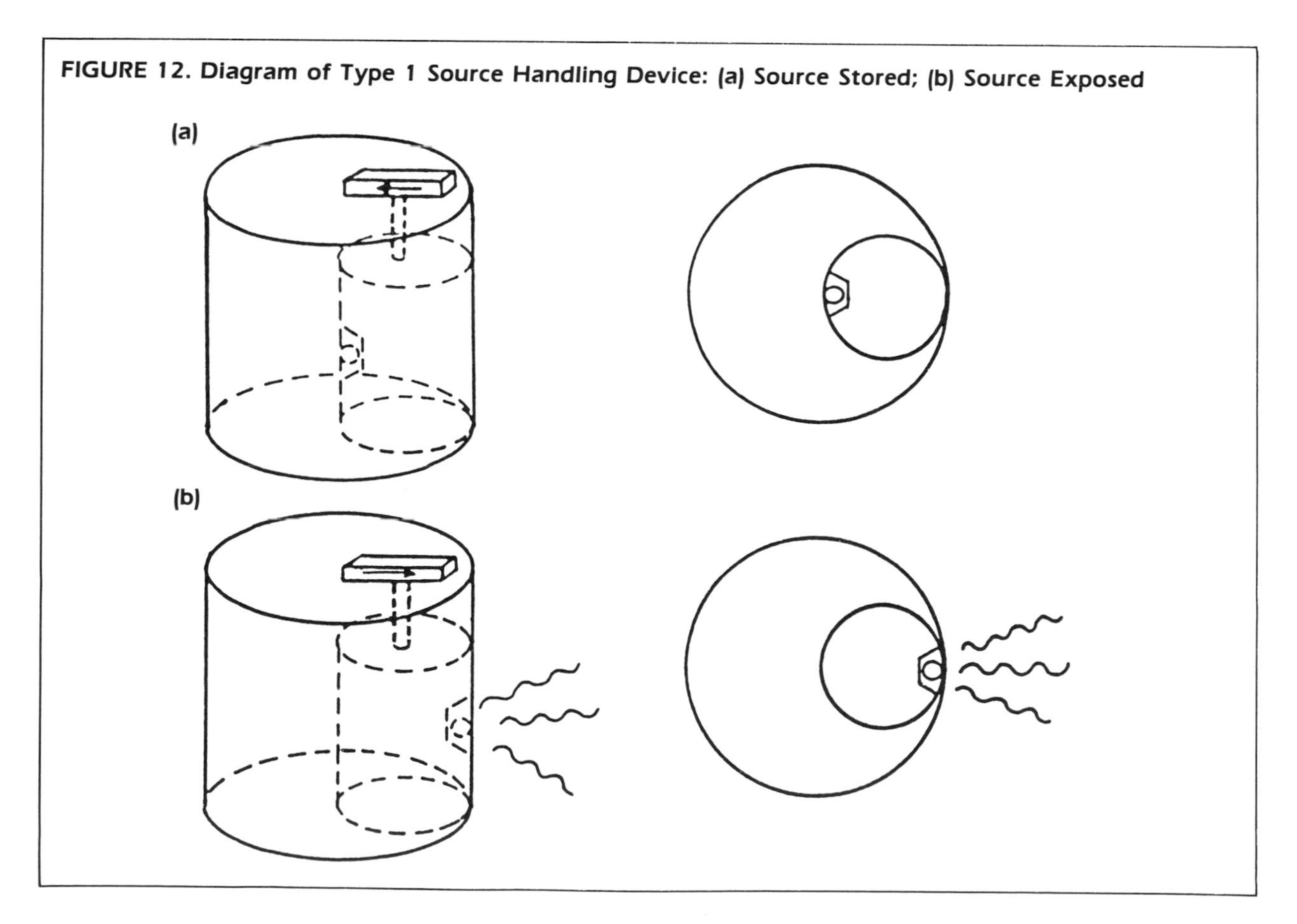

unless additional collimation is provided at the point of exposure. Both types have a great deal of utility in commercial radiography. Mechanical, electrical, hydraulic or pneumatic operation methods have been used. Experience shows the direct mechanical linkage to be inherently more fail-safe.

Manual Manipulation of Sources

The simplest equipment for making a radiographic exposure employs a source capsule handled by a rod or string. The rod or string is attached while the source is in its shield.

The rod generally is threaded for attachment to a threaded hole in the source capsule. Magnetized rods have also been used to remove the source capsule from its shield.

The string is tied to a loop on the source capsule and remains attached to the capsule during storage.

The string (often suspended from a light bamboo pole) or handling rods permit the operator to stay about 2 m (6 ft) from the unshielded source. Because the typical source for manual handling is 1 Ci (40 GBq) of cobalt-60 or 2 Ci (80 GBq) of iridium-192, the operator is exposed to gamma radiation fields up to 500 mR/hr (5 mSv/h) during source positioning. This limits the operator to ten or fewer exposures per day. Such handling, even for smaller sources, is either discouraged or prohibited at this time for routine radiographic exposures.

Remote Handling Equipment

Commercial field radiography currently requires the use of sources as much as 100 times larger than the manually handled sources. Radiography in laboratory or shop facilities may utilize sources of up to 1,000 Ci. Operator exposures from such sources would be intolerable if manual handling were attempted. Remote handling devices permit the radiographer to be two to ten times farther from the exposed source. With present-day equipment, these larger sources may be handled without exceeding exposure limits (which are one-third earlier limits) while making many more radiographic exposures per day.

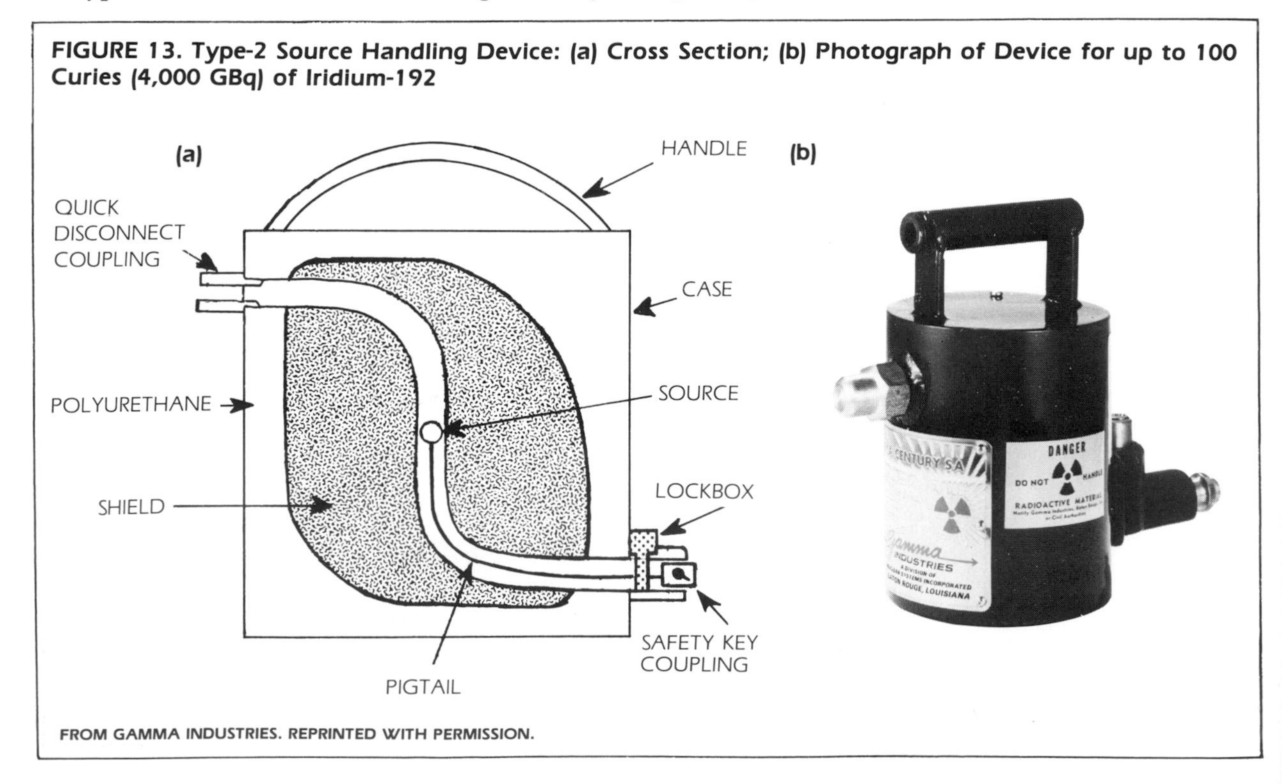

FIGURE 13. Type-2 Source Handling Device: (a) Cross Section; (b) Photograph of Device for up to 100 Curies (4,000 GBq) of Iridium-192

FROM GAMMA INDUSTRIES. REPRINTED WITH PERMISSION.

FIGURE 14. Diagram of Type-2 Source Handling Device Showing Operation

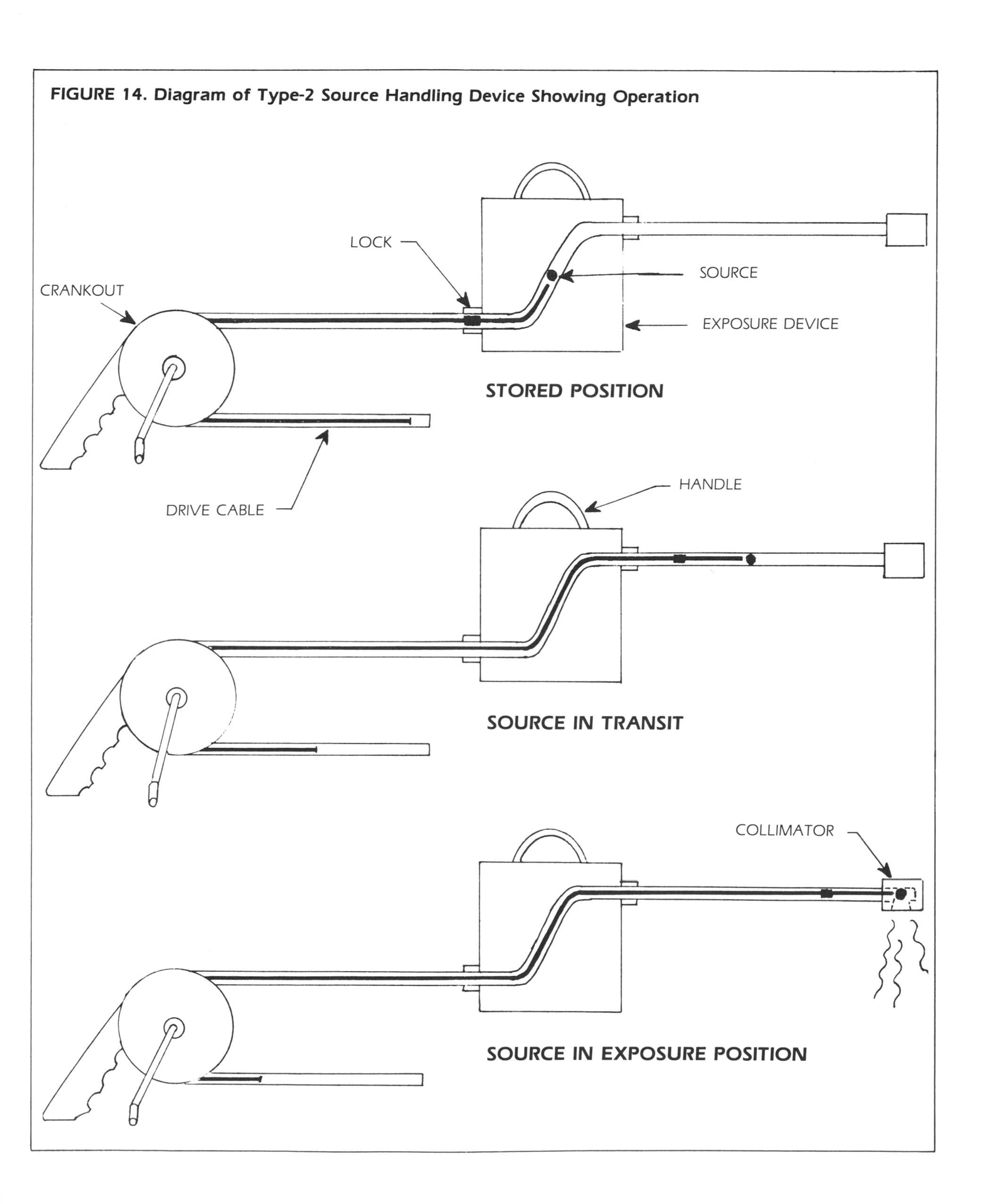

Type-1 Devices

These devices move the source capsule from the storage position in the shield to a position on the outside surface (see Fig. 12). The rotation of the eccentric cylinder carrying the source may be made manually (1) with the operator remaining behind the shield on the opposite side of the shield from where the source will appear, or (2) by turning a crank attached to the device by a metal drive cable. Shield material is usually depleted uranium (uranium with almost all the fissionable uranium-235 removed, leaving the uranium-238). Depleted uranium offers more shielding per unit mass than similar shields made from lead. A lead shield for 100 Ci of iridium-192 might weigh over 30 kg (66 lb); a uranium shield would be closer to 20 kg (44 lb). The

FIGURE 15. Guide-tube Collimators for Reducing Personnel Exposure: (a) Cross Section; (b) Photograph

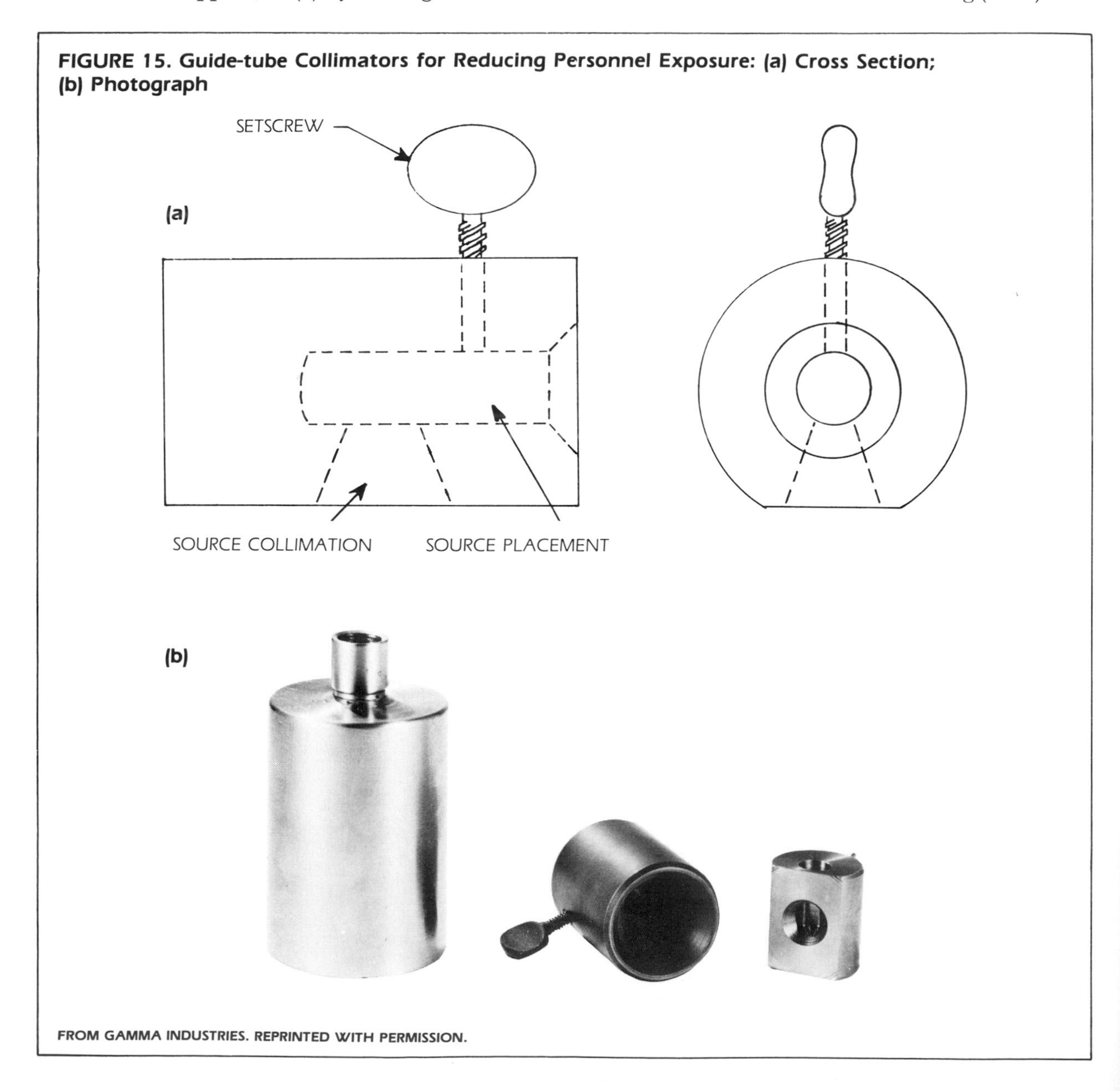

FROM GAMMA INDUSTRIES. REPRINTED WITH PERMISSION.

shield material is enclosed in either an aluminum or steel case. The source may be locked in the shielded or exposed position. These devices are frequently used for radiographing pipe.

Type-2 Devices

These devices permit the source capsule to be moved from the storage position inside the shield to an exposure location that may be many meters (yards) from the shield (see Figs. 13 and 14). The source is manufactured so that the capsule is attached to one end of a short piece of flexible metal cable. A special connector is attached to the end of the short cable, often called a pigtail. The connector end of the pigtail extends outside of the shield when the source capsule is centered in the shield. A drive is attached to the connector by another connector on the end of a flexible metal drive cable. The drive cable assembly includes a flexible, braided-metal tubular covering for the drive cable, a hand crank to move the cable, and a special fitting to attach the drive cable assembly to the shield. A locking mechanism is normally placed at the junction of the shield and the drive cable assembly to prevent unauthorized or accidental removal of the source capsule from its shielded position.

The source capsule moves through the shield (depleted uranium) in a smooth tube shaped to resemble a flattened letter S. The tube is zirconium or other similar metal that can withstand the casting temperature of the uranium. Another fitting at the exit end of the uranium-shield tube allows connection of a guide tube. The guide tube conducts the source capsule and drive cable to the exposure site. The guide tube may be made of flexible braided-metal tubing or of heavy-walled plastic tubing. The guide tube ends with a metal stop to position the source exactly where it needs to be to make the radiograph. A collimator (see Fig. 15) may be added to the stop at the end of the guide tube to provide a beam or circular band of radiation appropriate for the exposure. Such a collimator may reduce radiation intensity (outside of that needed for the radiograph) to less than 1/100 of the radiation without the collimator. Figures 16 through 22 show a variety of exposure devices for different sources and applications. Figure 23 illustrates field use of a portable exposure device.

FIGURE 16. Exposure Device: (a) Cross Section; (b) Photograph

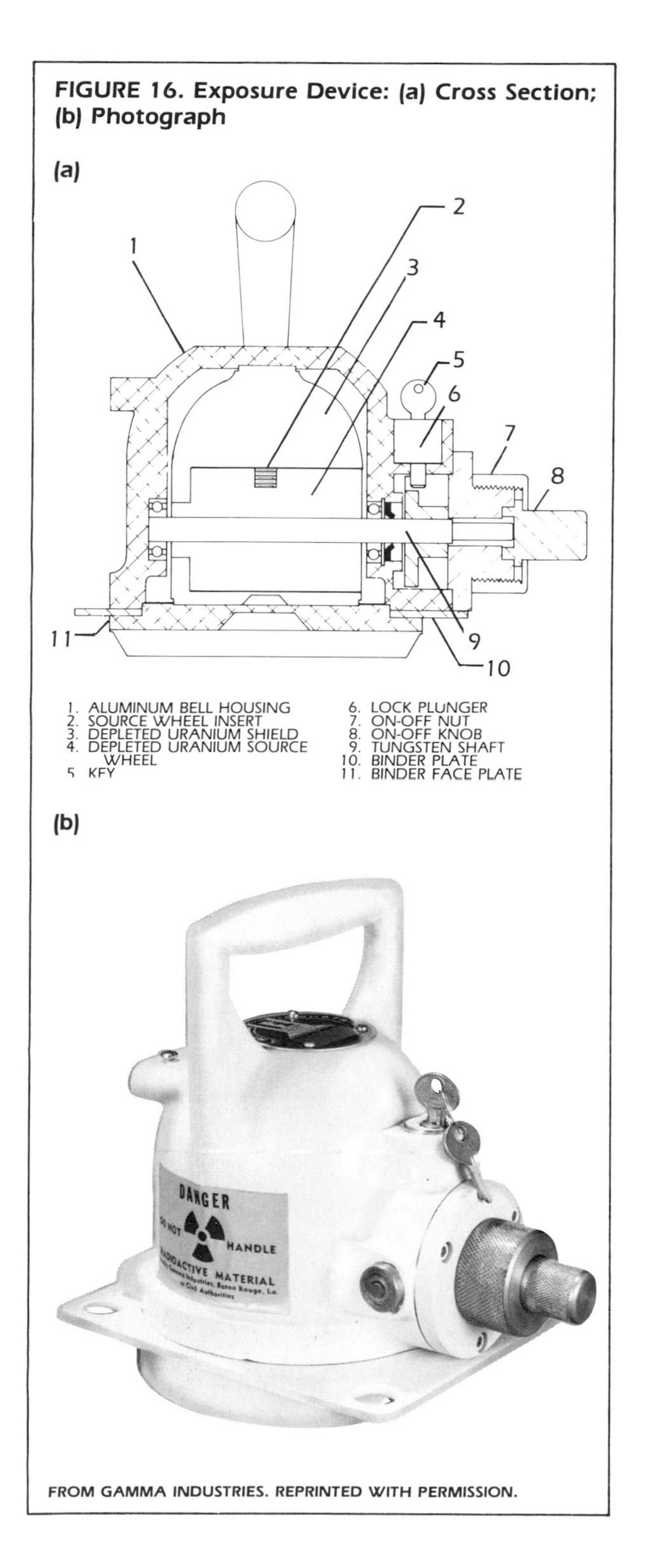

FROM GAMMA INDUSTRIES. REPRINTED WITH PERMISSION.

Reliability

Present-day radiographic exposure devices are quite reliable and are shielded better than earlier devices. This is important not only for the safety of the radiographer but for the general public as well. Unreliable items such as pneumatic drives, red-yellow-green lights to signal the position of the source (actually the position of the drive cable) and meters that count the turns of the hand crank have been properly abandoned. Novel devices, that prevent removal of the source capsule from its shielded position until correct connection between the drive cable and source pigtail is made, are good innovations to recent design. However, none of these can substitute for the proper use of a working survey meter by a trained radiographer.

FIGURE 17. Exposure Device for Up to 200 Curies of Iridium-192: (a) Cross Section; (b) Photograph

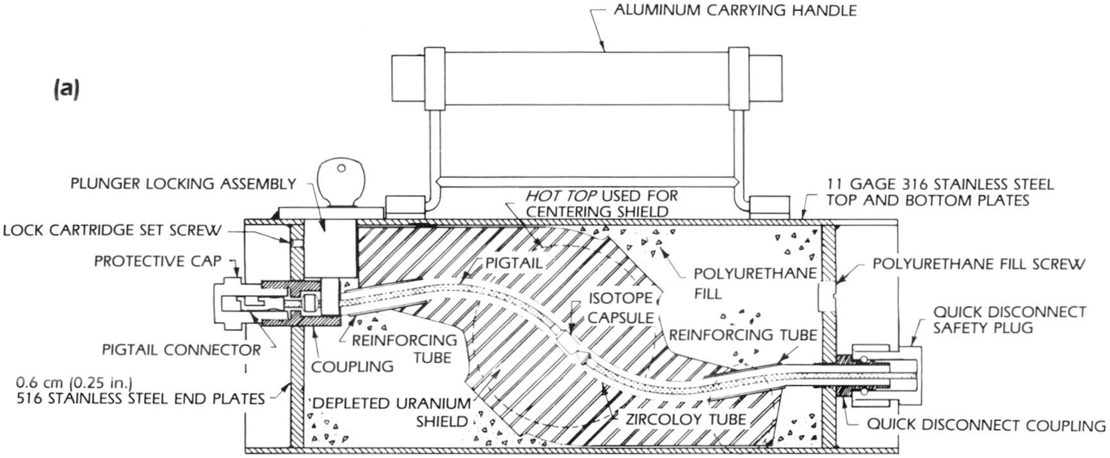

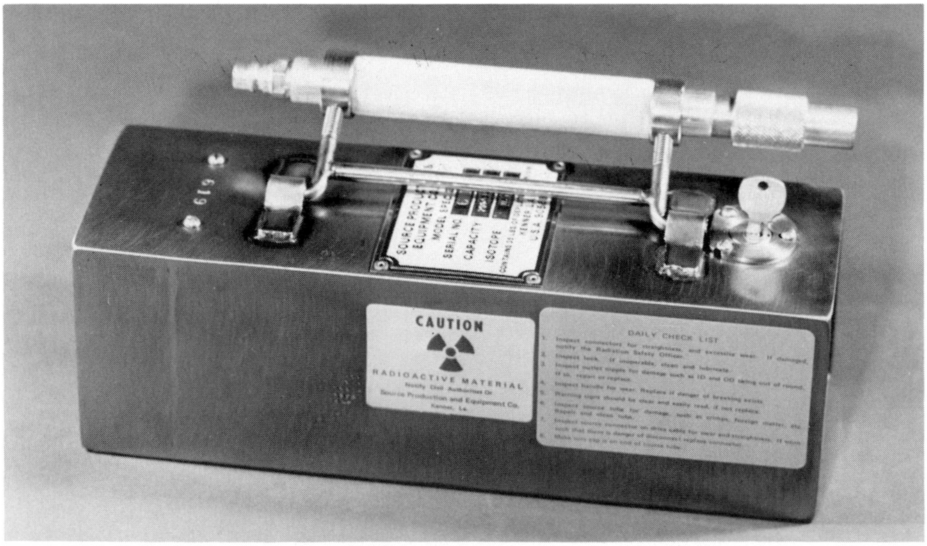

FROM SOURCE PRODUCTION AND EQUIPMENT COMPANY. REPRINTED WITH PERMISSION.

Safety Considerations

Proper attitude and understanding of the radiographic process by the operator are essential to the safe performance of radiography. Some other aspects, however, should be pointed out, including: time, distance and shielding. Obviously, the less time a radiographer spends in a radiation field, the lower the exposure will be. What is usually forgotten, or ignored, is long exposures at low exposure rates. For example, during transportation or storage of the shielded source, the radiographer may be exposed to as little as 2 mR/hr (20 μSv/h), which is generally ignored. However, if the time is long, such as eight hours, the exposure is more than 15% of a week's allowable average exposure.

Distance effects are described by the inverse square law; if the distance is increased, the exposure is decreased by the inverse of the distance squared. Double the distance gives one-quarter the exposure; ten times the distance gives 1/100 the exposure.

Shielding involves placing as much substance between the individual and the source as possible. Shielding works exponentially, so even small thicknesses are important. One of the important shields often omitted from use is the collimator described above (Fig. 15). Another shield is the one where the source is normally stored.

The adequacy of time, distance and shielding can only be established by use of an appropriate, working survey meter. Survey meters are discussed below.

Shielding Equipment

Because the radiographic source cannot be turned off like an X-ray machine, shielding must be provided whenever the source is not intentionally being exposed. This shielding must conform to regulations written for specific uses. In the US, regulations state that: (1) no more than 200 mR/hr (2,000 μSv/h) is permitted at any exterior surface of the shield; (2) no more than 10 mR/hr (100 μSv/h) at one meter is permitted from any exterior surface, for shields in which the source storage position is 10 cm (4 in.) or more from any exterior surface; or (3) no more than 50 mR/hr (500 μSv/h) is permited at 15 cm (6 in.) from any exterior surface of a smaller shield. When the radiographer is to remain in close proximity to these shields for long periods of time, a lower radiation level (better shielding) is necessary. It may be necessary to use a shield designed for a larger source. Again, collimators (Fig. 15) should be used whenever possible to reduce radiation exposures to personnel. Only a survey meter will provide information about the adequacy of the shielding.

FIGURE 18. Exposure Device for Up to 100 Curies of Iridium-192: (a) Note Department of Transportation Shipping Information; (b) Close-up of Crank-out Cable Attachment

(a)

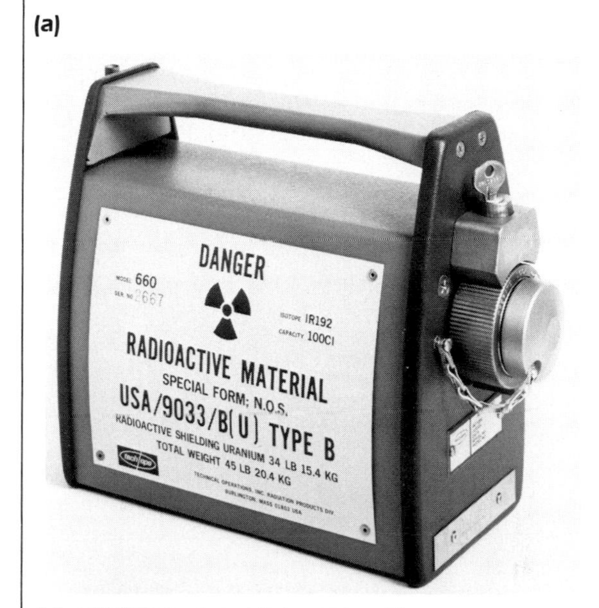

(b)

FROM TECH/OPS INCORPORATED. REPRINTED WITH PERMISSION.

Survey Meter Equipment

Use of a working, radiographic survey meter by a trained operator could reduce radiographic overexposures by 90%. Only such an instrument can tell an operator what the exposure rate is from the time the shielded source is obtained, through the working day, and until the source is replaced into storage. Because a human being cannot detect radiation without such a device, performing radiography without a survey meter, or ignoring the instrument when it is available, is inexcusable. United States' regulations require the use of an instrument that can operate in fields as low as 1 mR/hr and as high as 1,000 mR/hr. Additionally, the instrument must have been calibrated within 90 days of its use. Special survey meters must be used for X rays or neutrons and isotope radiography.

Dosimetry Equipment

In addition to the survey meter, which provides immediate information on radiation exposure, personnel dosimetry equipment must be used. Generally this will include at least one (preferably two) pocket dosimeters and either a film badge or a thermoluminescent dosimeter (TLD). These devices provide a total exposure reading for the operator. The pocket dosimeter can be read after each radiographic exposure or no less than daily. Records of these readings should be maintained. The film badge or TLD should be read monthly or on any occasion that an overexposure is suspected.

Procedures

Unless proper procedures are followed, radiographic equipment will not produce satisfactory

FIGURE 19. Exposure Device with Crank-out, Guide Tube, and Collimator Assemblies

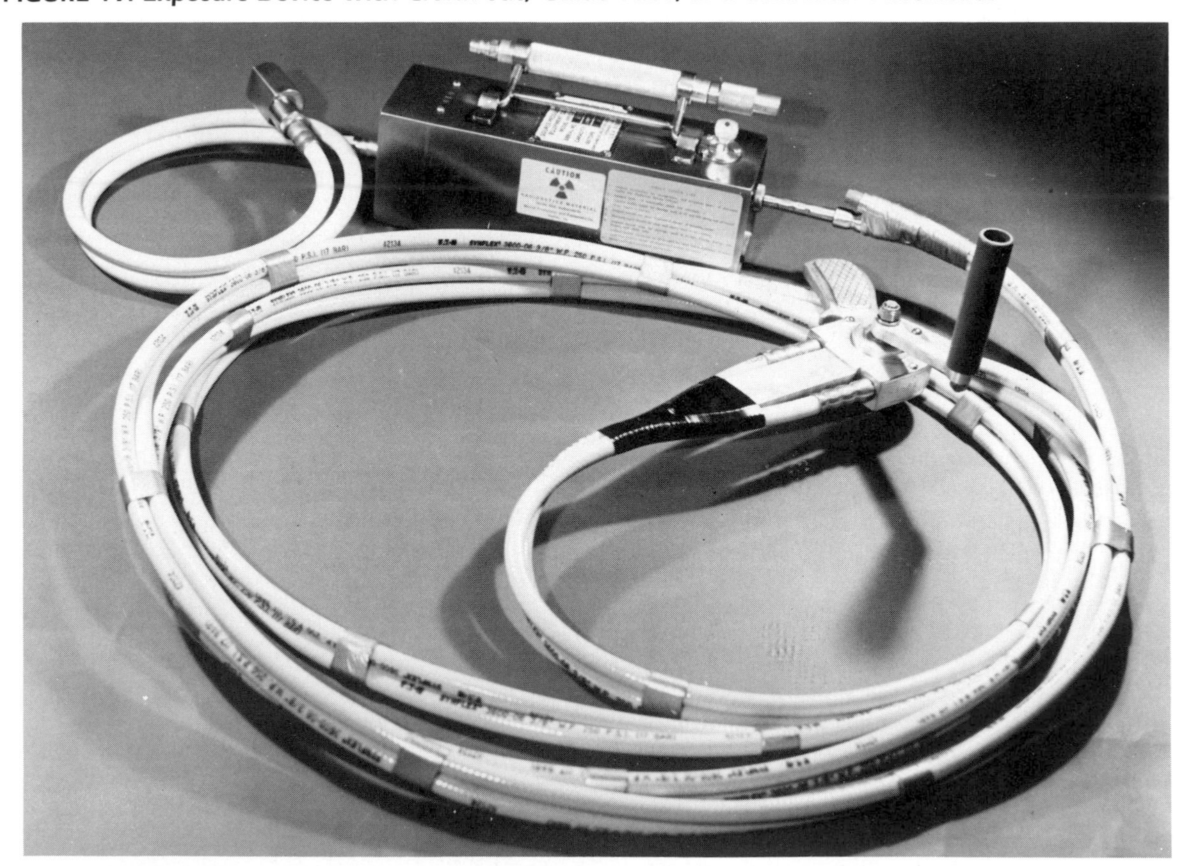

FROM SOURCE PRODUCTION AND EQUIPMENT COMPANY. REPRINTED WITH PERMISSION.

FIGURE 20. Exposure Device with Source Exchanger: (a) Exchanger [left] Closed, Exposure Device Fitted with Short Exchange Tube; (b) Exchanger Open and Attached, Ready for Source Transfer to Exposure Device; Note Old Source Pigtail in Right-hand Storage Position

(a)

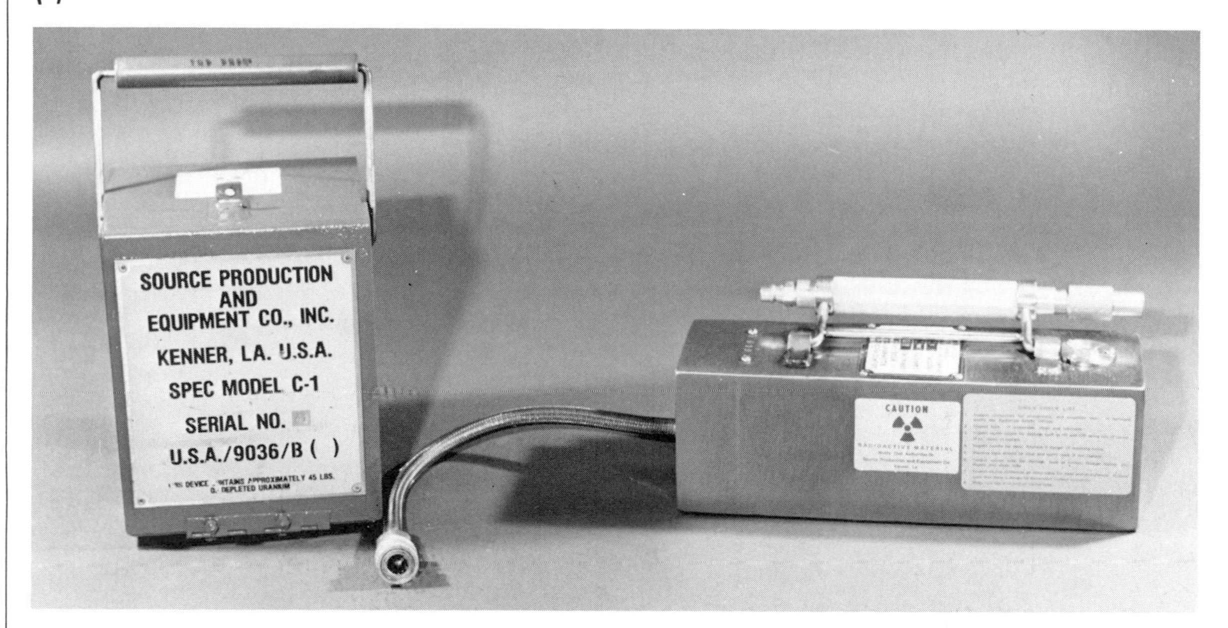

(b)

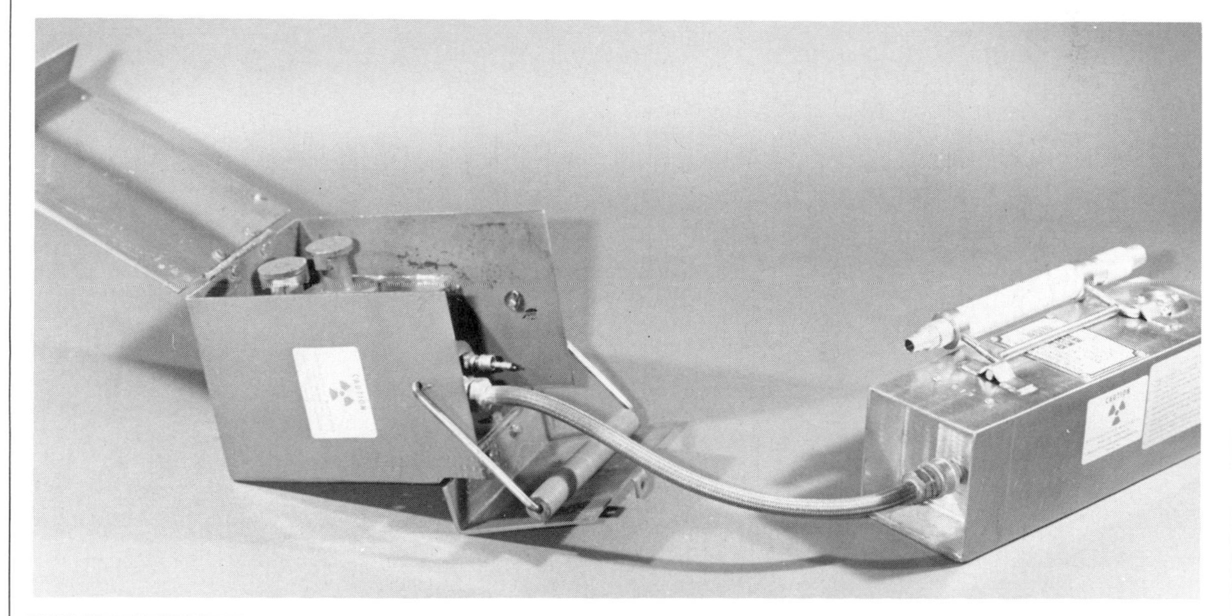

FROM SOURCE PRODUCTION AND EQUIPMENT COMPANY. REPRINTED WITH PERMISSION.

radiographs safely. Procedures related to safety include: (1) checking the radiographic exposure device with the survey meter at the beginning and end of the day; (2) using the survey meter during exposure of the source and whenever approaching the exposure device after an exposure; (3) placement of barricades to prevent public entry into exposure areas; (4) checking the radiation exposure rates at the barricade limits; and (5) periodic leak tests of the source capsule. Should the source capsule lose some of its radioactive material, serious overexposures to the radiographer and/or the public could occur. While the radiographic survey meter could detect severe leakage from the capsule, more sensitive means are necessary.

The possibility of leakage of radioactive material from a source capsule must be determined every six months following manufacture. The procedure involves locating the aperture in the shield nearest to the source, usually the tube opening which attaches to the guide tube. The aperture is wiped with a bit of moistened cotton or cloth attached to a small wooden stick. Water or a special solution may be used for moistening. The wipe is tested with a low background detector system that can detect less than 0.005 μCi of removable contamination. If more than 0.005 μCi of removable contamination is detected, the source is assumed to be leaking and must be destroyed. If such a source has been used for any length of time, radioactive contamination of the equipment and work areas may have occurred. Special survey meters must be used to detect this contamination. Finally, the radiographer should be provided with emergency procedures and access to experienced help in the event of an accident.

FIGURE 21. Portable Exposure Device for Up to 250 Curies of Cobalt-60

FROM TECH/OPS INCORPORATED. REPRINTED WITH PERMISSION.

FIGURE 22. Portable Exposure Device for Up to 100 Curies of Cobalt-60

FROM GAMMA INDUSTRIES. REPRINTED WITH PERMISSION.

FIGURE 23. Field Set-up for Pipeline Isotope Radiography, Showing Close-up of Crank-out Cable Connection and Guide Tube Placement

FROM TECH/OPS INCORPORATED. REPRINTED WITH PERMISSION.

PART 5

RADIOISOTOPES FOR OTHER TESTING METHODS

Other Testing Methods

While radiography is the testing method that involves the largest use of radioisotopes, other testing methods such as gages now consume large quantities of radioisotopic sources. The types of gages covered in this chapter are (1) gages using the transmission or backscatter of radiation; (2) gages using neutron interactions; and (3) instruments providing compositional information on specimens by the generation of characteristic X rays. Examples of less-used radioisotopic techniques not described here include Mossbauer resonant absorption measurements and radioisotope ionization detectors for gas chromatography.

Transmission Gages

Transmission gages utilize a collimated beam of radiation to inspect a specimen. Information concerning internal conditions, thickness, or density of the specimen results from measurement of the radiation transmitted through the specimen. This is similar to radiography except that the transmitted radiation detection is normally accomplished by an ion chamber, GM tube, or scintillation counter system. The counting system may be read either digitally or in analog (a meter). The source and detector are placed as close as possible to the specimen on opposite sides along either the diameter, chord, or some other straight line intersecting the region of interest. Figure 24 is a diagram of a simple transmission gage. More sophisticated designs can be used involving dual radiation beams through a specimen and a standard, or by chopping (modulating) the radiation beam alternately through a specimen and a standard, followed by demodulation of the detector signal.

Transmission Gage Theory

The theory supporting transmission gages using gamma radiation is identical to that for radiography:

$$I = I_o e^{-\mu t} \qquad \text{(Eq. 12)}$$

If the radiation is not well collimated, scattered radiation is detected and the equation becomes:

$$I = BI_o e^{-\mu t} \qquad \text{(Eq. 13)}$$

where B represents the increase in transmitted intensity from scatter. This build-up factor, or B, may be estimated by $(1 - \mu t)$. The absorption coefficient

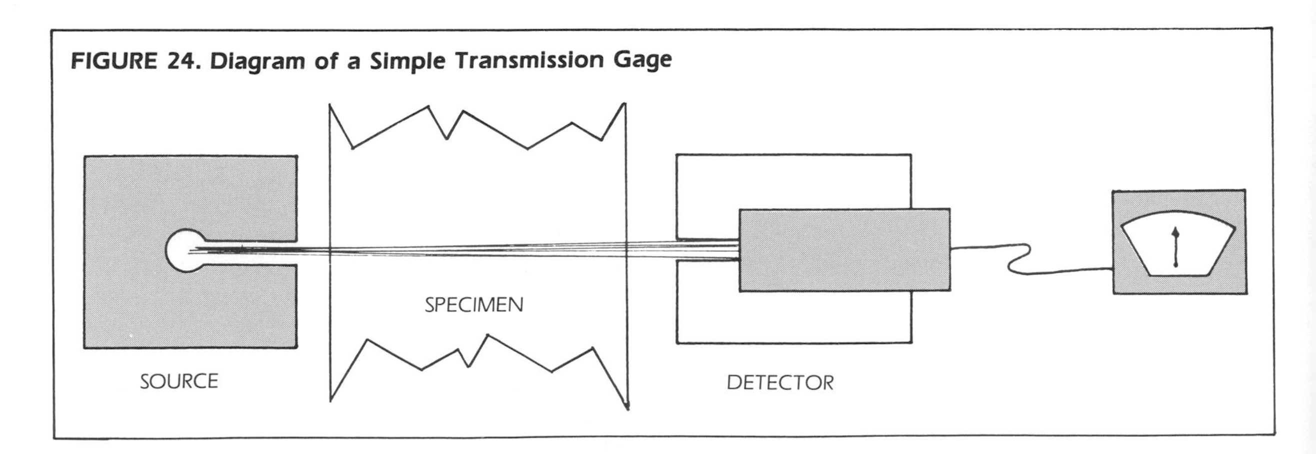

FIGURE 24. Diagram of a Simple Transmission Gage

μ is a constant for an isotope and an absorber. The selection of the best isotope or μ for a particular gaging application can be made by using the formula:

$$\mu t = 2 \qquad \text{(Eq. 14)}$$

This equation can be derived by setting the first derivative of eq. 12, with respect to μ, equal to zero and solving for μt. Probably no gamma emitter will have exactly the best μ, but a close approximation can often be obtained from the available radioisotopes. Table 3 contains a listing of the most frequently used gamma emitters, along with their more important properties.

Transmission gages, having beta emitting radioisotopes as sources, give a transmitted radiation intensity that can be expressed as an empirical relationship closely resembling gamma transmission:

$$I = I_o e^{-kx} \qquad \text{(Eq. 15)}$$

where k is a constant for the radioisotope, absorber, and gage configuration. Because betas are absorbed more easily than gammas, transmission gages using such sources are quite often employed when thin or low-density specimens are examined.

Transmission gages do not ordinarily use alpha sources because all of the alpha particles from a given source are either absorbed or transmitted by a given thickness of material. Alpha sources are used for measurement of gas pressure or for excitation of X rays from the surface of a specimen.

Backscatter Gages

Backscatter gages make use of the scatter of gamma and beta radiations from the surface layers of a specimen. As can be seen from Figure 25, a diagram of one type of backscatter gage, only one surface of the specimen need be available. Because of the nature of the scatter reactions of gamma and beta radiations with matter, the mathematical relationships for backscatter gages are usually empirical but often take the form of the following equations:

$$\left(\frac{I - I_o}{I_s - I_o}\right) = e^{-kx} \qquad \text{(Eq. 16)}$$

or

$$I = I_s\,(1 - e^{-kx}) + I_o \qquad \text{(Eq. 17)}$$

where I is the radiation intensity scattered from a thickness t; I_o is the radiation intensity scattered from zero specimen thickness (may be a thickness of supporting material such as a conveyor belt or substrate for a plating); I_s is the radiation intensity scattered from an infinitely thick layer of specimen (the thickness at which an increase in thickness produces no increase in scattered radiation); k is an empirical constant; and x is the thickness of the specimen in thickness units (cm) or density-thickness units (g/cm^2).

The main advantages of the backscatter gage are high sensitivity for small changes in thickness, one side access, and ability to distinguish thicknesses of one material from another.

Examples of the latter advantage are determination of silver or gold plating thickness on copper or the determination of paint or plastic coating thickness on steel or aluminum. Recent usage includes the measurement of copper thickness on printed circuit boards, as well as measurement of silver, gold, nickel or combination platings on copper.

Neutron Gages

Neutrons provide information about the hydrogen content of a specimen or information about certain elements that have exceptionally high neutron absorption cross sections (ability to absorb a neutron into the nucleus of the atom, giving an isotope of the target atom that is likely to be radioactive). Those gages that are used to measure quantities of high neutron cross section elements in a low neutron cross section matrix are related to neutron radiography just as gamma-ray gaging is related to gamma-ray radiography. Section 12, *Neutron Radiography*, provides more information on such properties and their measurements.

Hydrogen content is obtained differently. A source of high-energy or *fast* neutrons is used. The fast neutrons interact with hydrogen in the specimen so that the neutrons are slowed to kinetic energies that are related to the ambient temperature. These low energy neutrons are called *slow* or *thermal neutrons* and have a kinetic energy of 0.026 eV at 20 °C. The fast neutrons might have initial energies of 1 to 10 MeV. Hydrogen is the most effective element for this slowing or thermalization process. By measuring the thermal or backscattered neutrons, the hydrogen content of the specimen can be determined.

TABLE 3. Frequently Used Gamma Emitters

Isotope	Half-Life (years)	Emission[1] (MeV)	Use[2]	Application[3]	Source Size	Production[4]
^{241}Am	433	α 5.49-5.54 γ 0.0596	T,N,E	thin metal, moisture	μCi to 100 mCi	tU
^{14}C	5730	β 0.156	T,B	paper, plastic plating thickness	mCi	N(n,p)
^{109}Cd	1.24	ϵ,γ 0.0880 X 0.0221	T,E	thin metal	mCi	(n,γ)
^{242}Cf	2.63	α 6.12 spon fission	N	moisture composition	μCi to mCi	tU
^{60}Co	5.26	β 0.318 γ 1.17, 1.33	T	thick metal density	mCi to 1 Ci	(n,γ)
^{137}Cs	30.1	β 0.52 γ 0.662	T,B	thick metal density	mCi to 10 Ci	ff
^{55}Fe	2.7	X 0.0059	T,E	percent sulfur, exe	mCi	(n,γ) Mn(p,n)
^{63}Ni	100	β 0.0639	T	very thin sheet	mCi	(n,γ)
^{147}Pm	2.62	β 0.225 γ 0.121	T,B	paper, plastic plating	mCi	ff
^{239}Pu	24,400	α 5.10-5.16	N,E	moisture, composition	mCi to Ci	tU
^{145}Sm	0.9	ϵ,γ 0.0614	E	exe	mCi	(n,γ)
^{90}Sr	29	β 0.546	T,B	paper, plastic	mCi	ff
^{90}Y		β 2.28		thin metal		
^{204}Tl	3.78	β 0.763	T,B	paper, plastic plating	mCi	(n,γ)
^{170}Tm	0.35	β 0.968 β 0.884 X 0.052 γ 0.084	T	thin metal	mCi	(n,γ)

1. α = alpha rays
   β = beta rays
   γ = gamma rays
   X = X rays
   ϵ = electron capture

2. *T* = transmission gage
   B = backscatter gage
   N = neutron gage
   E = excitation of X-ray emission

3. exe = excitation of X-ray emission

4. (n,γ) = neutron capture by stable isotope one mass unit lighter
   (p,n) = accelerator produced by proton bombardment
   ff = fission fragment, produced by separation of elements left after fission of uranium or plutonium
   tU = transuranium element, made by capture of several neutrons by uranium-238 followed by beta decay

X-ray Emission Gages

When materials are bombarded with alpha, beta, X-ray or gamma radiation, some of the atoms present will emit characteristic X rays. Characteristic X rays are X rays whose energy is related to the atomic number of the element. The energy of the X ray emitted rises along with the atomic number (number of protons in the nucleus), from hydrogen as atomic number 1 to uranium as atomic number 92. As a result of this property, instruments are available that can give qualitative information (what elements are present) and/or quantitative information (concentration, thickness, or total amount) about a specimen being tested. If X rays are used to excite the emission of characteristic X rays from the sample, the method is called *X-ray fluorescence* (see Section 17, *X-ray Diffraction and Fluorescence*). If radiations other than X rays are used, the method may be called *X-ray emission spectrometry*, particularly if the detection of the characteristic X rays involves a detector that can separate the radiation by energy. Such detectors are NaI(Tl) scintillation and Si(Li) semiconductor detectors. At this time, the NaI(Tl) detectors are most often used in instruments operated in the field while Si(Li) detectors are most often used in laboratory instruments. Field use of the Si(Li) detectors, which have much better energy resolution than the NaI(Tl) detectors, is becoming more common.

Figure 26 is a diagram of an X-ray emission gage. Other designs can include filters to remove (1) undesirable excitation energies (especially when using a machine source for excitation); (2) unwanted X rays coming from the sample and secondary emission sources. Again, the excitation may be by alpha, beta, X-ray, or gamma radiation from radioisotopes or from machine-produced equivalents.

Figure 27 is a photograph of a commercial instrument designed for field use. The instrument can determine and report the alloy being tested in a few minutes (if that alloy and its composition are in the memory of the instrument). The composition of the alloy (elements and their percentages) may also be determined.

FIGURE 25. Diagram of a Simple Backscatter Gage

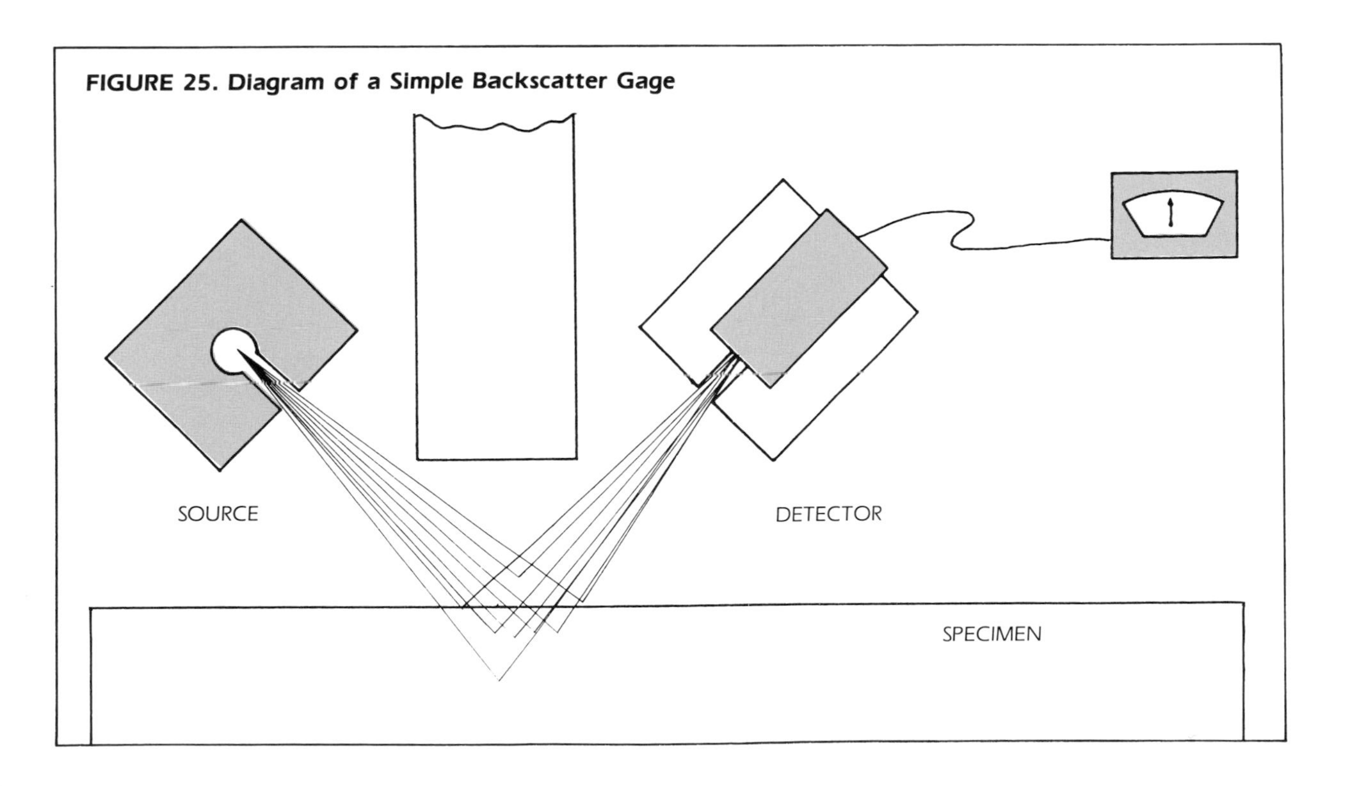

FIGURE 26. Diagram of an X-ray Emission Gage

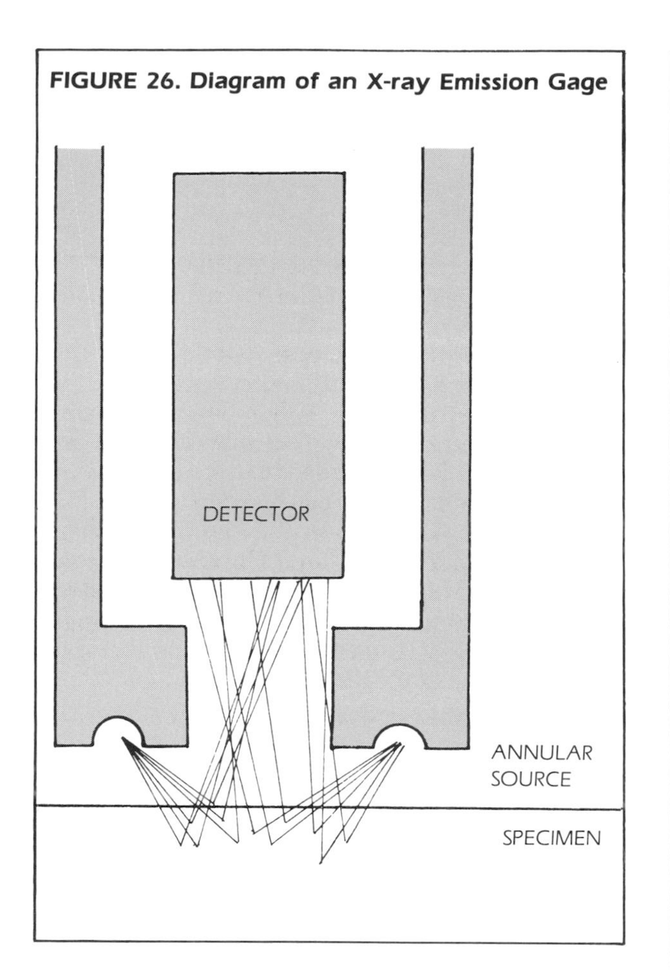

FIGURE 27. Commercial Field (Portable) Gage Using X-ray Emission: (a) for a Metal Bar; (b) for Weld Metal

(a)

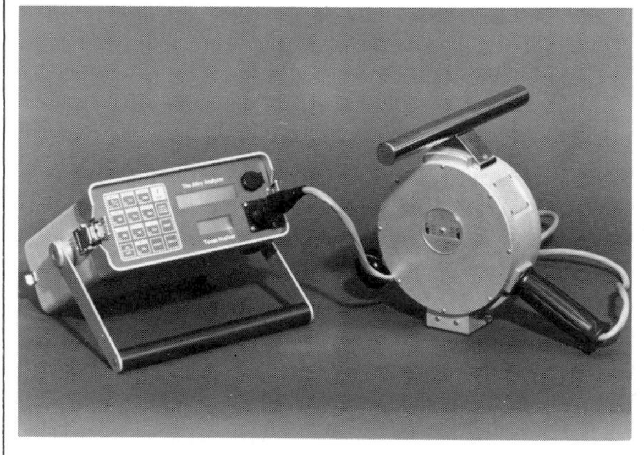

(b)

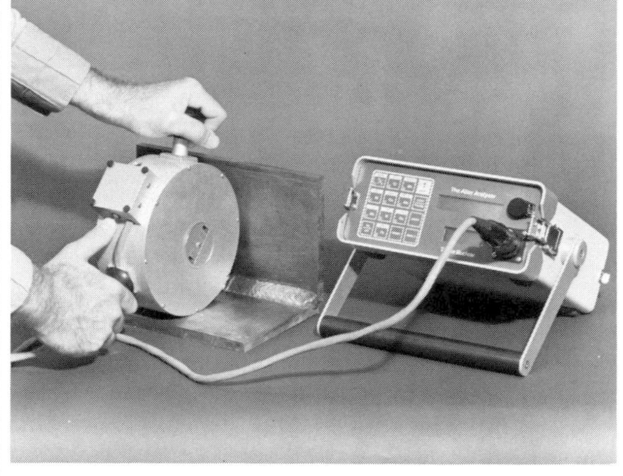

FROM TEXAS NUCLEAR. REPRINTED WITH PERMISSION.

PART 6

TRANSPORTATION OF RADIOGRAPHIC SOURCES

Introduction

Since the publication of the *NDT Handbook* first edition in 1959, regulations dealing with the transportation of radioactive material have changed considerably and probably will continue to change. The validity of the following text could change at any time because of these frequent revisions of regulations. Therefore, before transporting radioactive materials, carefully investigate the regulations for the specific materials and the areas through which they will move. Fines for failing to follow such regulations can be severe.

Sources of transport regulation information include:

1. US Code of Federal Regulations, Title 49, especially part 173;
   US Code of Federal Regulations, Title 10, Part 71;
2. International Atomic Energy Agency (IAEA) Regulations:
   Safety Series No. 6 (1979); and
   Safety Series No. 37 (1982).

For this discussion, only the more common types and quantities of radioisotopes will be considered for shipping in the United States. In 1983, the Department of Transportation (DOT) adopted regulations for the US that essentially conform to those issued by the IAEA in 1973.

Classification

The modern encapsulated sources manufactured for use in radiography can, with only few exceptions, qualify as a *special form* of material. First, they do not represent a hazard in terms of easy dispersion of the radioactive material. Second, they can meet a series of performance tests which include: (1) a free drop from 10 m (30 ft) upon an unyielding horizontal surface; (2) withstand an impact by 1.5 kg (3 lb) steel rod dropped from 102 cm (40 in.); (3) survive being heated in air at 800 °C (1,475 °F); and (4) do not leak upon immersion for 24 hours in water at room temperature.

The multicurie radiographic sources of cobalt-60, iridium-192 and cesium-137 qualify as *large quantity radioactive materials*. This puts them in the special transport designation, *type* A_1 *radioactive material*, which determines the type of container that must be used on common carriers. Special form in the A_1 group means that up to 20 Ci of iridium-192, 30 Ci of cesium-137 or 7 Ci of cobalt-60 may be carried in a Type A container; larger quantities must be carried in a Type B container.

Containers

Containers must have the usual features required for the shipment of any valuable material: a lock or seal; dimensions that exceed 10 cm (4 in.); prevention of loss of contents; and resistance to moderate heat.

Type A packaging must retain the contents and not become a less effective packaging if heated to 130 °C in direct sunlight, cooled to −40 °C, subjected to a pressure of 0.5 atmosphere, and vibrated as in normal transportation. Also, Type A packaging must pass consecutive application of at least two of the following tests: (1) a 30 minute water spray; (2) a free drop from 1.2 m (4 ft) onto an unyielding horizontal surface in such a way as to cause the most damage; (3) a free drop onto each corner of the package from 30.5 cm (12 in.); (4) an impact by a steel cylinder weighing 6 kg (13 lb) from 1 m (3 ft) into the most vulnerable surface; and (5) a compressive load of about five times the weight of the package or 2 psi times the cross sectional area of the package for 24 hours. Most radiographic exposure devices of recent manufacture qualify for Type A packaging.

Type B packaging requirements are more severe and most radiographic exposure devices require an overpack to be shipped by common carrier. Type B packaging requires that an accident not reduce the shielding or cause appreciable release of radioactive material after the following series of tests: (1) a free drop from over 9 m (30 ft) on to a flat unyielding horizontal surface to produce the most damage; (2) a free drop from 1 m (3 ft) onto a vertical steel spike; (3) exposure to 800 °C; and (4) immersion in at least 1 m (3 ft) of water for over 8 hours. Most manufacturers provide appropriate overpacks for shipping of the exposure devices to customers. The same overpack can be used to return the device to the manufacturer. However, the shipper must also make some important measurements concerning radiation levels around the package.

Radiation Levels

The package designed for shipment must not exceed 200 millirem (milliroentgens for gamma sources) per hour at any point on the external surface of the package and the transport index must not exceed 10. The *transport index* is the number placed on the package to designate the degree of control exercised by the carrier during transportation. The transport index indicates the largest dose rate (in mR/hr at one meter) from any accessible surface of the package. This transport index must be placed on the special shipping labels that are required for radioisotopic shipments.

Shipping Labels

A radioactive yellow-III label is required for shipment of most radiographic sources, even after extensive decay. A lower designation label can be used only if the radiation level on the surface of the package is less than 50 mR/hr and less than 1.0 mR/hr at 1 meter from the surfaces of the package. Any serious breach in these (or newer) requirements can result in heavy fines. Always refer to the most recent documents and regulations governing the transport of radioactive shipments.

REFERENCES

1. Evans, R.D. *The Atomic Nucleus*. New York: McGraw-Hill Book Co., Inc. (1955).
2. Reiffel, L. "Beta-ray Excited Low-energy X-ray Sources." *Nucleonics*. Vol. 13, No. 3 (1955): pp 22. Also, see: Kereiakes, J.G. and G.R. Kraft. "Thallium-204 X-radiography." *Nondestructive Testing*. Vol. 16, No. 6 (1958): pp 490.
3. Evans, R.D. and R.O. Evans. "Studies of Self-Absorption in Gamma-ray Sources." *Review of Modern Physics*. Vol. 20 (1948): pp 305.
4. Clarke, E.T. "Investigation of Isotopes for Aircraft Radiography." Wright Air Development Center Technical Report; ASTIA Document No. AD 118224 (1957): pp 56-440.
5. Johns, M.W. and S.V. Noble. "Disintegration of Iridium-192 and Iridium-194." *Physics Review*. Vol. 96 (1954): pp 1599.
6. West, R. "Low-Energy Gamma Ray Sources." *Nucleonics*. Vol. 11, No. 2 (1953): pp 20. Also, see: Halmshaw, R. "Thulium-170 for Industrial Radiography." *British Journal of Applied Physics*. Vol. 6 (1955): pp 8. Also, see: Clarke, E.T. "Gamma Radiography of Light Metals." *Nondestructive Testing*. Vol. 16 (1958): pp 265. Also, see: Carpenter, A.W. "Complete Portable Field X-ray Unit." Army Medical Research Laboratories Report No. 168 (1954).
7. Graham, R.L., J.L. Wolfson and R.E. Bell. "The Disintegration of Thulium-170." *Canadian Journal of Physics*. Vol. 30 (1952): pp 459.
8. *Catalog and Price List of Radioisotopes — Special Materials and Services*. Oak Ridge: US Atomic Energy Commission Radioisotopes Sales Department (1956): pp 144. Also, see: "Protection Against Radiations from Radium, Cobalt-60 and Cesium-137." *National Bureau of Standards (US) Handbook*. No. 54 (1954).
9. Harrington, E.L., H.E. Johns, A.P. Wiles and C. Garrett, "The Fundamental Action of Intensifying Screens in Gamma Radiography." *Canadian Journal of Research*. Vol. 28 (1948): pp 540.
10. O'Conner, D.T. and E.L. Criscuolo. "The Quality of Radiographic Inspection." *ASTM Bulletin*. No. 213 (1956): pp 53.
11. Halmshaw, R. "The Factors Involved in an Assessment of Radiographic Definition." *Journal of Photographic Sciences*. Vol. 3 (1955): pp 161.
12. Goldstein, H. and J.E. Wilkins, Jr. "Calculation of Penetration of Gamma Rays." US Atomic Energy Commission Report NYO-3075 (1954). Also, see: Frazier, P.M., C.R. Buchanan and G.W. Morgan. "Radiation Safety in Industrial Radiography with Radioisotopes." US Atomic Energy Commission Report 2967 (1954). Also, see: Ritz, V.H. "Broad and Narrow Beam Attenuation of Iridium-192 Gamma Rays in Concrete, Steel and Lead." *Nondestructive Testing*. Vol. 16, No. 3 (1958): pp 269.

BIBLIOGRAPHY

American Machinist. Vol. 210, SR-11-15 (November 1976).

Austin, J.C. and P. Richards. "Adapting the Universal Exposure Calculator for Radium Radiography to Radiography with Iridium-192." *Nondestructive Testing*. Vol. 14, No. 4 (1956): pp 16.

Berger, H. "Nuclear Methods for NDT." *National Bureau of Standards in Instrumentation Technology*. Vol. 23, No. 8 (1976): pp 45-50.

Berman, A.I. "Radioactivity Units and Radiography." *Nondestructive Testing*. Vol. 9, No. 2 (1950): pp 11.

Blair, J.S. "Iron and Steel Works Applications of Radioactive Isotopes for Radiography." *Iron Coal Trades Review*. Vol. 163 (1951): pp 1349, 1405.

Bukshpan, S. and D. Kedem. "Detection of Imperfections by Means of Narrow Beam Gamma Scattering." *Materials Evaluation*. Vol. 33, No. 10 (October 1975): pp 243.

Clack, B.N. "Natural and Artificial Sources for Gamma Radiography." *Engineer*. Vol. 194 (1957): pp 329.

Cosh, T.A. "An Exposure Calculator for Isotope Radiography." *Journal of Scientific Instrumentation*. Vol. 34 (1957): pp 329.

Doan, G.E. and S.S. Yound. "Gamma-ray Radiography." *ASME Proceedings*. Vol. 38, Pt. 2 (1938): pp 292.

Dutli, J.W. and G.M. Taylor. "Application of Cesium-137 to Industrial Radiography." *Nondestructive Testing*. Vol. 12, No. 2 (1954): pp 35.

Dutli, J.W., G.H. Tenney and J.E. Winthrow. "Radium, Tantalum-182 and Cobalt-60 in Industrial Radiography." *Nondestructive Testing*. Vol. 8, No. 3 (1949-50): pp 9.

"Neutron Radiography." *Engineering*. Vol. 211, No. 1 (April 1971).

Faulkenberry, B.H., R.H. Johnson and C.E. Cole. *Nucleonics*. Vol. 19, No. 4 (1961): pp 126.

Fletcher, L.S. *Welding Engineer* (June 1953).

Frazier, P.M., C.R. Buchanan and G.W. Morgan. "Radiation Safety in Industrial Radiography with Radioisotopes." US Atomic Energy Commission Report AECU 2967 (1954).

Gardner, R.P. and R.L. Ely, Jr. *Radioisotope Measurement Applications in Engineering*. New York, NY: Reinhold Publishing Corporation (1967).

Gezelius, R.A. and C.W. Briggs. *Radium for Industrial Radiography*. New York: Radium Chemical Co., Inc. (1946).

Gilbert, E. "Applications of Nondestructive Testing in the Petroleum Industry." *Nondestructive Testing*. Vol. 21, No. 4 (1963): pp 235.

Halmshaw, R. "Use and Scope of Iridium-192 for the Radiography of Steel." *British Journal of Applied Physics*. Vol. 5 (1954): pp 238.

"Thulium-170 for Industrial Radiography." *British Journal of Applied Physics*. Vol. 6 (1955): pp 8.

Handbook on Radiography, rev. ed. Ottawa: Atomic Energy of Canada, Ltd. (1950).

Hile, J. "Automatic Radiography with Cobalt-60." *Materials and Methods*. Vol. 40 (1954): pp 108.

Iddings, F.A. "Back to Basics." *Materials Evaluation*. Vol. 37, No. 11 (October 1979): pp 20.

"Take a Look at Nuclear Gages." *Instrument and Control Systems*. Vol. 49 (December 1976): pp 41-44.

Isenburger, H.R. "Exposure Charts for Cobalt-60 Radiography." *American Foundryman*. Vol. 18 (1950): pp 48.

Johns, H.E. and C. Garrett. "Sensitivity and Exposure Graphs for Radium Radiography." *Canadian Journal of Research.* Vol. 26A (1948): pp 292. Also, see: Johns, H.E. and C. Garrett. "Sensitivity and Exposure Graphs for Radium Radiography." *Nondestructive Testing.* Vol. 8, No. 3 (1949-50): pp 16.

Kahn, N.A., E.A. Imbembo and J. Bland. "A Universal Exposure Calculator for Radium Radiography and Its Application to Current Radiographic Films and Techniques." ASME Special Technical Publication 96 (1950): pp 41.

Karrer, C.A. "Safe and Economical Use of Isotopes in the Steel Industry." *Nondestructive Testing.* Vol. 13, No. 2 (1955): pp 29.

Kastner, J. "Units Used in Industrial Radiography to Describe Strength of Cobalt-60 Sources." *Nondestructive Testing.* Vol. 11, No. 1 (1952): pp 21.

Kiehle, W.D. "Radiography." *Nondestructive Testing.* Vol. 16, No. 4 (July-August 1958): pp 313.

Landalt, J.F. "A Technique for Placing Known Defects in Weldments." *Materials Evaluation.* Vol. 31, No. 10 (October 1973): pp 214.

Memorandum on Gamma-ray Sources for Radiography, rev. ed. London: Institute of Physics (1954).

Morrison, A. "Use of Radon for Industrial Radiography." *Canadian Journal of Research.* Vol. 23F (1945): pp. 413. Also, see: Morrison, A. "Use of Radon for Industrial Radiography." *Nondestructive Testing.* Vol. 6, No. 2 (1947): pp 24.

"Material Required to Carry Out Radiography with Cobalt-60 or Radium." National Research Council of Canada Report 2008 (1949).

"Exposures for Radium Radiography of Steel." *Metals Progress.* Vol. 57 (1950): pp 780.

"Exposures for Cobalt-60 Radiography of Steel." *Metals Progress.* Vol. 58 (1950): pp 80.

"Radiography with Cobalt-60." *Nucleonics.* Vol. 5, No. 6 (1949). Also, see: Morrison, A. "Radiography with Cobalt-60." *Nondestructive Testing.* Vol. 9, No. 4 (1951): pp 14.

Morrison, A. "Iridium 192 for Gamma-ray Radiography." *Nondestructive Testing.* Vol. 10, No. 1 (1951): pp 26.

Morrison, A. and E.M. Nodwell. "Radium Radiography of Thin Steel Section." ASTM Bulletin No. 127 (1944): pp 29.

"Methods and Limitations for In Service Inspection of Nuclear Power Plant." *Nuclear Engineering International.* (October 1976): pp 61-64.

O'Conner, D.T. and J.J. Hirschfield. "Some Aspects of Cobalt Radiography." *Nondestructive Testing.* Vol. 10, No. 1 (1951): pp 23; errata, Vol. 11, No. 1 (1952): pp 34.

Polansky, D., D.P. Case and E.L. Criscuolo. "The Investigation of Radioisotopes for the Inspection of Ship Welds." *Nondestructive Testing.* Vol. 17, No. 1 (1959): pp 21.

Pullin, V.E. "Radon, Its Place in Nondestructive Testing." *Welding.* Vol. 18 (1950): pp 166.

"Radiographic Inspection — An Adaptable Tool." *Quality Progress.* Vol. 7, No. 6 (June 1974).

Radioisotope Techniques. Vol. 2. London: H.M. Stationery Office (1952).

Reed, M.E. *Cobalt-60 Radiography in Industry.* Boston, MA: Tracerlab Inc. (1954).

Richardson, H.D. *Industrial Radiography Manual.* Superintendent of Documents Catalog. No. FS 5.284:84036 (March 1968).

Rigbey, J.V. and C.F. Baxter. "Iridium-192 in Industrial Radiography." *Nondestructive Testing.* Vol. 11, No. 1 (1952): pp 34.

Rummel, W.D. and B.E. Gregory. " 'Ghost Lack of Fusion' in Aluminum Alloy Butt Fusion Welds." *Materials Evaluation.* Vol. 23, No. 12 (1965): pp 568.

Schwinn, W.L. "Economics and Practical Applications of Cobalt-60 in Radiographic Inspection of Steel Weldments." ASME Special Technical Publication 112 (1951): pp 112.

Thompson, J.M. and P.A. Glenn. "Cesium Radioisotope — New Tool for Parts Inspection." *Iron Age.* Vol. 172, No. 11 (1953): pp 174.

"Interpreting Wild Radiographs." *Welding Engineer.* Vol. 56, No. 6 (June 1976).

West, R. "Low Energy Gamma Ray Sources." *Nucleonics.* Vol. 11, No. 2 (1953): pp 20.

SECTION 4

RADIATION DETECTION AND RECORDING

William B. Rivkin and Gerald Wicks, Health Physics Associates Ltd., Northbrook, IL

Material for Part 8 of this Section provided by:
Eastman Kodak Company

INTRODUCTION

Radiation such as heat, light and electromagnetism consists of waves and/or particles, produced electrically or emitted by radioactive nuclei. Emissions from radioactive nuclei, and radiation from that portion of the electomagnetic spectrum beyond the ultraviolet energies, can cause the ionization of atoms and molecules.

This *Nondestructive Testing Handbook* Section deals specifically with the detection and recording of such radiation. Ionizing radiation comprises: (1) charged particles such as alphas, betas, protons; (2) uncharged particles such as neutrons; and (3) electromagnetic radiation in the form of X and gamma rays.

PART 1

BASIC PRINCIPLES

Some forms of radiation, such as light and heat, can be detected by our senses; ionizing radiation, though, can only be detected by the aftereffect of its ionizing properties. If ionizing radiation does not interact with matter, its detection and measurement is not possible. For this reason, the detection process uses substances that respond to radiation, as part of a system for measuring the extent of that response.

The *ionization process* is used by a large class of detection systems, including: ion chambers, proportional chambers, Geiger Mueller counters and semiconductor devices (see Table 1).

Some systems depend upon the excitation and molecular dissociation that occur with ionization; these processes are useful in scintillation counters and chemical dosimeters. Though other types of detection systems exist, they are not generally used in radiation survey instruments.

TABLE 1. Effect of Ionization Used in Detection and Measurement of Ionizing Radiation

Effect	Type of Instrument	Detector
Electrical	Ionization chamber	Gas
	Proportional counter	Gas
	Geiger Mueller counter	Gas
	Solid state	Semiconductor
Chemical	Film emulsion	Photographic
	Chemical dosimeter	Solid or liquid
Light	Scintillation counter	Crystal or liquid
	Cerenkov counter	Crystal or liquid
Thermoluminescence	Thermoluminescent dosimeter	Crystal
Heat	Calorimeter	Solid or liquid

PART 2

GAS IONIZATION DETECTORS

Principles of Ionization

The mechanism most widely used in radiation survey applications is the ionization principle: charged particles producing ion pairs by direct interaction. These charged particles may (1) collide with electrons and remove them from their atoms, or (2) transfer energy to an electron by the interaction of their electric fields (see Fig. 1). If the energy transfer is not sufficient to completely remove an electron, the atom is left in a disturbed or excited state.

Gamma and X-ray photons interact with matter by photoelectric absorption, Compton scattering and pair production, each of which produces electrons and ions that may be collected and measured. The average energy expended in the creation of an ion pair, in air and most gases, is about 34 electron volts (eV).

The number of ion pairs produced per unit of path length is called *specific ionization*. Specific ionization is affected by the energy of the particle or photon and the nature of the ionized substance.

Ionization Chambers

In an ionization chamber, an electric field is applied across a volume of gas, between two electrodes. Often the chamber's geometry is cylindrical, a cylindrical cathode enclosing the gas and an axial, insulated rod anode (see Fig. 2).

Charged particles and/or photons pass through the chamber and ionize the enclosed gas. When an electric field is applied to the gas, ions drift along the electrical lines of force to produce an *ionization current*. Under normal conditions, electrons drift at speeds of about 10^6 centimeters per second. The drift velocity of ions is many orders of magnitude less.

When the electric field is increased slightly from zero, and a detector is placed in the constant radiation field, the collected ions still will be few in number because many re-combine. As the voltage is further increased, *recombination* yields to *ionization*, where all ions are collected (see Fig. 3).

Ion current chambers have a response magnitude proportional to the absorbed energy, and are therefore widely used for making dose measurements. When (1) recombination is negligible, (2) gas amplification does not occur and (3) all other charges are efficiently collected, then the steady state current produced is an accurate measurement of the rate at which ion pairs are formed within the gas. Measurement of this ionization current is the principle behind the *DC ion chamber*.

Ion chambers may be constructed of several different materials and, because radiation must penetrate the wall of the chamber in order to ionize the gas volume, chamber wall materials are chosen for the specific radiation energy being evaluated. When considering a particular instrument, the energy response curve should always be consulted (see Fig. 4). Some instruments may also have an angular

FIGURE 1. Ejection of Electron by Photoelectric Interaction

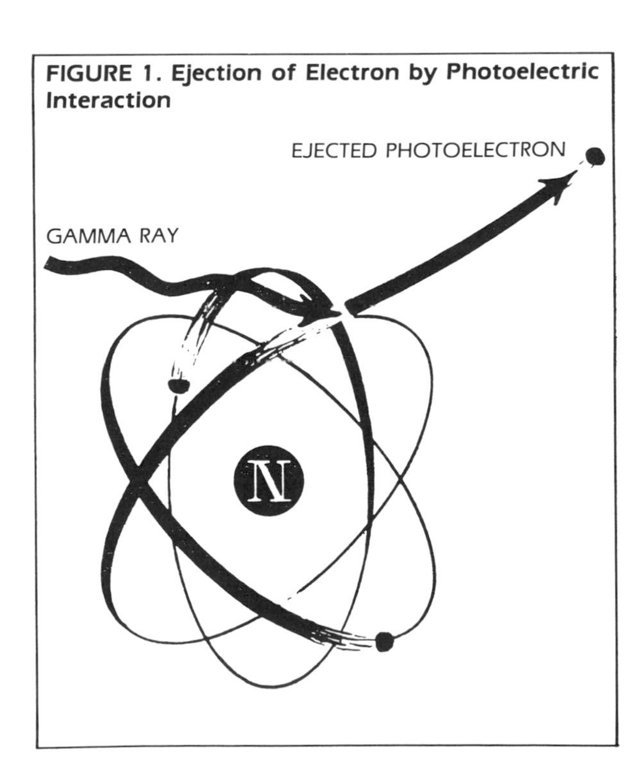

dependence (more sensitivity in some directions), which should be considered when making measurements. Radiofrequency-shielded ionization chambers are available for measurements made near high-level radiofrequency sources.

FIGURE 2. Basic Ionization Chamber with High-Value Resistance, *R*

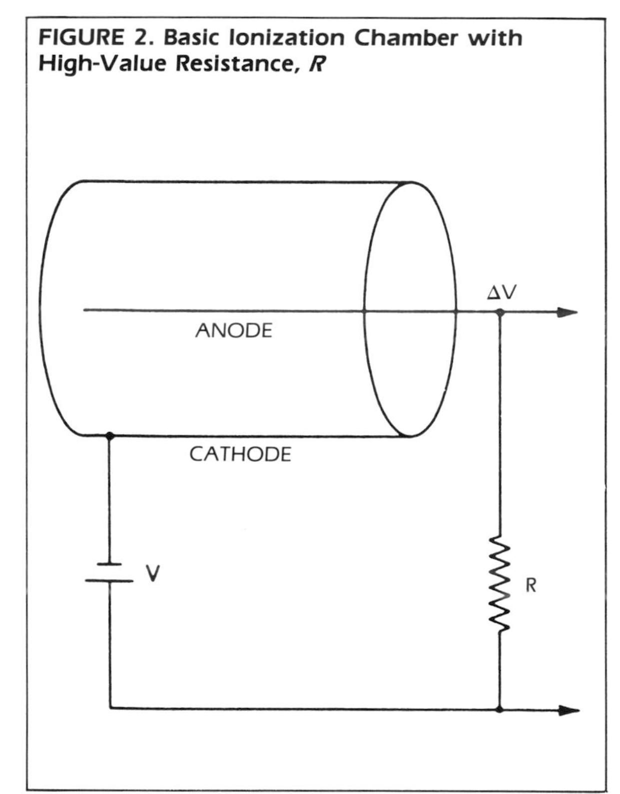

FIGURE 3. Pulse Size as a Function of Voltage in Gas Ion Chamber

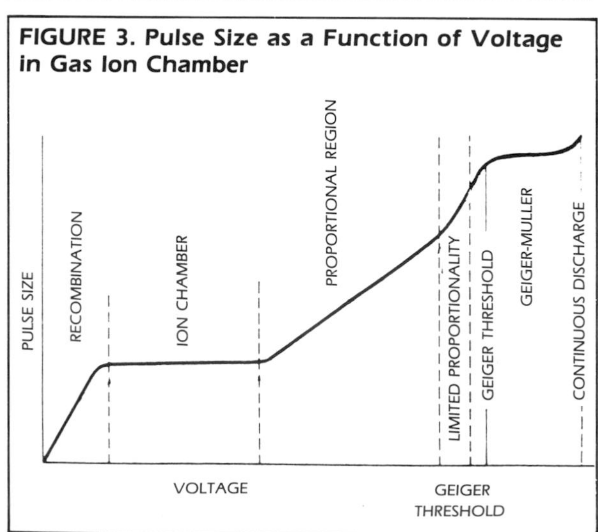

Output Current Measurement

The ionization current collected in the ion chamber flows through an external circuit for measurement. Although, in principle, an ammeter could be placed in the external circuit to read the ion current, in practice this is not normally done because the current is very small. A 440 cubic centimeter ion chamber typically produces about 4×10^{-15} A/μSv (4×10^{-14} A/mR), at standard temperature and pressure. A high-valued load resistor (on the order of 10^{10} ohms) is placed in the circuit, and the voltage drop across the resistor is measured with a sensitive

FIGURE 4. Energy and Directional Response of Typical Ion Chamber Survey Meters

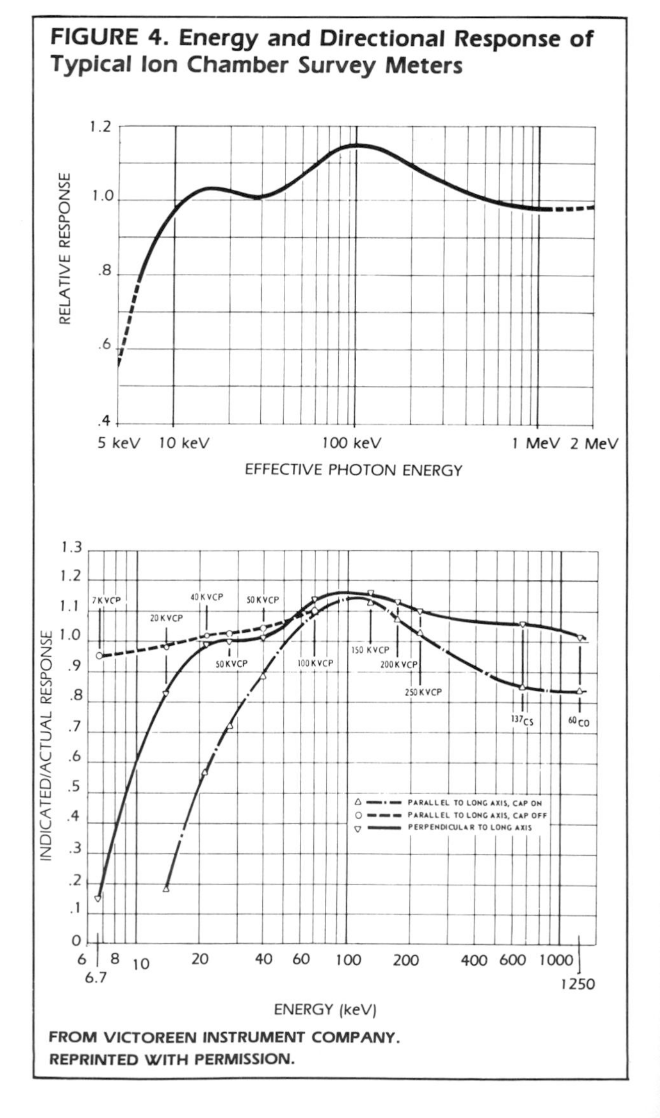

FROM VICTOREEN INSTRUMENT COMPANY. REPRINTED WITH PERMISSION.

electrometer. A metal oxide silicone field effect transistor (MOSFET) is used in some electometers. The MOSFET produces an input impedance on the order of 10^{15} ohms to amplify the collected current (see Fig. 5).

Vibrating Reed Electrometers

An alternative approach to ion current measurement is to convert the signal from DC to AC at an early stage. This allows a more stable amplification of the AC signal in subsequent operations. The conversion is accomplished in a dynamic capacitor or vibrating reed electrometer, by collecting the ion current across a resistive-capacitive circuit. The capacitance is then changed rapidly, compared to the time-constant of the circuit. The induced AC voltage is proportional to the ionization current (see Fig. 6).

Integrating Instruments

The instruments described above (see Fig. 7) are generally *rate meter* types; that is, they indicate the radiation at the time of exposure, and depending on

FIGURE 5. Operational Amplifier; Current Amplifier Configuration

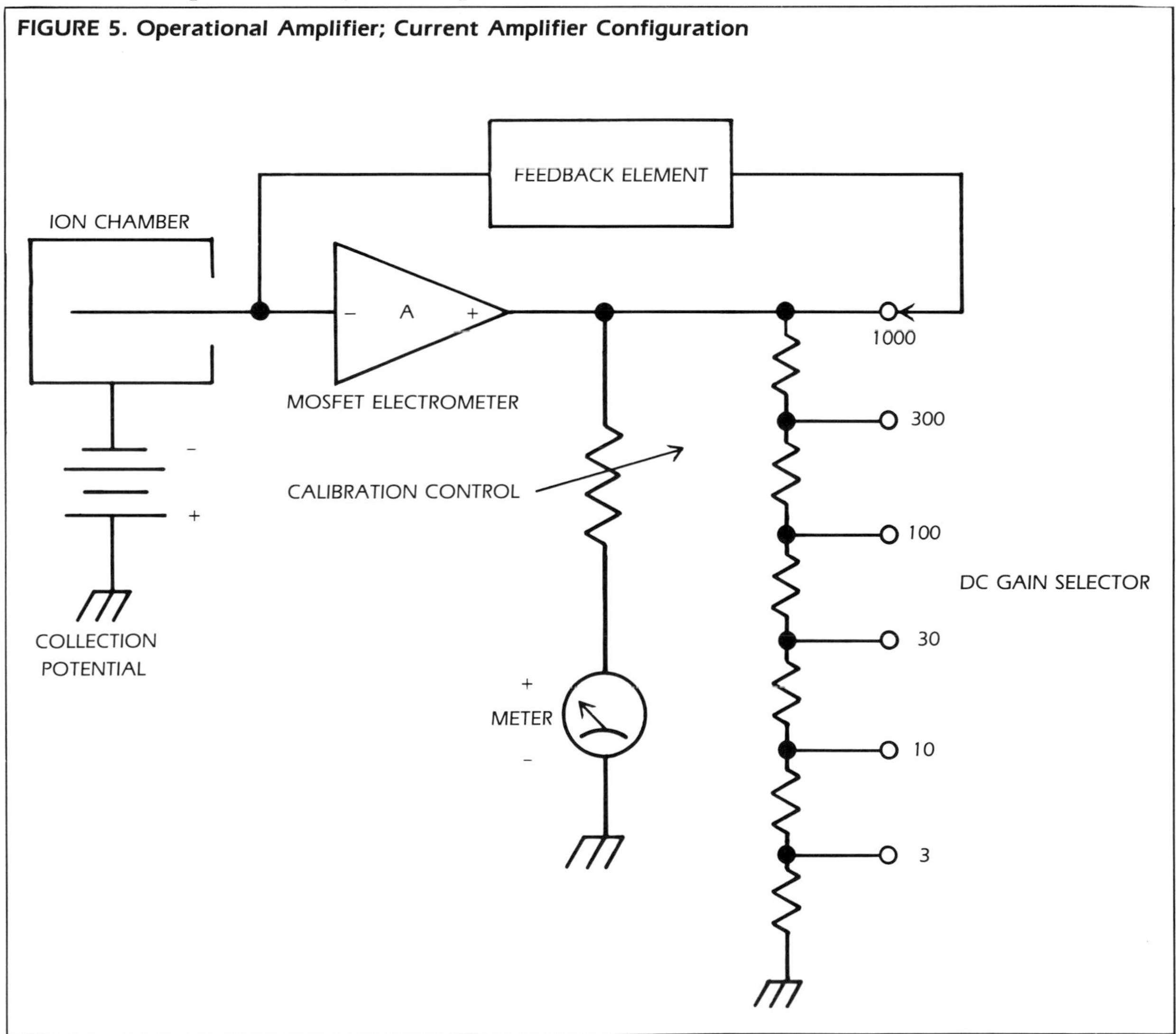

their time constants, will return to background levels as the source is removed.

Some instruments may have an integration switch that introduces a capacitor to the circuit to accumulate the charge. Leaving such an instrument at an operator's location will indicate the total amount of ionizing radiation that area has received, from the time the instrument is engaged.

FIGURE 6. Principle of the Vibrating Reed Electrometer; Oscillations of the Capacitance Induce an AC Voltage Which Is Proportional to the Steady-State Signal Current

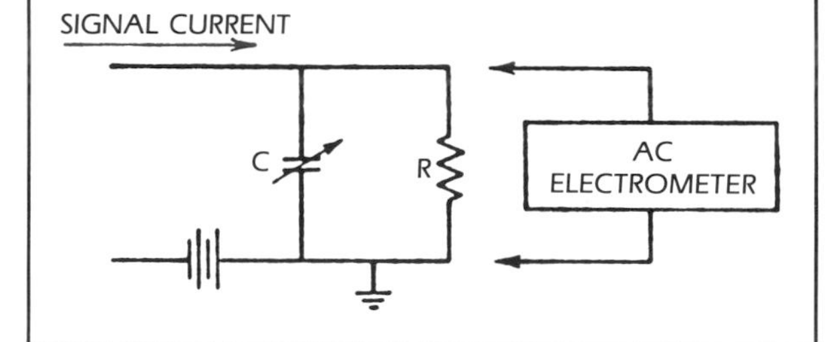

FIGURE 7. Examples of Ionization Chambers Located Externally on Survey Instruments; Lower Instruments Have Protective Caps Removed, Showing Thin Windows for Low Energy X-ray and/or Beta Detection

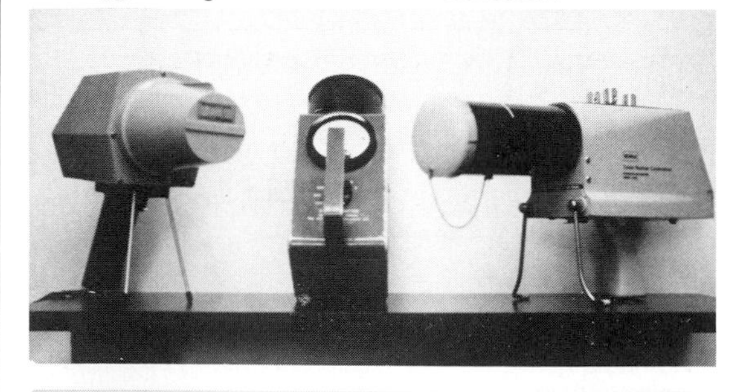

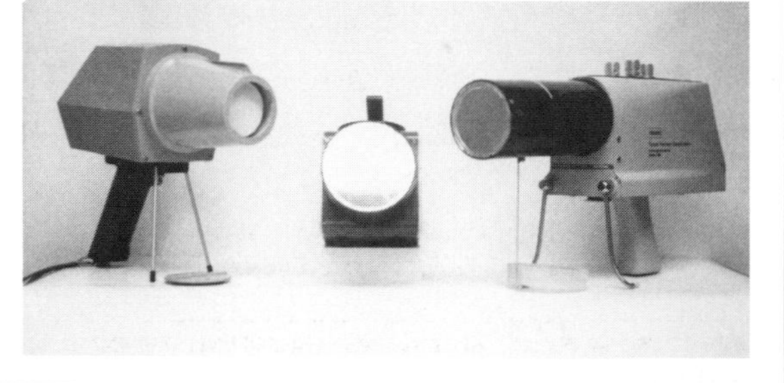

Personnel Monitoring Instruments

Pocket Chambers

Personnel monitoring instruments, some the size of a ball point pen, are usually the integrating type and contain an ionization chamber. One version, the pocket chamber, requires the application of an initial charge of 150-200 V by an external instrument. Zero dose is then indicated on a scale contained in the charging unit. Exposure of the chamber to ionization decreases the initial charge, and when reinserted into the charging unit, the reduced charge is translated to the level of exposure (see Fig. 8).

FIGURE 8. Cross Section of Quartz Fiber Pocket Dosimeter

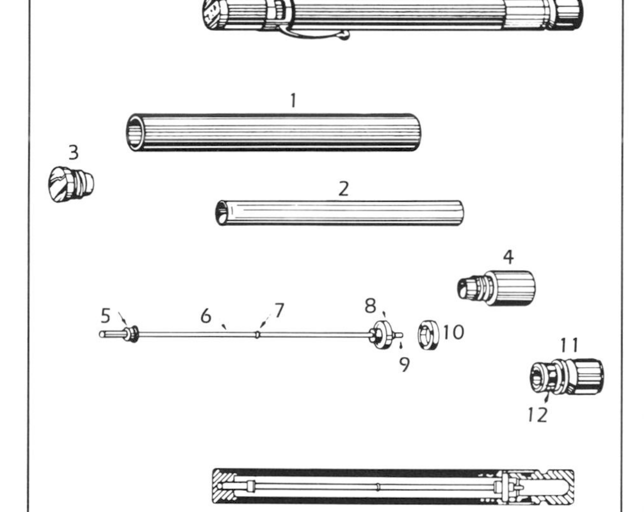

1. LOW ATOMIC NUMBER WALL
2. GRAPHITE-COATED PAPER SHELL
3. ALUMINUM TERMINAL HEAD
4. ALUMINUM TERMINAL SLEEVE
5. POLYSTYRENE SUPPORT BUSHING
6. CENTRAL ELECTRODE, GRAPHITE COATED
7. POLYETHYLENE INSULATING WASHER
8. POLYSTYRENE FIXED BUSHING
9. ELECTRODE CONTACT
10. RETAINING RING
11. ALUMINUM BASE CAP
12. POLYETHYLENE FRICTION BUSHING

FROM NATIONAL BUREAU OF STANDARDS.

Direct Reading Dosimeters

The direct reading dosimeter operates on the principle of the gold-leaf electroscope (Fig. 9). A quartz fiber is displaced electrostatically by charging it to a potential of about 200 V. An image of the fiber is focused on a scale and viewed through a lens at one end of the instrument. Radiation exposure of the dosimeter discharges the fiber, allowing it to return to its original position. Personnel dosimeters may have a full scale reading of 2 to 200 mSv (or 200 mR to several roentgens).

Chambers are available with thin walls for sensitivity to beta radiation over 1 MeV and may be coated on the inside with boron for neutron sensitivity.

Figure 10 demonstrates the energy response of self-reading pocket dosimeters. Table 2 lists performance specifications of dosimeters in general.

FIGURE 9. Cross Section of Pocket (Direct Reading) Ionization Chamber

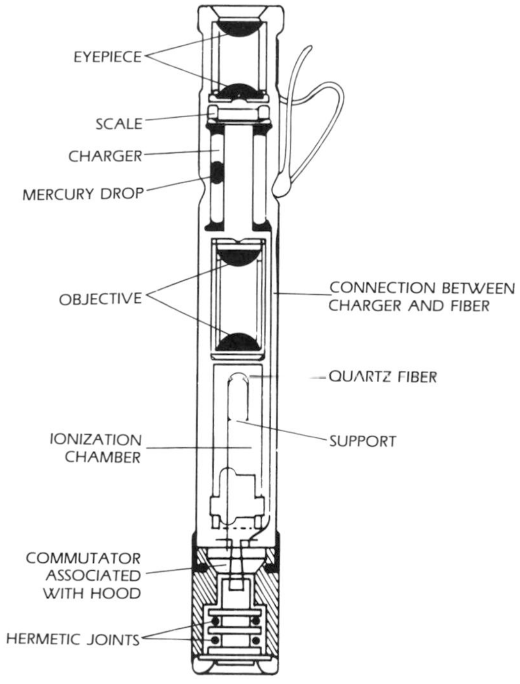

FROM NATIONAL BUREAU OF STANDARDS.

Proportional Counters

If the electric field in an ion chamber is raised above the ionization potential but below saturation potential, enough energy is imparted to the ions for production of secondary electrons by collision and gas amplification.

Operation at this electric potential overcomes the difficulty of the small currents in the ionization region yet takes advantage of pulse-size dependence for separating various ionizing energies. When an ionization chamber is operated in this region, it is called a *proportional counter.*

Because the size of the output pulse is determined by the number of electrons collected at the anode, the size of the output voltage pulse from a given detector is proportional to the voltage across the detector. By careful selection of gases and voltages, a properly designed proportional counter can detect alphas in the presence of betas, or higher-energy beta and gamma radiation in the presence of lower energies. Proportional counters are often used in X-ray diffraction applications.

TABLE 2. General Performance Specifications for Dosimeters

Characteristics	Performance Specifications
Accuracy	± 12% at 95% confidence
Energy dependence	± 10% over given range
Sensitivity adjustment	Sealed
Exterior surface	Smooth
Ruggedness	Withstands a 120 cm (48 in.) drop
Temperature	+ 50 °C to − 10 °C
Humidity	0% to 90%
Discharge	No more than 2% of full scale in 24 hours
Angular dependence	More than 70% at angles greater than 50° from direction of maximum response

FROM AMERICAN NATIONAL STANDARDS INSTITUTE. REPRINTED WITH PERMISSION.

FIGURE 10. Energy Dependence of the Response of Different Commercial Self-Reading Dosimeters

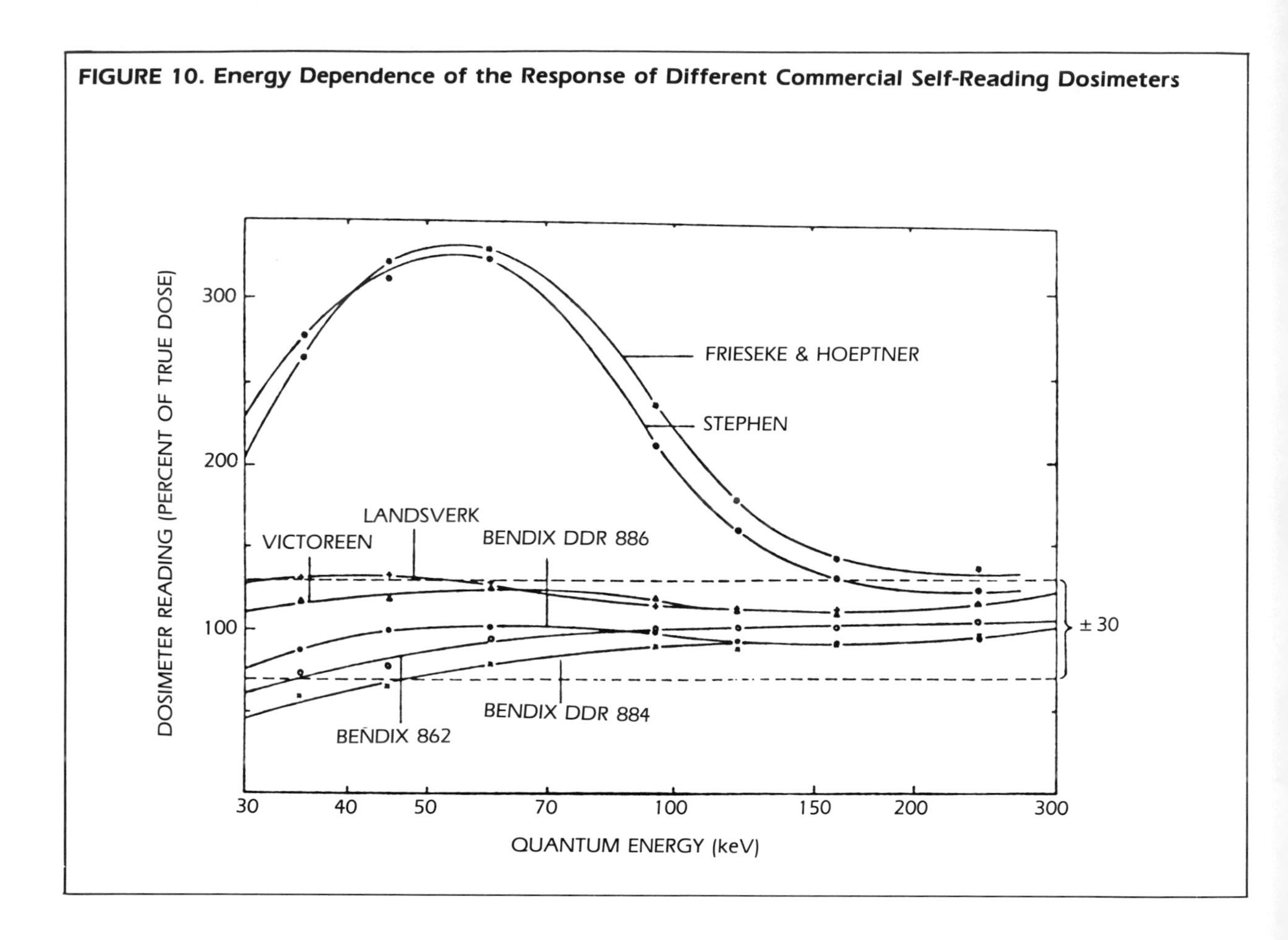

PART 3

GEIGER MUELLER COUNTERS

Operating Voltage Level

Increasing voltage beyond the proportional region (Fig. 3) will eventually cause the gas avalanche to extend along the entire length of the anode wire. When this happens, the end of the proportional region is reached and the Geiger Mueller region begins.

An instrument operating in this voltage range, using a sealed gas-filled detector, is referred to as a Geiger Mueller counter, a G-M counter or simply a Geiger tube. Introduced in 1928, the simplicity and low cost of this equipment has made it the most popular radiation detector in use today. G-M counters complement the ion chamber and proportional counter and comprise the third category of gas-filled detectors based on ionization.

Properties

Extension of the gas avalanche increases the gas amplification factors so that 10^9 to 10^{10} ion pairs are formed in the discharge. This results in an output pulse that is large enough (0.25 to 10 V) to require no sophisticated electronic amplification circuitry for readout. At this voltage, the size of all pulses, regardless of the nature of the ionization, is the same.

When operated in the Geiger region, a counter cannot distinguish among the several types of radiation and therefore is not useful for spectroscopy or for the detection of one energy event in the presence of another. An external shield is often used to filter out alpha and/or beta particles in the presence of gamma energies.

Resolving Time

As an ionizing event occurs in the counter, the avalanche of ions paralyzes the counter. The counter is then incapable of responding to another event until the discharge dissipates and proper potential is established. The time it takes to reestablish the electric field intensity is referred to as the *resolving time.* Average resolving time for a Geiger Mueller counter is approximately 100 μs, which must be corrected for, at high-level readings.

Resolving time (τ) of a counter may be determined by counting two sources independently (R_1 and R_2), then together ($R_{1,2}$). The background count is R_b.

$$\tau = \frac{R_1 + R_2 - R_{1,2} - R_b}{R_{1,2}^2 - R_1^2 - R_2^2} \qquad \text{(Eq. 1)}$$

Correct counting rate (R) can be calculated from observed counting rate (R_o) and resolving time (τ), in the following equation:

$$R = \frac{R_o}{1 - R_o\tau} \qquad \text{(Eq. 2)}$$

Dead Time

The relationship of resolving time to dead time and recovery is illustrated in Fig. 11. Resolving time may be a function of the detector alone or of the detector and its signal processing electronics. Its effect on the real counting rate depends on whether the system design is paralyzable or nonparalyzable.

Nonparalyzable Systems

In Fig. 12, a time scale is shown indicating six randomly spaced events in the detector.

At the bottom of the illustration is the corresponding dead time behavior of a detector assumed to be nonparalyzable. A fixed time τ follows each event that occurs during the live period of the detector. Events occurring during the dead time have no effect on the detector, which would record four counts from the six interactions.

Paralyzable Systems

The top line of Fig. 12 illustrates a paralyzable system. Resolving time τ follows each interaction, whether it is recorded or not. Events that occur during resolving time τ are not recorded and further extend the dead time by another period τ. The chart indicates only three recorded events from the six interactions. In this case, τ increases with increased number of interactions.

It can be demonstrated that with a paralyzable system (at increasingly higher interaction rates), the observed counting rates can actually decrease with an increased number of events. When using a counting system that may be paralyzable, extreme caution must be taken to ensure that low observed counting rates correspond to low interaction rates, rather than very high interaction rates with accompanying, long dead time. It is possible for a paralyzable system to record the first interaction and then be paralyzed, recording zero counts in high radiation fields.

Quenching

As positive ions are collected after a pulse, they give up their kinetic energy by striking the wall of the tube; ultraviolet photons and/or electrons are liberated, producing spurious counts. Prevention of such counts is called *quenching*.

FIGURE 11. Resolving Time, Dead Time and Recovery Time for G-M System

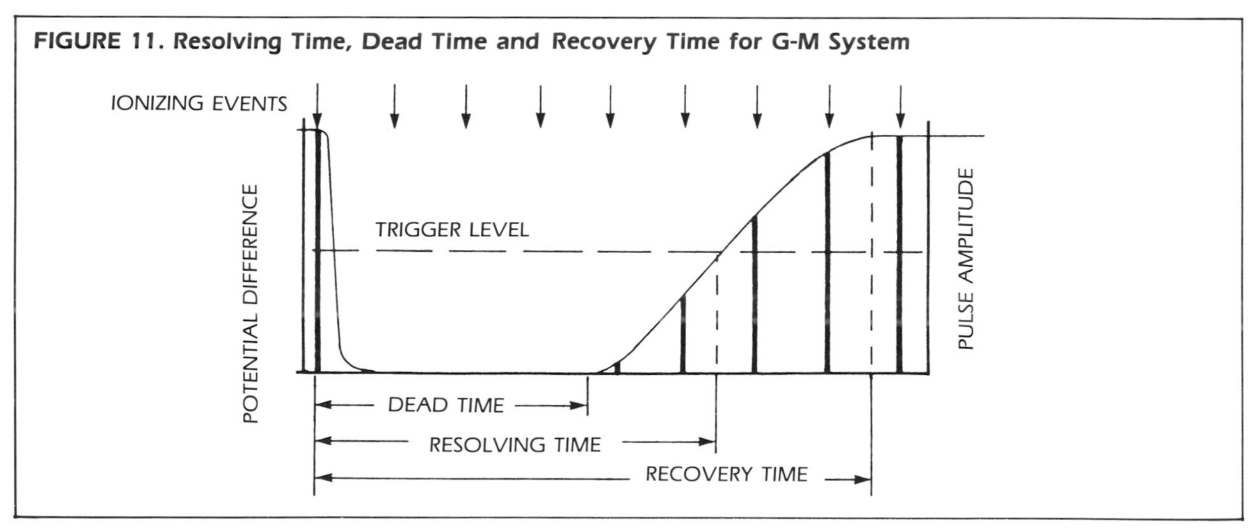

FIGURE 12. Processing of Detector Interactions in Paralyzable and Nonparalyzable Systems

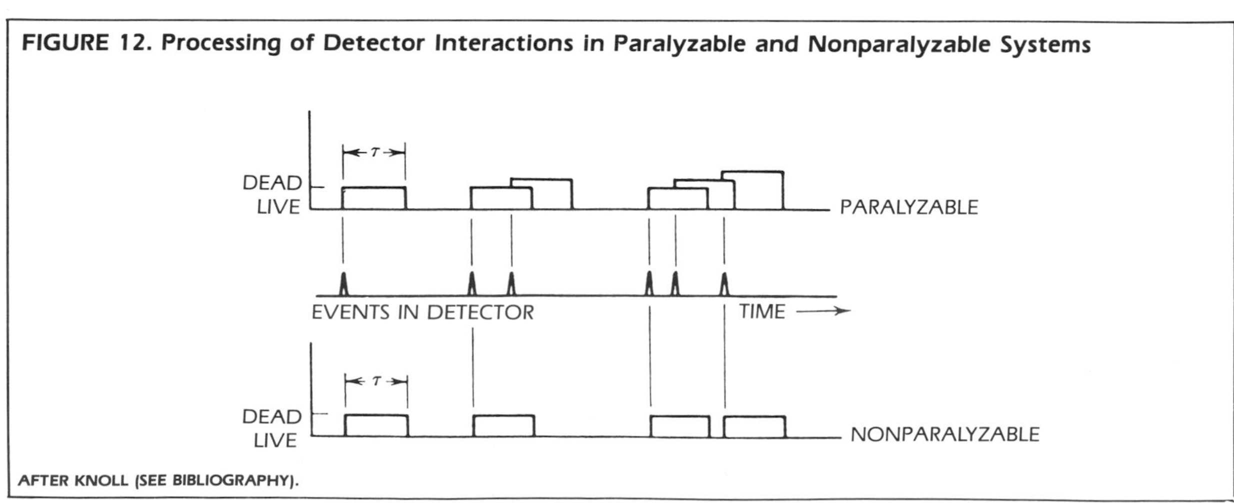

AFTER KNOLL (SEE BIBLIOGRAPHY).

Quenching may be accomplished electronically (by lowering the anode voltage after a pulse) or chemically (by using a self-quenching gas).

Electronic Quenching

Electronic quenching is accomplished by introducing a high value of resistance into the voltage circuit. This will drop the anode potential until all the positive ions have been collected.

Self-quenching Gas

A self-quenching gas is one that can absorb ultraviolet (UV) photons without becoming ionized. One method of using this characteristic is to introduce a small amount of organic vapor, such as alcohol or ether, into the tube. The energy from the UV photons is then dissipated by dissociating the gas molecule. Such a tube is useful only as long as it has a sufficient number of organic molecules to dissociate, generally about 10^8 counts.

To avoid the problem of limited lifetime, some tubes use halogens (chlorine or bromine) as the quench gas. The halogen molecules also dissociate in the quenching process but they are replenished by spontaneous recombination at a later time. Halogen quench tubes have an infinite lifetime and are preferred for extended use applications.

Reaction products of the discharge often produce contamination of the gas or deposition on the anode surface and generally limit the lifetimes of G-M tubes.

FIGURE 13. Assortment of G-M Counters Demonstrating Availability of Sizes and Shapes; Smallest Counter Shown is About 3 cm (1 in.) in Length

Design Variations

Geiger counters (see Fig. 13) are available in various shapes and sizes. The most common form is that of a cylinder with a central anode wire. If low-energy beta or alpha particles are to be counted, a unit with a thin entrance window (1 to 4 mg/cm^2) should be selected.

For surveying large surfaces, pancake or large window counters are available. High count rate instruments, greater than 500 μSv/h (50 mR/h), generally contain a small tube to minimize resolving time of the system; large volume detectors may require significant correction.

A G-M counter response to gamma rays occurs by way of gamma-ray interaction with the solid wall of the tube. The incident gamma ray interacts with the wall and produces a secondary electron which subsequently reaches the gas. The probability of gamma ray interaction generally increases with higher density wall material.

Personnel Monitors

Small G-M tubes are used in pocket-sized units for personnel monitoring. They generally emit a high-frequency chirp at a rate proportional to the subjected dose rate. The energy dependence curve for one such instrument is shown in Fig. 14.

Applications

Geiger Mueller counters are the most widely used, general purpose radiation survey instruments. It must be remembered that G-M counters, unlike current ionization chambers, read pulses (regardless of their energy or ionizing potential) and register in counts per minute. Some instruments have a scale calibrated in mR/hr, however this is an arbitrary scale calibrated with the radiation from radium-226, cesium-137 or some other energy (see Fig. 15).

A sensitivity versus energy table should always be consulted before making measurements with a G-M instrument.

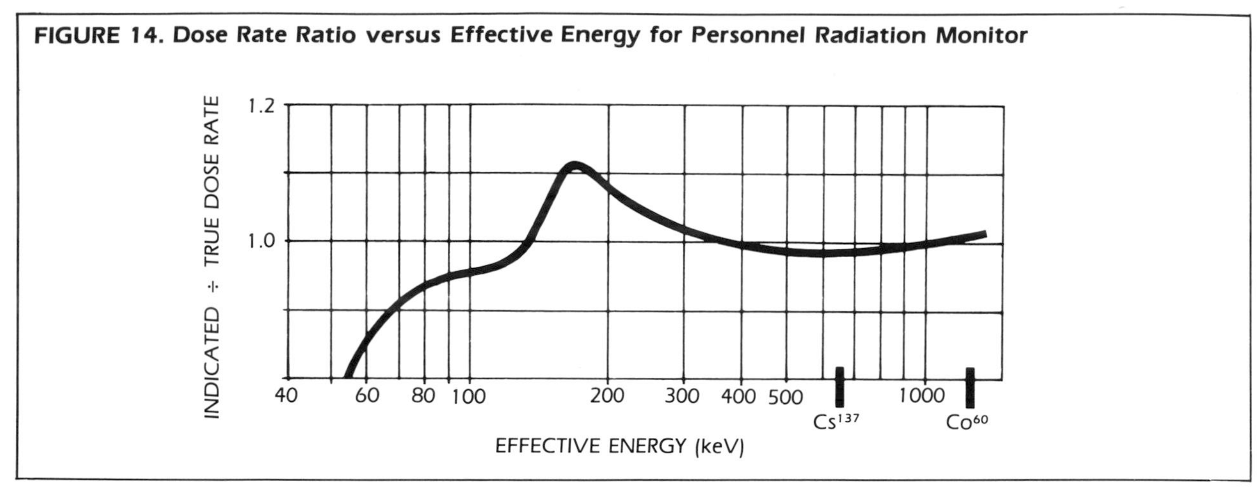

FIGURE 14. Dose Rate Ratio versus Effective Energy for Personnel Radiation Monitor

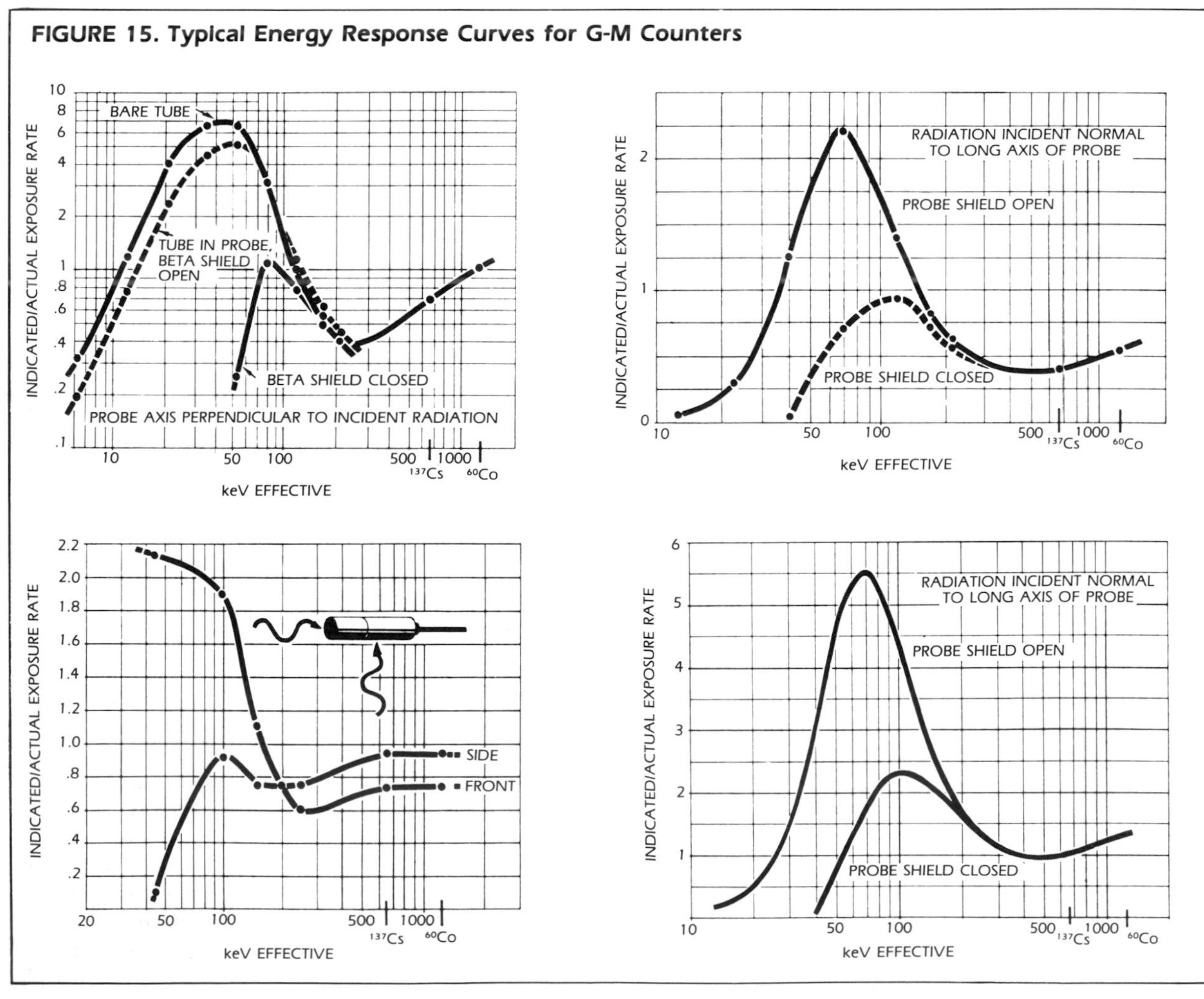

FIGURE 15. Typical Energy Response Curves for G-M Counters

PART 4

SCINTILLATION DETECTORS

Soon after the discovery of X rays and radioactivity, it was observed that certain materials emit visible light photons after interacting with ionizing radiation. These light photons appear to flash or sparkle and the materials are said to *scintillate*. Scintillators commonly used with radiation survey instruments are solid materials. Being denser than gases, these scintillators have greater detection efficiencies and are useful for low-level measurements. For gamma photons, scintillators have detection efficiencies 10^6 times greater than typical gas ionization chambers. Detections of alpha and beta particles, neutrons and gamma photons are possible with various scintillator systems (see Table 3).

Scintillation Process

Electrons, resulting from the interaction of radiation with matter, deposit their kinetic energy in matter by ionization and excitation. *Ionization* refers to the removal of an electron from an atom and *excitation* refers to the elevation of an electron's energy state. The return of excited electrons to their normal, lower energy state is called *de-excitation*. Scintillators excited by ionizing radiation return to lower energy states quickly and emit visible light during the de-excitation process. Radiation detection is possible by measuring the scintillator's light output (scc Fig. 16).

TABLE 3. List of Common Scintillators

Scintillator*	Chemical Symbol**	Radiation Type Detected
Sodium iodide	NaI(Tl)	gamma
Lithium iodide	LiI(Eu)	gamma, neutrons
Zinc sulfide	ZnS(Ag)	alpha
Bismuth germanate	$Bi_1Ge_3O_{12}$	gamma

*Many other scintillators are available but are not commonly used with radiation survey instruments.
**The parentheses indicate the impurity used as an activator.

Materials and Characteristics

Scintillation materials come in gaseous, liquid and solid forms. Organic liquids and solids, as well as inorganic gases and solids, are common scintillators. Organic, solid scintillators are available as crystals, plastics and gels. Inorganic solid scintillators are usually alkali-halide crystals. The scintillation process in inorganic materials requires the presence of small amounts of an impurity, or activator. Inorganic solid scintillators are commonly used with radiation survey instruments and are listed in Table 3.

Desirable Scintillator Characteristics

A useful and practical scintillator needs to have most of the characteristics listed below. Not all of

FIGURE 16. Energy Diagram of Scintillation Process

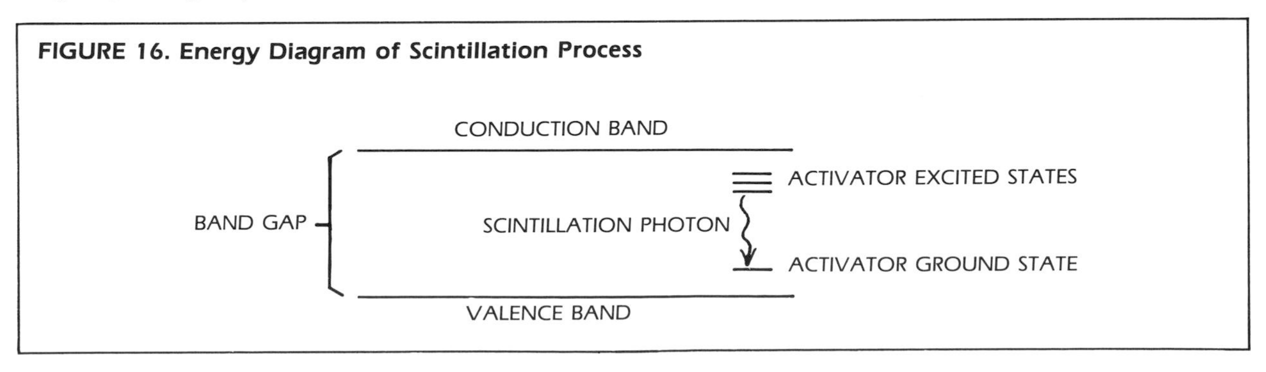

these characteristics are ideally satisfied by each scintillator and often a compromise is acceptable.

1. The scintillator should be of high density and large enough to ensure adequate interaction with the ionizing radiation.
2. Efficient conversion of the electron's kinetic energy into visible light is required and the light yield should be linearly related to the deposited electron kinetic energy.
3. The scintillator should be of good optical quality, transparent to its emitted light, and should have an index of refraction close to that of glass.
4. The wavelength of the emitted light should be appropriate for matching to a photomultiplier tube.

Photomultiplier Tubes

Before the advent of photomultiplier tubes (PMTs), scintillation light photons had to be visually counted. This limited the use and development of scintillators. In the 1940s, the PMT was developed and dramatically increased the use of scintillators, to the point where scintillators are preferred over other radiation detectors for many survey applications.

The PMT's function is to convert the scintillator's light output into an electrical pulse. The PMT is composed of a photosensitive layer, called the *photocathode*, and a number of electron multiplication structures called *dynodes*. Conversion of the scintillation light into photoelectrons is accomplished by the photocathodes through the photoelectric effect. To maximize the information contained in the scintillation light, the PMT photocathode should be matched to the scintillator; the scintillator and PMT should be optically coupled to minimize light losses.

Electron multiplication, or *gain*, is accomplished by positively charging the dynodes in successive stages, so that the total voltage applied to the PMT is around 1,000 V. Electrons emitted by the photocathode are focused toward the first dynode; more electrons are emitted than were initially incident on

FIGURE 17. Cutaway Drawing of a Photomultiplier Tube, Showing Crystal, Photocathode, Collecting Dynodes and the Voltage Divider Network

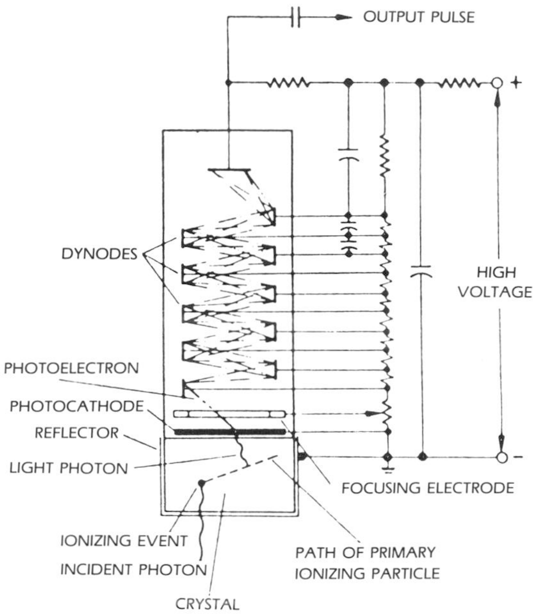

AFTER CEMBER (SEE BIBLIOGRAPHY).

the dynode. This is repeated at each dynode stage. The photocathode and dynodes are positioned in a glass-enclosed vacuum so that air molecules will not interfere with the pulsed electrical signal. The net result of the PMT may be an electron gain up to 10^{10} per emitted electron. Figure 17 illustrates the structure of a PMT.

System Electronics

Once the output pulse from a PMT is generated, it is amplified and analyzed. The pulse height, or amplitude, is proportional to the amount of energy deposited within the scintillator and can be correlated to a count rate or mR/hr scale when calibrated against a known energy source.

PART 5

THERMOLUMINESCENT DOSIMETRY

Thermoluminescence is the emission of light from previously irradiated materials after gentle heating. The radiation effect in thermoluminescent (TL) materials is similar to that observed in scintillators, except that light photon emission does not occur in TL materials until some heat energy is supplied (see Fig. 18). Measurement of the light photons emitted after heating permits correlation to the amount of ionizing radiation energy that was absorbed in the TL material. Thermoluminescent dosimetry (TLD) is possible for beta, gamma and neutron radiations, if the appropriate TL material is used.

LiF Properties

The most common TL phosphor used in gamma and neutron personnel dosimetry is lithium fluoride. Other TL phosphors are available for personnel dosimetry but, for various reasons, are not as well suited as LiF. The advantages of LiF include its: (1) usefulness over a wide dose range; (2) linear dose response; (3) near dose-rate independence; (4) reusability; (5) stability; (6) short readout time; and (7) near tissue-equivalence. Disadvantages include the loss of information after readout and lack of information about the incident radiation energy.

Interaction with Ionizing Radiation

Both gamma photons and neutrons produce ionization indirectly. Gamma photons interact with matter, releasing electrons that in turn cause ionization. LiF undergoes interaction with gamma photons and is therefore used in gamma dosimetry. Slow neutrons require the presence of LiF enriched with lithium-6 for detection of the (n, α) nuclear reaction. Fast neutron detection with LiF would only be possible if the fast neutrons were slowed down to thermal energies before reaching the LiF TLD. Nearly complete elimination of neutron response in LiF is possible with LiF enriched with lithium-7. In a mixed gamma and slow neutron field, distinction of gamma and neutron doses is possible by comparing the readings of two LiF TLDs with different lithium-6 contents.

FIGURE 18. Thermoluminescence Process

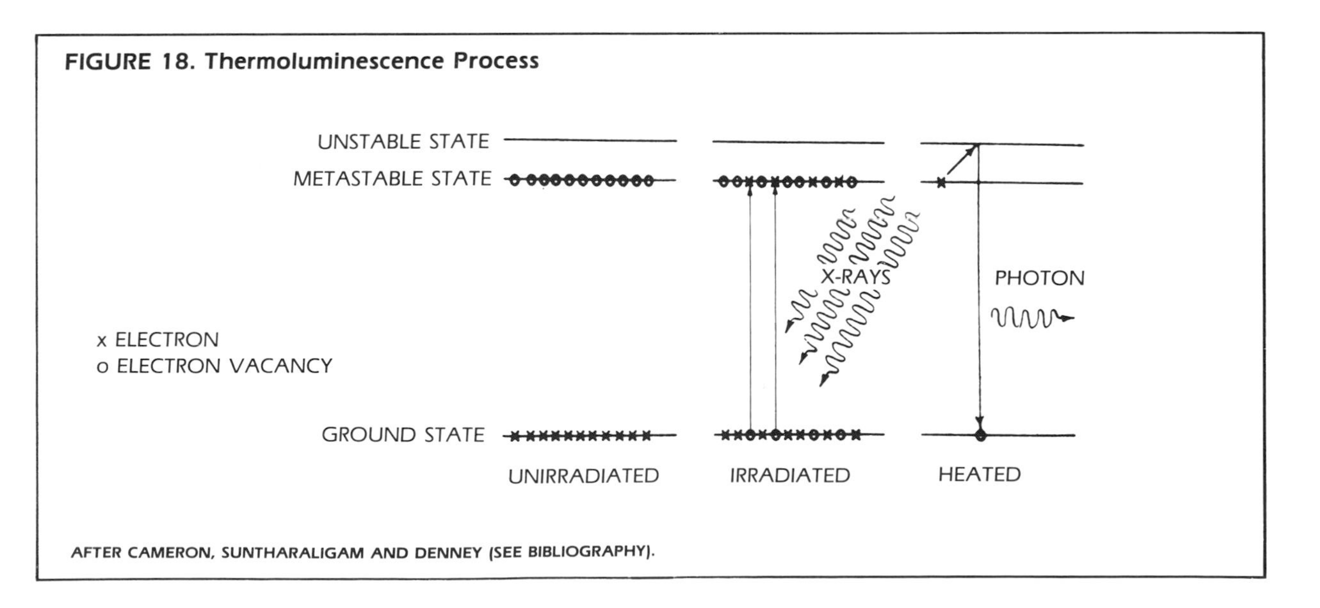

AFTER CAMERON, SUNTHARALIGAM AND DENNEY (SEE BIBLIOGRAPHY).

TLD Readout Systems

TLD readout systems are commonly made up of a sample holder, heating system, PMT (light detector), high voltage supply, signal amplifier and a recording instrument. The TLD sample is heated indirectly, using electrical resistance heat applied to a pan or planchette. The PMT converts the light output into an electronic pulse which is then amplified before recording. The recording instrument may be a plotter or any other type of instrument capable of measuring the amplified PMT output signal. A plot of the output signal versus time is equivalent to emitted light intensity versus heat and results in a *glow curve*. The area under the glow curve is proportional to the absorbed dose (see Fig. 19).

Uses of TLD include personnel dosimetry, medical dosimetry, environmental monitoring and archeological and geological dating.

FIGURE 19. Typical Glow Curve; Integrated Area under Curve is a Measure of Radiation Exposure

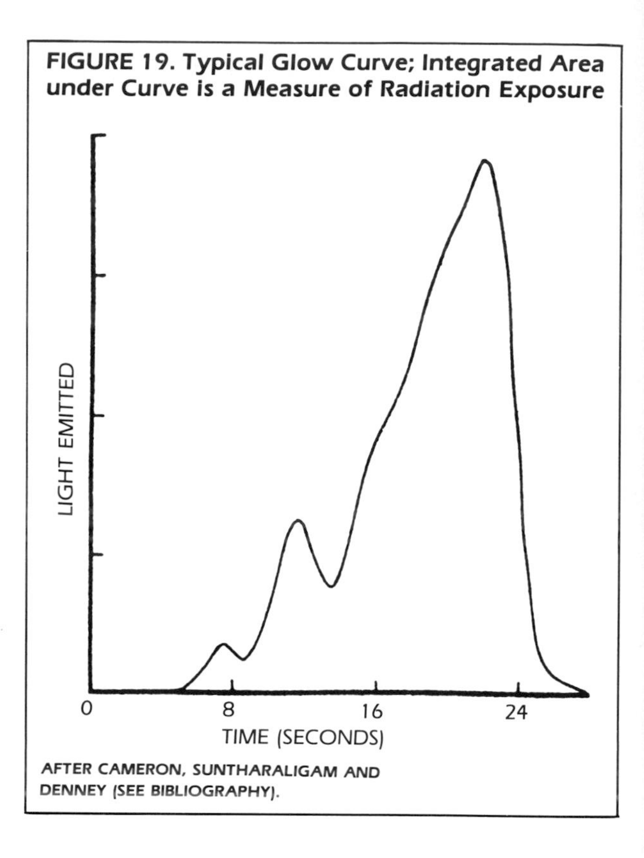

AFTER CAMERON, SUNTHARALIGAM AND DENNEY (SEE BIBLIOGRAPHY).

PART 6

NEUTRON DETECTION

Characteristics

The neutron is a part of the nucleus, has no charge, and is somewhat larger in mass than the proton. It is similar to the photon in that is has no charge and produces ionization indirectly; it is different from the photon because it is a nuclear particle and not a unit of electromagnetic energy. Because the neutron is an uncharged particle, its interactions with matter are different from those of charged particles or photons.

Ionization by neutrons is indirect: as a result of neutron interactions with matter, recoil nuclei, photons or charged particles are produced and these then interact with matter by various mechanisms that cause ionization.

Neutron Sources

Neutrons are classified according to their energies as shown in Table 4.

Some radionuclides (such as californium-252) may decay by spontaneous fission and emit neutrons with fission fragments, photons and electrons. Induced fission reactions, such as those occurring in a nuclear reactor with uranium, emit about 2.5 neutrons per fission. Fission neutrons range in energy from 0.025 eV to about 16 MeV. Other neutron sources are the result of various nuclear reactions and produce either a spectrum of neutron energies or monoenergetic neutrons. Common neutron producing nuclear reactions are the (γ, n), (α, n) (p, n), (d, n) and (α, 2n) reactions and may use radionuclide emissions or accelerated particles to initiate the reaction. Neutron radiography sometimes employs radionuclides that emit alpha or gamma photons and produce neutrons by (α, n) and (γ, n) reactions with various target materials.

TABLE 4. Neutron Classification

Class	Energy
Thermal	0.025 eV
Epithermal	1 eV
Slow	0.03 eV to 100 eV
Intermediate	100 eV to 10 keV
Fast	10 keV to 10 MeV
Relativistic	greater than 10 MeV

Neutron Detectors

There are several mechanisms and devices used to detect neutrons of various energies. Ionization chambers, proportional counters, scintillators, activation foils, track-etch detectors, film and nuclear emulsions, and thermoluminescent phosphors are some of the many devices used to detect neutrons. The main mechanisms used to detect neutrons in these devices are the (n, α), (n, p), (n, d), (n, f) and (n, γ) nuclear reactions.

Proportional Neutron Detectors

Many fast/slow neutron counters use proportional counting chambers filled with BF_3 gas. The interaction of thermal (slow) neutrons with boron gas releases an alpha particle of several MeV that is easily detected in the proportional mode. Fast neutrons are detected by a similar counter, in which thermal neutrons are absorbed in an external cadmium shield, while the fast neutrons that pass through the shield are thermalized in hydrogen-rich material and counted in the proportional chambers.

Scintillation

Scintillators containing lithium-6, boron-10 and hydrogenous plastics have been used as neutron detectors. Lithium-6 is used as LiI(Eu) and in lithium glasses to detect slow and fast neutrons. Scintillators loaded with boron-10 are used for slow neutron detection. Plastic scintillators with high hydrogen content are used in fast neutron detection and spectroscopy by measuring the energy deposited by recoil protons.

Activation Foils

Introducing certain materials to an incident neutron flux will result in these materials becoming radioactive. The process is called *activation* and gaining information about the incident neutron flux and energy is possible by analyzing the radiations emitted from the activated foil. Activation foils rely on (n, γ), (n, p), (n, α), (n, f) and other nuclear reactions to cause the activation. Selection of the proper activation foil can give a rough estimate of the neutron energy spectrum. In high neutron flux fields, where instruments would fail, activation foils are used as integrating detectors.

Miscellaneous Neutron Detectors

Track etch detectors, nuclear emulsions and film have all been used to detect neutrons. Various neutron interactions with the detector material or foils in intimate contact with the detectors allow these systems to operate as integrating dosimeters.

PART 7

SEMICONDUCTORS

Certain semiconductor crystals, when exposed to ionizing radiation, become conductors and may be used as radiation detectors. Semiconductors are most often used for low-level spectroscopic measurements of alpha particles, beta particles and gamma rays in laboratory settings and in X-ray diffraction equipment (see Table 5).

The most widely used semiconductor devices are diffused *p-n* junction, surface barrier, and lithium drifted detectors. Semiconductor detectors have found their broadest application in the field of spectroscopy, although lithium drifted detectors are also being used for gamma ray detection.

TABLE 5. Radiation Dector Types

Type of Radiation			
Charged Particle	γ-Ray	X-Ray	Type of Detector
X			Silicon surface barrier detectors
	X	X	Planar, pure Ge detectors: low-energy photon spectrometer for the energy range of 2 keV to 200 keV
	X		Coaxial Ge(Li) detectors
	X	X	Coaxial pure Ge detectors
		X	Si(Li) detector systems for X-ray detection exclusively below 30 keV

Diffused *p-n* Junction Detector

The diffused *p-n* junction detector obtains its name from its manufacturing process. A slice of *p*-type (electron depleted) silicon or germanium crystal, with a layer of *n*-type (electron rich) impurity (usually phosphorus) deposited on the surface, is heated to form a *p-n* junction just below the surface. The phosphorus may also be painted onto the silicon and made to diffuse into it by applying heat. Because the *n*-type material has an excess of electrons and the *p*-type has an excess of holes (*holes* may be thought of as unit positive charges), the natural action of the combined materials tends to align the electrons on one side of the junction and the holes on the other. Thus a difference of potential is built up across the junction.

By applying an external voltage to the crystal, of such polarity as to oppose the natural movement of electrons and holes (reverse bias), the potential barrier across the junction is increased and a depletion region is produced.

This depletion region is the sensitive part of the detector and is analogous to the gas volume in a gas ionization detector. Charged particles, upon entering the depletion region, produce electron-hole pairs analogous to the ion pairs produced in gas ionization chambers. Because an electric field exists in this region, the charge produced by the ionizing particle is collected, producing a pulse of current. The size of the pulse is proportional to the energy expended by the particle (see Fig. 20).

Surface Barrier Detectors

The operation of surface barrier and lithium drifted detectors is the same as for the *p-n* junction: a depletion region is produced, in which there exists an electric field. The method of producing the depletion region (as well as its dimension and location within the crystal) vary from one type of detector to another.

The operation of a surface barrier detector depends upon the surface conditions of the silicon or germanium. At the surface of a piece of pure crystal, an electric field exists such that both holes and electrons are excluded from a thin region near the surface. For *n*-type crystals, the field repels free electrons. If a metal is joined to the crystal, the free electrons are still repelled, but a concentration of holes is produced directly under the surface. If a reverse bias is then applied, a depletion region is produced (see Fig. 21).

Surface barrier detectors give better resolution for particle spectroscopy than *p-n* junctions, but wider depletion regions are possible with the latter. (The wider the depletion region, the higher the energy of particles which can be analyzed because a particle must expend all its energy in a depletion region.)

FIGURE 20. Cross Section of Diffused *p-n* Junction Detector

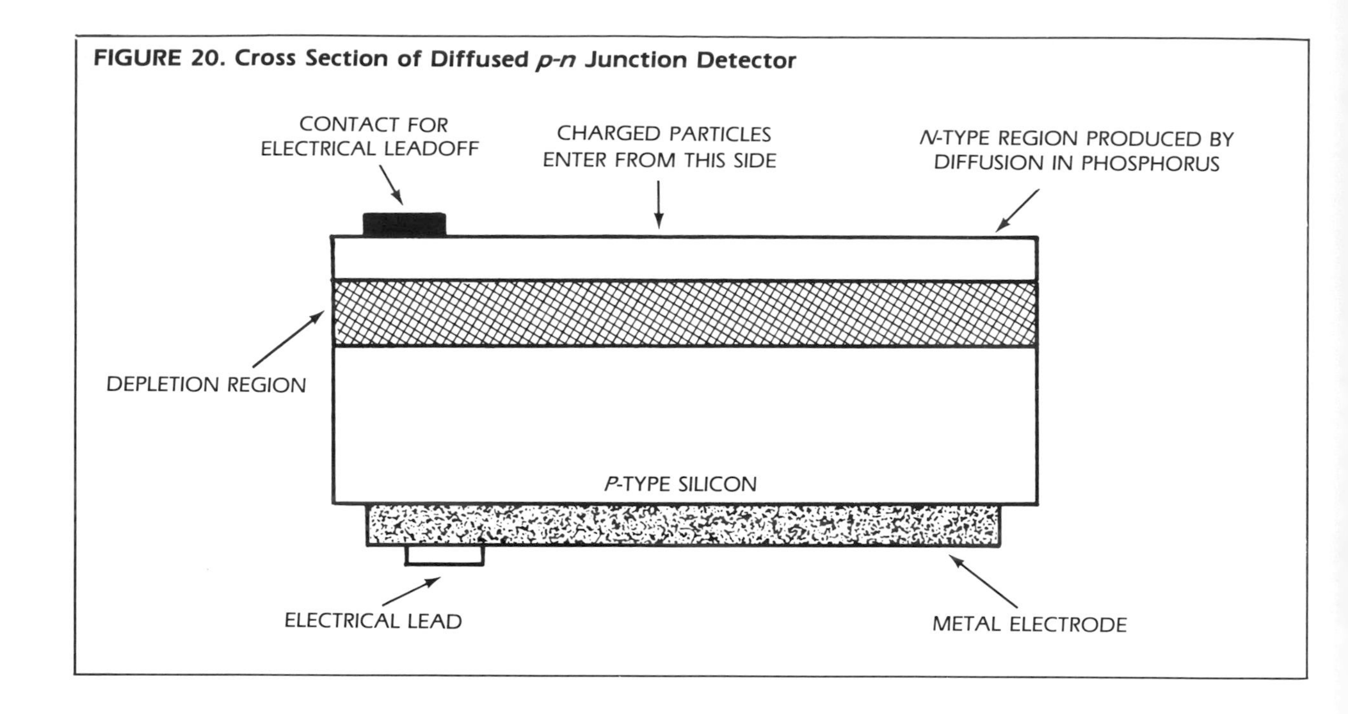

FIGURE 21. Cross Section of Surface Barrier Detector

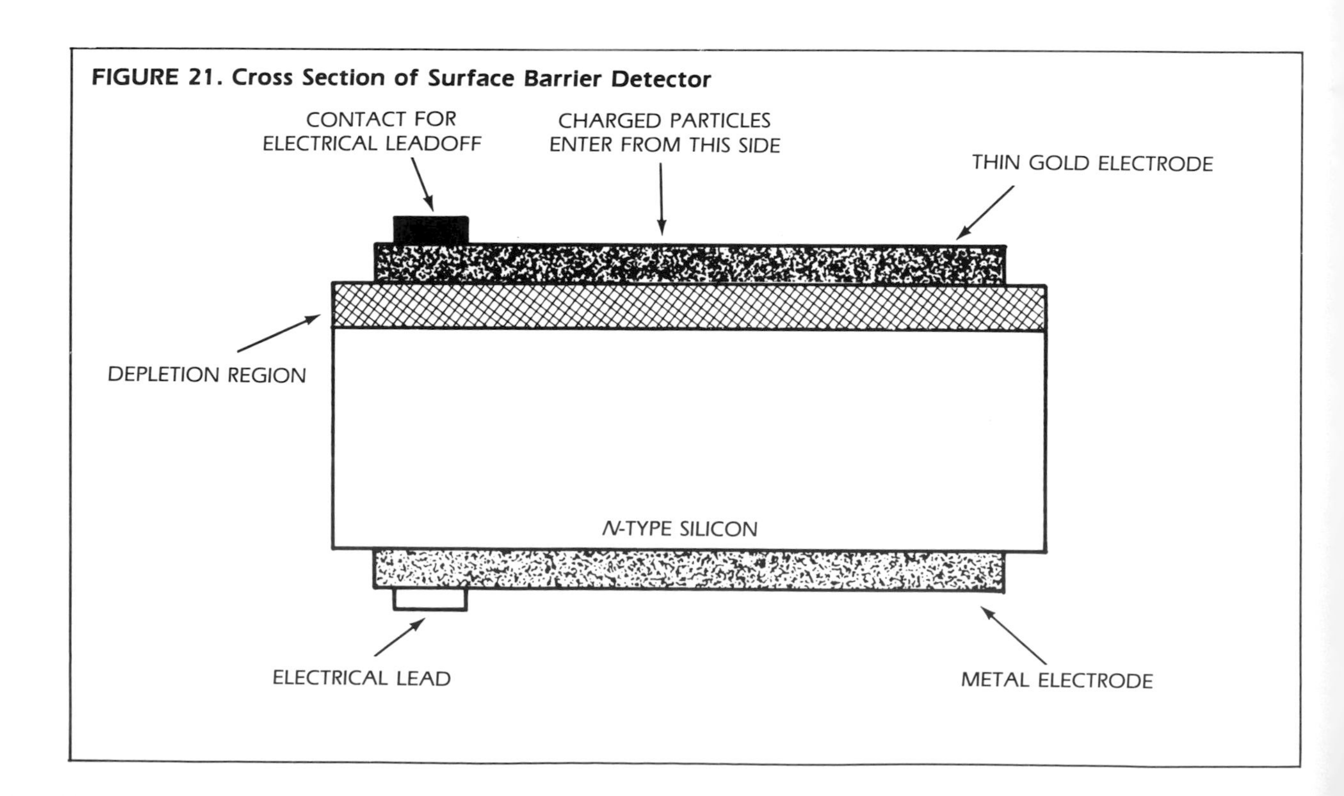

Lithium Drifted Detectors

The lithium drifted detector is produced by diffusing lithium into low resistivity *p*-type silicon or germanium. When heated under reverse bias, the lithium ions serve as an *n*-type donor. These ions drift into the silicon or germanium in such a way that a wide layer of the *p*-type material is compensated by the lithium, yielding an effective resistivity comparable to that of the intrinsic material (see Fig. 22). Wider depletion regions can be obtained with the lithium drift process than by any other method. Consequently, lithium drifted detectors are most useful in gamma spectroscopy work.

Silicon detectors can be operated at room temperatures, but exhibit low efficiency for gamma rays. Germanium detectors have higher gamma efficiencies, but must be operated at liquid nitrogen temperatures. For these reasons, coupled with the small sensitive volumes obtainable to date, semiconductor detectors have not received widespread application in radiation survey instruments.

FIGURE 22. Cross Section of Lithium Drifted Detector

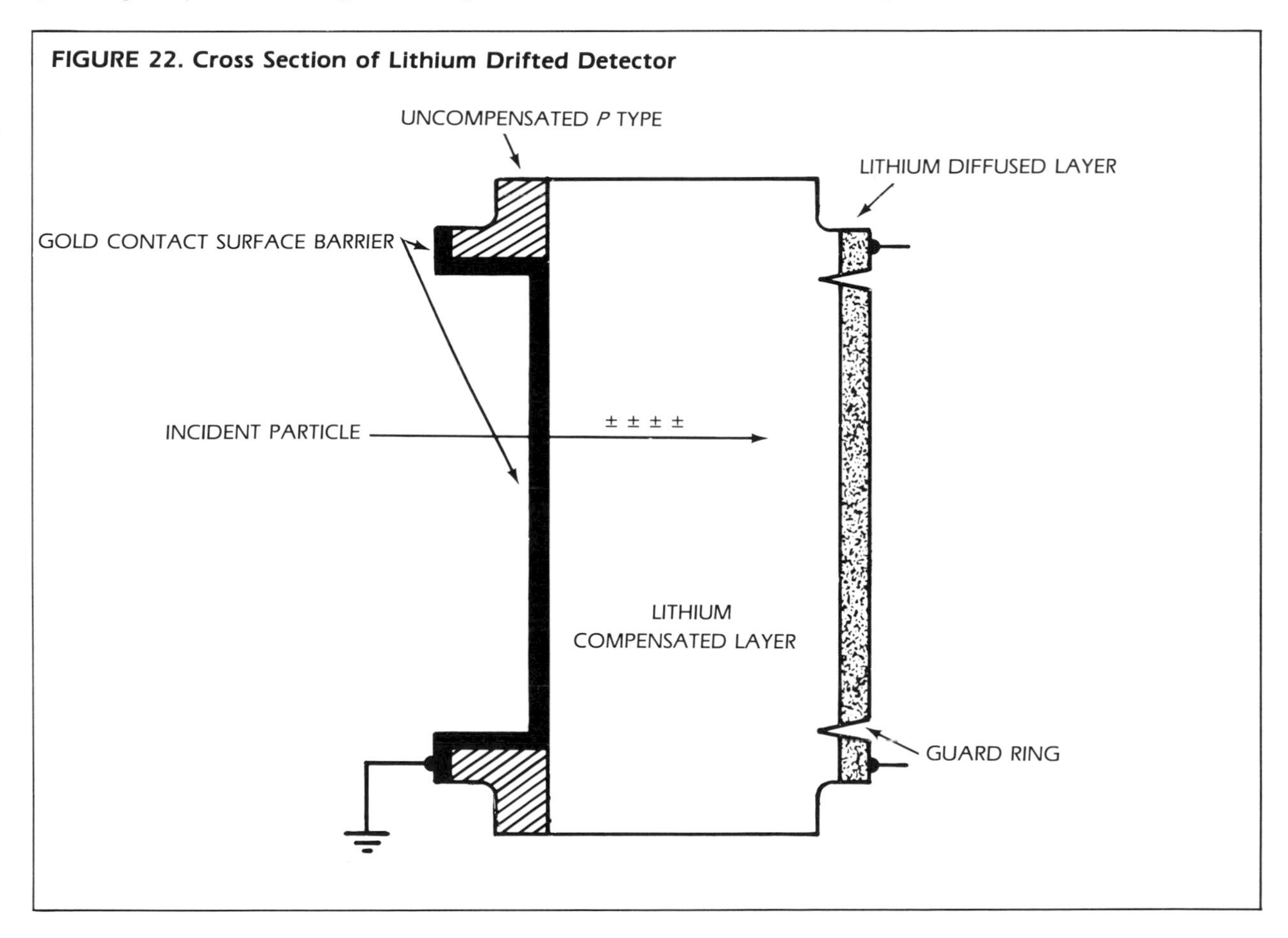

PART 8

RECORDING THE RADIOGRAPHIC LATENT IMAGE

Much of the following information is presented here in summary and is expanded in other sections of this Handbook volume. Consult the index to Volume 3 for more information on the radiographic latent image, its formation and processing.

Introduction

Throughout much of photography's history, the nature of the latent image was unknown. The first public announcement of Daguerre's photographic process was made in 1839, but it was not until 1938 that a satisfactory and coherent theory of photographic latent image formation was proposed. That theory has been undergoing refinement and modification ever since.

Some of the investigational difficulty arose because latent image formation is actually a very subtle change in the silver halide grain. The process may involve the absorption of only one or, at most, a few photons of radiation and this may affect only a few atoms out of some 10^9 or 10^{10} atoms in a typical photographic grain. Formation of the latent image, therefore, cannot be detected by direct physical or analytical chemical means.

A good deal *was* known about the latent image's physical nature. It was understood, for example, that the latent image was localized at certain discrete sites on the silver halide grain. If a photographic emulsion was exposed to light, developed, fixed and then examined under a microscope (see Fig. 23), the change of silver halide to metallic silver was visible at only a limited number of places on the crystal. Because small amounts of silver sulfide on the surface of the grain were known to be necessary for high photographic sensitivity, it seemed likely that the spots where the latent image formed were also concentrations of silver sulfide.

It was further known that the material of the latent image was, in all probability, silver. For one thing, chemical reactions that oxidized silver also destroyed the latent image. It was also a common observation that photographic materials given prolonged exposure to light darkened spontaneously, without the need for development. This darkening was known as the *print-out image*. The print-out image contained enough material to be identified

FIGURE 23. Localized Sites on the Grains

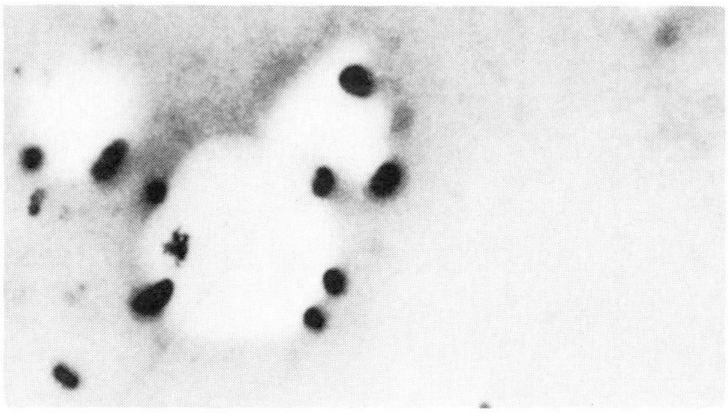

FROM EASTMAN KODAK COMPANY.

FIGURE 24. Localized Silver in a Print-out Image

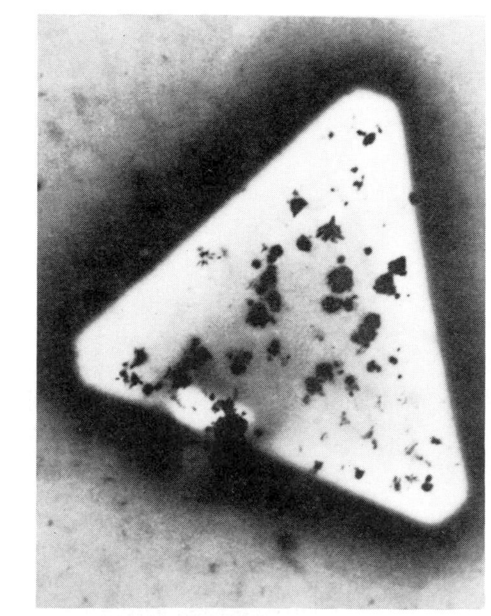

FROM EASTMAN KODAK COMPANY.

chemically as metallic silver. By microscopic examination, the silver of this image was also discovered to be localized at certain discrete areas of the grain (Fig. 24), just as the latent image.

Thus, the process that made an exposed photographic grain capable of transformation into metallic silver (by the mild reducing action of a developer) involved a concentration of silver atoms at one or more discrete sites on the photographic grain.

Any theory of latent image formation must account for the way that light photons, absorbed at random within the grain, can produce isolated aggregates of silver atoms. Most current theories of latent image formation are modifications of the mechanism proposed by R.W. Gurney and N.F. Mott in 1938. In order to understand the Gurney-Mott theory of the latent image, it is necessary to consider the structure of crystals, in particular, the structure of silver bromide crystals.

When solid silver bromide is formed, as in a photographic emulsion, the silver atoms each give up one orbital electron to a bromine atom. The silver atoms, lacking one negative charge, have an effective positive charge and are known as silver ions (Ag^+). The bromine atoms, on the other hand, have gained an electron and become bromine ions (Br^-). The plus and minus signs indicate, respectively, one fewer or one more electron than the number required for electrical neutrality of the atom.

A crystal of silver bromide is a regular, cubic array of silver and bromide ions, as shown in Fig. 25. It should be emphasized that the magnification used in the illustration is very high; the average grain in an industrial radiographic film may be about 0.001 mm (0.00004 in.) in diameter. Despite this minute size, the grain will contain several billion ions.

A crystal of silver bromide in a photographic emulsion is not perfect. First, within the crystal, there are silver ions that do not occupy the lattice positions shown in Fig. 25, but rather are in the spaces between. These are known as *interstitial silver ions* (see Fig. 26). The number of interstitial silver ions is small compared to the total number of ions in the crystal. In addition, there are distortions of the uniform crystal structure. These may be (1) foreign molecules, within or on the crystal, produced

FIGURE 25. A Silver Bromide Crystal is a Rectangular Array of Silver (Ag^+) and Bromide (Br^-) Ions

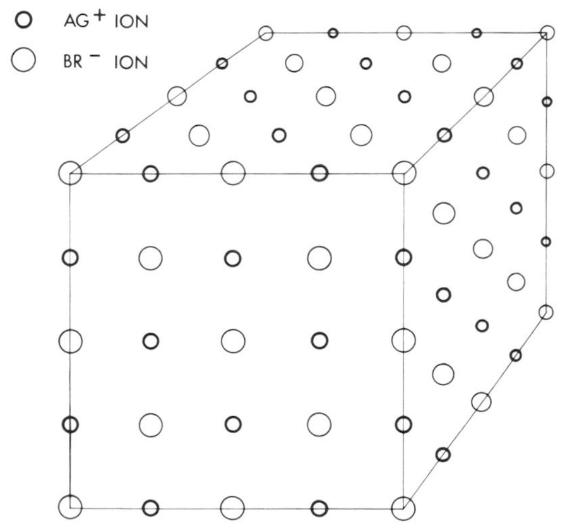

AG+ ION

BR − ION

FROM EASTMAN KODAK COMPANY.

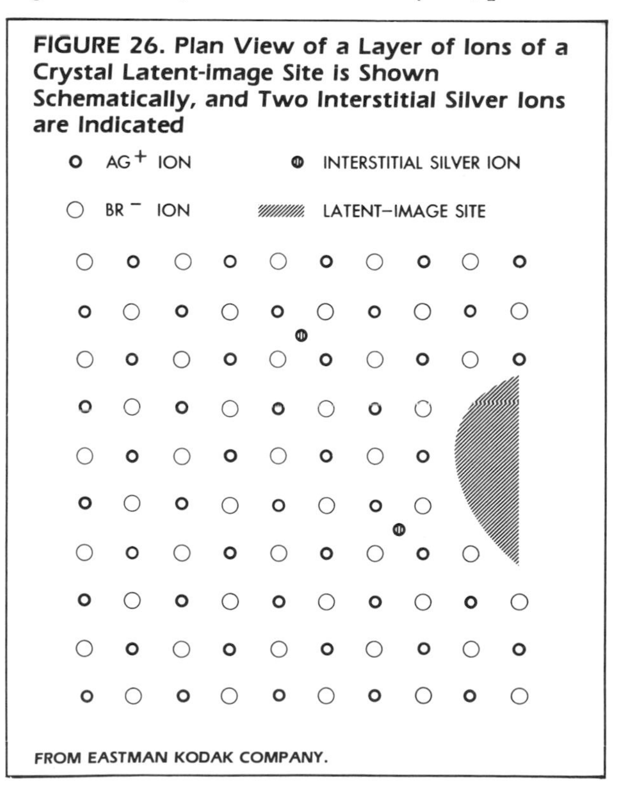

FIGURE 26. Plan View of a Layer of Ions of a Crystal Latent-image Site is Shown Schematically, and Two Interstitial Silver Ions are Indicated

FROM EASTMAN KODAK COMPANY.

by reactions with other components of the emulsion, or (2) distortions of the regular array of ions shown in Fig. 25. These anomolies are classed together and called *latent image sites.*

Radiographic Latent Images

In industrial radiography, the image-forming effects of X rays and gamma rays, rather than those of light, are of primary interest.

The agent that actually exposes a film grain (a silver bromide crystal in the emulsion) is not the X-ray photon itself, but rather the electrons (photoelectric and Compton) resulting from an absorption event.

The most striking difference between radiographic and visible-light exposures arises from the difference in the amounts of energy involved. The absorption of a single photon of *light* transfers a very small amount of energy to the crystal—only enough energy to free a single electron from a bromide (Br^-) ion. Several successive light photons are required to make a single grain developable (to produce within it, or on it, a stable latent image).

The passage of an *electron* through a grain can transmit hundreds of times more energy than the absorption of a light photon. Even though this energy is used inefficiently, the amount is sufficient to make the grain developable.

In fact, a photoelectron or Compton electron can have a fairly long path through a film emulsion and can render many grains developable. The number of grains exposed per photon interaction varies from one (for X radiation of about 10 keV) to 50 or more (for a 1 MeV photon).

For higher energy photons, there is low probability for a single interaction that transfers all the photons' energy. Most commonly, high photon energy is imparted to several electrons by successive Compton interactions. Also, high-energy electrons usually pass out of a film emulsion before all of their energy is transferred. For these reasons, there are, on the average, five to ten grains made developable per photon interaction at high energy.

For lower exposure values, each increment of energy exposes (on the average) the same number of grains. This, in turn, means that a curve of net density versus exposure is a straight line passing through the origin (Fig. 27). This curve is nonlinear only when the exposure is so great that appreciable energy is wasted on previously exposed grains. For commercially available fine-grain radiographic films, for example, the density versus exposure curve may be essentially linear up to densities of 2.0 or higher.

The fairly extensive straight-line relation between exposure and density is very useful for determining exposure values and for interpretation of densities observed on the resulting films.

If the curves shown in Fig. 27 are replotted as characteristic curves (density versus exposure), both characteristic curves are the same shape (see Fig. 28) and are separated along the log exposure axis. The similarity in toe shape has been experimentally observed for conventional processing and many commercial photographic materials.

Because a grain is completely exposed by the passage of an energetic electron, all radiographic exposures are, as far as the individual grain is concerned, extremely short. The actual time that an electron is within a grain depends upon the electron velocity, the grain dimensions and the squareness of

FIGURE 27. Typical Net Density Versus Exposure Curves for Direct X-ray Exposures

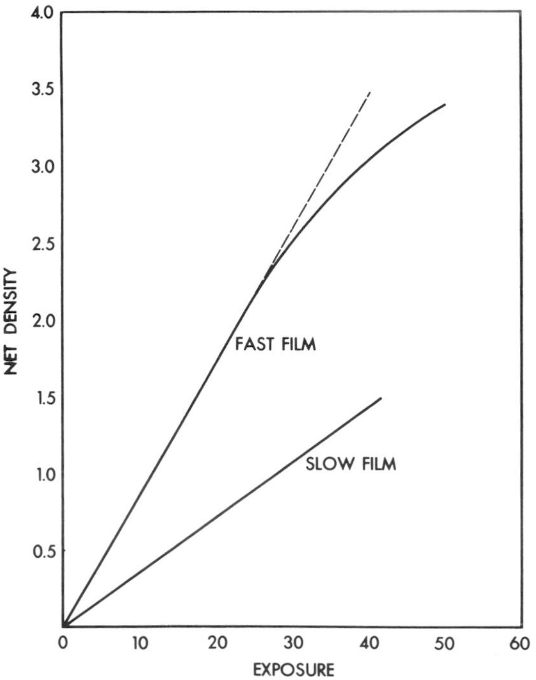

FROM EASTMAN KODAK COMPANY.

the hit. A time on the order of 10^{-13} seconds is representative. (In the case of light, the exposure time for a single grain is the interval between the arrival of the first photon and the arrival of the last photon required to produce a stable latent image.)

Development

Many materials discolor with exposure to light (some kinds of wood and human skin are examples) and could be used to record images. Most of these materials react to light exposure on a 1:1 basis: one photon of light alters one molecule or atom.

In the silver halide system of radiography, however, a few atoms of photolytically deposited silver can, *by development*, be made to trigger the subsequent chemical deposition of some 10^9 or 10^{10} additional silver atoms, resulting in an amplification factor on the order of 10^9 or greater. This amplification process can be performed at a time, and to a degree, convenient to the user and, with sufficient care, can be uniform and reproducible enough for quantitative radiation measurements.

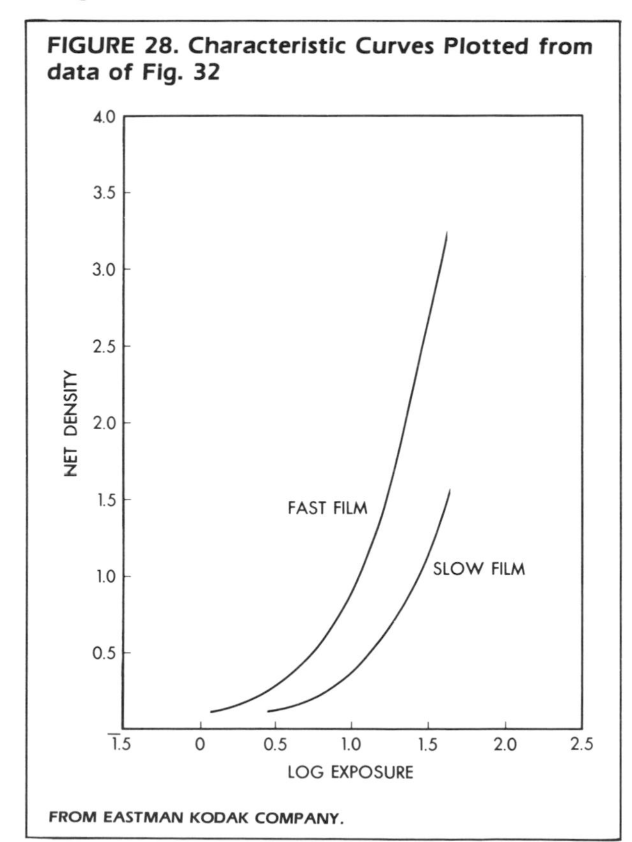

FIGURE 28. Characteristic Curves Plotted from data of Fig. 32

FROM EASTMAN KODAK COMPANY.

Development is essentially a chemical reduction in which silver halide is converted to metallic silver. In order to retain the photographic image, however, the reaction must be limited largely to those grains that contain a latent image; that is, to those grains that have received more than a prescribed minimum radiation exposure.

Compounds that can be used as photographic developing agents are those in which the reduction of silver halide to metallic silver is catalyzed (speeded up) by the presence of metallic silver in the latent image. Those compounds that reduce silver halide, in the absence of a catalytic effect by the latent image, are not suitable developing agents because they produce a uniform overall density on the processed film.

Many practical developing agents are relatively simple organic compounds (see Fig. 29) and their activity is strongly dependent on molecular structure and composition. The developing activity of a particular compound may often be predicted from a knowledge of its structure.

The simplest concept of the latent image's role in development is that it acts merely as an electron-conducting bridge, by which electrons from the

FIGURE 29. Configurations of Dihydroxybenzene, Showing How Developer Properties Depend upon Structure

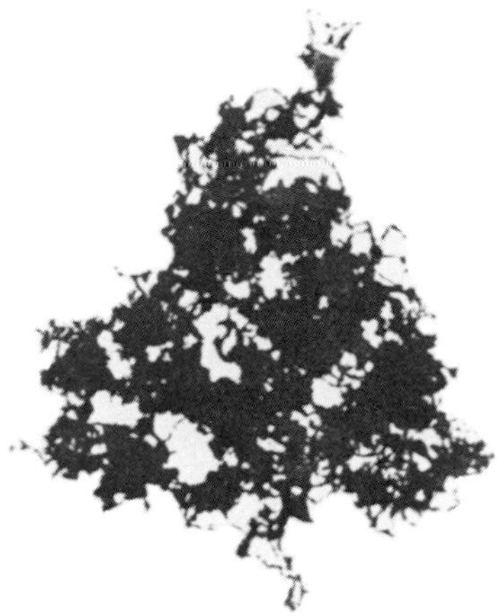

FROM EASTMAN KODAK COMPANY.

developing agent can reach the silver ions on the interior face of the latent image. Experiment has shown that this simple concept is inadequate for explaining the phenomena encountered in practical film development.

Absorption of the developing agent by the silver halide, or at the silver to silver-halide interface, is very important for determining the rate of direct or chemical development of most agents. The rate of development by hydroquinone, for example, appears to be relatively independent of the surface area of the silver and, instead, is governed by the extent of the silver to silver-halide interface.

The exact mechanisms of most developing agents are relatively complex and research on the subject is still very active. The broad outlines, however, are relatively clear.

A molecule of developing agent can easily give up an electron to an exposed silver bromide grain (one that carries a latent image), but not to an unexposed grain. This electron combines with a silver ion (Ag^{+}) in the crystal, neutralizing the positive charge and producing an atom of metallic silver. The process can be repeated many times until all the billions of silver ions in a photographic grain have been turned into metallic silver.

The development process has both similarities to, and differences from, the process of latent image formation. Both involve the union of a silver ion and an electron to produce an atom of metallic silver. In latent image formation, the electron is freed by the action of radiation and combines with a silver ion. In development, the electrons are supplied by a chemical electron donor and combine with the silver ions of the crystal lattice.

The physical shape of the developed silver has little relation to the shape of the silver halide grain from which it is derived. Very often the metallic silver has a tangled, filamentary form, the outer boundaries of which can extend far beyond the limits of the original silver halide grain. The mechanism for this filament formation is still in doubt. It is probably associated with another phenomenon, where filamentary silver is produced by vacuum deposition of silver atoms (in the vapor phase) onto suitable nuclei.

Contrast

The slope of the characteristic curve for X-ray film changes continuously along its length. It has been shown qualitatively that a density difference, corresponding to a difference in specimen thickness, is dependent on the region of the characteristic curve where the exposure falls. The steeper the slope of the curve in this region, the greater the density difference, and hence the greater the visibility of detail (assuming an illuminator bright enough so that a reasonable amount of light is transmitted through the radiograph to the eye of the observer.)

The slope of a curve at any particular point may be expressed as the slope of a straight line drawn tangent to the curve at that point. When applied to the characteristic curve of a photographic material, the slope of such a straight line is called the *gradient* of the material at that particular density. A typical

FIGURE 30. Characteristic Curve of a Typical Industrial Radiographic Film; Gradients Evaluated at Two Points on the Curve

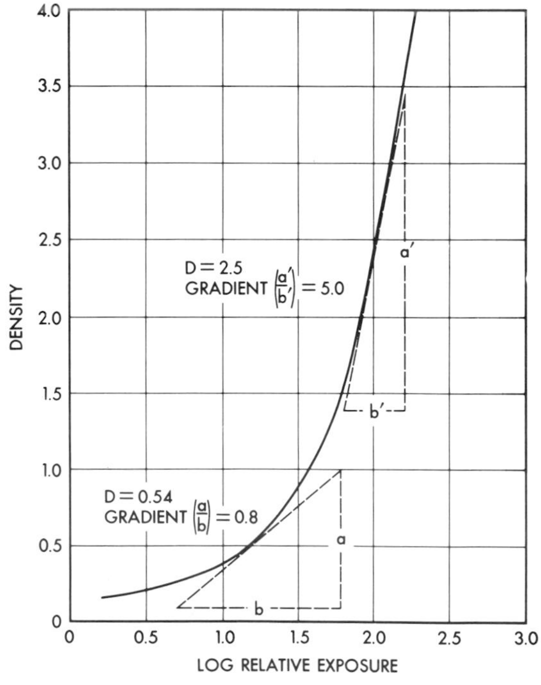

FROM EASTMAN KODAK COMPANY.

characteristic curve for a radiographic film is shown in Fig. 30. Tangents have been drawn at two points, and the corresponding gradients (ratios a/b and a'/b') have been evaluated. Note that the gradient varies from less than 1.0 in the toe to much greater than 1.0 in the high-density region.

Consider a specimen with two slightly different thicknesses that transmit slightly different radiation intensities to the film; there is a small difference in the logarithm of the relative exposure to the film in the two areas. Assume that, at a certain kilovoltage, the thinner section transmits 20 percent more radiation than the thicker section. The difference in logarithm of relative exposure ($\Delta \log E$) is 0.08, and is independent of the milliamperage, exposure time or source-film distance.

If this specimen is now radiographed with an exposure that puts the developed densities on the toe of the characteristic curve (where the gradient is 0.8), the X-ray intensity difference of 20 percent is represented by a density difference of 0.06 (see Fig. 31). If the exposure is such that the densities fall on the curve where the gradient is 5.0, the 20 percent intensity difference results in a density difference of 0.4.

In general, if the gradient of the characteristic curve is greater than 1.0, the intensity ratios, or subject contrasts, of the radiation emerging from the specimen are exaggerated in the radiographic reproduction; the higher the gradient, the greater the degree of exaggeration. Thus, at densities for which the gradient is greater than 1.0, the film acts as a contrast amplifier. Similarly, if the gradient is less than 1.0, subject contrasts are diminished in the radiographic reproduction.

A minimum density is often specified for radiographs. This is not because of any virtue in a particular density, but rather because of the gradient associated with that density; the minimum useful density is that density at which the minimum useful

FIGURE 31. Characteristic Curve of a Typical Industrial Radiographic Film; Density Differences Corresponding to a 20 percent Difference in Radiographic Exposure Evaluated for the Two Values of Gradient Illustrated in Fig. 30

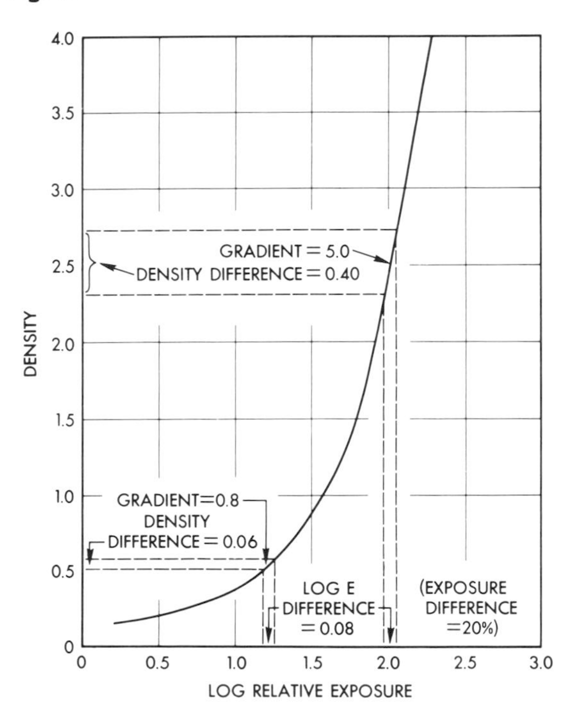

FROM EASTMAN KODAK COMPANY.

FIGURE 32. Gradient Versus Density Curves of Typical Industrial Radiographic Film

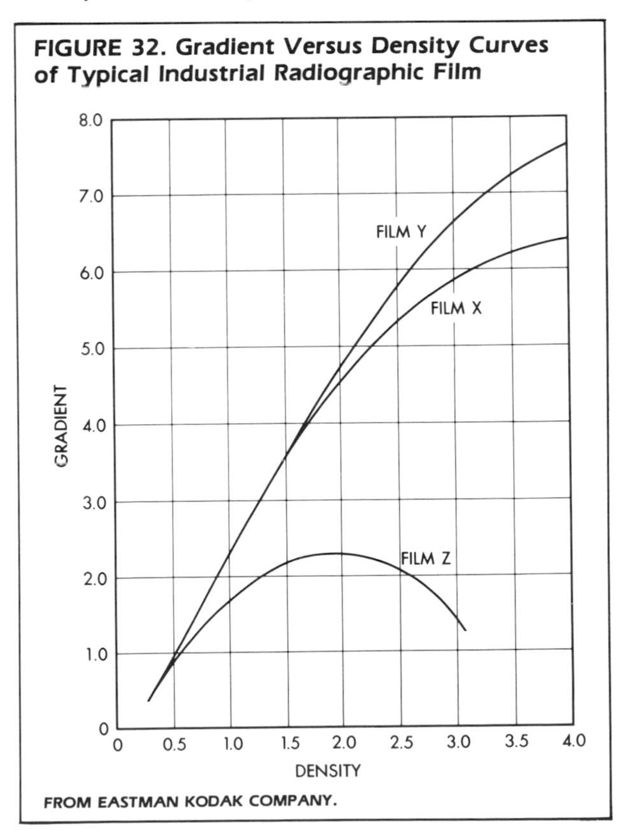

FROM EASTMAN KODAK COMPANY.

gradient is obtained. In general, gradients lower than 2.0 should be avoided whenever possible.

The ability of the film to amplify subject contrast is especially significant in radiography with very penetrating radiations which produce low subject contrast. Good radiographs depend on the enhancement of subject contrast by the film.

The gradients of radiographic film curves have been calculated, and are plotted in Fig. 32 against the density. The gradients of films *X* and *Y* increase continuously, up to the highest densities convenient for radiography.

The gradient versus density curve of film *Z* is different from the others in that the gradient increases, then becomes constant over the range of 1.5 to 2.5, beyond which it decreases. With this film, the greatest density difference (corresponding to a small difference in transmission of the specimen) is obtained in the middle range of densities. The maximum, as well as the minimum, useful density is governed by the minimum gradient that can be tolerated.

It is often useful to have a single number to indicate the contrast property of a film. This need is met by a quantity known as the *average gradient*, defined as the slope of a straight line joining two points of specified densities on the characteristic curve (see Table 6).

These two densities are often the maximum and minimum useful densities for a particular application. The average gradient indicates the average contrast properties of the film over this useful range; for a given film and development technique, the average gradient depends on the density range chosen. When high-intensity illuminators are available and high densities are used, the average gradient calculated for the density range 2.0 to 4.0 normally presents the contrast characteristics of the film fairly well. If high densities are not used, a density range of 0.5 to 2.5 is suitable. If intermediate densities are used, the average gradient can be calculated over another range of densities, 1.0 to 3.0, for example.

Experiments have shown that the shape of the characteristic curve is, for practical purposes, largely independent of the radiation wavelength (see Fig. 33 for the characteristic curves of typical industrial radiographic films). Therefore, a characteristic curve based on *any* radiation quality may be applied to exposures based on another quality, to the degree of accuracy usually required in practice; the same is true for values of gradient or average gradient derived from the curve.

TABLE 6. Average Gradient

Film	Density Range 0.5 to 2.5	Density Range 2.0 to 4.0
X	2.3	5.7
Y	2.6	6.3
Z	1.7	—

FIGURE 33. Characteristic Curve of a Typical Industrial Radiographic Film; Average Gradient Calculated over Two Density Ranges

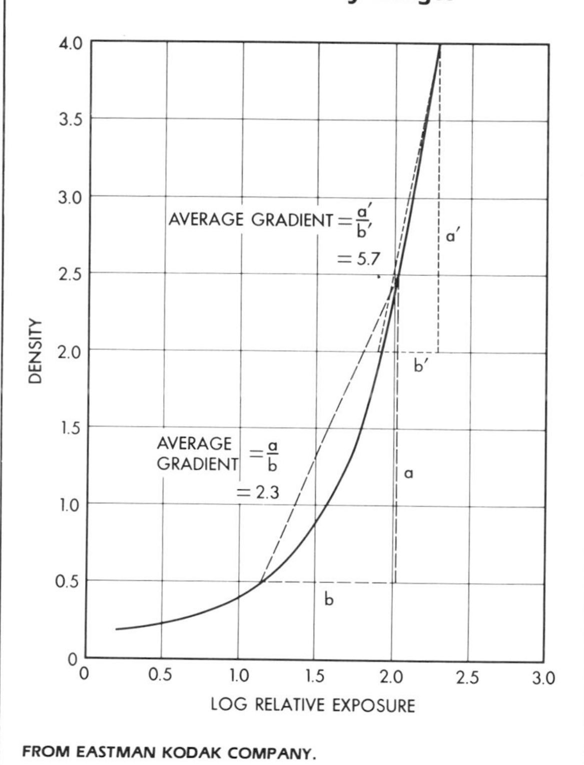

FROM EASTMAN KODAK COMPANY.

The influence of kilovoltage or gamma-ray quality on contrast in the radiograph, therefore, is the result of its effect on the subject contrast, and only very slightly, if at all, the result of any change in the contrast characteristics of the film.

Radiographic contrast can also be modified by choosing a film of different contrast, or by using a different density range with the same film. Contrast

TABLE 7. Densities Obtained Through 13 to 16 mm (0.5 to 0.6 in.) Steel Sections, Using an Exposure of 8 mA-min

kV	D_B	D_A	Radiographic Contrast	Relative Radiographic Contrast
120	0.50	0.27	0.23	20
140	1.20	0.67	0.53	46
160	2.32	1.30	1.02	88
180	3.48	2.32	1.16	100

is also affected by the degree of development, but in industrial radiography, films are developed to their maximum, or nearly their maximum, contrast.

In the early stages of development, both density and contrast increase quite rapidly with time of development. In manual processing, the minimum recommended development time gives most of the available density and contrast. With certain of the direct radiographic film types, somewhat higher speed and, in some cases, slightly more contrast are gained by extending the development; in no case should the maximum time recommended by the manufacturer be exceeded.

A special situation arises when, for technical or economic reasons, there is a maximum allowable exposure time. In such cases, an increase in kilovoltage increases the radiation intensity penetrating the specimen, and the film will contain a higher density. This may result in an increase in radiographic contrast.

Table 7 lists densities obtained through 13 to 16 mm (0.5 to 0.6 in.) sections, using an exposure of 8 milliampere-minutes. These data show that, when the exposure time is fixed, the density difference between the two sections increases; the visibility of detail is also improved as the kilovoltage is raised.

The improvement in detail visibility occurs in spite of the decrease in subject contrast (caused by the increase in kilovoltage), and is the direct result of using higher densities where the film gradient is higher. In this particular case, the film contrast increases (as a result of increased density) faster than the subject contrast decreases (as a result of increased kilovoltage).

Influence of Film Speed

It has been shown that the film contrast depends on the shape of the characteristic curve. The other significant value obtained from the characteristic curve is the relative speed which is governed by the location of the curve, along the log E axis, in relation to the curves of other films.

The spacing of the curves along the log E axis arises from differences in relative speed; the curves for the faster films lie toward the left, slower films toward the right. From these curves, relative exposures for producing a fixed photographic density can be determined. For some industrial radiographic purposes, a density of 1.5 is an appropriate level at which to compute relative speeds. However, the increasing trend toward high densities, with all radiographs viewed on high-intensity illuminators, makes a density of 2.5 more suitable for most industrial radiography. Relative speed values derived from characteristic curves, for two given density levels, are shown in Table 8, where film X has been assigned a relative speed of 100 at both densities. Note that the relative speeds computed are not the same; this is because of the differences in curve shape from one film to another.

Although the shape of the characteristic curve is practically independent of changes in radiation quality, the location of the curve along the log relative exposure axis, with respect to the curve of another film, does depend on radiation quality. Thus, if characteristic curves were prepared at a different kilovoltage, the curves would be differently spaced; that is, the films would have different speeds relative to the film that was chosen as a standard of reference.

TABLE 8. Relative Speed Values

	Density = 1.5		Density = 2.5	
Film	Relative Speed	Relative Exposure for D = 1.5	Relative Speed	Relative Exposure for D = 2.5
X	100	1.0	100	1.0
Y	24	4.2	26	3.9
Z	250	0.4	150	0.7

Density-exposure Relation

The most common way of expressing the relation between film response and radiation intensity is the characteristic curve (the relation between the density and the logarithm of the exposure). If density is plotted against relative exposure to X rays or gamma rays, in many cases there is a linear relation over a more or less limited density range (see Fig. 34). If net density (density above base density and fog), rather than gross density, is plotted against exposure, the straight line passes through the origin.

The linear relation cannot be assumed, however, but must be checked for the particular application, because of variations in film and processing conditions. The linear relation between density and exposure may be extremely useful in the interpretation of diffraction patterns and the evaluation of radiation monitoring films, provided that the limited linear range of the curve is considered.

Effect of Development Time on Speed and Contrast

Although the shape of the characteristic curve is relatively insensitive to changes in X- or gamma-ray quality, it is affected by changes in degree of development. Degree of development, in turn, depends on the type of developer, its temperature and its activity; the time of development increases the speed and contrast of any radiographic film. If, however, development is carried too far, the speed of the film, based on a certain net density, ceases to increase and may even decrease. In this case, fog increases and contrast may decrease.

Graininess

Graininess is defined as the visual impression of nonuniformity in the density of a radiographic (or photographic) image. With fast films exposed to high-kilovoltage radiation, graininess is easily visible with unaided vision; with slow films exposed to low-kilovoltage X rays, moderate magnification may be needed. In general, graininess increases with increasing film speed and with increasing radiation energy.

The clumps of developed silver responsible for the impression of graininess do not each arise from a single developed photographic grain. The particle of black metallic silver caused by the development of a single photographic grain in an industrial radiographic film is rarely larger than 0.001 mm (0.00004 in.) and is usually less.

The visual impression of graininess is caused by the random, statistical grouping of these individual silver particles. Each quantum (photon) of X radiation or gamma radiation absorbed in the film emulsion exposes one or more tiny crystals of silver bromide. These absorption events occur at random. Even in a uniform radiographic beam, the number of absorption events will differ from one small area of the film to the next, for purely statistical reasons. Thus, the exposed grains will be randomly distributed and their numbers will have a statistical variation from one area to the next.

With a very slow film, it might be necessary for 10,000 photons to be absorbed in a small area to produce a density of, for example, 1.0. With an extremely fast film it might require only 100 photons in the same area to produce the same density. When only a few photons are required to produce the density, the random positions of the absorption events

FIGURE 34. Density Versus Exposure Curve for Typical Industrial Radiographic Film Exposed to Direct X-rays or with Lead Screens

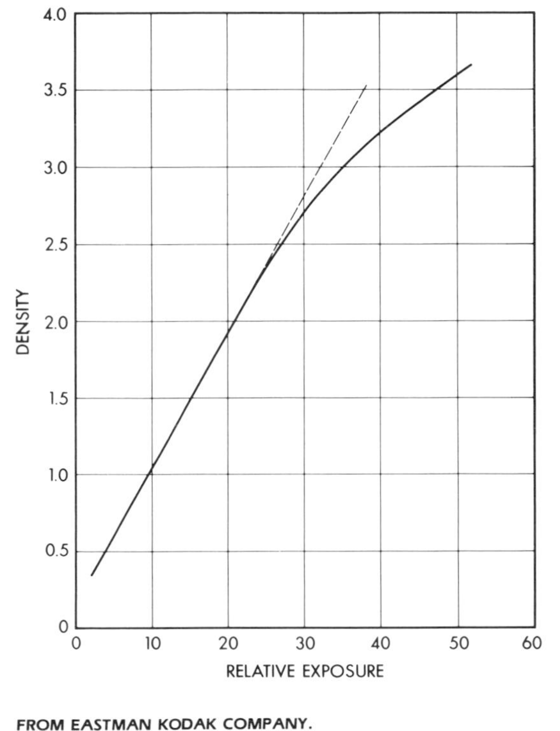

FROM EASTMAN KODAK COMPANY.

FIGURE 35. Typical X-ray Spectral Sensitivity Curve of a Radiographic Film, Showing Number of Roentgens Required to Produce a Density of 1.0 for Various Radiation Qualities

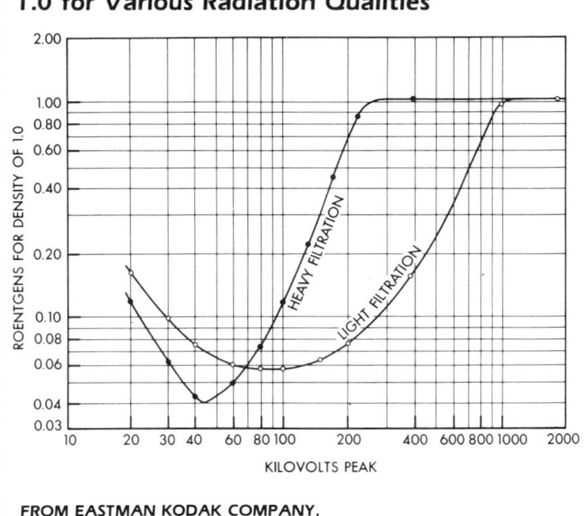

FROM EASTMAN KODAK COMPANY.

become visible in the processed film as film graininess. On the other hand, the more X-ray or gamma ray photons that are required, the less noticeable the graininess in the radiographic image, when all other conditions are equal.

In general, the silver bromide crystals in a slow film are smaller than those in a fast film, and this will produce less light-absorbing silver when they are exposed and developed. At low kilovoltages, one absorbed photon will expose one grain, of whatever size. Thus, more photons will have to be absorbed in the slower film than in the faster film to produce a particular density.

The increase in graininess with increasing kilovoltage can also be understood on this basis. At low kilovoltages, each absorbed photon exposes one photographic grain; at high kilovoltages, one photon will expose many grains. At high kilovoltages, then, fewer absorption events are required to produce a given density. Fewer absorption events, in turn, mean a greater relative deviation from the average, and hence greater graininess.

Screens

Although the above discussion is based on direct radiographic exposures, the concepts also apply to exposures made with lead screens. As stated earlier, the grains in a film emulsion are exposed by high-speed electrons. Silver bromide cannot distinguish between electrons from an absorption event within the film emulsion and those from a lead screen.

The quantum mottle observed in radiographs made with fluorescent intensifying screens has a statistical origin similar to that of film graininess. In this case, however, the number of photons absorbed in the screens is significant factor.

X-ray Spectral Sensitivity

The shape of the characteristic curve of a radiographic film is unaffected, for practical purposes, by the wavelength of the exposing X or gamma rays. However, the sensitivity of the film (the number of roentgens requried to produce a given density) is strongly affected by the wavelength of the exposing radiation.

Figure 35 shows the number of roentgens needed to produce a density of 1.0, for a particular radiographic film and specific processing conditions (exposures were made without screens).

The spectral sensitivity curves for all radiographic films have roughly the same features as the curves shown in Fig. 35. Details, among them the ratio of maximum to minimum sensitivity, differ with film type.

The spectral sensitivity of a film, or differences in spectral sensitivity between two films, need rarely be considered in industrial radiography. Usually such changes in sensitivity are automatically taken into account in the the preparation of exposure charts and tables of relative film speeds. The spectral sensitivity of a film is very important in radiation monitoring, because here an evaluation of the number of roentgens incident upon the film is required.

Reciprocity Law Failure

The Bunsen-Roscoe reciprocity law states that the density of a photochemical reaction is dependent only on the product of the radiation intensity and the duration of the exposure, and is independent of the absolute values of either quantity. Applied to radiography, this means that the developed density in a film depends only on the product of X-ray or gamma-ray intensity reaching the film and the time of exposure.

The reciprocity law is valid for direct X-ray or gamma-ray exposures, or those made with lead foil screens, over a range of radiation intensities and exposure times much greater than those normally used in practice. Reciprocity fails, however, for exposures to light and therefore for exposures using fluorescent intensifying screens. Figure 36 shows a conventional reciprocity curve.

The vertical axis in Figure 36 has been considerably expanded to make the curvature more apparent. The logarithms of the exposures that produce a given density are plotted against the logarithms of the individual intensities. It can be seen that, for a particular intensity, the exposure required to produce the given density is a minimum. It is for this intensity of light that the film is most efficient in its response.

FIGURE 36. Reciprocity Curve for Light Exposures; the Corresponding Curve for Direct Radiographic or Lead Screen Exposures Would be a Straight Line Parallel to the Log *I* Axis

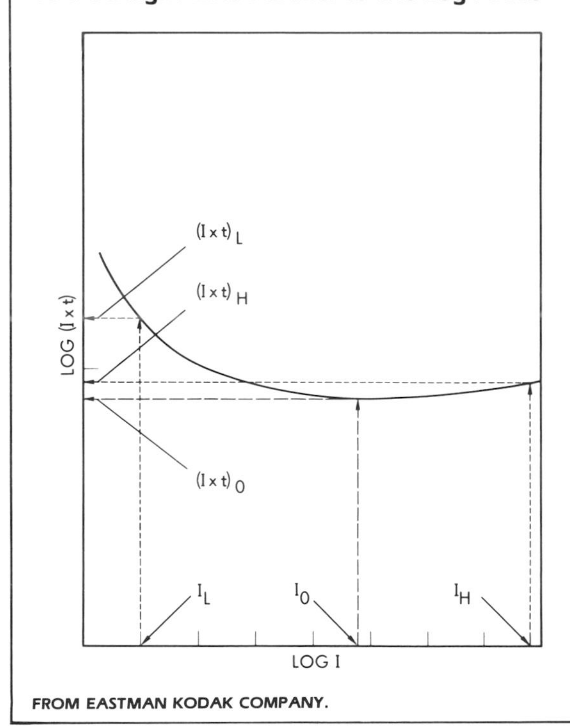

FROM EASTMAN KODAK COMPANY.

BIBLIOGRAPHY

1. Aerna, V. *Ionizing Radiation and Life.* St. Louis, MO: C.V. Mosby Co. (1971).
2. *Performance Specifications for Direct Reading and Indirect Reading Pocket Dosimeters for X and Gamma Radiation.* New York, NY: American National Standards Institute Report ANSI-N13.5 (1972).
3. *Radiation Protection Instrumentation Test and Calibration.* ANSI-N323 (1978).
4. *Specification of Portable X- or Gamma-Radiation Survey Instruments.* ANSI-1Y13.4 (1971).
5. *Luminescence Dosimetry.* Symposium Series No. 8. F.H. Attix, ed. Washington, DC: Atomic Energy Commission (1967).
6. Attix, F.H. and W.C. Roesch. *Radiation Dosimetry.* New York, NY: Academic Press. Vol. 2 (1966).
7. Cameron, J.R., N. Suntharalingam and G.N. Denney. *Thermoluminescent Dosimetry.* Madison, WI: University of Wisconsin Press (1968).
8. Cember, H. *Introduction to Health Physics,* 2nd edition. Pergamon Press (1983).
9. Hine, G. *Instrumentation in Nuclear Medicine.* New York, NY: Academic Press. Vol. 1 (1967).
10. *Personnel Dosimetry Systems for External Radiation Exposures.* Technical Report Series No. 109. Vienna: International Atomic Energy Agency (1970).
11. Knoll, G.F. *Radiation Detection and Measurement.* New York, NY: John Wiley and Sons (1979).
12. Lapp, R.E. and H.L. Andrews. *Nuclear Radiation Physics.* Englewood Cliffs, NJ: Prentice-Hall, Inc. (1972).
13. Moe, H.J. *Radiation Safety Technician Training Course.* Prepared for the US Atomic Energy Commission under contract W-31-109-Eng-38. Argonne, IL: Argonne National Laboratory (May 1972).
14. Morgan, K.Z. and J.E. Turner. *Principles of Radiation Protection.* New York, NY: John Wiley and Sons (1973).
15. *Instrumentation and Monitoring Methods for Radiation Protection.* NCRP Report 57. Washington, DC: National Council on Radiation Protection and Measurements (1978).
16. *A Handbook of Radioactivity Measurements Procedures.* NCRP Report 58 (1978).
17. Price W.J. *Nuclear Radiation Detection*, 2nd edition. New York, NY: McGraw-Hill Book Co. (1964).
18. Rivkin, W.B. *Personnel Monitoring Radiation Safety and Protection in Industrial Applications: Proceedings of a Symposium.* Washington, DC: Department of Health, Education and Welfare. DHEW Publication No. (FDA) 73-8012.
19. *Nuclear Medicine Physics, Instrumentation and Agents.* F. David Rollo, ed. St. Louis, MO: C.V. Mosby & Co.
20. Simmons, G.H. *A Training Manual for Nuclear Medicine Technologists.* Bureau of Radiological Health DMRG 70-3.

SECTION 5

FILM AND PAPER RADIOGRAPHY

Richard Quinn, Eastman Kodak Company, Rochester, NY (Parts 1-7)
J.C. Domanus, Risø National Laboratory, Roskilde, Denmark (Part 8)

Parts 1-7
Excerpted with permission from *Radiography In Modern Industry*, 4th edition

Part 8
Excerpted with permission from *Industrial Radiography on Radiographic Paper* (Risø Report 371)
Risø National Laboratory, Denmark (1977)
with additional material provided by the author

INTRODUCTION

Lawrence E. Bryant, Jr.
University of California
Los Alamos National Laboratory
Los Alamos, New Mexico, USA 87545

Radiography is one of the oldest and most widely used of nondestructive testing methods. Despite its established position, new developments are constantly modifying the radiographic techniques applied by industrial and scientific users, thereby producing technical or economic advantages, or both, over previous methods. This progressive trend continues with such special equipment and techniques as microfocus X-ray generators, portable linear accelerators, real-time and neutron radiography, imaging on paper, digital image analysis and image enhancement. Of course, all of us who are interested in radiography must begin our involvement with a certain basic knowledge, before progressing to and understanding current radiographic advances.

For many years, the Eastman Kodak Company has provided this basic knowledge in *Radiography in Modern Industry*, with a well-written text and illustrations that are clearly imprinted on our "radiographic minds." The American Society for Nondestructive Testing is indeed fortunate to have this material, excerpted from that book's 4th edition — in its usual illuminating style with some new material added to the good fortune of all radiographers.

Presented in this section are the basics of radiographic work, including: geometric principles; exposure variables and their relationships; radiation absorption in the specimen; radiographic screens; and radiographic film.

Also included are scattered radiation and its control, preparing exposure charts, the characteristic curve, radiographic image quality and detail visibility, and penetrameters.

Part 8 of this Handbook section discusses paper radiography; and, again, ASNT is fortunate to have it. It includes much practical information on characteristic curves, exposure latitude, relative speed, exposure charts, image quality, contrast, storage, and processing of radiography on paper.

This material was excerpted from the work of J.C. Domanus, one of the most widely published authorities on the industrial use of paper radiography.

PART 1

FILM EXPOSURE

Making a Radiograph

A *radiograph* is a photographic record produced by the passage of X-rays or gamma rays through an object onto a film (Fig. 1). When film is exposed to X-rays, gamma rays or light, an invisible change called a latent image is produced in the film emulsion. The areas so exposed become dark when the film is immersed in a developing solution, the degree of darkening depending on the amount of exposure. After development, the film is rinsed, preferably in a special bath, to stop development. The film is next put into a fixing bath, which dissolves the undarkened portions of the emulsion's sensitive salt. The film is washed to remove the fixer and dried so that it may be handled, interpreted and filed. The developing, fixing and washing of the exposed film may be done manually or in automated processing equipment.

The diagram in Fig. 1 shows the essential features in the exposure of a radiograph. The focal spot is a small area in the X-ray tube from which the radiation emanates. In gamma radiography, it is the capsule containing the radioactive material that is the source of radiation (for example, cobalt 60). In either case the radiation proceeds in straight lines to the object; some of the rays pass through and others are absorbed—the amount transmitted depending on the nature of the material and its thickness. For example, if the object is a steel casting having a void formed by a gas bubble, the void produces a reduction of the total thickness of steel to be penetrated. Hence, more radiation will pass through the section containing the void than through the surrounding metal. A dark spot, corresponding to the projected position of the void, will appear on the film when it is developed. Thus, a radiograph is a kind of shadow picture—the darker regions on the film representing the more penetrable parts of the object, and the lighter regions, those more opaque to gamma or X-radiation.

Industrial radiography is tremendously versatile. Radiographed objects range, in size, from microminiature electronic parts to mammoth missile components, in product composition through virtually every known material, and in manufactured form over an enormously wide variety of castings, weldments and assemblies. Radiographic examination has been applied to organic and inorganic materials, to solids, liquids, and even to gases. An industry's production of radiographs may vary from the occasional examination of one or several pieces to the examination of hundreds of specimens per hour. This wide range of applications has resulted in

FIGURE 1. Diagram of setup for making an industrial radiograph with X-rays.

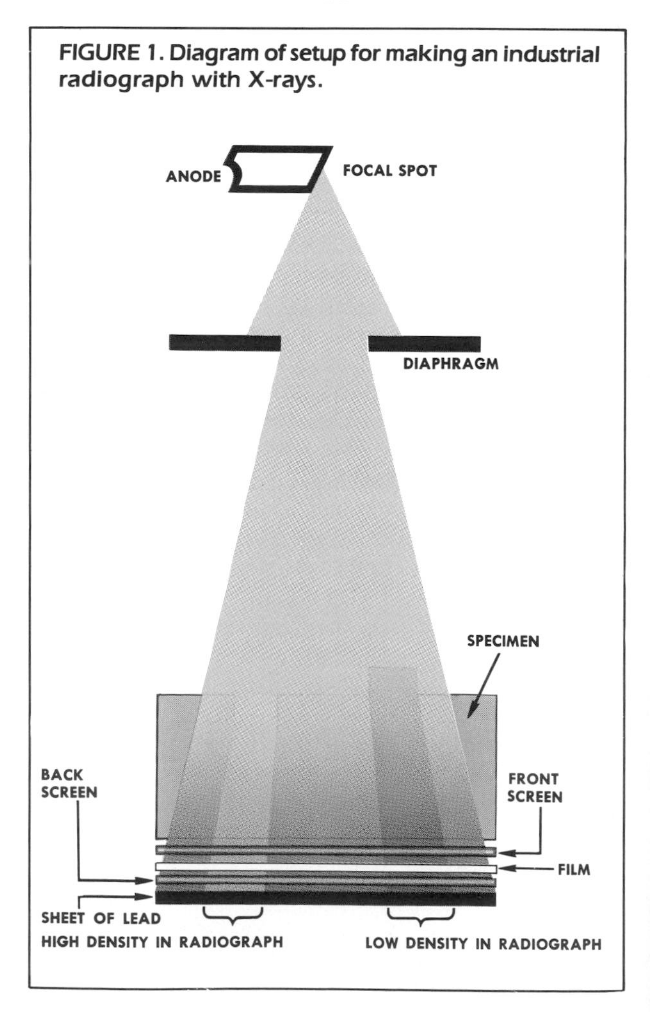

the establishment of independent, professional X-ray laboratories as well as radiographic departments within manufacturing plants. The radiographic inspection performed by industry is frequently monitored for quality by its customers—other manufacturers or governmental agencies—who use applicable specifications or codes, mutually agreed to by contract, and provided by several technical societies or other regulatory groups.

To meet the growing and changing demands of industry, research and development in the field of radiography are continually producing new sources of radiation such as neutron generators and radioactive isotopes; lighter, more powerful, more portable X-ray equipment as well as multimillion-volt X-ray machines designed to produce highly penetrating radiation; new and improved X-ray films and automatic film processors; and improved or specialized radiographic techniques. These factors, plus the activities of many dedicated people, broadly expand radiography's usefulness to industry.

Factors Governing Exposure

Generally speaking, the density of any radiographic image depends on the amount of radiation absorbed by the sensitive emulsion of the film. This amount of radiation in turn depends on several factors: the total amount of radiation emitted by the X-ray tube or gamma-ray source; the amount of radiation reaching the specimen; the proportion of this radiation that passes through the specimen; and the intensifying action of screens, if they are used.

Emission from X-ray Source

The total amount of radiation emitted by an X-ray tube depends on tube current (milliamperage), kilovoltage, and the time the tube is energized.

When other operating conditions are held constant, a change in milliamperage causes a change in the *intensity* (quantity of radiation leaving the X-ray generator per unit time) of the radiation emitted, the intensity being approximately proportional to the milliamperage. The high-voltage transformer saturation and voltage waveform can change with tube current, but a compensation factor is usually applied to minimize the effects of these changes. In normal industrial radiographic practice, the variation from exact proportionality is not serious and may usually be ignored.

Fig. 2 shows spectral emission curves for an X-ray tube operated at two different currents, the higher being twice the milliamperage of the lower. Therefore, *each wavelength* is twice as intense in one beam as in the other. Note that no wavelengths are present in one beam that are not present in the other. Hence, there is no change in X-ray quality or penetrating power.

As would be expected, the total amount of radiation emitted by an X-ray tube operating at a certain kilovoltage and milliamperage is directly proportional to the time the tube is energized.

Since the X-ray output is directly proportional to both milliamperage and time, it is directly proportional to their product. (This product is often referred to as the *exposure*.) Algebraically, this may be stated $E = MT$, where E is the exposure, M the tube current, and T the exposure time. The amount of radiation will remain constant if the exposure remains constant, no matter how the individual factors of tube current and exposure time are varied. This permits specifying X-ray exposures in terms of milliampere-minutes or milliampere-seconds, without stating the specific individual values of tube current and time.

FIGURE 2. Curves illustrating the effect of a change in milliamperage on the intensity of an X-ray beam (after Ulrey).

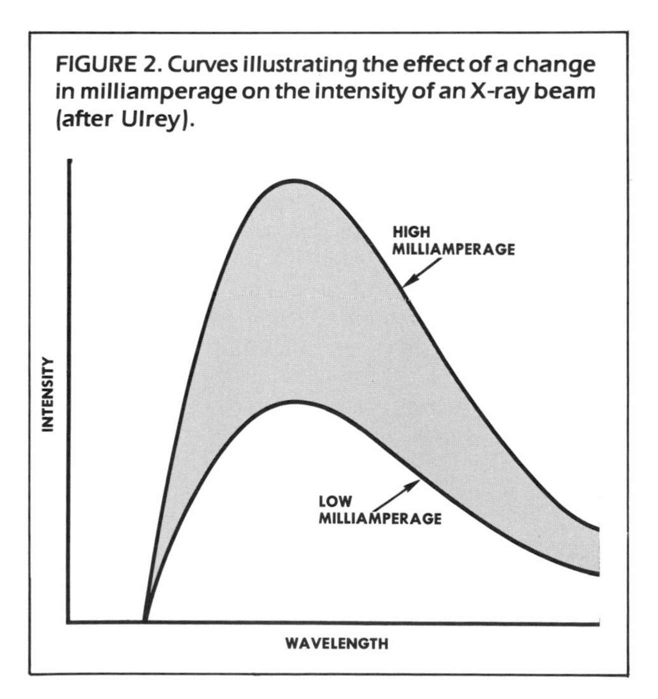

The kilovoltage applied to the X-ray tube affects not only the quality but also the intensity of the beam. As the kilovoltage is raised, X-rays of shorter wavelength, and hence of more penetrating power, are produced. Shown in Fig. 3 are spectral emission curves for an X-ray tube operated at two different kilovoltages but at the same milliamperage. Note that, in the higher-kilovoltage beam, there are some shorter wavelengths that are absent from the lower-kilovoltage beam. Further, all wavelengths present in the lower-kilovoltage beam are present in the more penetrating beam, and in greater amount. Thus, raising the kilovoltage increases both the penetration and the intensity of the radiation emitted from the tube.

Emission from Gamma-ray Source

The total amount of radiation emitted from a gamma-ray source during a radiographic exposure depends on the activity of the source (in curies or becquerels) and the time of exposure. For a particular radioactive isotope, the intensity of the radiation is approximately proportional to the activity of the source. If it were not for absorption of gamma rays within the radioactive material itself, this proportionality would be exact. In normal radiographic practice, the range of source sizes used in a particular location is small enough so that variations from exact proportionality are not serious and may usually be ignored.

Thus, the gamma-ray output is directly proportional to both activity of the source and time, and hence is directly proportional to their product. Analogously to the X-ray exposure, the gamma-ray exposure E may be stated $E = MT$, where M is the source activity and T is the exposure time; the amount of gamma radiation remains constant so long as the product of source activity and time remains constant. This permits specifying gamma-ray exposures in curie-hours or becquerel-seconds without stating specific values for source activity or time.

Since gamma-ray energy is fixed by the nature of the particular radioactive isotope, there is no variable to correspond to the kilovoltage factor encountered in X-radiography. The only way to change penetrating power when using gamma rays is to change the source, i.e., cobalt 60 in place of iridium 192.

FIGURE 3. Curves illustrating the effect of a change in kilovoltage on the composition and intensity of an X-ray beam (after Ulrey).

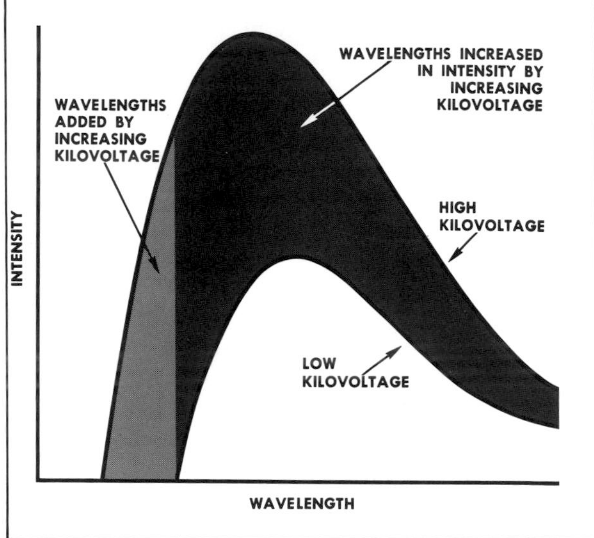

Geometric Principles

Since X-rays and gamma rays obey the common laws of light, their shadow formation may be simply explained in terms of light. It should be borne in mind that the analogy is not perfect since all objects are, to a greater or lesser degree, transparent to X-rays and gamma rays and since scattering presents greater problems in radiography than in optics. However, the same geometric laws of shadow formation hold for both light and penetrating radiation.

Suppose that, as in Fig. 4, there is light from a point L falling on a white card C, and that an opaque object O is interposed between the light source and the card. A shadow of the object will be formed on the surface of the card.

This shadow cast by the object will naturally show some *enlargement* because the object is not in contact with the card; the *degree of enlargement* will vary according to the relative distances of the object from the card and from the light source. The law governing the size of the shadow may be stated: *the diameter of the object is to the diameter of the shadow as the distance of the light from the object is to the distance of the light from the card.*

FIGURE 4. Illustrating the general geometric principles of shadow formation.

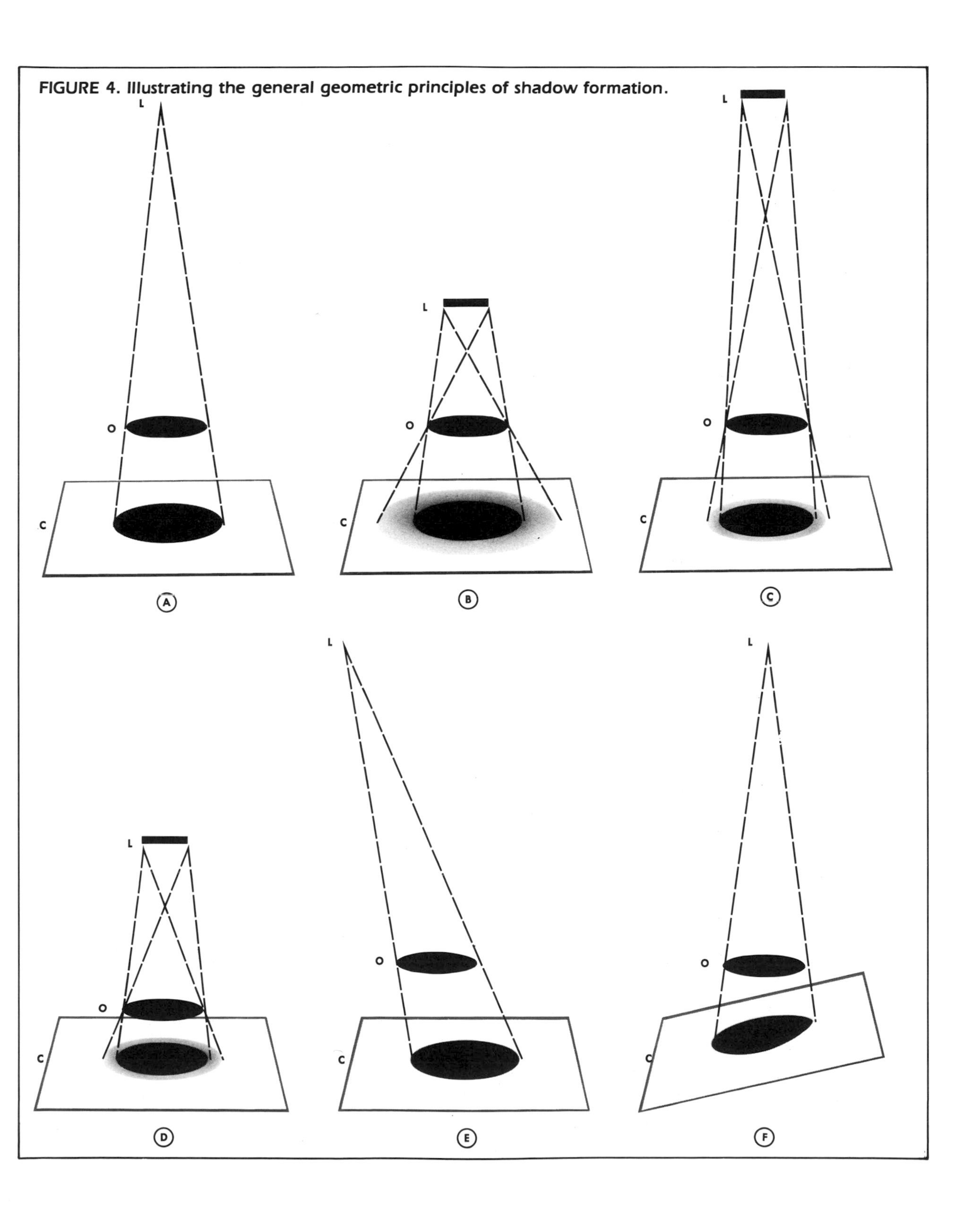

Mathematically, the degree of enlargement may be calculated by use of the following equations:

$$\frac{S_o}{S_i} = \frac{D_o}{D_i} \qquad \text{(Eq. 1)}$$

which may also be expressed as

$$S_o = S_i \left(\frac{D_o}{D_i}\right) \qquad \text{(Eq. 2)}$$

where S_o is the size of the object, S_i is the size of the shadow (or the radiographic image), D_o is the distance from source of radiation to object, and D_i the distance from the source of radiation to the recording surface (or radiographic film).

The degree of *sharpness* of any shadow depends on the size of the source of light and on the position of the object between the light and the card—whether nearer to or farther from one or the other. When the source of light is not a point but a small area, the shadows cast are not perfectly sharp (Fig. 4b-d) because each point in the source of light casts its own shadow of the object, and each of these overlapping shadows is slightly displaced from the others, producing an ill-defined image.

The *form* of the shadow may also differ according to the angle that the object makes with the incident light rays. Deviations from the true shape of the object as exhibited in its shadow image are referred to as distortion.

Fig. 4a to f shows the effect of changing the size of the source and of changing the relative positions of source, object and card. From an examination of these drawings, it will be seen that the following conditions must be fulfilled to produce the sharpest, truest shadow of the object:

1. The source of light should be small, that is, as nearly a point as can be obtained (compare Fig. 4a and 4c).
2. The source of light should be as far from the object as practical (compare Fig. 4b and 4c).
3. The recording surface should be as close to the object as possible (compare Fig. 4b and 4d).
4. The light rays should be directed perpendicularly to the recording surface (see Fig. 4a and 4e).
5. The plane of the object and the plane of the recording surface should be parallel (compare Fig. 4a and 4f).

Radiographic Shadows

The basic principles of shadow formation must be given primary consideration in order to assure satisfactory sharpness and freedom from distortion in the radiographic image. A certain degree of distortion will exist in every radiograph because some parts will always be farther from the film than others, the greatest magnification being evident in the images of those parts at the greatest distance from the recording surface.

Note, also, that there is no distortion of shape in Fig. 4e—a circular object having been rendered as a circular shadow. However, under circumstances similar to those shown in Fig. 4e, it is possible that spatial relations can be distorted. In Fig. 5 the two circular objects can be rendered either as two circles (Fig. 5a) or as a figure-eight-shaped shadow (Fig. 5b). It should be observed that both lobes of the figure eight have circular outlines.

Distortion cannot be eliminated entirely, but by the use of an appropriate source-film distance, it can be lessened to a point where it will not be objectionable in the radiographic image.

Application to Radiography

The application of the geometric principles of shadow formation to radiography leads to five general rules. Although these rules are stated in terms of radiography with X-rays, they also apply to gamma-ray radiography.

1. The focal spot should be as small as other considerations will allow, for there is a definite relation between the size of the focal spot of the X-ray tube and the *definition* in the radiograph. A large-focus tube, although capable of withstanding large loads, does not permit the delineation of as much detail as a small-focus tube.

Long source-film distances will aid in showing detail when a large-focus tube is employed, but it is advantageous to use the smallest focal spot permissible for the exposures required.

2. The distance between the anode and the material examined should always be as great as is practical. Comparatively long source-film distances should be used in the radiography of thick materials to minimize the fact that structures farthest from the film are less sharply recorded than those nearer to it. At long distances, radiographic definition is improved and the image is more nearly the actual size of the object.

Figures 6a-d shows the effects of source-film distance on image quality. As the source-film distance is decreased from 173 cm (68 in.) to 30 cm (12 in.) the image becomes more distorted until at 30 cm (12 in.) it is no longer a true representation of the casting. This is particularly evident at the edges of the casting where the distortion is greatest.

3. The film should be as close as possible to the object being radiographed. In practice, the film (in its cassette or exposure holder) is placed in contact with the object.

In Fig. 6b and e, the effects of object-film distance are evident. As the object-film dis-

FIGURE 5. Two circular objects can be rendered as two separate circles (A) or as two overlapping circles (B), depending upon the direction of the radiation.

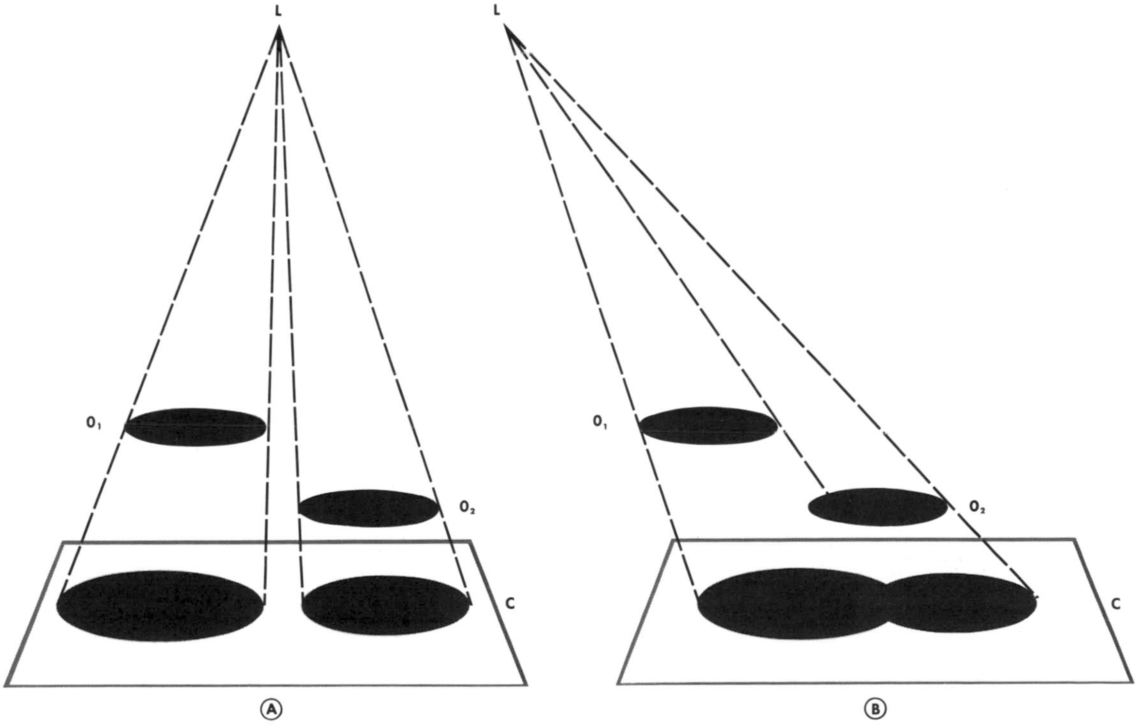

tance is increased from zero to 10 cm (4 in.), the image becomes larger and the definition begins to degrade. Again, this is especially evident at the edges of the chambers which are no longer sharp.

4. The central ray should be as nearly perpendicular to the film as possible to preserve spatial relations.
5. As far as the shape of the specimen will allow, the plane of maximum interest should be par-

FIGURE 6. The eight radiographs in this figure illustrate the effects on image quality when the geometric exposure factors are changed.

FIGURE 6a. Source-film distance: 173 cm (68 in.); object-film distance: 0.

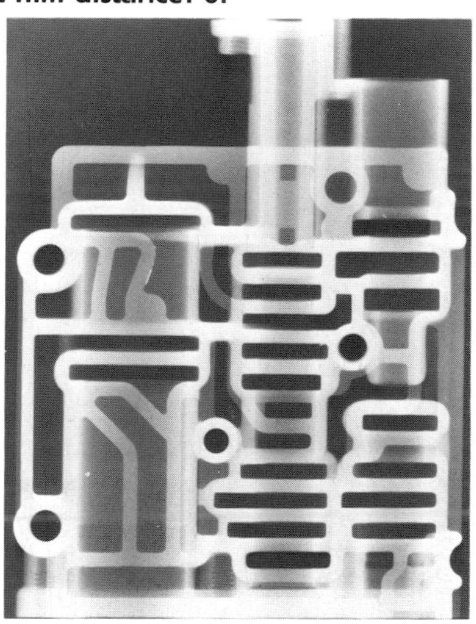

FIGURE 6b. Focal spot: 1.5 mm; object-film distance: 0.

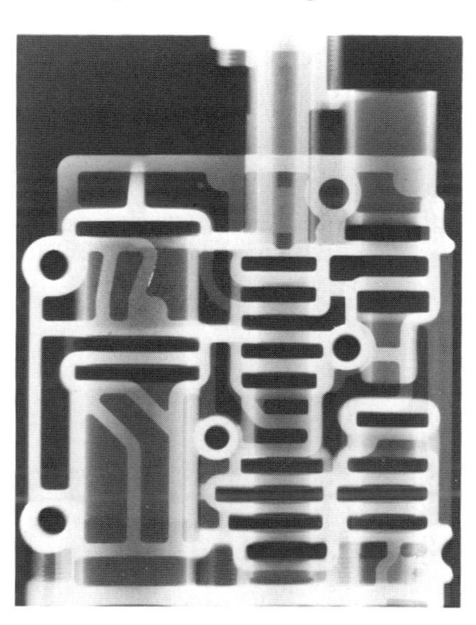

FIGURE 6e. Object-film distance: 10 cm (4 in.).

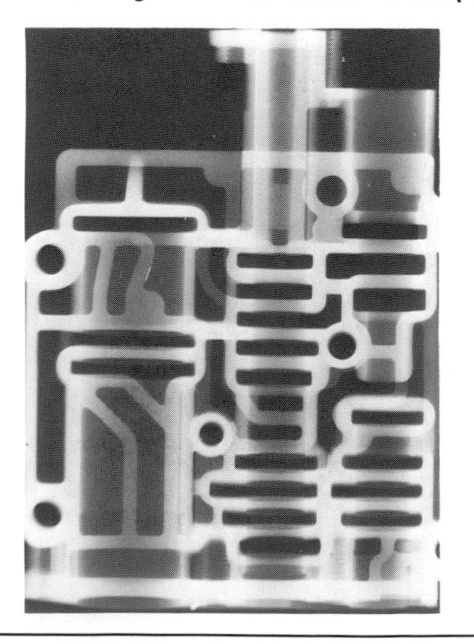

FIGURE 6f. Film-source angle: perpendicular; object-film angle: 45°.

allel to the plane of the film.

In Fig. 6f and g, the effects of object-film-source orientation are shown. When compared to Fig. 6b, the image in Fig. 6f is extremely distorted; although the film is perpendicular to the central ray, the casting is at a 45° angle to the film and spatial relationships are lost. As the film is rotated to be parallel with the casting (see Fig. 6g), the spatial relationships are maintained and the distortion is lessened.

FIGURE 6c. Intermediate size focal spot; intermediate source-film distance.

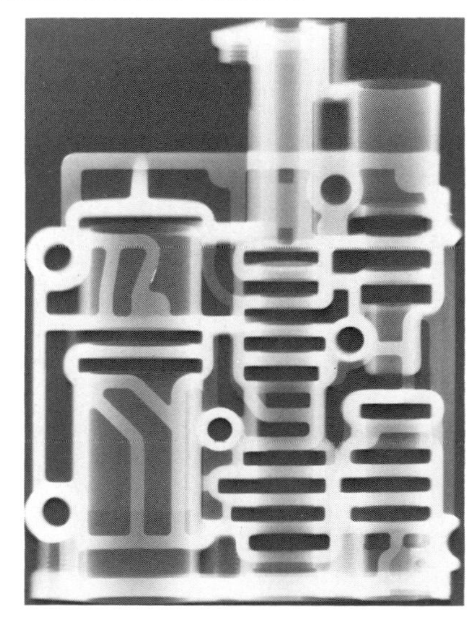

FIGURE 6d. Source-film distance: 30 cm (12 in.).

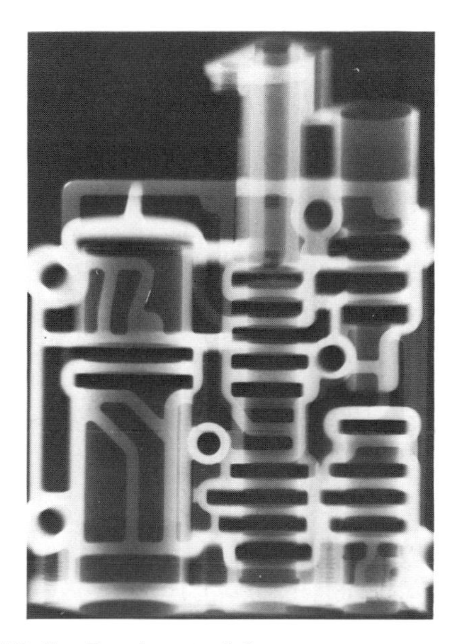

FIGURE 6g. Film-source angle: perpendicular; object-film angle: parallel.

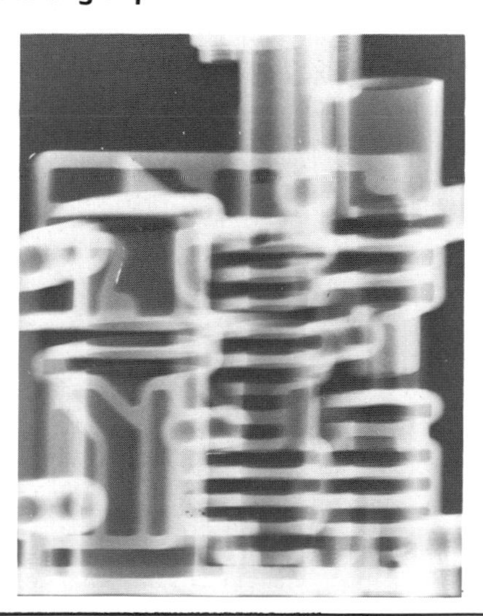

FIGURE 6h. Focal spot: 4.0 mm.

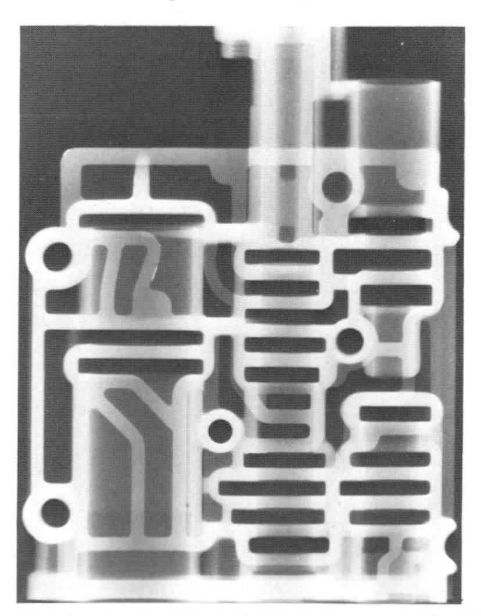

Calculation of Geometric Unsharpness

The width of the "fuzzy" boundary of the shadows in Fig. 4c and d is known as the *geometric unsharpness* (U_g). Since the geometric unsharpness can strongly affect the appearance of the radiographic image, it is frequently necessary to determine its magnitude. From the laws of similar triangles (see Fig. 7), it can be shown that:

$$\frac{U_g}{F} = \frac{t}{D_o} \quad \text{(Eq. 3)}$$

or

$$U_g = F\left(\frac{t}{D_o}\right), \quad \text{(Eq. 4)}$$

where U_g is the geometric unsharpness, F is the size of the radiation source, D_o is the source-object distance, and t is the object-film distance. Since the maximum unsharpness involved in any radiographic procedure is usually the significant quantity, the object-film distance (t) is usually taken as the distance from the *source side* of the specimen to the film.

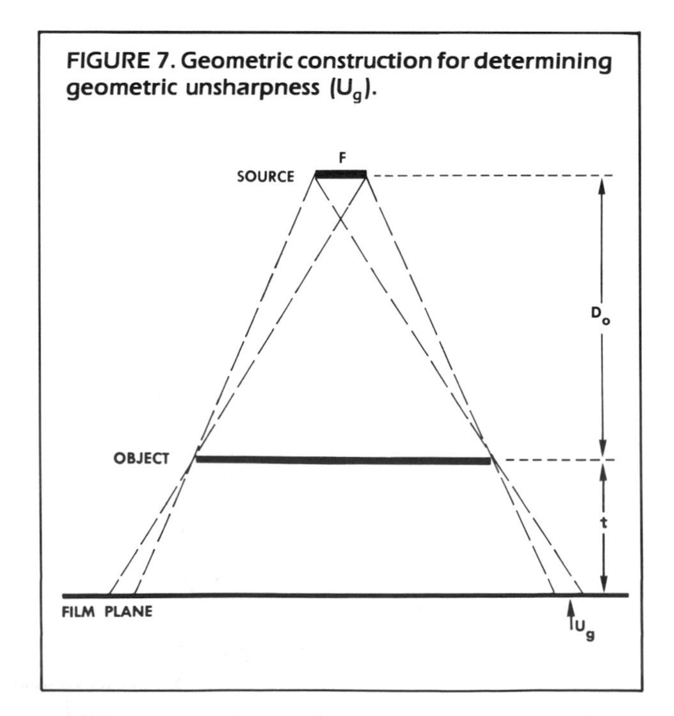

FIGURE 7. Geometric construction for determining geometric unsharpness (U_g).

D_o and t must be measured in the same units; inches are customary, but any other unit of length—say, centimeters—would also be satisfactory. So long as D_o and t are in the same units, eq. 3 or 4 will always give the geometric unsharpness (U_g) in whatever units were used to measure the dimensions of the source. The projected sizes of the focal spots of X-ray tubes are usually stated in millimeters, and U_g will also be in millimeters. If the source size is stated in inches, U_g will be in inches.

For rapid reference, graphs of the type shown in Fig. 8 can be prepared by the use of these equations. The graphs relate source-film distance, object-film distance and geometric unsharpness. Note that the lines of Fig. 8 are all straight. Therefore, for each

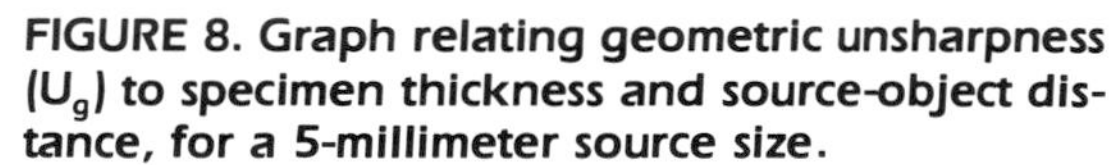

FIGURE 8. Graph relating geometric unsharpness (U_g) to specimen thickness and source-object distance, for a 5-millimeter source size.

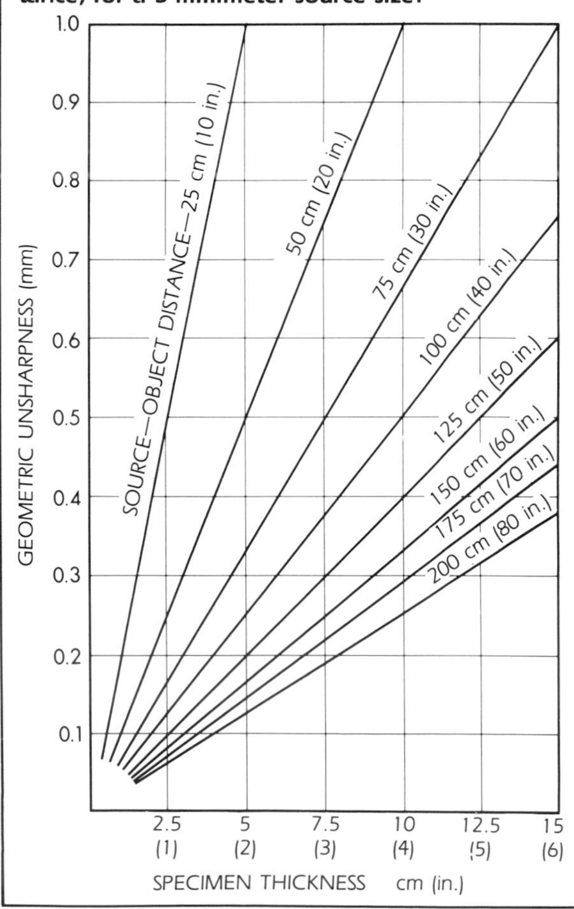

source-object distance, it is only necessary to calculate the value of U_g for a single specimen thickness, and then draw a straight line through the point so determined and the origin. It should be emphasized, however, that a separate graph of the type shown in Fig. 8 must be prepared for each size of source.

Geometric Enlargement

In most radiography, it is desirable to have the specimen and the film as close together as possible to minimize geometric unsharpness. An exception to this rule occurs when the source of radiation is extremely minute, that is, a small fraction of a millimeter, as in a betatron. In such a case, the film may be placed at a distance from the specimen, rather than in contact with it (see Fig. 9). Such an arrangement results in an enlarged radiograph without introducing objectionable geometric unsharpness. Enlargements of as much as three diameters by this technique have been found to be useful in the detection of structures otherwise invisible radiographically.

FIGURE 9. With a very small focal spot, an enlarged image can be obtained. The degree of enlargement depends upon the ratio of the source-film and source-specimen distances.

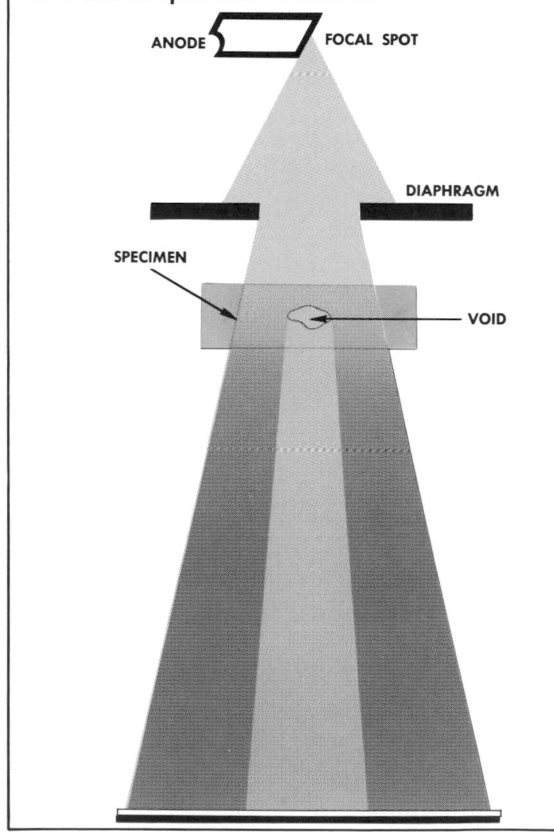

FIGURE 10. Schematic diagram illustrating the inverse square law.

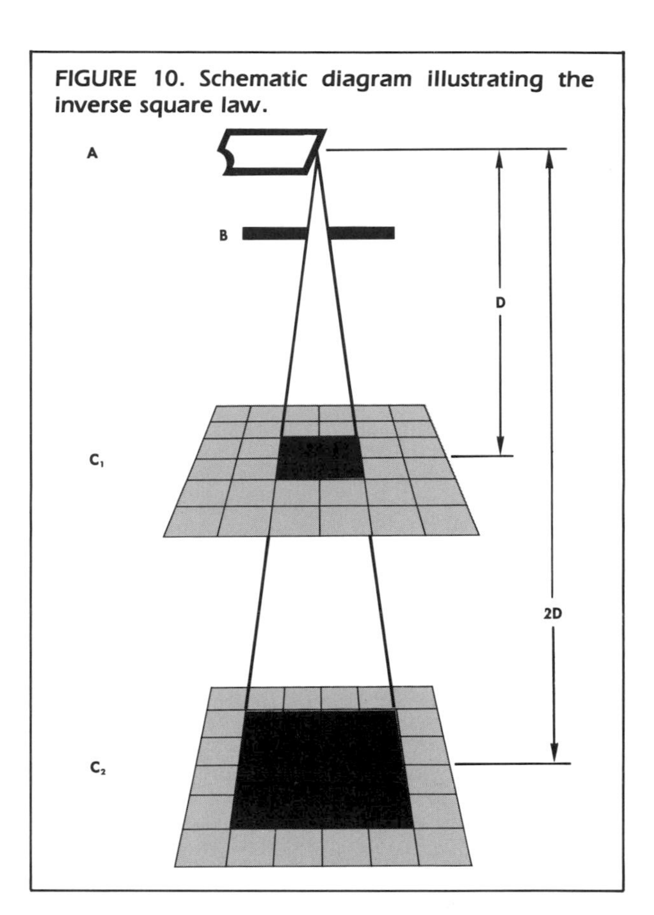

Inverse Square Law

When the X-ray tube output is held constant, or when a particular radioactive source is used, the radiation intensity reaching the specimen is governed by the distance between the tube (or source) and the specimen, varying inversely with the square of this distance. The explanation that follows is in terms of X-rays and light, but it applies to gamma rays as well.

Since X-rays conform to the laws of light, they diverge when they are emitted from the anode and cover an increasingly larger area with lessened intensity as they travel from their source. This principle is illustrated in Fig. 10. In this example, it is assumed that the intensity of the X-rays emitted at the anode A remains constant and that the X-rays passing through the aperture B cover an area of 4 cm^2 on reaching the recording surface C_1, which is 50 cm from the anode (distance D).

When the recording surface is moved 50 cm farther from the anode, to C_2, so that the distance ($2D$) from the anode is 1 m or twice its earlier value, the X-rays will cover 16 cm^2, an area four times as great as that at C_1. It follows, therefore, that the radiation per square unit on the surface of C_2 is only one-quarter of that at the level C_1. The exposure that would be adequate at C_1 must be increased four times in order to produce at C_2 a radiograph of equal density. In practice, this can be done by increasing the time or by increasing the milliamperage.

This inverse square law can be expressed algebraically as follows:

$$\frac{I_1}{I_2} = \frac{D_2^2}{D_1^2}, \qquad \text{(Eq. 5)}$$

where I_1 and I_2 are the intensities at the distances D_1 and D_2, respectively.

Relations of Milliamperage (Source Strength), Distance and Time

With a given kilovoltage of X-radiation or with the gamma radiation from a particular isotope, the three factors governing the exposure are the milliamperage (for X-rays) or source strength (for gamma rays), time, and source-film distance. The numerical relations among these three quantities are demonstrated below, using X-rays as an example. The same relations apply for gamma rays, provided the number of curies (or becquerels) in the source is substituted wherever milliamperage appears in an equation.

The necessary calculations for any changes in focus-film distance, D, milliamperage, M, or time, T, are matters of simple arithmetic and are illustrated in the following example. As noted earlier, kilovoltage changes cannot be calculated directly but must be obtained from the exposure chart of the equipment or the operator's logbook.

All of the equations shown on these pages can be solved easily for any of the variables by using one basic rule of mathematics: if a factor is moved across the equal sign (=), it shifts from the numerator to the denominator or vice versa.

$$\frac{A}{B} = \frac{C}{D} \qquad \text{(Eq. 6)}$$

To solve for B:

① $\frac{A}{B} = \frac{C}{D}$ ② $\frac{A}{1} = \frac{BC}{D}$

③ $AD = BC$ ④ $\frac{AD}{C} = B$

We can now solve for any unknown by:

1. Eliminating any factor that remains constant (has the same value and is in the same location on both sides of the equation).
2. Simplifying the equation by moving the unknown value so that it is alone on one side of the equation in the numerator.
3. Substituting the known values and solving the equation.

Milliamperage-Distance Relation

The milliamperage employed in any exposure technique should be in conformity with the manufacturer's rating of the X-ray tube. In most laboratories, however, a constant value of milliamperage is usually adopted for convenience.

Rule: *the milliamperage* (M) *required for a given exposure is directly proportional to the square of the focus-film distance* (D). The equation is expressed as follows:

$$M_1 : M_2 = D_1^2 : D_2^2 \qquad \text{(Eq. 7)}$$

or

$$\frac{M_1}{M_2} = \frac{D_1^2}{D_2^2} \qquad \text{(Eq. 8)}$$

Example: Suppose that with a given exposure time and kilovoltage, a properly exposed radiograph is obtained with 5 mA (M_1) at a distance of 50 cm (D_1), and that it is desired to increase the sharpness of detail in the image by increasing the focus-film distance to 100 cm (D_2). The correct milliamperage (M_2) to obtain the desired radiographic density at the increased distance (D_2) may be computed from the proportion:

$$5 : M_2 = 50^2 : 100^2$$

or

$$\frac{5}{M_2} = \frac{50^2}{100^2}$$

$$M_2 = 5 \times \frac{100^2}{50^2}$$

$$M_2 = 20 \text{ mA} \qquad \text{(Eq. 9)}$$

When very low kilovoltages, say 20 kV or less, are used, the X-ray intensity decreases with distance more rapidly than calculations based on the inverse square law would indicate because of absorption of the X-rays by the air. Most industrial radiography, however, is done with radiation so penetrating that the air absorption need not be considered. These comments also apply to the time-distance relations discussed below.

Time-Distance Relation

Rule: *The exposure time* (T) *required for a given exposure is directly proportional to the square of the focus-film distance* (D):

$$T_1 : T_2 = D_1^2 : D_2^2$$

or

$$\frac{T_1}{T_2} = \frac{D_1^2}{D_2^2} \qquad \text{(Eq. 10)}$$

To solve for either a new time (T_2) or a new distance (D_2), simply follow the steps shown in eq. 6.

Milliamperage-Time Relation

Rule: *The milliamperage* (M) *required for a given exposure is inversely proportional to the time* (T):

$$M_1 : M_2 = T_2 : T_1$$

or

$$\frac{M_1}{M_2} = \frac{T_2}{T_1} \qquad \text{(Eq. 11)}$$

Another way of expressing this is to say that for a given set of conditions (voltage, distance, etc.), the product of milliamperage and time is constant for the same photographic effect. Thus:

$$M_1T_1 = M_2T_2 = M_3T_3 = C \text{ (a constant)} \qquad \text{(Eq. 12)}$$

This is commonly referred to as the *reciprocity law*. (Important exceptions are discussed below.)

To solve for either a new time (T_2) or a new milliamperage (M_2), simply follow the steps shown in eq. 6.

Tabular Solution of Milliamperage-Time and Distance Problems

Problems of the types discussed above may also be solved by the use of a table similar to Table 1. The factor between the new and the old exposure time, milliamperage, or milliampere-minute (mA-min) value appears in the box at the intersection of the column for the new source-film distance and the row for the old source-film distance.

Suppose, for example, a properly exposed radiograph is produced with an exposure of 20 mA-min with a source-film distance of 76 cm (30 in.). The goal is to increase the source-film distance to 114 cm (45 in.) in order to decrease the geometric unsharpness in the radiograph. The factor appearing in the box at the intersection of the column for 45 in. (new source-film distance) and the row for 30 in. (old source-film distance) is 2.3. Therefore, the old milliampere-minute value (20) should be multiplied by 2.3 to give the new value: 46 mA-min.

TABLE 1. Milliamperage-Time and Distance Relationships*

OLD DISTANCE	NEW DISTANCE											
	25″	30″	35″	40″	45″	50″	55″	60″	65″	70″	75″	80″
25″	1.0	1.4	2.0	2.6	3.2	4.0	4.8	5.6	6.8	7.8	9.0	10.0
30″	0.70	1.0	1.4	1.8	2.3	2.8	3.4	4.0	4.8	5.4	6.3	7.1
35″	0.51	0.74	1.0	1.3	1.6	2.0	2.5	3.0	3.4	4.0	4.6	5.2
40″	0.39	0.56	0.77	1.0	1.3	1.6	1.9	2.2	2.6	3.1	3.5	4.0
45″	0.31	0.45	0.60	0.79	1.0	1.2	1.5	1.8	2.1	2.4	2.8	3.2
50″	0.25	0.36	0.49	0.64	0.81	1.0	1.2	1.4	1.7	2.0	2.2	2.6
55″	0.21	0.30	0.40	0.53	0.67	0.83	1.0	1.2	1.4	1.6	1.9	2.1
60″	0.17	0.25	0.34	0.44	0.56	0.69	0.84	1.0	1.2	1.4	1.6	1.8
65″	0.15	0.21	0.29	0.38	0.48	0.59	0.72	0.85	1.0	1.2	1.3	1.5
70″	0.13	0.18	0.25	0.33	0.41	0.51	0.62	0.74	0.86	1.0	1.1	1.3
75″	0.11	0.16	0.22	0.28	0.36	0.45	0.54	0.64	0.75	0.87	1.0	1.1
80″	0.10	0.14	0.19	0.25	0.32	0.39	0.47	0.56	0.66	0.77	0.88	1.0

*To convert to centimeters, multiply inch values by 2.54.

Note that some approximation is involved in the use of such a table, since the values in the boxes are rounded off to two significant digits. However, the errors involved are always less than 5 percent and, in general, are insignificant in actual practice. Also, a table of this type cannot include all source-film distances. However, in any one radiographic department, only a few source-film distances are used in the great bulk of the work, and a table of reasonable size can be made using only these few distances.

The Reciprocity Law

In the preceding text, it has been assumed that exact compensation for a decrease in the time of exposure can be made by increasing the milliamperage according to the relation $M_1T_1 = M_2T_2$. This may be written $MT = C$ and is an example of a general photochemical law: the same effect is produced for IT = constant, where I is intensity of the radiation and T is the time of exposure. This is called the *reciprocity law* and is true for direct X-ray and lead screen exposures. For exposures to light, it is not quite accurate and, since some radiographic exposures are made with the light from fluorescent intensifying screens, the law cannot be strictly applied.

Formally defined, the Bunsen-Roscoe reciprocity law states that the result of a photochemical reaction is dependent only on the *product* of radiation intensity (I) and the duration of the exposure (T), and is independent of absolute values of either quantity.

Errors that result from assuming the validity of the reciprocity law are usually so small that they are not noticeable in examples of the types given here. Departures may be apparent, however, if the intensity is changed by a factor of 4 or more. Since intensity may be changed by changing the source-film distance, failure of the reciprocity law may appear to be a violation of the inverse square law. Applications of the reciprocity law over a wide intensity range sometimes arise, and the relation between results and calculations may be misleading unless the possibility of reciprocity law failure is kept in mind. Failure of the reciprocity law means that the efficiency of a light-sensitive emulsion (in utilizing the light energy) depends on the light intensity.

Exposure Factor

The *exposure factor* is a quantity that combines milliamperage (X-rays) or source strength (gamma rays), time and distance. Numerically the exposure factor equals

$$\frac{\text{milliamperes} \times \text{time}}{\text{distance}^2} = \text{X-ray exposure factor} \tag{Eq. 13}$$

and

$$\frac{\text{curies} \times \text{time}}{\text{distance}^2} = \text{gamma ray exposure factor} \tag{Eq. 14}$$

Radiographic techniques are sometimes given in terms of kilovoltage and exposure factor, or radioactive isotope and exposure factor. In such a case, it is necessary merely to multiply the exposure factor by the square of the distance in order to find, for example, the milliampere-minutes or the curie-hours required.

Determination of Exposure Factors

X-rays

The focus-film distance is easy to establish by actual measurement, the milliamperage can conveniently be determined by the milliameter supplied with the X-ray machine, and the exposure time can be accurately controlled by a good time switch. The tube voltage, however, is difficult and inconvenient to measure accurately. Furthermore, designs of individual machines differ widely and may give X-ray outputs of a different quality and intensity even when operated at the nominal values of peak kilovoltage and milliamperage.

Consequently, although specified exposure techniques can be duplicated satisfactorily in the factors of focus-film distance, milliamperage, and exposure time, one apparatus may differ materially from another in the kilovoltage setting necessary to produce the same radiographic density. Because of this, the kilovoltage setting for a given technique should be determined by trial on each X-ray generator. In the preliminary tests, published exposure charts may be followed as an approximate guide. It is customary for equipment manufacturers to calibrate X-ray machines at the factory and to furnish suitable exposure charts. For the unusual problems that arise, it is desirable to record in a logbook all the data on exposure and techniques. In this way, operators will soon build up a source of information that will make them more competent to deal with difficult situations.

For developing trial exposures, a standardized technique should always be used. If this is done, any variation in the quality of the trial radiographs may then be attributed to the exposure alone. This method obviates many of the variable factors common to radiographic work.

Since an increase of kilovoltage produces a marked increase in X-ray output and penetration (see Fig. 3), it is necessary to maintain a close control of this factor in order to secure radiographs of uniform density. In many types of industrial radiography where it is desirable to maintain constant exposure conditions with regard to focus-film distance, milliamperage and exposure time, it is common practice to vary the kilovoltage in accordance with the thickness of the material to be examined in order to secure proper density in the radiographic image. Suppose, for example, it is desired to change from radiographing 4 cm (1.5 in.) steel to radiographing 5 cm (2 in.) steel. The 5 cm (2 in.) steel will require more than 10 times the exposure in milliampere-minutes at 170 kilovolts. However, increasing the kilovoltage to a little more than 200 will yield a comparable radiograph with the same milliampere-minutes. Thus, kilovoltage is an important variable because economic considerations often require that exposure times be kept within fairly narrow limits. It is desirable, as a rule, to *use as low a kilovoltage as other factors will permit.* In the case of certain high-voltage X-ray machines, the technique of choosing exposure conditions may be somewhat modified. For instance, the kilovoltage may be fixed rather than adjustable at the will of the operator, leaving only milliamperage, exposure time, film type and focus-film distance as variables.

Gamma Rays

With radioactive materials, the variable factors are more limited than with X-rays. Not only is the quality of the radiation fixed by the nature of the radiation emitter, but also the intensity is fixed by the amount of radioactive material in the particular source. The only variables under the control of operators, and the only quantities they need to determine, are the source-film distance, film type, and the exposure time. As in the case of X-radiography, it is desirable to develop trial exposures using the gamma-ray sources under standardized conditions and to record all data on exposures and techniques.

Contrast

In a radiograph, the various intensities transmitted by the specimen are rendered as different densities in the image. The density differences from one area to another constitute *radiographic contrast.* Any shadow or detail within the image is visible by reason of the contrast between it and its background of surrounding structures. Within appropriate limits, the greater the contrast or density differences in the radiograph, the more definitely various details will stand out. However, if overall contrast is increased too much, there is an actual loss in detail visibility in both the thick and the thin regions of the specimen.

Radiographic contrast is the result of both subject contrast and film contrast. Subject contrast is governed by the range of radiation intensities transmitted by the specimen. A flat sheet of homogeneous material of nearly uniform thickness would have very low subject contrast. Conversely, a specimen with large variations in thickness, which transmits a wide range of intensities, would have high subject contrast. Overall subject contrast could be defined as the ratio of the highest to the lowest radiation intensities falling on the film. Contrast is also affected by scattered radiation, removal of which increases subject contrast.

Choice of Film

Different films have different contrast characteristics. Thus, a film of high contrast may give a radiograph of relatively low contrast if the subject contrast is very low; conversely, a film of low contrast may give a radiograph of relatively high contrast if the subject contrast is very high. With any given specimen, the contrast of the radiograph will depend on the kilovoltage of the X-rays or the quality of the gamma rays, the contrast characteristics of the film, the type of screen, the density to which the radiograph is exposed, and the processing.

Radiographic Sensitivity

Radiographic sensitivity refers to the size of the smallest detail that can be seen in a radiograph, or to the ease with which the images of small details can be detected.

In the radiography of materials of approximately uniform thickness, where the range of transmitted X-ray intensities is small, a technique producing high contrast will satisfactorily render all portions of the area of interest, and the radiographic sensitivity will be greater than with a technique producing low contrast. If, however, the part radiographed transmits a wide range of X-ray intensities, then a technique producing lower contrast may be necessary in order to record detail (achieve radiographic sensitivity) in all regions of the part.

FIGURE 11. As kilovoltage increases, subject contrast decreases. More wavelenghts penetrate the subject in both thick and thin sections, thus reducing the overall difference in exposure between the two.

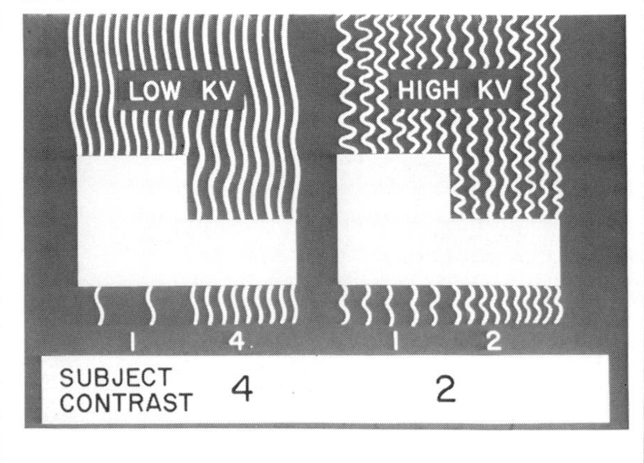

PART 2

ABSORPTION AND SCATTERING

Radiation Absorption in the Specimen

When X-rays or gamma rays strike an absorber (Fig. 12), some of the radiation is absorbed and another portion passes through undeviated. It is the intensity variation of the undeviated radiation from area to area in the specimen that forms the useful image in a radiograph. However, not all the radiation is either completely removed from the beam or transmitted. Some is deviated within the specimen from its original direction—that is, it is scattered and is nonimage-forming. This nonimage-forming scattered radiation, if not carefully controlled, will expose the film and thus tend to obscure the useful radiographic image. (Scattered radiation and the means for reducing its effects are discussed in detail later in this section.) Another portion of the original beam's energy is spent in liberating electrons from the absorber. The electrons from the specimen are unimportant radiographically; those from lead screens, as we discuss later, are very important.

FIGURE 12. Schematic diagram of some of the ways X- or gamma-ray energy is dissipated on passing through matter. Electrons from specimens are usually unimportant radiographically; those from lead foil screens are very important.

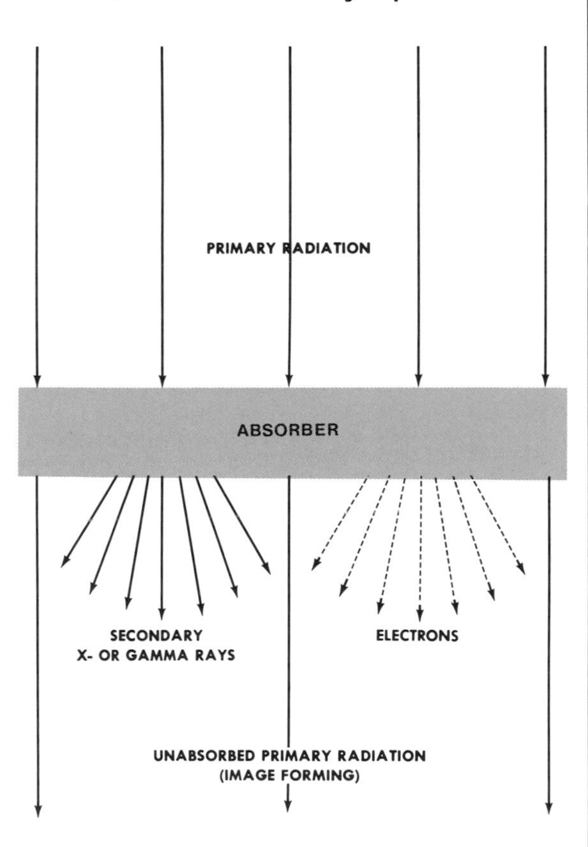

X-ray Equivalency

If industrial radiography were done with monoenergetic radiation, that is, with an X-ray beam containing a single wavelength, and if there were no scattering, the laws of absorption of X-rays by matter could be stated mathematically with great exactness. However, since a broad band of wavelengths is used and since considerable scattered radiation reaches the film, the laws must be given in a general way.

The X-ray absorption of a specimen depends on its thickness, on its density and, most important of all, on the atomic nature of the material. Comparing two specimens of similar composition, the thicker or the more dense will absorb more radiation, necessitating an increase in kilovoltage or exposure, or both, to produce the same photographic result. However, the atomic elements in a specimen usually exert a far greater effect upon X-ray absorption than either the thickness or the density. For example, lead is about 1.5 times as dense as ordinary steel, but at 220 kV, 0.25 cm (0.1 in.) of lead absorbs as much as 3.05 cm (1.2 in.) of steel. Brass is only about 1.1 times as dense as steel, yet, at 150 kV, the same exposure is required for 0.64 cm (0.25 in.) of brass as for 0.89 cm (0.35 in.) of steel. Table 2 gives approximate radiographic equivalence factors. It

TABLE 2. Approximate Radiographic Equivalence Factors

	X-rays								Gamma Rays			
Material	**50 kV**	**100 kV**	**150 kV**	**220 kV**	**400 kV**	**1000 kV**	**2000 kV**	**4 to 25 MeV**	**Ir 192**	**Cs 137**	**Co 60**	**Radium**
Magnesium	0.6	0.6	0.5	0.08								
Aluminum	1.0	1.0	0.12	0.18					0.35	0.35	0.35	0.40
2024 (aluminum) alloy	2.2	1.6	0.16	0.22					0.35	0.35	0.35	
Titanium			0.45	0.35								
Steel		12.	1.0	1.0	1.0	1.0	1.0	1.0	1.0	1.0	1.0	1.0
18-8 (steel) alloy		12.	1.0	1.0	1.0	1.0	1.0	1.0	1.0	1.0	1.0	1.0
Copper		18.	1.6	1.4	1.4			1.3	1.1	1.1	1.1	1.1
Zinc			1.4	1.3	1.3			1.2	1.1	1.0	1.0	1.0
Brass*			1.4*	1.3*	1.3*	1.2*	1.2*	1.2*	1.1*	1.1*	1.1*	1.1*
Inconel X alloy-coated		16.	1.4	1.3	1.3	1.3	1.3	1.3	1.3	1.3	1.3	1.3
Zirconium			2.3	2.0		1.0						
Lead			14.	12.		5.0	2.5	3.0	4.0	3.2	2.3	2.0
Uranium				25.				3.9	12.6	5.6	3.4	

Aluminum is the standard metal at 50kV and 100kV and **steel** at the higher voltages and gamma rays. The thickness of another metal is multiplied by the corresponding factor to obtain the approximate equivalent thickness of the standard metal. The exposure applying to this thickness of the standard metal is used.

Example: To radiograph 1.27 cm (0.5 inch) of copper at 220 kV, multiply 1.27 cm (0.5 inch) by the factor 1.4, obtaining an equivalent thickness of 1.78 cm (0.7 inch) of steel.

*Tin or lead alloyed in the brass will increase these factors.

should be emphasized that this table is approximate and is intended merely as a guide, since it is based on a compilation of data from many sources. In a particular instance, the exact value of the radiographic equivalence factor will depend on the quality of the X-radiation and the thickness of the specimen. It will be noted from this table that the relative absorptions of the different materials are not constant but change with kilovoltage, and that as the kilovoltage increases, the differences between all materials tend to become less. In other words, as kilovoltage is increased, the radiographic absorption of a material is less and less dependent on the atomic numbers of its constituents.

For X-rays generated at voltages more than 1000 kV and for materials not differing too greatly in atomic number (steel and copper, for example), the radiographic absorption for a given thickness of material is roughly proportional to the density of the material. However, even at high voltages or with penetrating gamma rays, the effect of composition on absorption cannot be ignored when dealing with materials that differ widely in atomic number. For instance, the absorption of lead for 1000 kV X-rays is about five times that of an equal thickness of steel, although its density is only 1½ times as great.

The kilovoltage governs the penetrating power of an X-ray beam and hence governs the intensity of the radiation passing through the specimen. It is not possible, however, to specify a simple relation between kilovoltage and X-ray intensity. Such factors as the thickness and the kind of material radiographed, the characteristics of the X-ray generating apparatus, and whether or not the film is used alone or with intensifying screens all exert considerable influence on this relation. The following example illustrates this point.

Data from a given exposure chart indicate that radiographs of equal density can be made of 1.3 cm (0.5 in.) steel with either of the following sets of exposure conditions:

80 kilovolts, 35 milliampere-minutes
120 kilovolts, 1.5 milliampere-minutes

In this case, a 50 percent increase in kilovoltage results in a 23-fold increase in photographically effective X-ray intensity.

Radiography of 5.08 cm (2 in.) aluminum can also be accomplished at these two kilovoltages. Equal densities will result with the following exposures:

80 kilovolts, 17 milliampere-minutes
120 kilovolts, 2.4 milliampere-minutes

In this case, the same increase in kilovoltage results in an increase in photographically effective X-ray intensity of only seven times. Many other examples can be found to illustrate the extreme variability of the effect of kilovoltage on X-ray intensity.

Gamma-ray Equivalency

Essentially the same considerations apply to gamma-ray absorption, since the radiations are of similar nature. It is true that some radioactive materials used in industrial radiography emit radiation that is monoenergetic, or almost so (for example, cobalt 60 and cesium 137). However, even with these sources, scattering is dependent on the size, shape, and composition of the specimen, which prevents the laws of absorption from being stated exactly. For those gamma-ray emitters (for example, iridium 192) that give off a number of discrete gamma-ray wavelengths extending over a wide energy range, the resemblance to the absorption of X-rays is even greater.

The gamma-ray absorption of a specimen depends on its thickness, density, and composition, as does its X-ray absorption. However, the most commonly used gamma-ray sources emit fairly penetrating radiations corresponding in their properties to high-voltage X-radiation. The radiographic equivalence factors in Table 2 show that the absorptions of the various materials for penetrating gamma rays are similar to their absorptions for high-voltage X-rays—that is, the absorptions of materials fairly close together in atomic number are roughly proportional to their densities. As with high-voltage X-rays, this is not true of materials, such as steel and lead, that differ widely in atomic number.

Scattered Radiation

When a beam of X-rays or gamma rays strikes any object, some of the radiation is absorbed, some is scattered, and some passes straight through. The electrons of the atoms constituting the object scatter radiation in all directions, much as light is dispersed by fog. The wavelengths of much of the radiation are increased by the scattering process, and hence the scatter is always somewhat "softer," or less penetrating, than the unscattered primary radiation. Any material—whether specimen, cassette, tabletop, walls or floor—that receives the direct radiation is a source of scattered radiation. Unless suitable measures are taken to reduce the effects of scatter, it will reduce contrast over the whole image or parts of it.

Scattering of radiation occurs, and is a problem, in radiography with both X-rays and gamma rays. In the text which follows, the discussion is in terms of X-rays, but the same general principles apply to gamma radiography.

In the radiography of thick materials, scattered radiation forms the greater percentage of the total radiation. For example, in the radiography of a 1.9 cm (0.75 in.) thickness of steel, the scattered radiation from the specimen is almost twice as intense as the primary radiation; in the radiography of 5 cm (2 in.) thick aluminum, the scattered radiation is two and a half times as great as the primary radiation. As may be expected, preventing scatter from reaching the film markedly improves the quality of the radiographic image.

As a rule, the greater portion of the scattered radiation affecting the film is from the specimen under examination (*A* in Fig. 13). However, any portion of the film holder or cassette that extends beyond the boundaries of the specimen and thereby receives direct radiation from the X-ray tube also becomes a source of scattered radiation which can

FIGURE 13. Sources of scattered radiation: (A) transmitted scatter; (B) scatter from cassette; (C) ''reflection'' scatter.

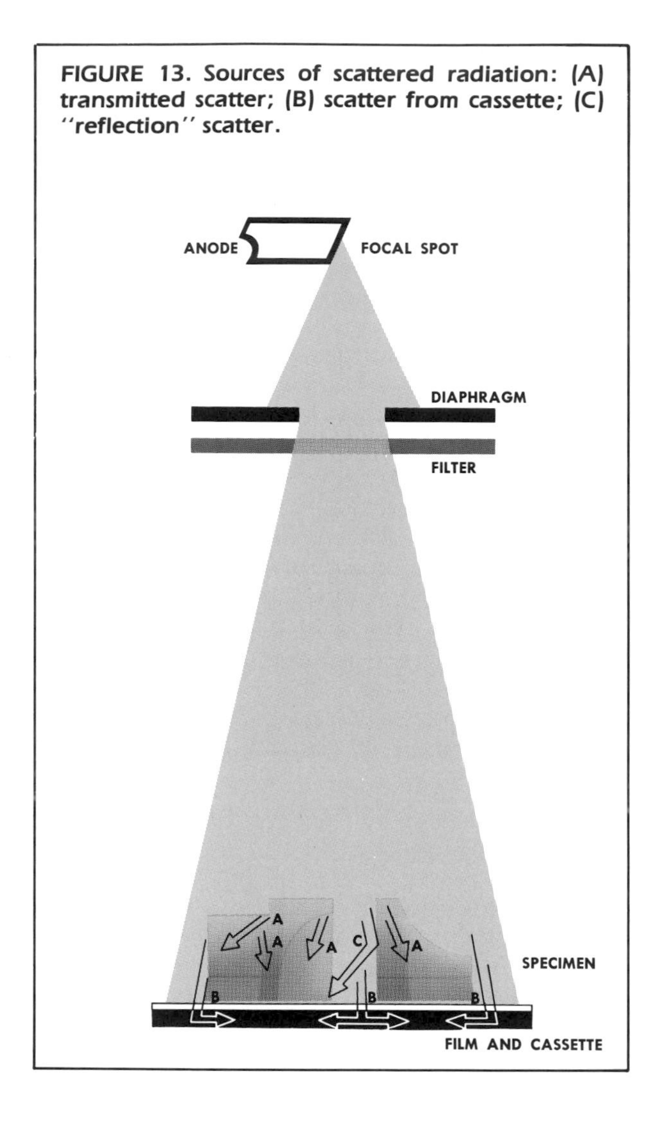

FIGURE 14. Intense backscattered radiation may originate in the floor or wall. Coning, masking or diaphragming should be employed. Backing the cassette with lead may give adequate protection.

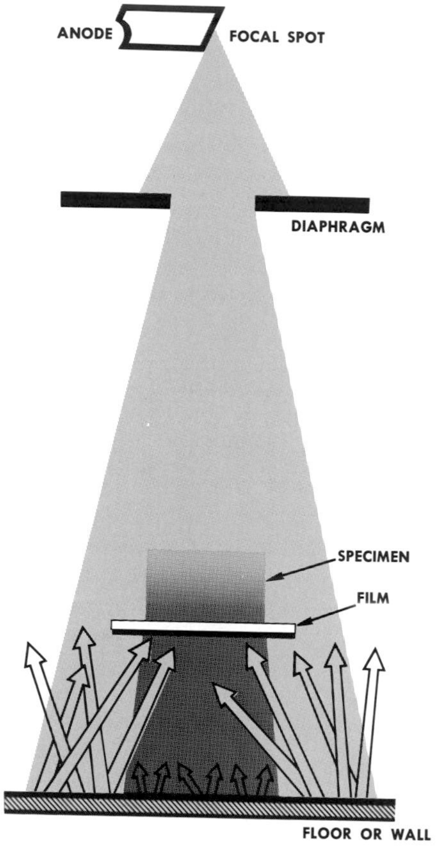

affect the film. The influence of this scatter is most noticeable just inside the borders of the image (*B* in Fig. 13). In a similar manner, primary radiation striking the film holder or cassette through a thin portion of the specimen will cause scattering into the shadows of the adjacent thicker portions. Such scatter is called *undercut*. Another source of scatter that may undercut a specimen is shown as *C* in Fig. 13. If a filter is used near the tube, this too will scatter X-rays. However, because of the distance from the film, scattering from this source is of negligible importance. Any other material, such as wall or floor, on the film side of the specimen may also scatter an appreciable quantity of X-rays back to the film, especially if the material receives the direct radiation from the X-ray tube or gamma-ray source (Fig. 14). This is referred to as *backscattered radiation*.

Reduction of Scatter

Although scattered radiation can never be completely eliminated, a number of means are available to reduce its effect. The various methods are discussed in terms of X-rays. Although most of the same principles apply to gamma-ray radiography, differences in application arise because of the highly penetrating radiation emitted by most common industrial gamma-ray sources. For example, a mask (see Fig. 15) for use with 200 kV X-rays could easily be light enough for convenient handling. A mask for use with cobalt 60 radiation, on the other hand, would be thick, heavy and probably cumbersome. In any event, with either X-rays or gamma rays, the means for reducing the effects of scattered radiation must be chosen on the basis of cost, convenience and effectiveness.

FIGURE 15. The combined use of metallic shot and a lead mask for lessening scattered radiation is conducive to good radiographic quality. If several round bars are to be radiographed, they may be separated along their lengths with lead strips held on edge by a wooden frame and the voids filled with fine shot.

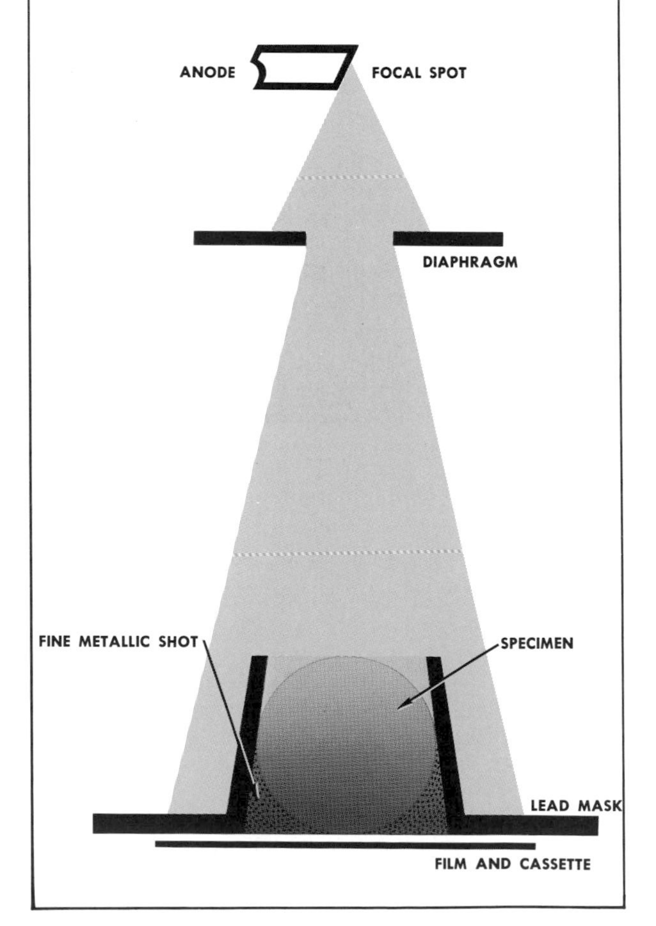

Lead Foil Screens

Lead screens, mounted in contact with the film, diminish the effect on the film of scattered radiation from all sources. They are beyond doubt the least expensive, most convenient, and most universally applicable means of combating the effects of scattered radiation. Lead screens lessen the scatter reaching the films regardless of whether the screens permit a decrease or necessitate an increase in the radiographic exposure. The nature of the action of lead screens is discussed more fully in Part 3 of this Section.

Many X-ray exposure holders incorporate a sheet of lead foil in the back for the specific purpose of protecting the film from backscatter. This lead will not serve as an intensifying screen; first, because it usually has a paper facing, and second, because it often is not lead of "radiographic quality." If intensifying screens are used with such holders, definite means must be provided to insure good contact.

X-ray film cassettes also are usually fitted with a sheet of lead foil in the back for protection against backscatter. Using such a cassette or film holder with gamma rays or with million-volt X-rays, the film should always be enclosed between double lead screens; otherwise, the secondary radiation from the lead backing is sufficient to penetrate the intervening felt or paper and cast a shadow of this material on the film, giving a granular or mottled appearance. This effect can also occur at voltages as low as 200 kV unless the film is enclosed between lead foil or fluorescent intensifying screens (see Fig. 21).

Masks and Diaphragms

Scattered radiation originating in matter outside the specimen is most serious for specimens which have high absorption for X-rays, because the scattering from external sources may be large compared to the primary image-forming radiation that reaches the film through the specimen. Often, the most satisfactory method of lessening this scatter is by the use of cutout diaphragms or some other form of mask mounted over or around the object radiographed. If many specimens of the same article are to be radiographed, it may be worthwhile to cut an opening of the same shape, but slightly smaller, in a sheet of lead and place this on the object. The lead serves to reduce the exposure in surrounding areas to a negligible value and to eliminate scattered radiation from this source. Since scatter also arises from the specimen itself, it is good practice, wherever possible, to limit the cross section of an X-ray beam to cover only the area of the specimen that is of interest in the examination.

For occasional pieces of work where a cutout diaphragm would not be economical, barium clay packed around the specimen will serve the same purpose. The clay should be thick enough so that the film density under the clay is somewhat less than that under the specimen. Otherwise, the clay itself contributes appreciable scattered radiation.

It may be found advantageous to place the object in aluminum or thin iron pans and to use a liquid absorber, provided the liquid chosen will not damage the specimen. A combined saturated solution of lead acetate and lead nitrate is satisfactory. This is prepared by dissolving approximately 3½ pounds of lead acetate in one gallon of hot water. When the lead acetate is in solution, about 3 pounds of lead nitrate may be added. (Warning! The lead solution is harmful if swallowed, harmful if inhaled. Wash thoroughly after handling. Use only with adequate ventilation.) Because of its high lead content this solution is a strong absorber of X-rays. In masking with liquids, care must be used to eliminate bubbles that may cling to the surface of the specimen.

One of the most satisfactory arrangements, combining effectiveness and convenience, is to surround the object with copper or steel shot having a diameter of about .025 cm (0.01 in.) or less (Fig. 15). This material "flows" and is very effective for filling cavities in irregular objects, such as castings, where a normal exposure for thick parts would result in an overexposure for thinner parts. Of course, it is preferable to make separate exposures for thick and thin parts, but this is not always practical.

In some cases, a lead diaphragm or lead cone on the tube head may be a convenient way to limit the area covered by the X-ray beam. Such lead diaphragms are particularly useful where the desired cross section of the beam is a simple geometric figure, such as a circle, square or rectangle.

Filters

In general, the use of filters is limited to radiography with X-rays. A simple metallic filter mounted in the X-ray beam near the X-ray tube (Fig. 16) may adequately serve the purpose of eliminating overexposure in the thin regions of the specimen and in the area surrounding the part. Such a filter is particularly useful for reducing scatter undercut in cases where a mask around the specimen is impractical or where the specimen would be injured by chemicals or shot. Of course, an increase in exposure or kilovoltage will be required to compensate for the additional absorption; but, in cases where the filter method is applicable, this is not serious unless the limit of the X-ray machine has been reached.

FIGURE 16. A filter placed near the X-ray tube reduces the subject contrast and eliminates much of the secondary radiation, which tends to obscure detail in the periphery of the specimen.

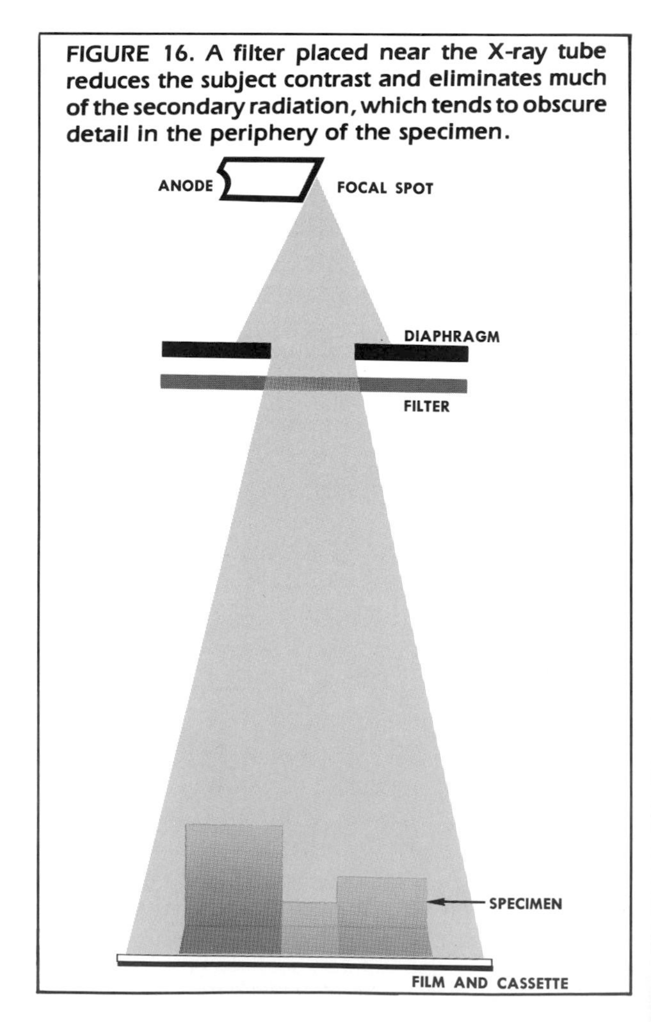

The underlying principle of the method is that the addition of the filter material causes a much greater change in the amount of radiation passing through the thin parts than through the thicker parts. Suppose the shape of a certain steel specimen is as shown in Fig. 16 and that the thicknesses are 0.6 cm (0.25 in.), 1.3 cm (0.5 in.), and 2.5 cm (1 in.). This specimen is radiographed first with no filter, and then with a filter near the tube.

Column 3 of Table 3 shows the percentage of the original X-ray intensity remaining after the addition of the filter, assuming both exposures were made at 180 kV. (These values were derived from actual exposure chart data.)

TABLE 3. Effect of Metallic Filter on X-ray Intensity

Region	Specimen Thickness cm (in.)	Percentage of Original X-ray Intensity Remaining After Addition of Filter
Outside specimen	0 (0)	less than 5%
Thin section	0.6 (.25)	about 30%
Medium section	1.3 (.50)	about 40%
Thick section	2.5 (1)	about 55%

Note that the greatest percentage change in X-ray intensity is under the thinner parts of the specimen and in the film area immediately surrounding it. The filter reduces by a large ratio the X-ray intensity passing through the thin sections or striking the cassette around the specimen, and hence reduces the undercut of scatter from these sources. Thus, in regions of strong undercut, the contrast is *increased* by the use of a filter since the only effect of the undercutting radiation is to obscure the desired image. In regions where the undercut is negligible, a filter has the effect of *decreasing* the contrast in the finished radiograph.

Although the highest possible contrast is often desired, there are certain instances in which too much contrast is a disadvantage. For example, it may be desired to render detail visible in all parts of a specimen having wide variations of thickness. If the exposure is made to give a usable density under the thin part, the thick region may be underexposed. If the exposure is adjusted to give a suitable density under the thick parts, the image of the thin sections may be grossly overexposed.

FIGURE 17. Curves illustrating the effect of a filter on the composition and intensity of an X-ray beam.

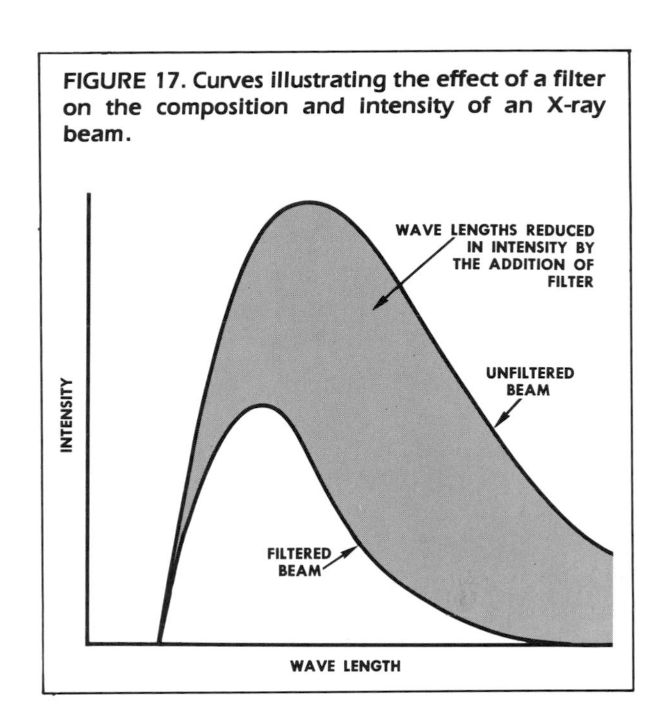

A filter reduces excessive subject contrast (and hence radiographic contrast) by *hardening* the radiation. The longer wavelengths do not penetrate the filter to as great an extent as do the shorter wavelengths. Therefore, the beam emerging from the filter contains a higher proportion of the more penetrating wavelengths. This is graphically illustrated in Fig. 17. In the sense that a more penetrating beam is produced, filtering is analogous to increasing the kilovoltage. However, it requires a comparatively large change in kilovoltage to change the hardness of an X-ray beam to the extent that will result from adding a small amount of filtration.

Although filtering reduces the total quantity of radiation, most of the wavelengths removed are those that would not penetrate the thicker portions of the specimen in any case. The radiation removed would only result in a high intensity in the regions around the specimen and under its thinner sections, with the attendant scattering, undercut and overexposure. The harder radiation obtained by filtering the X-ray beam produces a radiograph of lower contrast, permitting a wider range of specimen thicknesses to be recorded on a single film than would be possible otherwise.

A filter can act either to increase or to decrease the net contrast. The contrast and penetrameter visibility are *increased* by the removal of the scatter that undercuts the specimen and *decreased* by the hardening of the original beam. The nature of the individual specimen will determine which of these effects will predominate or whether both will occur in different parts of the same specimen.

The choice of a filter material should be made on the basis of availability and ease of handling. For the same filtering effect, the thickness of filter required is less for those materials having higher absorption. In many cases, copper or brass is the most useful, since filters of these materials will be thin enough to handle easily, yet not so thin as to be delicate (see Fig. 18).

Rules for filter thicknesses are difficult to formulate exactly because the amount of filtration required depends not only on the material and thickness range of the specimen but also on the distribution of material in the specimen and on the amount of scatter undercut to be eliminated. In the radiography of aluminum, a filter of copper about 4 percent of the greatest thickness of the specimen should prove the thickest necessary. With steel, a copper filter should ordinarily be about 20 percent, or a lead filter about 3 percent, of the greatest specimen thickness for the greatest useful filtration. The foregoing values are maximum values, and depending on circumstances, useful radiographs can often be made with far less filtration.

In radiography with X-rays up to at least 250 kV, the .0125 cm (0.005 in.) front lead screen customarily used is an effective filter for the scatter from the bulk of the specimen. Additional filtration between specimen and film only tends to contribute additional scatter from the filter itself. The scatter undercut can be decreased by adding an appropriate filter at the tube as mentioned before. Although the filter near the tube gives rise to scattered radiation, the scatter is emitted in all directions; and since the film is far from the filter, scatter reaching the film is of very low intensity.

Further advantages of placing the filter near the X-ray tube are that specimen-film distance is kept to a minimum and that scratches and dents in the filter are so blurred that their images are not apparent on the radiograph.

FIGURE 18. Maximum filter thickness for aluminum and steel.

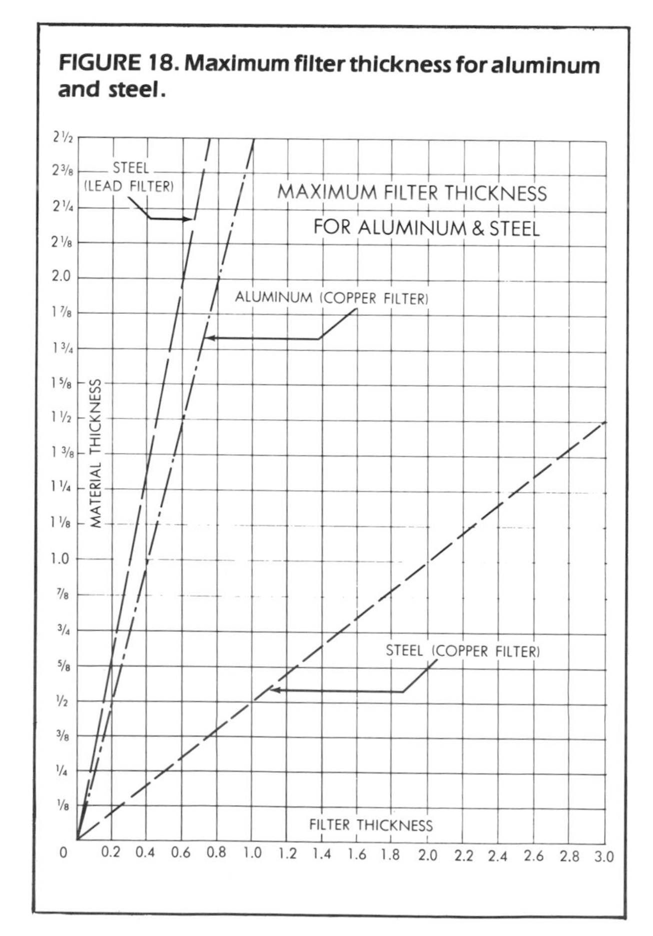

Grid Diaphragms

One of the most effective ways to reduce scattered radiation from an object being radiographed is through the use of a Potter-Bucky diaphragm. This apparatus (Fig. 19) consists of a moving grid, composed of lead strips held in position by intervening strips of a material transparent to X-rays. The lead strips are tilted, so that the plane of each is in line with the focal spot of the tube. The slots between the lead strips are several times as deep as they are wide. The parallel lead strips have the function of absorbing the very divergent scattered rays from the object being radiographed, so that most of the exposure is made by the primary rays emanating from the focal spot of the tube and passing between the lead strips. During the course of the exposure, the grid is moved, or oscillated, in a plane parallel to the film as shown by the black arrows in Fig. 19. Thus, the shadows of the lead strips are blurred to the point that they do not appear in the final radiograph.

Use of the Potter-Bucky diaphragm in industrial radiography complicates the technique to some extent and necessarily limits the flexibility of the arrangement of the X-ray tube, the specimen and the film. Grids can, however, be of great value in the radiography of beryllium more than 7.62 cm (3 in.) thick and in the examination of other low-absorption materials of moderate and great thicknesses. For these materials, kilovoltages in the medical radiographic range (approximately 50-150 kV) are used, and the medical forms of Potter-Bucky diaphragms are appropriate. Grid ratios (the ratio of height to width of the openings between the lead strips) of 12 or more are desirable.

The Potter-Bucky diaphragm is seldom used elsewhere in the industrial field, although special forms have been designed for the radiography of steel with voltages as high as 200 to 400 kV. These diaphragms are not used at higher voltages or with gamma rays because relatively thick lead strips would be needed to absorb the radiation scattered at these energies. This in turn would require a Potter-Bucky diaphragm, and the associated mechanism, of an uneconomical size and complexity.

FIGURE 19. Schematic diagram showing how the primary X-rays pass between the lead strips of the Potter-Bucky diaphragm. Most of the scattered X-rays are absorbed because they strike the sides of the strips.

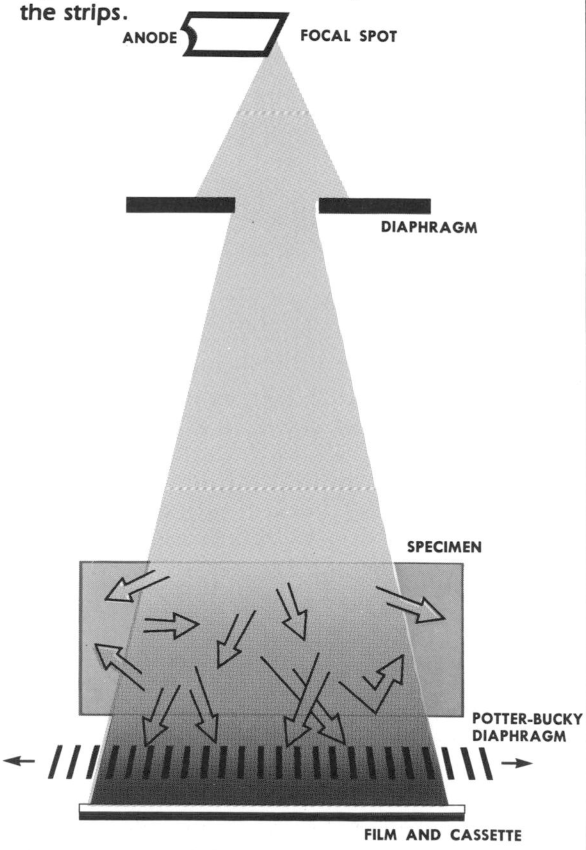

Mottling Caused by X-ray Diffraction

A special form of scattering caused by X-ray diffraction is encountered occasionally. It is most often observed in the radiography of fairly thin metallic specimens whose grain size is large enough to be an appreciable fraction of the part thickness. The radiographic appearance of this type of scattering is mottled and may be confused with the mottled appearance sometimes produced by porosity or segregation. It can be distinguished from these conditions by making two successive radiographs, with the specimen rotated slightly (1 to 5 degrees) between exposures, about an axis perpendicular to the central beam. A pattern caused by porosity or segregation will change only slightly; however, one caused by diffraction will show a marked change. The radiographs of some specimens will show a mottling from both effects, and careful observation is needed to differentiate between them.

Relatively large crystal or grain in a relatively thin specimen may in some cases reflect an appreciable portion of the X-ray energy falling on the specimen, much as if it were a small mirror. This will result in a light spot on the developed radiograph corresponding to the position of the particular crystal and may also produce a dark spot in another location if the diffracted, or reflected, beam strikes the film. Should this beam strike the film beneath a thick part of the specimen, the dark spot may be mistaken for a void in the thick section. This effect is not observed in most industrial radiography, because most specimens are composed of a multitude of very minute crystals or grains, variously oriented; hence, scatter by diffraction is essentially uniform over the film area. In addition, the directly transmitted beam usually reduces the contrast in the diffraction pattern to a point where it is no longer visible on the radiograph.

The mottling caused by diffraction can be reduced, and in some cases eliminated, by raising the kilovoltage and by using lead foil screens. The former is often of positive value even though the radiographic contrast is reduced. Since definite rules are difficult to formulate, both approaches should be tried in a new situation, or perhaps both used together.

It should be noted, however, that in some instances, the presence or absence of mottling caused by diffraction has been used as a rough indication of grain size and thus as a basis for the acceptance or the rejection of parts.

Scattering in 1- and 2-million Volt Radiography

Lead screens should always be used in this voltage range. The common thicknesses, 0.012 cm (0.005 in.) front and 0.025 cm (0.01 in.) back, are both satisfactory and convenient. Some users, however, find a 0.025 cm (0.01 in.) front screen of value because of its greater selective absorption of the scattered radiation from the specimen.

At these voltages filtration at the tube offers no improvement in radiographic quality. Filters at the film improve the radiograph in the examination of uniform sections but give poor quality at the edges of an image because of the undercut of scattered radiation from the filter itself. Hence, filtration should not be used in the radiography of specimens containing narrow bars, for example, no matter what the thickness of the bars in the direction of the primary radiation. Further, filtration should be used only where the film can be adequately protected against backscattered radiation.

Lead filters are most convenient for this voltage range. When used between specimen and film, filters are subject to mechanical damage. Care should be taken to reduce this to a minimum, lest filter defects be confused with structures in or on the specimen. In radiography with million-volt X-rays, specimens of uniform sections may be conveniently divided into three classes. Below 4 cm (1.5 in.) of steel, filtration affords little improvement in radiographic quality. Between 4 and 10 cm (1.5 and 4 in.) of steel, the thickest filter, up to 0.3 cm (.125 in.) lead, that allows a reasonable exposure time, may be used. Above 10 cm (4 in.) of steel, filter thicknesses may be increased to .63 cm (.25 in.) of lead, economic considerations permitting. It should be noted that in the radiography of extremely thick specimens with million-volt X-rays, fluorescent screens may be used to increase the photographic speed to a point where filters can be used without requiring excessive exposure time.

A very important point is to block off all radiation except the useful beam with heavy [1.3 to 2.5 cm (0.5 to 1 in.)] lead at the tubehead. Unless this is done, radiation striking the walls of the X-ray room will scatter back in such quantity as to seriously affect the quality of the radiograph. This will be especially noticeable if the specimen is thick or has parts projecting relatively far from the film.

PART 3
RADIOGRAPHIC SCREENS

Functions of Screens

Radiographic screens are employed to utilize more fully the X- or gamma-ray energy reaching the film. The physical principles underlying the action of both lead foil and fluorescent screens are discussed elsewhere and only the practical applications are discussed here.

When an X-ray or gamma-ray beam strikes a film, usually less than one percent of the energy is absorbed. Since the formation of the radiographic image is primarily governed by the absorbed radiation, more than 99 percent of the available energy in the beam performs no useful photographic work. Obviously, any means of more fully utilizing this wasted energy, without complicating the technical procedure, is highly desirable. Two types of radiographic screens are used to achieve this end—lead and fluorescent. Lead screens, in turn, take two different forms. One form is sheets of lead foil, usually mounted on cardboard or plastic, which are used in pairs in a conventional cassette or exposure holder. The other consists of a lead compound (usually an oxide), evenly coated on a thin support. The film is placed between the leaves of a folded sheet of this oxide-coated material with the oxide in contact with the film. The combination is supplied in a sealed, lightproof envelope.

Lead Foil Screens

For radiography in the range 150 to 400 kV, lead foil in direct contact with both sides of the film has a desirable effect on the quality of the radiograph. In radiography with gamma rays and with X-rays below 2,000 kV, the front lead foil need be only 0.01 to 0.015 cm (0.004 to 0.006 in.) thick; consequently, its absorption of the primary beam is not serious. The back screen should be thicker to reduce backscattered radiation. Such screens are available commercially. The choice of lead screen thicknesses for multimillion-volt radiography is much more complicated, and the manufacturers of the equipment should be consulted for their recommendations.

Effects of Lead Screens

Lead foil in direct contact with the film has three principal effects: (1) it increases the photographic action on the film, largely by reason of the electrons emitted and partly by the secondary radiation generated in the lead; (2) it absorbs the longer wavelength scattered radiation more than the primary; and (3) it intensifies the primary radiation more than the scattered radiation. The differential absorption of the secondary radiation and the differential intensification of the primary radiation result in diminishing the effect of scattered radiation, producing greater contrast and clarity in the radiographic image. This reduction in the effect of the scattered radiation decreases the total intensity of the radiation reaching the film and lessens the net intensification factor of the screens. The absorption of primary radiation by the front lead screen also diminishes the net intensifying effect; and, if the incident radiation does not have sufficient penetrating power, the actual exposure required may be even greater than without screens. At best, the exposure time is one-half to one-third of that without screens but the advantage of screens in reducing scattered radiation still holds.

The quality of the radiation necessary to obtain an appreciable intensification from lead foil screens depends on the type of film, the kilovoltage, and the thickness of the material through which the rays must pass (Fig. 20). In the radiography of aluminum, for example, using a 0.0125 cm (0.005 in.) front screen and a 0.025 cm (0.010 in.) back screen, the thickness of aluminum must be about 15 cm (6 in.) and the kilovoltage as high as 160 kV to secure any advantage in exposure time with lead screens. In the radiography of steel, lead screens begin to give appreciable intensification with thicknesses in the neighborhood of 0.63 cm (.25 in.), at voltages of 130 to 150 kV. In the radiography of 3 cm (1.25 in.) steel at about 200 kV, lead screens

FIGURE 20. Effects of kilovoltage on intensification properties of lead screens.

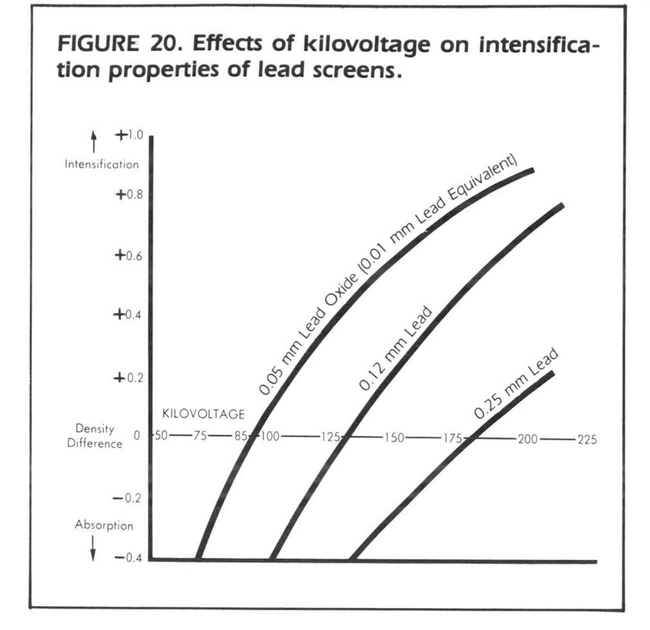

FIGURE 21. Upper area shows decreased density caused by paper between the lead screen and film. An electron shadow picture of the paper structure has also been introduced.

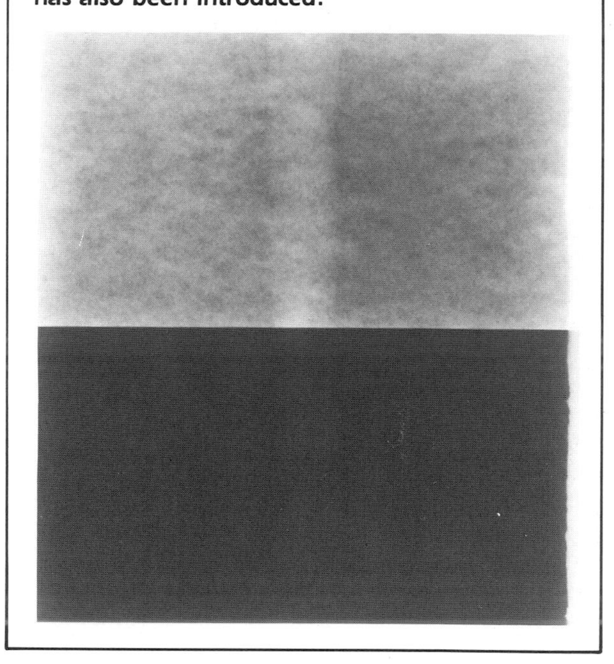

permit an exposure of about one-third of that without screens (intensification factor of 3). With cobalt 60 gamma rays, the intensification factor of lead screens is about 2. Lead foil screens, however, do not detrimentally affect the definition or graininess of the radiographic image to any material degree so long as the lead and the film are in intimate contact.

Lead foil screens diminish the effect of scattered radiation, particularly that which undercuts the object when the primary rays strike the portions of the film holder or cassette outside the area covered by the object.

Scattered radiation from the specimen itself is cut almost in half by lead screens, contributing to maximum clarity of detail in the radiograph; this advantage is obtained even under conditions where the lead screen makes an increase in exposure necessary.

In radiography with gamma rays or high-voltage X-rays, films loaded in metal cassettes without screens are likely to record the effect of secondary electrons generated in the lead-covered back of the cassette. These electrons, passing through the felt pad on the cassette cover, produce a mottled appearance because of the structure of the felt. Films loaded in the customary lead-backed cardboard exposure holder may also show the structure of the paper that lies between the lead and the film (Fig. 21). To avoid these effects, film should be enclosed between double lead screens, care being taken to ensure good contact between film and screens. Thus, lead foil screens are essential in practically all radiography with gamma rays or million-volt X-rays. If, for any reason, screens cannot be used with these radiations, a lightproof paper or cardboard holder with no metal backing should be used.

Contact between the film and the lead foil screens is essential to good radiographic quality. Areas lacking contact produce "fuzzy" images, as shown in Fig. 22.

Selection and Care of Lead Screens

Lead foil for screens must be selected with extreme care. Commercially pure lead is satisfactory. An alloy of 6 percent antimony and 94 percent lead, being harder and stiffer, has better resistance to wear and abrasion. Tin-coated lead foil should be avoided, since irregularities in the tin cause a variation in the intensifying factor of the screens, resulting in mottled radiographs. Minor blemishes do not affect the usefulness of the screen, but large "blisters" or cavities should be avoided.

Most of the intensifying action of a lead foil screen is caused by the electrons emitted under X-ray or gamma-ray excitation. Because electrons are readily absorbed even in thin or light materials, the surface must be kept free of grease and lint which will produce light marks on the radiograph. Small flakes of foreign material—for example, dandruff or tobacco—will likewise produce light spots on the completed radiograph. For this same reason, protective coatings on lead foil screens are not common. Any protective coating should be thin, to minimize the absorption of electrons and keep the intensification factor as high as possible, and uniform so that the intensification factor will be uniform. In addition, the coating should not produce

FIGURE 22. Good contact between film and lead foil screens gives a sharp image (left). Poor contact results in a fuzzy image (right).

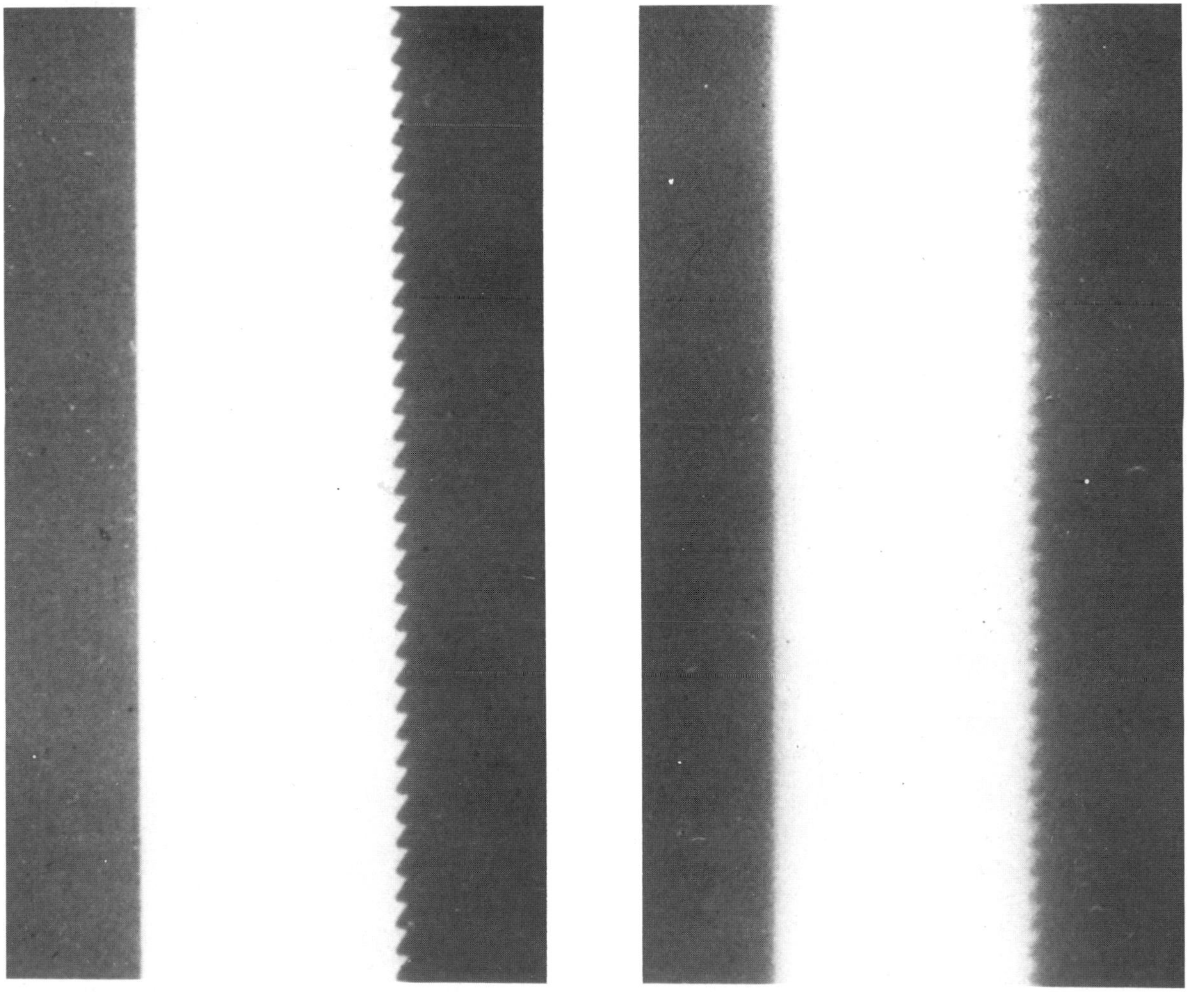

static electricity when rubbed against or placed in contact with film (see Fig. 23).

Deep scratches on lead foil screens, on the other hand, will produce dark lines on the radiograph (Fig. 24).

Grease and lint may be removed from the surface of lead foil screens with a mild household detergent or cleanser and a soft, lint-free cloth. If the cleanser is one that dries to a powder, care must be taken to remove all the powder and to prevent its being introduced into the cassette or exposure holder. The screens must be completely dry before use; otherwise, the film will stick to them. If more thorough cleaning is necessary, screens may be very gently rubbed with the finest grade of steel wool. If this is done carefully, the shallow scratches left by the steel wool will not produce dark lines in the radiograph.

FIGURE 23. Static marks resulting from poor film-handling technique. Static marks may also be treelike or branching.

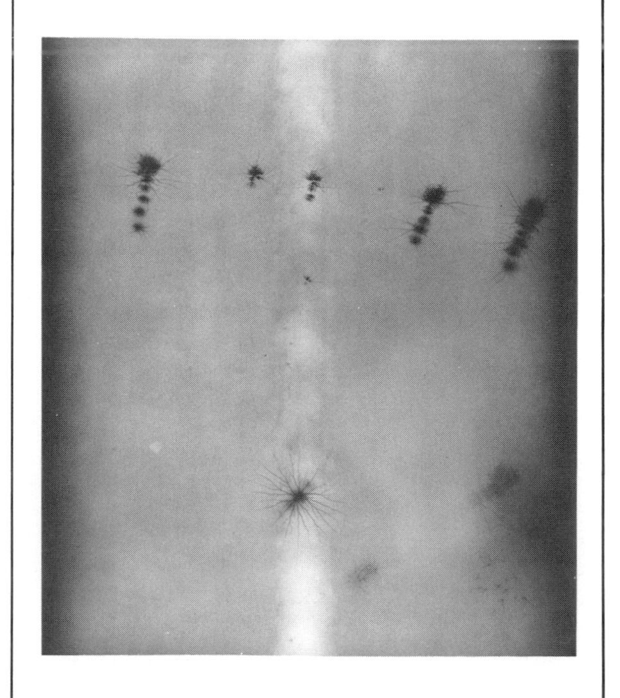

Films could be fogged if left between lead foil screens longer than is reasonably necessary, particularly under conditions of high temperature and humidity. When screens have been freshly cleaned with an abrasive, this effect will be increased; prolonged contact between film and screens should be delayed for 24 hours after cleaning.

Lead Oxide Screens

Films packaged with lead screens in the form of lead-oxide-coated paper, factory sealed in light-tight envelopes, have a number of advantages. One of these is convenience—the time-consuming task of loading cassettes and exposure holders is avoided, as are many of the artifacts that can arise from careless handling of film.

Another advantage is cleanliness. This is particularly important in the radiography of those specimens in which heavy inclusions are serious. Light material (such as hair, dandruff, tobacco ash) between the lead screen and the film will produce low-density indications on the radiograph. These can easily be confused with heavy inclusions in the specimen. The factory-sealed combination of film and lead oxide screens, manufactured under the conditions of extreme cleanliness necessary for all photographic materials, avoids all such difficulties and obviates much of the re-radiography that formerly was necessary. A further advantage is the flexibility of the packets, which makes them particularly valuable when film must be inserted into confined spaces.

Such packets may be used in the kilovoltage range of 100 to 300 kV. In many cases, the integral lead oxide screens will be found to have a somewhat higher intensification factor than conventional lead foil screens. They will, however, remove less scattered radiation because of the smaller effective thickness of lead in the lead oxide screen.

Scatter removal and backscatter protection, equivalent to that provided by conventional lead foil screens, can be provided by using conventional lead foil screens *external* to the envelope. Such screens can be protected on both surfaces with cardboard or plastic, and thus can be made immune to most of the accidents associated with handling. With this technique, the full function of lead foil screens can be retained while gaining the advantages of cleanliness, convenience and screen contact.

Fluorescent Screens

Certain chemicals fluoresce; that is, they have the ability to absorb X-rays and gamma rays and immediately emit light; the intensity of the emitted light depends on the intensity of the incident radiation. These phosphorescent materials can be used in radiography by first being finely powdered, mixed with a suitable binder, then coated in a thin, smooth layer on a special cardboard or plastic support.

For the exposure, film is clamped firmly between a pair of these fluorescent screens. The photographic effect on the film, then, is the sum of the effects of the X-rays and of the light emitted by the screens. A few examples will serve to illustrate the importance of intensifying screens in reducing exposure time. In medical radiography, the exposure is from 1/10 to 1/60 as much *with* fluorescent intensifying screens as without them. In other words, the *intensification factor* varies from 10 to 60, depending on the kilovoltage and the type of screen used. In the radiography of 1.3 cm (0.5 in.) steel at 150 kV, a factor as high as 125 has been observed and in radiography of 1.9 cm (0.75 in.) steel at 180 kV, factors of several hundred have been obtained experimentally.

FIGURE 24. The number of electrons emitted (per surface unit of the lead) is essentially uniform. More electrons can reach the film in the vicinity of a scratch, resulting in a dark line on the radiograph. (For illustrative clarity, electron paths have been shown as straight and parallel; actually, the electrons are emitted diffusely.)

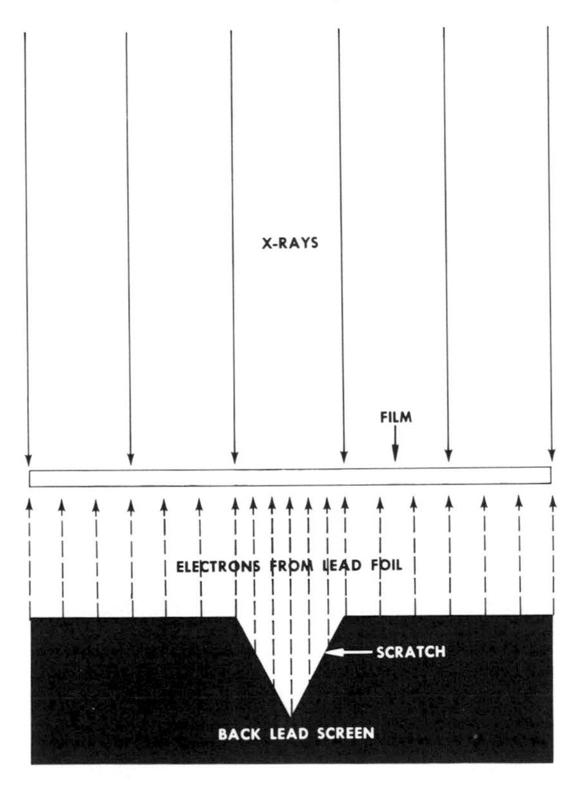

Under these latter conditions, the intensification factor has about reached its maximum, and it diminishes both for lower voltage and thinner steel and for higher voltage and thicker steel. Using cobalt 60 gamma rays for very thick steel, the factor may be 10 or less.

Despite their great effect in reducing exposure time, fluorescent screens are not widely used in industrial radiography. This is in part because they may give poor definition, compared to a radiograph made directly or with lead screens. The poorer definition can result from the spreading of the light emitted from the screens, as shown in Fig. 25. The light from any particular portion of the screen spreads out beyond the confines of the X-ray beam that excited the fluorescence. This spreading of light from the screens may account for the blurring of outlines in the radiograph.

The other reason fluorescent screens are seldom used in industrial radiography is because they may produce *screen mottle* on the finished radiograph. This mottle is characteristic in appearance, very much larger in scale and much softer in outline than the graininess associated with the film itself. It is not associated with the actual structure of the screen; that is, it is not a result of the size of the fluorescent crystals themselves or of any unevenness in their dispersion in the binder. Rather, screen mottle is associated with purely statistical variations in the numbers of absorbed X-ray photons, from one tiny area of the screen to the next. The fewer the number of X-ray photons involved, the stronger the appearance of the screen mottle. This explains, for example, why the screen mottle produced by a particular type of screen tends to become greater as the kilovoltage of the radiation increases. The higher the kilovoltage, the more energetic, on the average, are the X-ray photons. Therefore, on

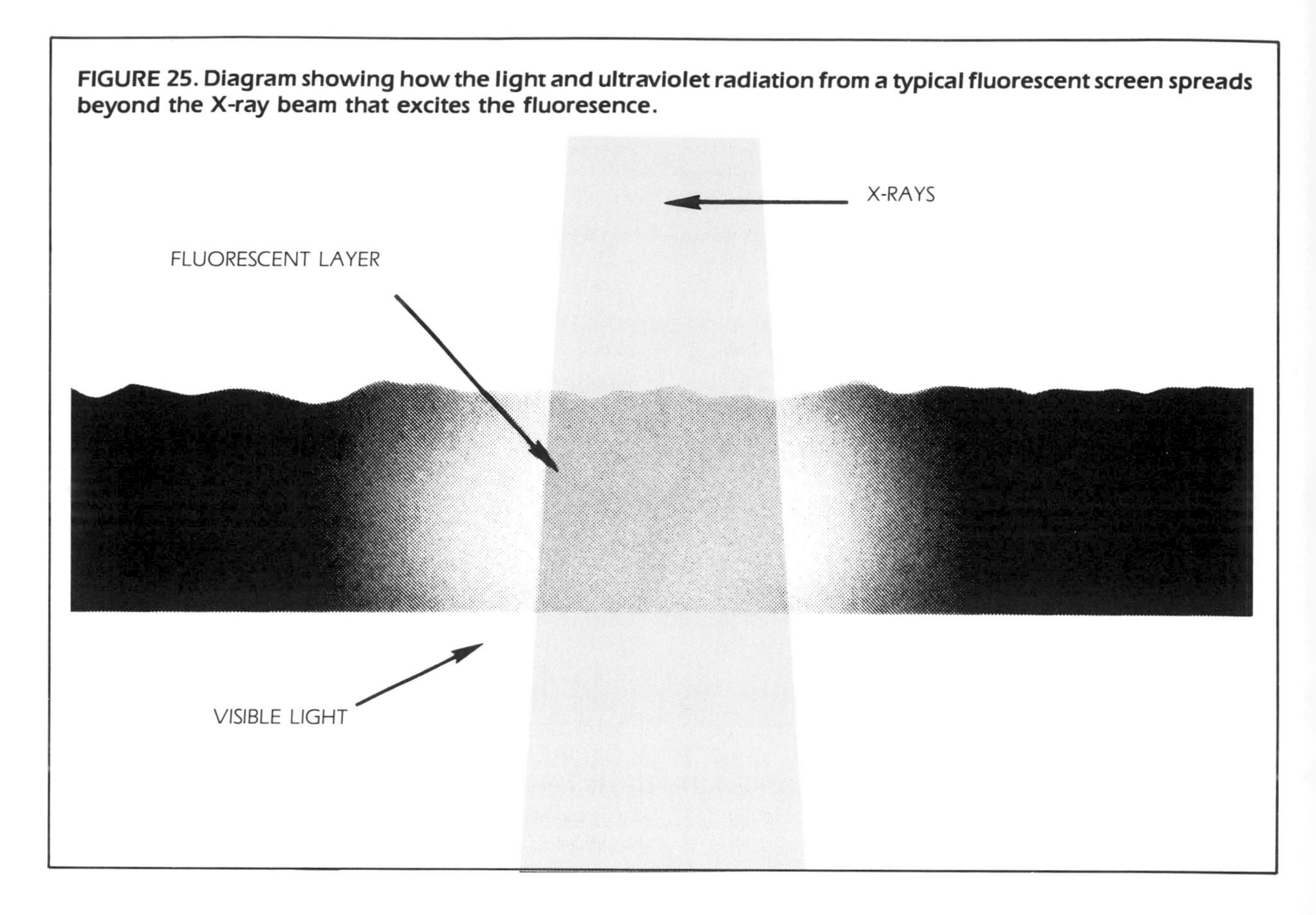

FIGURE 25. Diagram showing how the light and ultraviolet radiation from a typical fluorescent screen spreads beyond the X-ray beam that excites the fluoresence.

absorption in the screen, a larger "burst" of light is produced. The larger the bursts, the fewer that are needed to produce a given density and the greater is the purely statistical variation in the number of photons from one small area to the next.

Intensifying screens may be needed in the radiography of steel thicknesses greater than 5 cm (2 in.) at 250 kV, 7.5 cm (3 in.) at 400 kV, and 12.5 cm (5 in.) at 1,000 kV.

Fluorescent screens are not employed with gamma rays since, apart from the screen mottle, failure of the reciprocity law may result in relatively low intensification factors with the longer exposure times usually necessary in gamma-ray radiography. In the radiography of light metals, fluorescent screens are normally not used; but, should they be required, the best choice would be fluorescent screens of the slowest type compatible with an economical exposure time—if possible, those designed specifically for sharpness of definition in medical radiography.

At kilovoltages higher than those necessary to radiograph about 1.3 cm (0.5 in.) of steel, the fastest available screens are usually employed, since the major use of fluorescent intensifying screens is to minimize the exposure time.

There are a few radiographic situations which demand a speed higher than the fastest film designed for direct exposure (or exposure with lead screens), yet do not require the speed of film designed for use with fluorescent intensifying screens. In such cases a high-speed, direct-exposure film may be used with fluorescent screens. The speed of this combination will be intermediate between those of the two first-mentioned combinations. However, the contrast and the maximum density will be higher than that obtained with a film designed for fluorescent-screen exposure, and the screen mottle will be less because of the lower speed of the screen-film combination.

PART 4
INDUSTRIAL X-RAY FILMS

Modern X-ray films for general radiography consist of an emulsion (gelatin containing a radiation-sensitive silver compound) and a flexible, transparent base that sometimes contains a tint. Usually, the emulsion is coated on both sides of the base in layers about 0.0125 mm (0.0005 in.) thick (see Fig. 26 and 27). Putting emulsion on both sides of the base doubles the amount of radiation-sensitive silver compound, and thus increases the speed. At the same time, the emulsion layers are thin enough so that developing, fixing and drying can be accomplished in a reasonable time. However, some films for radiography in which the highest detail visibility is required have emulsion on only one side of the base.

When X-rays, gamma rays or light strike the grains of the sensitive silver compound in the emulsion, a change takes place in the physical structure of the grains. This change cannot be detected by ordinary physical methods. However, when the exposed film is treated with a chemical solution (called a developer), a reaction takes place, causing the formation of black, metallic silver. It is this silver, suspended in the gelatin on both sides of the base, that constitutes the image (see Fig. 28).

Although an image may be formed by light and other forms of radiation, as well as by gamma rays or X-rays, the properties of the latter two are of a distinct character, and, for this reason, the sensitive emulsion must be different from those used in other types of photography.

FIGURE 26. The silver bromide grains of an X-ray film emulsion (2,500 diameters). These grains have been dispersed to show their shape and relative sizes more clearly. In an actual coating, the crystals are much more closely packed.

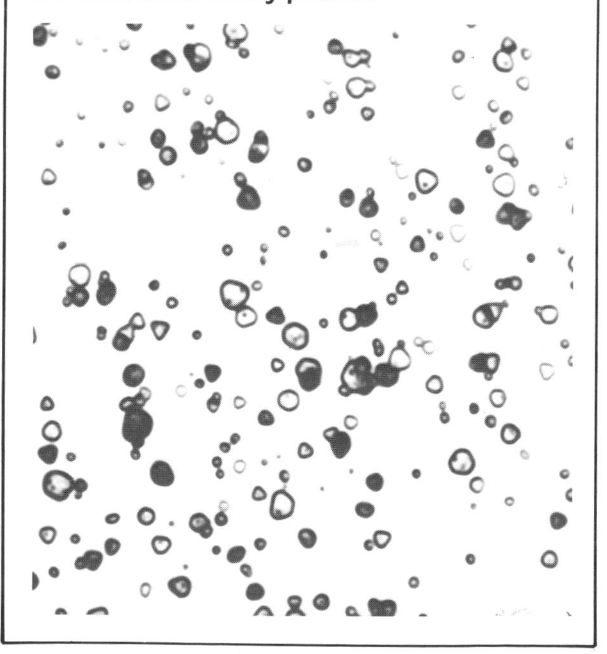

FIGURE 27. Cross section of the unprocessed emulsion on one side of an X-ray film. Note the large number of grains as compared to the developed grains of Fig. 28.

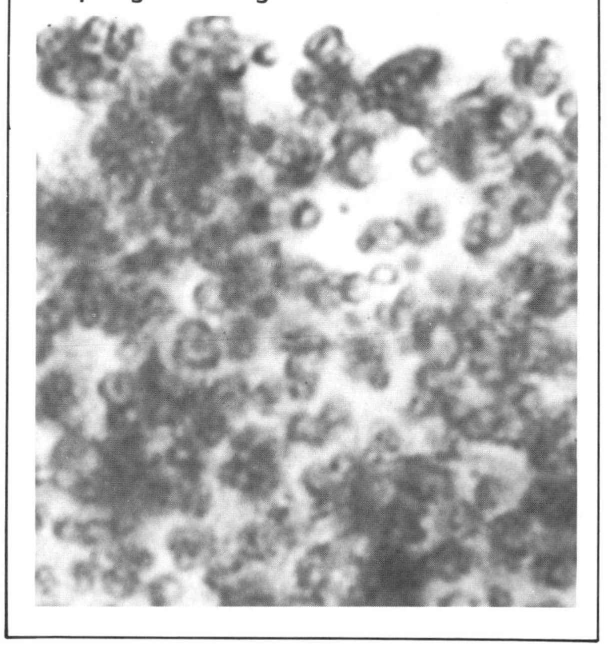

FIGURE 28. Cross section showing the distribution of the developed grains in an X-ray film emulsion exposed to give a moderate density.

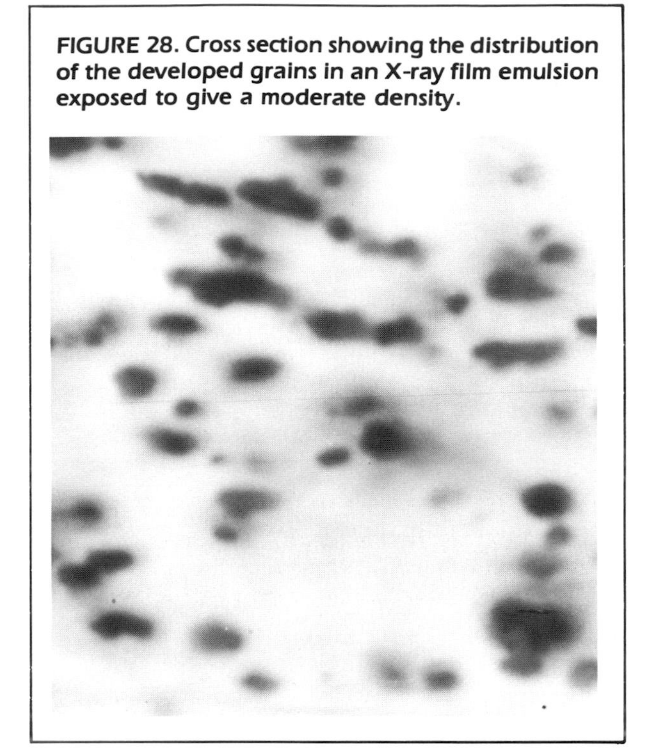

Selection of Films For Industrial Radiography

As pointed out at the beginning of this Section, industrial radiography now has many widely diverse applications. There are many considerations to be made in obtaining the best radiographic results, for example: (1) the composition, shape and size of the part being examined—and, in some cases, its weight and location as well; (2) the type of radiation used—whether X-rays from an X-ray machine or gamma rays from a radioactive material; (3) the kilovoltages available with the X-ray equipment; (4) the intensity of the gamma radiation; (5) the kind of information sought—whether it is simply an overall inspection or the critical examination of some especially important portion, characteristic or feature; and (6) the resulting relative emphasis on definition, contrast, density, and the time required for proper exposure. All of these factors are important in the determination of the most effective combination of radiographic technique and X-ray film.

The selection of a film for the radiography of any particular part depends on the thickness and material of the specimen and on the voltage range of the available X-ray machine. In addition, the choice is affected by the relative importance of high radiographic quality or short exposure time. Thus, an attempt must be made to balance these two opposing factors. As a consequence, it is not possible to present definite rules on the selection of a film. If high quality is the deciding factor, a slower and finer-grained film should be substituted for a faster one—for instance, for the radiography of steel up to 0.64 cm (.25 in.) thick at 120-150 kV, Film Y (Fig. 34) might be substituted for Film X. If short exposure times are essential, a faster film (or film-screen combination) can be used. For example, 4 cm (1.5 in.) steel might be radiographed at 200 kV using fluorescent screens with a film particularly sensitive to blue light, rather than a direct exposure film with lead screens.

Fig. 29 indicates the direction that these substitutions take. The direct-exposure films may be used with or without lead screens, depending on the kilovoltage and the thickness and shape of the specimen.

Fluorescent intensifying screens must be used in radiography requiring the highest possible photographic speed. The light emitted by the screens has a much greater photographic action than the X-rays either alone or combined with the emission from lead screens. To secure adequate exposure within a reasonable time, screen-type X-ray films sandwiched between fluorescent intensifying screens are often used in radiography of steel in thicknesses greater than about 5 cm (2 in.) at 250 kV, and more than 7.5 cm (3 in.) at 400 kV.

FIGURE 29. Change in choice of film, depending on relative emphasis on high speed or high radiographic quality.

IMPROVING QUALITY →

Screen-type film with fluorescent screens	Fast direct-exposure-type film	Slow direct-exposure-type film

← INCREASING SPEED

Photographic Density

Photographic density refers to the quantitative measure of film blackening. When no danger of confusion exists, photographic density is usually spoken of merely as *density*. Density is defined by the equation

$$D = \log \frac{I_o}{I_t} \tag{Eq. 15}$$

where D is density, I_o is the light intensity incident on the film and I_t is the light intensity transmitted.

Table 4 illustrates some relations between transmittance, percent transmittance, opacity and density. It shows that an increase in density of 0.3 reduces the light transmitted to one-half its former value. In general, since density is a logarithm, a certain *increase* in density always corresponds to the same *percentage decrease* in transmittance.

TABLE 4. Transmittance, Percent Transmittance, Opacity and Density Relationships

Transmittance	Percent Transmittance	Opacity	Density
$\frac{I_t}{I_o}$	$\frac{I_t}{I_o} \times 100$	$\frac{I_o}{I_t}$	$\log \frac{I_o}{I_t}$
1.00	100	1	0
0.50	50	2	0.3
0.25	25	4	0.6
0.10	10	10	1.0
0.01	1	100	2.0
0.001	0.1	1,000	3.0
0.0001	0.01	10,000	4.0

Densitometers

A densitometer is an instrument for measuring photographic densities. A number of different types, both visual and photoelectric, are available commercially. For purposes of practical industrial radiography there is no great premium on high accuracy in a densitometer. A much more important property is reliability, that the densitometer should reproduce readings from day to day.

X-ray Exposure Charts

An exposure chart is a graph showing the relation between material thickness, kilovoltage and exposure. In its most common form, an exposure chart resembles Fig. 30. These graphs are adequate for determining exposures in the radiography of uniform plates, but they serve only as rough guides for objects, such as complicated castings, having wide variations of thickness.

Exposure charts are usually available from manufacturers of X-ray equipment. Because, in general, such charts cannot be used for different X-ray machines unless suitable correction factors are applied, individual laboratories sometimes prepare their own.

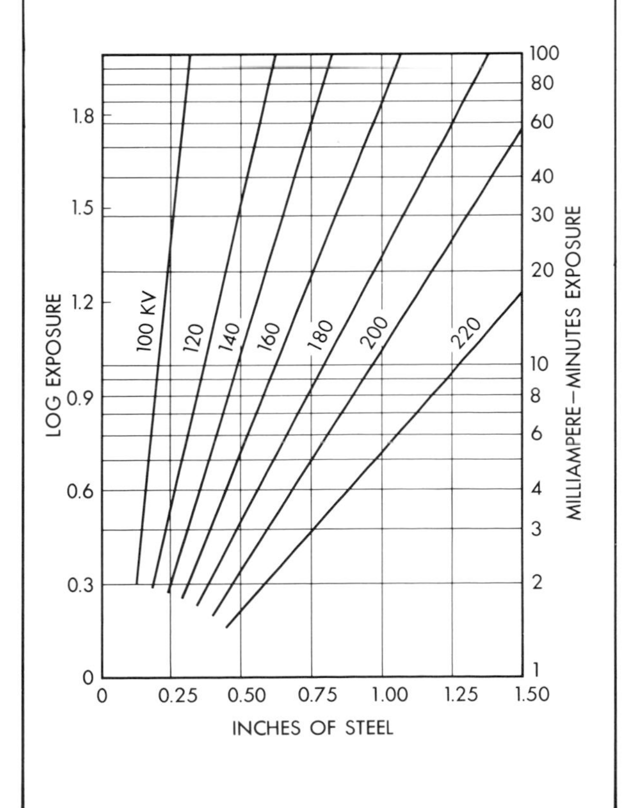

FIGURE 30. Typical exposure chart for steel. This chart may be applied to Film X (see Fig. 34), with lead foil screens, at a film density of 1.5. Source-film distance: 1.02 m (40 inches).

Preparing an Exposure Chart

A simple method for preparing an exposure chart is to make a series of radiographs of a pile of metal plates consisting of a number of steps. This step tablet, or stepped wedge, is radiographed at several different exposure times at each of a number of kilovoltages. The exposed films are all processed under conditions identical to those that will later be used for routine work. Each radiograph consists of a series of photographic densities corresponding to the X-ray intensities transmitted by the different thicknesses of metal. A certain density, for example 1.5, is selected as the basis for the preparation of the chart. Wherever this density occurs on the stepped-wedge radiographs, there are corresponding values of thickness, milliampere-minutes and kilovoltage. It is unlikely that many of the radiographs will contain a value of exactly 1.5 in density, but the correct thickness for this density can be found by interpolation between steps. Thickness and milliampere-minute values are plotted for the different kilovoltages in the manner shown in Fig. 30.

Another method, requiring fewer stepped-wedge exposures but more arithmetical manipulation, is to make one step-tablet exposure at each kilovoltage and to measure the densities in the processed stepped-wedge radiographs. The exposure that would have given the chosen density (in this case 1.5) under any particular thickness of the stepped wedge can then be determined from the characteristic curve of the film used. The values for thickness, kilovoltage and exposure are then plotted.

Note that thickness is on a linear scale, and that milliampere-minutes are on a linear scale. The logarithmic scale is not necessary, but it is very convenient because it compresses an otherwise long scale. A further advantage of the logarithmic exposure scale is that it usually allows the location of the points for any one kilovoltage to be well approximated by a straight line.

An exposure chart usually applies only to a single set of conditions, determined by:

1. The X-ray machine used;
2. A certain source-film distance;
3. A particular film type;
4. Processing conditions used;
5. The film density on which the chart is based;
6. The type of screens (if any) that are used.

Only if the conditions used in making the radiograph agree in all particulars with those used in preparation of the exposure chart can values of exposure be read directly from the chart. Any change requires the application of a correction factor. The correction factor applying to each of the conditions listed previously will be discussed separately.

1. It is sometimes difficult to find a correction factor to make an exposure chart prepared for one X-ray machine applicable to another. Different X-ray machines operating at the same nominal kilovoltage and milliamperage settings may give not only different intensities but also different qualities of radiation.
2. A change in source-film distance may be compensated for by the use of the inverse square law. Some exposure charts give exposures in terms of "exposure factor" rather than in terms of milliampere-minutes or milliampere-seconds. Charts of this type are readily applied to any value of source-film distance.
3. The use of a different type of film can be corrected by comparing the difference in the amount of exposure necessary to give the same density on both films (from relative exposure charts such as those shown in Fig. 35).

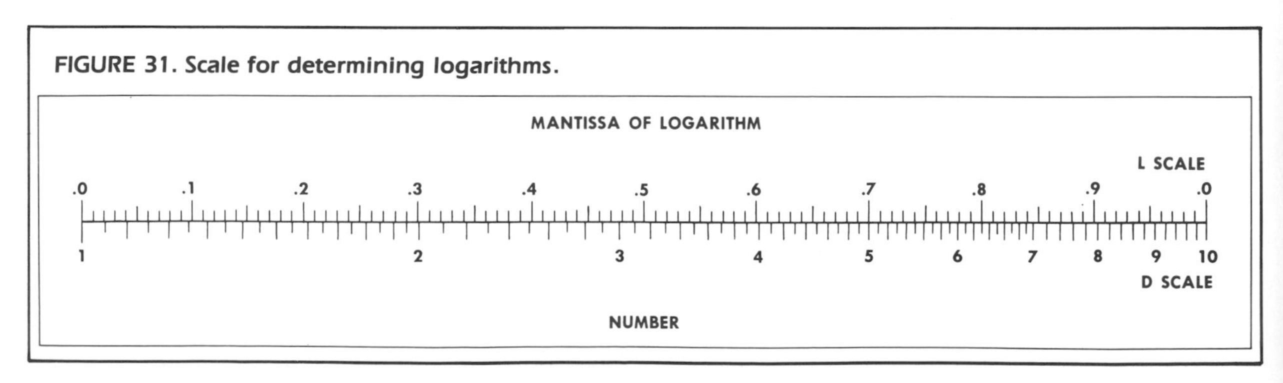

FIGURE 31. Scale for determining logarithms.

For example, to obtain a density of 1.5 using Film Y, 0.6 more exposure is required than for Film X.

This log exposure difference is found on the L scale (Fig. 31) and corresponds to an exposure factor of 3.99 on the D scale (read directly below the log E difference). Therefore, in order to obtain the same density on Film Y as on Film X, multiply the original exposure by 3.99 to get the new exposure. Conversely, if going from Film Y to Film X, divide the original exposure by 3.99 to obtain the new exposure.

These procedures can be used to change densities on a single film as well. Simply find the log E difference needed to obtain the new density on the film curve; read the corresponding exposure factor from the chart; then multiply to increase density or divide to decrease density.

4. A change in processing conditions causes a change in effective film speed. If the processing of the radiographs differs from that used for the exposures from which the chart was made, the correction factor must be found by experiment.
5. The chart gives exposures to produce a certain density. If a different density is required, the correction factor may be calculated from the film's characteristic curve.
6. If the type of screen is changed—for example, from lead foil to fluorescent—it is easier and more accurate to make a new exposure chart than to determine correction factors.

In some radiographic operations, the exposure time and the source-film distance are set by economic considerations or on the basis of previous experience and test radiographs. The tube current is, of course, limited by the design of the tube. This leaves as variables only the thickness of the specimen and the kilovoltage. When these conditions exist, the exposure chart may take a simplified form

FIGURE 32. Typical exposure chart for use when exposure and distance are held constant and kilovoltage is varied to conform to specimen thickness. Film X (see Fig. 34), exposed with lead foil screens to a density of 1.5. Source-film distance: 1.02 m (40 inches); exposure: 50 mA-min.

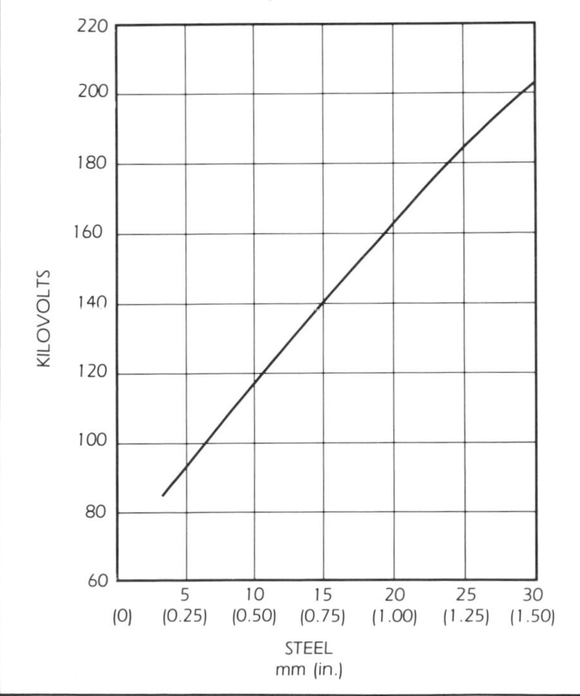

FIGURE 33. Typical gamma-ray exposure chart for iridium 192, based upon the use of Film X (see Fig. 34).

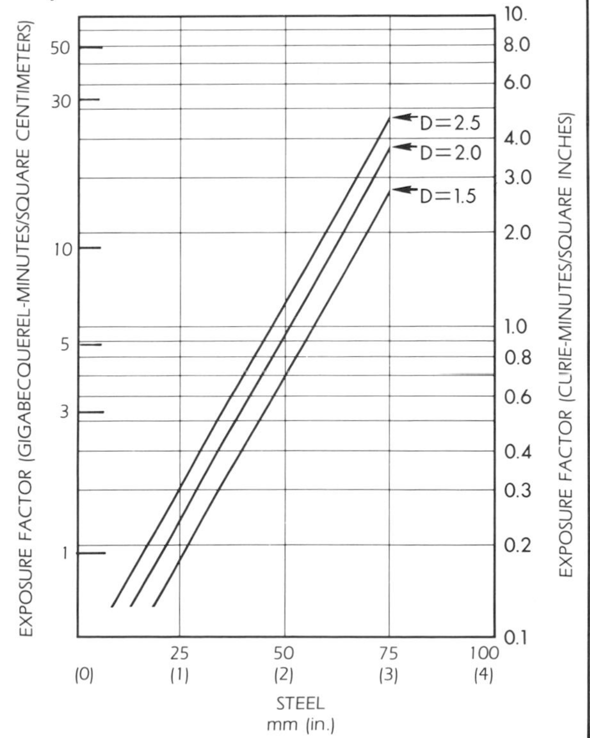

as shown in Fig. 32, which allows the kilovoltage for any particular specimen thickness to be chosen. Such a chart will be particularly useful when uniform sections must be radiographed in large numbers by relatively untrained persons. This type of exposure chart may be derived from a chart similar to Fig. 30 by following the horizontal line corresponding to the chosen milliampere-minute value and noting the thickness corresponding to this exposure for each kilovoltage. These thicknesses are then plotted against kilovoltage.

Gamma-ray Exposure Charts

A typical gamma-ray exposure chart is shown in Fig. 33. It is somewhat similar to Fig. 30; however, with gamma rays, there is no variable factor corresponding to the kilovoltage. Therefore, a gamma-ray exposure chart contains one line, or several parallel lines, each of which corresponds to a particular film type, film density, or source-film distance. Gamma-ray exposure guides are also available in the form of linear or circular slide rules. These contain scales on which the various factors of specimen thickness, source strength, and source-film distance can be set, and from which exposure time can be read directly.

The Characteristic Curve

The characteristic curve, sometimes referred to as the sensitometric curve or the H and D curve (after Hurter and Driffield who, in 1890, first used it), expresses the relation between the exposure applied to a photographic material and the resulting photographic density. The characteristic curves of three typical films, exposed between lead foil screens to X-rays, are given in Fig. 34. Such curves are obtained by giving a film a series of known exposures, determining the densities produced by these exposures, and then plotting density against the logarithm of relative exposure.

Relative exposure is used because there are no convenient units, suitable to all kilovoltages and scattering conditions, in which to express radiographic exposures. The exposures given a film are expressed in terms of some particular exposure, giving a relative scale. In practical radiography, this lack of units for X-ray intensity or quantity is no hindrance, as will be seen below. The use of the logarithm of the relative exposure, rather than the relative exposure itself, has a number of advantages. It compresses an otherwise long scale. Furthermore, in radiography, ratios of exposures or intensities are usually more significant than the exposures or the intensities themselves. Pairs of exposures having the same ratio will be separated by the same interval on the log relative exposure scale, no matter what their absolute value may be. Consider the pairs of exposures following:

FIGURE 34. Characteristic curves of three typical X-ray films, exposed between lead foil screens.

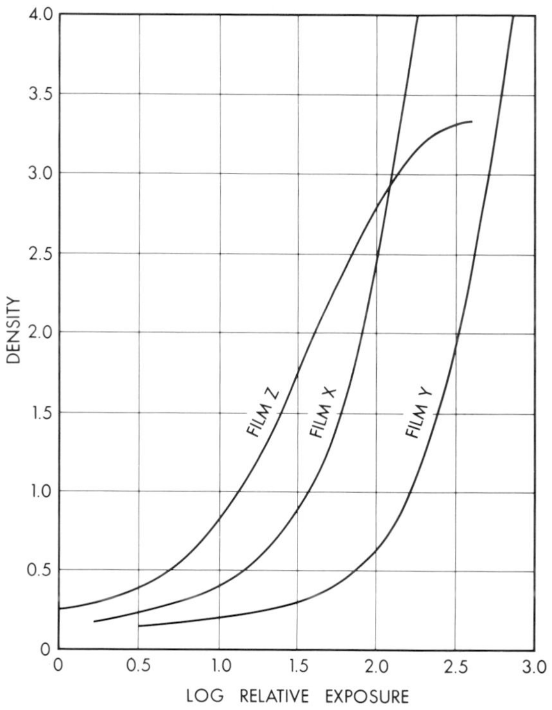

Relative Exposure	Log Relative Exposure	Interval In Log Relative Exposure
1	0.0	0.70
5	0.70	
2	0.30	0.70
10	1.00	
30	1.48	0.70
150	2.18	

This illustrates another useful property of the logarithmic scale; Fig. 31 shows that the antilogarithm of 0.75 is 5, which is the ratio of any pair of exposures. To find the ratio of *any* pair of exposures, it is necessary only to find the antilog of the log E (logarithm of relative exposure) interval between them. Conversely, the log exposure interval between any two exposures is determined by finding the logarithm of their ratio.

As can be seen in Fig. 34, the slope (or steepness) of the characteristic curves is continuously changing throughout the length of the curves. The effects of this change of slope on detail visibility are more completely explained later in this Section. It will suffice at this point to give a qualitative outline of these effects. For example, two slightly different thicknesses in the object radiographed transmit slightly different exposures to the film. These two exposures have a certain small log E interval between them; that is, they have a certain ratio. The difference in the densities corresponding to the two exposures depends on just where on the characteristic curve they fall; the steeper the slope of the curve, the greater is this density difference. For example, the curve of Film Z (Fig. 34) is steepest in its middle portion. This means that a certain log E interval in the middle of the curve corresponds to a greater density difference than the same log E interval at either end of the curve. In other words, the film contrast is greatest where the slope of the characteristic curve is greatest. For Film Z, as has been pointed out, the region of greatest slope is in the central part of the curve. For Films X and Y, however, the slope—and hence the film contrast—continuously increases throughout the useful density range. The curves of most industrial X-ray films are similar to those of Films X and Y.

Use of the Characteristic Curve

The characteristic curve can be used in solving quantitative problems arising in radiography, in the preparation of technique charts, and in radiographic research. Characteristic curves made under actual radiographic conditions *should* be used in solving practical problems. However, it is not always possible to produce characteristic curves in a radiography department, and curves prepared elsewhere must be used. Such curves prove adequate for many purposes although it must be remembered that the shape of the characteristic curve and the speed of a film relative to that of another depend strongly on developing conditions. The accuracy attained when using "ready-made" characteristic curves is governed largely by the similarity between the developing conditions used in producing the characteristic curves and those for the films whose densities are to be evaluated.

Quantitative use of characteristic curves are worked out in the following examples. Note that *D* is used for density and *log E* for logarithm of relative exposure.

Example 1: Suppose a radiograph made on Film Z (see Fig. 35) with an exposure of 12 mA-min has a density of 0.8 in the region of maximum interest. It is desired to increase the density to 2.0 for the sake of the increased contrast there available.

1. Log E at D = 2.0 is 1.62
2. Log E at D = 0.8 is 1.00
3. Difference in log E is 0.62
   Antilogarithm of this difference is 4.2

Therefore, the original exposure is multiplied by 4.2 giving 50 mA-min to produce a density of 2.0.

FIGURE 35. Characteristic curve of Film Z (see Fig. 34). Circled numerals correspond to items in text Example 1.

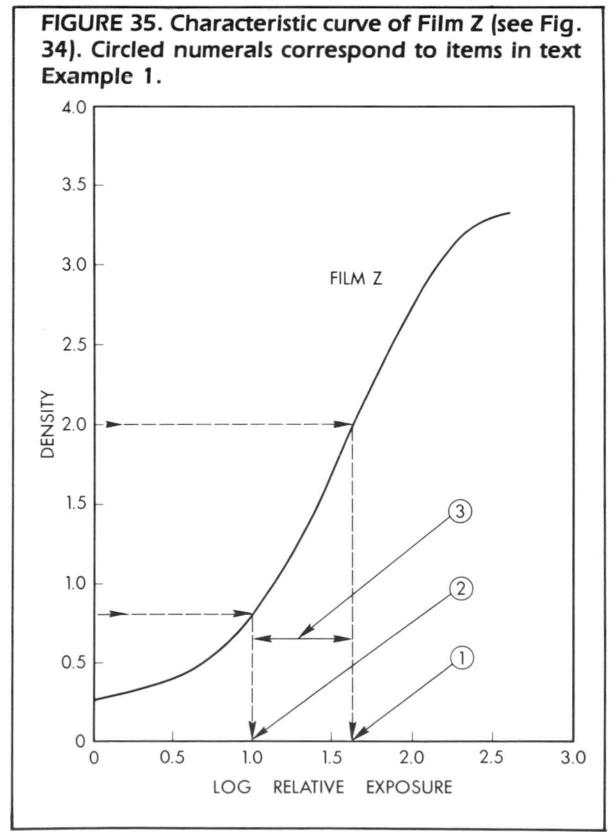

FIGURE 36. Characteristic curves of two X-ray films exposed with lead foil screens. Circled numerals correspond to items in text Example 2.

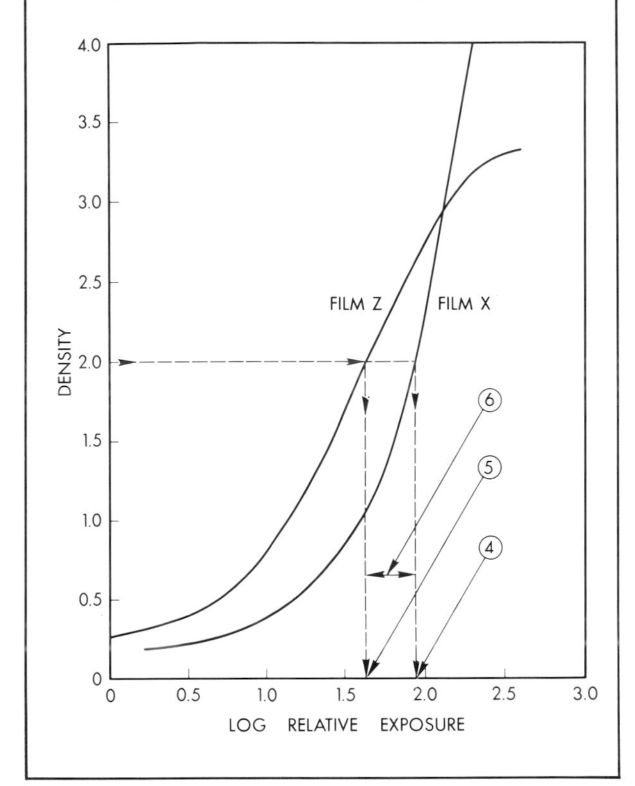

Example 2: Film X has a higher contrast than Film Z at $D = 2.0$ (see Fig. 36) and also a finer grain. Suppose that, for these reasons, it is desired to make the radiograph on Film X with a density of 2.0 in the same region of maximum interest.

4. Log E at D = 2.0 for Film X is 1.91
5. Log E at D = 2.0 for Film Z is 1.62
6. Difference in log E is 0.29
   Antilogarithm of this difference is 1.95

Therefore, the exposure for $D = 2.0$ on Film Z is multiplied by 1.95, for a density of 2.0 on Film X.

PART 5

RADIOGRAPHIC IMAGE QUALITY AND DETAIL VISIBILITY

Controlling Factors

The relationships of the various factors influencing radiographic sensitivity are shown in Table 5. Because the purpose of most radiographic inspections is to examine a specimen for inhomogeneity, a knowledge of the factors affecting the visibility of detail in the finished radiograph is essential. Table 5 shows the relationships of the various factors influencing image quality and radiographic sensitivity; following are a few important definitions.

Radiographic sensitivity is a general or qualitative term referring to the size of the smallest detail that can be seen in a radiograph, or to the ease with which the images of small details can be detected. Phrased differently, it is a reference to the amount of information in the radiograph. Note that radiographic sensitivity depends on the combined effects of two independent sets of factors: radiographic contrast (the density difference between a small detail and its surroundings) and definition (the abruptness and the "smoothness" of the density transition). See Fig. 37.

Radiographic contrast is the difference in density between two areas of a radiograph. It depends on both subject contrast and film contrast.

Subject contrast is the ratio of X-ray or gamma-ray intensities transmitted by two selected portions of a specimen. Subject contrast depends on the nature of the specimen, the energy (spectral composition, hardness, or wavelengths) of the radiation used, and the intensity and distribution of the scattered radiation, but is independent of time, milliamperage or source strength, distance, and the characteristics or treatment of the film.

Film contrast refers to the slope (steepness) of the characteristic curve of the film. It depends on the type of film, the processing it receives, and the density. It also depends on whether the film's exposure is direct, with lead screens or with fluorescent screens. Film contrast is independent, for most practical purposes, of the wavelengths and distribution of the radiation reaching the film, and hence is independent of subject contrast.

Definition refers to the sharpness of outline in the image. It depends on the types of screens and film used, the radiation energy (wavelengths, etc.), and the geometry of the radiographic setup.

TABLE 5. Factors Controlling Radiographic Sensitivity

Radiographic Contrast		Radiographic Definition	
Subject Contrast	**Film Contrast**	**Geometrical Factors**	**Graininess Factors**
Affected by: A. Thickness differences in specimen B. Radiation quality C. Scattered radiation	**Affected by:** A. Type of film B. Development time, temperature, and agitation C. Density D. Activity of the developer	**Affected by:** A. Focal-spot size B. Focus-film distance C. Specimen-film distance D. Abrupt thickness changes in specimen E. Screen-film contact	**Affected by:** A. Type of film B. Type of screen C. Radiation quality D. Development

Subject Contrast

Subject contrast decreases as the kilovoltage is increased. The decreasing slope (steepness) of the lines of the exposure chart (Fig. 30) as kilovoltage increases illustrates the reduction of subject contrast as the radiation becomes more penetrating. For example, consider a steel part containing two thicknesses, 2 and 2.5 cm (0.75 and 1 in.), which is radiographed first at 160 kV and then at 200 kV.

In Table 6, column 3 shows the exposure in milliampere-minutes required to reach a density of 1.5 through each thickness at each kilovoltage. These data are from the exposure chart, Fig. 30. It is apparent that the milliampere-minutes required to produce a given density at any kilovoltage are inversely proportional to the corresponding X-ray intensities passing through the different sections of the specimen. Column 4 gives these relative intensities for each kilovoltage. Column 5 gives the ratio of these intensities for each kilovoltage.

TABLE 6. Exposure of Steel Part Containing Two Thicknesses

kV	Thickness in. (cm)	Exposure to give D = 1.5 mA-min	Relative Intensity	Ratio of Intensities
160	0.75 (2)	18.5	3.8	3.8
	1.0 (2.5)	70.0	1.0	
200	0.75 (2)	4.9	14.3	2.5
	1.0 (2.5)	11.0	5.8	

Column 5 shows that, at 160 kV, the intensity of the X-rays passing through the 2 cm (0.75 in.) section is 3.8 times greater than that passing through the 2.5 cm (1 in.) section. At 200 kV, the radiation through the thinner portion is only 2.5 times that through the thicker. Thus, as the kilovoltage increases, the ratio of X-ray transmission of the two thicknesses decreases, indicating a lower subject contrast.

FIGURE 37. Advantage of higher radiographic contrast (left), is largely offset by poor definition. Despite lower contrast (right), better rendition of detail is obtained by improved definition.

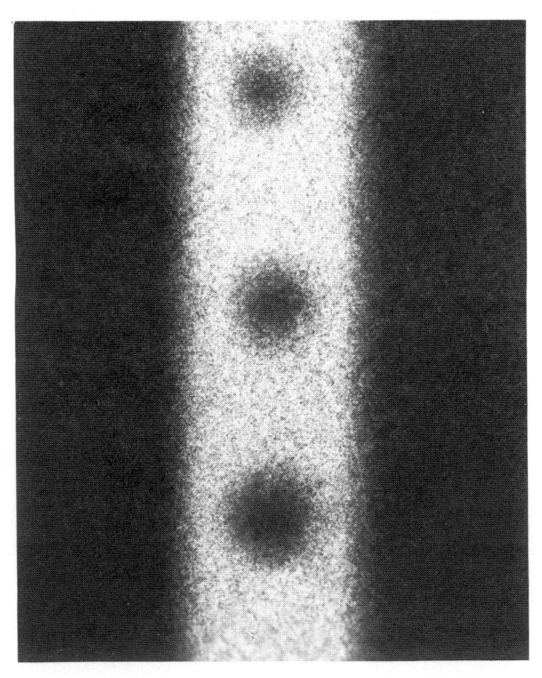

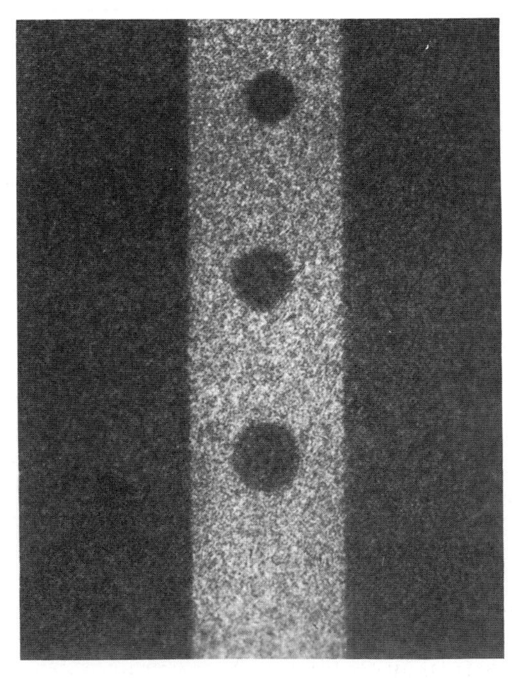

Film Contrast

The dependence of film contrast on density must be kept in mind when considering problems of radiographic sensitivity. In general, the contrast of radiographic films, except those designed for use with fluorescent screens, increases continuously with density in the usable density range. Therefore, for films that exhibit this continuous increase in contrast, the best density to use (or the upper limit of the density range) is the highest that can be conveniently viewed with the illuminators available. Adjustable high-intensity illuminators are commercially available and greatly increase the maximum density that can be viewed.

The use of high densities has the further advantage of increasing the range of radiation intensities which can be usefully recorded on a single film. In X-ray radiography, this in turn permits the use of lower kilovoltage, resulting in increased subject contrast and radiographic sensitivity.

The maximum contrast of screen-type films is at a density of about 2.0. Therefore, other things being equal, the greatest radiographic sensitivity will be obtained when the exposure is adjusted to give this density.

Film Graininess and Screen Mottle

The image on an X-ray film is formed by countless minute silver grains, the individual particles being so small that they are visible only under a microscope. However, these small particles are grouped together in relatively large masses which are visible to the naked eye or with a magnification of only a few diameters. These masses result in a visual impression called *graininess.*

All films exhibit graininess to a greater or lesser degree. In general, the slower films have lower graininess. Thus, Film Y (Fig. 34) would have a lower graininess than Film X.

The graininess of all films increases as the penetration of the radiation increases, although the rate of increase may be different for different films. The graininess of the images produced at high kilovoltages makes the slow, inherently fine-grain films especially useful in the million- and multimillion-volt range. When sufficient exposure can be given, fine-grain films are also useful with gamma rays.

The use of lead screens has no significant effect on film graininess. However, graininess is affected by processing conditions, being directly related to the degree of development. For instance, if development time is increased for the purpose of increasing film speed, the graininess of the resulting image is likewise increased. Conversely, a developer or developing technique that results in an appreciable decrease in graininess will also cause an appreciable loss in film speed. However, adjustments in development technique made to compensate for changes in temperature or activity of a developer will have little effect on graininess. Such adjustments are made to achieve the same degree of development as would be obtained in the fresh developer at a standard processing temperature, and therefore the graininess of the film will be essentially unaffected.

Another source of irregular density in uniformly exposed areas is the screen mottle encountered in radiography with fluorescent screens. The screen mottle increases markedly as hardness of the radiation increases. This is one of the factors that limits the use of fluorescent screens at high voltage and with gamma rays.

Penetrameters

A standard test piece is usually included in every radiograph as a check on the adequacy of the radiographic technique. The test piece is commonly referred to as a penetrameter in North America and an Image Quality Indicator (IQI) in Europe. The penetrameter (or IQI) is a simple geometric form made of the same material as, or a material similar to, the specimen being radiographed. It contains some small structures (holes, wires, etc.), the dimensions of which bear some numerical relation to the thickness of the part being tested. The image of the penetrameter on the radiograph is permanent evidence that the radiographic examination was conducted under proper conditions.

Codes or agreements between customer and vendor may specify the type of penetrameter, its dimensions, and how it is to be employed. Even if penetrameters are not specified, their use is advisable because they provide an effective check on the overall quality of the radiographic inspection.

Hole Type Penetrameters

The common penetrameter consists of a small rectangular piece of metal, containing several (usually three) holes, the diameter of which are related to the thickness of the penetrameter (Fig. 38).

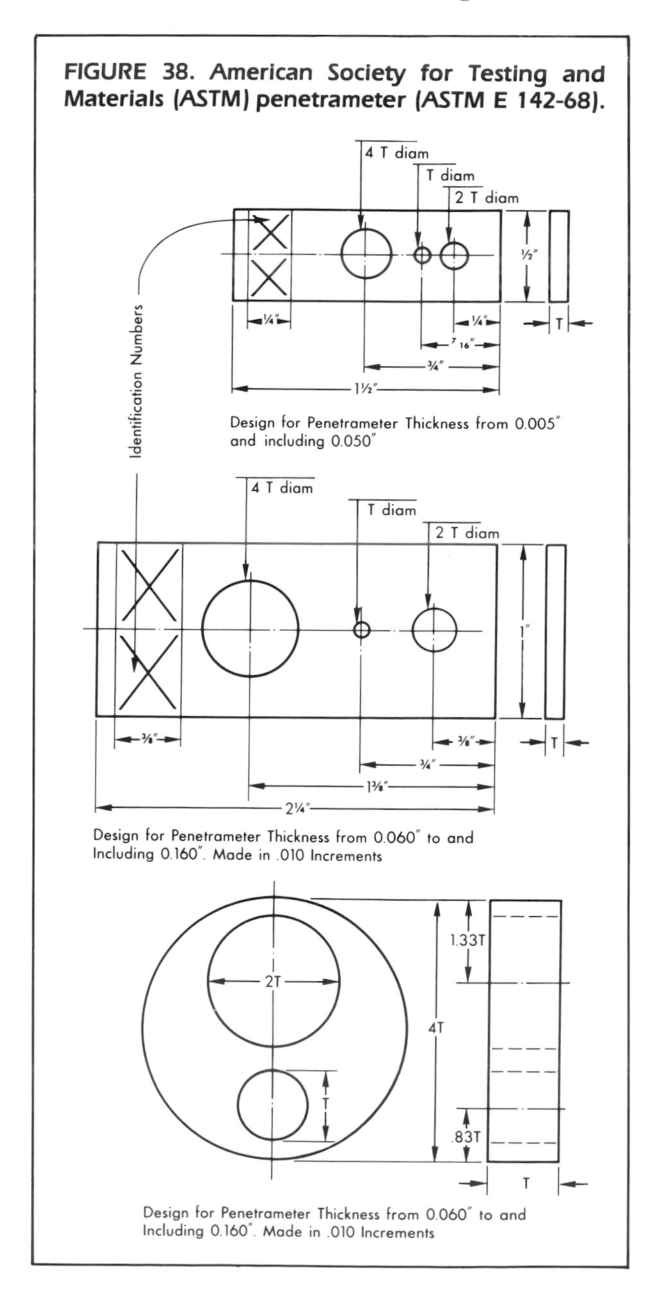

FIGURE 38. American Society for Testing and Materials (ASTM) penetrameter (ASTM E 142-68).

The ASTM (American Society for Testing and Materials) penetrameter contains three holes of diameters T, 2T, and 4T, where *T* is the thickness of the penetrameter. Because of the practical difficulties in drilling minute holes in thin materials, the minimum diameters of these three holes are 0.010, 0.020, and 0.040 in., respectively. These penetrameters may also have a slit similar to the ASME (American Society of Mechanical Engineers) penetrameter described below. Thick penetrameters of the hole type would be very large because of the diameter of the 4T hole. Therefore, penetrameters more than 0.180 in. thick are in the form of discs, the diameters of which are 4 times the thickness (4T) and which contain only two holes, of diameters T and 2T. Each penetrameter is identified by a lead number showing the thickness in thousandths of an inch.

The ASTM penetrameter permits the specification of a number of levels of radiographic sensitivity, depending on the requirements of the job. For example, the specifications may call for a radiographic sensitivity level of *2-2T*. The first symbol (2) indicates that the penetrameter shall be 2 percent of the thickness of the specimen; the second symbol (2T) indicates that the hole having a diameter twice the penetrameter thickness shall be visible on the finished radiograph. The quality level 2-2T is probably the one most commonly specified for routine radiography. However, critical components may require more rigid standards, and a level of 1-2T or 1-1T may be required. On the other hand, the radiography of less critical specimens may be satisfactory if a quality level of 2-4T or 4-4T is achieved. The more critical the radiographic examination—that is, the *higher* the level of radiographic sensitivity required—the *lower* the numerical designation for the quality level.

Some sections of the ASME Boiler and Pressure Vessel Code require a penetrameter similar (in general) to the ASTM penetrameter. It contains three holes, one of which is 2T in diameter, where *T* is the penetrameter thickness. Customarily, the other two holes are 3T and 4T in diameter, but other sizes may be used. Minimum hole size is 0.0625 in. Penetrameters 0.010 in., and less, in thickness also contain a slit 0.010 in. wide and 0.25 in. long. Each is identified by a lead number designating the thickness in thousandths of an inch.

Equivalent Penetrameter Sensitivity

Ideally, the penetrameter should be made of the same material as the specimen. However, this is sometimes impossible because of practical or economic difficulties. In such cases, the penetrameter may be made of a radiographically similar material—that is, a material having the same radiographic absorption as the specimen but which is better suited to the making of penetrameters. Tables of radiographically equivalent materials have been published, grouping materials with similar radiographic absorptions. In addition, a penetrameter made of a particular material may be used in the radiography of materials having *greater* radiographic absorption. In such a case, there is a certain penalty on the radiographic testers because they are setting more rigid radiographic quality standards for themselves than those which are actually required. This penalty is often outweighed by avoiding the problems of obtaining penetrameters for an unusual material.

In some cases the materials involved do not appear in published tabulations. Under these circumstances the comparative radiographic absorption of two materials may be determined experimentally. A block of the material under test and a block of the material proposed for penetrameters, equal in thickness to the part being examined, can be radiographed side by side on the same film with the technique to be used in practice. If the film density under the proposed penetrameter material is equal to or greater than the film density under the specimen material, that proposed material is suitable for fabrication of penetrameters.

In practically all cases, the penetrameter is placed on the source side of the specimen, in the least advantageous geometric position. In some instances, however, this location for the penetrameter is not feasible. An example would be the radiography of a circumferential weld in a long tubular structure, using a source position within the tube and film on the outer surface. In such a case a "film-side" penetrameter must be used. Some codes specify the film-side penetrameter that is equivalent to the source-side penetrameter normally required. When such a specification is not made, the required film-side penetrameter may be found experimentally. In the example above, a short section of tube of the same dimensions and materials as the item under test would be used in the experiment. The required penetrameter would be placed on the source side, and a range of penetrameters on the film side. If the source-side penetrameter indicated that the required radiographic sensitivity was being achieved, the image of the smallest visible hole in the film-side penetrameters would be used to determine the penetrameter and the hole size to be used on the production radiographs.

Sometimes the shape of the part being examined precludes placing the penetrameter on the part. When this occurs, the penetrameter may be placed on a block of radiographically similar material of the same thickness as the specimen. The block and the penetrameter should be placed as close as possible to the specimen.

FIGURE 39. DIN (German) penetrameter (German Standard DIN 54109).

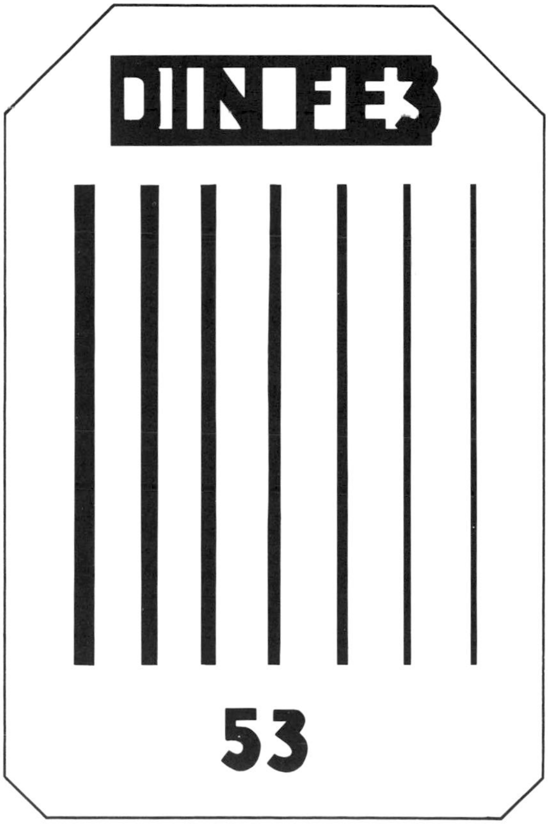

Wire Penetrameters

A number of other penetrameter designs are also in use. The German DIN (Deutsche Industrie-Norm) penetrameter (Fig. 39) is one that is widely used. It consists of a number of wires of various diameters sealed in a plastic envelope that carries the necessary identification symbols. The image quality is indicated by the thinnest wire visible on the radiograph. The system is such that only three penetrameters, each containing seven wires, can cover a very wide range of specimen thicknesses. Sets of DIN penetrameters are available in aluminum, copper and steel. Thus a total of nine penetrameters is sufficient for the radiography of a wide range of materials and thicknesses.

Comparison of Penetrameter Design

The hole type penetrameter is, in a sense, a *go, no-go* gauge; that is, it indicates whether or not a specified quality level has been attained but, in most cases, does not indicate whether the requirements have been exceeded, or by how much. The DIN penetrameter on the other hand is a series of seven penetrameters in a single unit. As such, it has the advantage that the radiographic quality level achieved can often be read directly from the processed radiograph.

Furthermore, the hole penetrameter can be made of any desired material but the wire penetrameter is made from only a few materials. A quality level of 2-2T may be specified for the radiography of, for example, commercially pure aluminum and 2024 aluminum alloy, even though these have appreciably different compositions and radiation absorptions. The hole penetrameter would, in each case, be made of the appropriate material. The wire penetrameters, however, are available in aluminum but not in 2024 alloy. To achieve the same quality of radiographic inspection for equal thicknesses of these two materials, it would be necessary to specify different wire diameters—that for 2024 alloy would probably have to be determined by experiment.

Special Penetrameters

Special penetrameters have been designed for certain classes of radiographic inspection. An example is the radiography of small electronic components in which some of the significant factors are the continuity of fine wires or the presence of tiny balls of solder. Special image quality indicators have been designed consisting of fine wires and small metallic spheres within a plastic block. The block is covered on top and bottom with steel approximately as thick as the case of the electronic component.

Penetrameters and Visibility of Discontinuities

It should be remembered that even if a certain hole in a penetrameter is visible on the radiograph, a cavity of the same diameter and thickness may not be visible. The penetrameter holes, having sharp boundaries, result in abrupt, though small, changes in metal thickness whereas a natural cavity having more or less rounded sides causes a gradual change. Therefore, the image of the penetrameter hole is sharper and more easily seen in the radiograph than is the image of the cavity. Similarly, a fine crack may be of considerable extent, but if the X-rays or gamma rays pass from source to film along the thickness of the crack, its image on the film may not be visible because of the very gradual transition in photographic density. Thus, a penetrameter is used to indicate the quality of the radiographic technique and not to measure the size of cavity which can be shown.

In the case of a wire image quality indicator of the DIN type, the visibility of a wire of a certain diameter does not assure that a discontinuity of the same cross section will be visible. The human eye perceives much more readily a long boundary than it does a short one, even if the density difference and the sharpness of the image are the same.

Viewing and Interpreting Radiographs

The examination of the finished radiograph should be made under conditions that favor the best visibility of detail combined with a maximum of comfort and a minimum of eye fatigue for the observer. To be satisfactory for use in viewing radiographs, an illuminator must fulfill two basic requirements. First, it must provide light of an intensity that will illuminate the areas of interest in the radiograph to their best advantage, free from glare. Second, it must diffuse the light evenly over the entire viewing area. The color of the light is of no optical consequence, but most observers prefer bluish white. An illuminator incorporating several fluorescent tubes meets this requirement and is often used for viewing industrial radiographs of moderate density.

For routine viewing of high densities, one of the commercial high-intensity illuminators should be used. These provide an adjustable light source, with a maximum intensity which allows viewing of densities of 4.0 or even higher.

Such a high-intensity illuminator is especially useful for the examination of radiographs having a wide range of densities corresponding to a wide range of thicknesses in the object. If the exposure was adequate for the greatest thickness in the specimen, the detail reproduced in other thicknesses can be visualized with illumination of sufficient intensity.

The contrast sensitivity of the human eye (the ability to distinguish small brightness differences) is greatest when the surroundings are of about the same brightness as the area of interest. Thus, to see the finest detail in a radiograph, the illuminator must be masked to avoid glare from bright light at the edges of the radiograph or transmitted by areas of low density. Subdued lighting, rather than total darkness, is preferable in the viewing room. The room illumination must be such that there are no troublesome reflections from the surface of the film under examination.

PART 6
FILM HANDLING AND STORAGE

X-ray film should always be handled carefully to avoid physical strains, such as pressure, creasing, buckling, friction, etc. The normal pressure applied in a cassette to provide good contacts is not enough to damage the film. However, when films are loaded in semiflexible holders and external clamping devices are used, care should be taken to insure that this pressure is uniform. If a film holder bears against a few high spots, such as those which occur on an unground weld, the pressure may be great enough to produce desensitized areas in the radiograph. Precaution is particularly important when using envelope-packed films.

Crimp marks or marks resulting from contact with fingers that are moist or contaminated with processing chemicals can be avoided if large films are grasped by the edges and allowed to hang free. A convenient supply of clean towels is an incentive to dry the hands often and well. Use of envelope-packed films avoids these problems until the envelope is opened for processing. Thereafter, of course, the usual care must be taken.

Another important precaution is to avoid drawing film rapidly from cartons, exposure holders or cassettes. Such care will materially help to eliminate objectionable circular or treelike black markings in the radiograph, the results of static electric discharges.

The interleaving paper should be removed before the film is loaded between either lead or fluorescent screens. When using exposure holders in direct exposure techniques, however, the paper should be left on the film for the added protection that it provides. At high voltage, direct-exposure techniques are subject to the problems mentioned earlier: electrons emitted by the lead backing of the cassette or exposure holder may reach the film through the intervening paper or felt and record an image of this material on the film. This effect is avoided by the use of lead or fluorescent screens. In the radiography of light metals, direct-exposure techniques are the rule, and the paper folder should be left on interleaved film when loading it in the exposure holder.

The ends of a length of roll film factory-packed in a paper sleeve should be sealed in the darkroom with black pressure-sensitive tape. The tape should extend beyond the edges of the strip to provide a positive lighttight seal.

Identifying Radiographs

Because of their high absorption, lead numbers or letters affixed to the film holder or subject furnish a simple means of identifying radiographs. They may also be used as reference marks to determine the location of discontinuities within the specimen. Such markers can be conveniently fastened to the film holder or object with adhesive tape. A code can be devised to minimize the amount of lettering needed. Lead letters are commercially available in a variety of sizes and styles. The thickness of the chosen letters should be great enough so that their image is clearly visible on exposures with the most penetrating radiation routinely used. Under some circumstances it may be necessary to put the lead letters on a radiation-absorbing block so that their image will not be "burned out." The block should be considerably larger than the legend itself.

Shipping of Unprocessed Films

If unprocessed film is to be shipped, the package should be carefully and conspicuously labeled, indicating the contents, so that the package may be segregated from any radioactive materials. It should further be noted that customs inspection of shipments crossing international boundaries sometimes includes fluoroscopic inspection. To avoid damage from this cause, packages, personal baggage, and the like containing unprocessed film should be plainly marked, and the attention of inspectors drawn to their sensitive contents.

Storage of Unprocessed Film

X-ray Film Storage

With X-rays generated up to 200 kV, it is feasible to use storage compartments lined with a sufficient thickness of lead to protect the film. At higher kilovoltages, protection becomes increasingly difficult; film should be protected not only by the radiation barrier for protection of personnel but also by increased distance from the source.

At 100 kV, a 3 mm (0.125 in.) thickness of lead should normally be adequate to protect film stored in a room adjacent to the X-ray room if the film is not in the line of the direct beam. At 200 kV, the lead thickness should be increased to 6 mm (0.25 in.).

With million-volt X-rays, films should be stored beyond the concrete or other protective wall at a distance at least five times farther from the X-ray tube than the area occupied by personnel. The storage period should not exceed the times recommended by the manufacturer.

Medical X-ray films should be stored at approximately 12 times the distance of the personnel from the million-volt X-ray tube, for a total storage period not exceeding two weeks.

It should be noted that the shielding requirements for films given in National Bureau of Standards Handbook 76 "Medical X-Ray Protection Up to Three Million Volts" and National Bureau of Standards Handbook 93 "Safety Standard to Non-Medical X-Ray and Sealed Gamma-Ray Sources, Part 1 General" are *not* adequate to protect the faster types of X-ray films in storage.

Storage Near Gamma Rays

When radioactive material is not in use, the lead container in which it is stored helps provide protection for film. In many cases, however, the container for a gamma-ray source will not provide satisfactory protection to stored X-ray film. In such cases, the emitter and stored film should be separated by a sufficient distance to prevent fogging. The conditions for the safe storage of X-ray film in the vicinity of gamma-ray emitters are given in Tables 7 and 8.

The Tables show the necessary emitter-film distances and thicknesses of lead that should surround the various gamma-ray emitters to provide protection of stored film. These recommendations allow for a slight but harmless degree of fog on films when stored for the recommended periods. The lead thicknesses and distances in the Tables are considered the tolerable minimum.

To apply the cobalt 60 Table to radium, values for source strength should be multiplied by 1.6 to give the *grams* of radium which will have the same gamma-ray output (and will require the same lead protection). This Table can be extended to larger or smaller source sizes very easily. The half-value layer, in lead, for the gamma rays of radium or cobalt 60 is about 1.3 cm (0.5 in.). Therefore, if the source strength is doubled or halved, the lead protection should be increased or decreased by 1.3 cm (0.5 in.).

The Table can also be adapted to storage times longer than those given in the tabulation. If, for example, film is to be stored in the vicinity of cobalt 60 for twice the recommended time, the protection recommendations for a source *twice* as large as the actual source should be followed.

Iridium 192 has a high absorption for its own gamma radiation. This means that the external radiation from a large source is lower, per curie of activity, than that from a small source. Therefore, protection requirements for an iridium 192 source should be based on the radiation output, in terms of roentgens per hour at a known distance. The values of source strength, in curies, are merely a rough guide and should be used only if the radiation output of the source is unknown. Table 8 can be extended to sources having higher or lower radiation outputs than those listed. The half-value layer of iridium-192 radiation in lead is about 0.48 cm (0.19 in.). Therefore, if the radiation output is doubled or halved, the lead thicknesses should be respectively increased or decreased by 0.48 cm (0.19 in.).

Tables 7 and 8 are based on the storage of a particular amount of radioactive material in a single protective lead container. The problem of protecting film from gamma radiation becomes more complicated when the film is exposed to radiation from several sources, each in its own housing. Assume that a radiographic source is stored under the conditions required by Table 7 or 8 [for example, a 50-curie cobalt 60 source, in a 16.5 cm (6.5 in.) lead container 30.5 m (100 feet) from the film storage]. This combination of lead and distance would adequately protect the film from the gamma radiation for the storage times given in the Tables.

TABLE 7. Cobalt 60 Storage Conditions for Film Protection

Source Strength (in Curies)	1	5	10	25	50	100
Distance from Film Storage (in Feet)	Lead Surrounding Source* (in Inches)**					
25	5.5	7.0	7.5	8.0	8.5	9.0
50	4.5	6.0	6.5	7.0	7.5	8.0
100	3.5	5.0	5.5	6.0	6.5	7.0
200	2.5	4.0	4.5	5.0	5.5	6.0
400	1.5	3.0	3.5	4.0	4.5	5.0

*Lead thicknesses rounded off to nearest half-value layer.
**To convert to centimeters, multiply by 2.54.

However, if a second source, identical with the first and in a similar container, is stored alongside the first, the radiation level at the film will be doubled. Obviously, then, if there are several sources in separate containers, the lead protection around each, or the distance from the sources to the film, must be increased over the values given in these Tables.

The simplest method of determining the film protection required for several sources is as follows: multiply the actual, total strength of the source in each container by the number of separate containers; then use these assumed source strengths to choose from the Tables the lead thicknesses and distances. Apply the values so found for the protection around each of the actual sources. For instance, assume that in a particular radiographic department there are two source containers, both at 30 m (100 feet) from the film storage area. One container holds 50 curies of cobalt 60 and the other an iridium 192 source whose output is 5 roentgens per hour at 1 meter (5 rhm). Since there are two sources, the 50 curies of cobalt 60 will require the protection needed for a single 100-curie source, and the iridium 192 source will need the protection of a single source whose output is 10 rhm. The thicknesses of lead needed are shown to be 18 cm (7.0 in.) for the 50 curies of cobalt 60 (Table 7) and 4.3 cm (1.7 in.) for the iridium 192, whose emission is 5 rhm (Table 8).

This method of determining the protective requirements for multiple sources is based on two facts. First, if several sources, say four, simultaneously irradiate stored film, the exposure contributed by each must be only one-quarter that permissible if the source were acting alone; in other words, the gamma-ray attenuation must be increased by a factor of four. Second, any combination of source strength, lead thickness, and distance given in Tables 7 and 8 results in the same gamma-ray dose rate (about 0.017 milliroentgens per hour) being delivered to the film location. Since the goal, in this example, is to reduce each source's radiation to one-quarter its original value, Tables 7 and 8 may still be used by simply setting up conditions for a source four times the actual strength.

Heat, Humidity and Fumes

During packaging, most X-ray films are enclosed in a moistureproof container which is hermetically sealed and then boxed. As long as the seal is unbroken, the film is protected against moisture and fumes. Because of the deleterious effect of heat, all films should be stored in a cool, dry place and ordered in such quantities that the supply of film on hand is renewed frequently.

Under no circumstances should opened boxes of film be left in a chemical storage room or in any location where there is leakage of illuminating gas

TABLE 8. Iridium 192 Storage Conditions for Film Protection

Output R/hr at 1 Meter	1	2	5	10	25	50
Source Strength (in Curies)	2	5	12.5	25	75	150
Distance from Film Storage (in Feet)	Lead Surrounding Source* (in Inches)**					
25	1.70	1.85	2.15	2.35	2.50	2.70
50	1.35	1.50	1.85	2.00	2.15	2.35
100	1.00	1.15	1.50	1.70	1.85	2.00
200	0.70	0.85	1.15	1.35	1.50	1.70
400	0.35	0.50	0.85	1.00	1.15	1.35

*Lead thicknesses rounded off to nearest half-value layer.
**To convert to centimeters, multiply by 2.54.

or any other types of gasses, or where there is a possibility of contact with formalin vapors, hydrogen sulfide, ammonia or hydrogen peroxide.

Packages of sheet film should be stored on edge, with the plane of the film in a vertical position. They should *not* be stacked with the boxes, horizontally, because the film in the bottom boxes can be damaged by the impact of stacking or by the weight of boxes above. In addition, storage of the boxes on edge makes it simpler to rotate the inventory and use the older films first.

Storage of Exposed and Processed Film

Archival Keeping Quality

Archival storage is a term commonly used to describe the keeping quality of X-ray film. It is defined by the American National Standards Institute (ANSI) as "those storage conditions suitable for the preservation of photographic film having permanent value." Archival storage is not defined *in years* by ANSI, but in terms of the thiosulfate content (residual fixer) permissible for storage of radiographs.

Although many factors affect the storage life of radiographs, one of the most important is the residual thiosulfate left in the radiograph after processing and drying. Determined by the methylene blue test, the maximum level is 2 micrograms per square centimeter on each side of coarse-grain X-ray films. For short-term storage requirements, the residual thiosulfate content can be at a higher level, but this level is not specified by ANSI.

Washing of the film after development and fixing, therefore, is most important. The methylene blue test and silver densitometric test are laboratory procedures performed on clear areas of the processed film.

Storage Suggestions

Regardless of the length of time a radiograph is to be kept, these suggestions should be followed to provide for maximum stability of the radiographic image.

1. Avoid storage in the presence of chemical fumes.

2. Avoid short-term cycling of temperature and humidity.

3. Place each radiograph in its own folder to prevent possible chemical contamination by the glue used in making the storage envelope (negative preserver). Several radiographs may be stored in a single storage envelope if each is in its own interleaving folder.

4. Never store unprotected radiographs in bright light or sunlight.

5. Avoid pressure damage caused by stacking a large number of radiographs in a single pile or by forcing more radiographs than can comfortably fit into a single file drawer or shelf.

The following ANSI documents contain recommendations useful in determining storage conditions, and may be obtained by contacting The American National Standards Institute, 1430 Broadway, New York, NY 10018.

1. ANSI PH1.41: *Specifications for Photographic Film for Archival Records, Silver Gelatin Type on Polyester Base.*
2. ANSI PH1.43: *Practice for Storage of Processed Safety Photographic Film.*
3. ANSI PH4.20: *Requirements for Photographic Filing Enclosures for Storing Processed Photographic Films, Plates, and Papers.*
4. ANSI PH4.8: *Methylene Blue Method for Measuring Thiosulfate and Silver Densitometric Method for Measuring Residual Chemicals in Film, Plates and Papers.*
5. ANSI N45.2.9: *Quality Assurance Records for Nuclear Power Plants, Requirements for Collection, Storage and Maintenance of.*

PART 7

INTRODUCTION TO PAPER RADIOGRAPHY

To many radiographers and interpreters the words "radiograph" and "film" are synonymous. However, a combination of factors, among them recurring silver shortages and rising costs of other nondestructive testing methods, has prompted increasing interest in the use of paper in industrial radiography. New developments in paper, screens and processing techniques have resulted from the realization that radiographs on paper have distinct advantages to offer the user who will consider them, not in the context of X-ray film, but as another recording medium, one which is to be viewed in an entirely different way.

Advantages of Paper Radiographs

What are some of the advantages of using paper in industrial radiography? For one, *rapid access.* A damp-dry radiograph can be put in the interpreter's hands in as little as 10 seconds after exposure. Moreover, radiographic paper, plus intensifying screens, plus proper exposure, equals *good image quality.* With direct exposure, the image has acceptable subject contrast combined with wide latitude. *Convenience and economy* also enter the picture. The paper processor is portable, requires no plumbing connections and, in addition, has a low operating cost.

Applications for Paper Radiography

Where can you use paper radiographs in your nondestructive testing program? The applications are numerous. For instance, there are many stages in production which may require radiographic inspection *before* code or specification radiographs are made. This is a primary use for paper radiography. Other applications exist in foundries where in-process X-ray procedures are used to monitor practices of a gating and risering system; checking core positions in wax patterns; detecting shrinkage flaws, porosity, dross, or cavities in castings; and monitoring root passes and subsequent weld passes for a variety of flaws.

Radiographs on paper also find application in checking circuit boards for proper assembly and absence of solder balls. In aircraft maintenance, paper radiographs can be used to inspect for water in honeycomb, foreign material in oil pumps, and for general survey work. Some other applications include on-site checking of pipelines, pressure vessels and weldments; inspection in industries such as food processing, wood products, tires, seeds, and titanium reprocessing; bomb detection; and many types of survey radiography.

Viewing Paper Radiographs

A correctly exposed, properly processed radiograph on paper is only part of the story (a detailed discussion of paper exposure and processing occurs in Part 8 of this Section). To be useful in providing information, the radiograph must be viewed, and viewing radiographs on the light-reflecting paper is entirely different from viewing radiographs on light-transmitting film. It is almost immediately apparent that some of the familiar methods of measurement and interpretation applicable to film are not relevant to the interpretation of paper radiographs.

Density—Transmission vs. Reflection

When electromagnetic radiation in the form of light, X-rays, or gamma rays reacts with the sensitive emulsion of X-ray film or radiographic paper, the emulsion will show a blackening after it has been processed. The degree of blackening is defined as *density.* Up to this point, radiographic paper acts identically to film, but beyond this point, differences appear.

Density Measurement

The density on transparent-based film is known as transmission density, D_T, and is defined as the logarithm of the ratio of the incident light intensity on the radiograph, I_O (from the illuminator), to the light intensity transmitted through the radiograph, I_T. The formula is:

$$D_T = \log \frac{I_O}{I_T} \quad \text{(Eq. 16)}$$

Since this formula applies only to light-transmitting images, it cannot be applied to an opaque-based imaging material such as radiographic paper. Therefore, a slightly different means of measuring density is necessary, and this is called *reflection density, D_R*. Reflection density is defined as the logarithm of the ratio of incident light intensity, I_O, to the reflected light intensity, I_R, from the image area. This formula is:

$$D_R = \log \frac{I_O}{I_R} \quad \text{(Eq. 17)}$$

So although the formula appears to be quite similar to that of transmission density, in practical application, reflection density measures the light reflected from the radiograph, not that which passes through. For example, reflection densities are measured by a reflection densitometer, and the familiar densitometer for measuring transmitted densities cannot be used.

Comparable Densities—Paper and Film

Interpreters could easily be led astray at this point by becoming involved in the purely objective relationship between reflection density and transmission density. They may theorize, for example, that under a given set of viewing conditions, a reflection density of 0.7 appears similar to a transmission density of 2.0. A more valuable relationship, however, is knowing which densities contain the same information if X-ray film and radiographic paper are used to record the same image.

A radiographic image on paper will contain the same important image details as a radiograph of the same subject on film (see Fig. 40). These details will be modified in density (and possibly in contrast) because the response is fundamentally different.

It must be recognized, of course, that the *total range* of recorded information will be less on a paper radiograph because the reflection density scale is shorter than its X-ray film counterpart. For example, a reflection density of 2.0 on a paper radiograph is so black that detail is completely obscured.

In addition, because of the opaque nature of the paper base, the method for viewing reflected densities of images on paper is different from the method for viewing transmitted densities. Although these differences exist and must be recognized, the similarities in practical usage between film and paper radiographs are striking. *Good practice indicates that the exposure given to radiographic paper can be adjusted until the necessary and desirable details of the image are distributed along the available density scale of the paper, within the constraints of optimum reflection viewing.*

FIGURE 40. The penetrameter is clearly visible in this paper radiograph of a weld in 0.64 cm (0.25 in.) steel. Exposure made at 120 kV, 10 mA, for 20 seconds at a focus-film distance of 90 cm (36 in.)

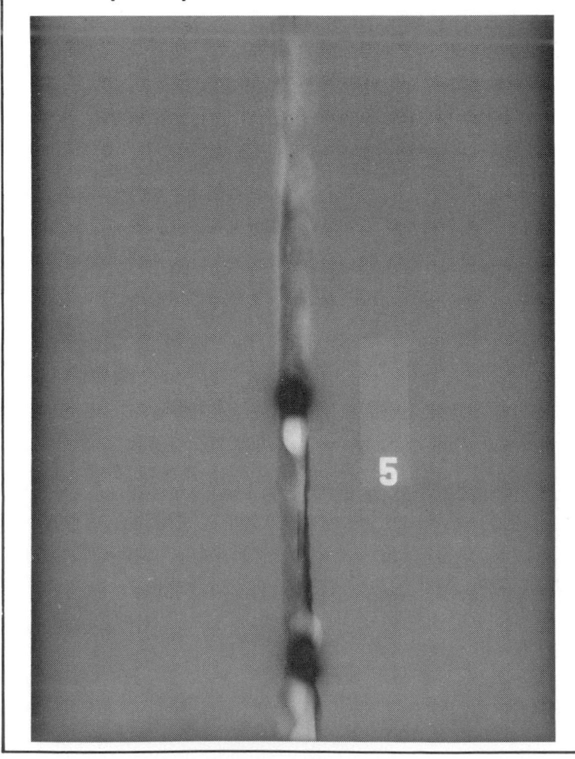

If this is done correctly, the important details will tend to be found in the mid-scale of subjective brightness provided by the density scale of the paper. This is strikingly similar to that of a film radiograph in which the details of a good image tend to be centered around the middle of the density scale (usually about 2.0). The center point, or aim point, then, is a significant factor for visualization of detail for both paper and film radiographs—even though the aim point may be a different value, and the densities may be reflected or transmitted.

Interpreting Paper Radiographs

Whether produced on film or on paper, the radiograph containing useful information must be viewed by an observer for the purpose of interpretation. The viewing process, therefore, is often a subjective interpretation based on the variety of densities presented in the radiograph. To perform this function, the eye must be capable of receiving the information contained in the image. Judgments, likewise, cannot be made if the details cannot be seen.

Viewing conditions are of utmost importance in the interpretation of radiographs. As a general rule, extraneous reflections from, and shadows over, the area of interest must be avoided, and the general room illumination should be such that it does not impose unnecessary eyestrain on the interpreter.

When following these guidelines in viewing *film* radiographs, the light transmitted through the radiograph should be sufficient only to reveal the recorded details. If the light is too bright, it will be blinding; if too dim, the details cannot be seen. The general room illumination should be at approximately the same level as that of the light intensity transmitted through the radiograph, to avoid shadows, reflections and undue eyestrain.

The natural tendency is to view radiographs on paper, like a photograph, in normal available light. For simple cursory examination this can be done, but since available light might be anything from bright sunlight to a single, dim, light bulb, some guidelines are necessary. It has been found from practical experience that radiographic sensitivity can be greatly enhanced if the following guidelines for viewing paper are observed.

1. As noted in the general rule, all extraneous shadows or reflections in the viewing area of the radiograph must be avoided. In fact, a darkened area, minimizing ambient lighting, is desirable.
2. Since radiographs on paper must be viewed in reflected light, several sources of reflected light have been used successfully. One method is the use of specular light (light focused from a mirror-like reflector), directed at an angle of approximately 30° to the surface of the radiograph, from the viewer's side so that reflected light does not bother the eyes. Light which comes from a slide projector is specular light. Other sources are the familiar high-intensity reading light, or a spotlight.

   Another type of light which has been found to be very effective is a circular magnifying glass illuminated around the periphery with a circular fluorescent bulb. When using this form of illumination, the paper radiograph should be inclined at an oblique angle to the light to produce the same specular lighting just discussed. These devices, found in drafting rooms as well as medical examining rooms, are usually mounted on some sort of adjustable stand and have the advantage of low power magnification—on the order of 3X to 5X. The magnification should be such that it does not distort the image. The magnifying glass, incidentally, will emphasize the fact that the graininess characteristics of paper radiographs are minimal.

Reflection density is different from transmission density. The difference is important in viewing and interpreting radiographs on paper, but presents no difficulty in procedure or visualization.

PART 8
INDUSTRIAL RADIOGRAPHY ON RADIOGRAPHIC PAPER

Introduction

This investigation was performed in order to study the quality of radiographic paper, after a review had been made of the rather scarce literature on the subject. Characteristic curves for papers* exposed with different intensifying screens** (in the low and intermediate voltage ranges) are reproduced. The relative speed, contrast and exposure latitude were calculated from these curves. The sharpness of the radiographic image as well as the influence of processing on speed and contrast are also discussed.

Characteristic Curves for Paper Radiography

The sensitometric properties of a radiographic material depend on the quality of the radiation reaching it. Taking this into account, characteristic curves were produced using radiation of a quality similar to that used in routine radiography (three X-ray machines of 50, 180 and 300 kV were employed, as well as all combinations of paper and screens). Note: throughout this text, data given for 50 kV and below are from a constant potential machine; above 50 kV, data are for self-rectified machines in peak kilovolts.

Low Voltage Range

To have the same radiation quality reaching the test papers as during actual radiography of a part (in this case, a fuel plate), a filter consisting of one of the parts (fuel plates) was placed at the X-ray tube window. Thus it produced no image.

The radiographic paper was exposed in rigid aluminum cassettes. All exposures were made at a focus-film distance (FFD) of 1 m; the resulting densities were plotted as a function of logarithm of exposure in milliampere-minutes (mAmin). See Fig. 41.

FIGURE 41. Characteristic curves of Paper A at 30 to 50 kV. Exposures made with X0, F1 and F2 screens; "0" indicates exposure without screen.

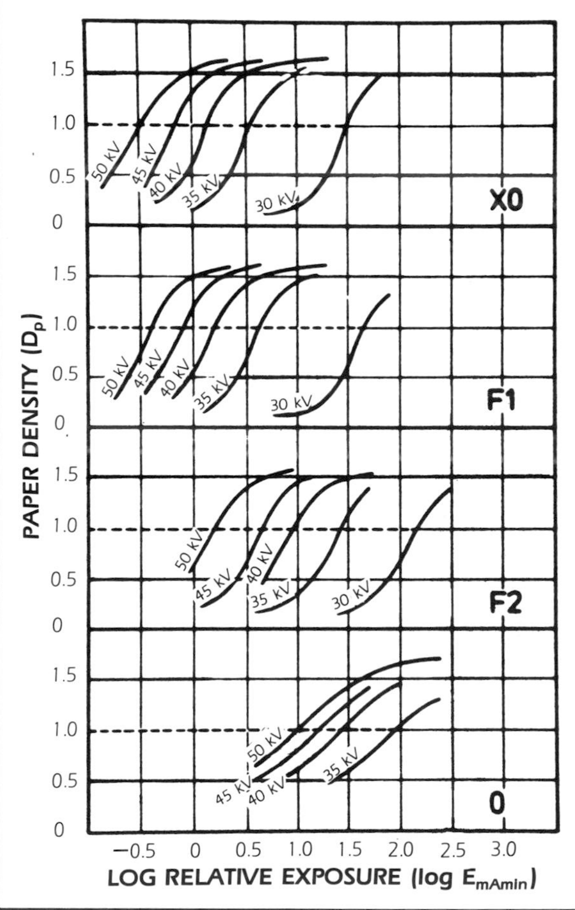

*In this Handbook Section, radiographic *Paper A* is KODAK INDUSTREX Instant 600 Paper; *Paper B* is KODAK INDUSTREX Rapid 620 Paper; and *Paper C* is KODAK INDUSTREX Rapid 700 Paper.

***Screen F1* is the KODAK INDUSTREX Intensifying Screen F1 ($BaSrSO_4$:Eu); high-speed, high-contrast recording capability for short exposures. *Screen F2* is the KODAK INDUSTREX Intensifying Screen F2 ($BaPbSO_4$:Eu); lower contrast and smaller intensification factor; yellow dye added for higher resolution image. *Screen XO* is an intensifying screen with performance characteristics similar to those of Screen F1.

Intermediate Voltage Range

To be able to study the sensitometric properties of radiographic paper at this range, characteristic curves were produced at 100 kV using a 30 mm aluminum filter at the X-ray tube window. It had been suggested that better quality paper radiographs could be obtained in this voltage range if a thin lead (Pb) filter were used on the cassette. This proved to be correct, and a 0.05 mm Pb filter was used both for the production of characteristic curves (see Fig. 42), as well as for routine radiography.

Relative Speed

The characteristic curves were produced by giving the optical density (D) as a function of the logarithm of exposures (in mAmin) at different voltages for a constant FFD of 1 m. From the curves, the relative paper speed may be derived by comparing the exposures necessary to obtain a constant density at each of the kilovoltages. For the sake of such comparison, a paper density of $D_p = 1.0$ was chosen. Table 9 lists the exposures necessary to reach $D_p = 1.0$ for different kilovoltages and paper-screen combinations, in the low voltage range. Table 10 is for the intermediate voltage range.

From these tables, *intensification factors* can be calculated for the different combinations of paper and screens. The factors are given in Table 11, as a quotient of the exposures needed to reach $D_p = 1.0$ for paper exposed without and with intensifying screens.

Contrast

Paper contrast (γ) can be figured from the characteristic curves, by measuring the angle (α) of the tangent to the curve (shown in Fig. 43).

Paper contrasts were calculated at different densities and are shown in Figs. 44 and 45.

As can be seen from the contrast curves, the maximum contrast is reached for paper densities between 0.9 and 1.1.

Exposure Latitude

Exposure latitude is the relation between the acceptable maximum and the acceptable minimum exposures for a specific radiographic paper or film. For both X-ray film and radiographic paper, the minimum exposure is limited to a point below which the radiographic quality is unacceptable. Density $D_{min} = 0.5$ can be considered as minimum density for paper radiographs.

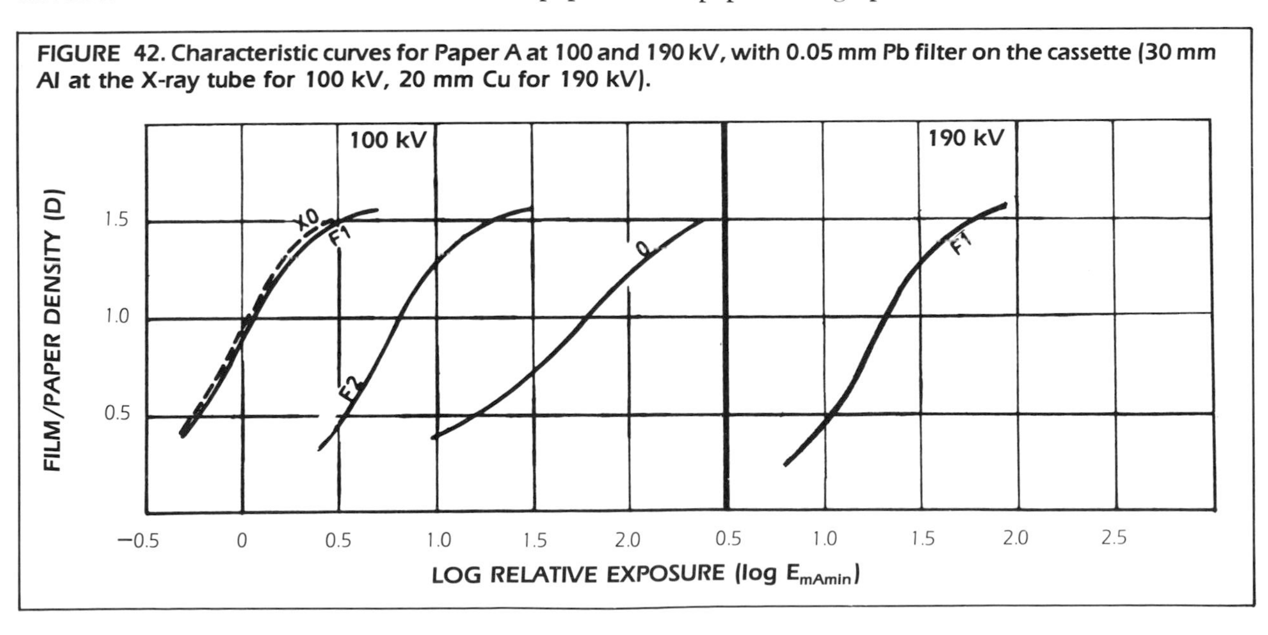

FIGURE 42. Characteristic curves for Paper A at 100 and 190 kV, with 0.05 mm Pb filter on the cassette (30 mm Al at the X-ray tube for 100 kV, 20 mm Cu for 190 kV).

The contrast of direct-exposure industrial X-ray film increases with its density; therefore, the maximum density will be limited only by the practical possibilities of viewing high film densities.

For radiographic paper, contrast decreases beyond a certain density, D_{max} = 1.3 can be considered the upper limit for paper density.

In calculating exposure latitude, densities between 0.5 and 1.3 are taken into consideration.

Fig. 46 shows how exposure latitude was calculated for a very specialized, low-energy technique: papers exposed at 10 kV, without intensifying screens. The results of similar calculations for various paper-screen combinations are shown in Tables 12 and 13.

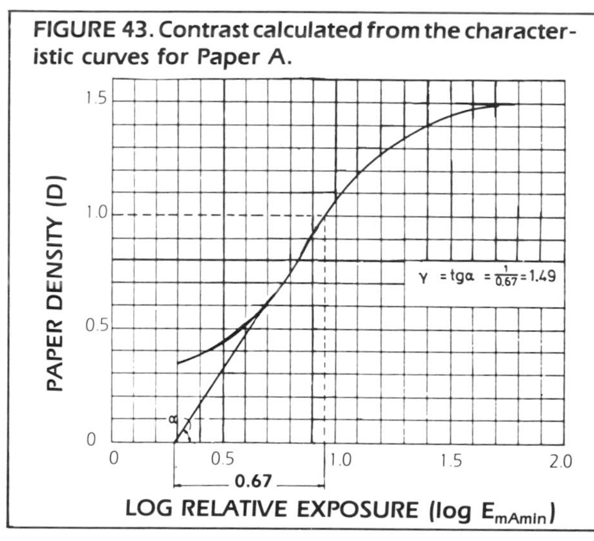

FIGURE 43. Contrast calculated from the characteristic curves for Paper A.

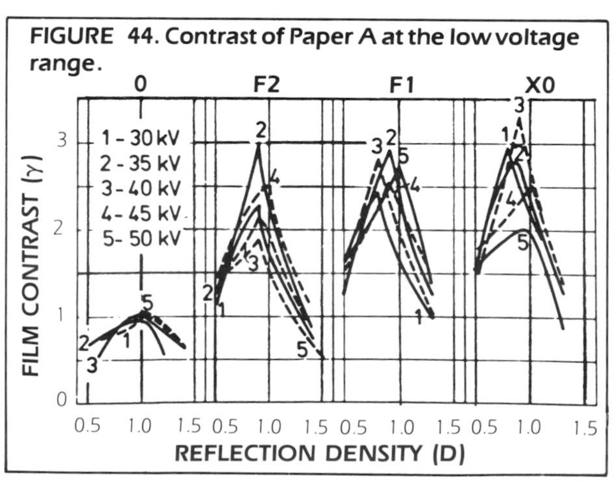

FIGURE 44. Contrast of Paper A at the low voltage range.

TABLE 9. Exposures in milliampere-minutes necessary to reach D_p = 1.0 in the low voltage range (FFD = 1 m). Paper A used at indicated voltages.

X-ray Energy	No screen	Intensifying Screen X0	F1	F2
30 kV	-	31.62	45.71	144.54
35 kV	39.81	3.47	4.07	26.30
40 kV	27.54	1.38	1.58	8.71
45 kV	15.85	0.71	0.83	4.47
50 kV	10.00	0.33	0.42	1.58

TABLE 10. Exposures in milliampere-minutes necessary to reach D_p = 1.0 in the intermediate voltage range (FFD = 1 m). Paper A used at indicated voltages.

X-ray Energy	No screen	Intensifying Screen X0	F1	F2
100 kV	60.26	1.10	1.01	6.46
190 kV	-	-	21.88	-
150 kV	-	0.10	-	-

TABLE 11. Intensification factors for different paper and screen combinations. Paper A used at indicated voltages.

X-ray Energy	Intensifying Screen X0	F1	F2
30 kV	-	-	-
35 kV	11.47	9.81	1.51
40 kV	19.96	17.43	3.16
45 kV	22.32	19.10	3.55
50 kV	30.30	23.81	6.33
100 kV	10.44	11.37	1.18

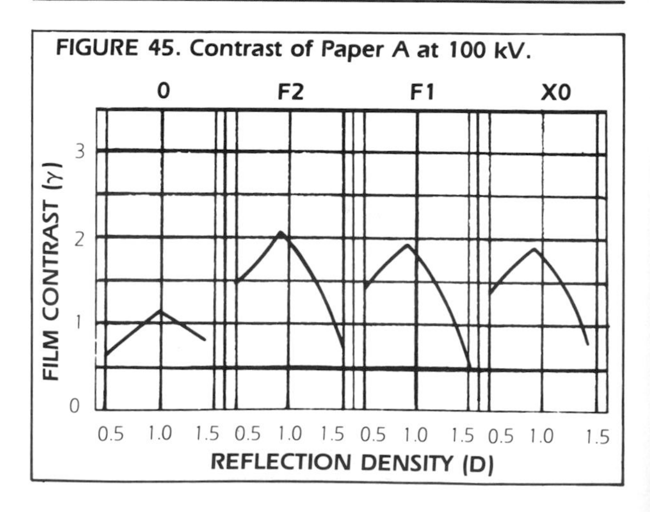

FIGURE 45. Contrast of Paper A at 100 kV.

TABLE 12. Exposure latitude (EL) and acceptable densities (D) in the low voltage range.

kV		30		35		40		45		50		10	
Paper	Screen	D	EL	D	EL	D	EL	D	EL	D	EL	D	EL
A	0	-	-	0.5-1.3	11.2	0.6-1.3	5.8	0.5-1.3	9.1	0.7-1.3	4.7	0.5-1.3	4.5
A	F2	0.5-1.3	3.3	0.5-1.3	3.0	0.5-1.3	3.1	0.5-1.3	2.8	0.6-1.3	5.0	-	-
A	F1	0.5-1.3	2.9	0.5-1.3	2.6	0.5-1.3	2.7	0.5-1.3	2.6	0.5-1.3	2.5	-	-
A	X0	0.5-1.3	2.3	0.5-1.3	2.6	0.5-1.3	2.2	0.5-1.3	2.4	0.5-1.3	3.2	-	-

FIGURE 46. Calculation of exposure latitude from the characteristic curves for Paper A.

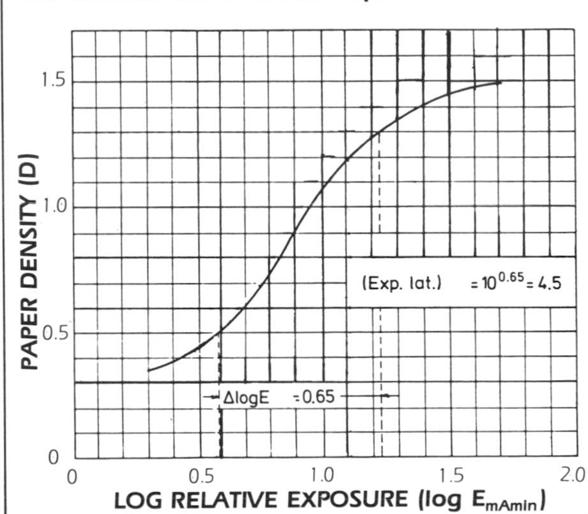

FIGURE 47. Paper density under different steps of the Al step wedge (exposed at 50 kV).

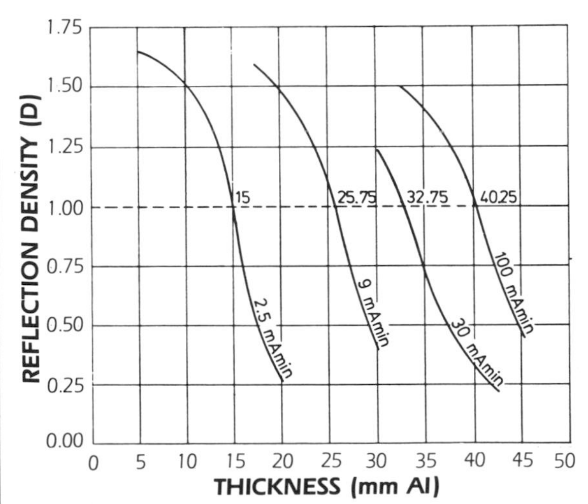

TABLE 13. Exposure latitude (EL) and acceptable densities (D) in the intermediate voltage range.

kV		100		190		150	
Paper	Screen	D	EL	D	EL	D	EL
A	0	0.5-1.3	7.9	-	-	-	-
A	F2	0.5-1.3	3.0	-	-	-	-
A	F1	0.5-1.3	3.2	0.5-1.3	3.2	-	-
A	X0	0.5-1.3	3.0	-	-	0.5-1.3	2.8

Exposure Charts

Radiograhic paper can be used with advantage in many fields of radiography, including the radiographic control of aluminum (Al) or steel (Fe) objects. Because objects made of these materials may differ in thickness, it is necessary to produce adequate exposure charts in order to avoid the time consuming trial-and-error method for finding adequate exposure conditions.

Aluminum Exposure Charts

Aluminum step wedges were exposed at different kilovoltages using different exposures (mAmin) at the same FFD = 1 m. For each setting and exposure, optical densities under all the Al steps were measured and plotted as a function of the step thickness (see Fig. 47).

Table 14 gives an example of such density readings (from which the curves in Fig. 47 were plotted). The readings were taken under different steps of an Al step wedge exposed at 50 kV (on Paper A) with X0 screen at FFD = 1 m.

Exposure charts should be plotted for paper density D_p = 1.0; intersections of the different thickness-density curves are tabulated at the bottom of Table 14. From these data, exposure charts for 50 kV can be produced in semilogarithmic scale. Repeating the procedure for other kilovoltages, exposure charts were produced for different paper-screen combinations.

For aluminum, exposure charts were produced for thicknesses up to 35 mm using soft X-rays and up to 70 mm using X-rays in the intermediate voltage range.

In the low voltage range, 25-50 kV were used in 5 kV increments (see Figs. 48-51).

In the intermediate voltage range, a 180 kV X-ray machine was used to produce exposure charts for aluminum, at FFD = 1 m and D_p = 1.0. A 0.05 mm lead filter was used on the cassette (see Figs. 52-54).

Steel Exposure Charts

Using a steel step wedge and the technique described above, exposure charts were produced for steel up to 70 mm using a 300 kV machine with voltages from 100-300 kV in 20 kV increments. A 0.05 lead filter was placed on the cassette; FFD = 1 m for D_p = 1.0 (Figs. 55 and 56.).

FIGURE 48. Exposure chart for Al at 50 kV.

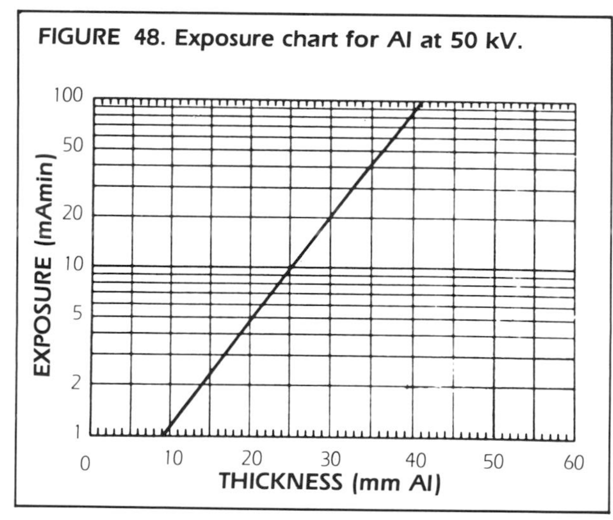

FIGURE 49. Exposure charts for Al for Paper A exposed with X0 screen.

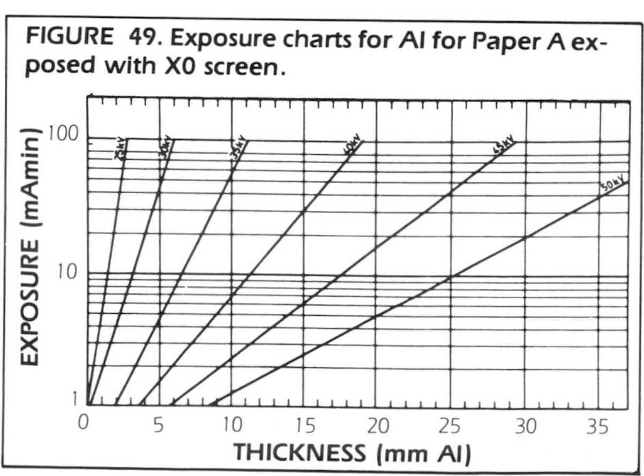

TABLE 14. Paper densities under different Al steps.

Step Wedge	Exposure in mAmin			
Thickness (mm)	2.5	9	30	100
5	1.64			
7.5	1.60			
10	1.53			
12.5	1.34			
15	0.95			
17.5	0.54	1.59		
20	0.28	1.50		
22.5		1.36		
25		1.07		
27.5		0.70		
30		0.40	1.23	
32.5			1.03	1.49
35			0.72	1.42
37.5			0.48	1.29
40			0.29	1.02
42.5			0.21	0.70
45				0.46
mm of Al for D_p = 1.0	15.00	25.75	32.75	40.25

FIGURE 50. Exposure charts for Al, Paper A with F1 screen.

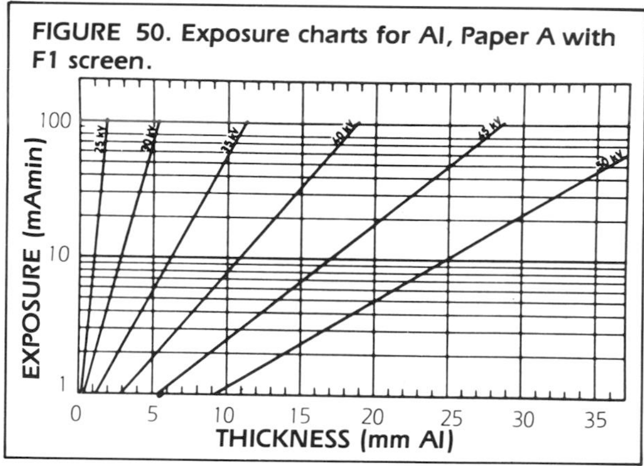

FIGURE 51. Exposure charts for Al, Paper A with F2 screen.

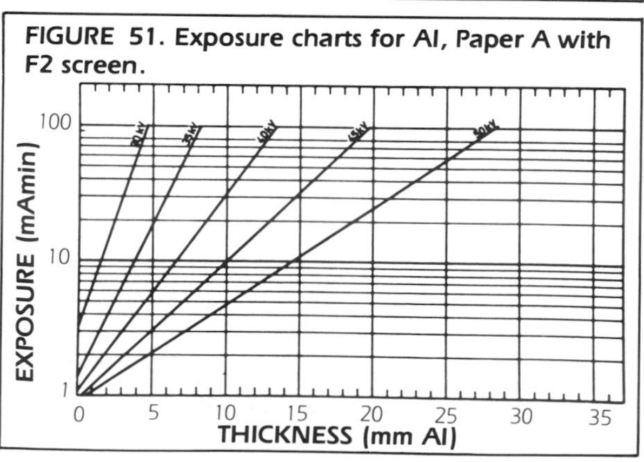

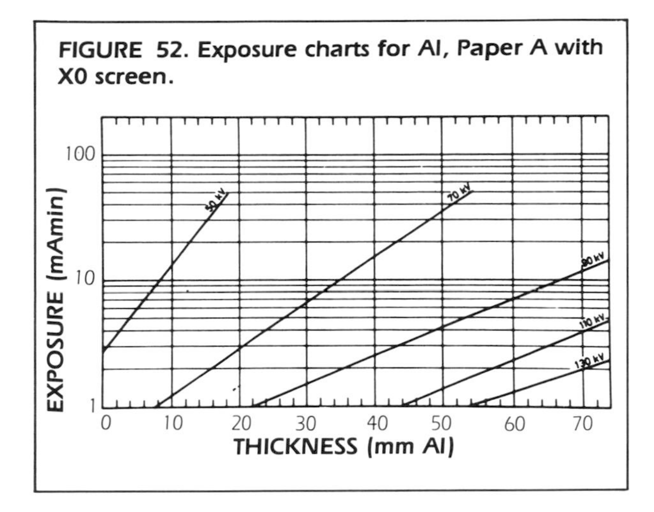

FIGURE 52. Exposure charts for Al, Paper A with X0 screen.

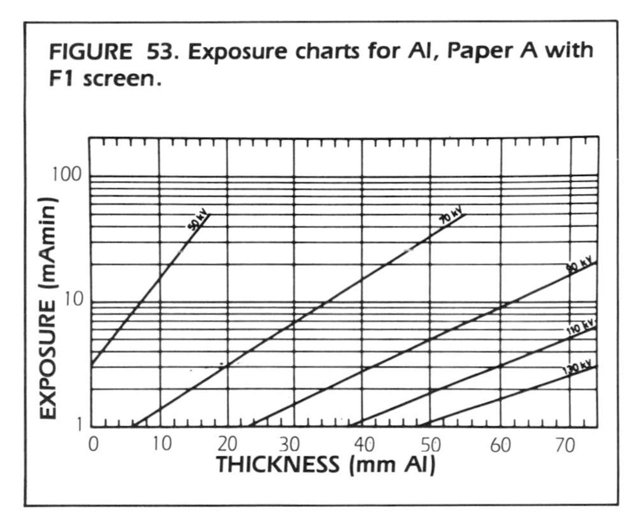

FIGURE 53. Exposure charts for Al, Paper A with F1 screen.

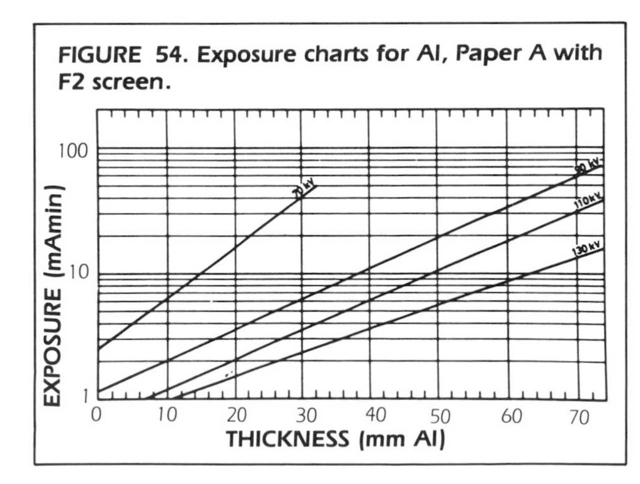

FIGURE 54. Exposure charts for Al, Paper A with F2 screen.

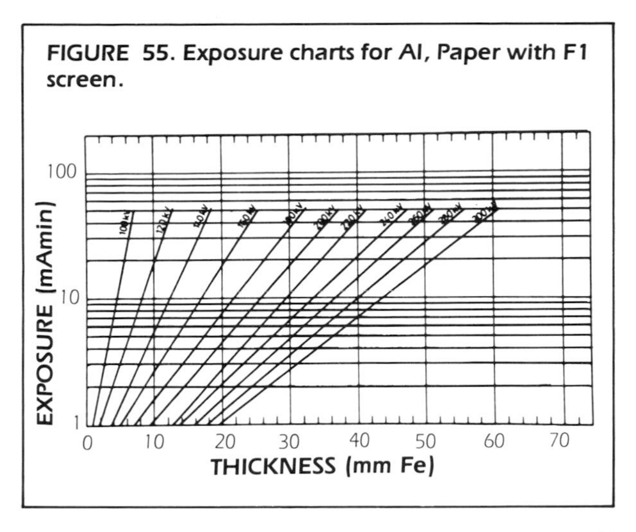

FIGURE 55. Exposure charts for Al, Paper with F1 screen.

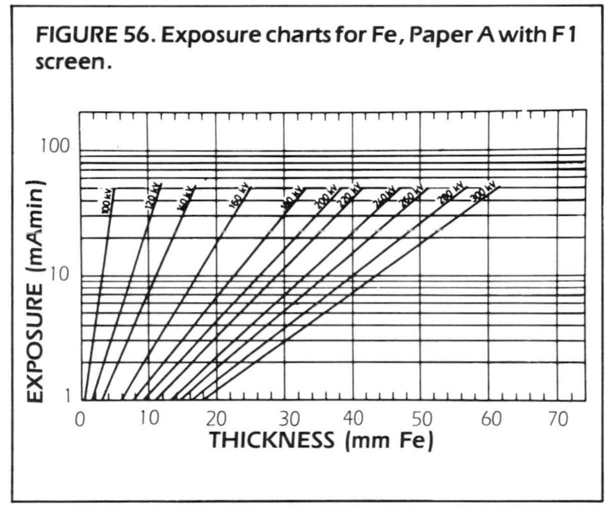

FIGURE 56. Exposure charts for Fe, Paper A with F1 screen.

Exposure Information for Different Paper Speeds

The fundamental procedures established in the earlier parts of this Section may be applied to other papers of different speeds (Papers B and C).

Exposure and specimen information are included in the individual figure headings. Note that F1 and F2 screens were used with both paper types.

Paper B was processed using standard stabilization equipment; Paper C was hand-processed in regular X-ray film processing tanks.

Characteristic Curves

FIGURE 57. Characteristic curves of Paper B at 25-50 kV. Exposures made with F1 and F2 screens; "0" indicates exposure without screen. Also listed are the thicknesses of the Al filters used at the X-ray tube.

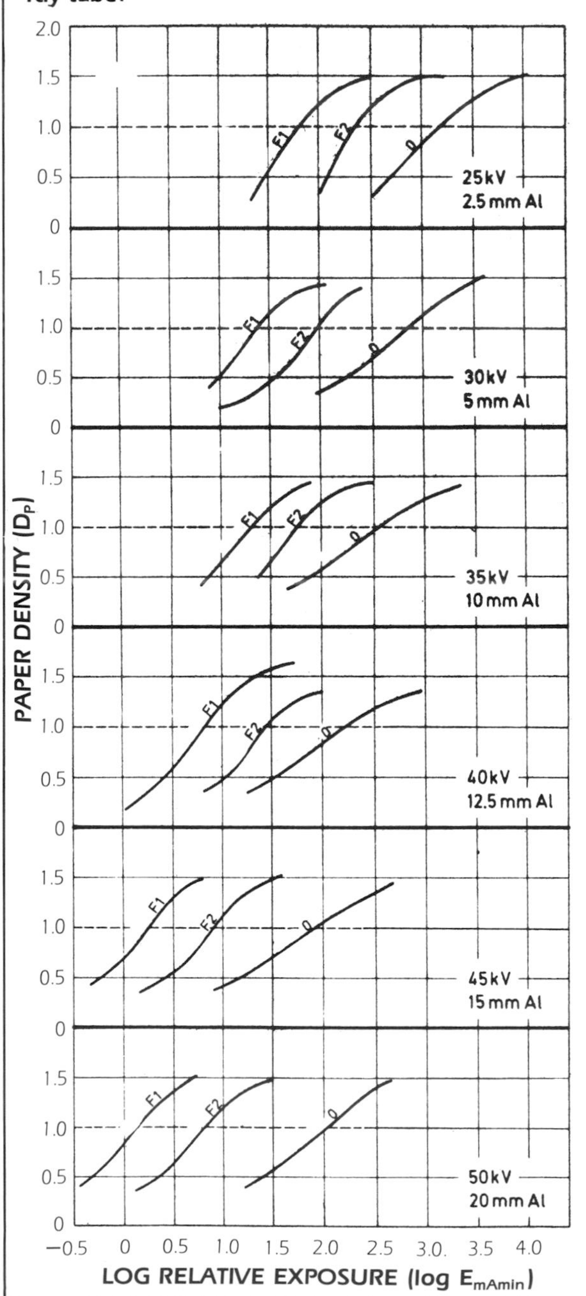

FIGURE 58. Characteristic curves of Paper C at 25-50 kV. Exposures made with F1 and F2 screens; "0" indicates exposure made without screen. Also listed are the thicknesses of the Al filters used at the X-ray tube.

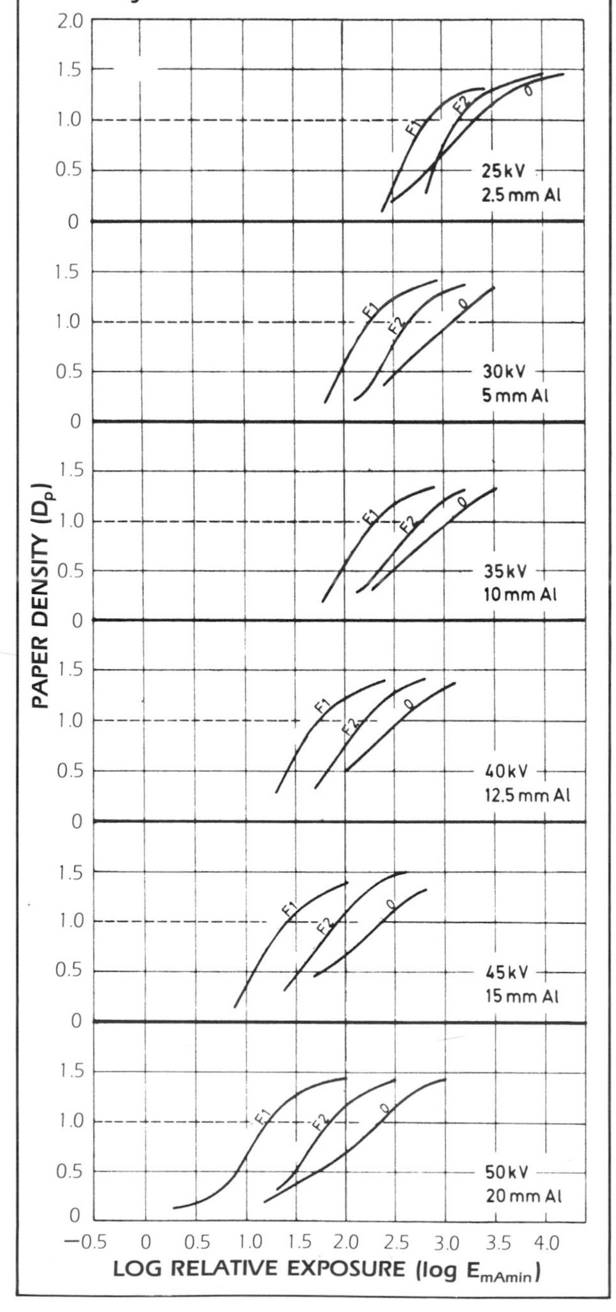

FIGURE 59. Characteristic curves of Paper B at (a) 100 kV, using a 180 kV X-ray machine, with a 30 mm aluminum filter, and (b) 190 kV, using a 300 kV machine with a 20 mm copper filter. Exposures made with F1 and F2 screens; "0" indicates exposure without screen. For comparison, the characteristic curve of KODAK INDUSTREX D X-ray film is also given.

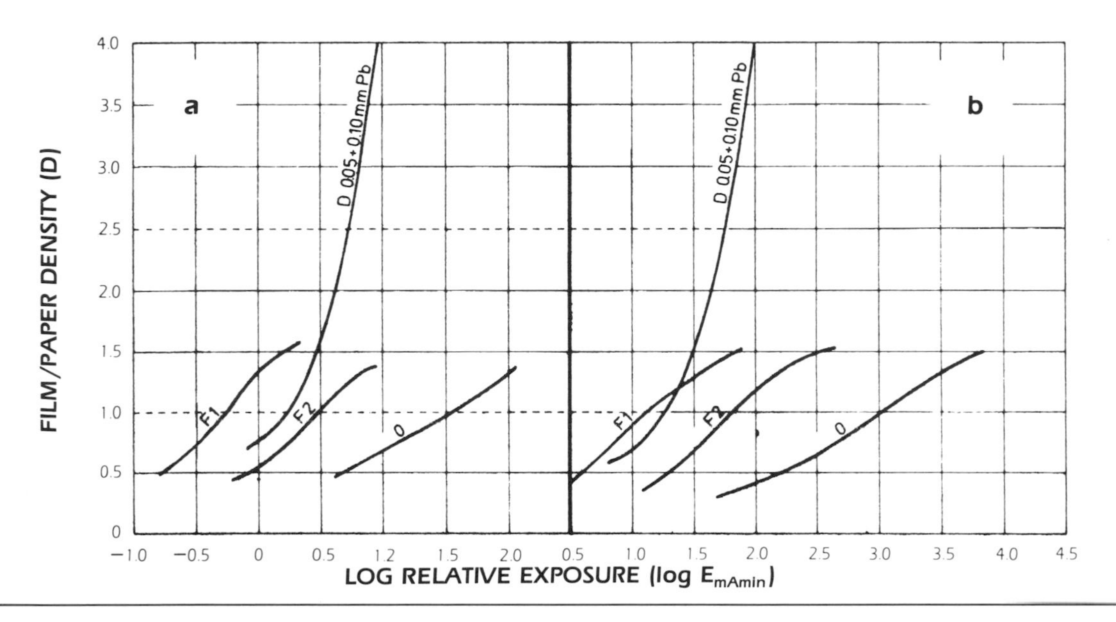

FIGURE 60. Characteristic curves of Paper Cat (a) 100 kV, using a 180 kV X-ray machine with a 30 mm aluminum filter, and (b) 190 kV, using a 300 kV machine with a 20 mm copper filter. Exposures made with F1 and F2 screens; "0" indicates exposure made without screen. For comparison, the characteristic curve of KODAK INDUSTREX D X-ray film is also given.

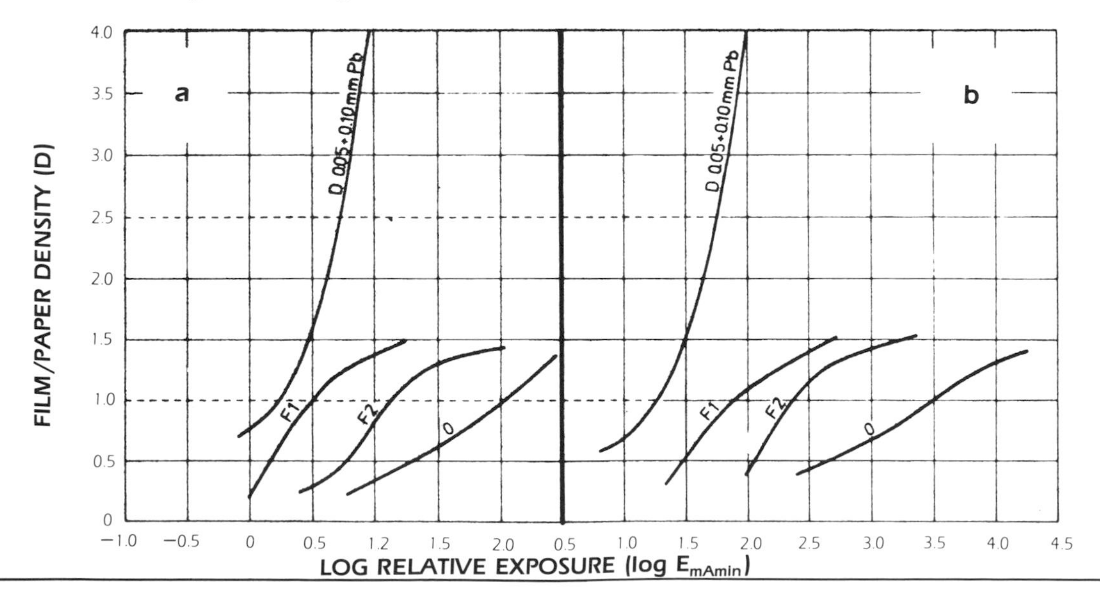

Exposure Charts

FIGURE 61. Exposure chart for aluminum: Paper B with F1 screen (solid lines) and Paper B with F2 screen (dashed lines). Exposures made at indicated voltages at FFD = 1 m, D_p = 1.0, using a 0.05 mm Pb filter.

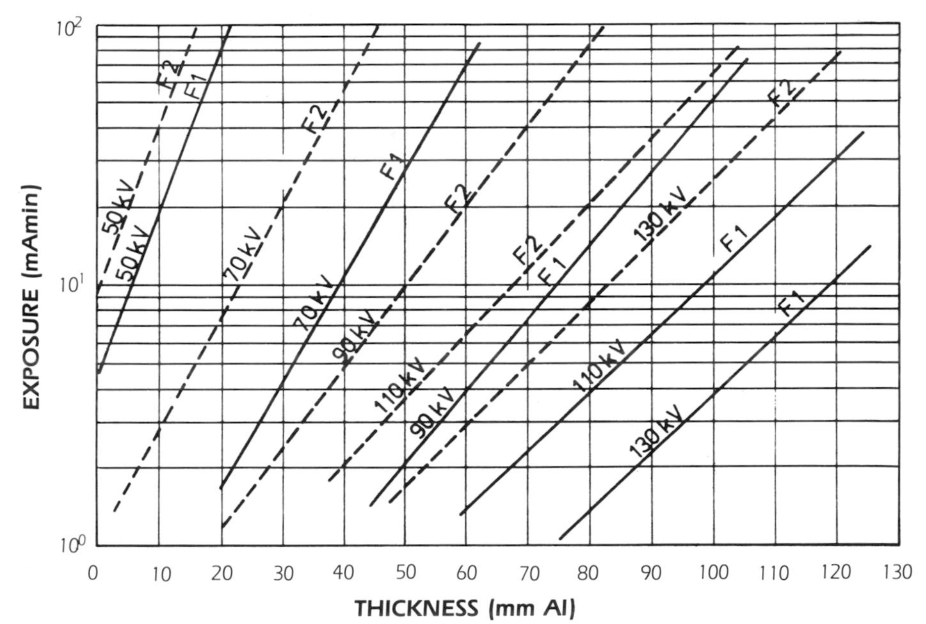

FIGURE 62. Same as Figure 61, at lower voltages. No lead filter used.

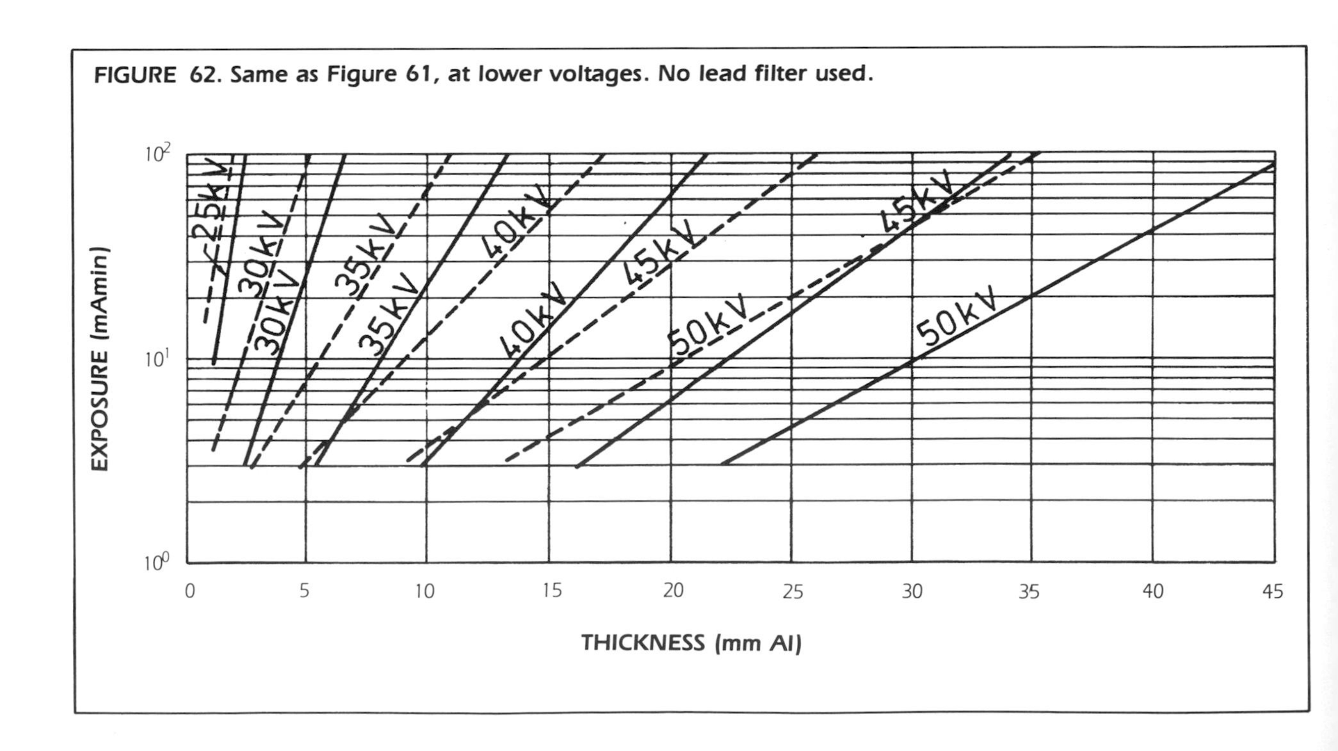

FIGURE 63. Exposure chart for aluminum: Paper C with F1 screen (solid lines) and Paper C with F2 screen (dashed lines). Exposures made at indicated voltages at FFD = 1 m, D_p = 1.0, using a 0.05 mm Pb filter.

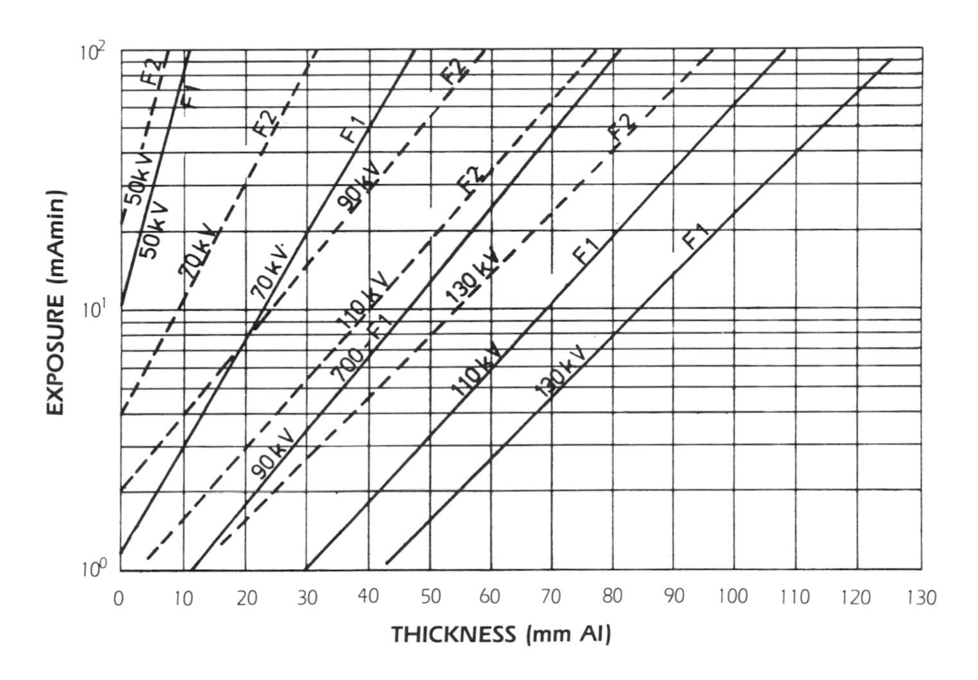

FIGURE 64. Same as Figure 63, at lower voltages. No lead filter used.

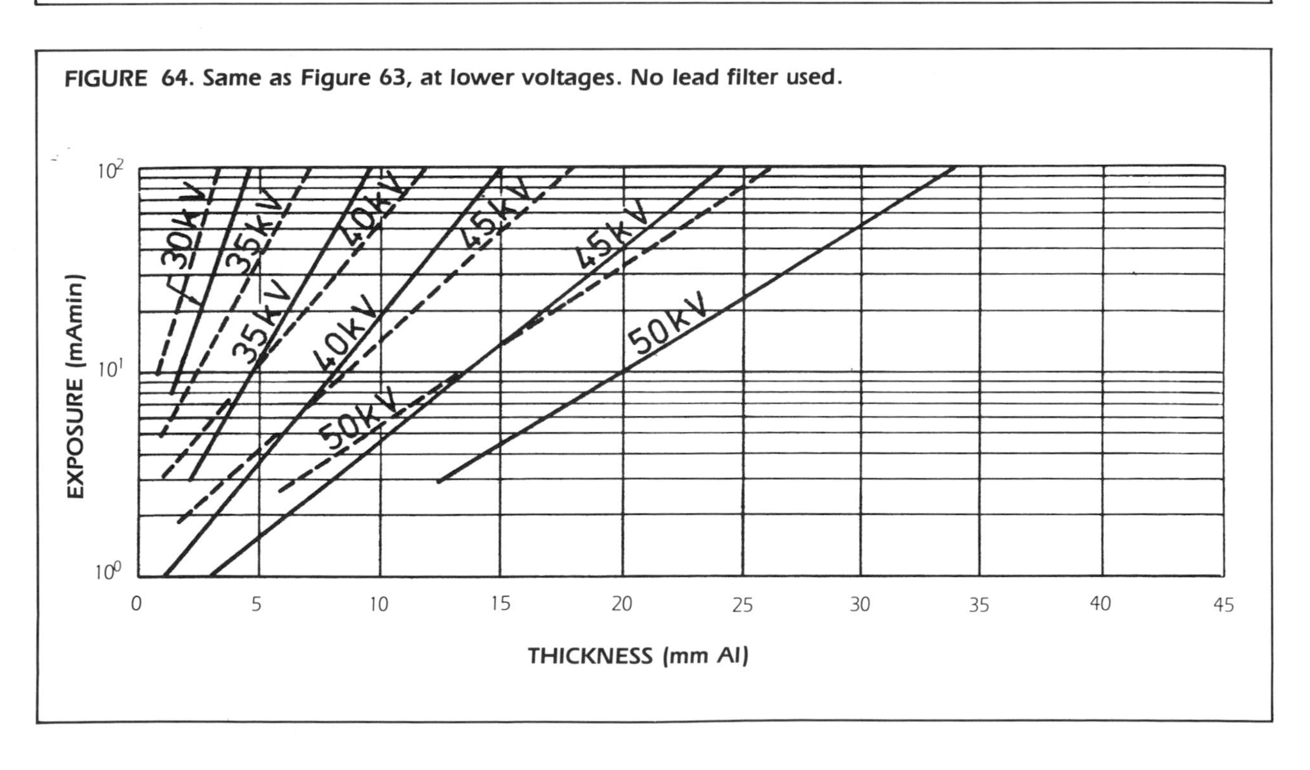

Sharpness of the Radiographic Image on Paper

The sharpness of a radiographic image depends on many factors: radiation source; focus size; focus-film and object-film distance; and grain size of the photographic emulsion. If fluorescent intensifying screens are used, sharpness also depends on good contact between the screen and the X-ray paper or film.

Even with correctly chosen exposure geometry (giving satisfactory geometric unsharpness), for the same paper-screen combination, one can get radiographs of varying sharpness because of poor contact between screen and paper.

As mentioned above, best results were obtained during radiography of certain specimens if a thin lead filter was present at the top of the cassette. In this case, a 0.05 mm Pb filter may be placed permanently between the front wall of the cassette and the intensifying screen. This filter can help improve contact between the paper and the screen.

Contact can be considerably improved in certain cassettes by inserting a sealed plastic bag containing a small amount of air. The bag is inflated just enough to ensure even distribution of pressure and is placed between the cassette lid and the paper. *The amount of air in the bags must be individually adjusted to each type and size of cassette.*

Processing of Radiographic Paper

Paper radiographics are processed using a stabilization method, which can be described in the following way.

Stabilization processing is a method of producing radiographs on paper much faster than is possible by conventional develop-stop-fix-wash processing. Exposed radiographic paper processed by stabilization makes good quality, ready-to-use radiographs available in a few seconds. These stabilized radiographs are not permanent because the chemical reactions within the emulsion have been stopped only temporarily. They will, however, last long enough to serve a number of practical purposes. In fact, stabilized radiographs often remain unchanged for many months if they are not exposed to strong light, high temperature, or excessive humidity.

Stabilization processing is a machine operation (see Fig. 65) in which radiographs on paper are processed in about 15 seconds and leave the processor in a slightly damp condition. They dry completely in a few minutes.

The main differences between stabilization processing and ordinary radiographic processing are in the speed of activation (development) and in the method of treating the unexposed light-sensitive silver halide left in the emulsion after development.

In conventional processing, the unused silver halide is dissolved by the fixer and traces of soluble silver salts left after fixing are removed by subsequent washing. The resulting radiographs are stable for long periods. In stabilization processing, however, the silver halide is converted to compounds which are only temporarily stable and the radiographs have a limited keeping time. Fixing and washing stabilized radiographs after they have served their initial purpose significantly improves their keeping characteristics.

In papers designed for stabilization processing, developing agents are incorporated in the paper emulsion. Development is achieved by applying an alkaline activator to the emulsion surface. The stabilizer is then applied to neutralize the activator and to convert any remaining silver halide to relatively stable, colorless compounds. Most photographic papers or X-ray films cannot be developed by this process because there is no developing agent present in either the emulsion or the processing chemicals. However, a stabilization paper with developing agents in the emulsion can be hand-processed in X-ray processing chemicals.

In stabilization processing, you exchange a measure of radiographic stability for the following advantages:

1. *Speed.* Stabilization processing is fast. Radiographs are ready for use in seconds.
2. *Simplicity.* The process is adaptable to uncomplicated mechanical systems.
3. *Space saving.* Darkroom space and plumbing needs are greatly reduced. In fact, some applications of the process do not require a darkroom.
4. *Water saving.* Stabilized radiographs do not require washing.
5. *Greater uniformity.* Mechanically processed radiographs have better day-to-day uniformity in density than those processed manually.

Important points for successful stabilization processing.

1. Correct exposure is essential because development time is constant.
2. Keep the processor clean. Follow the manufacturer's recommendations for cleaning and maintenance.
3. Don't overwork the chemical solutions. Observe the manufacturer's recommendations in regard to the capacity and renewal of solutions.
4. Make sure the processing trays in the machine are dry before loading them with chemicals. Some stabilization solutions are not compatible with water.
5. Avoid contamination of the activator with the stabilizer. Contamination results in chemical fog on the radiographs.
6. Avoid handling unprocessed paper after handling stabilized radiographs. The radiographs are impregnated with chemicals that easily mark or stain unprocessed material.
7. Do not wash stabilized radiographs unless they have been fixed in an ordinary fixing bath. Washing without fixing makes the radiographs sensitive to light.
8. Stabilized radiographs must not be heat dried. The combination of heat and moisture stains them an overall yellow-brown.
9. Because stabilized radiographs are impregnated with chemicals, do not file them in contact with processed X-ray film or other valuable material.

Long-Term Storage of Paper Radiographs

When long-term storage of paper radiographs is required, the following post-stabilization processing is recommended.

Fix the stabilized radiograph (time and temperature determined by the chemistry used). Wash at least 30 minutes in running water at 18-21 °C. Dry in a dust-free area on a matte dryer (if available). Otherwise, dry in a low heat dryer, mounted back-to-back on standard X-ray developing hangers.

Storing and Handling Paper Radiographs

The following rules are recommended for the storage and handling of radiographic paper.

The paper can be exposed to light transmitted by a suitable safelight with a brown or amber filter, using a 15-watt bulb at a distance of 1.2 m from the working surface.

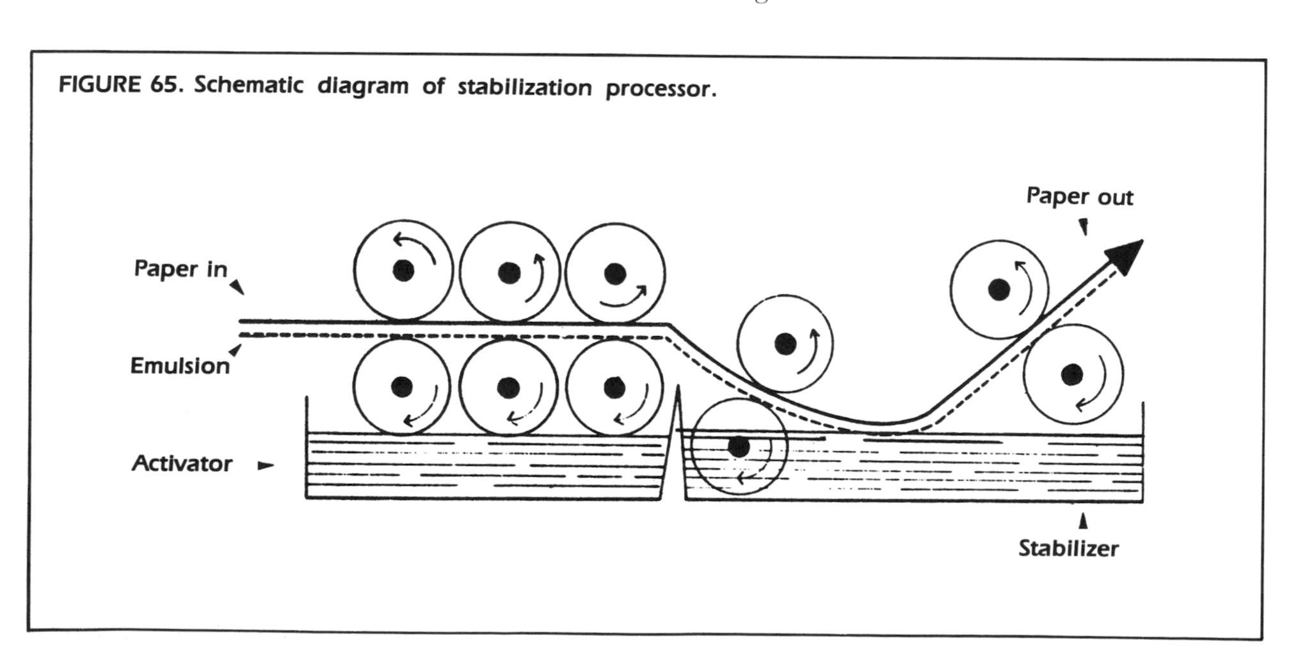

FIGURE 65. Schematic diagram of stabilization processor.

Before exposure and processing, store the paper under uniform conditions: 21-24 °C and 40-50 percent relative humidity are recommended.

After exposure and before processing, the paper may be stored for four days without any significant effect on speed or contrast.

After exposure and stabilization processing, examine the radiograph and dry thoroughly before storing. If the processed radiograph is to be kept longer than 6 to 10 weeks, fix and wash as recommended. Store the radiograph in a dry, dust-free place, away from harmful gases and chemicals.

After exposure and post-stabilization processing, radiographs have commercial keeping stability.

Radiographs are damp on leaving the processor, but at room temperatures they will dry completely, without curl, within a few minutes. Drying radiographs on hot drum machines is not recommended unless they have been fixed and washed. It has, however, proved possible to dry radiographs on a hot drum machine if the drum temperature does not exceed 50 °C.

Conclusions

Field of Application

Paper radiography can be used to examine the same objects as film radiography if a high sensitivity technique is not required. In general, the quality level required by most radiographic standards can be obtained (the 2-2T level require by ASTM, for example).

For paper radiographs, the best results can be obtained in the low and intermediate voltage ranges, though crude results at high voltages (up to 24 MeV) may also be useful.

For aluminum castings, very good results can be obtained up to 30 mm when using soft X-rays from a beryllium window tube at voltages up to 50 kV (though greater thicknesses can also be examined, up to 70 mm with voltages up to 130 kV).

For steel welds, paper radiography can be successfully used with voltages up to 120 kV and thicknesses up to 12 mm (or even 150 kV for 20 mm). Steel castings of larger thicknesses can be examined when high radiographic quality is not required.

Besides these two important fields of radiography (control of castings and weldings), paper radiography is suitable for the control of various assemblies, especially if they are composed of parts with different attenuation coefficients (metal in plastic, fibers in resin).

Paper radiography can also be used successfully for making a preliminary examination, during which the location of a defect can be determined and the correct kilovoltage found, after which the high sensitivity X-ray film technique is used. The above procedure is especially valuable because it considerably shortens the control process. Paper radiographs can be obtained, ready for use, in a far shorter time than film radiographs.

Paper radiography cannot be recommended when a high sensitivity technique is required (thicker steel welds for high pressure vessels, for example).

Equipment

The equipment necessary for the use of radiographic paper is simpler and less expensive than that for film radiography.

The same cassettes can be used with paper as are used with film. Instead of lead intensifying screens, used for most applications of film radiography, fluorescent intensifying screens are necessary for paper radiography. Because paper radiographs have only one light-sensitive surface, only one screen per cassette is needed.

Radiographic paper must be processed in special processors. These are simple to use, have smaller dimensions and are less expensive than film-processing tanks. A water supply is not required for radiographic paper processing. And paper can be processed in about 15 seconds, whereas hand-processing of X-ray films requires about 1 hour (in automatic processors this can be done in about 10 minutes).

The paper itself costs less than the X-ray film.

Other accessories, such as penetrameters, stepwedges or markers, are the same for both paper and film radiography. A reflection densitometer required to read the optical density of the paper costs about the same as a transmission densitometer needed for film.

Sensitometric Properties

The speed of radiographic paper, when used with intensifying screens, is very high, especially in the low voltage range. In the intermediate voltage range, radiographic paper shows much higher speed than X-ray film.

Comparing the contrast of film and paper, one must first observe that the contrast of industrial X-ray film increases constantly with the film density, whereas the contrast of radiographic paper reaches a maximum (usually at $D_p = 1.0$).

The contrast of X-ray film (measured at $D_f = 2.5$) is higher than the maximum contrast of the paper. The values for the maximum paper contrast (gamma) will be between 2 and 3; X-ray film shows a contrast that is twice as high (at $D_f = 2.5$).

Also, the exposure latitude of X-ray film is higher than that of paper radiographs.

Image Quality

In many applications, radiographic paper shows quality equal to radiographic film.

For aluminum and steel, paper radiographs do not show as high an image quality as that obtainable with X-ray film. Nevertheless, in most cases, a 2 percent IQI sensitivity can be achieved on paper.

General Conclusion

It can be said that radiographic paper is a valuable tool in the fields of industrial radiography and non-destructive testing. It can compete with X-ray film in many areas. It is especially useful when speed and low cost are required of the radiographic control, and less radiographic quality can be accepted.

SECTION 6

HIGH-ENERGY RADIOGRAPHY

B.J. Brunty, US Naval Weapons Station, Concord, CA

Parts of this chapter excerpted with permission from
"High-Energy X-ray Applications for Nondestructive Testing"
Varian Associates, Inc. (1982)

ACKNOWLEDGMENTS

The application of high-energy radiography as a method of nondestructive testing was a natural outgrowth of the use of low and medium-energy radiography. Much of the development of specialized procedures and techniques was accomplished in the 1950s and early 1960s during the time when intercontinental ballistic missiles were first being developed and tested. The Weapons Quality Engineering Center at the Naval Weapons Station, Concord, California, under the guidance of John H. Cusick, became heavily involved in the use of high-energy radiography for the inspection of these missiles. During this time, many procedures and techniques were developed, some of which form the basis for inspections being conducted today.

I am grateful to Russell Schonberg of Schonberg Radiation Incorporated, Charles H. Goldie of High Voltage Engineering Corporation and the late Howard Heffan for their contributions to this Section. I am very thankful to Larry Bryant of the Los Alamos National Laboratory for the coordination and editorial assistance he and his associates provided during preparation of this Section.

My gratitude is extended to Varian Associates, Incorporated, Palo Alto, California, for allowing the use of excerpts from the Linatron manual.

Appreciation is also extended to management personnel at the Naval Weapons Station, Concord, for their support in preparation of this Section. Special thanks goes to Rosalie M. Chiffoleau, Technical Writer/Editor who, by her long association with nondestructive testing and by her hard work and dedication, made this Section possible.

INTRODUCTION

Radiography using X-ray energies of one million electron volts (1 MeV) or greater is commonly considered to be in the high-energy range. The first commercial high-energy X-ray source was the 1 MeV resonant transformer introduced by General Electric in 1939. A few years later, a 2 MeV version, the Resotron, was fabricated. Though conceived eight years earlier, the Van de Graaff-style electrostatic generator also became commercially available around 1939, in the 1 and 2 MeV range. The 2 MeV generator was later changed to an upgraded version rated at 2.5 MeV. Some of these machines are still being used for radiography today. In the early 1940s, the betatron was developed, for energies from 15 to 25 MeV. The most recent addition to the list of high-energy X-ray machines is the electron linear accelerator, commonly referred to as the *linac.*

The original requirement for high-energy radiography was for inspection of thick metal castings and weldments; modern technologies in the nuclear field and in the rocket motor industries have, however, established requirements for higher energies and more X-ray output than those available in the earlier designs. The linac design adapts itself to these needs and generators are now available with energies up to 16 MeV, with radiation outputs up to 100 grays (10,000 rads) per minute when measured at one meter from the target.

PART 1

DESCRIPTION OF EQUIPMENT

Resonant Transformer X-ray Machines

The resonant transformer machine, with a multisection X-ray tube, was built when it became evident that more penetrating power was needed than could be generated with the conventional, single section X-ray tube. By resonating the high voltage circuit to the frequency of the AC power supply, the necessity for an iron core was eliminated. Additionally, by insulating the high voltage with Freon 12 or sulfur hexafluoride (SF-6), insulating oil was not needed. Elimination of the iron core and the insulating oil brought the size and weight of the machine into acceptable limits, allowing suspension and positioning by use of bridge cranes.

The resonant transformer X-ray machines, similar to Fig. 1, consist of an operating console, 180 Hz AC motor-generator set, motor controller, heat exchanger and a large steel tank. An air-core transformer is secured to the base of the steel tank by glass rods. A multisection X-ray tube is mounted coaxially with the transformer stack. A heated filament provides the free electrons in the tube and each section of the multisectional tube is connected to an appropriate tap on the transformer. Acceleration of the electrons through the tube is accomplished by uniform voltage distribution throughout the length of the tube. The electrons strike a water-cooled tungsten target at the end of the tube, thereby generating X-rays.

The majority of X-rays are generated in the forward direction; however, some of the beam is reflected from the target, and by opening ports around the lead collimator, a 360 degree circumferential beam can be extracted from the machine. The focal spot size can be as large as 10 millimeters (mm) in diameter in the transmitted direction. The apparent size of the target, when using the reflected beam, will depend upon the angle measured from the flat plane of the target. The radiation output from these machines is approximately 1.5 grays (150 rads) per minute when measured at one meter from the target.

FIGURE 1. General Electric 2 MeV Resotron

FROM US NAVY. REPRINTED WITH PERMISSION.

FIGURE 2. Van de Graaff 2.5 MeV Electrostatic Generator

FROM ARNOLD GREEN TESTING LABORATORIES, INC. REPRINTED WITH PERMISSION.

Electrostatic Generators

The Van de Graaff electrostatic generator shown in Fig. 2 has undergone some changes since its origin in the early 1930s. However, the basic principles of operation presented in Fig. 3 remain the same. The machine consists of a control console and a metal tank which houses power supplies, drive motors, an insulated charging belt, a generator column, high voltage terminals and electrical connections. An accelerator tube, which is highly evacuated, extends through the generator column. This metal tank is pressurized with nitrogen, carbon dioxide and sulfur hexafluoride to approximately 2600 kilopascals (375 pounds per sqaure inch) to prevent high voltage arcing. Within the tank, electron sprayers deposit electrons onto a moving belt which carries them to a terminal shell. Some of these electrons are emitted into the accelerator tube as beam current through a direct connection to the filament. The filament, when properly heated, emits electrons which are replaced from the terminal shell. The electrons flow through the accelerator tube at an energy determined by the terminal shell voltage. A water-cooled target is located at the end of the accelerator tube; when struck by the accelerated electrons, it produces X-rays. This design results in a direct current, constant potential X-ray machine. The machines currently manufactured are in the 2.5 MeV range, have a 2.5 mm target and produce approximately 1.7 grays (170 rads) per minute at one meter.

FIGURE 3. Operation of Electrostatic Generator

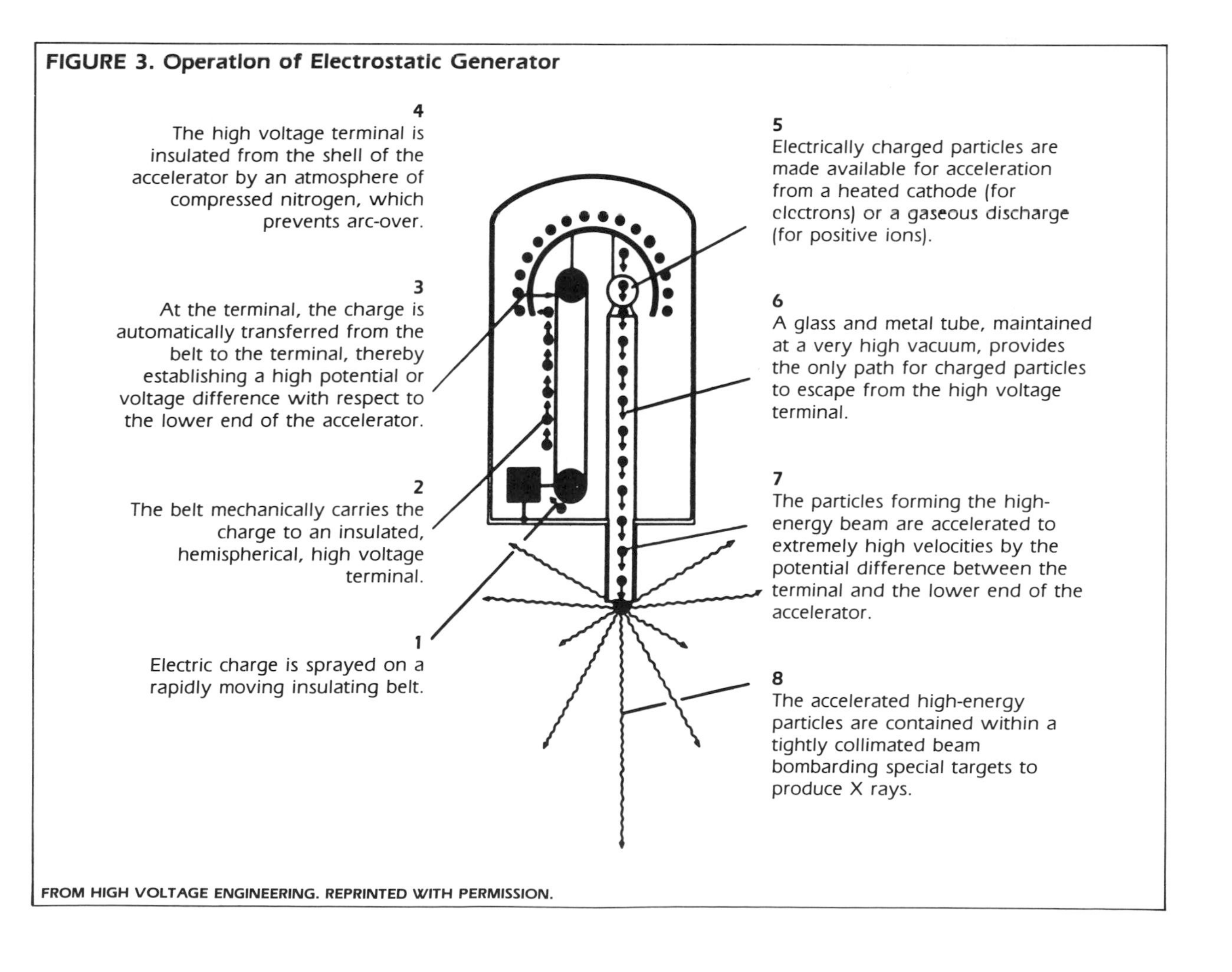

FROM HIGH VOLTAGE ENGINEERING. REPRINTED WITH PERMISSION.

Betatrons

As early as 1922, a patent was filed for an induction accelerator or *betatron* as it is now named. In the late 1920s, the principles of betatron operations were fully accepted; however, it was not until 1940-1941 that the first successful betatron was built. It produced 2.3 MeV electrons, and had an X-ray output equivalent to that from one gram of radium. The following year a 20 MeV betatron was built, then an 80 MeV model and finally in 1950 a 300 MeV machine was completed.

To accelerate electrons to high speed, the betatron, shown in Fig. 4, employs the magnetic induction effect used in a transformer. The primary winding in a transformer is connected to an AC voltage source which establishes a varying flux in an iron core. The secondary winding on this core has induced in it a voltage equal to the product of (1) the number of turns in the secondary winding and (2) the flux time rate of change. The resulting electric current is made up of the free electrons present in the wire. The betatron is essentially such a transformer, as shown in Fig. 5, except that, instead of wire, the secondary is a hollow circular tube. This tube, called a *doughnut*, is used to contain the electrons for many thousand revolutions.

The doughnut tube is usually made of porcelain and is coated on the inside with a conductive layer

FIGURE 4. Allis-Chalmers 25 MeV Betatron

FROM US NAVY. REPRINTED WITH PERMISSION.

FIGURE 5. Diagram of Betatron Generator: (a) Top View; (b) Cross Section

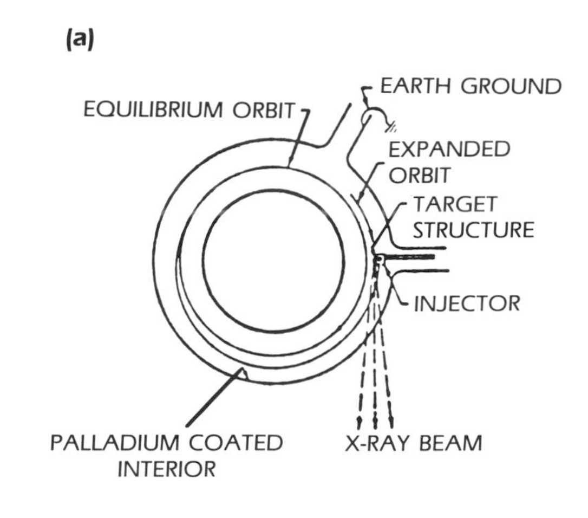

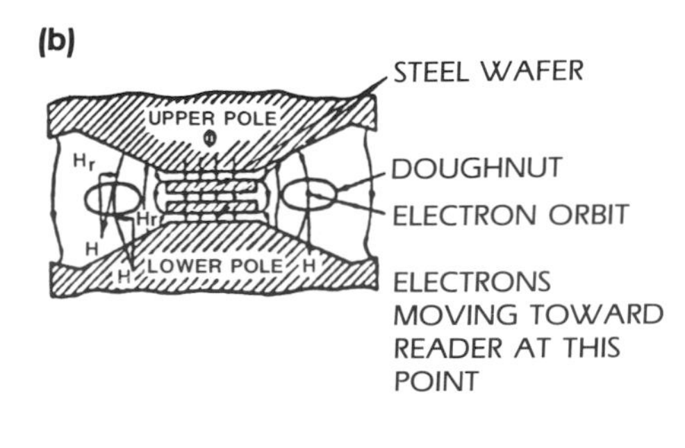

of palladium connected to a ground. The doughnut is placed between the poles of an electromagnet that produces a pulsating field. Electrons injected into the tube as the magnetic field increases will be accelerated in a circular path. The force acting on the particles is proportional to the rate of change of flux and the magnitude of the field. Since the electrons circle the orbit many times before striking the target, there is a large amount of energy gain. For example, in a 24 MeV betatron, the electrons circle the orbit about 350,000 times, travelling a distance of 400 kilometers (260 miles). The average voltage gain per turn at the orbit is about 70 volts, which gives approximately 24 MeV. As the electrons reach maximum energy, they are deflected by electrical pulse and caused to spiral outward until they strike the target. Betatrons are equipped with platinum wire targets with dimensions of 0.013 × 0.025 cm (0.005 × 0.010 in.) and have a radiation output of 1.5 grays (150 rads) per minute when measured at one meter from the target.

FIGURE 6. Varian Associates Model 2000 Linatron Linear Accelerator

FROM VARIAN ASSOCIATES, INC. REPRINTED WITH PERMISSION.

Electron Linear Accelerators

Electron linear accelerators, similar to the one shown in Fig. 6, are commonly referred to as linacs. Linacs accelerate electrons down a guide by means of radiofrequency (rf) voltages. These voltages are applied so that the electron reaches an acceleration point in the field at precisely the proper time. The accelerator guide consists of a series of cavities which causes gaps when the rf power is applied. The cavities have holes in each end which allow electrons to pass to the next cavity. When an electron is injected at the proper time it gains energy as it is accelerated across these gaps and out the other end of the cavity. When the rf power is phased properly, increased acceleration is achieved. Figure 7 shows the general arrangement of a linac's component parts.

For research purposes, heavier particles have been accelerated by linacs. An example of this type linac is the 20GEN, the two-mile long Stanford Linear Accelerator. Two general types of radiographic linacs are in use today. One employs the principle of the traveling wave and the other a standing wave method of acceleration.

Recent developments in linacs include increased radiation output capabilities, up to 100 grays (10,000 rads) per minute measured at one meter. Increased operating frequencies, up to 9,300 MHz, permit smaller, lighter-weight X-ray heads. One new configuration allows for the operation of the accelerator and collimator at a distance from the rf source; the source supplies power through a flexible waveguide. The total weight of the X-ray head is greatly reduced, permitting easy positioning for inspection of pipelines, valves and other test objects of limited accessibility. One such system is being used for in-service inspection of nuclear power plants.

For a summary of information on high-energy generators used less commonly in industrial radiography (synchrotron, cyclotron, etc.) see Volume 3, Section 2, "Conventional Electronic Radiation Sources."

FIGURE 7. General Arrangement of Linear Accelerator

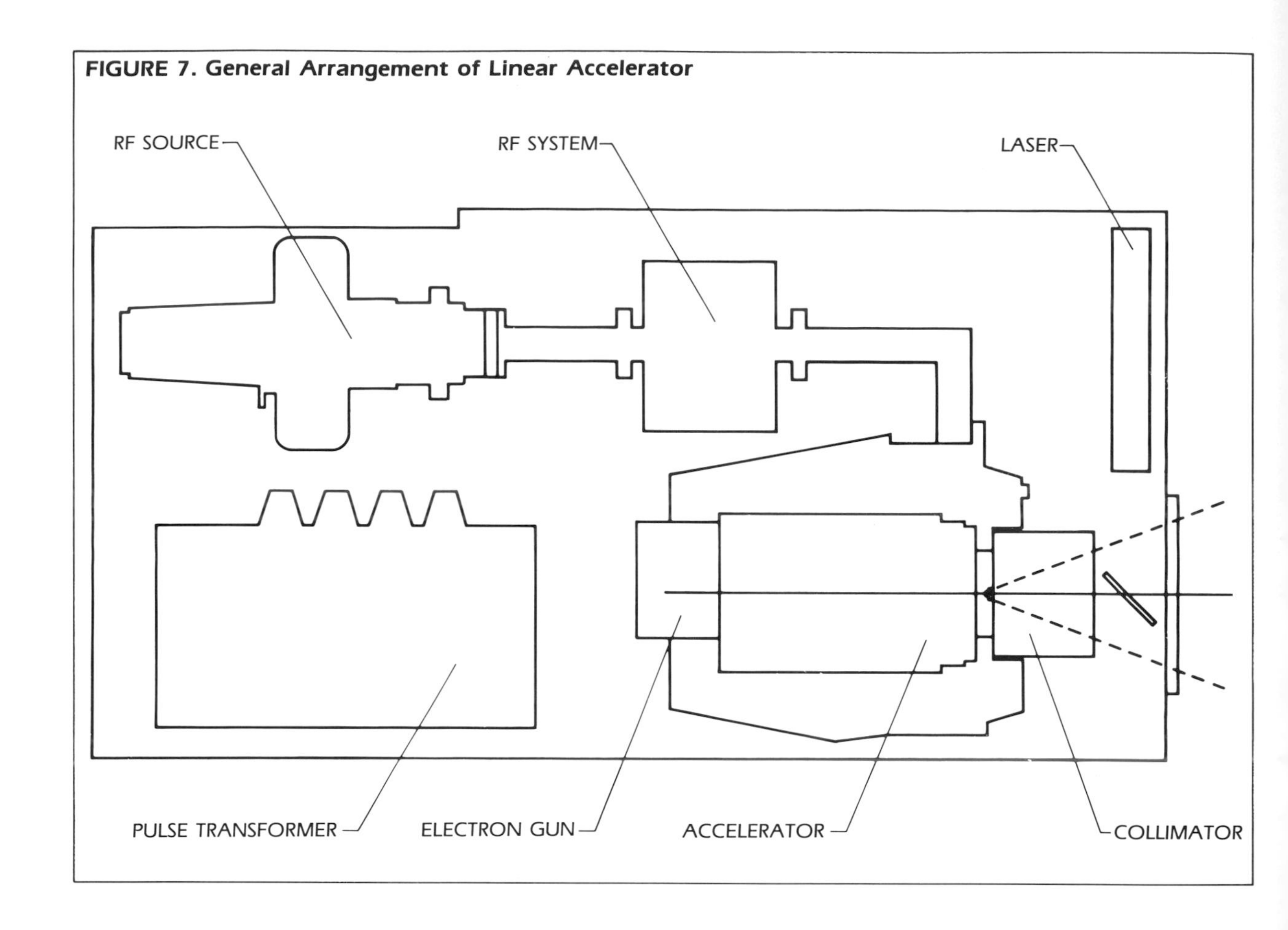

PART 2

CHARACTERISTICS OF HIGH-ENERGY RADIATION

Basic Principles

The basic principles of high-energy radiography are the same as those of low and medium-voltage X-radiography. Standard types of commercial X-ray film, with lead or other intensifying screens, are used to produce the radiographic image of the object being examined. The arrangement of the source, object and film, the shielding, blocking (masking), and other scatter reduction techniques and the use of penetrameters and identification numbers are all similar to methods used in radiography with other energies. The differences between high and low-energy radiography arise from several distinctive characteristics of a high-energy X-ray source, some of which prove to be advantageous.

Advantages

The major advantages of high-energy radiography are listed below.

1. The higher energy photons are more penetrating. Greater penetration means that radiography of thick sections is practical and economically feasible.
2. Large distance-over-thickness ratios (D/t) can be used with correspondingly low distortion.
3. Short exposure times and high production rates are possible.
4. The wide thickness latitude, good contrast and reduced amounts of high angle scatter reaching the film results in high quality radiographs, with excellent penetrameter sensitivity and good detail resolution.
5. Some machines have high output intensity, making possible the use of large focal-film distances, large areas of coverage and greater use of the low speed, fine grained and high contrast films.

Measurement of High-Energy Radiation

The resonant transformer machine and the Van de Graaff-style X-ray machines use clock timing mechanisms to determine exposure of the recording media, in a manner similar to low voltage X-ray machines. The betatron and linac machines measure ionizing radiation with ionization chambers.

Exposure from betatrons is normally measured in roentgens per minute at one meter from the target. The *roentgen* expresses the ionization produced by X-rays in air and is fundamentally a property of the beam, rather than a measure of the beam's effect on the irradiated object. This exposure in the SI system is *coulomb per kilogram*, giving the energy (in the form of ion pairs) in one kilogram of air.

Exposure from linacs is normally expressed in rads per minute at one meter from the target. The *rad* (radiation absorbed dose) is defined as the amount of energy imparted to matter, per unit mass of irradiated material, and is equal to 100 ergs per gram. The *gray* is the SI unit used to express exposure from betatrons and linacs; it equals 100 rads or one joule per kilogram.

To achieve calibration in rads, the radiation output of a linac is determined by measuring the charge produced by the X-ray beam in a given volume of air, using an ionization chamber dosimeter. Correction factors are then used to calculate the absorbed dose in a material. These ion chamber measurements are normally made at a specified depth in a water phantom or with the ion chamber surrounded by a plastic cylinder or equilibrium cap in order to achieve electronic equilibrium. This provides adequate buildup of the high-energy electrons which are set in motion by the X-rays. The equilibrium-cap thickness will range from one centimeter to five centimeters, depending upon the energy of the beam. For low atomic number materials, a roentgen measured in air is approximately equivalent to one rad of absorbed dose.

Target Characteristics

The intensity of the X-rays produced at the target is a function of the electron beam and the X-ray production efficiency of the target. *Target efficiency* is defined as the ratio of the total X-radiation power produced to the total power of the impinging electron beam. This efficiency depends on both target composition and geometry. The most efficient targets are made of materials with high atomic numbers (high Z elements). Of those elements, tungsten offers the best combined efficiency and physical properties, and is the primary material used in X-ray machine targets; however, gold, platinum and copper are also used.

Targets are normally constructed with a thickness slightly greater than the range of electrons in the material. Heat generated within the target must be dissipated to minimize erosion of the target material. Several methods are used to accomplish this. Water jackets are usually an integral part of the target assembly, allowing heat to be transferred to a heat exchanger and thereby dissipated. Rotating targets and targets that move under the electron beam are also used in conjunction with water jackets. These methods are designed to move new target material continually under the electron beam, thus improving the target's heat dissipating ability. Figure 8 shows a gold rotating target which has incurred some erosion from the electron beam.

FIGURE 8. Gold Rotating Target from Linac; Photograph Enlarged 3X

FROM US NAVY. REPRINTED WITH PERMISSION.

Focal Spot Size Determination

Because of the high energy and penetrating power of these machines, the pinhole technique for determining focal spot size cannot be used. There are, however, two commonly used methods for determining focal spot size and circularity.

The first method uses a special spot size camera with an optically aligned and welded assembly of photoetched plates. Each plate is about 25 cm (10 in.) thick and contains longitudinal holes 0.15 mm (0.006 in.) in diameter, on 0.25 mm (0.01 in.) centers. Dental film, instant film, or other equivalent film exposed through the spot size camera will yield a series of dots, corresponding to the holes in the plates, which are superimposed on the image of the X-ray spot. The spot size is determined directly by counting the dots covered by the spot.

The second technique uses an X-ray collimating assembly, which consists of alternating layers of thin lead foil and thin polyester sheets. The sandwich assembly, which is approximately 25 cm (10 in.) thick and two to five cm (1 to 2 in.) square, is aligned lengthwise in the X-ray beam, immediately next to the face of the X-ray head. An X-ray film is placed at the base of the assembly, and the first exposure is made. The stack is then rotated through 90 degrees without moving the film, and the second exposure is taken.

When developed, the film clearly shows the separations between layers of lead, where radiation from the target passed through the polyester film. The two views, 90 degrees apart, will image the focal spot, superimposed on a matrix of lines 90 degrees apart. The measurement is made in the same manner as before. With lead foil 0.13 mm (0.005 in.) thick and polyester film 0.13 mm (0.005 in.) thick, the uncertainty in the measurement is plus or minus 0.25 mm (0.01 in.) at the most. Use of thinner layers in the assembly diminishes the uncertainty in the measurement.

PART 3

RADIATION ATTENUATION AND SCATTERING

Radiation Attenuation

When an X-ray beam enters and passes through material, it becomes attenuated as the photons interact with nuclei and electrons of the object material. Three processes make up the total attenuation (see Volume 3, Section 1, "Radiation and Particle Physics").

Photoelectric Absorption

This is a process whereby a photon loses all of its energy to an atomic electron which in turn leaves the orbit of the atom and continues to move through the material at high speed. Photoelectric absorption occurs in steel, for example, with most of the low energy photons (0.1 MeV and less) which are present within the X-ray beam. The probability of photoelectric absorption decreases as energies increase above 0.1 MeV and rarely occurs in high-energy radiography.

Compton Scattering

This process occurs when the photon is deflected from its original path by an interaction with an atomic electron. The atomic electron is ejected from its orbital position and the photon continues to pass

FIGURE 9. Dominant X-ray Attenuation Process

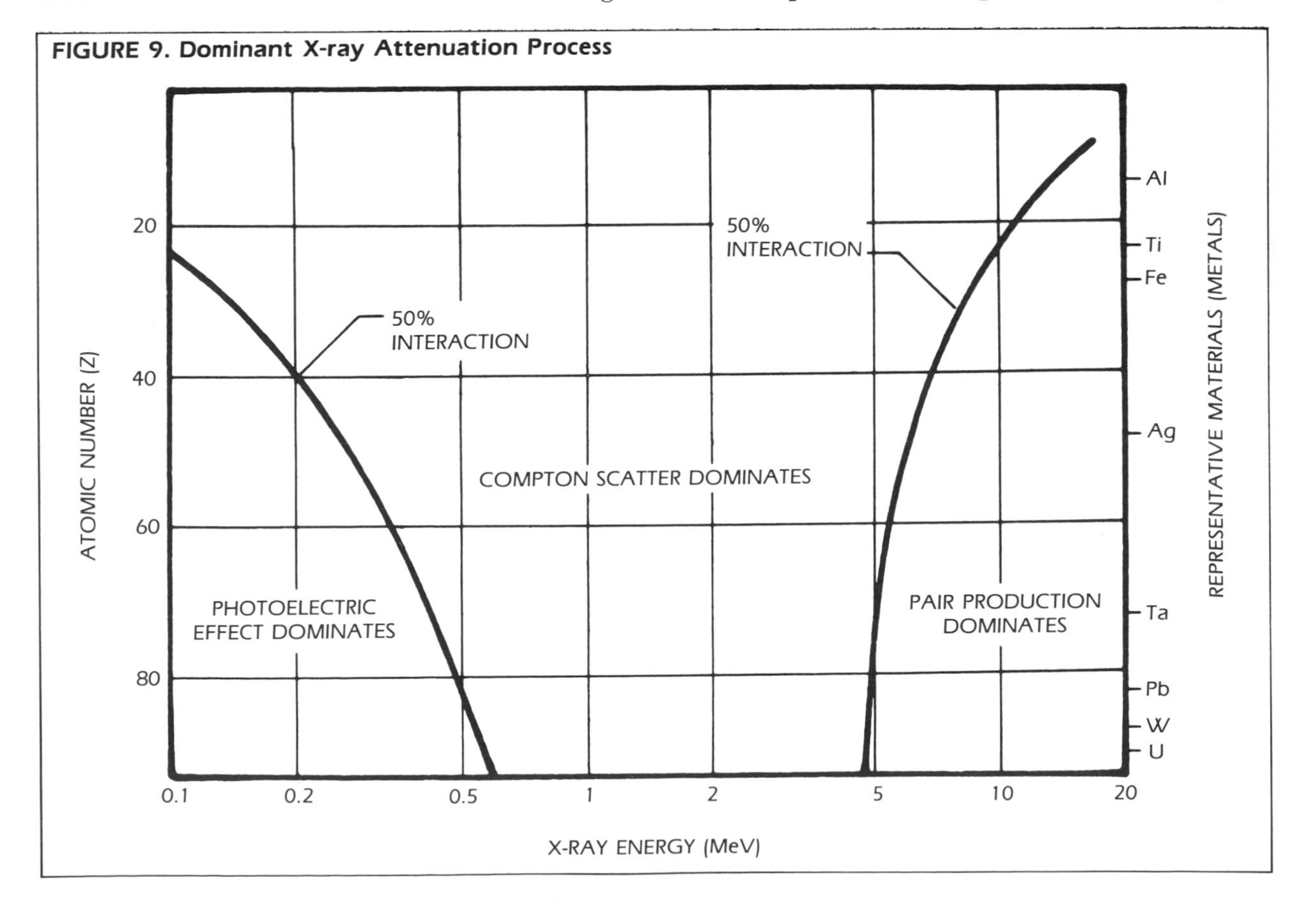

through the material in a new, deflected direction and at a reduced energy. For photon energies from 0.1 to 10 MeV, Compton scattering is the major attenuation process. Because many of the photons in a high-energy X-ray beam are in this energy range, a high intensity of scattered radiation can emanate from an object being radiographed, and indeed from every object that the beam happens to strike. In the back plane of the X-ray film, for example, the intensity of scatter from a wall or hardware bracket can easily reach values that equal the intensity of the transmitted primary rays.

Pair Production

Pair production is a process in which a photon is completely absorbed and an electron-positron particle pair is created. Pair production has a threshold energy of 1.02 MeV, and becomes significant in the attenuation of a high-energy X-ray beam when photons with energies above 4 MeV are sufficiently present.

Total Attenuation

The total attenuation of a high-energy X-ray beam is a combination of all three processes, plus some secondary processes (such as the generation of secondary X-rays within an object by the slowing-down process of the scattered electrons). Quantitatively, the amount of absorption and the total attenuation that occur when a high-energy X-ray beam passes through an object depend on the atomic number of the material, the density and thickness of the object, and the X-ray energies of the photons that make up the beam. Figure 9 is a plot of absorber atomic numbers versus electron beam energies, showing the range of absorber atomic number and photon beam energy where each of these three occurrences dominates the attenuation process.

Half-value Layers

The X-ray attenuation of a material is often defined with the *half-value layer* (HVL), or the thickness of material that reduces the intensity of the transmitted X-rays by a factor of two. HVL is related to the linear attenuation coefficient, μ, by:

$$\mathrm{HVL} = \frac{0.693}{\mu} \qquad \text{(Eq. 1)}$$

The unfiltered X-rays emitted by the target of an X-ray machine contain photons with a range of energies, up to a maximum determined by the incident electron energy. This collection of photons makes up the energy spectrum of the X-ray beam.

After the X-rays are produced, the target, the cooling jacket, other materials between the target and the exterior of the X-ray machine, the air in the space between the X-ray machine and the detector, and even the walls of the chamber used to measure the radiation, all absorb some of the emitted photons. Because the low-energy photons are always preferentially absorbed by every material, the emerging X-ray beam changes to a somewhat filtered and modified beam.

Narrow Beam

Two methods may be utilized when determining half-value layers. One is termed *narrow beam geometry* where the absorber material is placed next to the exit cone of the X-ray machine and the detecting media is placed at some distance from the material. This method helps prevent forward scatter from reaching the detector and allows for the evaluation of the transmitted primary beam. *Narrow beam* has been defined as that condition where scatter reaching the X-ray film or detector is only that scattered radiation within angles less than 0.01 steradian. The first two half-value layers will be thinner than subsequent HVLs, because less material is needed to reduce the intensity of a beam containing large numbers of lower energy photons. Thus, the first two layers modify the beam, filter it and make it harder.

Broad Beam

The second method of determining HVLs is called *broad beam geometry* and more nearly represents radiographic conditions: the material is located a specified distance from the machine and the recording media is placed directly behind the material. Again, the first two HVLs preferentially filter out the low-energy photons. Under these conditions, however, scattered radiation from the material contributes to the exposure of the recording media. After about two HVLs, the attenuation reaches equilibrium and a constant value of the HVLs results. Because of the conditions described above, it is normal radiographic practice not to use the first two HVLs

TABLE 1. Typical Broad Beam Half-value Layers

Material (Density)	Typical Half-value Layer*						
	1 MeV	2 MeV	4 MeV	6 MeV	8 MeV	10 MeV	16 MeV
Tungsten (18 g/cm³)							
HVL (cm)	0.55	0.90	1.15	1.20	1.20	1.20	1.15
HVL (in.)	0.21	0.36	0.45	0.48	0.48	0.48	0.45
Lead (11.3 g/cm³)							
HVL (cm)	0.75	1.25	1.60	1.70	1.70	1.70	1.65
HVL (in.)	0.30	0.49	0.63	0.67	0.67	0.67	0.65
Steel (7.85 g/cm³)							
HVL (cm)	1.60	2.00	2.50	2.80	3.00	3.20	3.30
HVL (in.)	0.63	0.79	1.00	1.10	1.20	1.25	1.30
Aluminum (2.70 g/cm³)							
HVL (cm)	3.90	5.40	7.50	8.90	9.60	10.00	11.00
HVL (in.)	1.50	2.10	2.90	3.50	3.80	3.90	4.30
Concrete (2.35 g/cm³)							
HVL (cm)	4.50	6.20	8.60	10.20	11.00	11.50	12.70
HVL (in.)	1.80	2.40	3.40	4.00	4.30	4.50	5.00
Solid Propellant (1.7 g/cm³)							
HVL (cm)	6.10	8.40	11.60	13.80	14.90	16.50	20.40
HVL (in.)	2.40	3.30	4.60	5.40	5.90	6.50	8.00
Lucite (1.2 g/cm³)							
HVL (cm)	10.50	12.10	16.80	19.90	21.50	23.80	29.50
HVL (in.)	4.10	4.80	6.60	7.80	8.50	9.40	11.60

*Values measured by film techniques may vary somewhat depending upon actual material characteristics, scatter control and other factors.

in determining exposure techniques. Typical broad beam half-value layers are shown in Table 1.

Energy Quality

Since the linear attenuation coefficient and the HVL have definite values for each material and for each photon energy, these quantities are also used to express the quality of energy, or energy makeup, of the beam from an X-ray generator. As previously mentioned, practical radiography setups use broad beam radiation; that is, scatter is present in the exposure. In that arrangement, the demonstrated or measured HVL thickness at a given generator energy setting may vary with each setup, depending on the amount of scatter that the film or detector receives. The slope of the exposure curve and the contrast and latitude achieved in a step block exposure are indicators of the HVL and the effect of scatter. In high-energy X-radiography, the types and thicknesses of the test materials determine to a large extent the generator energy that should be used. The broad-beam HVL is a useful material index for the radiographer to use in the energy selection, since it is related directly to exposure time. Figures 10 through 13 illustrate HVL as a function of incident electron energy for steel, rocket propellant, lead and concrete. These values represent equilibrium half-value layers.

Scatter Radiation

Scatter will be present in every high-energy radiographic application. Because this scatter can be as intense as the primary beam or, in certain circumstances, can be even greater than the primary beam, it is very important that it be controlled. When scatter is not minimized, film contrast will be reduced with a corresponding reduction of image quality.

Latitude

A common task in high-energy installations is the radiography of objects with varying shapes and thicknesses. A single film exposure can cover a range

FIGURE 10. Half-value Layer for Steel as a Function of Energy

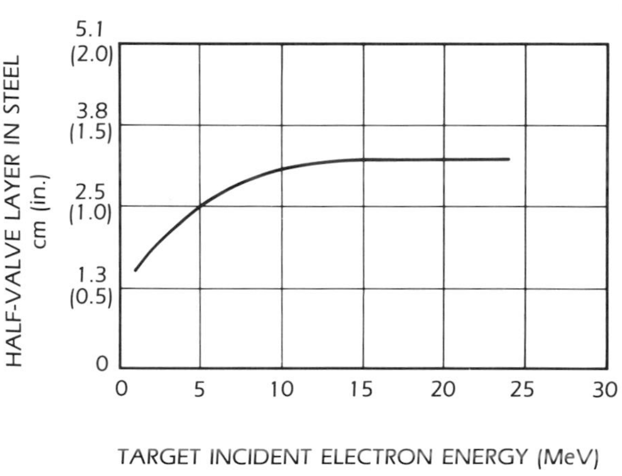

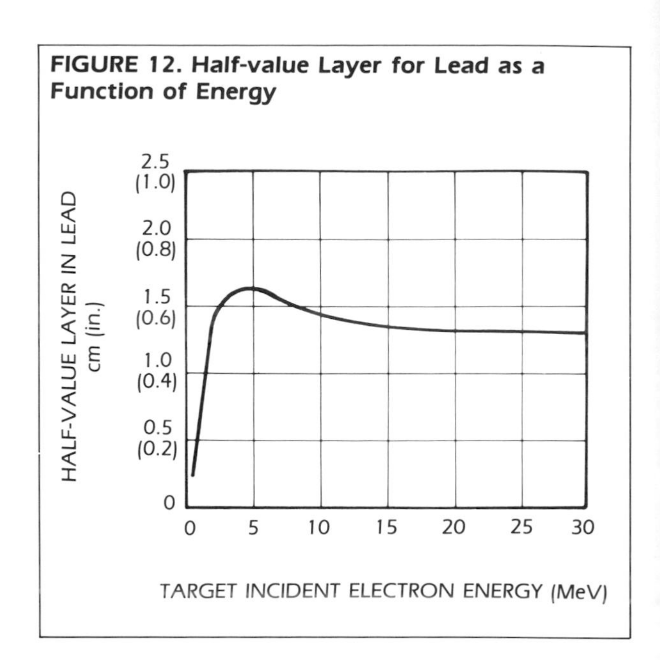

FIGURE 12. Half-value Layer for Lead as a Function of Energy

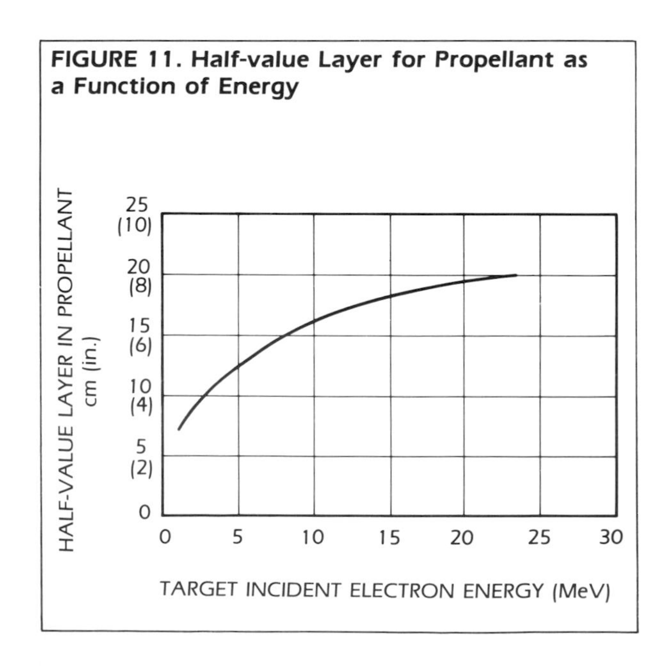

FIGURE 11. Half-value Layer for Propellant as a Function of Energy

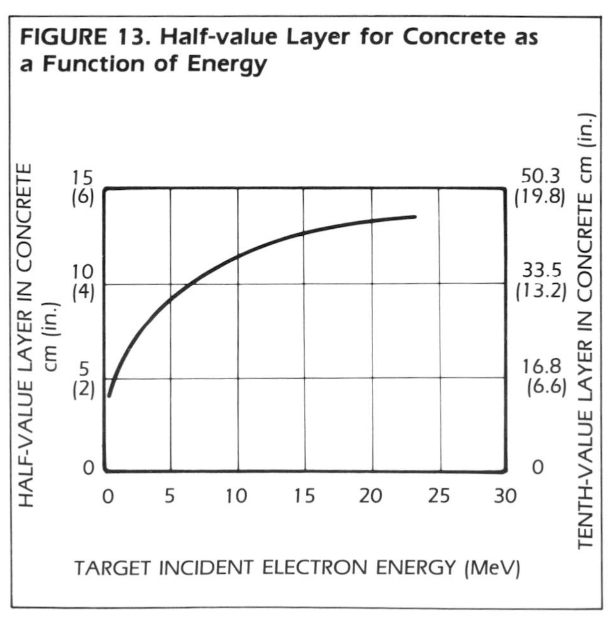

FIGURE 13. Half-value Layer for Concrete as a Function of Energy

of useful film densities where sensitivity and interpretability are accurate and valid; the thickness range that corresponds to the range of useful densities is called the *latitude of exposure.* Latitude depends on the film gradient, or contrast, and on the attenuation of the material. Naturally, when two films of different speeds are used to image the same object in one exposure, the latitudes of the films are summed, to expand the total latitude for the exposure.

When a wedge-shaped object is used to generate the exposure curve, the points on the wedge image, corresponding to the minimum and maximum film densities allowed by the appropriate specifications, will provide data for determining latitude. Figure 14 shows a plot of latitude in steel for several energies.

FIGURE 14. Latitude Ranges for Exposures of Steel; Circled Numbers are H&D Units

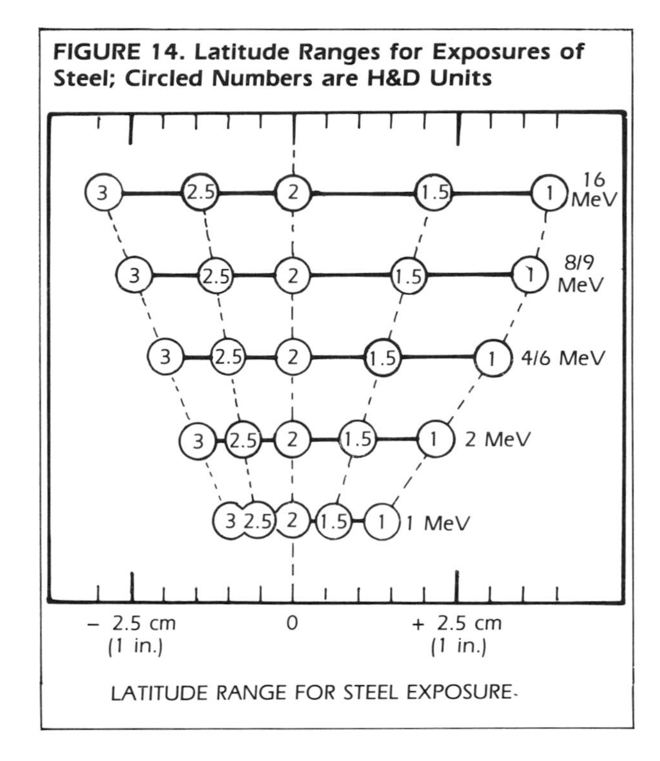

PART 4

BEAMING AND FIELD FLATNESS

Beaming

In high-energy X-ray machines, electrons reach speeds approaching that of light. Most of the electrons continue to travel in the forward direction after their initial interactions with the target atoms. The deflection angle of scatter tends to be small, and decreases as the energy of the incident electrons increases. Because high-energy interactions during electron penetration produce high-energy X-ray photons, the direction of these emitted photons, like that of the scattered electrons, is also predominantly

FIGURE 15. X-ray Intensity Distributions for Varian Linatron Models 200 and 400

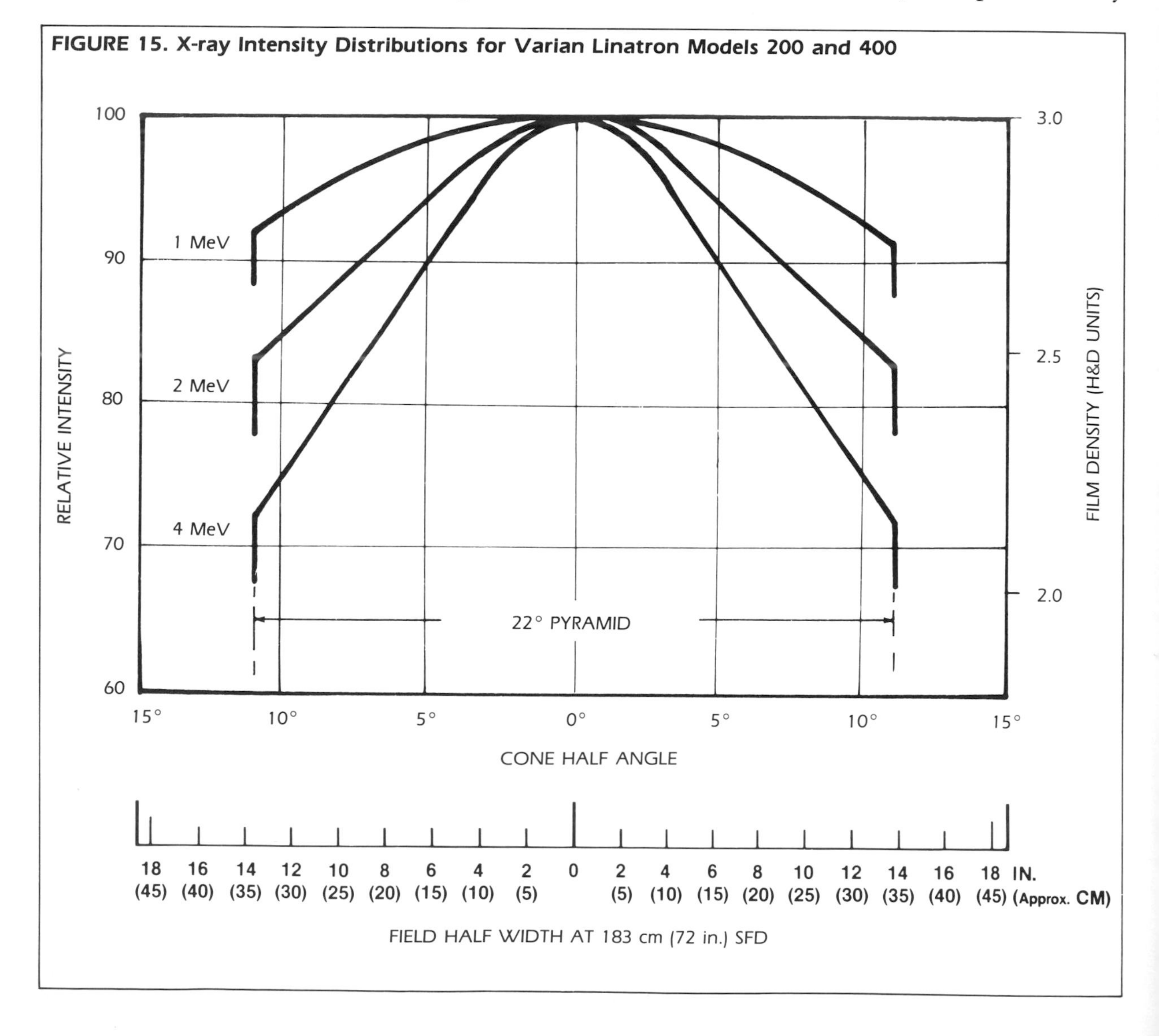

forward. Thus, the radiation intensity across the X-ray field is not uniform or *flat;* this effect is termed *beaming* and it increases with increased energy.

Target Thickness

When targets are slightly thicker than the electron path length of the most energetic electron, a proportionally larger number of lower energy photons is produced, when compared to the number of photons produced by a thin target. This thicker target lessens the beaming effect, broadens field coverage and increases sensitivity in inspection of thin and low-density materials.

Compensators

In the very high-energy linacs and betatrons, the intensity of the X-ray beam is so much greater at its centerline than at small angles off-center that a compensator or field flattener may be employed to reduce the centerline intensity and produce a more uniform intensity across the field. These compensators are usually made of aluminum. They are designed to be proportionally thicker in the center of the beam to compensate for the higher beam intensity of the central ray; they are smoothly tapered and become thinner at the edges of the radiation field. As a consequence of the differential X-ray attenuation produced by the thickness variations, the resulting beam profile is flatter. In some machines, the field flattener is located in the beam in such a way that it attenuates the X-rays after the output has been measured by the ion chamber. In such cases, adjustments must be made to achieve proper radiographic exposure.

Field Flattening

For the radiographer who requires a uniform field intensity in order to obtain a uniform exposure across the radiograph, beaming can present a problem, and may become a controlling factor in applied radiography. As long as exposure times do not become excessively long, a lower X-ray energy source (with its flatter beam) can provide the more uniform field. Increasing the source-to-film distance reduces the effect of the beaming for a fixed film

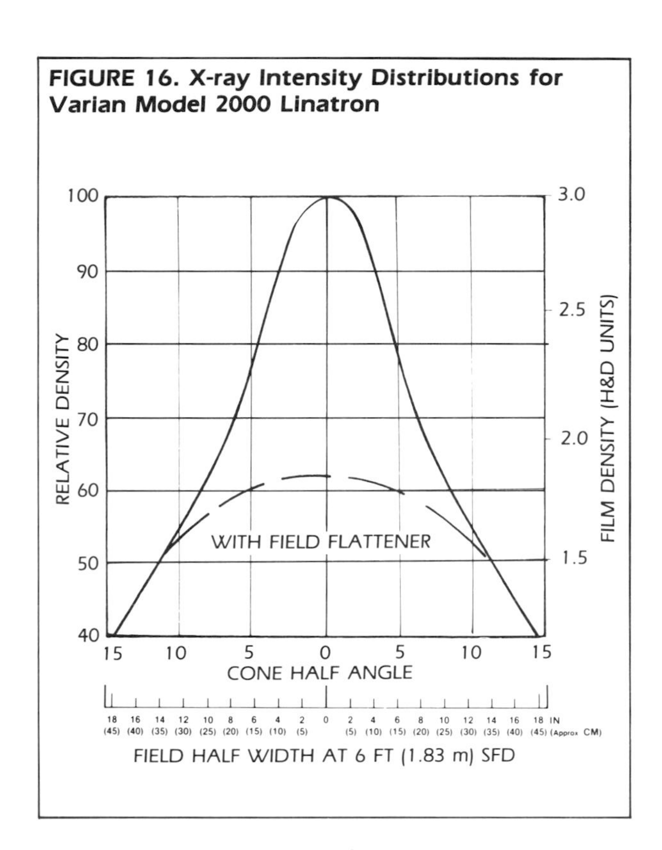

FIGURE 16. X-ray Intensity Distributions for Varian Model 2000 Linatron

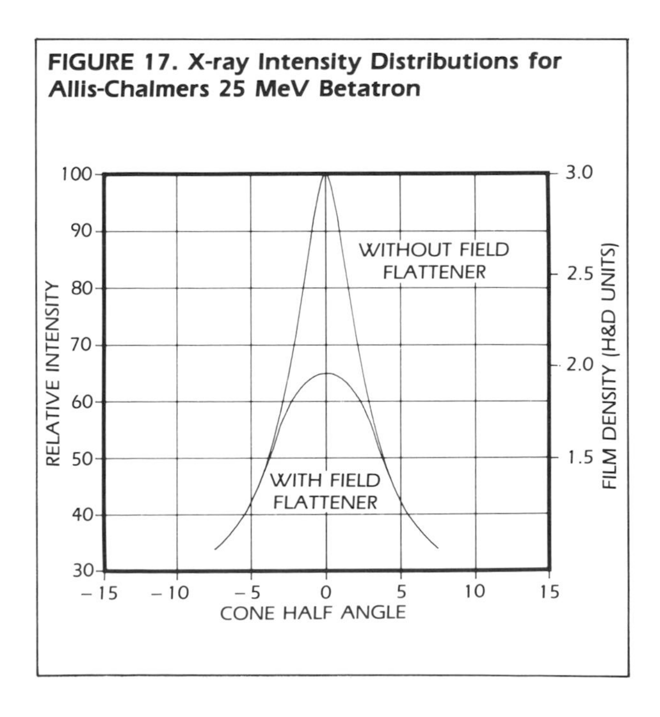

FIGURE 17. X-ray Intensity Distributions for Allis-Chalmers 25 MeV Betatron

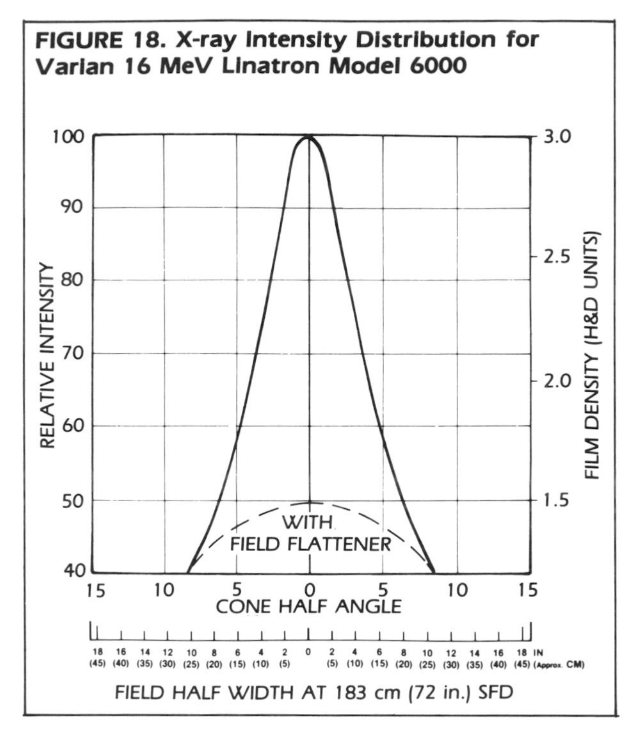

FIGURE 18. X-ray Intensity Distribution for Varian 16 MeV Linatron Model 6000

TABLE 2. Physical Density Ratios for Common Materials

Multiply by factors below to convert steel exposures to approximate exposures for these materials:	
Zirconium	0.83
Copper	1.13
Molybdenum	1.3
Silver	1.33
Lead	1.44
Uranium or Tungsten	2.38
Gold	2.45
Multiply by factors below to convert propellant exposures to approximate exposures for these materials:	
Sodium	0.57
Lucite	0.69
Magnesium	1.02
Carbon	1.3
Concrete	1.38
Aluminum	1.6
Titanium	2.67

size. Use of a higher energy, a more powerful source with compensator, and a large source-to-film distance, permits the use of larger film areas, makes it possible to reduce overall inspection time, and increases production rates.

The beaming characteristics of an X-ray machine can be useful in some cases. For example, when large, solid cylinders are radiographed diametrically, it can be advantageous to have a more intense beam in the center than at the outer diameter of the cylinder. Large solid-propellant rocket motor radiography is an example of how the use of optimum energy and field uniformity yields economical inspection. Special field flatteners may be constructed for special applications. When large numbers of items are radiographed with thickness variations greater than the combined latitude capability of the machine and film, a specially shaped compensator can be constructed to flatten the radiographic field. An example of this application is the radiography of large caliber artillery shells.

Intensity Distribution

Figure 15 shows plots of X-ray intensity distribution across the uncompensated beams of the 1 MeV, 2 MeV and 4 MeV Varian Linatrons. Figure 16 gives beam profiles for compensated and uncompensated 8 MeV Linatrons. Figure 17 shows a plot of X-ray intensities for an Allis-Chalmers 25 MeV betatron with and without a field flattener. Figure 18 shows plots for the 16 MeV Linatron, compensated and uncompensated. These plots, which illustrate the beaming effect and the degree of field flatness, can be used to determine variations in final film density from the center to the edge of a film, or to the edge of the field, if a film covers the entire field.

For example, when radiographing an object of constant thickness with a 35.5 × 43 cm (14 × 17 in.) film centered in the beam, and with a 183 cm (72 in.) source-to-film distance (SFD), the corner of the film will be 28 cm (11 in.) from the central ray (CR), representing an angle of 8.7 degrees. By using the field flatness charts, the relative intensity and therefore the film density can be predicted for the outer edges of the film.

Exposure Curves

Table 2 lists the physical density ratio conversion factors for converting a steel or solid propellant exposure curve into curves for a wide variety of materials.

PART 5

DEFINITION AND UNSHARPNESS

Definition

It is usually important for radiographs to exhibit the best possible definition, the highest possible contrast and the least amount of unsharpness. *Definition* can be defined as the ability to detect image edges on a radiograph and is related to the unsharpness. The amount of unsharpness is significant in high-energy radiography. Total image quality needs definition and high contrast to assure good radiographic sensitivity.

Unsharpness

In general radiography, three sources or kinds of *unsharpness* have been identified: geometric, inherent (of the film/screen) and scatter. Of these, the two most important are: U_f, inherent unsharpness of the film and screens; and U_g, geometric unsharpness due to the size of focal spot and the thickness and arrangement of the object. U_f is usually the major element of unsharpness in high-energy X-radiography. It increases with increasing radiation energy and film grain size; it is a function of screen material and thickness and is affected by the film's developing temperature, chemistry and processing technique. The values of U_f in Table 3 can be expected in high-energy radiography with lead screens, fine-grained film (speed index = 4, at density = 2.0), and automatic processing.

U_g is a linear unsharpness, described by the diagram in Fig. 19a and the expression:

$$U_g = \frac{F}{D/t} \qquad \text{(Eq. 2)}$$

where F is the focal spot size; D is the distance from the focal spot to the front surface of the object; and t is the thickness of the object or the distance from the front surface of the object to the film.

FIGURE 19. Geometric Unsharpness in Radiographs: (a) Diagram of Unsharpness; (b) Total Unsharpness as a Function of Focal Spot Size

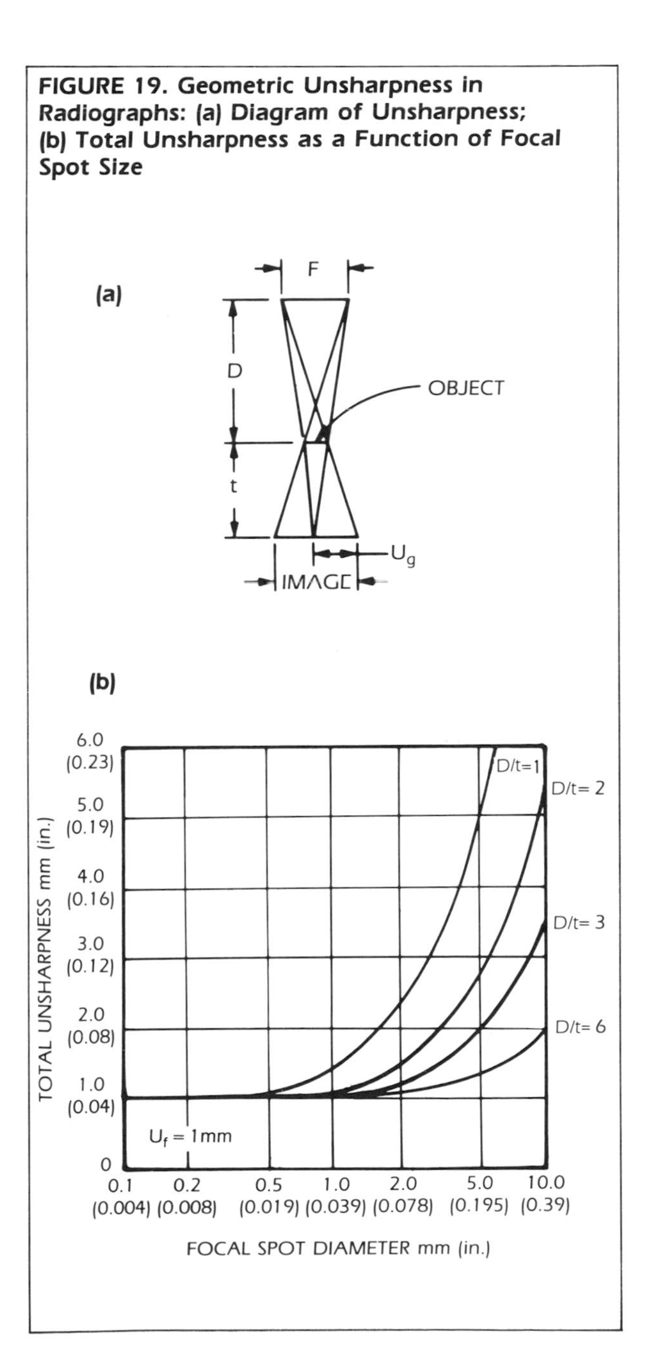

The total unsharpness (U_{TOT}) results from the combined effect of the inherent, scatter and geometric unsharpnesses, and can be expressed as the cube root of the sum of the cube of each, as:

$$U_{TOT} = (U_f^3 + U_g^3 + U_s^3)^{1/3} \qquad \text{(Eq. 3)}$$

where U_s is the unsharpness due to scatter.

In practice, however, the total unsharpness is reduced to include only the two most prominent causes, inherent and geometrical unsharpness. In this instance, the expression for total unsharpness will be:

$$U_{TOT} = (U_f^2 + U_g^2)^{1/2} \qquad \text{(Eq. 4)}$$

TABLE 3. Film Unsharpness for High-Energy Radiography

Energy (MeV)	U_f (mm)
1	0.15
2	0.3
4	0.4
8	0.6
10	0.8
16	1.0

PART 6

SENSITIVITY AND IMAGE QUALITY

Sensitivity

Most procedures, specifications and standards require that radiographic inspections demonstrate 2% sensitivity, based on the visibility of penetrameter wires or holes. However, most high-energy X-ray machines can achieve 1% sensitivity through a wide range of material thicknesses if proper precautions are taken with techniques. In determining machine capability, four kinds of sensitivity are commonly evaluated: thickness, wire penetrameter, plaque penetrameter and radiographic sensitivity.

Thickness Sensitivity

Thickness sensitivity, sometimes referred to as contrast sensitivity, refers to the ability of the radiographic inspection to demonstrate a thickness step by the visibility of the density difference the step produces on the film. With a reasonably sharp step edge in the image, the minimum density difference a trained eye can reliably perceive under good viewing conditions is approximately 0.006 density units. Depending on the energy of the source and the contrast achieved by the film and screens, this minimum perceptible density difference varies with energy and object thickness.

FIGURE 20. Wire IQI Sensitivity for Steel; Fine Grained Film, Lead Screen, Low Scatter

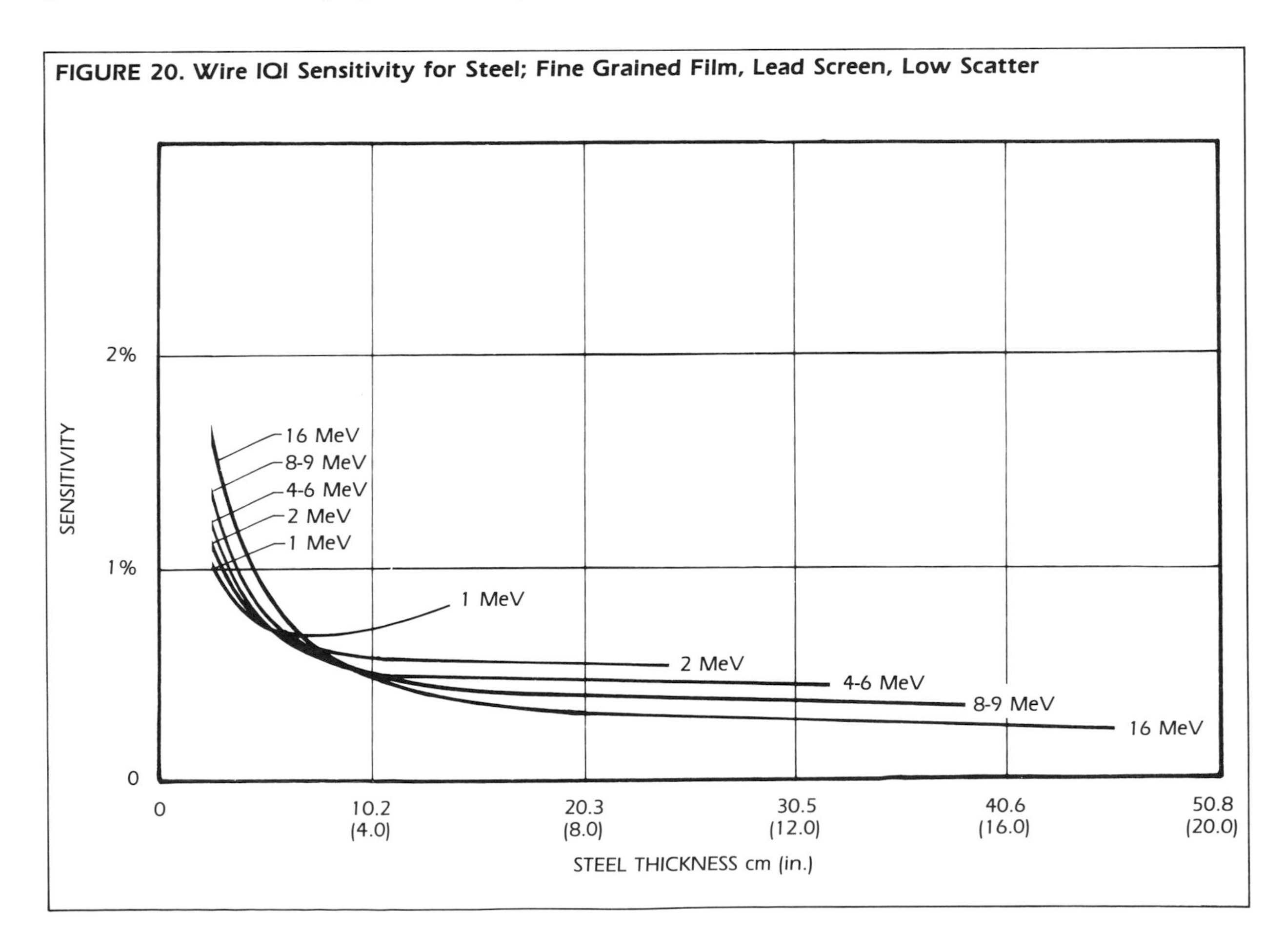

Wire Penetrameter Sensitivity

Wire penetrameters are in general use in Europe and in some applications in the United States. The German DIN wires come in 16 sizes, from 3.2 mm to 0.10 mm, in geometric progressions. Compared to discerning the image of a thickness step, wires are a little more difficult to see in a high-energy radiograph because of unsharpness factors and the graininess of the film. Wire sensitivities for several linacs are plotted in Fig. 20; here wire sensitivity is defined as the wire diameter divided by the object thickness. The figure shows, for example, that all machines should demonstrate better than 1% sensitivity above 3 cm (1.0 in.) of steel, and should achieve better than 0.5% for steel thicknesses above 15 cm (6.0 in.).

Drilled Hole Plaque Penetrameter Sensitivity

The drilled hole plaque has been the standard American penetrameter for all radiography. Drilled hole penetrameter sensitivities for several linacs are plotted in Fig. 21.

Intensifying Screens

Thickness

The intensification with a specific screen and for a given application can be determined experimentally by making exposures with and without the screen. With narrow beam geometry, the increase in film density with a screen is entirely due to the screen's intensifying effect. Figure 22 shows the maximum intensification obtained at several energies

FIGURE 21. Drilled Hole Plaque IQI Sensitivity for Steel; Fine Grained Film, Lead Screens, Low Scatter

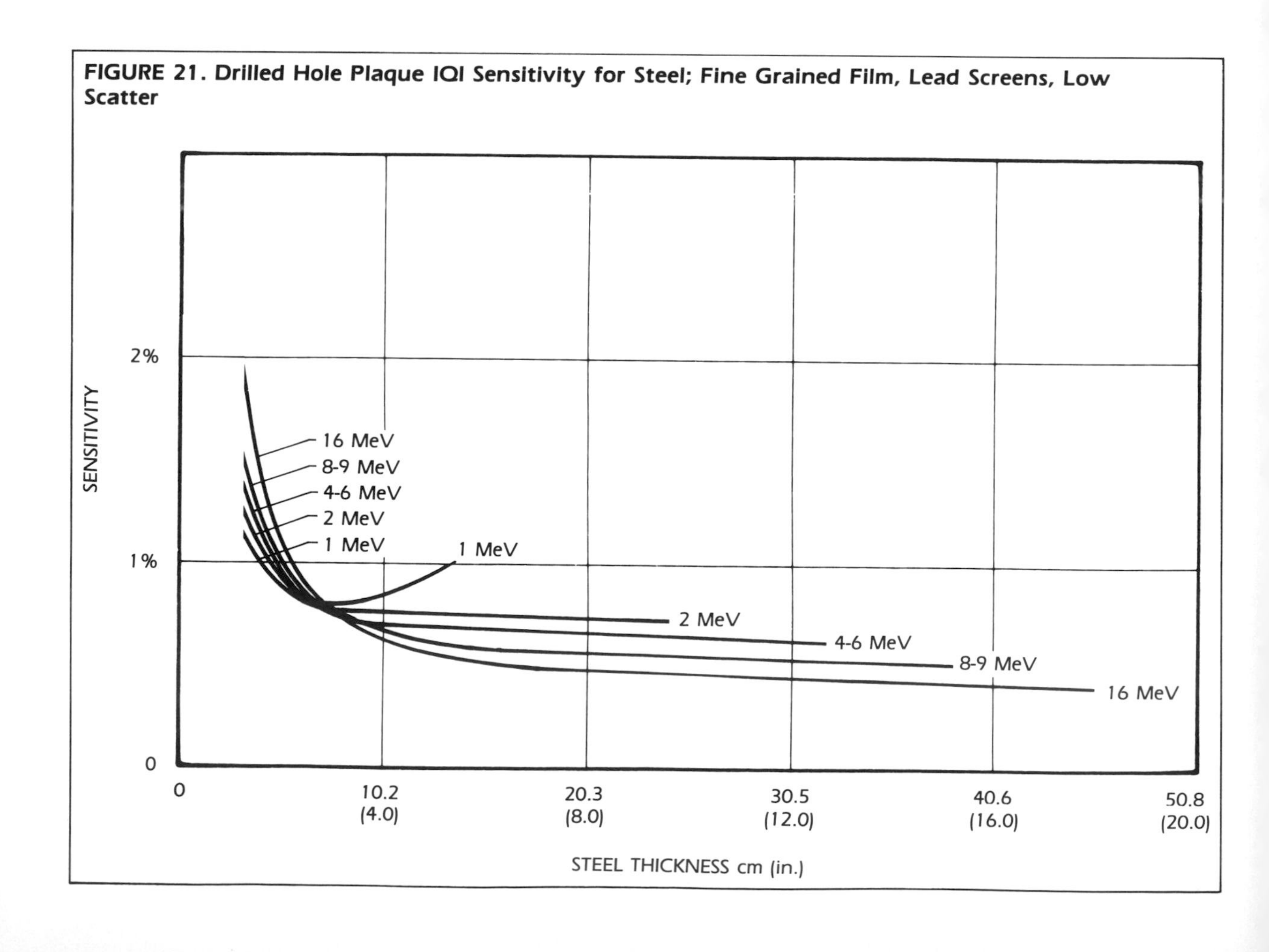

TABLE 4. Maximum Intensification

MeV	Front Lead Thickness for Maximum Intensification cm (in.)	Front Lead Thickness for Optimum Image cm (in.)
1	0.012 (0.005)	0.005 - 0.013 (0.002-0.005)
2	0.025 (0.010)	0.013 - 0.025 (0.005-0.010)
4	0.051 (0.020)	0.025 (0.010)
8	0.102 (0.040)	0.051 - 0.076 (0.020-0.030)
16	0.152 (0.060)	0.076 - 0.127 (0.030-0.050)

from the approximate lead screen thickness listed in Table 4.

In broad beam geometry, which is the normal radiographic arrangement, the filtering of the lead screen determines the relative amount of scatter that reaches the film. The net density on the film is the combined effect of filtering and intensification. As indicated in Fig. 22, intensification for narrow beam, low front-scatter conditions reaches maximum values for each energy. At 2 MeV, for example, a film density increase of about 0.8 density units would be observed for a 0.025 cm (0.010 in.) thick lead screen, when compared to an exposure without a front screen. That corresponds to a film exposure increase of about 110%. If normal amounts of front scatter impinge on the film holder and screen, as in a routine radiographic arrangement, an apparent intensification from 20% to 40% might occur, compared with an exposure without a front screen.

High sensitivity in high-energy radiography requires sharp images and high contrast. Image contrast is affected by the screen and film responses; therefore, screens must be selected and used carefully in order to obtain optimum results at each energy and for each application. Table 5 suggests lead screen and filter thicknesses that typically achieve optimum image quality at the various energies, for radiographic setups with low-scatter and high-scatter conditions. In practice, it may be desirable for the front and back screens to be of equal thickness to minimize handling difficulties.

Metal-phosphor Screens

Composite screens (consisting of a layer of fluorescent salt phosphor over a metal foil) have recently been developed to the point that they can sometimes be used in high-energy radiography. Composite screens can shorten exposure times beyond those used with lead screens and can also obtain degrees of contrast beyond those from lead screens. These composite screens can be used with ordinary X-ray film, and do not require special screen film. The metal foil, which can be tantalum, tungsten or lead, serves as the source of electrons which strike the phosphor, causing visible light photons to radiate and expose the film. Composite screens of lead and calcium tungstate are available commercially; rare earth phosphors such as gadolinium oxysulfide are a more recent addition.

The shorter exposure times obtained with these screens make it possible to use thin phosphors and thus obtain sharper images. The higher speed screens usually have a thicker phosphor layer. Typically, metal foil thicknesses ranging from 0.025 to 0.15 cm (0.01 to 0.06 in.) and phosphor thicknesses less than 0.05 cm (0.02 in.) provide the optimum range of speed, contrast and image sharpness. Table 6 lists some relative speeds for various types of fluorescent screens. Fluorescent screens must be well maintained, carefully handled and kept clean. Dust particles leave images (white spots) on the film. Also, there are reports of screen deterioration with age.

FIGURE 22. Front Lead Screen Intensification, 1 to 16 MeV X-rays

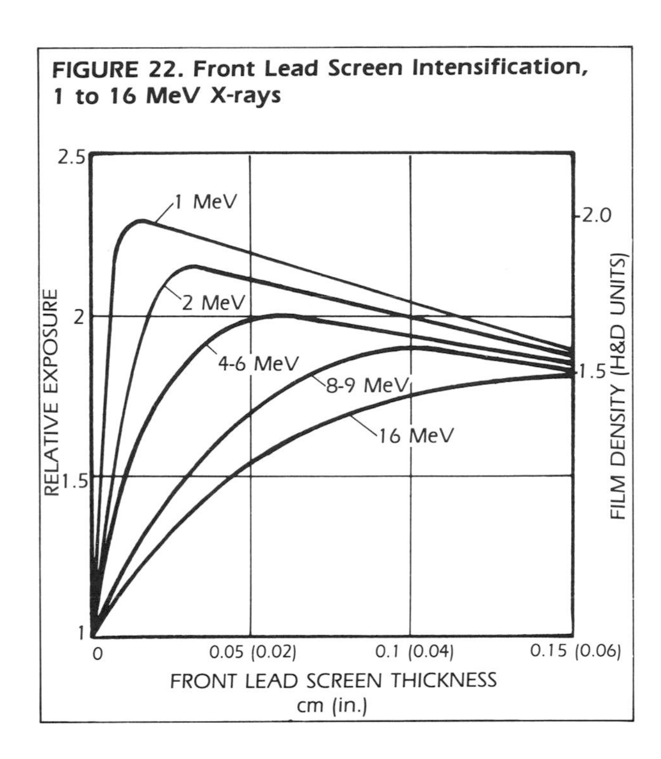

TABLE 5. Suggested Screen and Filter Thickness

Energy (MeV)	Radiographic Conditions	Screen Thickness cm (in.)		Remarks
		Front	Back	
1-4	Flat object, low scatter, up to 10 cm (4 in.) steel or equal	0.025 (0.010)	0.025 (0.010)	backing lead as needed*
	Complex object, high scatter, or thick object over 10 cm (4 in.) of steel or equal	0.051 (0.020)	0.025 (0.010)	0.076 cm (0.030 in.) lead or composite object filter may improve sensitivity**
6-10	Flat object, low scatter, up to 12.7 cm (5 in.) steel or equal	0.051 (0.020)	0.025 (0.010)	backing lead as needed*
	Complex object, high scatter, or thick object over 12.7 cm (5 in.) of steel or equal	0.076 (0.030)	0.025 (0.010)	0.076 cm (0.030 in.) lead or composite object filter may improve sensitivity**
12-25	Flat object, low scatter, up to 15.25 cm (6 in.) steel or equal	0.076 (0.030)	0.025 (0.010)	backing lead as needed*
	Complex object, high scatter, or thick object over 15.25 cm (6 in.) steel or equal.	0.127 (0.050)	0.025 (0.010)	0.318 cm (0.125 in.) lead or composite filter may improve sensitivity**

* Back lead filters may be placed behind film holder if there is a large amount of backscatter present; 0.64 cm (0.25 in.) thickness should be adequate for all cases.
** Object filters are placed between the object and the film holder to reduce the effect of secondary or scatter radiation or scatter generated by the object.

TABLE 6. Relative Intensification Index of Screens*

Screen (Front and Back)	X-ray Energy (MeV)			
	1-2	4-6	8-10	16
Lead (optimum thickness)	1	1	1	1
Lead and tungstate salt composites				
Very high speed	0.14	0.20	0.33	0.50
High speed	0.33	0.50	0.67	0.77
High definition	0.60	0.70	0.8	1.0
Lead and gadolinium oxysulfide composites				
High speed	0.20	0.33	0.50	0.67
High definition	0.55	0.67	0.80	1.0

*The intensification index is arranged with decreasing numbers for faster screens, corresponding in proportion to the exposure in rads needed to produce a specified density on film.

PART 7

X-RAY FILM CHARACTERISTICS AND RADIOGRAPHIC VARIABLES

Film Speed

The sensitometric properties (the relative speed) of commercial X-ray film can be established by exposing each film in a real radiographic arrangement, until it will produce a film darkening, when developed, of a specified density on the Hurter-Driffield (H&D) scale. If all other variables are held constant, a relative film speed index can be assigned to each film, with a numerical value that is proportional to the exposure (in rads or grays) needed to obtain the 2.0 film density. The data for high-energy radiography presented in Table 7 were obtained using this procedure.

Other X-ray film characteristics (contrast, response curves, characteristic curves, gradient and the use of multifilm techniques) are substantially the same for all radiographic energy levels and discussions of these topics are included elsewhere in this Volume. Following is a discussion of the radiographic variables which most directly affect the quality of a high-energy radiograph.

Collimation

Collimation is the elimination of unwanted portions of the X-ray beam at the source by the use of shielding. Most high-energy X-ray machines have collimators built into the exit opening of the machine. These are sometimes made from lead, tungsten or depleted uranium. Some are fixed in size and shape and some can be adjusted, manually or remotely by electric motors. Remotely adjustable collimators are excellent for use with real-time imaging systems, where the varying size of the opening can be seen on the imaging screen. In radiography, collimators should be of the size and shape needed to cover only the area being radiographed.

Magnification and Image Sharpness

Magnification and unsharpness occur to images of internal conditions as they are projected onto film. This is caused by the fact that the image recording medium is located some distance from the internal condition. Certainly, images of defects that occur on the film side of the object have very little projection or magnification and therefore are usually sharper. Because of the physics of image formation with a nonpoint source of radiation, the sharpness of all images depends upon the amount of projection involved. Some images of small voids that are located well up in the front (source side) of the object may be totally unsharp and only visible on the radiograph because of high contrast on the film. Within the practical limits permitted by focal spot size and other factors, moving the film away from the object will reduce the intensity of the scatter radiation originating from within the object and may improve sensitivity for the case where the film/screen unsharpness is greater than the geometrical unsharpness.

Projection radiography with the resultant magnification can intentionally be used to advantage when the factors controlling geometrical unsharpness are correct. The betatron, because of its small focal spot size, is well suited for projection radiography. With such a small focal spot, high magnification is possible. This can enlarge small detail which might otherwise be unseen or too small to measure.

This projection technique can also be used for radiography of highly radioactive materials. It decreases the fogging effect of the radioactive object and allows the image-forming radiation to contribute the majority of film darkening. In this case, the beneficial effect is not the magnification but the distance from radioactive object to film and its effect of minimizing fog and therefore improving contrast.

TABLE 7. Relative Speeds of Industrial X-ray Film at High X-ray Energies with Lead Screens and Automatic Processing

Film Type	Film Density (H&D Units) 1.5	2.0	2.5	3.0	3.5
	Film Speed Index				
DuPont NDT 75	0.35	0.5	0.7	0.9	1.2
Kodak AA	0.7	1	1.3	1.6	2
Gevaert D-7	0.7	1	1.3	1.6	2
DuPont NDT 70	0.7	1	1.3	1.6	2
DuPont NDT 65	0.8	1.1	1.4	1.7	2.1
DuPont NDT 55	1.2	1.8	2.4	2.8	3.4
Kodak T	1.4	2	2.5	3	3.7
Gevaert D-4	2.2	3	4	5	5.6
Kodak M	2.7	4	5	6	7
DuPont NDT 45	5	7	8.7	10	11
Gevaert D-2	5.5	7	8.7	10	11
Kodak R	7	10	12.6	15	17

Blocking

Blocking (masking) is the elimination of unwanted portions of the X-ray beam (and scatter) at the object through the use of lead bricks, shot-filled bags and other shielding materials. Blocking is needed less often in high-energy X-radiography than in the low and medium kilovoltage range, but it may be required when large thickness differences in an object permit production of high-intensity extraneous scatter, or when the object is several HVLs thick and is smaller than the radiation field.

A radiographer can judge the effectiveness of the blocking in a particular setup by observing the film density in the area of the radiograph over which the lead shielding projects. If that area is of very low density, which indicates that it has received minimal radiation, it is a good indication that the desired imaging area also received little scatter.

A second test of the effectiveness of scatter reduction by blocking is to compare the exposure and density for the particular application with standard exposures and densities from an exposure curve or other reference. If a shorter exposure of the material gave a higher film density than other well-blocked exposures, it is very likely that scatter has contributed to the exposure.

Filters

Absorbing plates or assemblies placed between the object and the film holder may be necessary to achieve the best image quality when:

1. the object shape, position or composition produces extensive scatter;
2. large object thickness produces scatter greatly in excess of the primary radiation;
3. the object is smaller than the radiation field;
4. large thickness differences exist in the object; and
5. high-speed photoelectrons produced in the object may reach the film holder.

PART 8

RADIOGRAPHIC APPLICATIONS

Discussion of Requirements

Practices in the high-energy radiography of castings, welds, propellants and explosives, assemblies, and some special applications are described. The information presented is of a general nature but can be used in preparing specific operating procedures for a high-energy radiographic facility.

Safety

With the advent of machines which are capable of higher and higher radiation output, radiation safety practices have become a matter of great importance. Some machines are capable of delivering a potentially lethal dose of radiation to the body in a few seconds. Some are powered by radiofrequency generators and when turned on for maintenance or repair, even with the electron generating filament disconnected from the system, can generate X-rays due to acceleration of free electrons within the system. Likewise, machines which utilize the principle of a revolving belt can generate X-rays when the belt is turning, even though the electron sprayers have been disconnected from the system.

Some X-ray generators utilize internally installed collimators made from depleted uranium which may be a source of radiation and require licensing by the Nuclear Regulatory Commission. Other collimators manufactured from lead can become a source of neutron production when placed in a high-energy radiation field (see Section 12, "Neutron Radiography"). These neutrons can be a personnel hazard as well as a detriment to film quality. Precautions must be exercised to attenuate these neutrons to protect operating personnel. Normal industrial safety hazards such as electrical shock, moving equipment, noise and tripping hazards are always present in the work place and must be addressed. Additionally, facilities which inspect propellants and explosives are faced with the hazards inherent in working with these materials. To optimize safety, each facility must be studied thoroughly for radiation and other hazards; a safety and operating procedure must be written, identifying hazards and demonstrating methods of operation which prevent injury to personnel.

Radiation Precautions

There are two major forms of hazardous radiation generated by high-energy radiographic sources: photons and neutrons.

As previously stated, the generation of neutrons is an unwanted side effect, which complicates shielding requirements. However, because this Section focuses on bremsstrahlung X-rays, the discussion on shielding will emphasize only materials which provide X-ray absorption.

Shielded rooms for high voltage X-ray sources are usually constructed to provide the required personnel protection as first priority, facility utilization requirements as second priority and costs as third.

Personnel protection can be provided by shielding either the work area (housing the X-ray source) or shielding the control area (housing the operating personnel) and isolating access to the X-ray source. In locations where space is not at a premium, an inexpensive method of isolation is accomplished by placing fences at sufficiently great distances to reduce dose rates to acceptable levels. The control area containing unexposed film and all operating personnel can be modest in size, thereby reducing the cost of shielding.

When shielding of the work area is required, the first steps to consider are: (1) the directions in which the primary beam can be pointed, and (2) the access requirements.

Personnel access must be studied to determine whether a labyrinth or direct opening door is the most useful. A labyrinth shield is not suitable if large parts must be moved into the work area; if this is the case, swinging or rolling doors should be considered. When making a decision on work area access, it is helpful to remember that any photon which has impinged on at least three surfaces will be attenuated to a safe level of radiation.

From an economic point, it is usually best to use concrete as the main shielding material. Arguments can be made for other materials, especially if there is a potential need to relocate the facility. With high-energy, high output machines, it is not uncommon to have required concrete shield wall thicknesses in excess of 152 cm (60 in.) for the direct beam and 91 cm (36 in.) for scattered radiation; the cost of such shielding is often high.

The attenuating ability of most shield materials is directly related to the material density. Of the common shield materials, lead is the most absorbing at 11.35 grams per cubic centimeter. While lead is absorbent, it has some less desirable features, such as cold flow (sag) and the resulting need for substantial skeleton structures to support it. It is usually necessary to laminate lead with steel sheet or plywood exterior to reduce the chance of mechanical damage from fork lifts, etc. Lead is also subject to melting in the event of a building fire. On the other hand, it takes up the least space of any common shielding material, and with a 5% antimony alloy, it is stiff enough to support its own weight.

Concrete, with a density of about 2.3 g/cm³, is about five times less dense than lead; it can, however, be rather inexpensively constructed to fit customer needs.

In some instances, it is possible to use topographic

FIGURE 23. Typical High Energy Radiographic Installation

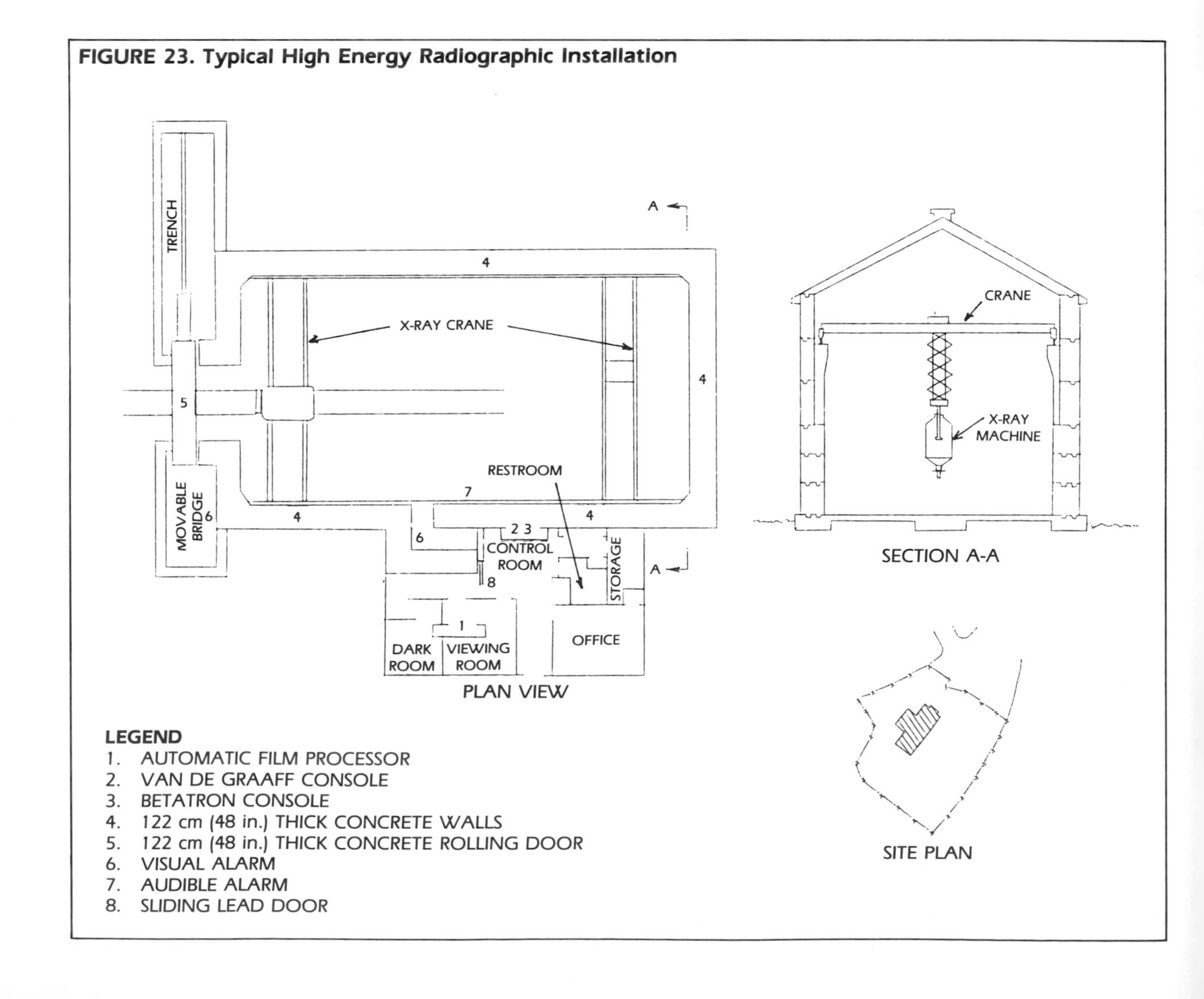

characteristics to advantage by placing the radiographic facility on a hillside, excavating the space needed and then partially burying the top and sides, allowing for the access wall. This can result in a major cost reduction without any sacrifice in safety or accessibility.

Other shield materials include iron, with a density of 7.8 g/cm^3 and water with a density of 1.0 g/cm^3. Iron is most often used for doors and the other moveable portions of the enclosure, to simplify structural support requirements. Water is most often used where there is a need to reduce the neutron flux from a high-energy facility. In some cases, a flooded roof is used to reduce sky-shine. In other cases, water is used to line a labyrinth to absorb scattered neutrons. In general, water is not effective for photon absorption because of its low density.

Electrical Shock Precautions

High-energy X-ray machines operate on incoming power with AC voltages of 208 V, 220 V or 440 V. Additionally, extremely high voltages exist in capacitors, pulse-forming networks and transformers located in several places within these machines. Such voltages can range from 3000 to 60,000 V and because these voltages can be lethal, utmost caution must be exercised when repairing or servicing the equipment. Most machines are interlocked with the designed intent of preventing accidental contact with high voltages. Interlocks should never be bypassed. Capacitors will remain charged after the incoming voltages have been disconnected. Shorting sticks should be used to discharge these high voltage circuits before the operator touches any circuit component.

General Requirements

When considering high-energy radiography, one must be aware of the equipment, building and personnel requirements which constitute the entire inspection system. Requirements for a typical high-energy radiographic installation such as the one shown in Fig. 23 can be extensive.

It is extremely important to keep careful records of all radiographic work. The daily log should show the names of the radiographers and the identification of each part radiographed. In addition, the log should contain the following:

1. date and time of shipment and identity of material;
2. receiving inspection findings, noting any unusual conditions associated with handling or shipping damage;
3. number of exposures and film made on the object;
4. identification of specification(s) and procedure that apply to the job; if the exposure plan is not on record at the X-ray facility, it should be sketched; and
5. report number and date, if applicable.

FIGURE 24. Typical Rocket Motor Configurations: (a) Longitudinal Section; (b) Transverse Sections

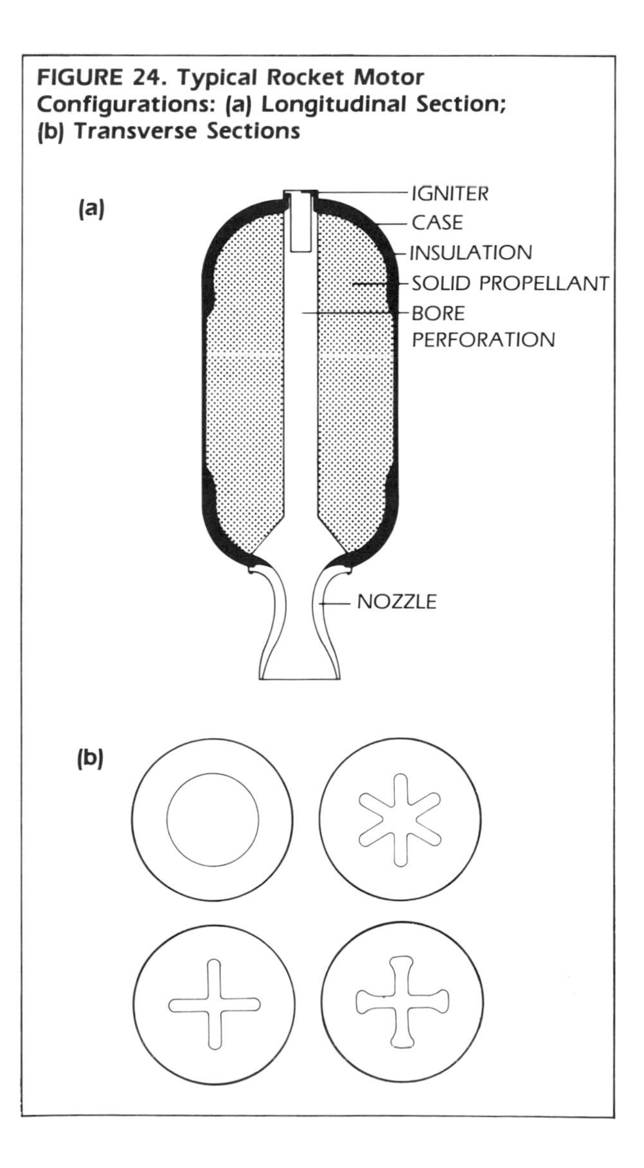

In addition to the job information, a daily log should be maintained containing information on radiation hazards and operation of all equipment, including safety devices, interlocks, warnings, etc. All malfunctions should be dated and identified. Checks of all the equipment after repairs or adjustments should be logged.

Exposure Room Equipment

In addition to the high-energy X-ray machine and its manipulation system, an X-ray facility may require the following:

1. materials handling crane, forklift, or other means to move and set up the parts that require X-ray examination;
2. electrical power outlets;
3. air compressor or compressed air line;
4. grounding system leading to an outside earth ground;
5. waist-high exposure table; and
6. film cassette support and holders.

Accessory Equipment

Along with this general equipment, a facility designed to examine objects such as large rocket motors or other similar assemblies will require yokes, belt slings, roller support frames, motorized turntables, and tangential shielding supports. Other accessories and supplies for radiography are listed below:

1. image quality indicators (IQI) applicable to the standards and the type of material being radiographed;
2. deep-block lead numbers and letters;
3. film holders (all sizes), rigid cardboard, soft plastic and vacuum cassettes;
4. lead screens from 0.025 to 0.076 cm (0.010 to 0.030 in.) thick, assorted sizes, some unbacked and some with cardboard backing;
5. lead sheets 0.32 to 1.27 cm (0.125 to 0.5 in.) thick, in all cassette sizes;
6. lead or copper shot in canvas bags;
7. lead bricks;
8. calipers and other wall-thickness gages;
9. survey meters for radiation protection surveys; also *R*-meters, pocket dosimeters, film badges, warning lights, signs and other safety equipment;
10. technique charts for each material to be radiographed;
11. well-developed operating procedure manuals for all equipment;
12. film processing equipment, temperature-controlled, automatic if the volume of work requires it;
13. film viewing equipment with variable intensity viewers;
14. magnifying glass, 3× to 5×, for film reading; and
15. densitometer with capabilities up to density 4.0 H&D.

Film Processing Facilities

Since a high-energy radiographic facility is often in a separate building it may be necessary to have film processing capability included in this remote location.

Radiography of Castings or Welds

High-energy radiography of castings or welds has a major advantage over lower energy radiography: its increased latitude and decreased wide-angle scatter, which usually combine to produce a very read-

FIGURE 25. Cross Section of Front End of Solid Propellant Rocket Motor, Showing Construction Details for Radiographic Examination

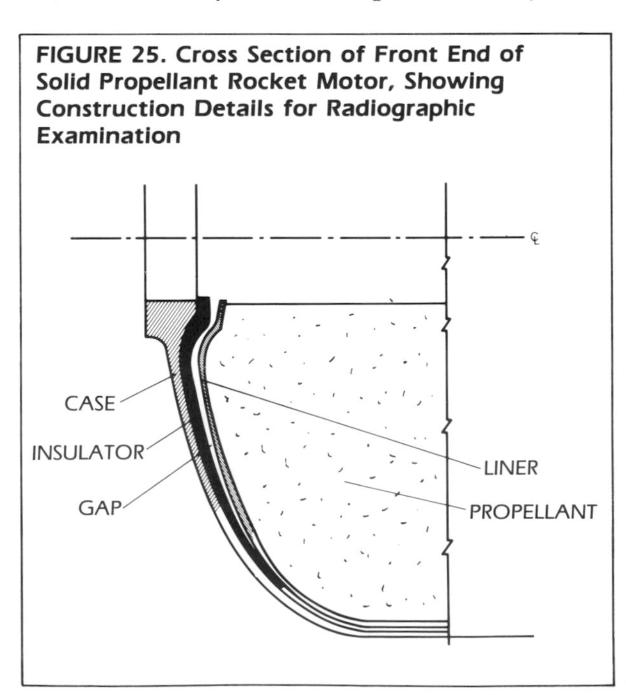

able film. However, high-energy X-rays do produce radiographs with less contrast. Thus, the imaging of fine cracks and filamentary porosity can be difficult.

Double Wall Radiography

When applied to valve bodies, for example, high-energy X-rays can penetrate both outside walls and provide the necessary sensitivity for demonstrating internal section defects, especially in the critical valve seat areas. Exposures are made by placing the film behind the valve body and beaming into the object at the best angle to project the critical internal section onto the film. This same double wall radiography is used on welded pipe and fixtures where the internal surface is not accessible for film placement. Blocking, filtering or other scatter reduction techniques should be used to ensure maximum sensitivity; an adequate *D/t* ratio is needed to minimize distortion and undesirable enlargement.

Radiography of Propellants

Rocket Motors

Solid propellant rocket motors are made with diameters of 5 cm (2 in.) or less to 305 cm (120 in.) or more, with a variety of bore configurations, some of which are shown in Fig. 24. These configurations have a designed burning surface area that produces a predictable pressure/flight curve. If this area is increased by the presence of a crack, void or separation, overpressurization of the case can occur, thus causing a malfunction.

Essentially, a solid propellant rocket motor consists of a rigid case, an internally bonded insulator and liner, and the solid propellant. These arrangements are shown in Fig. 25. One or more nozzles at the back end complete the basic motor. The case may be made of wound and epoxy-bonded glass or other fiber material, high strength steel, or titanium. The insulator and liner are often made of an asbestos and rubber composition.

In general, the propellant in large motors is adhesively bonded to the liner to provide structural support and to restrict the burning to the bore surfaces. There are many types of propellant; the two most common are a rubbery mixture of an organic fuel/binder and an oxidizer, and a more rigid double-base compound made with plasticized nitroglycerine. By the nature of their design, solid propellant rocket motors provide low subject contrast when radiographed; therefore, every precaution must be taken to increase the radiographic contrast. This contrast can be greatly enhanced by selecting the

FIGURE 26. Support Equipment for Rocket Motor Radiograpy; (a) Roller Support Stand for Horizontal Inspection; (b) Turntable for Vertical Inspection

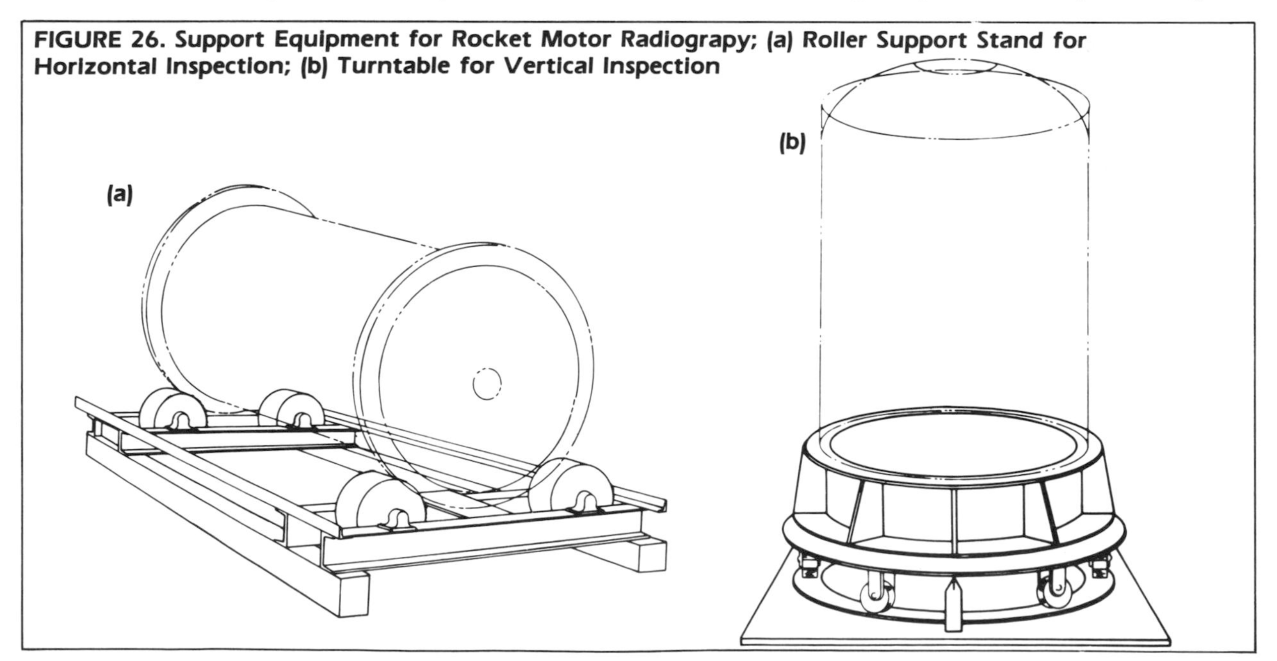

proper X-ray energy for radiography and by using digital video processing.

As a rule, radiography of the propellant grain and the peripheral bonded regions, including the front and back domes, is required. Radiographs of the grain should reveal the presence of cracks, voids, nonuniform mixing, lack of bond between internal interfaces, bore deformation, foreign material, excessive porosity, low-density volumes, and other defects that are characteristic of the propellant type and the manufacturing process.

Tangential radiography of peripheral areas reveals liner-propellant bond failures, deterioration or separation of the liner-propellant bond, insulation failure, peripheral porosity, and other defective or substandard conditions. Radiographs are also used to judge the quality of rocket motor repair work.

Motor Attitude and Support Equipment

Radiography of many solid propellant rocket motors is best accomplished with the motor horizontal on a roller support stand, as illustrated in Fig. 26. Of course, the structural strength of the motor must allow it to be placed in the horizontal position, and rings must be attached to the motor to support it and permit its rotation. With the motor arranged in this manner, the X-ray source can be positioned above the area to be tested, and film for the grain exposures can be placed on a shelf below. This is a very convenient production arrangement. It is easily changed to accommodate larger or smaller motors; it makes the bore accessible; and it allows tangential shields and film holders to be placed along both sides of the motor for tangential exposures.

FIGURE 27. Visualization of Crack-like Defects, as a Function of Propellant Thickness and Crack Angle, for $\Delta x/x = 0.2\%$

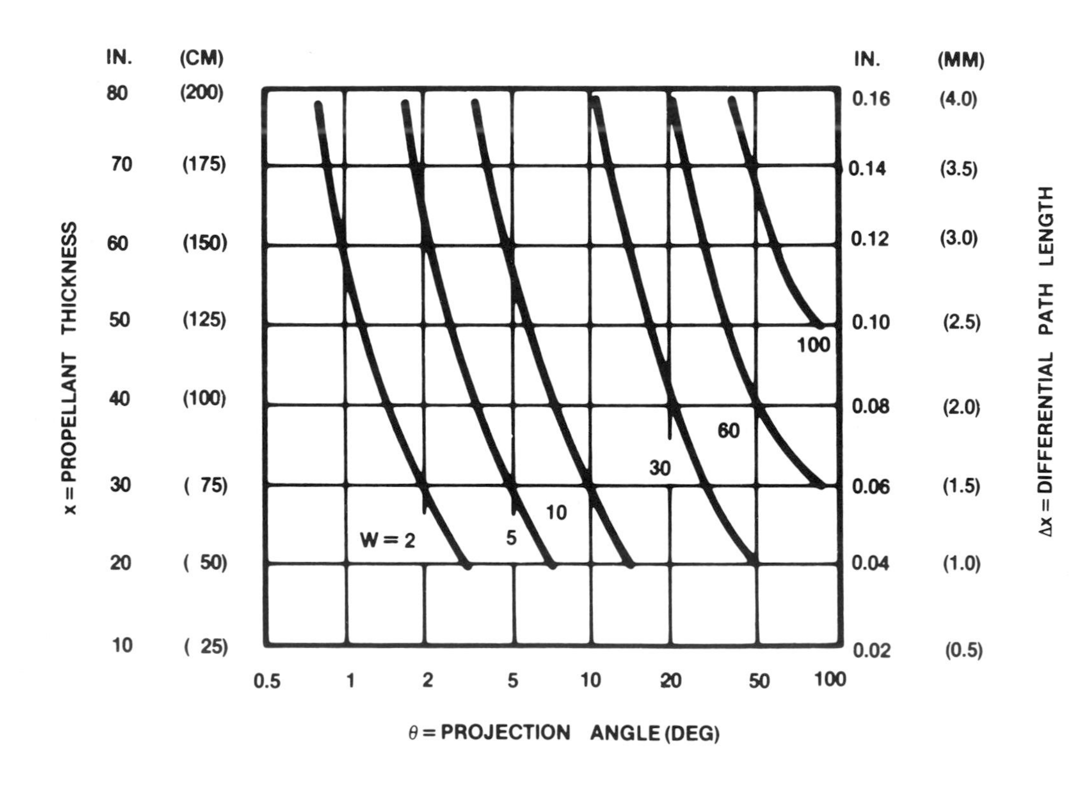

Radiography must be accomplished with the motor in the vertical attitude when the structural strength of the motor case will not support the weight of the motor in the horizontal attitude, or when this placement is convenient because of motor production techniques. This type of radiography requires the use of a turntable as shown in Fig. 26. For large motors these tables are usually powered, and in some facilities they are placed on an elevator to permit making longitudinal changes in the exposure locations. In addition, adapter rings, handling yokes, belt slings, and other special motor-handling equipment may be required.

Rocket Motor Radiographic Procedures

Radiography of the grain of a solid propellant rocket motor is usually performed in accordance with an established layout and exposure plan that is designed to afford complete coverage of the grain, including all the regions that are known to be critical. In a typical plan, the cylinder is divided into a number of segments that correspond to the film centers. These segments may be exposed individually, or they may be grouped into exposure regions, with two, four or more films exposed simultaneously. In all cases, in addition to specifying the film locations, the plans should show the exact location and direction of the central ray. A significant amount of film overlap is used in most plans to ensure complete coverage. The overlap also allows some of the propellant to be exposed a second time at a different angle, which offers an added opportunity to show cracks that may be present but at an angle that would not allow detection on the first exposure. The dome and tangential areas require separate exposures. Plaque or wire-type image quality indicators should be placed on the front (source side) of the motor, so that at least one image appears on a film in each exposure. Plaque-type IQI thickness sensitivity of 1% is routinely demonstrated.

Source-to-film Distance

The source-to-film distance should be as large as practical, although a D/t ratio of 3:1 produces acceptable radiographs through propellant grains. When calculating the film coverage for a grain exposure layout, the angle from the central ray to the edge of the film should not exceed 15 degrees.

Screens

Lead front and back screens (or equivalent screens of other materials such as tungsten, copper, tantalum, or fluorometallics), 0.05 cm (20 mils) thick, should be used. In some instances, objects are present near the path of the X-rays, causing forward scatter to be generated; cassette filters may be required to control the effects of this scatter, or additional lead may be added to the front screens.

FIGURE 28. Radiograph Showing Propellant-to-Insulation Separation in Dome Area of Rocket Motor

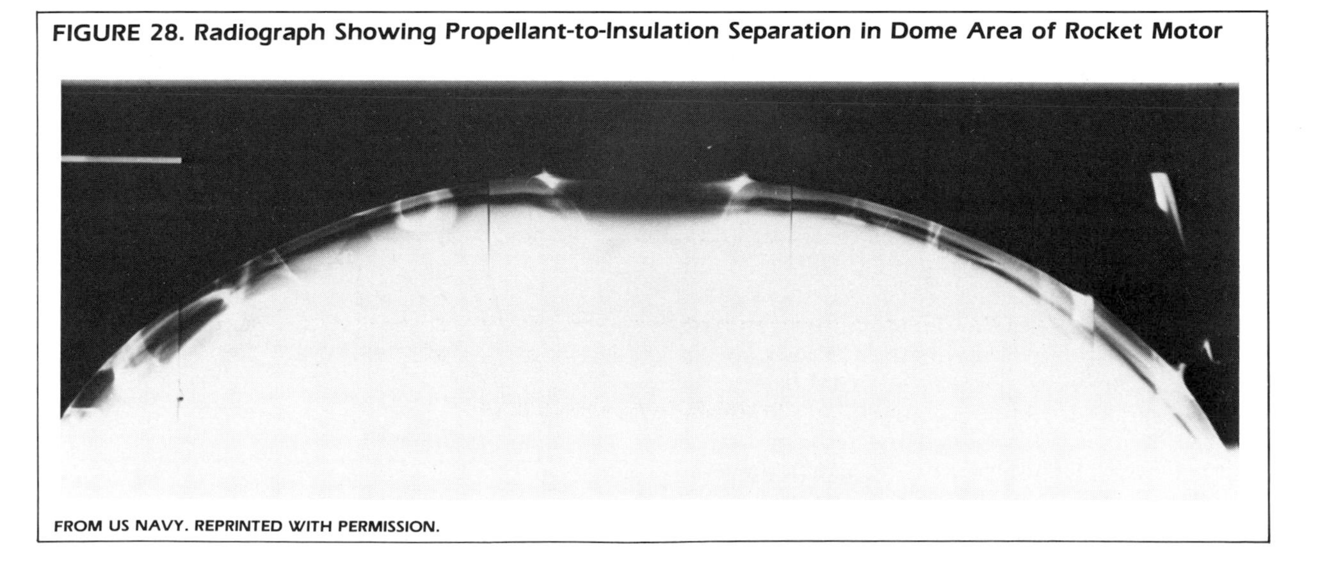

FROM US NAVY. REPRINTED WITH PERMISSION.

The film holders should be placed as close to the back of the motor as possible for the through-body exposures, and ideally they should be enclosed in an evacuated envelope to ensure film-screen contact. If a vacuum system is not available, then film-screen contact should be maintained by spring-pressure cassettes or other mechanical means.

Film Density

Ideal film densities, when testing propellants, are from 1.5 to 3.0. When thicknesses are greater than those that will render these densities, the multiple (two-speed) film technique should be used. Although film exposure curves will normally indicate film combinations that can be used for various thicknesses, some experimental exposures are usually required to finalize film selection and exposure. Sometimes a thin lead foil is used as a separator between films when optimum intensification and definition are required. Most often the intensifying effect of adjacent film, without an interleaving screen, provides the necessary extended latitude.

Grain Radiography Coverage Requirements

With optimum radiographic production techniques, contrast sensitivities of better than 1% can be achieved with most high-energy X-ray machines. Some machines are capable of demonstrating 0.2% contrast sensitivity. At that sensitivity, defects such as voids and gross density variations in the solid propellant are readily detected. When these are the only concern, the simplest radiographic procedure (e.g., one exposure through the grain or two exposures, one at 0 degrees and the second at 90 degrees) may be used. However, cracks and separations are also of great concern in rocket motors. Cracks with widths of 0.02 to 0.15 mm (0.001 to 0.005 in.) are detected only when they are aligned with the radiation source. Most propellants which crack will open to a width of 0.2 to 0.8 mm (0.01 to 0.03 in.) minimum. At this width, cracks can readily be detected at almost any angle to the radiation, through large thicknesses of propellant. Figure 27 illustrates the width of crack-like defects that can be imaged through various thicknesses of propellant, at various projection angles, with an assumed contrast sensitivity of 0.2%. This sensitivity level can be achieved if optimum laboratory techniques are used.

Reference is made to the approximate relationship $\Delta x = W/\Theta$, where Δx is the change in propellant thickness, Θ is the projection angle and W is the crack width. It can be seen from Fig. 27 that the ratio W/Θ increases as propellant thickness increases. This means that as the propellant thickness increases the same width crack will be shown at smaller projection angles (see Table 8).

To obtain the required crack detectability in rocket grains, the number of exposures must be increased as the diameter increases. When the technique being used produces contrast sensitivity other than $\Delta x/x = 0.2\%$, the procedures must be modified to obtain equivalent defect detectability.

Tangential Radiography

Propellant-to-insulation separations caused by failure of the adhesive bond are considered critical defects wherever they occur in a rocket motor, and separations in the domes are viewed as more serious than those that may occur in the cylinder. Figure 28

FIGURE 29. Probability of Detecting Propellant-to-Insulation Separation Versus Number of Exposures for Various Circumferential Lengths of Separation (L_c) for 137 cm (54 in.) Diameter (D) Rocket Motor, $P = L_c/\pi DN$

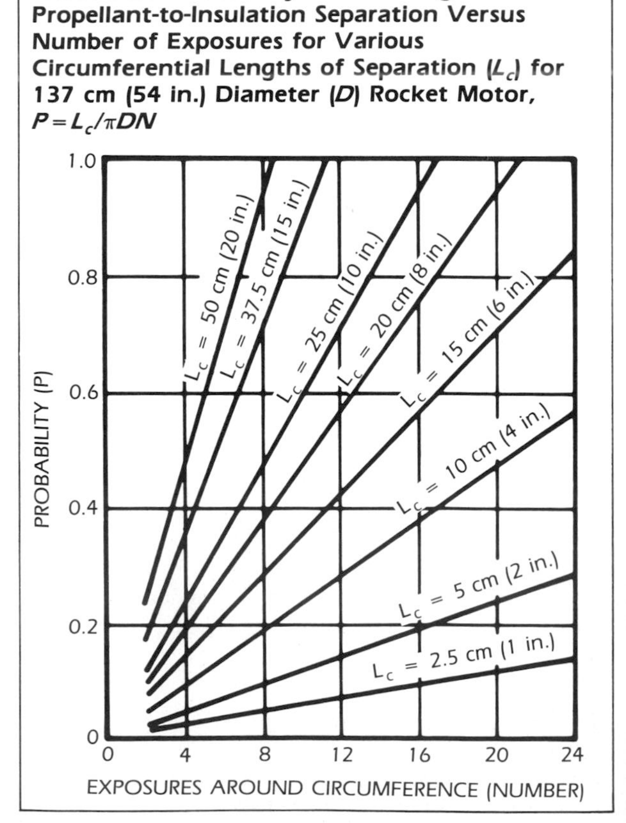

illustrates propellant-to-insulation separation in the dome area of a large solid propellant rocket motor. With the proper technique, separations as small as 0.05 mm (0.002 in.) can be shown on the radiograph when the separation occurs at the tangential point. Tangential radiography is performed by placing the areas of inspection at nearly right angles to the X-ray beam. The X-rays forming the outermost part of the cone of radiation will pass through the interfaces formed by successive diameters of the propellant, insulator and case. By placing the film against the motor case, below the inspection area, a radiographic image of the tangential points of successive interfaces is formed. The detectability of a separation depends upon the number of tangential exposures, as illustrated in Fig. 29. For example, 12 tangential exposures evenly distributed around a 137 cm (54 in.) diameter motor have 50% probability of showing an area of separation that extends 17.75 cm (7 in.) around the periphery and 100% probability of showing an area extending 38 cm (15 in.) or more around the periphery.

Radiographic Coverage

Radiography of the peripheral areas of the dome and cylindrical areas of a solid propellant rocket motor requires an exposure plan similar to that for the grain. In fact, the same layout and marking may be used for the tangential exposures. The central ray positions depend on the size of the motor and on the particular source to be used. When the X-ray machine has high output intensities and a large radiation cone, the grain exposure with simultaneous bilateral tangential exposures can be made by directing the central ray radially. With a less powerful source or one with a small cone of radiation, only one side of the motor can be exposed at a time. In general there is no need to use the multifilm technique, since one film may have sufficient latitude to show the critical areas inside the case. As with all rocket motor radiography, some experimental exposures may be needed to finalize the optimum technique.

Forward scatter generated when taking tangential radiographs with a high-energy X-ray machine will have a very low angle to the primary beam; therefore, tangential shielding may not be required. When exposures with very high contrast sensitivity are required, a lead shield can be added to the outer surface of the motor at the tangential point of the

TABLE 8. Ratio of Crack Width to Projection Angle for Various Propellant Thicknesses

Propellant Thickness (cm)	Ratio of Crack Width to Projection Angle (mm per angular degree)
75 -100	$W/\Theta \sim 0.025$
100 - 130	$W/\Theta \sim 0.040$
130 - 150	$W/\Theta \sim 0.050$

FIGURE 30. Radiograph of an Explosive-loaded, Fused 5 in. (12.7 cm) Projectile

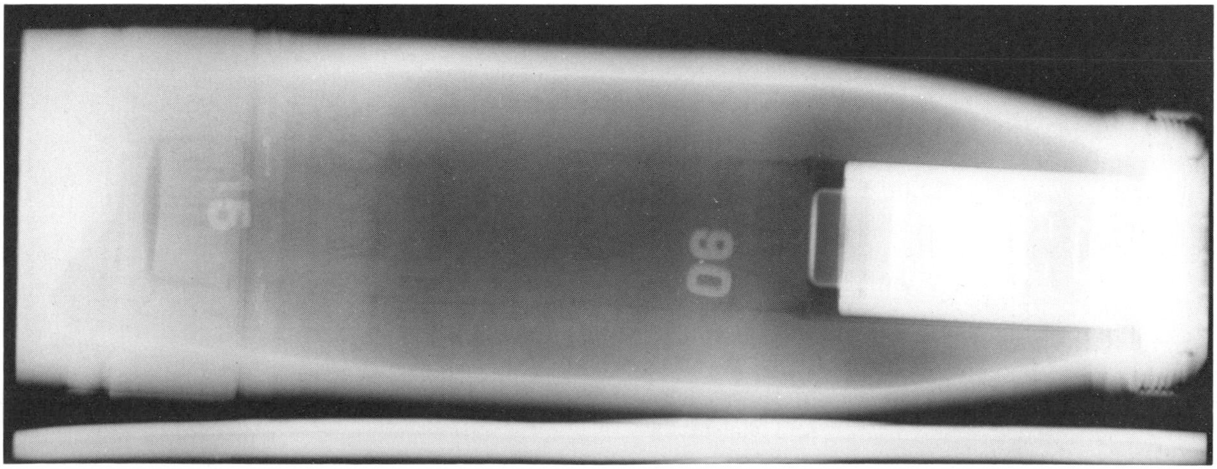

FROM US NAVY. REPRINTED WITH PERMISSION.

area being covered. The attenuation of the shield should be approximately the same as the average attenuation of the chord at the propellant interface. With known thicknesses of the case and insulation, the total equivalent chord length, and a good approximation of the X-ray attenuation in the case, insulation and propellant can be calculated from simple geometrical relations. This calculation should also be made as the basis for choosing the thickness of the image quality indicator that will be used.

Wire penetrameters and plaque-type penetrameters with slits built into them and with thicknesses often less than 1% of the total equivalent chord at the propellant interface, are routinely employed in tangential radiography. Lead screens and filters should be used as in the grain exposures. Because the curvature of a motor prevents the placement of film close to the tangent point, the interface image is a projection. For this reason, the largest possible source-to-film distance should be used, consistent with reasonable exposure times.

Radiography of Explosives

Explosive Projectiles and Warheads

Explosives such as projectiles and warheads also require radiographic inspection for manufacturing defects and for defects that occur as a result of storage and handling. Warheads aboard aircraft that are subjected to repeated arrested landings on carriers can sustain substantial forces on crucial suspension and bearing points. Projectiles can become damaged as a result of the extreme handling and storage environments to which they must be subjected in remote sites around the world.

Complexity

High-energy radiography is used to examine and recertify most of these explosive items. In doing so, some or all the complexities of the various types of radiography (i.e., casting, welding and assembly testing) must be addressed.

Some of these items are manufactured by pressing granulated powder into the containing vessel. Some, however, are made by casting the explosive compound into the vessel. In this case, voids, cracks, shrinkage and piping can be present just as in the case of cast metals. Each item has outer metal parts which can be welded, forged or extruded. Additionally, most of these items have fusing or other types of detonating devices which must be examined while assembled in the explosive device. Figure 30 is a radiograph of a projectile showing some of the conditions found in this type ordnance material.

Radiography of Assemblies

Assemblies such as jet engines, gas turbines, valves, nuclear fuel elements and explosive devices (bombs and fuses) are frequently radiographed with high-energy X-rays to show internal conditions or dimensions. These assemblies may have material thicknesses that vary by several HVLs at adjacent regions when projected on the film. Also, many assemblies can have material and assembly characteristics that produce forward scatter, which obscures the sharpness of the radiographic image. In some instances, such as jet aircraft engines or gas turbines, in-motion radiographic techniques are utilized to determine dynamic dimensions between mating surfaces, gas seals, etc. Thus it is difficult to prescribe radiographic techniques that are universally applicable to all assembly radiography. In each case, as with other types of radiography, some experimental radiographs must be taken before the technique can be finalized.

FIGURE 31. Minac-3 Linear Accelerator with Shrinkac X-ray Head

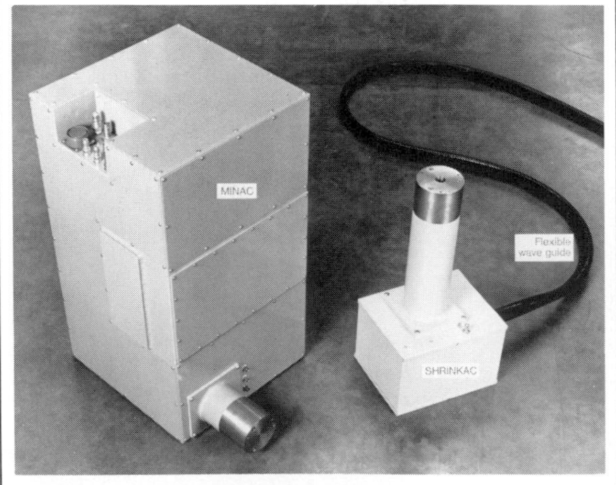

FROM SCHONBERG RADIATION, INC. REPRINTED WITH PERMISSION.

Radiographic Practices

The following radiographic practices should be observed when establishing these techniques:

1. to minimize the effects of forward scatter, use the highest available X-ray energy;
2. use object-film filters and screen-film combinations that produce the highest contrast and best sharpness;
3. use large *D/t* ratios to avoid distortion;
4. for radiography of a radioactive object, place the film at a distance from the object sufficient to reduce fogging of the film by the object's radiation; also use an object-film filter in back of the object to further reduce object emanation;
5. use multifilm techniques where thickness variations exceed the range of a single film.

Exposure times for each film may be obtained from established exposure curves when the average thickness of the objects in the area of penetration is known.

Special Applications

Mobile Applications

With the advent of smaller X-ray machines in the high-energy class, it has become advantageous to transport equipment to field locations. In such cases, these machines produce a high-energy, high-intensity output source with small targets, providing good definition and sensitivity. The military has a program for recertifying suspended ammunition by use of transportable high-energy X-ray equipment, whereby the ammunition can remain in stock at the storage location, whether it be in the United States or in a foreign country, and the radiation inspection facility is brought to the storage site. This is not only more cost-effective, but allows the stock to remain available, if required.

Figure 31 shows the Minac-3 linear accelerator with the Shrinkac X-ray head. This arrangement allows for remote positioning of the X-ray head at a distance from the control console and power supply. The configuration provides high-energy X-rays

FIGURE 32. Minac-3/Shrinkac Arrangement for Radiography

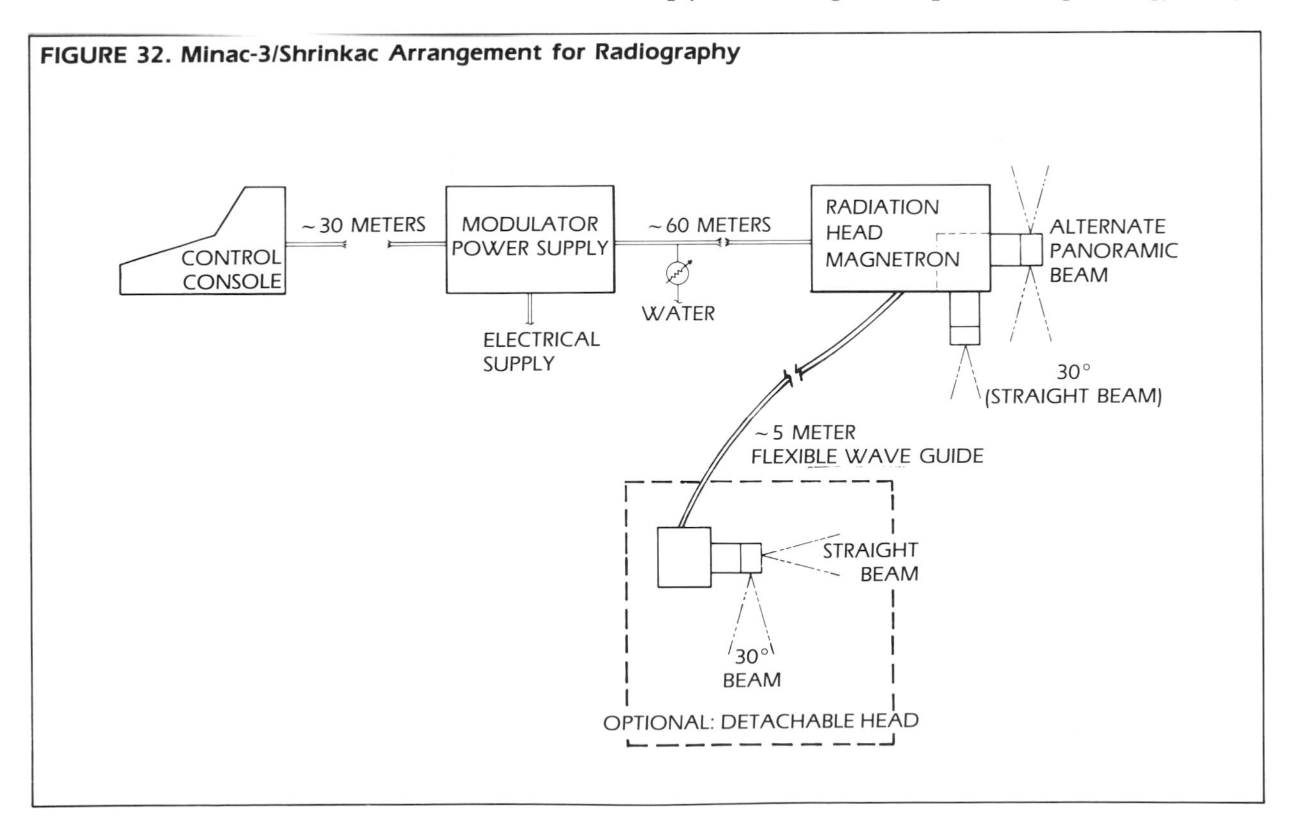

to inspect items that are located in confined places. Figure 32 shows in schematic form the arrangement of the Minac-3/Shrinkac concept.

An example of field site radiography using high-energy equipment is the inspection of welded pipe assemblies in a nuclear power plant. As high-energy machines get smaller and more flexible, it is evident that they will be used more for field work such as for pipeline inspection and the like. As this occurs, radiation safety will become more and more difficult to maintain. Safety procedures must be complete and exacting to ensure safety for the operating personnel and the general public.

CONCLUSION

A great amount of radiography has been accomplished using the upper limits of older equipment capability. Modern, smaller, light-weight machines producing high-energy and high-intensity X-ray output have eliminated the constraints of the older machines.

Modern manufacturing technology has presented requirements for radiography of large assemblies and structures which cannot be moved to inspection facilities and which cannot be inspected adequately using radioactive sources. High-energy X-ray machines, that can be transported to the test site, have provided a means of accomplishing these inspections.

When utilized in electronic imaging radiography (real-time radiography), high-energy X-ray machines provide instant imaging of thick, high-density parts.

These and other features of high-energy radiography demonstrate the advantages of its use in nondestructive testing and assure continual progress in the applications of the method.

PART 9

HIGH-ENERGY EXPOSURE CURVES

FIGURE 33. Typical Linatron Exposure Curves for Steel

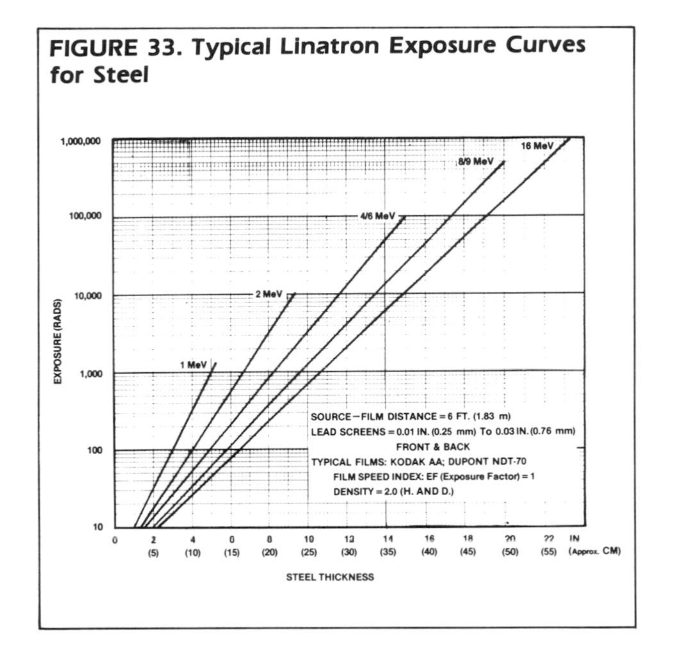

FIGURE 34. 1 MeV Linatron Exposure Curves for Steel

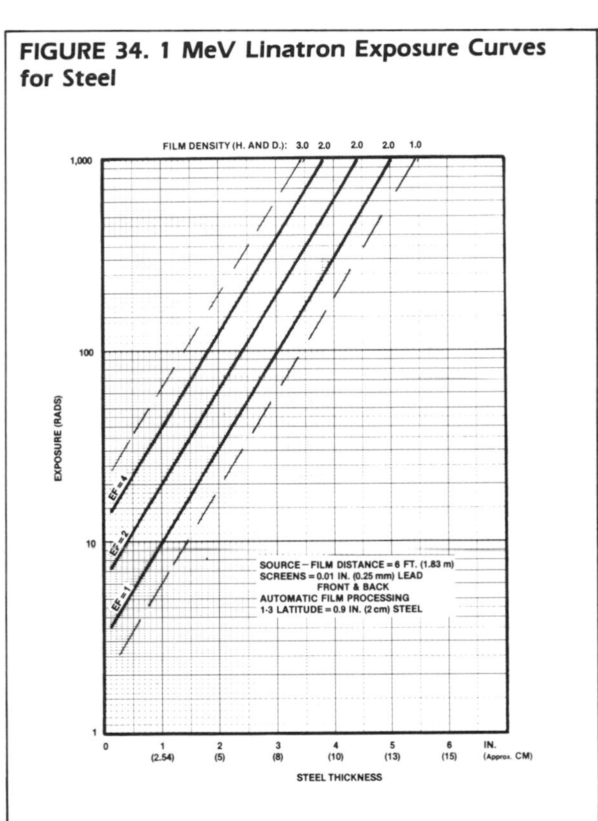

FIGURE 35. 2 MeV Linatron Exposure Curves for Steel

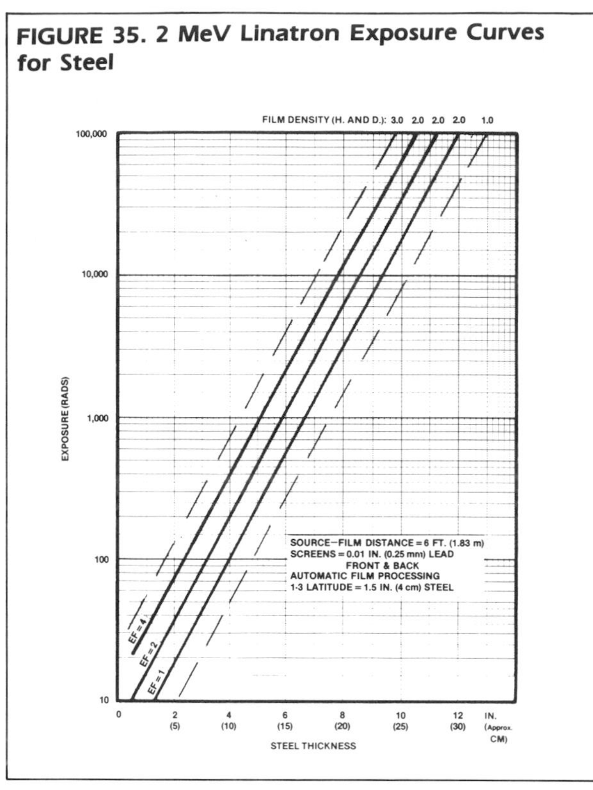

FIGURE 36. 2 MeV Resotron Exposure Curve for Steel

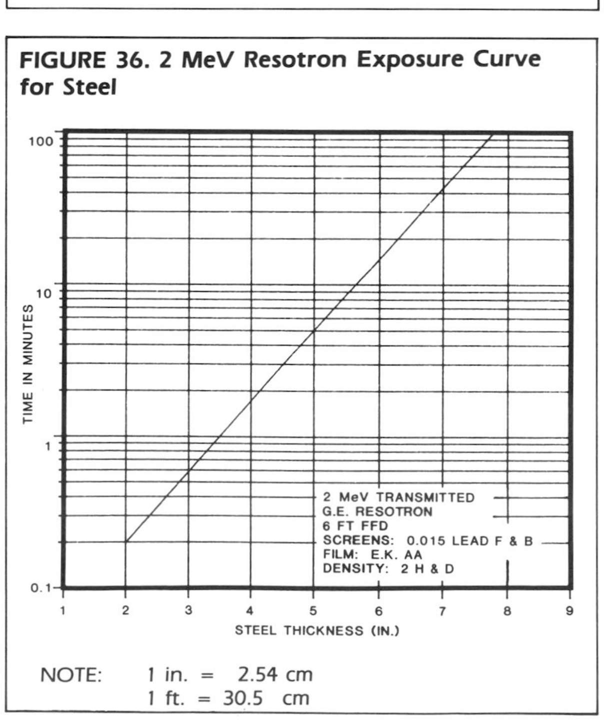

NOTE: 1 in. = 2.54 cm
1 ft. = 30.5 cm

FIGURE 37. 2.5 MeV Van de Graaff Exposure Curve for Steel

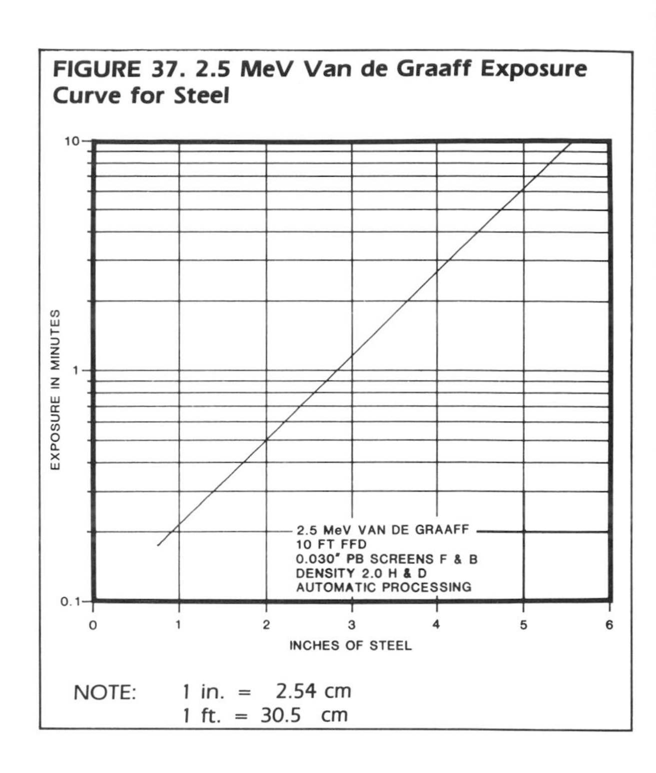

NOTE: 1 in. = 2.54 cm
1 ft. = 30.5 cm

FIGURE 38. 4 MeV Linatron Exposure Curves for Steel

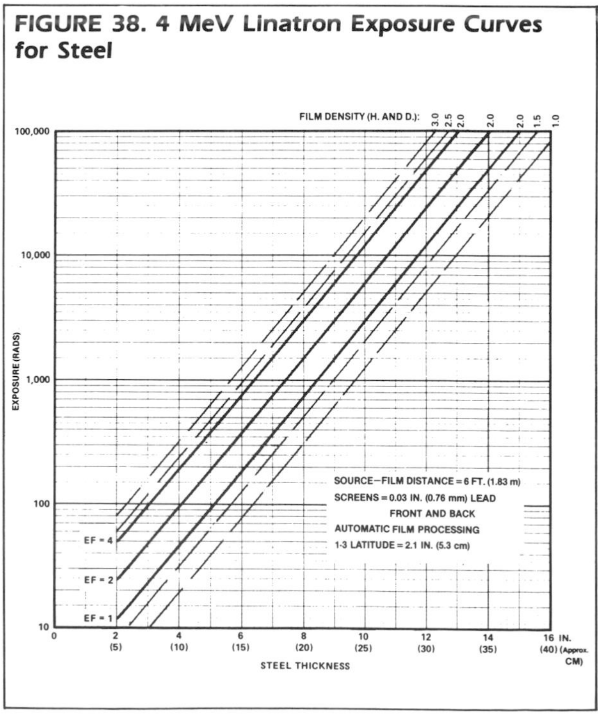

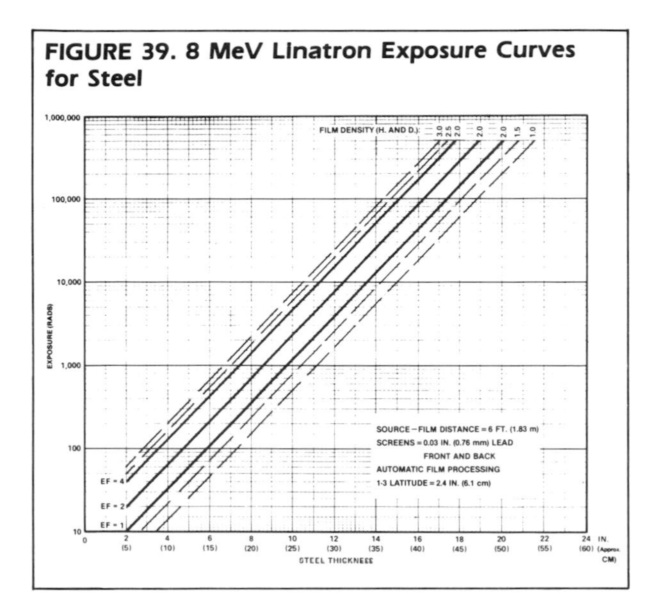

FIGURE 39. 8 MeV Linatron Exposure Curves for Steel

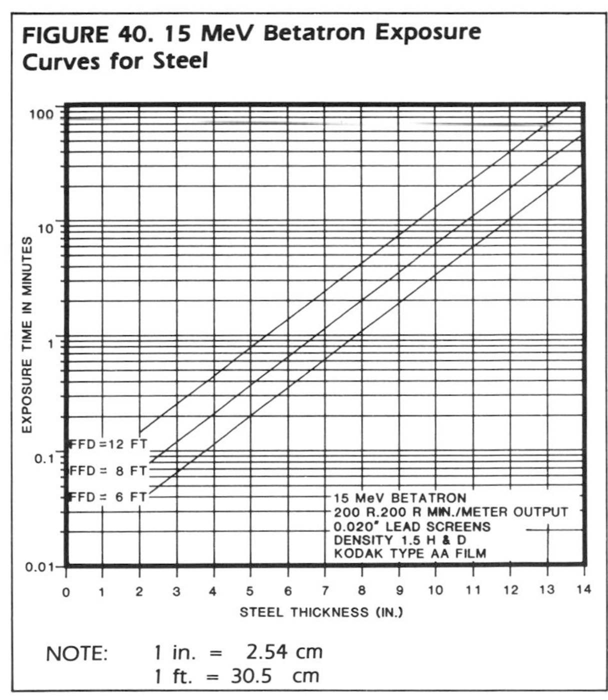

FIGURE 40. 15 MeV Betatron Exposure Curves for Steel

NOTE: 1 in. = 2.54 cm
1 ft. = 30.5 cm

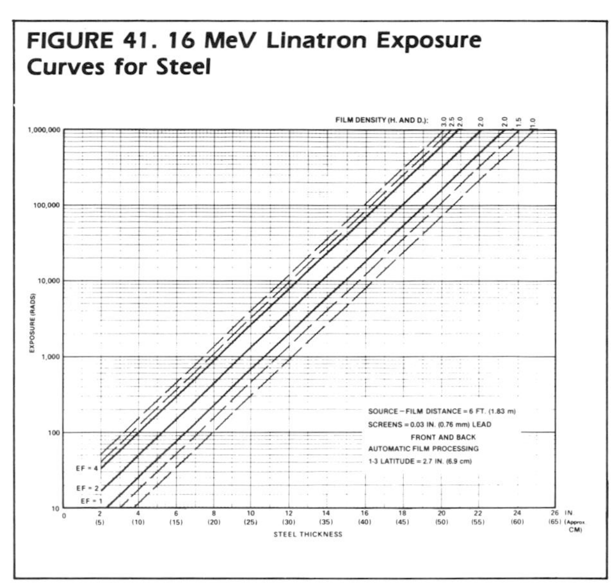

FIGURE 41. 16 MeV Linatron Exposure Curves for Steel

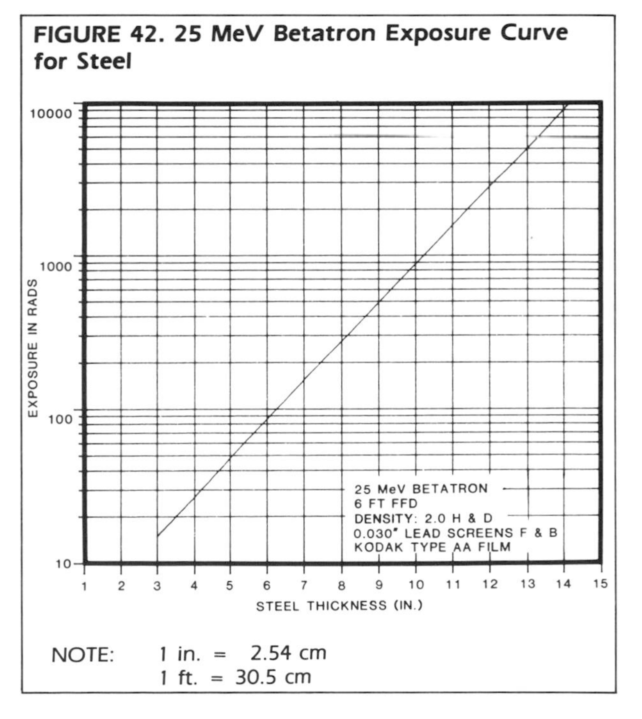

FIGURE 42. 25 MeV Betatron Exposure Curve for Steel

NOTE: 1 in. = 2.54 cm
1 ft. = 30.5 cm

FIGURE 43. Typical Linatron Exposure Curves for Solid Propellant

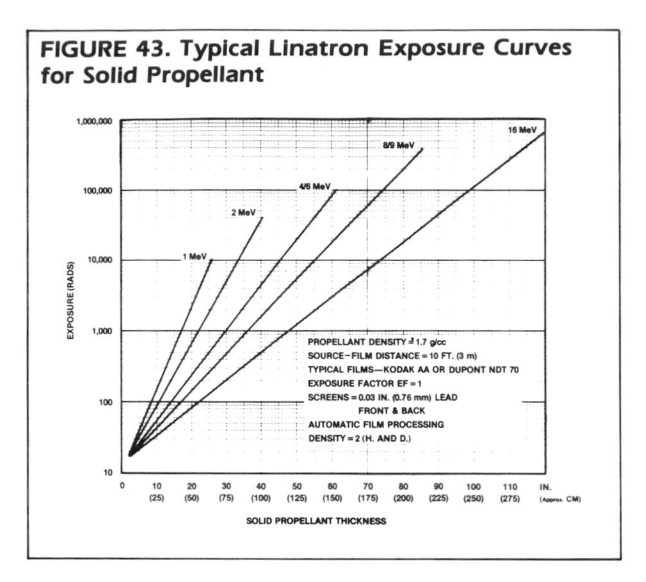

FIGURE 44. 1 MeV Linatron Exposure Curves for Solid Propellant

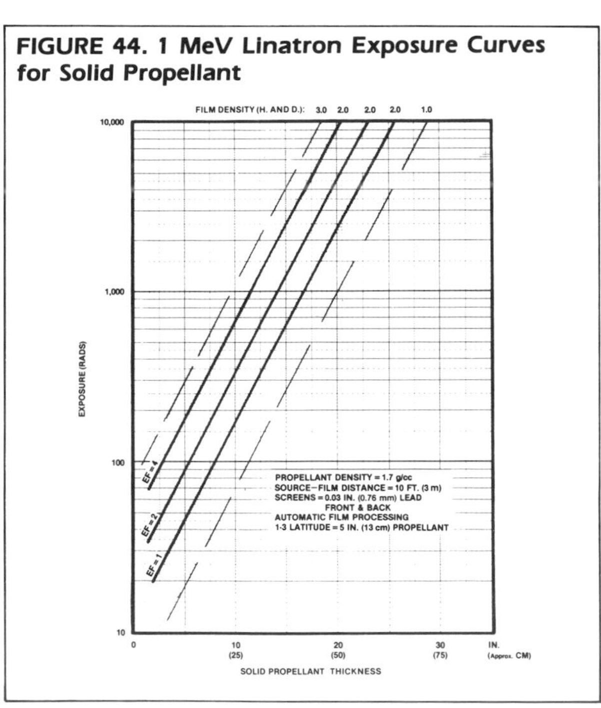

FIGURE 45. 2 MeV Linatron Exposure Curves for Solid Propellant

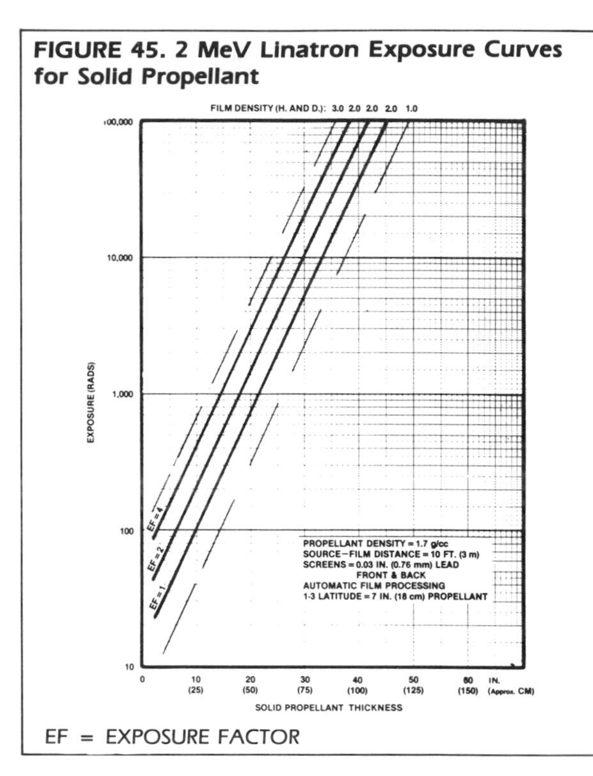

EF = EXPOSURE FACTOR

FIGURE 46. 4 MeV Linatron Exposure Curves for Solid Propellant

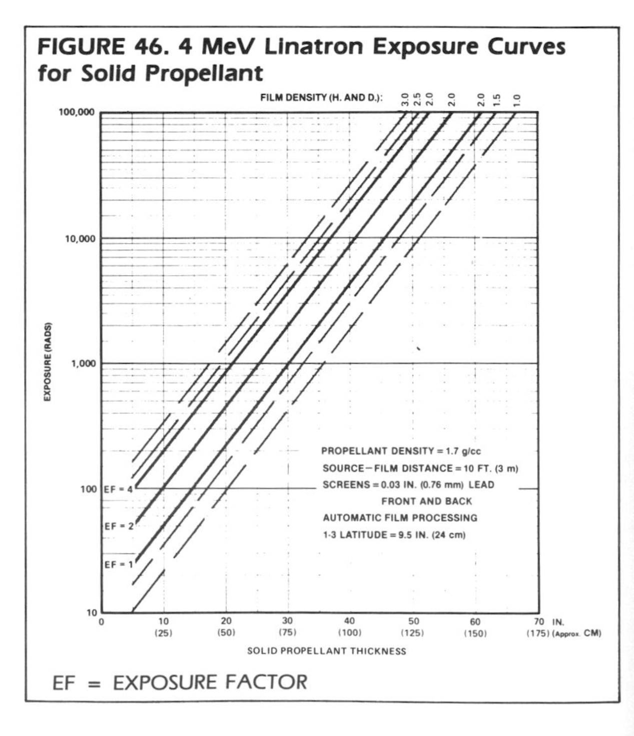

EF = EXPOSURE FACTOR

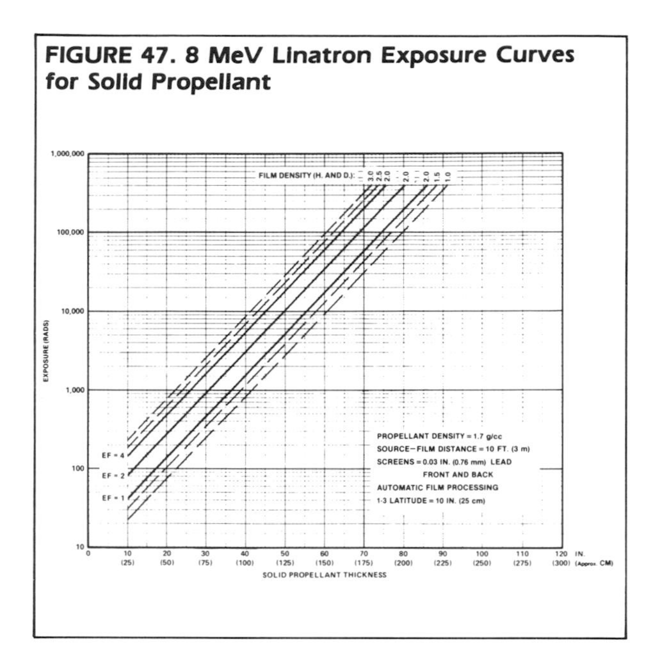

FIGURE 47. 8 MeV Linatron Exposure Curves for Solid Propellant

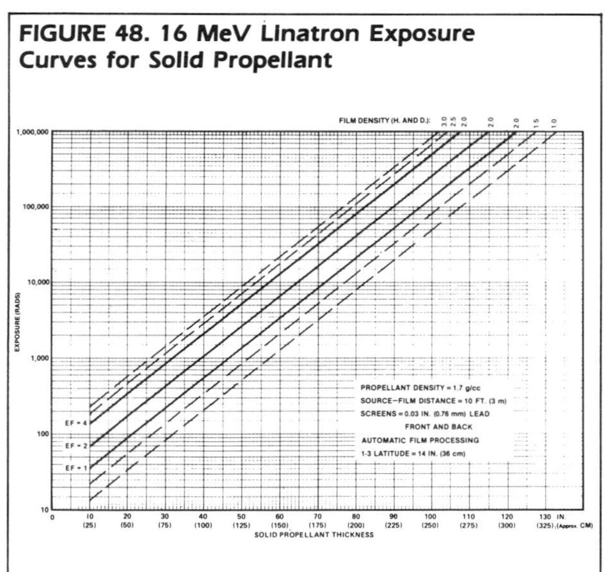

FIGURE 48. 16 MeV Linatron Exposure Curves for Solid Propellant

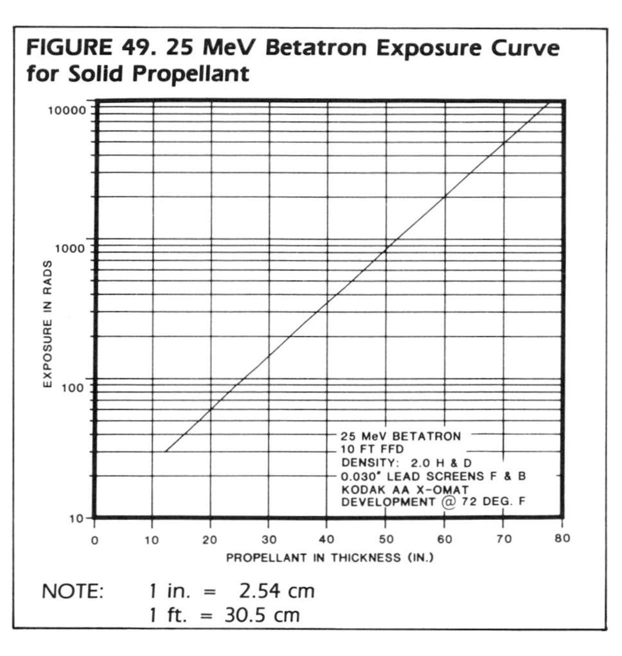

FIGURE 49. 25 MeV Betatron Exposure Curve for Solid Propellant

NOTE: 1 in. = 2.54 cm
1 ft. = 30.5 cm

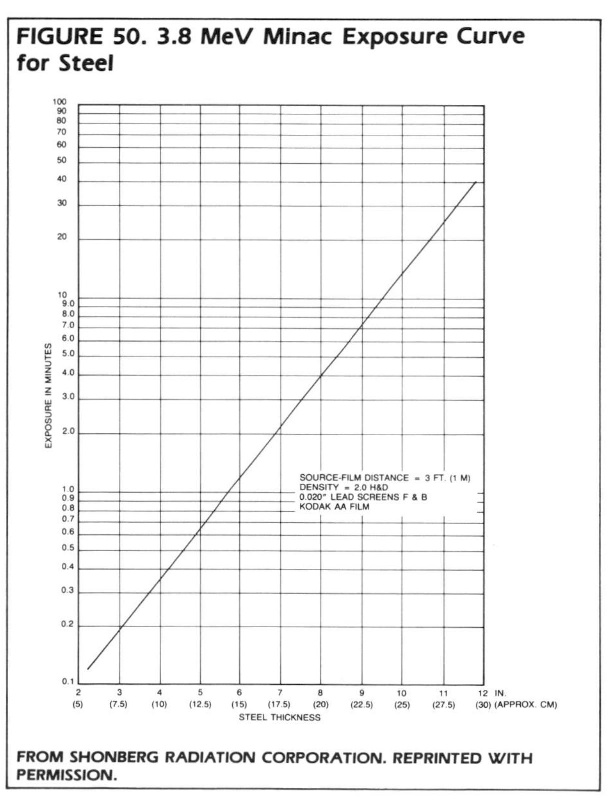

FIGURE 50. 3.8 MeV Minac Exposure Curve for Steel

FROM SHONBERG RADIATION CORPORATION. REPRINTED WITH PERMISSION.

BIBLIOGRAPHY

1. Morgan, R.H. and K.E. Corrigan, eds. *Handbook of Radiology*. Chicago: The Year Book Publishers (1955).

2. Wilshire W.J., ed. *A Further Handbook of Industrial Radiology*. London: Edward Arnold & Co. (1957).

3. McMaster, R.C., ed. *Nondestructive Testing Handbook*, 1st ed. New York: Ronald Press (1960).

4. Hogarth C.A. and J. Blitz. *Techniques of Nondestructive Testing*. London: Butterworth & Co. (1960).

5. McGonnagle, W.J. *Nondestructive Testing*. New York: McGraw Hill Book Co. (1961).

6. Halmshaw, R., ed. *Physics of Industrial Radiology*. New York: American Elsevier Publishing Co. (1966).

7. *Radiography in Modern Industry*. Rochester, N.Y.: Eastman Kodak Company.

8. Halmshaw, R. *Industrial Radiology Techniques*. London: Wykeham Publishing Co. (1971).

9. *High-Energy X-ray Applications for Nondestructive Testing*. RAD-1936. Palo Alto, CA: Varian Associates, Inc. (1982).

10. Pollitt, C.C. "Radiography with High-energy Radiation." *Journal of the British Steel Castings Research Association*. No. 65 (1962).

11. O'Connor, D.T., E.L. Criscuolo and A.L. Pace. *10-MEV X-ray Technique*. Special Technical Publication No. 96. Philadelphia, PA: American Society for Testing and Materials.

12. Halmshaw, R. and C.C. Pollitt, "Radiology with High-energy X-rays." *Progress in Nondestructive Testing*, Vol. 2. London: Heywood & Co. (1959).

13. Cusick, J.H. and J. Haimson. "10-MeV Rotating Target Linear Accelerator for Radiography of Large Rocket Motors." *Proceedings, Missiles & Rockets Symposium*. Concord, CA: U.S. Naval Ammunition Depot (1961).

14. Haimson, J. "Radiography of Large Missiles with the Linear Electron Accelerator." *Nondestructive Testing* (1963).

15. Bly, J.H. and E.A. Burrill. *High-energy Radiography in the 6- to 30- MeV Range*. Special Technical Publication No. 278. Philadelphia, PA: American Society for Testing and Materials (1959).

16. McMaster, R.C., ed. "High Voltage Radiography." *Nondestructive Testing Handbook*, 1st ed. New York: Ronald Press (1960).

17. *Metals Handbook*. Vol. 11, 8th ed. Metals Park, OH: American Society for Metals (1978).

SECTION 7

RADIOGRAPHIC LATENT IMAGE PROCESSING

William E.J. McKinney, E.I. DuPont de Nemours & Co., Wilmington, DE

INTRODUCTION

Most radiographers are highly skilled, motivated, and generally interested in the challenges of creating an image on film. Much training goes into being able to select the correct exposure. However, the image or exposure is useless until it is developed. This invisible image is called a *latent image*. It is through the chemical process called development that the hidden (latent) image is transformed into the useful visible image. For a radiographer, or an industrial lab in general, to know only about latent image formation and not visible image formation is to know only half of the technology called radiography. We must be knowledgeable and skilled in both areas if we are to control the efficiency, economics, and the quality we are responsible for.

The basic steps in processing are development (latent image into visible image), fixation (stop development and remove all remaining undeveloped crystals and unexposed crystals), washing (remove fixer to provide archival quality), and drying. All of the chemical reaction steps are controlled by elements of time (immersion time in solution), temperature (of the solution) and activity (replenishment, agitation, moisture). Time, temperature and activity, in turn, depend on various sytems: transport (time factor), temperature control, replenishment, circulation/filtration (agitation, uniformity of chemicals), electrical, and dryer systems. These six electromechanical systems constitute the processor (manual or automatic). In processing, there are a total of seven systems and the six supporting electromechanical processor systems are based on the needs of the seventh system—chemistry (developer, fixer, and wash).

Though the developer has its own time/temperature/activity relationship (as do the fixer and wash), it is a fact that one of the controlling factors of the developer is the fixer. And if the fixer is not washed out properly the film may be damaged. Thus the chemistry system includes developer, fixer and wash.

How do we know if the processor is right? This really means how do we know if processing is correct; will I get a good radiograph? If a technically accurate exposure (exposed radiograph with a latent image) is put into a processor, will it come out okay? Will it be free of artifacts and have the correct density and contrast? What if you are not sure your exposure technique is optimal or whether the quality achieved on the visible film is the result of bad exposure, bad processing, or both? The answer to all of these questions, which are quite common in industrial radiography, is in two parts. First, there is no single value greater than correct exposure with full development. Over- or under-development or exposure is inappropriate and inefficient. The processing completes what the exposure started; it cannot add information. Second, the sum total of the radiographer's efforts is to produce a useful visible image, whose density levels and contrast may be measured. To monitor and control processing and the total visible image production we use *sensitometry*, which may be defined as the quantitative measure of the film's response to exposure and development. The total value of the visible image is the result of exposure *and* development. To know only how to make exposures is to know only half of the technology required.

PART 1
THE CHEMISTRY OF RADIOGRAPHY

The Latent Image

When a radiographic film is exposed to a radiation energy source, it forms what is called a *latent image*. When the film is processed in chemicals, a visible image appears. This is, in its simplest terms, the chemistry of radiography. But, because the chemistry actually allows radiography to exist, it is most important that it be better understood. When we speak of radiographic chemistry we mean the total concept of the chemical constituents and mechanisms of film, processing chemistries, and the reactions that occur during exposure.

Radiographic film is composed of base and emulsion. Processing chemistries include developer, fixer, and the wash water. To help understand the total concept, we will construct a theoretical sheet of film, expose it, and process it from the standpoint of chemistry.

The Chemistry of Film

Film Base

Dimethyl terephthalate + Ethylene glycol = polyethylene terephthalate. The modern base, a plastic base, is made of polyethylene terephthalate. Ethylene terephthalate is an ester; thus the name polyester. This type of base (innovated by Du Pont as Cronar® Polyester film base) has become the accepted standard today. Its strength, improved clarity, superior transport characteristics, stability, and the fact that it does not absorb water are but a few of its important features.

Polyester bases require an adhesive so that the emulsion will adhere properly to the smooth surface. The adhesive is applied to both sides of the base as a substrate layer. The tint, composed of a delicate balance of many dyes, is usually found as an integral part of the base.

Film Emulsion

Once the film base is made ready to receive the emulsion, the emulsion is applied to both sides of the base. The emulsion is composed of a silver halide recording media and a binder of gelatin manufactured from collagen. Collagen is a naturally occurring fibrous protein and is a major component of animal skin, bone, and certain tissues. Collagen is treated with lime or an acid that breaks down the protein into a very pure gelatin. The gelatin has a great affinity for water—that is, it can absorb great quantities of water by swelling—and this is very important in film processing.

To the gelatin is added a sensitized silver halide. Silver halide is usually silver bromide. Other useful members of the halide group are chlorine and iodine. The halide might also be a combination such as chlorobromide or iodobromide.

Silver bromide is formed in this way:

(Eq. 1)

$$\underset{\text{silver}}{2Ag^{\circ}} + \underset{\substack{\text{nitric}\\ \text{acid}}}{2NHO_3} \longrightarrow \underset{\substack{\text{silver}\\ \text{nitrate}}}{2AgNO_3} + H_2$$

$$\underset{\substack{\text{silver}\\ \text{nitrate}}}{AgNO_3} + \underset{\substack{\text{potassium}\\ \text{bromide}}}{KBr} \longrightarrow \underset{\substack{\text{silver}\\ \text{bromide}}}{AgBr\downarrow} + \underset{\substack{\text{potassium}\\ \text{nitrate}}}{KNO_3}$$

The silver bromide is sensitized with a sulfur compound and mixed into the gelatin. Several washing operations follow until the emulsion is ready to be coated onto the base. And, of course, all of these steps must be carried out in total darkness.

Exposing a Film

Latent Image Formation

(Eq. 2)

$$\underset{\text{silver bromide}}{AgBr} + \underset{\text{energy}}{h\nu} \longrightarrow \underset{\text{latent image}}{AgBr + Ag_i^{\circ}}$$

Gurney and Mott developed a theory that is the accepted basis for explaining image formation. In the above formula, we see that the latent image is composed of metallic silver and that the crystalline silver bromide is undisturbed. Gurney and Mott found that crystals (silver bromide) sensitized with a foreign sulfur compound were easier to expose. They called these sensitizers *sensitivity specks.* At the moment of exposure the energy of exposure initiates an autocatalytic (self-completing) reaction. The crystal is coated with an excess of bromide ions containing excess electrons. At exposure, some of these electrons are released and are trapped at the sensitivity specks—now termed *sensitivity sites.* The bromine becomes gas and is absorbed in the gelatin. Since the sensitivity site contains numerous electrons, it is of a negative value and exerts a magnetic pull on silver ions floating in the crystal lattice structure. This unbonded silver, which needs an electron(s), is termed *interstitial silver* (Ag_i^+) Thus the migration phase results in the silver ion moving to the sensitivity site and picking up the electron(s). If the exposure of this single crystal is accurate, then at least five atoms of silver (Ag_i°) will deposit and thereby constitute a *development site.* Without this site the crystal will not develop.

The Chemistry of Processing

After the exposure has been made, and before development, both exposed and unexposed silver bromide crystals exist within the film emulsion. This is the *latent image*. The exposed crystals will be made visible as black metallic silver by reducing the structural silver bromide to simple metallic silver and by clearing away the *un*exposed crystals. This is the basis of chemical processing and we can see just how important its role is in the field of radiography. As we discuss processing chemistry, keep in mind the fundamental rule of radiography: "Processing completes what the exposure only started."

In our discussion we will often be using the following terms: *speed* is the film's response to exposure, or its sensitivity; *D. Max.* is the maximum density for the maximum exposure; *D. Min.* is the minimum density for the minimum exposure; and *contrast* is a difference in densities for a range of exposures.

Development

The film emulsion is now composed of two types of crystals: unexposed and exposed. The developer selectively seeks out the exposed crystals containing a development site made up of five atoms of interstitial silver and converts them to black metallic silver. The entire crystal becomes metallic silver and now contains *1,000,000,000* atoms of silver. This amplification factor of about 10^9 is the result of the oxidation-reduction reaction whereby the developer is consumed (oxidation) and the crystal is reduced from a compound to a simple element (reduction).

We may show this sequence of events with eq. 3 and 4:

(Eq. 3)

$$\underset{\text{silver halide salt crystal}}{AgX} + \underset{\text{photon of energy}}{h\nu} \longrightarrow \underset{\text{latent image}}{AgX + Ag_i^{\circ}}$$

X=Br=bromine, Cl=chlorine, I=iodine or hybrids

Radiation: X-ray, gamma, light, heat

Ag_i°=interstitial silver

(Eq. 4)

$$\underbrace{AgX + \underset{\text{5 atoms}}{Ag_i^{\circ}}}_{\text{latent image}} + Developer \rightarrow e^- \xrightarrow{\text{conditions: time, temperature, activity}} \underbrace{\underset{10^9\ \text{atoms}}{Ag^{\circ}}}_{\text{visible image}} + Developer\ (\text{oxidized})$$

TABLE 1. Developer Components

Chemical	General Function	Specific Function
Phenidone	Reducer	Quickly produces gray tones
Hydroquinone	Reducer	Slowly produces blacks
Sodium carbonate	Activator	Provides alkaline media; swells emulsion
Potassium bromide	Restrainer	Prevents reduction of unexposed crystals
Sodium sulfite	Preservative	Maintains chemical balance
Water	Solvent	Dissolves chemicals
Gluteraldehyde	Hardener	Permits transport of films by controlling swelling

This reaction is controlled, as are all chemical reactions, by elements of time, temperature, and activity, which will be discussed later. To keep the developer chemical strength (activity) at a constant level we use a replenishment system (manual or automatic).

The typical constituents of a radiographic developer can be seen in Table 1.

The primary function of the developer is to reduce silver ions to black metallic silver. However, there are five criteria for a modern developing agent; that it:

1. Provides a reducing agent for silver ions; that is, a source of electrons to reduce silver ions (Ag^+) to black metallic silver (Ag°);
2. Provides reduction of the exposed silver halide in preference over the unexposed crystals;
3. Is water soluble or soluble in an alkaline media;
4. Is reasonably stable and resistant to aerial oxidation;
5. Should yield colorless, soluble oxidation products.

Reducing Agents. Developers composed of metol (Elon) and hydroquinone are referred to as *MQ developers.* Modern developers are composed of phenidone and hydroquinone and are called *PQ developers.*

The basic reaction might be written:

$$\text{AgBr (exposed)} + \text{developer} \longrightarrow \quad \text{(Eq. 5)}$$

$$\text{Ag}^\circ + \underset{\text{(oxidized)}}{\text{developer}} + \underset{\text{hydrobromic acid}}{\text{HBr}}$$

It is important to notice that the developer is oxidized. Oxidized developer becomes a deep brown color and this indicates exhaustion. Since the rate of development is pH-dependent, pH is standardized with buffers against the effect of different water supplies and working conditions. *Buffering* means that the formulas are designed so that additional hydrogen or hydroxyl groups cause an internal rearrangement that prevents any appreciable alteration of pH.

The single most important function of the developer is the action of the reducing agents. The reducing agent or developing agent supplies the electrons necessary to enable the essential reaction of development to occur. As in other branches of chemistry, radiographic or photographic, there must be compromises. Therefore, such reducers as chlor-hydroquinone, which is more soluble in an alkaline solution than hydroquinone, are not used in large quantities due to their great expense. Due to their concentrated stability, they are used as minor components. Metol (Elon) is the standard in radiographic chemistry. It is rapid-working and produces high emulsion speeds with low contrast and density. It is, however, affected adversely by bromide accumulation.

Phenidone has replaced metol as the partner to hydroquinone because it is more stable in the highly alkaline solutions used in today's faster processing. But the more important reason is that phenidone shows superior superadditivity over metol, in combination with hydroquinone.

Superadditivity. **The superadditive nature of developer agents is a peculiar phenomenon whereby** the two basic reducers produce a greater effect when used together than when their individual values are added together. This is also called a *synergistic effect.* Phenidone, as mentioned above, attacks the exposed silver halide crystal very rapidly, resulting in deposits of black metallic silver in the grey tones of the image. This action results in low contrast and density and contributes to high emulsion speeds. The sensitometric curve would appear as shown in Fig. 1.

Hydroquinone, on the other hand, works very slowly in performing its function of developing the

FIGURE 1. Phenidone Response

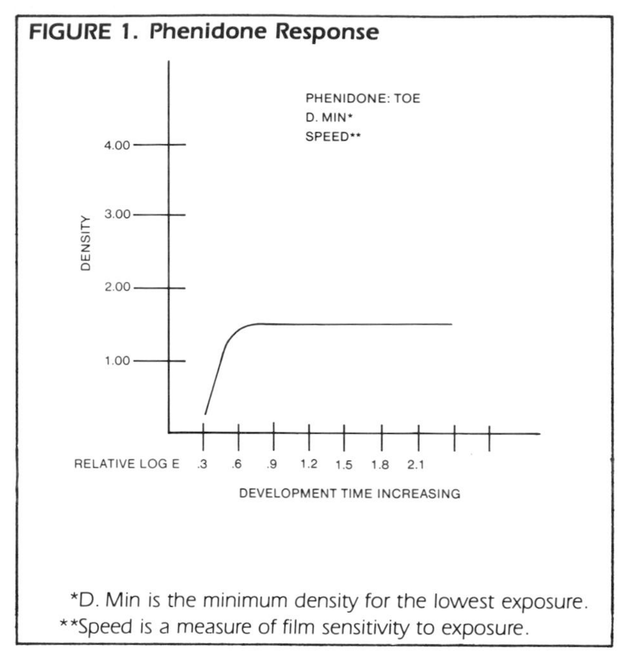

*D. Min is the minimum density for the lowest exposure.
**Speed is a measure of film sensitivity to exposure.

FIGURE 2. Hydroquinone Response

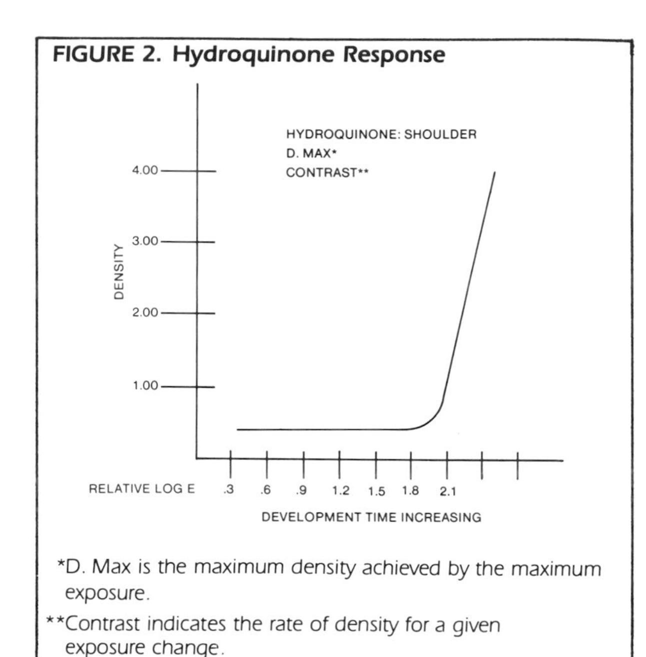

*D. Max is the maximum density achieved by the maximum exposure.
**Contrast indicates the rate of density for a given exposure change.

black or dense areas of the image. It strives to produce high contrast images utilizing maximum density without extracting maximum emulsion speed. The characteristic curve is shown in Fig. 2.

If we simply add the phenidone and the hydroquinone curves together we can see they would balance each other, forming the resultant curves, shown in Fig. 3. However, when phenidone and hydroquinone work together, one in the presence of the other, there is a synergistic effect in which the total effect is greater than the sum of the individual parts — there is an unexpected super-additive effect. The characteristic curve then would be as shown in Fig. 4. The shape of this sensitometric curve is similar to a typical relative log square-root-of-2 step wedge exposure for a moderate speed, high contrast, high maximum density radiographic film processed in a phenidone/hydroquinone (or PQ) developer. All of radiography as it is presently known revolves around the point that the development process has made the hidden image visible in a specific way. How well the reducers work together is important to the entire processing system.

Consider, if you will, the result of oxidized reducers. The reducing agents are relatively susceptible to oxidation and oxidation stops development. When developer is exposed to air, then aerial oxidation occurs. Simply mixing the concentrated portions in the wrong sequence can expose them to a chemical imbalance that will result in oxidation. Let's look specifically at how hydroquinone works.

Hydroquinone + O_2 → Semi-Quinone + O_2 → Quinone

As can be seen in the above reaction, hydroquinone is oxidized to quinone giving two electrons, which stabilize atoms of silver, and also hydrogen peroxide. For every atom of silver built into the image, a bromide ion is released. The *development rate* (the rate of reaction of the reducing agents) can increase as the bromide ions accumulate, depending on the class of film used. Bromide levels can also affect the pH of the solution, which can further alter the reaction rate. For every silver atom built into the image, a bromide ion is released. *Bromide depression* occurs when excess bromine is presented to the reducers so that they link with the bromine indiscriminately, resulting in an effect similar to underdevelopment. Bromide depression can occur due to under-replenishment, processing

FIGURE 3. Anticipated Composite Curves

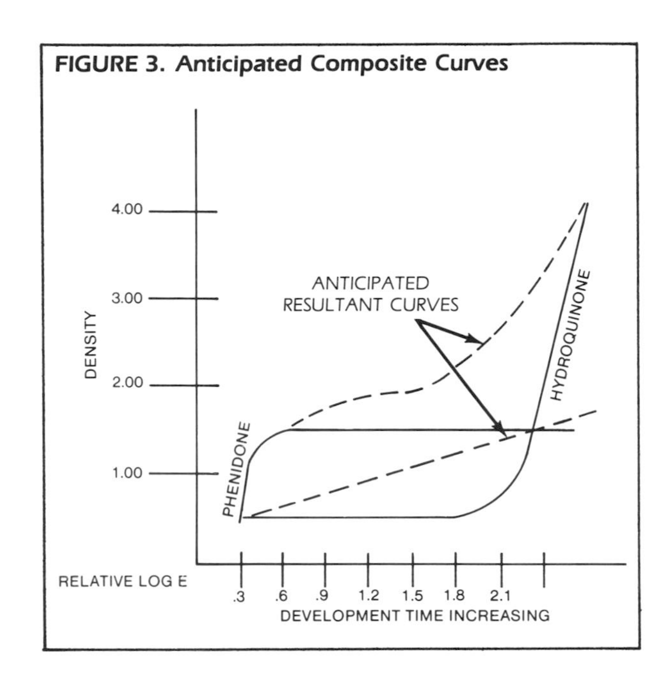

FIGURE 4. Actual Curve Showing Superadditivity

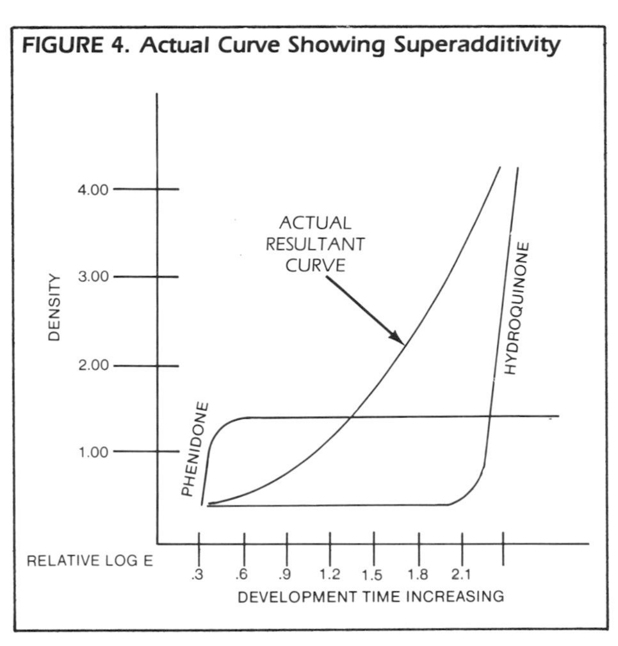

large volumes of roll film, or as a result of chemical deterioration due to improper mixing, storage, or contamination. Modern PQ (phenidone-hydroquinone) developers are much less susceptible to bromide depression and this allows longer working life. MQ (metol-hydroquinone) developers are often affected by bromide depression.

The function of the developing agents is to supply the electrons necessary for the essential reaction of development to occur. This function is accelerated or activated by an alkaline medium of sodium carbonate. The *accelerator* increases the ability of the development agents to donate electrons. The normal pH is between 9.8 and 11.5. Working solutions might be pH 10.5, but the replenishment solution is higher at about pH 11.0. In general, measuring pH is difficult because of the high levels and influence of temperature and is therefore only a relative indicator of chemical quality. The best test of overall quality is a sensitometric test.

Reaction of Alkaline Phenidone with Silver Ions. Oxidation will break down the products in the absence of restrainers or in long storage. Oxidation products generally stain the gelatin and form a sludge with a low degree of solubility.

Restrainer. The *restrainer* is a soluble bromide, such as inorganic potassium bromide (KBr). It is also referred to as an *anti-foggant* or an *anti-fogging agent.* The restrainer is usually found in the developer chemistry, but anti-foggants are also found in the emulsion. Common organic anti-fogging agents are Benzotriazole, 6-nitrobenzimidazole, and phenyl mercaptotetrazole (PMT).

The development of fog is similar to the development of the latent image, but is more dependent on the bromide content of the developer. Fog results when the unexposed silver halide crystals are attacked by the developing agents. Of course, here we are speaking only of development fog. Soluble bromide—that is, potassium bromide—reduces the rate of fog formation by protecting the unexposed crystals. This creates a development selectivity whereby the rate of fog formation is reduced or retarded to greater extent than the development of the latent image. The use of a starter solution allows fresh chemicals to be used immediately because it simulates this chemical balance.

Preservative. The standard preservative is sodium sulfite. Its primary function is to retard aerial oxidation of the developer agent. The preservative is not just one chemical but generally a host of chemicals. Other reagents that might be included in a formula are crysteine, stannous salts, dihydropyrogallol, and ascorbic acid. Any one of these reagents might be used as a preservative just for hydroquinone. These are also called *buffering agents*, and are both abundant in variety and found in small amounts. In addition to aiding and controlling the developer agent reactions under normal conditions, they also retard the influences of oxidation and different solvent conditions. The general solvent is tap water and it varies in pH and general hardness depending on the city.

Solvent. Water is the solvent, and it comprises over 80 percent of the developer solution. Water should be of "drinking quality" with a carbonate hardness of between 40 and 150 parts per million. Metals in water can accelerate developer oxidation and result in high fog. Calcium bicarbonate reacts with sodium carbonate and sodium sulfite producing a precipitate of calcium carbonate or calcium sulfite.

$$\underset{\text{calcium bicarbonate}}{Ca(HCO_3)_2} + \underset{\text{sodium carbonate}}{Na_2CO_3} \rightarrow \underset{\text{calcium carbonate}}{CaCO_3 \downarrow} + \underset{\text{sodium bicarbonate}}{2NaHCO_3} \quad \text{(Eq. 6)}$$

This precipitate will coat processor tank walls, heater thermocouples, and the heat exchanger tubes and act as an insulator, thus affecting the ultimate development. The metals in water are usually copper and iron. Other compounds are magnesium bicarbonates, magnesium sulfate, and calcium sulfate. But, one might ask, why use tap water instead of distilled, demineralized, or soft water? Besides the fact that certain aspects of tap water are helpful, it is also less expensive and more convenient. In addition, distilled or soft water tends to soften the gelatin in the emulsion too much. To make it possible to use any water supply, sequestering agents are added, to lock up or tie up the complexes of calcium and magnesium in more soluble complexes so they cannot react with the carbonate and sulfite ions. A widely known organic sequester is E.D.T.A. or ethylenediamine tetraacetic acid.

Hardener. Automatic processors lack a first rinse or an acid stop bath primarily to conserve space, yet it is still desirable to control development action to prevent fog. Thus a hardener agent (any aldehyde compound) is added. Formaldehyde is quite good except for its objectionable fumes; glutaraldehyde is most often used. Its primary function is to keep the gelatin from swelling too much. This chemical is acidophillic (likes acid) and thus begins to deplete at mixing. Two to three weeks after mixing the developer, processed films may become overdeveloped, unclear, wet, have high hypo retention, and physical artifacts may occur.

Temperature Influence on Developer Action. Developing agents are temperature dependent, resulting in temperature coefficients. There is about a ± 0.05 pH change per 10°C. Sensitometrically the optimal developer temperature occurs when it produces the maximum or a specific gamma (contrast) level. *Optimal* means achieving the best levels of speed, D.Max, and D.Min. for maximum contrast. Deviation in either direction due to temperature change will generally result in lower contrast (see Fig. 5).

Agitation. Agitation increases both the rate of development and the rate of reduction. To clarify, development rate is increased because agitation permits a constant mixing of the solution and aids in washing bromine and the oxidized developers out of the emulsion. Agitation aids reduction by constantly swirling the reducing agents in and around the silver halide crystal lattice.

FIGURE 5. Gamma Versus Temperature Response Curve

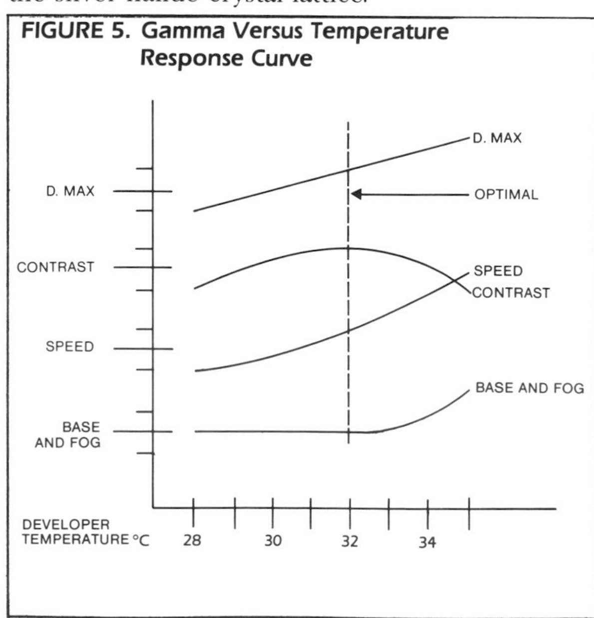

When replenishment systems are used, agitation helps keep the stronger replenishment solution properly mixed into the working solution. Agitation also helps in the filtration of reaction by-products, mostly gelatin, by circulating the chemicals through a filter. Finally, it keeps the temperature uniform.

Replenishment. There are two generally accepted methods of replenishment: the exhaustion or batch method and the regulated method. The batch method means that replenishment of the developer occurs after the whole batch is exhausted. The regulated method replaces a small amount of chemistry for that which is removed through carryover within the emulsion. Basically the classical figure for carryover of the developer within the swollen emulsion is about 2 ounces (60 ml) per 14 × 17 in. of industrial Class II film. (Note: use of the 14 × 17 in. measurement is a common industry practice. For reference, the metric size is 35 × 43 cm.)

Those desiring consistent processing should use only the regulated system. Chemically defined, *replenishment* is only a replacement of quantity, of volume, a maintenance of a preset amount. *Regeneration* is the second function of an adequate replenishment system and its job is to insure consistent activity by a replacement of spent chemicals. It is the purpose of developer regeneration to insure that the characteristics of the finished radiograph—its speed, contrast level, fog level, and maximum density—remain substantially constant.

A good replenishment/regeneration system will prolong the life of chemistries, aid in the maintenance of consistent quality and may lead to improved sensitometric quality. The proper replenishment/regeneration system means that chemistry need never be changed in a processor. Changing out chemistry once a month or every 2 to 6 weeks should be by choice and convenience and never because the activity has been lost.

Starter Solution. This is an acid solution (pH 2 approximately) containing bromides that is added to fresh developer each time the automatic processor is filled. Between 2.5 and 3.2 oz per gal (20 and 25 mL per L) of developer are added to the processor, depending upon the manufacturer. Each manufacturer's brand of starter should be used with the corresponding brand of developer. Starter is not normally added to the replenishment chemistry.

Starter gets its name from the fact it is used when we start up a fresh batch of developer. Its acid nature primarily deactivates the developer to help control fog. Its bromides are added to simulate used developer and thereby provide consistent, reproducible quality batch to batch. For Class I films, which benefit from higher bromide levels, the starter both lowers pH and increases the development rate.

The developer chemistry manufacturer provides guidance on the amount of time and temperature to use with its product. The manufacturer bases his recommendations on the assumption that all instructions have been carefully followed, including the addition of the correct amount of the correct brand of starter.

Faults from Developer. Types of faults: too much or too little density (toe, straight line, shoulder areas); too much or too little contrast.

TABLE 2. Faults from Developer

Underdevelopment = loss of density = loss of contrast		Overdevelopment = gain of density = loss of contrast
low	temperature	high
fast	time-transport rate	slow
low	solution level	—
over	replenishment Class I	under
under	replenishment Class II	over
under	replenishment Class III	over
under	replenishment Class IV	over
low	agitation	—
high	chemical-oxidation	—
high	contamination	high

TABLE 3. Fixer Components

Chemical	General Function	Specific Function
Ammonium Thiosulfate	Clearing Agent	Clears away unexposed, undeveloped AgBr
Aluminum Chloride	Hardener	Shrinks and hardens the emulsion
Acetic Acid	Activator	Acid media—neutralizes developer
Sodium Sulfite	Preservative	Maintains chemical balance
Water	Solvent	Dissolves chemicals

Automatic versus Manual Processing and Chemistry. Automatic developers contain gluteraldehyde as a hardening agent to control emusion (gelatin) swelling. Since manual developers have no hardeners the gelatin carries out more of the developer volume.

Manual processing uses a short-stop, first rinse, or acid bath between the developer and fixer to stop development or prevent excess developer from carrying into the fixer and diluting or contaminating it (to prolong the fixer life).

In automatic processing, in addition to developer hardener, the processor uses squeegee rollers to remove excess developer and an automatic replenishment system to sustain both volume and activity levels of all chemicals. Of course the processor has no short stop and this reduces the overall size by approximately 20 percent.

Automatic developers can generally operate at higher temperatures than manual developers.

Fixers are generally the same for both automatic and manual processing.

Manual Acid Short Stop. The acid stop bath, normally 2 to 3 percent acetic acid solution, functions in several ways: it neutralizes alkaline developer by rapidly lowering the pH to the point where development stops; it helps prevent aerial oxidation of the developer agent, which otherwise could form staining products; it dissolves or retards the formation of calcium scum; and preserves the acidity of the fixer and helps control gelatin swelling. Some commonly used agents are acetic acid, citric acid, diglycolic acid, and sodium bisulfite.

The rate of neutralization for the acid stop bath of the fixer depends on:

1. Nature and thickness of emulsion
2. pH value of stop bath and/or fixer
3. Total acidity of the stop bath and/or fixer
4. Agitation
5. Developer alkalinity
6. Developer pH
7. Type of developer agents used
8. Age (a function of replenishment)
9. Temperature.

Fixer

Standard fixers are composed of the chemicals listed in Table 3.

Fixing Agent. The function of the fixing agent is to form soluble stable complexes of silver salts that can be removed readily from the emulsion. Fixing agents should have no effect on the emulsion binder or on the already developed silver. Thiosulfate, in the form of sodium or ammonium salts, is the usual fixing agent. Sodium thiosulfate is best known as *hypo.* However, all of the terms hypo, fixer, clearing agent, fixing agent, and thiosulfate are generally held to be synonomous. The basic reaction between thiosulfate and silver halide is that of dissolving and carrying away the undeveloped silver. Thiosulfate can, however, attack the developed silver if the pH is decreased (moved toward a neutral or basic pH). Thus, replenishment is important to the fixer in regeneration of chemical strengths. The developer carryover into the fixer replaces what fixer is carried out, but it also reduces the pH slightly. If left within the emulsion, thiosulfate reacts with silver particles to form silver sulfide (Ag_2S), which has a characteristic objectionable yellow-brown stain. This is referred to as ***residual hypo*** or ***hypo retention*** and will be discussed later.

Hardener. The hardener shrinks and hardens the emulsion. Aluminum chloride is most frequently used but actually any aluminum compound, such as potassium alum or chrome alum, will work. More specifically the hardener objectives can be stated as follows: (1) to increase resistance to abrasion; (2) to minimize water absorption by the gelatin (this reduces drying time); and (3) to reduce swelling to permit roller transport.

Activator. Acetic acid provides acid media of about pH 4.0 and aids in the hardening of the emulsion. However, the most important function is the neutralizing of developer carryover and of the developer trapped within the emulsion. The reducers of the developer require high basic or alkaline media in which to react and they will continue to react, even after the film is removed from the developer solution, until they are neutralized. Because a very small portion of the fixer (acid) will neutralize or at least lower the pH of a larger volume of developer, greater care is required when mixing chemistries so that contamination of the developer with fixer does not occur.

Acetic acid is usually used because it is a "weak acid." It achieves good buffering and a slightly acid medium permits the use of aluminium hardeners.

Preservative. Sodium sulphite is also the preservative for the fixer, but its general function is to prolong the life of thiosulfate in the fixer by reacting with free sulphur in the presence of the activator to regenerate the thiosulfate complex.

$$\underset{\text{sodium sulphite}}{Na_2SO_3} + \underset{\text{sulphur}}{S} \longrightarrow \underset{\text{sodium thiosulphate}}{Na_2S_2O_3} \qquad \text{(Eq. 7)}$$

Solvent. Water is again the solvent and as with the developer it need be only of drinking quality.

Rate of Fixation. The rate of fixation is primarily dependent on: (1) the diffusion rate of the fixing agent into the emulsion; (2) the solubility of the silver halide grains; and (3) the diffusion rate of the complex silver ions out of the emulsion. Thus it can be seen that adequate agitation and replenishment are important to proper fixation.

The rate of fixation is the amount of time required to totally "fix" the emulsion, including clearing of all unexposed silver from the emulsion and hardening of the emulsion. In general it is said that the fixer clears and fixes. The rate is determined by the rule-of-thumb principle: the total fixing time is twice the clearing time. A simple clearing time test might be: using a 70 mm × 30 mm (3 in. × 1 in.) strip of fresh unexposed, unprocessed film, place a drop of fixer on both sides of the film, wait for 10 seconds, then dunk the strip into the fixer, agitating gently, watching for the spot to disappear. The clearing time is when the overall film is as clear as the spots, which had a head start. Additional time will not make it any clearer. Clearing time is critical for industrial films, especially in automatic processors where immersion time is fixed. Normally films will clear in 20 °C (70 °F) fixer in 20 to 60 seconds depending on brand and class of film.

Faults from Fixer include:

1. Rise in pH—decreased hardening: wet films, poorer archival quality;
2. Dichroic stain—reaction of developer and silver loaded fixer;
3. Streaks—nonuniform removal (neutralization) of the developer;
4. Precipitation resulting from too low pH;
5. Brown stain (produced by the formation of hydroquinone monosulfanate) from electrolytic oxidation of carryover developer; sulfite content low.

Water

Wash water is a photographic processing chemical. Its purpose is to dilute or wash out the residual **fixer chemicals; its action is to swell the emulsion; and the rate is usually 3 gal/min (11 L/min).**

Wash Principle. Washing steps are included in photographic processing to remove reagents that might adversely affect later operations—and at the end of processing to eliminate all soluble compounds that might impair the stability of the film.

Purpose of the Wash Water. Water removes fixing salts contaminated with dissolved silver compounds in the form of complexes with the thiosulfate.

Failure to remove these silver compounds eventually causes stain in the highlights and the unexposed areas, whereas the presence of thiosulfate, its oxidation product *tetrathionate*, and other *polythionates* will, with time, cause slow sulfiding of the image. This stain is silver sulfide (Ag_2S) and is called *hypo retention stain*. The rate of diffusion of thiosulfate from emulsion is affected by:

1. Amount of silver image present
2. pH of the fix
3. Type of thiosulfate
4. Degree of fixer exhaustion
5. Temperature of wash
6. Agitation rate
7. Rate of flow of water
8. Design of wash apparatus

Washing System. In the *counter-current principle*, the water enters at the point where the films exit—the films leave uncontaminated water. One thousand square feet of film will deposit about four troy ounces of silver in a stagnant water tank. Agitation is normally supplied as a function of the water volume (replenishment flow rate) and directly affects efficiency.

Hypo Retention. Hypo retention is the amount of residual hypo or thiosulfate remaining in the emulsion after the film is processed. Hypo retention levels will vary with different brands of film. The type of processor, processing cycle, and the situation of the chemistries have influence on hypo retention levels. The amount of residual hypo, which affects the archival qualities of the radiograph, is measured in micrograms of thiosulfate per square inch of film ($\mu g/in.^2$) or micrograms per square centimeter ($\mu g/cm^2$). Twenty-five micrograms per square inch (4 $\mu g/cm^2$) is the ANSI upper limit of retained thiosulfate for prolonged storage (in excess of five years). A retention level higher than this may cause a general brown stain to appear on the film. Film with a level of 500 will usually last only one year before stain appears and the film becomes legally useless.

Hypo retention tests, requiring the normal processing of an unexposed film should be made twice a year. Write down the processing conditions (time, temperature, chemical age, processor number, date, and so on) and submit to your technical representative to have an analytical test made.

Hypo estimator kits are available from your X-ray dealer and are used on a daily basis to indicate a general "go/no go" status. These are convenient and very useful but are only estimators. It is important to have an analytical test made periodically, and to compare the test results to your use of the estimator.

However, the most important aspect of silver sulfide formation is the storage conditions. Films held in long-term storage require the same ambient conditions as fresh film: 21 °C (70 °F) or cooler, and 60 percent or less relative humidity (RH). Even with low hypo retention levels, improper storage (e.g., 32 °C [90 °F], 90 percent RH) can result in unwanted stain. If you must keep films, then you must keep them without stain. (See information on Archival Quality.)

Water's "Mechanical" Function. Water is required primarily to wash the fixer out of the film; this is its chemical function, discussed previously.

Mechanically, the wash water is either the source of heat for the developer solution (manual and a few automatic processors) or the primary developer temperature stabilizer in automatic processors. In manual processors, the developer and fixer tanks sit in a larger tank filled with circulating, temperature-selected water. The water controls the temperature of the other chemistries. In automatic processors, the wash water flows through a heat exchanger at about 3 °C (5 °F) less than the desired developer temperature. The cooler water and the warmer developer are in proximity with a common steel wall. The cooler water picks up heat from the developer, causing the developer thermostat and heater to respond more rapidly and thereby provide greater stability. The fixer tank in an automatic unit is usually heated by the developer on one side and cooled by the wash on the other side. The wash water tank also provides an insulation barrier between the hot dryer section and the chemical section.

Conclusion

Chemistry necessitates reaction controls, such as the time-temperature method of processing. Filtration, circulation, pumping, metering, replenishment system, emulsion characteristics, transport systems, aerial oxidation, contamination, and chemistry aging are all various aspects of the chemistry system in the processing of radiographs. It is these things which greatly influence the processing of radiographs to obtain optimal informational integrity. It has been rightly stated that "radiography begins and ends in the darkroom" and that "processing completes what the exposure started."

The only real difference between manual and automatic processing chemistries is the developer hardener, and the only real difference in the two methods is the increased degree of consistency derived from the machine. Manual processing can be as fast as automatic, but with the human operator controlling time, agitation, replenishment, and other factors there are many chances for variables. On the other hand, the automatic processor, although consistent, is not entirely automatic and may produce consistently good or consistently bad product depending on the knowledge of and control by the operator. These machines cannot analyze chemicals or compensate for time/temperature variations; the technology exists but is not in present equipment.

In summary, processing is a chemical reaction that converts the latent image into the visible image through an oxidation/reduction reaction that amplifies the latent image into a useful, or usable, visible image at a rate of 10^9 times per crystal. Unexposed and undeveloped crystals are removed in the fixer. Wash water removes remaining residual chemicals and the film is ready to be dried.

PART 2
THE DARKROOM

Darkroom Technique

Principle

The radiographic darkroom is two things: a scientific laboratory and a *dark* room. A darkroom, where the lighting is kept at a very low level with special filters, must be constantly inspected to insure that it is indeed dark. The reason is that the X-ray film is sensitive to light and will turn black. X-ray or radiographic film can be affected by heat, light, humidity, static electricity, pressure, chemical fumes, and/or radiation. Since so many things can alter radiographic film the darkroom must be considered a scientific laboratory. As a laboratory, certain dictates of common sense are required to establish and maintain a desired level of quality. All variables that can alter the scientific processes within your darkroom must be known and eliminated. A routine system of checking for the existence of these variables must be made.

To reinforce the idea that the darkroom is indeed a scientific laboratory, even though it exists in the dark, let's consider just two points. The first point is that in radiography we strive to make excellent-clarity radiographs. The most common cause of unsatisfactory radiographs is fog, a non-informational density or blackness from silver deposits that occur in the wrong place and mask over the visibility of detail. As mentioned above, many forms of energy cause radiographic film to become black. Once radiation exposure has been made the radiographic film becomes at least *twice* as sensitive to all types of energy, and, thus, extreme care is required in working within the darkroom laboratory. The second point is that the darkroom laboratory exists because processing in a very precise manner is required to change the image formed by the exposure (the latent image) into the useful visible image. Processing is an exact science based on a scientific principle called the *time/temperature* method of processing. This time/temperature principle is based on a controlled level of chemical activity, which is monitored by the technician. Processing is completely vital to radiography and must be performed completely.

The darkroom laboratory should be, by its descriptive name, light tight and should have all of the requirements and equipment of a laboratory. Most laboratories are well ventilated, well organized, clean, pleasant, and safe places to work. This then should be a starting point as we consider darkroom techniques.

Design

The basic requirement for designing a darkroom is usually available space. It is most unfortunate when the darkroom is considered so unimportant that it is crowded into a former closet or basement area. However, this attitude is gradually changing and expert thought is being given to the proper design of a darkroom. Any darkroom must be designed so that there is a smooth and orderly work flow pattern.

The layout of a darkroom is generally considered to be either for centralized processing or decentralized (dispersion) processing. Centralized processing has, until recently, been the most advantageous system. However, some large industrial facilities have found that, with the convenience of automatic processors, dispersion processing and darkroom location are better suited to the needs and requirements of increased workloads. Incidentally, dispersion processing has been in existence for a very long time as has the idea of portable darkrooms (and processing). It is only now that the idea can be fully exploited.

Darkroom layout should first be designed for convenience and safety. Consideration must be given to saving steps and time for the darkroom personnel, since darkroom efficiency is directly related to exposure planning efficiency. Since the darkroom is a laboratory, every applicable safety standard must be followed. Separate the darkroom into a wet and a dry area and keep these areas as far apart as possible. Keep surfaces dry and clean.

There should be adequate ventilation to provide a sufficient supply of fresh, clean air. Dust is very destructive in the darkroom because it scratches films, salt screens, and equipment, resulting in permanent damage. Metal filings (carried by radiographers' hair or clothes into the darkroom) can adversely affect the developer and cause artifacts on the film.

Near the darkroom should be a "viewing room," which is referred to as the "lightroom," in which processed films are sorted and organized, and some supplies are stored. The most important aspect of the lightroom area is the availability of view boxes. Films can be reviewed rapidly in the lightroom for consultation or directive purposes rather than waiting for a more formal reading to be made and recorded. Wherever possible, wet readings, if used, should be made in the lightroom so that normal darkroom functions are not disturbed.

Equipment and Practice

Maintenance

It is recommended practice to sponge off routinely the outside and inside of a manual solution tank cover. Always replace tank covers when solutions are not in use to minimize oxidation, contamination, and dust. Dust sticks to a wet surface, so always wipe up spills as they occur and buff surfaces dry. Periodically wipe down walls and shelves (including side walls of shelves). Make sure the room is light tight, free of strong chemical fumes, and radiation protected through the establishment and continuation of a regular maintenance schedule. Time, money, and effort are saved through a regular preventive maintenance program requiring a few minutes a day.

Inspect the darkroom at the beginning and end of each shift or work day. Clean up and put things in their places. Make sure adequate supplies are on hand for each day's workload.

Every darkroom should have a mop handy for floors, and sponges for cleaning walls, surfaces, and the processing equipment. A source of hot water is necessary for cleaning, lintless rags for wiping surfaces dry, a calibrated bi-metallic or electronic thermometer for checking temperature, and nonmetallic scouring pads for removing chemical encrustations. Do not use soaps or detergent around the processing solutions. Protective waxes can be applied to the *exterior* surface. Spare safelight bulbs, laboratory brushes, beakers, funnels, graduates, and carrying buckets are all very helpful supplies. Keep everything in a given place so that it is easy to locate — even in the dark.

Darkroom Lighting

For general darkroom lighting, either direct or indirect sources of light are satisfactory. White or light-colored walls and tested ceiling safelight fixtures give good overall illumination. Direct-type safelights may be located over the loading bench and processing tanks or the processors.

Safelight Testing Procedure

All illuminators should be tested thoroughly and frequently to avoid light fog. This testing procedure is suggested: make a very low intensity exposure to produce an approximate density 0.50. Unload the film in total darkness and place it under a mask under one safelight. Turn on the safelight. Uncover sections of the film at one minute intervals until a maximum exposure of ten minutes has been given. Turn off the safelight. Develop the film normally but in total darkness. Process and inspect the film. The time required to produce an increased trace of fog indicates the time limit for the safelight fixture.

Extraneous light in the darkroom is just as bad as stray X-radiation and must be eliminated. Possible sources of white-light leaks are doors, windows, keyholes, ventilators, joints in walls, and partitions. To monitor monthly for stray light, enter the darkroom and allow your eyes to adjust for 15 minutes. Move around looking for light leaks. Look high and low. Make sure *all* lights are on in adjacent rooms. Correct any leaks and reinspect. Keep records.

Safelights

The highest sensitivity of X-ray film is in the blue region of the spectrum. Therefore, safelights should be made with amber or red filters. Filters especially designed for X-ray darkrooms are available from your X-ray dealer. The following precautions should be taken:

1. Use white frosted bulbs, 7½ to 15 watts, at 1.2 m (48 in.) distance.
2. Examine filters frequently for cracks or chips that let white light leak from the safelight.
3. Make sure that the filter lens is correctly installed following the replacement of the bulb.
4. Observe for fade, density change, or color change on a regular basis.
5. Safelights are helpful and important in the darkroom provided they are properly maintained. Only one person should work with them (including bulb changes) to avoid accidents. Do not keep adding more light to a darkroom or it will become a lightroom!

Unwanted Radiation

Because X-ray films are highly sensitive, they must be protected from accidental exposure to energized X-ray tubes (generators) or radioactive materials (isotopes).

If fogging of film occurs, the storage room, if located near sources of radiation, should be checked for possible stray radiation coming from radium, radon needles, radioactive isotopes, X-ray tubes, or other sources. It is advisable to perform this test every six months as a precaution. The following is a simple, inexpensive test.

Attach a small coin or equivalent penetrameter with adhesive tape to each of several X-ray films (use fastest speed) in plastic bags or cardboard holders (day-pack works very well) and place them on the bin and on the walls of the room in which films are stored. The coin is toward each possible source of radiation. After two weeks, develop the films. If an image of the coin appears on any of them, radiation may be reaching the stored films and should be eliminated. Another method is to use normal radiation testing devices such as a personnel film badge or an ion chamber device. In the latter case, tests must be made during full exposure. The use of film as a testing device provides indication of accumulated dose, if any.

Ventilation

It is important that the darkroom be well ventilated. Ventilation provides comfort to the darkroom personnel and makes the darkroom a better place to work. Ventilation helps to maintain proper ambient (room) temperature and relative conditions vital to the proper storage of film. It also helps insure against artifacts from static electricity, handling, or moisture. In addition, adequate ventilation is needed to keep harmful chemical fumes from building up and affecting either the darkroom personnel or the radiographic film.

Cleaning Tanks

Corrosion seldom occurs when the tanks are full of normal chemical solutions and are kept *clean*.

Deposits often form on the walls of the developer tanks because of the action between mineral salts dissolved in the water and carbonate in the developing solutions. These deposits can be removed by using commercially prepared stainless steel tank cleaner. Follow the directions of the manufacturer, being sure to rinse the tank walls with fresh water. Wipe the tank out with a clean cloth or cellulose sponge.

Clean the exterior stainless steel before any deposits can attack the surface. Wipe with a cloth and warm soapy water then rinse, making sure no soap deposits get into the chemistry. Once a week, use a stainless steel cleaner according to the directions on the lable.

Always give special attention, when you clean, to welds and corners where deposits can cling.

If deposits are heavy, remove the worst of them with fiber brush, or plastic cleaning pad, then polish with a stainless steel cleaner. If an abrasive is required, use a very fine sandpaper.

Caution: Never use metallic abrasives, steel wool, or wire brushes, as they can contaminate the surface of stainless steel. Any foreign metal particles will cause corrosion and may contaminate chemicals. Do not use commerical steel wool pads or strong detergents, since these are hard on the stainless steel and could react unfavorably with the chemistry.

Cleaning Illuminators

Quite frequently, good radiographs will appear dull because they are viewed on faulty illuminators. Illuminators are faulty when the glass plate is dirty and/or bulbs of different wattage, age, color, or size are used. Old interior paint that is dull, dusty, or rusty will cause the radiograph to appear dull. Use a regular photographic exposure meter to test your illuminator. At ASA 100 a standard viewbox should register an exposure value (EV) of 13, which indicates 500 foot-candles of power, on direct contact reading. EV14 is 1000 foot-candles, EV12 is 250 foot-candles. Test high-intensity viewers at a fixed viewing of 300-350 mm, (12-14 in.) distance and calibrate the control knob according to brightness level. Identical radiographic studies should be viewed at the same intensity. Installation of an amperage meter to control uniform output would be warranted.

Wash the outside of the viewing glass plate every day. Once a month wash out the inside of the view box. Always use matched bulbs. When conditions indicate paint deterioration, unplug the illuminator, remove the front diffuser and wash it. Then paint the inside with a good, durable white enamel finish.

Tank Stoppers

In darkrooms where developer and fixer tanks are emptied into the wash-water compartment by the removal of a stopper, it is better that this stopper be pushed into the hole from the inside of the tank than from the wash-water side. This method uses the weight of the chemical solutions to tighten the stopper, thus minimizing the danger of premature leaking or emptying of the tank.

Avoiding Static

There are two ways to avoid markings on X-ray films. One is to prevent the generation of static electricity; the other is to cause such charges, once generated, to leak off gradually rather than to discharge rapidly, which is what causes damage.

The most successful procedure is to keep a high relative humidity in the surrounding atmosphere. An accurate instrument for measuring relative humidity, called a *psychrometer*, is a valuable addition to any radiographic darkroom. Periodic checks on prevailing darkroom humidity enable one to take special precautions necessary to minimize the generation and discharge of static electricity. The relative humidity in the darkroom should be between 40 and 60 percent. There are several types of electric humidifiers on the market that are suitable. The relatively new ion generators may be used in place of humidifiers and are as effective without needing water.

The following precautions will be of assistance in overcoming the most common causes of static:

1. If you are using X-ray film that has interleaving paper, handle film gently. Let the interleaving paper fall away from the film and place film in the cassette gently, without sliding it over the screen.
2. Following X-ray exposure, the cassette or holder should be opened slowly and the film removed carefully. The reason for careful handling is that the film is more than twice as susceptible to an energy source once it is exposed. Thus, a film will react to much smaller electrical discharges after exposure.
3. Move slowly when handling the film.
4. Make sure everything is grounded.
5. Use X-ray anti-static salt screen cleaner regularly.
6. Avoid the use of static-generating synthetic clothing.

Grounding. Electrically ground the metal top of the film loading bench, film bin, X-ray table, pass boxes, and other equipment such as processors.

In the darkroom, avoid nonconductive floor covering (rubber tile), hard floor waxes on concrete, rubber- and plastic-soled shoes, intensifying salt screens with worn surfaces, and using a dry cloth to clean intensifying screens. A camel-hair brush or vacuum cleaner should be used for dusting, and a lint-free cloth and screen cleaner with antistatic solution should be used for washing intensifying salt screen surfaces. In periods of low ambient humidity (winter time or northern climates), when static is prone to occur, antistatic solution can be applied to intensifying screens as added protection against static discharges.

Common Static Markings. Three kinds of static markings are illustrated in Fig. 6. Crown and tree, the first two types of static markings, are considered to be results of heavier electric discharges. They can be generated by very rapid motions, such as occur when film is removed from interleaving paper, when interleaving paper breaks contact with the film, or when the film is touched by the fingers.

Smudge static markings may result from photographic exposure to visible light produced by sparks that occur in the air next to the film surface. Smudge static is considered to be produced when relatively low potential discharges occur over a large area.

FIGURE 6. Static Electricity Markings

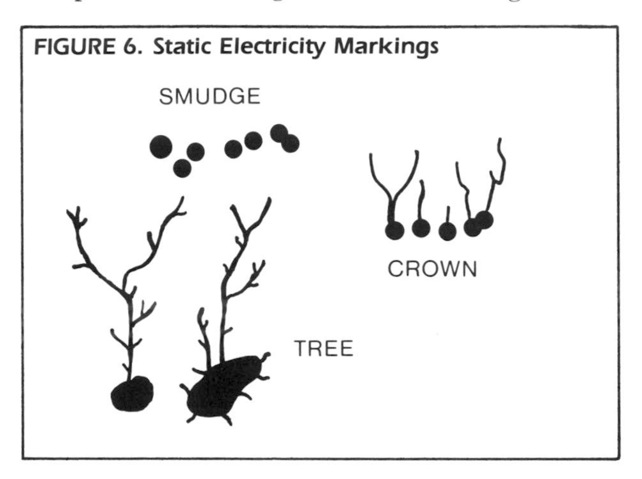

Color Conditioning

Surfaces above the working area should be finished with a white or light-tinted paint, in a semigloss or flat finish, because a gloss finish sometimes reflects light in an objectionable manner. An easily cleaned surface is a primary concern.

Storing X-Ray Film

Recommended storage conditions for all types of X-ray films are temperatures between 18 and 24 °C (65 and 75 °F) and 40 to 60 percent relative humidity.

Usually most radiographic facilities will have two storage areas. One area is for long-term supplies and another, usually the darkroom, is for short-term needs. In either case, it is required that a stock rotation plan be instituted. The plan is quite simple in that as new film arrives, it is placed on the right side of the supply. As the film is needed, it is removed from the left side. This is called a FIFO system: first-in-first-out. To assist in rotating film boxes, remember that all film boxes have an emulsion number and an expiration date on the end label. A system many people use is to write on each box the date when it is received. Whatever system you use, be sure to have a system, keep records, and always use the oldest film first. Film must be stored and inventoried by expiration date, film brand, type (class) and/or speed.

Film boxes should be stored on their edges. This distributes the weight and helps protect the film from pressure marks.

When using one box at a time, or when there is no film bin, always be sure to fold over or close the bag and to replace the lid after each film is removed. Film is packaged in hermetically sealed light-tight and moisture-proof pouches—for example, black polyethylene or aluminized plastic. Once the pouch is opened to expose the fresh film there is still enought bag remaining to fold over to make the bag light-tight again. This feature is not to eliminate the use of the box top or film bin, but it is an added safeguard.

Cassettes should not be stored for prolonged periods of time loaded with film. Load cassettes and holders with fresh film before each use.

PART 3

MANUAL PROCESSING TECHNIQUE

Principle

Processing completes what the exposure started. After the exposure is made, the film is removed from the cassette and placed on a hanger. The film is processed. This is the simplicity of radiographic processing, but there is obviously more to forming a visible image radiograph than is stated above.

There is indeed an exactness to manual processing. This is a controlled scientific process in which something in one form (the latent image) is converted into another form (the visible image). No matter how superior your exposure techniques are, if you permit anything less than optimal processing technique then you will obtain a visible image that is less than optimal.

Equipment And Practice

Thermometers

In developing X-ray films by the time and temperature system, an accurate thermometer is of the utmost importance. Service thermometers should be checked at regular intervals against a thermometer whose accuracy is known. They should be graduated in degrees or, better, in half degrees.

Never use glass thermometers containing mercury or iodine. If the glass breaks, they can become hazardous to personnel and developer. A thermometer should be read while it is inserted in the thoroughly mixed solution. To avoid parallax, the thermometer should be held so that an imaginary line from the eye forms a right angle to the axis of the thermometer. To further avoid this problem, a bimetallic or electronic thermometer might be considered.

Mixing Solutions

Make sure the solution tanks are clean. Carefully read the instructions on the chemical container. Mix as recommended at the suggested temperature to insure satisfactory performance. Avoid high temperatures, inclusion of air, and contamination during preparation. Stir thoroughly but not too violently or too long.

If the efficiency of the processing solutions is below par, investigate these four factors:

1. Are the tanks really clean?
2. Are solutions properly mixed? Did you mix in the right sequence? Did you over agitate?
3. Are temperature and timer accurate?
4. Are your exposure techniques accurate?

Water for Processing

The water temperature for mixing should be ±3 °C (±5 °F) of the manufacturer's recommendation.

In spite of the variety of impurities in water supplies, most city water is pure enough for photographic processes. The rule of thumb is to use water of drinking quality.

If you plan to use well water rather than city water, first be sure to make a thorough check of your local conditions. Water hardness analyses (and other chemical analyses) are required and are available from water service companies.

Deep wells are stable in temperature and chemical content throughout the year; they are also normally cold enough to require heating for photographic use. Shallow wells, on the contrary, are subject to seasonal fluctuations in both temperature and chemical content. Since heating is much cheaper than chilling, this alone usually justifies the cost of a deep well versus a shallow well.

Water Impurities. Normally, city water is chemically acceptable for photoprocessing. Some potential sources of impurities should still be kept in mind, however. Two impurities, sulfide and iron, are the most serious. Sulfide causes photographic fog. Up to 0.1 parts per million (ppm) is tolerable, and it is rarely a problem (usually it is found in surface water or shallow wells). If you can smell sulfur odors, the water cannot be used.

Iron over 0.1 ppm can cause trouble in some photographic processes, but up to 10 ppm is permissible in other processes. It causes stain on film and prints, and over a period of time will rust the tanks. Iron will also react with other metal ions, causing pits on paper and film. More common in water from deep wells, iron can be removed with a green sand filter.

Seasonal changes in river water are common, and municipal treatment of water may change seasonally or through a change in the source of supply. Some areas can change abruptly. If you suspect a change, contact your local water department for a water analysis and advise them that you are working with photographic chemicals.

There are five methods of purifying water: (1) distillation; (2) boiling; (3) filtration; (4) chemical treatment; (5) ion exchange. For processing, filtration is the most practical method; ion exchange is a possibility where filtration is insufficient. Filtration of the hot water, cold water, and tempered water lines is important to providing clean water. If you use a mixing valve, then check to see if it has strainers at the point where incoming lines attach. If there are strainers, take them out and clean them periodically.

Water Hardness. Water that is either too hard or too soft is undesirable for processing. Hardness is essentially a measure of the calcium and magnesium ions present in water. It is usually expressed as so many parts per million of calcium carbonate; water is considered moderately hard if it contains between 40 and 150 ppm. Note: Grains per U.S. gallon (avoirdupois) $\times$ 17.118 = ppm.

Water harder than 150 ppm may leave hard-to-remove scum deposits on drying radiographs. Calcium and magnesium also build up scale in water lines, processing tanks, and other equipment. Water over 150 ppm may have to be softened, if deposits are being left on the radiographs.

Water softer than 50 ppm can allow the emulsion to over soften and may reticulate (gelatin cracks in going from hot to cold solutions).

Combating Algae. Algae are carried into wash tanks by the water supply and the air, and grow rapidly when combined with the gelatin washed from the film in combination with moisture and heat. In extreme conditions, the algae form streamers that break off occasionally and are deposited on the film or paper, leaving spots that are almost impossible to remove.

Algae accumulation can be prevented by draining the tanks daily and allowing them to dry out overnight. Cleaning the tanks with a fiber brush may be necessary about once a week. Filtering incoming water or running water continuously are ineffectual and expensive. Adding algaecide to the water at the end of each day will retard growth. Use 25 mL/L (3 oz/gal).

Algae can be removed from the walls of the water compartment by several methods, and repeated treatments generally are required. Several photographically neutral algaecides are available for this purpose; consult a pharmacist, photo supply, X-ray dealer or local swimming pool chemicals dealer. A very efficient and inexpensive method is to use diluted acid or diluted commercial laundry bleach (5% sodium hypochlorite) to clean the tanks and retard algae growth.

Foaming. Foaming occurs when there is a sudden release of air in the water, caused by either: (1) very high water supply pressure (when the pressure is released, the trapped air expands, causing foam); or (2) cold water (cold water can hold more air than hot water). When the cold water mixes with hot in a mixing valve, there is a sudden increase of pressure in the air in the cold water—resulting in bubbles.

For relief, install a pressure reducer ahead of the mixing valve. If the problem is serious, you may have to install a tank between the mixing valve and the wash tanks to allow the expanded air to bleed off.

Working in the Darkroom

The workflow in the darkroom is illustrated in Fig. 7.

Contamination

When mixing solution, great care should be taken to prevent the fine powder particles (powdered formulas) from blowing, or concentrates (liquid formulas) from spilling around the darkroom. They are injurious to screens and films. Fine particles of developer coming in contact with an undeveloped film can produce black spots; particles of fixer can produce white spots. Developing solutions can cause permanent stains on salt intensifying screens.

FIGURE 7. Manual Processing Darkroom Workflow

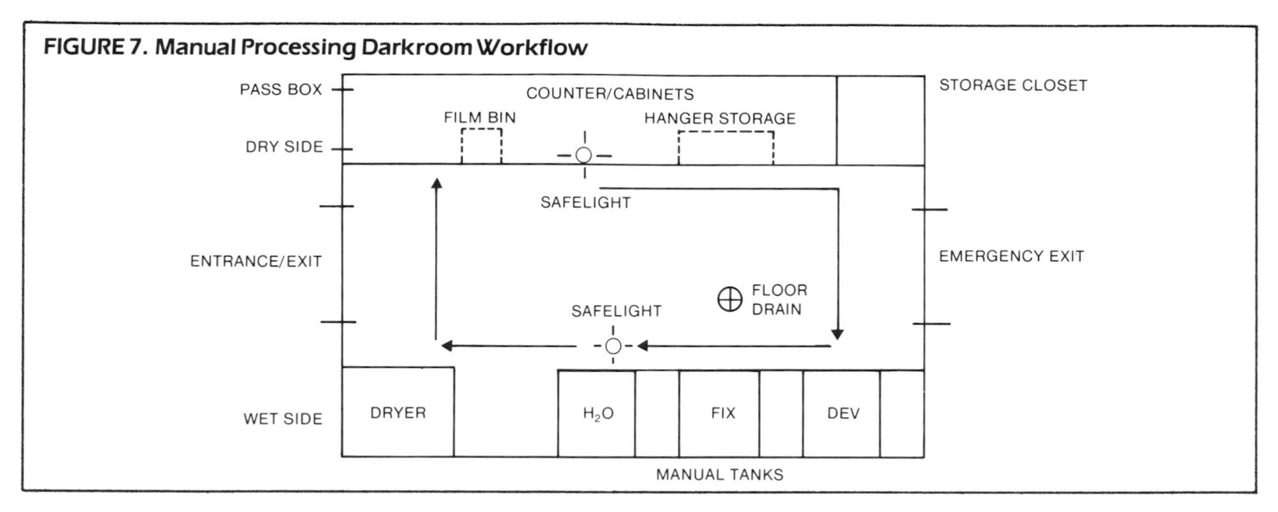

Therefore, it is always best to mix and store chemicals outside the darkroom.

Before processing films, it is wise to remove any scum, lint, dust, or other foreign matter that may have collected on the surface of the processing solutions. A piece of cheesecloth mounted on a 20 × 25 cm (8 × 10 in) film hanger or an absorbent paper towel will do the job. Any scum or foam will adhere to the cheesecloth or paper towel as it is drawn across the surface of the solution.

Time/Temperature Method

The time/temperature method provides the controlled basis for obtaining consistent, optimal radiographic information. This scientific method also permits us to alter the processing cycle to suit our specific needs and requirements.

Regardless of the optimal time/temperature recommendations for a given chemistry and film, it must be remembered that everyone has different likes and desires. With this in mind, consider that faster processing, higher or lower contrast, greater speed, less density, and other considerations can be altered through the judicious use of the time/temperature method. At a given temperature, longer times will overdevelop the film, increasing density. At a given time, higher temperatures will cause overdevelopment fogging (increased noninformational density) of the film. This does not mean that "sight" development is advocated. No one should use anything but the specific time/temperature method (see Fig. 8).

FIGURE 8. Example of a Manufacturer's Recommendation. Taken from "Radiographer's Reference," Wilmington, DE: E.I. DuPont de Nemours & Co., Inc. (1981).

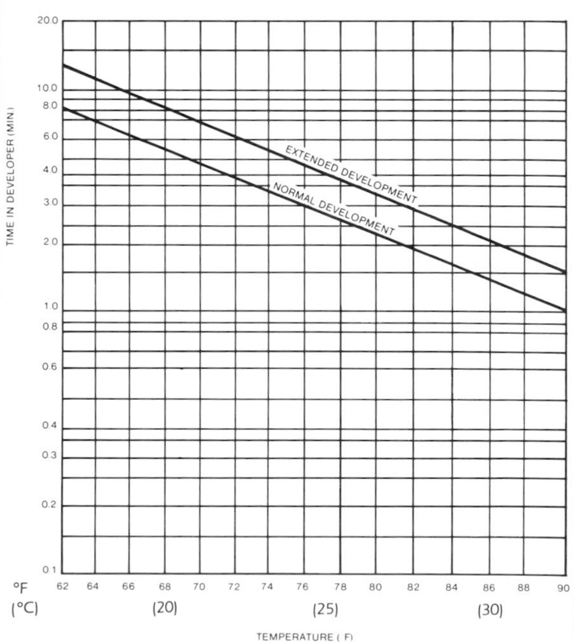

Developers

Optimal radiographs require correct development. The developing time to be followed differs with the type of film being processed, the type of developer employed, and the temperature of the processing solutions.

Developers and replenishers are available in both powdered and liquid concentrated form. They are supplied in convenient sizes for making anything from 3.8 to 760 L (1 to 200 gal) of working solution. Powdered developer and replenisher are packed in airtight cans. Each can has (1) a small inner container (the reducing agents), and (2) the bulk chemicals. Developer oxidizes rapidly and thereby loses its efficiency. To preserve the developer, keep it tightly covered when not in use.

In mixing fresh solutions, it is important to know the exact capacity of the mixing tank, because solutions that are too concentrated or too diluted will not give uniform processing results. To determine the capacity of your rectangular processing tank, multiply the inside length by the width by the depth in inches and divide the result by 231 to get the tank capacity in gallons. Multiply the length by the width by the depth in centimeters and divide by 1,000 to get the capacity in liters. The capacity of cylindrical tanks may be calculated, but it is more accurately determined by filling the tank using a calibrated bucket. Similarly, if the solution tank insert has bulged at the sides, it is necessary to determine the tank volume by adding known quantities of water.

To obtain even, overall development, the developer must be stirred before each use daily. Stir the solutions before checking the temperature. Churn-type mixers are recommended and they should be washed after each use. Do not use developer and fixer mixers interchangeably. Never mix violently or for long periods. In use, the temperature of the developer should never be allowed to fall below 13 °C (55 °F). Below this temperature, the reducing agents form a cloudy precipitate and fail to function properly.

By following the directions printed on the container, the necessity for cooling the solution before use can be eliminated. The chemicals will go into solution easily. Never add ice to developer merely to reduce its temperature. Doing so causes dilution, which weakens the solution. However, when dilution of fresh solution is required, clean ice may be used for the twofold purpose of dilution (to the extent needed to fill the tank) and of bringing the solution to the required temperature. (When doing this, remember that ice has a greater volume than water.) Keeping a liter or two of developer replenisher in a refrigerator during summer months provides an easy way of cooling the developer without dilution.

In removing films from the developer, do not let them drain back into the tank for more than two seconds. The final drippings are oxidized so badly that they contaminate and weaken the solution remaining in the tank. Excessive draining time also may lead to streaks or fog in the film.

Radiographs require just the right amount of contrast. Too much reduces the range of densities covered by a single exposure; the thinner parts become too dark and the thicker ones too light. Conversely, insufficient contrast, though it affords more latitude in exposure, lessens the total differentiation, thereby obscuring fine details. To achieve an ideal degree of contrast, a given type of x-ray film should be developed in the same brand of solutions using the time/temperature and replenishment characteristics supplied by the manufacturer.

Compensating for Developer Exhaustion

Replenisher Technique. Replenishers replace the reducing agents as they are exhausted, and if added correctly eliminate the need for adjusting developing time to maintain constant density and contrast over the useful life of the developer. Replenishers should be mixed as directed on the label and should be added frequently and in small amounts to maintain the developer level; if an amount greater than 10 percent of the developer tank volume is added at one time, fog may increase. The remaining replenisher should be kept in a tightly stoppered bottle or in a plastic jug or tank with a floating lid and dust cover.

Because developer exhaustion depends upon the type of film emulsion and the film density as well as the film area, the quantity of replenisher per film will not be constant. As a rule, about 90 mL (3 oz) of replenisher are required for each 35 × 43 cm (14 × 17 in.) film processed.

The developer tank should be drained and thoroughly cleaned when replenisher has been added in an amount equal to five times the volume of the tank. For example, if we were using a 38-liter (10-gallon) developing tank, 190 liters (50 gallons) of replenisher would be added before the tank is drained and cleaned.

Non-replenisher Technique. The correct developing time when replenisher is not used depends upon the degree of developer exhaustion as well as the temperature of the solution. The most accurate way to measure exhaustion is to keep a record of the number of films processed. An alternative method, although less accurate, is to increase the developing time at periodic intervals.

For example, if the tank full of developer normally lasts four weeks, developing time at a constant temperature of 20 °C (68 °F) may be five minutes during the first two weeks, six minutes during the third week, and seven minutes during the fourth week. This system is based upon the assumption that approximately the same number of films are developed each day.

In working with other sizes, the following equivalents can be used:

- Three 8 × 10 inch (20 × 25 cm) films equal one 14 × 17 inch film
- Two 10 × 12 inch (25 × 30 cm) films equal one 14 × 17 inch film
- One 11 × 14 inch (28 × 35 cm) film and one 8 × 10 inch film equal one 14 × 17 inch film.

Air Bells

When a film is placed in the developer, the film hangers should be agitated gently for a few seconds to remove air bubbles (referred to as bells) from the film. Defects resulting from air bells usually appear as small, round, clear spots because the bubble of air prevents the development of the emulsion beneath it. Undeveloped silver, upon complete fixation, is dissolved, leaving a transparent circular area.

Film hangers, when suspended in the developer, should be spaced at least one centimeter (approximately half an inch) apart. Agitate for a minimum of five seconds at the beginning, middle, and end of the time period.

Quality Control

Quality is defined as the sensitometry of the visible image. Maintenance of the time/temperature method of processing depends on some system of quality control. Quality control kits are available to give you experience and to help in setting up your own program. A simple quality control test is to use the same cassette or film holder and a step wedge with known exposure, film, and processing techniques. Establish a "control" film and routinely make test films that can be compared to the control film. Keep exact records, as these will become your personal teaching and record file.

Sensitometry. The study and measurement of relationships between exposure, processing conditions, and film response to exposure is known as sensitometry. The properties of a film that affect or govern the relationships are known as sensitometric properties. Sensitometry uses many terms that have specific meanings; therefore, we must have some common understanding of the definitions of these terms.

Density. Density and adequate differences in density (contrast) are considered the most important of all properties in the radiograph. Proper densities and adequate contrast make visible the structural details within the image of the object.

Radiographic density has been defined as the amount of film blackening that is the result of metallic silver deposits remaining on the film after exposure and processing. A useful method of measuring the amount of film blackening is to determine the manner and degree to which it interferes with a beam of light passing through a radiograph. The amount of light absorbed by the film is measured in terms of density by a densitometer.

Sensitometrically, density is defined as the common logarithm of the ratio of the amount of light striking one side of the radiograph compared to the amount of light that passes out the other side. When the metallic silver in the emulsion allows one tenth of the light to pass through, this ratio is 10:1. The common logarithm of 10 equals 1 and the silver deposit is said to have a density of 1. Density can now be defined by the equation:

$$D = \log \frac{I_I}{I_O} \qquad \text{(Eq. 8)}$$

where D is density, I_I is intensity of light incident to the film (also termed I_o, original), and I_o is intensity of light output through the film (also termed I_t, transmitted).

The data obtained by sensitometric procedures are usually plotted in the form of graphs. Fig. 9 shows a typical characteristic curve of an X-ray film exposed with intensifying salt screens. The portion of the curve designated as the "toe" demonstrates the nonlinear response of the emulsion to relatively small amounts of radiant energy. With uniform increases in exposure, the density builds up slowly until the linear response "straight line" portion of the curve is reached. Along this straight line portion, the density increases uniformly with the logarithm of the exposure until the nonlinear "shoulder" of the curve is reached. The shoulder is not produced with direct exposure to industrial films; it is produced when salt screens are used. Additional exposure results in smaller increases in density to a point where additional exposure does not produce greater density.

Contrast. Contrast by definition is the difference between two densities. As a radiograph is viewed on an illuminator, the difference in brightness of the various parts of the image is called *radiographic contrast.* This is the product of two distinct factors: (1) film contrast—inherent in the film and influenced by the developing process; and (2) subject contrast—a result of differential absorption of radiation by the subject. Although radiographic contrast can be altered by changing one or both of these factors, it is good practice to standardize the film and processing procedure and to control radiographic contrast by changing subject contrast. This can be done easily by adjusting the kilovoltage or general exposure technique to alter the quality of radiation.

Sensitometrically, contrast refers to the slope or steepness of the characteristic curve of the film; this is also called gamma.

Gamma. Gamma is the slope or steepness of the straight line portion of the characteristic curve. In plotting a characteristic curve *density* (a logarithmic value) is most often plotted against log relative exposure.

Exposure is defined as the intensity multiplied by the time; it can be expressed either in absolute exposure units (ergs/cm^2 or J/m^2 of X-radiation) or in relative units. Relative exposure is much more convenient and equally useful for us. In radiography we think of exposure in terms of milliampere-seconds (mAs) or milliampere-minutes (mAm). Then if the mAs is doubled, the exposure is doubled, kVp remaining constant.

FIGURE 9. H&D Curves

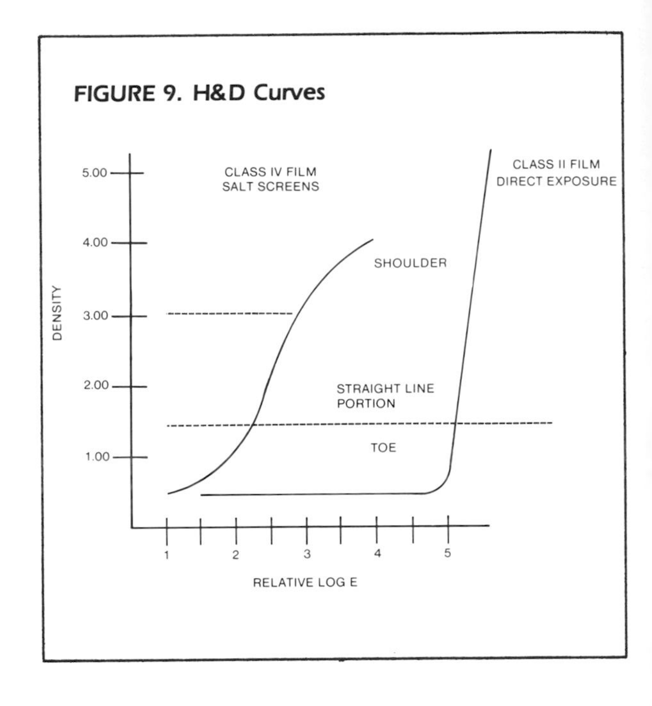

If the kilovoltage remains constant, the ratios of the exposure reaching the film through two different regions of the subject are always the same, no matter what the values of milliamperage, time, or focal-film distance may be. For example, two exposures, one of which is twice the other, will always be separated by 0.3 on the logarithmic exposure scale (the logarithm of 2 is 0.3).

Speed. It has been determined that the contrast of a film is indicated by the *shape* of the characteristic curve. Speed is indicated by the *location* of the curve along the exposure axis. The "faster" film will lie toward the left of the graph. In Fig. 10, film A is faster than film B, but it has the same contrast. The separation of the films is a measure of the speed difference. Film A requires less exposure to achieve a speed point of density 2.00 above base plus fog. Film B requires more exposure, so we say it responds slower or is less fast.

The convenience of using relative exposure also applies to speed. The speed of one film can be expressed on a basis relative to another film when one is made the standard of comparison. This reference film can be assigned any arbitrary speed value, such as 100. If another film requires only half

FIGURE 10. Speed Shifts

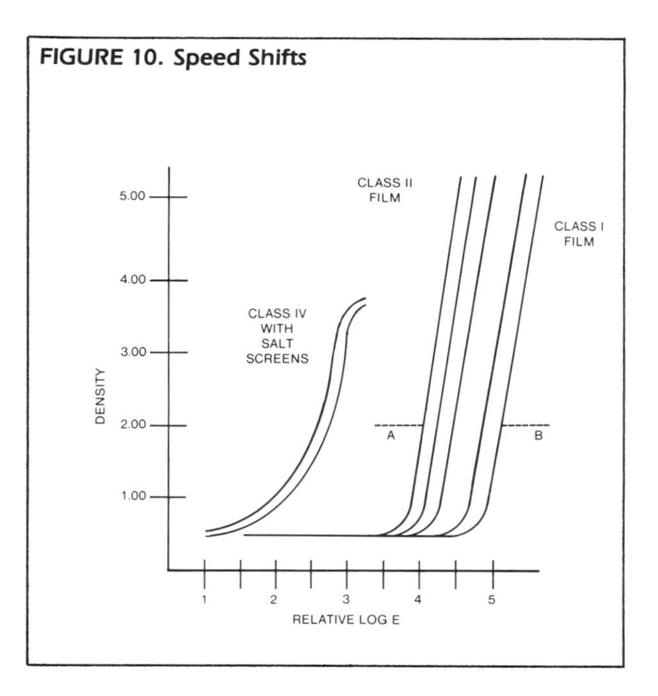

the exposure to reach the same density as the reference film, the faster film will have a *relative speed* of 200.

A density of 2.00 above base plus fog has been designated as the density to compute film speed. This density has been chosen because it represents the minimum useful density range for much of radiography. For lower density films, speeds are often calculated at the lower density of 1.00.

The establishment of a quality control program is discussed further in Part 4.

Fixer

Failure to fix a film sufficiently results in its discoloring with age. A good rule of thumb for determining the minimum fixing time is as follows: After a film has cleared, leave it in the fixer an additional two times as long as it took to clear. For example, if it takes one minute for a film to clear, then it should be left in the fixer at least two more minutes, or a total period of three minutes.

To prepare fixing baths from liquid concentrate fixer, it is essential that directions on the container be followed. Attention is paid only to the temperature of the developer, because fixer temperature is less critical; however, at high temperatures, it is important that the processing baths be maintained at about the same temperature. Changes in temperature cause the gelatin of the emulsion to swell and contract. When the temperatures of two baths differ excessively, this change in the gelatin takes place so abruptly that unevenness is likely to result. The effect produced on the film is known as "reticulation" (the gelatin breaks, producing fine cracks).

Agitate the film briefly after placing it in the fixer. This helps avoid streaks and stains on the finished radiograph. Remember, excessive fixing necessitates an increase in washing time and may etch away some silver.

Constant agitation of a film in a fixing bath accelerates the action of the solution. When one is anxious to examine a radiograph, this technique used in conjunction with a high-speed developing technique materially shortens the time required to process a film for visual inspection.

It is important not to turn on the white light until the film has been in the fixer for one minute. Cloudy areas or streaks may be caused on partially fixed film by turning on white lights too soon. A longer time is necessary if the fixer is old or partially exhausted, particularly if a white light of high intensity (such as a viewing illuminator) is used.

Fixing baths should be changed frequently. It is false economy to use fixer solution beyond its normal life, for exhausted fixer eventually may cause discoloration to the radiographic image. The time necessary to clear a film is a reliable gauge of fixer exhaustion. It is based on a clearing time test. If a film normally clears in 20-60 seconds (depending on type) at about 20 °C (68 °F), then this becomes the standard. Longer times indicate exhaustion. Always use the fastest speed film for tests (Class II). Dip a corner in the fixer until clear. Time until a second patch (immerse further) is equally clear. Further information about the clearing time test may be found in Part 1.

Precautions for minimizing fixing troubles:

1. Maintain the fixer at proper acidity to minimize sludging and to get maximum hardening.
2. Agitate the film for 15 seconds after immersion in the fixer to avoid streaks and stains.
3. Change fixer baths frequently.

Archival Quality. The importance of adequate fixing to archival keeping quality cannot be overemphasized. Poorly fixed films will not deteriorate until many months after processing. The lighter portion of a radiograph may become yellow and the image may tend to fade. This delayed action may be traced back to the nature of the fixing process, which is believed to consist of two steps. Undeveloped silver halide is first converted to silver-thiosulfate complexes, $Ag(S_2O_3)=$, that possess only slight solubility; the reaction then proceeds to form soluble compounds that are removed by washing.

Inadequate fixing leaves small amounts of these complexes that cannot be removed by normal washing. In time, these residues break down and react with the silver image to form silver sulfide (Ag_2S). This sulfide is usually yellow in color, although it may range from pink to brown. It cannot be easily removed from the radiograph. Extremely minute amounts of residual thiosulfates are sufficient to cause serious deterioration. Therefore, the radiograph may appear transparent and at the same time be inadequately fixed. For this reason, *close* attention to recommended fixing technique is mandatory.

Hypo Displacer

Washing time can be shortened by using a hypo displacer (a 5 percent solution of sodium sulfite is effective). After removal from the fixing bath, films should be given a 5-second water rinse (with agitation) then placed in the hypo displacer for 1 to 2 minutes with moderate agitation. Follow this with a 5-minute wash.

Washing

Radiographs must be washed thoroughly to prevent discoloration with age and to assure preservation of the image.

Proper washing time for X-ray films depends principally upon the frequency with which the water is changed. Running water at a temperature of 20 °C (68 °F) is recommended. The rate of flow should be adjusted so that the water in the tank changes completely at least 10 times each hour. Under these conditions, Class II films should be washed at least 20 minutes, the time being computed from the moment the last of the group of films is placed in the wash tank. Class I films may be washed 10 minutes. Of course these times vary with brands of film and the quality of the chemicals. To avoid excessive swelling of the emulsion, film should not be left in the wash tank overnight.

Archival keeping quality is strongly dependent on (1) the concentration of thiosulfate (fixer) remaining in the radiograph after processing, and (2) storage temperature and humidity. Thiosulfate (residual hypo) content is best measured quantitatively by an analytical test. Your local technical representative can arrange to have this test run for you.

Drying

Avoid temperatures above 120 °F in drying and maintain a steady volume of air across the film surface. An excessively low relative humidity within the dryer may cause water-spot drying marks and streaks on the film surface.

To accelerate drying:

1. Use fresh fixer or raise the acidity of the old fixer to the proper level to insure optimum hardening. A correctly hardened film drys faster.
2. Swab film with a moist fine-grain cellulose sponge, absorbent cotton, or a chamois. This technique also can be used to eliminate water marks and streaks that occasionally occur while drying film.
3. Use a wetting agent bath to decrease drying time and to help eliminate water-spot drying marks.

High-Speed Processing Technique

When immediate examination of a radiograph is necessary, film can be rapid-processed with little sacrifice of quality. The following three techniques are recommended:

1. If it does not exceed the capacity of the X-ray equipment, double the exposure and reduce the developing time by one-half.
2. Develop for 60 seconds at 32 °C (90 °F), agitating frequently and vigorously. Rinse for 10 seconds in water at the same temperature or immerse in an acid stop bath for the same period. Fix until clear (approximately 20 to 60 seconds). After viewing the film, replace in fixer and complete processing in a routine manner.
3. Consider newly introduced films that permit faster processing and/or reduction in exposure through the use of salt screens. Processing may be as fast as 3 minutes total.

Rinsing. If an acid short stop is not used, the film should be rinsed carefully for 15 to 20 seconds upon removal from the developer, preferably in a bath of running water at about 20 °C (68 °F). A shorter rinse may be used if the film is agitated or if warmer water is used.

The tops of the film hangers should be placed below the surface of the rinse water, so that they will be rinsed between developing and fixing. Adequate rinsing is important and should not be overlooked, because it reduces carry-over of the alkaline developer, thus preserving the acidity and the efficiency of the fixer. Inadequate rinsing also can cause stain or streaks on the finished radiograph.

Stop Bath. An acid stop bath, rather than a water rinse, is recommended between development and fixation. The use of such a bath, consisting of a 2 to 3 percent acetic acid solution, serves two purposes. First, and most important, it neutralizes the alkali of the developer and thereby protects the acidity of the fixing bath. Second, it stops development quickly.

Carry-over of alkali developer eventually reduces the acidity of the stop bath to a point where it no longer protects the fixer or stops development of the film. The pH of the stop bath should be maintained at approximately 2.6 and should not exceed pH 4.0. The pH of stop bath and fixer may be measured with pHydrion paper #325 range 3.0 to 5.5. Further information may be found in Part 1.

Table 4 shows two formulas for acetic acid stop baths.

TABLE 4. Acetic Acid Stop Bath

Glacial acetic acid	1 part: 12 mL or 1.6 oz.
Water to make	80 parts: 1 L or 1 gal
or	
28% acetic acid	1 part: 50 mL or 6 oz
Water to make	20 parts: 1 L or 1 gal

Hot Weather Processing

Through the summer months, darkrooms and chemical solutions frequently get warmer, than normal. For best results, maintain the developer, fixer, and wash water at the same temperature. Ice should not be placed in the solution, because excessive dilution will result as the ice melts. Although processing films in hot solution is not recommended, satisfactory radiographs can be produced in solution up to 35 °C (95 °F). Water temperatures can shoot up to dangerous heights even in an air-conditioned darkroom. Prolonged washing at high temperatures may damage film—so keep washing to a minimum if your water is too warm. Automatic water mixers will require watching too—they cannot keep water any cooler than the temperature of the cold-water supply.

Restrainer

With temperatures up to 24 °C (73 °F), no extra precautions are needed. However, when temperatures range between 27 and 35 °C (80 and 90 °F), restrainer can be added to the developer.

A restrainer for developing solutions is made up of 18 g of sodium bicarbonate per liter (2.5 oz/gal) of diluted developer, or 4.5 g/L of concentrated solution.

Weigh out the total amount of the proper restrainer needed for your full tank of developer, dissolve it in about 200 mL (6 oz) of warm water (27 °C, 80 °F) and then add the resulting solution to the developer. Stir the mixture thoroughly.

Rinse

The gelatin in the emulsion swells more in warm solutions and absorbs more developer. Therefore longer rinsing times are required at higher temperatures. Poorly rinsed films carry more alkali into the fixer and thereby reduce the speed and hardening action of the fixer.

Fixing at High Temperatures

A fixing bath that contains an acid hardener minimizes the tendency of the emulsion to "frill" during the final washing. Even when rinsing is done carefully, the fixer acidity declines with use. The addition of fixer replenisher will maintain pH 4.5 and the fixer's hardening ability.

Washing Film in Hot Weather

In the summer, excessive washing should be avoided. Prolonged immersion in warm water may cause the emulsion to frill. To determine the correct rate of water flow, measure the time required to refill the tank after removing a given quantity of water and adjust the flow so that water in the tank changes at least 10 times each hour.

Drying Film in Hot Weather

The high relative humidity generally prevailing in hot weather increases the time required to dry an X-ray film. Three of the factors that affect drying time are (1) the degree to which the film has been hardened in the fixer; (2) the length of time it was washed; and (3) the water-absorbing property of the gelatin used to make the emulsion. Methods of controlling the first two factors have been described previously. Control of the third depends on the type of film used. Faster processible film is recommended especially because it absorbs a minimum of water.

Overnight Cooling

In laboratories where 10-20 L (3-5 gal) solution tanks are used, the following recommendations may prove useful. Before closing the laboratory for the day, remove 4 L (1 gal) of developer and 4 L (1 gal) of fixer and place them in separate labeled glass containers. Store them in a refrigerator overnight and in the morning add the chilled solutions to the warm solution to bring the working temperature closer to normal. Make certain the bottles are dedicated and correctly labeled.

Safety

Developer, with its hydroquinone and alkalinity, forms a very hazardous solution. Always use good ventilation when mixing chemicals. A nose filter or respirator is suggested when powdered chemicals are mixed. Goggles are required by OSHA to protect the eyes. In addition an eye-wash station is required. If developer gets into the eye the worker must begin washing within 15 seconds and continue washing for 15 minutes minimum. One hour washing is preferred.

While the developer is most hazardous, care must be taken when working with fixers also. In addition to the above, rubber gloves will protect the skin and keep chemicals out of cuts; rubberized or plastic aprons will protect the worker and clothing.

Other Helpful Information

Clearing Film Base

Liquid laundry bleach will dissolve the gelatin and produce a clean, clear sheet of base plastic. Warm solutions work faster. Enzymes could be used but are very hazardous.

Removing Stains from Clothing

Silver stains on white cotton or synthetic fibers or work clothes, caused by spilling of developer or fixer on the cloth, can be removed by rinsing the fabric in plain water immediately after it has been stained. If the spilled chemicals have dried out or have gone unnoticed for any length of time, however, the stained fabric must be treated with a solution made as follows:

Sodium hypochlorite bleach,
liquid laundry bleach (5% solution) 15 mL (½ oz)
Acetic acid, 5% (vinegar) ..... 15 mL (½ oz)
Water about 38 °C (100°F) .... 4 L (1 gal)

Soak the stained portion of the fabric in this solution for 5 to 10 minutes. Then soak the stained portion in *fresh* fixer. Rinse in plain water and dry. Small stains can be removed by touching the stain with iodine, then with fresh hypo, followed by a plain water rinse, as explained under Iodine-Hypo Cleaner.

Caution: Do not follow these techniques with colored fabrics. Use of the ***iodine-hypo cleaner*** is not recommended with nylon because the resulting iodine stains are difficult to remove. The sodium hypochlorite-acetic acid treatment is satisfactory with nylon, however.

Stain removers are available to remove developer stains from clothing.

Iodine-Hypo Cleaner for Clothes

Spots caused by X-ray processing solutions can be removed from white cotton fabrics as follows: first, soak the area for a short time in warm, fresh water. This will remove most developer stains. If the spots remain, moisten with a solution of 1 part tincture of iodine and 1 part water. After a minute or two, wet the material with *fresh, unused* fixer. Then rinse thoroughly in water. Repeat if necessary.

Removing Developer Stains from Hands

Developer solutions should be considered hazardous and hands should not be routinely submerged in them. Rubber gloves should be worn. When manually processing or working on automatic processor rack components where contact is prevalent, be sure to rinse hands frequently in fresh water. In manual processing make sure hands are rinsed as you rinse the film or fix it.

To remove developer stains from hands and fingers, prepare the following solution for washing (sequentially), then wash with soap and water.

Solution A
Potassium permanganate .... 7.5 g (¼ oz)
Water .... 1 L (32 oz)

Solution B
Sodium bisulfite .... 500 g (16 oz)
Water .... 1 L (32 oz)

PART 4

AUTOMATIC PROCESSING

Introduction

Automatic processing is a chemical reaction performed in a machine; it is often spoken of as if the machine were the important part. However, since the introduction in 1957 of what we consider to be a modern roller transport processor, we have yet to achieve complete automation. What has been produced is a machine that provides improved consistency over manual processing. But humans control the machine as they control manual processing, only less frequently. Keep in mind that "automatic" processing produces consistent quality; the quality will be consistently good or bad depending on the operator.

There are three distinct advantages to automatic processing: consistent quality, improved quality, and ecomomy of time and labor. Some people will question the middle advantage, but given a well-functioning, properly adjusted machine, film after film you will have better quality from the machine than from a person "sight-developing." As a result, processors have found their way into small and large laboratories, trailers and the backs of pickup trucks.

Automatic processing developer chemistry has a hardener that is extremely important in controlling the amount of emulsion swell. An over-swollen film can result in overdevelopment fog, uncleared films, poorer archival quality, wet films, and increased transport problems. Automatic chemistry is usually replenished automatically to sustain volume and activity.

In an institution or laboratory all the processors may be located in one central darkroom/processing area or in several areas. The major advantage of centralization is that one darkroom person can feed several processors, saving space and manpower. The main advantage of dispersed or decentralized processing is that the processor can be placed in different areas, such as production, quality control, and research, to reduce downtime and the confusion of intermixed films. The major disadvantage is increased space and manpower. It is a good idea with dispersed processing, to centralize the mixing and distribution of chemicals to various locations and to return spent fixer to a central collection/recovery area.

Daylight Installation. A new concept related to processing is the "daylight" approach. The system dispenses films into hard cassettes. After exposure the film is fed directly into a processor, sitting in a lighted area, by means of an unloading, feeding device. Film is never touched by humans until after processing. Processors can be placed anywhere there are utilities. There is no darkroom or associated personnel.

Darkroom Installation. In a conventional darkroom, the processor may be installed totally inside the darkroom. The best method is to put the bulk of the unit outside the darkroom to reduce heat inside. Some people put the bulk of the unit inside the darkroom so that any jams can be cleared in the dark. The disadvantage of this system is that all service requiring white light necessitates closing down the darkroom. An alternative would be to put the bulk of the unit in the outer room, but construct the room so that it can become a darkroom.

All processors should have a minimum of 600 mm (24 in.) access on a side. If this is not possible the processor can be made portable on wheels or skids with quick disconnects to allow easy service. Near the processor there should be a floor sink for washing racks or the entire processor may be placed over a large grill work with a drain below.

Automatic Feeders. Automatic feeders are available for some makes of processors, which allow a stack of films to be placed in the feeder and the lid closed; the operator can do other jobs as the feeder automatically feeds films. But feeders require adjustment and periodic monitoring. In addition, it may take twice as long to feed one sheet of film automatically as it would manually, because of the cycle time and delay of the mechanism. This is primarily true at the faster processing times (in excess of seven minutes). But still, these units can be very useful.

FIGURE 11. Automatic Processor Darkroom Work Flow

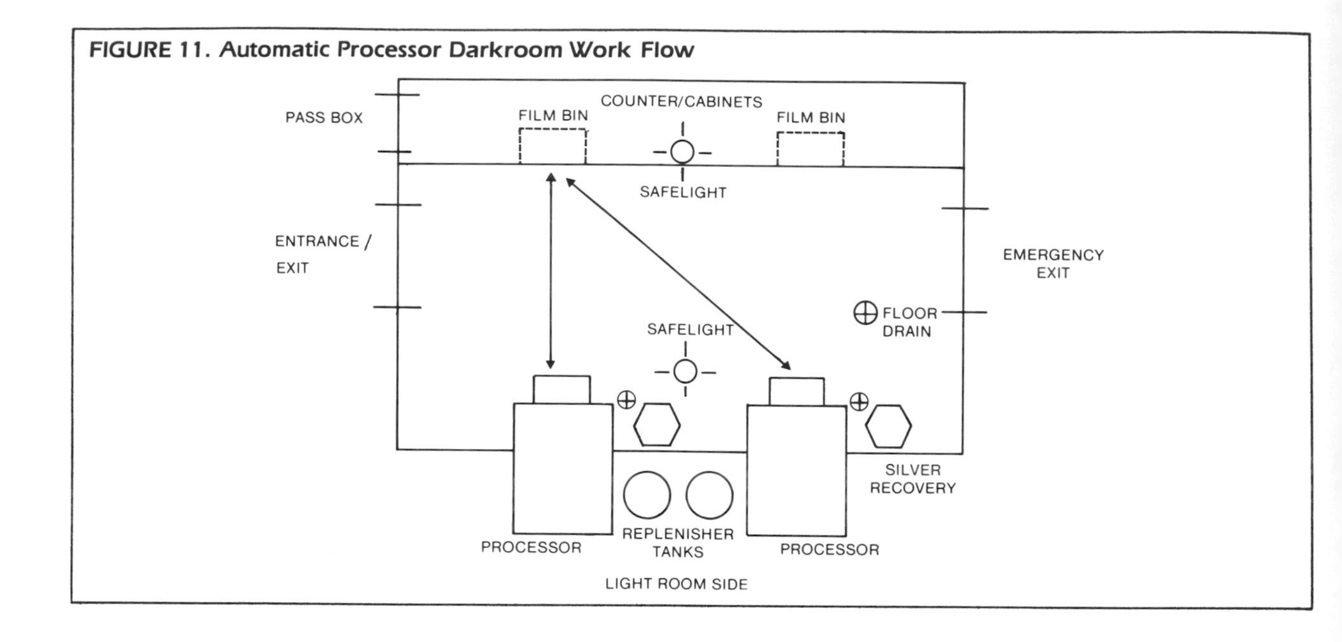

Darkroom Workflow

An automatic processing darkroom shares many of the features of a conventional darkroom, but there are significant differences. No provision need be made for hangers or hanger storage. Deep tanks for developing and fixing solutions and for washing and drying may be eliminated, since these operations take place within the automatic processor; however, deep tanks may be retained for training and emergency use. Only the input end of the processor needs to be located within the darkroom. Dual processor installation is convenient. To handle peak loads, both machines are operated simultaneously; one suffices at other times.

Either the entire processor can be within the darkroom, or all of it except the catch bin (see Fig. 11).

Understanding the Systems

Understanding the mechanization and principles involved in an automatic processor is made easier by recognizing that there are different systems and subsystems that work together to produce the end result. If a processor fails, the problem can be isolated to one of the systems, then pinpointed and corrected. In this way, fewer mistakes are made, the fundamental mechanics are easier to learn and repair is surer.

All automatic processors have these systems: transport (to move the film); electrical (to energize the equipment); tempering (to insure proper chemical temperature); replenishment (to maintain and extend the life of the chemistry); chemistry (to develop the radiographic image); circulation/filtration (to keep chemistry clean, thoroughly mixed and agitated); and air/dryer (to dry the film).

Transport System

The transport system consists of the feed tray (with or without an automatic feeder), the entrance rollers, the entrance-to-developer crossover, the developer rack, developer-to-fixer crossover, the fixer rack, the fixer-to-wash crossover, the wash rack, the wash-to-dryer crossover, the dryer rack, and the catch bin.

Some processors have a minimum number of rollers and rely on guide plates and guide "shoes" to direct the film from one set of rollers to the next. In most processors, guide plates are fixed while the guide shoes, made of plastic or metal, are usually adjustable and direct the film around a bend.

Adjustable guide shoes should be open to catch film on the ingress side and closed on the egress side to direct the film into the next roller assembly. Make sure the guide shoe does not gouge the film and that it is parallel to the rollers.

In general, the transport system is composed of horizontal or vertical "plane" assemblies and turn-around assemblies. The rollers can either be paired or staggered (see Fig. 12).

A fractional horsepower electrical motor drives the entire assembly. The drive shaft of the motor can be connected to the processor drive shaft directly, by a drive chain or by gears. The processor drive shaft transfers its power to the various rack assemblies through worm gear drives. The racks, in turn, can be powered by a direct gear, a vertical

As an example, one particular system uses a No. 25 roller chain in the transport system, which is an ASA Standard single roller chain (non-roller type) with a pitch of 0.25 in., width of 0.125 in., bushing diameter of 0.130 in., and average failure load of 890 lbs.

Sprockets are numbered with an alpha-numeric code, such as 25B32 or K2536. The number 25 is the type of pitch (0.25 in.) of the teeth and this also indicates that No. 25 (non-roller) chain is required. This pitch number is always the first number set. The letter refers to a manufacturer and can come first or in the middle of the two number sets. The second number set refers to the number of teeth. Gears are numbered similarly, such as 16Y32, with the 16 meaning 16° diametral pitch and the 32 indicating the number of teeth. Gears run on gears; sprockets run on chains.

Electrical System

Electricity flows from a wall-mounted circuit breaker (as required by National Electric Code) and is conducted through switches and relays to motors and heaters. The motors used in processors are:

1. Drive motor. (These are also classified as gear motors.) Generally, there are three types of drive motors in common use: capacitor starting, capacitor running, and split-phase. They contain starter windings and are generally universal/reversible in the direction of rotation.
2. Circulation pump motor. This motor has starter windings and is started by means of a starter relay.
3. Replenishment pump motor. This is a simple shaded-pole motor that does not require an external capacitor or starting relay.

Electrical switches used in the processor are generally single actuator (throw), double-pole type (STDP). Microswitches (meaning small movement is required to activate) are used at the entrance detector assembly. They perform two functions. When the switch plunger is in the up position, a circuit is closed to operate the replenishment pump. When the switch is open (plunger down), a timer is activated to prevent films from being fed too rapidly.

Resistance devices (heaters) operate on a basic principle: resistance to electrical flow generates heat. The developer and dryer resistance elements are designed to impose loads and give off heat.

FIGURE 12. Rack Components

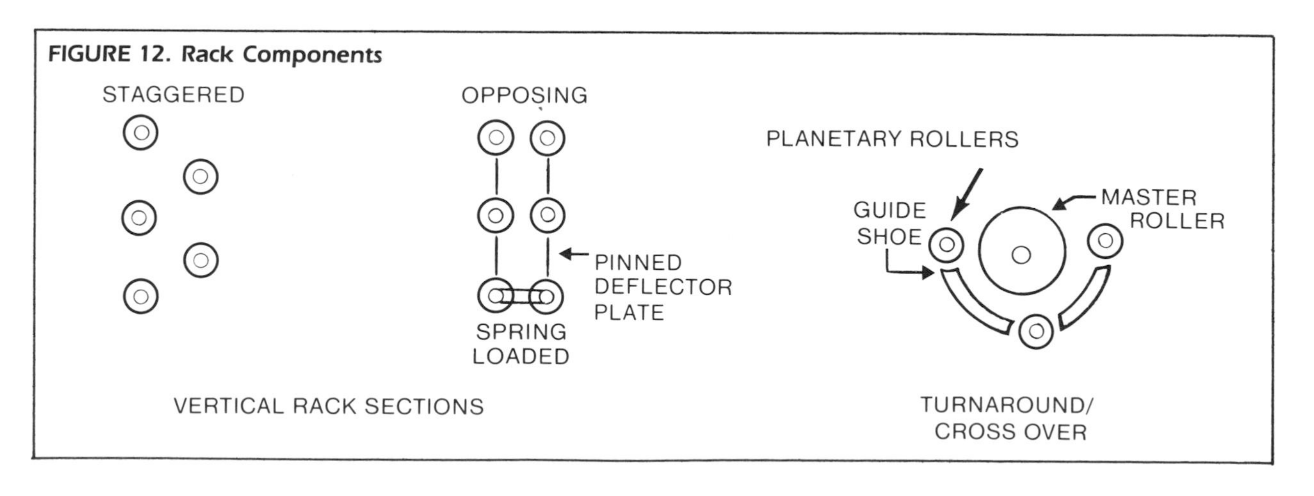

When trying to learn the electrical system, follow the diagrams and compare the various symbols with the actual parts and wiring on the processor.

Tempering System (Temperature Control)

Tempering of chemistry is important to the consistency of radiographs. The time/temperature relationship requires relatively constant temperatures. Use of a mixing valve for the wash water allows control of the wash water temperature to within 3 °C (5 °F) of the developer temperature. The mixing valve is a "fail-safe-to-hot" device when properly adjusted and maintained. This means if only hot water is present the valve will turn off.

Temperature fluctuations directly affect film density. Thus, great care is needed to control temperature, but under controlled conditions it can be used to alter density.

In addition to flowing to the wash tank, the tempered water flows through a heat exchanger where it helps stabilize the developer temperature. It then flows through a water jacket or heat exchanger to warm the fixer solution. The fixer is further heated through the steel tank wall by proximity to the developer solution.

Replenishment System

This is a mechanical system that meters a unit of chemistry to the processing tanks to compensate for loss of volume and activity. The replenisher pump is activated by a film detector switch that either turns the pump on or activates a timer which then turns on the pump to meter a unit of replenisher.

Chemistry System

The chemistry has considerable latitude in which to operate as a constant relative to the time/temperature relationship. However, it is sensitive to the proper operation of the replenishment and the circulation/filtration systems.

The chemistry affects the film contrast, density, resolution, graininess, fog levels, drying, and hypo retention levels.

Chemical contamination is the area of greatest concern to proper operation. Whenever a change of chemistry is made, flush all parts of the chemical system—the entire replenisher system, circulation/filtration system, and all tubing—with warm water. Systems cleaners are very strong chemicals requiring several thorough rinsings after use. Avoid soaking racks in cleaners.

Another important aspect of chemistry is thorough and complete mixing according to the instructions. All mixing equipment should be very clean. The temperature of the water used to mix the chemistry should not be warmer than recommended in the instructions. Due care should be exercised in the pouring and mixing of the chemicals. Avoid excessive stirring.

A detailed explanation of the chemistry is provided in this section under "Automatic Processor Chemistry."

Circulation/Filtration System

Circulation is responsible for chemistry (developer, fixer, and water) uniformity and for agitation to increase the effectiveness of the chemistry. Agitation of the developer and fixer is accomplished with circulation pumps that move the chemistry at a rate of about 10-20 L (3-5 gal) per minute. A few processors also recirculate the water. Wash water flows in the bottom of the tank and out the top, overflowing a weir, or dam. The overflow then mixes with the developer and fixer overflow to dilute them as they are drained off.

Circulation can fail because of chemical deposits in the pump housing, burn-out of the pump motor from lack of lubrication or running without solution, or because of clogged filters.

The filtration system is a part of the overall attempt to keep the chemistry in its best condition for uniformity and reactivity. Filters and built-in strainers on the hot and cold lines before the water mixing valve remove particles that could damage the mixing valve. Filters should be installed before the replenisher pump to protect it. The developer filter found on most processors removes gelatin. Several industrial processors will also have a fixer filter.

Air/Dryer System

Room air is drawn over a heater and directed to the films through air tubes. After passing over and drying the films, the warm air is exhausted from the room. A small portion of the warm air is circulated back to the blower to heat the incoming air.

Air flow depends on several factors. First, it depends on the blower speed and cleanliness. Blower speed can be varied by changing sheaves (pulleys) and drive belts. Cleanliness is best assured by frequently cleaning the blower "squirrel" cage and all of the air-dryer tubes.

Second, air flow is dependent on the exhaust system. If it is necessary to use no more than five elbows, use a 100 mm (4 in.) diameter exhaust duct.

Time/Temperature Relationships

Figure 13 demonstrates the relationship or balance between the time factor (immersion in the developer) and the temperature factor (of the developer). Note that these data are based on a specified chemical brand and type, which influences the relationship, as does the brand and type of processor.

FIGURE 13. Developer Average Time-Temperature Relationship

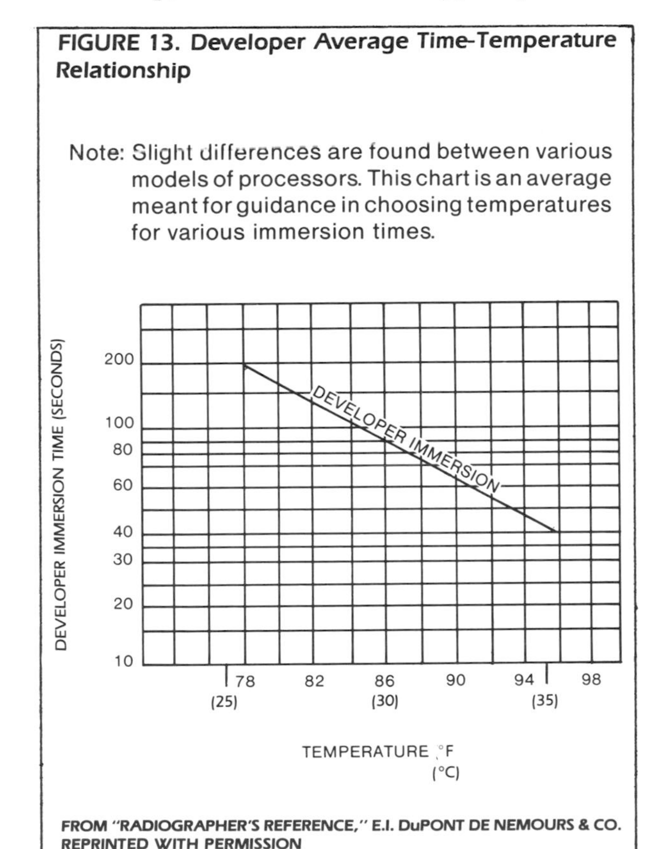

FROM "RADIOGRAPHER'S REFERENCE," E.I. DuPONT DE NEMOURS & CO. REPRINTED WITH PERMISSION

Automatic Processor Chemistry

Processing Chemicals

Processing chemicals for automatic equipment differ from deep-tank chemicals. Never attempt to use conventional chemicals in automatic processors. Although the basic chemical reactions are similar, automatic processing chemicals are especially formulated for high-speed roller operation.

Modern automatic processing chemicals make possible maximum ease of use and uniformity of finished radiographs, regardless of the make of processor. One hundred percent liquid chemicals eliminate tedious and messy mixing of powders. In addition, they can be mixed with tap water throughout the range of normal room temperatures; careful control of temperature is not required.

While all chemicals operate efficiently with any make or speed of film, they generally provide best results with the same brand of film. In addition, using companion products causes fewer variables and is better understood by a manufacturer.

All ingredients, plus full instructions, are included in each package of developer/replenisher, fixer/replenisher, and starter solution. These chemicals are prepared in strict conformity with basic formulas known to produce excellent results. When properly mixed and cared for, they operate efficiently over long periods of time. Properly measured drinking-quality water for mixing (not necessarily distilled), clean tanks, tight covers, and freedom from contact with corrosive materials are vitally important to maintaining performance.

Always flush the entire processor (replenishment tanks, processing tanks, filter chamber, pumps, and all tubing) with warm water for several minutes whenever a different chemistry is to be used. This includes switching from one brand of chemistry to another or after using a "systems cleaner" chemical. The slightest amount of contaminant will result in the chemistry not working properly. The best and safest cleaning method is to use plenty of warm (38 °C; 110 °F) rinse water. Avoid abrasives such as scouring powder, which could react unfavorably with the developing solution and scratch equipment surfaces.

A more detailed discussion of processing chemicals may be found in Part 1.

TABLE 5. Automatic Developer Components

Chemical	General Function	Specific Function
Phenidone	Reducer	Quickly produces gray tones
Hydroquinone	Reducer	Slowly produces blacks
Sodium carbonate	Activator	Provides alkaline media
Potassium bromide	Restrainer	Prevents development of unexposed crystals
Sodium sulfite	Preservative	Maintains chemical balance
Water	Solvent	Dissolves chemicals
Gluteraldehyde	Hardener	Permits transport of films

Developer

Phenidone is most often used in automatic processor chemistry. It is called a reducer because it attacks the exposed silver bromide crystals in the emulsion and reduces or removes the bromide ion leaving only the silver atom. Phenidone is fast acting and attacks only those exposed crystals found in the gray tonal areas of a radiograph. This action promotes maximum film speed.

Hydroquinone is the standard reducer of the exposed silver halide crystal in the black areas of the radiograph. It works slowly to produce the dense blacks on the radiograph. Hydroquinone controls the contrast and maximum density (D. Max.).

Gluteraldehyde is a hardener found in the developer used in automatic processors where it is necessary to retard the swelling and softening action of the chemistry upon the emulsion so that the film will transport. Many times when damp films begin to appear in the receiver bin of an automatic processor, the cause can be traced to the depletion of hardener in the developer rather than a failure of the fixer hardener.

Developer Starter. Always use the starter solution of the same make as the developer. Follow instructions exactly to insure proper sensitometric values and consistent and reproducible quality.

Replenishment of the Developer. As the film leaves the developer solution, the developer solution that is partially trapped within the emulsion is carried out of the tank into the fixer tank. The amount removed will vary depending on conditions, but the rule of thumb is that 60-90 mL (2-3 oz) are removed by carry-over per 35 × 43 cm (14 × 17 in.) film. This volume is replaced to maintain the volume in the tank. Also, it is necessary to control the proper bromide level. Properly replenished developer will contain about 12-16 g/L of bromide for Class I films and 10-12 g/L for Class II films. Use the higher level for mixed types.

Most replenishment systems in automatic processors are activated when a film is placed in the entrance rollers. Again, automatic replenishment is automatic *if* there is no air in the replenishment lines, if there is replenisher in the replenisher tanks and/or if the replenishment in-line filters are not clogged. Also, it is automatic if the microswitch over the top entrance roller has power, is in good working order, and is adjusted to close whenever a film pushes the top entrance roller up against it. And of course, replenishment is automatic only *if* all of these are checked regularly.[1]

Over and under-replenishment affect the radiograph as shown in Figs. 14-17.

Fixer

The clearing agent, ammonium thiosulfate, clears all of the *unexposed* and undeveloped silver halide crystals from the emulsion.

By providing an acid media of about pH 4.0, acetic acid aids in the hardening of the emulsion. The reducers of the developer require a highly basic (or alkaline) media in which to react; they will continue to react even after the film is removed from the developer solution and until they are neutralized. Because a very small portion of the fixer (acid) will neutralize—or at least lower the pH

TABLE 6. Automatic Fixer Components

Chemical	General Function	Specific Function
Ammonium thiosulfate	Clearing agent	Clears away exposed AgBr
Potassium alum	Hardener	Shrinks and hardens the emulsion
Acetic acid	Activator	Acid media—neutralizes and stops developer action
Sodium sulfite	Preservative	Maintains chemical balance
Water	Solvent	Dissolves chemicals

of—a large volume of developer, great care is required when mixing chemistries so that contamination of the developer with fixer does not occur.

***Replenishment of the Fixer*. This is usually 120-150 mL (4-5 oz) of fresh solution for each 14 × 17 in.** film. Replenishment is not made to maintain chemical volume, but rather to maintain a relative level of pH and hardener content. Since the volume remains relatively constant due to carry-out of solutions, there will be a gradual buildup in silver content. Replenishing increases the volume and causes an overflow to carry out some of the used fixer. Correct replenishment will sustain the fixer at **a concentration of silver at about 10 g/L (1.2 ± 0.2 troy oz/gal). Silver estimating paper is available** from dealers; it is used like pH paper and is useful for silver recovery work and/or quality control work to insure the most efficient replenishment rate.

Hypo Retention (Archival Quality)

Hypo retention (the amount of residual hypo or thiosulfate remaining in the emulsion after the film is processed) is related to the archival quality of the film. Any thiosulfate left in the film emulsion will react with the air and silver, resulting in a brown stain that makes the film useless (not of archival quality). **The American National Standards Institute suggests less than 4μg/cm^2 (25 μg/in.^{2}) of retained thiosulfate for a film to last 5 years and not form a general brown stain. Deteriorated developer and fixer hardeners and storage of films in warm, humid areas will increase stain (silver sulfide) production. Film companies will perform an analysis if supplied an unexposed, processed film, and hypo estimating kits are also available for this purpose.**

FIGURE 14. Effect of High Developer Replenishment Rate on ASTM Class I Film

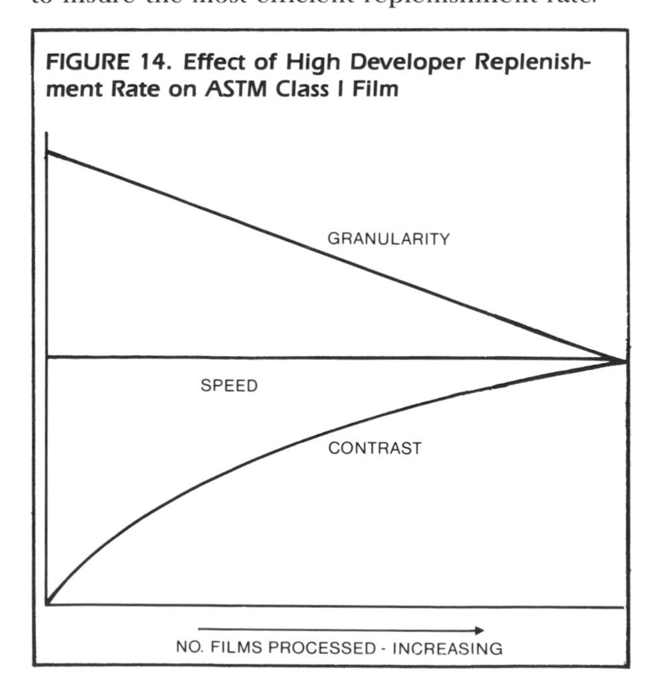

FIGURE 15. Effect of High Developer Replenishment Rate on ASTM Class II Film

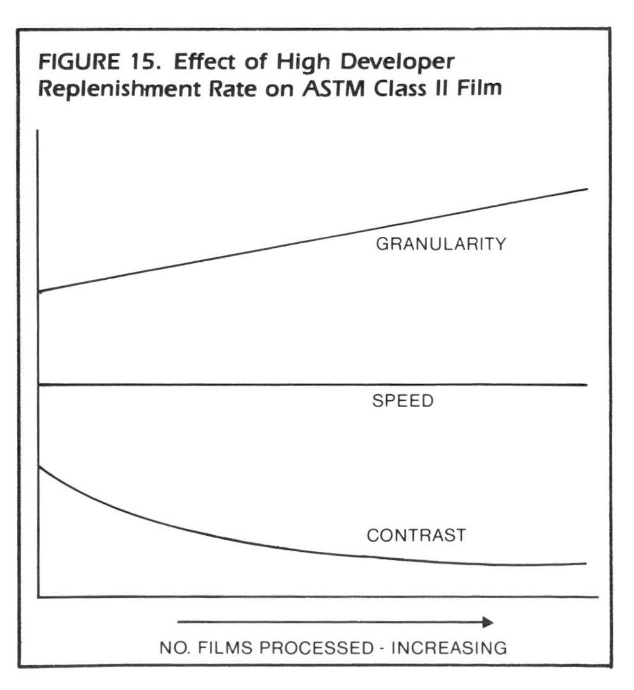

FIGURE 16. Effect of Low Developer Replenishment Rate on ASTM Class I Film

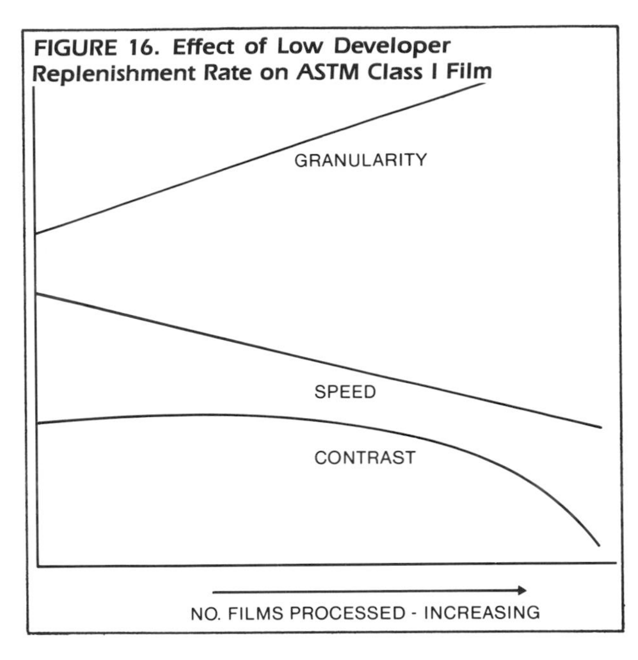

FIGURE 17. Effect of Low Replenishment Rate on ASTM Class II Film

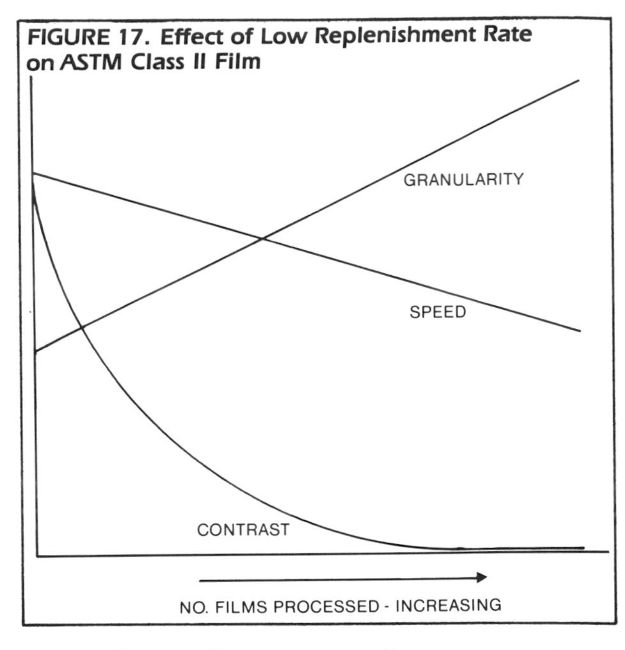

Water

In processing, water is one of the chemicals. Its primary function is to dilute and flush out the thiosulfate clearing agent. The water should flow from the bottom of the tank upward over the film and out an overflow drain at a rate of about 15 L/min (3 gal/min). (An exception to this rule is in those processors that have water circulation pumps to provide agitation; they use about 4 L/min [1 gal/min] with full washing.) The water should be approximately 3 °C (5 °F) cooler than the developer. Actual amounts depend on manufacturers' recommendations. The filtered, tempered water then helps to stabilize the temperatures of the chemistry and is still sufficiently warm to dissolve the fixer out of the emulsion.

When to Change Solutions

Every processor will have its own processing loads and degree of cleanliness and operator efficiency. Solutions will need to be changed in a properly maintained processor with adequate replenishment rates about once a month, on the average. Processors with smaller tank capacities generally require more frequent changes. A unit installed in a mobile field unit, where it is operated under abnormal conditions or used intermittently, requires more frequent changes than a processor in the main department. Higher-volume units accumulate gelatin debris. Lower-volume units subject the developer to excessive aerial oxidation, which weakens it and lowers its activity. A correct replenishment rate to sustain chemical activity is different from that required to fight aerial oxidation.

The institution that utilizes a daily sensitometric quality control system will be able to observe when it is time to change chemistries or to make corrections to sustain activity. Developers should be changed by choice—such as at cleaning time—and not by necessity.

Silver Recovery

Introduction

The scarcity and high price of silver make its recovery from fixing baths important for ecological, environmental, and economic reasons. Fixer contains about 40 percent of the original silver in the film. In 1974, a law was passed acknowledging the toxicity of dissolved silver and banning it from being placed in drains. Thus, in addition to the value and scarcity of silver, it is a violation of the Toxic Substances Act to put silver into drains. Even the smallest user should treat the used fixer to remove silver; the money made will pay for the effort.

Recovery Methods

The methods of recovery are chemical (which includes precipitation and metallic replacement) and electrolytic. Metallic replacement is simplest but requires a low volume continuously. Electrolysis is recommended for higher-volume processors.

Metallic Replacement. If a fixer solution containing silver ions is brought into contact with a metal, the less-noble metal (such as steel wool, zinc, copper, or steel turnings) is replaced by the silver. The silverless fixer cannot be reused.

Metallic replacement units, also called buckets or cartridges, contain steel wool or zinc screen and are usually used in tandem. As the acid fixer breaks down the less-noble steel, the more-noble silver metal precipitates as metallic silver. A sludge of iron oxide and silver forms in the bottom of the container.

This method is both inexpensive and efficient; it can remove 60 to 95 percent of the silver. One pound of steel wool will collect 3-4 pounds of silver. However, efficiency is based on a slow, steady, continuous flow of silver-laden fixer. Efficiency is about perfect for the first 25 percent of the unit's life expectancy (100-200 gal) and then often drops to only 30 percent effective. Also, the sludge produced by this method is expensive to ship and refine.

Chemical Precipitation. Precipitation is a chemical reaction that separates the silver from the solution in an insoluble, solid form. This type of unit, particularly those that use sodium sulfide and zinc chloride, produces toxic and volatile fumes, and so should be avoided. The units using sodium borohydride are very efficient and safe, but require constant pH adjustment by a technician.

Electrolytic Recovery. Silver is recovered by passing an electric current (dc) through the silver-laden fixer. Electrolytic recovery systems (called cells) are classified as agitated or nonagitated, high or low current density units. The two terminals (electrodes) are a positively charged anode (usually graphite) and a negatively charged cathode of stainless steel. Positively charged silver ions (Ag^{+}), are attracted to the cathode, where they plate out as metallic silver (Ag°), called flake. This is the most efficient method for medium or larger installations.

Efficiency of Silver Recovery Equipment

During processing, the developer converts the latent image-bearing silver halide crystals into a visible image—black metallic silver. Those crystals exposed, but not developed, and those not exposed are washed out in the fixer. In industrial radiography the silver is dissolved into the fixer at a usual rate of one troy ounce per gallon (about 10 grams per liter) of fixer. Silver recovery or reclamation is the process of converting the silver to metallic silver. Understanding the factors that control the efficiency of this operation will help in understanding and upgrading existing systems or in generating specifications for new systems.

Dwell Time. Sufficient time is required for the reaction to occur. Electrolytic units (cells) are rated in troy ounce per hour capacity. Buckets are rated in cubic centimeters per minute or gallons per hour of fixer flow. Exceeding these limits will result in silver going through the unit and down the drain.

Agitation. Buckets provide agitation by flowing the fixer over the many wire filaments. Electrolytic cells use pumps or impellers. Greater physical agitation increases the unit's efficiency in producing metallic silver and allows higher plating currents to be used. However, agitation should not be so violent as to cause splatter, spill-over, or excess evaporation.

Surface Area. The larger the surface area the higher the plating current can be in a cell. In any unit, increased area increases recovery rate.

Edge Effect. The edge effect is related to surface area and electron flow efficiency; the more edges and/or surface area, the better the efficiency.

Electrolysis. As the acid fixer enters the bucket the steel wool is attacked, producing chemical electrolysis. The steel becomes oxidized to iron oxide, and the silver in solution becomes metal. Cells contain an anode and a cathode, a rectifier, and a transformer to pass direct current through the solution. The higher the current, the greater the efficiency. Too high a current (usually when little silver is present) can break down the fixer, this is called sulfurization and is to be avoided.

Maintenance. All units require records, regular inspection, and regular maintenance to ensure proper use. Buckets can clog, back up, or leak. Cells can become too loaded with silver, short out, blow a fuse, or burn up an agitation pump. The amperage should be automatically or manually adjusted according to the film (silver) volume during the day.

Centralization. Centralized recovery is the most efficient system where three or more units are involved. A holding tank feeds a single cell a continuous supply for optimal efficiency.

Sizes of Recovery Units

It is important to choose a size to meet the needs of each processor, since they will probably vary in work load. Centralized recovery eliminates this problem, and one cell can usually handle several processors. When trying to determine the proper size or capacity of recovery unit for your individual needs, the following guidelines and examples should be used. Some steps require the gathering of data and others involve calculation.

- Normal silver content in a processor with stable chemistry is 8 g/L (1.0 troy oz/gal); measure using silver estimating paper.
- Average film feeding rate per busy 2-hour period is 100 35 × 43 cm (14 × 17 in.) or equivalent; monitor over several weeks.
- Average film feed rate per hour is 50 14 × 17 in.
- Replenishment rate per 14 × 17 in. is 130 mL; monitor weekly.
- Total replenishment per hour (50 films average) is 50 × 130 = 6500 mL (Converted to gallons at approximately 4000 mL per gallon = 1.6 gallons.)
- Each 1.6 gallons contains 1.0 troy ounce. Therefore, maximum silver recovery rate is 1.6 troy ounces per hour for which a 2.0 troy ounce per hour cell would be required. If a bucket is used it should be rated at a maximum of 300 mL per minute, or well within its capability.
- In either cells or buckets, when a processor fixer tank is drained, it must not be drained rapidly through the unit or it will exceed the recovery rate due to a lack of dwell time. A 10-gallon processor tank may contain 10 or more troy ounces of silver.

General Recommendations. Silver estimating paper, which indicates the relative amount of silver in used fixer, can be readily purchased from most silver reclaimers. The test strips are used just like pH paper strips. Industrial radiography usually operates at a level of about 10 g/L (1.2 ± 0.2 troy oz/gal).

An even simpler method of determining if there is any silver in the solution is to put a brightly polished copper tube in the solution. Any silver in the used fixer will quickly adhere to the copper tube and give it a gray color.

Purchase tailing or central electrolytic units according to the determined troy ounce per hour capacity based on your calculations. Collect the fixer in the processor at cleaning time. If a silver recovery unit malfunctions, disconnect it or isolate it so that it cannot ruin radiographs being processed.

Types of Installation

Tailing units may be buckets or cells placed individually or in tandem on each processor. Usually buckets are used in tandem, with the second becoming the first after every 100-300 gallons of fixer, depending on brand and size. Properly sized cells should not require a tailing unit.

Recirculating cells take the fixer from the processor, remove most of the silver, and return the fixer to the processor. This reduces fixer consumption and lost silver by carry-out into the wash. The major disadvantages are increased cost of the cell and potentially higher hypo-retention levels because of reduced fixer efficiency. Fixer can only be recirculated from cells—never from buckets or precipitation.

Setting Up Silver Recovery Units

The following precautions should be observed when installing or operating any silver recovery units:

1. Make sure that fixer overflow to the silver recovery unit is a continuous downward flow.
2. Clean the standpipe on metal exchange units regularly.
3. Ensure against air locks in electrolytic units.
4. Be sure there is an air break (electrolytic break) between the solution in the electrolytic recovery

unit and the incoming fixer. Without one, there is danger of plating silver in the processor fixer tank.

5. Use the highest amperage possible for optimal recovery — but keep it short of sulphurization (characterized by yellow color and smell of rotten eggs). High amperage produces soft, black silver; low amperage produces hard, shiny silver but of approximately equal quality. The higher amperage helps to ensure removal of the majority of silver.

Scrap Film

Films that are to be discarded also have value for their silver content. About 60 percent of the original silver remains in the film to form the visible image. Both waste film and outdated records are worth a considerable amount of money and should be sold for silver content.

Security and Selling Recovered Silver

In radiography, business economics revolve around the cost of producing a visible image versus the value or price of the product. The single largest budget item is manpower, which must be used efficiently.

Radiographic film is expensive. It must be kept in the best possible condition and protected against abuse and theft, whether in its fresh or used form, through inventory controls such as records, policy, procedures, and security. Such programs will pay for themselves in improved earnings.

Silver should be recovered from used fixer and used films to reduce the cost of the original fresh stock. Because a substantial value is represented by recovered silver, it is important to impose inventory and security controls.

Security. Fresh green (unexposed) film and black or scrap (processed or discarded green) film represent money that is easily transported or lost. Both conditions of x-ray film require inventory controls including records, proper storage conditions, and security.

Inventory controls include records by type, size, quantity, and emulsion numbers of all products ordered, received, moved in and out of storage and the darkroom, and date of use. Scrap film should be weighed weekly (or daily if the volume warrants), inventoried, and placed in a secure place (such as where fresh product is locked up). Silver flake from electrolytic recovery units should be dried, weighed, inventoried, and locked up monthly or as needed. If buckets are used they must be periodically removed, tagged, inventoried, and stored in a secure place. For example, a single pound of silver flake the size of a small ball, represents about $150 at a silver price of $10 per troy ounce. A small 5-gallon bucket may contain 150 troy ounces of silver, which represents $1500. (Dollar amounts in this text are presented for the purpose of illustration and are based on a theoretical price of $10 per troy ounce. Due to historical fluctuations in silver prices, calculations should be redetermined as necessary.)

Security is included in inventory, in that all product must be signed in to and out of a record book. All fresh product, scrap film and silver should be under lock and key with controlled access.

Balance sheet analysis should be applied to all inventory, since specific relationships can be expected and these may be worked with to insure or improve business economics. Your records will allow you to determine current inventory status, inventory turnover rates, usage of a certain size film, and will allow you to analyze a repeat/reject program, identifying causes for scrap and improving efficiency. Film purchases must be cross-checked against production and scrap film records to ensure that all film dollars are accounted for.

Storage conditions for fresh green film [21 °C (70 °F) or cooler, 60 percent RH, on box edge, with a first-in-first-out (FIFO) system] are well recognized. One company had no record of what was ordered versus what was delivered; X-ray had no inventory, storage security policies or procedures of its own. It was suggested that controls would cost less than the 10 percent estimated minimum loss of their $500,000 film budget. Certainly if good product is damaged due to poor storage or security conditions then a radiographer's work may have to be repeated, which escalates the costs.

Selling Silver. Valuable silver is found on black and green scrap films as well as in silver flake or sludge from recovery units. Some general guidelines apply to selling film and silver:

- *Stockpile* for at least a year if possible, and for as long as seems practical. Purchasers (dealers or refiners) will pay more for larger quantities.
- *Charges* may be a flat fee and/or a percentage.

Percentage charges should change with volume to be competitive with a flat fee approach.

- *Deal direct* with the largest refiner you can find to eliminate middleman profits that reduce your income.
- *Sample* the marketplace by sending out small accurately weighed identical samples to several purchasers. Determine who pays the most (actual bottom line) and/or provides the best service (turnaround, help, records). Five samples will produce five results.
- *Package* carefully for shipping. Many purchasers provide containers, labels and instructions. For instance, silver flake should be labeled "chemical sample" rather than "scrap silver." Silver flake is chemically stable; it can be labeled "inert metal."
- *Pricing* is an area that can be very confusing. The price quoted is usually the spot price for fabricated product (99.95 % pure) of the commodities market listing ("Cash Prices") in the daily *Wall Street Journal (WSJ)*. However, this is the selling price of pure, assayed product—not scrap of unknown value. Thus, one would expect to receive less than the quoted price, allowing for profit and expenses of the purchaser. The seller must ask himself, "How can I receive $15 per troy ounce from a dealer who will sell to a refiner who will refine it and sell it for $15?" The *WSJ* provides the common denominator. The dealer and refiner make their profit from charging handling, service, record, assay, shipping, insurance, and other hidden or stated fees.

For instance, if you sell 10 pounds of silver, you will be quoted a price of perhaps "10 times the *WSJ* price " for that day. This means price "per pound." There are 14.583 troy ounces per U.S. standard pound (avoirdupois). For each of your 10 pounds, the dealer will pay the *WSJ* price for 10 ounces and keep 4.583 ounces, or about 30 percent, as his charge. If you sell 100 pounds, you might get 11 or 12 times the *WSJ* price. In this case, there are usually no other charges.

Prices for scrap film are based on pounds of film, not silver. A pound of scrap film is mostly the weight of the base material. Green scrap film should be removed and sold separately or used for processor cleaning. Older films may contain more silver and should bring a higher price.

Of all the silver in the film, forty percent goes into the fixer and sixty percent remains on the film. Ninety percent of the silver from the fixer and seventy percent of the film silver may be recovered, giving a total of about 75 percent of the original silver that is recoverable. Considering that the film costs are approximately five times the price of silver, only 10 to 15 percent of the film's list price may be recovered. If your department buys $100,000 worth of film per year, up to $15,000 might be recovered.

Specific Suggestions

Silver flake is derived from scraping off the collection plate of an electrolytic silver recovery unit (cell). There is little significance whether it is silver colored and hard (result of low current levels) or black and soft (result of high current levels). A properly sized cell will collect 90-95 percent of the silver, which will be 95 percent pure. Dry the silver before weighing. (If weighed wet, deduct 5 percent of the weight for trapped moisture).

Silver sludge from buckets should be shipped in solution in the bucket. Draining the fixer exposes the sludge to air and an exothermic reaction produces heat. If possible, the sludge should be dried in a large open pan and then shipped, because the sludge damages the refiner's crucibles and he charges more to handle sludge than flake. One refiner who "sells" buckets provides them free if they are returned. The refiner charges 15-20 percent for use of the bucket and for refining service.

Operating Procedures

Processor Start Up

1. Remove all easily removable side covers and panels.
2. Make sure solution levels are correct.
3. Turn on main isolator (circuit breaker) and wash water.
4. Turn switches on one at a time.
5. Adjust water temperature and flow rate. These may vary with time of day and may depend on use of water elsewhere.
6. While temperatures rise to pre-set working level, inspect the machine. Temperature will rise about one-half degree per minute.

Look for leaks, solution circulation, water flow, rollers turning properly, crossovers seated properly.

Listen to motors, gear system, drive system, pumps and blower. Listen for grinding sounds, squeaks, sounds of slippage.

Notice if strong chemical fumes rise when top cover is removed. If so, make sure the cover is left open at night to permit fumes to escape. Smell for burning insulation, because start-up creates the greatest electrical load. Smell for dust that will be moved around as heat builds up. If processor is sufficiently dirty, you will smell the dust actually burning.

Feel crossovers to make sure they are properly seated and aligned. Run a finger along rollers to detect chemical deposits. Place a hand on motor mounts. Feel for abnormal heat build-up and vibrations. Feel tension on blower fan belt (if so designed) when at rest. Normally there should be about 6 mm (0.25 in.) deflection at the center between the two sheaves (pulleys). Check dryer roller assembly drive belt, if so equipped, for tension and wear. Check while components are warm.

7. Run at least two clean sheets of 14 × 17 in. scrap green film to clean rollers of deposits resulting from standing. Make sure replenisher pump works. Check films as they fall into the catch bin for grit on surface and wetness.
8. Processor should now be ready for a normal day's work. Using a bimetallic or electronic thermometer, check solution temperatures. Make any final adjustments in thermostat and/or thermometer.
9. Replace all covers. Make sure they are fully closed. Run a sensitometric quality control strip and measure results. If okay, begin processing films. This entire procedure takes approximately 15 minutes, the time it takes for solutions to come up to working temperatures. The result will be increased life and uniformity from your processor and less downtime for repairs.
10. Note date and time and list any observations in a log book. Record sensitometric results.
11. Wipe counters, work surfaces, processor feed tray and receiver bin with a damp, lintless cloth.

Running

Darkroom personnel should periodically check all operating temperatures during the work day. They should be constantly aware of how this machinery is functioning. At any indication of trouble, the operator must know how many films are in the machine so that when the last film is finished, he can take immediate action to verify the problem, locate it, and correct it.

It is very important for the processor operator to take a few seconds every hour or so to check for signs of possible trouble. Inspect the replenisher tanks for sufficient chemistry and for a steady reduction in the level. Inspect films for proper drying; uniformity of drying patterns, completeness of drying, and proper transportation are important. Also, inspect the film for the general chemical activity based on sensitometry, transport artifacts, and general cleanliness of the radiographic image and of the surface. Inspect fixer overflow occasionally. The silver recovery equipment should be checked for proper flow of fixer and for proper current in electrolytic units.

These checks are a part of the preventive maintenance program and should be recorded in the processor log book. Awareness and involvement will reduce retakes and downtime, increase productivity, and improve economics.

Standby

Several processors are equipped with automatic or semiautomatic controls that place the processor in a "standby" state whenever films are not fed within a certain period of time. Basically, these systems turn off the drive system and the fixer circulation pump while maintaining the chemistry temperature near the working level. The dryer may be turned off or lowered to a new setting.

The standby feature may be useful for saving water and electricity, and reducing noise and heat, but may increase wear on the processor and cause deterioration of chemistry. Standby is a convenience rather than a quality control device. Whenever possible the heat should be removed from the developer to prolong its life.

Shutdown

1. Turn each switch off, one at a time.
2. Cut the electrical power at the main isolator (circuit breaker) and turn the water off (use the shut-off valve, not the mixing valve).
3. Remove all easily removable external panels and inspect equipment for leaks.
4. Remove all crossovers to a sink, inspect, and wash with hot water. A plastic cleaning pad can be used lightly on hard rollers only. Rinse the rollers (and crossover guide plates) thoroughly. Leave crossovers out of the processor overnight to prevent chemical fume buildup. Protect them from dust.
5. Next, with a damp sponge or damp, lintless cloth, wipe off all processor parts, including sides, support bars, tubing, motor housings, wire, and the inside of all panels. The reasons for this wipe down are simple: a spotless processor is easier to keep clean and a clean surface is less prone to accumulate dirt. But more important, a clean processor will permit easier location of leaks and other problems.
6. As you clean, inspect the equipment; if any problems are discovered, correct them immediately. Install film supplies and chemicals for replenisher tanks in preparation for the following day.

   Note: enter into the processor log book the details of any problems found, adjustments made, and the amount and code number of any chemistry mixed along with general observations and recommendations.
7. Replace panels. Leave top cover slightly open to allow ventilation. Internal fume control hoods/splash guards should be removed and be placed on top or with the crossovers.

Mixing Chemistry

Start with clean equipment and read the instructions. Use safety eye protection and follow basic safety practices. Most radiographic chemistries contain several component parts, which must be mixed sequentially to prevent improper reactions.

Fill the mixing tank with a measured volume of water at the temperature recommended in the instruction sheet. Violent or prolonged agitation is to be avoided. Use a mixing paddle made of, or coated with, chemically inert plastic.

Add each component in proper sequence and mix properly in between. When the chemistry is completely mixed, cover with a floating lid and a dust cover. The floating lid is especially important for the developer because it reduces oxidation; in the fixer tank it controls fumes.

Adding Chemistry to the Processor

To avoid contamination, always start with the fixer first, because a small portion—less than one percent by volume—of this acidic chemistry will contaminate a large quantity of the developer. Should spillage occur, the empty developer tank can easily be rinsed. If your processor does not have splash guards, you can make them from scrap X-ray films.

If it is suspected that some fixer splashed into the developer tank, do not take a chance; flush the tank with warm water. If the processor has a developer filter, remove it and replace with a new one.

Since developer solution filters take a long time to fill up, soak them in the developer replenisher tank or in water for about half an hour before they are needed. Then place the filter into its compartment with the lid on loosely. Place the splash guard in position and then transfer the developer solution. When the tank is about half full, add the starter solution. For each liter of developer required in the developer processing tank, add 18-24 mL (2.5-3.2 oz per US gal) of starter solution. This is the *only* time that the starter solution is used. Use the same brand starter as the developer and use the specified quantity.

Next, add the remainder of the required amount of developer. Install the developer rack slowly to avoid overflow. As the rack is lowered, the level will rise in the tank and in the filter compartment where the trapped air is being forced out. As the rack is installed watch the filter lid for signs of leaks, which will indicate that all air has been removed. When all air has been removed, the lid and/or air release valve can be tightened. Plastic filter compartment lids should be put on with fingers only—never with a wrench. Metal lids with screw-type clamps should be made very tight.

Finally, clean the work area and equipment and put the equipment away. Then install the crossovers and prepare the processor for operation. This installation of fresh chemistry should be entered in the processor log book.

Cleanup and Inspection

Operating instructions for each processor also include suggestions for cleaning and inspection. In general, these procedures require little time and trouble, but this does not minimize their importance. They contribute substantially to efficient operation and to maintenance of optimum film quality.

Solution Services

In many localities, specialized organizations handle the mixing and maintenance of chemicals for automatic processing equipment. These services include routine inspection, cleaning and refilling. Although the services are provided by professionals, it is still good policy for each X-ray department to have its own in-house specialists and to perform routine, daily inspection.

Below is a suggested approach to a negotiated contract with an outside dealer to provide adequate service. The same contract might be used as a basis for a job description for an in-house processing specialist or operator.

SOLUTION - SERVICE MAINTENANCE CONTRACT

Chemicals

To be delivered (premixed or concentrated) and installed (as needed or at least twice monthly) quantity of specified brand, type, and volume.

Developer added to clean processor will have starter solution added according to the manufacturer's recommendations. The volume to be measured accurately and total volume recorded each time by the service person. Correct brand of starter must be used.

Installation of chemicals: service person will remove floating lid carefully and discard any trapped oxidized chemicals to a sink or other receptacle. Oxidized developer must not be dumped back into tank. Floating lids will be used and must be replaced. Containers must be wiped free of loose dirt and dust prior to pouring to keep debris out of replenishment tanks. Used containers must be rinsed with fresh water and recapped. Service person must record volume in tank at start, volume added, color and clarity of solutions, and observations of dirt and crystal formation in the tank or tubing. Service person will remove all empty containers or place in a designated area.

Service personnel will use accepted safety equipment and procedures as specified by OSHA when working with chemicals or components of processor.

Service person, in installing chemicals to clean processor, will install fixer first using splash guards. Any fixer splashed or spilled in the developer tank will be flushed out prior to adding developer.

Racks will be reinstalled to prevent fixer spill into the developer or general chemical overflow.

After racks are reinstalled service person will inspect for proper seating. If solution levels are low, tanks must be topped off.

Replace developer filter and water filters, clean replenishment line filter and mixing valve strainer.

All chemical spills, drips, and so on must be immediately cleaned from inside processor, and especially from the exterior of tanks and area around the processor.

If at any time service person is uncertain about chemical problems, designated staff member should be consulted.

Service person will record all chemicals delivered, installed, and stored. Date, time spent, time of day, and signature should be recorded.

Processor

Using accepted (OSHA) safety equipment service person will (twice) monthly remove all racks, including crossovers, dryer tubes, and rollers.

Systems cleaners must not be used on rollers at any time.

Clean rollers using only warm water, 40 °C (100 °F), and plastic cleaning pad or sponge, especially for resilient rollers (including inside).

Inspect all roller surfaces circumferentially for nicks or heaving (bumps) by feel. Inspect all rollers for distortion. Inspect all rollers for gear wear, stud hole/stud wear and play, and/or sprocket wear.

Inspect, adjust, align racks for squareness and drive chains.

Inspect ladder chain and sprockets for alignment and tension.

Inspect main drive shaft, internal and external bearings.

Check alignment side to side and end to end of all turnarounds.

Clean all accessible surfaces inside and outside of processor cabinet including cabinet panels.

Clean and lubricate (if possible and using dry sprays) microswitch mechanisms.

Lubricate main drive shaft bearings, worm gears, and chains after cleaning where possible.

Lubricate all pumps where possible.

Test action of all switches.

Monitor heating, agitation, replenishment, and transport components in run mode.

Check replenishment rates, temperatures in all solutions, and transport rate in run mode.

Clean dust away from motor fans and wipe off all motor housings.

Check alignment of main drive chain.

Check all tubings, fittings, valves, and clamps. Annually (and whenever a pump or component is removed for service) replace all tubing. All replacement clamps must be screw clamps of 316 stainless steel. Inexpensive "O" ring spring clamps are to be gradually eliminated; wear safety glasses and use correct pliers for removing these clamps; keep other personnel away during clamp removal.

Inspect insulation on wiring where accessible, check terminals, relays, and other accessible components visually.

Make adjustments to all out-of-adjustment components and record component, finding, and corrective action.

(Optional) Replace all obviously broken or worn parts. Record finding and action. Deliver old part to specified staff member.

Prepare a list of parts to be ordered for immediate installation and/or at next service call and submit to specified staff member.

Document completely, on a check list or in a report, all calls, time spent, reason, observations, and recommendations.

Processing

Before and after each service call, a sensitometric test film will be processed and inspected relative to a master film to indicate a base line of conditions and to provide guidance. These films will benefit dealer and institution and should be submitted to the institution.

General

This contract covers one or all of the following areas of purchased and agreed-upon expertise.

1. Solution service—defined as chemical delivery, mixing, installation, inspection, counseling, testing, and quality control.
2. Processor preventive maintenance inspection—defined at routine inspection to locate problem parts so they can be replaced.
3. Processor preventive maintenance—defined as cleaning and disassembly to inspect and subsequently replace damaged parts or those predicted to fail.

This contract covers twice-monthly solution service and maintenance on listed processors and additional (or less) on other listed units.

It is estimated that a proper "call" on a single processor should take at least one hour of continuous work, with two hours being the average.

This contract (optional) includes 24-hour emergency service.

This contract is let on the basis of service by a trained person (named). Training must include safety, chemistry, and sensitometry as well as electromechanical systems. This includes emergency service and other times when the designated person is unavailable.

The service person and his supervisor will quarterly (adjustable) meet with staff members to review performance of equipment and make recommendations.

Service person will routinely discuss with operators, designated staff members, and technical supervisors what they each observe regarding the processor, processing, and film quality, including artifacts. This should be done informally before and after each call. If a call is performed at night the service person should leave a report for the institution.

It is vital that open communication exists—that everyone work together to produce consistent high quality radiographs with the greatest economy of money and radiation. This then should be the primary tenent and judge of performance of this contract.

Processor Quailty Control

Processing of a radiograph is done in an automated processor to achieve consistent quality. The processing system is a chemical process with specific conditions of time and temperature based on a given chemical activity. The sum result of processing is the "goodness" of the quality of radiographs produced.

Processing quality control is a procedure of monitoring to see if, and to what extent, there is consistency. Consistency is necessary before "quality" can be improved upon, because the variables of processing must be identified, their degree of fluctuation and cycling patterns noted, and the limits of acceptability established. Once the uncontrolled variables are identified, the best control measure will be more apparent. Sometimes, the variable may have to be compensated for, minimized, or eliminated. Also, in this process of identifying and controlling the variables, indications are often provided as to the best way to consistently achieve optimum quality.

The basic procedure is to select one processor, identify "quality," identify and control all of the variables of the processing system, and establish consistency. Next, select a second processor and match it to the first in quality and consistency.

Then add a third processor. Continue until the entire facility is controlled and each processor produces radiographs of consistent quality within acceptable limits. The variables in each processor are about the same, but they are independent of one another and will deviate differently. For this reason one should not attempt to control all processors at the same time. Even the matching of many processors at one time is difficult; it entails changing some variables, but the wrong variable might be chosen or be placed in a situation of greater fluctuation.

Sensitometry

Processing is the vital link between the latent image and the visible image radiograph; careful control of the many factors involved is essential. Establishing processor quality control to maximize uniformity is desirable.

Control strips are processed periodically during a day and read on a densitometer (see Fig. 18). The changes in density levels of the exposure are plotted and/or the characteristic curve is generated. Speed and contrast can be determined. This system is the professional approach to quality control.

The matched exposure radiographs are easily duplicated from day to day. Check the processor, note the conditions mentioned below, and process the control strip. This control strip becomes your master and subsequent daily strips are simply matched to it. Should deviations from the master strip occur, interpretation, investigation, and corrective actions are required. The master test strip should be processed under optimum conditions of chemistry and processor performance. (See further discussion of sensitometry under "Manual Processing.")

Archival Quality

Archival quality **is the retention of radiographs, unchanged by stain, for a specified period of time. How long radiographs will last in storage depends on the amount of thiosulfate left in the emulsion after processing. The thiosulfate in the fixer may not be adequately removed for any of several reasons. Fixer is often called "hypo" and the life of stored films is said to be dependent on "hypo retention." Retained hypo reduces the life of the film by reacting with silver and air to turn first a pale green and later a deep orange brown. This coloring is detrimental to radiographic interpretation and may make radiographs valueless for legal, scientific, teaching, or reference purposes.**

Hypo retention, or the amount of residual thiosulfate, is determined by a chemical reaction test which, with the aid of a densitometer, indicates the amount of thiosulfate in micrograms per square centimeter or square inch. Proper storage of processed films in a cool, dry environment does not improve the hypo level but does retard the chemical reaction that causes the stain.

Testing for Residual Chemicals

The methylene blue test described in the ANSI PH4.8-1971 document quantifies, by a relatively lengthy wet chemical analysis, the amount of residual thiosulfate remaining after processing. The fact that this test measures thiosulfate *only* limits it to films that have been processed within a week or two of testing. Thiosulfate decomposes into polythionates and other compounds with time but the simpler and faster silver densitometric method described as an alternate in the same document converts *all* residual compounds into a visible stain. This stain approximates the maximum amount of discoloration that would be generated during storage.

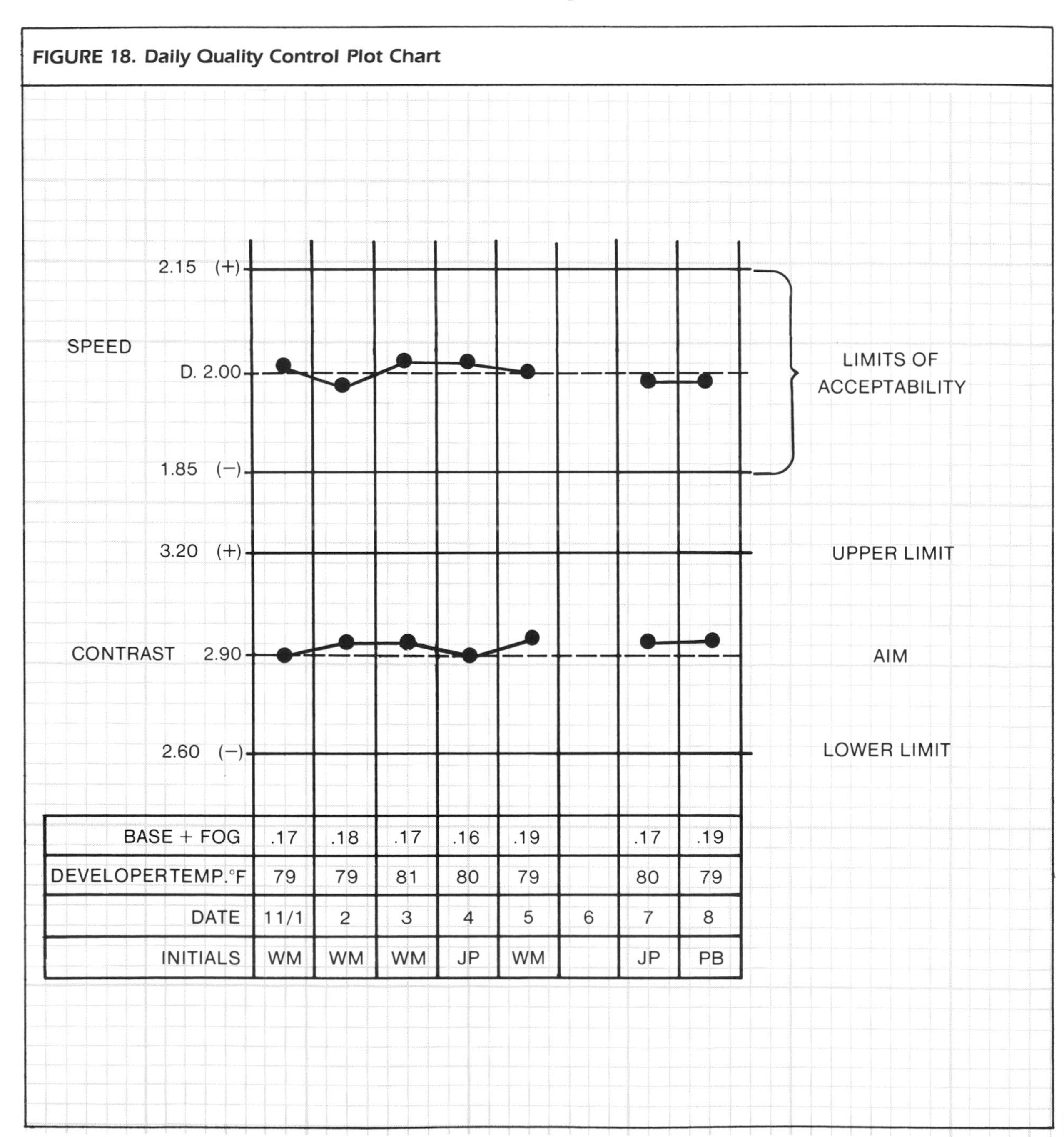

FIGURE 18. Daily Quality Control Plot Chart

Spot Testing Procedure

Dissolve 10 g (0.35 oz) of silver nitrate in a solution containing 30 mL (1 oz) glacial acetic acid in 750 mL (25 oz) de-ionized or distilled water. Dilute to 1 L (34 oz) and store in a brown bottle away from strong light. *CAUTION:* Discard the test solution if it has darkened or discolored in the bottle.

Once daily for machine-processed radiographs, or on each batch of manually washed radiographs that are removed from the water to dry, perform a spot check by placing a drop of the washing test solution on a fully dried *lowest* density area. If no discoloration appears in two minutes, the radiographs should be archival and should not discolor in storage.

If a stain appears, it approximates the maximum amount of discoloration that would be generated on *that side* of the radiograph during storage. Since such staining indicates incomplete washing, the radiographs should be re-washed and re-tested. To accelerate re-washing, the radiographs can be agitated in a 5 percent solution of sodium sulfite (50 **g/L of water or 6.7 oz/gal) for 15-30 seconds to completely neutralize the residual fixer and then re-washed briefly and re-tested.**

If large numbers of radiographs have to be re-washed, the fixer tank in an automatic processor can be filled with the sodium sulfite solution described above and with the developer-fix crossover assembly removed, the radiographs can be hand fed into the down side of the fixer rack into the sodium sulfite, then automatically through washing and drying.

When *less* than archival quality is acceptable, the following table lists the various net stain densities that might be considered acceptable for the described years of storage:

TABLE 7. Stain Densities Versus Storage Years

Net Stain Density	Storage Years
0.50 or over	under 1
0.25	1-3
0.15	3-10
0.06	over 10 (archival)

To find the net stain density, the complete silver densitometric test described in ANSI PH4.8-1971 must be used on an unexposed but fully processed piece of film that has the *highest relative speed.* This is usually the film with the heaviest coating of emulsion, which is usually the slowest to wash.

The storage year estimates were based on natural aging stain data taken from radiographs stored in envelopes in office conditions, air conditioned and heated, for approximately ten years. The natural aging stain densities were *approximately one-third* of the original stain test densities and might be generalized to: density+natural aging=net stain density found by ANSI test × storage years ÷ 30, provided the radiographs are stored under similar conditions.

Since warm, moist, moving air accelerates staining, we ran accelerated aging tests at 83 percent relative humidity and 43 °C (110 °F) in moving air and found that staining increased rapidly in the first two weeks then leveled off after four weeks.[1]

TABLE 8. Accelerated Aging-Net Stain Densities

	Elapsed days										Max. Net
Film	1	5	6	7	8	14	16	21	29	36	Stain
1	0.07	0.14	0.15	0.15	0.15	0.18	0.26	0.31	0.31	0.33	0.41
2	0.09	0.18	0.19	0.20	0.21	0.33	0.35	0.37	0.41	0.41	0.83
3	0.05	0.06	0.07	0.07	0.07	0.09	0.09	0.09	0.12	0.13	0.20
4	0.19	0.36	0.42	0.44	0.48	0.91	1.06	1.10	1.10	1.10	1.97
5	0.13	0.29	0.32	0.33	0.37	0.58	0.58	0.59	0.59	0.60	1.13
6	0.10	0.14	0.14	0.14	0.14	0.23	0.26	0.29	0.32	0.33	0.62
7	0.07	0.09	0.09	0.10	0.10	0.12	0.12	0.13	0.15	0.18	0.29
8	0.02	0.02	0.02	0.03	0.03	0.04	0.04	0.04	0.04	0.05	0.06
9	0.01	0.01	0.01	0.01	0.01	0.01	0.02	0.02	0.03	0.03	0.03

Other Helpful Information

Keep exact records on your processor on a daily basis.

When feeding film into a processor, do not put fingers on the film because they will cause pressure marks. Place fingers behind or on the trailing edge, using both hands, and guide the film in this manner. Always line up the film along the guide plates to the side of the feed tray to insure proper alignment. Should the film fail to transport properly and migrate toward the outer edges of a roller assembly, try to isolate the cause by remembering that a cocked film will change direction every time it goes around a master (sun) roller.

Check for proper roller tension by holding several rollers and attempting to turn the drive sprocket. There should be no slippage. Also, without any restriction on the rollers, they should turn freely as a unit.

To form tubing, where called for, insert a length of heavy electrical cord the same size as the inside diameter of the tubing. Bend this assembly very tight and heat all sides with a match or hot air blower. The cord keeps the inner walls from collapsing. Maintain the tight bend until the tube is cool and then remove the cord. Repeat this procedure until the tubing has achieved the desired bend.

As you work on the electrical and plumbing areas of a processor, take the time to label the wires and tubes. This will assist you while you work and speed the location of parts in the future. It is also an aid for teaching.

The thermometer gauges usually are adjustable by very carefully and gently prying off the front bezel (the retainer ring that holds the glass). Insert a small screwdriver into the stem slot and rotate the pointer to the new position. Replace the bezel.

When removing tubing that has solution in it, take great care that the solution does not spray or splash your face and eyes. If the tubing is long enough or if it is to be discarded, and no other way is available for releasing pressure and draining the line, then use a knife to slit the tubing on the side away from you. *Always wear safety glasses.*

When removing "O" ring spring tubing clamps, use hose clamp pliers. These pliers are inexpensive and have holes in the end to securely grip the clamp. The clamp is made of spring steel—it can be dangerous if it slips or explodes. Do not reuse the clamps but replace them with screw clamps.

On certain metering pumps, the rubber poppet valves on occasion will work free of the plastic housing and lodge crosswise in the line. This usually occurs when the nozzle works loose and is indicated by a replenishment rate that is approximately one-half of normal. The poppet valve should have a small retainer lip at its apex.

When a processor is upgraded or converted to a different processing cycle, save the discarded parts for future emergencies.

Should the replenisher system fail, simply add the required replenishment manually on an hourly basis.

Keep a week's supply of chemicals and film on hand in the event of weather or delivery problems.

Establish a small supply of parts. All processors have several interchangeable parts: sprockets, gears, rollers, bushings. Refer to manufacturer's recommended parts lists for specific recommendations of spare parts. Good preventive maintenance programs, including cleaning and lubrication, afford the operator the advantage of spotting potential trouble areas and allowing time to order replacement parts. Initially, any parts ordered (except costly motors) should be double the quantity immediately required. A replacement part similar to the original part will *probably* fail after a similar length of time. Also, this is a practical way to build a parts inventory.

To set microswitch clearances, use a strip of polyester-based film about 150 mm long × 50 mm wide (6 in. × 2 in.). This film is 0.2 mm (0.007 in.) thick and is easy to use.

Processing Roll Film

Because of the quantity of film processed, and because roll film can range in size from 16 mm to 43 mm (17 in.) wide, an adjustment to the replenisher system should be made. A replenisher pump cut-off switch can stop excess "charging" of the chemistry and the resultant variations in chemistry activity. It is relatively easy for an electrician to hook up a toggle switch for this purpose. Smaller format film may require replenishment for half or a third of the time. Larger format film may require additional chemicals added by hand or through the use of an auxiliary pump.

When processing roll film, smaller and thinner films need to be hooked to a "leader." Some suggested splicing techniques are:

1. Tape the film to be processed to a leader film such as a standard 14 × 17 inch film using acetate-based pressure-sensitive tape. Tape the film securely to the leader, remembering that the area under the tape will not be developed. Several 16 and 35 mm film widths can be taped to a single 17-inch wide (430 mm) film.
2. Fold a sheet, 250 mm (10 in.) wide, of thin film (4 mils or less) over the leading edge of the film to be fed and insert it into the rollers. The increased thickness permits the film to be pulled through the processor.
3. Cut slots in a standard 14 × 17 inch polyester-based film and feed the roll film into the front slot. Then double the roll film back and out the rear slot, locking the roll of film in place.

There probably are many other very good systems. However, avoid adhesive tape that cannot withstand high temperatures or chemicals, and do not use staples. Staples can affect the chemistry and gouge the rollers.

PART 5

AUTOMATIC RADIOGRAPHIC FILM PROCESSORS

Advantages of Automatic Processing

While manual processing techniques will always be needed to meet emergencies and to develop radiographs at locations other than the darkroom, most X-ray departments now depend increasingly on automatic processing. Without automatic processing, in fact, radiography could no longer keep up with ever-increasing demands for volume, speed, and uniformity. Depending on the design of the machine, new processors can handle up to 100 films per hour. Processing time is just 5 to 10 percent of that needed for manual processing of conventional industrial films. Radiographic uniformity is assured because all variables—temperature, time, concentration of processing chemicals—are automatically controlled.

It is because of this increased efficiency, uniformity, and speed that quicker handling of workloads is possible. Efficiency will also result in better part handling and better scheduling.

Types of Equipment

Industrial Processors

From the many makes and models available, the radiographer can choose the machine or combination of machines that best fits his requirements of capacity, floor space and workflow. In most processors, a series of rollers carries films at a constant rate through the four basic processing stages—developing, fixing, washing, and drying. At the end of the cycle the machine delivers a complete product ready for interpretation.

Medical for Industrial Use

Many older medical radiographic film processors can be used for industrial radiography. It is important, however, to seek out those units which originally operated at 3 to 5 minutes dry to dry or longer. Equally important is to seek a unit with a developer tank capacity of 20 L (5 gal) or larger. Preferably the racks should be equal length, but if not, they usually can be reworked easily.

OUTLINE OF PROCEDURE FOR INITIATING AN AUTOMATIC PROCESSING QUALITY CONTROL PROGRAM

I. Preparation of Processor
   - A. Clean the processor very thoroughly. Flush out all lines including the replenishment tubing coming from the replenishing tank.
   - B. Using a checklist, inspect all components of the processor. Inspect alignment and wear points.
   - C. Correct any mechanical problems. Make note of and order new parts where indicated. Record this information.
   - D. Lubricate all necessary points.
   - E. Mix and install chemistry. Make note of chemistry brand, type, order code or other identification, and serial number. Also note amount and brand of starter solution added to developer.
   - F. Install new filter (if used in processor). Note make, type, and size in micrometers.
   - G. Reassemble processor and turn unit on.
   - H. While processor is "coming up" to operating condition, clean outside of processor, especially feed tray and receiver bin.

II. "Operational" Conditions
   - A. When the processor indicates normal functions, make note of the following specifications and record. (Begin with current levels of temperature, etc.)
     1. Developer immersion time: leading edge into developer chemistry until leading edge exits the developer.
     2. Fixer immersion time
     3. Wash immersion time
     4. Dryer time
     5. Dry-to-dry time for 35 cm (14 in.) length, in the direction of transport
     6. Exit time for 35 cm (14 in.) length, in the direction of transport (this added to "dry-to-dry" provides "dry-to-drop")
     7. Inches per minute rate

8. Developer temperature: determine by using bi-metallic or electronic thermometer calibrated on processor.
9. Fixer temperature
10. Wash temperature (calibrate mixing valve thermometer)
11. Dryer temperature
12. Replenishment rate (per 14 × 17 in.) fed 17 in. across and with replenishment tank half full of developer/fixer.
13. Wash water flow rate
14. Ambient conditions

B. Make sure that unit is clean and properly functioning, and that temperature and speed are stabilized by feeding six unexposed 14 × 17 in. films through processor. Carefully inspect last film for artifacts.
C. Recheck film feeding rate, developer temperature and record.
D. Observe and comment on chemical circulation and blower functions. Also note activity of any indicator lights.
E. Make sure no person alters any temperatures or rates in the processor.

III. Quality Control Test
A. With the process at a controlled "known" state, test films can be processed.
B. In the darkroom, from a preselected box of film, expose a film using a step wedge, penetrameter and/or sample part and immediately process along the drive side of the processor.
C. Repeat as above with a film processed along the nondrive side of the processor.
D. Process several normal or routine films.
E. After processing, re-check developer temperature for stability and record. While feeding the last film, the entrance time can be monitored and recorded.

IV. Interpretation
A. The two sensitometric strips should be compared to verify that equal results are obtained on either side of the processor.
B. The routine films should be reviewed to see if they truly represent "average" or "normal" films insofar as penetrometer, density, contrast, and artifacts are concerned.
C. The next step:
1. If the two sensitometric films do not match or if the routine films are unacceptable, make corrections and repeat test procedure.
2. If the test films match and the routine films are acceptable, then the test films become "Master" films and should be so labelled and filed. File in separate places in case one film is lost. Record routine film identification numbers should it be necessary to retrieve these films for comparison.

D. The "Master" control film now represents average or normal films under the known conditions of processing.
E. This should be filed with all available information.

V. Continuous Monitoring for Consistency
A. On a regular basis, such as three times daily or during "shifts", but at least daily, another film should be sensitometrically exposed and processed.
B. The test film is compared to the "Master" control film.
C. Next step:
1. If the films do not match but are within acceptable limits, make note of this by recording density shift. If convenient, recheck all specifications such as temperature.
2. If films do not match and are beyond acceptable limits, then a complete check must immediately be made of all processing conditions to identify which variable (or variables) is out of limits. Record cause and effect relationship. process another test film to verify that acceptable processing has been regained.

D. Periodically, such as weekly, all records must be reviewed, updated, and trends noted.

VI. Expansion of Program
A. Control test films should be processed before and after any major changes in conditions such as when chemistry is changed, or a drive or pump motor replaced.

B. Processing should be consistent. The routine processing of a controlled exposure film will indicate the degree of consistency as well as help to identify variables and the extent that they are individually inconsistent. Before a variable can be controlled, it must be identified.

C. When consistent quality processing has been achieved in one processor for a period of one or more months, then consideration should be given to insuring that optimal quality is being provided. Should a different definition of quality be produced, then the "Master" control film must be replaced with a new film to reflect the new specifications of processing.

D. When one processor is proven to be producing consistent quality, then work should begin on establishing the consistency of a second processor relative to itself and relative to the first processor. If two processing systems can be held consistent, then the third processor should be added. In each case, the "Master" film is the controlling factor to which other test films should be matched.

PART 6

PREVENTIVE MAINTENANCE

Introduction

Like an automobile, an automatic processor runs best when maintenance procedures are followed carefully. And like the newer automobiles, the newer processors are more sophisticated and require a minimum of maintenance. Most processors need lubrication only at infrequent intervals, such as when solutions are being changed. But reasonable attention to these minimal needs will often avoid the need for a service call and attendant downtime.

In order to work on a processor, it should be clean and it should be properly lubricated. These words of advice are often put aside by the average processor owner. But those processor owners and operators who have had to wade through years of dirt before correcting a problem during a critical work period can appreciate the importance of a clean processor.

Below are some guides to aid you in working with each system. This listing and approach supports and extends that found in your processor manual.

Transport System

Check the level and alignment of the feed tray. Keep the entrance rollers free of dirt and free to move up and down. Check the alignment of the main drive motor, drive sprocket (or gear), drive chain, and driven sprocket (or gear). The chain should be loose enough for a 6 mm (0.25 in.) deflection at its center. Make sure the motor is securely and tightly mounted. Next, check the processor's main drive shaft for proper alignment, cleanliness and lubrication. By disassembly and visual inspection, check the bearings for signs of wear and the shaft for wear at the bearing point. Make sure roller rack assemblies are supported by the support bar and not by the gears and drive shaft. This test is done by pushing down on the racks and listening for the motor to slow down, or by removing the racks, one at a time, and listening for the motor to speed up. Make sure that all roller rack drive mechanisms are properly aligned and that bushings (when normally out of solution) are lubricated. Watch for signs of wear on all gears, such as teeth wearing down or cracks at the hub due to stress and/or heat. Place roller assemblies on a flat, level surface and make sure the assembly is square. Tighten all spacer bars to maintain alignment. Where a ladder chain drives the rack assembly, check for wear on the teeth and make sure the chain is relatively parallel to the side plates. The hooks of a ladder chain pull down on the cross bar and face outward.

Make sure dryer rollers properly contact the drive belt or drive shaft and that primary drive shafts are free-turning and well lubricated (avoid lubricants in the actual dryer chamber). Keep tension on the dryer drive belt, if so equipped, and look for signs of flat spots on the drive gears. Check for warped dryer rollers. Keep roller friction gears and the catch bin clean.

Proper lubrication of the transport system is very important. A small amount of lubricant should be applied on a regular schedule rather than a large amount infrequently. Too much lubricant can contaminate chemistry, create a dirty mess, or damage films. All exposed roller assembly gears, the processor drive shaft, and other gears not in chemistry solutions should be coated with a *very* light layer of heavy grease. Smaller parts, such as the drive chain, can be lubricated with a cotton-tipped swab. Disposable oilers with needles will place a drop of oil where required. Ideally, the processor will be cleaned thoroughly each week to remove all used and dirty lubricant, and new lubricant will be applied. Glycerin is a good lubricant for bushings or bearing assemblies that are over or near chemistry because it mixes with the chemistry better than regular petroleum-based products. Always wipe off all excess lubricant, since only a slight film is required.

Electrical System

Install and use an external wall-mounted mains/isolator (circuit breaker) with proper current rating (refer to National Electric Code). Never, for any reason, substitute pennies or copper tubes for fuses. Keep one spare fuse on hand for each type of fuse used and replace as needed. When working on the electrical system, turn off the main isolator (circuit breaker) and switches, and test with a pocket tester to make sure there is no current flow. For increased safety, lock out the circuit breaker and put a sign on it. If fuses blow more than once in a relatively short period, do not keep replacing fuses; call in an electrician to isolate the cause. This will save the expense of a new motor or new wiring and is the reason for using fuses.

Make sure all electrical connections—including plugs — are covered. Connections should be wrapped with electrical tape. Take whatever measures are required to keep moisture from electrical power. A variety of protective sprays and resins can be applied, and even grease is useful.

Whenever working on the electrical system with the power off, check that all connections are tight and inspect for signs of insulation deterioration. Should a motor stop working without apparent reason, check the switches and starter relays involved for proper operation. Most control panel switches are double switches; should one switch become defective (most switches are repairable) simply locate open (unused) terminals on other switches.

Electrical problems are almost nonexistent. Those that do occur are generally the result of improper cleaning, alignment, and/or lubrication. Keep in mind that in an emergency, most electrical equipment can be easily rewired to keep your processor running, perhaps by using different motors or other equipment. However, such measures are only temporary and any changes made must be placed on file, indicated on the processor in a conspicuous manner, and noted on the electrical schematic and/or wiring diagram. Motors with burned out starter windings or starting devices can be manually started.

Tempering System

Do not use the mixing valve to turn water on and off, as this tends to wear out the seals faster. Instead, use the hot and cold main valves or install a shutoff valve on the tempered water line to the processor. If your valve is equipped with strainers, turn off the main water supply to the mixing valve and remove and clean the strainers about once a month, depending on the local water conditions.

Whenever tanks are cleaned out, fill them with warm (40 °C, 100 °F) water and allow the water to circulate through the heat exchanger to clean it of debris. Remember to fill all tanks.

Observe and record heat-up time and the approximate on-off cycle time of the thermostats (developer and dryer) by monitoring the indicator light. Do this when the equipment is new and periodically thereafter. Verify that thermometers are calibrated and working.

Chemistry System

Use clean equipment when mixing fresh solution and follow exactly the mixing instructions found on or in the package. Change the developer filter regularly. Avoid contamination. Ventilate the tanks when not in use. Use dust covers and floating covers in the replenishment tanks. Check replenishment rates at least once a week. Make sure all tubing lines are free of debris and kinks, and that their connections are secure. Avoid pumping air through the developer. Where possible, do not store chemistry in under-processor replenisher tanks because of the high temperatures generated by the air/dryer system. Make sure that all drain and overflow lines are fully open to prevent backup. Check operating temperatures with a bimetallic or electronic thermometer. Keep the water temperature about 3 °C (5 °F) below the developer temperature. And finally, establish sensitometric quality control procedures to keep a daily check on chemical activity for optimum performance. Keep all metal grinding dust out of the chemicals—which means out of the area entirely (darkroom and light room).

Air/Dryer System

Keep the air tubes (all of them) clean. Keep them "free" (easy to remove) and their holders clean. Never twist a screwdriver into air tube slots in an attempt to increase the air flow. This will only cause hot and cold spots by disturbing the air flow pattern. Check to see that the air slot is uniform. Inspect the blower fan belt(s) periodically for correct tension (13 mm or 0.5 in. deflection at the center between the blower and motor pulleys) and for signs of wear. Depending on local conditions, about once a year the blower should be inspected for dirt buildup on fins. Periodically clean the heater element, as well as any accessible areas of the plenum, with a vacuum cleaner.

Use larger 100 mm (4 in.) exhaust tubing with as few elbows as possible and make sure connections are tight. Check for any back pressure at the external end of the exhaust tube. Some processors have damper valves in the exhaust system that can accidentally be closed and, therefore, should be periodically checked.

Keep the processor area at a comfortable working temperature (20 °C, 70 °F) with proper humidity (60 percent RH). Too high or low a temperature and/or humidity is uncomfortable to work in and it retards the efficiency of the air/dryer system. Do not forget to wash or replace all air filters as they become dirty. Prevent any liquids (water, chemicals or lubricants) from coming in contact with any part of the air/dryer system as they will be blown around and deposited on the radiographs.

Use the least amount of heat to dry eight to twelve 14 × 17 in. films fed one after the other. Start at 40 °C (100 °F) and work upward.

Circulation/Filtration System

Circulation

Replace all tubing once a year or as needed, such as when work is performed on a pump head. Look for constrictions and hardening of tubes. Keep circulation pump motors lubricated (if required) and inspect the pumps themselves, when chemistry is changed, for chemical deposit buildup. Each day at start-up and shut-down remove the crossovers and visually inspect each tank for proper solution circulation. Some processors have line restriction "down stream" from the pump; this must be transferred when the tubing is changed. If a "pump" fails, inspect to see if you need a new motor, coupler (drive magnet or shaft) or pump (impeller). Most often only the pump portion, which is a relatively inexpensive part, is bad.

Filtration

Replace the developer filter frequently in accordance with the manufacturer's recommendations, workload, and visual inspection. For increased accuracy, keep records on filter replacement rates. Keep water filters (which protect the mixing valve mechanisms) new and clean for greatest efficiency. Clean out the strainers (if provided) in the water mixing valve assembly about once a month. Clean out the replenisher line filters (which protect the pump and valves) whenever chemistry is changed.

Upon installation of filters, remove all air so that solutions flow properly. This is done for the developer filter by bleeding the air off as the solution level rises. Keep screen-type re-useable filters wet or put them in running water before installing. All filter compartment lids and tubing clamps must be tight to prevent leaks and/or air intake.

Replenishment System

Keep microswitches working properly. Measure the pumping rate weekly when the replenisher tanks are half full, to eliminate the effects of the variability of pressure. When measuring replenisher flow, hold the solution outlet and container at about the level of the solution in the processor tanks. The fixer replenisher is approximately double that of the developer. Make sure all connections are tight, especially the nozzles on metering pumps. Check all lines for chemical deposits or debris such as lint. Keep the replenisher filter clean. Use the minimum amount of hose from the pump to the replenisher tanks to prevent kinks. Store replenisher chemicals, preferably outside the processor, in tanks that have floating covers (may be solid plastic or mini-float balls). If the replenisher tanks are elevated off the floor, place a spring around the tubing where it connects at the bulk head fitting to keep a kink from forming or the fitting from breaking. It is better, although space consuming, to use separate replenisher tanks for each processor so that, should a problem arise, only one processor is affected. It is also generally better to use smaller replenishment tanks that are refilled once a day rather than larger tanks that are refilled once a week.

Weekly, check the waterflow rate at different times of the day on different days to insure minimum water replenisher flow rates. If you do not have a flow meter, then drain the tank and time how long it takes to fill up and overflow for a specific setting of the water valve (usually wide open). Consult the processor specifications for tank capacity.

Cleaning

A clean processor is easier to maintain, lasts longer, and produces higher quality radiographs with greater uniformity than a dirty processor. The best cleaning agents are warm water (about 40 °C or 100 °F) and hard, conscientious work. Roller assemblies tend to build up chemical deposits on both the inside and outside of rollers; they are best cleaned by soaking in warm water. Soaking should be followed by the light scouring of hard rollers with a plastic cleaning pad; avoid pads with soap or metal content. Scouring should be done carefully to avoid cutting the roller material. Do not scour soft rollers, because scouring cuts through the rollers' gelatin glaze causing scratches that hold debris such as metallic silver, lint, dust. Feel the roller for cleanliness.

After cleaning, rinse all parts again with warm water.

The feed tray is often neglected in cleaning procedures, but it picks up oily fingerprints and dust. Therefore, clean the feed tray several times daily with a dry rag to remove dust and oil. Once a day, or as needed, wash the surface with warm water. A small amount of soap can be included to act as an anti-static agent. Wash out the receiver bin at the beginning of each day. (Note: a "day" is a single work shift.)

System cleaners, or chemical cleaners, are available for cleaning processors. But these cleaners, because they are strong chemicals, require extreme care in handling and considerable flushing afterward to prevent contamination of new chemistry. Remember that when you drain the processor tanks you probably have not drained the entire circulation system where contamination can be trapped.

If the processor is very dirty, use system cleaner to remove scale and deposits from the tanks and circulation lines. Never soak rack assemblies in cleaners. If a cleaner is to be used on rollers (not the cleaner's intended purpose) pour a weak solution over them. Wear protective equipment (goggles, gloves, and apron).

To avoid contamination of the new developer, be sure to thoroughly flush all cleaner out of the tanks, lines, and heat exchanger. If you use a powdered system cleaner, mix it in a separate container and then transfer it to the processor. If powder is dumped directly into the processor, undissolved chemicals can get into heat exchangers and transfer lines where they remain to cause trouble later. The crystals can destroy pump parts.

About 7 mL of sodium hypochlorite bleach per liter (1 oz per US gal) of water can also be used as a strong developer or wash tank cleaning agent. Keep this solution out of contact with fixer. Remember to use care and to flush thoroughly afterwards to remove all traces.

To clean a processor completely, drain off all chemistry (keeping silver-laden fixer for silver recovery) and remove and discard the developer filter. Close the drains and fill all tanks, including the replenishment tank, when possible, with warm (40 °C, 100 °F) water. Turn the processor on, activate the microswitches and allow the warm water to flow throughout all systems. Drain all lines, remove the racks, and replenisher tanks. Repeat several times.

Other suggested cleaning includes:

1. Approximately once a year, replace the tubing lines. These lines are difficult to keep clean and they eventually harden, discolor and form restrictions in the line. When hoses are replaced, use new connectors and screw-type clamps. Transfer line restrictors and install in the proper direction. Pump inlets are always in the center of the pump housing.
2. Remove and inspect the replenisher line filter weekly to insure that the small screen-type strainer is not clogged.
3. Clean any chemical deposits from sprockets, gears, drive shaft, worm gears, and drive chains. A mild solvent or soap can be used in these areas with a toothbrush for getting into tight areas. Do not forget to lubricate afterwards. Do not use solvents or soaps on rollers or any components that will contact chemicals.
4. Air tubes are often neglected in cleaning, especially those that are hard to remove. However, for proper air flow these tubes, all of them, must be removed on a regular basis and flushed with warm water. At the same time the roller support mounts should be cleaned and inspected. Annually remove the dryer blower and clean the

fins with a small brush and water. The blower is balanced; you may find weights, which should not be disturbed, or drilled holes, which should be cleaned.

5. All surfaces should be wiped daily to remove dust, chemicals from splashing, and general dirt. This helps reduce the collecting of dirt and facilitates locating the source of problems such as leaks. Also, chemicals collected on most surfaces will gradually eat at the paint and metal resulting in costly replacements. Tanks corrode from spills left on the outside surface.

Depending on the workload and the processing cycle speed, it may be necessary to remove the rack and disassemble it to facilitate proper cleaning of some of the rollers. Steam has been found to work well and does not require disassembly of the racks. If steam is not readily available, check a local automobile car washing facility or garage that advertises steam cleaning. Do not use soap or detergent. Larger institutions have found ultrasonic rack cleaning units to be useful.

Caring for Stainless Steel

The main thing to note about "stainless" steel is that it is not necessarily stainless. Stainless steel derives its characteristics from an invisible, self-healing film of oxide. Anything that penetrates or destroys the oxide film is likely to damage the steel surface.

There are many types of stainless steel; they differ in the ratio of chromium, manganese, and nickel or by the amount of molybdenum or columbium they contain. One kind — Type 316 — has proved most satisfactory for photographic work, but even it is not impervious to chemical attack.

Screws, paper clips and other non-stainless steel parts will cause the stainless steel to corrode if contact is maintained for any length of time. Because of the metal particles it leaves behind, steel wool should not be used to clean stainless steel.

If stainless steel becomes stained, rusted, or pitted, clean it with a 2-5 percent solution of oxalic acid. Scrub the contaminated area thoroughly with a plastic or fiber-bristled brush. Then rinse the area well with water, and treat it with a 20 percent solution of nitric acid to restore its protective film. This is called "passivation". Where pin holes have developed, the area can be epoxied. Rinse these treated areas to remove all traces of acid before installing chemicals. Consider rinsing the developer tank several times with a gallon or two of developer.

Before you put your new processor into service, clean all its stainless steel surfaces. Clean weekly thereafter. To prepare new surfaces, or to keep old surfaces looking new, scrub with cleaners specially designed for stainless steel.

At the end of the processing day, turn the water off and place the required quantity of algaecide in the tank. Liquid laundry bleaches can be used to clean the tanks and retard algae growth. A simple trick is to always leave the wash drain partially open (about one liter per minute outflows) so that the wash tank always drains dry whenever the water is turned off. Since normal water flow is substantially greater, the tank will fill and function normally with the drain partially open.

Lubrication

Lubrication reduces friction, which in turn reduces the load on a system. The reduced load requires less energy and, therefore, less heat. Heat will reduce the service life of motors in the automatic processor.

Over-lubrication is possible and it also causes problems. In most cases, a small amount of lubricant is sufficient. A drop of oil or a smear of grease will protect the part from dirt and reduce the load on the motors. But because it is often difficult to determine when a motor or bearing requires lubrication, it is wise to establish a weekly or monthly lubrication checklist.

The drive motor, unless otherwise recommended by the manufacturer, should have one small drop of oil (20-30 weight nondetergent dielectric oil) per week in each oil cap. The drive motor gear reducer assembly should have new grease packing once a year. The drive chain should be coated with a light grease. When the lubricant gets dirty and gritty, clean it off and relubricate. Always remove excess lubricants.

Processor drive shafts will have from two to eight bearing blocks for support and proper alignment. Each bearing should have a drop of oil a week. Drive shaft worm gears and their matched gears should have a very light coat of grease. Only a very light application is needed, since friction usually is not as much of a problem as is dirt collecting in the teeth of the drive mechanism, causing a build up and load increase.

Some processors have small support bushings or sleeves around the shaft where the gears are driven by the main shaft. These bushings collect chemical deposits that corrode the shaft, create pits and "load" the entire drive system. Glycerin should be applied with a disposable syringe (100 mL) and needle (#18 with the point removed) for exact placement of the lubricant with minimum chance of spill. If the bushing is frozen, it will be necessary to disassemble the parts and sandpaper the shaft to remove corrosion.

Microswitches must be kept free but, at the same time, lubrication must be avoided.

Circulation pump motors, such as the vertical "sump" type found on some commonly used processors, have oil holes at the top and bottom of the motor (in the end plates). This particular motor fails most often due to lack of lubrication in one or both of the bearings. Newer machines using this motor will have oil tubes running up the side of the motor from the lower hole. However, while this is an obvious lubrication point, do not overlook the oil hole on top. A drop of oil once a week in each hole will prolong motor life.

Any time a motor burns out or must be replaced for some other reason, take the time to take it apart. Look for oil or grease pads around the bushings. Check for signs of wear and corrosion resulting from lack of lubrication. Also inspect for corrosion of the rotor due to chemical fumes.

Lubrication records showing the date, part lubricated, and lubricant used should be kept in the processor log book.

Correct Hose Installation

Sharp bends in hose lines (tubing) should be avoided. Bends tend to shorten hose life, increase pressure drop, retard swivel action, increase pump load and raise the temperature of the fluid. To avoid strain on hose and connections, the hose should be long enough to allow for hose flexing in locations where it is used.

Any flexible polyvinyl chloride (PVC) tubing may be used. It comes in clear and colors (use red for acid fixer and blue for alkaline developer).

TABLE 9. Recommended Minimum Hose Length for 180° Bends

Hose ID	Minimum Length
5 mm (0.25 in.)	380 mm (15 in.)
10 mm (0.375 in.)	430 mm (17 in.)
15 mm (0.50 in.)	480 mm (19 in.)
20 mm (0.75 in.)	610 mm (24 in.)
25 mm (1 in.)	760 mm (30 in.)

NOTE: Hose size is generally based on the inside diameter of the hose. It is preferable to use 3 mm (0.125 in.) wall thickness for greatest strength. Also, rigid plastic (P.V.C.) elbows may be used. Joining tubing sections or connecting to a fitting is best done by making a butt-joint. A small piece of the next larger size tubing is slipped over both parts and is secured with screw clamps.

Processing Maintenance Schedules

Daily

1. Start up:
   - Check all solution levels
   - Check replenishment tank quantity
   - Inspect transport running
   - Work microswitches
   - Clean feed tray
   - Check for leaks, general inspection
   - Wash crossovers
   - Check chemical color
   - Record observations, work, recommendations
2. At Run Status (beginning of shift):
   - Feed several green clean-up films
   - Run a sensitometric test film and make a visual comparison to a standard. If they do not match acceptably, check processor time and temperatures. Correct condition to produce acceptable results.
   - Record observations, work, conclusions, including sensitometric plot.
3. Mid-Day (middle of shift):
   - Clean feed tray
   - Inspect general parameters (replenishment, temperature, feed time, etc.)
   - Run sensitometric control, make visual comparison; correct if needed
   - Record observations

4. After periods of one hour or more on standby:
   Clean feed tray
   Run several green clean-up films and/or clean crossovers
   Run sensitometric strip
   Record

5. Running serial roll of strip film:
   Run sensitometric strip prior. If okay, process work film
   Run sensitometric film after to verify consistent processing or to indicate a need for change prior to next run
   Record

6. Shut Down (end of shift):
   Run sensitometric strip. Compare to standard.
   Record.
   Remove and wash crossovers
   Wipe out the inside of the processor removing all splatter, spills, etc.
   Inspect generally
   Leave fume controls, splash guards, and crossovers on top of processor. Cover with a towel.
   Never soak rollers or racks in system cleaners

7. Other—General:
   Train operators to be aware of and responsive throughout the day to sounds, smell, and vibrations of processor, the way film feeds and any unusual occurences. Establish direct and easy communication and recording of these observations.
   Restrict processor monitoring, sensitometric evaluation, jam clearing, etc., to one person per area or processor
   Sensitometric tests are used to show if unit is performing correctly within predetermined limits. If the processing is out of limits or tending toward out of limits, corrective action must be taken or unacceptable radiographs will result.
   Sensitometric data plots can be posted near processors
   At all times wear safety goggles and other protective equipment. Use safe working procedures, such as posting signs, and restricting non-workers from area during work, at all times.
   Never soak rollers or racks in system cleaners.

Weekly

1. Remove the developer rack, hose down with hot water, inspect, and replace
2. Check replenishment rate for accuracy (the volume with tank full, half full, empty) and reproducibility (same volume with same volume of replenisher tank at the same pump setting)
3. Check to see if replenishment pump, thermostat, and mixing valve settings have been altered
4. Verify or recalibrate all thermometers, including the one on the mixing valve, using a calibrated bimetallic stem thermometer
5. Check temperatures of all solutions with calibrated thermometers
6. Check developer immersion time and feed rate (this allows calculation of transport rate)
7. Lubricate main drive shaft bearings, dryer drive mechanism, and motors with fittings with light oil
8. Clean and grease worm gears on main drive shaft
9. Clean and relubricate main drive chain
10. Check alignment and tension of main drive chain and dryer belt
11. Review weekly sensitometry trends. If a densitometer is available, read, plot, calculate, and record sensitometric values: D. Max, contrast, speed, base and fog

Record all observations and work performed

Monthly

1. Twice Monthly
   Remove deep racks and scrub with hot water and plastic cleaning pad. Inspect and replace.
   Clean up all spills and splatters under tanks
   Record observations and work
2. Once Monthly
   Clean dryer air tubes, rollers, and compartment
   Change out all chemicals
   Change developer and water filter; clean replenishment and mixing valve strainers
   Scrub tanks, rinse, fill with warm water, and circulate for five minutes. Drain and rinse.
   Make sure all fluid is out of pumps
   Make any repairs necessary
   Refill with fresh chemicals
   Record observations and work
   Review records for month with staff. Training could also be included

Quarterly

1. Drain all chemicals from processor *and replenishment tank*, scrub out, rinse, fill with warm water and pump
2. Calibrate (or verify) markings on replenishment tanks from empty to completely filled
3. Check processor, support bars, racks, feed tray for level
4. Record

Biannually

1. Use system cleaner to clean heat exchanger in developer section only if a problem is anticipated. Fill tank, add cleaner, circulate for five minutes, drain, flush three times.

Annually (Spread work throughout the year)

1. Disassemble each rack and crossover completely, inspect and replace parts. Especially inspect stud holes, studs, bearing, and shafts for signs of wear.
2. Disassemble main drive shaft and components, clean and inspect bearings
3. Replace all tubing
4. Record
5. Review all records, develop program for following year

PART 7

TOOLS

SUGGESTED LIST OF TOOLS FOR WORKING ON AUTOMATIC PROCESSORS

Electrical

1. VOM (volt/ohm meter)
2. Assorted solderless connectors
3. Pocket probe electrical tester
4. All-purpose electrical repair tool (Alternate: wire stripper only)
5. Electrical tape—plastic roll
6. Soldering gun kit
7. Fuse puller, 12 cm (5 in.)

Pliers

8. 12 cm (5 in.) diagonal side cutter
9. 15 cm (6 in.) slip joint
10. 15 cm (6 in.) long nose
11. 12 cm (5 in.) bent long nose
12. Adjustable vise grip
13. "O" ring hose clamp

Wrenches

14. Socket wrench set, ⅜ in. drive
15. Socket wrench universal joint
16. Open-end wrench set (6 pieces)
17. Adjustable wrench 10 cm (4 in.)
18. Adjustable wrench 15 cm (6 in.)
19. Allen wrench set (14 pieces)

Screwdrivers

20. Screwdrivers set (7 pieces)
21. Screw starter and magnet

Drills

22. Variable speed drill
23. Drill bit set
24. Drill bit and tap set

Miscellaneous

25. Drilling oil, penetrating oil—small can
26. Flashlight
27. Flashlight batteries "D" size
28. Tape measure, steel
29. Small pocket knife
30. Ignition/carburetor tool kit (small wrench set)
31. Pocket stop-watch
32. Light household oil, small can
33. Plastic scouring pads
34. Hemostats, 200 mm (8 in.) or larger
35. Disposable syringe and hypodermic needle or similar oiler
36. Silver estimating paper
37. Bimetallic thermometer
38. Stepped wedge, penetrameter, other image quality devices
39. Coveralls, rubber gloves
40. Safety goggles with side shields
41. Processor log book
42. #25 chain connector
43. #25 chain master links
44. Screw-type hose clamps
45. Assorted small stainless steel machine screws, nuts, bolts, washers
46. Machine level
47. Hammer, ball peen, 200 g (8 oz)
48. Assorted fuses for mains isolator
49. Lint-free rags
50. Permanent ink marking pen
51. Work cart
52. Small part pickup tool
53. Offset screwdriver
54. Brushes
55. Other!
56. Tool Case

PART 8

PROBLEM SOLVING GUIDE

Transport Problems

Sheet Film Jams

1. Turn off the drive motor or the power. Look inside and try to save as many films as possible.
2. Remove the crossover rack ahead of the pile-up.
3. Remove the films at this point. Put them in a tray of water to keep them from sticking together.
4. Clear the film jam. If you have to remove a solution rack, turn off the circulation pump for that solution. *Caution:* If you must remove the fixer rack be careful not to drip fixer into the developer. Remove the rack where the jam occurred, and examine it for the cause of the problem.
5. Feed the removed films into the rack nearest the point of jamming, to complete the processing cycle.

Film Twisting or Turning in Processing Section

1. Be sure the crossovers are seated properly.
2. Be sure the crossovers and solution racks are square.
3. Check for burrs on the guide plates in the racks.
4. Check for stretched springs on the racks, which would lead to uneven roller pressure.

Film Jams in the Processing Section

1. See causes under "Film Twisting or Turning" above.
2. Be sure two films were not fed one on top of the other.
3. Films shorter than 100 mm (4 in.) cannot be reliably fed without a leader tab.
4. Check for badly stretched roller springs on the racks.
5. Check for broken or bound end bearings on the rack rollers.
6. Check for broken teeth on the rack gears.
7. Be sure all gears are meshed and not just riding on top of the teeth.
8. Check for worn gears.
9. Check for chemical deposits on the racks.
10. Check for warped rollers.
11. Check for a temperature or chemical problem that is causing film to be tacky.
12. Check for loose guide plate.

Film Jams in the Dryer

1. Be sure the exit crossover is seated properly.
2. Be sure the dryer rollers are seated properly.
3. If the dryer drive has stopped, check the connecting gears.
4. Check for any conditions causing the film not to dry properly—dryer temperature too low, exhaust inefficiency, low replenishment, solution fix replenishment too low, temperatures too high, high ambient humidity.
5. Make sure air tubes face the correct direction.

Lengthwise Scratches on the Film

1. Check for encrusted chemical deposits on the racks—particularly at or above solution level and feed tray.
2. Check for burrs on the guide plates in the racks.
3. Check for a roller not turning because of broken gear teeth, or some other cause.
4. Check for reversed or out of position guide shoes.

Pressure Marks on Films

1. Check for foreign material or rough spots on the rollers.
2. Check for roller that is warped or too tight.
3. Consider developer hardener failure; replace the developer.

Excessive Gear Damage

If the sun gear on the drive end of the racks gets "chewed up":

1. The rack may have been dropped into the tank too hard, bending the indexing pins. With a hammer, carefully drive down the pins until they are

parallel with the rest of the rack.
2. If the plastic and steel gears do not mesh, bend the rail slightly. There should be very little play in the gears.
3. Be sure the gears are kept clean.

Electrical Problems

Motor Starting Relays

The circulation pump motors are started by current-type motor starting relays. When voltage is applied to the motor (by closing the heat or cool switch), the first rush of current through the motor main winding and starting relay coil closes the normally open (N.O.) relay contacts. This allows current to flow in the motor starter winding and starts the motor. As the motor speed increases, the current drops in the main winding and relay coil, opening the relay contacts to disconnect the starter winding. The motor is then in its normal condition.

If defective, the relay should be replaced as a unit. When you install the new relay, see the wiring diagrams for color coding of the wires and position.

Developer Switch Motor Does Not Operate with Heat or Cool Switch Depressed

1. If one switch activates the pump motor but the other does not, either the latter switch is defective or there is an error in the wiring to it.
2. If neither switch activates the motor, turn the developer thermostat all the way down, to energize the cool solenoid valve. If the solenoid is not energized, trace the power supply to the machine and to the heat switches.
3. However, if the solenoid in step 2 is energized, trace the wiring to the motor and motor starter.
4. If the wiring appears correct, replace the motor starter. If this still does not solve the problem, replace the motor.

Fixer Pump Motor Does Not Operate with Circulation-Replenishment Switch Depressed

1. Feed a sheet of film into the machine. If the replenisher pump motors are activated, the power supply to the circulation-replenishment switch is correct. Check the wiring to the fixer circulation pump motor and motor starter. If the wiring seems correct, replace the motor starter. If this still does not solve the problem, replace the motor.
2. However, if the replenisher pump motors in step 1 are not energized, trace the power supply to the machine as far as the circulation-replenishment switch.

Replenisher Pump Does Not Run

If a replenisher pump does not run when the circulation-replenishment switch is on and a replenisher microswitch is depressed and

1. If only one pump motor does not run, trace the wiring to that motor. If the wiring seems correct, replace the motor.
2. If neither motor runs, trace the wiring to the coil of relay K4 (if included) and to its N.O. contacts. If the wiring seems correct, be sure the replenisher microswitch is being fully depressed. If it is not, readjust it.
3. If the microswitch is being activated, jumper across its terminals. If this energizes relay K4, replace the switch; if not, replace the relay.

Drive Motor Does Not Run with Drive Switch Depressed

If there is power to the rest of the machine, check the wiring to the switch and drive motor to see if it is correct. If it seems correct, check the switch and drive motor.

Tempering Problems

Developer Temperature Fluctuates

1. Check switch and fuses for the immersion heaters.
2. See "Circulation/Filtration Problems" for possible causes of impaired circulation.
3. Be sure the incoming cold water to the heat exchanger is adequate and at least 3 °C (5 °F) colder than the desired water temperature.
4. Check developer thermometer for function and accuracy.
5. Replace the developer thermostat.
6. Replace the thermocouple sensing element.
7. Replace the immersion heater(s).
8. Check the solenoid valve where used.

Solenoid Valve Leaks

1. Disassemble the valve as outlined above and clean the valve seat and ball. Use crocus or emery cloth to remove any scale or rust deposits on the seating surface.
2. If the seating surface is damaged, or if the plunger or spring is worn, replace the valve.

Solenoid Valve Does Not Open or Close Properly

1. Disassemble the valve and check for foreign matter.
2. Be sure the plunger is not binding in the tube.
3. Check the action of the spring.

Film

Problems with film and chemistry, and the solution to these problems, will vary with the material you are using. For most of the following problems, therefore, you should also consult your technical sales representative(s).

Incorrect Film Density

1. Check the mixing procedure for the developer and replenisher to be sure they conform to the procedure supplied with the particular batch of chemicals you are using. Mixing instructions can change from batch to batch, and the only instructions that should be used are those enclosed with the chemicals.
2. Check the replenishment rate. If the change in density has been gradual, you may be over- or under-replenishing.
3. Check the causes of over- or under-replenishment listed under "Common Replenishment System Problems."
4. Do not use the developer beyond the time recommended by the manufacturer.
5. Be sure the developer has not been contaminated with fixer of foreign chemicals.
6. Be sure the developer temperature is as recommended by the film manufacturer.
7. Make sure the processing tank is at the correct level.
8. Make sure there is adequate agitation.

Mottling. Mottling of the film is most often caused by too high a developer concentration. Check with your developer supplier and/or perform a specific gravity test using a hydrometer.

Streaking. Streaks on the film are most often caused by weak developer and/or weak fixer. Check the replenishment rate; change the solutions if necessary. See also the information on drying streaks under "Common Air/Dryer Problems."

Yellow Smudges. Yellow smudges are often caused by exhausted fixer.

Film Does Not Clear. If the film does not clear, the fixer concentration is probably too low, is contaminated, is old or appears so because the developer hardener fails to control swelling.

Pi-Lines. Pi-Lines on the film are normal with fresh chemicals, and should disappear after a short run. Run "green" films to clean rollers.

Dirt on the Film. Dirt on the film usually comes from the water supply. If you have a filter in the water supply line, change the cartridge; if you are not using a filter, it would be wise to install one. Too little agitation or flow of any solution will contribute to dirt accumulation. Check for algae in the wash tank. Check for dust in the dryer.

Drying

Films Stick to the Dryer Rollers

If films stick to the rollers in the dryer, they have not been properly hardened. There are several possible causes:

1. The chemicals may not be designed for roller processing or may be incompatible with the brand of film being used.
2. The fixer may be exhausted or under-replenished.
3. The developer may be diluted, under-replenished, or old.
4. The solution temperatures may be too high for the film.
5. The solutions may be contaminated, or may have been improperly mixed.
6. Solution circulation may be inadequate. Check causes under "Common Circulation/Filtration System Problems."
7. The wash may be inadequate. Check the causes of inadequate wash water flow in the mixing valve instruction manual.
8. There may be a dryer malfunction.

Processor Trouble-Shooting Checklists

Replenishment System

Components:

1. Detector switch (microswitch)
2. Detector assembly: airflow, rollers
3. Main switch, fuses, relays
4. Pump: head, motor, gearbox
5. Lines: tubing, fittings
6. Gauges: flow indicators, meters
7. Tanks: outside processor
8. Check valves
9. Needle valves
10. Filters

Replenishment Problems Are Caused by:

1. Detector switch malfunction
2. Detector switch adjustment
3. Detector assembly malfunction
4. Cleanliness of detector assembly, switch
5. Electrical supply
6. Pump setting
7. Pump accuracy
8. Pump reproducibility
9. Pump leak
10. Pump malfunction
11. Lines kinked, blocked
12. Lines—air leak, air block
13. Lines—solution leak
14. Lines—fittings damaged
15. Gauges not calibrated
16. Filters or strainers clogged
17. Tanks not calibrated
18. Check valves stuck
19. Check valves deteriorated
20. Check valves installed incorrectly
21. Adjustment valve failure
22. Replenish tanks empty
23. Frequency of mix
24. Absence of floating lids
25. Water supply problems

TABLE 10. Replenishment Trouble-shooting Guide

Replenishment Problems	Common Cause*	Or Check These Causes*
Chemistry volume low	6-8	1-24
Chemistry volume high	6-8, 18	1, 2, 4, 6-8, 15, 17-20, 23
Chemistry activity low	6-8	1-24
Chemistry activity high	6-8, 18	1, 2, 6-8, 15, 17-20, 23
Increased density	6-8	1-4, 6-8, 15, 17-21, 23
Unclear film	6-8	1-24
Unwashed film (mixing valve)	6-8	1-24
Undry film	25	1-24
Scratches, abrasions	6-8	1-24
Film jams	6-8	1-24
Decreased density	6-8	1-24
Chemical breakdown	6-8	1-24
High consumption	6, 24	1-24
Leaks	13, 14	9, 11-14, 16, 12-21
Variable sensitometry	7, 8	1, 4, 7, 8, 17

*Numbers refer to items in preceding list.

Electrical System

Components

1. Power supply
2. Lock-out box, fuses, terminals
3. Processor switches
4. Processor relays
5. Processor terminals
6. Processor fuses, circuit breakers
7. Motor protectors
8. Motors: pump, drive
9. Heaters: developer, dryer
10. Indicator lights
11. Thermostats: developer, dryer
12. Conductors

Electrical Problems Are Caused by

1. No supply at source
2. Half normal supply (1 line)
3. Wrong phase
4. Variable supply
5. Line voltage fluctuation
6. Loaded line (elevator, etc., on same line)
7. Fuses burned
8. Fuses corroded
9. Fuses broken
10. Wrong size fuse
11. Circuit breaker blown (open)
12. Circuit breaker broken
13. Relay stuck closed
14. Relay stuck open
15. Burned contacts
16. Welded contacts
17. Bent contacts, rocker plates
18. Broken switch open
19. Broken switch closed
20. Jammed switch open
21. Jammed switch closed
22. Motor starter switch defective
23. Motor starter windings burned, shorted
24. Motor running windings burned, shorted
25. Heater element burned, shorted
26. Thermostat switch malfunction
27. Indicator light burned out
28. Loose connections
29. Short circuit in conductors
30. Corroded connectors
31. Heat
32. Lack of lubrication

TABLE 11. Electrical System Trouble-shooting Guide

Electrical Problems	Common Cause*	Or Check These Causes*
Nothing runs	1–3	1–3, 6, 11, 12, 14, 15, 17, 18, 20, 29, 30
Drive motor won't start	22, 23	1–9, 11, 12, 14, 15, 17, 18, 20, 22, 23, 28–32
Drive motor starts, won't run	24	24
Drive motor manual start, will run	22, 23	1–9, 11, 12, 15, 17, 18, 20, 22, 23, 28, 30, 32
Pump motor won't start	22, 23	1–9, 11, 12, 14, 17, 20, 22, 23, 28–31, 33
Pump motor won't run	24	24
Motor speed changes	4, 32	3–6, 24, 28, 32
Motors fail annually	2–6	2–6, 10, 30–32
No circulation	22, 26, 32	1–9, 11, 12, 14, 15, 17, 18, 20, 22–24, 26, 28–32
No heat	7, 25, 26	1–9, 11, 12, 14, 15, 17, 18, 20, 25–30
Too much heat	10, 13, 26	3–5, 10, 13, 16, 17, 19, 21, 26, 27
Too little heat	26	2–9, 11, 12, 14, 26–31
Insufficient replenishment	14	2–9, 14, 17, 24, 28–32
Heater won't turn off	13	10–13, 16, 17, 19, 21, 26, 29
Motor won't turn off	13, 16, 21	13, 16, 17, 19, 21, 29
Insulation burned	6, 13	3–6, 10, 29–31
Insulation crumbling	6, 13	3–6, 10, 29–31

*Numbers refer to items in preceding list.

Circulation/Filtration System

Circulation Components
1. Electrical power supply
2. Switch, fuses, relays
3. Pump motor starter
4. Pump motor
5. Heat exchanger
6. Tubing, fittings

Filtration Components
1. Filter type
2. Filter size

Circulation/Filtration Problems Are Caused by
1. Line voltage fluctuations
2. Pumps not lubricated
3. Failure of pump motor thermostat
4. Burned motor starter
5. Burned motor starter windings
6. Dirty magnets, shafts
7. Worn bearings
8. Loose pump/motor linkage
9. Warped pump head
10. Leaking pump head
11. Leaking tubing
12. Leaking fittings
13. Air blockage of pumps
14. Air blockage of filters
15. Blocked tubing
16. Improper flow (i.e., backwards)
17. Filter clogged
18. Filter caked
19. Filter channels
20. Filter size too small
21. Filter size too large
22. Filter missing
23. Heat exchanger leaking
24. Heat exchanger clogged
25. Dirty pump
26. Defective switch

TABLE 12. Circulation/Filtration Trouble-shooting Guide

Circulation/ Filtration Problems	Common Cause	Or Check These Causes
No circulation	4, 17	2–18, 20, 25–27
Reduced flow	26	1–18, 20, 24–27
Variable flow	1–3	1–3, 6–9, 26
Pump not running	2, 3	1–5, 7, 9, 13–18, 20, 25–27
Pump okay, no flow	13, 14	10–18, 20, 25
Pump motor won't start	3, 4	1–5, 27
Pump runs after manual start	4, 5	4, 5
Very loud, noisy pump	6, 26	6–9, 13, 26
Leaks at pump	9	9–12, 16
General leaks	10	10–18, 20
Foam	13, 14	9–14, 19, 21, 23
Increased temperature of chemistry	25, 17	9–18, 20, 25
Frequent filter changes (i.e., weekly)	20	17, 18, 20, 24–27
Lower density	2, 3	1–18, 20, 24–27
Uncleared films	2, 3	1–18, 20, 24–27
Reduced or no replenishment	17	11, 12, 14, 17, 18
Dirt on films	21	17, 19, 21, 26

*Numbers refer to items in preceding list.

Temperature Control System

Components

1. Thermostat
2. Thermometers
3. Switch, fuses, relays
4. Water supply
5. Electrical power supply
6. Heaters
7. Indicator lights
8. Circulation pump
9. Mixing valve (if used)
10. Heat exchanger

Temperature Problems Are Caused By

1. Electrical power fluctuations
2. Water supply fluctuations
3. Loss of cold water
4. Loss of hot water
5. Loss of volume
6. Loss of pressure
7. Mixing valve malfunction
8. Clogged filters, strainers
9. Clogged heat exchanger
10. Clogged tubing
11. Thermostat stuck open
12. Thermostat stuck closed
13. Thermostat uncalibrated
14. Thermometer uncalibrated
15. Thermometer broken
16. Heater broken
17. Heater relay stuck open
18. Heater relay stuck closed
19. Indicator light burned out
20. No circulation
21. Switch or fuse failure

TABLE 13. Tempering System Trouble-shooting Guide

Temperature Problems	Common Cause*	Or Check These Causes*
Decreased temperature/decreased density	11	1, 2, 4–7, 11, 13–17, 19–21
Increased temperature/increased density	3, 18	1–3, 5–10, 12–15, 18–21
Unclear films	3 or 4	3–7
Unwashed films	3 or 4	1–15
Undried films	17	11, 13–17, 19–21
Streaks on films	20	8–10, 20
Unequal development	8, 9, 20	1–10, 20
Unequal clearing	10, 20	1–10, 20
Unequal washing	7, 9, 20	1–10, 20
Unequal drying	16, 20	1–10, 16, 20
Temperature fluctuations	1, 2	1, 2, 7–10, 13, 14, 20
Developer gets hotter, heater off	2	2, 3, 5–7, 19
Developer does not heat	2, 21	1, 2, 4–11, 13–17, 19–21
Indicator light cycles on/off slowly	20	8–10, 13, 20
Hot water heats developer	7	2, 3, 5–10

*Numbers refer to items in preceding list.

Dryer System

Components

1. Electrical supply
2. Main switch, fuses, relays
3. Blower/motor
4. Heater(s)
5. Thermostat
6. Thermometer
7. Safety thermostat
8. Indicator light
9. Air tubes
10. Exhaust
11. Sheaves (pulleys, belts)
12. Receiver bin

Dryer Problems Are Caused By

1. Electrical supply fluctuations
2. Insufficient electricity
3. Broken (open) switch, relay fuse
4. Broken (closed) switch, relay fuse
5. Stuck switch, relay
6. Blower fins clogged with dirt
7. Blower bearings worn, broken
8. Blower drive belt misaligned
9. Blower drive belt broken
10. Motor starter windings burned
11. Motor starter relay broken
12. Motor running windings burned
13. Motor circuit breaker open
14. Heater burned open
15. One of heaters burned open
16. Thermostat stuck closed
17. Thermostat stuck open
18. Safety thermostat broken or open
19. Too high setting on safety thermostat
20. Thermostat not calibrated
21. Thermometer not functioning
22. Indicator light broken
23. Air tubes dirty
24. Air tubes installed wrong
25. Exhaust tube too small
26. Exhaust tube blocked
27. Exhaust too small
28. Exhaust too great
29. Room intake air too cool
30. Rollers dirty
31. Rollers allow film slippage
32. Squeegee roller function
33. Developer depleted
34. Fixer depleted
35. No wash water
36. Transport speed too fast
37. Ambient conditions

TABLE 14. Dryer Trouble-shooting Guide

Drying Problems	Common Cause*	Other Frequent Causes*
Overall wet, cool, soft films	3, 17, 19, 33, 34	1–3, 5–15, 17, 21, 23–27
Same as above on trailing edge	15, 33, 34	1, 2, 6–8, 15, 19–21, 23–27, 29–34, 36
Film cool, barely dry	15, 19, 20, 33	1, 2, 6–8, 15, 19–21, 23–34
Film hot, damp	19, 20, 37	6–10, 15, 19–21, 23–27, 30–34, 36, 37
Overall mist, warm hard films	27	6, 25, 28, 37
Same as above on trailing edge	27	6, 25–28, 37
6th to 10th film becomes damp	27, 33	17, 19–22, 25–30, 33, 34
14 × 17 in. dry; 8 × 10 in. damp	31	31
8 × 10 in. dry; 14 × 17 in. damp	3, 17, 19, 33, 34	1–3, 5–15, 17–21, 23–27
Serial films become damper	27, 33	19–22, 25–30, 33, 34, 37
Marginal drying, high heat	19, 20, 25, 28	15, 18–28, 30–37
Water spots	32, 33	6, 10, 14, 15, 17–21, 30–34
Cross hatch pattern	6, 30, 33	6, 10, 15, 17–21, 23, 24, 30–34
Wave-form pattern	19, 20	4–6, 10, 15, 16, 19–27
Baked, glossy surface	19, 20	4, 16, 19, 20–22
"Light" area streaks	23, 24	23, 24
Jams	31, 33	3, 6–14, 17–37
Film cocking	31, 33	31, 33
Dirt on films	30	23, 24, 30, 33, 34
No heat	3, 14	1–3, 5, 9–12, 17–22, 28, 29, 37
No air	9. 10, 11	1–3, 5–13, 23, 24

*Numbers refer to items in preceding list.

Transport System

Roller Subsystem Components
(film-handling problems)

1. Rollers: size, hardness, driven, bearings
2. Guide shoes: adjustable, non-adjustable
3. Deflector plates, wires, bars
4. Face plates
5. Tie bars, support bars
6. Feed tray

Motor Subsystem Components
(speed, drive problems)

1. Electrical power supply gears
2. Gears, sprockets, pulleys
3. Chains, belts
4. Bearings
5. Shafts

Transport Problems Are Caused By

1. Misaligned gears, sprockets
2. Misaligned chains, belts
3. Misaligned turnarounds, crossovers
4. Misaligned racks
5. Misaligned guideshoes, deflector plates
6. Misaligned drive shafts
7. Misaligned support bars
8. Misaligned feed tray
9. Worn, broken gears, sprockets
10. Worn, broken chains, belts
11. Worn, broken bearings, shafts
12. Tension springs too tight, too loose
13. Roller gear-ends loose
14. Dryer drive alignment
15. Roller studs binding
16. Time-delay-on-feed too short
17. Dirty rollers
18. Fuses, breakers, protectors
19. Slipping gears, sprockets
20. Slipping chains, belts
21. Broken electrical conductor, terminal, connector
22. Damaged switch
23. Electrical power source
24. Drive motor gears damaged
25. Drive motor starter circuit damaged
26. Drive motor run circuit damaged
27. Damaged rollers
28. Damaged drive shafts
29. Chemical failure: developer, fixer, hardeners
30. Film characteristics: size, type
31. Lubrication
32. Wrong parts used
33. Cleanliness
34. Film-feeding procedure

TABLE 15. Transport Trouble-shooting Guide

Transport Problems	Common Cause*	Or Check These Causes*
Film jams	1–8	1–17, 19, 20, 27–34
Abrasions	1–4	3–5, 7–15, 17, 19, 20, 27–33
Overlapped films	16, 34	3–5, 7–16, 19, 20, 27, 28, 30, 32, 34
Cocked films	8, 34	3–5, 8, 11–15, 17, 27, 28, 31–34
Detector switch activation	8, 22	1, 2, 5–13, 15, 17, 19, 20, 22, 27, 28
Unusual noise	1, 2	1, 2, 4, 6, 7, 9–15, 19–20, 24, 27, 28, 31–34
Film dropping through dryer	14, 32	1, 2, 4, 6, 9–15, 19–20, 27, 28, 30, 32
Film retained in dryer	29	1, 2, 4, 7, 9–12, 14–17, 19, 20, 27–33
Increased density	27, 31	1, 3–5, 9–15, 19, 20, 27–29, 31, 33
Decreased density	29	4, 14, 23, 29–33
Rapid gear, sprocket wear	31, 33	1–7, 14, 15, 17, 19, 20, 27–29, 31, 33
Component not turning	1, 9	1–7, 9–15, 18–28, 31–33
Scratches	5, 29	1–15, 19, 20, 27–34
Gelatin pick-off, deposit	17, 29	3–7, 9–15, 17, 19, 20, 27–30, 32, 33
Pressure marks	4	3–7, 9–13, 15, 17, 19, 20, 27, 29, 33, 34

*Numbers refer to items in preceding list.

PART 9

OTHER SYSTEMS

Paper Processing (Stabilization Process)

Paper processing is usually associated with paper "prints" made from negatives or in the processing of chart recording paper that has a photographic image rather than an ink image. These systems require conventional processing.

In radiography, paper processing involves using a sheet of paper coated with an emulsion on one side. Some or all of the developer agents are contained in the emulsion layer. After exposure, the paper is processed in what is termed "stabilization processing." The significance of the stabilization process is that it is very fast, producing an almost dry film in as little as ten seconds. The speed is accomplished by eliminating short stop, wash, and dry steps.

Although stabilization processing usually involves paper, film is also made for this type of system. In industrial radiography, paper is most commonly used because the white paper base produces an image that is more black and white. Also, since the paper is not transparent, viewing and interpretation are a little different. The primary advantages are lower cost and faster speed. The primary disadvantages are decreased image quality and a less permanent record.

There are three basic stabilization processing systems:

1. Monobath, where the developer is in the emulsion layer and the film is passed through a single alkaline fixing bath (called a "fixavator").
2. Two-step (the most common), where the paper contains only the reducing agents. The film is fed into an activator solution that is highly alkaline, and then into a stabilizing bath.
3. Three-step processing, where films or paper have no developer agents in the emulsion. The first bath applies the developer agents, the second bath the activator, and the third bath the stabilizer.

The two-step system, which uses the "stabilizer" from which the overall process gets its name, is most common. In previous chapters, we have discussed how the fixer works to neutralize the developer, clear away unexposed and exposed (undeveloped) crystals, and shrink and harden the emulsion. In stabilization processing, fixer is not used. Stabilizers do not clear away unexposed crystals; rather they form new complexes that render crystals insensitive to light so they remain unexposed, undeveloped, and "hidden" in the gelatin.

Stabilization inactivates the silver halide crystals with its fast-acting complexing compounds. These compounds are usually of the thiocyanate type, although several other types exist. Perhaps most common is ammonium thiocyanate (sulfocyanide of ammonium); $5NH_4NCS \bullet AgNCS$ is the complex formed. Another popular stabilizer is bisulfite-hypo, which uses sodium thiosulfate ("hypo").

Processing

Processing in trays involves the following three-step sequence:

1. Develop (activator) EKD-72 (2:1)
   4-10 seconds 38 °C (100 °F)
2. Stop bath, acetic acid, 5%
   2 seconds 38 °C (100 °F)
3. Stabilizer (bisulfite-hypo)
   4-10 seconds 38 °C (100 °F)

Tray processing may also be done at room temperature with different solutions and greater control of moisture pickup.

"Hand processing" differs from tray in that, instead of placing the paper in trays to absorb excess amounts of chemicals, the chemicals are "brushed on" using a sponge. This action provides agitation and keeps solution pickup to a minimum, which is very important to the stabilization process. Since the emulsion layer is only on one side and the base is paper, it is very important to coat only the emulsion side. Excess moisture must be kept out of the paper and from the nonemulsion side. Hand processing usually takes a total of 20 seconds, involving only an activator (5-10 seconds) and stabilizer (10-15 seconds).

Stabilization processors are very small and compact and can easily sit on a darkroom counter. There is no plumbing connection as water is not used. The processor usually has an activator tank and a stabilizer tank. There is a transport system with squeegee rollers at each tank exit. Processing is usually done at room temperature, so that the only electrical power requirement is for the transport. The exposed paper is turned over and fed into the processor emulsion-side down. The activator activates the developer reducing agent in the emulsion. There is very little chemical absorption. The excess surface moisture is squeegeed off. The next step is the stabilizer which converts the unstable silver halide crystals to more stable complexes. The excess chemicals are squeegeed off. The paper exits almost dry and ready for reading.

Archival Quality

In the stabilization process wash water is not used because the water would dilute and break down the stabilizers and the stabilized silver complexes. Thus the conclusion and concern is that the image is not permanent. Though it lacks archival quality, an image will last for days, weeks, or months before it starts to deteriorate. Deterioration depends on the type of stabilizer used; thiocyanates usually result in image fade and thiosulfates usually result in common hypo retention stain (yellow to brown). For most applications, the paper is read and discarded.

Usually, if the paper film is to be read and held for a short period of time (up to one month), no further processing is needed. However, bright illumination, high heat, and/or high humidity can cause deterioration and must be avoided. Also, since the paper is "unwashed" in the conventional sense and contains chemicals, these images must not be stored next to or in proximity to conventional radiographs.

To make paper-stabilization images "permanent" (this is a relative factor, but will increase the life to years) additional processing is required: Using normal x-ray automatic processor type fixer in trays:

Fix for 3-4 minutes
Wash for 5 minutes
Dry

Using photographic products:

Fix in a compatible quick fix for 2 minutes
Hypo eliminator for 3 minutes
Wash for 20 minutes
Dry

If a conventional automatic X-ray processor is available on a longer 10- to 14-minute cycle, it would be possible to experiment by passing the paper through fix, wash, and dryer sections. A leader will probable be required.

Remember, for archival quality in stabilization processing (the "no wash" processing) the film must be further fixed, then washed. Washing the film immediately after stabilization with fixer will destroy the complexes and ruin the film. Thus it is "no wash" because wash is not needed and must not be used. Also, once the film or paper is fixed, adequate washing is required to remove all of the fixer or it will deteriorate the image.

Papers used in this system have much less gelatin, which makes them much harder than conventional radiographic films. These two features reduce the amount of moisture pickup and retention and permit the "almost dry" characteristics of stabilization processing.

Because antifoggants and other ingredients may be placed in either the emulsion or in the activator, the general rule is to use the paper and activator from the same manufacturer.

Other "Paper" Systems

Electrophotography is a "paper system" that results in a positive or negative image as a nontransparency. This is called the "dry" process. X-rays bombard a special cassette containing a plate coated with electrical charges. The plate is usually selenium bonded to an aluminum backing plate. The electrical charge (electrostatic) is applied in the "darkroom" and placed in a cassette. Exposure energy alters the charged layer pattern as the X-rays alter the insulating characteristics of the selenium and the electrical charges leak through to the aluminum plate, which is grounded. Thus the "pattern" of the part is formed in charged areas left on the plate. To make a useful image, the plate is placed in a "processor," where it is sprayed with charged particles that adhere to the charged parts of the plate. The image may be "flip flopped" electronically from a negative mode to a positive mode.

For a permanent record, the image of particles is transferred to paper that has light adhesive layer.

The paper is fed into the processor pressed against the plate image, heat is applied to fuse the particles, and the image is transferred to the paper. A dry, positive nontransparency is produced. The plate is cleaned off to be reused.

This system depends on the phenomenon called photoconductivity. There may be some uses (such as edge-enhancement, dry processing, and/or cost savings as silver prices escalate) for this process in industrial radiography. However, the disadvantages are that a special processor and special cassettes are required, a different type image is produced and the hardware is expensive.

Glass Plate Processing

Introduction

Glass plates are seldom used today because flexible bases such as cellulose triacetate and polyester are easier to handle and may be used with automatic processing systems. The inflexible glass plate must be optically clear, flat, free of color and bubbles.

There is the advantage, however, of being able to coat the glass with various "custom" emulsions for a variety of reasons, e.g. to form "glass pictures," for nuclear tracking and for ultra high resolution imaging such as microradiography. In each case the glass is coated with an emulsion layer. Great care is required to make sure that gelatin will not frill at the edges and that the layer is uniform in thickness. Since image quality may be poorer within 10 mm (0.375 in.) of the outside edges, the useful image area should be restricted. Proper and careful curing and storage of the coated plate is important.

Glass Pictures

A paper mask (negative) of an image is chemically treated with sensitized gum and dusted with bitumen dust. This *resist mask* is placed against a warm glass plate to which the bitumen dust sticks. The glass is exposed to hydrofluoric vapors, which etch the glass in all unprotected areas. After etching is complete the bitumen dust is washed off in a petroleum solvent, leaving the photographic image.

The etched image is a positive of the original negative (or negative of an original positive) but acts as a negative (positive) because of the way light is scattered through the image. The image is an integral part of the glass. Other images can be superimposed with further etching or by using dyes, stains, or regular emulsion layers.

Another entirely different type of glass picture results from the use of colloidal silver and other photosensitive metals inside the glass structure, in suspension. The glass plate is covered with a negative and is exposed to a strong source of ultraviolet light. The metal form is altered so it will precipitate when further processed by heating. The image is integral to the glass, but is inside the glass rather than on the surface.

Glass Plates for Track Emulsions

Nuclear track emulsion layers usually are grouped as 100 μm and less or 200 μm and thicker. The thinner (100 μm or less) emulsions layers are often coated on glass plates to provide rigidity and ease of handling. Because the gelatin coating is thinner, conventional glass plate processing may be used. The thicker (200-600 μm) emulsions are generally self-supporting or are handled by means of small rods or pegs. The very thick gelatin layer requires special processing techniques, which are discussed below.

Processing Nuclear Track Glass Plates. Conventional emulsions of less than 10 μm coated on glass plates are processed in trays in a conventional manner using conventional fine-grain developers. The glass plate is placed emulsion-side-up in a tray containing developer at room temperature (20 °C, 70 °F) for about 5 minutes. Agitation is important and may be accomplished by sliding the plate or moving the solution. Moving the solution by lifting one corner of the tray slightly or by using a small paddle is preferred and easier than moving the glass plate. If the plate is moved it should sit on small glass (or stainless steel) rods, which act as rollers for easy movement. The plate should be pushed from the edge rather than grasped with tongs or the fingers, since these can damage the soft gelatin. As in all manual processing, as the solution temperature goes up, time is reduced and vice versa.

The next step is 10 seconds or so in a stop bath for first rinse followed by 10-15 minutes in fixer solution at room temperature. This can be followed by a hypo eliminator for 30 seconds prior to washing. Washing may be as long as an hour to ensure good archival quality. After washing, a wetting agent may be used to promote uniform drying. Sometimes a squeegee or sponge is used to remove surface moisture. The plate should be placed upright to allow moisture to drain off.

Thicker emulsions used in nuclear tracking, as noted above, are divided into two categories: up to 100 μm and over 200 μm (up to 600 μm). For emulsions in the thinner group, up to 100 μm, conventional processing is used. The gelatin layers may be free from support during exposure, but should be mounted on glass for processing. During processing the gelatin softens and can distort, which increases the risk of physical and informational damage. The plate (glass with single emulsion layer) will be processed "conventionally" but with a different approach. The first step is a pre-soak in water at room temperature (about 20 °C, 70 °F) for 15-30 minutes to swell up the gelatin and aid developer entry. Next the plate is placed in a tray of fine-grain developer diluted 5 to 10 times more than normal. The excess dilution will reduce the activity so that the developer works more slowly. It must be understood that the gelatin layer is so thick it is difficult to get the developer throughout the thickness to develop all exposed crystals fully. Time is required along with swollen gelatin. Develop for one hour at 20 °C (70 °F), this will vary with emulsion type and thickness. The plate should be gently agitated every five minutes. Place the plate in a 1 percent acetic acid stop bath solution for 10 minutes. If silver deposits on the surface, it may be scraped or rubbed off, but be careful not to gouge the gelatin. Fix the plate for 2 hours in regular fixer solution diluted with twice to three times normal water (50-30 percent dilution). Hypo eliminator may be used. Wash for 4-6 hours. Agitate gently in each step. A wetting agent may be applied. Drying should be very gradual, using a gentle air flow and gradual, even humidity reduction. Usually this is not a problem, since the humidity is generally high in the darkroom during processing times and tapers off as the moisture is removed. However, although special control of the humidity may not be required, make sure that moisture is not removed rapidly through the effects of a dehumidifier or air conditioner, or leaving the darkroom door open to a very low humidity room.

For the thicker emulsions of 200-600 μm the real problem is slow, uniform development of the very thick gelatin layer. This is further complicated because in nuclear tracking there are generally several layers sandwiched together and a nuclear emission might track through one layer into another. After processing it is important to be able to track or map the track of a nuclear particle. If the emulsion is distorted during processing, the information is distorted or destroyed. There are two accepted techniques to prevent distortion: the two-solution method proposed by Blau and DeFelice and the temperature cycle method first suggested by Dilworth, Ochialini, and Payne. This latter method is the most common today and involves these steps:

1. Mount the gelatin layer (often called pellicles) on glass plates and air dry at room temperature.
2. Soak the dried plate in distilled water starting at 20 °C and gradually lower to 5 °C over 2½ hour soak period.
3. Replace the water with Amidol developer at 5 °C and allow a 2½ hour soak.
4. Remove plate, blot surface gently and "develop" at room temperature for one hour. Temperature elevation is about 1 °C per minute. The developer inside the gelatin is sufficient to complete development. The top temperature may be as high as 30 °C, which will reduce the overall time.
5. Development is ceased by reducing the temperature to 5 °C and by adding a stop bath solution of 0.5 percent acetic acid for 1-2 hours in accordance with the gelatin thickness, 600 μm taking at least two hours. "Time varies as the square of the emulsion thickness" according to L.F.A. Mason.
6. Place plate in 30 percent sodium thiosulfate solution (preferred over ammonium thiosulfate) at 5°C for 2-4 days with regular agitation. The fixer must be replenished continually to keep the silver concentration at about a 5 g/L level. Clearing may be finished in 2 days, but total fixing should continue for at least 50 percent longer.
7. Washing is accomplished by slowly diluting (3 percent of total tray volume per hour) the fixer with water at 5 °C. This will require 15-20 hours and is followed by 30 hours of regular washing at 5 °C.
8. Excess water is removed by soaking in various ratios of glycerine/alcohol/water baths ending with a bath containing 5 percent glycerine, 70 percent alcohol and 25 percent water.
9. Air dry at room temperature (20-22 °C) with 55-65 percent RH for 15 hours.

In summation, the purpose of this one-week process is to insure complete processing and to prevent distortion. An alternative to the temperature cycle method is the "constant low temperature" method, where the plate and processing are all at 0 °C and development is a very slow process requiring four times longer or at least 10 hours for thinner (100 μm) emulsions.

Developers for nuclear emulsions are based on Amidol, because its pH is very close to that of gelatin and so promotes entry. Mason lists these two:

TABLE 16. Developers for Nuclear Emulsions

	Bristol	Brussels
Amidol		
(2, 4-diaminophenoldihydrochloride)	3.25 g	4.50 g
Sodium sulfite, Amhy.	7.20 g	17.50 g
Sodium metabisulphite	0.90 g	—
Potassium bromide	0.87 g	0.80 g
Boric acid	—	35 g
Add distilled water to make	1 L	1 L
pH of developer	6.6	6.4

Ultra High Resolution Imaging[2]

These glass plates are coated with a 6 μm emulsion layer sensitive to green light or blue or green light at 525-550 nm of wavelength. These plates offer resolution of 2000 lines per mm, very high contrast, and very fine grain (RMS Granularity L6). The type 2 film is half the photographic speed of the regular HRP. Plates are offered in standard sizes of 2 × 2 to 8 × 10. However, as noted above, the image area should avoid the 1 cm (0.375 in.) on outer edges where there will be some distortion.

Processing High Resolution Glass Plates[2,3]

These plates may be processed in three ways: Standard, negative format; Ultra clean, negative format; Standard, reversal format. In all cases processing is conducted in trays in a very clean environment. It is very important to wipe away all dust and to keep such debris from becoming a part of the image.

Standard, Negative Processing

1. Developer should be diluted 1:2 or 1:4. Below are the developer time/temperature relationships. For other than normal temperatures keep **all solutions within ± 3 °C (5 °F) of the developer temperature.**

TABLE 17. Developer Time/Temperature Relationships for High Resolution Glass Plates

Time	Temperature		Developer
2 minutes	29.5 °C	85 °F	HRP 1:4 or D-19
3 minutes	25.0	77	
4 minutes	23	73	
5 minutes	20.0	68	(normal)
1½ minutes	29.5	85	HRP 1:2
2 minutes	26	79	
3 minutes	23	72	
3¾ minutes	20.0	68	(normal)

In the various time/temperature relationships above, for a given temperature such as 20 °C (68 °F), increasing the time causes photographic speed shifts without loss of contrast except in the underdeveloped state.

2. Agitate continuously during development
3. Rinse: 30 seconds at about 20 °C (68 °F)
4. Fix: 1 minute at about 20 °C (68 °F) with frequent agitation; avoid prolonged fixing.
5. Wash: 5-10 minutes at about 20 °C (68 °F)
6. Dry in dust-free atmosphere

Ultra-Clean, Negative Processing. To insure extreme cleanliness this procedure calls for ultrasonic cleaning in various solvents prior to exposure. Sprays are used under pressure during processing. All of this requires special equipment, procedures, and additional hazards that must be carefully considered. All solutions are continuously filtered to 0.5 μm and/or discarded after each use. Surface debris must be skimmed off as it appears. Instead of skimming manually the trays may be continuously flooded to float out debris. Distilled and/or deionized water must be used.

Spraying requires less than 70 kPa (10 psi) at 50 mm (2 in.) from the plate. "Spray" means a flat flood rather than atomized droplets.

All of the following steps are conducted at 20 °C (68 °F).

1. Presoak plate in water for 1 min.
2. Spray rinse with water for 15 sec.
3. Develop for 4.5 min.
4. Spray rinse with developer for 15 seconds
5. Rinse in water for 30 seconds

6. Spray rinse with water for 15 seconds
7. Fix for 45 seconds
8. Spray rinse the fixer for 15 seconds
9. Wash in water for 4 minutes
10. Spray rinse with water for 15 seconds
11. (If the plate has backing material that is to be removed, proceed to the next step. If there is no backing material or if it is to be left on, then treat the next step as an extension of the wash times above). All temperatures are 20 °C (68 °F).
12. Remove backing in 0.05N (normal) sodium hydroxide for 30 seconds
13. Spray rinse with water for 15 seconds at 8 in.
14. Wash in water for 4 minutes at 8 in.
15. Spray rinse with water for 15 seconds at 8 in.
16. Soak in alcohol drying bath, 50 percent solution for 15 seconds
17. Soak in alcohol drying baths, 75 percent solution for 15 seconds
18. Soak in alcohol drying bath, 100 percent solution for 15 seconds
19. Dry in clean atmosphere for 10 minutes

Total time is about 28 minutes from dry to dry.

The alcohol drying bath acts to dehydrate the gelatin and reduce tackiness, but the process must be gradual to prevent physical shock and distortion; thus the concentrations increase. These baths are made of ethanol or methanol and water.

Reversal Plate Processing. Instead of working with a negative it is possible to reverse the image into the positive phase chemically. A positive image can be used to make several duplicates where copies are required. The plate should be overexposed by 2½ times normal exposure. All solution temperatures are 20 °C (68 °F).

1. Develop (first Developer), agitate 3 minutes
2. Wash in running water 2 minutes
3. Bleach, agitate 2 minutes
4. Rinse in running water 30 seconds
5. Clearing bath, agitate 3 minutes
6. Wash in running water 4 minutes
   Re-expose during this cycle using an enlarger lamp or roomlights to produce 30-40 foot candles, for 30 seconds.
7. Develop (second Developer), 1:4, 5 minutes
8. Fix 2 minutes
9. Dry the film in a clean atmosphere at 43 °C (110 °F).

Instant Processing

Instant images are nontransparency images on paper, the same as the typical amateur product. In radiography, of course, the exposure is made with direct X-rays or indirectly with X-rays exciting salt-intensifying screens. The film packet contains a receptor film (negative), a pod of chemicals, and a print film (positive). After exposure, the film is pulled out of its holder through rollers which open the pod and spread the chemicals. After about 10 seconds, the two films are separated. The final film is viewed with reflective light. This system, technically referred to as "diffusion transfer reversal process," in effect employs a monobath processing system. This is also called an "instant" system because it will produce a dry image faster than any other system short of television.

The major advantages are for rapid reading, to aid in setting up parts for conventional radiography, and technique development. This system also eliminates conventional processing. The major disadvantages are the nontransparency type image, reduced image quality, and increased expense. The image is in the positive mode rather than negative, so that reading and interpretation are more difficult. In this regard, the radiographer must remember that if the positive is too light it is the result of the negative being too dark from overexposing; a too dark positive image results from underexposure (to the negative). Exposures, thus, are adjusted accordingly.

For archival quality some prints require a coating with a colorless "fixer" that comes in a special applicator. The print may be used as is, but tends to curl; the remedy is to mount it. A negative may be made from the positive print as desired, but special equipment is required. The original negative must be discarded.

In the fall of 1981, a new system was introduced by Polaroid. It has two components: a light-sensitive negative in a light-tight envelope and a transparent, blue-tinted polyester image sheet. The cassette contains a rare earth intensifying salt screen, which allows for very short exposures or a general reduction in energy. The negative part retains a conventional X-ray image, the polyester film presents a positive image, although it can be placed on a conventional viewbox. This sheet must be coated with a typical print coater.

The entire system is portable and the processor, although electric, can be manually operated.

REFERENCES

1. Becker, G.L., E.I. Du Pont de Nemours & Co., Inc. "Assuring NDT Radiographic Processing." 38th National ASNT Convention (October 1978).

2. Garfield, J.F. "Apparatus and a Laboratory for Processing Thick Nuclear Track Emulsions." *Photographic Science and Engineering.* Vol. 2 (August 1958).

3. "Kodak High Resolution Plate," and "Kodak High Resolution Plate, Type 2" No. P-226. Rochester, NY: Eastman Kodak Company.

Reference Texts

Very few texts or research papers deal specifically with industrial radiographic processing, since radiography is a specialized area of photographic science. All of the references listed contain sections useful to an increased understanding of this specialty. Recommended primary sources are indicated by an asterisk(*).

Basic Chemistry of Photographic Processing, Parts 1 and 2. Self-teaching texts. Rochester, NY: Eastman Kodak.

Bruce, H.F. *Your Guide to Photography.* New York: Barnes and Noble.

Control Techniques in Film Processing. Society of Motion Picture and Television Engineers (SMPTE), 19606 E. 41st St., New York, NY 10017.

Darkroom Technique Guide. Wilmington, DE: E.I. Du Pont de Nemours & Co.

Eaton, G.T. *Photographic Chemistry.* New York: Morgan and Morgan (1957).

Gray, J.E. *Photographic Processing, Quality Control and the Evaluation of Photographic Materials,* Vol. 2. Rockville MD: Bureau of Radiologic Health (1977). HEW Publication 77-8018.

Gray, R.H., ed. *Applied Processing Practice and Techniques.* Washington, D.C.: Society of Photographic Science and Engineering (1968).

Gregg, D.C. *Principles of Chemistry.* Boston: Allyn and Bacon (1963).

Grob, B. *Introduction to Electronics I.* New York: McGraw-Hill (1959).

Haist, G. *Modern Photographic Processing,* Vols. 1 and 2. New York: Wiley Interscience (1978).

Handbook of Photography. Society of Photographic Scientists and Engineers, 1330 Massachusetts Ave., N.W., Suite 204, Washington, D.C. 20005.

Haus, A., ed. *The Physics of Medical Imaging* (1979). American Institute of Physics, 335 E. 45th St., New York 10017.

Herz, R.H. *The Photographic Action of Ionizing Radiation.* New York: Wiley Interscience (1969).

Hillson, P. *Photography.* New York: Doubleday Science Series.

James, T.H., and G.C. Higgins. *Fundamentals of Photographic Theory.* New York: Morgan and Morgan (1960).

John, D.H.O. *Radiographic Processing.* London: Focal (1967).

* Kirby, C.C., W.E.J. McKinney, and T.T. Thompson. *Radiographic Processing and Film Quality Control.* Chicago: American Society of Radiologic Technologists (1975). Designed to be used with tapes in following entry, but may be used separately.

* McKinney, W.E.J. *Radiographic Processing and Film Quality Control.* (16 half-hour video tapes) Produced by the U.S. Veterans Administration. Available from VA information centers or American Society of Radiologic Technologists, Chicago, IL.

Marcus, A. *Basic Electricity.* Englewood Cliffs, NJ: Prentice-Hall (1958).

* Mason, L.F.A. *Photographic Processing Chemistry.* New York: Focal Library (1975).

Mees, C.E.K. *The Theory of the Photographic Process.* New York: Macmillan (1954).

* Mees, C.E.K., and T.H. James. *The Theory of the Photographic Process*, 3d ed. New York: Macmillan (1971).

Mitchell, J.W., ed. *Fundamental Mechanisms of Photographic Sensitivity.* London: Butterworths Science Publications Ltd. (1951).

Pauling, L. *College Chemistry.* San Francisco: Freeman (1964).

Recovering Silver from Photographic Materials. Rochester, NY: Eastman Kodak, J10.

Sensitometric Properties of X-Ray Film. Rochester, NY: Eastman Kodak (1963).

Silver Recovery. Wilmington, DE: E.I. Du Pont de Nemours & Co., Inc.

Stiles, E. *Handbook for Total Quality Assurance*, Vol. 26. Waterford, CT: National Foreman's Institute, The Complete Management Library.

Todd, H.N. *Photographic Sensitometry:* A Self-Teaching Text. New York: Wiley Interscience (1977).

* Todd, H.N., and R.D. Zakia. *Photographic Sensitometry.* New York: Morgan and Morgan (1969).

Reference Journals and Proceedings

Journal of the American Water Works Association
Journal of Applied Photographic Engineering (SPSE)
Journal of Applied Physics
Journal of Colloid Science
Journal of Optical Society of America
Journal of Photographic Engineering
Journal of Photographic Science
Journal of Photographic Society of America
Journal of Physical Chemistry
Journal of Society of Motion Picture and Television Engineers (SMPTE)
Journal of the Society of Photographic Instrumentation Engineers (SPIE): Newsletter
Journal of the Society of Photographic Scientists and Engineers (SPSE)

Technical Magazines

American Photographer
Industrial Photography
Materials Evaluation
Photo Methods for Industry
Photographic Processing
Photographic Science and Engineering
Photographic Science and Technology
Photographic Scientific Engineering
Physics Today
Quality Control
Radiography
Scientific Industrial Photography
Technical Photography

SECTION 8

RADIOGRAPHIC INTERPRETATION

Charles Hellier, Brand Examination Services and Testing Company, Essex, CT
Sam Wenk, Southwest Research Institute, San Antonio, TX

ACKNOWLEDGMENTS

The authors and editorial staff gratefully acknowledge the contributions to this section by R.R. Hardison of Newport News Shipbuilding & Drydock Co., and his associates, L.S. Morris, D.L. Isenhour and R.D. Wallace; and to Gary Yonemura of the National Bureau of Standards. Appreciation is also expressed to Eastman Kodak Co., Electric Power Research Institute and the Southwest Research Institute for permission to use illustrative material.

Special recognition is due Carlton H. Hastings, NDT pioneer, past president and honorary member of ASNT, who is quoted in Part 2 of this Section.

INTRODUCTION

Radiographic interpretation is the art of extracting the maximum information from a radiographic image. This requires subjective judgment by the interpreter and is influenced by the interpreter's knowledge of:

1. the characteristics of the radiation source and its energy level(s) with respect to the material begin examined;
2. the characteristics of the recording media in response to the selected radiation source and its energy level(s);
3. the processing of the recording media with respect to resultant image quality;
4. the product form (object) being radiographed;
5. the possible and most probable types of discontinuities that may occur in the test object; and,
6. the possible variations of the discontinuities' images as a function of radiographic geometry, and other factors.

Visual acuity is essential to the interpreter, but this too is strongly influenced by the individual's knowledge and experience, as well as viewing conditions. Because radiographic interpreters have varying levels of knowledge and experience, training becomes an important factor for improving the agreement level between interpreters.

In a program conducted by a research laboratory,[1,2] a comparison was made among five certified film interpreters who were trained by a master-apprentice program. These five certified film interpreters reviewed 350 radiographs and reached agreement on 238 radiographs, or disagreed 32 percent of the time.

The results of this research were then incorporated into a unified training program, using discontinuity categories from the welding process. Subsequently, a procedure was developed wherein nine certified film interpreters trained under the unified training program were compared to nine certified film interpreters trained under the master-apprentice program. Using 96 radiographs, the master-apprentice group disagreed 44 percent of the time; the unified training group disagreed only 17 percent of the time.

In a similar study of medical radiology,[3] the reproducibility of a tuberculosis diagnosis was examined. This study revealed an average disagreement in one out of three cases, or 67 percent agreement. On a second independent reading of the same radiographs, a physician would disagree with his previous diagnosis in an average of one out of five cases, or 80 percent agreement.

It is evident that even under the best possible circumstances of training and experience, qualified film interpreters are not likely to reach agreement more than 90 to 95 percent of the time. Therefore, in all applications where quality of the final product is important for safety and/or reliability, a minimum of two qualified interpreters should evaluate, and pass judgment on, the radiographs.

Reference radiographs are a valuable training and interpretation aid. An in-house library of radiographs and accompanying photographs of discontinuity macrosections are also recommended.

PART 1
FUNDAMENTALS

The five essential features of radiographic nondestructive tests are:

1. supplying a suitable form and distribution of energy from an external source to the test object (X-ray or gamma ray source);
2. modification of the energy distribution within the test object as a result of its discontinuities or variations in the material properties which correlate with serviceability (variations in X-ray attenuation and/or changes in thickness);
3. detection of the change in energy by a sensitive detector (X-ray film, fluorescent screens, etc.);
4. recording of the energy change from the detector in a form useful for interpretation (X-ray film, fluorescent screens, etc.); and
5. interpretation of the radiographic image for compliance with applicable codes and acceptance standards (radiographic interpreter).

Specifying Nondestructive Tests[4,5]

Nondestructive tests must be designed and specified for validity and reliability in each individual application. The tests are specific to the problem; there is no such thing as a general nondestructive test applicable to every kind of material, part or structure, nor to all their functions or operating conditions. Instead, each nondestructive test must be based upon a thorough understanding of the nature and function of the part being tested and of the conditions of its service.

These fundamentals translate back to the basic experience and knowledge qualities that a radiographic interpreter should possess. Specific radiographic procedures must be prepared and adhered to in both the production and interpretation of the resultant radiographic image. These procedures are based on applicable specifications, codes and standards and the interpreter must be thoroughly familiar with their requirements in order to properly assess the image and product quality.

The first step of the film review (also applicable to paper radiographs and real-time images) requires the assessment of the radiographic quality. This includes determining if the radiograph (1) has all the required identification information; (2) is free from artifacts which could mask discontinuities; (3) has the correct penetrameter and quality level; and (4) has the correct station markers. If the radiograph is unsatisfactory, it should be rejected.

The next step is to assess the product quality in the radiographic area of interest. Here is where individual visual acuity and experience, as well as an understanding of the radiographic process, become the dominating factors.

Because the eye is sensitive to moving objects, moving the radiograph back and forth often helps to visualize small details; angulating the film or the angle of view diminishes the effect of low contrast. Keeping the viewing area relatively small provides for better observation of fine detail. A good magnifying glass is also an aid in making judgments of indications. On the other hand, large radiographs for

FIGURE 1. Schematic Diagram of the Effective (or Projected) Focal Spot of an X-ray Tube

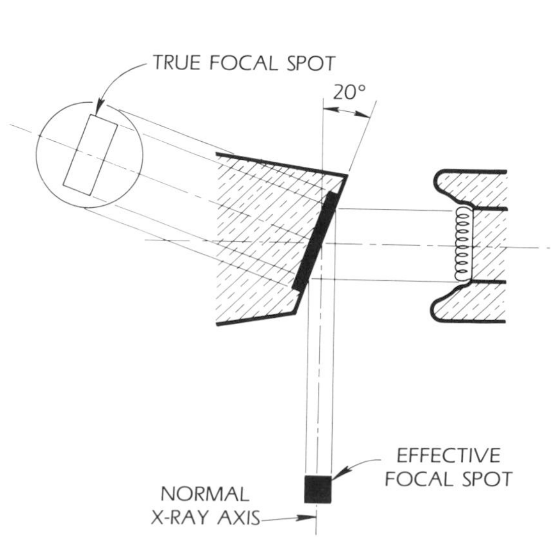

TRUE FOCAL SPOT
20°
EFFECTIVE FOCAL SPOT
NORMAL X-RAY AXIS

assembly conditions require large area viewing. A visual examination of the test object should be conducted if there appears to be a possible surface indication. It may be necessary to re-radiograph for verification or to obtain better resolution, perhaps with a change in geometry if the discontinuity is unfavorably oriented or is a significant distance from the center of the film (see Figs. 1 and 2).[6]

FIGURE 2. Diagram Showing Change of Shape and Size of Projected X-ray Focal Spot as a Function of Position in X-ray Field

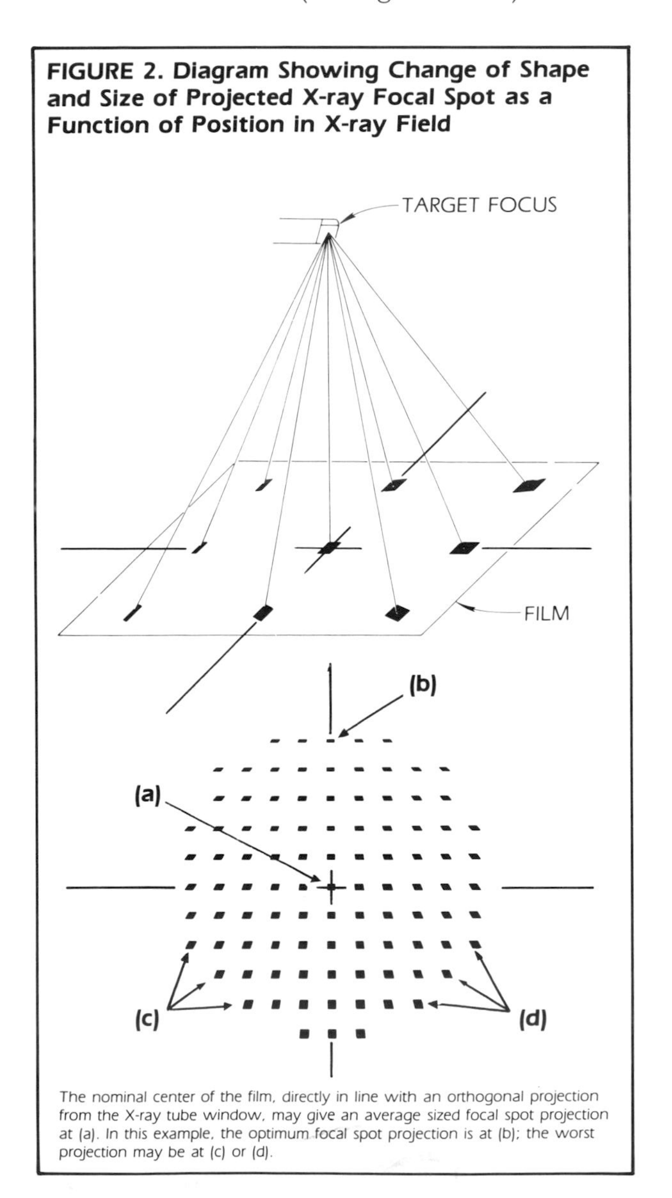

The nominal center of the film, directly in line with an orthogonal projection from the X-ray tube window, may give an average sized focal spot projection at (a). In this example, the optimum focal spot projection is at (b); the worst projection may be at (c) or (d).

PART 2

STANDARDS, CODES AND SPECIFICATIONS

All radiography (except research and development) should be performed in accordance with written procedures developed from applicable standards, codes or specifications, as required by contractual agreement. This means that the radiographic interpreter must have both a working knowledge of and ready access to pertinent documents to verify the required quality level of (1) the radiography and (2) the product.

However, radiographic personnel should understand that specified quality levels may vary depending on the specifications in effect, and that all radiographic quality levels are minimum requirements which can be exceeded.

Product quality levels should be based on the service (use) of the component being examined, even though this is not always addressed in the governing code or specification. Ideally, the product quality level should be established by appropriate personnel in conjunction with the radiographic specialist, thus providing the maximum degree of inspectability and ensuring that the most critical discontinuities will be detected.

All parties involved in the design, fabrication and examination of a product must remember that the only known flaw-free, homogeneous material is a single crystal. Carlton H. Hastings succinctly defined *material* as a "collection of defects, with acceptable material being a [fortunate] arrangement of defects and rejectable material being an unfortunate arrangement of defects."[7] The message is clear: regardless of the radiographic technique used, there can never be the assurance of a component totally free of discontinuities. Hence, a thorough understanding of radiography's limitations is essential for choosing the optimum techniques to achieve the best possible radiographic quality level.

PART 3

VISUAL ACUITY

Visual acuity is vital to the first step in the three-step interpretation process: (1) detection; (2) interpretation; and (3) evaluation. Individual visual acuity can and does vary from day to day depending on physiological and psychological factors. An annual visual acuity examination cannot test this daily fluctuation's influence on interpretation. Recognizing this encourages the use of a daily visual acuity test. Standardizing the sensory capacity of the human eye should be based on the circumstances of the viewing; we must know what the eye is expected to see. In radiographic inspections, the physical measure of interest for visual acuity testing is the discontinuity as displayed on the film, regardless of how much it may differ from the actual discontinuity in the test object. The visual acuity test described below[8] was developed using microdensitometric scans of discontinuities taken directly from actual radiographs.

Samples of the acuity test slides are presented in Fig. 3. The optotype (acuity test target) is a thin line darker than the background. Line orientation serves two important functions: (1) it increases the number of possible responses (two with no orientation, four with orientation); and, (2) it includes astigmatic effects. The four orientations are horizontal *(H)*, vertical *(V)*, oblique-right *(R)*, and oblique-left *(L)*.

The recommended parameters for this test are summarized in Table 1. The background luminance (brightness) of the test chart is kept constant at 85 ± 5 cd/m² or 25 ± 1.5 fL (see footnote to Table 1 for definitions of these units). Three contrast levels and line widths are recommended. The length of the lines is kept constant at 107 minutes of arc. These angular measures are based on a viewing distance of 40 cm (16 in.). Two levels of line sharpness are included; one with a sharp edge similar to most optotypes used in visual acuity testing, and one with a blurred edge conforming to many actual radiographs.

These slides are designed for self-testing as well as testing by designated examiners. The front of the slides contain only the optotypes (see Fig. 3) with all necessary information given on the reverse side. This procedure assumes that for self-testing the examinee will look at the side giving the correct response *(H, V, R* or *L)* only after evaluating the target orientation.

TABLE 1. Visual Acuity Test Parameters

Variable	Number of Conditions	Conditions
Figure-ground	1	Dark on light
Background luminance* [cd/m² (ftL)]	1	85 ± 5 (25 ± 1.5)
Contrast	3	0.1, 0.3, 0.85
Line width (minutes)	3	0.75, 1.0, 1.5
Line length (minutes)	1	107
Viewing distance [cm (in.)]	1	40 (16)
Blur	2	Sharp, blurred
Line orientation	4	Perpendicular/horizontal oblique right/left
Light source (viewer)	1	Incandescent, fluorescent
Total (combinations)	72	

*The international unit for luminance is candela per square meter (cd/m²) and is sometimes called the *nit*. The English unit for luminance is the footLambert (fL) and is equal to 3.426 cd/m².

FIGURE 3. Examples of Reference Acuity Tests Showing Line Orientations and Dimensions

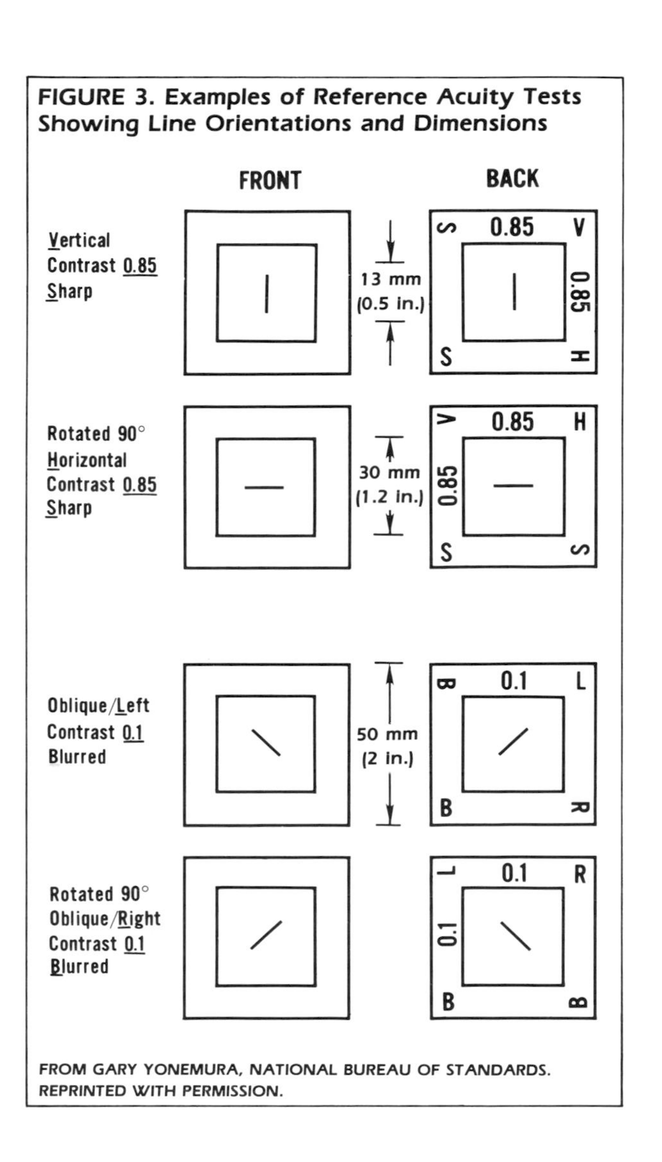

FROM GARY YONEMURA, NATIONAL BUREAU OF STANDARDS. REPRINTED WITH PERMISSION.

PART 4

RADIOGRAPHIC VIEWING CONDITIONS AND EQUIPMENT

Viewing conditions are of utmost importance. The examination of the finished radiograph should be done under conditions which afford maximum visibility of detail together with a maximum of comfort and a minimum of fatigue for the interpreter. Subdued lighting in the viewing area is preferable to total darkness. The room lighting must be arranged so that there are no reflections from the surface of the film (or paper) being interpreted. Adequate table surface must be provided on either side of the viewer to accommodate film and to provide a writing surface for recording the interpretation. Quick and easy access to a densitometer, reference radiographs, applicable codes, standards and specifications should also be provided. In addition, it is important for the film interpreter to be free of distractions, including telephone and receptionist duties, in order to maintain concentration. These conditions apply to both film and paper radiography.

For real-time radiography, the same general conditions apply, but can vary depending on the specific system being used. Direct viewing generally requires dark adaptation; use of a remote viewing system with a video presentation allows individual control of brightness and contrast for maximum visual acuity.

If the interpretation of the radiographic image is to be meaningful, it is essential that proper viewing equipment, in good working condition, be used. If slight density variations in the radiographs are not observed, rejectable conditions may go unnoticed. In many cases, various types of discontinuities are barely distinguishable even with the use of optimized techniques and fine-grained film. In order to optimize the interpreter's ability to properly evaluate the radiographic image, ideal viewing conditions and suitable equipment are absolutely necessary.

High Intensity Illuminators

A radiograph that meets the density requirements of current codes and specifications will permit only a small fraction of the incident light to pass through it. The density of a radiographic film is logarithmic, expressed in equation form as:

$$Density = \log\left(\frac{I_o}{I_t}\right) \qquad \text{(Eq. 1)}$$

where *Density* is the degree of blackness resulting from radiographic exposure; I_o is the incident light intensity (from the high intensity illuminator or densitometer); and I_t is the light transmitted through a specific region of the radiograph.

If a film is perfectly clear, the density will be 0.

$$Density = \log\left(\frac{100}{100}\right) = \log(1) = 0$$

A film that permits 1% of the incident light to be transmitted will have a density of 2.0.

Following the same procedure, it can be seen that a film density of 3.0 permits only .1% of the incident light to pass through and a density of 4.0, a mere 0.01%.

Typically, radiographic density requirements through the area of interest range between 2.0 (1% light transmission) and 4.0 (.01% light transmission); this explains the need for a source of high intensity viewing light.

There are many types and styles of high intensity illuminators, although they are generally classified into four groups: (1) spot viewers; (2) strip film viewers; (3) area viewers; and (4) combination spot and area viewers.

Spot viewers provide a limited field of illumination, typically 7.5 to 10 cm (3 to 4 in.) in diameter. These viewers are usually the most portable and least expensive.

The strip film viewer (Fig. 4) permits interpretation of strip film including 9 × 43, 13 × 43, 10 × 25 and 13 × 18 cm (3.5 × 17, 4.5 × 17, 4 × 10 and 5 × 7 in.) and the 35 mm or 70 mm sizes. The viewing area is rectangular and the area of illumination may be adjusted to conform to the film dimension by employing metal or cardboard masks.

The area viewers are designed to accomodate large films up to 35 × 43 cm (14 × 17 in.) The illumination is generally provided by fluorescent lights or a bank of photo-flood bulbs. The fluorescent light intensity may not have suitable brightness to permit effective examination through the higher densities and this could result in a serious limitation. The combination spot and area viewers (Fig. 5) provide the interpreter with spot capability while allowing the viewing of a large area of film. A switch determines which light source will be activated.

FIGURE 4. High Intensity Illuminator Designed for Viewing Strip Film

FIGURE 5. High Intensity Combination Illuminator with Iris Diaphragm Spot Viewer and Large Viewer

X-RAY PRODUCTS. REPRINTED WITH PERMISSION.

Heat

Since light of high intensity also generates significant amounts of heat, it is necessary that the illuminator have a means of dissipating or diverting the heat to avoid damaging the radiographic film while viewing. Light sources in typical illuminators consist of one or more photo-flood bulbs. Other light sources such as flood lights and tungsten halogen bulbs are also used.

Diffusion

To eliminate variation in the intensity of the light, it is also important that the light be diffused over the area used for viewing. This is accomplished with a diffusing glass, usually positioned between the light source and the viewing area, or with a white plastic screen at the front of the viewer.

Intensity Control

Another essential feature of the illuminator is the variable intensity control. This permits subdued intensity when viewing lower densities, and maximum intensity as required for the high density portions of the radiograph.

Masks

Masks can be extremely helpful when attempting to evaluate a small portion of a larger radiograph, or when the radiograph is physically small. The objective is to illuminate that portion of the radiograph identified as the area of interest, while masking other light from the eyes of the interpreter. Some spot viewers are equipped with an iris diaphragm which permits the spot size to be varied with the simple adjustment of a lever. This feature is especially helpful when small areas or fine details must be examined.

Precautions

The illuminator's front glass or screen touches the film and should always be clean and free of blemishes on both sides. Scratches, nicks, dirt, or other imperfections on the front glass or screen will cast shadows on the radiograph, causing unnecessary images.

Another precaution will help minimize film scratches. The front of the viewer should be carefully examined to ensure that there are no sharp edges or other obstructions; these could cause scratches to the sensitive surface of the radiograph as it is moved or positioned on the viewer.

Magnifiers

Normally, radiographs can be effectively evaluated without the aid of magnification devices. There may be occasions, however, when such devices are helpful. For example, if the article being radiographed contains very small variations or consists of minute components, magnification may be essential. This application will generally require the use of fine-grained film which can be suitably magnified. Some of the coarser grained films are difficult to view with magnification because the graininess is also enlarged; this can make discernment of slight density changes impossible.

There is a wide assortment of magnifiers appropriate for the evaluation of radiographs. The most common is the handheld magnifying glass, available in many shapes, sizes and powers. For convenience, a gooseneck magnifier may be employed. Because this magnifier is self-standing and attached to a weighted metal base, it leaves the interpreter's hands free during use. One device that offers magnification *and* measuring capabilities is a comparator with an etched glass reticle (Fig. 6).

If any form of magnification is employed, it should be done with caution and limited to only those applications where it is necessary.

Other Viewing Accessories

Additional accessories that aid the interpreter and should be available in the film reading area, include (but are not limited to):

1. supply of wax pencils for marking the film;
2. rulers (the most appropriate would be clear, flexible plastic);
3. a small flashlight to reflect light off the radiographic film to assist in the identification of surface artifacts such as scratches, roller marks, dirt, etc.;
4. gloves, usually cotton or nylon, to minimize direct contact between the film and the fingers of the interpreter;
5. charts, tables and other technical aids that will assist in the prompt establishment of density range (see example in Table 2), determination of geometric unsharpness, and other data relating to the applicable codes or specifications.

Viewing Paper Radiographs

Typically, paper radiographs are reviewed under normal lighting conditions utilizing reflective white light. It is essential that the light be of suitable intensity and, in some cases, positioned at an angle to prevent glare. Various lighting sources have been successful, including the high intensity reading lamps and specular lights (light focused from a reflector). Magnifiers containing fluorescent bulbs also provide an effective means of evaluating the paper radiograph, while magnifying the image. High magnification (above 5×) is not usually beneficial due to the normal graininess and lack of sharpness inherent in the paper radiograph. See Section 5, *Film and Paper Radiography* for more information.

FIGURE 6. Comparator With Etched Glass Reticle

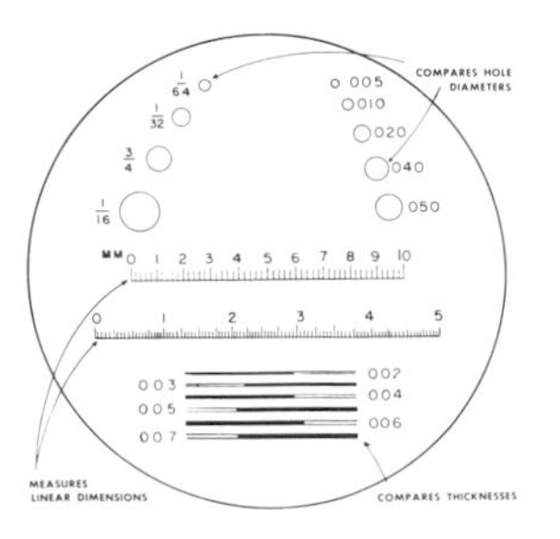

EDMUND SCIENTIFIC PRODUCTS. REPRINTED WITH PERMISSION.

TABLE 2. Typical Density Range Table: Based on +30% and −15% Densities

Density Through Penetrameter	Maximum (+30%)	Minimum (−15%)
1.5	1.95	1.28
1.6	2.08	1.36
1.7	2.21	1.45
1.8	2.34	1.53
1.9	2.47	1.62
2.0	2.60	1.70
2.1	2.73	1.79
2.2	2.86	1.87
2.3	2.99	1.96
2.4	3.12	2.04
2.5	3.25	2.13
2.6	3.38	2.21
2.7	3.51	2.30
2.8	3.64	2.38
2.9	3.77	2.47
3.0	3.90	2.55
3.1	4.03	2.64
3.2	4.16	2.72
3.3	4.29	2.81
3.4	4.42	2.89
3.5	4.55	2.98
3.6	4.68	3.06
3.7	4.81	3.15
3.8	4.94	3.23
3.9	5.07	3.32
4.0	5.20	3.40

Densities less than 2.0 and more than 4.0 are considered unacceptable by some codes or specifications.

PART 5

DENSITOMETERS

The densitometer is an instrument that measures film density (see Figs. 7 and 8). Before the invention of portable densitometers, densities were estimated by comparing the radiographic density to a comparator strip. The strip contained a series of densities established by cumbersome and unwieldy early densitometers. Many of these early radiographic density determinations were simple, visual estimates.

The operation of present-day densitometers is quite simple. After calibration, using a density strip with known values for a number of different densities, the radiograph is positioned between the light source, usually located at the base of the densitometer, and the head, which contains a photomultiplier. If the transmitted light intensity decreases as a result of increased radiographic film density, less light reaches the photosensitive surface in the head and the voltage output from the photomultiplier (to the meter or digital display) will indicate a higher density reading. Conversely, as more light passes through a lower density region of the radiographic film and interacts with the photosensitive surface in the head, a lower density is indicated on the meter or digital display.

An aperture is installed near the light source to establish the precise amount of light passing through the film. Changing apertures requires recalibration.

Procedure

The first step in the proper use of the densitometer is warm-up. Most instruments now contain solid-state circuitry and warm-up time is minimal. It is good practice to wait at least five minutes after the densitometer has been turned on before taking density readings. This provides ample time for electronic stabilization.

FIGURE 7. Digital Transmission Densitometer

MACBETH PROCESS MEASUREMENT. REPRINTED WITH PERMISSION.

FIGURE 8. Battery Powered Densitometer

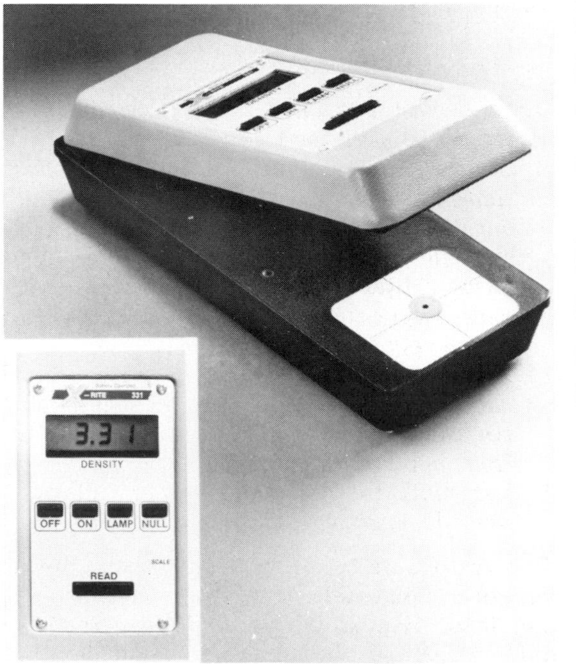

X-RITE CORPORATION. REPRINTED WITH PERMISSION.

The next step is the most important one. No matter how simple the densitometer may appear to be, it is necessary to calibrate. This is accomplished using a calibrated density strip. Because different densitometers have different controls and procedures for calibration, the specific instruction manuals should be consulted. After calibration is accomplished, a series of readings for a number of density steps should be taken using the calibrated strip. This should be repeated frequently during the use of the densitometer to detect electrical shifts or inadvertent changes to the controls.

It is good practice to record calibration readings in a daily log book. Some codes and specifications require the use of a master density strip traceable to a standards organization. The master strip can be used to calibrate other density strips that are typically used for daily calibration. As the daily calibration strips wear out and become damaged due to use, new ones can be prepared by comparison to the master strip.

After calibration, the densitometer is ready to use.

Precautions

Several precautions should be kept in mind:

1. the densitometer is a sensitive electronic instrument and must be treated with care;
2. the densitometer must be kept clean at all times; the aperture, glass portions of the head and the reflective mirror (if used) should be cleaned with care using a cotton swab moistened with alcohol;
3. *never* take density readings if the film is not completely dry;
4. when replacing the bulb, exercise extreme care; make sure the densitometer is unplugged and take time to remove smudges resulting from handling;
5. keep both the daily and master calibration strips in a protective cover or envelope.

It is reasonable to expect readings with an accuracy of ± 0.02 when the densitometer is properly maintained. Repeatability should generally fall within ± 0.01. If the readings vary from these tolerances, the equipment should be checked for possible corrective action.

Density of Paper Radiographs

Density readings of radiographic film are made using a transmission densitometer. Light cannot be effectively transmitted through paper radiographs, therefore the density must be measured with a reflection densitometer.

Reflection density can be determined using eq. 2:

$$D_R = \log \left(\frac{I_o}{I_R}\right) \qquad \text{(Eq. 2)}$$

where D_R is the reflection density; I_o is the incident light intensity; and I_R is the reflected light intensity.

While this equation is similar to the one used to determine the transmission density in radiographic film, density readings of paper radiographs are achieved by measuring reflected light. There are a number of commercially available reflection densitometers and several transmission densitometers that also have the ability to read reflected densities.

Scanning Microdensitometers

Densitometry for conventional radiographs is done with the transmission densitometer. This instrument is generally suitable for assuring compliance with radiographic technique requirements. However, it may not provide sufficient information for certain specialized radiographic analyses. The transmission densitometer is limited, in certain respects, by its relatively large aperture and by its inability to automatically scan a film or produce a permanent record. These limitations dramatically affect the accuracy of relative density determinations, especially if the area of interest on a film is small (two or three millimeters). The scanning microdensitometer (SMD), which is also called a recording microdensitometer, was designed to overcome these limitations.

The SMD automatically scans a predetermined area on a film and produces a graphic depiction of the density changes occurring in the scan path. The accuracy of the SMD is greatly enhanced by its adjustable aperture, which may be set for openings as small as three microns (thus the term *microdensitometer*). The SMD concept is based on the synchronous combination of an elaborate densitometry system and a compatible scanning/recording system.

Description of Equipment and Operation

The principle of operation for conventional SMD equipment (see Fig. 9) is based on a true double-beam light system, in which two beams, emanating from a single light source, are switched alternately to a single photomultiplier. One of the light beams is directed, through a series of prisms and mirrors, to the aperture that actually scans the film; the other light beam is directed to an aperture that sends it through a mobile, calibrated *gray wedge*.

Any differences in light intensity are automatically corrected so that both apertures transmit the same light quantity. During a scan, the film is placed on an automatically propelled carriage, which transports the film across the aperture's light beam (the aperture remains stationary). As the film traverses the light beam, continuous density readings are transmitted to a computer that feeds these readings to the gray wedge portion of the apparatus.

The mobile gray wedge (which is calibrated based on degree of density change per centimeter) will shift its position so that the density through the gray wedge matches the density of the film being scanned. The mobile gray wedge is mechanically attached to a recording pen assembly; the recording pen is in contact with graph paper that is mounted on a graph carriage moving at the same rate as the film carriage. The end result of this system is a graphic depiction of the density changes occurring in the scan path of the film (see Figs. 10 and 11). The film density change from base metal to shim is denoted by ΔD_1. The film density change from base metal to total weld thickness is denoted by ΔD_2. These values are shown in the cross-sectional drawing of Fig. 10 and the scanning microdensitometry graphs of Fig. 11.

Advantages and Limitations of Scanning Microdensitometry

The SMD was designed to overcome the scanning, recording and accuracy limitations of conventional densitometry equipment. With these limitations eliminated, a broad array of information can be derived from the SMD.

The instrument provides numerous means by which accuracy can be enhanced and/or optimized. The aperture opening may be set as small as 3 micrometers (0.00012 in.) to provide information associated with film grain dispersion and grain size. The ratio arm will allow for graph-to-scan path ratios of 1:1, 2:1, 5:1 and so on, which is very beneficial for scanning small areas. The ratio arm setting can also be used to optimize accuracy, provided other equipment adjustments are set accordingly. The scan-graph itself can be incorporated into radiographic records to demonstrate verification of dimensional tolerances or adherence to density tolerances. The SMD unit can be a very useful and cost-effective radiographic tool; however, its limitations should not be overlooked.

The major limitation of this equipment is assuring that it will produce interpretable results. Some graph peaks are signals (relevant) and others are noise (nonrelevant). The ability to distinguish between signal and noise is highly dependent on the aperture opening and its relation to the film being

FIGURE 9. Double-beam Microdensitometry Schematic

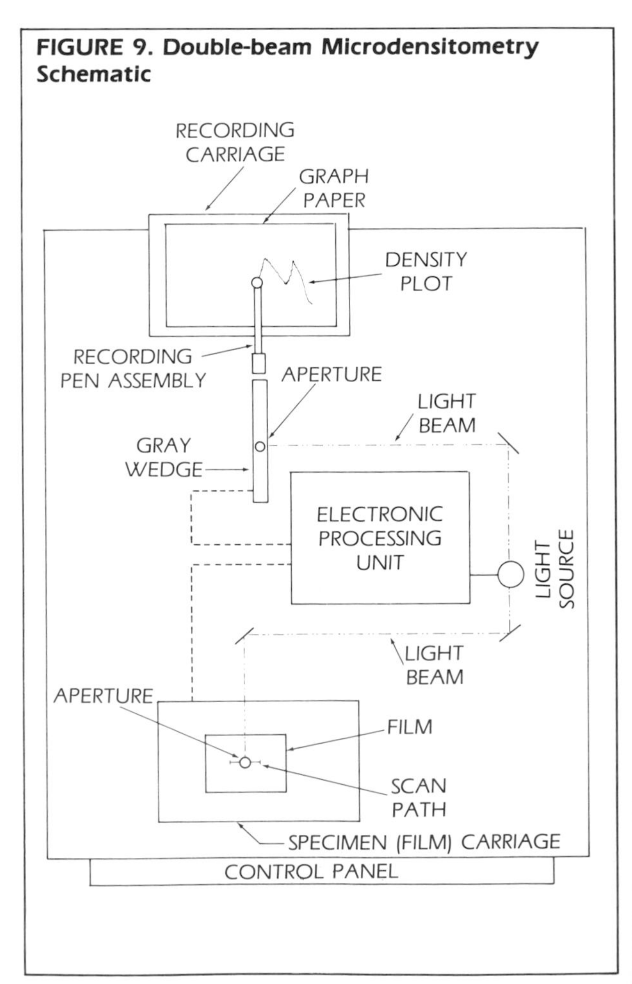

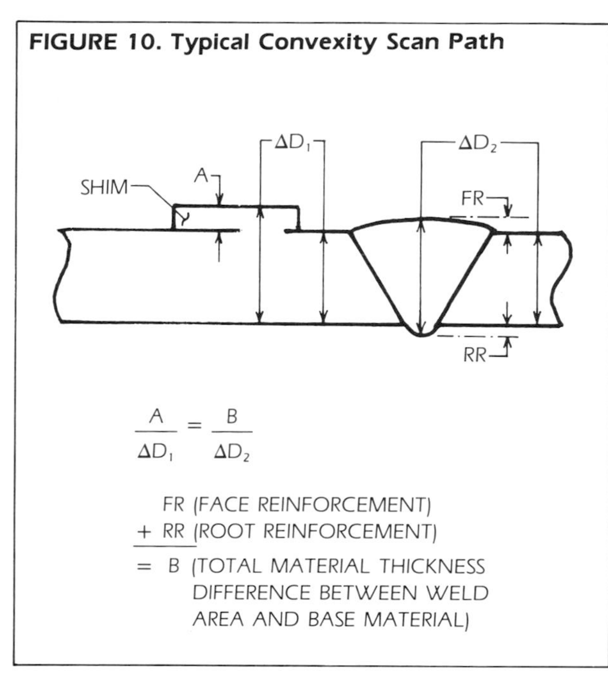

FIGURE 10. Typical Convexity Scan Path

scanned. If a very grainy, high-contrast film is scanned, a large aperature of 90 micrometers (0.0036 in.) should be used. Otherwise, the signal-to-noise ratio of the scan-graph will make interpretation difficult. Conversely, if a fine grained film is scanned, a small aperture opening is appropriate.

The SMD operator should be thoroughly familiar with the variables of the equipment so that the scanning technique can be optimized, based on the objective of the scan and the data-producing capabilities of the film.

Applications

Scanning microdensitometry equipment can be very useful for certain industrial radiography applications. Among these applications are X-ray focal spot measurements and determination of total radiographic unsharpness. A principle application of the SMD in industrial radiography is verification of dimensional tolerances of questionable piping weld root conditions.

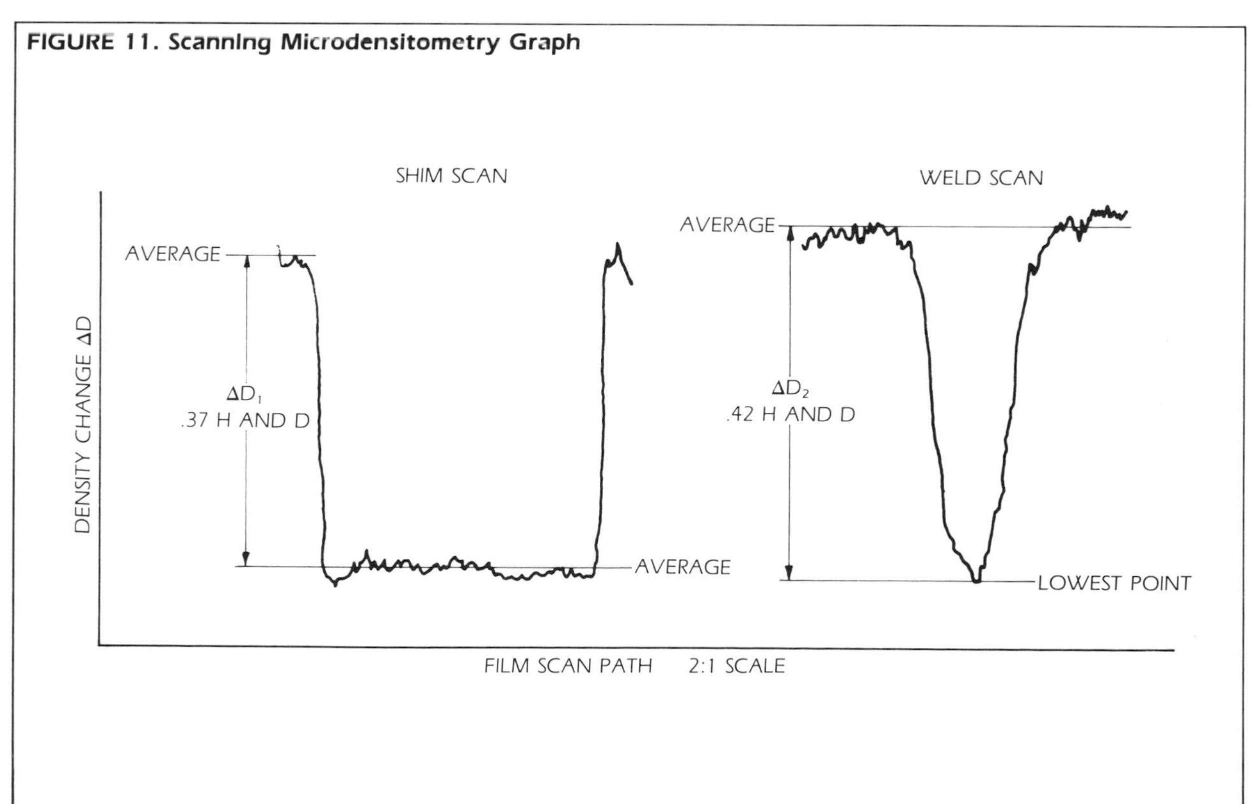

FIGURE 11. Scanning Microdensitometry Graph

The ideal method for verifying the dimensional tolerances of a given weld root condition is by performing a visual inspection and actually measuring the condition involved. Ideal situations are infrequent in pipe radiography, however, and the majority of pipe weld joints are inaccessible for visual inspection of the weld root. In a large portion of these situations, an additional radiograph, showing a profile view of the questionable condition, will provide the information necessary to support or determine the film interpreter's judgment. Additional radiography is normally an effective means for determining the accept/reject status of a radiograph; however, the additional time, cost and material required with this method frequently make scanning microdensitometry more suitable.

The SMD graph (see Fig. 11) allows conversion of density differences to material thickness differences (provided there is an item of known thickness on the radiograph, such as a shim). This, in turn, allows the X-ray film interpreter to determine the degree of the weld condition without additional radiography.

The scanning microdensitometer is also used for determining adherence to dimensional tolerances of assemblies such as nuclear fuel elements, ordnance items, and other assemblies where hidden component tolerances are critical. Where physical density variations are a matter of concern, the scanning densitometer is useful for controlling the density of matrix and composite materials.

In an unusual application, a microdensitometer employing a double-light beam source was used to measure the stress in rock specimens when mine rock anchor bolts of various types were inserted.[9]

The scanning microdensitometer will generally transfer certain information from a radiograph to a medium (a graph) that can be understood by nonradiographic personnel. It should be noted, however, that some degree of interpretation is necessary in order to fully understand the information provided by the microscan graph. Therefore, the microscan graph should only be interpreted by knowledgeable and qualified personnel.

The limitations of SMD systems must be realized. The actual scanning technique must be devised based on the objective of the resultant graph.

PART 6

RADIOGRAPHIC INTERPRETATION REPORTING

When reporting and documenting the results of radiographic film interpretation, complete and accurate information must accompany the radiographs. The reason for this is subsequent customer review, and possibly regulatory agency review; these later reviews may not occur until long after the completion of the radiography and acceptance by the fabricator/supplier. Lack of explanatory information and documentation can result in costly delays for resolving apparent or suspect indications of the radiographs.

Suppose, for example, that there is a surface defect in a casting mold and this results in a number of castings which have the same defect. The castings are subsequently radiographed and the radiographs reveal the same indication. The condition of the mold is well known to the initial film interpreter and might be omitted from the readers sheet. Later reviewers will not have this basic information and must then develop it. This generally requires reconstruction of the shooting sketch and visual examination of the casting, frequently a time-consuming task, particularly if the shooting sketch does not adequately identify reference points and the indication is on an inside surface. To further complicate matters, the casting may be unavailable for routine visual examination. Documentation needed to minimize confusion during interpretation includes, but is not limited to, the following items:

1. The contract or purchase order should clearly delineate the applicable codes, standards, specifications and procedures, including acceptance criteria and personnel qualification requirements. Exceptions to codes, standards, or specification requirements, if any, should also be noted.
2. Required quality levels and techniques as referenced in the applicable codes, multifilm techniques if used, section thicknesses, penetrameter selection and placement for each thickness range covered.
3. General exposure technique(s) utilized:
   a. shooting sketches, including film coverage and identification;
   b. kV, mA, target-to-film distance and target size (for X rays); source type and curie strength, source-to-film-distance and physical source size (for gamma rays);
   c. film type(s), intensifying screens used;
   d. calculated geometric unsharpness;
   e. blocking and masking;
   f. manual or automatic processing;
   g. quality level required and obtained;
   h. required and obtained film density.
4. Repairs should be documented so that the ultimate reviewer knows the cause and corrective action as an aid to interpretation. Radiographs taken after repair should be so indicated. Also, indications that are determined to be surface conditions on the test object should be recorded as such, together with any corrective action. If not radiographed after corrective action, that fact should be noted.
5. Disposition of each radiograph should be noted. All indications which are within the allowable acceptance criteria should be classified and sized (i.e., *Station No. 7, SLAG, ¼ inch long)* and entered on the readers sheet.

PART 7

RADIOGRAPHIC ARTIFACTS

Indication Description

Because most *nonrelevant indications* can be related to their actual causes, this category of indications is comparatively easy to interpret. False and actual discontinuity indications will be presented here in order to provide guidance for the radiographic film interpreter.

The interpretation of radiographs is not a precise science. As mentioned earlier in this chapter, even those qualified film interpreters with years of experience will often disagree on the nature of discontinuities and their disposition. The descriptions and illustrations[10] contained in this chapter may be used as a general guideline to help identify similar indications encountered during the interpretation process.

False Indications (Film Artifacts)

The radiographic process is very intolerant of dirt and careless handling of the recording media. Violations of good darkroom practice in film loading, unloading and processing will result in artifacts that must be recognized for what they are, not what they may appear to be.

Erroneous interpretations may be made as the result of not recognizing artifacts. Emulsion scratches are a common cause of such misinterpretation. These and most other artifacts are quickly recognizable by viewing both surfaces of the film with reflected light.

The double-film technique is one of the most effective steps in recognizing artifacts, by simply comparing the area of interest on both films. If the indication is on one film and not the other, is not in the same place, or has changed in appearance, it is an artifact.

There are many different types of artifacts, some of which can be confused with actual discontinuities. It is extremely important to identify these false indications and to note their presence in the film interpreter's report (sometimes referred to as the *readers sheet*). Figure 12 is a typical readers sheet for weld interpretation; Fig. 13 is for castings. In some cases the existence of artifacts in the area of interest may require re-radiography. It is therefore important to take every possible step to minimize artifacts.

Artifacts Caused Prior to Processing

Film Scratches

Radiographic film emulsion is quite sensitive and scratches can be caused by most abrasive materials; fingernails and rough handling during loading or unloading are examples. Film scratches can be identified by reflecting light at an angle to the film surface.

Crimp Marks

Crimp marks are caused by bending the film abruptly, usually when loading and unloading the film holder. If the film is crimped prior to exposure, it will produce a crescent-shaped indication that is lighter in density than the adjacent film density (see Fig. 14). If crimped after exposure, the film will produce an indication that is darker than the adjacent film density.

Pressure Marks

Pressure marks are caused by severe localized applications of pressure to the film. For example, a part may be dropped on the film holder during setup. This will produce an artifact on the processed film (see Fig. 15).

Static Marks

Static charges may develop when the radiographic film is handled roughly or moved rapidly during loading or unloading the film holder. It may also be caused by rapid removal of the paper wrapper used an an interleaf. The appearance of *static marks* will range from branch-like, jagged dark lines to irregular, abrupt dark spots (see Fig. 16).

FIGURE 12. Typical Radiographic Inspection Readers Sheet for Welds

COMPANY OR LABORATORY NAME
ADDRESS PHONE NUMBER

CUSTOMER ID ____
WELD ID/LOCATION ____
PART ID/LOCATION ____

RT REPORT NO. ____ DATE ____

PART DATA

WELD THICKNESS ____
BASE MATERIAL THICKNESS ____
MATERIAL TYPE ____
WELDING TECHNIQUE ____
WELD STATUS ____
WELD JOINT CONFIGURATION ____
DIAMETER OR WELD LENGTH ____
APPLICABLE CODE/STANDARD ____
RT PROCEDURE ____

TECHNIQUE DATA

PENETRAMETER ID/NUMBER ____
SHIM THICKNESS ____
FILM TYPE ____
FILM SIZE ____
SINGLE ____ MULTIPLE (NO.) ____
kV ____ OR SOURCE STRENGTH ____
RADIATION SOURCE-TYPE ____
TARGET OR SOURCE SIZE (EFFECTIVE) ____
SFD ____ UNSHARPNESS ____
FILM PROCESSING: MANUAL ____ AUTOMATIC ____

SCREENS

TYPE ____
FRONT THICKNESS ____
BACK THICKNESS ____
CENTER THICKNESS ____

SHOOTING SKETCH

SHOW SOURCE/TUBE LOCATION, DIRECTION OF RADIATION, POSITION AND LOCATION OF PART, FILM, PENETRAMETER, SHIM AND LEAD NUMBER/LETTER ID.

INTERPRETATION DATA

VIEWED SINGLE VIEWED SUPERIMPOSED

PART NUMBER	WELD ID	STATION	ACC.	REJ.	DISCONTINUITY CODE	QUALITY LEVEL	DENSITY	ARTIFACT	REMARKS

DISCONTINUITY CODE

POR - POROSITY
S.I. - SLAG INCLUSION
S.L. - SLAG LINE
IP - INCOMPLETE PENETRATION
LOF - LACK OF FUSION
UC-EX - UNDERCUT OUTSIDE
UC-IN - UNDERCUT INSIDE
CONC - CONCAVITY
CONV - CONVEXITY
Hi.Lo - HIGH LOW
BT - BURN-THROUGH
IC - ICICLES
W - TUNGSTEN INCLUSION
CR(L) - LONGITUDINAL CRACK
CR(T) - TRANSVERSE CRACK
CR(R) - ROOT CRACK
CR(C) - CRATER CRACK
AB - ARC BURN
UI - UNCONSUMED INSERT
OR - OXIDIZED ROOT
SURF - SURFACE

REVIEWED BY ____ NAME ____ LEVEL ____ DATE ____

CUSTOMER OR CUSTOMER REPRESENTATIVE ____ NAME ____ LEVEL ____ DATE ____

FIGURE 13. Typical Radiographic Inspection Readers Sheet for Castings

COMPANY OR LABORATORY NAME
ADDRESS PHONE NUMBER

CUSTOMER ID ______
CASTING ID/LOCATION ______
PART ID/LOCATION ______

RT REPORT NO. ______ DATE ______

PART DATA

MATERIAL THICKNESS ______
MATERIAL TYPE ______
APPLICABLE CODE/STANDARD ______
RT PROCEDURE ______

TECHNIQUE DATA

PENETRAMETER ID/NUMBER ______
SHIM/BLOCK THICKNESS ______
FILM TYPE ______
SINGLE ______ MULTIPLE (NO.) ______
kV ______ OR SOURCE STRENGTH ______
RADIATION SOURCE-TYPE ______
TARGET OR SOURCE SIZE (EFFECTIVE) ______
SFD ______ UNSHARPNESS ______
FILM PROCESSING: MANUAL ______ AUTOMATIC ______

SCREENS

TYPE ______
FRONT THICKNESS ______
BACK THICKNESS ______
CENTER THICKNESS ______

SHOOTING SKETCH

SHOW SOURCE TUBE LOCATION, DIRECTION OF RADIATION, POSITION AND LOCATION OF PART, FILM, PENETRAMETER, SHIM AND LEAD NUMBER LETTER ID

INTERPRETATION DATA

VIEWED SINGLE VIEWED SUPERIMPOSED

PART NUMBER	CASTING ID	STATION	ACC.	REJ.	DISCONTINUITY CODE	QUALITY LEVEL	DENSITY	ARTIFACT	REMARKS

DISCONTINUITY CODE

MP - MICROPOROSITY
POR - POROSITY
WP - WORM HOLE POROSITY
BH - BLOW HOLE
SI - SAND INCLUSION
SL - SLAG INCLUSION
DI - DENSE INCLUSION
SS - SPONGE SHRINKAGE
MS - MICROSHRINKAGE
SH - SHRINKAGE
DR - DROSS
HT - HOT TEAR
UC - UNFUSED CHAPLET
MR - MISRUN
CS - CORE SHIFT
SEG - SEGREGATION
SURF - SURFACE

REVIEWED BY ______ NAME ______ LEVEL ______ DATE ______

CUSTOMER OR CUSTOMER REPRESENTATIVE ______ NAME ______ LEVEL ______ DATE ______

Screen Marks

Scratches and other blemishes in a lead screen will become intensified and can create significant indications on the film image. This may be especially noticeable when the film holder containing the lead screens is bent to accommodate part configuration. Dirt on fluorescent screens will interfere with light transmission to the film and a light area will result after the film is processed. Dirt on lead screens interferes with electron bombardment of the film and also produces a light area in the image. Screens should have a unique serial number inscribed in a corner to identify these problems and to make it easier to locate the faulty screen (see Fig. 17).

Small bits of foreign material (such as lint, tobacco, paper or dandruff) between the film and fluorescent or lead screens will cause light spots in the processed film. To minimize false indications from screens, it is imperative that they be absolutely clean, smooth, free of imperfections and foreign matter. A further word of caution: if a plastic protective coating is used, make sure it is removed prior to using new screens.

Fog

Fog is a slight, overall exposure effect caused when unprocessed film is exposed to low levels of radiation, to high humidity and/or temperature, or to a safelight that is above acceptable levels of intensity. Information regarding safelight intensity limits can be obtained from the film manufacturer.

Light Exposure

Inadvertent exposure of the unprocessed film will result if the film box is opened in a darkroom that is bright, or if the film is placed in an exposure holder that is not totally light-tight. This results in severe overexposure (see Fig. 18) and exposure holders should be examined regularly to eliminate the problem.

Finger Marks

Marks such as fingerprints are normally easy to recognize. They may be darker or lighter images on the film.

FIGURE 14. Crimp Marks Resulting From Poor Handling of Individual Sheet of Film (a) Before Exposure and (b) After Exposure

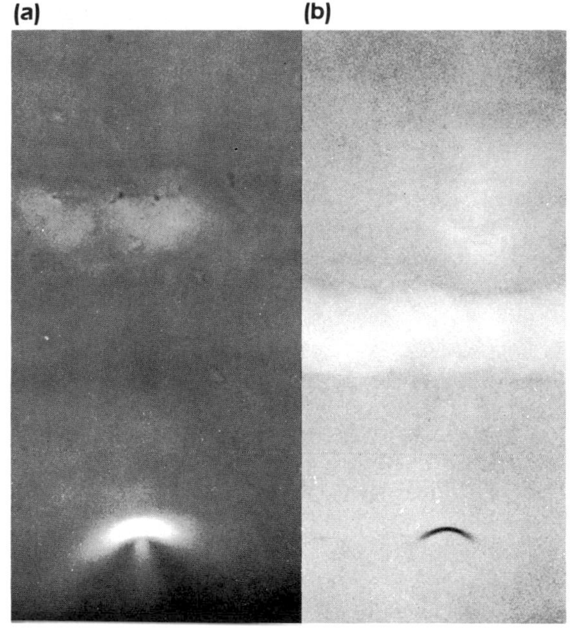

FIGURE 15. Pressure Mark Caused Before Exposure, Visible as Low Density

FIGURE 16. Static Marks Resulting from Poor Film-handling Technique

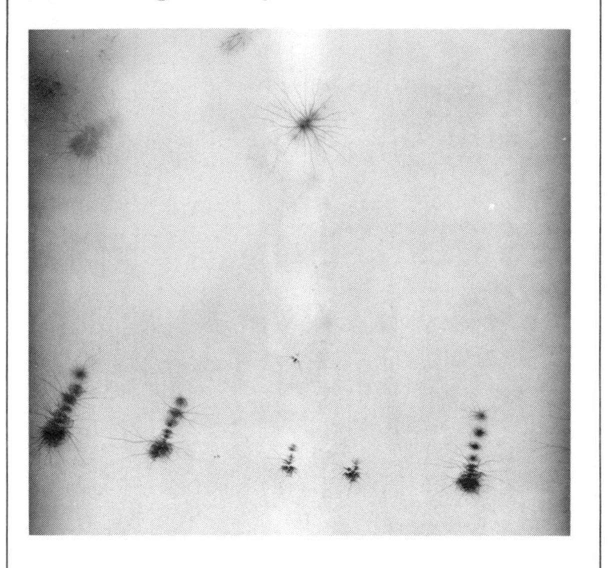

EASTMAN KODAK COMPANY. REPRINTED WITH PERMISSION.

FIGURE 17. The Words "Front and Back" Scratched in the Surface of Front and Back Lead Foil Screens Before Radiography of a 2.5 cm (1 in.) Welded Steel Plate; Hairs Placed Between the Respective Screens and Film Visible as Light Marks Preceding the Scribed Words

EASTMAN KODAK COMPANY. REPRINTED WITH PERMISSION.

FIGURE 18. Light Leaks

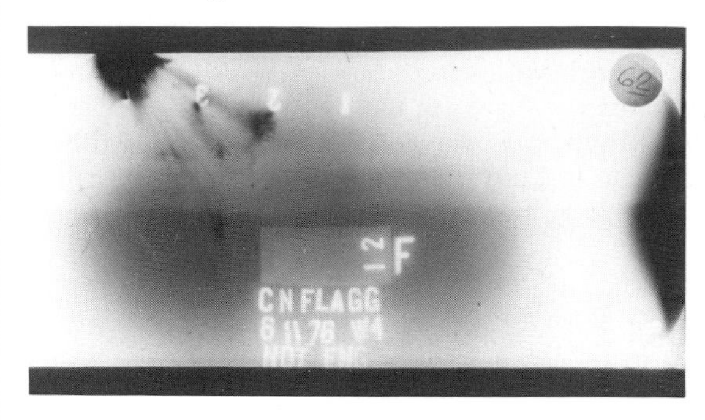

Artifacts Caused During Processing

Chemical Streaks

During manual processing, streaks on the film may result if chemicals from prior processing are not adequately removed from the hanger clips (see Fig. 19). Overall film streaking may also result when the film is placed directly into a water rinse without first placing it into the stop bath solution. Developer carry-over into the fixer may cause an overall streaking condition. A further cause of streaking is insufficient agitation of the film hanger during the development step.

Spotting

If fixer solution (Fig. 20) comes in contact with the film prior to development, light areas or spots will result. If drops of developer or water inadvertently reach the film prior to placing it into the developer, dark spots can result (Fig. 21).

Another spotting condition may occur from water droplets on the film surface. During the drying process, these droplets take longer to dry and leave a distinct circular pattern on the film surface.

Delay Streaks

These are uneven streaks in the direction of film movement through an automatic processor. A delay in feeding successive films may result in the drying of solutions on the processor rollers. Cleaning the exposed rollers with a damp cloth should eliminate *delay streaks*.

Air Bells

Air bells are caused by air bubbles clinging to the surface of the film when it is immersed in the developer, causing light spots on the film image. If the film hanger is tapped abruptly against the side of the tank then properly agitated, the air bubbles should become dislodged (see Fig. 22).

Dirt

If dirt or other contaminants accumulate on the surface of the developer, stop bath or fixer, a noticeably dirty pattern will probably appear on the film. If the rinse water is not adequately replenished, it can also cause a similar problem, especially if the water coming into the wash tank is dirty and filtration is not used. The condition can be verified by observing the surface of the film in reflected light.

Pi Lines

These lines run across the film, perpendicular to the direction of rolling, when an automatic processor is used. They occur at regularly spaced intervals, 3.14-times the roller diameter. This condition is apparently caused by a slight deposit of chemicals on the rollers by the leading edge of the film (see Fig. 23).

Pressure Marks

Pressure marks may be caused by a buildup of foreign matter on rollers in an automatic processor or by inadequate clearances between rollers. Rollers should be thoroughly cleaned and properly adjusted to minimize this condition (see Fig. 24).

Kissing

Film that comes in contact with other film, especially in the developer during manual processing, will result in a severe blotch in the area of contact.

FIGURE 19. Streaking Caused by Inadequately Cleaned Film Hangers

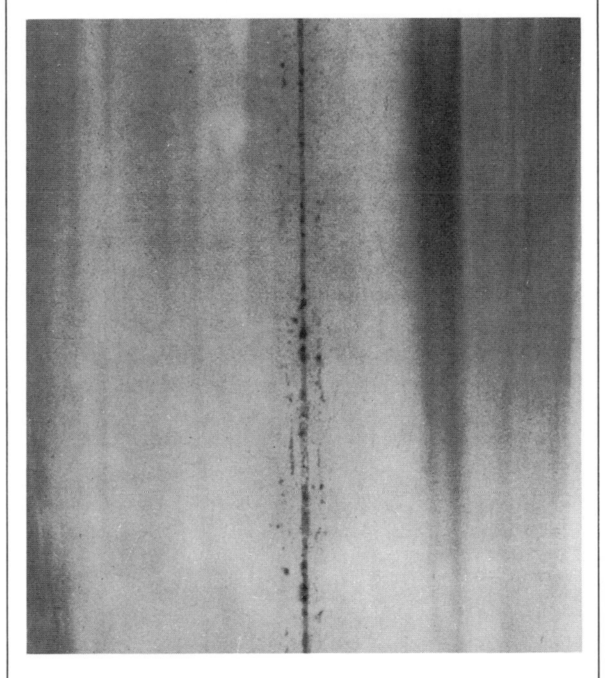

EASTMAN KODAK COMPANY. REPRINTED WITH PERMISSION.

FIGURE 20. Light Spots Caused by (a) Stop Bath or (b) Fixer Splashed on Film Before Development

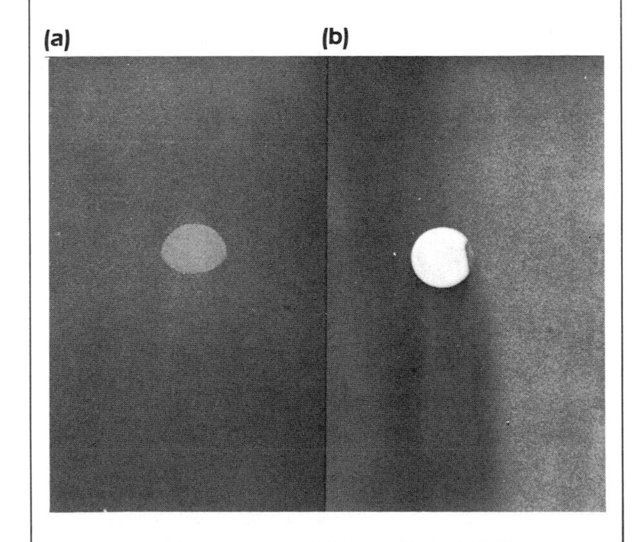

EASTMAN KODAK COMPANY. REPRINTED WITH PERMISSION.

FIGURE 21. Dark Spots Caused by (a) Water or (b) Developer Splashed on Film Before Development

(a) (b)

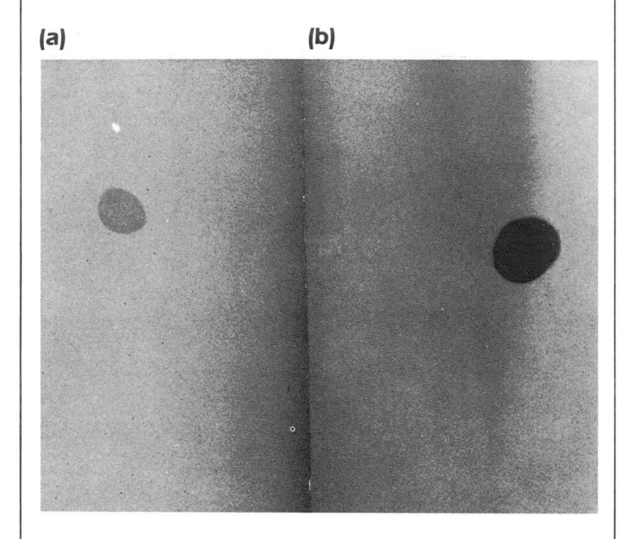

FIGURE 22. Surface Deposits Caused by Contaminated Wash Water in an Automatic Processor

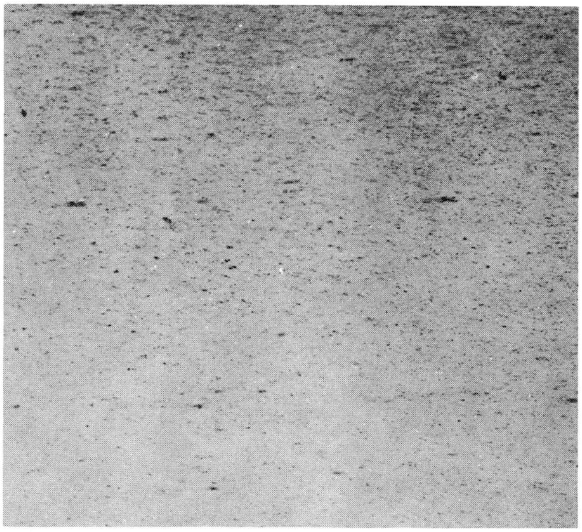

FIGURE 23. Pi Line

FIGURE 24. Pressure Marks Caused by Foreign Matter on Rollers or Improper Roller Clearance

Artifacts Caused After Processing

Scratches

Scratches result from rough handling. Even after processing, the emulsion is sensitive to all types of abrasion and care should be taken to minimize damage to the emulsion.

Fingerprints

These occur when handling the film during interpretation. To prevent fingerprints, radiographs should not be handled unless cotton or nylon gloves are worn.

Real-time Artifacts

Real-time radiographic artifacts are also operator dependent and must be recognized. They are caused primarily by electronic noise generated in video systems and can be corrected by filtering. Dust on the lens surface is another common cause of real-time artifacts. When using image enhancement techniques on radiographs, as described in Sections 14 and 15 of this volume, a very careful examination of the film should be made to identify all artifacts prior to enhancement. Otherwise, the artifacts will also be enhanced and could possibly be difficult to identify in subsequent evaluations. This is also true when radiographs are duplicated or microfilmed.

PART 8

DISCONTINUITY INDICATIONS

Discontinuity Indications for Welds

The various discontinuities found in weldments are illustrated and described here strictly as representative conditions. Cross-sectional photographs or sketches are also shown. These examples are for illustrative purposes; actual discontinuities vary in shape, size and severity.

Porosity

These are voids that result from gas being entrapped as the weld metal solidifies (see Fig. 25). Porosity is generally spherical but may be elongated. In some cases, porosity may appear to have a tail as a result of the gas attempting to escape or move while the weld metal is still in the liquid state. Porosity is often uniformly scattered to different degrees of severity but may also appear as a cluster where there is a concentration of pores in a relatively small area. *Linear porosity* is a condition which involves a number of pores that are aligned and separated by a distance usually stipulated in the acceptance standards. *Piping porosity* is severely elongated gas holes that are well defined and may vary in length from very short to as long as 38 cm (15 in.) or more. This type of porosity is sometimes referred to as *wormhole porosity*. *Hollow bead* (see Fig. 26) is an elongated gas void that is usually centrally oriented in the root pass and may also extend for a significant length.

In general, porosity is not considered to be a critical discontinuity unless it (1) is present in large quantities; (2) contains sharp tails; or (3) is aligned in significant numbers in a relatively short distance. The severity of piping and wormhole porosity or hollow bead conditions is generally determined by length and amount.

Slag

Referred to as *inclusions*, these disk-shaped indications are caused by nonmetallic materials entrapped in the weld metal, between weld passes or between the weld pass and the base material (see Fig. 27). Slag inclusions occur in all shapes and sizes but can be generally categorized as an inclusion (short, isolated piece) or as a slag line (relatively narrow but having length). Slag inclusions are evaluated based on size, quantity and length.

Dense Inclusions

This is a metallic inclusion that has a material density greater than the density of the weld metal (see Fig. 28). The most common dense inclusion is the *tungsten inclusion* which results from pieces of the tungsten electrode breaking off and becoming entrapped in the weld metal. These inclusions are readily seen in a radiograph, appearing as light spots on the film.

Incomplete Penetration

Incomplete penetration is an area of nonfusion in the root area (see Fig. 29). This may result from inadequate heat while the root pass is being deposited. It may also be caused by faulty joint design, improper fit-up or problems with the welding procedure. This condition is considered more severe than the porosity/slag type discontinuity because it is more of a *stress riser*. Incomplete penetration is usually easy to detect and identify radiographically due to its location in the weld and its relatively straight, well-defined image.

Lack of Fusion

Lack of fusion is an area of nonadhesion between successive weld passes or between a weld pass and the side wall of the base material (see Fig. 30). It is primarily the result of improper welding techniques or poor joint design. Many lack-of-fusion conditions are relatively narrow, and in some cases angularly oriented, so that this discontinuity is not always readily detected by radiography. When it is observed, it may not be clearly defined but will have a tell-tale linear alignment, running in the same direction that the weld was deposited.

FIGURE 25. Porosity: (a) Photomacrograph; (b) Radiographic Image

(a)

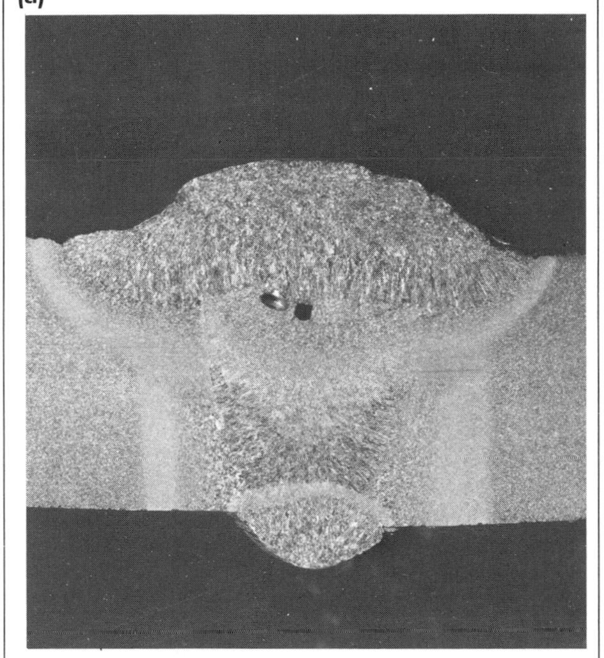

(b)

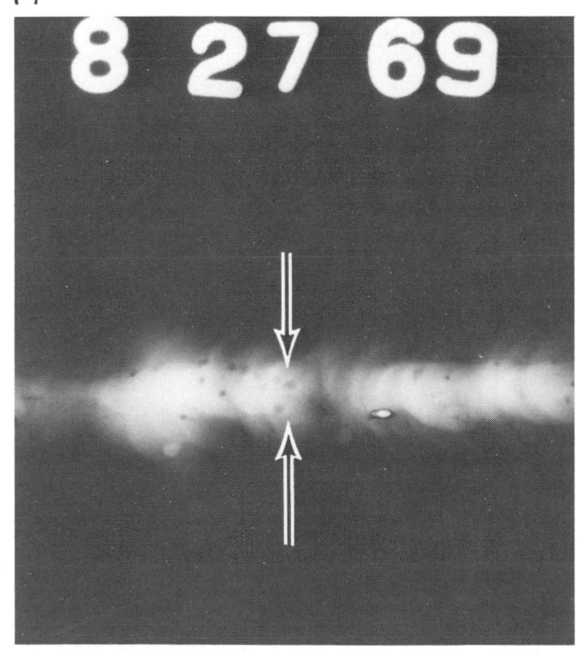

SOUTHWEST RESEARCH INSTITUTE. REPRINTED WITH PERMISSION.

FIGURE 26. Hollow Bead: (a) Photomacrograph; (b) Radiographic Image

(a)

(b)

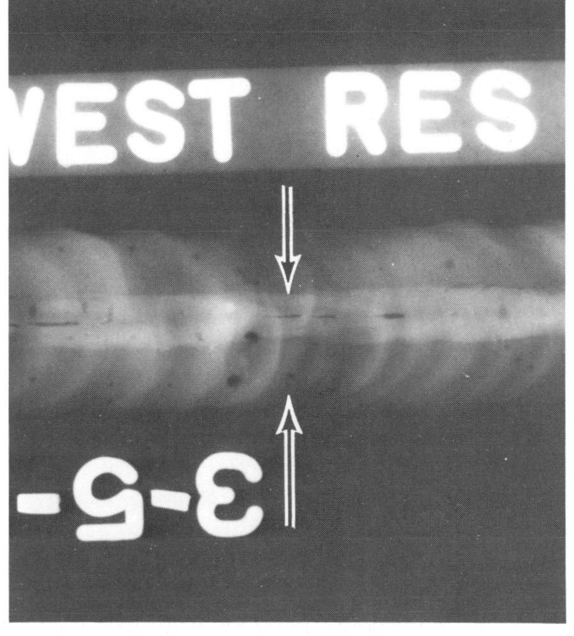

SOUTHWEST RESEARCH INSTITUTE. REPRINTED WITH PERMISSION.

FIGURE 27. Slag Inclusion: (a) Photomacrograph at 4×; (b) Radiographic Image at 1×

(a)

(b)

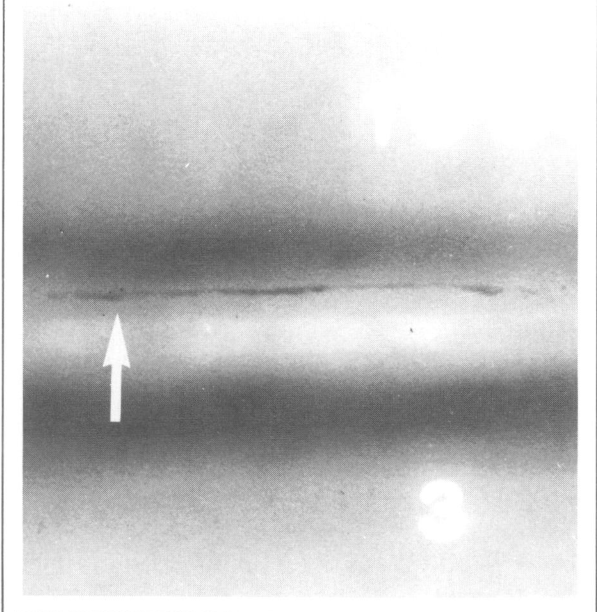

ELECTRIC POWER RESEARCH INSTITUTE. REPRINTED WITH PERMISSION.

FIGURE 28. Tungsten Inclusion: (a) Photomacrograph at 4×; (b) Radiographic Image at 1×

(a)

(b)

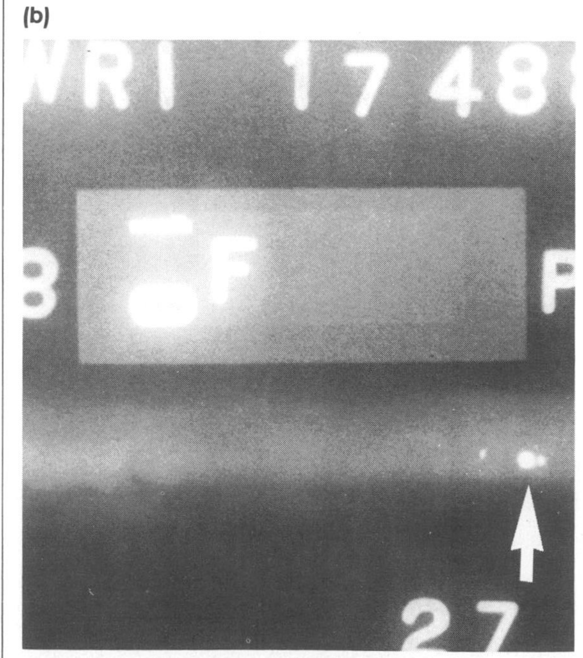

ELECTRIC POWER RESEARCH INSTITUTE. REPRINTED WITH PERMISSION.

FIGURE 29. Incomplete Penetration: (a) Photomacrograph; (b) Radiographic Image

(a)

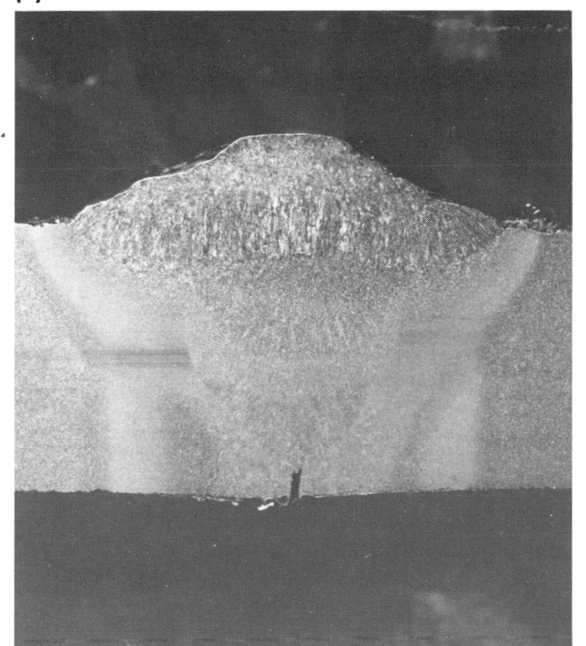

(b)

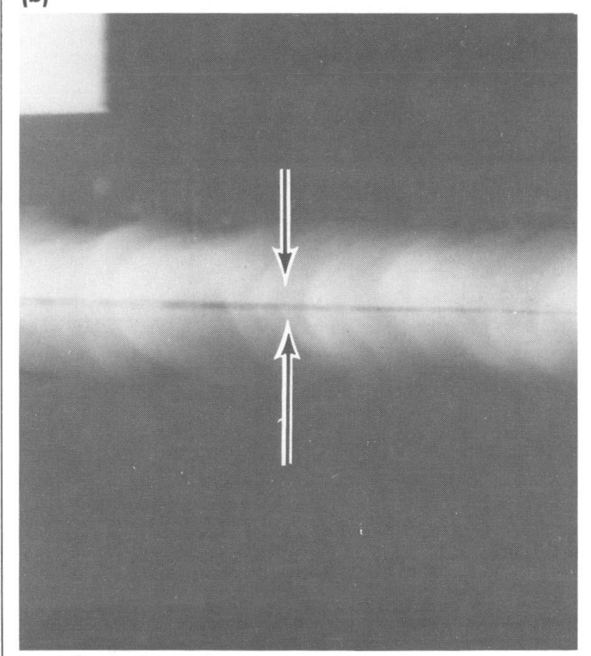

SOUTHWEST RESEARCH INSTITUTE. REPRINTED WITH PERMISSION.

FIGURE 30. Lack of Fusion: (a) Photomacrograph at 4×; (b) Radiographic Image at 1×

(a)

(b)

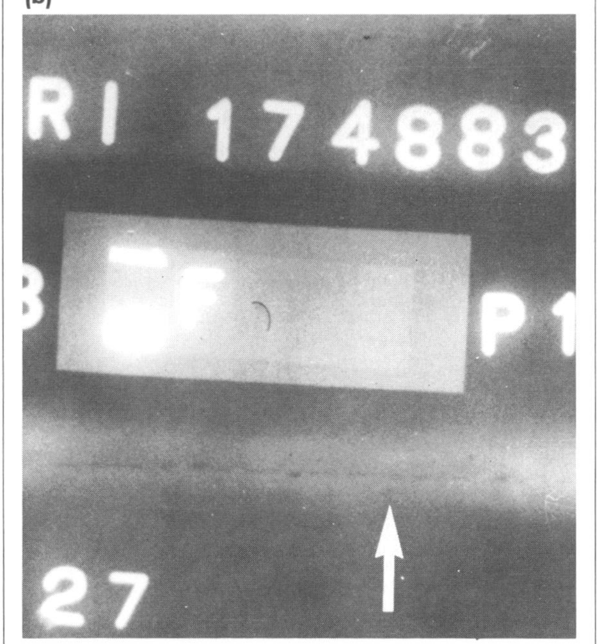

ELECTRIC POWER RESEARCH INSTITUTE. REPRINTED WITH PERMISSION.

Underfill

Underfill is a condition where the weld joint is not completely filled, as evidenced by a depression or lack of weld metal at the face of the weld. This condition is readily observed by an increase in the film density in the weld area; the extent should be confirmed by physical measurement.

Undercut

Undercut is generally described as a groove or depression located at the junction of the weld and base material (the fusion zone) on the weld surface (see Fig. 31). This depression is caused by a melting away of the base metal during the welding process. Undercut can be readily seen and identified on a radiograph but the extent should be measured physically, if possible. Generally, undercut is not considered to be a serious condition if it is relatively shallow (within specification requirements) and not sharp.

Overlap

Overlap is an extension of unfused weld metal beyond the fusion zone. In many cases the overlap forms a tight stress riser notch and is not easily seen in the radiograph. It is generally considered severe when it is detected and confirmed visually.

Excessive Penetration

This is sometimes referred to as *convexity* and results from excessive heat input while the root pass is being deposited (see Fig. 32). The reinforcement of the root becomes excessive and, in some cases, results in a corner or notch condition on the inside surface at the toe of the weld. When excessive penetration occurs in short or intermittent droplets, it may be referred to as *icicles* and is usually accompanied by a burn-through area which lacks weld metal (see Fig. 33).

Concavity

Concavity is a concave condition in the root pass face which results from insufficient heat input while depositing the root pass (see Fig. 34). Concavity causes a dimensional change in the thickness of the weld which may then be less than the required thickness. Because the condition is usually a gradual dimensional change, it shows as a slight and gradual density change in the radiograph. The extent should be determined by physical measurement but may be estimated by density measurements.

Hi-Lo

Hi-Lo is an expression used to describe a misalignment in pipe welds which results in an offset union of the two sections being welded (see Fig. 35).

Cracks

Cracks are fractures or ruptures of the weld metal occurring when the stresses in a localized area exceed the weld metal's ultimate tensile strength. *Hot cracks* occur as tears while the weld metal is in the plastic condition and *cold cracks*, sometimes referred to as *delayed cracks*, occur after the weld metal has cooled. There are a number of crack types associated with weldments.

A *longitudinal crack* (see Fig. 36) is oriented along the length, or approximately parallel to the longitudinal axis, of the weld.

Transverse cracks (see Fig. 37) are approximately perpendicular to the longitudinal axis of the weld.

Underbead cracks form in the heat-affected zone and are usually short but may also be an extensive network.

Toe cracks begin at the toe of the weld and propagate along the plane of highest stress.

Root cracks (see Fig. 38) are longitudinal cracks located in the root pass.

Crater cracks are usually star-shaped patterns which occur in the crater (a depression at the end of a weld bead).

FIGURE 31. Undercut on the Outside Diameter: (a) Photomacrograph at 4×; (b) Radiographic Image at 1×

(a)

(b)

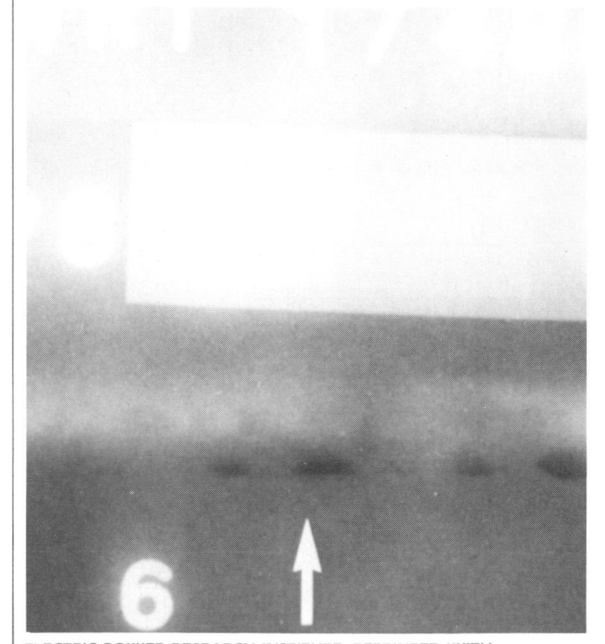

ELECTRIC POWER RESEARCH INSTITUTE. REPRINTED WITH PERMISSION.

FIGURE 32. Excessive Penetration: (a) Photomacrograph at 4×; (b) Radiographic Image at 1×

(a)

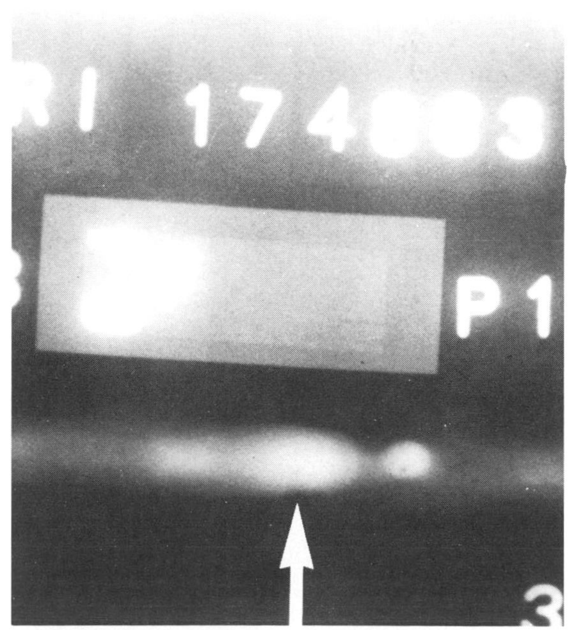

ELECTRIC POWER RESEARCH INSTITUTE. REPRINTED WITH PERMISSION.

FIGURE 33. Burn-through Area: (a) Photomacrograph; (b) Radiographic Image

(a)

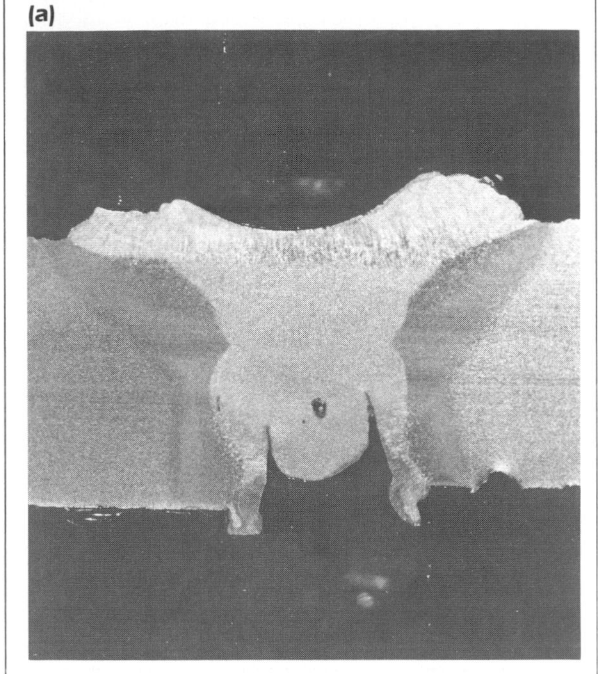

(b)

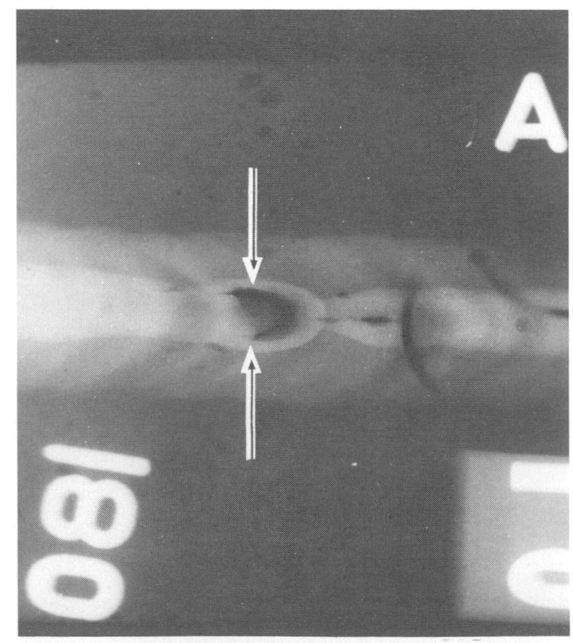

SOUTHWEST RESEARCH INSTITUTE. REPRINTED WITH PERMISSION.

FIGURE 34. Concave Root Surface: (a) Photomacrograph at 4×; (b) Radiographic Image at 1×

(a)

(b)

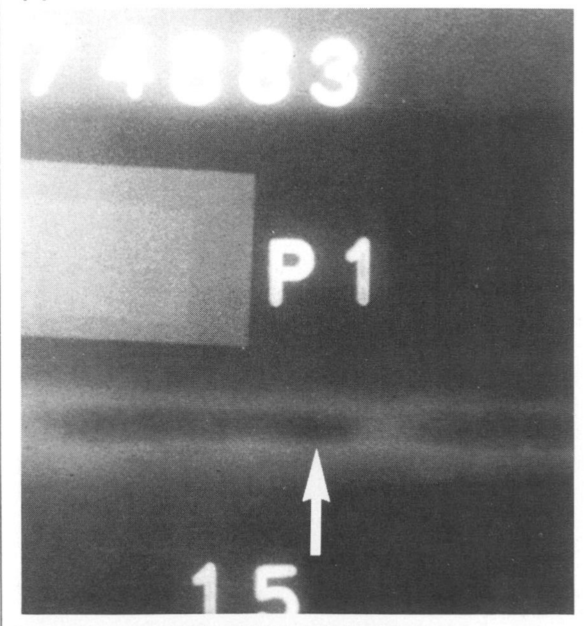

ELECTRIC POWER RESEARCH INSTITUTE. REPRINTED WITH PERMISSION.

FIGURE 35. Hi-Lo Condition, also Called Mismatch: (a) Photomacrograph at 4x; (b) Radiographic Image at 1x

(a)

(b)

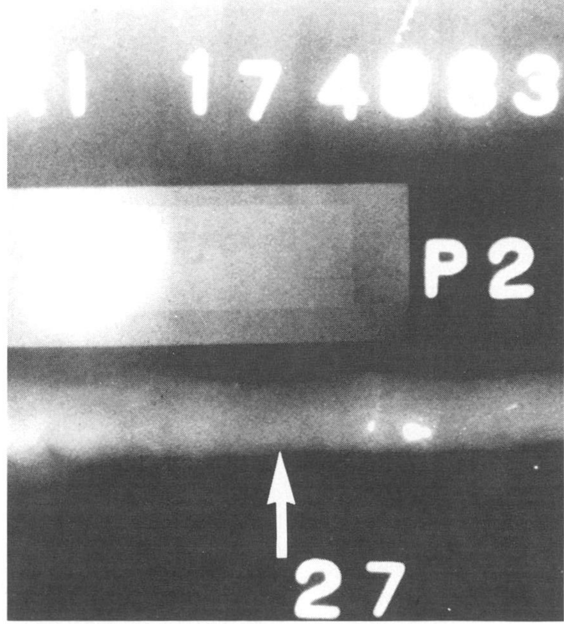

ELECTRIC POWER RESEARCH INSTITUTE. REPRINTED WITH PERMISSION.

FIGURE 36. Crack: (a) Photomacrograph at 4×; (b) Radiographic Image (Collimated Source) at 1×; (c) Radiographic Image at 1×, Same Conditions as (b) without Collimation

(a)

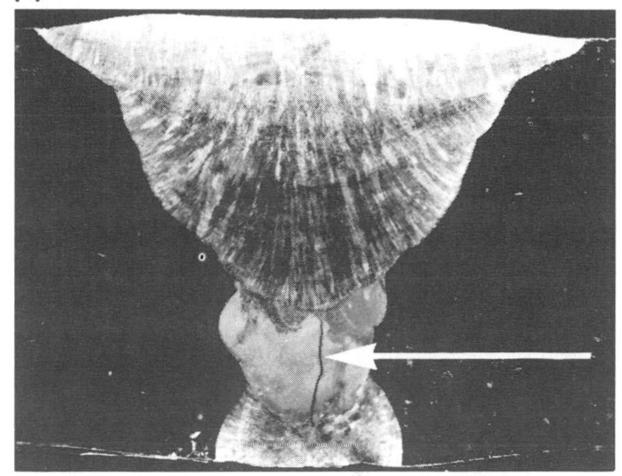

(b)

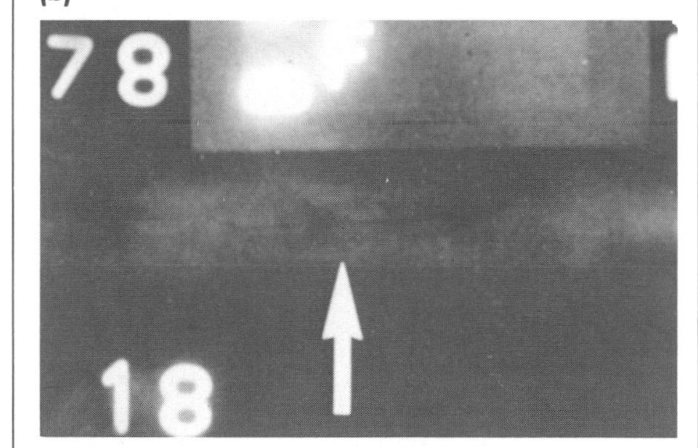

(c)

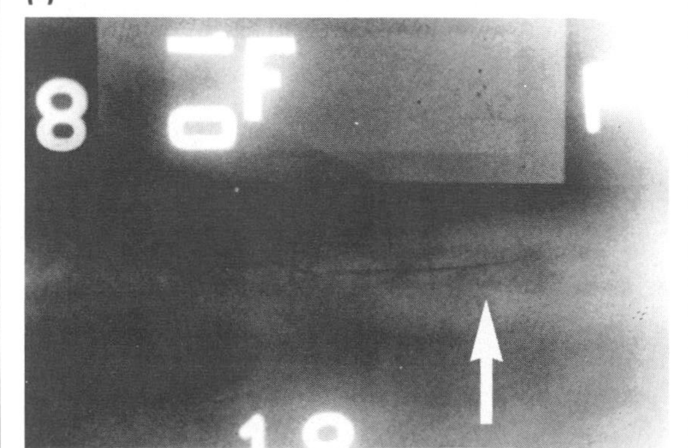

ELECTRIC POWER RESEARCH INSTITUTE. REPRINTED WITH PERMISSION.

FIGURE 37. Transverse Crack: (a) Photomacrograph; (b) Radiographic Image

(a)

(b)

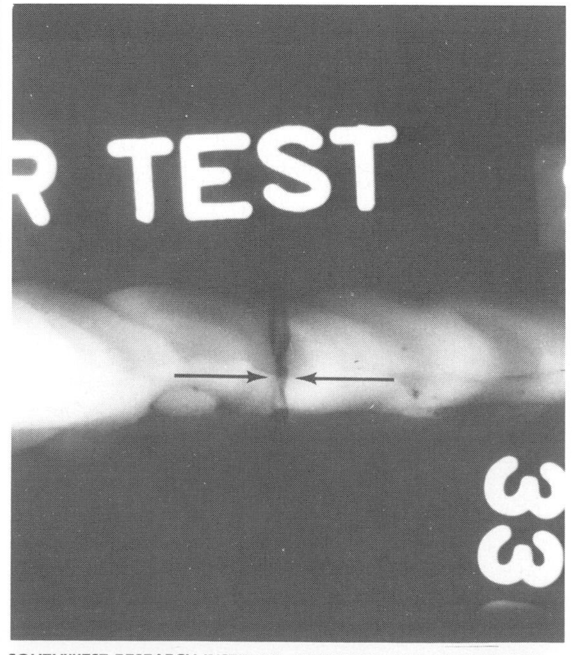

SOUTHWEST RESEARCH INSTITUTE. REPRINTED WITH PERMISSION.

FIGURE 38. Crack Adjacent to Root: (a) Photomacrograph; (b) Radiographic Image

(a)

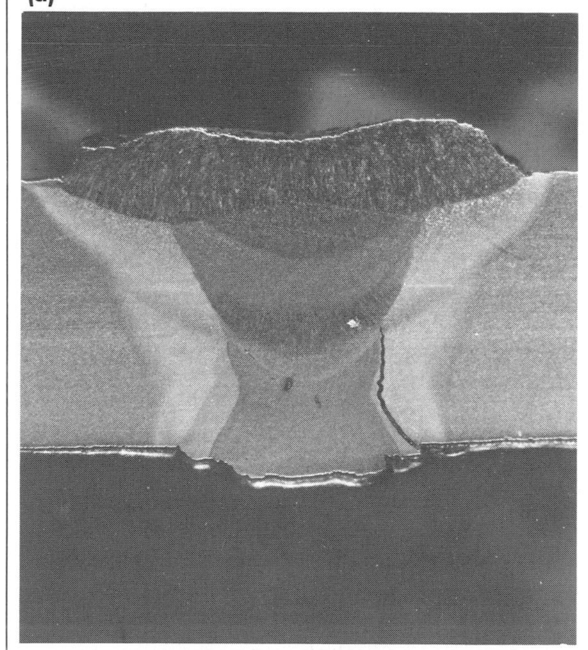

(b)

SOUTHWEST RESEARCH INSTITUTE. REPRINTED WITH PERMISSION.

Void-type Discontinuity Indications for Castings

Casting discontinuities, as with weld discontinuities, will vary in shape, size and appearance depending upon many variables, including material type, mold design, casting process, casting size and foundry control. The examples used to illustrate the various discontinuities found in castings are typical and are not intended for any purpose other than guidance.

Porosity occurs when gas is formed, generally while the molten metal is being poured into the mold, and becomes trapped during solidification. The gas may initiate from the molten metal (air entrapped by the turbulence of the pouring process) or from gas given off by the mold material.

Microporosity

Microporosity is an extremely fine network of gas that appears to be localized. In many cases, these voids are so small they cannot be seen as individual pores (see Fig. 39). This condition is fairly common in aluminum and magnesium alloy castings.

Porosity

Porosity can be individually identified and defined in the radiograph as distinct, globular, gas voids (see Fig. 40). They will vary in size and concentration and these characteristics are used for classification of porosity. Such voids may be present at the surface of the casting or throughout the cross section.

Gas Voids

The most serious gas voids are referred to as *gas holes* (see Fig. 41), *wormhole porosity*, or *blow holes*. A larger, darker (film density) porosity condition is called a gas hole to distinguish it as a more severe condition compared to typical porosity. Wormhole porosity is so named because of its likeness to a wormhole. The shape is caused by the tendency of entrapped gas to escape during solidification and this, in turn, occurs because the gas is considerably lighter in material density than the base metal. During its escape attempt, the gas forms a tail-like linear pattern resembling wormholes.

The most severe gas voids are called blow holes: severe, well-defined cavities that occur when the hot, molten metal is deposited into a mold containing moisture or other impurities. The extremely hot metal causes the moisture to rapidly change to steam which develops a series of linear voids extending into the metal from the surface.

FIGURE 39. Microporosity

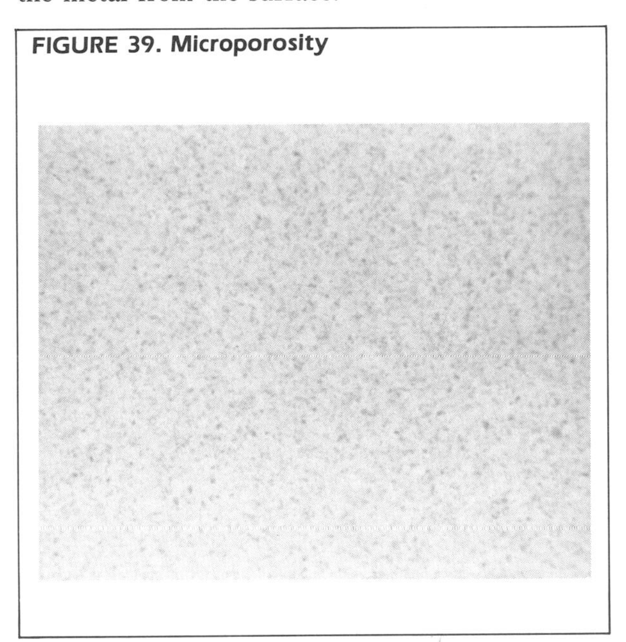

FIGURE 40. Porosity

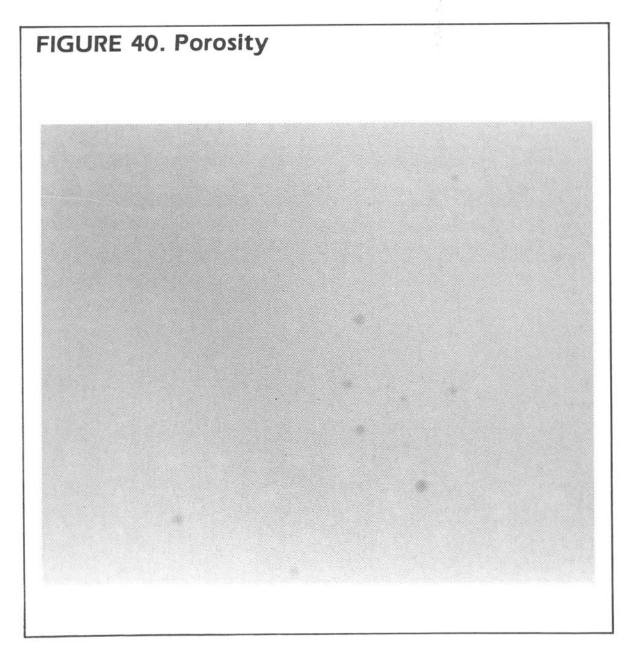

FIGURE 41. Gas Holes

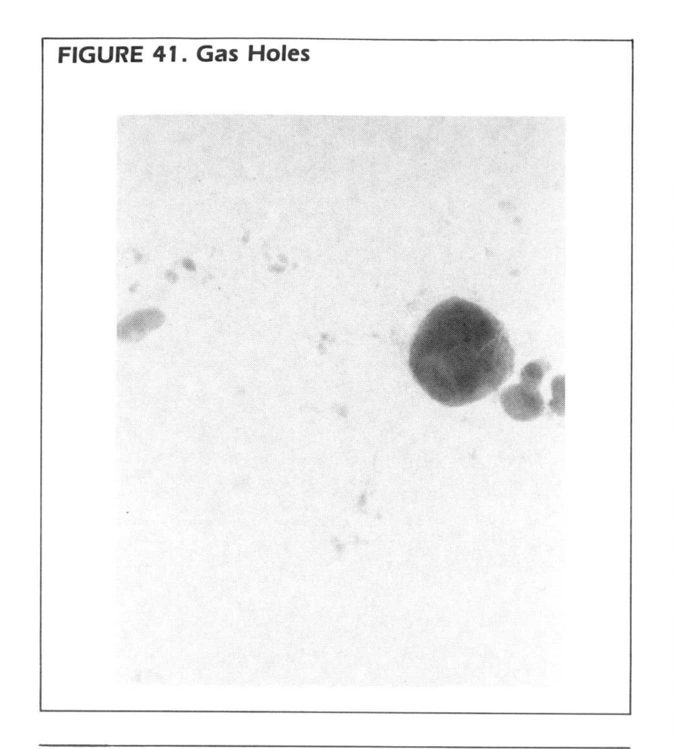

FIGURE 42. Slag Inclusions

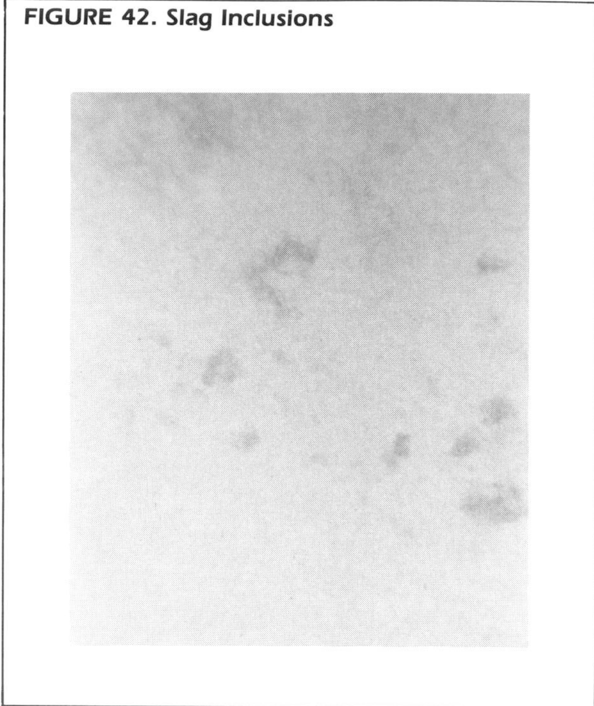

Inclusion Indications for Castings

There are many inclusion-type discontinuities that may be encountered during the radiographic examination of a casting.

Sand and Slag Inclusions

Sand inclusions are pieces of sand which have broken off the sand mold. Radiographically, they resemble a pocket of sand with a granular appearance if observed closely.

Slag inclusions (see Fig. 42) are impurities introduced into the mold with the molten metal. They may also be the result of oxide or impurities that did not rise to the surface prior to metal solidification.

Dross is sometimes referred to as the scum of the melt. Dross may become entrapped, resulting in a general zone of impurities. Dross is usually irregular compared to slag and may be accompanied by gas voids.

Dense inclusions (Fig. 43) can result from the inadvertent addition of more dense objects (such as core wire, bits of metal or other high density metals) to the molten cast metal. These dense inclusions will result in a lighter area of film density in the radiograph.

FIGURE 43. Dense Inclusions

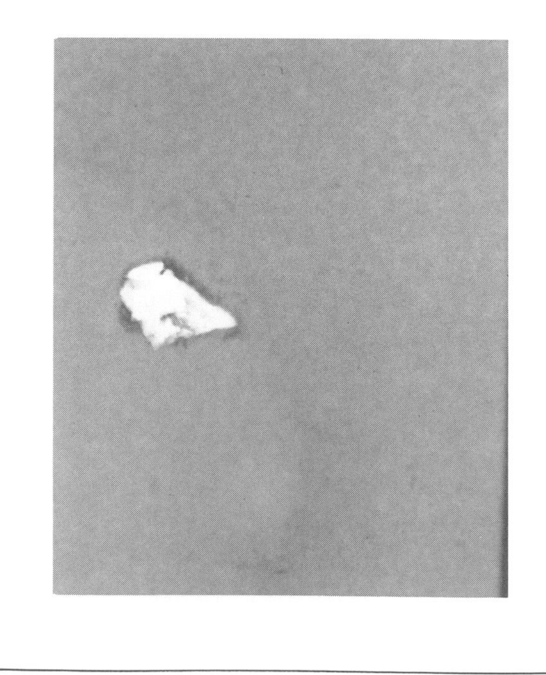

Crack-type Discontinuities for Castings

Shrinkage

This discontinuity results from localized contraction of the cast metal as it solidifies and cools. In more severe cases, shrinkage may also exhibit a cavity area accompanied with a zone of fine, branch-like indications.

Microshrinkage (Fig. 44) is a zone of a very minute shrinkage, usually resulting in a pattern exhibiting a slight density increase in the radiograph. Individual linear indications are not defined because this condition generally consists of many small, tight, shrinkage branches that may overlap and superimpose.

Sponge shrinkage (Fig. 45) is a system of small interconnected areas of shrinkage, somewhat coarser than microshrinkage and resembling sponge. It is not usually an easily defined condition unless it exists in relatively thin sections. In thicker sections, it may exhibit the same patchy pattern as microshrinkage.

Other shrinkage (see Fig. 46) discontinuity patterns will range from minor branch-like zones to major, abrupt linear indications resembling a tree with a trunk and many branches.

Hot tears (Fig. 47) are a series of well-defined cracks or ruptures caused by high stresses which develop as the casting is cooling. In a radiograph, the hot tear appears as a dark, jagged linear indication, sometimes intermittent. A hot tear is certainly the most severe discontinuity of the shrinkage family.

Cracks (Fig. 48) are usually better defined than hot tears, and are normally confined to a single linear indication.

FIGURE 44. Microshrinkage

FIGURE 45. Sponge Shrinkage

FIGURE 46. Shrinkage

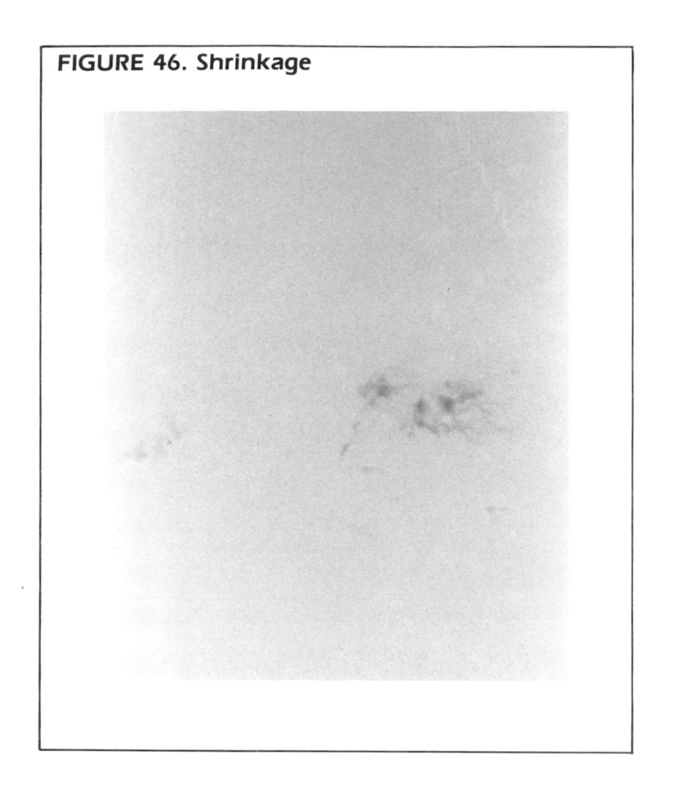

FIGURE 47. Hot Tears

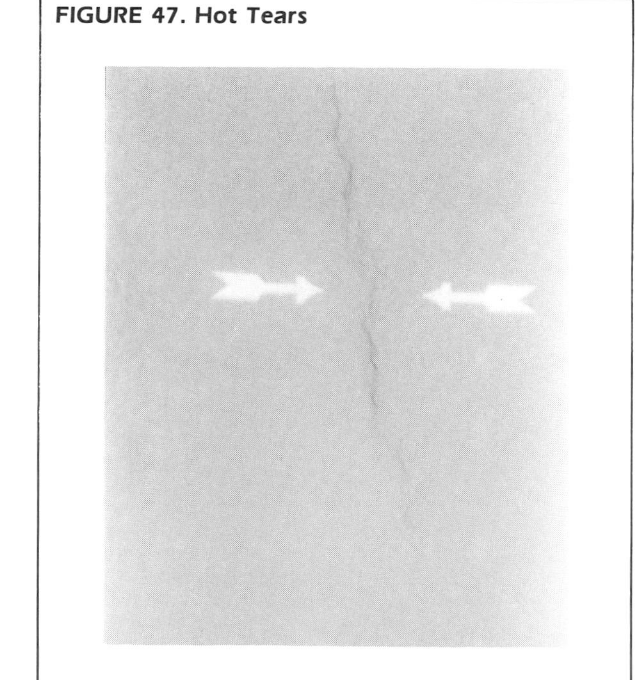

FIGURE 48. Cracks

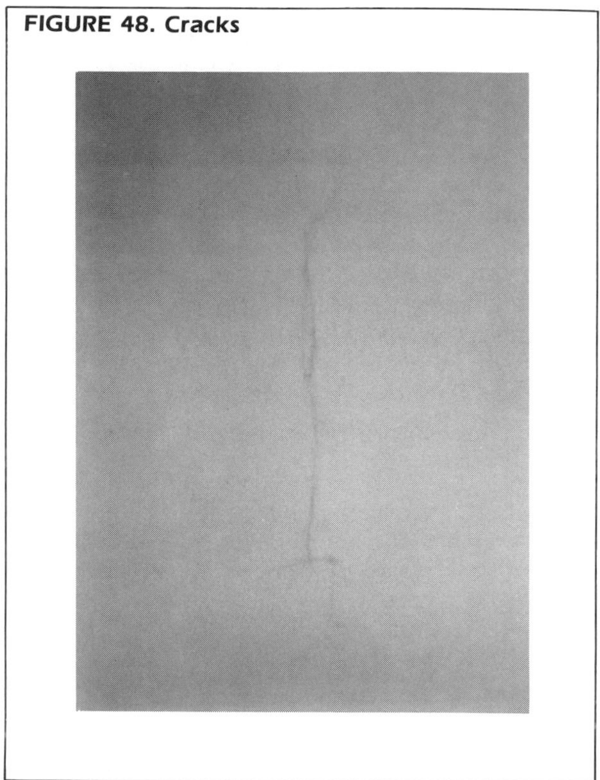

Other Casting Discontinuities

When two bodies of molten metal fail to metallurgically fuse in a casting, the discontinuity that results is referred to as a *cold shut*. Radiographically, it is not easily detected unless part of the unfused surface plane is parallel to the radiation beam. If it is detected, it will appear as a faint linear indication with slightly higher density.

Chaplets are the metal devices used to support the core inside the mold. Chaplets are usually made of the same material as the casting and generally will be consumed when the molten metal comes in contact with them. If this does not occur, or if only part of the chaplet melts, the condition that results is referred to as an *unfused chaplet* (see Fig. 49). Radiographically, these appear as a characteristic circular shape with only slight density differences between it and the cast material.

The mold core is designed to provide a specific shape and size cavity within the casting. If the core shifts or moves during pouring, the resulting condition is called a *core shift* (see Fig. 50). It is readily observed when evaluating the radiograph unless the movement is slight.

A *misrun* (see Fig. 51) is a void or cavity that occurs when the mold is not completely filled with molten metal. This may be caused by trapped gas which is not properly vented, solidification of a portion of the casting before the mold is completely filled, or too little molten metal deposited in the mold. In any event, the misrun is easily identified, visually as well as radiographically.

Segregation is considered more of a metallurgical problem than a discontinuity. It is a condition that results from concentrations of any of the alloy's constituents; the concentrations may be localized or present throughout the casting in bands. In a radiograph, the condition may appear as a slight or significant overall film density pattern, depending upon the degree of segregation and the density of the segregated material.

FIGURE 49. Unfused Chaplet

FIGURE 50. Core Shift

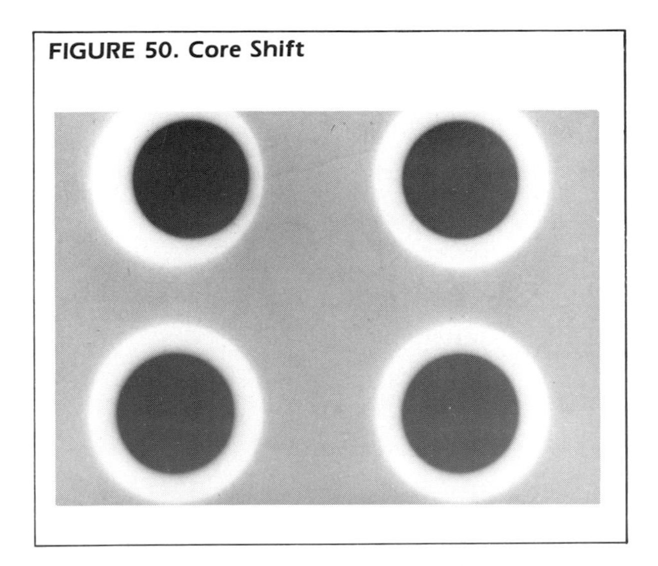

FIGURE 51. Misrun

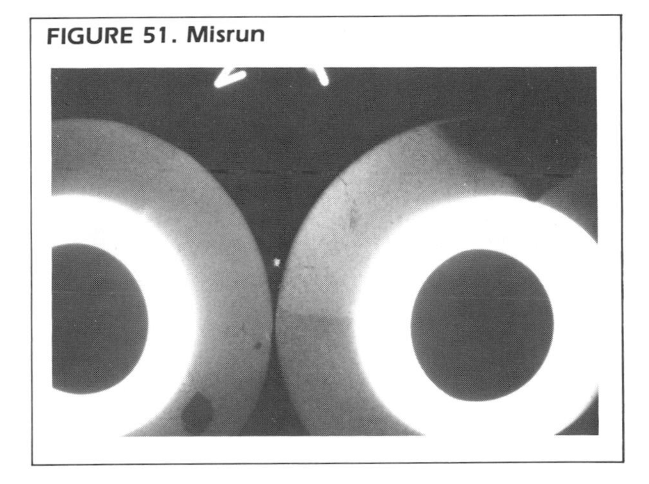

Conclusion

There is a multitude of structures, assemblies, materials and components that can be effectively radiographed. Interpretation, if it is to be meaningful must only be attempted with a complete understanding of the following:

1. part dimensions and configuration;
2. material type;
3. radiographic technique used;
4. processing used with object;
5. applicable code;
6. acceptance standard; and
7. other information desired from the examination.

The key to successful interpretation, after all other variables are optimized, rests with the individual doing the interpretation. Judgment must be based on complete knowledge of the radiographic process and a thorough understanding of the test object, coupled with extensive radiographic interpretation experience and training.

REFERENCES

1. Megling, R.C. and M.L. Abrams. *Relative Roles of Experience/Learning and Visual Factors on Radiographic Inspector Performance.* Research Report SRR73-22. San Diego, CA: Naval Personnel and Training Research Laboratory (June 1973).

2. Berock, J.F., R.G. Wells and M.L. Abrams. *Development and Validation of an Experimental Radiographic Reading Training Program.* Report AD-782-332. San Diego, CA: Navy Personnel Research and Development Center (June 1974).

3. Lusted, L.B. "Signal Detectability and Medical Decision-Making." *Science.* Vol. 171 (March 1971): pp 1217-1219.

4. McMaster, R.C. and S.A. Wenk. *A Basic Guide for Management's Choice of Nondestructive Tests.* Special Technical Publication No. 112. Philadelphia, PA: American Society for Testing and Materials (1951).

5. *Nondestructive Testing Handbook*, 1st edition. Vol. 1, Secs. 1 and 4 (1959).

6. McClung, R.W. "An Introspective View of Nondestructive Testing" (1974 ASNT Lester Honor Lecture). *Materials Evaluation.* Vol. 33, No. 2 (February 1975): pp 16a-19a, 43a-45a.

7. Hastings, C.H. "Nondestructive Testing as an Aid to Fracture Prevention Mechanics." *Journal of the Franklin Institute.* Vol. 290, No. 6 (December 1970).

8. Yonemura, G.T. Report NBS-TW 1143. Washington, DC: National Bureau of Standards (June 1981).

9. Foster, B.E., S.D. Snyder, R.W. McClung and W.J. Godzinsky. "Development of High Voltage Radiography and Dual Microdensitometric Techniques for Evaluating Stressed Rock Specimens." *Materials Evaluation.* Vol. 31, No. 11 (November 1973): pp 229-236.

10. *Radiography in Modern Industry*, 3rd edition. Rochester, NY: Eastman Kodak Co. (1969): pp 147-153.

SECTION 9

RADIOGRAPHIC CONTROL OF WELDS

Herb Chapman, Canadian Welding Bureau, Toronto, Ontario

ACKNOWLEDGMENTS

Since the original version of this chapter was published in the 1959 edition of the *Nondestructive Testing Handbook*, significant advances have been made in the art and science of radiography. The testing of weldments, though, has maintained its place as a prime candidate for this NDT technique.

This section has been rewritten both to expand the basic information on welds and their radiographic examination and to integrate this material with the section on the interpretation of radiographs. There is no atlas of discontinuities and their related images on radiographic film, but some examples of radiography have been included to cover special conditions in which information on the welding is an integral part of the inspection. In particular, a few typical radiographs of pipe welds, courtesy of the Electric Power Research Institute, are presented in Part 9 (excerpted with permission from *NDE Characteristics of Pipe Weld Defects*, No. NP-1590-SR©1980). Part 10 of this Section contains reproductions of films supplied by Pat Walsh of Ontario Hydro and Wayne Carter of Stasuk Testing and Inspection Ltd. and reproduced primarily by Lester Northeast of Dupont Canada.

An attempt has been made to incorporate information on advanced equipment for radiographing welds as part of an extended inspection system for reactor cores. For this testing purpose, equipment has been developed that is extremely powerful yet remarkably compact and precise. In the study of reactor components, the availability of such high-power X-ray generators (and, consequently, short exposure times) is an important advance, mainly because the presence of film near a nuclear reactor results in unwanted exposure to the film when long radiographic exposures are required.

The planning of this chapter was done by Lawrence E. Bryant, Los Alamos National Laboratory, who separated the topic of radiography into its many subdivisions, of which this effort is a part. For each of these subdivisions, topical headings and subheadings were made available to the author as a framework.

Two other individuals should be recognized here for their help in reviewing the manuscript, namely, Bill Havercroft of Ottawa and Len Baxter of Montreal. Bill Havercroft was a contributor to the original *Nondestructive Testing Handbook*, and to him special thanks are due for his continuing, friendly interest in this work.

INTRODUCTION

Radiographic examination applied to a new product is intended to provide assurance that the product will be clear of significant discontinuities. In the industrial world where products are made, or worked on, by people, it can be considered normal to have something go wrong. This is the case for all industrial processes, from the historically established trades to modern production systems.

Some of the very earliest manufactured products still available for study are exemplified by carved stone and cast metals. Among the latter, the precious metals are most in evidence. These metallic components were intended to serve simple decorative functions. Their inspection or quality control was in the hands of the craftsman. With products such as rings, brooches and pectorals, so long as they had intrinsic beauty, they would be acceptable. They may and often do contain very significant internal flaws as revealed by modern NDT methods. A *significant flaw* is one that could affect the serviceability of the component.

There are many beautiful gold artifacts in existence that are considered priceless, but contain very large voids and inclusions of flux or fueling materials. It is these specimens that have survived an original inspection, and for the use intended, they were most certainly *adequate to the purpose*. Undoubtedly, the artisan made many of these items more than once, perhaps as commanded by a demanding superior or by pride of workmanship. With gold, rejects would simply be remelted and recast, and we would have no way of knowing how many times the metal had been recycled before being locked into the shape that now survives. The variability of the casting process makes some reworking almost routine, and even today castings may be scheduled for repair; the term used for this repair is *conditioning*. This variability, which may not be apparent until a product is near completion, applies also to fusion welding.

The quality of welds may be partly determined visually; there are general requirements of workmanship that are most easily met by having the welds *appear* correct. The primary reason for concern with welds, however, is that they must be sound throughout or as sound as their intended use demands. In determining the requirements for intended use, when some minor discontinuities might be present, radiography has very special advantages. The radiographic process provides good information on the precise nature of a defect. Some discontinuities are not inherently hazardous so this ability to confirm their identity with some assurance becomes very important. For example, cases of isolated spherical porosity are seldom considered serious in themselves because they do not constitute stress risers within the component. It is important however that they be recognized; their presence could indicate that the welding process has gone awry and this could presage more serious difficulties.

In the hierarchy of quality functions, radiography comes at the functional peak. A welding procedure would usually be regarded as a proven routine, though certain unexpected events may occur. For instance, a gradual long-term change could develop in a welding process, such as drift in an instrument; there might be no external indication that the process had been altered. Radiography can contribute very meaningfully here because of the detailed record it provides of the internal condition of a weld.

Radiography is useful in welding technique development. In addition, there are many codes which require that radiography be employed to some degree in the final inspection steps (see refs. 1 through 6, 11 and 13). In some cases, when welds become hidden in complex assemblies, inspection must occur fairly early in the component's manufacturing cycle.

PART 1

MECHANICS OF WELD RADIOGRAPHY

Most weldments consist of two pieces of metal joined in a way that satisfies a specification, a drawing, or some other means of stating a requirement. In industry, welded joints are most often secured by fusion welding.

Butt Joints with Square Groove

Figure 1 illustrates the fundamental joints encountered in industry. These could be fastened with screws, pegs or rivets and would maintain their identity as joints. In this chapter, we will concentrate on joints fastened by welds; the types of welds involved are illustrated in Figs. 2 and 4 through 8.

The basic type of welded assembly is the *square butt joint* in which the original square-cut faces are butted or prespaced as shown in Fig. 1a. When there is some space left between the faces, which is the common form of assembly, this space is referred to as a *groove*. In the case of butt joints, the groove is square; its shape involves only right angles. Figure 2a also demonstrates this angular relationship.

It is possible to make a weld in a square-cut joint with the joint faces fitted tightly together (with no groove). Such a tight joint can lead to the entrapment of nonweld materials in the weld metal because the joint is not open to let these materials float away. The chances for this kind of difficulty are much reduced when only parent metal is present in the fluid state, as for example, when clean stainless steel faces are fused in an inert atmosphere. If such material is relatively thin, on the order of 1.5 to 3 mm (0.06 to 0.12 in.), it is possible to make a fully acceptable weld by fusing the edges with the arc from a tungsten electrode. If the components have been properly cleaned, the joint faces do not contribute foreign material to the fusion; likewise assurance is required that the electrode, which is nonconsumable, is free of contaminants, and that the gaseous shield is free of material that could contribute inclusions.

An inert atmosphere is needed because oxygen or other gases react with the base metal to produce oxides. Most oxides are ceramic in nature and many are not fluid at the melting point of their primary metal. Thus their presence in a joint often produces entrapped solids that may be considered defects. It is also important that no moisture be present in the protective atmosphere because the arc breaks down the moisture into its constituents, oxygen and hydrogen. In its free state, the oxygen combines with the base metal to form oxides and the liberated hydrogen can cause embrittlement and cracking in the weld metal.

FIGURE 1. Joint Types: (a) Butt Joint; (b) T-joint; (c) Corner Joint; (d) Lap Joint; and (e) Edge Joint

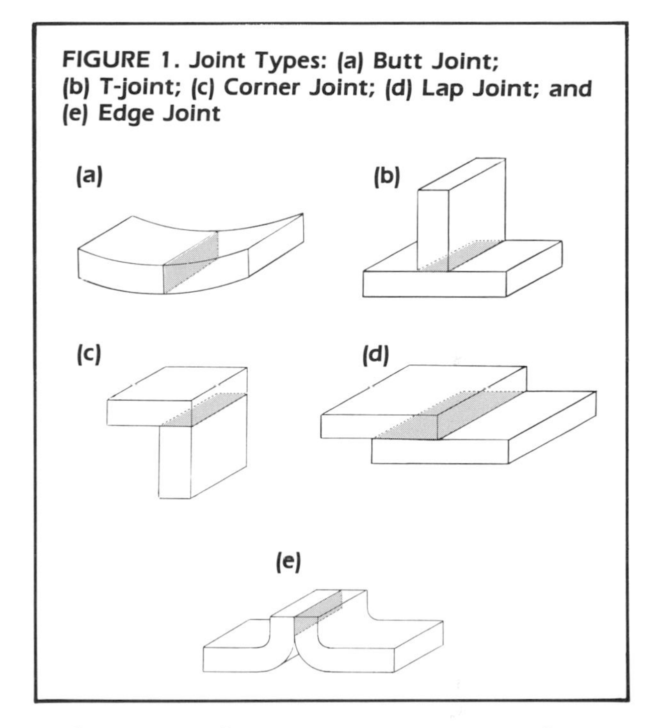

The end result of this is that a square groove joint made without the addition of filler metal can contain discontinuities typical of the joint shape and of the welding process used. In a radiograph, these discontinuities occur in very straight lines when gross quantities of entrapped solids are present. In as much as they usually will have lower mass than the metal involved, they will appear as darker zones in the radiographic image.

Why is the square butt weld often run with zero gap? This is a question of economics. The appeal of this type of joint is based on its simplicity: the weld is made in one pass and no filler material is required when the faces are tightly fitted. There may be a slight depression along the center line of the weld on both surfaces, but the major dimensions of this valley will ordinarily be far under any formal limit. On a radiographic film, the image should be shadow-free except that the melted base metal appears slightly darker.

Butt Joints with Prepared Groove

Single Bevel Joints

Another type of butt joint has a prepared groove. Figure 2 shows ten welded assemblies made up of a single type of *joint*, the butt joint, and a single type of *weld*, the groove weld. The shape of the groove serves to classify the set still further. In Figs. 2a, c, e, g and i, the groove is limited to one side of the joint, the *top* side as drawn. These make up the *single groove* set, with groove shapes as follows: (1) *square*, because of 90° rectangular shape (Fig. 2a); (2) *bevel*, with one angled face (Fig. 2c); (3) *V*, a pair of angled surfaces facing one another (Fig. 2e); (4) *J*, one face shaped to assure economy in the amount of weld metal (Fig. 2g); and (5) *U*, a pair of J-surfaces facing one another (Fig. 2i). Figures 2b, d, f, h and j comprise repetitions of the previous five, but doubled through the thickness. The double joint is used mainly for thick material, but can be specified for thinner members when it is desired to have a better balance of material on the two sides, to minimize distortion.

FIGURE 2. Butt Joints with Groove Welds: (a) Square Groove Single Weld; (b) Square Groove Double Weld; (c) Single Bevel Groove; (d) Double Bevel Groove; (e) Single V-groove; (f) Double V-groove; (g) Single J-groove; (h) Double J-groove; (i) Single U-groove; and (j) Double U-groove

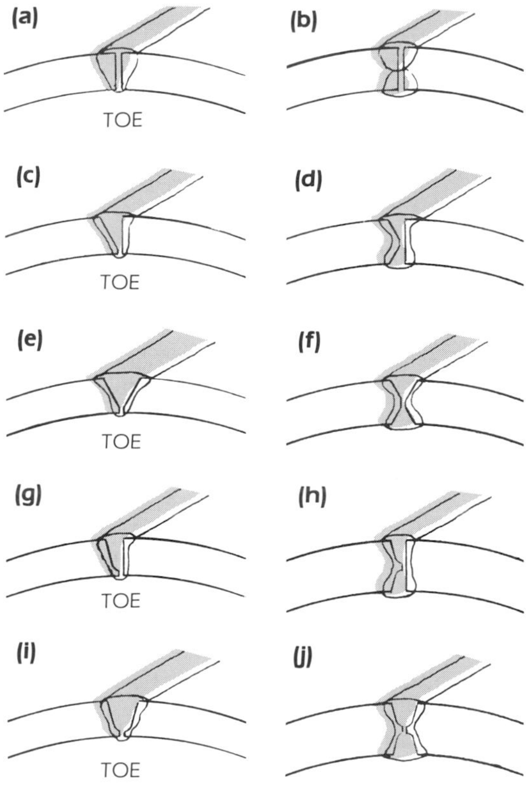

These grooves can be symmetrical or asymmetrical. The groove form is a necessity for joints in thicker materials where there must be access for maintaining an arc and allowing metal to be deposited under controlled conditions. Usually, economic choice will determine whether the V-groove or the bevel groove is used. Note that the smaller bevel groove requires less electrode material to make the joint. Less weld metal means less distortion. In these simple joints, metal is added from one side only; when distortion occurs, the assembly bends in the direction shown in Fig. 3.

In Fig. 2, there are flat zones visible at the toes of the joints. This configuration is common, giving some stability to the bottom corner. If the corner were a sharp V, the pointed area would melt away under the arc and excessive penetration could result. This small zone at the root is actually a square groove joint of modest dimensions. When a series of weld passes is required to complete a joint, the first pass, in which this small, square butt is involved, can become very critical. Often this early pass will be cut away after the weld is completed, and a new root weld will be made from the back side. The radiography of such a joint should not be done until all these steps are completed.

There are cases, as in pipe welding, where it is very difficult to properly clean out and reweld the

FIGURE 3. Potential Distortion with Welding Done from One Side Only

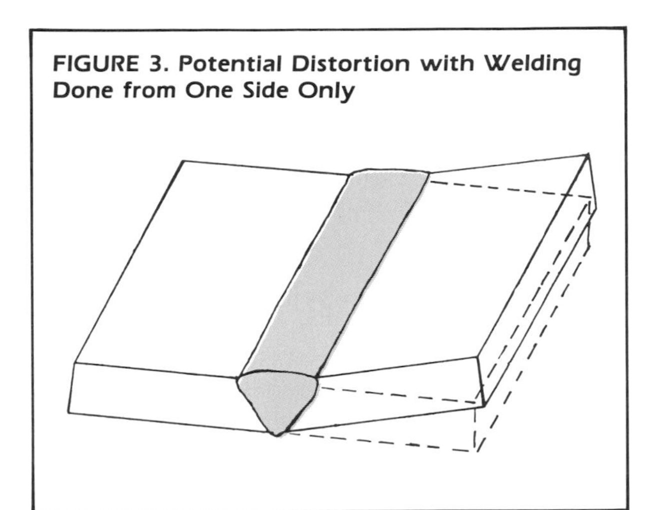

single V groove or bevel joint from the back side. This kind of difficulty usually contributes to the production of poor welds.

A refinement in one-side welded joints is to use a specially shaped filler that fits neatly into the groove from the back side. This filler is then melted into the first pass made from the open side. The filler is called a *consumable insert* because it is consumed during the root pass of the weld. Radiographic inspection of such a joint includes checking that complete fusion has occurred at the inner corners. In the welding trade, these are known as *EB* or *electric boat* inserts.

Double Bevel Joints

In order of increasing complexity, the next form of joint is the *double-V* or *double-bevel* preparation shown in Figs. 2d and 2f. These are made up of successive mirror images of the single bevel joints and are used when the material is too thick to be welded from one side without serious distortion. A root pass is again performed, but this time it is done in the mid-section of the weld. This mid-section is a small, square groove, butt joint, and requires checking, as above, for the similar geometry.

In general, any difficulties in this type of joint tend to be concentrated around the mid-section. Discontinuities will be further from the film and this effect will reduce image sharpness. The condition is not necessarily critical because the mid-section of a weld usually is near the neutral axis of the shape involved; this is the zone where neither tension nor compression occurs under bending. When fitness-for-purpose criteria are applied, fairly large discontinuities can be tolerated in the mid-section.

In the case of the bevel weld preparation, there is a vertical face involved, when the weld is to be made in the flat position (Figs. 2c, 2d and 9a). This configuration can lead to some difficulties in welding. When the square face is oriented vertically, and weld metal is being deposited, there could be problems in developing a uniform fusion layer on this face. The resulting discontinuity is referred to as *incomplete fusion*, and its potential effects are described later under *Discontinuities in Welds*. When it is feasible to position the joint so that the square face is horizontal during welding (Fig. 9b), this difficulty may be avoided.

J-Groove Joints

There is another group of joints for groove welding, those using the *J-groove* (Figs. 2g and 2h). This configuration is used in the same situations as the bevel and V preparations, with the J shape replacing the bevel or V. This is a technique used to permit welding in thicker materials while requiring less filler metal than the angled preparation. Using less filler metal has two positive effects: (1) costs will be reduced; and (2) there is less chance of distortion from weld metal shrinkage because the top of the groove will be much narrower than with the bevel. The side of the J approaches a vertical orientation when welds are in the flat position. Because of the vertical face, this joint is susceptible to the discontinuities typical of the bevel groove, described above.

U-Groove Joints

The last of the regular groove-weld joints is the *U-groove* (Figs. 2i and 2j), consisting of two opposing J-grooves. As with the other configurations, the U-groove may use the single or double mode. It is intended to ensure access for the welding of thicker sections, while minimizing the amount of filler metal involved, and will be subject to the same sort of discontinuities as the J-groove, plus the possibility of fusion problems in the root area.

FIGURE 4. T-joint with Fillet Welds

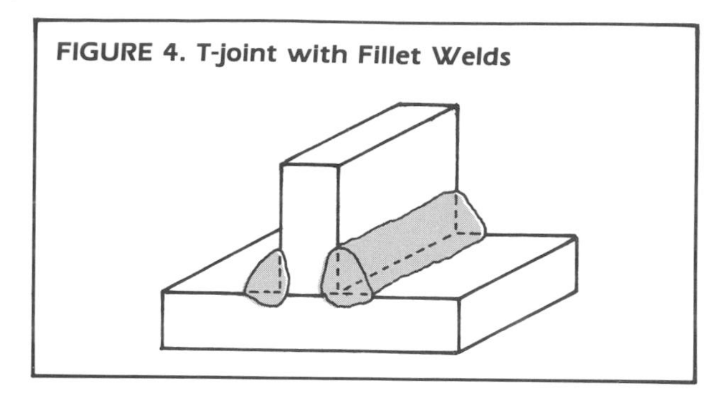

FIGURE 5. T-joints with Groove Welds: (a) Single Bevel Groove; (b) Double Bevel Groove

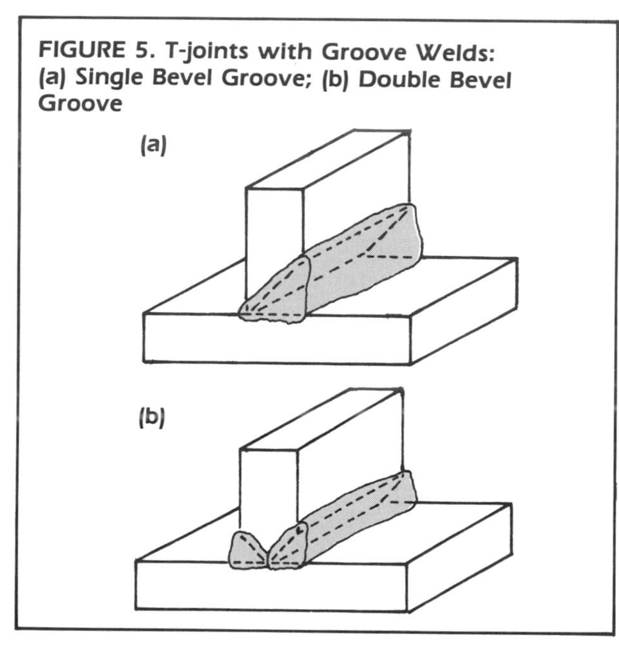

FIGURE 6. Corner Joint with Single Bevel Groove

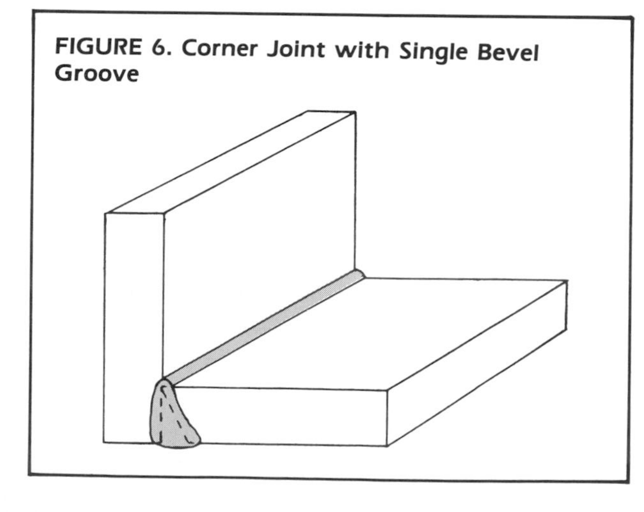

Other Joints and Welds

There are some further joints of more complex geometry which are not as well suited to radiographic inspection as the butt joint.

The *T-joint*, for example, may be assembled with one or two *fillet welds* (Fig. 4). This is the weld most commonly used with the T-joint. It is usually the most economic assembly because no special preparation (machining) is required. The T-joint with fillet welds is not an easy assembly to inspect because it is not easy to set up for reliable radiography. In its most rudimentary form, a single fillet weld, confined to one side, would be a candidate for radiography. While this weld may be radiographed by shooting from above (with respect to the orientation of Fig. 4), such welds should not be subjected to refined methods of inspection; the reason is that an unfused zone exists in the original joint interface, and this thin gap looks very much like incomplete penetration (see Part 5 of this section for more information on fillet weld radiography).

More refined models of the T-joint are shown in Figs. 5a and 5b, where prepared welds are illustrated. Some options available to the weld designer are shown in this figure. Other forms of preparation (shown in Fig. 2) might also be used. The welds shown would be amenable to radiography by positioning the film under the base plate. Joints such as this may be expected to carry dynamic loads in service, such as in bridges.

Another fundamental joint, the *corner joint*, may be bonded by several kinds of weld. A totally welded corner, with a prepared single bevel weld groove, is shown in Fig. 6.

This is one of several ways used for fastening the corner joint. The simplest method is a fillet weld on the inside corner, such as those fillets shown in Fig. 4. Any of the other single groove shapes in Fig. 2 could be used here as well.

For static service (buildings not subject to variable wind loads, for example) the simple fillet weld could be used on the inside corner.

A fourth complex joint is the *lap joint*. This is usually assembled with a pair of fillet welds as shown in Fig. 7. This unbalanced-looking joint is a natural for the fillet weld. There would be no point in cutting any groove shapes as a preparation. A similar joint can be used to join two extended plates by butting them and adding a pad bonded by fillet welds; the plate junction is made by including one of the groove welds associated with the butt joint.

A final edge joint requires the simplest form of weld. As shown in Figs. 8a, b and c, this joint is ordinarily secured by melting existing flanges. Note that the configuration in Fig. 8a could be welded on the reverse side, using the natural groove there; it would then be classed among the groove welds.

FIGURE 7. Lap Joint with Double Fillet Welds

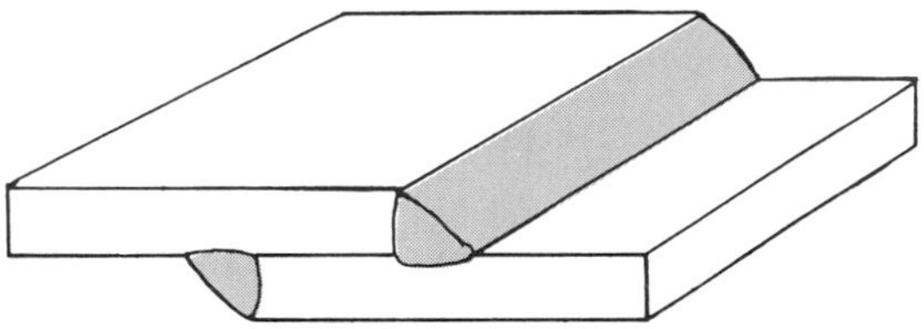

FIGURE 8. Edge Joints: (a) Normal Fused Edge; (b) Thin Sheet before Welding; (c) Thin Sheet after Welding with Flange Consumed

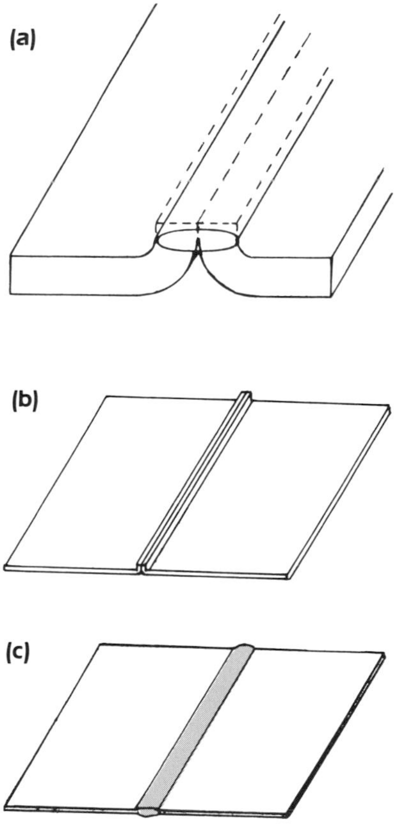

Welding Position

There are six recognized welding positions. Discontinuities associated with gravity, with fluidity, and with the skill of the welder can occur in at least four of these positions: flat, horizontal, vertical and overhead. In Fig. 9, the welds are shown lined up with the horizon, or at right angles to it. Actually the various positions can rotate or tip through a range of about 60° and still be described as labeled in Fig. 9.

The welding positions for groove welds in pipe are shown in Fig. 10. Except for Fig. 10a, in which the pipe is rolled while the weld is made from the top, these pipe welds are difficult for the welder. The T-connections or the skewed joints in the Y-configuration and the K-configuration (see Fig. 11) are even more challenging, but are not suited to inspection by the radiographic method.

The *flat position* (Figs. 9a and 10a) is generally considered the easiest to work in, but this is only true for symmetrical welds of the square groove and V-groove types. The bevel and the J-groove for the flat position are much more difficult. When planning for radiography, it would be useful to know the actual welding position so that the X-ray beam can be appropriately directed.

The *horizontal position* (Figs. 9b and 10b) is next in welding difficulty. When the joint has a single bevel, or a single J — and the cut face is horizontal, or near horizontal — it is considered much less troublesome to weld. A welder fully skilled in this position can produce welds of high quality, but if the groove should be more open, the horizontal platform would no longer exist and the chances for error are enhanced.

The third regular position is the *vertical position* (Fig. 9c). In this case, gravity acts along the groove and the groove shape does not have nearly the same importance. The shape does matter, however, when considering accessibility to the prepared faces. This proviso applies for all positions. All prepared faces must be properly fused with the filler metal. Angles must not be so sharp as to make the groove too narrow; this would interfere with the manipulation of the electrode.

The fourth position in this sequence is the *overhead* (Fig. 9d) and this is by far the most difficult welding to perform. Again, however, fully skilled welders can produce acceptable welds in this position. The effects of gravity differ from the previous cases because a double effect prevails here. First,

FIGURE 9. Welding Positions for Groove Welds in Plate: (a) Flat 1G; (b) Horizontal 2G; (c) Vertical 3G; (d) Overhead 4G

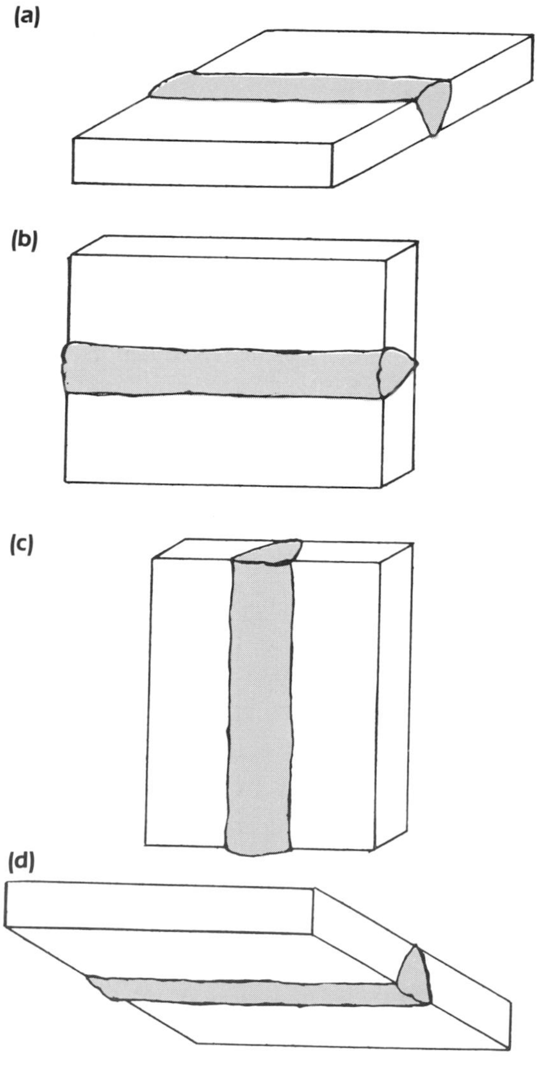

(a)

(b)

(c)

(d)

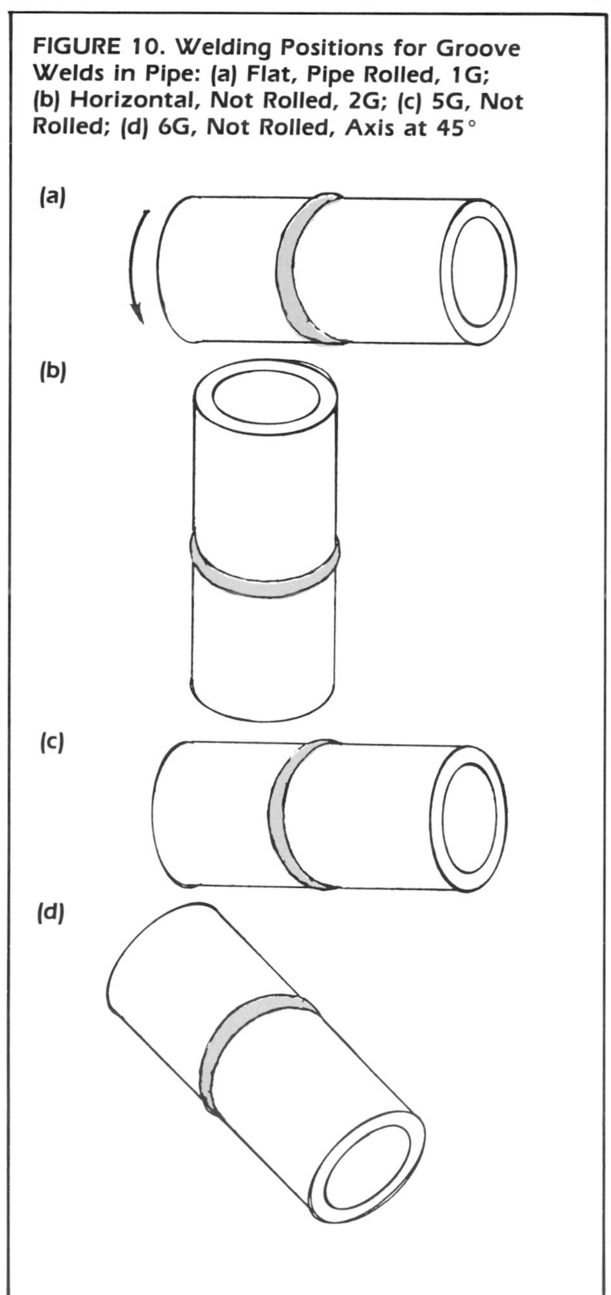

FIGURE 10. Welding Positions for Groove Welds in Pipe: (a) Flat, Pipe Rolled, 1G; (b) Horizontal, Not Rolled, 2G; (c) 5G, Not Rolled; (d) 6G, Not Rolled, Axis at 45°

there is a tendency for the weld metal to slump, and this results in irregular welds. Secondly, when flux is involved (which is most of the time), it first tends to float up and out of the weld metal and into recesses of the joint; the welder must overcome this tendency by ensuring that thrust from the arc (*arc blow*) displaces the molten flux before it can harden in place.

The above four positions are basic and apply to groove welds and fillet welds, as summarized in Figs. 9 and 10 for groove welds. The letter *G* following the position number stands for *groove*. In the case of fillet welds, the position designations would be *1F* through *4F*.

In the welding of piping, another degree of freedom exists in that sometimes pipe can be rotated; then a joint which might have required all the positional skills can be completed *downhand* or essentially in the flat position. The basic set is shown in Fig. 10 where the designations are *1G Rotated* (flat), *2G* (horizontal), *5G* (combining the four basic positions) and *6G* (oblique). The term *rolled* applies for only one case.

The illustrations presented here represent test models to evaluate a welder's skill. They also represent the welding positions that apply to actual fabrications. In the workplace, conditions can be more extreme. There can be difficulties with access to joints or to portions of joints. If it has been difficult for the welder to get access to a weld, it will be equally difficult for the radiographer.

Standards for welding now often include requirements covering accessibility in joints that are to serve in static or dynamic roles. The American Bureau of Shipping's *Rules for Building and Classing Steel Vessels* (see ref. 3) includes (1) clause 30.21.3, requiring that pipeline joints welded on board are to be in proper positions; and (2) clause 30.49.2, assigning qualification tests for welders, and referring to figure 30.13 which indicates that test welds in pipe must be 300 mm from two walls and 150 mm from the floor.

However, such standards usually do not include specific requirements to facilitate radiography or the other NDT methods.

In considering these welding positions, it is not to be inferred that the radiographic practice or requirement would vary to suit the various positions. The welds are designed to carry loads, static or cyclic, and the inspection to be done must be based on the implied service of the weld, not on the position at the time of welding.

FIGURE 11. Tubular Configurations: (a) T-connection; (b) Y-connection; (c) K-connection

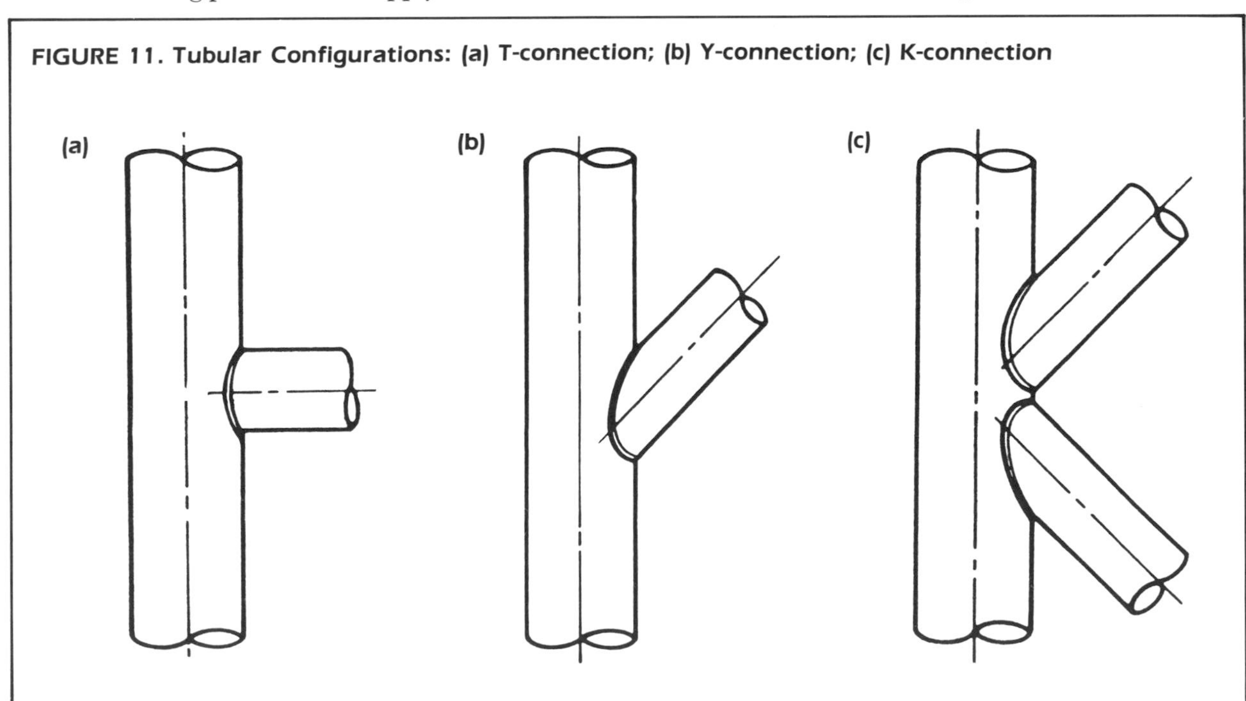

PART 2

WELDMENT MATERIAL AND THICKNESS

Material Density and Atomic Number

The material involved in a weldment influences radiography mainly through its density and, more importantly, its atomic number. The complete set of regular metals, from lithium to osmium, exhibit densities from 0.53 to 22.57 grams per cubic centimeter (g/cm^3). The penetrating ability of the radiation is a direct function of the density of the material being examined. Fortunately, however, radiographic equipment designed to examine commercial alloys does not have to cover the complete range just quoted.

Welded materials ordinarily range in density from 1.78 g/cm^3 for magnesium to 8.57 for niobium. Comparisons of density alone are not sufficient when comparing materials; the absorption of ionizing radiation by a specimen is also affected by the atomic structure of the material.

In general, materials of high atomic number and/or increased thickness require a relatively high radiation energy for radiography. There are practical limits involved here because neither the energy of the radiation source nor the time allowed for an exposure can be extended indefinitely.

The first limitation arises with transformer-type high voltage sources; the insulation materials needed to withstand extreme voltages either do not exist or would be prohibitively expensive. There are other means of attaining the higher energy needed, but these voltage generators are often too expensive for routine work with weldments.

The second limitation, the time required for an exposure, may become impractical. Excessively long exposures may imply an inappropriately low radiographic energy.

Material Thickness

The thickness of the material under examination has an influence that acts in parallel with the specimen density. As the thickness is increased, the energy of the radiation source must be raised. A limit occurs because more energetic components of the radiation can move through light materials with minimal attenuation. As a result, the detector can be overexposed to primary radiation.

Another effect related to the thickness of the material has to do with how close the zone of interest is to the plane of the detector. This distance has a direct influence on the sharpness of the shadow generated by the radiation beam.

The geometrical unsharpness is also controlled by the size of the radiation source. In X-ray devices, it is the size of the focal spot; in a gamma-ray source, it is the physical size of the isotope.

When butt joints are examined, one zone in the weld could be of more interest than another. It is possible to ensure that the detector is positioned so that the potential discontinuities are closer to the film. This applies to the single-V or bevel groove, when the root of the joint is virtually on one surface. This portion of any groove weld with root passes is the most susceptible to trouble. With double groove welds, however, the root zone is at or near the mid-section and the above positioning technique will not work (see Figs. 2d, 2f, 2h and 2j).

PART 3

RADIATION SOURCES FOR RADIOGRAPHY OF WELDS

Equipment that could be used for the radiography of weldments is usually one of two types: (1) standard X-ray generators; and (2) gamma-ray sources.

Standard X-ray Generators

All X-ray units are high voltage devices. The standard equipment recommended for weldments ranges from 50 kilovolts-peak (kVp) to 25 million electron volts (MeV). This range covers weldment materials with densities of 1.8 to 8.6 g/cm³. There are limitations in thicknesses that could be adequately covered, due to a combination of the attenuation coefficients and thicknesses. Figure 12 presents some thickness limits for X-ray machines both in the regular voltage ranges and for high-energy X-ray generators.

Radioactive Isotopes

The exposure time required for radiography with isotopes depends on the sources' specific activity, and to some degree on the dimensions of the isotope capsule. This activity level is measured in *curies* (Ci).

The shape of the isotopic pellet also influences its usefulness in radiography. Only the atoms near the outside of the pellet emit at the full rate. For those in the interior, the radiation they emit must first penetrate part of the pellet itself, and because the

FIGURE 12. X-ray Potential and General Thickness Limits for Steel

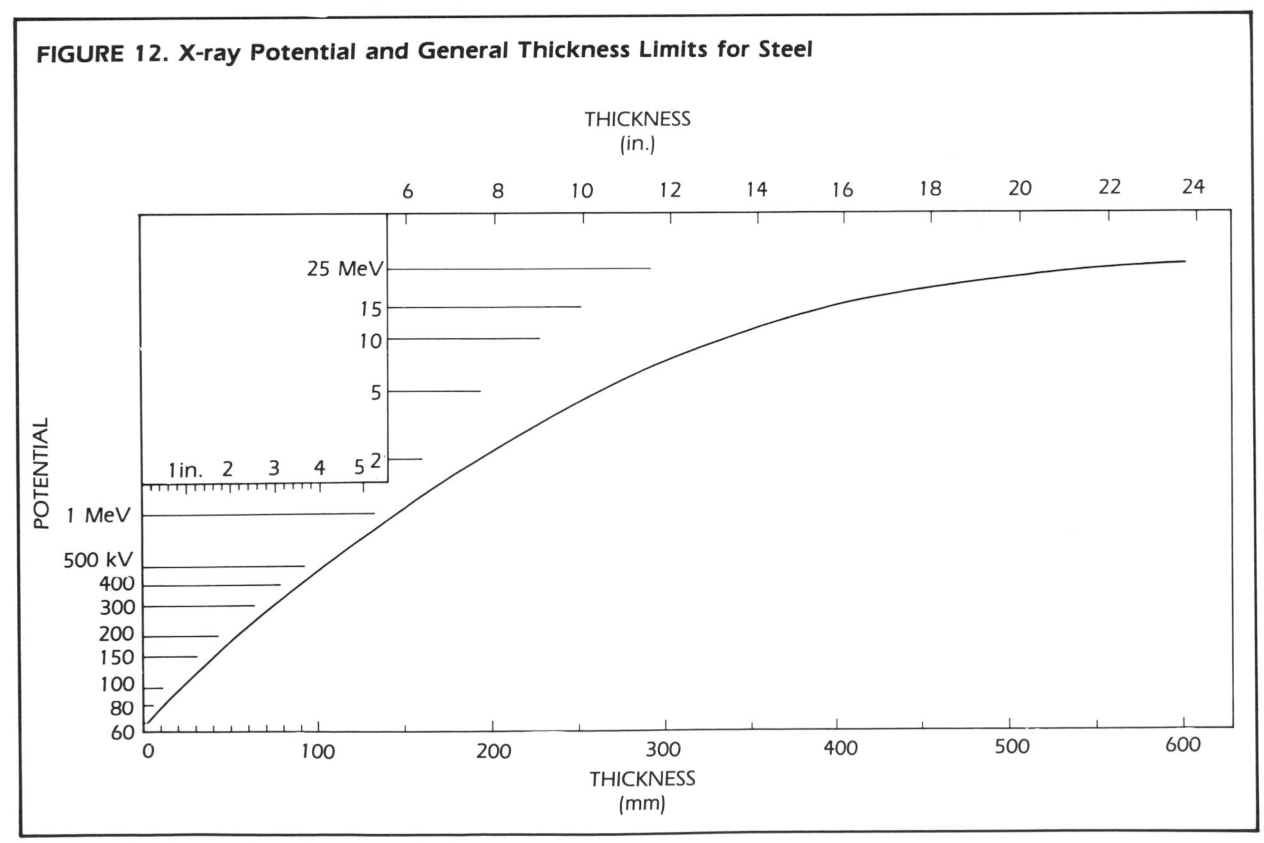

FIGURE 13. Shape Factor for a Radioactive Isotope: (a) Point Source; (b) Source on End; (c) Source on Side

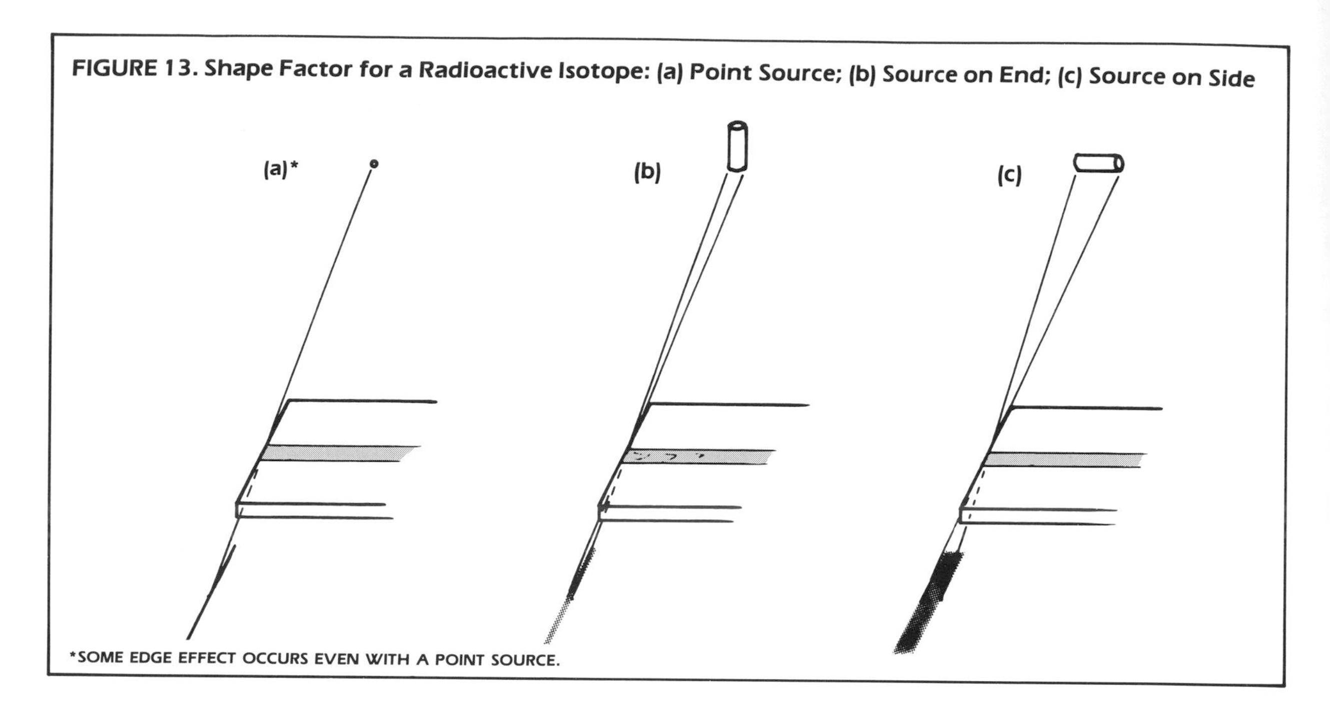

materials used are very dense and of medium to high atomic number, a loss in intensity ensues. Thus, the more active pellets are cylindrical with a relatively long axis.

Figures 13a, b and c show how the sharpness of a weldment shadow is a function of the size of the source of radiation (shape factor). In the case of larger sources, every point on the surface acts as a single source and generates its own portion of the shadow. The larger the cross-sectional area, the more diffuse the shadow; for sharp radiographs, smaller sources are better.

Figure 14a, with sources of equal net power, demonstrates the geometry effect shown in Fig. 13. The smaller, more intense source will yield a sharper image for an equivalent exposure time.

Figure 14b again has sources of different specific activity, but this time of the same size. They will generate equivalent images, with the exposure time being the only difference.

FIGURE 14. Isotopic Exposure of Radiograph: (a) Geometric Effect with Equivalent Exposure; (b) Exposure Altered with Equivalent Geometry

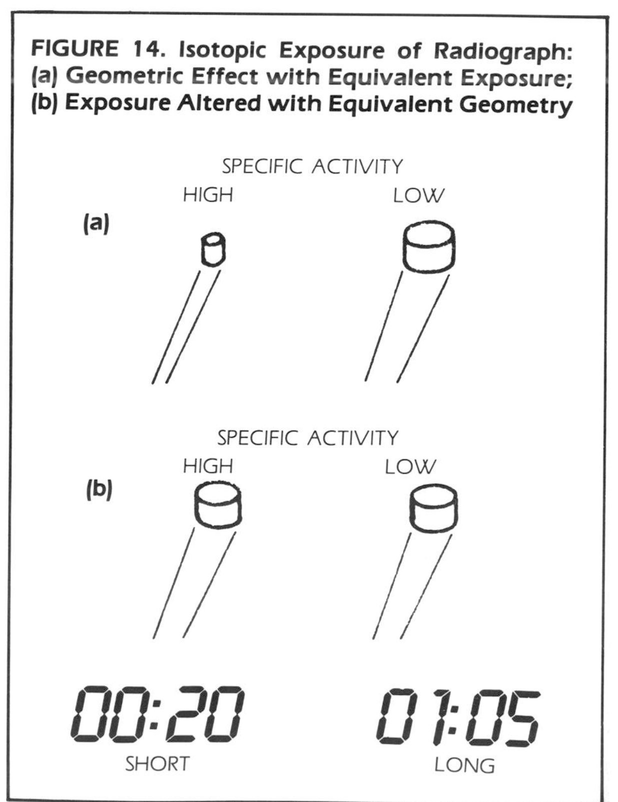

Frequency Distribution

The energy span for radiation sources is based on a frequency distribution, but is customarily presented in terms of wavelength, the inverse of frequency. The electromagnetic spectrum extends from very long wavelength radio waves, which can be several miles in length, to very short and energetic X-rays. This is a continuous spectrum, with visible light occurring near the middle of an extremely wide band (see Fig. 15). Moving toward the shorter wavelengths, visible light gives way to ultraviolet, though there is no sharp boundary. Continuing toward shorter wavelengths, the ultraviolet graduates into soft X-rays. These longer wavelength X-rays have low penetrating ability. Radiation in this range is not used for industrial radiography but does find use in forensic science. This low energy zone is used for X-ray diffraction analysis where it yields information on inter-atomic spacings in metals and nonmetals. Tube voltages on the order of 25 kV to 50 kV can be used for thin welds. At somewhat shorter wavelengths comes the medical and surveillance X-ray band. The tube voltages involved here are typically from 45 kV to 100 kV. Overlapping into this section is the functional range for the industrial radiography of light metals (magnesium and aluminum alloys) and this extends to about 220 kV for heavier sections.

FIGURE 15. Electromagnetic Spectrum

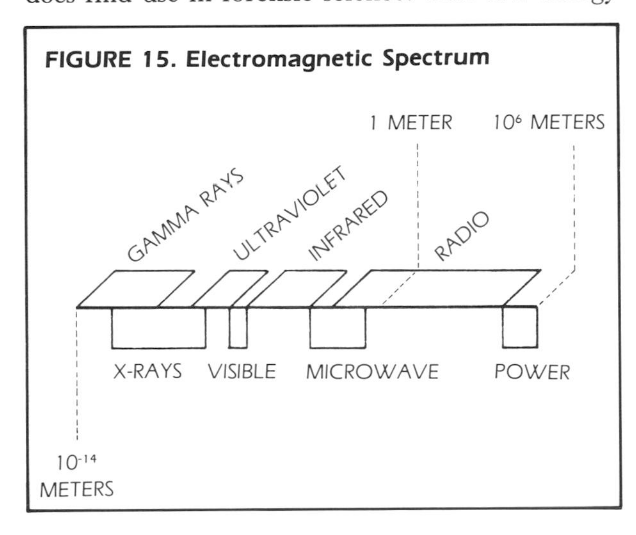

The next range, up to 25 MeV, incorporates steel and other metals of similar radiographic behavior; it overlaps the aluminum/magnesium range, down to about 100 kV. The lower end of the range will function for thinner weldments joined by resistance, spot and seam welding.

Figure 16 illustrates that the radiation from radium-226 is at the highest energy, but cobalt-60 exhibits the highest practical energy content. (Figure 16 is a simplified illustration. For more specific information, see Section 3, *Isotope Radiation Sources*). Virtually all the useful radiation from cobalt-60 occurs in a narrow, high-energy band. This effect is the opposite of having a wide frequency band, as occurs with X radiation.

As a result, if a thin weldment, for example, of aluminum or steel, is irradiated with cobalt-60, it is saturated and a film detector shows no detail even if there are significant changes within the weld. By contrast, wide spectrum radiation suffers more absorption and this induces some variation in the image of a thinner part. A radiographic film exposed to cobalt-60 may have a very flat look, because the nearly single wavelength has passed uniformly through the material. A different effect can be ex-

FIGURE 16. Major Radioactive Isotopes Used in Industrial Radiography

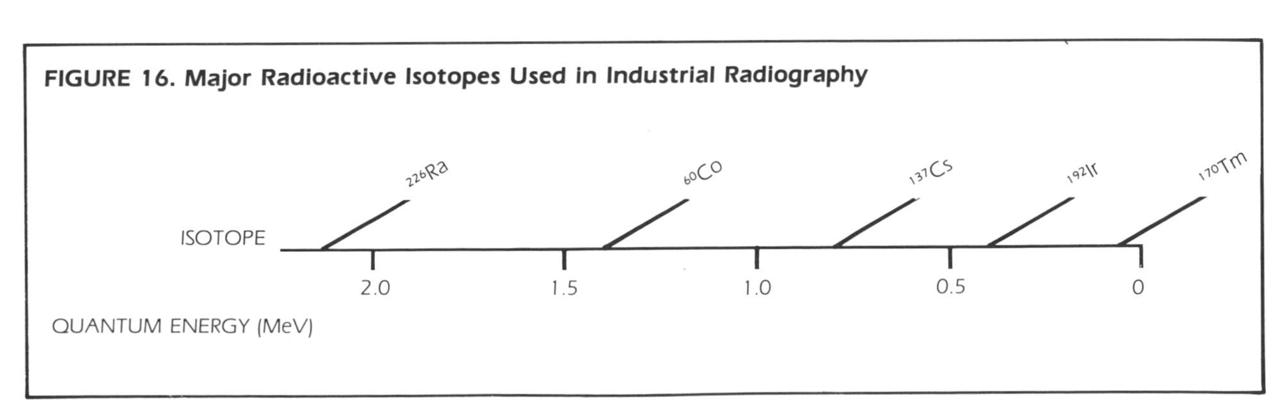

TABLE 1. Thickness Ranges In Steel Corresponding to Radiation Energy

Maximum Radiographic Energy	Effective Penetration Range in Steel
150 kV	Up to 16 mm (0.625 in.)
250 kV	Up to 40 mm (1.5 in.)
400 kV	Up to 60 mm (2.5 in.)
Iridium-192	25 to 75 mm (1.0 to 3.0 in.)
1,000 kV (1 MV)	6 to 90 mm (0.25 to 3.5 in.)
Cobalt-60	40 to 220 mm (1.5 to 8.5 in.)
2.0 MeV	6 to 250 mm (0.25 to 10.0 in.)
4.5 MeV	25 to 300 mm (1.00 to 12.0 in.)
7.5 MeV	60 to 450 mm (2.25 to 18.0 in.)
20.0 MeV	75 to 600 mm (3.00 to 24.0 in.)

pected when the material is of high atomic number, such as tantalum or uranium.

The other commercial isotopes are less penetrating than cobalt and will accommodate thinner weldments. Table 1 shows some thickness ranges in steel, corresponding to the respective radiation energies. Note that upper and lower limits are presented for each. The higher energy gamma ray source isotopes are not suitable for critical applications involving light alloy weldments.

PART 4

RECORDING MEDIA

Film

An important factor influencing the choice of film is the total cost of the operation. Slow, fine-grained film costs roughly the same as other film types but it can end up costing much more because of the longer exposures it demands. The time effect can be twofold: (1) radiographic personnel are required for a longer time; and (2) the working areas of fabricating shops may have to be shut down for safety reasons. With the more dangerous radiation sources (the isotopes) it is customary to operate hazardous equipment on off-shifts. This procedure can be very safe but it can add materially to the interval between the actual exposure and the evaluation of the component, when the film reading must occur on the following day.

With film, there is also a choice to be made for the type of radiation being employed: high or low energy. High voltage X-rays and some of the isotopes penetrate materials more easily and may reduce contrast. The information available from film is contained in its optical density variations. If this *variation* is reduced by the quality of the radiation, whether the film is light or dark, the information content also decreases. The choice of film is then governed to a degree by the type of radiation; for example, high contrast film can help to compensate for relatively low subject contrast.

Another film parameter to consider is the kind of material being radiographed. The lighter metals require radiation such as a low voltage X-ray source or an isotope that emits less penetrating radiation (thulium-170). For thin sections of light metals, only the lower energy X-ray beam will suffice; the useful isotopes do not emit enough lower energy radiation to match that available from low energy X-ray generators. Thus, for thin sections of light metals, the film is chosen to respond to lower energy radiation.

Paper

Radiographic paper, like film, has a silver-halide emulsion, but on paper the emulsion is present in just a single layer. The image formed must be viewed by reflection rather than by transmission. For convenience and rapid sorting, however, the paper system has definite advantages, not the least of which is cost. Acceptance codes tend to refer to film as the reference medium and this means that paper radiographs would not match a formal requirement. Paper does have a real place, however, in quality surveillance systems where the cost of the inspection operation is an important factor.

PART 5

TECHNIQUE DEVELOPMENT

Sensitivity

The choice of a radiographic technique is usually based on sensitivity. A weld zone will always be different in structure and density from the parent material and ordinarily there is no need for these differences to be highlighted or reported as significant information. Thus, the sensitivity of a proposed technique should not approach absolute.

Figure 17 presents sensitivity information in terms of percentages, with 2% applying to fairly critical projects and 4% to less critical work. As an example of scaling, in a steel section 10 cm (4 in.) thick, a 2% sensitivity would correspond to 2 mm (0.08 in.). This latter figure is roughly the total thickness of a typical spot-welded assembly.

Setup for Exposure of Various Weld Types

Following are examples showing how the radiography of specific weldments could be handled.

Butt Joint

Consider a butt joint in 7.5 cm (3.0 in.) steel, as might occur in the wall of a pressure vessel. The welding technique should be in accordance with the American Society of Mechanical Engineers *Pressure Vessel Code*. For this example, Section VIII, Part I (see ref. 1) will be chosen.

To ensure weld quality, the designer will probably have the joint prepared for welding from two sides, with the initial welding pass made at the midsection, which becomes the root of the joint. This weld preparation is called *U-preparation*; the groove shape consists of two J's facing one another. Such a configuration permits access to the root for welding and requires much less material to fill than a V-shaped opening. Good practice requires that the backside of the root pass be ground or gouged, leaving a groove for the first pass on the second side. Such a weld usually requires at least five passes on each side. These would be sequenced to minimize stress and distortion, but this does not influence the radiographic method.

If the initial root pass and the first pass on the second side have been performed properly (and confirmed by visual examination at the time) then there should be no radiographic indications associated with the root of the joint. The most likely areas for difficulty are on the steeply sloping sides of the groove, and it is for these zones that the radiographic test is designed. The radiography will require two exposures, with the beam directed in line with the preparation angle of the groove. The symmetry of the joint will ensure that one shot on either side of the weld can be used to sample the two flanks (Fig. 18).

Figure 19 further illustrates these conditions. The film will be evaluated according to the ASME Section VIII requirements. The actual length of the weld being sampled affects the exposure geometry; note that with a flat joint, the distance from the source to the weld will be at a minimum along the central axis of the radiation beam.

The beam direction has a further effect on the test because as it shifts from the 90° orientation, a discontinuity's projected shape and position change. The visibility of the penetrameter will serve to show whether the distortion significantly affects the image. The specified film density is broad enough to accommodate the variation arising from this projection effect. In some cases, the finished weld bead must be ground to make the weld region flush. In the case chosen here, grinding is not required; instead, a maximum bead height is indicated. If this were not taken into account, the extra bead metal could lead to some underexposure of the film; exposure parameters must be based on the total thickness.

Joining Varying Thicknesses

Another common variation occurs when two members of differing thicknesses are fastened by a butt joint. For such a case, the weld area is a tapered zone. While there are limits set for the angle of such a taper, the difference between the two thicknesses has a critical effect on the radiographic process. Accommodating such a difference may require a special arrangement, such as the use of two film speeds for a single exposure. By this method, the thinner steel section will be captured on slower film and the thicker section on faster film. To evaluate such a radiograph, the two films must be viewed together; otherwise, a further, synthetic tapered effect is superimposed on the image of the tapered metal.

Joining Different Materials

Only one type of material is involved in the previous example, but another variable can arise in a welded joint when more than one type of material is involved. A typical case of this sort involves joints that will be exposed to significant magnetic fields, such as in large electrical equipment. These configurations often include two parts fabricated from steel joined by a nonmagnetic weld of stainless steel. The weld serves to break the continuity of the magnetic field that occurs when the component is in service. To the radiographer, there will be a noticeable change in radiographic density between the steel and the stainless steel. It is important that the form of the joint be known so that the change in density will not be interpreted as a change in thickness.

Lap Joints

The butt joint with a groove weld is the simplest welding arrangement and the one yielding the easiest interpretation because of the general uniformity of the assembly. A less uniform weldment is the *lap joint* secured by two fillet welds.

FIGURE 17. X-ray Potential and Steel Thicknesses for Two Levels of Sensitivity

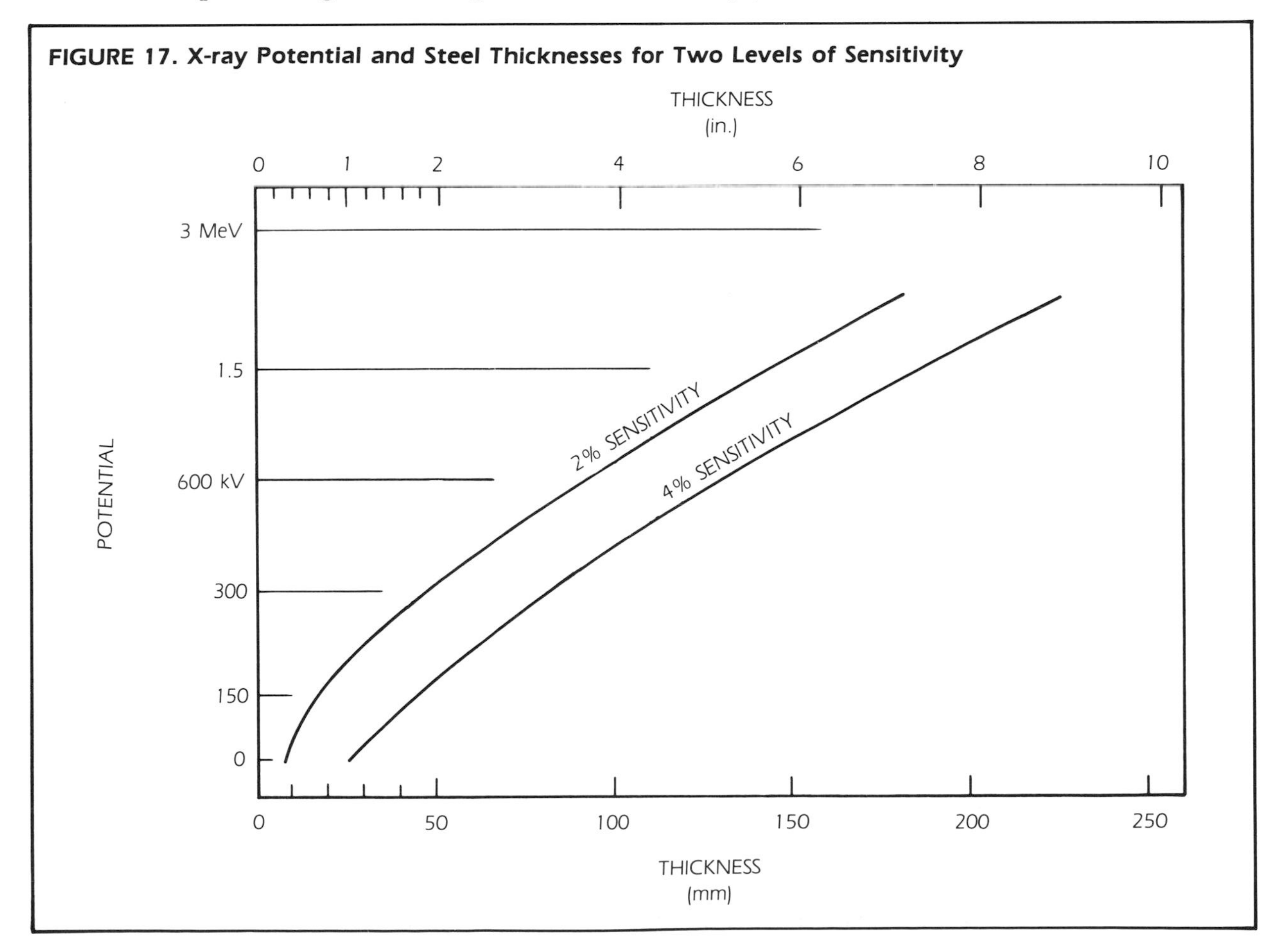

This type of joint is used on vessels fabricated according to ASME VIII specifications; if the requirement were to Part 2, with an opening already fitted with a nozzle, a pipe might be attached using a lap joint. Being integral with its vessel, such a joint is subject to loading and will have some low strength requirements. The assurance requirements for this joint could include verifying that sufficient weld metal is present in the joint.

It is possible to make a fillet weld that, from the outside, appears full-sized but is actually hollow. A radiographic requirement is imposed to ensure that the bond has been made over the complete edge of the applied pipe. The simplest technique for this is to shoot the radiograph through the weld area.

Because the weld is triangular, there is no means of directing a beam so that a constant thickness is examined. Also, the film plane is not close to the weld zone. The interpretation of the film from this seemingly simple joint is quite complex because there is a large film density gradient over a small dimension (the projected width of the weld). In this case then, one could expect only to confirm the presence of relatively large discontinuities. This is acceptable because the use of a simple lap joint indicates that the joint is not expected to meet high strength requirements. The radiographic test, therefore, is in agreement with the projected service.

T-Joints

A further increase in complexity occurs with the T-joint, of which there are two types: one with full penetration, making a completely welded assembly, and the simpler case with fillet welds in the corners. When there are only two fillets involved (or even one) the radiographic assessment becomes very similar to that for the lap joint: a varying thickness of material is presented to the radiation beam, and the film plane is separated from the weld metal by the thickness of the lower plate. Again, however, if radiography is required, the sensitivity does not have to be of an extremely high order because a joint of this type is only partially welded and would never be used in a critical application.

A more refined T-joint involves a groove weld, or welds, rather than simple fillets. The vertical member is prepared from one or both sides. The complete weld is made through the thickness of the web. As shown in Figs. 5a and 5b, such a weld could be radiographed, but a distorted image will result and the effective sensitivity is significantly reduced because of the base material thickness.

There is still some variation in the effective thickness across the weld, but this is much less than for the fillet weld. Note that a moderate fillet is usually present in a practical T-joint, but that the extra material is cosmetic and does not contribute to the strength of the joint.

The weld in a prepared T-joint can be considered fully load-bearing and may therefore have performance requirements that will justify a sensitive radiographic technique.

Corner Joints

As with the lap joint and the simple T-joint, the corner joint may be assembled with a minimum of welding using a fillet weld in the corner. This weld is not used when full loading is required, and thus does not require refined radiography. If radiography is used, the joint has one special advantage: there could be a preferred direction for the beam that would not involve the unwelded portion of the joint (see Fig. 20). There is still a variable thickness

FIGURE 18. Radiography of Weld Joint Based on Double U-groove; Both Source Positions in Line with Prepared Faces

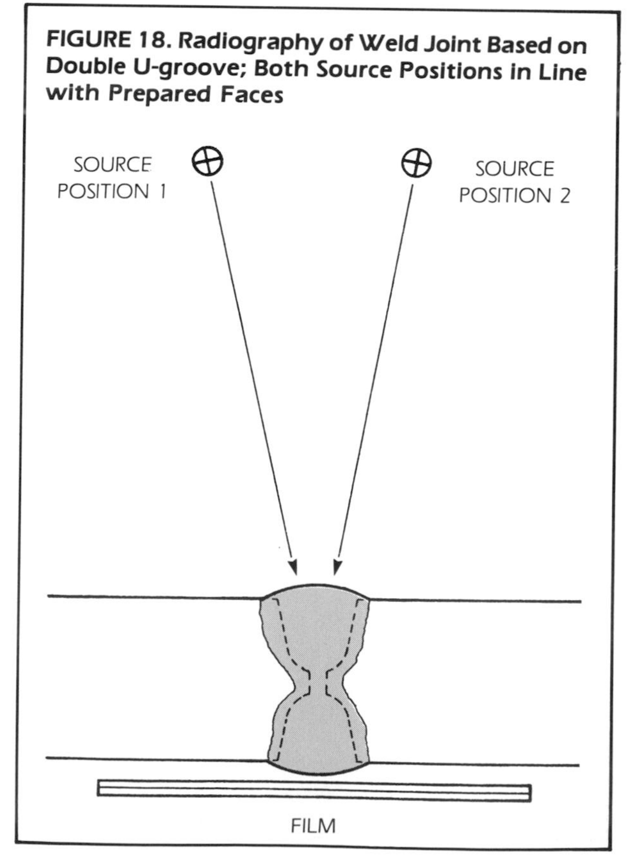

presented to the radiation, but this effect may be minimized by having the beam come in at about 20° from the vertical.

The unwelded portion presents a large open notch that limits the usefulness of such a simple assembly. The corner joint may also be prepared as a groove weld, just as for the T-joint. For a groove weld, the preparation usually involves work on one member of the joint. When welded, the corner must be solid metal. The weld is amenable to radiography because the film and the weld metal are virtually in contact.

Fillet Welds

The fillet weld is commonly used with the lap joint, the corner joint and the T-joint; the fillet weld may be used in conjunction with a groove weld in a corner or T-joint.

For a 90° T-joint with a symmetrical fillet weld, the shortest path through the weld is that bisecting the 90° angle, giving a radiation beam angle of 45°. If the film holder is positioned in contact with the flange of the T-joint (the top of the *T*), then the radiation beam must pass through a relatively large thickness of metal.

Another approach to the radiography of fillet welds is to position the film on the weld side of the joint. In this case, the thickness of the film holder will have some considerable influence on the radiographic resolution, but this limitation only applies when the film holder is flat. Flexible holders are available that can be bent to conform to the weldment and thereby bring the film closer to the weld bead. The extra metal in the radiation path does *not* fall between the weld zone and the film, and resolution will usually be improved. In this set-up, the beam is at right angles to the weld surface, and this is the best arrangement for detecting centerline cracking. A similar arrangement is useful for T-joints in which the web (the stem of the *T*) is prepared for welding.

Other arrangements may be used to compensate for the uneven geometry of the fillet weld. One example is the introduction of metal wedges, prepared with shapes complementary to the fillet shape, and fitted between the film and the radiation source. The wedges may be in contact with the weld sur-

FIGURE 19. Factors Determining the Image Quality on a 7.5 cm (3.0 in.) Weldment

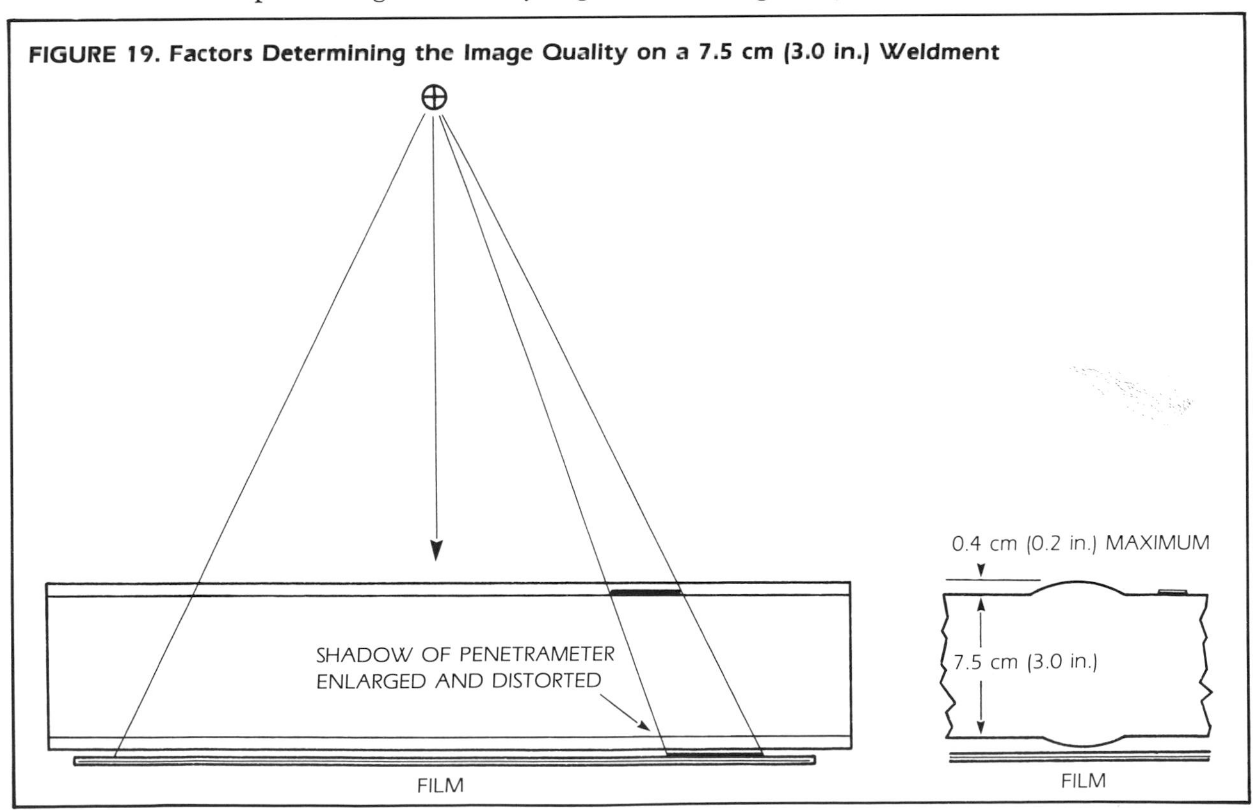

face or on the opposite side, where they should be in contact with the film holder. This set-up can be used with any of the three joint types using the fillet weld.

Among the special illustrations in Part 10 of this Section is one referred to as a *socket weld*, which is another, special form of fillet weld. In that example, the component was small enough to be radiographed through its entire cross section. Such small components are often rotated in order to radiograph other zones of the fillet weld.

Edge Joints

The edge joint is illustrated in Fig. 8. This form of joint is usually intended for sealing and is not considered structural. When the weld is completed, there is a severe notch on one side that reaches to the root of the weld and precludes any significant bending or tensile loading. The edge joint has an evident deficiency in load-bearing capacity, but may be radiographically inspected when sealing is a crucial factor. The total geometry of the joint discourages radiography: although the weld is fairly thin, it is mounted on legs that act as a thick section when viewed from directly above. When sealing a cylindrical object with an edge weld, tangential radiography is frequently employed.

A form of the edge joint made with flanges could be readily checked radiographically (Figs. 8a and 8c). If made from thin material, when fairly sharp bending might be assumed in the weld preparation, a careful weld could be designed to use up all the projected material and melt through to the back, rather like the consumable insert. The resulting weld is somewhat thicker than the sheet and is suited to the radiographic examination process. Such welds are used on articles made of precious metal, where the amount of material is always a great part of the total cost.

FIGURE 20. Radiography of Fillet Weld on Corner Joint

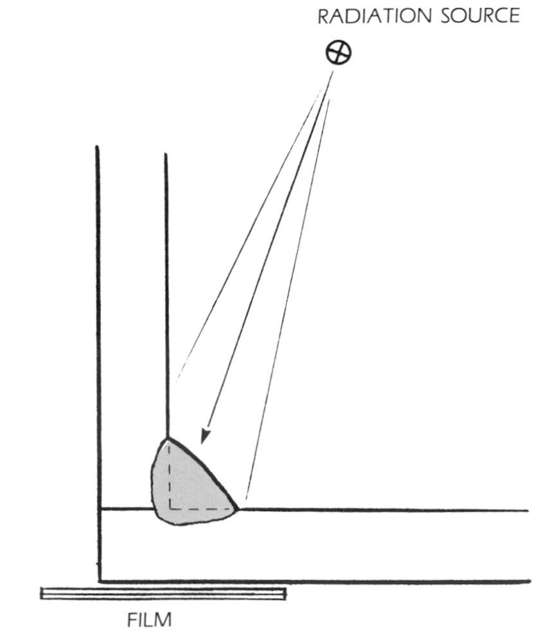

RADIATION SOURCE

FILM

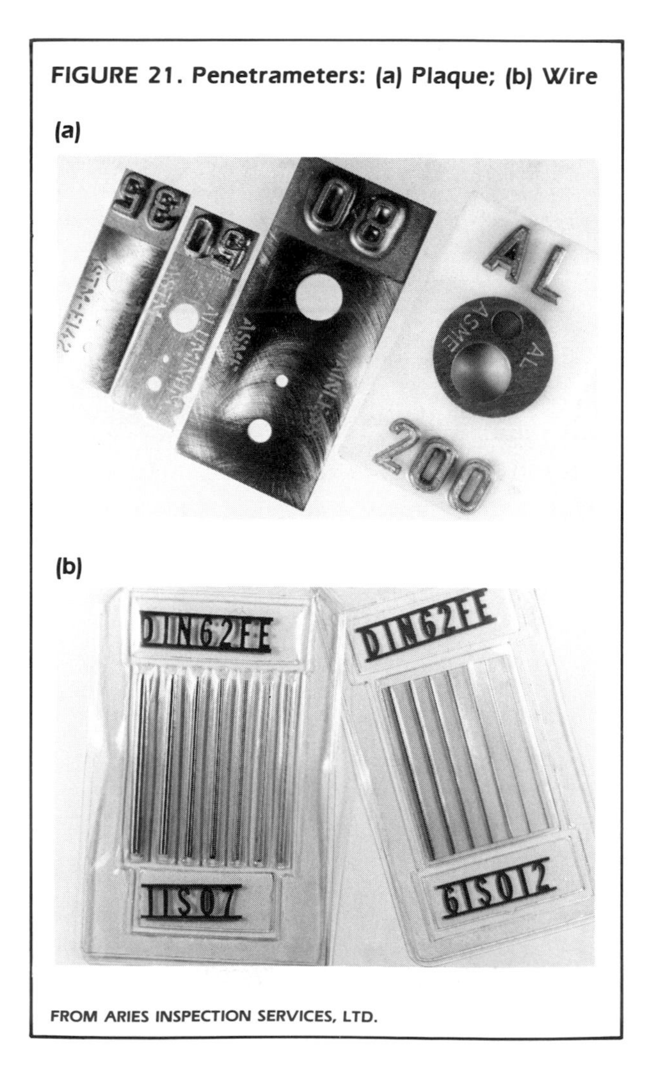

FIGURE 21. Penetrameters: (a) Plaque; (b) Wire

FROM ARIES INSPECTION SERVICES, LTD.

Use of Penetrameters

There are three basic penetrameters for indicating the quality of a radiograph. In North America, quality is most commonly judged by checking for the discernibility of (1) certain small holes in a thin plaque, and (2) the outline of the plaque itself. At one end of the plaque there are lead characters that identify the material of the plaque and its thickness (see Fig. 21a).

The plaque-type penetrameter first functions in a thickness mode, the thickness being a percentage of the weldment thickness. A second aspect of this sensitivity indicator is related to the actual hole diameter. Customarily, because many codes are so phrased, the plaque thickness (T) will be 2% of the weld thickness, and visibility on the radiograph of a hole whose diameter is two times the plaque thickness (2T) will be required.

Penetrameters that are made to American Society for Testing Materials Standard E142, *Standard Method for Controlling Quality of Radiographic Testing* (see ref. 8), will incorporate the 2T hole and two others, a 1T diameter hole and a 4T diameter hole. For more critical applications, a contract could require that the smallest hole be used. Alternatively, plaques could be prepared with other versions of the hole diameters. The 2% level, however, has the advantage of international recognition because many codes refer to the 2% level.

FIGURE 22. Mil-Std-453 Type Penetrameter

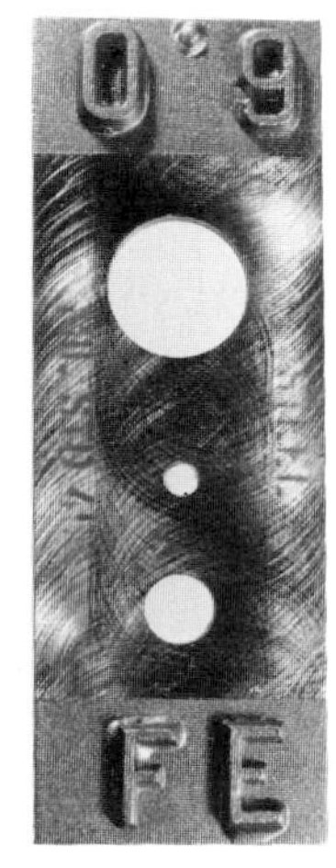

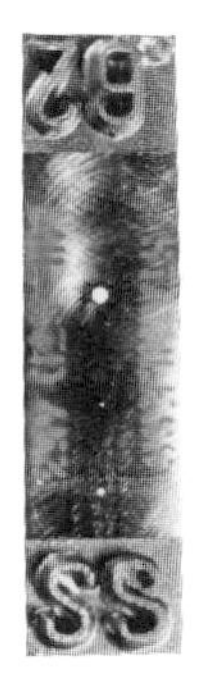

The next most common form of penetrameter is the wire type (Fig. 21b). The wires are arranged by diameter and are all of the same length. The wire material is chosen to match the material in the weldment. In use, a specification will state the smallest diameter that should be visible on a finished radiograph. The wire package is of a standard design, and information about the wire set is presented in lead characters.

Customarily it would be stated that at least two-thirds of the significant wire length should be resolved on the film. See Table 2 for some details on the design of the wire set; this is adapted from ASTM E747-80, *Standard Method for Controlling Quality of Radiographic Testing Using Wire Penetrameters* (see ref. 9).

A third major form of penetrameter portrays the material thicknesses and identity on each device, with the 1T, 2T and 4T holes (Fig. 22). This is referred to as 'the Mil-Std-453 model' (Military Standard 00453-B, *Inspection, Radiographic*).[7]

Purpose of Penetrameters

All forms of the penetrameter have one basic purpose: displaying a measure of the radiographic quality and thereby some measure of the quality of the radiographic process. It may at first appear confusing to be concerned about the quality of a quality control system. The concern is real, however, because serious judgments are often made on the basis of shadow images seen on a radiographic film.

Some of the most critical radiographic test objects are pressure-containing zones in nuclear reactors. These zones comprise the main pressure vessels, heat exchangers, pipe fittings, valves and piping, and are linked with one another by welds. Because of the extremes in pressure and temperature these components must withstand, they are built from very thick metal. They are, however, not much thicker than they need to be to maintain the security of the contents; no large safety factors are involved. Exceeding minimum standards could make these components so thick and heavy that the material and fabrication costs would be prohibitive. The aim is to ensure that the assembly is of high enough quality that it can function in complete safety. The welds involved are expected to be as sound as the materials being joined. The materials are subjected to quality assurance systems throughout their manufacture and the welded joints become equally important. When radiography is used to determine the quality of the welds, the inspection system itself logically becomes subject to question, and to similar control.

Area Coverage

The area to be covered by the film is a major concern. For one thing, the film size must be adequate for the size of the weld zone involved. The film must also be large enough to contain information on radiographic quality (the penetrameter and any shims must be visible). Larger film sizes may be required when complex film identification systems are involved (see *Identifying Welds and Films* below).

The area to be covered is directly affected by the geometry of the arrangement. As shown in Fig. 19, the radiation source images precisely only when the object and the film are normal (at right angles) to the beam. With flat workpieces, this becomes a definite limitation on the length of a weld that can be covered in a single exposure. Often a specification will incorporate a limit here; for example, the American Welding Society Standard D1.1 *Structural Welding Code* (see ref. 2), clause 6.10.5.3, 1983 ed., demands a maximum of 26.5 degrees half-angle for such a case. This angle of spread is illustrated in Fig. 19.

Such exposure angles have three important effects and all tend to degrade the image. First, the true shape of an indication will not be presented because the beam will generate a shadow that is elongated in the plane common to the weld metal and the axis of the radiation beam. The second effect has to do with the position of an image compared with the actual position of the discontinuity that generates the image. An indication on the film will always be projected away from the true position of the discontinuity causing the image. Like the distortion effect above, this can become quite severe if limits are not imposed. The third effect is that the radiation must pass through greater material thickness away from the beam axis, and the resulting film density and image quality may be lower than the standard requirement.

Specifications usually incorporate clauses covering exposure geometry and may even go a step further in demanding that the penetrameters be placed at the outboard end of the joint, where the effect is intensified.

Screens

For steel weldments, use of a lead foil screen between the film and the radiation source is usually required. The screen improves the image quality by filtering out longer wavelength radiation which causes general fogging. A companion screen, fitted on the back side of the film, also has an intensifying effect; additionally, it reduces backscattered radiation.

The lead screen is not used for light alloy weldments because, with aluminum and other light metals, the softer radiation must reach the film unimpeded to yield an adequate contrast range through the material. A screen on the back side of the film, however, acts positively and is often used. In this position, it contributes to latent image exposure (1) by electron intensification, and (2) by blocking secondary radiation (backscatter). Such radiation is always lower in energy when compared to the primary beam; if not stopped, it could cause general film fogging. The ASME VIII, Sec. 1 specifications[1] require that a lead character (*B*) be present on the back of the film holder as evidence that a screen was used. If a significant amount of backscattered radiation impinges on the wrong side of the film holder, the *B* is imaged.

Section Thickness

The quality of any radiography is very much dependent on the object's effective section thickness. A simple geometric concept governs here; the further the zone of interest is from the film or other detector, the less distinct the radiographic shadow will be. This is a very real effect. The result is a moderate loss of resolution when a thick weld is considered, and the effect can be very pronounced when complex shapes are involved, such as some forms of corner and T-joints. The penetrameter image becomes very important as a monitor, and the radiographic procedure must make it clear that the penetrameter should not be positioned where it would yield misinformation. If the section thickness is such that the image quality is in doubt, a greater focal distance, and perhaps even a higher energy and greater intensity source of radiation should be used.

The ultimate sensitivity is not expected to be constant over a wide range of weldment thicknesses. The sensitivity parameter is presented as a percentage of the object thickness (see Table 2 for a comparison of plaque and wire penetrameter sizes). Specifications that present acceptance limits for radiography also present the scaling of the acceptable indication sizes.

With the wire penetrameter, the parallel relationship holds (Table 2); there is no target hole to consider because the wire covers the two functions of indicating (1) a change in thickness traversed by the beam and (2) the spatial resolution of an image. Thus, while a single pore of 1 mm (0.4 in.) diameter could be acceptable on a 50 mm (2 in.) section thickness, it does not follow that a 40 mm (1.5 in.) pore would be acceptable on a 150 mm (6 in.) section. Such a large flaw indicates that there must have been some special difficulty with the welding process.

TABLE 2. Comparison of Plaque and Wire Penetrameter Sizes

Plaque Thickness[1]		Wire Diameter[2]	
(in.)	(mm)	(in.)	(mm)
0.005	0.13	0.005	0.13
0.006	0.16	0.006	0.16
0.008	0.20	0.008	0.20
0.009	0.23	0.010	0.25
0.010	0.25	0.013	0.33
0.011	0.28	0.016	0.40
0.012	0.31	0.020	0.51
0.020	0.51	0.025	0.64
0.100	2.54	0.032	0.81
0.150	3.81	0.040	1.02

1. Plaque thicknesses are an arbitrary set.
2. Wire diameters are in a geometric series, each multiplied by 10th root of 10 (1.2589).

Identifying Welds and Films

Because a radiograph is a permanent reference of a weldment's condition, it is usually required that the finished film be traceable to the specific joint. This need is generally met with a marking system consisting of lead characters that image clearly on film through the thickness range involved. Records made at the time of the radiography must adequately show how to match the film: (1) to the weld; (2) to the location on the weld; and (3) to the position of the film and the beam direction of the original exposure, as evidence that coverage was complete. Careful marking and recordkeeping are necessary, especially if checks are made later to determine whether some condition is developing into a significant discontinuity.

Devices that mark by imprinting an image from the radiation must be considerably more dense than the material under study. Lead is commonly used for overprinting on steel, and such numbers, letters and symbols are available in many sizes and thicknesses. On very thick welds, even quite thick lead markers can virtually disappear on the film; mounting them on a block of steel or sheet of lead will keep the image from burning out during exposure. A similar case exists when a weld specimen is smaller in area than the film. If the numbers are thin and placed directly on the film, their image could disappear; they should be positioned on a separate lead shim or on a shim of thickness similar to that of the weld material.

In ASTM Standard E94, *Recommended Practice for Radiographic Testing* (see Clause 15, "Identification and Location Markers on Radiographs"), there are suggestions for the type of information that should be retrievable from film marking systems.[15]

Users of lead markers are reminded that the placement of these barriers to radiation can be quite important. They must, of course, be located so that the film will image them, but it is equally important that they do not intrude on a significant portion of the weld. On thick welds, extra care in positioning is required with markers that are near the end, or near the edge, of the film. Here, beam divergence could result in part of the identification image being lost. There might be a need to reshoot films that were perfectly adequate representations of the weld, but were poorly identified.

There are also possibilities for confusion when some of the symmetrical letters are used and care is needed when choosing the actual symbols. Some of the simply symmetrical letters are *A*, *T*, *V*, *W* and *Y*; *O* and *Z* are doubly symmetrical.

Of course the written record of the radiography must include sufficient information to ensure that the shorthand used on the film can be deciphered.

Notes and drawings must be prepared and preserved with care. It is not unusual to retain records on processes and inspection results for many years.

In large radiographic projects, a few simple identifying numbers are not sufficient for permanent reference; on occasion the film identity scheme could require as much film as the weld zone itself. There are refined coding systems that print information on the film using visible light as the exposure medium. These systems have the disadvantage of being indirect compared to the identity scheme using lead characters exposed simultaneously on the radiograph.

For some welds, the area of interest is not limited to the weld metal. For example, a form of cracking can occur adjacent to a weld, rather than directly in the weld. The identifying symbols must be positioned accordingly, so as not to obscure a possible discontinuity.

With light materials, identifying the film with lead characters is quite easy due to the difference in density between aluminum and lead.

Film Handling

All of the standard precautions for film handling apply when radiographing welds. Extreme care should be taken in loading, unloading and handling the film holder. Cleanliness is essential in the use of screens, and during the processing cycle.

During exposure, the film seems safe in its cassette, yet there can be mishandling at this stage too. A special case for welds occurs when a high weld bead, coupled with a heavy load, imposes a pattern (pressure mark) on the film, one that will show after processing.

PART 6

IN-HOUSE AND FIELD RADIOGRAPHY OF WELDS

A very common specification for sensitivity is the 2% level. There are, though, situations in which 2% would be considered totally inadequate. Spot welds in electronic devices and some of the welded joints common to nuclear fuel packages are examples. Test objects such as these are relatively small and are commonly handled in a permanent radiography location.

A sure way of increasing the precision of the radiographic arrangement is to significantly extend the source-to-film distance. In the field, this is not always possible. The small linear accelerator shown in Fig. 23 introduces new possibilities here; however, it is assuredly a special case. Another small, regular X-ray generator is shown in Fig. 24a, with a heavier companion in Fig. 24b.

Field work does have some advantages, especially when there is a small zone of interest in a large assembly; examples include aircraft, installed pressure vessels and bridges. When radiography is performed for such products, field work becomes a necessity.

Handling devices and fixtures become very important items to consider when doing field X-ray work. Even the smallest commercial tube of reasonable potential, say 200 kV, is sufficiently massive to require careful mounting. The structure being studied may not be rigid enough to support an X-ray device and completely separate auxiliary equipment has to be considered.

FIGURE 23. 3.5 MeV Minac Linear Accelerator Head

FROM SCHONBERG RADIATION CORPORATION.

In true field work, the entire radiographic process, including film processing, will be completed away from a permanent base. The equipment available for this purpose is entirely adequate to the task. Radiographic field equipment has become so efficient that it is now commonly found as part of some permanent testing laboratories.

FIGURE 24. Typical X-ray Units Used in Weld Radiography

(a)

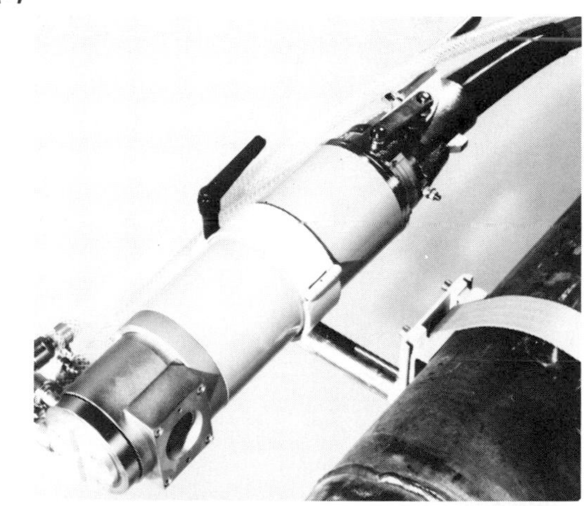

(b)

FROM SEIFERT X-RAY CORPORATION.

PART 7

DISCONTINUITIES IN WELDS

While many of the following discontinuities occur in several types of welded joints, the differences in joint geometry produce differences in the placement and orientation of the discontinuities. Thus, the radiographic procedure for imaging one discontinuity might not be useful for another.

Weld discontinuities that can be found by radiography comprise mainly: porosity; slag; incomplete penetration; lack of fusion; cracks; underfill; undercut; and arc strikes.

Some weld discontinuities that are not commonly found by radiography include: underbead cracks, lamellar tearing; lamination; and delamination.

Porosity

Porosity is a condition in which the metal has voids containing gases. These gases are dissolved in the metal while it is hot and then separate as it cools. The solubility of gases in metals can be very high at high temperatures.

The gases result from chemical and physical reactions that take place during welding. A certain amount of gas is always generated in standard welding but is usually not detectable radiographically. At times, however, excessive gas is produced, leading to the discontinuity called *porosity*.

Most porosity is spherical in shape. Porosity can take on other shapes related to some welding conditions; it can also be distributed in specific ways that are again related to welding condition.

Porosity can be found almost anywhere throughout the weld metal, though it would not normally be present in the heat-affected zone of the material.

The restrictions on porosity in welds tend to be rather lenient. The reason is that the condition does not seriously weaken a welded joint. There are two aspects contributing to this apparent lenience.

First, spherical porosity does not act as a stress riser, meaning that it does not behave like a sharp notch which would indeed weaken a joint. Secondly, the strength of weld metal is customarily greater than the nominal strength of the material being joined. Welding electrodes, with their special coatings, are designed to deposit a relatively high strength material, one with a very fine grain structure as a result of an extremely rapid cooling rate. Most weldments then can accomodate a fair amount of porosity, especially when the service conditions are predictable.

Common Types of Porosity

Uniformly distributed, spherical porosity indicates that the cause of the difficulty has been averaged over the entire weld.

Clustered spherical porosity occurs in zones where there has been a perturbation in the welding process, especially when welding has been stopped and started up again; in such cases the pores would be spherical.

Spherical porosity in a linear grouping is confined to the earlier stages of the weld, often following a joint edge. It is usually indicative of contaminated material.

Spherical porosity is usually easy to identify in a radiographic image, although it is possible to have inclusions that are similarly shaped.

Piping

Piping is a term applied to porosity of a tear-drop shape; it usually has a linear distribution. The condition is easily recognized in a radiograph because of its streamlined shape and customary grouping.

Herring Bone Porosity

Herring bone porosity is a special case, occuring in automatic arc welding processes. It has a tear drop shape, similar to piping porosity, and is distributed in linear fashion with a high degree of regularity. Herring bone porosity is caused by contamination, generally the presence of air. In the gas metal arc process, a special atmosphere provided by the equipment helps avoid this discontinuity.

Slag

Inclusions of slag or other nonmetallic solid materials may be trapped in or adjacent to weld metal. This form of discontinuity, like porosity, is consistently found by the radiographic process. It can occur in near-spherical, isolated forms, but the tendency is for inclusions to accumulate linearly. Most inclusions consist of residual slag from the electrode coatings in welding processes that use flux. The trapped material tends to occur in the welding direction, and therefore appears in a linear arrangement.

In building up a weld, a channel effect occurs on both sides of the bead. If through a fault in the process some slag is trapped, there could be two affected zones, separated by the weld bead width. When such parallel slag lines occur they are called *wagon tracks.*

In multipass welding, the slag covering will remain on the as-welded workpiece, just as it does on completed welds, and measures must be taken to ensure its removal.

If this interbead slag is not completely removed before the next pass is made, inclusions will result. In a radiograph, this type of slag is not distinguishable from the material trapped during its initial fusion.

All the welding processes involving flux can exhibit slag inclusions. Just as for porosity, the automatic processes, when they go awry, tend to exhibit long, regular discontinuities.

The spherical type of slag inclusion may often be distinguished from porosity because there is a major density difference between the solid slag and the open void. Should both appear in the same image, confirmation is easier. A major slag inclusion may also have a fine structure that can be resolved in a radiograph.

The limits on slag are more stringent than for porosity because the linear form of the inclusion can act as a stress riser. The interpreter is usually directed by the applicable code in determining the allowable length and diameter of such inclusions.

An intermittent slag line is actually a single slag deposit with gaps. The accumulated length of the broken run plus the gap space may be used to define the total discontinuity length, in some specifications.

Incomplete Penetration

Incomplete penetration is a condition that may or may not be defective. One definition states "inadequate penetration of the joint constitutes penetration which is less than specified." Thus it is possible to have an incompletely bonded (or incompletely filled) joint that would still be allowed by specification. Should such a weld be radiographed, clear geometric indications of the unwelded zone are obtained. Normally, welds are expected to occupy all of the available cross sections; incomplete penetration must be monitored and measured to determine limits.

On a radiograph, this indication is usually a linear, dark zone. In the case of joints welded from one side only, the image will be well defined, assuming the detector is on the root side. When incomplete penetration occurs in thicker joints that are welded from both sides, the outlines will not be sharp. This is because (1) the zone of interest is further from the detector, and/or (2) the discontinuity is a smaller percentage of the total specimen thickness.

The effect of this condition on the weldment can be very serious; a pair of sharp notches may be present, extending for a considerable distance. The joint would be weak in bending across the weld, especially for repeating loads or cycles. In these cases, specifications such as AWS D1.1 (see ref. 2) impose strict limits on the length permitted.

Lack of Fusion

Lack of fusion is a discontinuity characterized by unbonded zones in a weld, where weld metal is cast in-place but not bonded to a prepared surface or to a previous bead. The shape may then be precisely or roughly linear. In its usual form, the discontinuity exhibits a gap over some of its width, gradually closing to zero. The typical radiographic appearance shows evidence of tapering (the image would be longitudinal and shaded across its width). Note that when lack of fusion comprises only weld metal in complete contact with another material, the thickness of the discontinuity could be near zero. In this arrangement, the flaw would not be detectable radiographically unless the beam were directed along the unfused zone. In ordinary weld sampling, with the beam directed at the weld bead, the result would be negative. If the radiographic requirements are designed to detect lack of fusion, a requirement for the appropriate beam direction (Fig. 18) would be included.

Lack of fusion can occur in all types of fusion welds. The limits for lack of fusion are generally equivalent to those for incomplete penetration, and specifications often group the two conditions together, classed as *fusion discontinuities*.

Cracks

A crack may be defined as a split, exhibiting a sharp tip and typically a very small opening. Its detection by radiography is a function of the orientation of the beam to the crack. As in the case of incomplete fusion, cracks are ordinarily detectable only by ensuring that the beam direction is in line with the crack. Typically, when cracking is suspected but cannot be confirmed, more radiography will be required to confirm the indication. The typical crack, when clear in a radiograph, will appear as a narrow, irregular line.

Cracks detected by radiography are always considered discontinuities and rejection follows, with repair being an option. They are the most serious form of discontinuity because their sharp tip acts as a severe stress riser. For cyclic loading, cracks are always dangerous. Note that in interpreting radiographs, the very presence of a crack is sufficient evidence for rejection; there is no guarantee that the actual limits of a crack will match the visible length of its radiographic image. Furthermore, portions of cracks can be closed very tightly because of weld shrinkage stresses, and no tightly closed crack would be detectable.

Hot Cracks

A form of cracking, referred to as *hot cracking* or *hot tearing*, is radiographically detectable. This discontinuity originates during or just after the hardening of the metal when shrinkage is excessive. When narrow, deep welds are made, the weld metal does not stay molten long enough to fill the normal shrinkage spaces.

Hot cracks have some visible width and are therefore more easily detected with radiography.

Crater Crack

A variation on the hot crack is the *crater crack*. These occur if the electrode is removed too soon at the point where a weld is terminated. Crater cracks usually consist of sets of radial cracks and are recognizable in a radiograph by this typical pattern.

Other Separations

Other cracks may develop adjacent to the weld metal, in the heat-affected zone. Such cracks, which are associated with hydrogen in the metal, may appear up to several hours after the completion of the weld. They will not be open and will generally not be oriented in a predictable direction. The radiographic method should not be used for their detection.

Other serious forms of separation which are not ordinarily detectable with radiography include:

1. the underbead crack (a cold crack sloped from the plate surface);
2. lamellar tearing (separation within the plane of the plate); and
3. delamination (the opening of an in-plate separation).

Dense Inclusion

Dense inclusions are usually particles from the tungsten electrodes that become deposited in weld metal through faulty procedures. The particles may be spherical and appear much denser than the material being welded, typically aluminum alloys and stainless steels. The very high density serves as positive identification.

Tungsten inclusions are not considered especially harmful and are usually evaluated in conjunction with porosity. The condition does indicate some processing or procedural problem, and should thus be monitored.

Undercut

Undercut refers to a groove melted into base metal directly adjacent to a weld bead. Oridinarily it will be evident visually. When undercut occurs in assemblies, the original surface may not be accessible and the need for radiographic interpretation may arise.

Undercut is a processing fault and may be repaired by adding an extra, narrow weld bead.

The condition is generally recognizable because the plate thickness will exhibit a definite tone on a radiograph, while adjacent to the weld there will be a darker area where the plate thickness has actually been reduced.

In some standards, a certain amount of undercut is permitted. In more critical cases, no undercut is permitted, and a radiographic determination leads to rejection.

Arc Strikes

Arc strikes are discontinuities that result from establishing the welding arc in zones other than a weld. They consist of remelted metal and portions of electrode metal in unscheduled places. Their potential danger arises from steep changes in metal properties that develop when a material such as steel has been subjected to very rapid heating and cooling. Excessive hardness can result, leading to possible fracture during service.

The condition is identified by its position (away from the weld metal) and by a small patch of extra thickness that is often intermittent but linear.

Arc strikes are usually cause for rejection on critical weldments because of the possible effect on service life. Repairs are possible and would generally involve grinding through the thin affected layer, with fine grinding as a finishing operation; no further welding would be scheduled.

A collection of radiographs showing some special conditions for weldments is presented in Part 10 of this Section, *Special Radiographic Images of Welds.*

PART 8

CODES, STANDARDS AND SPECIFICATIONS FOR RADIOGRAPHY OF WELDS

There is much published material on the radiographic inspection method, its controls and limits of acceptance. This section will necessarily be limited to those publications that concern or contain specific information on the radiographic method as it applies to weldments. The welds in question are joining methods for metal products such as the following:

1. pressure vessels, boilers and heat exchangers;
2. buildings, industrial structures and bridges;
3. ships and marine structures;
4. transmission pipelines;
5. industrial pipe;
6. storage tanks;
7. rail vehicles;
8. machinery;
9. aircraft and spacecraft; and
10. road vehicles.

A *specification* is a document that states in some detail the set of requirements associated with the radiographic testing method. The source of such a specification is usually the buyer of the product. Instead of composing a complex technical document, such a buyer could choose a particular standard document that adequately covers the particular method.

A *standard* is a published specification, test method, classification or practice that has been prepared by an issuing body. To satisfy the needs of a contract, a standard or portions of a standard can function as a specification.

A *code* is a collection of related standards and specifications, often applying to a particular product line. An example is the ASME *Boiler and Pressure Vessel Code* which consists of many specifications covering pressure vessels, their manufacture and inspection, their licensing, and their in-service inspection; it incorporates many ASTM standards. Another example is the *National Building Code of Canada*, incorporating CSA design standards, CSA welded construction requirements and construction safety measures to ensure public safety in buildings.

The ASTM standards do not presume to set test limits for radiography becaue these vary according to the product. The ASME code contains such limits, as does the AWS *Structural Welding Code*, and the American Bureau of Shipping (ABS) *Rules for Building and Classing Steel Vessels.* Piping and pipelines are covered by American National Standards Institute B31.1, *Code for Pressure Piping* and American Petroleum Institute (API) 1104, *Standard for Welding Pipelines and Related Facilities.* For storage tanks refer to American Waterworks Association D100, *Standard for Welded Steel Elevated Tanks, Standpipes and Reservoirs for Water Storage,* and API 650, *Standard for Welded Steel Tanks for Oil Storage.* Requirements for rail vehicles refer to the AWS Standard D1.1 already quoted. For aircraft and aerospace, reference standards are more specialized, as issued by NASA and by the Department of Defense. Machinery is typically incorporated into existing standards. A typical standard that could apply to road vehicles is AWS D14.3, *Welding on Earth Moving and Construction Equipment.* (See the reference lising at the end of this chapter for more information on these standards.)

Radiographic Requirements for Welds in Selected Standards

Although there is some commonality in the standards cited for the radiographic control of welds, there is variation in the acceptance limits. For the method and control, respectively, it is normal to see a reference to ASTM E94-77, *Recommended Practice for Radiography Testing,*[15] and ASTM E142-77, *Standard Method for Controlling Quality of Radiographic Testing.*[8] The acceptance limits are, not unnaturally, associated with the type of product involved, so those products that are deemed more critical will carry more stringent requirements.

The acceptance criteria are usually expressed in graded words and numbers, occasionally with a graphical presentation as a backup for estimating the extent of porosity. Where porosity charts are included, there is also an absolute limit based on total area and individual diameters.

PART 9

RADIOGRAPHY OF PIPE WELD DISCONTINUITIES

Three selected pipe weld discontinuities are reproduced here. With each photograph there is an iridium-192 radiograph showing the identical discontinuity in projection. Discontinuities were deliberately introduced into the specimens, which were based on 250 mm (10 in.), schedule-80 stainless steel pipe, designation Type 316. The complete report, from which these illustrations are reprinted,[16] also includes a comparison of ultrasonic testing indications, with photographs of the response signals.

Additional illustrations of pipe welding discontinuities can be found in Section 8, *Radiographic Interpretation*.

Arc Strikes

Arc strikes (Fig. 25) are disturbances left on the surface of the base metal where a welder has momentarily touched an arc-welding electrode to start the arc.

Arc strikes can cause failure to the affected material. These failures initiate at the abnormal structural conditions produced by the arc strike. The careful welder strikes the arc in the joint where the base metal will be melted by penetration, as the operation progresses beyond the striking point. The welder may also use a scrap of metal as a starting tab, or a high-frequency arc starter. Severe metallurgical conditions can exist when the careless welder makes an arc strike on the surface of the base metal adjacent to the weld and then quickly moves the electrode into the joint to perform the welding operation. Arc strikes often harbor minute cracks, porosity and hard zones.

Drop-through

Weld drop-through (Fig. 26) is an undesirable sagging or surface irregularity at the weld root, usually encountered when the welding temperature is near the solidifying temperature of the base metal. This can be caused by over-heating with rapid diffusion or alloying between the filler metal and the base metal. Drop-through is characterized by excessive root bead width and a slumping of weld metal on the back side of the weld.

Unconsumed Insert

An unconsumed insert (Fig. 27) results from preplaced filler metal that is not completely melted and fused in the root joint. This condition is caused by low welding current, improper weaving procedure, improper joint design and improper welding speed. Considerable welder technique and skill must be developed to ensure high quality root beads when using inserts with the gas tungsten arc welding process. Proper welding parameters and sufficient skill of the welder will produce melting and fusion of the insert and the side walls of the joint preparation. This results in a satisfactory root bead profile.

FIGURE 25. Arc Strike: (a) Photomacrograph at 2×; (b) Radiograph at 1×

(a)

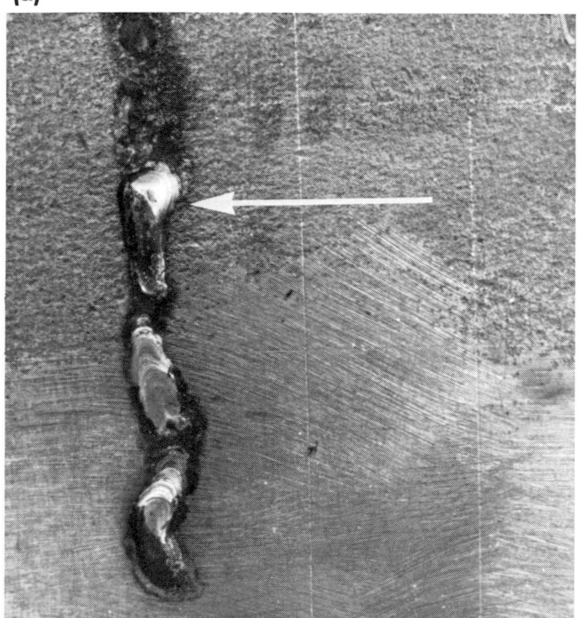

(b)

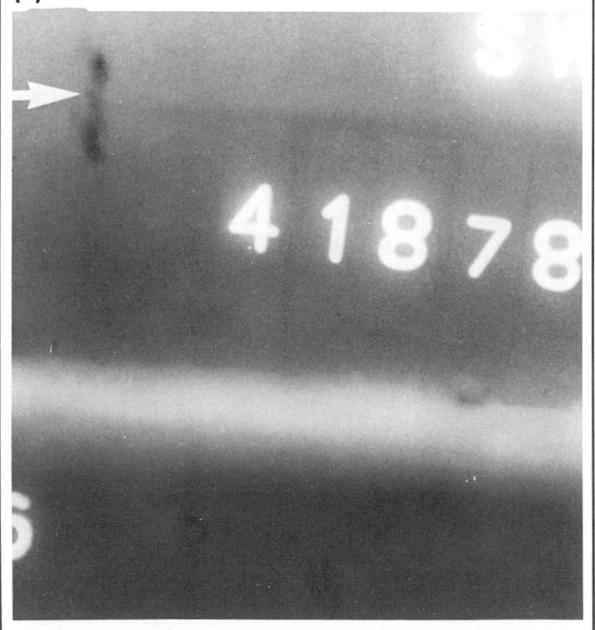

FROM ELECTRIC POWER RESEARCH INSTITUTE. REPRINTED WITH PERMISSION.

FIGURE 26. Drop-through: (a) Photomacrograph at 4×; (b) Radiograph at 1×

(a)

(b)

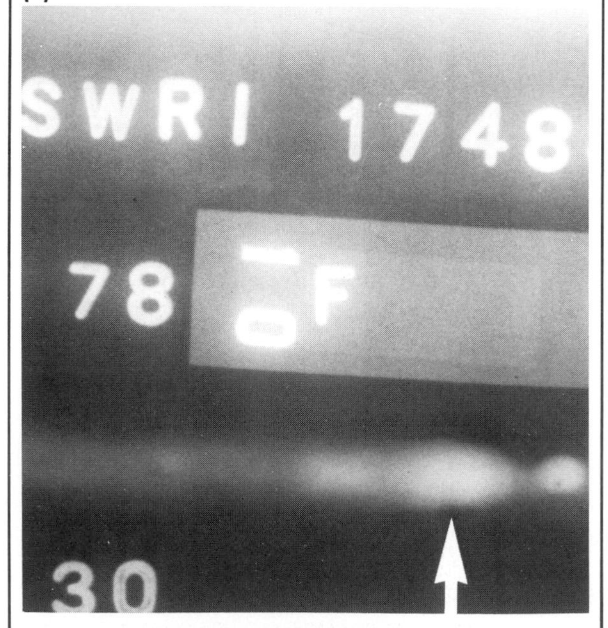

FROM ELECTRIC POWER RESEARCH INSTITUTE. REPRINTED WITH PERMISSION.

FIGURE 27. Unconsumed Insert:
(a) Photomacrograph at 4×; (b) Radiograph at 1×

(a)

(b)

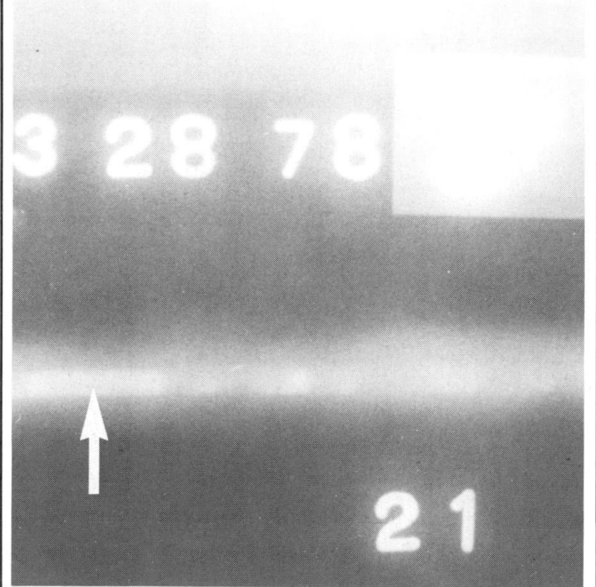

FROM ELECTRIC POWER RESEARCH INSTITUTE. REPRINTED WITH PERMISSION.

PART 10

SPECIAL RADIOGRAPHIC IMAGES OF WELDS

A search was made for radiographs showing weld conditions that were unusual, or that had some other special merit.

The films also had to be reproducible for publication. These two criteria, *special* and *reproducible*, quickly narrowed the set down to the images presented here.

Concerning the reproducibility of radiographs for the print medium, it is worthwhile to note that most good radiographs do not reproduce well. The reason for this is that the original film bears a double emulsion layer. If the radiograph is good it will look quite dark to the eye because the darker films (as noted in Section 5, *Film and Paper Radiography*) will cover a wider tonal range and will therefore contain more information than lighter films. Such radiographs cover too wide a density range for the capacity of the single emulsion photographic film.

The subject matter in these radiographs varies. The images do have two things in common: the objects are made of steel or stainless steel; and the source of radiation for each was iridium-192.

Thermit Process

A well-established joining method for railway track is the *thermit process*, in which molten iron is cast into place around a rail junction. There are no electrodes or arc involved, the heat being supplied by a thermal reaction in a crucible arrangement.

In Fig. 28, a simple shooting sketch and a radiograph demonstrate the special geometry of this process. The image shows some porosity in the joint zone in the web section of the rail. The object portrayed was not considered rejectable because the porosity is not associated with the working portion of the rail. The finished joint has been ground smooth over much of its area; a reinforcement remains at the joint zone of the web.

The following data describe the test object in Fig. 28: standard crane rail: weight = 150 kg/m (105 lb. per foot); height = 130 mm (5 in.); foot width = 130 mm (5 in.); and web thickness = 25 mm (1 in.).

Socket Welds

A *socket weld* (Fig. 29) is a fillet weld used to seal a tube inserted into a fitting. The case in Fig. 29 shows a rejectable condition at the root of the fillet, incomplete penetration. Pipe joints with this condition would ordinarily be of good mechanical strength because the weld metal section exceeds the wall thickness of the tube. For nuclear service, however, the requirements are more stringent because of the extensive work required should subsequent repairs become necessary in a radiation field. The case shown is rejectable because of the implied service.

FIGURE 28. Radiograph of a Thermit Weld; Film Holder in Contact with Vertical Side of Head and Reaching to Foot; (a) Shooting Sketch; (b) Radiograph

(a)

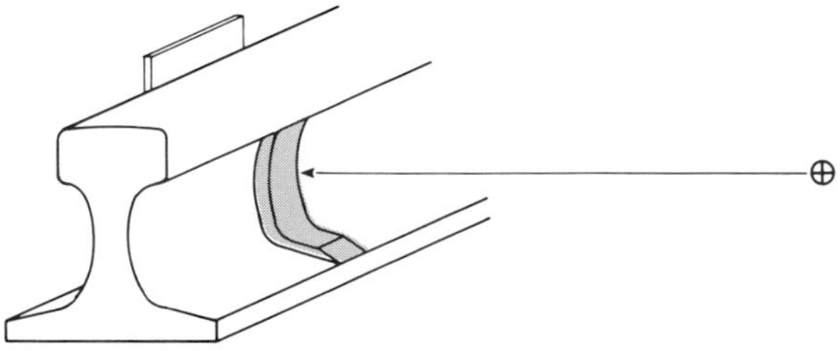

(b)

O 21256
M2636Y

FIGURE 29. Radiograph of Socket Weld

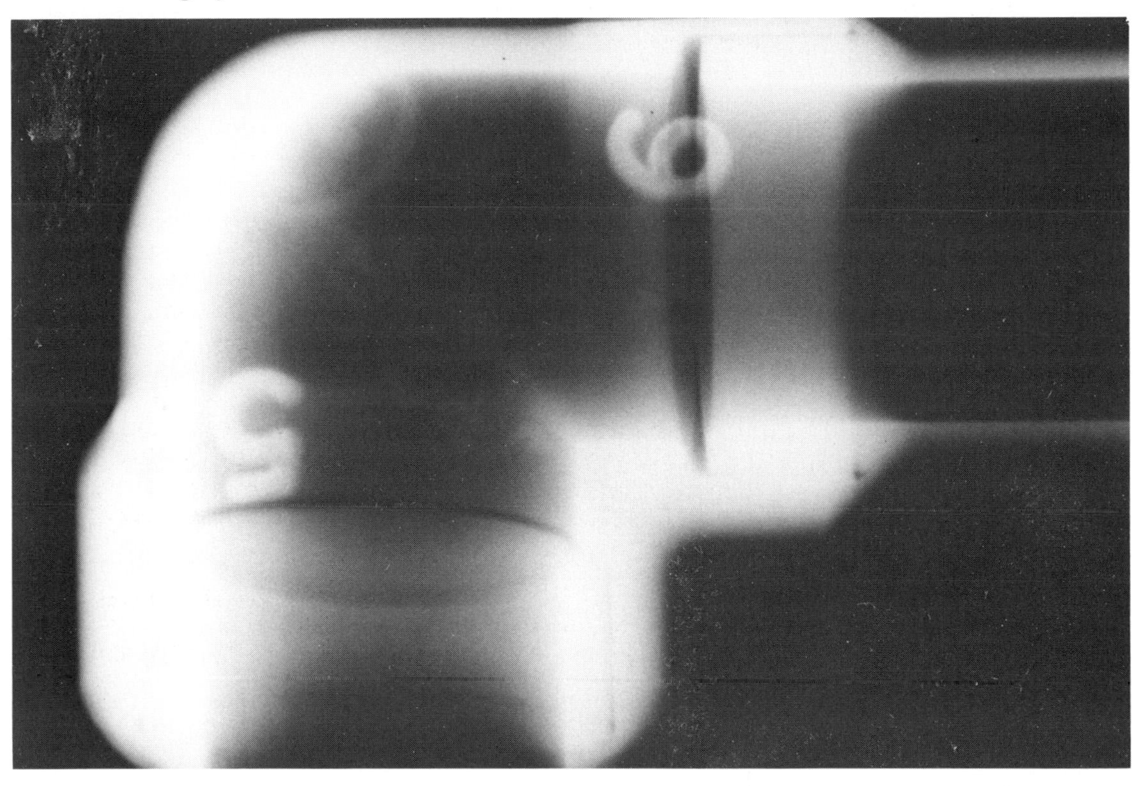

Data for this test object include: tubing diameter = 40 mm (1.5 in.); tubing wall thickness = 5 mm (0.2 in.); and source to film distance = 300 mm (12 in.)

Inadvertent Indication

Figure 30 is a case where a radiograph was intended to check weld quality but also exposed serious shrinkage in a valve casting. While some porosity had developed in the weld joining a pipe to the casting, the shrinkage effect was in a position of overriding importance. In this case the radiographic procedure required that the source be in contact with the outer surface of the pipe, with the film on the remote side (*contact shot*). This technique is used with large objects, particularly pipe, when there is no access to the interior. Because the beam passes through a two-wall thickness, it is filtered to a degree; any features in the metal on the source side are not imaged as sharply. The discontinuities visible in the weld are discussed in Section 8, *Radiographic Interpretation*. The test object in Fig. 30 has the following dimensions: diameter = 600 mm (24 in.); pipe wall thickness = 10 mm (0.375 in.) and valve wall thickness = >50 mm (>2 in.)

Overlap of Film

The circumferential joint in a large pipe has been radiographed for Fig. 31. The effect demonstrated, a very noticeable light zone over much of the film, is a result of film overlapping. This is done when large welds are radiographed to show that the complete weld has been imaged. The light area visible in Fig. 31 indicates the presence of a second length of film, in contact with the pipe. The film images the light zone at normal density. The two radiographs are then matched to show that the features of interest are visible in both films. This procedure is a special case of film identification, an extension

FIGURE 30. Radiograph of Shrinkage in Casting Adjacent to Weld

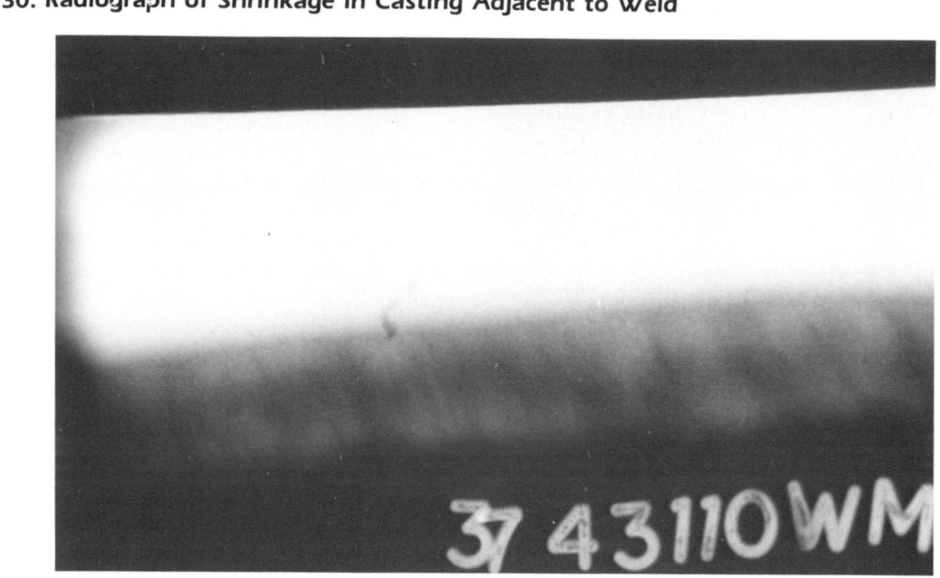

of the subject matter in Part 5 of this Section, *Technique Development*.

Data for Fig. 31 include: contact shot; diameter = 500 mm (20 in.); and wall thickness = 10 mm (0.37 in.).

Drop-through

The condition illustrated in Fig. 32 is also seen in Fig. 26. The condition shown here is much more severe, as evidenced by the weld material hanging inside the pipe wall. The droplet shape is easily identified, together with the cavity remaining after the material was dislodged. Figure 32 shows a pipe with: diameter = 150 mm (6 in.); and wall thickness = 11 mm (0.43 in.).

Stainless Steel Pipe

Figure 33 illustrates a major blowhole with associated incomplete penetration in a large stainless steel pipe. The bead shape, which changes from a smooth to a rippled form, further shows that there was a perturbation in the welding process, such as a change in heat input or speed. The test object data include: contact shot; diameter = 400 mm (16 in.); and wall thickness = 10 mm (0.37 in.).

FIGURE 31. Radiograph Showing Film Overlap

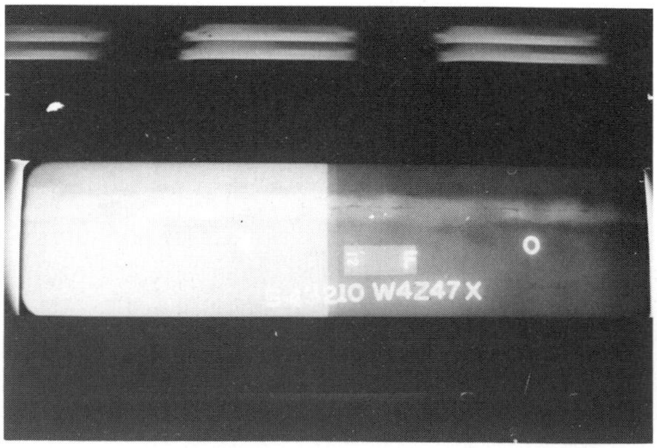

FIGURE 32. Radiograph of Severe Case of Drop-through

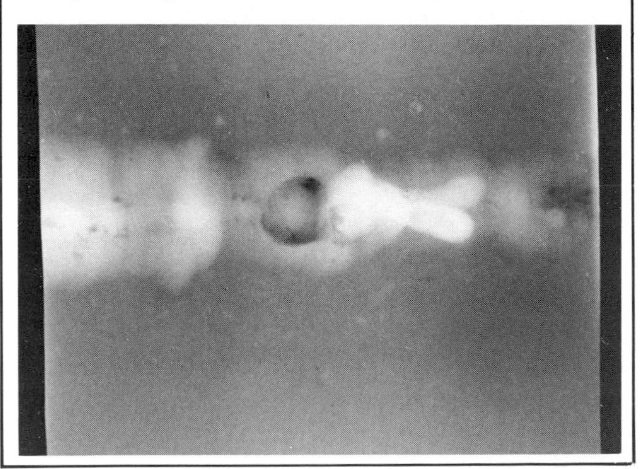

Incomplete Penetration

Figure 34 portrays what is probably a case of 100 percent incomplete penetration around a butt joint in a pipe. The film shows about 80 percent of the pipe diameter; and a very uniform indication extends along the weld centerline. It seems to disappear toward the edges of the film, but this is the section in which poor projection geometry prevails, so that the effective wall thickness, as seen by the beam, increases very rapidly.

The film demonstrates the loss in image definition as the optical conditions degrade. This figure may be compared with Figs. 13, 14 and 19 in this Section. Data for the pipe include: diameter = 150 mm (6 in.); and wall thickness = 11 mm (0.43 in.).

FIGURE 33. Radiograph Showing Blowhole and Associated Incomplete Penetration

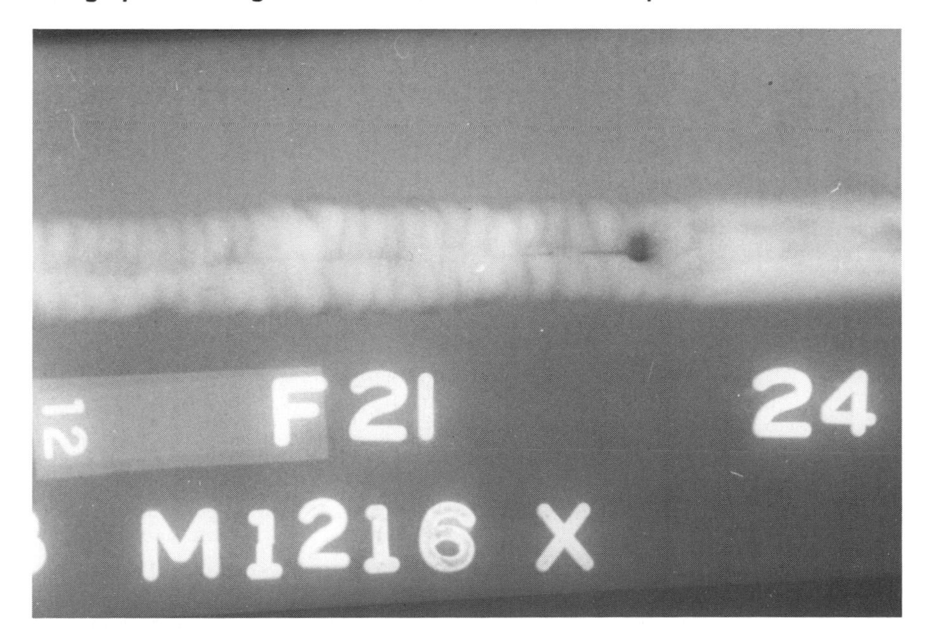

FIGURE 34. Radiograph Showing an Indication of Continuous Incomplete Penetration

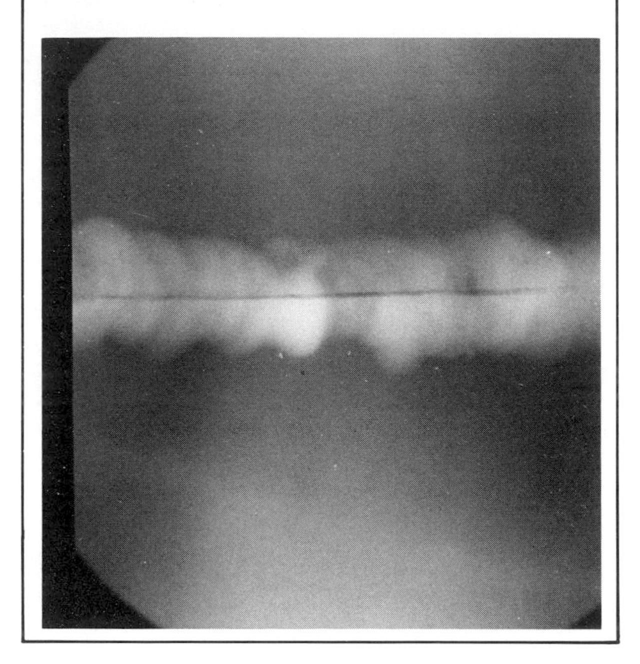

Conclusion

Weldments are often subjected to high pressure, high temperature and cyclic loading. These stresses are usually strongest in structures that can least tolerate the possibility of failure (nuclear reactors, for example). Because of the variety of welding techniques and the variety of joined materials, nondestructive testing of welds is not a simple matter.

The radiographic testing of welds, then, is a vitally important task that is often difficult to accomplish. The intent of this Handbook section has been to describe the welding process in some detail, and to verify the specific ways that radiography can be used in the quality control of this critical joining method.

REFERENCES

1. ASME *Boiler and Pressure Vessel Code.* Section III, "Nuclear Components"; Section V, "Nondestructive Evaluation"; Section VIII, "Pressure Vessels"; Section XI, "Rules for Inservice Inspection of Nuclear Power Plant Components." New York, NY: American Society of Mechanical Engineers (1983): pp 4, 23, 32 and 48.

2. AWS *Structural Welding Code.* Standard D1.1. Miami, FL: American Welding Society (1982): pp 4, 31, 42 and 48.

3. ABS *Rules for Building and Classing Steel Vessels.* New York, NY: American Bureau of Shipping (1983): pp 4, 13 and 48.

4. ABS *Rules for Nondestructive Inspection of Hull Welds.* New York, NY: American Bureau of Shipping (1975): pp 4.

5. API *Standard for Welding Pipelines and Related Facilities.* Standard 1104. Washington, DC: American Petroleum Institute (1983): pp 4 and 48.

6. CSA *Welded Steel Construction.* Standard W59. Toronto, Ontario: Canadian Standards Association (1982): pp 4.

7. *Inspection, Radiographic.* Mil-Std-00453B. Washington, DC: Department of Defense (1975): pp 30.

8. ASTM *Standard Method for Controlling Quality of Radiographic Testing.* Standard E-142. Philadelphia, PA: American Society for Testing Materials (1977): pp 29 and 49.

9. ASTM *Standard Method for Controlling Quality of Radiographic Testing Using Wire Penetrameters.* Standard E-747. Philadelphia, PA: American Society for Testing Materials (1980): pp 30 and 34.

10. *National Bulding Code of Canada.* Ottawa, Ontario (1980): pp 48.

11. ANSI *Code for Pressure Piping.* Standard B31.1. New York, NY: American National Standards Institute (1983): pp 4 and 48.

12. AWWA *Standard for Welded Steel Elevated Tanks, Standpipes and Reservoirs for Water Storage.* Standard D100. Denver, CO: American Waterworks Association (1979): pp 48.

13. API *Standard for Welded Steel Tanks for Oil Storage.* Standard 650. Washington, DC: American Petroleum Institute (1980): pp 4 and 48.

14. AWS *Welding on Earth Moving and Construction Equipment.* Standard D14.3. Miami, FL: American Welding Society (1977): pp 48.

15. ASTM *Recommended Practice for Radiographic Testing.* Standard E-94. Philadelphia, PA: American Society for Testing Materials (1977): pp 34 and 49.

16. *NDE Characteristics of Pipe Weld Defects.* Report No. NP-1590-SR. Palo Alto, CA: Electric Power Research Institute (1980): pp 42.

SECTION 10

RADIOGRAPHIC CONTROL OF CASTINGS

Solomon Goldspiel, P.E., Brooklyn, NY

INTRODUCTION

Casting of metals is a fabrication method chosen primarily because it produces complex shapes with a minimum of finished weight. In addition, casting permits the manufacture of a desired product with fewer components, thus minimizing the amount of joining (by fasteners or welds). Possible casting problems are often associated with the mold, the molten metal or patterns in the cast object; these must be properly considered to avoid difficulties and to ensure required quality levels. To provide a justification for nondestructive testing of castings, with emphasis on the radiographic technique, these problems are summarized in this chapter. Special thanks to Vaughn Jensen, Hercules, Inc.; Robert Koenke, X-ray Products, Corp., XRP Laboratories; David Walshe, ITT Grinnell Corporation; Arnold Greene Testing Laboratories, Natick, MA; and Lawrence E. Bryant, Los Alamos National Laboratory, for their contributions to this chapter.

Casting Defects

Pouring of castings is simplified when the melt fluidity is high. This is achieved by superheating the material above its melting range to a point where associated gas absorption is kept to a reasonable minimum; this is especially important for ordinary-air molten practices. Failure to superheat creates the possibility of gas-pocket formation in some portions of the casting. Gas pockets tend to occur in the heavier cast sections while gas porosity occurs under an initial layer of solidified metal near the mold and/or core walls. Large pockets of gas are obviously detrimental to the strength of the casting unless it is designed to allow pocket formation only in portions which are discarded during finishing.

Subsurface gas is unacceptable because finish machining may produce surface openings which act as locations of stress concentration. Subsurface porosity occurs because the gas takes some time to move toward the mold and core walls and, in that time, abstraction of heat causes metal skins to form, preventing the gas from escaping. The extent of subsurface porosity varies with mold type. Sand molds, depending on their degree of dryness, have a greater tendency to produce solid metal skins near mold walls, when compared to shell and plastic (precision casting) molds. Possible sources of such defects should be eliminated during the casting process.

PART 1

BASIC PRINCIPLES OF CASTING RADIOGRAPHY

The major goals of casting inspection are (1) to determine whether castings are sound, serviceable, and whether they are in compliance with strength, pressure and other requirements over their expected life span; and (2) to help provide product control of the various factors and operations involved in production. The latter is especially true when the number of castings is high and the production of pilot castings is possible and practical.

Radiography, with film as its recording medium, is one of the most effective NDT methods for quality control of castings.

Radiography utilizes penetrating ionizing radiation and the method provides permanent reference data. The film, in principle, is a record of the total thickness of radiation-absorbing material displayed in a single plane; or a planar projection of conditions prevailing in the three-dimensional space of the item tested. Other common nondestructive testing methods are affected by the metallographic structure and degree of working to which the metal has been subjected; generally these considerations have no effect on the radiation transmission or subsequent legibility of the radiographic indications. The only exception is the indication called mottling, as described in some detail later. Hence radiography is the most effective nondestructive testing method for castings, which generally are studied for presence of volumetric defect types. As is true of any important engineering fabrication where stress concentrations may cause trouble, surface methods such as magnetic particle or liquid penetrant testing should supplement radiography since discontinuities such as surface cracks are more difficult to determine radiographically.

To be effective, the radiographic method requires proper interpretation of resulting films. Interpretation, in turn, demands a familiarity with the casting process. Important things to consider include: the method of casting; mold and core materials and design; melting and pouring temperatures; cooling rates and time; and possible interaction of the particular metallographic structure with the test radiation.

The Casting Process

The casting shape and method often dictate how gates and risers are handled. *Gates* are the means for introducing the melt (molten metal) into the mold. *Risers* are appendages added to the shape being cast; ideally, material shrinkage will be located in the risers, which are cut off and discarded during finishing operations.

The positioning of gates and risers depends on the shape actually produced, the solidification characteristics of the particular alloy being used, and the manner in which the mold is split into its lower (drag) and upper (cope) portions. Castings with flanges and protrusions (called bosses) must be designed to provide adequate access of molten metal even to the object's thickest portions. Risering must also allow for shrinkages characteristically occurring in all common casting metals and alloys.

Heating and Cooling

Prior to introducing the metal into the mold, melting allows slag, dross and other nonmetallics to float to the surface where they are skimmed off. The melt must be sufficiently superheated to permit the slag to rise to the tops of risers where it does no harm. At the same time, the superheating must not be so high as to promote excessive gas absorption from air and from reactions with the mold and core materials.

The control of cooling rates produces the optimum quality metal in the casting. Cooling is affected by the casting's geometry, mold and core material properties and by the location of risers which, in effect, also act as heat reservoirs. The mold and core materials affect the speed of heat removal during solidification and the possible formation of a solid metal skin which tends to lock in gases attempting to escape from the interior liquid metal as it solidifies.

PART 2

GENERAL TECHNIQUES OF CASTING RADIOGRAPHY

Radiation Sources Used

Gamma Radiation

Gamma ray sources, especially as used in casting radiography, have the following advantages over X-rays: (1) simple apparatus; (2) compactness; (3) independence from external power; (4) ability to provide simultaneous testing of many objects and complete circumferences of large cylindrical objects; (5) they are useable when access to the interior of the object is difficult; and (6) when testing must occur in confined spaces. Radium, a natural gamma ray source, came into use in the late 1930s; it has since been replaced by man-made isotopes (see Table 1).

The use of X-rays, however, is essential for some applications. Low (kilovolt) X-ray energies are needed for obtaining required radiographic sensitivities in light metals (such as aluminum) and in thin material thicknesses of steel. High energy (megavolt) X-rays are necessary for penetration of steel thicknesses in excess of 20 cm (8 in.). As with all nondestructive testing, casting inspection is best done with the method that produces the desired results for the specific application.

Gamma ray sources, unlike X-ray machines, emit penetrating radiation having only one, or a few, discrete wavelengths. Sources are commonly specified by the energy of the individual quantum (using units of *electron volts*) rather than by wavelength. Thus a gamma ray with an energy of 1.25 MeV is equivalent in wavelength and penetrating power to the most penetrating radiation emitted by an X-ray tube operating at 1.2 MeV. The total penetrating power is about equal to that of a 3 MeV X-ray machine which emits a spectrum or wide range of energies.

The wavelengths (or energies) emitted by a gamma ray source depend only on the nature of the emitter and are not variable at the will of the operator, as X-rays are.

Important gamma source characterisitics include: (1) the curie (or, in SI, becquerel) value and specific activity; (2) half-life; (3) energy of quanta; (4) dosage rate; (5) application thickness limit. The intensity of gamma radiation depends on the number of radioactive atoms that disintegrate per second in the source. For small or moderate sources, the radiation intensity is proportional to the source activity

TABLE 1. Gamma-ray Sources Used in Industrial Radiography

Radioactive Element	Energy (MeV)	Half-Life	Dosage Rate R/hr/Ci at 1 meter	Metallurgical Applications
Radium	0.24 - 2.20*	1620 years	0.84	1/2 to 5 in. (10-120 mm) steel or equivalent
Cobalt-60	1.17 and 1.33	5.3 years	1.35	1 to 8 in. (20-200 mm) steel or equivalent
Iridium-192	0.137 to 0.651**	75 days	0.55	1/2 to 3 in. (10-80 mm) steel or equivalent
Cesium-137	0.66†	33 years	0.39	1/2 to 3 1/2 in. (10-90 mm) steel or equivalent
Thulium-170	0.048 and 0.054††	127 days	0.003	1/32 to 1/2 in. (1-12 mm) steel or equivalent

* Infrequently used; has eleven principal gamma rays.

** Has at least twelve gamma rays; principal ones at 0.310, 0.470 and 0.600 keV.

† Usual form used is CsCl, a soluble powder calling for special precautions to prevent leakage, such as double encapsulation in silver brazed type 316 stainless steel.

†† Produced only in limited intensities for desired energies, so use is not widespread. Has excellent portability, however, and may be used at 2% sensitivity down to 12 mm (0.5 in.) aluminum and for light nonmetals.

(in curies or becquerels). The proportionality fails for large sources or for those emitting low energy gamma rays due to self-absorption.

The *specific activity*, expressed in curies per gram or cubic centimeter (or becquerels per gram), is important because it influences geometrical unsharpness of the images. Higher specific activity allows shorter source-to-film distances without a loss of image sharpness.

Gamma-ray dosage is expressed in roentgens per hour at one meter per curie or, in SI, millisieverts per hour at one meter per gigabecquerel. The roentgen is the quantity of radiation which produces one electrostatic unit of electricity for every 0.001293 grams of air ions. Gamma-ray sources lose activity with time, the rate depending on the material *half-life* shown in Table 1. Knowing the half-life of the isotope allows preparation of *decay curves* as illustrated for iridium-192 in Fig. 1. This graph of activity versus time permits revised calculation of exposure time as the activity of the gamma-ray source diminishes.

X Radiation

X-rays are made by using electrical energy to produce electrons that are accelerated to very high velocities. X-rays are emitted by deceleration of the electrons when they strike a target, which for industrial radiography is usually made of tungsten. The higher the voltage of the applied energy, the greater the speed of the electrons striking the focal spot. The result is a decrease in wavelength of the X-rays emitted, with a simultaneous increase in penetrating power and intensity. Thus, unlike gamma-ray sources, X-ray machine radiations may be varied at the will of the operator within the range of the equipment used.

The various X-ray machines commercially available may be very roughly classified according

FIGURE 1. Decay Curves for Iridium-192: (a) Linear Plot; (b) Logarithmic Plot

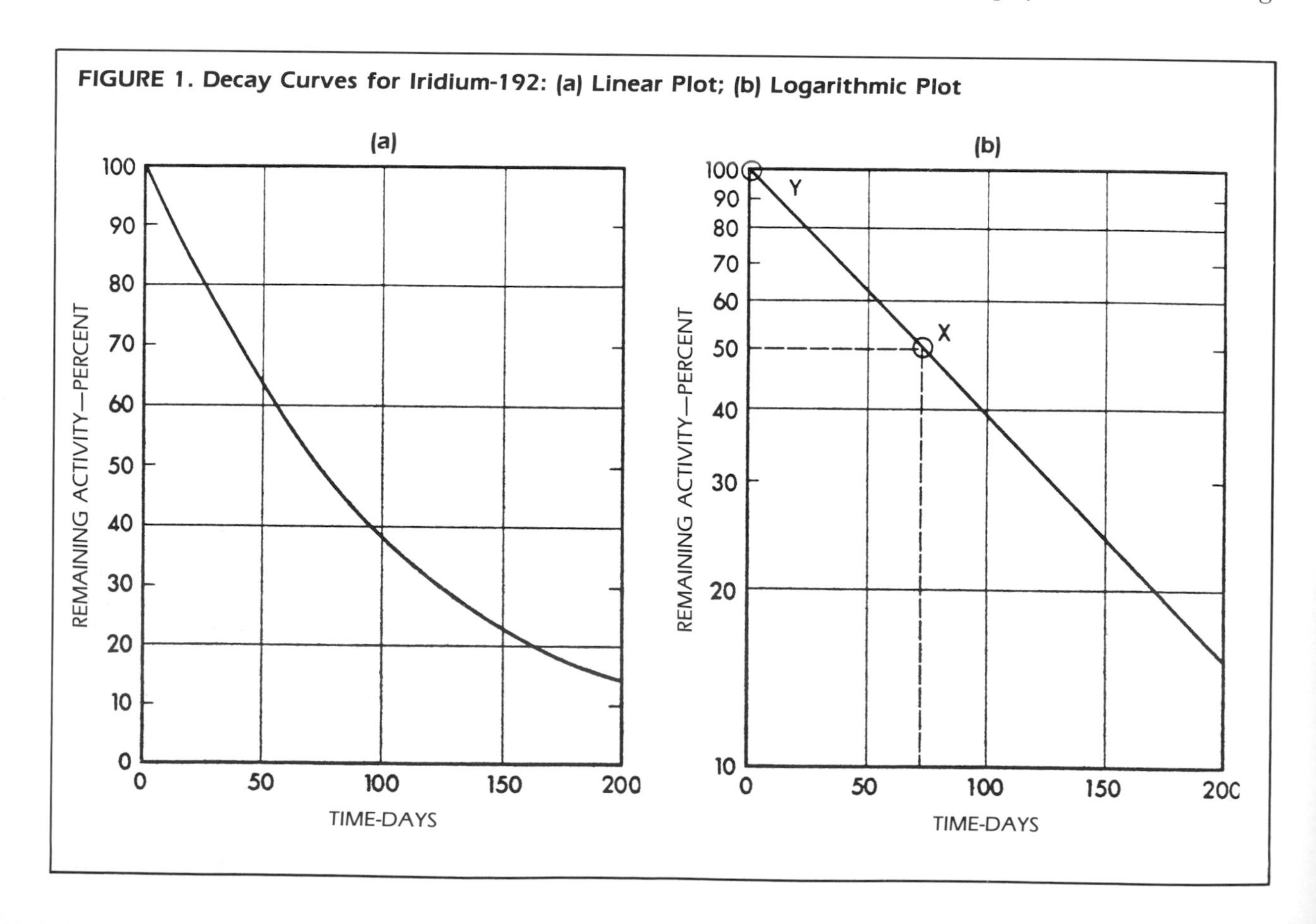

to their maximum voltage. Table 2 is a generalized guide for typical X-ray machines by voltage ranges and applications.[2] The table must be used with the understanding that particular machines differ in their specifications from model to model.

While most commonly used X-ray machines are less mobile than radioisotopes and depend on availability of electrical current, they are available in portable designs. In addition, those with rod-anode tubes (in which the X-ray beam axis is perpendicular to the electron stream) can be used to produce radiographs with circumferential coverage of appropriate items. The major advantages of X-rays are their safety; their ability to generate higher intensities of penetrating radiation; and their variability in radiation intensity (quantity) and energy (quality) as determined by the operator and the application.

Specification Requirements

Radiography of castings indicates the quality of the product within the limits of the inspection method. Hence, the contract between producer and consumer should, for the mutual benefit of each, refer to time-proven specifications.

Among the most widely used specifications are those of The American Society for Testing and Materials (ASTM) with emphasis on method and interpretation of radiographic indications;[3] and those of the American Society of Mechanical Engineers (ASME) with emphasis on product quality and in-process weld repairs.[4] During the preparation of radiographic test requirements, it is essential to recognize that radiographic coverage has its limitations, especially when inspecting castings with complex geometries or part configurations. Geometry or part configurations that do not allow complete coverage with normal radiographic methods should be acknowledged by supplier and user before the start of work. The use of supplementary nondestructive testing methods should be clearly agreed upon from the outset. It should be noted that the radiographic method is particularly effective for discontinuities which displace a volume of cast material. Cracks and planar discontinuities which do not displace an appreciable volume of material may not be detected by radiography unless the radiation is favorably oriented. Additional nondestructive test methods, and criteria for acceptability, must be agreed upon by supplier and user.

TABLE 2. Typical X-ray Machines and Their Applications

Maximum Voltage (kV)	Screens	Metallurgical Applications and Approximate Thickness Limits*
50	None	Thin sections of most metals**
150	None or lead foil None or lead foil Fluorescent	5 in. (125 mm) aluminum or equivalent 1 in. (25 mm) steel or equivalent 1 1/2 in. (40 mm) steel or equivalent
250	Lead foil Fluorescent	2 in. (50 mm) steel or equivalent 3 in. (75 mm) steel or equivalent
400	Lead foil Fluorescent	3 in. (75 mm) steel or equivalent 4 in. (100 mm) steel or equivalent
1000	Lead foil Fluorescent	5 in. (125 mm) steel or equivalent 8 in. (200 mm) steel or equivalent
2000	Lead foil	8 in. (200 mm) steel or equivalent
8 to 25 (MeV)	Lead foil	16 in. (400 mm) steel or equivalent

* Also used for important nonmetallics such as moderate thicknesses of graphite and small electronic components, wood, plastics, etc.
** Lower limit depends on particular machine and how the secondary voltage may be adjusted.

Radiographic Setup

Radiographic coverage is determined by the casting geometry, especially those portions to which gates and risers are connected during casting; cylindrical portions; flanges; bosses; and, portions that are inaccessible to radiation and/or film. The inaccessible portions always call for special considerations in nondestructive testing contract dealings. All radiographic testing (especially of portions critical to use or loading) requires the making of *radiographic shooting sketches* (RSS) and the compilation of associated data. Recommended RSS types (see Figs. 2 and 3) are covered by ASTM specification.

Shooting sketches may take into consideration such aspects as simultaneous coverage of cylindrical portions, completely or by sectors; and single or double wall shots when IDs are relatively small, around 10 cm (4 in.) or less. With isotope radiography, it is often convenient to examine a number of castings simultaneously. In all cases, special consideration must be given to proper identification of the films. This is usually done using lead letters attached to the object under test. Because sources are never actually ideal point-sources, the source-to-film distance should be such that it limits geometric unsharpness, U_g, with reasonably economical exposure times. This unsharpness is given by:

$$U_g = Ft/D \qquad \text{(Eq. 1)}$$

where F is the effective focal spot size, t is the object-to-film distance; and D is the source-to-object distance.

Penetrameters

Plaque-type Penetrameters[5]

A relatively simple way to determine whether the radiographic testing procedure has met the required quality level (even without detailed consideration of the many factors involved) is by the proper choice and use of penetrameters. The penetrameter (also referred to as the *image quality indicator* or *IQI*) is simply a metal plaque with holes. Its material is chosen to have radiation absorption characteristics close to those of the material under test. Its thickness and hole sizes are predetermined percentages of the section thickness to be radiographed. The plaque, as most commonly used in casting radiography, is 2% of the section thickness and its three holes have diameters of 2, 4 and 8% of the section thickness. Where casting thickness and/or exposures vary appreciably, more than one penetrameter must be used to indicate film sensitivity.

Penetrameters are usually placed on the portion radiographed. If this is not possible, because of curvature of the part or possible interference with radiographic legibility, the penetrameters are placed as close as possible to the portion radiographed, on blocks of suitable thickness and as close to the casting as possible. In addition, penetrameters are usually placed on the source side of the casting. When they are placed on the film side, the comparability of penetrameter size to the one required for the source side must be suitably demonstrated. Figure 4 shows an actual radiograph with both the commonly used ASTM plaque penetrameter and the wire-type (DIN) penetrameter discussed below.

The radiograph in Fig. 4 shows a plaque-type penetrameter with a thickness (T) equal to 2% of the casting's wall thickness, and holes with diameters of 1T, 2T and 4T. The wire-type penetrameter has six wires 25 mm (1 in.) long with diameters ranging from 0.25 to 0.8 mm (0.010 to 0.032 in.). For the particular section thickness radiographed, 20 mm (.75 in.), equivalency of the two penetrameter types is demonstrated when all six wires of the wire-type and the 2-2T hole are visible. In this illustration, the thickness and hole diameter required are about 0.4 mm (0.015 in.) each.

Wire Penetrameters[6]

Wire penetrameters, originally introduced in Germany and known as DIN types, have been recently adopted in the U.S. as an alternative means of radiographic quality control (see Fig. 4). When it is used, the size must be equivalent to the customarily specified plaque type. Table 3 shows equivalence data which, in this instance, compares wire penetrameters to plaque types for the 2-2T level. This designation signifies that radiographs must show, as a minimum, a plaque thickness (T) within 2% and a hole diameter (2T) within 4% of the section thickness radiographed.

FIGURE 2. Sample of Radiographic Shooting Sketch (RSS) Showing Film Placement and Identification

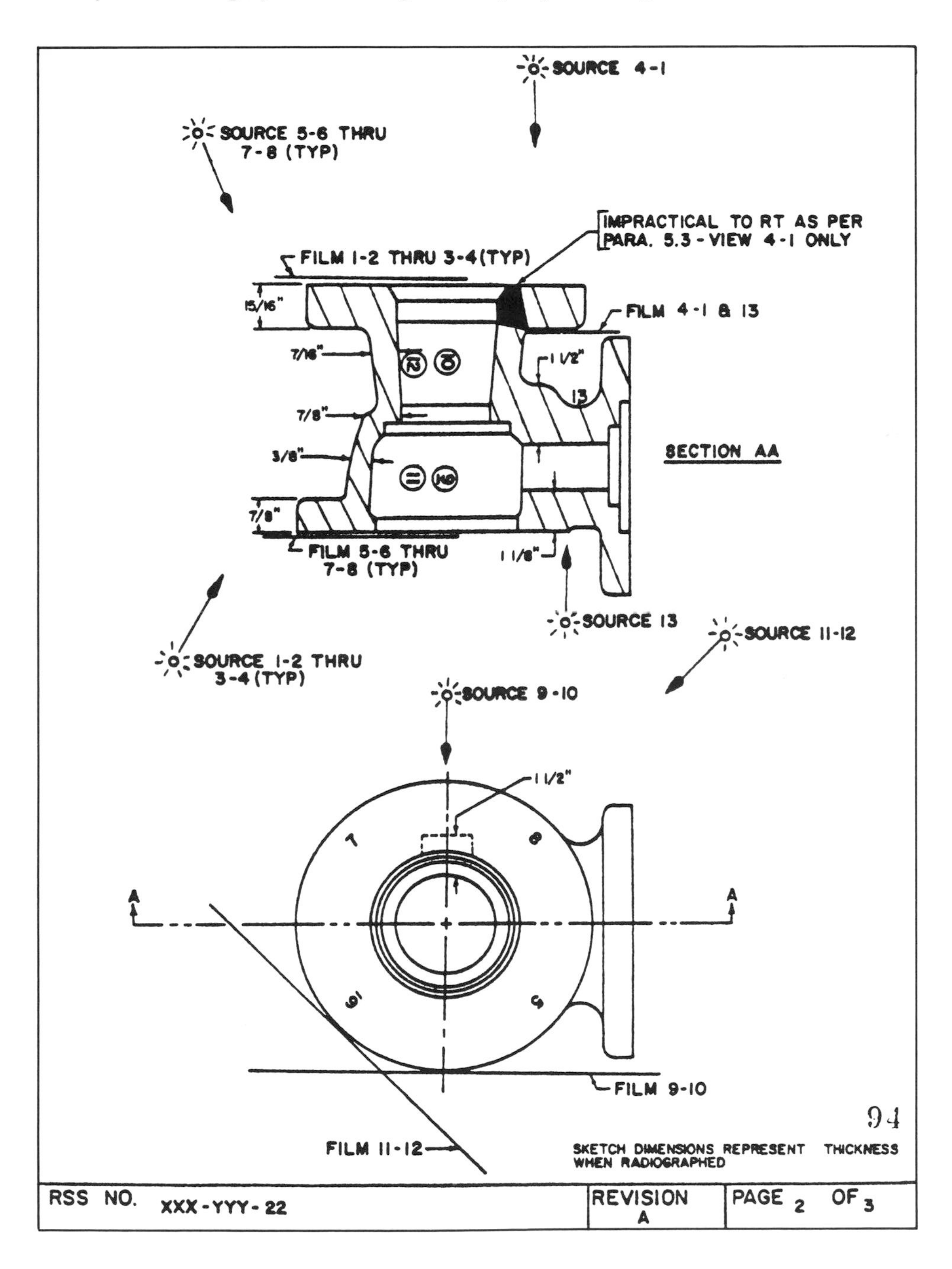

FROM THE AMERICAN SOCIETY FOR TESTING AND MATERIALS. REPRINTED WITH PERMISSION.

FIGURE 3. Data Tabulation for Radiographic Standard Shooting Sketch (RSS)

GENERAL INFORMATION	CASTING IDENTIFICATION		
COMPANY PREPARING RSS	DRAWING NO.	REVISION	PIECE NO.
COMPANY PERFORMING RT	DESCRIPTION BODY		
FOUNDRY CASTING IDENTIFICATION METHOD STAMPED [X] ETCHED [ ] AT RT LOCATION 9-10	MATERIAL NI-CU	MATERIAL SPEC.	
SURFACE CONDITION WHEN RT'D AS CAST [ ] ROUGH MACH'D [X] FINISH MACH'D [ ]	PATTERN NO.	HEAT NO.	

RSS APPROVAL			
SUPPLIER		CUSTOMER	
1.	DATE		DATE
2.	DATE	2.	DATE

SAMPLE

RT PARAMETERS								
VIEWS	1-2 thru 4-1	5-6 thru 7-8	9-10	11-12	13			
SOURCE TYPE	IRID 192							
FINISHED THICKNESS	13/16"	3/4"	5/8"-2-1/8	5/8"-2-3/8	3/4"-2-3/8			
THICKNESS WHEN RT'D	15/16"	7/8"	3/4"-2-5/16	3/4"-2-1/2	7/8"-2-1/2			
PENETRAMETER(S)	17	17	15 - 45	15 - 50	17 - 50			
SOURCE TO FILM DISTANCE	30"							
FILM TYPE	1		1 & 2					
FILM SIZE	5 X 7		8 X 10					
QUALITY LEVEL	2-2T							
ACCEPTANCE STANDARD	ASTM E-272							
SEVERITY LEVEL	2							

NOTES	REVISIONS					
	REV.	DESCRIPTION	APPROVAL SUPPLIER BY	SUPPLIER DATE	CUSTOMER BY	CUSTOMER DATE
	A	ORIGINAL ISSUE	—	—	—	—

RSS NO. XXX-YYY-22	REVISION A	PAGE 1 OF 3

Defect Detection and Radiographic Sensitivity

It must be remembered that a penetrameter is used to indicate the quality level of the radiographic technique and not necessarily to provide a measurement of the size of minimum discontinuity which can be shown in the object. Thus, if required penetrameter details are visible in the radiograph, there is no certainty that an equivalent flaw in the casting will be revealed. This is because the penetrameter holes have sharp boundaries while natural casting holes of the same size may have boundaries that are more or less rounded, with sides gradually merging into surrounding casting portions. Hence, the hole of the penetrameter may be readily discerned, even though its density differs only slightly from that of the surrounding casting area.

Similar considerations apply to linear or crack-like indications. If the plane of the linear discontinuity is inclined away from the beam (by at least 7 degrees), the crack may not be visible on the radiograph because of the relatively gradual transition of densities in the image. Similarly, the visibility of a wire penetrameter does not guarantee that a casting discontinuity of the same cross section will actually be visible. The human eye discerns a long boundary more readily than it does a short one, even if the density increase and image sharpness are the same. Nevertheless, it is true that the probability of flaw detection in the object radiographed (to assure the required quality) is related to the discernibility of penetrameter features, when all other factors are the same.

FIGURE 4. Plaque and Wire-type Penetrameters, Shown in Radiograph of 1.9 cm (0.75 in.) Casting Portion

FROM ARNOLD GREENE TESTING LABORATORIES, NATICK, MA.

TABLE 3. Wire Penetrameter Sizes Equivalent to 2-2T Hole-Type Levels

Minimum Specimen Thickness in. (mm)	Wire Diameter in. (mm)
0.25 (6.35)	0.0032 (0.08)*
0.313 (7.95)	0.004 (0.1)*
0.375 (9.5)	0.005 (0.13)*
0.500 (12.7)	0.0063 (0.16)
0.625 (15.9)	0.008 (0.2)
0.750 (19.1)	0.010 (0.25)
0.875 (22.2)	0.013 (0.33)
1.00 (25.4)	0.016 (0.4)
1.25 (31.8)	0.020 (0.51)
1.5 (38.1)	0.025 (0.64)
1.75 (44.4)	0.032 (0.81)
2 (50.8)	0.040 (1.02)
2.5 (63.5)	0.050 (1.27)
3 (76.2)	0.063 (1.6)
3.5 (88.9)	0.080 (2.03)
4 (102)	0.100 (2.5)
4.5 (114)	0.126 (3.2)
5 (127)	0.160 (4.06)

* Wire diameters for use with specimens less than 1.27 cm (0.5 in.) thickness do not represent the true 2-2T level. They follow the same relationship as the hole type.

FROM THE AMERICAN SOCIETY FOR TESTING AND MATERIALS. REPRINTED WITH PERMISSION.

Special Considerations for Radiography of Castings

Mold Type

As castings progress from sand mold-type to shell mold, permanent mold, investment, precision, plastic mold and die castings, the radiographic procedure must be changed to accommodate more castings with more complex shapes and thinner sections. All these factors provide justification for using one or more pilot runs designed to improve yields and minimize or eliminate systematic flaws.

Mold type also determines the amount and frequency of radiographic testing. For sand castings, sections are relatively thicker with rougher skins; there is more allowance for machining, especially for surfaces joined to other system components. Surface roughness and sections with substantially thicker-than-finished dimensions increase the difficulties in radiographic interpretation.

Exposures must be made in such a way that penetrameter sensitivity, as dictated by finished section thicknesses, is not compromised. This is usually done by using penetrameters based on finished rather than rough-wall thicknesses. As foundry methods improve, precision and die cast sections become thinner and smoother; the interpretation of radiographs is improved and radiation energies must be reduced. Of course, in the more precise casting methods, cost of all production steps must be watched. Systematic flaws must be determined using pilot runs of sufficient number to assure the required quality levels. It should be noted that smoother surfaces also tend to considerably reduce random flaws.

Individual sand castings generally require more radiography since the possibility of non-systematic (random) flaws is larger than for the more precise casting methods. Important sand castings require individual radiography, especially of critical portions, to locate both systematic and random flaws. It may be pointed out that systematic flaws are generally associated with the casting details (gates, risers, junctions of heavy to thin portions, etc.). Random flaws may be due to accidental conditions (local gas due to mold moisture; local stresses causing incidence of linear flaws, etc.).

Testing Alloy Castings

When testing alloys, the major factors affecting the radiography are alloy density and the major ele-

TABLE 4. Common Alloy Casting Densities vs. Radiographic Sources Used and Available ASTM Reference Radiograph Documents

Major Element or Alloy Type	Density Range (g/cm^3)	Available ASTM Radiographic Sources Commonly Used	Reference Radiographs
Magnesium	1.79 - 1.86	X-rays*	E155 and E505
Aluminum	2.57 - 2.95	X-rays*	E155 and E505
Titanium	4.43 - 4.65	X-rays and Ir-192	**
Cast iron	5.54 - 7.48	X-rays, Ir-192 and Co-60	E802 plus applicable steel documents
Zinc	6.6 - 6.7	X-rays, Ir-192 and Co-60	None: use available documents for closest density
Carbon steels	7.81 - 7.84	X-rays, Ir-192, Co-60	E192, E446, E186 and E280
Stainless steels	7.53 - 7.75	X-rays, Ir-192, Co-60	E192, E446, E186 and E280
Aluminum bronze	7.50 - 7.80	X-rays, Ir-192, Co-60	E272
Manganese bronze	7.7 - 8.3	X-rays, Ir-192, Co-60	E272
Silicon bronze	8.30	X-rays, Ir-192, Co-60	E272
Tin bronze	8.7 - 8.8	X-rays, Ir-192, Co-60	E310
Navy bronze	8.7	X-rays, Ir-192, Co-60	E272
Nickel silver	8.85 - 8.95	X-rays, Ir-192, Co-60	E272
Tantalum	16.6	Cs-137	None

*Up to about 300 kVp

**Depending on thickness involved, use applicable documents for aluminum or steel including source type.

ment's atomic number; these determine the energy levels needed in the radiation sources. It is well known that alloys may react with the atmosphere or mold material. The general solidification peculiarities of alloys may affect the indication types which are discernible in their radiographs. Industrial casting alloys fall into the following major types based on atomic number: (1) light metals (including magnesium, aluminum and titanium); (2) intermediate alloy types (including zinc, cast iron, steels, brasses and bronzes); and (3) heavy metals (including lead and tantalum). The densities of some of the most common alloys are shown in Table 4.

Of the heavy alloys, tantalum may be used as an example for a typical casting application. Tantalum castings are used in acid-resistant chemical equipment such as heat exchangers, centrifugal pumps and valves. This is because the element has a combination of characteristics not found in many refractory metals. These include: ease of fabrication; low ductile-to-brittle transition temperature; and high melting point.

Tantalum oxidizes in air above 299 °C (570 °F). Because of its high density, it requires longer exposures or higher radiation energy than needed for less dense materials. Thus a section 0.75 cm (0.30 in.) thick requires an exposure of about half an hour with a medium strength cesium-137 source when fast industrial films are used with intensifying screens.

Standards

ASTM standards are widely used in radiographic testing of castings. The standards were written (with the help of producers, consumers, government and educational institutions) for radiographic practices[3,7] and radiograph quality control.[5,6,8,9] Following the standards ensures the production of radiographs that can be meaningfully read by representatives of both producer and consumer interests.

Of course, the reading leads to an interpretation of the casting's soundness, and interpretation requires some sort of standardized guideline.

ASTM's work on reference radiograph documents (beginning in 1950 and continuing to the present) has led to the development of standards that encourage relatively unbiased determination of casting quality for use in meeting contractually required acceptance criteria.

Table 5 lists most of the ASTM reference radiograph documents.[10] Included is information on fabrication material of the hardware used; section thickness ranges; discontinuity types represented; and numbers of discontinuities shown in graded types.

Reference radiographs for steel castings were first issued by the U.S. Navy Bureau of Engineering in 1938 (see ref. 10) as "Gamma Ray Radiographic Standards for Steam Pressure Service". These were reissued in 1942 by the Bureau of Ships as "Reference Radiographic Standards for Steel Castings" and adopted in 1952 by ASTM as E71, "Reference Radiographs for Steel Castings". This document contained copies of relatively thin-walled casting radiographs; the hardware from which they were produced however was not maintained. (It may be pointed out that all current ASTM reference radiograph illustrations are backed by hardware, the exceptions being E310 and E272 for which some castings were lost.) Document E71 was used for many years until replaced by suitable ASTM documents.

ASTM has also developed improved methods for mass production and monitoring of document illustrations, either in the form of actual radiographs or photographic copies of originally selected radiographs. Its recent work involves reference radiographs for titanium alloys.

Since Table 5 was compiled, documents for ductile and gray iron castings[11,12] have also been produced. Another document now available discusses how images change as certain radiographic parameters vary.[9]

TABLE 5. ASTM Casting and Weld Reference Radiograph Documents Developed and Published through Spring 1982

Material and Fabrication	Designation and Year Issued	Section Thickness Range (in.)[a]	Discontinuity Types Illustrated	Number Shown
Aluminum and magnesium castings	E 155 (1960)	1/4 to 2	*aluminum:* gas holes, two thicknesses; shrinkage cavity; shrinkage, sponge, two thicknesses; foreign material, less and more dense, two thicknesses	8 each
			magnesium: gas holes, two thicknesses; microshrinkage, sponge and feathery, two thicknesses; foreign material, less and more dense, two thicknesses; reacted sand inclusions; segregation, eutectic and gravity	8 each
			ungraded illustrations: pipe, shrinkage, flow line, hot tear	1 each
Aluminum and magnesium die castings	E 505 (1974)	1/8 to 1	porosity; cold fill; shrinkage	4 each
			ungraded illustration: foreign material	1 each
High-strength copper base and nickel-copper alloy castings	E 272 (1965)	up to 6[b, c]	gas porosity, two thicknesses; sand inclusions, two thicknesses; dross inclusions, two thicknesses; shrinkage, linear;[c] shrinkage linear;[c] shrinkage feathery;[b] shrinkage spongy[c]	5 each
Tin bronze castings	E 310 (1966)	up to 2[d]	porosity; shrinkage, linear, feathery, spongy	5 each
			ungraded illustrations: hot tear; chaplet	1 each
Steel thin precision castings	E 192 (1962)	1/8 to 3/4[e]	gas holes; shrinkage, spongy, dendritic, filamentary	8 each
			ungraded illustrations: discrete discontinuities, six types; mold defects, four types; diffraction pattern, two types	1 each
Steel castings	E 446 (1972)	up to 2[f]	gas porosity; sand and slag inclusions; shrinkage, four types	4 each
			ungraded illustrations: crack, hot tear, insert, mottling	1 each
Steel castings, heavy walled	E 186 (1962)	2 to 4 1/2[g]	gas porosity; sand and slag inclusions; shrinkage, three types	5 each
			ungraded illustrations: crack, hot tear, insert	1 each
Steel castings, heavy walled	E 280 (1965)	4 1/2 to 12[h]	gas porosity; sand and slag inclusions; shrinkage, three types	5 each
			ungraded illustrations: crack, hot tear, insert	1 each

TABLE 5. Continued

Material and Fabrication	Designation and Year Issued	Section Thickness Range (in.)[a]	Discontinuity Types Illustrated	Number Shown
Steel fusion welds	E 390, Vol. 1 (1969)	up to 1/4[i]	porosity, three types; tungsten inclusions	5 each
			ungraded illustrations: linear porosity; lack of fusion; worm holes; burn through; icicles, transverse crack; crater crack, undercut	1 each
	E 390, Vol. 2	1/4 to 3[j]	porosity, four types; slag inclusions; incomplete penetration; lack of fusion	5 each
			ungraded illustrations: worm hole; burn through; icicles; longitudinal and transverse crack; crater crack; undercut	1 each
	E 390, Vol. 3 (1969)	3 to 8[k]	porosity, three types; slag inclusions; incomplete penetration; lack of fusion	5 each
			ungraded illustrations: transverse and crater cracks	1 each
Gray iron castings	E 802 (1982)	up to 4	feathery shrinkage	7 each

[a] 1 in. = 25 mm
[b] Radiation sources: low-voltage X-rays up to 2 in. (50 mm) and 2 MeV X-ray or Co-60 for 2 to 6 in. (50 to 150 mm).
[c] Radiation sources: 2 MeV X-rays or Co-60.
[d] Radiation sources: low-voltage X-rays or Ir-192.
[e] Radiation sources: 130 to 250 kVp X-rays.
[f] Radiation sources: moderate-voltage X-rays; 1-MeV X-rays or Ir-192; 2-30 MeV X-rays.
[g] Radiation sources: 1 MeV X-rays or Ir-192; 2 MeV X-rays or Co-60.
[h] Radiation sources: 2 MeV X-rays or Co-60; 4-30 MeV X-rays.
[i] Radiation sources: 90-150 kVp X-rays.
[j] Radiation sources: X-rays to 2 MeV or Co-60.
[k] Radiation sources: 2 MeV X-rays or Co-60.

PART 3

RADIOGRAPHIC INDICATIONS FOR CASTINGS

The major objective of radiographic testing of castings is the disclosure of discontinuities (with emphasis on volumetric types) which adversely affect the strength of the product. These discontinuities, of course, are related to casting process deficiencies which, if properly understood, can lead to accurate accept-reject decisions as well as to suitable corrective measures. Following is a brief description of the most common discontinuity types included in existing reference radiograph documents (in graded types or as single illustrations).

Gas porosity is a form of spherical or elliptical holes within the cast metal. These are usually due to occluded gas in the melt which had no chance to rise and escape through the casting top or its risers. Gas porosity often comes from the atmosphere or from mold-metal interactions. Overheating of melt and excessive moisture in molds and/or cores tends to promote this flaw type (see Figs. 5, 6 and 7).

Sand inclusions and *dross* are nonmetallic oxides, appearing on the radiograph as irregular, dark blotches. These come from disintegrated portions of

FIGURE 5. Example of Gas Holes in 0.64 cm (0.25 in.) Aluminum Casting

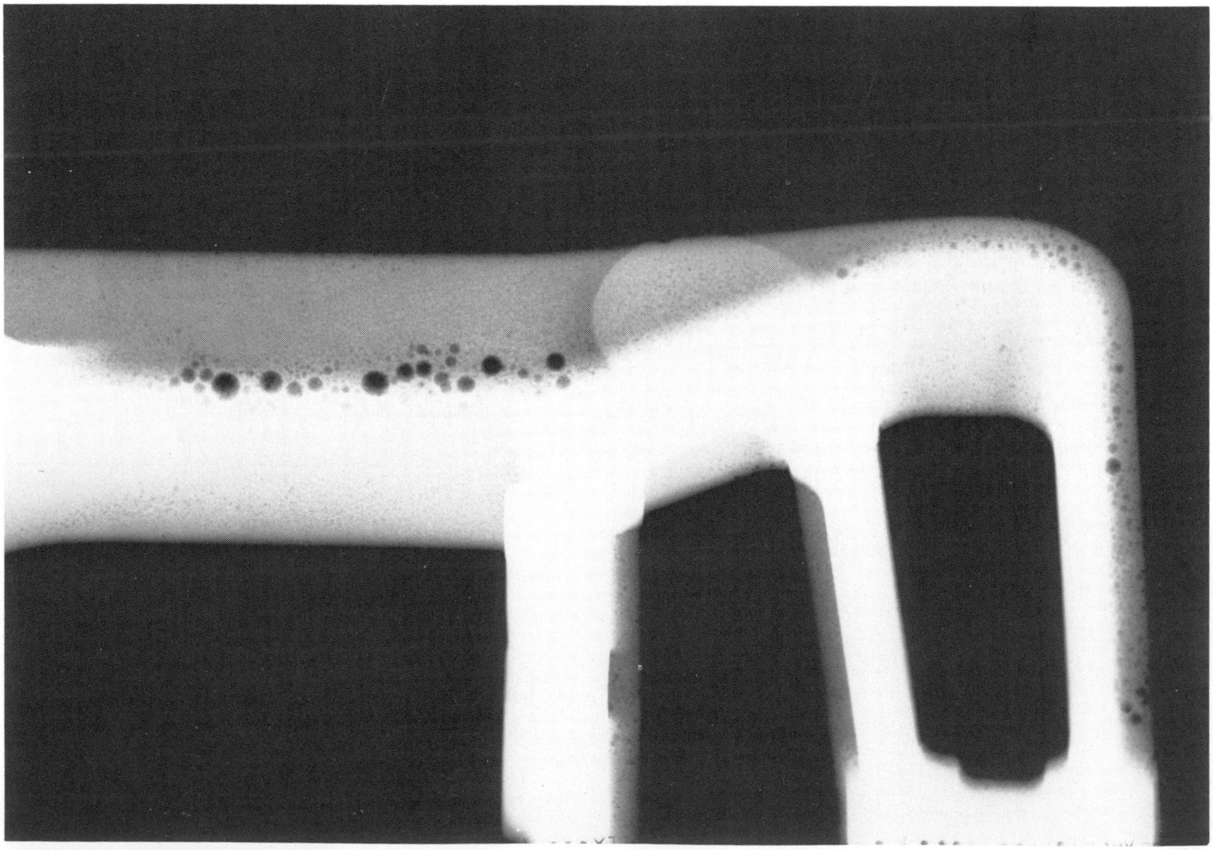

FROM X-RAY PRODUCTS CORPORATION.

mold or core walls and/or from oxides (formed in the melt) which have not been skimmed off prior to introduction of the metal into the mold gates. Careful control of the melt, proper holding time in the ladle and skimming of the melt during pouring will minimize or obviate this source of trouble.

Shrinkage is a form of discontinuity which appears as dark spots on the radiograph. Shrinkage assumes various forms but in all cases it occurs because molten metal shrinks as it solidifies, in all portions of the final casting. Shrinkage is avoided by making sure that the volume of the casting is adequately fed by risers which sacrificially retain the shrinkage.

Shrinkage can be recognized in a number of characteristic but varying appearances on radiographs. There are at least four types: (1) cavity; (2) dendritic; (3) filamentary; and (4) sponge types. Some documents designate these types by numbers, without actual names, to avoid possible misunderstanding.

Cavity shrinkage appears as areas with distinct jagged boundaries. It may be produced when metal solidifies between two original streams of melt, coming from opposite directions to join a common front; cavity shrinkage usually occurs at a time when the melt has almost reached solidification temperature and there is no source of supplementary liquid to

FIGURE 6. Example of Elongated Gas Porosity in 0.64 cm (0.25 in.) Aluminum Casting

FROM X-RAY PRODUCTS CORPORATION.

feed possible cavities.

Dendritic shrinkage is a distribution of very fine lines or small elongated cavities that may vary in density and are usually unconnected.

Filamentary shrinkage usually occurs as a continuous structure of connected lines or branches of variable length, width and density, or occasionally as a network.

Sponge shrinkage shows itself as areas of lacy texture with diffuse outlines, generally toward the mid-thickness of heavier casting sections. Sponge shrinkage may be dendritic or filamentary shrinkage; filamentary sponge shrinkage appears more blurred because it is projected through the relatively thick coating between the discontinuities and the film surface (see Fig. 8).

Cracks are thin (straight or jagged) linearly disposed discontinuities which occur after the melt has solidified. They generally appear singly and originate at casting surfaces.

Cold shuts generally appear on or near a surface of cast metal as a result of two streams of liquid meeting and failing to unite. They may appear on a radiograph as cracks or seams with smooth or rounded edges (see Fig. 9).

Inclusions are nonmetallic materials in a supposedly solid metallic matrix. They may be less or more dense than the matrix alloy and will appear on the radiograph, respectively, as darker or lighter indications. The latter type is more common in light metal castings (see Figs. 10 and 11).

Core shift shows itself as a variation in section thickness, usually on radiographic views representing diametrically opposite portions of cylindrical casting portions.

Hot tears are linearly disposed indications which represent fractures formed in a metal during solidification because of hindered contraction. The latter may occur due to overly hard (completely unyielding) mold or core walls. The effect of hot tears, as a stress concentration, is similar to that of an ordinary crack; hot tears are usually systematic flaws. If flaws are identified as hot tears in larger runs of a casting type, they may call for explicit improvements in technique.

Misruns appear on the radiograph as prominent dense areas of variable dimensions with a definite smooth outline. They are mostly random in occurrence and not readily eliminated by specific remedial actions in the process.

FIGURE 7. Example of Round Gas Porosity in 0.64 cm (0.25 in.) Aluminum Casting

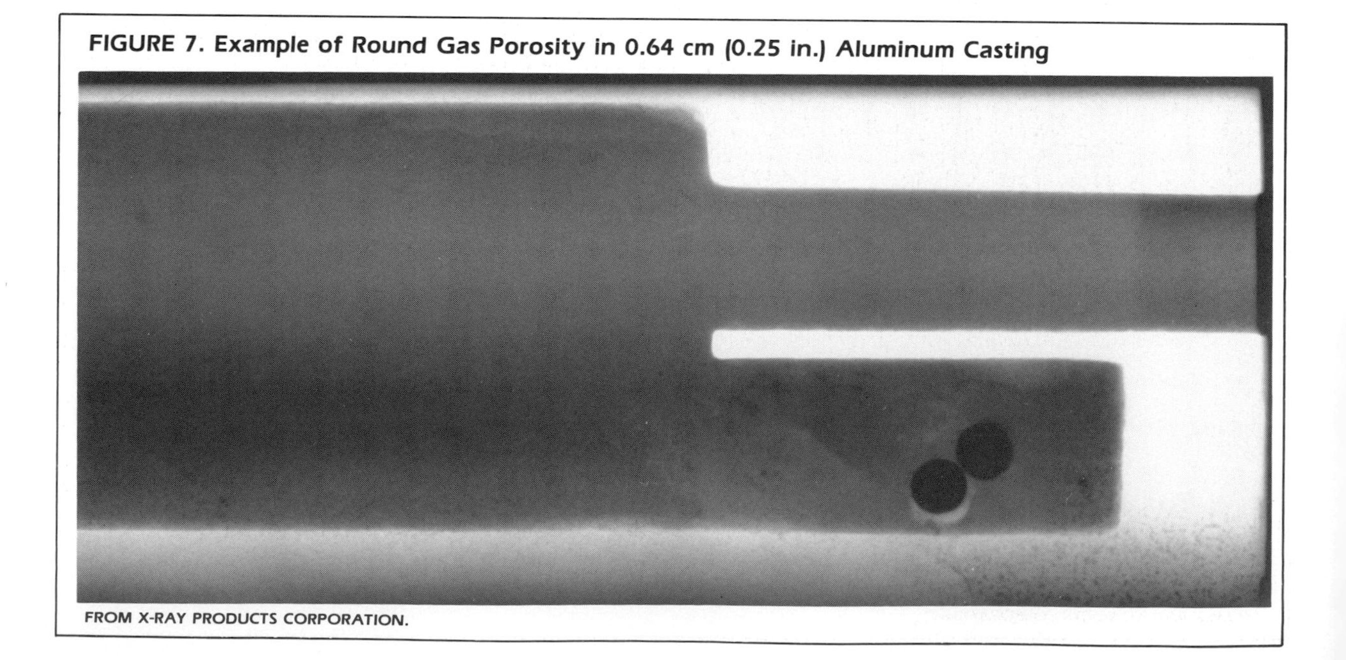

FROM X-RAY PRODUCTS CORPORATION.

Mottling is a radiographic indication which appears as an indistinct area of more or less dense images. The condition is a diffraction effect which occurs on relatively vague, thin-section radiographs, most often with austenitic stainless steel. Mottling is caused by interaction of the object's grain boundary material with low-energy X-rays (300 kV or lower). Inexperienced interpreters may incorrectly consider mottling as indications of unacceptable casting flaws. Even experienced interpreters often have to check the condition by re-radiography from slightly different source-film angles. Shifts in mottling are then very pronounced, while true casting discontinuities change only slightly in appearance.

Radiographic Indications for Casting Repair Welds

Most common alloy castings require welding either in upgrading from defective conditions or in joining to other system parts. It is mainly for reasons of casting repair that these descriptions of the more common weld defects are given here. The terms appear as indication types in ASTM E390, which is cited among the casting reference radiograph documents of Table 5. For additional information, see the *Nondestructive Testing Handbook*, Volume 3, Section 9 on the "Radiographic Control of Welds."

Slag is nonmetallic solid material entrapped in weld metal or between weld material and base metal. Radiographically, slag may appear in various shapes, from long narrow indications to short wide indications, and in various densities, from gray to very dark.

Porosity is a series of rounded gas pockets or voids in the weld metal, and is generally cylindrical or elliptical in shape.

Undercut is a groove melted in the base metal at the edge of a weld and left unfilled by weld metal. It represents a stress concentration which often must be corrected, and appears as a dark indication at the toe of a weld.

FIGURE 8. Example of Sponge Shrinkage in 1.27 cm (0.5 in.) Aluminum Casting

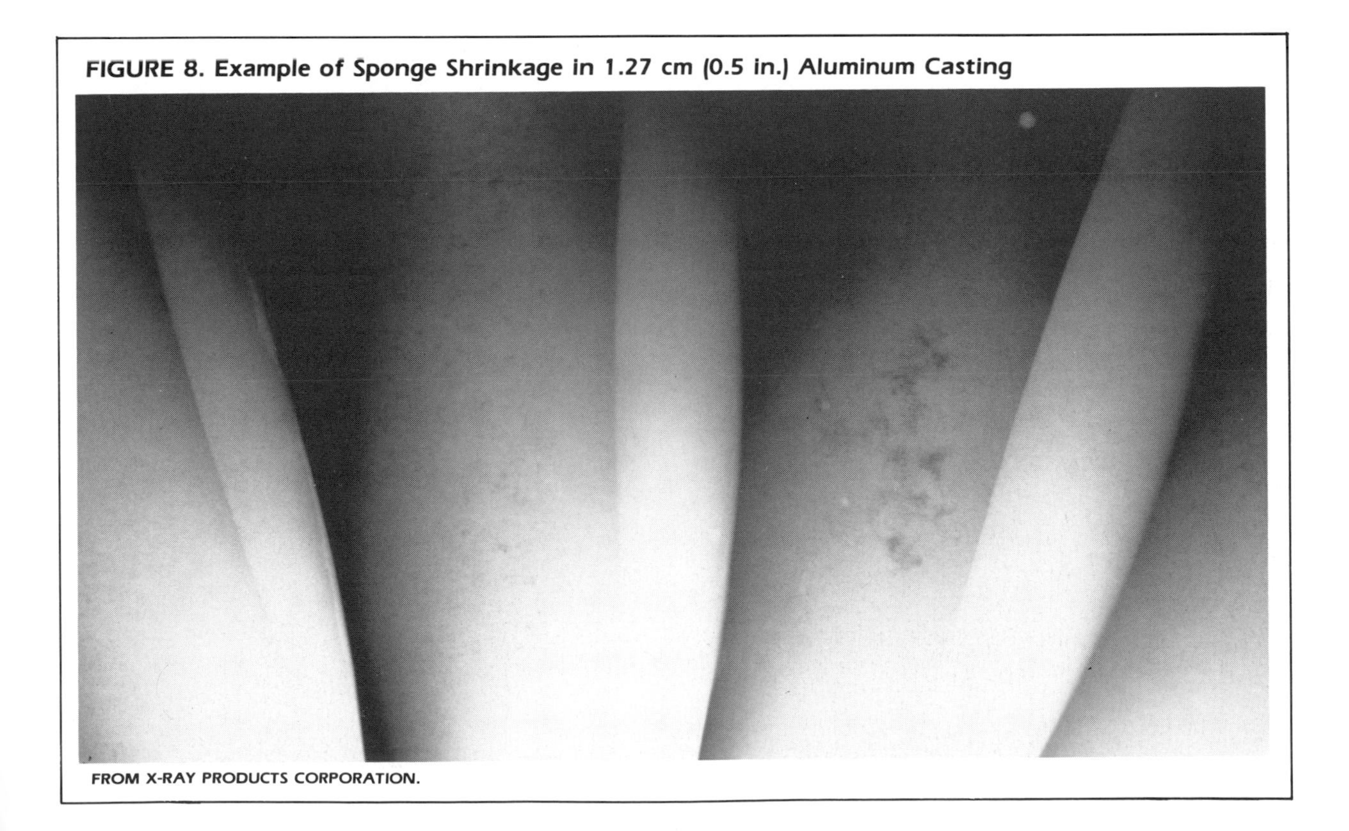

FROM X-RAY PRODUCTS CORPORATION.

Incomplete penetration, as the name implies, is a lack of weld penetration through the thickness of the joint (or penetration which is less than specified). It is located at the center of a weld and is a wide, linear indication.

Incomplete fusion is lack of complete fusion of some portions of the metal in a weld joint with adjacent metal; either base or previously deposited weld metal. On a radiograph, this appears as a long, sharp linear indication, occurring at the center line of the weld joint or at the fusion line.

Melt-through is a convex or concave irregularity (on the surface of backing ring, strip, fused root or adjacent base metal) resulting from complete melting of a localized region but without development of a void or open hole. On a radiograph, melt-through generally appears as a round or elliptical indication.

Burn-through is a void or open hole into a backing ring, strip, fused root or adjacent base metal.

Arc strike is an indication from a localized heat-affected zone or a change in surface contour of a

FIGURE 9. Example of a Cold Shut in 0.32 cm (0.125 in.) Aluminum Casting

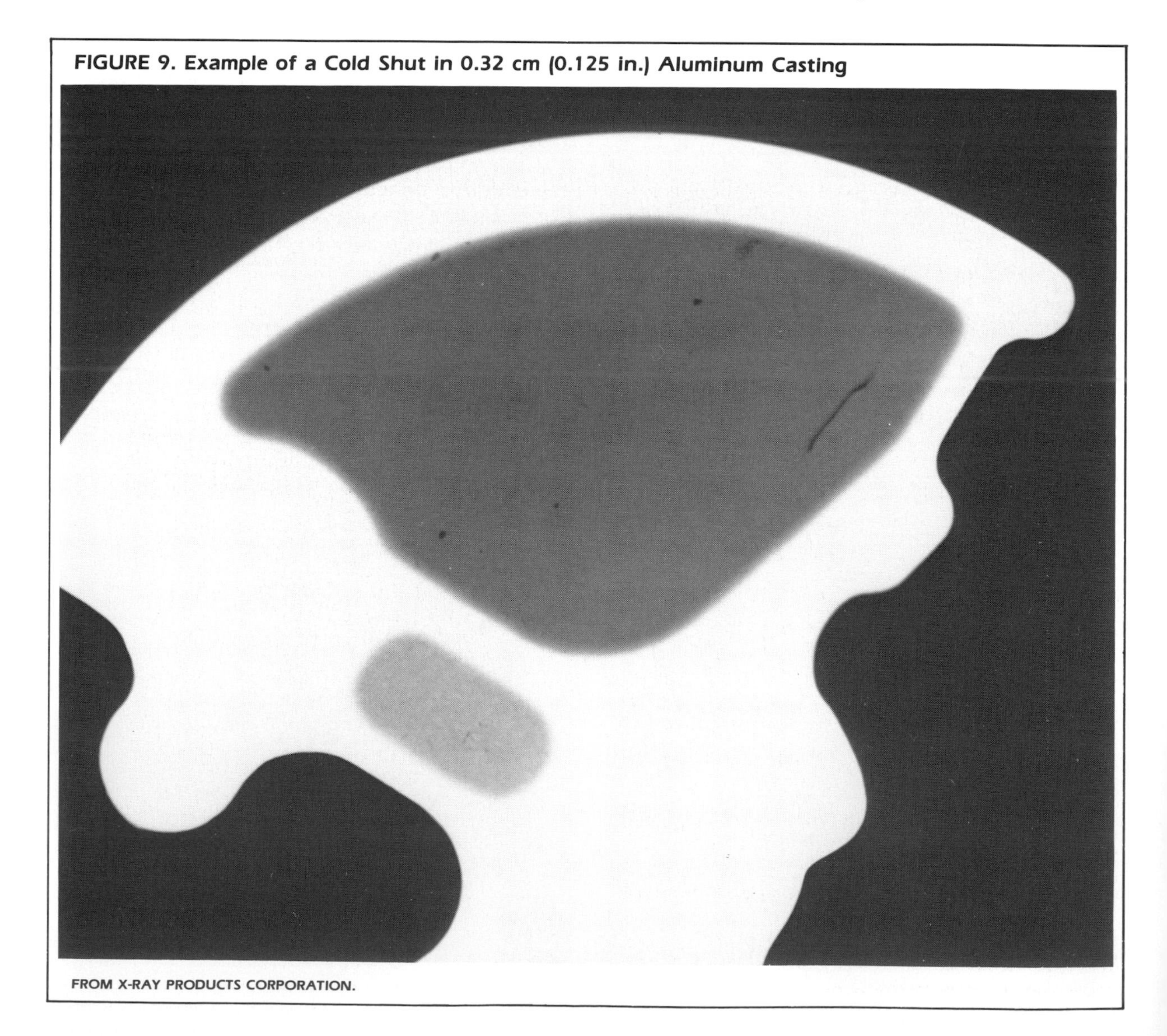

FROM X-RAY PRODUCTS CORPORATION.

finished weld or adjacent base metal. Arc strikes are caused by the heat generated when electrical energy passes between surfaces of the finished weld or base metal and the current source.

Weld spatter occurs in arc or gas welding as metal particles which are expelled during welding and which do not form part of the actual weld; weld spatter appears as many small, light cylindrical indications on a radiograph.

Tungsten inclusion is usually more dense than base-metal particles. Tungsten inclusions appear as mostly linear, very light radiographic images; accept/reject decisions for this defect are generally based on the slag criteria.

FIGURE 10. Example of an Inclusion Less Dense than Surrounding Material in 0.32 cm (0.125 in.) Aluminum Casting

FROM X-RAY PRODUCTS CORPORATION.

Oxidation is the condition of a surface which is heated during welding, resulting in oxide formation on the surface, due to partial or complete lack of purge of the weld atmosphere. Also called *sugaring*.

Root edge condition shows the penetration of weld metal into the backing ring or into the clearance between backing ring or strip and the base metal. It appears in radiographs as a sharply defined film density transition.

Root undercut appears as an intermittent or continuous groove in the internal surface of the base metal, backing ring or strip along the edge of the weld root.

FIGURE 11. Example of an Inclusion More Dense than Surrounding Material in 0.64 cm (0.25 in.) Aluminum Casting

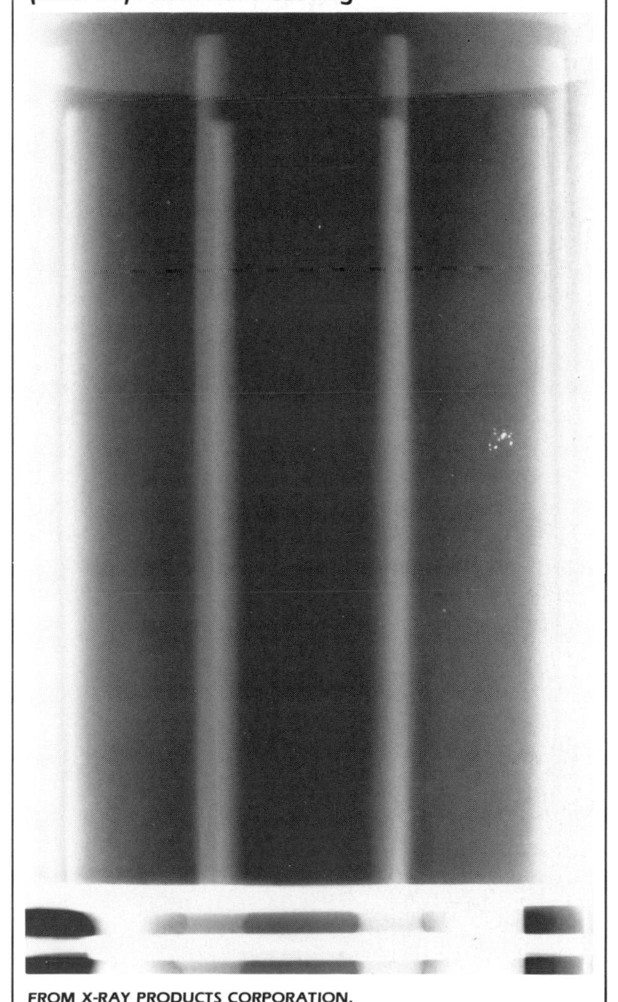

FROM X-RAY PRODUCTS CORPORATION.

PART 4

RADIOGRAPHIC TESTING AND PROCESS SCHEDULING

It is important to determine when radiographic testing should be done, with respect to required heat treatment, necessary repair welding and rough or finish machining. The timing of such tests is determined either by mutual agreement between producer and user, or by the appropriate, specified code.

If a casting could possibly be rejected, further steps in its manufacture and testing should be minimized or eliminated. In addition, it is also known that heat treatment, especially drastic temperature changes, may cause aggravation of some discontinuity types.

On the other hand, the closer the casting surfaces are to their finished condition, the easier it is to read radiographic films and disclose linear surface flaws; these tend to act as stress concentrations and could be the most damaging defects in-service. All other things being equal, however, radiographic testing is performed as early in the process as possible, to permit necessary weld repairs. Consequently, radiographic interpreters must have the actual casting available for their inspection in order to make allowance for surface irregularities.

In addition, experience has shown that castings should be examined in their final form with surface NDT methods, usually magnetic particle or liquid penetrant testing. The coverage and test frequency (if many castings of a single type are involved) is subject to mutual agreement between producer and user, with a logical statistical basis for the number and exact locations to be tested.

Factors Influencing Choice of Acceptance Criteria

The advent of reference radiographs has done much to improve the measurement of casting quality, especially as applied to important technological applications. It must be remembered that reference radiographs are sets of illustrations (especially of the graded severity discontinuity types), which cover a wide range of attainable casting quality levels. There is no recommendation of acceptability criteria for particular applications.

The ASME Boiler and Pressure Vessel Code[4] does point out accept-reject criteria for various graded steel casting discontinuity types. Briefly, accept-reject criteria for castings should be based on the following considerations: (1) alloy type; (2) section thickness; (3) pressure (including temperature and superheat when steam is involved); (4) service stress; (5) presence of impact and vibration; (6) fatigue; (7) exposure to penetrating radiation; (8) accessibility for maintenance and replacement during expected life; and (9) alloy solidification peculiarities, if any. At best, however, the acceptance criteria are largely qualitative.

PART 5

IMPROVEMENT OF CASTING RADIOGRAPHY

There is room for improvement in the effectiveness of radiography as a testing method for castings; for example, the relationship between radiographic indications and change of mechanical strength properties should be thoroughly investigated. Availability of quantitative data, relating static tensile parameters and impact-bend to severity of radiographic indications, would promote more objective determinations of acceptance; such data could be used in conventional design calculations. While it is recognized that establishment of these relationships is costly and their use is by no means easy, the trend in recent years has been in this direction.

Work of this type, but related to different materials and thickness ranges has been reported[12-17] and should be considerably extended to supply data for thicker sections of currently used materials and for materials coming into use in modern technological applications. The relatively high cost of such development work and the importance of its being reported in an unbiased way frequently causes it to be undertaken by government, technical societies and educational institutions. Thus, work of this type is sporadic and depends to a large degree on availability of required funding.

To illustrate the possible advantages of correlating radiographic discontinuity indications with destructive test results, some data are presented in Table 6.[16] This summarizes work on Class B steel castings (complying with MIL-S-15083) and manganese-nickel-aluminum bronze alloy #2 (complying with MIL-B-21230), and presents the following conclusions.

For steel castings:

1. Visual gradation of radiographic indications yields severity levels which are inversely proportional to ultimate strength, for all defect types except inclusions.
2. The deterioration in ultimate strength of steel in the annealed condition may be estimated from the severity of radiographic indications. The deterioration is most severe for linearly disposed discontinuity types such as shrinkage.
3. The yield point is practically unrelated to the severity of radiographic indications within the range studied, except for the most severe levels of the linear types.
4. Ductility of cast steel is affected drastically by soundness and cleanliness. Hence, where elongation is important in the design, the quality of steel castings is dictated by this parameter.
5. While radiographs of castings should by no means be used to replace tensile testing, the static tensile strength of particular portions may be assessed with considerable reliability by use of the relationships shown in Table 6.

For manganese-nickel-aluminum bronze castings,[16] similarly obtained but with fewer data, the table shows that:

1. The general trends for steel are verified.
2. Slopes for most curves are somewhat smaller than for steel, possibly due to nonequivalence of the severity levels of the reference radiographs of the two materials; and relatively lower notch sensitivity of the bronze.

An attempt to develop correlations between radiographic severity levels (of various discontinuity types) and flexural resistance (for the materials just cited) gave negative results.[16] Of all discontinuity types considered, gas porosity in steel castings appears to be most detrimental to the ability of the material to resist flexural loading. For all discontinuity types, location with respect to the tension surface is more important than extent; this is shown in plan-view radiographs when compared to reference radiographs in a conventional manner. The most important conclusion from this correlation work is that performance of castings in flexure cannot be assessed reliably with radiography alone, as it is usually practiced. Performance for static tensile loading can be predicted with considerable reliability.

Future Developments

Industrial radiography has been in use since 1938, and in that time its practice has achieved many things. In addition to helping improve quality in production, casting radiography has made worthwhile contributions to fabrication, while increasing product reliability at reasonable costs. Future developments should include accurate information on casting positions for which various radiographic indications are obtained, especially those indications for which ordinary destructive tests rarely, if ever, give needed information. Helpful additions would include correlation data on discontinuity significance in terms of service performance.

TABLE 6. Statistical Correlation Data Between Tensile Properties and Severity of Representative Types of Radiographic Indications in 7.6 cm (3 in.) Thick Steel and Manganese-Nickel-Aluminum Bronze Plate Castings[16]

		Tensile 1000 psi*		Yield 1000 psi*		Percent Elongation (4 in. Gage Length)*	
Discontinuity	Material	Slope**	95% Tolerance Limit	Slope	95% Tolerance Limit	Slope	95% Tolerance Limit
Gas porosity A	MIL-S-15083 Class B Steel	−3.28	±5.2	−0.43	†	−3.65	±5.0
Inclusions B	MIL-S-15083 Class B Steel	−0.03	†	−0.03	†	−1.36	†
Linear shrinkages Ca	MIL-S-15083 Class B Steel	−8.11	±6.2	1.76	±2.3	−3.38	±4.8
Dendritic shrinkage Cb	MIL-S-15083 Class B Steel	−8.11	±9.2	−0.69	±3.0	−1.22	±3.8
Worm hole shrinkage Cc	MIL-S-15083 Class B Steel	−7.60	±5.4	−1.43	±2.2	−3.46	±5.6
Hot tears D	MIL-S-15803 Class B Steel	−8.06	±6.8	−1.23	±2.0	−4.40	±5.8
Chill inserts Ea	MIL-S-15083 Class B Steel	−2.58	±5.2	−0.08	†	−2.59	±4.3
Chaplet inserts Eb	MIL-S-15083 Class B Steel	−4.93	±5.4	−0.061	±1.9	−3.26	±3.8
Gas porosity A	Mn-Ni-Al Bronze MIL-B-21250A, Alloy #2	−4.07	±6.4	††	††	†	†
Sand inclusions B	Mn-Ni-Al Bronze MIL-B-21250A, Alloy #2	−0.06	±7.2	††	††	†	†
Dross inclusions Bb	Mn-Ni-Al Bronze MIL-B-21250A, Alloy #2	−3.85	±9.0	††	††	−1.20	±3.9
Linear shrinkage Ca	Mn-Ni-Al Bronze MIL-B-21250A, Alloy #2	−3.62	±8.4	††	††	−1.10	±2.6
Spongy shrinkage Cd	Mn-Ni-Al Bronze MIL-B-21250A, Alloy #2	−3.58	±10.1	††	††	−0.93	±3.0

* 1000 psi = 6.89×10^3 kilopascals (kPa); 1 in. = 2.54 cm.
**The slope is the deterioration per grade of severity and is given by the equation $Y = a - bX$, where Y is the property, X the severity of indication and a the average value for substantially sound plates.
† No significant relationship indicated.
††Data not taken.

PART 6

SPECIAL APPLICATIONS OF CASTING RADIOGRAPHY

Flash Radiography in Casting Control

The stop-motion capabilities of flash radiography and the sequential recording capability of cineradiography have become useful techniques in the area of casting quality control. Flash cineradiography may be used to study the actual pouring of castings, showing a mold's gating characteristics and locating possible casting defects early in the process.

This kind of radiographic control is especially effective when the number of castings is high and the use of pilot runs is practical. Slow flash cineradiography systems are used to study test runs and locate possible defects, such as those caused by improper flow of the melt within the mold. Hence, flash radiography of pilot castings is an important technique for improving yields and minimizing system flaws. Flash radiography and flash cineradiography may also be used to study the welding stages of casting repair.

FIGURE 12. Setup for Experimental Flash Radiography of Lead Casting

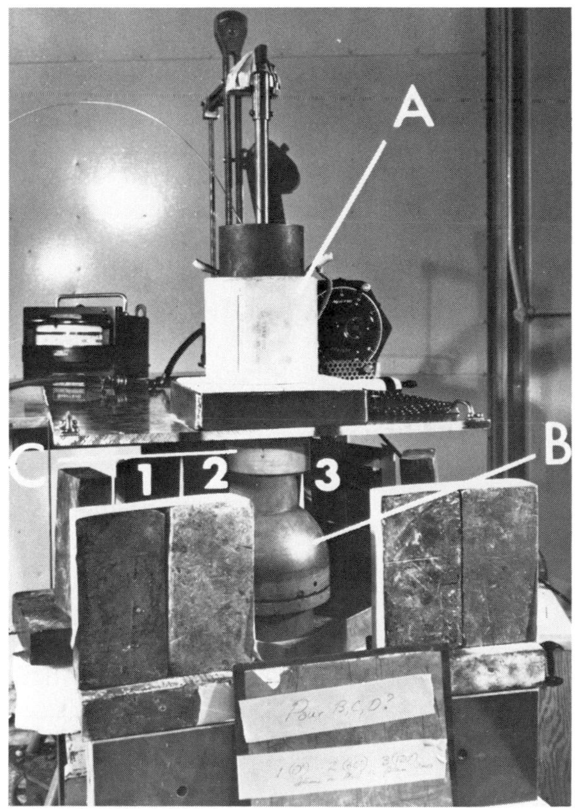

FROM THE LOS ALAMOS NATIONAL LABORATORY. REPRINTED WITH PERMISSION.

FIGURE 13. Pouring of a Lead Casting Imaged with Flash Radiography; X-ray Pulse Occurred 50 Milliseconds after Melt Interrupted Trigger Circuit

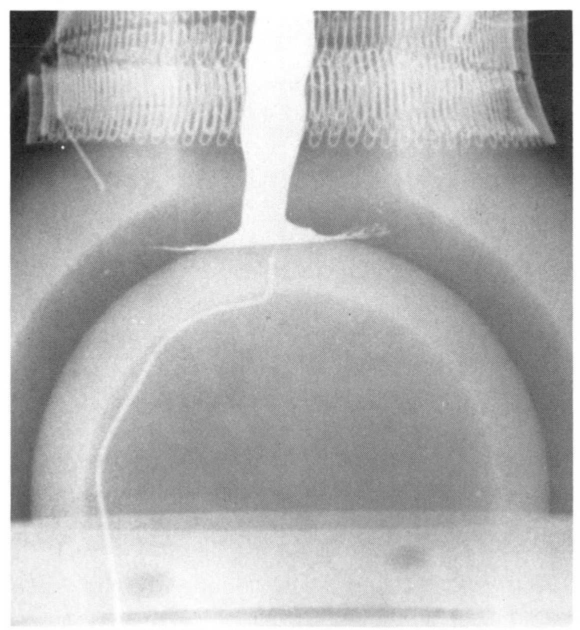

FROM THE LOS ALAMOS NATIONAL LABORATORY. REPRINTED WITH PERMISSION.

Experimental Example of Casting Flash Radiography

The following application was conducted at the Los Alamos National Laboratory[18] and it illustrates the kinds of images and information that can be obtained from a flash radiographic study of the casting process. Flash X-ray equipment is used here to image molten lead poured into a graphite mold.

The radiographic setup includes three flash X-ray tube heads as shown in Fig. 12. Atop the work table (arrow A) is an electrically heated crucible used to melt the lead. Underneath the table is the hemispherical mold (arrow B). On top of the mold is a ring (arrow C) containing a narrow, beamed light source and a light-detecting photocell.

The light beam and photocell are the trigger circuit. The falling, molten lead interrupts the light beam, giving a *zero time* electrical signal, which for this experiment is then sent to the three trigger amplifiers. A delay time may be chosen at each trigger amplifier and, in this case, times of 50, 75 and 100 milliseconds were chosen in order to view three sequential phases of the casting process.

Behind the mold are three lighttight film packages (labeled 1, 2 and 3). These contain a high-sensitivity X-ray film; positioned on either side of the films are fluorescent intensifying screens. Although the X-ray pulses used in flash radiography are of significant energy and of extremely brief duration, their total quantity of radiation is very low. The intensification from the fluorescent screens is needed to darken the films enough for interpretation.

Lead bricks are positioned in front of and to either side of the graphite mold. The lead serves a masking or collimating function, allowing only the radiation from one X-ray tube to expose its intended film.

Figures 13, 14 and 15 show the X-ray images at 50, 75 and 100 milliseconds from the time the lead

FIGURE 14. Pouring of a Lead Casting Imaged with Flash Radiography; X-ray Pulse Occurred 75 Milliseconds after Melt Interrupted Trigger Circuit; 25 Milliseconds Elapsed Time from Image in Figure 13

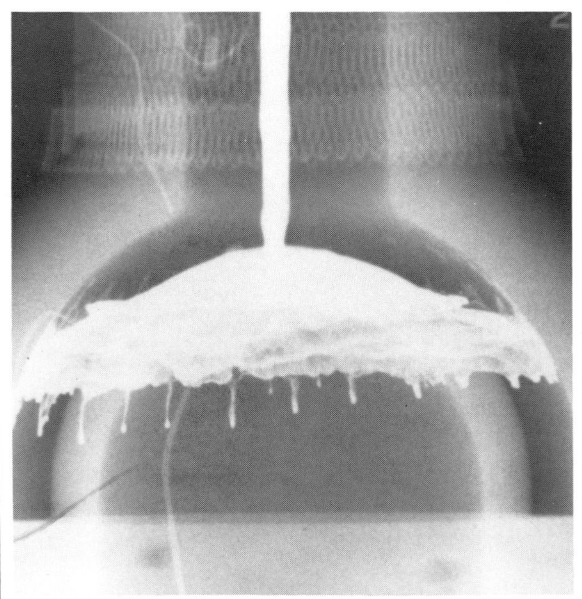

FROM THE LOS ALAMOS NATIONAL LABORATORY. REPRINTED WITH PERMISSION.

FIGURE 15. Pouring of a Lead Casting Imaged with Flash Radiography; X-ray Pulse Occurred 100 Milliseconds after Melt Interrupted Trigger Circuit; 25 Milliseconds Elapsed Time from Image in Figure 14

FROM THE LOS ALAMOS NATIONAL LABORATORY. REPRINTED WITH PERMISSION.

interrupted the light beam until the X-ray pulses occurred.

In Fig. 13, the falling molten lead (the white, irregular column at the top) is just beginning to spray horizontally after striking the inner, polar region of the graphite mold. In Fig. 14 (25 milliseconds later), the molten lead continues to fall on the polar region and the sideward spray has reached the outer wall of the mold. Rivulets of molten metal can be seen running both *up* and *down* the side walls. Finally, in Fig. 15 (at 100 milliseconds) the mold is partially filled. Note that one portion of the bottom rim is just beginning to receive the melt.

Such flash radiographic tests provide unique stop-motion images of the casting process; images may also be produced in sequence at very short time intervals, using cineradiography techniques. This kind of nondestructive testing is now being used to improve casting techniques, mold designs and the quality of cast metal parts. For more information on flash radiography, see *Nondestructive Testing Handbook* Volume 3, Section 11.

PART 7

TYPICAL PROBLEMS ENCOUNTERED IN CASTING RADIOGRAPHY

The important decisions in casting radiography involve setup, actual radiographic procedures and interpretation. Each of these may at times become problems in actual practice. The intent of this section is to consolidate the discussion of casting control problems and to indicate various solutions.

The Radiographic Source

The choice of radiographic source is based on many considerations. Often the source is determined by availability. If this is the case, care must be taken to assure that the penetrameter sensitivity required by contractual agreement can be met. The usual limits for most commonly used radiation sources, as far as metal section thickness is concerned, are cited in Tables 1 and 2. When deviation from these limits is considered for available sources, trial shots with carefully chosen compensating parameters should be made to determine whether the required sensitivity can be achieved. These parameters include finer grained film, larger source-to-film distance for improved sharpness, and proper base density.

When a choice of source can be made, the preferred one is determined by a combination of factors, including: section thickness; the ability to produce simultaneous, complete coverage of a cylindrical casting portion; and the desirability for simultaneous radiography of several castings at one time. A reasonably small-sized gamma-ray source is often the best choice, especially when portability is desired and the radiography is scheduled for raw castings (prior to surface preparations, necessary machining and repair or assembly welding).

Radiographic Coverage

Contrary to a common misconception, there is no such thing as 100 percent radiographic coverage for all castings. To make sure that no coverage problems arise between producer and purchaser, it is essential to follow proper and early planning of the radiography. Decisions on the radiographic techniques, especially for castings produced in considerable quantities, can best be made by the use of shooting sketches and associated tabular data, as illustrated in Figs. 2 and 3.

In the control of castings, radiographic coverage is recognized to be a problem with certain configurations of the mold and core and these include: portions connecting flanges with bodies; and, transition portions between relatively thin bodies and heavy bosses, especially when source location (with respect to film) is limited by casting geometry and details. Thus, for example, the blackened portion in the casting of Fig. 2 cannot be radiographed properly. Placement of the source on the inside of the cylinder is prohibited by the size of the inside diameter; placement of the source on the outside is made practically impossible by the limited space between the other flange and the portion of interest.

Radiographic Scheduling

Scheduling of radiography is also an important aspect which should be agreed upon by all concerned parties, prior to actual production. This is especially true if the casting requires considerable machining and welding. Radiography performed early in the manufacturing process has the advantage of saving further production expenses on castings which may ultimately be rejected or may require extensive repair and associated, unplanned heat treatments. Where an appreciable number of a particular casting is produced, pilot castings may yield valuable information for corrective action, even if they are radiographed in the as-cast condition (with gates and risers attached and section thicknesses considerably greater than finished dimensions). Note, however, that required penetrameter sensitivities are often based on the finished, rather than on the as-cast, thickness.

Once major problems in the pilot castings(s) have been suitably solved, future castings can be radiographed in the nearly finished and weld-repaired condition, thereby obtaining the best penetrameter sensitivity and optimized film sensitivity. The interpretation of a pilot radiograph often includes consideration of how the casting process might have caused a given flaw. Thus, radiographic interpretation is not simply a search for casting defects; it also allows the discovery and eventual elimination of defect-causing conditions within the casting process.

Radiographic Interpretation

Proper reading of films demands that the casting be available for reference purposes. Regardless of when the radiography is performed, visual inspection of the part may help decide whether indications on the radiograph are true internal flaws or indications caused by surface finish or other conditions.

Knowledge of gate and riser locations, if the casting is radiographed after their removal, can also help film interpretation. Shrinkage is more likely to occur near risers, if they are of small volume or if the melt solidifies quickly. Gas inclusions are more likely to be found near gates due to the influx of melt when other portions have partially solidified. Interior casting surfaces may at times cause false indications on radiographs due to surface or near-surface conditions of the wall; moisture in the core may introduce gas into the melt before it has time to move upward and out through the risers. At other times, brittle components of the core surface may be carried by the melt stream into the casting wall and may then result in near-surface density differences; these also could be incorrectly interpreted as defects.

The hot tear is another discontinuity whose positive recognition is sometimes difficult but nonetheless important. This defect is linear in nature and has all the adverse effects of a crack. It occurs when the melt is nearly solidified and is the result of an applied force in excess of what the just-solidified metal could withstand at considerably higher temperatures. This is the clue for determining the accuracy of a *hot tear* interpretation.

There must exist in the casting, as confined by mold and core, a source of suitably directed stress to make this type of defect possible. In a cylindrical casting, for example, a relatively unyielding core may cause an outward force as the metal shrinks and thus produce a hot tear. In the same casting shape, a linear indication perpendicular to the axis usually cannot be a hot tear, because there is no opportunity for forces in the axial direction to act during solidification. In other words, when interpreting a linear discontinuity as a hot tear, it must be verified that the geometry and mold configuration are capable of providing appropriate stresses during the last stages of solidification.

Choice of Reference Radiographs

There is at present a wide range of reference documents for various alloy types and section thicknesses, though not all alloys are represented. Because of this, a decision must be made by producer and user on the document mutually considered appropriate for judging discontinuities. For example, ASTM document E-310, based on leaded bronze hardware, should not be used for bronzes that tend to solidify more rapidly (those containing little or no lead). Titanium alloy castings, as another example, are currently judged by aluminum or steel reference radiographs, pending completion of titanium references.

Conclusion

Typical problems in the radiographic control of castings are difficult to enumerate because of the wide variety of casting materials, configurations and methods. To assure that problems are kept to an absolute minimum, the radiographic testing of castings, in all its aspects, should be specified in contracts, purchase orders, product specifications and drawings.

PART 8

GLOSSARY OF CASTING TERMINOLOGY[19]

age hardening: A process of aging that increases hardness and strength, but that ordinarily decreases ductility. Age hardening usually follows rapid cooling from solution heat treatment temperatures. Also known as *precipitation hardening.*

air dried: Refers to air drying of a core or mold without application of heat.

air injection machine: A type of die casting machine in which air pressure acts directly on the surface of molten metal in a closed pot (gooseneck) and forces the metal into a die.

anneal: To subject a casting to a temperature of 315 to 426°C (600 to 800°F) and then cool it slowly to increase ductility and relieve stresses.

annealing: Any treatment of metal at high temperature for the purpose of softening and removing residual stresses.

arbor: A bar or mandrel on which a core is built.

back draft: A reverse taper on the pattern which prevents its removal from the mold.

backing board: A second bottom board where molds are opened.

baked core: One which has been heated or baked until it is thoroughly dry.

baked permeability: The property of a molded mass of sand heated at a temperature above 110°C (230°F) until dry and cooled to room temperature to permit passage of gases.

barium clay: A molding clay containing barium, used to eliminate or reduce the amount of scattered or secondary radiation reaching the film.

basin: A cavity on top of the cope into which metal is poured before it enters the sprue. Also called *pouring basin.*

batch: An amount of core or mold sand or other material prepared at one time.

bead: A half-round cavity in a mold, or a half-round projection or molding on a casting.

bedding a core: Placing an irregularly shaped core on a bed of sand for drying.

bed-in: A method of ramming the drag mold without rolling over it.

bentonite: A plastic, adhesive type of clay that swells when wet. It is derived from decomposed volcanic ash, and is used for bonding molding sand.

binder: A material used to hold the grains of sand together in molds or cores. It may be cereal, oil, clay or natural or synthetic resins.

bleed: Refers to molten metal oozing out of casting. It is stripped or removed from the mold before complete solidification.

blended sand: A mixture of sands of different grain sizes and clay content that is needed to produce a sand possessing more suitable characteristics for foundry use.

blind riser: An internal riser that does not reach to the exterior of the mold.

blowhole: A hole in a casting or a weld caused by gas entrapped during solidification.

body core: Main core.

bond: A cohesive material used to bind sand.

bond clay: Any clay suitable for use as a bonding material in molding sand.

bond strength: The degree of cohesiveness that the bonding agent exhibits in holding sand grains together.

book mold: A split mold hinged like a book.

bottom board: The board or plate on which the mold rests.

bottom pour mold: A mold gated at the bottom.

branch gates: Gates leading into a casting cavity from a single runner and sprue.

bridging: Premature solidification of metal across a mold section before the metal below or beyond solidifies.

Brinell Hardness: A measure of the hardness of a metal, as determined by pressing a hard steel ball into the smooth surface under standard conditions. For aluminum, the steel ball is 10 millimeters in diameter and total load is 500 kilograms. Results are calculated as the ratio of applied load to total surface area of indentation and are referred to in terms of Brinell Hardness Number or BHN.

buckle: An indentation in the casting, resulting from expansion of the sand.

bumper: A machine used for packing molding sand in a flask by repeated jarring or jolting.

bumping: Ramming sand in a flask by repeated jarring and jolting.

burnt-in sand: A defect consisting of a mixture of sand and metal cohering to the surface of the casting.

cast structure: The internal physical structure of a casting evidenced by shape and orientation of crystals and segregation of impurities.

casting: (1) An object at or near finished shape obtained by solidification of a substance in a mold; (2) pouring molten metal into a mold to produce an object of desired shape.

casting shrinkage: (1) Liquid shrinkage — the reduction in volume of liquid metal as it cools to the liquidus; (2) solidification shrinkage — the reduction in volume of metal from the beginning to ending of solidification; (3) solid shrinkage — the reduction in volume of metal from the solidus to room temperature; (4) total shrinkage — the sum of liquid, solidification and solid shrinkages.

casting strains: Strains in a casting caused by casting stresses that develop as the casting cools.

casting stresses: Stresses set up in a casting because of geometry and casting shrinkage.

cast-weld assembly: An assembly formed by welding one casting to another.

cavity: The die impression that gives a casting its external shape.

centrifugal casting: A casting made in a mold (sand, plaster, or permanent mold) which rotates while the metal solidifies under the pressure developed by centrifugal force.

chaplet: A metal support used to hold a core in place on a mold.

chill: (1) A metal insert imbedded in the surface of a sand mold or core or placed in a mold cavity to increase the cooling rate at that point; (2) white iron occurring on a gray iron casting, such as the chill in the wedge test.

chill (to): To cool rapidly.

chipping: (1) Removing seams and other surface defects in metals manually with chisel or gouge or by a continuous machine, before further processing; (2) similarly, removing excessive, though not defective, metal.

chuck: A small bar set between crossbars to hold sand in the cope.

cire perdue process: See *lost wax process.*

cold chamber machine: A die-casting machine where the metal chamber or plunger are not heated.

cold shut: (1) A discontinuity that appears on the surface of a cast metal as a result of two streams of liquid meeting and failing to unite; (2) a portion of the surface of a forging that is separated, in part, from the main body of metal by oxide.

columnar structure: A coarse structure of parallel columns of grains, having the long axis perpendicular to the casting surface.

combination die (die casting): A die having two or more different cavities for different castings.

continuous annealling furnace: A furnace in which castings are annealed, or heat treated, by being passed through different heat zones kept at constant temperatures.
continuous casting: A casting technique in which an ingot, billet, tube, or other shape, is continuously solidified while it is being poured, so that its length is not determined by mold dimensions.
cooling stresses: Residual stresses resulting from nonuniform distribution of temperature during cooling.
cope: The upper or topmost section of a flask, mold, or pattern.
core: (1) A specially formed material inserted in a mold to shape the interior of another part of a casting which cannot be shaped as easily by the pattern; (2) in a ferrous alloy, the inner portion that is softer than the outer portion.
core blower: A machine for making foundry cores, using compressed air to blow and pack the sand into the core box.
core pin: A core, usually a circular section having some taper.
core plate: A plate on which a green core is baked.
core wash: A liquid with which cores are painted to produce smoother surfaces on the casting.
coupon: A piece of metal from which a test speciman is to be prepared, often an extra piece, as on a casting or forging.
cover-half: The stationary half of a die-casting die.
crush: A casting defect caused by a partial destruction of the mold before the metal was poured.
crushing: The pushing out of shape of a sand core or sand mold when two parts of the mold do not fit properly where they meet.
daubing: The act of filling cracks in cores.
deburring: Removing burrs, sharp edges or fins from metal parts by filing, grinding or rolling the work in a barrel with abrasives suspended in a suitable liquid medium. Sometimes called *burring.*
degasifier: A substance that can be added to molten metal to remove soluble gases which might otherwise be occluded or entrapped in the metal during solidification.
degassing: Removing gases from liquids or solids.
dendrite: A crystal that has a tree-like branching pattern being most evident in cast metals slowly cooled through the solidification range.
deoxidizer: A substance that can be added to molten metal to remove either free or combined oxygen.
deoxidizing: (1) The removal of oxygen from molten metals by use of suitable deoxidizers; (2) sometimes refers to the removal of undesirable elements other than oxygen by the introduction of elements or compounds that readily react with them; (3) in metal finishing, the removal of oxide films from metal surfaces by chemical or electrochemical reaction.
dewaxing: Removing the expendable wax pattern from an investment mold by heat or solvent.
die casting: (1) A casting made in a die; (2) a casting process where molten metal is forced under high pressure into the cavity of a metal mold.
directional solidification: The solidification of molten metal in a casting in such manner that feed metal is always available for that portion that is just solidifying.
drag: The bottom section of a flask, mold, or pattern.
drop: A defect in a casting due to a portion of the sand dropping from the cope or other overhanging section of the mold.
drop out: The falling away of green sand from the walls of a mold cavity when the mold is closed.
dross: The scum that forms on the surface of molten metals largely because of oxidation but sometimes because of the rising of impurities to the surface.
facing: Any material applied in a wet or dry condition to the face of a mold or core to improve the surface of the casting.
facing sand: A special sand mixture placed against the pattern to produce a satisfactory casting surface.
feeder: A reservoir of molten metal connected to, but not a part of, the casting; designed to remain liquid while the casting is solidifying. It is located so that it will feed liquid metal to the larger portions of the casting which are the last to solidify.
flask: A metal or wood frame used for making and holding a sand mold. The upper part is called the *cope;* the lower, the *drag.*
fluidity: The ability of molten metal to flow readily; usually measured by the length of a standard spiral casting.
foreign materials: They may appear as isolated, irregular or elongated variations in radiographic film density, not corresponding to variations in thickness of material or to cavities. May be sand, slag, oxide or dross metal or any material included in the material being examined.
foundry: A commerical establishment or building where metal castings are produced.
fracture: A break, rupture or crack large enough to cause a full or partial partition of a casting.
free carbon: The part of the total carbon in steel or cast iron that is present in the elemental form as graphite or temper carbon.
gaggers: The metal supports that reinforce sand in the cope.
gas holes: Holes created by a gas escaping from molten metal; appear in radiographs as round or elongated, smooth-edged dark spots, occurring individually, in clusters, or distributed throughout a casting.
gas porosity: Refers to porous sections in metal that appear in radiographs as round or elongated dark spots corresponding to minute voids; usually distributed through the entire casting.
gate: The portion of the runner in the mold through which molten metal enters the mold cavity. Sometimes the generic term is applied to the entire network of connecting channels which conducts metal into mold cavity. Also called *ingate.*
gated pattern: A pattern designed to include gating in the mold.
gooseneck: The pressure vessel or metal injection pump in an air-injection type of casting machine.
grain refiner: Any material, usually a metal from a special group, added to a liquid metal or alloy to produce a finer grain in the hardened metal.
green core: One that has not been baked.
green sand: Core sand intended for use in a damp state.
gross porosity: Pores, gas holes or globular voids that are larger and in greater number than obtained in good practice.
growth: The expansion of a casting because of aging.
heat treatment: Heating and cooling a solid metal or alloy in such a way as to obtain desired conditions or properties. Heating for the sole purpose of hot working is excluded from this definition.
hindered contraction: The condition when the geometry will not permit a casting to contract in certain regions in keeping with the coefficient of expansion of the metal being cast.
horn gate: A curved gate shaped like a horn and arranged to permit entry of molten metal at the bottom of a casting cavity.
hot cracks: Appear as ragged dark lines of variable width and numerous branches. They have no definite line of continuity and may exist in groups. They may originate internally or at the surface.
hot spot: The point of retarded solidification caused by an increased mass of metal at the juncture of two sections. It frequently results in shrinkage and inferior mechanical properties at this location.
hot tear: A fracture formed in a metal during solidification because of hindered contraction.
impregnation: The treatment of porous castings with a sealing medium to stop pressure leaks.
impurities: Elements or compounds whose presence in a material is undesired.
inclusion: Any foreign matter contained in welds or castings.
ingate: Same as *gate.*
inverse segregation: Segregation in cast metal in which an excess of lower-melting constituents occurs in the earlier freezing portions, apparently the result of liquid metal entering cavities developed in the earlier solidified metal.
investment casting: (1) Casting metal into a mold produced by surrounding (investing) an expendable pattern with a refractory slurry that sets at room temperature after which the wax, plastic or frozen mercury pattern is removed through the use of heat.

Also called *precision casting*, or *lost-wax process;* (2) a casting made by the process.

investment compound: A mixture of graded refractory filler, a binder and a liquid vehicle, used to make molds for investment castings.

investment molding: A method of molding by using a pattern of wax, plastic, or other material that is invested (surrounded) by a molding medium in slurry or liquid form. After the molding medium has solidified, the pattern is removed by subjecting the mold to heat, leaving a cavity to catch the molten metal. Also called *lost wax process* or *precision molding*.

joint: The part of the mold where the cope and cheek, cope and drag, or cheek and drag come together.

loose piece: A core positioned near, but not fastened to, a die and arranged to be ejected with the casting. The loose piece may be removed and used repeatedly for the same purpose. Also, it is similarly used in or on patterns, core boxes and permanent molds.

lost-wax process: An investment casting process in which a wax pattern is used.

macroshrinkage: A casting defect, detectable at magnifications not exceeding ten diameters, consisting of voids in the form of stringers shorter than shrinkage cracks. This defect results from contraction during solidification where there is not an adequate opportunity to supply filler material to compensate for the shrinkage. It is usually associated with abrupt changes in section size.

malleable cast iron: A cast iron made by a prolonged anneal of white cast iron in which decarburization or graphitization, or both, take place to eliminate some or all of the cementite. The graphite is in the form of temper carbon.

match plate: A plate of metal or other material on which patterns for metal castings are mounted or formed as an integral part so as to facilitate the molding operation. The pattern is divided along its parting plane by the plate.

match plate pattern: A sand-molding pattern, partly on the cope side and partly on the drag side of the plate that forms the parting between the cope and drag sections of the molding flask. Permanent forms for runners, gates, sprue and riser locations, and sometimes complete risers, are included. Such a pattern usually is made of aluminum and is used extensively with molding machines.

microshrinkage: A casting defect, not detectable at magnifications lower than ten diameters, consisting of interdendritic voids. This defect results from contraction during solidification where there is not an adequate opportunity to supply filler material to compensate for shrinkage. Alloys with a wide range in solidification temperature are particularly susceptible.

misrun: A casting not fully formed, resulting from the metal solidifying before the mold is filled.

mold: A form or cavity into which molten metal is poured to produce a desired shape. Molds may be made of sand, plaster or metal and frequently require the use of cores and inserts for special applications.

mold jacket: Wood or metal form that is slipped over a sand mold for support during pouring.

mold wash: An aqueous or alcohol emulsion or suspension of various materials used to coat the surface of a mold cavity.

molding machine: A machine for making sand molds by mechanically compacting sand around a pattern.

nodular cast iron: A cast iron that has been treated while molten with a master alloy containing an element such as magnesium or cerium to give primary graphite in the spherultic form.

normalizing: Heating a ferrous alloy to a suitable temperature above the transformation range and then cooling in air to a temperature substantially below the transformation range.

open-sand casting: Any casting made in a mold that has no cope or other covering.

parting line: The mark left on the casting where the die halves meet; also, the surface between the cover and ejector portions of the die.

parting sand: Fine sand for dusting on sand mold surfaces that are to be separated.

pattern: A form of wood, metal or other material, around which a molding material is placed to make a mold for casting metals.

peeling: (1) The dropping away of sand from the casting during shakeout; (2) the detaching of one layer of a coating from another or from the basic metal, because of poor adherence.

permanent mold: A metal mold (other than an ingot mold) of two or more parts that is used repeatedly for the production of many castings of the same form. Liquid metal is poured in by gravity.

permeability: The characteristics of molding materials which allow gases to pass through them. Permeability Number is determined by a standard test.

pinhole porosity: Numerous small holes distributed throughout the cast metal.

plaster molding: Molding, wherein a gypsum-bonded aggregate flour in the form of a water slurry is poured over a pattern, permitted to harden, and after removal of the pattern, thoroughly dried. The technique is used to make smooth nonferrous castings of accurate size.

plunger machines: Die casting machines having a plunger in continuous contact with molten metal.

porosity: Fine holes or pores within a metal.

pouring: Transferring molten metal from a furnace or a ladle to a mold.

pouring basin: A basin on top of a mold to receive the molten metal before it enters the sprue or downgate.

pull cracks: In a casting, cracks that are caused by residual stresses produced during cooling, and that result from the shape of the object.

ramoff: A casting defect resulting from the movement of sand away from pattern because of improper ramming.

recarburize: (1) To increase the carbon content of molten cast iron or steel by adding carbonaceous material, high-carbon pig iron or a high-carbon alloy; (2) to carburize a metal part to return surface carbon lost in processing.

riser: A reservoir of molten metal connected to the casting to provide additional metal to the casting, required as the result of shrinkage before and during solidification.

runner: (1) A channel through which molten metal flows from one receptacle to another; (2) the portion of the gate assembly that connects the downgate sprue or riser with the casting; (3) parts of patterns and finished castings corresponding to the described portion of the gate assembly.

runner box: A distribution box that divides the molten metal into several streams before it enters the mold cavity.

runout: (1) The unintentional escape of molten metal from a mold, crucible or furnace; (2) the defect in a casting caused by the escape of metal from the mold.

sand: A granular material resulting from the disintegration of rock. Foundry sands are mainly silica. *Bank sands* are found in sedimentary deposits and contain less than 5% clay. *Dune sand* occurs in wind blown deposits near large bodies of water and is very high in silica content. *Moulding sand* contains more than 5% clay; usually between 10 and 20%. *Silica sand* is a granular material containing at least 95% silica and often more than 99%. *Sand core* is nearly pure silica. *Miscellaneous sand* includes zircon, olivine, calcium carbonate, lava and titanium minerals.

scab: A defect consisting of a flat volume of metal joined to a casting through a small area. It is usually set in a depression, a flat side being separated from the metal of the casting by a thin layer of sand.

scarfing: Cutting surface areas of metal objects, ordinarily by using a gas torch. The operation permits surface defects to be cut from ingots, billets or the edges of plate that is to be beveled for butt welding.

scrap: (1) Defective product not suitable for sale; (2) discarded metallic material from whatever source that may be reclaimed through melting and refining.

sealing: (1) Closing pores in anodic coatings to render them less absorbent; (2) plugging leaks in a casting by introducing thermosetting plastics into porous areas and subsequently setting the plastic with heat.

seam: (1) On the surface of metal, an unwelded fold or lap which appears as a crack, usually resulting from a defect obtained in casting or in working; (2) mechanical or welded joints.

segregation: Nonuniform distribution of alloying elements, impurities or microphases.

semipermanent mold: A permanent mold in which sand or plastic cores are used.

shakeout: Remove castings from a sand mold.

shell core: A shell-molded sand core.

shell molding: Forming a mold from thermosetting resin-bonded sand mixtures brought in contact with preheated (149 to 260 °C [300 to 500 °F]) metal patterns, resulting in a firm shell with a cavity corresponding to the outline of a pattern.

shift: A casting defect caused by mismatch of cope and drag or of cores and mold.

shot peening: Cold working the surface of a metal by metal-shot impingement.

shrink mark: A surface depression on a casting that sometimes occurs next to a thick section that cools more slowly than adjacent sections.

shrinkage cavities: Cavities in castings caused by lack of sufficient molten metal as the casting cools.

shrinkage cracks: Hot tears associated with shrinkage cavities.

shrinkage porosity (nonferrous alloys): A localized lacy, or honeycombed, darkened area on a radiographic film that indicates porous metal.

skim gate: A gating arrangement designed to prevent the passage of slag and other undesirable material into the casting.

skimmer: A tool for removing scum, slag and dross from the surface of molten metal.

skin: A thin outside metal layer, not formed by bonding as in cladding or electroplating, that differs in composition, structure or other characteristics from the main mass of metal.

slag: A nonmetallic product resulting from the mutual dissolution of flux and nonmetallic impurities in smelting and refining operations.

slag inclusions: Nonmetallic solid material entrapped in weld metal or between weld metal and base metal.

slag lines: Elongated cavities containing slag or other foreign matter.

slide: Part of a die generally arranged to move parallel to the parting line, the inner end forming a part of the die cavity wall and involving one or more undercuts and sometimes including a core or cores.

slush casting: A casting made by pouring an alloy into a metal mold, allowing it to remain long enough to form a thin shell, and then pouring out the remaining liquid.

snap flask: A hinged flask that is removed from the mold after the mold is made.

soaking: Prolonged holding at a selected temperature.

soldiers: Wooden blocks or sticks used to reinforce bodies of sand in the cope. They usually overhang the mold cavity.

solidification shrinkage: The decrease in volume of a metal during solidification.

solution heat treatment: A heat treatment which causes the hardening constituent of an alloy to go into solid solution, followed by a quench to retain it temporarily in a supersaturated solution state at lower temperatures.

spheroidizing: Heating and cooling to produce a spheroidal or globular form of carbide in steel.

split gate: A gate having the sprue axis in the die parting.

sprue: (1) The channel that connects the pouring basin with the runner; (2) sometimes used to mean all gates, risers, runners and similar scrap. Also called *downsprue* or *downgate*.

stress relieving: Heating to a suitable temperature, holding long enough to reduce residual stresses and then cooling slowly enough to minimize the development of new residual stresses.

temper: (1) In heat treatment, reheating hardened steel or hardened cast iron to some temperature below the eutectoid temperature for the purpose of decreasing the hardness and increasing the toughness. The process also is sometime applied to normalized steel; (2) in tool steels, temper is sometimes used, but inadvisedly, to denote the carbon content; (3) in nonferrous alloys and in some ferrous alloys (steels that cannot be hardened by heat treatment), the hardness and strength produced by mechanical or thermal treatment, or both and characterized by a certain structure, mechanical properties, or reduction in area during cold working.

temper brittleness: Brittleness that results when certain steels are held within, or are cooled slowly through, a certain range of temperature below the transformation range. The brittleness is revealed by notched-bar impact tests at or below room temperature.

tempering: Reheating a quench-hardened or normalized ferrous alloy to a temperature below the transformation range and then cooling at any rate desired.

tie rod: A bar used in a casting machine to hold dies locked against pressure and, in general, also to serve as a way along which the movable die platen slides.

toggle: The linkage in a casting machine employed to multiply pressure mechanically in locking the dies; also, linkage used for core locking and withdrawal in a die.

trimming: (1) In forging or die casting, removing the parting-line flash and gates from the part by shearing; (2) in castings, the removal of gates, risers and fins.

unit die: A die block that contains several cavity inserts for making different kinds of die castings.

upset: A frame used to deepen either the cope or drag.

vacuum melting: Melting in a vacuum to prevent contamination from air, as well as to remove gases already dissolved in the metal; the solidification may also be carried out in a vacuum or at low pressure.

vent: A small opening in a mold for the escape of gases.

wash: A coating applied to the face of a mold prior to casting.

water line: A tube or other passage through which water is circulated to cool a casting die.

weak sand: Refers to sand that will not hold together when used to make a mold.

williams riser: An atmospheric riser.

zircon sand: A highly absorptive material used as a blocking or masking medium for drilled holes, slots and highly irregular geometric parts to reduce or eliminate scattered radiation during radiography.

REFERENCES

1. McMaster, R.C., ed. *Nondestructive Testing Handbook.* New York, NY: Ronald Press Co. (1959).
2. *Radiography in Modern Industry.* Rochester, NY: Eastman Kodak Co. (1969).
3. "Standard Method for Radiographic Testing of Metal Castings." Philadelphia, PA: American Society for Testing and Materials (1983).
4. "Examination of Steel Castings." *ASME Boiler and Pressure Vessel Code*, Section VIII, Division I, Pressure Vessels, Appendix VII. New York: American Society of Mechanical Engineers.
5. "Standard Method for Controlling Quality of Radiographic Testing." *Annual Book of ASTM Standards.* ASTM Specification E142-77, Part 11 (1982): pp 193-202.
6. "Standard Method for Controlling Quality of Radiographic Testing Using Wire Penetrameters." *Annual Book of ASTM Standards.* ASTM Specification E747-80, Part 11 (1982): pp 674-680.
7. "Standard Recommended Practices for Radiographic Testing." *Annual Book of ASTM Standards.* ASTM Specification E94-77, Part 11 (1982): pp 118-133.
8. "Standard Method for Determining Relative Image Quality Response of Industrial Radiographic Film." *Annual Book of ASTM Standards.* ASTM Specification E746-80, Part 11 (1982): pp 668-673.
9. "Obtainable ASTM Equivalent Penetrameter Sensitivity for Radiography of Steel Plates 1/4 to 2 in. Thick with X-Rays and 1 to 6 in. Thick with Cobalt-60." *Annual Book of ASTM Standards.* ASTM Specification E592-77, Part 11 (1982): pp 597-602.
10. Goldspiel, S. and W.N. Roy. "Reference Radiographs." *Standardization News* (November 1982).
11. "Standard Reference Radiographs for Ductile Iron Castings." *Annual Book of ASTM Standards.* ASTM Specification E689-79, Part 11 (1982): pp 612-614.
12. Goldspiel, S. "Development of Radiographic Standards for Castings." Tokyo and Osaka, Japan: Third International Conference on Nondestructive Testing (1960).
13. Materk, L.J. "Correlation of Radiographically Observed Flaws with Tensile Properties of Stainless Steel Castings." *Materials Research and Standards.* Vol. 2, No. 8 (August 1962).
14. Briggs, C.W. "Significance of Discontinuities in Steel Castings on Basis of Destructive Testing." *Materials Research and Standards.* Vol. 3, No. 6 (June 1963).
15. Goldspiel, S. "New Approach to Nondestructive Testing Demanded by the Deep Diving Submersible Program." New London, CT: ASME Underwater Technology Conference (May 1965).
16. Goldspiel, S. "Development of Radiographic Standards for Castings." *Materials Research and Standards* (July 1969).
17. Goldspiel, S. "Status of Reference Radiographs." *Nondestructive Testing Standards.* Harold Berger, ed. ASTM STP 624 (1977).
18. Bryant, L.E. "Portable Flash X-ray at the Los Alamos Scientific Laboratory." LASL-78-19 (May 1978).
19. *NDT Terminology.* E.I. du Pont de Nemours & Company, Photo Products Department, Wilmington, DE.

SECTION 11

FLASH RADIOGRAPHY

Francis Charbonnier, Hewlett-Packard McMinnville Division, McMinnville, OR

INTRODUCTION

Flash radiography is a special type of radiography which is used to produce a single stop-motion image or a series of sequential images of high speed phenomena. In conventional radiography, the subject is motionless during exposure. The exposure time can therefore be lengthened as necessary to obtain proper film exposure once the tube-to-film distance, the focal spot size and the tube voltage have been adjusted for optimum sharpness (lateral resolution) and contrast (depth resolution). In flash radiography, the stop-motion requirement places an upper limit on the exposure time, i.e. on the duration of the X-ray pulse or the duration of X-ray detector activation. This limit depends on the velocity of the object being radiographed; for instance, millisecond exposures may be adequate to stop motion in vibration studies, whereas submicrosecond exposures are generally required for ballistic or shockwave studies and subnanosecond exposures may be required for extremely high speed or extremely short duration events such as nuclear fuel pellet implosion.

Real-time radiography, using X-ray image intensifiers and cameras or television display systems, represents still another type of radiography that produces an essentially continuous display of a dynamic event. In real-time, however, the event's rate of motion or change must be sufficiently slow to allow millisecond exposures at a frame rate on the order of 30 to 60 frames per second.

Flash radiography provides one added element, time resolution or stop motion, which is not present in conventional radiography. This added capability, however, carries some limitations:

1. The exposure time and radiation intensity must generally be preset, and there is no opportunity to change them while the exposure is in progress.
2. Relatively high voltages are required to achieve useful X-ray intensities during a very short pulse. Hence, image contrast is more limited than in conventional radiography.
3. Even at high voltages, the total radiation intensity per pulse is relatively low, typically an incident dose of 1 to 50 milliroentgens (mR) at 3 meters from the X-ray source; the intensity is much less after penetration of a thick object. Consequently very fast film/screen combinations must be used, resulting in a loss of sharpness and an increase in quantum noise.
4. Because the target cannot be cooled effectively during the submicrosecond pulse, the X-ray target must be physically large (one to several millimeters) to absorb the electron beam energy. The large focal spot places an additional limitation on the sharpness of the radiographic image.
5. During the observation of very violent events, such as large explosive detonations or impact phenomena, suitable shields and a substantial physical distance from the event must be used to protect the film from damage. This further degrades the contrast and sharpness of the image.

The ability to control and optimize technique factors is much more limited in flash radiography than it is in conventional static radiography, and the same image quality cannot generally be attained. However, in many situations involving high speed events, flash radiography represents the only available technique for imaging or observation, and its limitations are tolerated because of its unique ability to freeze motion and provide time-resolved information.

History and General Principles

The general principles which govern the production and the imaging characteristics of X-rays are identical for conventional static radiography and flash radiography. Subjects such as the energy and intensity of X-ray brehmsstrahlung and characteristic radiation (as a function of tube voltage and target material), or X-ray penetration and image contrast (as a function of tube voltage), or the relation of image sharpness to geometry and focal spot size, have been widely discussed in the literature.

This discussion of flash radiography makes extensive references to several, more highly detailed publications. The excellent book *Flash Radiography* by Jamet and Thomer,[1] and the *Proceedings of the Flash Radiography Symposium*,[2] given at the 36th National Fall Conference of the American Society for Nondestructive Testing (ASNT), are particularly useful general references. Several recent developments in cine flash radiography are covered in the *Proceedings of the Paris 1981 First European Conference on Cineradiography*,[3] published by the Society of Photo-Optical Instrumentation Engineers (SPIE).

The first flash radiographic studies were reported in 1938 by Kingdon and Tanis in the US[4] and by Steenbeck in Germany[5] who independently produced high intensity microsecond pulses of X-rays, using mercury vapor discharge tubes powered by capacitors charged to 100-150 kV. These gas discharge tubes had very low efficiency. Muehlenpfordt contributed a significant improvement by developing continuously pumped vacuum discharge tubes, at about 10^{-4} torr (10 millipascals) pressure, using a conical tungsten anode and a stainless steel ring-shaped cathode. This simple but very effective tube design overcame the inability to focus the electron discharge and succeeded in producing a controlled, relatively small focal spot with reasonable pulsed heat absorption capability; it remains in wide-spread use today. The design was further developed by Schaafs[6,7] using, first, a single capacitor, then later a Marx-Surge generator to extend the high voltage to 400 kV. This design was first applied extensively in the early 1940s by Schardin at the Ballistic Research Institute in Berlin. Flash X-ray systems were developed in the US by Slack and Ehrke in 1941,[8] and subsequently improved by Criscuolo.[9]

Dyke and co-workers achieved further improvements in the efficiency and reproducibility of flash X-ray tubes by using multiple-needle field-emission cathodes in high vacuum sealed-off tubes.[10] These tubes, together with improved coaxial Marx-Surge generator designs, resulted in the development of technically improved and commercially manufactured flash X-ray systems[11] which are now in use throughout the world.

PART 1

FLASH X-RAY SYSTEMS

Methods of X-ray Generation

In X-ray tubes for conventional radiography, a thermionic cathode is used to produce an electron beam that is accelerated and focused to strike a small spot on a metal plane target. This basic mechanism of generating X-rays by impact of high energy electrons on a metal target is also used in flash radiography. However, since thermionic cathodes are not capable of producing the very high peak current densities and total currents required for flash radiography, different electron sources must be used. These sources do not allow effective focusing of the electron beam, so different X-ray tube and target geometries must be designed for achieving the necessary confined focal spot.

Gas Discharge Tubes

Gas discharges were first used to accelerate electrons into the target and produce X-rays. Gas discharges are capable of producing very large electron currents. However, electrons lose energy by collision or ionization and the average energy of the electrons striking the target is considerably less than the voltage applied to the tube; a severe loss in X-ray generating efficiency and X-ray hardness is the result. For these reasons gas discharge tubes are no longer used in flash X-ray systems.

Field Emission

Field emission is a process in which electrons are emitted into high vacuum by applying an extremely high electric field (30-100 MV/cm) at the surface of a cold metal cathode, generally made of tungsten. The cathode is electrolytically etched into a very sharp needle so that the high electric field necessary for emission can be concentrated at the tip of the needle and can be produced at a reasonable applied voltage. The electric field applied at the cathode surface thins and lowers the potential energy barrier at the surface and, when the applied electric field is sufficiently high, electrons at or slightly below Fermi energy can tunnel through the surface energy barrier with a probability (described by quantum mechanical theory) that increases exponentially with field strength. Thus field emission draws from the almost limitless supply of free electrons in the atomic conduction bands, and is capable of producing extremely high current densities, in the 10^7 to 10^8 amperes per square centimeter (A/cm^2) range.[12] If the applied electric field is increased further, the emitted current density becomes so large that the cathode over-heats and evaporates; this results in a vacuum arc which causes a larger increase in emitted current.[13] With sufficiently high applied fields, the transition from field emission to a vacuum arc can occur in a fraction of a nanosecond. Field emission cathodes for flash X-ray tubes generally consist of a large number of nearly identical needles arranged in linear or circular arrays around the conical target; space charge effects at the cathode, and in the cathode/anode gap, help control and equalize the current emission from the various needles.

Vacuum Discharges

Vacuum discharges occur in a residual gas pressure low enough so that the electron mean free path is many times larger than the gap spacing between the electrodes; under these conditions, the avalanche breakdown and electron-gas molecule collisions are essentially eliminated. The discharge occurs when the applied voltage exceeds a certain threshold, and may be explained by two alternate theories: *field emission*[14] or *clump initiation*.[15]

In the field emission-initiation theory, the electric field at the cathode becomes large enough to cause field emission, then a vacuum arc (as previously described), at very sharp microprotrusions randomly distributed on the cathode surface. The local field enhancement at the tips of these microprotrusions, which depends on the shape but not on the size of the protrusions, can be as high as 50 to 100 times. This explains why a vacuum discharge may occur at gross electric fields (assuming a perfectly smooth cathode surface) on the order of 10^6 volts per centimeter, i.e. 50-100 times smaller than the local fields required to produce a field emission-initiated vacuum arc. The field emission-initiation hypothesis is more likely to apply under conditions of high vacuum, very short duration voltage pulses, relatively close cathode-anode spacing and high strength electrode materials such as tungsten.

The other discharge initiation theory is the microparticle or clump theory which assumes that microparticles, on the surface of either electrode, acquire an induced electrostatic charge under high field stress, become detached from the electrode by electrostatic repulsion and are then accelerated to the other electrode. Here the impact causes local evaporation and formation of a plasma, i.e. a large source of electrons and ions, which immediately leads to a vacuum discharge. For parallel plane electrodes, the clump theory predicts that the threshold voltage for vacuum discharge is proportional to the square root of the gap spacing, a relationship which is often observed in voltage breakdown experiments at high voltages. The clump initiation hypothesis generally applies for long pulse durations, for large gap spacings and for electrode materials of relatively low tensile strength, such as aluminum.

In a vacuum discharge tube, a finite time is required for the discharge to develop even though, in the field emission-initiation mode, the discharge initiation time can be extremely short at sufficiently high voltages. The high density plasma formed by cathode evaporation provides a source of electrons which are accelerated to the anode and produce X-rays. The plasma front propagates across the gap at a velocity which depends on gap field and cathode material and is generally on the order of 10^7 to 10^8 cm/s. As the plasma propagates, the effective cathode-anode gap spacing is reduced and the tube impedance decreases, causing an increase in tube current and a decrease in tube voltage. Thus the X-ray hardness decreases during the discharge. If the voltage pulse is long enough, the plasma front may reach the anode, causing gap closure and effectively short-circuiting the tube and ending X-ray production. Gap closure is undesirable and is avoided by proper choice of voltage, pulse duration and gap spacing.

Once initiated, a vacuum discharge is terminated only by removing the voltage applied to the tube, and there is then a finite recovery time required for the plasma to recombine and the metal vapor in the gap to condense and disappear. This recovery time, which depends on prior discharge current and energy as well as gap spacing and electrode material, sets the minimum time interval between successive X-ray pulses from a given discharge tube; it is typically on the order of microseconds.

Vacuum gaps are sometimes designed with an added trigger electrode located close to the cathode. A very short high-voltage pulse applied to the trigger electrode will initiate a discharge at the cathode and produce the necessary plasma, which is then directed to the anode where a longer voltage pulse is applied. This type of tube is called a *triggered gap triode discharge tube* and is particularly useful when very soft X-rays are desired; soft X-rays require anode voltages which are too low for fast, reliable discharge initiation in a diode tube.

Vacuum discharges are widely used in flash X-ray tubes and, depending on the applied voltage and tube geometry, either the field emission or the clump initiation mechanism may predominate. Vacuum discharges are capable of generating peak currents as large or larger than those generated by the multiple-needle field emission cathodes described earlier. The field emission cathodes generally produce better control and reproducibility of the discharge delay time and the X-ray output intensity from pulse to pulse.

Flash X-Ray Tubes

Flash X-ray tube designs are based on the basic mechanisms and principles just discussed. Figure 1 illustrates a typical design for a high vacuum sealed-off field emission flash X-ray tube, operating at relatively high voltages (100 to 2000 kV); Fig. 2 shows the design of a special purpose continuously pumped vacuum-discharge triode tube intended to produce the high intensity, low energy, copper K_α characteristic radiation needed for flash X-ray diffraction studies.

The simple sealed-off tube of Fig. 1 uses a multiple-needle field emission cathode consisting of six linear arrays surrounding a conical target made of tungsten. The focal spot and resolution characteristics of this flash X-ray source have been discussed.[16] The sharpness of the X-ray image is determined by the base diameter of the cone, the effective focal spot diameter being about two-thirds of the cone base diameter. The pulsed heat absorption capability of the target is proportional to the lateral area of the cone (and also to the electron range into the target).

For a given resolution, the X-ray intensity can be increased by using a cone with a small half angle (θ). If θ becomes too small, X-rays are reabsorbed more heavily in the target and the effective X-ray beam coverage is reduced. In practice, a θ value on the order of 7 degrees is found to be about optimum for flash radiography. The maximum X-ray intensity increases very rapidly with increasing voltage, both because the electron range increases rapidly with voltage and because the efficiency of X-ray generation also increases with voltage; hence it is difficult to achieve large X-ray intensities at low voltages without excessive evaporation and rapid destruction of the target.

Figure 2 illustrates a different design intended for a specific use: the generation of intense copper K_α radiation for application to flash X-ray diffraction. This tube was designed and built at Institut St. Louis (ISL).[1] The tube is a triode, to permit lower voltage operation; the cathode and trigger electrodes are close-spaced, concentric rings with very sharp edges. The target is made of copper to produce the desired K_α characteristic radiation energy, of approximately 8 keV. The tube is operated at a target voltage of only 30 kV to reduce the intensity of high-energy bremsstrahlung background radiation, though the voltage pulse applied to the trigger electrode may be substantially higher than 30 kV to ensure a prompt, intense discharge. A polyester film window is used to permit extraction of the very low energy characteristic X-rays with only slight attenuation. These design characteristics, particularly the polyester window, require that the tube be continuously pumped to a residual pressure on the order of 1 mPa (10^{-5} torr) by means of a conventional or a small ion pump.

Pulsed High Voltage Sources for Flash Radiography

A number of techniques have been used to generate the pulsed high voltage and to accommodate the high current, low impedance characteristics of flash X-ray tubes.

The oldest and simplest technique is to charge a capacitor to a high voltage, then discharge the capacitor through a low-impedance pressurized-gas triggered spark gap. This technique is limited to 100 to 150 kV, in practice.

Capacitive energy storage at relatively low voltage (15 to 100 kV) followed by voltage multiplication is routinely used to generate high voltages. The voltage multiplication is achieved by means of a pulse transformer or a Marx-Surge generator. Pulse transformers are suitable for output voltages up to 400 kV. Marx-Surge generators can be used over a

FIGURE 1. High Voltage Sealed Off Field Emission Flash X-ray Tube

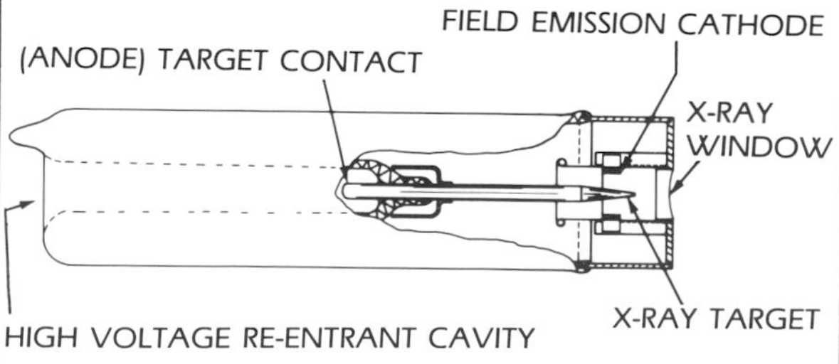

FIELD EMISSION CATHODE
(ANODE) TARGET CONTACT
X-RAY WINDOW
HIGH VOLTAGE RE-ENTRANT CAVITY
X-RAY TARGET

HEWLETT-PACKARD. REPRINTED WITH PERMISSION.

FIGURE 2. Continuously Pumped Low Voltage Flash X-ray Tube Made for the Production of Copper Kα Radiation

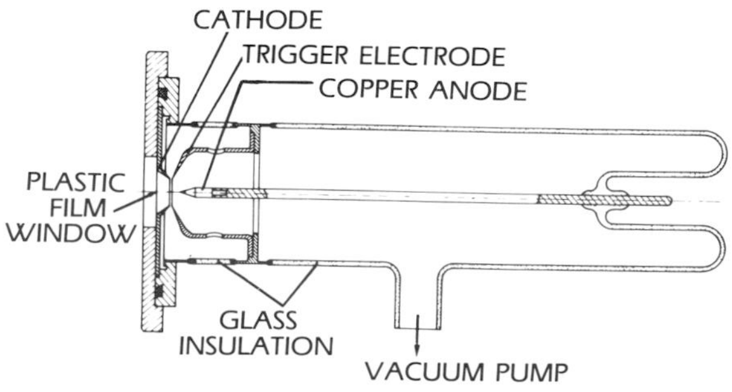

CATHODE
TRIGGER ELECTRODE
COPPER ANODE
PLASTIC FILM WINDOW
GLASS INSULATION
VACUUM PUMP

AFTER JAMET, INSTITUT ST. LOUIS. ELSEVIER SCIENTIFIC PUBLISHING © 1976. REPRINTED WITH PERMISSION.

FIGURE 3. Schematic of Marx-Surge High Voltage Pulse Generator

450 kV
FLASH X-RAY SYSTEM

SYSTEM SPECIFICATIONS:
15 modules
Output energy 140 joules
6000 amps
TUBE OUTPUT:
20 mR (200 μGy) dose at one meter
25 ns pulse duration
Pulse to pulse repeatability (RMS) 5%

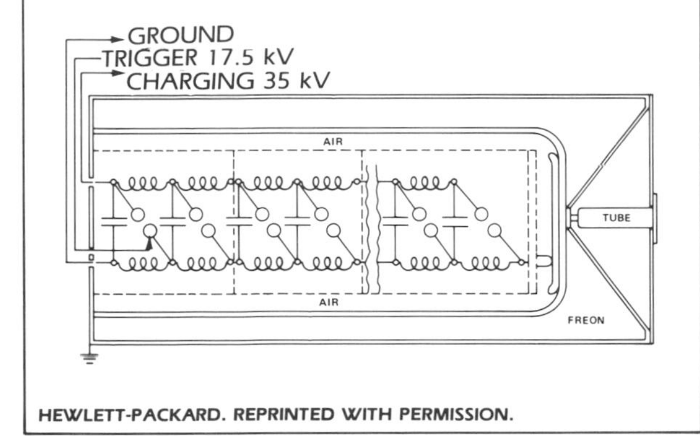

HEWLETT-PACKARD. REPRINTED WITH PERMISSION.

much broader range of voltages, up to several megavolts, and are more commonly used in present flash X-ray systems.

Marx-Surge Generators

Figure 3 shows a schematic diagram of a Marx-Surge generator[17] in its simplest form. A bank of N capacitors (of individual capacitance C) is charged in parallel to a DC voltage V_o (typically 15 to 100 kV) then discharged in series by means of cross-connected spark gaps. The open circuit output voltage is then NV_o. The output voltage waveform, into a resistive load R, decays exponentially as shown in eq. 1.

$$V = NV_o e^{-\frac{NT}{RC}} \qquad \text{(Eq. 1)}$$

For flash radiography applications, the capacitor in each stage of the Marx-Surge generator is replaced by a pulse-forming network to produce a more effective, more nearly rectangular output waveform. The spark gap housing, and sometimes the whole pulser, is contained in a pressure vessel filled with the appropriate gas (dry air or CO_2) and the gap spacing and gas pressure are adjusted to optimize the output discharge characteristics. The first gap contains a trigger electrode which is used to initiate the output discharge. The spark gaps are generally located in close proximity and in direct sight of each other so that the ionizing and ultraviolet radiation from one gap contributes to the breakdown of the following gap, yielding the fast rise-time and low time jitter (uncertainty in discharge timing) which are desired for flash radiographic applications.

Blumlein Line Generators

The rise time of the voltage pulses delivered by a Marx-Surge generator is not easily reduced below 5 ns. The difficulty in designing a device with sufficiently low inductance puts a lower limit of approximately 15 ns on the X-ray pulse length. The design also limits the maximum current intensity and the minimum characteristic impedance of the pulser, particularly for high voltage systems which contain a large number of stages. For these reasons, more complex high voltage-generating circuits are used for applications which require output pulses of very short duration, very high current or very low impedance. The most common of these circuits is the Blumlein line generator,[18] pulse-charged by a Marx-Surge generator.

A Blumlein generator generally consists of three coaxial cylinders which behave as two transmission lines connected through a resistive load (the flash X-ray tube). Figure 4 illustrates a specific embodiment used in a 600 kV, 3 ns flash radiographic or electron beam system. The three coaxial electrodes are labeled A, B and C. When the Marx-Surge generator is fired, the rapidly rising positive output voltage V_1 causes the series spark gap $SG1$ to break down, thereby pulse charging the middle electrode B to a voltage V_2 approximately equal to V_1. The rise time in V_2 is on the order of 15 ns and the inductance L is chosen low enough to keep the center electrode A close to ground. As V_2 approaches V_1, the radial spark gaps $SG2$ become over volted (and irradiated with ultraviolet from $SG1$) and they spontaneously fire, initiating the discharge of the Blumlein from left to right. The inductance L is large enough to allow V_3 to rise and a voltage pulse appears across the field emission tube, having a duration equal to the 2-way propagation time of the discharge in the Blumlein line (approximately 3 ns). The Blumlein and the tube are designed to have approximately equal impedance, about 60 ohms, yielding a 600 kV, 10,000 A, 3 ns electron or X-ray pulse.

Electron Accelerators

Linear accelerators are sometimes used to produce very high electron energies for flash radiographic applications requiring moderately short (approximately 0.1 to 10 microseconds) repetitive pulses of very high energy X-rays.

FIGURE 4. Schematic of Two-Stage High Voltage Blumlein Pulse Generator (Pulse-Charged by Marx-Surge Generator) For the Generation of Low Impedance Very Short Duration Pulses

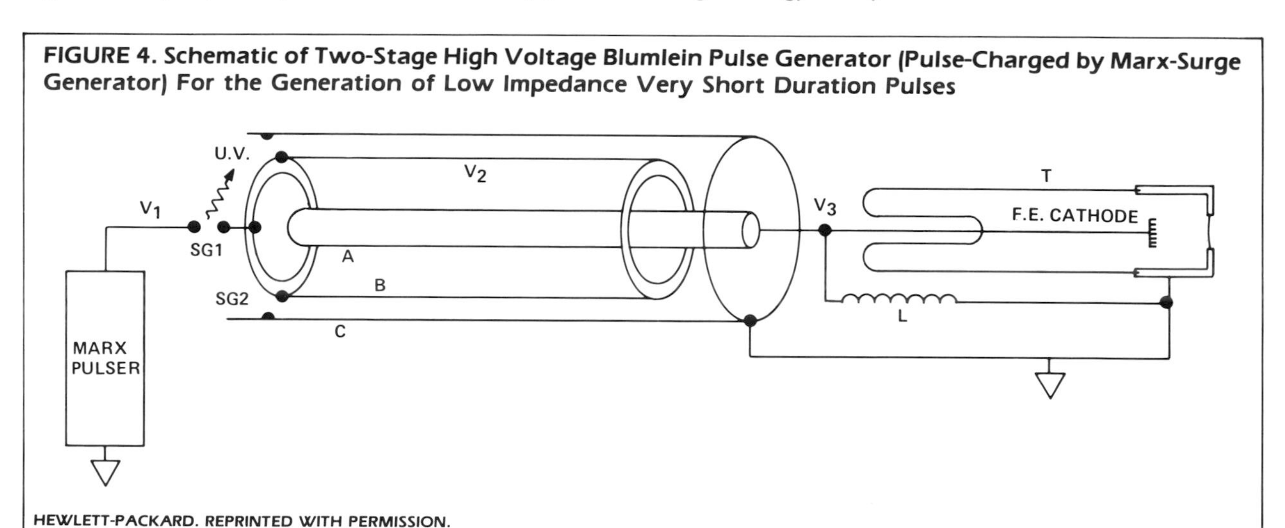

HEWLETT-PACKARD. REPRINTED WITH PERMISSION.

FIGURE 5. Schematic Diagram of PHERMEX Machine. Very High Energy High Intensity Flash X-ray System

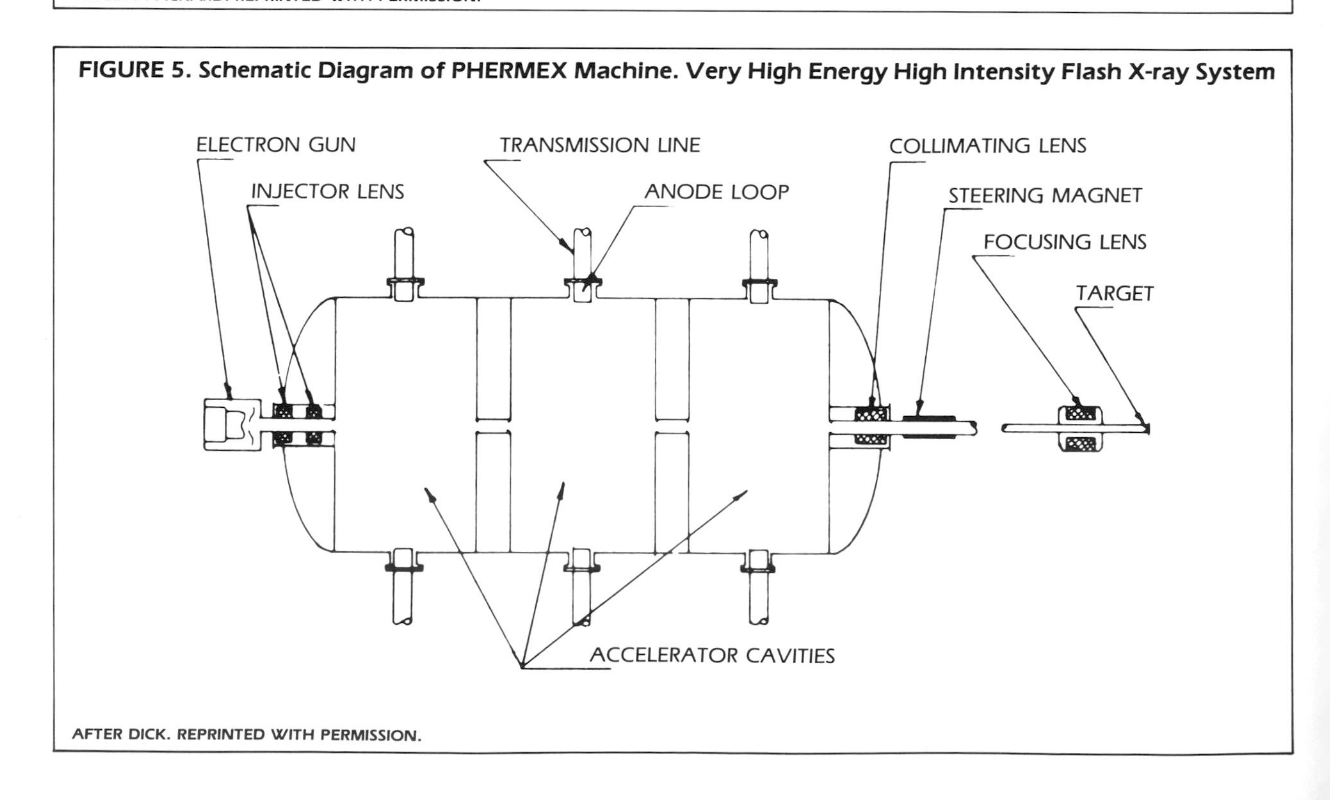

AFTER DICK. REPRINTED WITH PERMISSION.

A unique design for a very high energy flash radiography system is PHERMEX (Pulsed High Energy Radiographic Machine Emitting X-rays) at the Los Alamos National Laboratory.[19] PHERMEX, built in 1965, is shown schematically in Fig. 5. Three large cylindrical cavities, resonant at 50 MHz, are arranged in series and excited by input radiofrequency energy at 50 MHz. Standing waves in the TM010 mode develop and increase in amplitude during the excitation period. When the standing wave amplitude reaches its maximum (approximately 5 MV/cm axial field), a high current pulsed electron gun produces an intense, relatively low-energy (500 keV) electron beam. The beam is injected into the RF cavities and accelerated to 27 MeV. The emerging electron beam is focused onto a transmission X-ray target, producing a very high energy X-ray source only 1 mm in diameter. PHERMEX is ideally suited for flash radiographic applications demanding very high penetration and high image quality.

Commercial Flash X-ray Systems

Commercial flash X-ray systems are available over a range of voltages (from approximately 100 to 2,000 kV) and with a wide range of output capability. The two main companies manufacturing such systems are Hewlett-Packard in the US[20] and Scanditronix in Sweden.[21] Tables 1 and 2 summarize the published characteristics of systems from these companies. Commercial flash X-ray systems are also available from the USSR[22] and more recently from Japan[23] and from the UK.[24]

The different available outputs provide additional flexibility and imaging capability. Soft X-ray output, achieved with special flash X-ray tubes using a thin beryllium or polyester film window, allows much higher X-ray dose and contrast. Soft X-rays are useful in the observation of low density media or in flash X-ray diffraction studies. Remote

TABLE 1. Flash X-ray Systems from Hewlett-Packard[20]

Nominal peak voltage (kV)	150	300	450	1000	2300
Nominal pulse duration (ns)	70	30	30	30	30
Available Outputs					
Standard X-rays	X	X	X	X	X
Soft X-rays	X	X	X		
Remote tubehead	X	X	X	X	
Dual tubeheads (stero)	X	X	X		
Electron beam		X	X	X	
Standard X-ray Output					
Shadowgraph distance (meters)	4	6	10	20	40
Penetration in aluminum at 2 meters (mm)	20	60	100	170	240
Penetration in steel at 2 meters (mm)	—	10	230	50	80
X-ray dose per pulse at 1 meter in mR (μGy)	1.6 (16)	6 (60)	20 (200)	55 (550)	500 (5000)

TABLE 2. Flash X-ray Systems from Scanditronix[21]

Nominal peak voltage (kV)	150	300	450	600	1000	1200
Nominal pulse duration (ns)	20	20	20	20	20	20
Remote tubehead	X	X	X			
Dual tubehead	X	X	X			
Dose per pulse at 1 meter in mR (μGy)		8 (80)	20 (200)		50 (500)	
Penetration in steel at 2 meters (mm)		10	18		35	

tubeheads, connected to the pulser by high voltage coaxial cables, are particularly useful in ballistic or explosive studies where the event is very violent; use of a remote tubehead makes it easier to protect the high voltage pulser from damage. It is also possible, with appropriate higher impedance tubes, to divide the output of the pulser into two remote tubes, allowing synchronized orthogonal views for three dimensional reconstruction of an event. Finally, reversal of the high voltage pulse polarity and replacement of the flash X-ray tube by a flash electron beam tube makes it possible to produce a high power short-duration electron beam instead of a pulsed X-ray beam. High intensity pulsed electron beams can be used for special imaging applications (betagraphy), for irradiation of materials, for pulse radiolysis or for gas laser pumping. The intense high energy X-ray beams needed for radiation-effects studies may also be produced more efficiently by generating a high energy electron beam (up to several MeV) and converting it to X-rays in a thin external transmission X-ray target; this can replace the method of using a conventional flash X-ray tube with an internal cone reflection X-ray target. Flash electron beam tubes use the same electron generation techniques (field emission or vacuum discharge cathodes) as do flash X-ray tubes, and the high energy pulsed electron beam is extracted from the tube through a thin vacuum window usually made of titanium or polyester film.

The characteristics of some systems which have been used for flash radiography and special applications, and which are generally mobile or commercially available, are summarized in Table 3.

A number of other flash X-ray systems have also been designed, built and used, particularly for very high penetration studies, for a variety of ballistic and diffraction applications,[1] and for soft X-ray and characteristic X-ray generation for flash X-ray diffraction studies.[27]

TABLE 3. Special Flash Radiography Systems[25]

X-ray Energy	Type	Designer-Builder	Date Built	Dose per Pulse at 1 Meter*	Nominal Pulse Duration	Focal Spot Size (mm)	Main Applications
85 keV (see ref. 26)	Blumlein	Lawrence Livermore National Laboratory	~1967	30 mR	50 ns	1.1	Flash X-ray diffraction and low density media
100 keV (see ref. 27)	Capacitor/Coaxial Lines	Ernst Mach Inst.	1978		60 ns		Flash X-ray diffraction and low density media
3 MeV (see ref. 28)	Marx/Blumlein (SWARF)	AWRE, U.K.	1973	50 R	60 ns		Studies requiring very high penetration at large distance
7 MeV (see ref. 29)	Marx/Blumlein "Pulserad 1480"	Physics International Co.	1974	500 R	150 ns	5	Studies requiring very high penetration at large distance
7 MeV (see ref. 30)	Marx/Blumlein "Grec"	CEA Vaujours	~1980	300 R	80 ns		Studies requiring very high penetration at large distance
10 MeV (see ref. 31)	Linear Accelerator	Los Alamos National Laboratory		2500 R/min up to 240 pps	4 μs		Cineradiography of rocket motors
27 MeV (see ref. 17)	50 MHz Linear Accelerator in Stored energy mode "Phermex"	Los Alamos National Laboratory	1965	15 R at 100 R at	40 ns 200 ns		Large explosive - metal systems
50 MeV (see ref. 32)	Linear Accelerator "Artemis"	CEA Vaujours	~1982	20R at 150 R at	50 ns 15 μs		Large explosive - metal systems
80 MeV	Linear Accelerator	Lawrence Livermore National Laboratory	~1955	60R	300 ns		Large explosive - metal systems

*1 mR = 10 μGy
1 R = 10 mGy

PART 2

FLASH RADIOGRAPHY TECHNIQUES

Film Recording

Flash radiographic images are most often recorded on film. Usually, the X-ray intensity available at the film is severely limited by the small output of flash X-ray systems (this is due to the very short pulse duration: 10-1000 ns) and by the need to provide a large distance between the flash X-ray tube and the event under study (both to reduce geometric blur and to protect the tube). Hence, very fast film/screen combinations are used to get adequate film exposures. In such combinations, a dual emulsion light-sensitive film is placed in close contact between two fluorescent screens which absorb and convert to light a portion of the incident X-rays, each screen exposing primarily the emulsion facing it. The spectral sensitivity of the film is chosen to match the light output of the screen. The speed of very fast film/screen combinations can be 100 times larger than the speed of non-screen films directly exposed to X-rays, yielding adequate image film densities at X-ray exposures of 0.5 to 10 μGy (0.05 to 1 mR), depending on X-ray energy and film/screen selection.

The general photographic density response (Hurter-Driffield or *H and D* curve) of these dual emulsion films has a typical S shape as illustrated in Fig. 6. The film density range is more limited than that of films used in conventional radiography (maximum density approximately 2.5 vs 4) and the slope of the H and D curve (gamma factor) is small at both low and high exposure, resulting in very low image contrast. Hence, careful adjustment of film exposure is necessary to achieve image quality and contrast.

The choice of screen depends on the optimum trade-off between speed and resolution. Thicker screens have more output but reduced resolution. High speed films are generally chosen, though sometimes a lower film speed is used to avoid excessive quantum noise in the image.

Until 1975 the standard film/screen combinations used for flash radiography generally included thick calcium tungstate screens (emitting light predominantly between 350 and 500 nanometers) matched to blue-sensitive dual-emulsion films. Since that time, faster and more intense rare earth screens, emitting in the green or blue spectral range, and appropriate matching films sensitive to the longer wavelengths in the green region, have become widely used in medical radiography and have been tried for flash radiography. Rare earth screens are not as effective as calcium tungstate screens in absorbing very high energy X-rays. Consequently the rare earth screen combinations have proven useful at lower flash radiography voltages (100 to 300 kV) but not at higher voltages.[33]

Electro-Optical Systems

Electro-optical systems are increasingly used in flash radiography for one of several desired goals: preservation of the image in a destructive event; high brightness gain in order to record very faint images; or separation of successive images at high frame rates in cine flash radiography.

FIGURE 6. Film Density and Contrast Factor Curves Typical of Dual Emulsion Films Used with Intensifying Screens for Flash Radiography

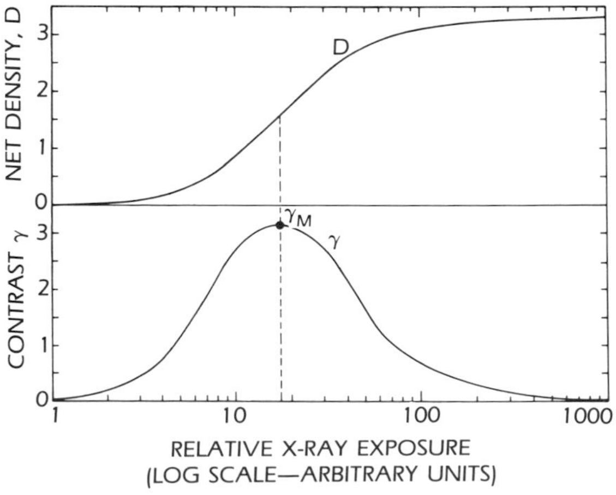

The system illustrated in Fig. 7 achieves the first two stated goals. In the study of large explosive events (see ref. 34), it was impossible to use direct film recording because film located close to the object would have been destroyed by the event. The system consists of a fluorescent screen, two mirrors, a lens, an electromagnetic shutter, a light image intensifier and a standard camera. The fluorescent screen is destroyed by the explosion but not before its image has been captured, amplified and recorded. Because the fluorescent screen will be destroyed anyway, it can be placed much closer to the object, with a consequent gain in image sharpness; the screen also does not have to be placed in a protective cassette, allowing a consequent gain in image exposure and contrast. With a high gain light image intensifier (10^5 to 10^6 photon gain) images can be recorded at much lower exposure, though the image will generally be grainy due to increased quantum noise. Image contrast is also reduced because the fluorescent screen and the light sensitive film used in the camera have a more limited density range and a lower contrast factor (gamma) than the dual emulsion films generally used for direct recording in flash radiography.

This technique was extended[35] to allow the recording of up to 6 separate images of a high speed event at frame rates up to 10^5 frames per second, using the arrangement shown in Fig. 8. The fluorescent screen material, gadolinium oxysulfide with praseodymium doping, was chosen for its combination of high light output (approximately 10% conversion efficiency) and very fast decay (approximately 10 microseconds) which avoids mixing of successive images. The fluorescent screen image is relayed by a mirror, then captured by six lenses, each coupled to a gated image intensifier, with the final images recorded on instant film. This constitutes a very powerful flash cineradiography system for applications requiring high penetration and a small number of images at very high frame rates.

Still another approach for recording a limited number of X-ray images at extremely high frame rates uses an image converter camera to acquire and separate the images. In an early implementation,[36] a linear accelerator was used as the X-ray source (10-20 MeV, 0.1-30 microsecond pulses); the detector system consisted of an external scintillator screen coupled by mirrors to an image converter camera capable of producing nine images at 300,000 frames/second. More recently, a similar system has been developed, the key element of which is a magnetic focus, magnetic deflection image converter camera which produces on its output screen up to nine 24 × 30 mm images at frame rates of 10,000 to 250,000 frames/second (see ref. 37).

FIGURE 7. High Intensity Flash X-ray System with Remote Detection Consisting of Fluorescent Screen, Mirrors, Image Intensifier and Camera, for Observation of Destructive Events

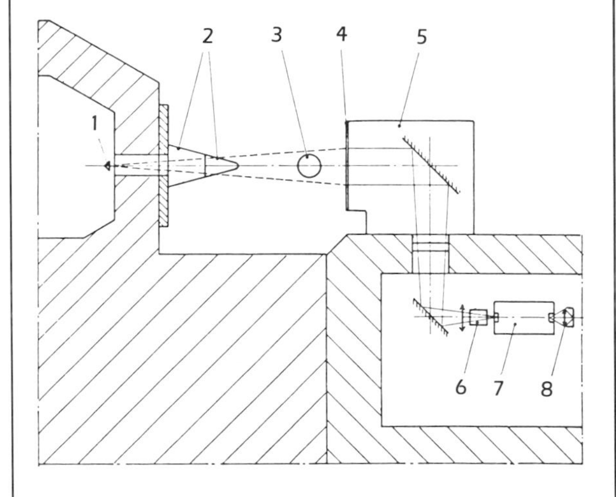

1. X-ray source
2. X-ray output and protection
3. Object radiographed
4. Fluorescent screen
5. Light-tight box
6. Electromagnetic shutter
7. Intensifier tube
8. Optical system (for recording output screen image)

A different imaging system has been described[38] in which the image is acquired by an X-ray sensitive vidicon using a beryllium input window to enhance sensitivity to low energy X-rays. The vidicon output is amplified and displayed on a television screen; it can also be stored for subsequent recall or image processing. This system can be used to record small size images either singly or, in flash cineradiography, at frame rates compatible with the television system capability.

Another system was used for biomedical flash cineradiography studies of impact or crash injury.[39] The X-ray source, which will be discussed in the next section, uses a single Marx-Surge generator and tube assembly capable of producing 350 kV pulses at rates up to 1,000 pulses per second, with a maximum pulse train of 60-100. The recording system consists of an X-ray image intensifier coupled to a 16 mm pin-registered framing camera. The system is characterized by high brightness gain; the image noise is less than that of the previous systems using an external fluorescent screen and a light image intensifier. The input image size is limited to 30-40 cm (12-16 in.) diameter by the X-ray image intensifier. Most present X-ray image intensifiers offer several modes of operation with different input image diameters (typically 15, 23 and 30 cm) for optimum fit to each specific study. Several X-ray image intensifiers have been used, including high performance X-ray image intensifiers developed specially to meet the requirement of high speed flash cineradiography.[40] One model uses a special cesium iodide input screen which is relatively thick and is processed for very fast decay. As a result, the system can operate at frequencies up to 10^6 per second without excessive image stacking, and can retain acceptable quantum detection efficiency when used with flash X-ray sources in the 200 to 2,000 kV range.

Triggering Methods

Successful flash radiography generally demands that the instant of the flash radiograph be precisely known and precisely controlled. This requires the use of reliable and precise sensors and trigger generators, and also requires that the flash X-ray generator (and, when applicable, the gated detector) respond to the trigger with minimum delay and very

FIGURE 8. Block Diagram of Multiple Tube/Detector Flash Cineradiography System

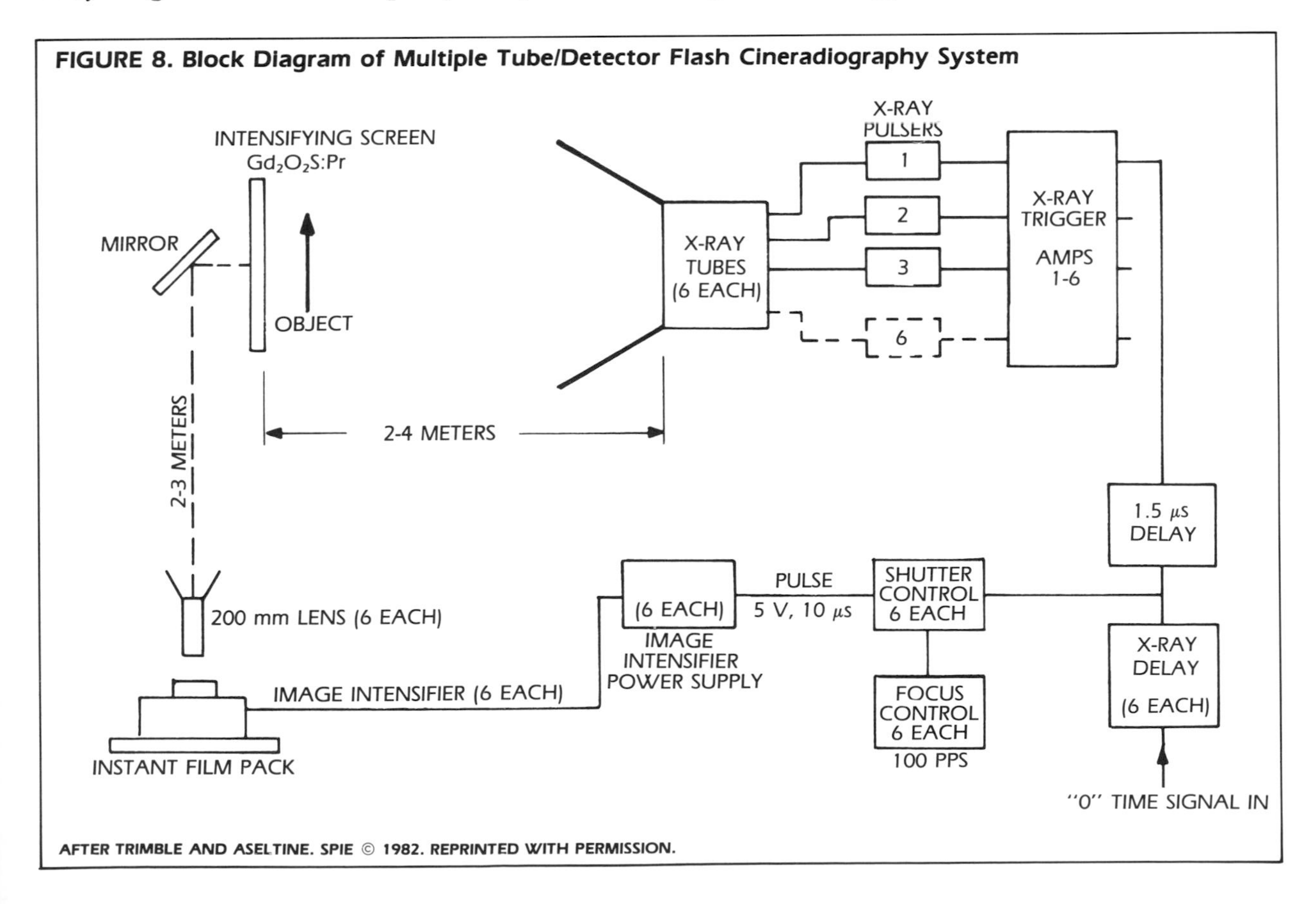

AFTER TRIMBLE AND ASELTINE. SPIE © 1982. REPRINTED WITH PERMISSION.

low jitter. Internal delays of 100 to 300 ns, and jitters of 10 to 50 ns, are typical of present well-operated flash X-ray systems; lower jitter (down to 2 ns) has been achieved for special applications.

There is a great variety of trigger sensors and generators, each suited to a particular application. One of the most common trigger sensors in ballistic studies is the *make* circuit, consisting of 2 copper foils separated by thin polyester film; a voltage on the order of 400 V is applied between the two copper foils. When the projectile strikes the sensor, it short-circuits the copper foils, generating an electrical input pulse signal to the X-ray trigger generator. Break circuits using a wire or a conducting paint line on a paper sheet are also often used, though they may give erroneous signals under some conditions. Other trigger circuits include photoelectric devices, X-ray beam interruption sensors, magnetic loop triggers, acoustic or pressure transducer triggers, microwave doppler techniques and capacitance triggers. A description of these various triggering methods is presented in ref. 41.

Protection of Equipment and Film

Many important applications of flash radiography involve very destructive events such as detonations, ballistics, explosives or impact phenomena which expose the equipment and film to blast waves, overpressures (or negative pressures), ground shocks, and fragments; this makes film and equipment protection essential. The risk of damage is decreased by maximizing the distance from the event. This is easier to achieve for the equipment than for the film because the relatively large X-ray source size requires the film to be much closer to the object to avoid excessive image unsharpness.

The film and intensifying screens are placed inside a commercial film holder which is inserted in a steel frame. The frame is designed with a front plate made from a material possessing the highest possible strength and the lowest possible absorption to X-rays. Magnesium alloys or plastics such as polystyrene are generally used for low X-ray energies; aluminum alloys or even steel are used for high X-ray energies. It is generally necessary to use several plates of protective material, separated by buffer spaces filled with air, felt or compressible foam, to minimize pressure marks on film. The film holder can be shock-mounted inside the protective frame and the frame itself is loosely mounted (suspended from a tripod or placed inside a barrel) so that the whole assembly can recoil upon impact, dissipating some of the blast energy and reducing peak stress on the film. The event under study is sometimes so violent that the film cannot be reliably protected and alternate image techniques must be used: for example, the fluorescent screen/mirrors/image intensifier and camera system discussed previously.

The X-ray tube is protected by a heavy armor plate with a central aperture just large enough to accommodate the useful X-ray beam. A lighter shield, such as aluminum, is placed over the aperture to stop blast waves and small fragments. A cone shaped shield, as shown in Fig. 7, is more effective than a piece of armor plate. A remote tubehead can be used, at voltages up to 600 kV, to remove the high voltage pulser to a more distant underground location so that only the tubehead is at risk. The pulser is generally shock-mounted to reduce damage from ground-transmitted shocks. The flash X-ray system control console is often located in a remote bunker, as much as 100 to 1,000 meters away from the pulser and tube.

PART 3

FLASH CINERADIOGRAPHY

Many applications require a sequence of images rather than a single image. The technique of recording the radiographic sequence, with a movie camera, is called *cineradiography*. The total number of images required, the time interval between successive images and the image size vary greatly from one application to another. Cineradiography techniques suited to several specific needs will be discussed here.

Small Number of Images at Very High Frame Rates (10^4 to 10^8 Frames per Second)

Ballistic events usually last only a fraction of a millisecond and image frame rates of 10^5 to 10^6 per second are needed to record these events; individual X-ray pulse durations for flash cineradiography are generally below 0.1 microseconds. Nuclear fusion studies require much shorter pulses and higher frame rates; special techniques are required to generate X-ray pulses and to record and separate images at these very high rates.

Multiple Tube, Multiple Pulser Techniques

An elementary though very costly approach to flash cineradiography is a system containing several flash radiography tubes, powered by separate high-voltage pulsers and triggered in sequence. With this technique, the time interval between successive images can be made as short or as long as desired, the time intervals between successive images may be different, and the maximum X-ray intensity can be obtained for each pulse. A special three-channel flash X-ray system, optimized for ion beam pellet implosion studies, has been described.[42] It achieves exceptional spatial and temporal resolution as well as exceptionally high frame rates. The individual X-ray sources are only 0.1 mm in diameter, using replaceable targets in continuously pumped tubes. The duration of the individual X-ray pulses is three nanoseconds; the time separation between successive pulses can be as short as two nanoseconds; and the timing of these three pulses can be controlled to an accuracy of one nanosecond.

FIGURE 9. Multiple Electrode Soft X-ray Tube for Cineradiography: Six Small Triode Discharge Tubes Arranged in a Single Vacuum Vessel

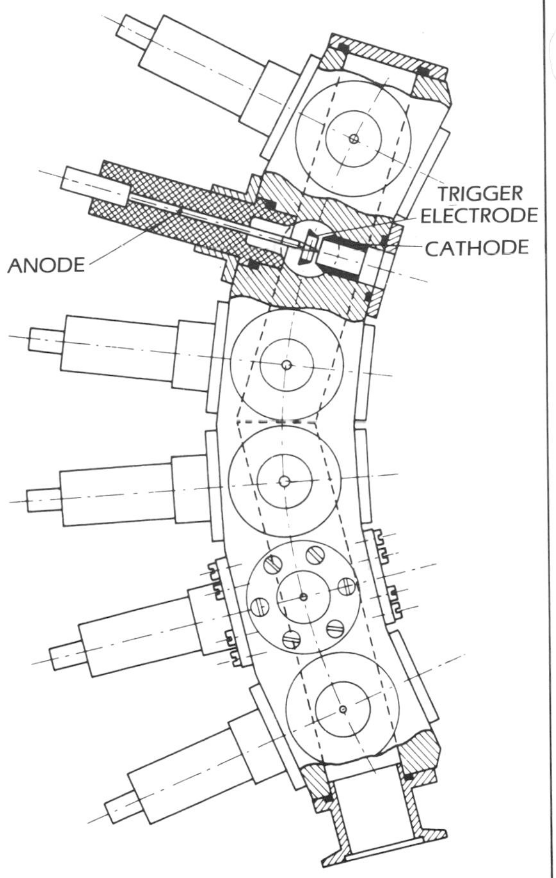

AFTER JAMET. ELSEVIER SCIENTIFIC PUBLISHING © 1976. REPRINTED WITH PERMISSION.

Multiple tubes must be used in applications, such as external ballistics, where the trajectory and stability of an object is observed over a large distance. In this case, the successive images are automatically separated and are recorded on separate film.

Multiple tubes can be used (and image separation achieved) in a radial tube array as illustrated in Fig. 10. This arrangement is well-suited to the study of events with symmetry of revolution. The tubes can be used either singly or in synchronized pairs, where each pair yields three-dimensional image reconstruction (stereoradiography).

The high bulk and cost of a multiple channel flash X-ray system severely limits the number of images which can be practically obtained: 10 or 12 at most. Another disadvantage for some applications is that the use of separate tubes results in the separation of the X-ray sources, causing parallax and complicating interpretation of the image sequence.

Due to the large size of the high voltage pulsers, excessive parallax is caused if the flash X-ray tubes are located inside the pulsers; hence remote tubeheads are generally used, enabling the X-ray sources to be located closer to each other. A further reduction in parallax can be achieved by using a multiple tube tubehead (*cluster tubehead*).[35] Up to nine tubes of the same or different voltages, connected to nine pulsers through individual high voltage cables, are located inside a single high voltage housing. This can also be achieved by locating several discharge systems (cathode, trigger electrode and X-ray target) inside a single vacuum vessel (see Fig. 9).[43] Except for the multiple tube arrangements shown in Figs. 9 and 10, where image separation is automatically achieved by proper X-ray beam collimation, image separation normally requires special arrangements, separate lenses and cameras[35] or image converter cameras[36,37] as previously discussed.

Single Tube, Multiple Pulser Techniques

Single tube, multiple pulser techniques have been developed to eliminate parallax while achieving frame rates exceeding the pulse repetition capability of existing high voltage pulsers.

The first design for such a system[44] is shown in Fig. 11. Several high voltage pulsers are connected to a single flash X-ray tube through high voltage isolation diodes. These diodes are low forward impedance vacuum discharge diodes possessing two critical characteristics: (1) they are capable of passing a large forward current with a low voltage drop (when their associated pulser is fired), and (2) they

FIGURE 10. Multiple Flash X-ray Tubes and Collimators in Radial Configuration

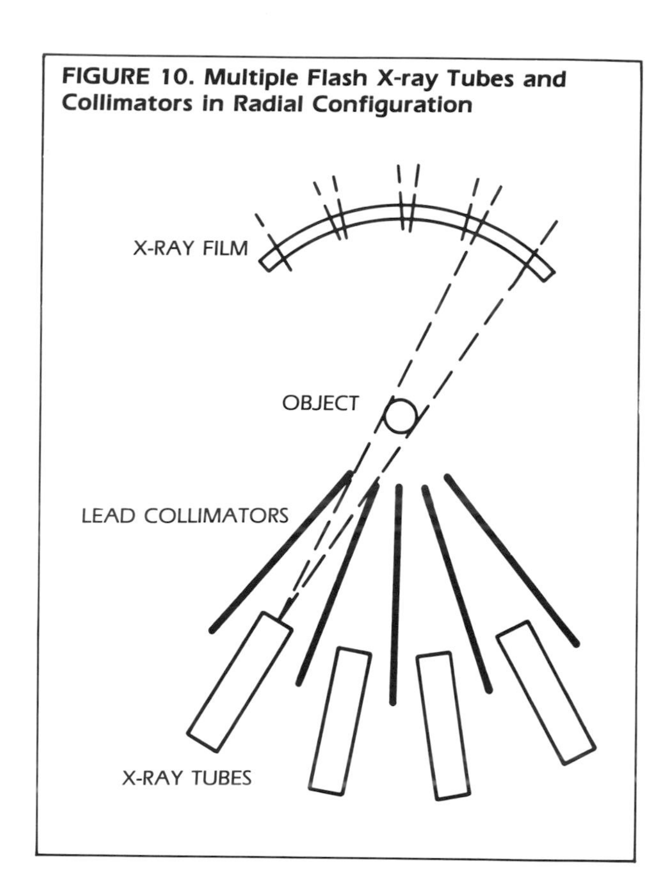

FIGURE 11. Multiple Pulser, Single Flash X-ray Tube System Designed to Produce Short Sequences of Flash Radiographs at Very High Frame Rates

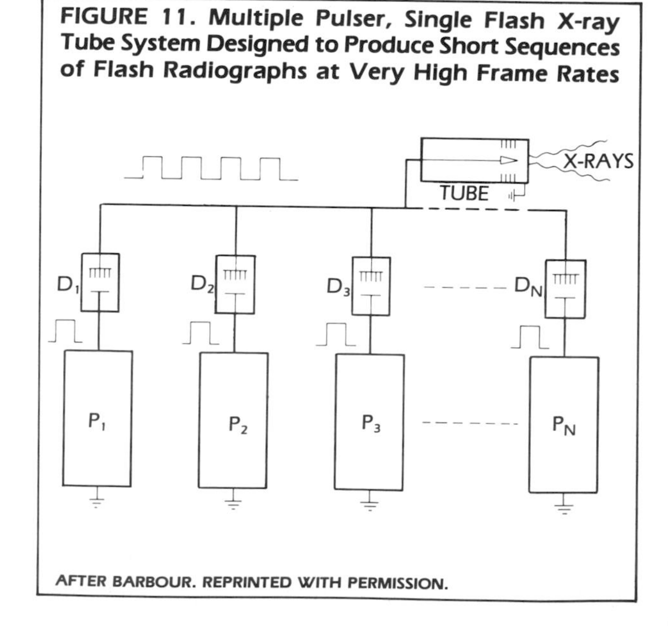

AFTER BARBOUR. REPRINTED WITH PERMISSION.

can hold off the pulser voltage in the reverse direction when another pulser is fired. For instance, with a system consisting of 150 kV, 100 ohm pulsers and a single 100 ohm flash X-ray tube, several 25 ohm forward impedance diodes will transfer 120 kV, 1500 A pulses to the tube. The maximum pulse rate is approximately 10^5 pulses per second and is limited by the recovery of the cine diodes and the recovery of the flash X-ray tube; the total number of pulses is limited by the cost of the system and by the heat capacity of the X-ray target. This system design is limited to relatively low-voltage, high-impedance pulsers. At higher voltages the cine diode must have large gap spacing to hold off the reverse voltage and will necessarily have much higher forward impedance. The voltage applied to the tube, V_t is

$$V_t = V_o \frac{R_t}{R_d + R_t} \qquad \text{(Eq. 2)}$$

where V_o is the voltage applied to the tube in the absence of a cine diode; R_d is the forward impedance of the cine diode; and R_t is the impedance of the pulser and the flash X-ray tube. Hence, if R_d is large, the voltage loss is large and the X-ray output falls drastically.

Based on a sophisticated flash X-ray tube, another system[45] is capable of producing four X-ray pulses from a single large anode, at rates up to 4×10^7 pulses per second. The system is illustrated in Fig. 12. Four 500 kV Mark-Surge generators with negative voltage output are connected directly to the four cathodes of a single target flash X-ray tube. Normally the recovery time following a discharge in the X-ray tube limits the pulse rate to approximately 10^5 pulses per second. However, the addition of deflectors to the X-ray target divides the tube into four separate chambers and confines each discharge to one chamber; in this way, a discharge can be initiated in one of the chambers before metal vapors and ions from a previous discharge in another chamber have recombined. This special tube design, though very complex to build, could conceivably be extended to more than four chambers. Only one-fourth of the anode is involved in each discharge, so that strictly speaking the four X-ray sources do not coincide but are separated by approximately 20 mm. The very large target easily absorbs the energy of the four pulsers. A small focal spot (2.5 mm) is achieved by using a very small cathode-anode spacing, though this forces a low tube impedance and a short pulse duration (25 ns) in order to avoid gap closure.

Large Number of Images at Low to Moderately High Frame Rates (1 to 10^5 Frames per Second)

A repetitively pulsed single pulser and tube combination is generally used for these applications. The frame rate may be limited by (1) the high voltage pulser (e.g., by the recovery time of the pressurized gas spark gaps) or (2) by the maximum speed of the film (which sets the maximum frame rate at which images of a given format can be separated). Due to

FIGURE 12. Novel Flash X-ray Tube with Multiple Discharge Chambers and Single Anode; for Cineradiography at Extremely High Frame Rates with Minimum Parallax

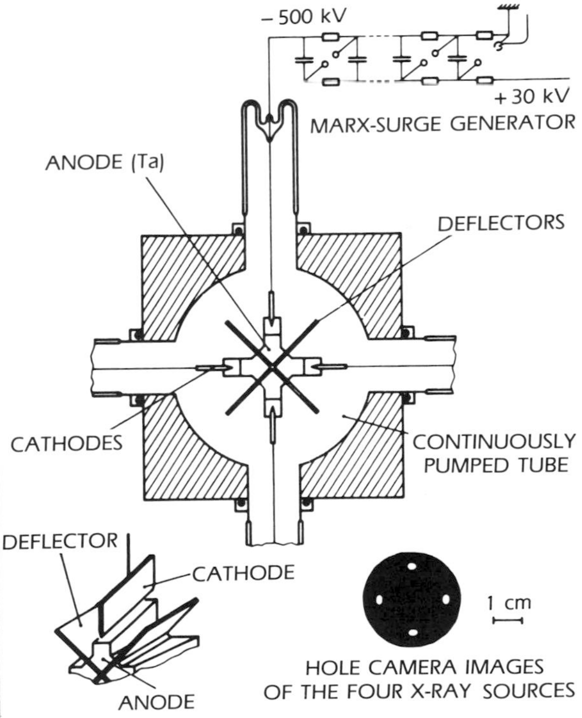

AFTER JAMET. SPIE © 1983. REPRINTED WITH PERMISSION.

the extremely short duration of the individual X-ray pulses, the film usually need not be stopped for each image but can be in continuous motion.

Figure 13 shows a very simple arrangement[46] suitable for high voltage cine flash radiography with a small number of large format images at lower frame rates (for example, a total of eight 35 × 43 cm [14 × 17 in.] images at 10 to 30 images per second). A typical application would be for the dynamic study of casting processes or rocket motors. The film drum speed is continuously variable and it determines the frame rate; a microswitch on the film drum senses the film position and triggers the flash X-ray system when each successive film cassette reaches the right position. Much higher frame rates are required for other applications, limiting the images to smaller format or requiring the use of different technologies. A series of up to sixty low voltage (30 kV) X-ray pulses at rates up to 5,000 pulses per second has been produced.[47] The key to this development was a special pressurized-gas spark gap designed for extremely fast recovery. For this purpose, hydrogen gas was used because of the high mobility of hydrogen ions; the gap consisted of a series of very closely spaced metallic discs to speed up cooling of the plasma and quenching of the discharge. A more sophisticated system was subsequently designed, capable of achieving frame rates up to 12,000 frames per second and voltages up to 120 kV.[48] The X-ray images were recorded by a high speed continuous motion drum camera with a maximum film speed of approximately 100 meters per second; the maximum frame rate of 12,000 frames per second could be used only for images less than 8 mm high.

Extensive work has also been done on powerful flash X-ray systems capable of producing very long trains of pulses at very high frame rates, from moderate voltages up to 300 kV.[49,50] These systems use the different technology shown in Fig. 14. A capacitor in parallel with a heavy duty quenching spark gap produces a train of 10 joule, 60 kV pulses at a controlled rate. These pulses serve as the input to the primary of a pulse transformer. The transformer output, at 200 to 300 kV, is applied to the flash X-ray tube through another quenching gap used to speed up the deionization time of the X-ray tube. The X-ray tube is continuously pumped, demountable, and the target and X-ray windows can be replaced as often as necessary. The imaging system consists of direct film recording on a drum camera or, as shown in Fig. 14, an X-ray image intensifier and a high speed camera. With this system, up to 2,000 frames at 500 frames per second, or up

FIGURE 13. Slow Cineradiography System with High Voltage and Large Image Size Capability

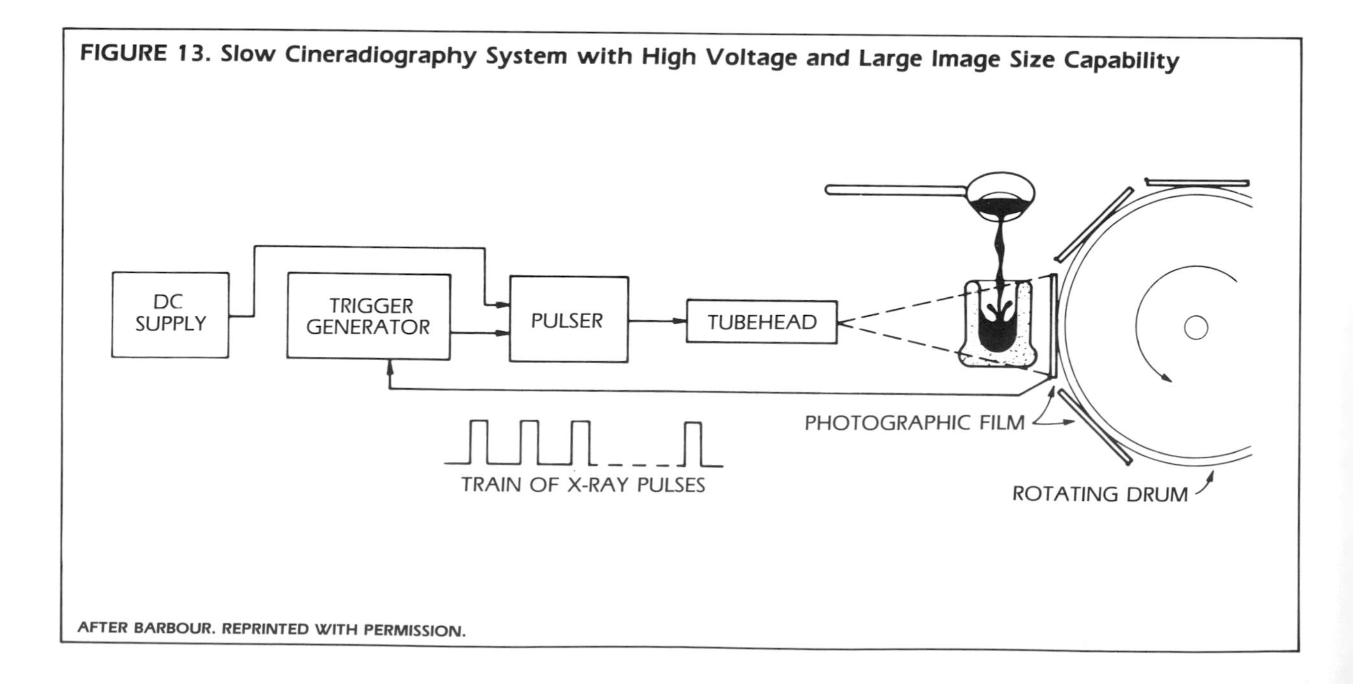

AFTER BARBOUR. REPRINTED WITH PERMISSION.

to 300 frames at 40,000 frames per second, have been produced.

Illustrated in Fig. 15 is a system to produce high energy and high voltage X-ray pulses for biomedical or crash injury studies.[39] A large capacitor stores 8,000 joules of energy at 8 kV and recharges the high voltage pulser (Marx-Surge generator) through a resonant charging circuit to increase energy efficiency. A high voltage, short persistence X-ray image intensifier acquires the low intensity X-ray images and produces small, bright, light images which are recorded by the camera. A pin-registered, high-speed framing camera is used to achieve greater optical efficiency and image quality. The camera can be set at the desired speed, from 24 to 1,000 frames per second, and produces a shutter correlation pulse which is used to trigger and synchronize the flash X-ray system. The X-ray pulse intensity and the imaging system sensitivity are sufficient to produce well-exposed images of the chest or skull of a human or a large primate at distances of 2 to 3 meters. With a high vacuum sealed-off field emission flash X-ray tube, the maximum number of pulses is limited by the heat capacity of the X-ray target to 60 to 100 pulses at 1,000 pulses per second. The frame rate is limited to 1,000 frames per second by the camera. The Marx-Surge generator, which uses carbon dioxide instead of air in the spark gaps, would be capable of operating at up to 1,200 to 1,500 pulses per second, and special fast decay X-ray image intensifiers would eliminate stacking at even much higher frame rates. The camera therefore is the limiting factor in achieving higher frame rates.

Linear accelerators are natural X-ray sources for very high-voltage flash radiography with relatively long pulse duration (0.5 to 10 microsecond) and moderate frame rates. Several such systems were

FIGURE 14. "Strobokin" Flash Cineradiography System Designed to Produce Very Long Pulse Trains at Moderately High Frame Rates

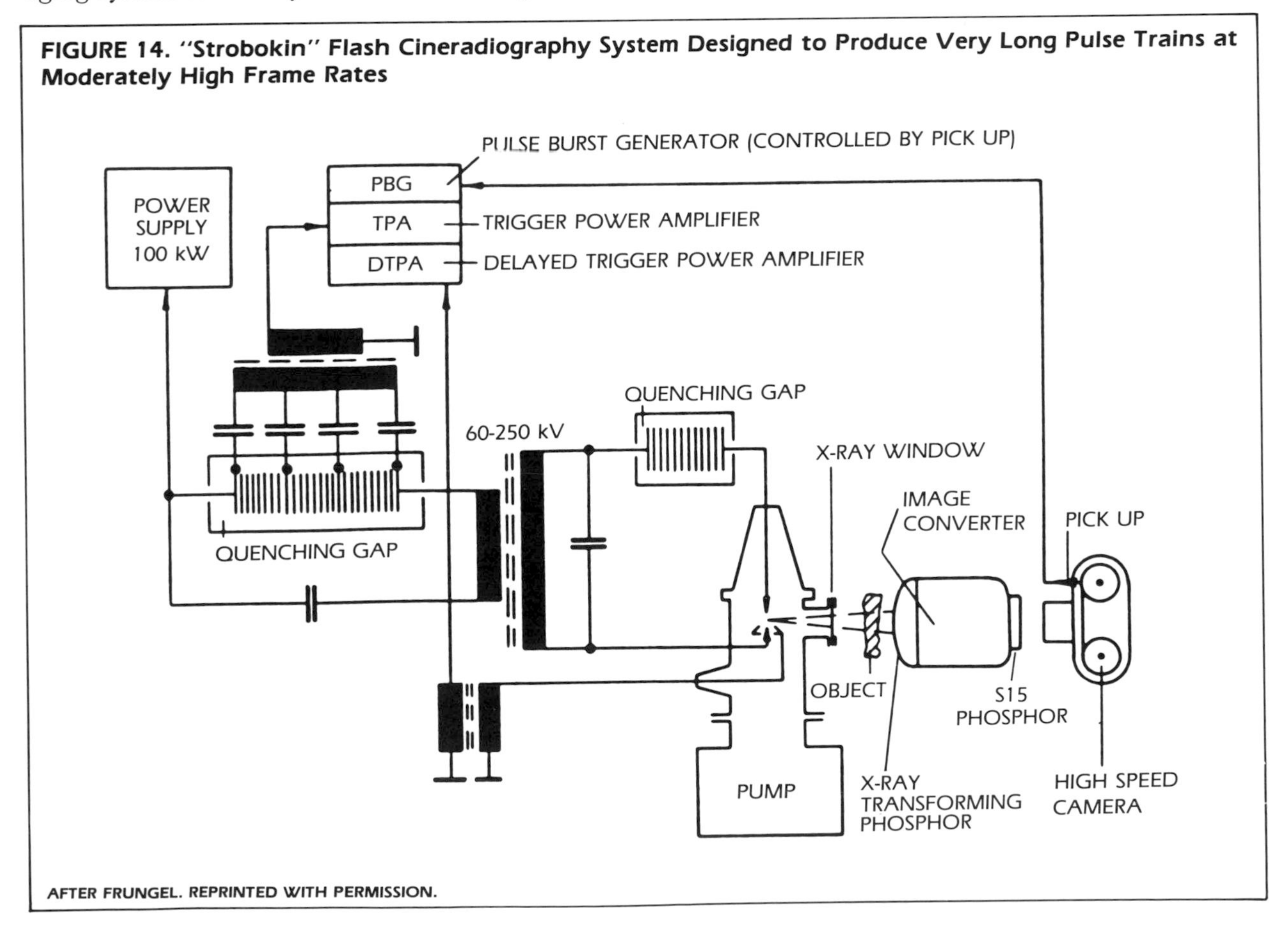

AFTER FRUNGEL. REPRINTED WITH PERMISSION.

listed in Table 3 and discussed in the accompanying text. Betatrons have also been used as very high-voltage impedance X-ray sources, particularly in the USSR, where they are more readily available than linear accelerators.[51]

Finally, cine flash radiography can also be done using a continuous radiation source (X-ray system or radioisotope) and an intermittent imaging system with a gated X-ray image intensifier and/or a shutter camera. One such system[52] uses a high-gain high energy X-ray image intensifier and a rotating prism camera. The system operates at rates up to 10,000 frames per second, with exposure times down to 10 microseconds, by gating the X-ray image intensifier with high voltage pulses initiated by the camera. Image stacking was observed above 5,000 frames per second. A system being tested at Los Alamos National Laboratory combines a moderate or a high voltage continuous X-ray source (150 kV system, or 2 MeV Van de Graaff accelerator) with an X-ray image intensifier and a high framing-rate video system capable of recording, storing and later displaying frames at rates up to 2,000 per second for full frames and 12,000 per second for partial frames. Results are encouraging and the video system has a significant advantage: images can be examined immediately, avoiding the delays of film processing and permitting optimization of radiographic technique and image quality at the beginning of the study. Generally, systems using continuous X-ray sources are appropriate for the study of relatively slow events, allowing the use of relatively long (0.1 to 10 milliseconds) exposure times; continuous X-ray sources have limited penetration at short exposure times (10 to 100 microseconds) and are therefore not useful for very fast events; the X-ray intensity produced by a continuous X-ray source is grossly inadequate when the exposure must be limited to a few microseconds or less to freeze motion.

FIGURE 15. High Voltage (350 kV) High Penetration Flash Cineradiography System with 1000 Frames per Second Capability

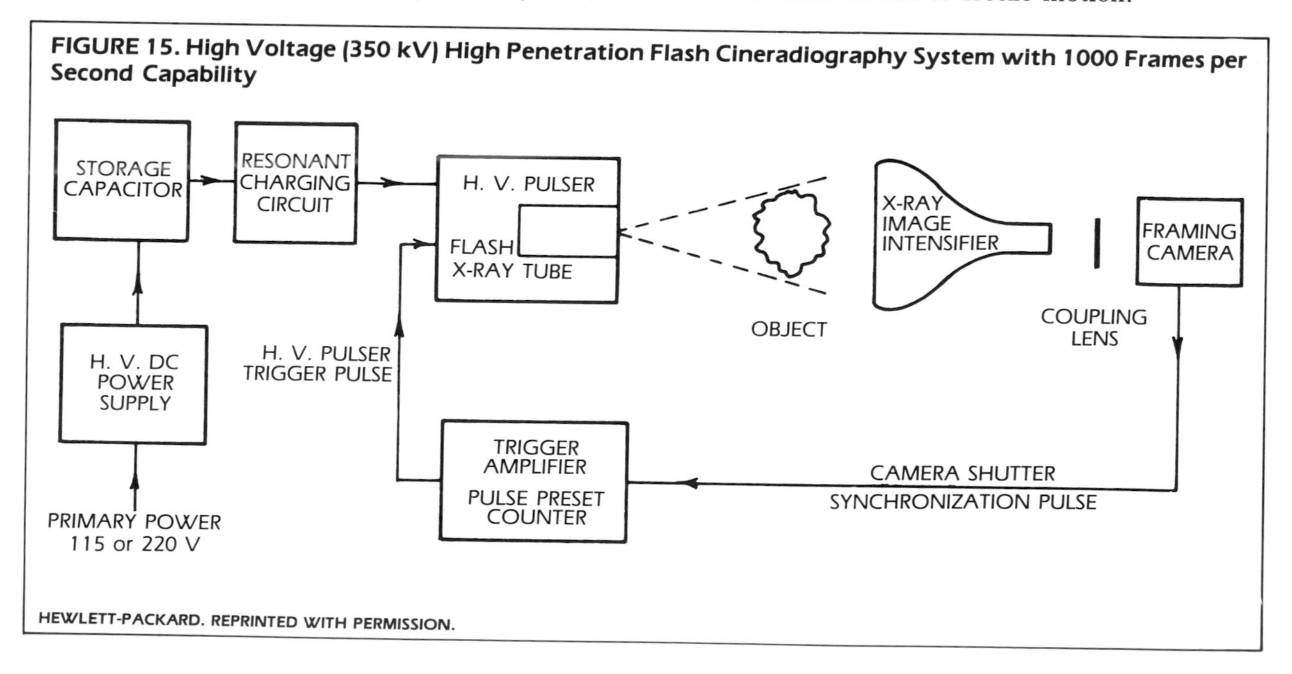

HEWLETT-PACKARD. REPRINTED WITH PERMISSION.

PART 4

SPECIAL HIGH SPEED RADIATION IMAGING TECHNIQUES

Soft X-rays

Occasionally the object under study contains a very wide range of thicknesses and detail size. In such cases soft (long wavelength) X-ray techniques and dual film recording can yield substantially more information. The principle of the technique is shown schematically in Fig. 16 and a possible application[54] is illustrated in Fig. 17. For soft X-ray applications, the flash X-ray tube is provided with a thin beryllium window that is transparent to the intense low energy (8-30 keV) X-rays which the tungsten target produces even at high tube voltages. The first film provides a high contrast, high resolution image of very small detail and is generally a medium or fine grain industrial film in a paper pack. Behind it is a conventional cassette with dual emulsion film and two intensifying screens; this produces a complementary image of the denser portions of the object.

FIGURE 16. Dual Film Technique for Obtaining Simultaneous Soft and Hard Flash X-ray Images

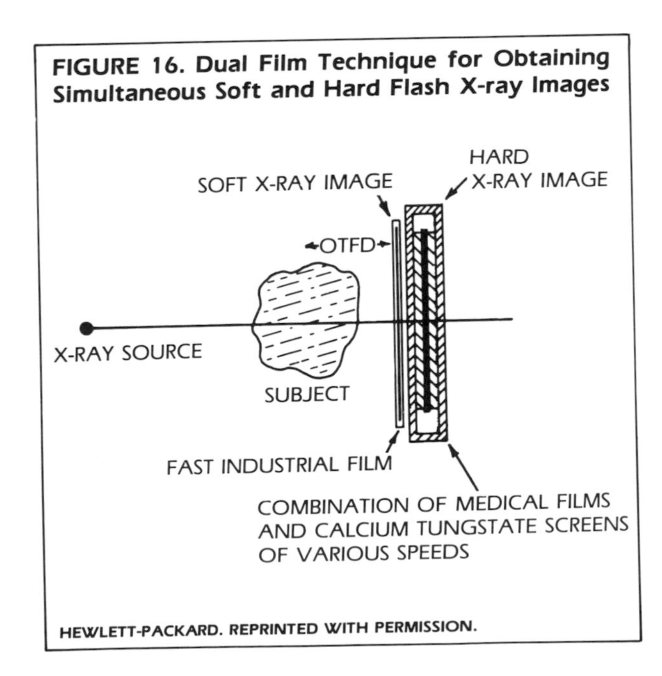

HEWLETT-PACKARD. REPRINTED WITH PERMISSION.

Very Low Voltage Flash X-rays

Sometimes the object of interest consists entirely of thin, low density portions. In such cases, the presence of high energy X-rays is undesirable and a very low tube voltage is needed. Standard flash X-ray diode tubes do not perform reliably below 50 kV but this problem can be solved by using a triode flash X-ray tube; a trigger electrode is pulsed briefly at the high voltage (50 kV) necessary to initiate the vacuum discharge; the X-ray target (anode) is held at a lower voltage (20 kV) for the duration of the X-ray pulse. The X-ray tube window may be made of beryllium in sealed-off tubes or of polyester film in continuously pumped tubes. This technique has been fully described in the literature.[1, 55]

FIGURE 17. Simultaneous Soft and Hard X-ray Images of Bullet Impact on Aluminum Bar

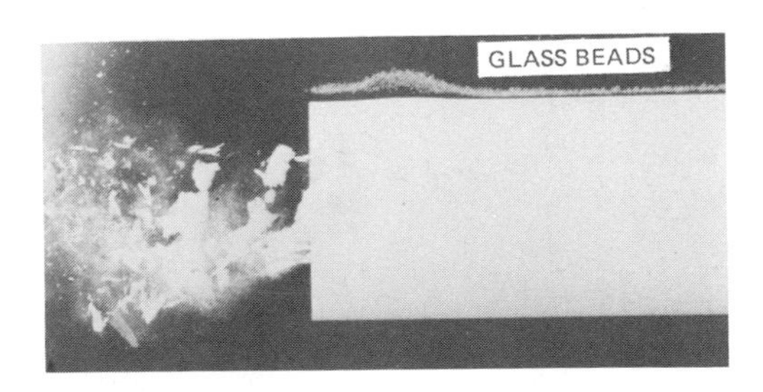

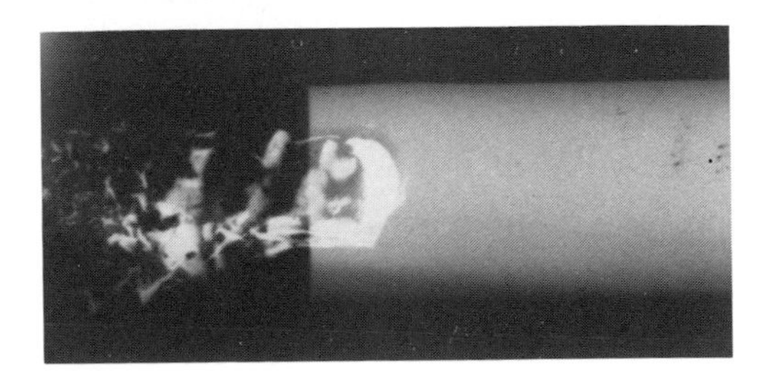

HEWLETT-PACKARD. REPRINTED WITH PERMISSION.

Multiple Film Cassettes

When a high voltage flash X-ray system is used to image a very dense object, several film/screen packages can be stacked in a single cassette to enhance penetration and contrast. This technique is useful because at very high voltages the X-rays reaching the cassette are sufficiently penetrating that the front pair of intensifying screens may absorb only 10 to 20 percent of the incident X-rays. A second film/screen package located just behind the first package, in the same cassette, will absorb 10-20 percent of the remaining X-rays and produce a slightly weaker image on the second film. The images on the two films are in precise time and spatial registration. When the two films are superimposed on a view box, the combined image will offer a wider range of density and a higher contrast than would be feasible with a single film. The effect of the technique is to increase the penetration capability of a given system by almost one half-value layer (for example, 8 millimeters of steel for a 600 kV system) and to essentially double the contrast. At very high voltage, a third film/screen package may be added to further enhance the effect. Bright lighting may be required for good observation of the darker portions of the film.

Flash Betagraphy

Flash betagraphy is a technique in which a pulsed electron beam from a small (apertured) electron source is used to produce a projection image of very

FIGURE 18. Simultaneous Flash Betagraph and Radiograph of Bullet Penetrating Two Copper Plates

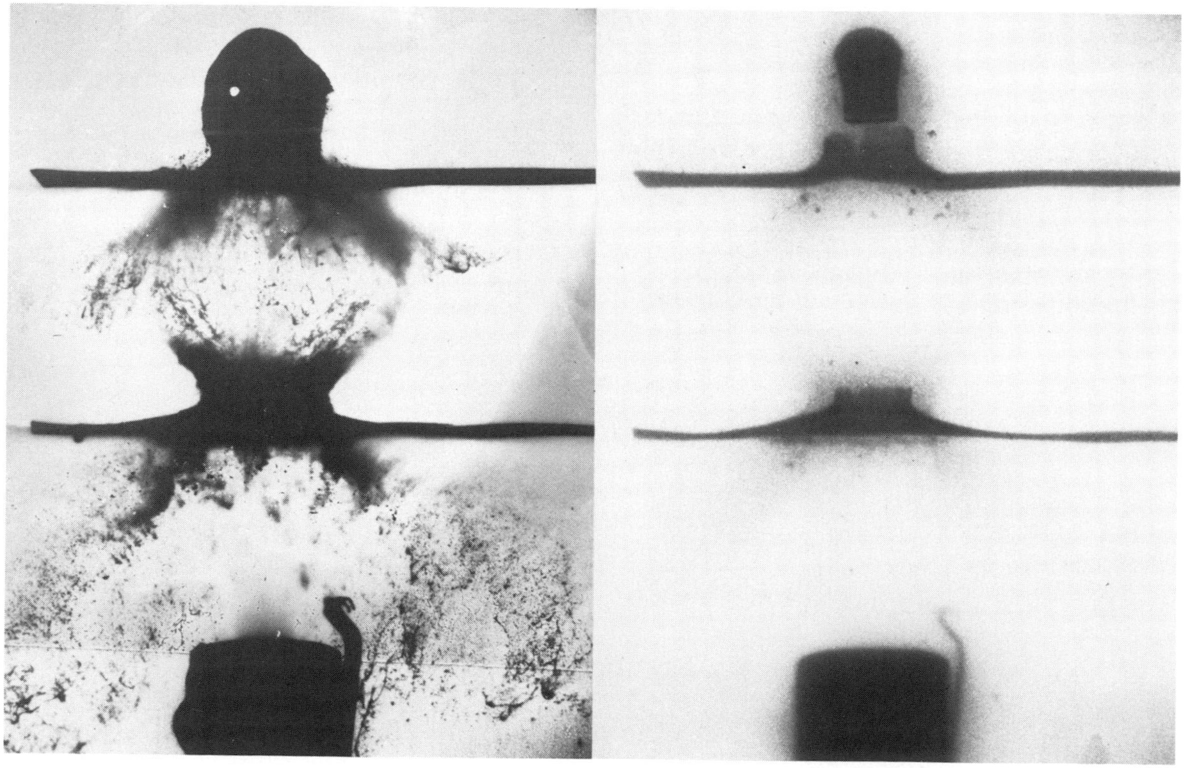

AFTER BARBOUR. REPRINTED WITH PERMISSION.

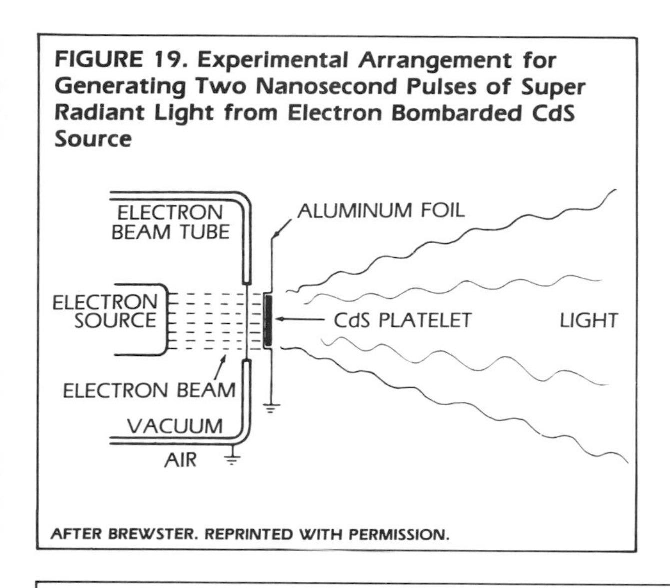

FIGURE 19. Experimental Arrangement for Generating Two Nanosecond Pulses of Super Radiant Light from Electron Bombarded CdS Source

AFTER BREWSTER. REPRINTED WITH PERMISSION.

small particles in near vacuum.[56] The image is normally recorded on a fine or extra-fine grain film (without screens) in a tight black polyester package. A very thin particle casts a shadow on film not by absorbing the high energy electrons but by deflecting them (scattering). Scattering cross sections are so high that exeeedingly thin and low density particles (down to a few microns), which would be transparent even to soft X-rays, will yield high contrast betagraphic images. The technique is illustrated in Fig. 18 which shows simultaneous flash radiography and betagraphy images of a bullet impacting on two plates. Despite its very high contrast capability, betagraphy is a very difficult and limited technique which can only be used in near vacuum (pressure less than 0.1 torr); residual air would cause prohibitive scattering of the electrons and destroys the imaging process. Flash betagraphy has been used

FIGURE 20. Experimental Setup Used to Produce Photo in Fig. 21: Shock Tube (about 9 × 9 cm) is Helium-Driven to Produce Dynamic Pressure of About 7MPa (1000 psi) in Flow Behind Shock Wave

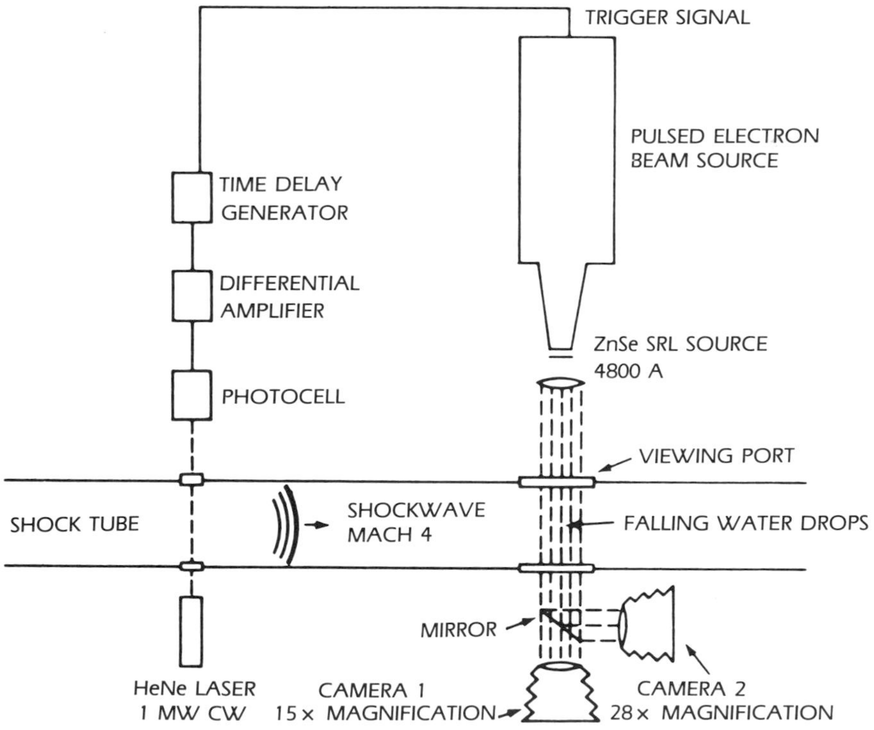

TRIGGER SIGNAL
PULSED ELECTRON
BEAM SOURCE
TIME DELAY
GENERATOR
DIFFERENTIAL
AMPLIFIER
ZnSe SRL SOURCE
4800 A
PHOTOCELL
VIEWING PORT
SHOCK TUBE
SHOCKWAVE
MACH 4
FALLING WATER DROPS
MIRROR
HeNe LASER
1 MW CW
CAMERA 1
15 × MAGNIFICATION
CAMERA 2
28 × MAGNIFICATION

AFTER AESCHLIMANN. REPRINTED WITH PERMISSION.

to image fine debris and gas clouds[56] and to study simulated micrometeorite impact.[1]

Super Radiant Light

Super radiant light (SRL)[57] for high speed photography is produced by bombarding a selected semiconductor material, such as cadmium sulfide, with a short, very high intensity pulse of energetic electrons from a field-emission tube. This is illustrated in Fig. 19. If the electron current density is sufficiently high (typically greater than 100 A/cm^2), an electron population inversion can be created between the valence and conduction bands of the semiconductor, and light output is greatly amplified by stimulated emission. There is no afterglow, and the SRL pulse duration is slightly shorter than the exciting electron pulse (for example, two nanoseconds when a three nanosecond pulsed electron beam is used).

Excellent stop motion can be achieved with this technique. The SRL output light is essentially monochromatic, with a quantum energy approximately equal to the energy band gap of the semiconductor. Wavelengths from 3400 Å to 7400 Å (covering the

FIGURE 21. Interaction of Falling Water Droplets with Shock Wave as Photographed by Experimental Setup Diagrammed in Fig. 20.

AFTER AESCHLIMANN. REPRINTED WITH PERMISSION.

visible and near ultraviolet portions of the spectrum) can be produced by appropriate choice of the semiconductor material. The SRL pulse has characteristics (short duration, precise timing, near monochromaticity, and noncoherence) that make it more useful than spark sources or lasers for certain applications.

Figures 20 and 21 illustrate an application of SRL for the observation of a high-speed shock wave moving at Mach 4 and interacting with falling water drops 0.5 mm in diameter. The need for a high-intensity, short-duration light source that could be precisely synchronized with the shock wave was indicated in this case. A diagram of the setup is shown in Fig. 20. A helium neon laser beam was used to sense the arrival of the shock wave and trigger the pulsed electron beam source after a preselected time delay. The interaction phenomenon was illuminated by super radiant light emitted from a single zinc selenide source and was recorded simultaneously at two magnifications by two cameras, as shown in Fig. 20. A semitransparent mirror presented the image to both cameras simultaneously. One of the resulting shadowgraphs is shown in Fig. 21.

SRL can also be used for black and white or color display of dynamic stress, using a birefringent (double refracting) material to produce stress-dependent rotation of the polarization plane.

PART 5

APPLICATIONS OF FLASH RADIOGRAPHY

Flash radiography has been used in a large number of diverse studies since the 1940s, and only a summary of the major applications can be given here.

Ballistics

Ballistics, the study of the dynamics and flight characteristics of projectiles, is one of the main applications of flash radiography. Ballistics is commonly divided into four areas: internal ballistics, intermediate ballistics, external ballistics and terminal ballistics.

Internal Ballistics

Internal ballistics deals with the study of explosive combustion, acceleration of the projectile and motion of various mechanical elements while the projectile is still inside the gun barrel. Internal ballistics generally requires a high voltage flash X-ray system to penetrate the gun barrel which, for large caliber guns, may be made of steel several inches thick. Protection problems are generally not severe and the X-ray tube and film can be located close to the gun barrel, enhancing penetration and sharpness. Figure 22 is a familiar illustration showing a

FIGURE 22. 600 kV Flash Radiograph of Colt 45; Note Bullet Near Gun Muzzle

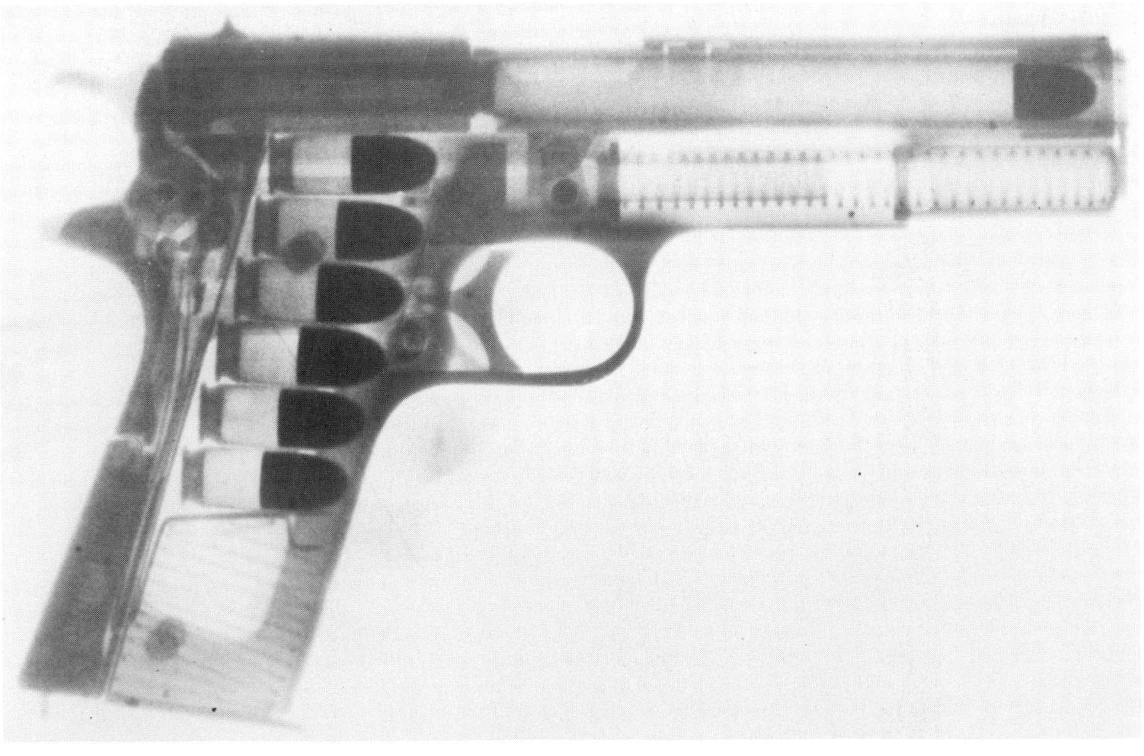

HEWLETT-PACKARD. REPRINTED WITH PERMISSION.

Colt 45 being fired with the bullet just ready to emerge from the barrel. A 600 kV flash X-ray system was used for optimum penetration. The ability to image internal mechanisms and to freeze motion is well illustrated by the sharp outline of the bullet near the end of the barrel. Figure 23 is another example of internal ballistics. The Figure shows two superimposed sequential flash radiographs of an experimental cartridge containing 16 separate projectiles fired simultaneously in an M-79 grenade launcher.[58] The second flash radiograph, taken approximately one millisecond after the first, shows the projectiles further down the tube and reveals significant differences in individual velocity.

FIGURE 23. Two Interior Ballistic Radiographs of Experimental Cartridge Containing 16 Projectiles

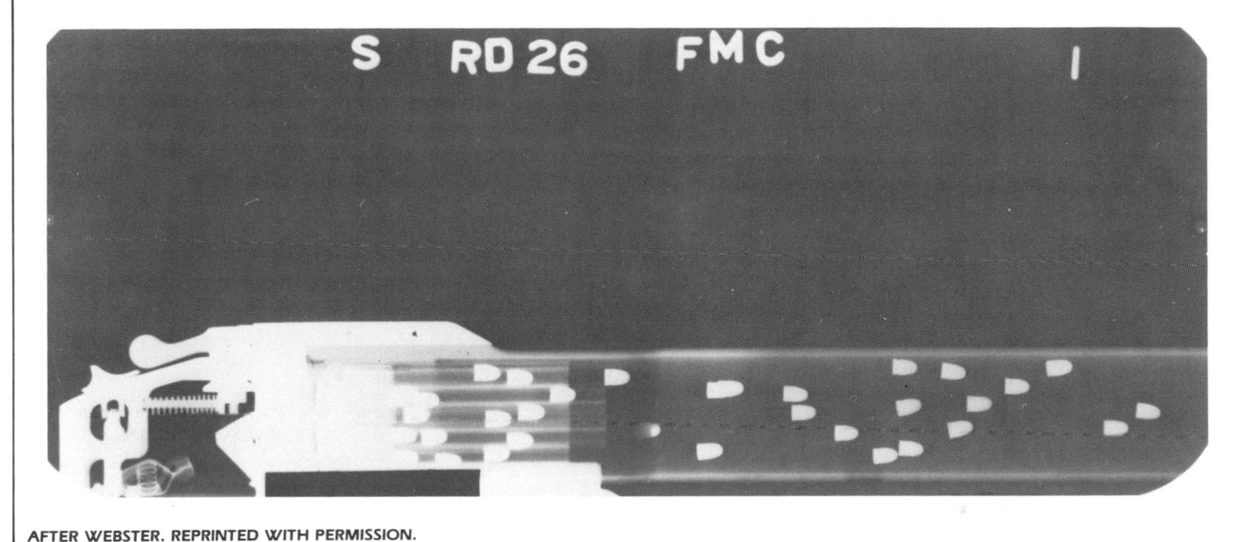

AFTER WEBSTER. REPRINTED WITH PERMISSION.

FIGURE 24. Soft X-ray Flash Photograph of a 20 mm Projectile Near the Gun Muzzle

AFTER JAMET. ELSEVIER SCIENTIFIC PUBLISHING © 1976. REPRINTED WITH PERMISSION.

Intermediate Ballistics

Intermediate ballistics is the study of projectiles just after they emerge from the gun barrel. It is one of the more difficult phases to observe because the muzzle flash makes light photography ineffective; the strong muzzle blast creates a hazard for the X-ray tube and the film, which must be located a significant distance away; accurate timing of the event is difficult to achieve because sensors at the gun muzzle can be triggered prematurely by explosive gases preceding the projectile. Figure 24 illustrates the feasibility of imaging the gases which surround the projectile.[1] This radiograph is of interest in part

FIGURE 25. Flash Stereoradiography Sequence of Model Being Launched in Aeroballistic Tunnel

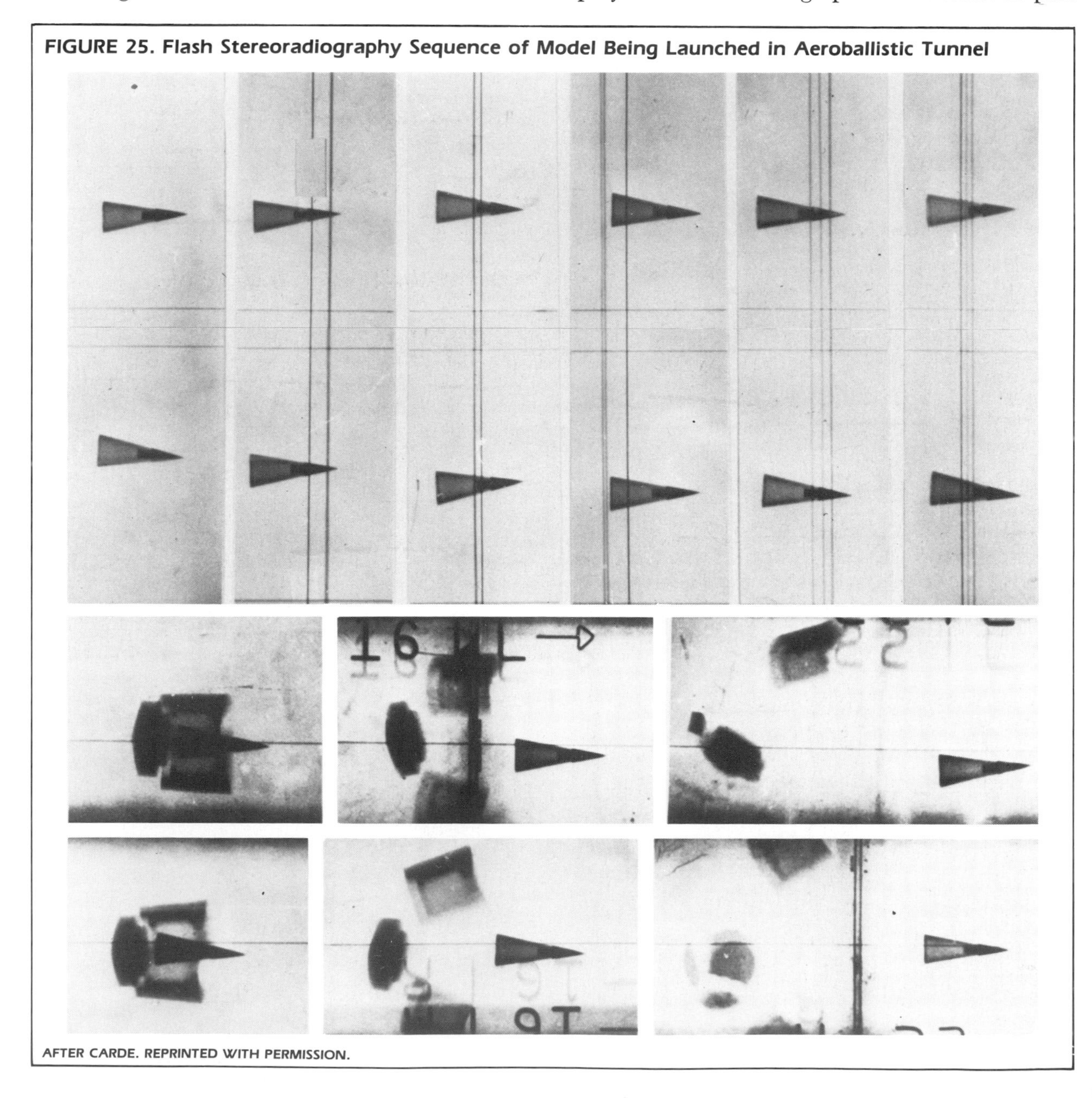

AFTER CARDE. REPRINTED WITH PERMISSION.

because of the special technique used to visualize the powder gases: soft X-rays were used and the image was recorded on X-ray film without intensifying screens; even then, the stagnation powder gas ring surrounding the body of the projectile would not have produced detectable contrast without the addition (to the gunpowder) of 10% barium nitrate which greatly enhanced X-ray absorption.

Flash radiography is also commonly used in intermediate ballistics to observe (1) the separation of sabots and (2) the deployment of fins in projectiles of complex shape which must be housed in a sabot for acceleration in the gun barrel or which require fins to enhance the aerodynamic stability of their trajectories. Figure 25 illustrates the separation of the sabot and the subsequent flight of a high velocity conical projectile. A series of stereo pairs allowed reconstruction of the projectile trajectory and attitude in three dimensions.

External Ballistics

External ballistics is the study of the flight characteristics of projectiles, and is usually the easiest stage to radiograph unless the trajectory is very erratic. Flash radiography can also be used during this

FIGURE 26. Observation of Ablative Effects in Free Flight (Aluminum Sphere, 8 mm Diameter, Shot by a Light Gas Gun at 3,500 ms^{-1} in Air): (a) Photography in Visible Light; (b) X-ray Flash Photograph of Same Sphere at Rest; and (c) X-ray Flash Photograph of Same Sphere in Flight.

(a)

(b)

(c)

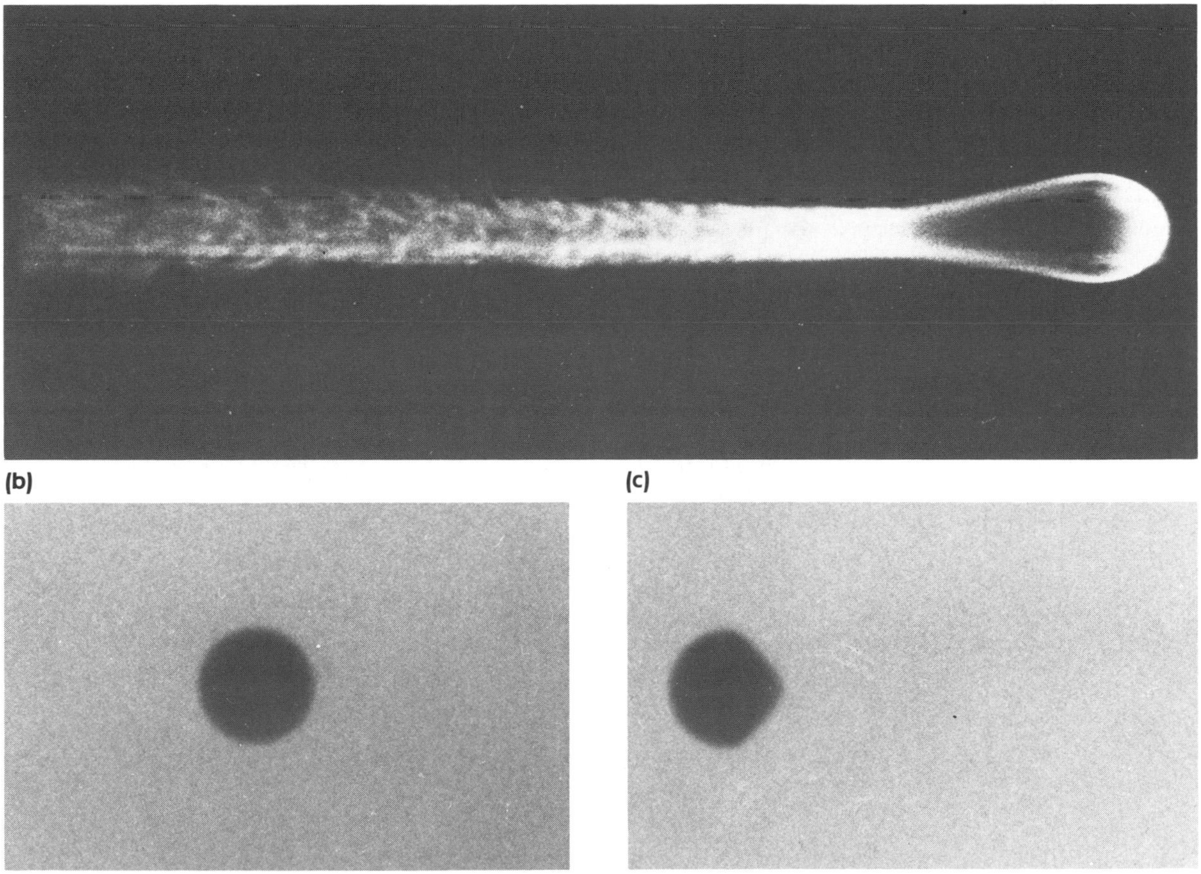

AFTER JAMET. ELSEVIER SCIENTIFIC PUBLISHING © 1976. REPRINTED WITH PERMISSION.

stage to verify displacement of parts inside the projectile. For example, a fuse which was disarmed (for safety purposes before and during firing) must be armed in flight so that the projectile will explode on impact. Flash radiography is also used extensively in external ballistics to study the stability, drag, deceleration or ablation of various objects moving at extremely high velocities through the rarified gas inside a pressurized aeroballistic range; this is done to simulate and study the reentry of space vehicles into the atmosphere. Figure 26 illustrates this type of ablative study[1] by showing a light photograph and a flash radiograph of an 8 mm diameter aluminum sphere fired by a light gas gun into air at a velocity of 3500 meters per second. The light photograph clearly shows the trail of incandescent debris and

FIGURE 27. Penetration of 20 mm High-Explosive Incendiary Projectile Through 3.2 mm (0.125 in.) Steel Plate

Approximately 20 millionths of a second after impact.

The nose of the shell has passed through the plate. There has been no shift of the detonating parts of the fuse.

The shell has started to swell in the region of the bourrelet. The booster has not yet detonated.

Approximately 35 millionths of a second after impact.

The shell has swelled more than the shell pictured above. The maximum swelling is still in the region of the bourrelet. The booster has detonated. The ring of petals about the shell can be seen.

The shell body has swelled to almost twice normal diameter. The petals have constrained the bourrelet region of the shell from swelling as rapidly as the body of the shell. The base of the shell has swelled very little.

Approximately 60 millionths of a second after impact.

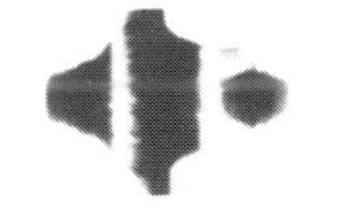

The shell has ruptured. The base of the shell is starting to mushroom out. The constraining influence of the petals is evident.

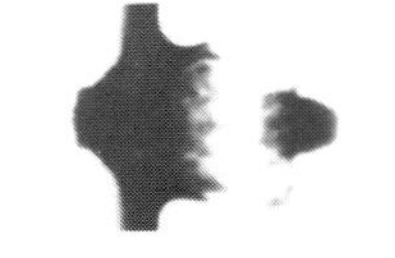

The base of the shell has mushroomed out even further. The nose and some of the shell fragments have passed out of the constraining influence of the petals.

Approximately 100 millionths of a second after impact.

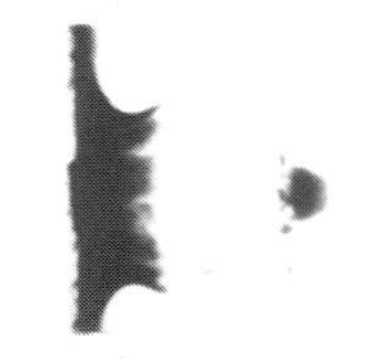

The fragments of the base of the shell are passing through the hole. The petals have curled back to produce a hole approximately 6 cm (2.5 in.) in diameter.

The shell fragments are almost all completely through the plate. The petals have continued to be curled back to produce a hole approximately 15 cm (5 in.) in diameter.

AFTER WEBSTER. REPRINTED WITH PERMISSION.

FIGURE 28. Sequence of Three Flash Radiographs Recorded on a Single Film; Multiple Pulser/Single Tube System as Shown in Fig. 11 Was Used

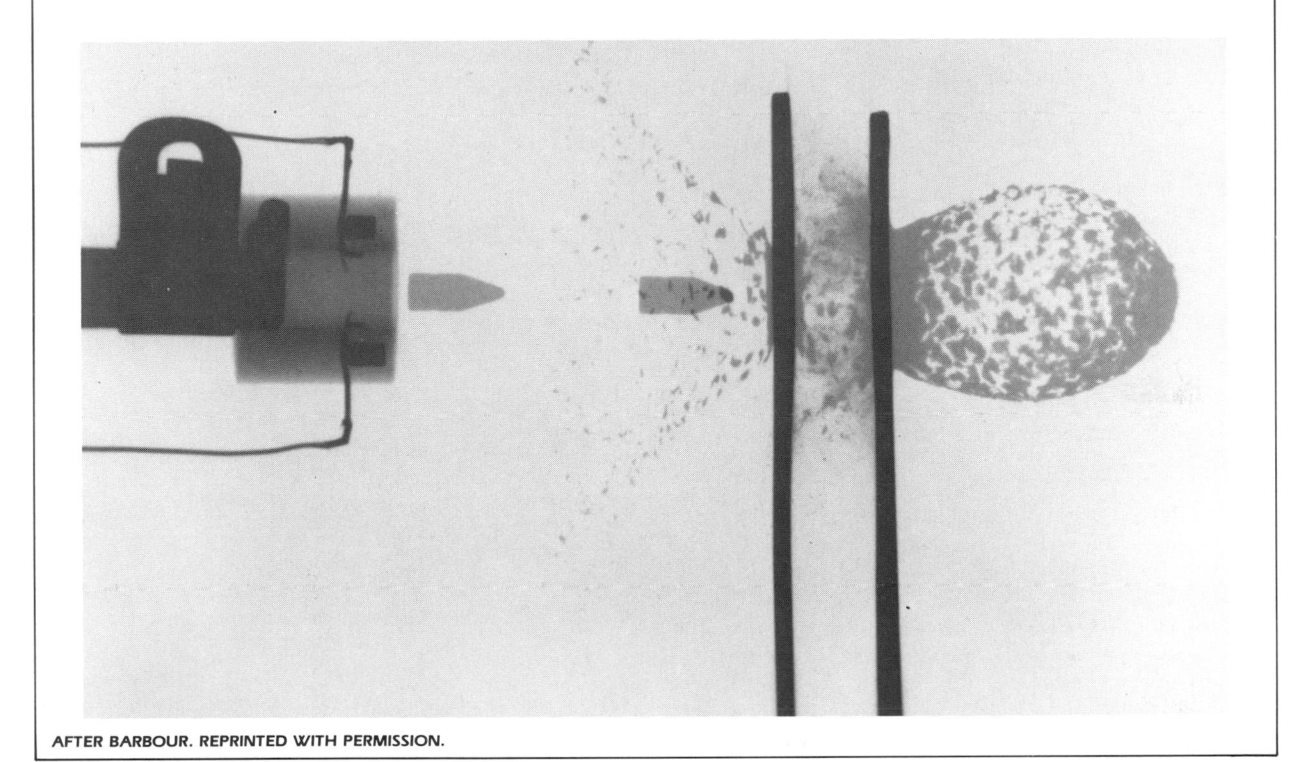

AFTER BARBOUR. REPRINTED WITH PERMISSION.

the luminosity due to ionization of the air, but only the flash radiograph can show the profile of the eroded sphere in flight.

Terminal Ballistics

This is the study of the impact of different projectiles on various targets. It is perhaps the most difficult to observe of all the phases of ballistics. Light flash photography is generally not practical because the impact produces intense light, smoke and debris. Flash X-ray tube and film protection is particularly difficult because of blast waves, fragments and ground shocks, and the point of impact may be uncertain unless the projectile trajectory is precisely controlled. On the other hand, the flash radiograph can generally be timed precisely at the desired instant by locating a sensor on the back of the target and setting the desired delay on the trigger amplifier which fires the flash X-ray system. Figure 27 presents a sequence of flash radiographs[58] showing the penetration of armor plate by a 20 mm projectile. The event was highly reproducible and each picture in the sequence was obtained from a different firing, with gradually increasing delays from the time of impact. Figure 28 illustrates a complete three-frame sequence radiograph obtained with three flash X-ray pulsers connected to a single flash X-ray tube through cine diodes, a technique discussed earlier and illustrated in Fig. 11. The first two flash radiographs show the bullet in flight in two successive positions, allowing determination of the bullet attitude and velocity. The third flash radiograph was taken after the bullet had penetrated two lead plates and shows the fragmentation pattern.

FIGURE 29. Series of Shaped Charge Radiographs: v_j = 9000 ms^{-1}

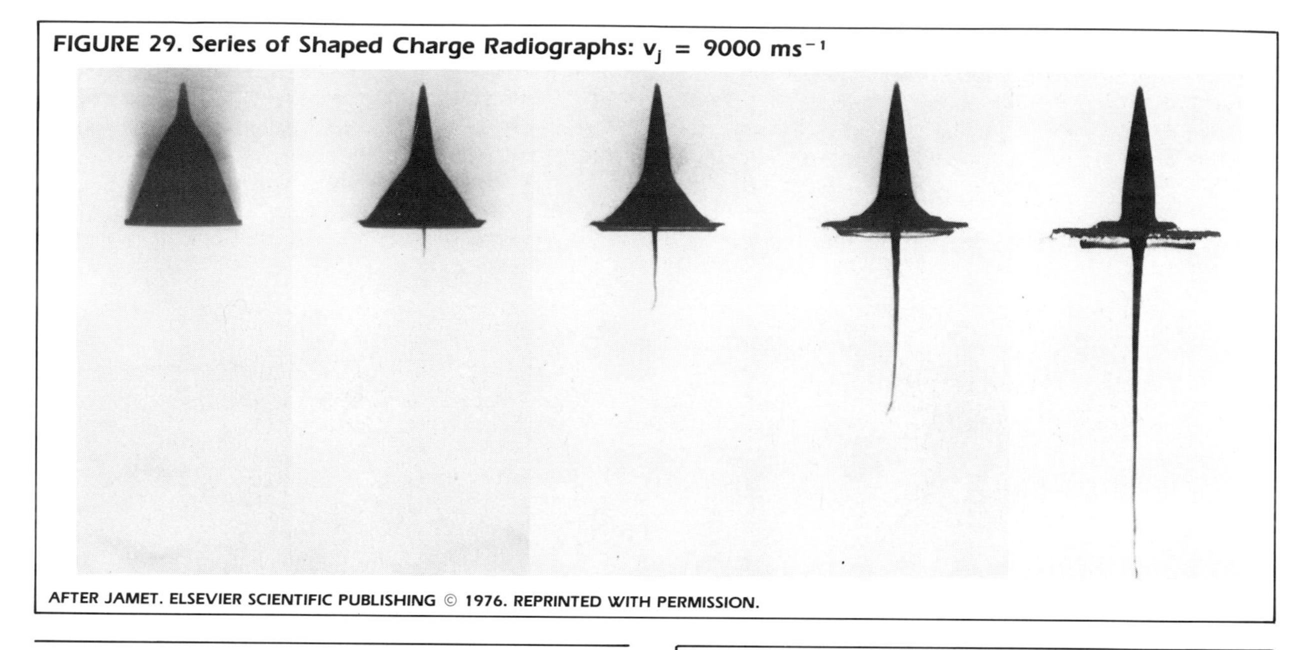

AFTER JAMET. ELSEVIER SCIENTIFIC PUBLISHING © 1976. REPRINTED WITH PERMISSION.

Detonation Phenomena

Flash radiography is perhaps the most effective technique for observation of detonics processes (the formation, propagation velocity and intensity of detonation waves, and related effects such as formation and propagation of compression or shock waves in solid, liquid or gaseous media). Shaped charges provide a good illustration of this type of study. Figure 29 shows a sequence of flash radiographs[1] of the formation of a shaped charge jet. The sequence was compiled from individual 20 ns exposures, using the radial tube/film arrangement shown in Fig. 10.

Figure 30 illustrates the penetration of a copper shaped charge jet into an aluminum target, clearly showing the cavity produced, the zone of compressed material and the shockwave propagating outward from the cavity.

FIGURE 30. Penetration of a Shaped Charge Jet in a Duraluminum Target

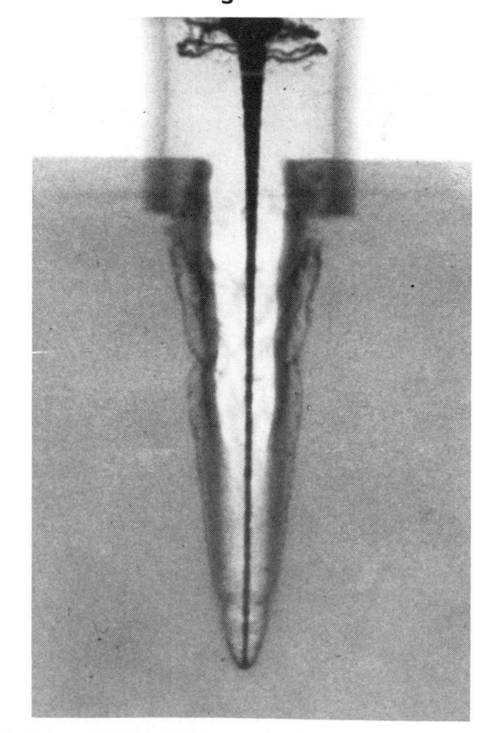

AFTER JAMET. ELSEVIER SCIENTIFIC PUBLISHING © 1976. REPRINTED WITH PERMISSION.

Industrial Applications

Industrial applications are not yet as widespread as applications to ballistics and detonics. A dynamic study of liquid-filled high voltage power switches (10 kV, 600 A) was made to investigate arc initiation and quenching during switch opening.[59]

Arc welding[60] and electron beam welding[61] have also been studied by flash radiography and flash cineradiography. Another application is to the metal casting process, as illustrated in Fig. 31. This sequence of four radiographs was obtained with the 600 kV slow cineradiography system shown in Fig. 13. The four radiographs illustrate successful phases during a test pour of a multiple cavity shell mold for steel fittings, and shows the gating characteristics as well as casting defects due to premature partial filling of the upper cavities (arrows).

High voltage cine flash radiography has also been

FIGURE 31. Applications of High Voltage Slow Cine Flash Radiography System to Study Casting Process: Sequential Radiographs of the Filling of a Multi-Cavity Shell Mold

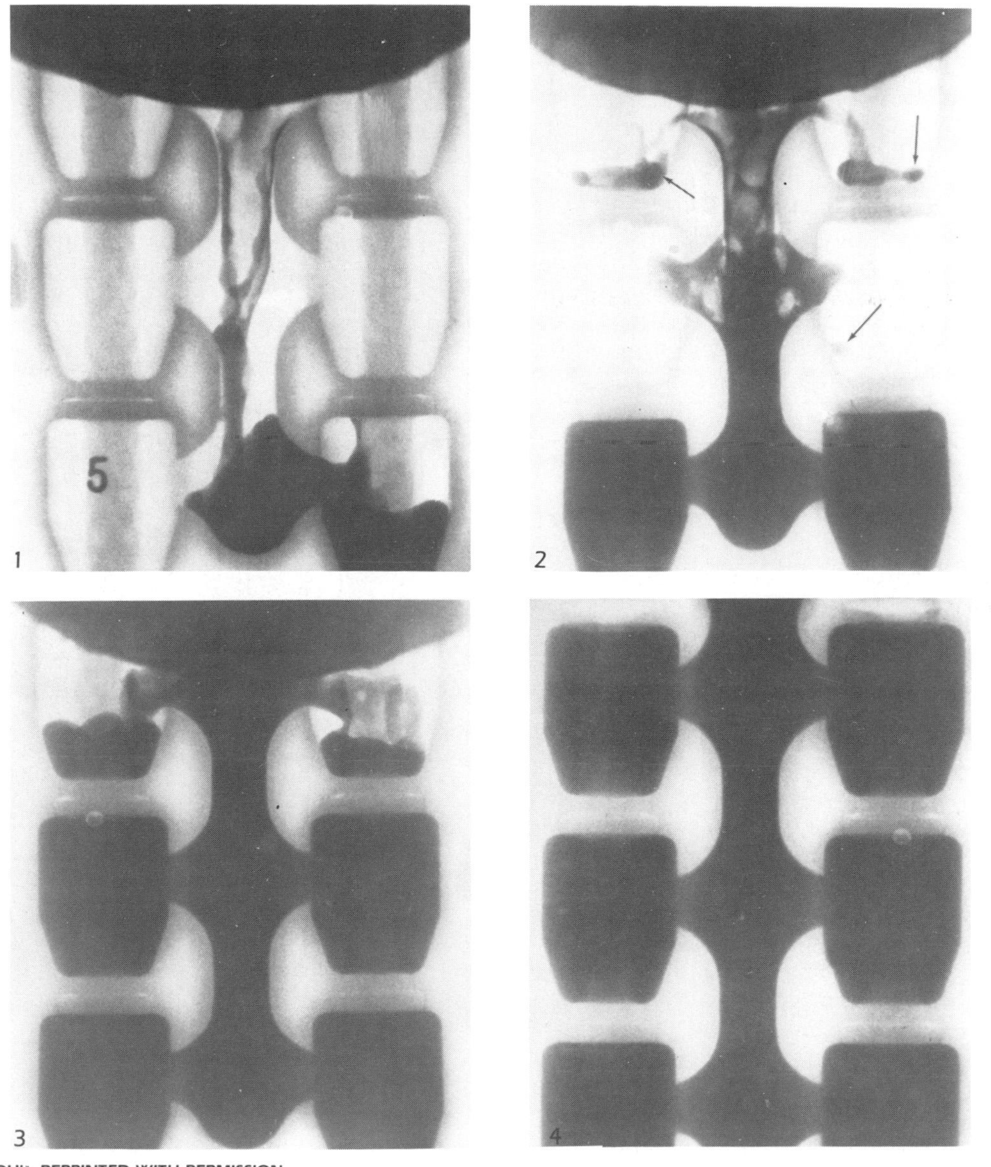

AFTER BARBOUR. REPRINTED WITH PERMISSION.

used successfully to study dynamic conditions inside gas turbines and jet engines.[62] The main purpose of such studies is to obtain accurate engineering data on clearances between the rotating and static parts of gas turbines, at all steady state or transient running conditions, between cold static and maximum power. A 2 MV flash X-ray system capable of delivering five pulses in 15 seconds was used in early experiments on large jet engines. Better results were subsequently obtained using a linear accelerator as the pulsed X-ray source for an X-ray image intensifier coupled to a 16 mm rotating prism camera for image recording. These studies led to the development of a multispectral imaging system[62] which can operate at very high voltage (8 MV) or at relatively low voltage depending on the application. The high voltage configuration uses a linear accelerator as the X-ray source and the special high energy X-ray image intensifier previously discussed (also, see ref. 40). The low voltage configuration uses an industrial DC X-ray system (360 kV, 10 mA or 200 kV, 70 mA) and a gated X-ray image intensifier.

Biomedical Applications

Flash radiography has been used to study internal organ displacement under the effect of strong acceleration, and for studies of crash injuries,[63] as illustrated in Fig. 32. The flash cineradiography system shown in Fig. 15 was used to study the displacement of internal organs under impact, and particularly brain displacement under lateral head impact. The motion was made visible to X-rays by injecting a contrast medium into the carotid arteries a fraction of a second before impact, so that the major blood vessels in the brain could be followed after impact and used to measure displacements. Another high speed flash cineradiography system has been developed for biomechanics impact research.[64] The system consists of a DC X-ray source, a fluorescent screen, a gated light image intensifier and a 35 mm rotating prism camera capable of 2500 frames per second. Because of the DC X-ray source, the penetration capability of the system is very limited at high frame rates.

Nuclear Technology

High speed radiography has been applied to nuclear technology, particularly to the imaging of stainless steel-clad fuel elements. Another important application is the investigation of fuel-coolant interactions in safety studies of liquid sodium-cooled fast breeder reactors. Both the flash X-ray system (250-350 kVp, 1,000 pulses per second) and the DC X-ray source[65] (420 kV, 15 mA) have been used in these studies. The imaging system consists of an X-ray image intensifier (gated when used with the DC X-ray source) and a 16 mm pin-registered framing camera or rotating prism camera. Both systems have produced useful cineradiographs of simulated safety failures (fuel-coolant interactions) in steel pipe test sections.

FIGURE 32. Application of High Voltage Fast Cineradiography System to Study of Crash Injury: Sequential Radiographs During Lateral Head Impact. Blood Vessels Made Visible by Injection of Contrast Medium In the Carotids

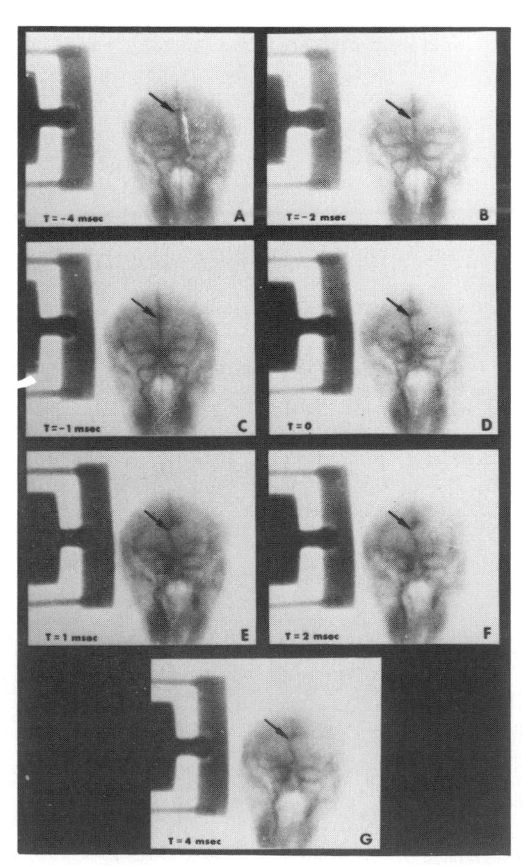

AFTER SHATSKY. REPRINTED WITH PERMISSION.

X-ray Backlighting for Pellet Implosion Studies

One of the elements of controlled nuclear fusion research involves the study of pellet implosion to produce high density, high temperature plasmas. The pellets are small hollow glass, plastic or metal spheres filled with a deuterium-tritium gas mixture under pressure. Bombardment by intense, short pulse ion beams or, more commonly, laser beams causes the pellet to implode, generating the very high densities, temperatures and pressures required to initiate thermonuclear fusion reactions. Flash radiography of imploding pellets, using an external X-ray source (backlighting), is exceptionally difficult because it requires extremely short pulses, very precise timing and very high resolution. The three-channel high voltage system described earlier[42] meets these requirements and has been used successfully for flash radiography of high density metal pellets. Very low energy X-rays are required for glass and plastic pellets and it is possible to produce the necessary flash X-ray source characteristics by bombarding a metal target with a very short duration, very high power, sharply focused laser beam; this evaporates the target and produces a plasma which radiates X-rays. The X-rays emitted are predominantly L shell characteristic radiation from the target and the X-ray energy is thereby controlled through choice of the target material. In this manner a 0.1 mm, 1.5 keV, 50 picosecond duration plasma source has been produced[66] by bombarding a brass target with a 50 picosecond, 1 joule glass laser pulsed beam. The technique was improved by using two separate laser-bombarded targets (copper and molybdenum) to produce two X-ray microsources 250 picoseconds apart and at energies of 1.2 and 2.6 keV.[67] By using a small copper wire instead of a plane target, a point X-ray source only 6 micrometers in diameter was produced.

FIGURE 33. Flash X-ray Transmission Patterns Obtained from an Exploding Gold Foil: (a) Before Explosion; (b) 1.3 μs After Discharge; (c) 1.5 μs After Discharge; and (d) 1.8 μs After Discharge

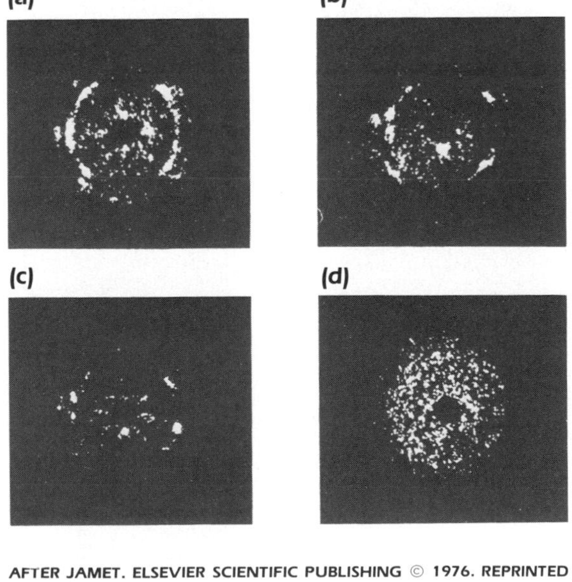

(a) (b)

(c) (d)

AFTER JAMET. ELSEVIER SCIENTIFIC PUBLISHING © 1976. REPRINTED WITH PERMISSION.

Flash X-ray Diffraction

X-rays are used to record diffraction patterns that yield detailed information about the crystal structure, texture, residual stresses and defect distribu-

FIGURE 34. Laue Transmission X-ray Diffraction Pattern from a Sapphire Crystal Using a Single 70 ns Pulse of 150 kV X Radiation from a Tungsten Target Tube; the Photograph Was Recorded Directly from the Image Intensifier Output onto Instant Film

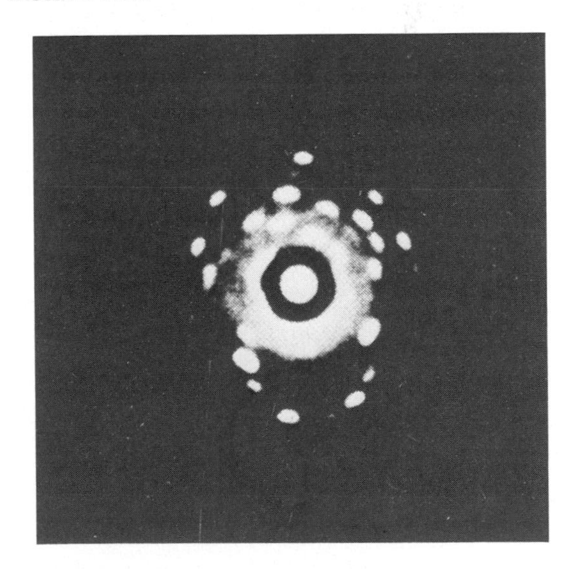

AFTER GREEN. REPRINTED WITH PERMISSION.

tion of various materials. X-ray diffraction studies normally use fine beams of low energy, continuous or monochromatic X rays and require the long exposure times which restrict such studies to static situations. However, it is of great interest to obtain information about the crystal structure of materials under very high dynamic stresses or other rapidly changing conditions which could produce significant changes in the crystalline state. Intensive efforts have been made to produce the very high intensity, short pulse, low energy X-ray beams required for flash X-ray diffraction with very short exposure times. These efforts have been at least partly successful, and useful flash X-ray diffraction patterns have been obtained, though the quality of the patterns is generally not as good as that obtained with long exposures

FIGURE 35. Comparison of Debye-Scherrer Diffraction Patterns of a 20 μm Thick Copper Foil. Taken with a Molybdenum Anode Tube and Different Pulse Generators; Patterns Recorded with Instant Film Cassettes: (a) Capacitor Discharge Pulser; (b) Blumlein Transmission Line; (c) Cable Loop Generator, 3 Loops; and (d) Cable Loop Generator, 7 Loops

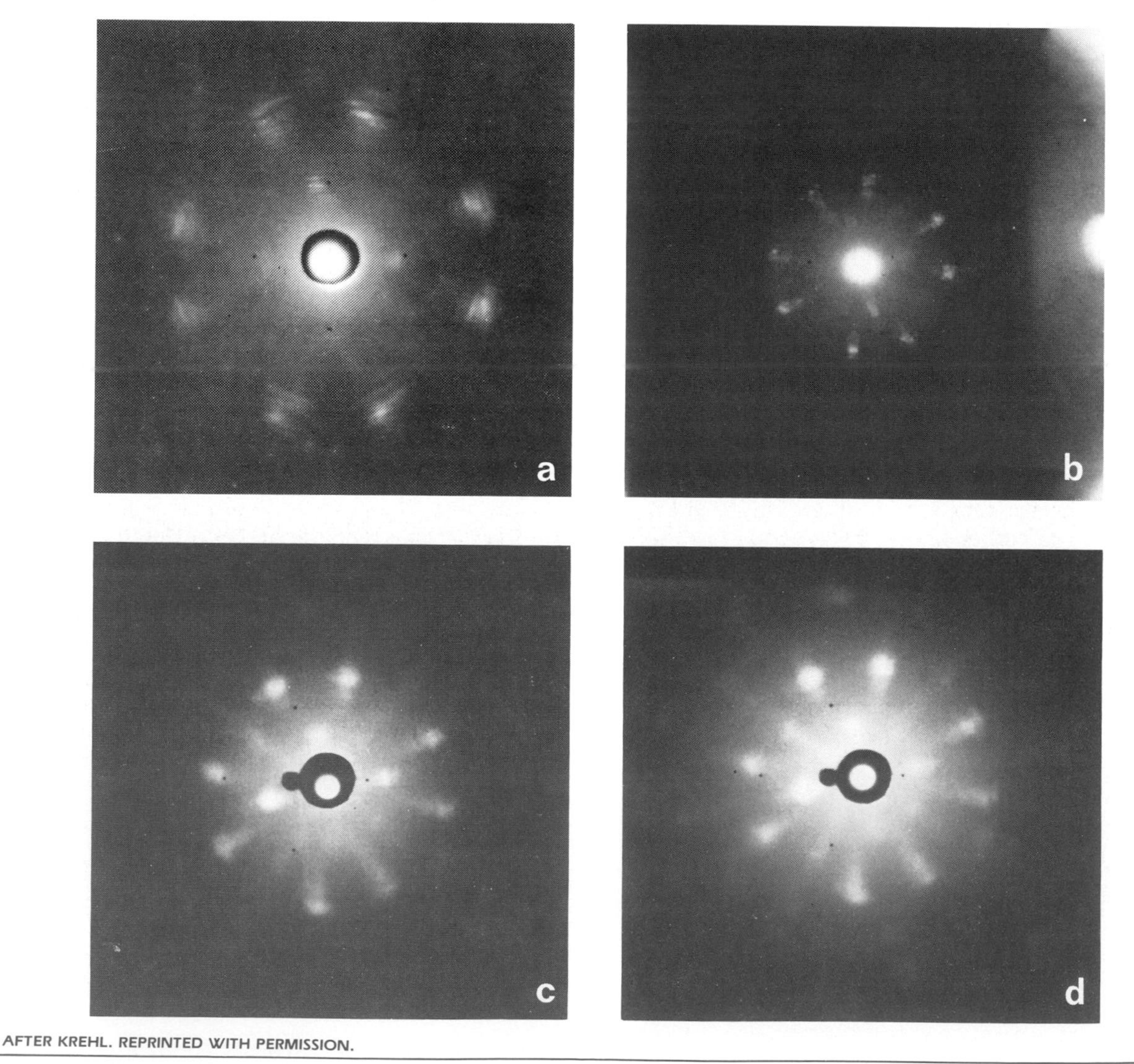

AFTER KREHL. REPRINTED WITH PERMISSION.

under static conditions. As early as 1942, it was possible to photograpically record Laue diffraction patterns of single aluminum crystals using a pulsed X-ray generator (60 kV capacitor discharge, 1 millisecond duration). In the 1950s, successful recordings were made of single crystal and powder patterns with exposure times of approximately 1 microsecond.

In 1967, a very effective source for flash X-ray diffraction was designed, using a low voltage Blumlein design to produce very intense, short duration, low energy X-ray beams (85 kV, 40 kA, 30 ns). The source was used to produce flash X-ray patterns of a variety of materials under shock wave compression.[71, 72]

In 1972, several reports were made at the 10th International Congress on High Speed Photography and Photonics in Nice, on independent flash X-ray diffraction studies with 20-50 ns time resolution. One report detailed the use of a fluorescent screen, lens-coupled to a multiple stage light image intensifier tube to enhance the intensity of the patterns. Figure 33 shows transmission powder patterns of an exploding gold foil 5 micrometers thick.[73]

In 1973, a commercial flash X-ray system (with a beryllium window tube) coupled with an X-ray image intensifier detection system was used to record good quality transmission Laue patterns of single crystals, as illustrated in Fig. 34.[74] Powder patterns of polycrystalline aggregates were also produced with this system. In 1975, a flash X-ray tube (using a molybdenum target to produce high intensity molybdenum characteristic K_α radiation at 17 keV) was used to obtain simultaneous flash radiographs and transmission Laue patterns of an aluminum shaped charge jet.[75] Pattern analysis showed that the jet was composed of cold-worked, particulate solid with a size distribution ranging from 0.01 to 1 mm.

A still more powerful and flexible source for flash X-ray diffraction has been built.[76] The pulser consists of several (up to 10) low impedance coaxial high voltage cables connected in parallel. The number of cable loops in parallel determines the pulser impedance, which can be reduced below 1 ohm. A special continuously pumped flash X-ray tube, with replaceable electrodes and a beryllium window, was designed to match the pulser. Figure 35 illustrates the quality of the Debye-Scherrer diffraction patterns which can be obtained using this system.

CONCLUSION

It can be said, in summary, that successful flash radiography demands the good practices of conventional radiography as well as its own specialized practices (triggering and protection from potentially violent test environments, for example).

Flash X-ray equipment and techniques are now available for obtaining single or sequential images. A variety of flash X-ray generators have been built to provide different degrees of penetration, intensity, portability and pulse duration. Additional flexibility results from the fact that flash X-ray image recording may be done on X-ray film with the help of fluorescent screens or with image converter camera/photographic film techniques.

The flash X-ray method is taking its place as a valuable diagnostic tool in the study of high speed dynamic events; compared with photographic techniques, flash radiography is essential when desired detail is internal or where detail is obscured by light, smoke or debris.

Useful applications of the flash radiographic method are constantly expanding and presently include such diverse uses as: ballistics; detonics; biomedical, mechanical and nuclear applications; and dynamic diffraction studies. All of these factors combine to make flash radiography a practical and increasingly valuable testing technique.

REFERENCES

1. Jamet, F., and G. Thomer. *Flash Radiography*. Elsevier Publishing Company (1976).

2. Bryant, L.E., ed. *Proceedings of the Flash Radiography Symposium*. Houston, TX: American Society for Nondestructive Testing National Conference (1976)

3. Marilleau, J., ed. *Proceedings of the First European Conference on Cineradiography with Photons or Particles*, Paris, 1981. Society of Photo-Optical Instrumentation Engineers (1983).

4. Kingdon, K.H., and H.E. Tanis. *Physical Review*. Vol. 53 (1938): pp 128.

5. Steenbeck, M. *Wissenschaftlische Veroffnete Siemenswerke*. Vol. 17 (1938): pp 363-380.

6. Schaaffs, W. *Zeitschrift fur Angevandte Physik*. Vol. 1 (1949): pp 462.

7. Schaaffs, W. *Ergebnisse Exakten Naturwissenschaften*. Vol. 28 (1954): p 1.

8. Slack, C.M., and L.F. Ehrke. *Journal of Applied Physics*. Vol. 12 (1941): pp 165. Also, "Microsecond Radiography and its Development." American Society for Testing and Materials Bulletin (1948): pp 59-72.

9. Criscuolo, E.L., and D.T. O'Connor. *Review of Scientific Instrumentation*. Vol. 24 (1953): pp 944.

10. Dyke, W.P. "X-ray Tubes with Field Emission Cold Cathodes for Flash Radiography." *Encyclopedia of X Rays and Gamma Rays*. Reinhold Publishing Company (1963).

11. Dyke, W.P. *Scientific American*. Vol. 210 (1964): pp 108.

12. Dyke, W.P., and J.K. Trolan. *Physical Review*. Vol. 89 (1953): pp 799-808.

13. Dyke, W.P., et al. *Physical Review*. Vol. 91 (1953): pp 1043-1057.

14. Charbonnier, F.M., et al. *Journal of Applied Physics*. Vol. 38 (1967): pp 627-640.

15. Cranberg. L. *Journal of Applied Physics*. Vol. 23 (1952): pp 23.

16. Charbonnier, F.M., et al. *Radiology*. Vol. 117 (1975): pp 165-172.

17. Marx, E. *E.T.Z.* Vol. 45 (1924): pp 652.

18. Blumlein, A.D. British Patent No. 589,127 (1947).

19. Dick, R.D. "PHERMEX." *Proceedings of the Flash Radiography Symposium*, L.E. Bryant, ed. American Society for Nondestructive Testing (1976): pp 43-58.

20. Brewster, J.L. "Current Flash X-ray Equipment." *Proceedings of the Flash Radiography Symposium*, L.E. Bryant, ed. American Society for Nondestructive Testing (1976): pp 15-22.

21. Mattson, A. "Flash X-ray Systems." *Proceedings of the First European Conference on Cineradiography with Photons of Particles*, J. Marilleau, ed. Society of Photo-Optical Instrumentation Engineers (1983): pp 270-274.

22. Bichenkov, E.I., et al. "Soft X-ray Unit PIR 100" and "X-ray Unit PIR-600M." Moscow: Fourteenth International Congress on High Speed Photography and Photonics (ICHSPP) (1980).

23. Uyemura, T. "Recent Developments in High Speed Photography in Japan and China." San Diego: Fifteenth ICHSPP (1982).

24. Bracher, R.J. "Technology for Flash Radiography of Automobile Tires in Motion." San Diego: Fifteenth ICHSPP (1982): pp 798.

25. Bryant, L.E. "Status of Flash Radiography in the USA and Future Possibilities." Tokyo: Thirteenth ICHSPP (1978): pp 401-404.

26. Bahl, K.L., and H.C. Vantine. "A Flash Radiographic Technique Applied to Fuel Injection Sprays." *Proceedings of the Flash Radiography Symposium.* L.E. Bryant, ed. American Society for Nondestructive Testing (1976): pp 193-200.

27. Krehl, P. "Low Impedance High Intensity Flash Soft X-ray Machine." Tokyo: Thirteenth ICHSPP (1978): pp 409-413.

28. Gilbert, J.F., and Carrick. *British Journal of Nondestructive Testing.* Vol. 16 (1974): pp 65.

29. D'A Champney, P., and P.W. Spence. "Pulsed Electron Accelerators for Radiography and Irradiation." *Proceedings of the Flash Radiography Symposium.* L.E. Bryant, ed. American Society for Nondestructive Testing (1976): pp 23-42.

30. Buchet, J., et al. "Generateur De Radiographie Eclair a Haute Energie 'Grec'. *Proceedings of the First European Conference on Cineradiography with Photons or Particles*, J. Marilleau, ed. Society of Photo-Optical Instrumentation Engineers (1983): pp 240-245.

31. Bryant, L., et al. "High Energy Cineradiographic Experiment." Tokyo: Thirteenth ICHSPP (1978): pp 431-435.

32. Hauducoeur, A., et al. "Installation de Cineradiographie a Haute Energie 'Artemis'." *Proceedings of the First European Conference on Cineradiography with Photons or Particles*, J. Marilleau, ed. Society of Photo-Optical Instrumentation Engineers (1983): pp 256-261.

33. Byrant, L., et. al. "X-ray Film/Screen Study for Flash Radiography." Houston: ASNT National Conference (1980).

34. Bergon, J.C., and J. Constant. "Application of Electro-optical Recording Equipment to Flash X-ray Studies of Explosive Processes." Denver: Ninth ICHSPP (1970): pp 278-282.

35. Trimble, J.J., and C. Aseltine. "Flash X-ray Cineradiography at 100,000 Frames per Second." San Diego: Fifteenth ICHSPP (1982): pp 688-695.

36. Viguier, P., and G. Bourdarot. "High-Speed Cineradiography at High Energy." Denver: Ninth ICHSPP (1970): pp 315-320.

37. Bracher, R.J. "Development of a Nine Frame Image Converter Camera for Cineradiography at 250,000 Frames per Second." San Diego: Fifteenth ICHSPP (1982): pp 797.

38. Baikov, A.P., et al. "X-ray Television System for Pulse Image Visualization." Denver: Ninth ICHSPP (1970): pp 283-286.

39. Charbonnier, F.M., et al. "Design of High Voltage High Frame Rate Cine Flash Radiography System." London: Eleventh ICHSPP (1974): pp 348-355.

40. Driard, B. "X-ray Image Intensifiers for High Speed Cineradiography." Tokyo: Thirteenth ICHSPP (1978): pp 423-427.

41. Bryant, L.E. "Flash Radiographic Techniques: Exposure, Recording, Triggering, and Film Protection." *Proceedings of the Flash Radiography Symposium*, L.E. Bryant, ed. American Society for Nondestructive Testing (1976): pp 59-73.

42. Chang, J., et al. "Three-Frame FXR Systems for Ion Beam Pellet Implosion Studies." San Diego: Fifteenth ICHSPP (1982): pp 696-699.

43. Thomer, G., and F. Jamet. "A Tube with Six Sources for Soft X-ray Flash Radiography." Stockholm: Eighth ICHSPP (1968): pp 256-258.

44. Barbour, J.P., et al. "Cineradiography at 10^5 Frames Per Second with a Single X-ray Tube." Zurich: Seventh ICHSPP (1964).

45. Jamet, F., and F. Hatterer. "Flash X-ray Cinematography at Frame Rates up to 4×10^7 Images per Second — Applications to Terminal Ballistics." *Proceedings of the First European Conference on Cineradiography with Photons or Particles*, J. Marilleau, ed. Society of Photo-Optical Instrumentation Engineers (1983): pp 164-167.

46. Barbour, J.P., et al. "Cineradiography Using Field Emission Technology." Stockholm: Eighth ICHSPP (1968): pp 237-239.

47. Thomer, G., and A. Stenzel. "Studies of X-ray Cineradiography with Soft X-rays." Darmstadt: Fourth ICHSPP (1958).

48. Thomer, G., and A. Stenzel. "X-ray Flash Cinematography up to 12,000 Images per Second." Washington: Fifth ICHSPP (1960): pp 173-175.

49. Frungel, F., et al. "High Speed X-ray Flash Cinematography of Small Objects." Washington: Fifth ICHSPP (1960): pp 170-172.

50. Frungel, F. "Repetitive Sub Microsecond Light and X-ray Flash Techniques and Applications in Precision Analysis of Motion and Flows." Tokyo: Thirteenth ICHSPP (1978): pp. 37-48.

51. Moscalev, V.A., et. al. "High Speed Cineradiography by High Current Betatrons." Tokyo: Thirteenth ICHSPP (1978): pp 415-418.

52. Huston, A.E., and P.A.E. Stewart. "Cineradiography with Continuous X-ray Sources." Toronto: Twelfth ICHSPP (1976).

53. Bryant, L.E. Private communication: to be published.

54. Charbonnier, F., et. al. "New Tubes and Techniques for Flash X-ray Diffraction and High Contrast Radiography." Nice: Tenth ICHSPP (1972): pp 356-361.

55. Bracher, R.J., and H.F. Swift. "Low Voltage Flash Radiography." Houston: ASNT National Conference (1980).

56. Brewster, J.L., et al. "Three Nanosecond Exposures with X rays, Electrons and Light." Stockholm: Eighth ICHSPP (1968): pp 240-244.

57. Brewster, J.L., et al. "A New Technique for Ultra-Bright Nanosecond Flash Light Generation." Denver: Ninth ICHSPP (1970): pp 303-309.

58. Webster, E.A. "Flash X-ray Studies of Ballistic Phenomena at Frankford Arsenal." *Proceedings of the Flash Radiography Symposium*, L.E. Bryant, ed. American Society for Nondestructive Testing (1976). Also, San Diego: Fifteenth ICHSPP (1982): pp 682-687.

59. Schaaffs, *Messtechnik.* Vol. 5, No. 70 (1970): pp 85-92.

60. Stenzel, A., and G. Thomer. Journal of Society of Motion Picture and Television Engineers (SMPTE). Vol. 70 (1961): pp 18-20.

61. Bryant, L.E. "Flash Radiography of Electron Beam Welding." *Materials Evaluation.* Vol. 29, No. 10 (1971): pp 237-240.

62. Pullen, D., and P. Stewart. "Some Applications of Cineradiography to Gas Turbines." *Proceedings of the First European Conference on Cineradiography with Photons or Particles*, J. Marilleau, ed. Society of Photo-Optical Instrumentation Engineers (1983): pp 40-49.

63. Shatsky, S.A. "Flash X-ray Cinematography During Impact Injury." Seventeenth Stapp Car Crash Conference (1974).

64. Bender, M., et al. "High Speed Cineradiographic System for Biomechanics Impacts Research." *Proceedings of the First European Conference on Cineradiography with Photons or Particles*, J. Marilleau, ed. Society of Photo-Optical Instrumentation Engineers (1983): pp 130-135.

65. Will, H., et al. "Investigation of Material Movement Within Steel Test Sections by Means of High Speed X-ray Photography." *Proceedings of the First European Conference on Cineradiography with Photons or Particles*, J. Marilleau, ed. Society of Photo-Optical Instrumentation Engineers (1983): pp 16-21.

66. Launspach, J. "Imagery of Laser Imploded Targets in the X-ray Domain." *Proceedings of the First European Conference on Cineradiography with Photons or Particles*, J. Marilleau, ed. Society of Photo-Optical Instrumentation Engineers (1983): pp 101-109.

67. Yamanaka, C., et al. "Ultrashort X-ray Probe Backlighting for Implosion Fusion Experiments." San Diego: Fifteenth ICHSPP (1982): pp 783-788.

68. Tsukerman, V., and A. Avdeenko. *Zh. Tekhn. Fiz.* Vol. 12 (1942): pp 185-196.

69. Schall, R. "Feinstrukturaufnahmen in Ultrakurzen Zeiten Mit Dem Rontgenblitzrohr." *Zeitschrift fur Angevandte Physik.* (1950): pp 83-88.

70. Schaaffs, W. "Eine Rontgenblitzrohre Zur Erzeugung Von Rontgeninterferenzen in Einer Mikrosekunde." *Zeitschrift fur Angevandte Physik.* Vol. 8 (1956): pp 299-302.

71. Q. Johnson, et al. "X-ray Diffraction Experiments in Nanosecond Time Intervals." *Nature.* Vol. 213 (1967): pp 1114-1115.

72. Johnson, Q., and A. Mitchell. "First X-ray Diffraction Evidence for Transition During Shock-Wave Compression." *Physical Review.* Vol. 29 (1972): pp 1369-1371.

73. Jamet. F. "Recording of X-ray Diffraction Patterns Using Flash X-rays and an Image Intensifier." Journal of the Society of Motion Picture and Televsion Engineers. Vol. 80 (1971): pp 900-901.

74. Dantzig, J., and R.E. Green. "Flash X-ray Diffraction Systems." *Advances in X-ray Analysis.* Vol. 16. New York: Plenum Press (1973): pp 229-241.

75. Green. R.E. "First X-ray Diffraction Photograph of a Shaped Charge Jet." *Review of Scientific Instrumentation.* Vol. 46 (1975): pp 1257-1261.

76. Krehl, P. "A New Low Impedance High Intensity Flash Soft X-ray Machine." Tokyo: Thirteenth ICHSPP (1978): pp 409-413.

SECTION 12

NEUTRON RADIOGRAPHY

Harold Berger, Industrial Quality, Inc., Gaithersburg, MD
Darrell C. Cutforth, Argonne National Laboratory, Idaho Falls, ID
Donald A. Garrett, National Bureau of Standards, Washington, DC
Jerry Haskins, Lawrence Livermore National Laboratory, Livermore, CA
Frank Iddings, Louisiana State University, Baton Rouge, LA
Richard L. Newacheck, Aerotest Operations, Inc., San Ramon, CA

INTRODUCTION

Lawrence E. Bryant, Jr.
University of California
Los Alamos National Laboratory
Los Alamos, New Mexico 87545

Neutron radiography is a valuable nondestructive testing technique which ideally complements conventional radiography. The first publications covering neutron radiography found it convenient to compare a neutron radiograph, side-by-side, with an X-ray image, to point out the benefits of the neutron image. This sometimes gave the impression that the two techniques were competitive; happily, this has turned out *not* to be the case. Neutron radiography is now a widely accepted, specialized testing technique. Sometimes an object can be most thoroughly analyzed radiographically with both neutrons *and* X (or gamma) rays.

This Section of the NDT Handbook is a useful reference to practicing neutron radiographers and can serve as an introduction for students or conventional radiographers as well. The text includes helpful discussions on: neutron sources; moderation; collimation; techniques for neutron radiography; neutron imaging methods; and reference material concerning regulatory control, neutron radiography standards and cross sections. A discussion of applications is also included, with some well-illustrated examples.

All of the authors for this Section deserve the gratitude of the technical community, the Handbook Editor and the Handbook Coordinator. Special thanks go to Harry Berger, Industrial Quality, Inc., for his service as primary author and contact with the publications staff.

The authors in turn extend their thanks to John P. Barton (N-Ray Engineering) for his helpful review of the chapter, to Roger A. Morris (Los Alamos National Laboratory) for permission to use the table of thermal neutron cross sections and attenuation coefficients, and to the American Society for Testing and Materials for permission to reprint the table of thermal neutron cross sections and attenuation coefficients (Table 9), and the Figure on half-value layers (Fig. 10).

PART 1

PRINCIPLES OF NEUTRON RADIOGRAPHY

Development

Radiography with thermal neutrons can be traced to the mid-1930s, shortly after the discovery of the neutron. Research work in this field has carried through to the present time, with a significant increase in developmental activity since 1960. Commercial interest in thermal neutron radiography began in the mid-1960s.

Principles

Neutron radiography extends the ability to image the internal structure of a specimen, beyond what can be accomplished with photon (X and gamma) radiation. Similarities, as well as obvious differences, exist when neutron radiography is compared to photon radiographic techniques. Similarities include the ability to produce a visual record of changes in density, thickness and composition of a specimen. Indeed, the neutron radiograph can look very much like a photon radiograph.

Advantages

It is the differences between the techniques which provide the advantages of neutron radiography over photon radiography. The major difference is the way in which neutrons are removed from the inspection beam by the specimen. Neutrons interact only with the nuclei of the atoms in the specimen; the neutrons may be scattered or absorbed by the atomic nuclei. Because the neutron interactions involve nuclei rather than the numerous orbital electrons, marked differences between the transmission of neutrons and the transmission of photons through a specimen may take place.

Neutron Transmission

Mathematically the relationship for neutron transmission looks much like that for photons, but the variation of the action site (electron orbits or nucleus) produces large differences in the amount of transmitted beam.

For photons:

$$I = I_o\, e^{-\mu_x t} \qquad \text{(Eq. 1)}$$

For neutrons:

$$I = I_o\, e^{-N\sigma t} = I_o e^{-\mu_n t} \qquad \text{(Eq. 2)}$$

Where I is the transmitted intensity; I_o is the incident intensity; μ_x is the linear attenuation coefficient for photons; t is the thickness of specimen in the beam path; N is the number of atoms per cubic centimeter; σ is the neutron cross section of the particular material or isotope (a probability or effective area); and, μ_n is the linear attenuation coefficient for neutrons ($N\sigma$).

Figure 1 provides a comparison of the change in attenuation with increasing atomic number for X-rays (125 kV) and thermal (0.025 eV) neutrons. Such comparisons indicate some of the advantages of using neutrons for radiography. One advantage is found in the imaging of certain low atomic number materials in some high atomic number matrices. Photon radiography works best for the opposite circumstances. Neutrons can image a high cross section element in a low cross section matrix (an element's cross section is its total probability per atom for scattering or absorbing a unit of applied energy), such as cadmium in tin or in lead. Even changes in the isotopic composition of some elements can be imaged because cross sections of the isotopes may be different.

Radioactive Objects

An additional advantage of neutron radiography is its ability to radiograph specimens that are intense sources of photons (radioactive specimens). The neutrons transmitted through a radioactive specimen will strike a metal detection foil such as indium, dysprosium or gold, rather than a converter screen with film. Atomic nuclei in the metal screen absorb neutrons to produce short-lived radioactive isotopes. After removal from the neutron beam, the decay of radioisotopes in the screen exposes a film, giving an *autoradiograph* of the specimen.

Imaging

Neutron radiographs usually are recorded on conventional X-ray film. Although neutrons have little direct effect on film, many techniques have been devised to convert neutrons into radiations that will expose a film or produce light for a real-time imaging system. These converter screens are similar to the intensifying screens used in photon radiography.

Disadvantages

Disadvantages of neutron radiography include the high cost of the sources, the relatively large size of the most practical neutron source assemblies, and the personnel protection and safeguard problems associated with neutrons. These disadvantages combine to yield a major limitation of the technique; no really portable, inexpensive system is available. Nevertheless, equipment to produce neutron radiographs is available; in special circumstances, the unique information provided by neutron radiography outweighs the disadvantages.

Applications

Such critical areas as the inspection of aerospace components, explosives, adhesive components, nuclear control materials and nuclear fuel are examples of applications making use of neutron radiography's advantages. Neutron radiography is also useful for the detection of corrosion (particularly in aircraft structures) and for locating areas of water entrapment and hydrogen embrittlement.

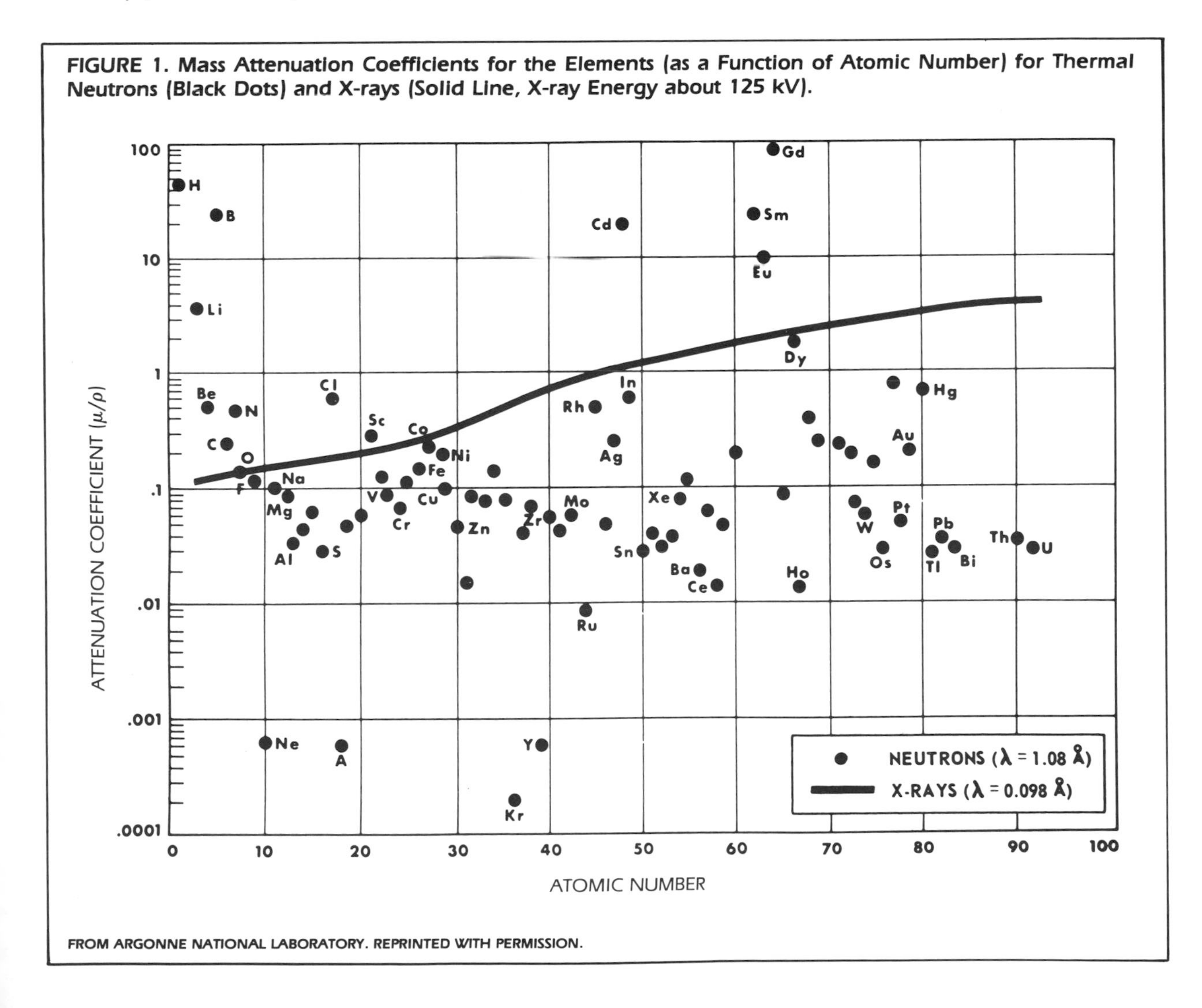

FIGURE 1. Mass Attenuation Coefficients for the Elements (as a Function of Atomic Number) for Thermal Neutrons (Black Dots) and X-rays (Solid Line, X-ray Energy about 125 kV).

FROM ARGONNE NATIONAL LABORATORY. REPRINTED WITH PERMISSION.

PART 2

EQUIPMENT AND PROCEDURES

Neutron Sources

The two main constituents of the nucleus are the proton and the neutron. The force between any pair of these particles is strong, attractive, and of very short range; stable nuclei must have an external source of energy supplied for a separation of particles to occur. Any nucleus can be disrupted if adequate energy is supplied, but several light materials as well as several heavy materials can be made to yield free neutrons by supplying only moderate amounts of energy to their nuclei. A very limited number of materials emit neutrons during the spontaneous disruption of their nuclei. The general types of reactions used for neutron production are outlined below.

Fission

When a neutron enters a nucleus, the new (compound) nucleus gains an energy equal to the sum of the binding energy and the kinetic energy of the neutron. For some heavy isotopes, the addition of the binding energy of a neutron is sufficient to cause instability leading to nuclear fission. There can be, on the average, more than one free neutron produced per neutron absorbed, in assemblies containing fissionable materials. This net gain in free neutrons makes a nuclear chain reaction possible and is the basis for the more prolific neutron sources in general use. Isotopes most commonly used in fission reactors are uranium-235 and plutonium-239.

Positive-Ion Bombardment

Neutrons can be readily obtained from the action of energetic positive ions such as protons or alpha particles on light materials such as deuterium, tritium, beryllium, lithium or boron. The source of positive ions can be either a radioactive isotope or a particle accelerator.

Photoneutrons

The necessary energy for neutron emission can be supplied to the nuclei in the form of intermediate to high energy photons. Target materials which are of special interest for photoneutron production are beryllium-9 and possibly hydrogen-2 for low energy photons (1.6 to ~ 9 MeV) and uranium for high energy photons (E ~ 9 MeV or higher).

Spontaneous Fission

A few materials fission spontaneously, that is, without the input of neutrons from outside the material. A practical spontaneous fission source for neutron radiography is californium-252.

Neutron Energies

As with all penetrating radiation, many different energies are available for use in radiography. Neutron energy ranges which are potentially useful for radiography include thermal, epithermal or resonance, cold and fast. Table 1 describes these various energy ranges and discusses their usefulness.

Thermal Neutrons

Most neutron radiography has been done with thermal neutrons. The reason is that useful and interesting attenuation characteristics are found in the thermal region. In addition, imaging of thermal neutron beams is relatively straightforward and efficient. A source for thermal-neutron radiography must include moderator material to slow down, or thermalize, the source's fast neutrons. Fortunately, moderator materials can be incorporated into most source assemblies. The energy distribution in a beam extracted from a weakly absorbing slowing-down medium is such that thermal-neutron radiography can be done, if the intensity of the beam is sufficient. Actually, such beams contain a wide range of neutron energies, but the imaging process usually provides excellent discrimination against higher energy neutrons.

Epithermal Neutrons

There are applications in which thermal neutrons do not penetrate the inspection object sufficiently for meaningful neutron radiography. Epithermal neutrons are sometimes used to satisfy test requirements for these applications. The energy range for neutrons in this category can vary, but a commonly used interval is 0.5 to 10 keV. The neutron beam may be monoenergetic; a multi-energy beam may be used if proper care is exercised in image detection. The most extensive use of epithermal neutron radiography has been the inspection of highly enriched nuclear reactor fuel specimens. A comparison that demonstrates the improved neutron transmission for epithermal neutrons through enriched reactor fuel is shown in Fig. 2. Similarly improved transmission can be obtained with other materials. For example, epithermal neutrons can be used for inspecting thickness of hydrogenous materials greater than can be inspected with thermal neutrons. Very high sensitivities to a particular material or isotope can be obtained if the neutron radiography is accomplished at the energy of a resonance, by using a detector with a resonance reaction or a time-of-flight separation of energies (exposing the detector only to those neutrons which travel from the source in a specific elapsed time).

Cold Neutrons

Very low energy neutrons can offer advantages for some specialized inspections; the penetrating ability of neutrons can be greatly enhanced for

TABLE 1. Neutrons Classified According to Energy

Term	Comments	Energy Range
Slow		0.00 eV to 10^3 eV
Cold	Materials possess high cross-sections at these energies, which decrease the transparency of most materials but also increase the efficiency of detection. A particular advantage is the reduced scatter in materials at energies below the Bragg cutoff.	Less than 0.01 eV
Thermal	Produced by slowing down of fast neutrons until the average energy of the neutron is equal to that of the medium. Thermal neutrons provide good discriminatory capability between different materials; sources are readily available.	0.01 eV to 0.5 eV
Epithermal	Produced at energies greater than thermal, e.g. fission energies, and surrounded by a moderator. Neutrons are slowed down until they have energies in thermal equilibrium with the moderator molecules. At any location where thermal equilibrium has not been achieved the distribution of neutron velocities will contain velocities that exceed that permitted by a Maxwellian distribution of the moderator temperature. Such neutrons are referred to as epithermal	0.5 eV to 10^4 eV
Resonance	Certain nuclei exhibit strong absorption characteristics at well-defined energies called resonance absorptions. Neutrons in these specific energy ranges are referred to as resonance neutrons and provide excellent discrimination for particular materials by working at energies of resonance. Greater transmission and less scatter occur in samples containing materials such as hydrogen and enriched reactor fuel materials.	1 eV to 10^2 eV
Fast	Fast neutrons provide good penetration. Good point sources of fast neutrons are available. At the lower energy end of the spectrum fast neutron radiography may be able to perform many inspections performed with thermal neutrons, but with a panoramic technique. Poor material discrimination occurs, however, because the cross-sections tend to be small and similar.	10^3 eV to 20 MeV
Relativistic		>20 MeV

some radiographic specimens by taking advantage of the reduced scatter at neutron energies below the Bragg cutoff (the point where an energy's wavelength, compared to the specimen's atomic spacing, becomes sufficiently long to prohibit diffraction). Specifically, iron becomes more transparent at a neutron energy of about 0.005 eV because of reduced scatter. In fact, the use of cold neutrons allows radiographic inspection of iron specimens in the thickness range of 10 to 15 cm. Another application for cold neutrons involves taking advantage of the high absorption cross sections in many materials. This may allow the imaging of small concentrations of materials, too small to be imaged well with thermal neutrons. The efficiency of detectors also increases in the cold energy region.

Fast Neutrons

The primary advantages of fast neutrons for radiography are their excellent penetrating qualities and their point emission. Interference from scattering and limited detector response combine to restrict the practical applications of fast neutron radiography. Also, the similarity of attenuation cross sections for most materials for fast neutrons places a significant restriction on contrasting ability.

Most neutrons are born in the fast region; this means that moderator materials are to be avoided (or at least restricted) if radiography is to be done with high-energy neutrons. The availability of *point neutron sources* has been a significant factor in attracting investigation into radiography with fast neutrons.

FIGURE 2. Neutron Radiographs of Highly Radioactive, Enriched, Mixed Oxide Reactor Fuel (Pellets about 6 mm Diameter), after Irradiation to about 10 Percent Burnup; (a) Thermal Neutron Transfer Radiograph; (b) Indium Transfer Radiograph, Taken with Indium Resonance Neutrons and Cadmium Filter, Shows Greater Transmission Through the Fuel.

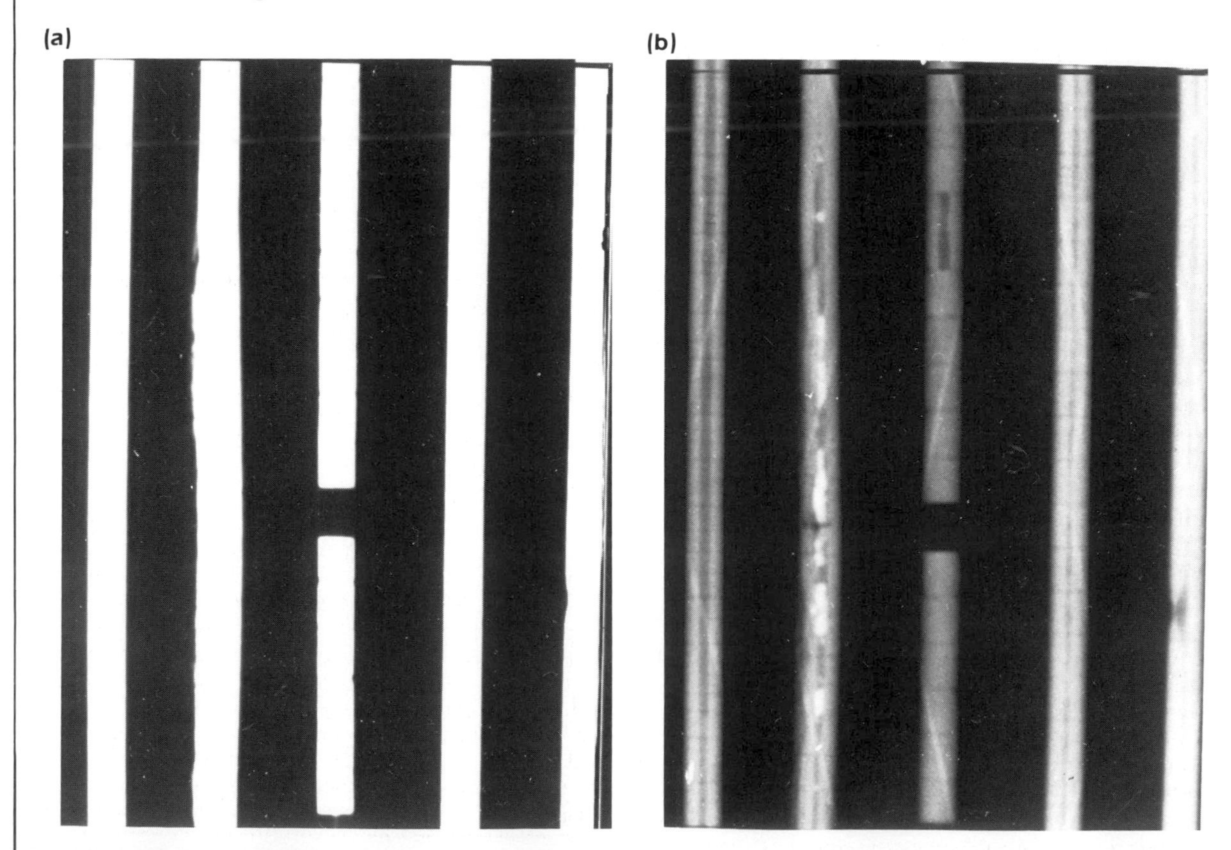

FROM ARGONNE NATIONAL LABORATORY. REPRINTED WITH PERMISSION.

Moderation

In general, liberated neutrons have considerably more kinetic energy than the atoms or molecules of the host material; this energy may be dissipated through numerous collisions with nuclei in the host material. The transformation of fast neutrons to slow neutrons is achieved by a moderating material (moderator). Its presence produces a slowing down of the fast neutrons by elastic scattering collisions (between the moderator nuclei and the neutrons) until the average kinetic energy of the neutrons is the same as that of the moderator nuclei.

Moderators

A complete description of the suitability of a substance as a moderator requires information on its scattering cross section, the average loss in neutron energy per elastic collision, the number of scattering centers per unit volume, and the absorption cross section. Two parameters which provide a measure of the efficiency of a material as a moderator are the *slowing down power* and the *moderating ratio*. The slowing down power represents the average decrease in the logarithm of the neutron energy per unit of path length. The moderating ratio is the ratio of the slowing down power to the macroscopic absorption cross section (Σ, or the number of atoms times the atomic cross section value). The slowing power and the moderating ratio for several moderators are listed in Table 2.

Water and other hydrogenous materials have relatively small moderating ratios because of the large absorption cross section for hydrogen; however a large slowing-down power makes water, oil and plastic materials attractive for small, compact neutron systems.

TABLE 2. Slowing Down Properties of Moderators

Moderator	Slowing Down Power	Moderating Ratio
Water and other hydrogenous materials	1.3	60
Heavy Water	0.18	~6,000-20,000*
Beryllium	0.16	135
Graphite	0.06	175

*The absorption cross section in heavy water is greatly affected by light water contamination.

Thermalization Factors

Thermalization factors have been determined experimentally for many neutron sources moderated in water. The *thermalization factor* is the ratio of the total 4π (all directions) fast neutron yield from the source in neutrons per second (n/s) divided by the peak thermal neutron flux in the surrounding moderator, in neutrons per square centimeter seconds ($ncm^{-2}s^{-1}$). Table 3 gives experimental values for several sources in a water moderator. Neutron sources that are physically small (allowing efficient moderation to take place) and which yield relatively low energy neutrons (that are easy to slow down) have favorably low thermalization factors. A schematic arrangement for source moderation of a thermal neutron beam is shown in Fig. 3.

Neutron Collimation

Beam extraction and neutron collimation form an important part of neutron radiography's source technique. Early facilities used collimators designed to render the beam parallel (single tube or multiple tube), but nearly all of the newer facilities use divergent collimators.

TABLE 3. Some Thermalization Factors Measured in a Water Moderator

Source	Thermalization Factor Measured	Thermalization Factor Rounded*
D(d,n) at 150 keV	196	200
T(d,n) at 150 keV	645	650
Be(p,n) at 2.8 MeV	38	40
Be(d,n) at 2.8 MeV	298	300
^{252}Cf (10 mCi or 370 MBq)	78	100
^{241}Am-Be (1 Ci or 37 GBq)	200	200
^{124}Sb-Be (10 mCi or 370 MBq)	46	50

* Rounded-up figures are quoted to draw attention to the variability of thermalization factors in practice due to thermal neutron absorption in the target holder or source. Sb-Be and large Am-Be sources are likely to suffer particularly in this respect. Care was taken to minimize flux distortions in these measurements apart from the beryllium-target reactions where a heavy stainless-steel target holder was used in anticipation of high-flux work later.

AFTER HAWKESWORTH (SEE REF. 2).

Collimator Designs

Many types of collimators have been proposed and used. The principle of the point source, the parallel-wall collimator, and the divergent collimator are illustrated in Fig. 4. The divergent collimator is widely used. Although the divergent collimator appears similar to point-source geometry, it is generally used to extract a beam from a relatively large moderator assembly. Therefore, structures (walls) are required to limit the uncollimated background radiation reaching the imaging plane. Limiting the background radiation is generally as important as geometric collimation for obtaining good quality neutron radiographs.

Divergent Collimators

The primary advantage of the divergent collimator is that a uniform beam can be projected easily over a large inspection area. Other significant advantages include the fact that a uniform flux is not required over a large volume at the inlet end and the fact that the inlet imposes minimal perturbation to the source assembly. Some distortion occurs at the edges of the beam coverage from a divergent collimator because the neutron paths are radial rather than parallel. However, the beam distortion generally does not present problems except in a few special applications.

FIGURE 3. Moderator Surrounding a Fast Neutron Source (left). A Straight Collimator Is Shown with the Neutron Beam Moving Left-to-Right Toward the Object with Detector.

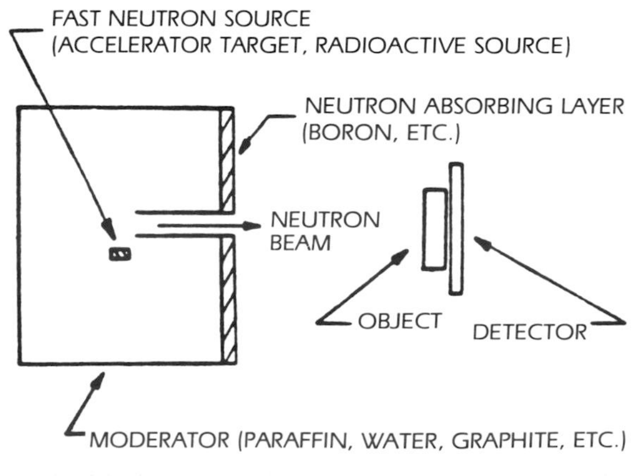

FAST NEUTRON SOURCE (ACCELERATOR TARGET, RADIOACTIVE SOURCE)
NEUTRON ABSORBING LAYER (BORON, ETC.)
NEUTRON BEAM
OBJECT
DETECTOR
MODERATOR (PARAFFIN, WATER, GRAPHITE, ETC.)

L/D Ratio

The important geometric factors for a neutron collimator are the total length (L) from inlet aperture to detector and the effective dimension of the inlet of the collimator (D). This information is usually expressed as the *L/D ratio.* These parameters determine the angular divergence of the beam and, to a large extent, the neutron intensity at the inspection plane. An efficient collimator requires that scattering from structural components and penetration of collimator walls by uncollimated background radiations be minimized.

FIGURE 4. Neutron Collimators: (a) Point Source for Fast-Neutron Radiography; (b) Parallel-Wall Collimator; (c) Divergent Collimator.

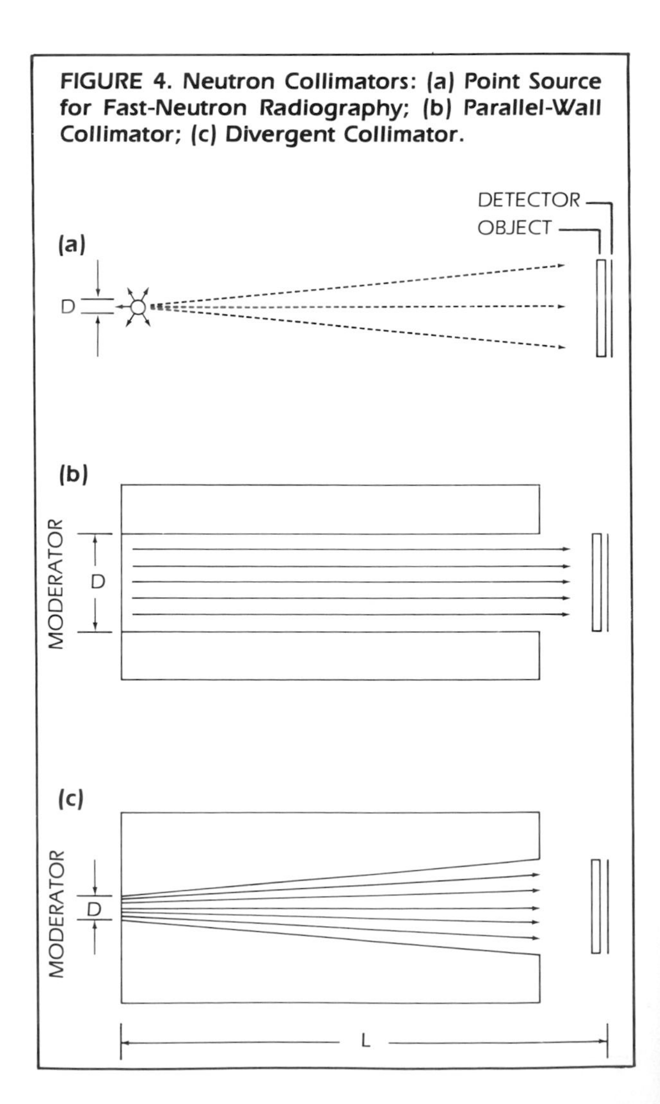

Unsharpness

As in X-radiography, the geometric unsharpness can be calculated for neutron radiography, using the values L and D. The value D is, in effect, the focal spot size and the value L is the source-to-film distance. Therefore, the geometric unsharpness, U_g, is as follows:

$$U_g = D\left(\frac{t}{L\text{-}t}\right) \quad \text{(Eq. 3)}$$

where t is the object thickness or the separation of the object plane of interest from the detector. In cases where t is small compared to L, as would be the case in many practical situations,

$$U_g = \frac{t}{L/D} \quad \text{(Eq. 4)}$$

Neutron Intensity

The values L and D are also related to the neutron intensity needed for radiography. For a collimator with dimensions L and D and an entrance port at a point in the moderator where thermal neutron flux is Φ, the intensity at the collimator's exit can be approximated by the following relationship (for a round aperature of diameter D):

$$I \cong \frac{\Phi}{16\,(L/D)^2} \quad \text{(Eq. 5)}$$

Backscatter

It is good practice to contain all radiation that passes through the imaging plane in a beam catcher. Radiation backscatter can seriously degrade the quality of any radiograph; a well designed beam catcher will keep backscatter from reaching the imaging device.

Types of Neutron Sources

Many sources have been used for neutron radiography and gaging. These include reactor, accelerator, radioactive and subcritical assembly sources. A summary of the general characteristics of these various sources is given in Table 4. The limits of available intensity, as well as the other observations in the Table, have been generalized to give an indication of what may normally be expected from each type of source.

TABLE 4. Average Characteristics of Thermal-Sources

Type of Source	Typical Radiographic Intensity*	Resolution**	Exposure Time	Characteristics
Radioisotope	10^1 to 10^4	Poor to Medium	Long	Stable operation, medium investment cost, possibly portable.
Accelerator	10^3 to 10^6	Medium	Average	On-off operation, medium cost, possibly mobile.
Subcritical Assembly	10^4 to 10^6	Good	Average	Stable operation, medium to high investment cost, mobility difficult.
Nuclear reactor	10^5 to 10^8	Excellent	Short	Stable operation, medium to high investment cost, mobility difficult.

*Neutrons per square centimeter per second.
**These classifications are relative.

Nuclear Reactor

The majority of practical neutron radiography has been done using a nuclear reactor as the source. The main reasons for this are listed below.

1. Reactors are prolific sources of neutrons, even when operating at low to medium power levels.
2. Most reactor beams are rich in thermal neutrons.
3. Most of the early research in, and the applications of, neutron radiography were associated with the reactor community.
4. Neutron radiography can be essentially a by-product of many reactor operations.
5. Organizations that first offered neutron radiography as a commercial service used reactors as their neutron sources.
6. Most present applications have not required portability from the neutron source.
7. Experience with reactors has shown long, relatively trouble-free operation.

Some desirable characteristics of a reactor for radiography are low cost per neutron; high available thermal neutron flux for beam extraction; some capability for beam tailoring; small physical size; and low power output. The two most formidable problems associated with procuring a special-purpose reactor are overall cost, plus the need to satisfy the requirements and regulations imposed by various regulatory agencies. The capital cost of a reactor facility is high compared to most other pieces of nondestructive test equipment. Even a small reactor would require several trained people to operate it and perform the radiography.

Accelerators

Numerous nuclear reactions can be used to produce neutrons from accelerated charged particles. Free neutrons can be produced by positive-ion bombardment of selected materials with accelerating potentials in the 100 keV to many MeV energy range. Some specific reactions for positive-ion bombardment are H-3 (d, n) He-4, H-2 (d, n) He-3, Li-7 (p, n) Be-7 and Be-9 (d, n) B-10.

(d,T) Source

A reaction that has received much attention for neutron radiography applications is H-3 (d,n) He-4, often abbreviated as a (d,T) or deuteron-tritium source. The source is attractive because a relatively high yield can be achieved from low bombarding energies (about 150 keV). Operation at yields of about 10^{11} n/s can be expected from such machines; higher yields have been achieved. However, target/tube lifetimes are limited at high neutron output, and the emitted neutrons are very energetic ($\sim$14 MeV). A practical thermalization factor for 14 MeV neutrons is typically 1000.

High Energy Machines

Several compact accelerators are commercially available. These are capable of accelerating positive ions to energies between 1 and 30 MeV; examples include Van de Graaff accelerators and cyclotrons. Many of these machines are versatile in that the accelerated particle can be a proton, a deuteron or another positive ion. However, sophisticated equipment is required to supply accelerating potentials of this magnitude, and the capital investment for such equipment is substantial. Good radiographic results have been reported for 15-30 minute exposures using a Van de Graaff accelerator operating at 3 MeV, 280 μA with the Be-9 (d,n) B-10 reaction.

(X,n) Sources

Electron accelerators can be used as neutron sources by irradiating a suitable neutron-yielding material with bremsstrahlung radiation, produced when energetic electrons strike a target of high atomic number. Any nuclide (nuclear species) can be made to undergo photodisintegration with photons of adequate energy; for radiographic purposes, beryllium and uranium appear to be the most useful materials. Beryllium has a low threshold energy for photoneutron production ($\sim$1.66 MeV) and is normally used with photon energies up to 10 MeV. The threshold energy for photoneutron production in uranium is $\sim$9 MeV; the yield rises quite rapidly as photon energy increases. The yield from a uranium target is considerably higher than the yield from a beryllium target for energies in excess of 15 MeV. The higher energy machines with uranium targets could produce high flux intensities ($>10^{11}$ ncm^{-2}s^{-1}) and offer potential for many practical applications.

Although a linear accelerator has the disadvantage of high X-ray background, it has the advantage of serving a dual role; it can be used for neutron radiography and for high-energy X-radiography. Changeover time from one radiation to the other could be as low as a few hours.

Plasma Sources

Radiography with fast neutrons from an electromagnetic plasma accelerator has also proven to be feasible. Up to 10^{10} neutrons can be produced in 100 nanoseconds (ns) from a machine powered by a capacitor bank of only 30 kilojoules (kJ). Up to 10^{12} neutrons can be produced from a large machine with a capacitor bank of 400 kJ. Such a device could allow radiographic recording of fast events without blur due to object motion.

Radioactive Sources

Many isotopic neutron sources make use of either the (α,n) or the (γ,n) reaction for neutron production. These sources have been used for many years for a variety of applications and they have the desirable features of being reliable and at least semi-portable. However, the thermal-neutron intensities that can be achieved from such isotopic source assemblies tend to be low, especially when compared to an operating nuclear reactor. Spontaneous fission of transplutonium elements is another neutron-producing reaction that has received considerable interest for neutron radiographic applications. Some pertinent data for several isotopic sources are tabulated in Table 5.

(γ,n) Sources

The (γ,n) reaction will probably not find much use for general neutron radiographic applications. The main reason for this is the high gamma-ray background associated with such sources. The antimony-beryllium combination has had some application in inspection of irradiated reactor fuel; in this application, the gamma radiation from the neutron source is not a problem because the high radiation level of the specimen itself requires a hot-cell (highly shielded) operation.

(α,n) Sources

Sources that make use of the (α,n) reaction provide marginal neutron intensity for most practical neutron radiography. These sources can be used if long exposure times are suitable. Also, (α,n) reactions produce adequate intensity for many gaging applications where high-efficiency neutron detectors can be used.

Cf-252

On the basis of technical performance, spontaneous fission from californium-252 is the most attractive isotopic source for neutron radiography. Its neutron intensity is limited by economic rather than technical considerations, and the gamma background is low enough to allow direct-exposure radiography. Californium-252 has been studied widely and several investigators report the production of useful radiographs when using this material as the neutron source. The quality of the radiographs has generally been below the quality obtained from reactors but this can be improved if long exposure times are acceptable. The source is well suited for applications which require moderate resolution, an in-house system, or a source which can be moved.

Subcritical Flux Boosters

By defintion, it is not possible to maintain a chain reaction in a subcritical reactor assembly. However, a subcritical reactor gives a neutron amplification of $1/(1-k_{eff})$, where k_{eff} is the effective multiplication constant for the system. Thus, an isotopic or accelerator source, driving a subcritical assembly, could supply more neutrons for radiography than could be obtained from the same source in a purely moderating medium. A subcritical assembly used as a flux booster for neutron radiography should be relatively inexpensive, small and moveable, and constructed so that there is little chance of achieving an unwanted criticality. Work on subcriticals indicates that the flux advantage in the assemblies, compared to the neutron source surrounded by a pure moderator, is small unless k_{eff} of the system is close to unity. Subcritical systems have provided an increase in available flux for neutron radiography of about 30 times.

Application Considerations of Sources

The fundamental figure of merit for a radiography source is the maximum thermal neutron flux available for beam extraction. This available source intensity is then apportioned to beam collimation or intensity, depending on test requirements. With a given source assembly, beam intensity (speed of exposure) can be increased only by sacrificing collimation (resolution) and vice versa.

Neutron Intensity

Very few, if any, practical applications of neutron radiography can tolerate a detector exposure of less than 10^5 n/cm^2. This means that even with poor collimation, several seconds would be required to produce a fast-film radiograph of a thin specimen when the flux inside the source assembly is 10^8 to 10^9 ncm^{-2}s^{-1}. Neutron sources with intensities lower than this will require very long exposures for radiographic inspection. Practical high-resolution neutron radiography requires an available thermal neutron flux before collimation of 10^{10} ncm^{-2}s^{-1} or more.

TABLE 5. Some Radioactive Sources for Neutron Radiography and Gaging

Source	Reaction	Half-Life	Average Neutron Energy (MeV)	Neutron Yield (n/s-g)	Gamma Dose* (rads/hr at 1m)	Gamma Ray Energy (MeV)	Comments
Sb-124—Be	(γ,n)	60 days	0.024	2.7×10^9	4.5×10^4	1.7	Short half-life and high gamma background, available in high intensity sources, low neutron energy is an advantage for thermalization.
Po-210—Be	(α,n)	138 days	4.3	1.28×10^{10}	2	0.8	Short half-life, low gamma background
Pu-238—Be	(α,n)	89 years	~4	4.7×10^7	0.4	0.1	High cost, long half-life
Am-241—Be	(α,n)	458 years	~4	1×10^7	2.5	0.06	Easily shielded gamma output, long half-life, high cost
Am-241—Cm-242—Be	(α,n)	163 days	~4	1.2×10^9 (80% Am, 20% Cm)	Low	0.04 0.06	Increased radiation yield over Am-241—Be for relatively little more cost but with a short half-life
Cm-242—Be	(α,n)	163 days	~4	1.4×10^{10}	0.3	0.04	High yield source, but half-life is short
Cm-244—Be	(α,n)	18.1 years	~4	2.4×10^8	0.2	0.04	Long half-life, low gamma background are attractive. Source can also be used as spontaneous fission source, with about half the neutron yield. Because Cm-244 is produced in nuclear fuel, this radioisotope could be widely available as a by-product material
Cf-252	Spontaneous fission	2.65 years	2.3	3×10^{12}	2.9	0.04 0.1	Very high yield source, present cost high, projected future cost makes it attractive, small size and low energy are advantages for moderation

*The gamma ray dose is normalized to a neutron yield of 5×10^{10} n/s; 100 rads = 1 gray.

Beam Collimation

The collimator L/D ratio should normally be 10 or greater to give useful radiographs. Collimator L/D ratios between 50 and 500 are recommended for most applications where thick objects are to be inspected (see unsharpness equations 3 and 4).

Interference Radiation

Controlling unwanted interference radiation is a more complex problem for neutron radiography than for X-ray or gamma ray radiography. The usual problem, which includes scattering from all structural materials (including back-scattering), must be evaluated and controlled. However, neutron radiography is also subject to interference from neutrons which penetrate the collimator walls, secondary radiation from neutron capture in structural materials, and from gamma-ray contamination in the primary neutron beam. Direct exposure imaging techniques are normally useful only in cases where the neutron radiographic beam has a relatively low component of electromagnetic radiation. A high gamma-ray background can greatly reduce the radiographic contrast of low atomic number materials in the inspection object. For medium speed X-ray films, direct metal screen-film image detectors give about the same exposure from 10^5 n/cm^2 as from 10 micrograys (1 milliroentgen) of cobalt-60 gamma radiation. The exact ratio of neutron and gamma-ray response depends on the film and screen used.

Beam Tailoring

A neutron beam extracted directly from a source assembly without altering its basic content is called a *raw* or *unfiltered neutron beam*. Neutron radiographs are usually made with such beams because they give satisfactory results for most applications and they require the least effort to produce. Specially tailored neutron beams should be considered for neutron radiography only when unfiltered beams do not give the desired results.

Gamma Filtration

Generally, the gamma rays in a neutron beam do not cause serious interference problems in neutron radiographs. However, some neutron sources have inherently high gamma background because of special materials used in the assembly. Because some applications require a very low gamma background, gamma-ray filtration in the neutron beam should be considered. The most commonly used material for gamma filtration is bismuth; lead can also be used. The required thickness of the gamma filter is strongly dependent on the intensity and energy of the photons in the beam.

Foil Filters

Some materials have special characteristics in the behavior of their neutron cross sections; these can cause a dramatic change in neutron beam characteristics even with a small thickness of the material between the source and the image detector. Such foil filters are sometimes used to selectively remove neutrons from the beam in a particular energy interval. The most common use is a threshold material which removes most of the neutrons below a certain energy. Cadmium, gadolinium or samarium filters, for example, are used to remove thermal neutrons when the beam is being used for epithermal neutron radiography. In principle, several materials could be used in combination to allow a narrow band of neutrons to be transmitted through the filter pack. In practice, a multifilter technique would be justified only for very special applications.

Window Filters

For some elements, an inherent interference cancels much of the potential scattering on the low energy side of the resonance. This low cross section segment will act as a window for neutrons in that energy range. If a sizable thickness of a material which has this characteristic is placed in the neutron beam, neutrons with the energy of the window will pass through, while neutrons of other energies are attenuated. Several materials exhibit this effect; examples include iron, scandium and silicon.

Spectral Shift from Changing Moderator Temperature

Changing the temperature of the neutron moderator material can have a significant effect on the average neutron energy extracted from that material. For example, increased intensity of cold neutrons can be obtained by using cryogenic moderators rather than moderators at room temperature. Also, the epithermal neutron content in a beam can be enhanced considerably with a hot moderator (greater than 300 °K).

Time of Flight

It is possible to use time-of-flight techniques to select a narrow neutron energy increment for radiography. The technique requires a pulsed neutron source and an image detector system which can be gated on and off in a very short time. This method can be used to analyze materials by observing resonances in the neutron spectra; time-of-flight can also be used to make a system sensitive to a particular material by working at a resonance of that material.

Other Techniques

A neutron crystal spectrometer is capable of producing a diffracted monoenergetic neutron beam essentially gamma-free and usually well collimated. Aluminum, sodium chloride, or beryllium crystals are used for neutrons in the 0.02 to 10 eV energy range. The energy of the neutron beam depends upon the plane spacings of the crystal and the angle of the crystal with respect to the main beam. Beams from crystal spectrometers are generally limited to small sizes and low intensities.

Neutron Image Detectors

All the methods used to detect X-ray images can be used to detect neutron images: film, radiographic paper, real-time fluoroscopy, and so on. In addition, some imaging methods not useful for X-rays can be used witn neutrons: activation transfer and track-etch methods, for example. Various imaging methods are useful for different energies of neutrons. Since most work has been accomplished with thermal neutrons, that energy range will be emphasized here.

Film Methods

Film techniques involve two different approaches. In one, the direct exposure method, the film is actually present in the neutron beam during the exposure. In the other, the transfer method, the film need not be exposed to the neutrons; the film exposure is made by autoradiography of a radioactive, image-carrying metal screen. The two techniques are illustrated in Fig. 5. Conversion or intensifying screens are used with both techniques. For the direct exposure method these screens increase the

FIGURE 5. Diagrams for Neutron Radiography with Film Using (a) Direct Exposure Method, and (b) Transfer Exposure Method.

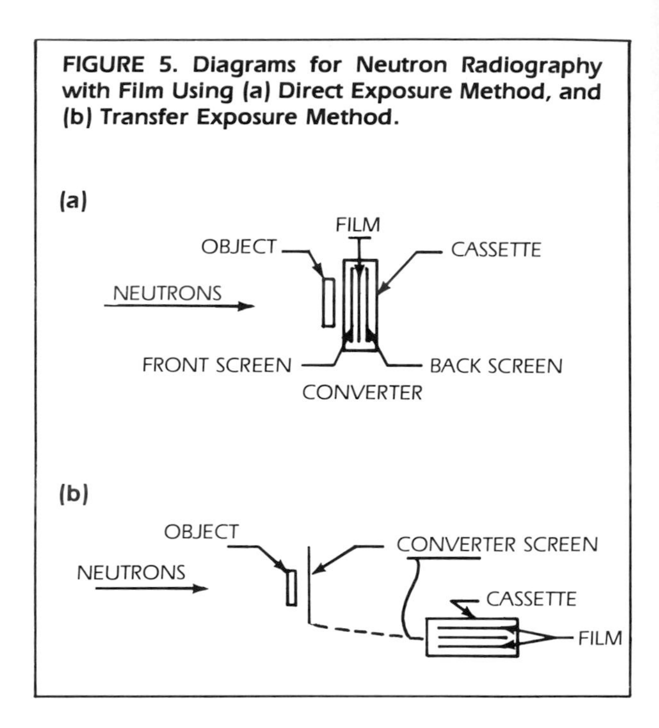

detector response by the emission of radiation to which the adjacent film is sensitive. Direct sensitivity of film to neutrons is relatively low. For the transfer method, the screens are chosen from materials that tend to become radioactive upon thermal neutron exposure. The properties of some useful conversion screen materials for thermal neutron imaging by both techniques are given in Table 6.

Transfer Method

In the transfer method, the film is not present in the image beam; this means the film will not be exposed to gamma radiation (from either a radioactive specimen, reactions between neutrons and objects in the beam path, or from those gamma rays present in the beam itself.) The transfer technique, therefore, has been widely used to inspect radioactive materials such as irradiated reactor fuels. These neutron radiographs show none of the gamma radiation fogging that would appear in conventional X-radiographs. The most common materials for transfer neutron radiography are indium (54 minute half-life) and dysprosium (140 minute half-life); thicknesses used are usually 125 to 250 μm (0.005 to 0.010 in.) Since transfer neutron radiographs really involve two exposures, one in the neutron

TABLE 6. Properties of Some Thermal Neutron Radiography Conversion Materials

Material	Useful Reactions	Cross Section for Thermal Neutrons (barns)	Half-Life
Lithium	$^{6}Li(n,\alpha)^{3}H$	910	prompt
Boron	$^{10}B(n,\alpha)^{7}Li$	3,830	prompt
Rhodium	$^{103}Rh(n)^{104m}Rh$	11	4.5 min
	$^{103}Rh(n)^{104}Rh$	139	42 s
Silver	$^{107}Ag(n)^{108}Ag$	35	2.3 min
	$^{109}Ag(n)^{110}Ag$	91	24 s
Cadmium	$^{113}Cd(n,\gamma)^{114}Cd$	20,000	prompt
Indium	$^{115}In(n)^{116}In$	157	54 min
	$^{115}In(n)^{116m}In$	42	14 s
Samarium	$^{149}Sm(n,\gamma)^{150}Sm$	41,000	prompt
	$^{152}Sm(n)^{153}Sm$	210	47 h
Europium	$^{151}Eu(n)^{152m}Eu$	3,000	9.2 h
Gadolinium	$^{155}Gd(n,\gamma)^{156}Gd$	61,000	prompt
	$^{157}Gd(n,\gamma)^{158}Gd$	254,000	prompt
Dyprosium	$^{164}Dy(n)^{165m}Dy$	2,200	1.25 min
	$^{164}Dy(n)^{165}Dy$	800	140 min
Gold	$^{197}Au(n)^{198}Au$	98.8	2.7 days

1 barn = $10^{-28}m^2$

beam and the other an autoradiograph exposure, these radiographs usually require more total time than the direct exposure approach. Another disadvantage is that there is a lower limit of neutron intensity in which this type of exposure will work. The activation and decay of the radioactivity each have a half-life. There is very little activity or decay radiation to be gained after about three half-lives, when 87.5 percent of the total activity or decay will have been accomplished. Therefore, even for fast films, 10^4 $ncm^{-2}s^{-1}$ is about the minimum useful intensity for transfer neutron radiography with a high cross section, reasonable half-life conversion material such as dysprosium.

Transfer Method Examples

Transfer radiographs are mostly used with high intensity neutron sources, such as reactors, and for inspections involving radioactive material, namely nuclear reactor fuel (see Fig. 2 for examples). For transfer to a medium speed X-ray film, for several half-lives, typical neutron exposures with indium would be 5 to 10 minutes and with dysprosium 2 to 5 minutes for a thermal neutron intensity of 10^7 $ncm^{-2}s^{-1}$. Spatial resolution capabilities of such neutron radiographs are on the order of 50 μm. The contrast sensitivity is such that one can detect thickness differences of one percent in 25 mm uranium or steel objects. This contrast is somewhat better than that normally achieved with direct exposure neutron radiographs because the contrast of the transfer radiographs is not reduced by secondary gamma radiation.

Gadolinium Direct Exposures

The most widely used detection method for industrial neutron radiography is the direct exposure technique with a gadolinium conversion screen. A typical arrangement is a single 25 μm (0.001 in.) thick gadolinium metal foil in a back screen configuration. In this case it is desirable to use a single-emulsion film (as opposed to normal, double-emulsion X-ray film). The low-energy internal conversion electrons emitted from the gadolinium upon neutron bombardment essentially expose only the emulsion facing the gadolinium. The elimination of the second film emulsion, therefore, reduces the detector response to thermal neutrons only slightly while substantially reducing the detector response to the gamma or X-radiation components of the radiation beam. Slow speed X-ray films with single emulsions are available from several manufacturers.

Exposures, Contrast, Resolution

A typical thermal neutron exposure for a slow, single-emulsion, X-ray film and a single gadolinium conversion screen is about 3×10^9 n/cm². This can be reduced to about 10^8 n/cm² for a fast X-ray film. In addition, some increase in speed can be obtained by using double gadolinium screens (a 6 μm front screen and a 50 μm thick back screen combination has provided good results) or a rhodium-gadolinium converter screen combination with double emulsion films. The single gadolinium conversion method has provided excellent spatial resolution in thermal neutron radiographs. An experimental spatial resolution value of 10 μm has been reported, in reasonable agreement with theoretical analyses. The contrast capability of the direct exposure method is usually somewhat poorer than that of the transfer method, because fogging radiation is often present. A typical thickness sensitivity for 25 mm steel or uranium objects is 2 percent.

Scintillators

Because scintillator-film techniques provide much faster results than gadolinium conversion screens (by factors as large as 100), there has been much interest in that detection method. The scintillators usually involve materials that undergo a prompt reaction with the neutrons, and an associated phosphor material. Common scintillator constituents are alpha emitters such as boron and lithium combined with a phosphor such as ZnS(Ag) or scintillating glasses; now widely used is the rare earth phosphor gadolinium oxysulfide. Combined with a fast light-sensitive film, scintillators can provide useful thermal neutron images with total exposures in the range of 10^5 to 10^6 n/cm^2. The capabilities of scintillator-film neutron radiography vary appreciably with screen and film characteristics. In general terms, a spatial resolution of 50 to 100 μm can be achieved. Contrast for metal specimens such as steel and uranium is on the order of 2 to 4 percent.

Reciprocity-Law Failure

Scintillator neutron radiographs, like other light detection systems employing film, are subject to reciprocity law failure. The same total exposure for a scintillator-film detector at one intensity will not necessarily produce the same film density at another intensity, as would be true for a lead screen X-radiograph or a gadolinium direct exposure neutron radiograph. The light exposure system is rate-dependent. For neutron scintillators, it has been shown that the effective exposure for constant film density is:

$$E = It^p \qquad \text{(Eq. 6)}$$

where I is the neutron intensity; t is the exposure time over the range of 1 to 10^3 s; and the exponent p is the Schwartzchild index, which is always other than 1, and in this case is equal to 0.74. In an experimental case with a B-10 based scintillator and a fast, light sensitive X-ray film (the reciprocity law failure is dependent on the film), a neutron exposure at 3×10^5 ncm^{-2}s^{-1} yielded a film exposure four to five times higher than an identical total exposure at an intensity value of 3×10^3 ncm^{-2}s^{-1}. The effect of reciprocity law failure complicates exposure calculations at very short (0.001 s) or very long (more than 10 s) exposures. Workers in other light detection fields have also encountered these reciprocity problems. One solution is to cool the detector to minimize reciprocity law failures at long exposure times. This has been verified for neutron radiography. Substantial improvements were found if detectors were cooled to dry ice temperature during the neutron exposure.

Neutron-Gamma Response

One important property of direct exposure methods is the relative response of the detector to neutron and gamma or X-radiation. A transfer neutron radiograph has essentially no response to gamma rays. For direct exposures, however, the film is in the beam and will offer some response to gamma radiation. Depending on what information is sought from such a radiograph, the gamma image will probably reduce the contrast obtained with a true neutron radiograph. In most cases a gamma-ray component in a neutron radiograph is something to be avoided. Of the direct exposure methods just discussed, neutron scintillator techniques employing a ZnS phosphor (with an alpha emitter) provide the best neutron-gamma response ratio. Typically for these scintillators, an exposure of 10^4 n/cm^2 will yield about the same film density as an exposure of 10 micrograys (1 mR) of cobalt-60 gamma radiation; the exact ratio is dependent on the film and other variables. The metal conversion screen methods typically have more response to gamma rays by a factor of ten. The glass scintillators and the rare earth scintillators provide a relative neutron-gamma response between these two extremes.

Film Response, Summary

A comparison of the two general classes of film detection methods shows that transfer techniques yield high contrast images with no gamma response. Direct exposure methods, on the other hand, provide much faster results and have yielded much better spatial resolution. Although most of this discussion related to X-ray film, other films and photographic materials, including instant film, X-ray sensitive paper, and light-sensitive film and paper can also be used for these photographic neutron detection methods.

Real-Time Neutron Image Detection

For neutron radiography, as for other forms of radiography, it is sometimes desirable to observe a dynamic event, or to view many objects passing by a detection station. Real-time radiographic systems provide this capability. Most of the systems depend on the fact that neutron scintillators yield light when irradiated with thermal neutrons. The resultant light can be amplified by image intensifiers and/or detected by television cameras. The television display can provide real-time viewing of the neutron images at locations removed from the radiation area. Several such systems have been assembled and tested for thermal neutron response. Systems have included scintillator screens viewed with light detectors such as image intensifiers and TV cameras and integral image intensifier tubes made for neutron imaging. Scintillating screens of ZnS(Ag) + (Li-6)F have been used. Most neutron real-time systems now make use of gadolinium oxysulfide scintillators. At high neutron intensity (10^5 to 10^7 $ncm^{-2}s^{-1}$, these systems show a high-contrast spatial resolution of about 0.25 to 0.5 mm. Contrast sensitivity for steel and uranium samples is in the 2 to 4 percent range. Object motion of several meters per minute can be followed without objectionable blur. At lower neutron intensities where image statistics are less favorable, these characteristics degrade somewhat. For example, in experimental tests at an intensity about 10^4 $ncm^{-2}s^{-1}$ (a value equivalent to less than 10^3 n/cm^2 per television frame), the single frame contrast observed for a 13 mm thickness of steel was 16 percent. Electronic integration, (adding successive frames) can be used to minimize statistical fluctuations and improve low intensity real-time images.

Other Detectors for Thermal Neutron Radiography

In addition to film and real-time detectors, many other image detectors have been used for thermal neutrons. These include: radiographic paper; instant film; xeroradiography; multi-wire spark counters; proportional counters; thermoluminescent detectors; pressurized gas cell methods (similar to ionography); point detectors such as scintillation counters; and gas cell detectors scanned across the image. One other detector now used routinely is the track-etch detector.

Track-Etch Detectors

The tracks caused by radiation damage in a dielectric (a material which is electrically insulating, or which can sustain an electric field with a minimum loss of power) can be chemically etched in a preferential manner so the tracks become visible. A collection of many tracks can form a visual image similar to the collection of dots used to make a newspaper picture. For neutron work, most track-etch applications have involved plastics and neutron converters that emit alpha particles (such as boron or lithium) or fission fragments (uranium converters). The advantages of the track-etch approach include no light sensitivity (for ease of handling), linear response, no sensitivity to X-rays, and an indefinite time to accumulate an image (no saturation effects such as those that occur with transfer detection). The insensitivity to X-rays and the indefinite exposure time, coupled with the excellent spatial resolution for the track-etch method (25 μm), place it in competition with the transfer detection method for inspecting radioactive material. Track-etch detectors have been applied to the inspection of irradiated reactor fuel.

Track-Etch Results

The track-etch method has been investigated mostly with several cellulose nitrate and polycarbonate plastics. Results reported have shown a spatial resolution of 25 μm and a sensitivity to thickness variations (in 25 mm thick uranium) of 1 percent. These results were obtained for both polycarbonate plastics with uranium-235 foils and for cellulose nitrate with a (Li-6)F conversion screen; exposures were in the 10^8 to 10^9 n/cm^2 range. The most sensitive materials required an exposure of only 2×10^7 n/cm^2. The cellulose nitrate was etched for 4 min in 6.5 N sodium hydroxide solution at 55 °C; the polycarbonate was etched in the same solution at 70 °C for 40 min.

Image Detectors for Other Neutron Energies/Cold Neutrons

The same general techniques used for thermal neutron image detection apply to other neutron energies. However, the imaging work done with neutrons outside the thermal energy region has not been as extensive as that done with thermal neutrons. For cold neutrons, detector exposure requirements will decrease somewhat compared to those cited for thermal neutrons. Detection work had been reported with film methods, both direct exposure and transfer, real-time and track-etch methods. Conversion materials were the same as used for thermal neutrons.

Epithermal Neutrons

Work with neutrons in the epithermal or resonance energy region (approximately 0.5 eV to 10 keV energy) has been concentrated mainly on the lower end of the energy spectrum. Although neutron cross sections tend to decrease as the neutron energy increases above the thermal energy range, there are also large resonances that occur in this energy region. Therefore, detectors such as indium, with a large activation resonance at 1.46 eV, have been widely used for these neutrons. Gold, with a resonance at 4.9 eV, is another potentially useful detector. A common detection method has been to filter the neutron beam with cadmium or gadolinium (to remove thermal neutrons) and to then use a transfer detection method with a conversion screen such as indium. This technique has been used to examine fast reactor fuels because the higher energy provides greater neutron transmission through enriched U-235 and plutonium materials (see Fig. 2 for an example). Greater transmission (with less scatter) for radiography of hydrogenous objects can be accomplished in a similar manner.

Resonance Neutrons

Another approach for neutrons in this energy range involves the use of a neutron pulse and a time-of-flight scheme to permit detection of neutrons at a particular energy. Neutron pulses have been provided by reactors and accelerators and detected by a variety of methods. Time-gated detectors such as position sensitive proportional counters, scintillators with time-gated image intensifiers, and photomultiplier tubes have been used.

Fast Neutrons

The other energy range of interest for neutron radiography is fast neutrons, energies above 10 keV. Neutron cross sections at these energies tend to be low so that detectors generally require greater neutron exposures than do thermal neutron radiographs. The detection approaches are similar to those used for thermal neutrons, but different conversion materials are normally used. Direct exposure has been obtained with fluorescent screens made for X-radiography; the screens are combined with light-sensitive X-ray films. The X-ray phosphor screens work primarily because neutron interactions with hydrogen, in the plastic binder or in the plastic or paper backing, yield protons that stimulate light from the phosphor. Fast X-ray phosphor screens and film provide neutron radiographs of MeV energy neutrons with total exposures on the order of 10^7 to 10^8 n/cm^2. X-ray film without screens also responds to fast neutrons but requires exposures of 4×10^8 n/cm^2 to more than 10^9 n/cm^2. The film exposures can be reduced slightly by adding conversion screens of paraffin, plastic, tantalum or lead. Fast neutron radiographs made by these techniques display changes in metal specimen thickness in the range of 3 to 6 percent and have demonstrated a spatial resolution better than 1 mm. Transfer detection is also used for fast neutrons. Some useful materials for transfer conversion screens are summarized in Table 7.

TABLE 7. Characteristics of Some Transfer-Detection Materials for Fast Neutron Radiography

Material	Reaction	Abundance of Target Isotope in Normal Material (percent)	Threshold Neutron Energy for Reaction (MeV)	Half-Life	Reaction Cross Section (millibarns) 1 MeV	3 MeV	14 MeV
Aluminum	$^{27}Al(n,\alpha)^{24}Na$	100	6.1	15 h	—	—	120
Phosphorus	$^{31}P(n,p)^{31}Si$	100	1.8	2.65 h	—	80	140
Sulfur	$^{32}S(n,p)^{32}P$	95.02	1.7	14.3 days	—	150	220
Copper	$^{63}Cu(n,2n)^{62}Cu$	69.1	11	10 min	—	—	460
Gallium	$^{69}Ga(n,2n)^{68}Ga$	60.2	10.2	67.5 min	—	—	1070
Gallium	$^{71}Ga(n,2n)^{70}Ga$	39.8	9.2	21 min	—	—	2180
Rhodium	$^{103}Rh(n,n')^{103m}Rh$	100	<0.1	57 min	450	800	—
Cadmium	$^{111}Cd(n,n')^{111m}Cd$	12.7	0.3	48.7 min	130	300	—
Indium	$^{115}In(n,n')^{115m}In$	95.7	0.5	4.5 h	55	340	—
Antimony	$^{121}Sb(n,2n)^{120}Sb$	57.2	9.3	16.5 min	—	—	1180

It should be noted that the neutron reactions cited have threshold energies. The reaction will not occur for energies lower than the threshold. This limits the response of the detector to a portion of the neutrons that may be available, thereby tending to increase the required exposure. On the other hand, since scattered neutrons are of lower energy, the threshold response can be valuable in eliminating radiographic detection of scattered neutrons. As an example, copper screens are often used to detect (d-T) accelerator-produced neutrons having an energy of 14 MeV. Since the threshold for the 10-minute half-life reaction is 11 MeV, a copper screen exposure of 30 minutes or less will essentially detect only primary neutrons from the source. A typical exposure for that type of radiograph would require 10^{10} n/cm^2, for transfer to a fast X-ray film.

Detector Discussion

Image detectors for neutrons have been shown to provide a broad capability in terms of exposure requirements, spatial resolution and contrast. It must be recognized that the image properties in an actual radiographic system will depend on several factors in addition to the detector properties. The thickness of the radiographic object and the geometry of the imaging system will have a significant influence on the spatial resolution obtained. Similarly, scattered and secondary radiation that reaches the detector will strongly influence the contrast. The contrast capability discussed for the various detectors has usually been described in terms of a percentage thickness change that can be observed in a metal specimen. This is a common method for describing the capability of an X-radiographic system. The use of that method for describing contrast in neutron radiography, therefore, has merit by permitting X-radiographic comparisons. On the other hand, neutron radiography seldom offers an advantage for inspecting a homogeneous material. Its significant advantage is in differentiating between materials or isotopes. That should be kept in mind when thinking about neutron contrast, neutron radiographic test pieces, or specifications.

PART 3

APPLICATIONS

General

General applications for neutron radiography include inspections of nuclear materials, explosive devices, turbine blades, electronic packages and miscellaneous assemblies including aerospace structures (metallic honeycomb and composite components), valves and other assemblies. Industrial applications generally involve the detection of a particular material in an assembly containing two or more materials. Examples include detection of (1) residual ceramic core in an investment-cast turbine blade, (2) corrosion in a metallic assembly, (3) water in honeycomb, (4) explosives in a metallic assembly, or (5) a rubber "O" ring in a valve. Nuclear applications depend on the capability of neutron radiography to yield good, low background radiographs of highly radioactive material (by such methods as transfer or track-etch), to penetrate fairly heavy assemblies and to discriminate between isotopes (e.g. between U-235 and U-238 or Cd-113 and other cadmium isotopes).

Explosives

Explosives or pyrotechnic devices account for a large part of neutron radiographic applications. Small explosive charges in metallic assemblies (such as lead, shaped-charge lines or steel explosive bolts) can be detected, measured and assessed in terms of density, uniformity, foreign material, etc. Many of these pyrotechnic devices are relatively small assemblies containing metal, explosive and, in some cases, plastic components. An X-radiograph shows the metallic parts very well. The neutron radiograph shows the low atomic number materials, including explosives, plastic and adhesives. Together, the two radiographic methods provide a relatively complete inspection. Information sought often includes: the presence or absence of the explosive; breaks in the explosive train; density; uniformity; and foreign material. Figure 6 shows an explosive bolt as radiographed by both neutrons and X-rays.

Turbine Blades

Turbine blades made by the investment-cast process are inspected by neutrons to be sure there is no residual ceramic core left in the internal cooling passages after the leaching process. The nickel alloy used in most turbine blades is highly attenuating for X-rays but reasonably transparent to neutrons. A few ceramic materials have reasonable attenuation for neutrons. In many applications of this type, the neutron attenuation of the material to be found (the ceramic) is assured by adding a 1 to 2 percent of gadolinia. The addition of this high neutron cross section tracer assures that even small traces of residual ceramic can be observed. Figure 7 is a neutron radiograph of turbine blades, showing residual ceramic.

Electronic Devices

Electronic devices such as relays are inspected by neutron radiography to detect foreign materials, such as cloth or paper, that might interfere with the operation. An example of a neutron radiograph of a small relay is shown in Fig. 8. If there were pieces of hydrogenous foreign matter in the case, pieces that might prevent the relays from making good contact, the neutron radiograph would show them.

Assemblies

Various assemblies and components are usefully inspected by neutron radiography. Rubber "O" rings in metallic assemblies can be detected to be sure the ring is present and properly seated (see Fig. 9); this is often a difficult task for X-radiography because of the relatively high attenuation of the surrounding metal. Honeycomb assemblies for aerospace and other applications can be inspected with neutrons to show adhesives during manufacture or repair, or to show the presence of water during service. Neutron radiographs of composite assemblies show distribution of adhesives or resins.

Metallic assemblies can be inspected to show corrosion. This is used primarily for aerospace applications where the metal may be aluminum and the corrosion is an hydroxide which can be detected by neutrons.

Contrast Agents

Contrast agents can be used, as in the case of the ceramic core in turbine blades, to show areas that might be difficult to detect otherwise. Similiarly,

FIGURE 6. Radiographs of Explosive Bolt (about 5 cm in Height): (a) Thermal Neutron Radiograph; (b) X-radiograph. Neutron Radiograph Shows Explosive Material (Salt and Pepper-like Image) Through the Stainless Steel Threaded Region in Upper Part of Bolt; Also Shows Paper (Upper White Line), Plastic (Mid-Region) and Epoxy (Lower part).

(a)

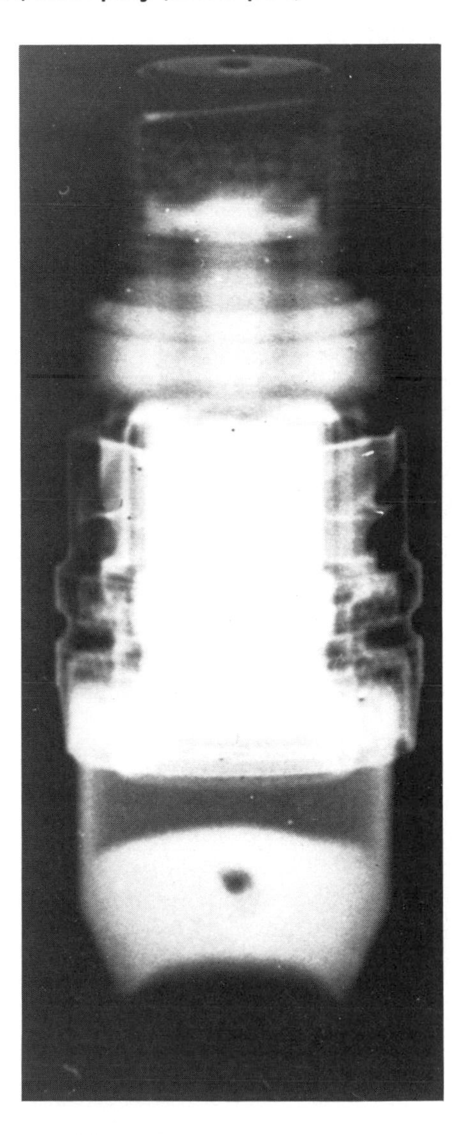

(b)

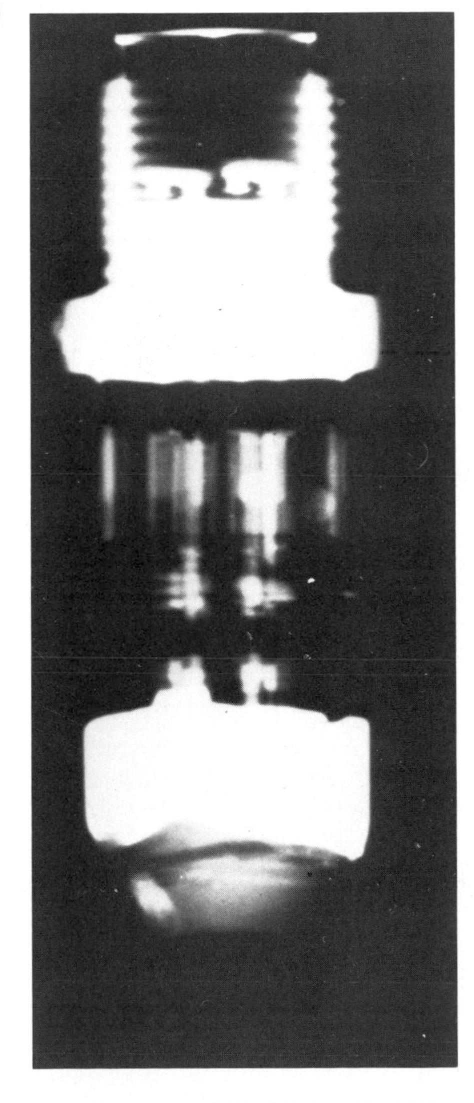

liquid penetrants, sometimes doped with additional neutron attenuating material such as gadolinium, boron or lithium, can enter surface discontinuities and serve as a contrast agent to outline delaminations or cracks.

Metallurgy

Metallurgical studies have been made with neutron radiography to observe the distribution of high neutron cross section alloying agents such as cadmium. Hydriding of materials such as zirconium or titanium can be detected by neutron radiography. Distribution of other agents such as boron or lithium can also be observed. Often radiographs of thin specimens are taken on fine grain film and enlarged to show small details.

Nuclear Industry

Nuclear applications depend on different capabilities of neutron radiography. Highly radioactive materials, such as irradiated nuclear fuel, can be inspected by neutron radiography. Such radiographs can be used to measure dimensions, to show the condition of the fuel, to observe coolant leakage or hydriding and to observe isotopic distributions. New nuclear fuel assemblies have been neutron

FIGURE 7. Thermal Neutron Radiograph of Investment-Cast Turbine Blade. Residual Ceramic Core Is Detected (Sharp White Image at Left Center).

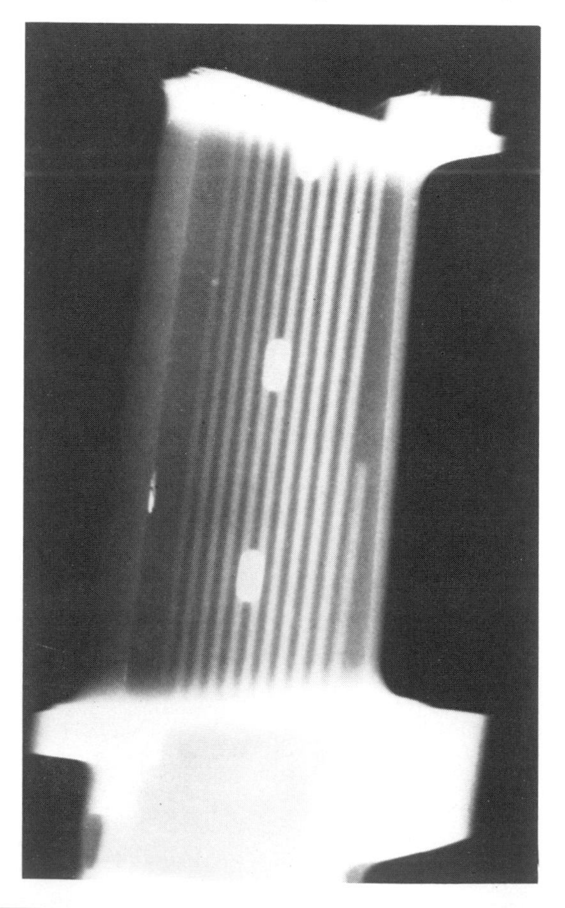

FIGURE 8. Thermal Neutron Radiographs of Electric Relays.

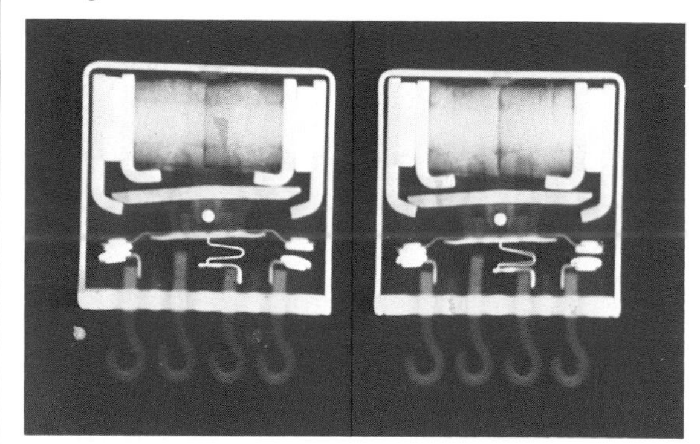

FROM AEROTEST OPERATIONS, INC. REPRINTED WITH PERMISSION.

FIGURE 9. Thermal Neutron Radiograph of Assembly Showing Two Rubber "O" Rings (Arrows) That Have Twisted Out of Grooves.

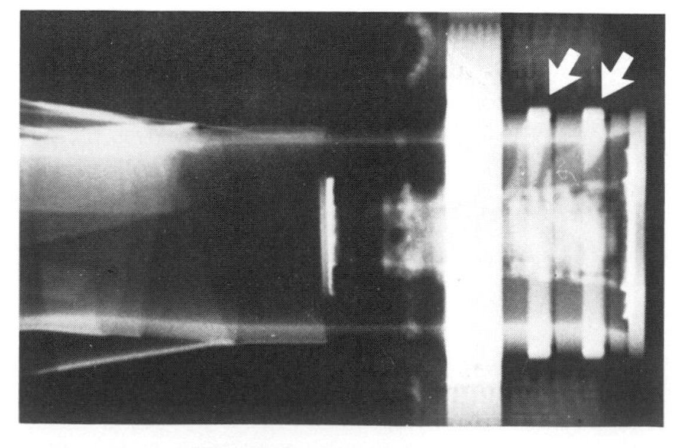

FROM AEROTEST OPERATIONS, INC. REPRINTED WITH PERMISSION.

radiographed to show the fuel condition and to detect the presence of foreign material in assemblies. Reactor control assemblies have been neutron radiographed to show nonuniform distribution, both before and after reactor service. Poison elements, which are used to control neutron flux in a reactor, can be neutron radiographed to show their distribution.

Application Summary

In industrial applications, neutrons are used to show a neutron-attenuating material such as an adhesive, rubber, plastic, fluid, or explosive in an assembly. Neutrons are used in these cases because X-radiographic capability is limited when viewing light materials behind metal. These observations can be made both in time exposures with film and paper or in real-time using fluoroscopic, television methods. For example, the flow of oil in an operating engine has been observed dynamically using real-time neutron radiography to determine the time for the lubricant to arrive at the upper parts of the engine, to detect oil blockages, and so on. The nuclear applications are by nature concentrated in nuclear centers around the world and fill the need to examine nuclear fuel and to control material before and after irradiation.

PART 4

NEUTRON RADIOGRAPHY STANDARDS, RECOMMENDED PRACTICES AND CONTROL

Personnel Qualification

Personnel qualification in neutron radiography is discussed by the American Society for Nondestructive Testing in its Recommended Practice No. SNT-TC-1A. This document covers seven NDT disciplines and includes recommended training, experience and test questions for each of those disciplines. The document is a guideline for the certification of NDT personnel.

System Performance Standards

Standards for neutron radiography system performance are written by the American Society for Testing and Materials (ASTM), subcommittee E07.05 on neutron radiography and gaging. There are three ASTM standards applicable to neutron radiography, as listed below.

1. E545 - Standard Method for DETERMINING IMAGE QUALITY IN THERMAL NEUTRON RADIOGRAPHIC TESTING
2. E748 - Standard Practices for THERMAL NEUTRON RADIOGRAPHY OF MATERIALS.
3. E803 - Standard Method for DETERMINING THE L/D RATIO OF NEUTRON RADIOGRAPHY BEAMS.

These documents deal with standardization from two viewpoints. E748 provides the basis for good working practices which lead to the production of high quality radiographs. E545 and E803 provide detailed methods for measuring the quality obtained from good facility design and good working practice. In both E545 and E803 the results can be quantified and assigned a numerical figure of merit.

Regulatory Control

The federal government and many states exercise control of radiation equipment, personnel and procedures.

Licensing Requirements

The Atomic Energy Act of 1954, as amended, establishes a regulatory framework for ensuring health and safety of the public. This is codifed in Title 10, Chapter I, Code of Federal Regulations (CFR). The Act is quite specific on delineating those sources of radiation which are subject to control by the Nuclear Regulatory Commission (formerly Atomic Energy Commission). Those sources of radiation subject to NRC regulatory control are as follows:

1. Special nuclear material: uranium enriched in the isotope 233 or 235 or plutonium.
2. Source material: uranium or thorium.
3. Production facilities: equipment or devices capable of producing significant quantities of special nuclear materials.
4. Utilization facilities: equipment or devices capable of making use of significant quantities of special nuclear material.
5. Byproduct material: radioactive material (except special nuclear material) made radioactive by the utilization or production of special nuclear material.

Section 274 of the Atomic Energy Act was enacted in 1959. Among other things, this section provided a statutory means by which the NRC can relinquish to the states a part of its regulatory authority. The NRC retains control over production and utilization facilities and large quantities of special nuclear material. Certain other areas are reserved to NRC as stated in 10CFR 150.15 (a).

TABLE 8. Principle Regulation Requirements for Neutron Radiography

Type of Radiation Source	10 CFR 30 or Agreement State Equivalent	10 CFR 34 or Agreement State Equivalent	10 CFR 70 or Agreement State Equivalent	10 CFR 73	10 CFR 50	10 CFR 20 or State Equivalent	State Registration	Special State Requirements
1. Reactor	—	—	—	—	X	X	—	—
2. Byproduct material	X	X*	—	—	—	X	X	—
3. Special nuclear material	—	—	X	X	—	X	X	—
4. Naturally occurring or accelerator-produced radioactive material	X**	X**	—	—	—	X	X†	—
5. Radiation machine	—	—	—	—	—	X	X	X
6. Accelerator	—	—	—	—	—	X	X	X††

*Tritium targets for neutron generators not subject to Part 34.
**Agreement states and some non-agreement states.
†Non-agreement states may register or license.
††Not all states have special requirements.

States

Authority over source material, byproduct material, and small quantities of special nuclear material can be borne by individual states. In addition, the states have regulatory programs for naturally occurring radioactive materials, accelerator-produced radioactive materials, and all radiation-producing machines (for example, X-ray machines and particle accelerators). Some states license all radioactive materials and they register all machines. Table 8 shows a summary of the regulatory requirements for the various types of sources.

Personnel Protection

All NRC licenses must comply with the requirements of 10CFR Parts 19 and 20. Part 19 contains requirements for the posting of notices to workers, providing radiation safety instructions to workers, notification to NRC of the occurrence of incidents, worker rights during NRC inspections. Part 20 contains the standards for protection against radiation. This part includes the standards for permissible doses for individuals, permissible levels of radiation in unrestricted areas, survey requirements, personnel monitoring requirements, requirements for precautionary signs and labels, procedures for picking up, receiving, and opening of packages, storage of licensed materials, maintaining certain records, notification to NRC of the occurrence of incidents, and a requirement for reporting certain personnel monitoring results. Part 34 deals specifically with radiography. It deals with personnel training, use of survey meters, leak tests for radiographic sources and other matters relating to radiographic use of radioactive sources.

PART 5

NEUTRON CROSS SECTIONS AND ATTENUATION

Neutron cross sections are defined in Part 1 of this Section. Values for thermal neutrons for many materials (elements) are given in Table 9 (see Bibliography item 8 for a more extensive compilation). Generally, neutron cross sections decrease with increasing neutron energy; exceptions include resonances, as mentioned earlier. Cross section values can be used to calculate the attenuation coefficients and the neutron transmission as shown in eqs. 1 and 2. For compound inspection materials, the method for calculating the linear attenuation coefficient is shown following Table 9.

TABLE 9. Thermal Neutron Linear Attenuation Coefficients Using Average Scattering and 2200 m/s Absorption Cross Sections for the Naturally Occurring Elements

Element		Cross Section (barns)		Linear Attenuation Coefficient (cm^{-1})
Atomic No.	Symbol	Scattering	Absorption*	
1	H_2	38.0	0.332	gas
2	He	0.8	. . .	gas
3	Li	1.4	71.0	3.36
4	Be	7.1	0.010	0.88
5	B	4.4	755	99
6	C	4.8	0.003	0.541
7	N_2	10	1.88	gas
8	O_2	4.2	0	gas
9	F_2	3.9	0.01	gas
10	Ne	2.4	2.8	gas
11	Na	4.0	0.536	0.115
12	Mg	3.6	0.063	0.158
13	Al	1.4	0.23	0.0984
14	Si	1.7	0.16	0.0965
15	P	5.0	0.20	0.184
16	S	1.1	0.52	0.0591
17	Cl_2	16.0	33.6	gas
18	A	1.5	0.66	gas
19	K	1.5	2.07	0.047
20	Ca	3.2	0.44	0.0849
21	Sc	24.0	24.0	1.609
22	Ti	4.0	5.8	0.555
23	V	5.0	4.9	0.698
24	Cr	3.0	3.1	0.509
25	Mn	2.3	13.2	1.224
26	Fe	11.0	2.53	1.149
27	Co	7.0	37.0	4.01
28	Ni	17.5	4.8	2.04
29	Cu	7.2	3.8	0.931
30	Zn	3.6	1.1	0.309
31	Ga	4.0	2.8	0.347
32	Ge	3.0	2.45	0.242
33	As	6.0	4.3	0.475
34	Se	11.0	12.3	0.856
35	Br	6.0	6.7	0.263
36	Kr	7.2	31	gas
37	Rb	5.5	0.73	0.0673
38	Sr	10.0	1.21	0.201
39	Y	8.0	1.31	0.347
40	Zr	8.0	0.180	0.346
41	Nb	5.0	1.15	0.341
42	Mo	7.0	2.7	0.621
43	Tc	5.0	22.0	density unknown

TABLE 9. Continued

Element		Cross Section (barns)		Linear Attenuation Coefficient (cm^{-1})
Atomic No.	Symbol	Scattering	Absorption*	
44	Ru	6.0	2.56	0.615
45	Rh	5.0	156.0	11.70
46	Pd	3.6	8.0	0.746
47	Ag	6.0	63.0	4.05
48	Cd	7.0	2450	113.5
49	In	2.2	196	7.60
50	Sn	4.0	0.625	0.171
51	Sb	4.3	5.7	0.370
52	Te	5.0	4.7	0.286
53	I	3.6	7.0	0.248
54	Xe	4.3	. . .	gas
55	Cs	7.0	29.0	0.306
56	Ba	8.1	1.2	0.143
57	La	9.3	9.3	0.496
58	Ce	2.8	0.73	0.102
59	Pr	4.0	11.6	0.434
60	Nd	16.0	46.0	1.785
61	Pm	. . .	60	density unknown
62	Sm	. . .	5600	173.0
63	Eu	8.0	4300	89.1
64	Gd	. . .	46 000	1405.0
65	Tb	. . .	46.0	1.455
66	Dy	100	950	33.3
67	Ho	. . .	65	2.08
68	Er	7.8 (coh)	173	5.94
69	Tm	7	127	4.46
70	Yb	12	37	1.195
71	Lu	. . .	112	3.75
72	Hf	8	105	5.07
73	Ta	5	0.03	0.278
74	W	5	19.2	1.53
75	Re	14	86	6.64
76	Os	15.2 (coh)	15.3	2.17
77	Ir	. . .	440	30.9
78	Pa	10	8.8	1.244
79	Au	9.3	98.8	6.39
80	Hg	20	380	16.3
81	Tl	14	3.4	0.607
82	Pb	11	0.17	0.368
83	Bi	9	0.034	0.258
84	Po	. . .	. . .	. . .
85	At	. . .	. . .	density unknown
86	Rn	. . .	. . .	gas
87	Fr	. . .	. . .	density unknown
88	Ra	. . .	130	1.69
89	Ac	. . .	510	density unknown
90	Th	. . .	1500 (fission)	44.1
91	Pa	. . .	1500 (fission)	60.4
92	U	. . .	7.68 (includes fission)	0.788
93	Np	. . .	900 (fission)	density unknown
94	Pu	. . .	160 (fission)	7.96

*All cross-section values are most probable values with no implied accuracy.

If the material under inspection contains only one element, then the linear attenuation coefficient is:

$$\mu = \rho \frac{N\sigma}{A} \qquad \text{(Eq. 7)}$$

where μ is the attenuation coefficient (cm^{-1}); ρ is the material density (g/cm^3); N is Avogadro's number (6.023×10^{23} atoms/mole); σ is the total cross section in cm^2; and A is the gram atomic weight of material in g/mol.

If on the other hand, the material under inspection contains several elements, or is in the form of a compound, then the linear absorption coefficient is:

$$\mu = \rho \frac{N}{M}(\nu_1\sigma_1 + \nu_2\sigma_2 + \ldots \nu_i\sigma_i) \qquad \text{(Eq. 8)}$$

where μ is the linear attenuation coefficients of the compound (cm^{-1}); ρ is the compound density (g/cm^3); N is Avogradro's number; M is the molecular weight of the compound; ν_i is the number of absorbing atoms of ith kind per compound molecule; and, σ_i is the total cross section of the ith atom (cm^2).

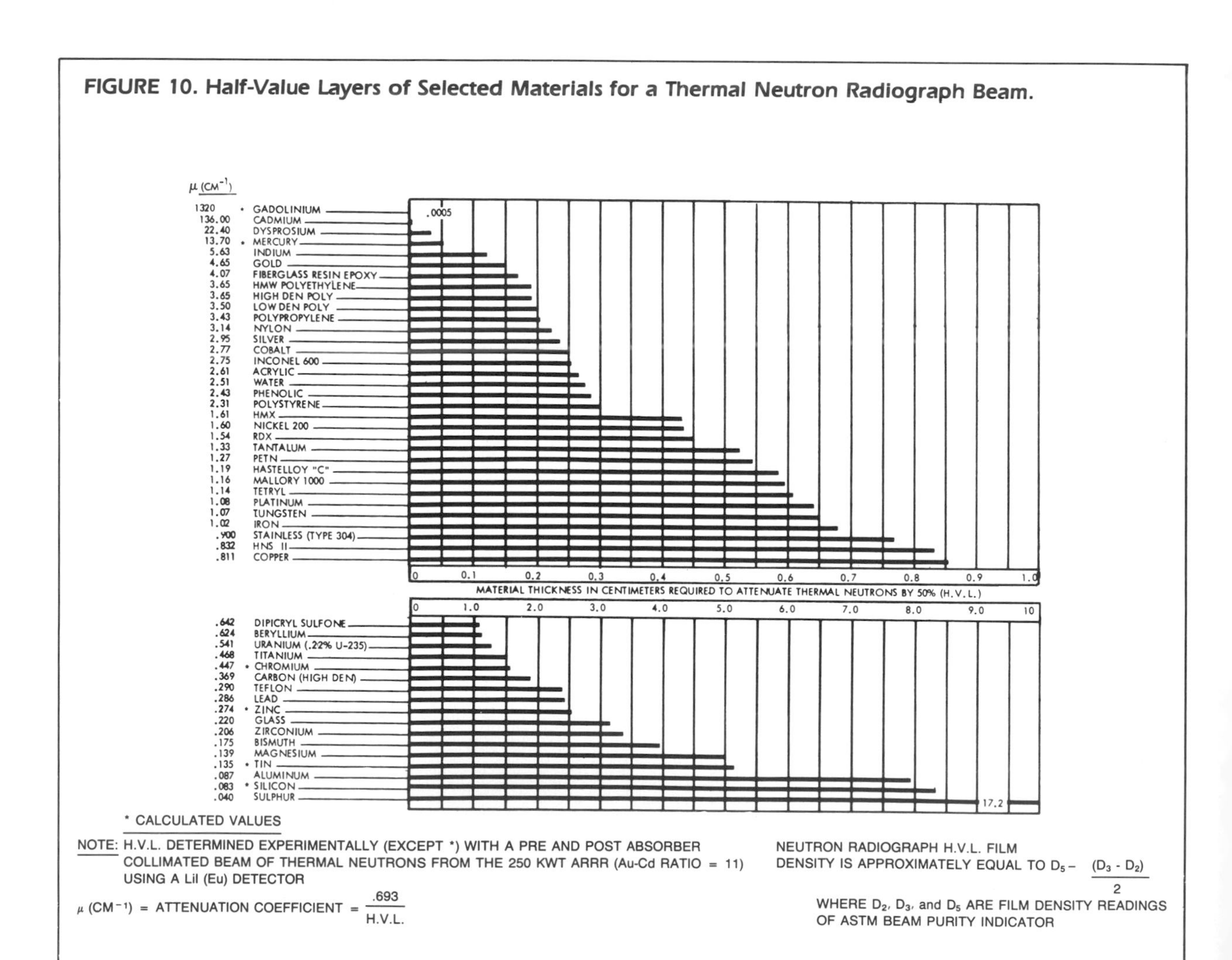

FIGURE 10. Half-Value Layers of Selected Materials for a Thermal Neutron Radiograph Beam.

* CALCULATED VALUES

NOTE: H.V.L. DETERMINED EXPERIMENTALLY (EXCEPT *) WITH A PRE AND POST ABSORBER COLLIMATED BEAM OF THERMAL NEUTRONS FROM THE 250 KWT ARRR (Au-Cd RATIO = 11) USING A LiI (Eu) DETECTOR

μ (CM^{-1}) = ATTENUATION COEFFICIENT = $\frac{.693}{\text{H.V.L.}}$

NEUTRON RADIOGRAPH H.V.L. FILM DENSITY IS APPROXIMATELY EQUAL TO $D_5 - \frac{(D_3 - D_2)}{2}$

WHERE D_2, D_3, and D_5 ARE FILM DENSITY READINGS OF ASTM BEAM PURITY INDICATOR

As an example, consider the calculations of the linear attenuation coefficient, μ, for the compound polyethylene $(CH_2)_N$:

$$\mu = \rho \frac{N}{M}(\nu_c\sigma_c + \nu_H\sigma_H)$$

where:

$\rho = 0.91\ g/cm^3$
$N = 6.023 \times 10^{23}$ atoms/mol
$M = 14.0268$ g/mol
$\nu_c = 1$
$\sigma_c = 4.803 \times 10^{-24}\ cm^2$
$\nu_H = 2$
$\sigma_H = 38.332 \times 10^{-24}\ cm^2$

Thus, the linear attenuation coefficient is:

$$\mu = \frac{0.91(6.023 \times 10^{23})}{14.0268}[(1)\ (4.803) + (2)\ (38.332)]10^{-24}$$

$$\mu = \frac{0.390}{10}\ (81.457)$$

$$\mu = 0.390\ (8.1457)$$

$$\mu = 3.177\ cm^{-1}$$

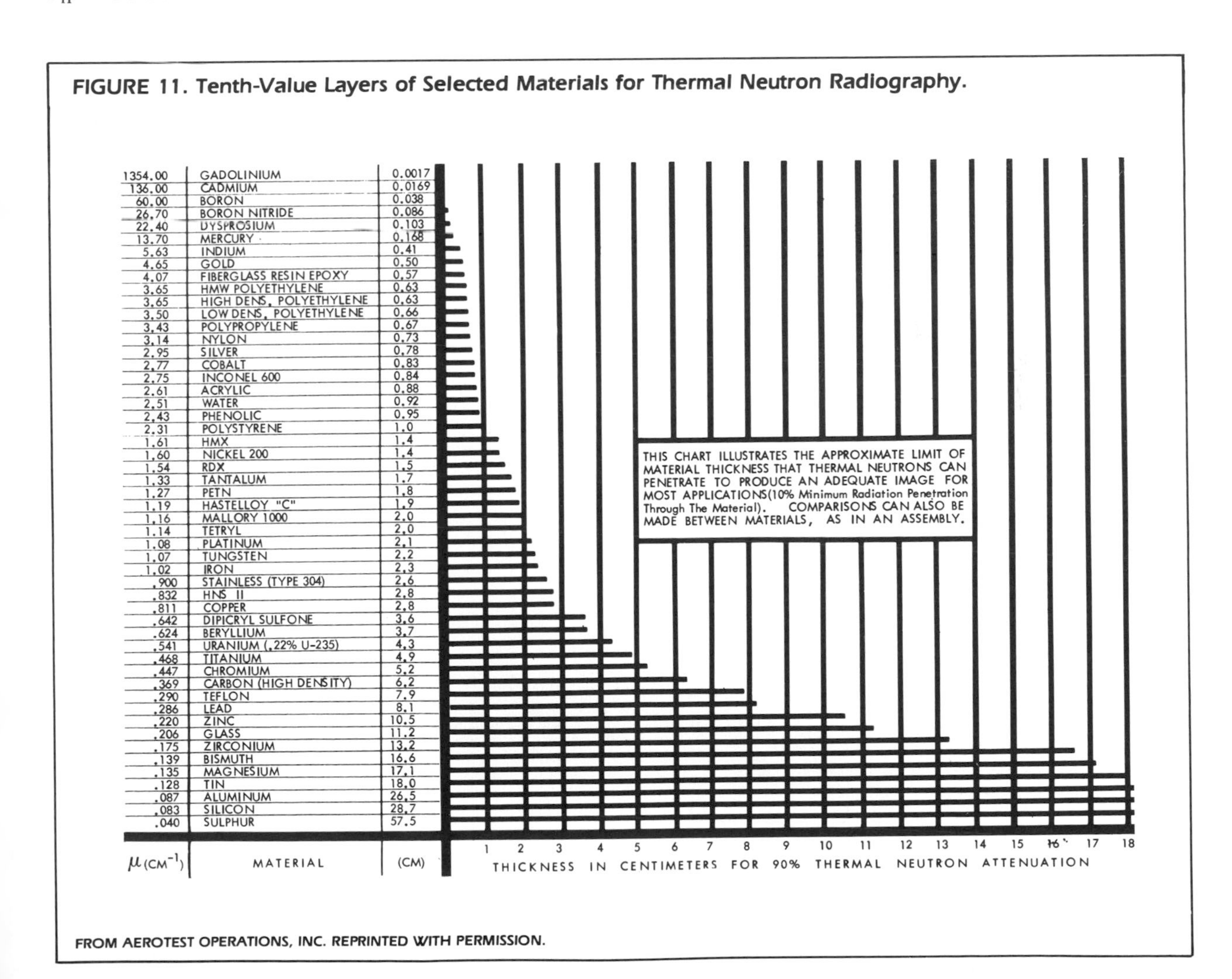

FIGURE 11. Tenth-Value Layers of Selected Materials for Thermal Neutron Radiography.

FROM AEROTEST OPERATIONS, INC. REPRINTED WITH PERMISSION.

Half-Value Layers

An important concept for radiography is the *half-value layer* (HVL); that is, the thickness of material that will reduce the radiation intensity by a factor of two. A plot of half-value layers for a practical thermal neutron radiographic beam is given in Fig. 10. This information can be used to estimate the transmission and detectability of various materials combined with others.

Tenth-Value Layers

There will be a thickness of material that is sufficiently thick that little of the neutron beam penetrates. In Fig. 11, thicknesses of material that will transmit only 10% of an incident thermal neutron radiographic beam are plotted. This thickness represents about the limit that should be attempted in normal thermal neutron radiography. Variations in neutron energy should be considered for thicknesses greater than those shown in Fig. 11.

CONCLUSION

Neutron radiography is a valuable method for nondestructive testing. The attenuation differences between X-rays and neutrons make these two radiographic methods, to a large degree, complementary. Figure 6 in this Section is an illustration of how the two methods provide a more complete inspection when used together. The neutrons in this example show light materials such as the explosive, plastic and epoxy components, while the X-radiograph shows the metallic components.

Neutrons offer sensitivity to different isotopes and can also be very useful for inspecting highly radioactive material. These two characteristics offer advantages particularly to the nuclear industry. Other areas of application include aerospace, the military and transportation industries.

The neutron radiographic technique is relatively expensive, but it can be used to perform inspections that present problems for other NDT methods. When used for these unique inspections, neutron radiography is a cost-effective nondestructive testing technique.

BIBLIOGRAPHY

1. Annual ASTM Book of Standards, Vol. 03.03, "Metallography; Nondestructive Tests." American Society for Testing and Materials, Philadelphia.

2. *Atomic Energy Review,* Neutron Radiography Issue. Vol. 15, No. 2, International Atomic Energy Agency, Vienna (1977).

3. Barton, J.P. and P. von der Hardt, eds. "Neutron Radiography—Proceedings of the First World Conference." D. Reidel Publ. Co., Dordrecht, Holland and Boston (1983).

4. Berger, H., *Neutron Radiography—Methods, Capabilities and Applications.* Elsevier, Amsterdam (1965).

5. Berger, H., ed. "Practical Applications of Neutron Radiography and Gaging." ASTM Special Technical Publication 586, American Society for Testing and Materials, Philadelphia (1976).

6. Hawkesworth, M., ed. "Radiography With Neutrons." British Nuclear Energy Society, London (1975).

7. Herz, R. *The Photographic Action of Ionizing Radiations.* Wiley-Interscience, New York (1969).

8. Hughes, D.J. and R.B. Schwartz. *Neutron Cross Sections.* Report BNL-325 (second edition and supplements). Brookhaven National Laboratory, Upton, N.Y. (1958-1973; supplements issued periodically).

9. Metals Handbook, Vol. 11. *Nondestructive Inspection and Quality Control.* American Society for Metals, Metals Park, Ohio (1976).

10. Morgan, K.Z. and J.E. Turner. *Principles of Radiation Protection.* Wiley, New York (1967).

11. NCRP Report No. 38. "Protection Against Neutron Radiation." National Council on Radiation Protection, Washington, DC (1971).

12. NBS Handbook 72. "Measurement of Neutron Flux and Spectra for Physical and Biological Applications." National Bureau of Standards, Washington, DC (1960).

13. Tyufyakov, N.D. and A.S. Shtan. "Principles of Neutron Radiography." Translated from Russian and reprinted by Amerind Publ. Co., New Delhi, India (1979).

14. von der Hardt, P. and H. Rottger, eds. *Neutron Radiography Handbook.* D. Reidel Publ. Co., Dordrecht, Holland, Boston and London (1981).

SECTION 13

IMPLEMENTATION OF NEUTRON RADIOGRAPHY

John P. Barton, Consultant, La Jolla, CA

ACKNOWLEDGMENTS

This Section of the *Nondestructive Testing Handbook* contains illustrations from more than twenty neutron radiography centers in five countries. The author wishes to offer special thanks to those individuals who so promptly provided artwork and who, in many cases, reviewed the chapter's text.

INTRODUCTION

Neutrons interact with matter in ways more complex than X rays. This is an advantage because of the resulting diversity of application possibilities.

When considering neutron radiography, it usually does little good to contact a local NDT supplier. Systems for this form of testing are not bought off the shelf. Managers considering implementation of neutron radiography should ask "What has been put into practice already?" This section helps answer this vital question.

PART 1

FUNDAMENTALS OF IMPLEMENTATION

A comprehensive review of neutron radiographic principles is provided in Section 12. This Section focuses on implementation and provides guidance on available systems.

Approaches to Implementation

The options for using neutron radiography (NR) include:

1. trial neutron radiographs at a commercial center;
2. trial neutron radiographs at specialized non-commercial facilities;
3. development contract; or
4. procurement of a custom-designed system.

Step 1: Commercial Services

For a list of current NR service centers, a suggested source is: The American Society for Nondestructive Testing (Penetrating Radiation Committee).

A typical NR service center can provide a high-quality radiograph using thermal neutrons. Turnaround time is normally a day or two; an overnight delivery service is often used. Costs are similar to custom X-radiography.

Suggested questions for the service center include:

1. types of NR available (thermal, epithermal, cold, etc.);
2. imaging method (film, converter, real-time, etc.);
3. radiographic image quality (ASTM E545 or alternate);
4. collimation available (ASTM E803 or alternate);
5. orientations available (beam vertical, horizontal);
6. part limitations (size, explosive content, etc.);
7. induced radioactivity (typically insignificant)
8. radiation damage (typically insignificant);
9. part identification; and
10. NR interpretation assistance.

If routine neutron radiography service can be established, consider the preparation of specifications, and note that: (1) thermal neutron radiographs can vary significantly in technique and quality (specify both); (2) ASTM standards for NR quality control should often be supplemented by reference

FIGURE 1. Comparison of (a) N-ray and (b) X-ray as Aid to Interpretation; Example is a Full-size Motorcycle

(a)

(b)

FROM ROCKWELL CORPORATION. REPRINTED WITH PERMISSION.

standards matching the parts and defects of concern; and (3) correct interpretation of the neutron radiograph can be as important as correct performance of the radiography (specify interpretation criteria).

Step 2: Specialized Neutron Radiography Centers

If commercial service NR centers cannot meet the neutron radiography requirements, consider existing centers with specialized equipment. There are, for example, about 70 research reactors in operation in the United States (350 worldwide).[1] There are also numerous accelerator facilities offering continuous or pulsed neutron sources.[2] Information on the range of existing systems is given in Parts 2 through 5 of this Section.

Step 3: Technique Development

Consider next whether your application has sufficient potential to warrant a development project. Outlines of some successful development projects are given in Part 6.

Step 4: Custom Systems

Consider also the range of custom-designed systems already available for implementation. Justification of an in-house neutron radiography system should include analysis of required performance, throughput, logistics, safety and economics.

FIGURE 2. Comparison of Radiographs Exposed with Different N-ray Energies (Example is an Explosive Bridge Wire Igniter): (a) Thermal Neutron Image; (b) Cold Neutron Image

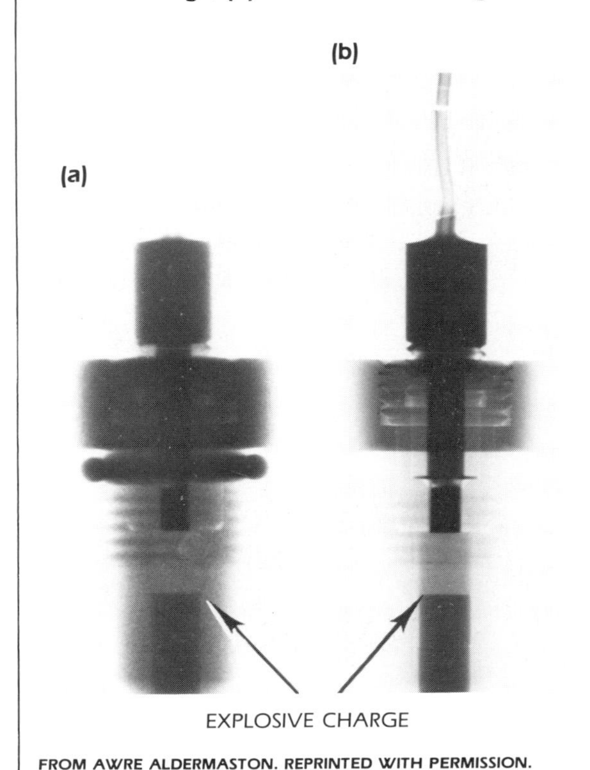

FROM AWRE ALDERMASTON. REPRINTED WITH PERMISSION.

FIGURE 3. Comparison of N-rays (a) Before and (b) After Processing; Example is Moisture Globules in Aluminum Honeycomb Panel, Later Dried

(a)

(b)

FROM US AIR FORCE. REPRINTED WITH PERMISSION.

FIGURE 4. Comparison of (a) Drawing and (b) N-ray of the Part as an Aid to Interpretation; Also, Comparison of N-rays Showing Helium-3 Gaseous Penetrant Applied to (c) Serviceable Unit and (d) Defective Unit; Part is an Electric Bridge Wire Squid

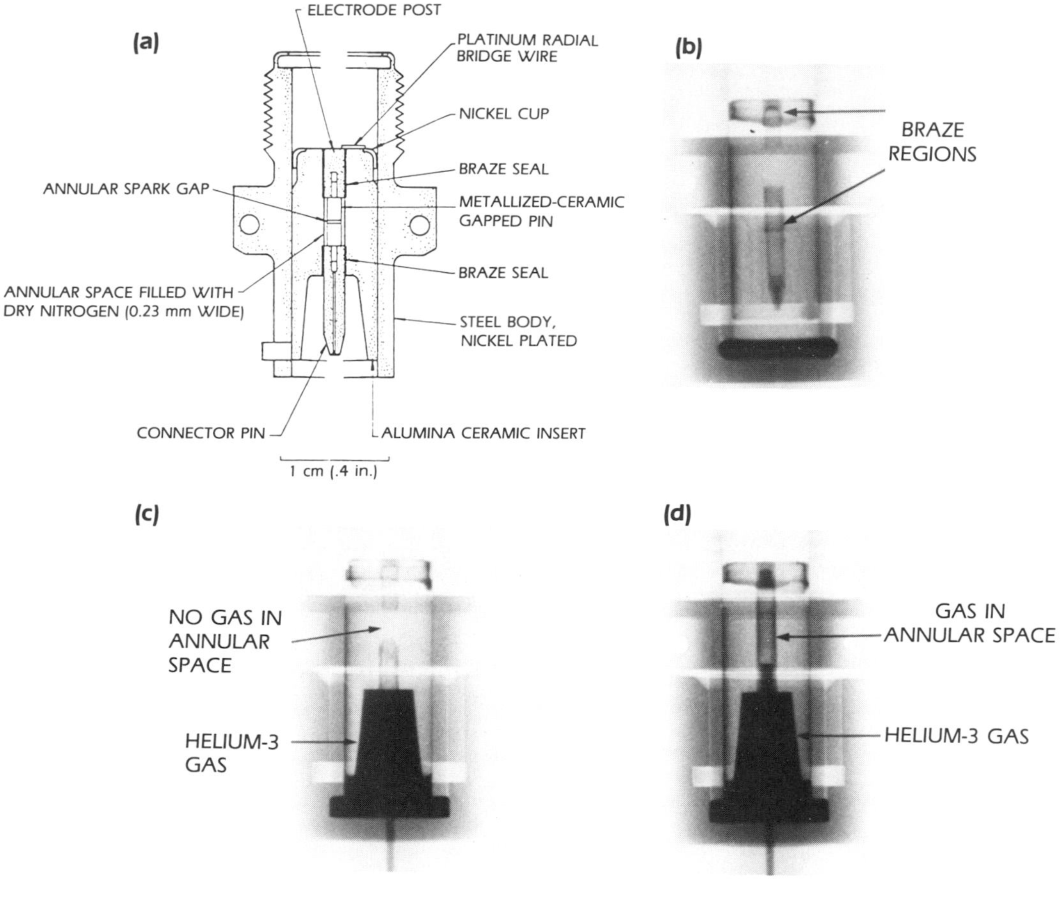

FROM AWRE ALDERMASTON. REPRINTED WITH PERMISSION.

Approaches to Interpretation

Tools to assist in interpretation of neutron radiographs include:

1. attenuation data (see Section 12, Part 5);
2. reference objects (with known defects);
3. reference radiographs (custom developed);
4. complementary nondestructive testing (nonradiographic);
5. X-ray comparison (Fig. 1);
6. N-ray comparisons at different energies (Fig. 2);
7. N-ray comparisons before and after changes (Fig. 3); and
8. tracers or penetrants (Fig. 4).

Once a neutron radiographic technique is established, interpretation of results is normally not difficult. It is, however, important. For instruction courses on NR interpretation, contact one of the service centers or equipment supply centers.[3]

PART 2

IMPLEMENTATION OF LOW-YIELD SYSTEMS

Low-yield systems fall into three categories: stationary, maneuverable and mobile. A stationary system can be located close to the application site. A maneuverable system is one which, fixed within a building, can be moved to scan an object such as an aircraft. A mobile system can be switched off and moved to an entirely different location upon demand. Source types used for low-yield systems are the *d,T* accelerator, and the californium-252 isotopic source.

The thermal neutron flux available from the standard *d,T* accelerator equals that from 7 mg of californium-252. Rapid scintillator-film combinations or electronic imaging can produce useful images of modest quality, typically in a few minutes. At a low collimator ratio of 20:1, such systems require up to 24 hours for the high-quality film exposure (3×10^9 n/cm^2) that will be used as a reference for medium- and high-yield systems. Overnight exposures are practical with californium-252 systems.

d,T Accelerator Systems

An available neutron radiography system is shown in Figs. 5 and 6. The power supply and cooling system are on mobile platforms. The accelerator head is near the center of a sphere of moderator material from which the collimator assembly can be seen protruding. Radiation safety for this mobile system is provided by barriers and interlocks to ensure sufficient distance between source and all personnel during operation.

The neutron-generating tube is inside the permanently sealed SF_6 pressurized accelerator head (Fig. 6a), which is movable on long cables. The tube itself consists of an ion source, titanium tritide target and focusing electrodes. In operation, a mixed beam of deuteron and tritium ions are accelerated to continuously replenish the target. Yields of 7×10^{10} neutrons per second (n/s) with less than 50% deterioration after 200 hours are claimed for this generator. The high-voltage power supplies are generated using a reliable full-wave voltage doubler in the cylindrical tank, which is pressured with SF_6 (Fig. 6b, right). The cooling system (Fig. 6b, left) provides a closed-loop water circuit to cool the target, and a freon circuit to cool the ion source. The control unit for the accelerator (Fig. 6c) contains all instrumentation necessary for remote operation. Electrical requirements for this mobile unit are standard 115 and 230 volts.

Californium-252 Systems

A stationary californium-252 neutron radiography system is shown in Fig. 7. It consists of a solid moderator, a collimator, and a surrounding shield that accounts for the majority of the size. Since the first deliveries of californium-252 based systems in 1974, continuing development has produced dramatic improvements in image quality. By using sufficient time for exposures, it is normal to obtain low collimation radiographs that are otherwise as high in image quality as some of the best reactor-produced radiographs.

Construction of the californium-252 source is illustrated in Figs. 8 and 9. The source material, which is small, is doubly encapsulated in stainless steel or in combinations of steel, platinum and zirconium. These sources are tested to withstand severe crush and fire tests, so the hazard is no greater than that of any typical gamma-ray source widely used in industrial radiography.

Comparisons

The *d,T* source can be switched off like an X-ray source. If the *d,T* source is to be highly utilized, budget considerations should include the cost of replacement tubes which are typically 10% of the accelerator cost for each few hundred hours of operation. Because of the tritium content (about 10 curies), implementation of a *d,T* source must be preceded by application for a license, as for californium-252.

A limited number of californium-252 sources (by-products of other irradiation programs) could be purchased or loaned, in quantities about ten times more abundant in thermal neutrons than the available d,T generators. Other properties of californium-252 sources are (1) small size; (2) the ability to moderate the neutrons compactly with less weight than the d,T source; and (3) the ability to shield the neutrons more easily. Purchase costs of intense sources are very high and, whether loaned or purchased, there will be encapsulation and shipment costs affected by the sources' 2.5 year half-life.

FIGURE 5. Mobile *d,T* NR System

FROM VOUGHT CORPORATION. REPRINTED WITH PERMISSION.

FIGURE 6. Components of Mobile *d,T* NR System: (a) *d,T* Source Head, Typically on 6 m (20 ft) Cables; (b) Cooling Unit (Left) and Power Supply (Right); and (c) Control Unit

(a)

(b)

(c)

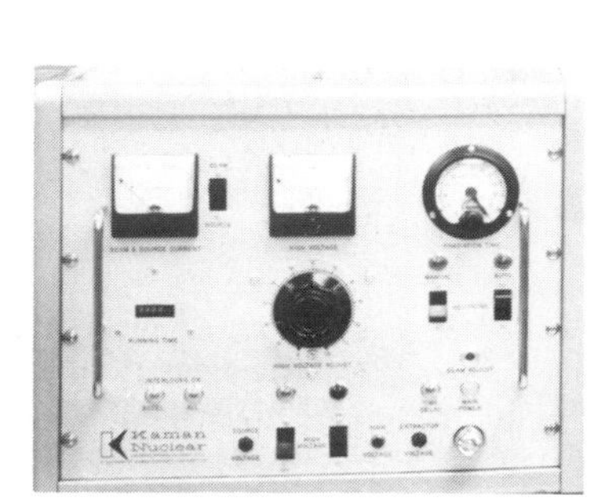

FROM KAMAN SCIENCES CORPORATION. REPRINTED WITH PERMISSION.

FIGURE 7. Stationary Californium-252 NR System

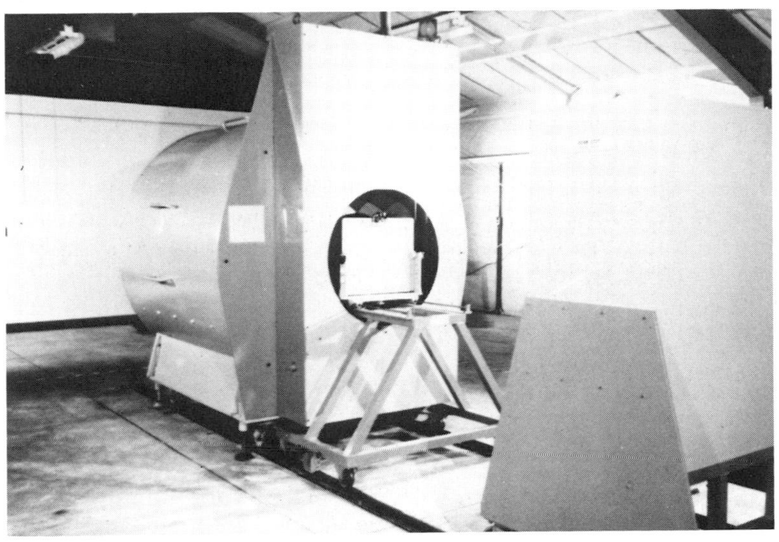

FROM KENNEDY SPACE CENTER. REPRINTED WITH PERMISSION.

FIGURE 8. Californium-252 Sources Compared in Size to Postage Stamp

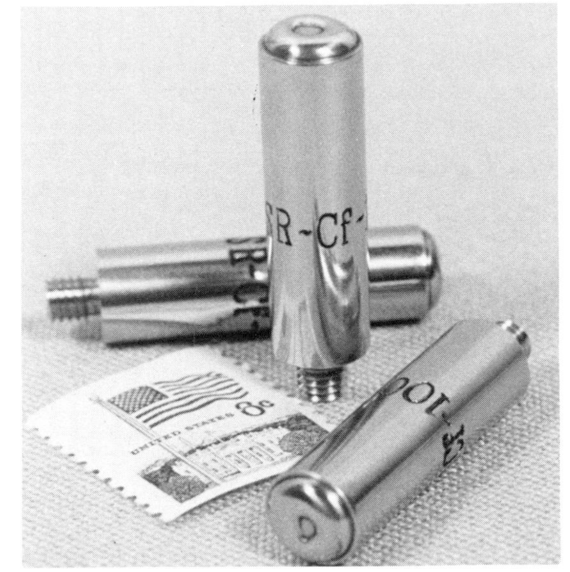

FROM SAVANNAH RIVER LABORATORY. REPRINTED WITH PERMISSION.

FIGURE 9. Cross Section of Californium-252 Source, Doubly Encapsulated

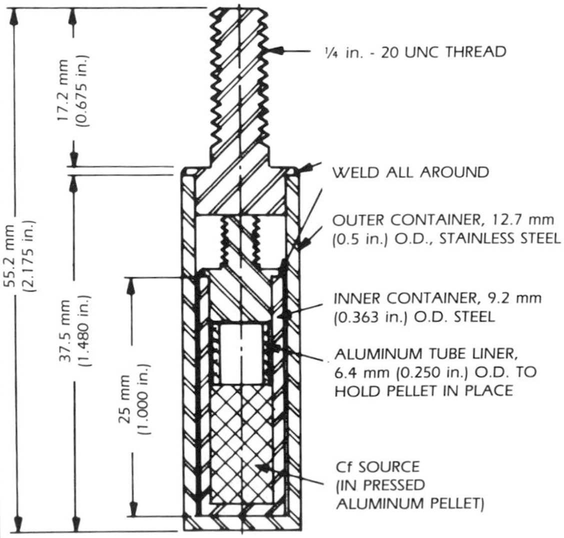

17.2 mm (0.675 in.)
55.2 mm (2.175 in.)
37.5 mm (1.480 in.)
25 mm (1.000 in.)
¼ in. - 20 UNC THREAD
WELD ALL AROUND
OUTER CONTAINER, 12.7 mm (0.5 in.) O.D., STAINLESS STEEL
INNER CONTAINER, 9.2 mm (0.363 in.) O.D. STEEL
ALUMINUM TUBE LINER, 6.4 mm (0.250 in.) O.D. TO HOLD PELLET IN PLACE
Cf SOURCE (IN PRESSED ALUMINUM PELLET)

FROM OAK RIDGE NATIONAL LABORATORY. REPRINTED WITH PERMISSION.

PART 3

IMPLEMENTATION OF MEDIUM-YIELD SYSTEMS

Medium-yield systems successfully implemented for routine application of neutron radiography include: the subcritical neutron multiplier; the Van de Graaff accelerator; and the mini-reactor. These are designed to meet specific needs for in-house neutron radiography. By use of widely divergent beams, the number of films simultaneously exposed can partly compensate for exposure times that are long compared with X-ray or high-yield N-ray systems.

Subcritical Multiplier

A subcritical multiplier (see Figs. 10 and 11) is similar to a standard plate-type research reactor, but is designed to operate just below the level at which a chain reaction can be sustained. Advantages, compared with a research reactor, include lower cost, minimal licensing and operator requirements. Two types of subcritical multipliers have been manufactured for neutron radiography: one with solid plastic moderator, and one with water moderator. The latter type, which permits higher power levels, is described below.

FIGURE 10. Elevation of a Subcritical Multiplier System

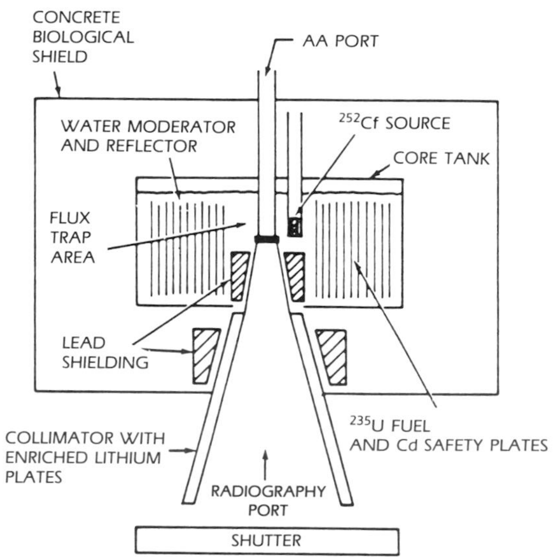

CONCRETE BIOLOGICAL SHIELD
AA PORT
WATER MODERATOR AND REFLECTOR
^{252}Cf SOURCE
CORE TANK
FLUX TRAP AREA
LEAD SHIELDING
COLLIMATOR WITH ENRICHED LITHIUM PLATES
^{235}U FUEL AND Cd SAFETY PLATES
RADIOGRAPHY PORT
SHUTTER

FROM IRT CORPORATION. REPRINTED WITH PERMISSION.

Subcritical Multiplier with Water Moderator

In this system, the fuel is aluminum-clad metallic plates with 93% uranium-235. The neutron source, 40 mg californium-252, is placed in a central flux trap. Other components include aluminum-clad cadmium safety shutdown blades, power monitors, control system, water pump, filter, demineralizer, tank, frame to hold fuel, and shielding.

Because the system is small, it can be positioned to suit the neutron radiography application. One such facility houses an above-ground multiplier, and a beam extracted from the flux trap vertically downward (Fig. 10). Delicate items can be placed on top

FIGURE 11. Subcritical Multiplier Facility with Control Room and Tower

FROM MONSANTO CORPORATION. REPRINTED WITH PERMISSION.

of film cassettes for radiography without risk of damage. By changing the table height, collimator ratios may be selected between 15:1 and 100:1.

With a source of 40 mg californium-252 in the multiplier, the power is 130 watts, and the peak flux is 7×10^9 $ncm^{-2}s^{-1}$. Using L:D = 100:1, the exposure time for high-quality, single-emulsion film (3×10^9 n/cm^2) is about 16 hours. Using L:D = 20:1, the exposure time is about 30 minutes. Four films of 36 × 43 cm (14 × 17 in.) can be exposed simultaneously. The operator needs to be present only for about ten minutes at startup, and ten minutes at shutdown. Otherwise, the equipment operates unattended and can be routinely set for overnight exposures.

Van de Graaff System

A Van de Graaff system designed for neutron radiography is illustrated in Fig. 12. Deuterons (3 MeV, 280 μA) are accelerated onto a disk-shaped, water-cooled beryllium metal target. Neutrons in the range of 2 to 6 MeV are emitted preferentially in the forward direction, and moderated in water. The 4π (solid angle) yield of 5×10^{11} n/s produces a peak thermal neutron flux of 2×10^9 $ncm^{-2}s^{-1}$. At a collimator ratio of 36:1, the typical exposure time for high-quality film (3×10^9 n/cm^2) is about two hours.

The principle of the Van de Graaff machine is illustrated in Fig. 13. A rotating belt transports charge from a supply to a high-voltage terminal. An ion source within the terminal is fed deuterium gas from an external reservoir. A radiofrequency system ionizes the gas and positive ions are extracted into the accelerator tube. The terminal voltage of about 3 MV is distributed by a resistor chain over about 80 gaps forming the accelerator tube, all of which is enclosed in a pressure vessel filled with insulating gas (N_2 and CO_2 at 20 atmospheres or 2 MPa).

The particle beam is extracted along flight tubes. It bombards the water-cooled beryllium target in the center of the water moderator tank, which also serves as a partial shield. The accelerator tank for the 3 MeV machine measures 5.2 m (17 ft) in length and 1.5 m (5 ft) in diameter. The weight is 6,100 kg (13,500 lb). The dimensions of the water tank

FIGURE 12. Elevation of a Van de Graaff Accelerator System

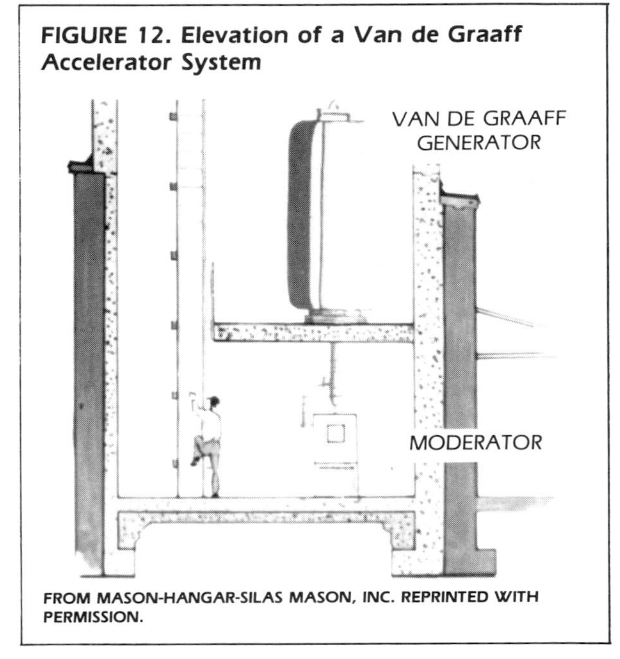

FROM MASON-HANGAR-SILAS MASON, INC. REPRINTED WITH PERMISSION.

FIGURE 13. Cross Section Showing Van de Graaff Principle

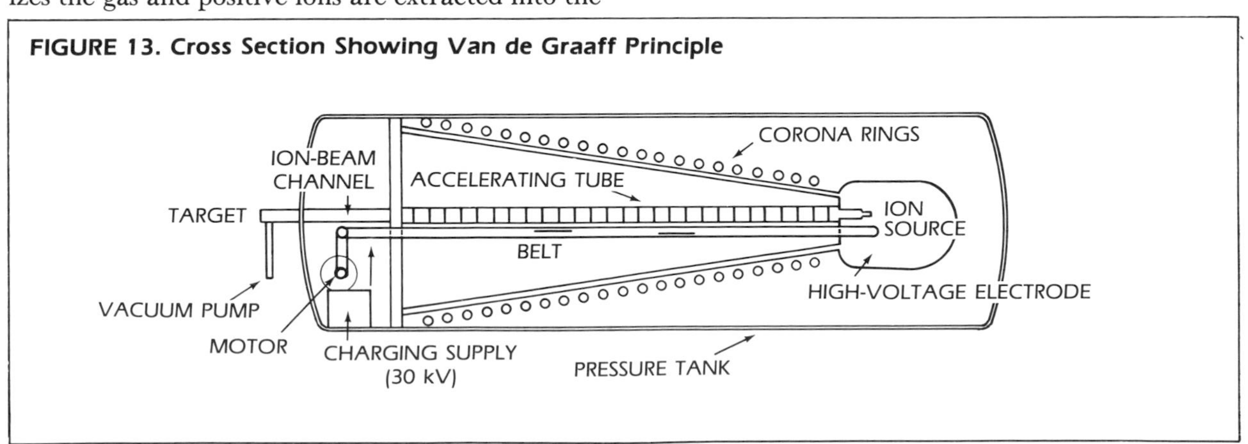

FIGURE 14. Mirene Mini-reactor for NR

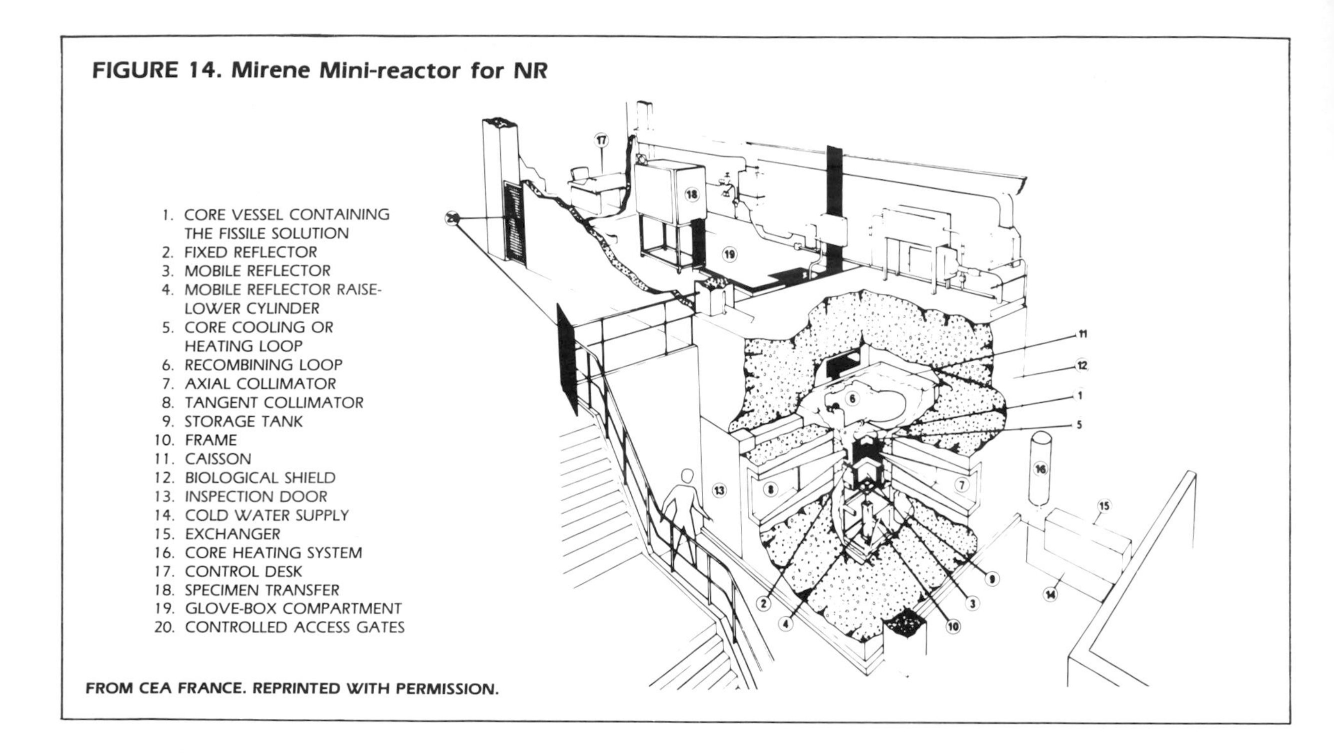

FROM CEA FRANCE. REPRINTED WITH PERMISSION.

FIGURE 15. Mini-reactor Time Behavior for One NR Exposure

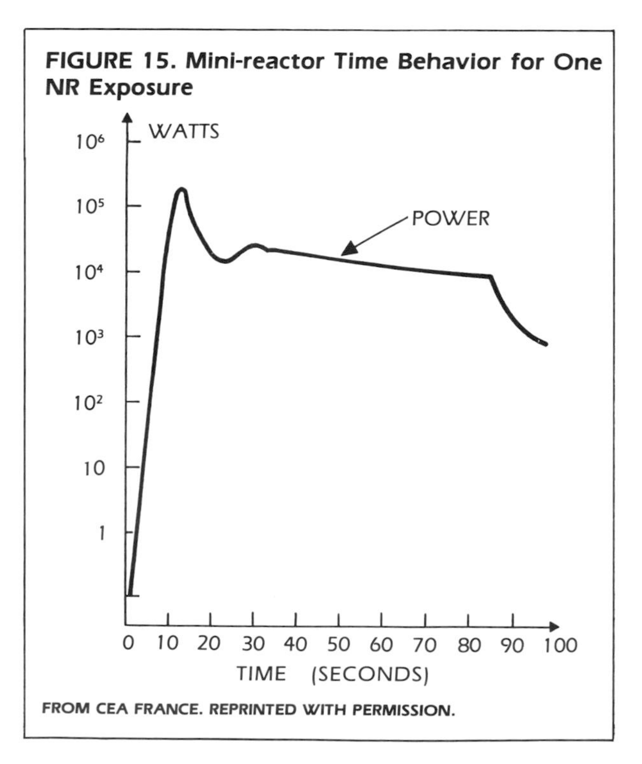

FROM CEA FRANCE. REPRINTED WITH PERMISSION.

are approximately 1 m (3 ft) on each side. Neutron beams can be extracted through three horizontal beam collimators.

The vertical configuration of the system shown in Fig. 12 is not essential. Horizontal tank geometry should be considered for ease of maintenance. The major part of the accelerator can be outside the neutron shield, with the deuteron flight tube passing through a small hole in the shield wall. X-ray shielding will be needed around the reverse electrode at the rear end of the tank.

Unlike reactors, subcritical multipliers or d,T accelerators, the Van de Graaff accelerators utilize no radioactive source material and sometimes require less stringent license processes.

The Mini-Reactor

The French Commissariat a' L'Energie Atomique has developed the Mirene, a simple, inexpensive reactor for NR. It is capable of taking up to four exposures per hour at collimator ratios of about 20:1 using high-quality imaging (3×10^9 n/cm²).

The reactor (Fig. 14) uses a solution of uranyl nitrate containing 93% enriched uranium. The core, containing this solution, is a steel tank about 30 cm (12 in.) in diameter. Other components include fixed side reflectors, movable lower reflector with air-jacks, cooling system, expansion chamber, recombination system for gases released in the operation, collimators and shields.

The reactor operates by upward movement of the lower reflector, which produces a brief burst of power. The system then shuts itself down slowly under the influence of a negative temperature coefficient. Reverse motion of the lower reflector serves to shut the system down completely. A typical pulse (Fig. 15) has a duration of 100 seconds and a peak power of 170 kW.

PART 4

IMPLEMENTATION OF HIGH-YIELD SYSTEMS

High-yield NR systems can provide over 100 direct exposure neutron radiographs per day, each with a collimator ratio of 100:1 or better, and with high imaging quality (3×10^9 n/cm^2). They can provide over 20 indirect neutron radiographs per day using overnight image transfers.

Source Choice

These systems typically use small, nuclear reactors. Such reactors are usually considered research reactors, though they may be used exclusively for application work. In directories, reactors are normally classified by their maximum power, in kilowatts- or megawatts-thermal [MW(t)]. For neutron radiography, however, the most important parameter is peak neutron flux[1].

For thermal neutron radiography, a flux in the range of 10^{11} to 10^{12} ncm^{-2}s^{-1} is desirable at the collimator input. At a collimator ratio of 100:1, this flux produces directed neutron currents in the range of 10^6 to 10^7 ncm^{-2}s^{-1} incident on the object.

High-yield NR system designs have used both (1) existing reactors, and (2) reactors planned specially for neutron radiography.

The choice between an existing reactor and a specially installed reactor will depend primarily on required throughput, need for special design, and logistics. Other considerations include procurement options and safety review.

FIGURE 16. Typical NR Reactor, Cutaway View

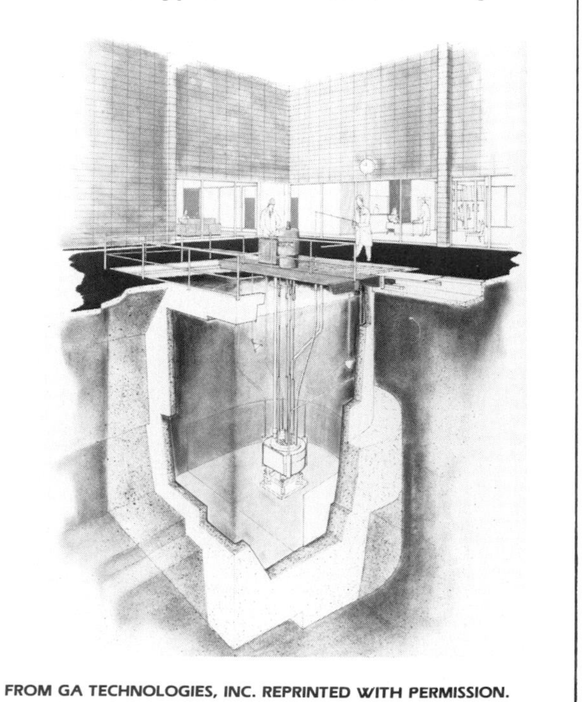

FROM GA TECHNOLOGIES, INC. REPRINTED WITH PERMISSION.

Procurement

Precedent exists for acquisition of new reactors or transfer of surplus reactor components. Reactor fuel frequently has been obtained by universities and national laboratories on loan from the US government

FIGURE 17. Typical Low-cost Reactor Room

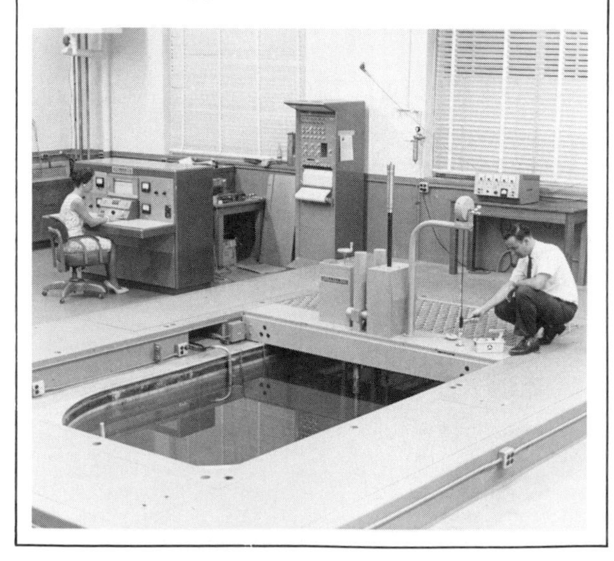

and an initial fuel load typically lasts many years without replenishment. Concerning costs, many reactors have demonstrated useful lives well in excess of 20 years. The largest expense over this life is normally the radiography staffing and film costs (not reactor staff or capital cost). Maintenance and fuel costs are comparatively low. Decommissioning costs must be anticipated, but actual costs can be significantly less than published projections.[4]

Safety

The Department of Energy has stringent review procedures which are applied both to its own research reactors and to those of the Department of Defense, under Section 91b of the 1954 Atomic Energy Act, as amended.

A non-government institution wishing to install a new research reactor must apply to the Nuclear Regulatory Commission (NRC), which will prepare an environmental impact assessment and review all safety aspects, including a preliminary safety analysis report (prior to construction), and a final safety analysis report (prior to operation).

Environmental impact appraisals for numerous small research reactors licensed by the NRC are available in public records. Because of the low power at which these reactors operate, and the consequent low inventory of radioactive fission products, the NRC has issued a generic finding of no significant impact under either normal or accident conditions for research reactors of less than 2 MW(t).

The reactor type selected for some NR systems[5] offers characteristics of inherent safety and standardized design. Over 60 such reactors are in operation worldwide, including 27 in the US. Most are in city centers, and one has operated for over 20 years in the basement of a six-story hospital.

Safety consultants, hardware suppliers, and other services needed for implementation of a small reactor system are available.[6]

FIGURE 18. Reactor Core

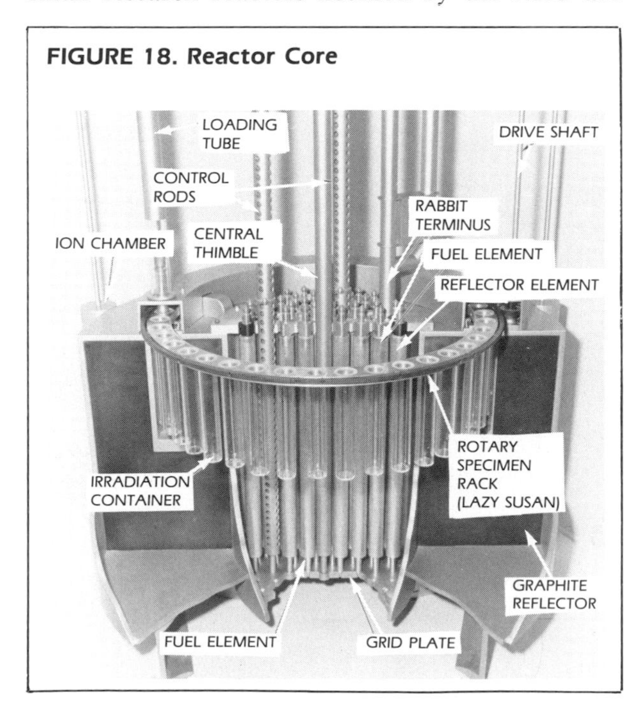

System Description

The reactor type used most extensively for neutron radiography is illustrated in Figs. 16 through 19. It has four basic subsystems: core, coolant, control and shielding. The core, situated near the base of a water tank about 7 m (24 ft) deep, consists of about 70 fuel rods, each about 3.8 cm (1.5 in.) in diameter, held by grid plates to form a 61 cm (24 in.) cylindrical array (Fig. 18). Each fuel rod consists of uranium-zirconium-hydride, clad in stainless steel. The uranium enrichment is just less than 20%, so that stringent theft safeguards are not required.

Cooling is by natural convection, using water at normal pressure, which is circulated in a closed loop through a small heat exchanger. The control system

FIGURE 19. Control Console

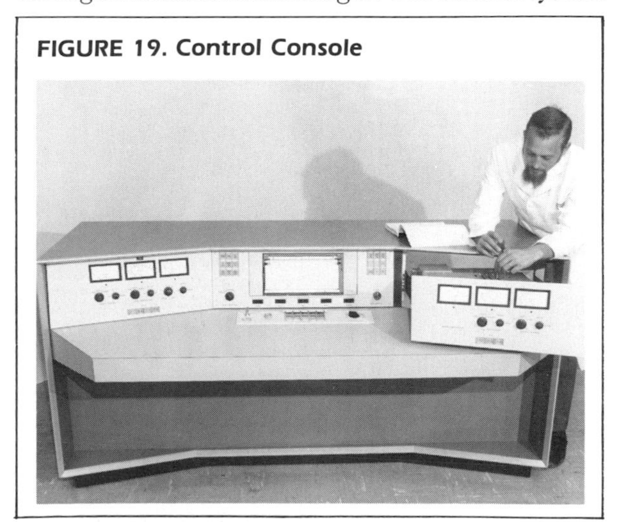

consists of boron-loaded rods suspended from drive mechanisms at the top of the tank, flux monitors immediately around the core, and a solid-state control panel (Fig. 19).

Shielding is most inexpensively provided by situating a reactor below ground level. The water provides shielding in the upward direction.

Beam Geometries

Beams can be extracted tangentially or radially, and vertically or horizontally (Fig. 20). The tangential beam gives a good thermal-neutron to gamma-ray ratio for direct exposure image quality. A radial beam, penetrating toward the core center, can give improved flux and epithermal neutron content,

FIGURE 20. Beam Geometry: (a) Vertical Beam Extraction for Simple Object Positioning; (b) Horizontal Beam Extraction for Multiple Beam Space; (c) Radial Beam for Maximum Epithermal Neutron Intensity; and (d) Tangential Beam for Optimum Direct Imaging Quality

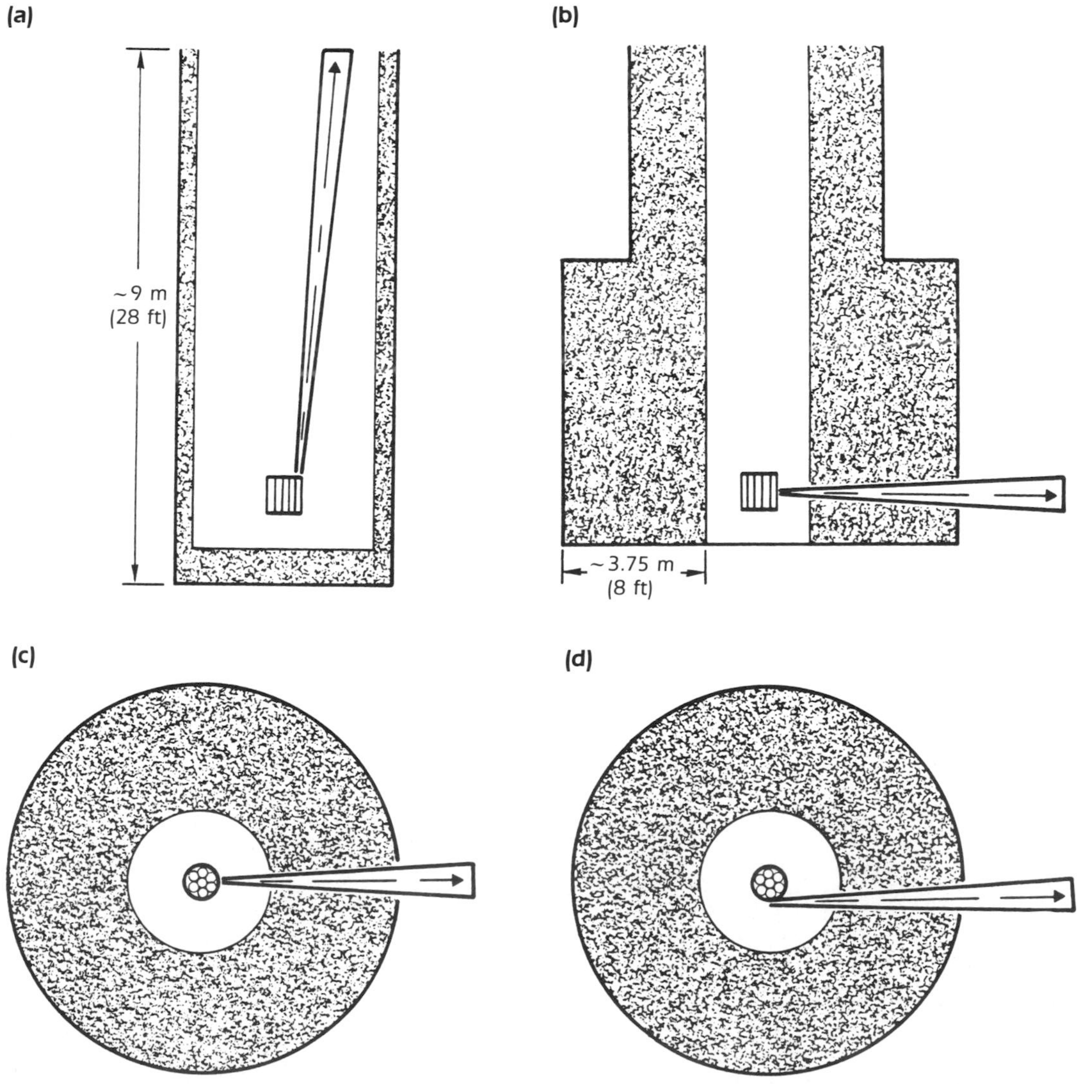

(a)

~9 m
(28 ft)

(b)

~3.75 m
(8 ft)

(c)

(d)

which is sometimes valuable for penetration of thick objects, including nuclear fuel. Vertical beam extraction provides for convenient placement of objects on a horizontal exposure plane. Horizontal beam geometry requires a more expensive aboveground reactor, but provides the potential for expanded working space on multiple beams. Beams inclined at about 45 degrees from vertical combine the advantages of vertical and horizontal extraction.

Typical Systems

High-yield systems available for neutron radiography are illustrated in Figs. 21 through 24. The systems in Figs. 21, 22 and 23 are adaptations of previously existing reactors. The arrangement of Fig. 22 is of interest because it uses a fuel-free zone in the center of the core to provide optimum flux and neutron-to-gamma ratio. The arrangement of Fig. 23 provides three beams of different neutron energies, available on a single reactor. Figure 24 shows a reactor installed especially for NR.

FIGURE 21. Typical NR Service Center for Non-nuclear Applications

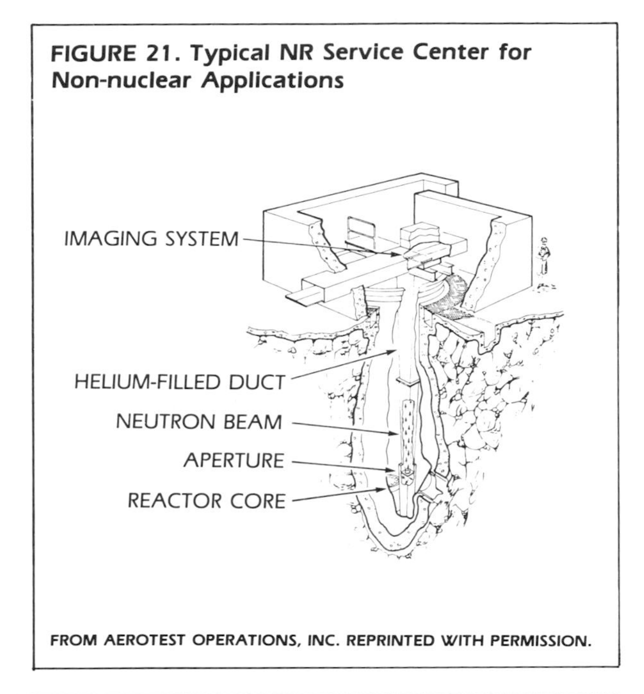

FROM AEROTEST OPERATIONS, INC. REPRINTED WITH PERMISSION.

FIGURE 22. Typical NR Service Center for Nuclear and Non-nuclear Applications

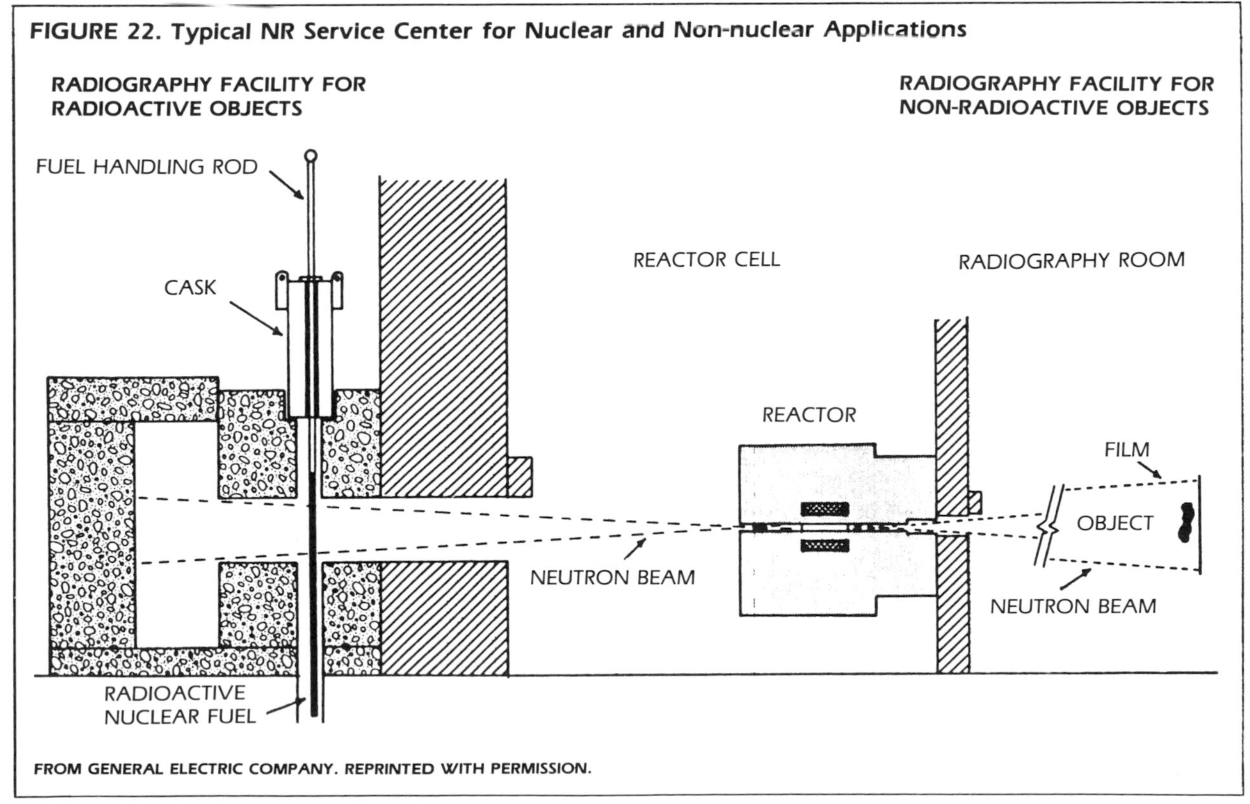

FROM GENERAL ELECTRIC COMPANY. REPRINTED WITH PERMISSION.

FIGURE 23. Reactor Providing Different Types of Beam

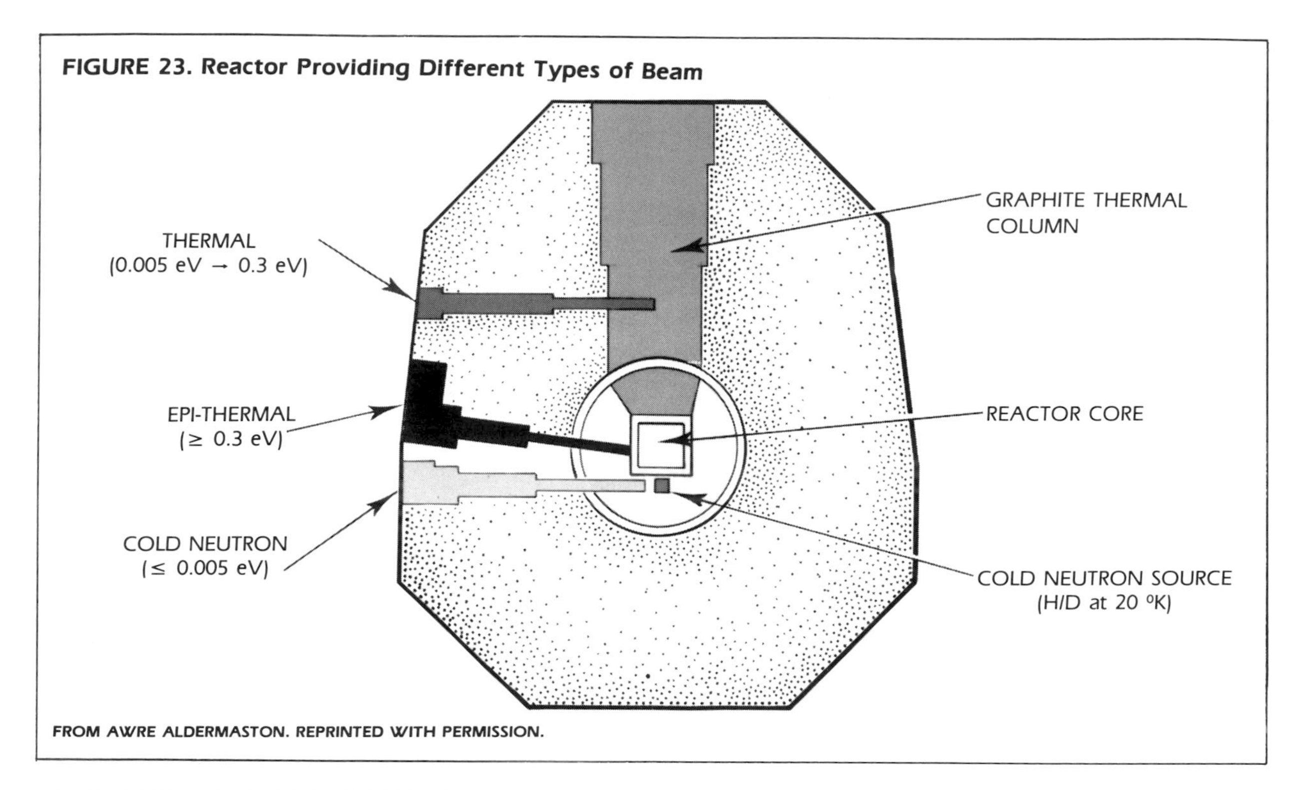

FROM AWRE ALDERMASTON. REPRINTED WITH PERMISSION.

FIGURE 24. Reactor Installed for Inspection of Nuclear Fuel

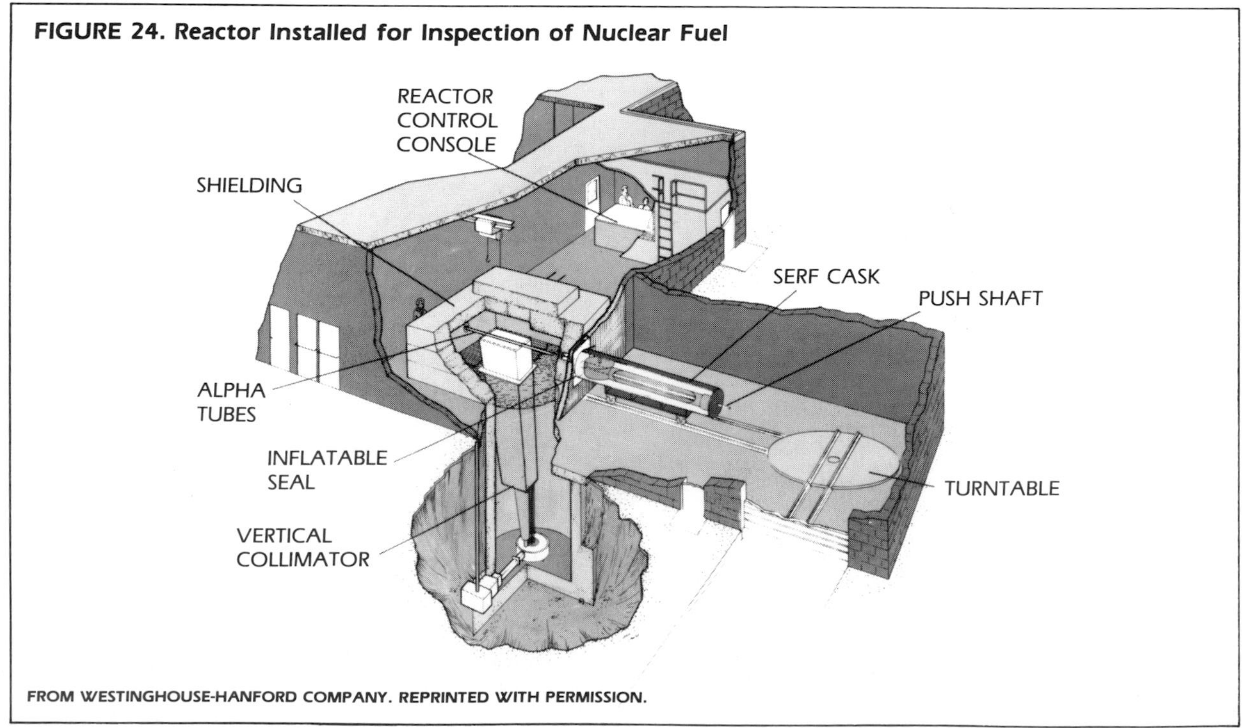

FROM WESTINGHOUSE-HANFORD COMPANY. REPRINTED WITH PERMISSION.

PART 5

NUCLEAR INDUSTRY SYSTEMS

The particular value of NR for nuclear fuel inspection is evident from the frequency of literature citation, and the capital invested in NR systems.[7] In the US, both government centers and commercial service centers offer facilities for nuclear fuel inspection.[3,5] Implementation of nuclear fuel NR, therefore, can be achieved by shipping samples to existing facilities, or by consideration of an in-house system.

System Descriptions

Although some reactor systems have been installed specifically for inspection of new fuel, in most cases the ability to inspect irradiated fuel is essential.

Figures 25 through 27 show three approaches.

FIGURE 25. Cask and Pedestal for Nuclear Fuel NR

FROM ARGONNE NATIONAL LABORATORY. REPRINTED WITH PERMISSION.

The dry system (Fig. 25) uses a horizontal beam extracted from a reactor into a pedestal to which a shielded fuel transfer cask can be mated. The pedestal serves as the fuel exposure position, the image screen positioner, and the beam stop.

Pool-type systems, illustrated in Fig. 26, use water as the radiation shield. In many cases the objects inspected are under irradiation tests in the reactor itself, and therefore can be periodically radiographed and returned to the core without extraction from the water shield. Such collimators, with reliable means of sealing and drying specimens under water, have been sold internationally.

The hot cell system, shown in Fig. 27, enables fuel to be radiographed without extraction from the inert gas atmosphere. In this case, the small reactor is adjacent to the hot cell complex. The drawing shows the reactor (left), the beam emergent horizontally, and a vertical specimen tube that opens to the hot cell complex at the top.

Automated System

Details of a highly automated dry system are shown in Fig. 28. The beam, extracted horizontally from a 700 kW pool reactor, is of limited diameter

FIGURE 26. Collimator for Underwater System

FROM NERF PETTEN. REPRINTED WITH PERMISSION.

due to safety considerations of the tank and shield penetration. The beam can be switched off by rotation of the inner collimator section. The fuel to be inspected, typically a 3.7 m (12 ft) high rack containing five side-by-side rods of power reactor fuel, is movable in a vertical channel 7.3 m (24 ft) long, that intersects the beam. At the exposure position, the beam size is 10 × 15 cm (4 × 6 in.) and the intensity is 2.5×10^7 $ncm^{-2}s^{-1}$. An automated control system moves the fuel rack through the beam in consecutive steps, and at each step a track etch film is exposed using a continuous strip to cover the full fuel-rod length. Multi-stage development of the latent image permits wide latitude analysis from object regions of low absorption to regions of high absorption. Dimensional measurements are obtained directly from the negative (nitrocellulose) film using optical instruments. High-contrast images are obtained at selected points in the negative development process using silver halide films and an enlarger to avoid scattered light.

Transportable Systems

Transportable neutron radiography system designs are available for use in the spent fuel storage pool of power reactors or in hot cells. They use low-yield neutron sources (californium-252 or Sb-Be) and compensate by deployment of very low collimation ratios and long image integration times made possible by track etch imaging. Such systems are suitable for inspection of reactor control blades, as well as fuel.

Neutron Hodoscope

This specialized system provides spatial and time-dependent information on the fast neutrons emitted by fissile material or scattered by objects in a nuclear reactor after neutron bombardment.

The fuel sample is typically enclosed in thick metal containers and situated in the center of a test reactor, where it is subjected to simulated accident conditions.

The hodoscope (Fig. 29) consists of a massive collimator with hundreds of channels through which flow the neutrons from the object. The viewing area of some hodoscopes is 7 × 120 cm (3 × 50 in.). An array of fast neutron detectors measures the neutron intensity corresponding to each channel, and its corresponding region of the object.

FIGURE 27. Hot Cell Fuel Inspection System

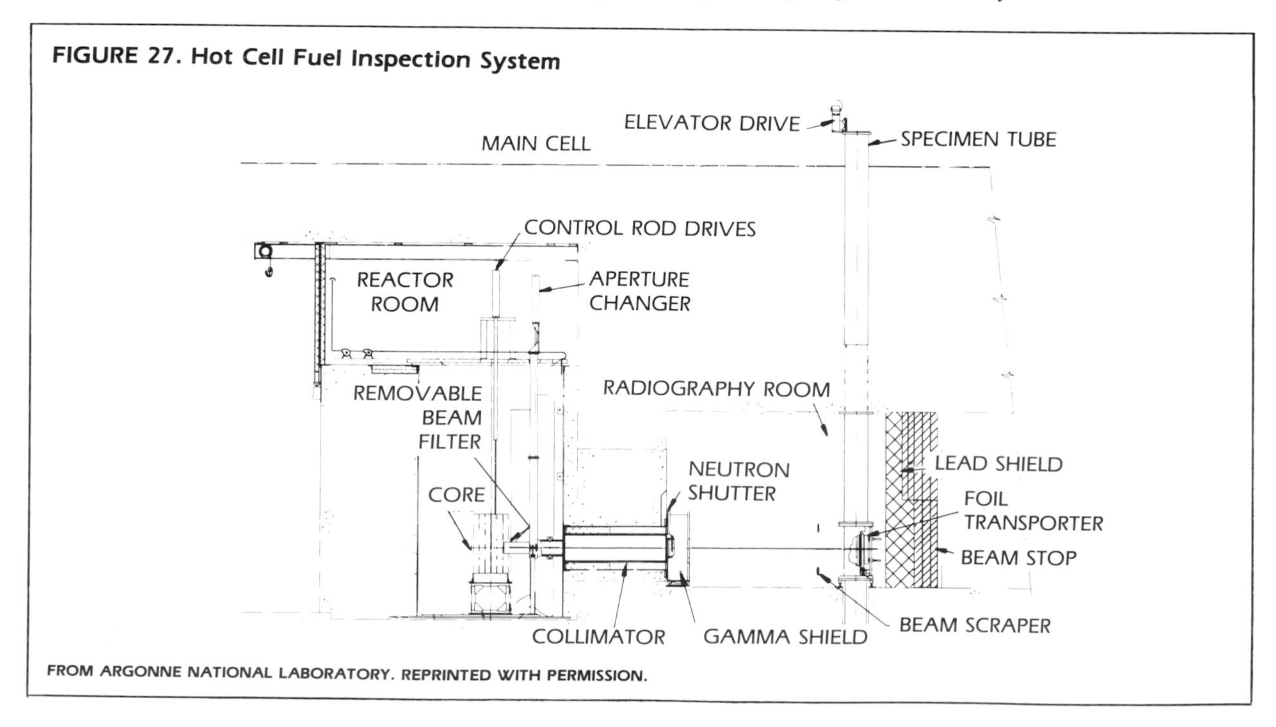

FROM ARGONNE NATIONAL LABORATORY. REPRINTED WITH PERMISSION.

Time resolution on the order of 1 millisecond (ms), for durations of many seconds, and fuel displacement resolutions of 0.2 mm horizontal and several millimeters vertical are experienced in the cineradiographic mode of operation.

In addition to performance of transient tests, the hodoscope is used to inspect the test object before the experiment and also after the experiment; this last inspection is made before possible disturbance of the sample during its removal from the reactor. For this radiographic scan, the test reactor is operated at a low power to generate neutron emission from the test fuel. The collimator is then scanned horizontally and vertically across the fuel, a technique that improves resolution. By rotation of the object, a form of tomography can be accomplished.

Reference Items

Nuclear fuel reference specimens have been made to meet specific needs. Some, for example, contain pellets of different enrichment, and different agglomeration sizes of PuO_2 in mixed oxide fuel. A reference nuclear fuel rod (see Fig. 30) is available as a standard for dimensional measurements on light water reactor fuels.[8]

Figure 31 shows a selection from an extensive collection of neutron radiographs of nuclear fuel.[9] Readers are referred to the fuel development literature for further examples of neutron radiographs.

FIGURE 28. Automated System for Nuclear Fuel Inspection

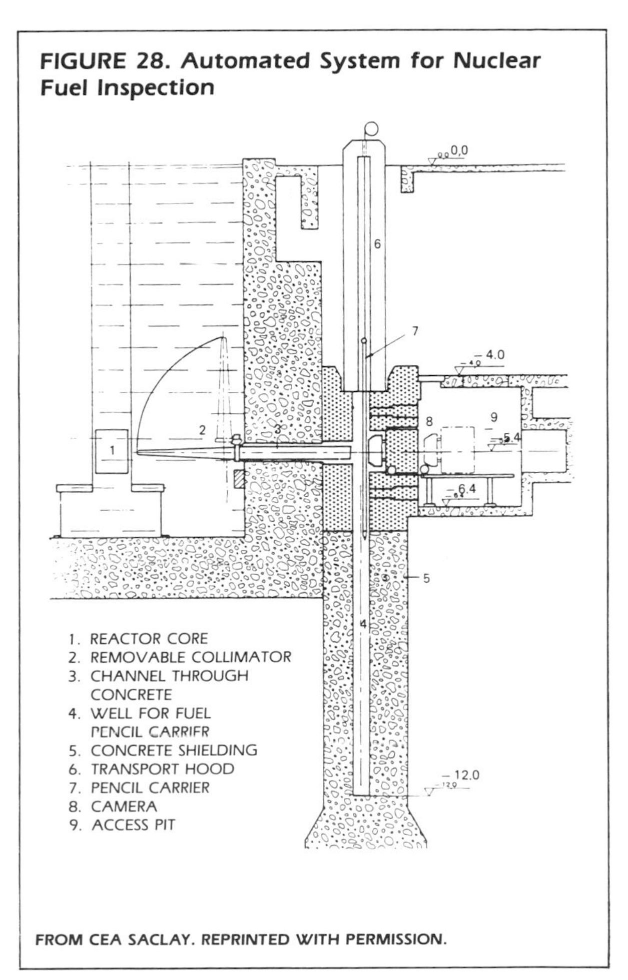

FROM CEA SACLAY. REPRINTED WITH PERMISSION.

FIGURE 29. Neutron Hodoscope

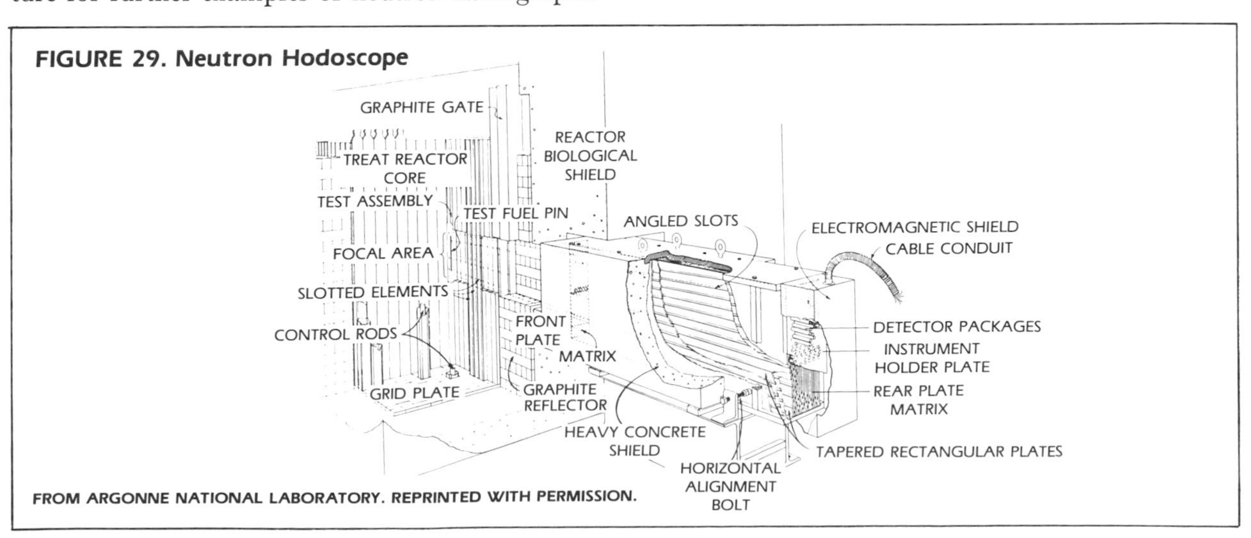

FROM ARGONNE NATIONAL LABORATORY. REPRINTED WITH PERMISSION.

FIGURE 30. Reference Nuclear Fuel Pin

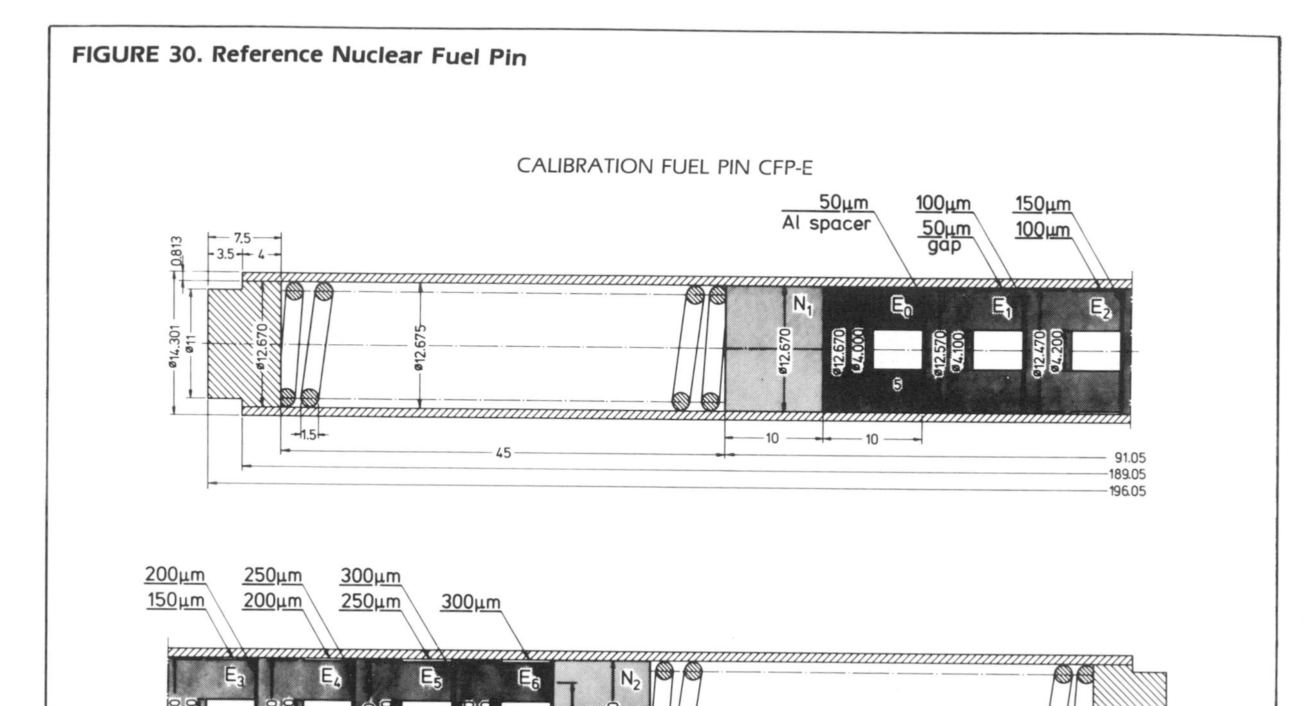

FROM RISØ NATIONAL LABORATORY. REPRINTED WITH PERMISSION.

FIGURE 31. Neutron Radiographs of Fuel: (a) Longitudinal Cracks in Pellets; (b) Missing Chips in Compacted Fuels; (c) Inclusions of Pu in Pellets; (d) Accumulation of Pu in Central Void; (e) Deformed Cladding; and (f) Hydrides in Cladding

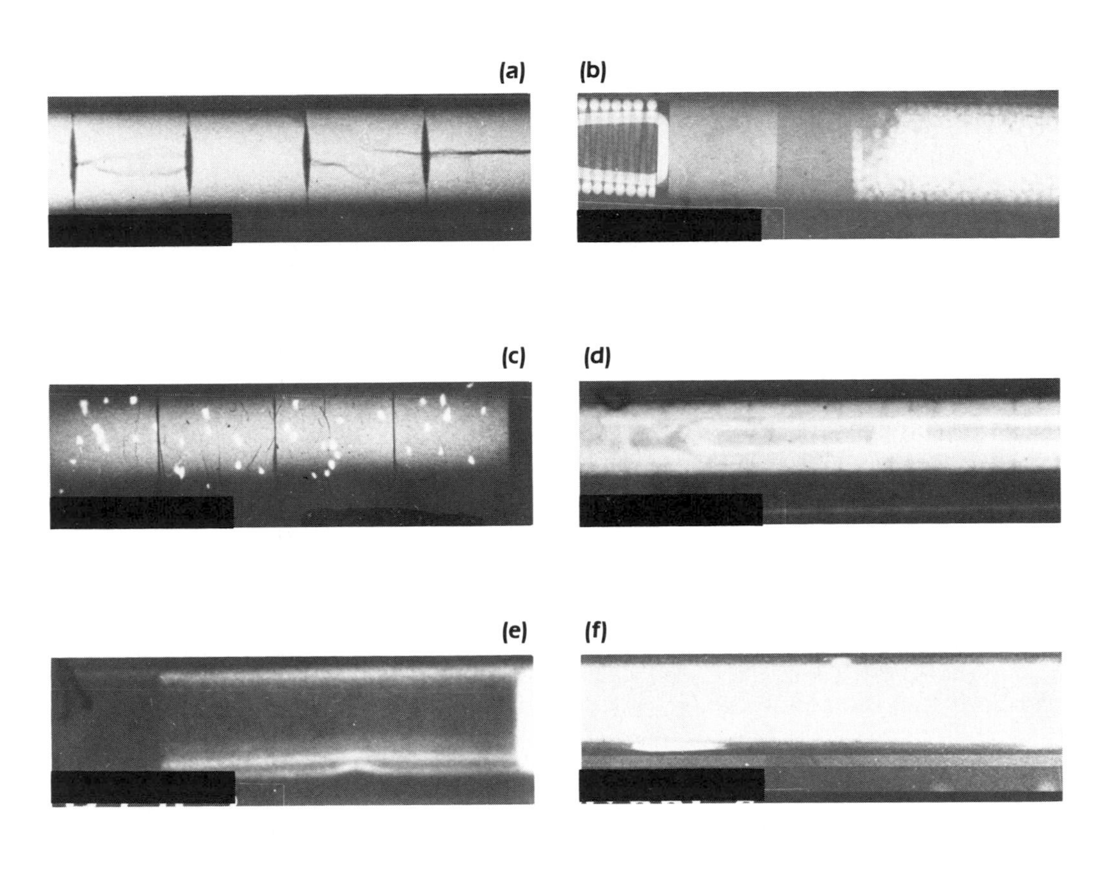

FROM RISØ NATIONAL LABORATORY. REPRINTED WITH PERMISSION.

PART 6

SPECIAL SYSTEMS

Static Cold Neutron Radiography

Previous parts of this section refer primarily to thermal neutron radiography. In this section, a cold neutron radiography system is described. This design has been used to examine large quantities of refractory metal for hidden impurities. In terms of object throughput, it has been one of the largest single neutron radiographic applications to date.

Principle

The principle of cold neutron radiography is illustrated in Figs. 32, 33 and Table 1. By using low-energy (cold) neutrons, the penetration of crystalline materials can be significantly improved. The neutron energy spectrum is lowered, in this case, by use of a refrigerated volume of moderator.

System Description

The equipment, designed and operated specifically for neutron radiography,[10] uses a one-liter volume of solid methane positioned at the edge of a 1.5 MW reactor and cooled by two helium circuit refrigerators to below 50 °K (Fig. 33). The cooled moderator is insulated by a vacuum jacket. The thermal neutron flux at the position of the cooled moderator in undisturbed geometry is 2×10^{13} $ncm^{-2}s^{-1}$. To minimize gamma-ray heating of the cooled moderator, a lead shield containing a single crystal bismuth filter separates the reactor from the moderator. Other components of the system include a cooled beryllium crystal to scatter out high-energy neutrons before entering the collimator, and an argon-purged collimator.

TABLE 1. Relative Neutron Attenuation Coefficients

Element	Thermal Neutrons	Cold Neutrons
Beryllium	0.861	0.055
Silicon	0.092	0.050
Iron	1.160	0.568
Nickel	1.980	1.350
Zirconium	0.340	0.047
Lead	0.370	0.049
Bismuth	0.250	0.126

FROM GA TECHNOLOGIES, INC. REPRINTED WITH PERMISSION.

FIGURE 32. Attenuation of Materials for Thermal and Cold Neutrons

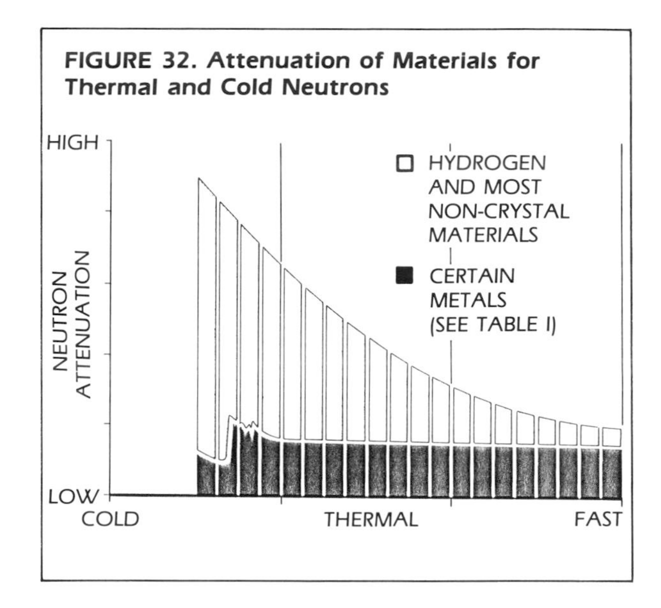

FIGURE 33. Cold Neutron Radiography System as Used for Routine Inspection of Metals

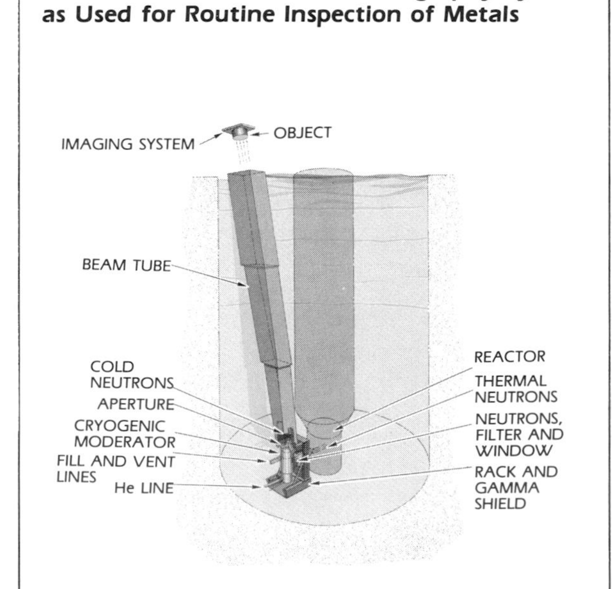

FROM GA TECHNOLOGIES, INC. REPRINTED WITH PERMISSION.

In-motion Cold Neutron Radiography

This second example of a special system can provide engineering information such as in-motion neutron radiographic video recordings of aircraft and automobile engines, operating in test conditions.

Unique information on the flow of lubricant and fuel can be revealed using this system. The neutron energy is selected by filters in the beam, rather than by a cryogenically cooled moderator volume, as in the previous system. The increased opacity of thin lubricant films to cold neutrons is as important as the increased transmission capability of crystalline metals.

System Description

The layout of this system is shown in Fig. 34. The source is a 26 MW, multipurpose research reactor. The beam, extracted radially, is about 15 cm (6 in.) in diameter at the reactor end, and diverges to 30 cm (12 in.) diameter at the exposure position, about 25 m (80 ft) away. The filters, positioned in the beam tube, consist of about 30 cm (12 in.) thickness of polycrystalline beryllium and 23 cm (9 in.) of single-crystal bismuth. Both filters are cooled by liquid nitrogen to improve their transmission of cold neutrons. The flight tube is filled with helium to avoid losses by air scatter.

Imaging systems used include: (1) a Delcalix image intensifier that makes use of Bouwers optics to efficiently collect light from a $ZnSLi_6F$, or gadolinium oxysulphide (GOS) screen; (2) a Thomson-CSF image intensifier, with 23 cm (9 in.) input diameter, precoated with GOS; or (3) a modular imaging system using $ZnSLi_6F$ screen, and a low light level, silicon-intensified TV camera.

Performance

The neutron flux at the imaging plane is 3×10^5 $ncm^{-2}s^{-1}$ and the measured collimator ratio is 300:1. This flux has been shown capable of penetrating several engine types, providing true real-time imaging. By frame averaging, the quality of image improves up to about 5×10^6 n/cm^2, at which level the statistical limitations become dominated by the other system limitations. Examples of cold neutron in-motion radiographs are shown in Fig. 35.

High-speed Motion Thermal Neutron Radiography

A system has been developed to perform neutron radiographic analysis of dynamic events having a duration of several milliseconds. The high frame rate, up to 10,000 frames per second, can be used to study the behavior of pyrotechnical devices during the firing cycle. In addition to developmental engineering studies, the design can be used for fundamental studies such as two-phase flow.

System Description

The neutron source is a reactor capable of providing a pulse with a peak power of 3,000 MW and a full width at half maximum of 9 ms. A tangential beam is extracted to optimize neutron-to-gamma ratio, and a low collimator ratio of 30:1 is selected to further maximize the flux (4×10^{11} $ncm^{-2}s^{-1}$ at the object).

The diagram in Fig. 36 shows a steel rifle barrel with shell, firing breech and bullet trap set in place to examine the ballistic cycle of a propellant. Immediately on the source side of the object is a massive, finely adjusted beam mask to reduce scatter, and on the far side is a $ZnSLi_6F$ scintillator screen optimized for transmission mode neutron radiography, as opposed to the more common reflection mode.

The light generated in the scintillator is transmitted through fiber-optic coupling to a two-stage, 40 mm diameter image intensifier, providing a gain of about 300 and a resolution of 35 line pairs per millimeter. The photocathode is Type S-20, and the output screen is P-11 to meet requirements of low persistence and spectral matching. A rotating-prism, high-speed camera is used to record the output image with selected film types in the form of 16 mm width, 38 meter rolls (8,000 frames per second maximum) or 120 meter rolls (11,000 frames per second maximum). The shutter mechanism of the camera reduces the actual exposure time per frame by a factor of 2.5 so that at 10,000 frames per second the exposure time is 40 μs. The entire imaging system is carefully insulated from shock or vibration caused by the firing.

A sodium iodide scintillator gamma-ray detector positioned in the beam enables the pulse delay time to be measured and subsequently preset. Synchronization between the pulsing of the reactor, the speed curve of the camera, and the firing event is provided by electrical circuits.

Performance

Figure 37a shows a portion of a high-speed motion neutron radiograph of a stationary test object. The original negative film strip contains high-quality image information. Figure 37b illustrates a neutron radiographic series taken at 5,000 frames per second through a thick steel breech with a blank cartridge inside. Analysis of the original negatives reveals various features of the burnup cycle. Tagging of propellant grains with gadolinium oxide permits grain movement and time of burnup to be identified.

Neutron Computerized Tomography

Computerized tomography (CT) uses radiographic information from many angles, and computes cross sections of the object in planes perpendicular to the axis of rotation. In medical applications of X-ray CT, the ability to discern small density differences is of prime importance. In neutron radiography of nuclear fuel, CT systems have become valuable for a different reason: the ability to view a complex, high-contrast object as if it had been sliced in a plane perpendicular to the axis of rotation, thereby avoiding the interference present in standard radiography.

System Description

Figure 38 illustrates a process used extensively for CT neutron radiography of neutron fuel. The beam is optimized for maximum epithermal neutron content by partially penetrating the core region of the source reactor. The object to be radiographed is supported in the beam by a mechanism which provides for step rotation with accurate setting of orientation. Neutron radiographs are taken at a series of orientations. Indium foil activation transfer imaging is used because of the high resonance integral cross section for epithermal neutrons. Cadmium or gadolinium filters are used to eliminate unwanted scatter, and fission caused by thermal neutrons. Although indium has a large resonance at 1.4 eV, this does not completely dominate the epithermal reaction rate.

A microdensitometer with digital readout is used to provide traces across a selected plane for the set of films. Standardization against a reference wedge compensates for variations between films.

For reconstruction of the tomographs, the digital radiographic data is fed to a computer. Both convolution methods (which are analytical) and algebraic methods (which are iterative) have been used. The iterative algebraic methods are better suited to high-contrast objects, especially when relatively few angles of orientation can be radiographed in an acceptable time.

Typical applications have included inspection of light water reactor fuel assemblies (encased in zircalloy) and fast reactor fuel assemblies (encased in steel). The number of orientations has varied from 18 (each 10 degrees) to 90 (each 2 degrees).

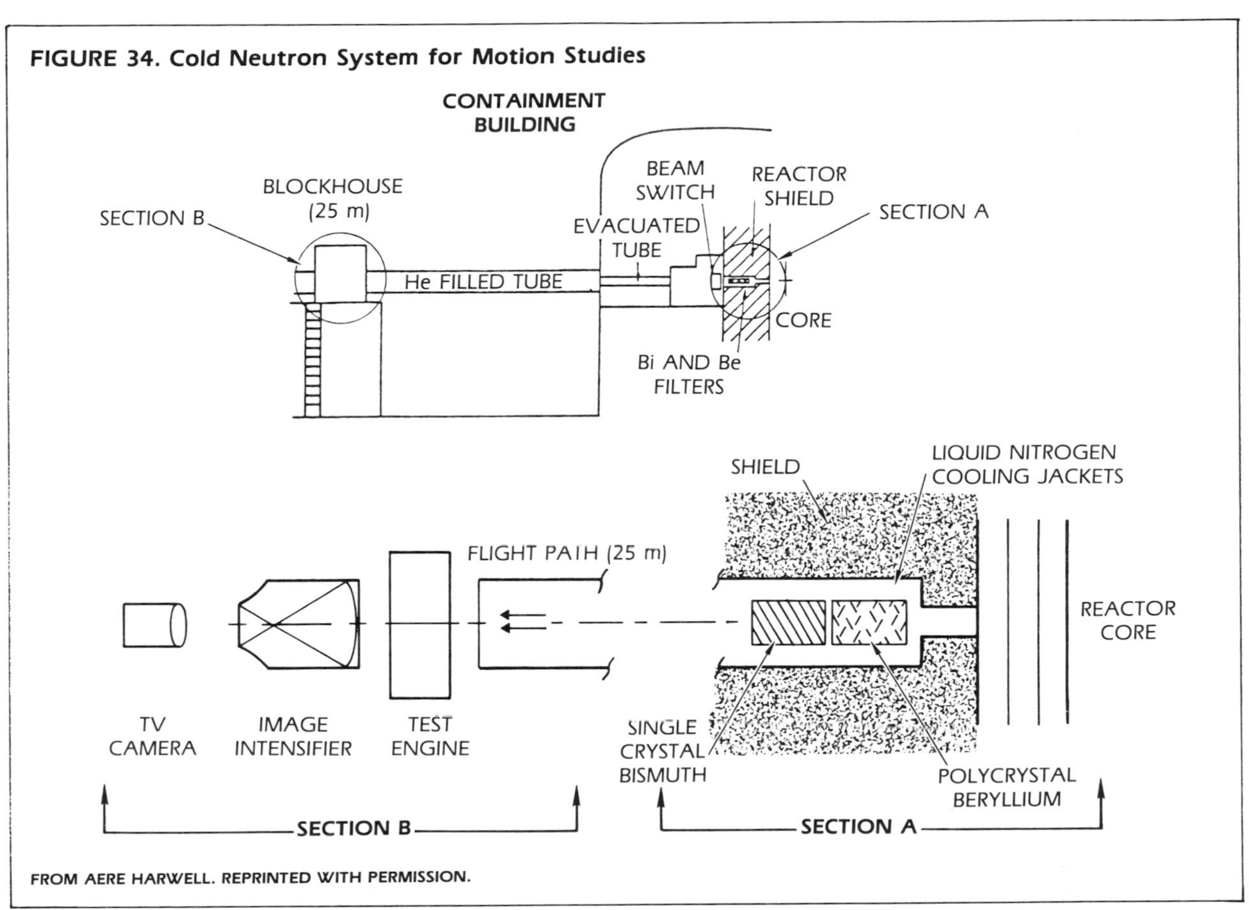

FIGURE 34. Cold Neutron System for Motion Studies

FROM AERE HARWELL. REPRINTED WITH PERMISSION.

FIGURE 35. Frames from Real-time Studies of Operating Aircraft Engine

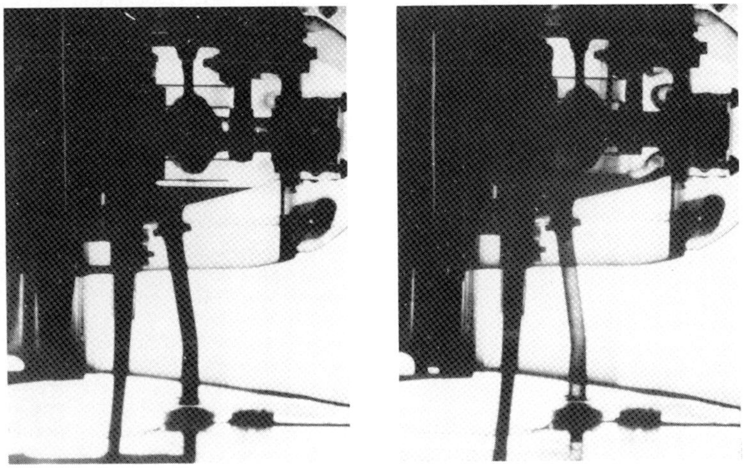

FROM ROLLS ROYCE LTD. REPRINTED WITH PERMISSION.

FIGURE 36. Thermal Neutron System for High-speed Events

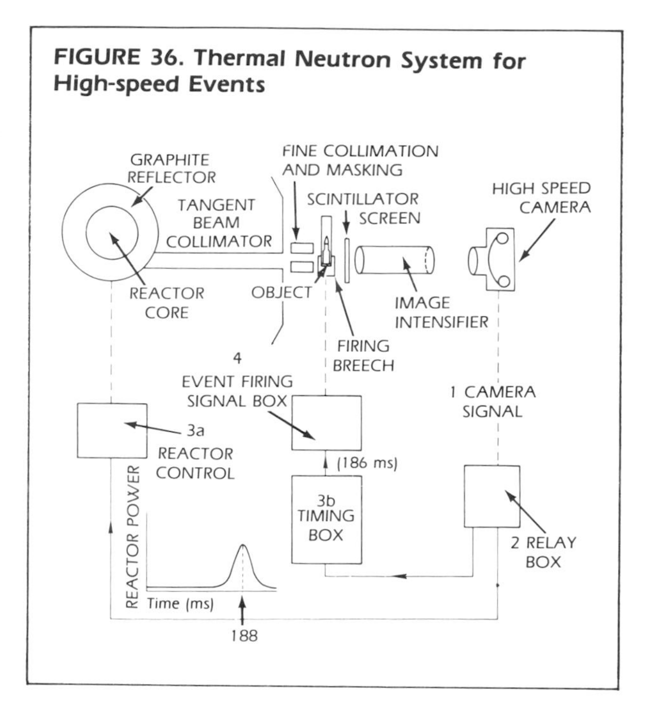

FIGURE 37. Thermal Neutron Radiographs: (a) Stationary Test Object Radiographed at 6,000 Frames per Second; and (b) Neutron Radiographs Taken at 5,000 Frames per Second Showing Burnup Cycle of Propellant Grains of Rifle Shell Inside Steel Barrel

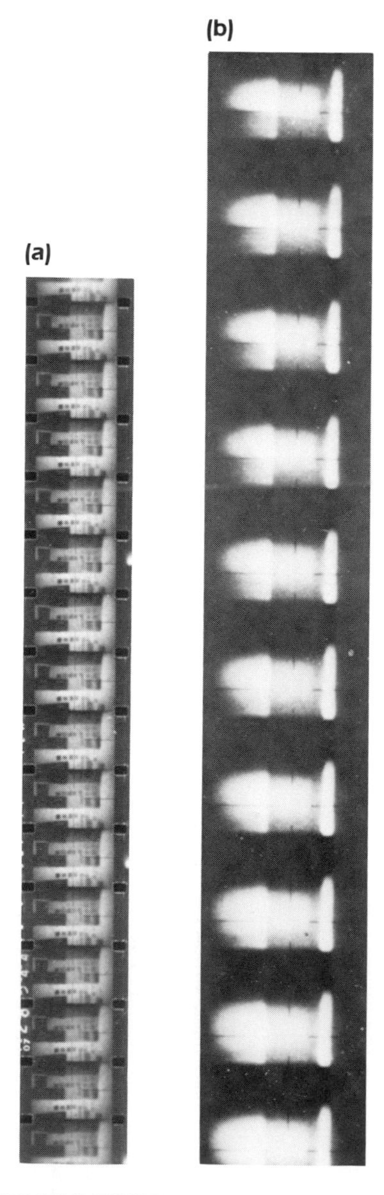

FROM OREGON STATE UNIVERSITY. REPRINTED WITH PERMISSION.

FIGURE 38. Computerized Tomography System For Nuclear Fuel Inspection

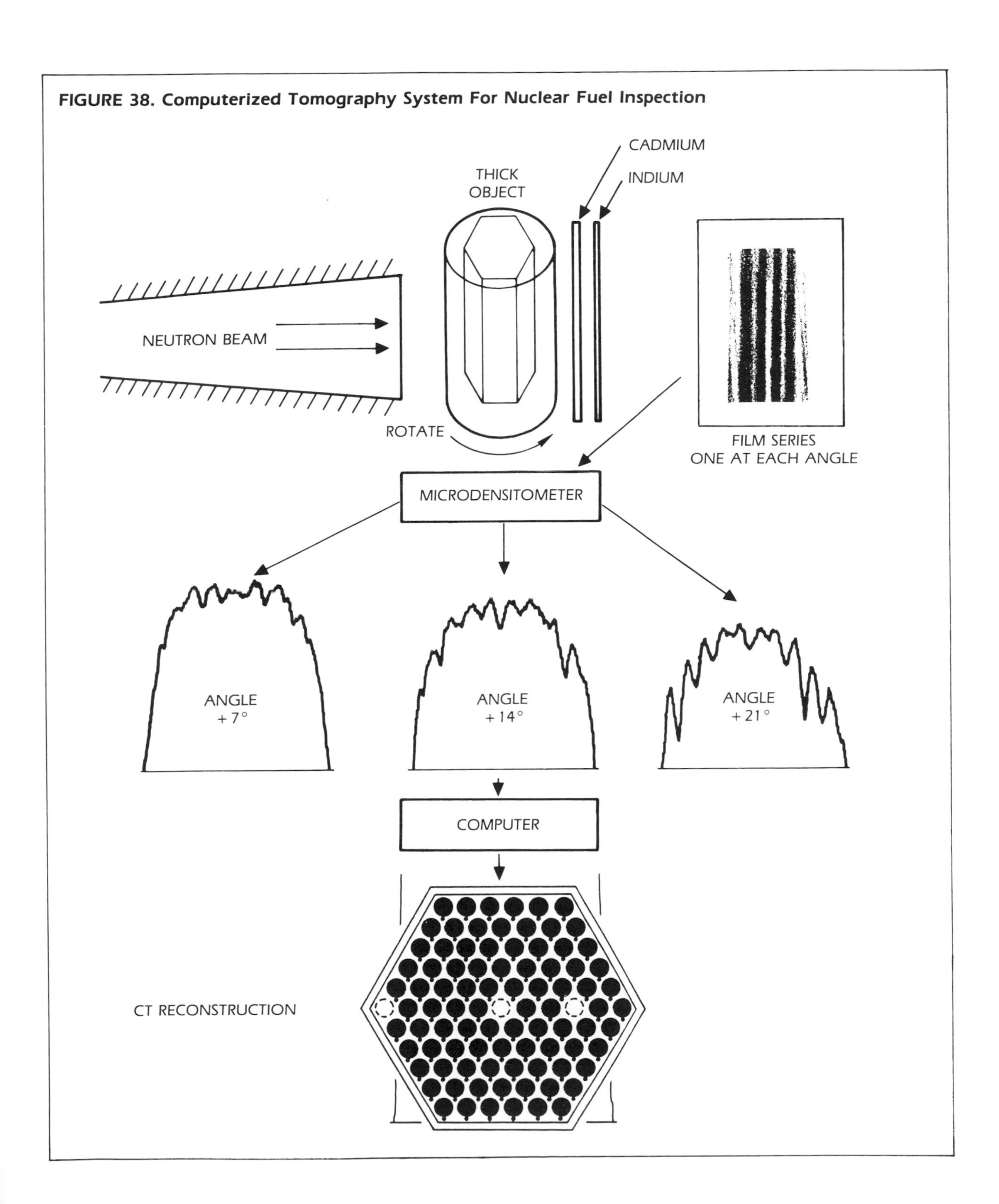

REFERENCES

1. *Research, Training, Test and Production Reactors Directory.* La Grange Park, IL: American Nuclear Society (1983).

2. *Neutron Sources for Basic Physics and Applications.* S. Cierjacks, ed. Oxford: Pergamon Press (1983).

3. *Neutron Radiography Newsletter.* American Society for Nondestructive Testing. Vols. 1-15 (1978). Continued as the *International Neutron Radiography Newsletter.* British Journal of Nondestructive Testing (1984).

4. *Technology, Safety and Costs of Decommissioning Reference Nuclear Research and Test Reactors.* NUREG/CR-1756. US Nuclear Regulatory Commission (1982).

5. *Neutron Radiography, Proceedings of the First World Conference.* J.P. Barton and P. von der Hardt, eds. D. Reidel Publishing Company (1983).

6. *Nuclear News, Buyer's Guide.* La Grange Park, IL: American Nuclear Society.

7. "Neutron Radiographic Inspection of Nuclear Fuels." *Atomic Energy Review,* A.M. Ross, ed. Vol. 15, No. 2. Vienna: International Atomic Energy Agency (1977).

8. *Neutron Radiography Handbook.* P. von der Hardt and H. Rottger, eds. Chapter 3. D. Reidel Publishing Company (1981).

9. *Reference Neutron Radiography of Nuclear Fuel.* J.C. Domanus, ed. D. Reidel Publishing Company (1984).

10. *Seventh Biennial US TRIGA User Conference Proceedings.* Section 4-1. San Diego, CA: GA Technologies, Inc. (1980).

SECTION 14

REAL-TIME RADIOGRAPHY

Richard Bossi, Lawrence Livermore National Laboratory, Livermore, CA
Charles Oien, Sandia National Laboratories, Livermore, CA
Paul Mengers, Quantex Corporation, Sunnyvale, CA

INTRODUCTION

Lawrence E. Bryant, Jr.
University of California
Los Alamos National Laboratory
Los Alamos, New Mexico, USA 87545

Real-time radiography has been around for a long time, as is so humorously illustrated in the first figure of this Section. For many years, real-time radiography was known as *fluoroscopy;* in fact, for those of us middle aged or older, we may have encountered it before any other nondestructive testing technique — watching our feet squirm in new shoes on the fluoroscopic system at our friendly shoe department.

As is often the case in NDT, real-time radiography has benefited from advances that were originally intended for the medical field; i.e., the development of the X-ray sensitive image intensifier in the 1950s brought a major advance to real-time radiography.

By the mid 1970s, many diverse technologies were converging on this technique, producing even more rapid progress. These advances included improved resolution fluors and high energy X-ray sensitivity for image intensifiers; digital video processing for image enhancement; microfocal X-ray generators for high definition imaging; highly automated handling systems for rapid inspection; and computer control of these various subsystems.

This section of the Nondestructive Testing Handbook covers all of these developments, as well as the basic principles of real-time radiography.

Because of the rapidly evolving technology, our authors have had an extremely demanding job in covering such a broad field. Credit is also due to our technical reviewers: Dr. Robert Buchanan, Lockheed Missiles and Space Company, Research Laboratory, Mountain View, California; Charles Arney, Real-Time Radiography Consultant, Mountain View, California; Frank Patricelli, Science Applications, Inc., San Diego, California; and Roy Braley, Lawrence Livermore National Laboratory, Livermore, California.

PART 1

FUNDAMENTALS OF REAL-TIME RADIOGRAPHY

Principles

Real-time radiography is a nondestructive testing method that uses penetrating radiation to produce images which are viewed concurrent with the irradiation. In the case of dynamic systems, real-time radiography allows radiographic interpretation to be performed simultaneously with the progress of the event. The arrangement of the radiation source, object and image plane is similar to conventional radiography.

The most important process in real-time radiography is the conversion of radiation to light by means of a fluorescent screen. The light signal may then be observed directly, amplified and/or converted to a video signal for presentation on a television monitor and subsequent recording.

Fluoroscopy is the original term used to describe the direct viewing of fluorescent screens. *Real-time radiography* or *real-time imaging* are the terms now used to describe industrial imaging systems.

Real-time radiography is often applied to objects on assembly lines for rapid inspection. Remote adjustment of the object position allows inspectors the freedom to review details of interest or to move on to other locations. Accept-or-reject decisions may be made immediately without the delay or expense of film development.

For dynamic events, real-time radiography is typically used in the range of several seconds to several minutes, though long-time events (on the order of hours) also can be monitored. Short-time events may be imaged in real time and replayed at slower rates; television frame rates of 30 frames per second can cover events in the fraction of a second range. Even flash radiographic events can be imaged using frame-grabbing techniques.

FIGURE 1. European Customs Fluoroscope Inspection (1897)

Background

Real-time radiography has its roots in the discovery of X-rays; Roentgen, for example, used phosphor screens for X-ray detection. Figure 1 shows the inspection of baggage at the Brussels railroad station in 1897.

Barium platinocyanide, willemite, calcium tungstate, cadmium zinc sulphide, and cesium iodide are a few of the phosphors developed for fluoroscopy systems used over the years in medicine and industry. Typical industrial applications in the first half of this century included the examination of golf balls, cables, candy bars, shoes, light alloy castings, and packages.[19, 34]

Modern fluoroscopy systems operate on the same principles as earlier systems but with considerable improvement in the radiation sources, object handling, fluorescent screen performance, and personnel radiation protection. The low brightness level of fluorescent screens is still a significant disadvantage in fluoroscopy systems and requires the inspector to function in a darkened environment for best results.

Advances in the electronics industry have greatly aided real-time radiography. Image amplifiers and television systems were first introduced around 1950. Visual acuity improvements, no dark-adaptation time, and image contrast improvements are all important technical advances that image amplifiers have provided. The use of television systems has advanced the safety of real-time radiography by allowing the inspector to be further removed from the object and the radiation field. The digital electronic advances of the late 1960s and 1970s have increased the possibilities of image enhancement and information storage. Remote operations with sophisticated image enhancement capability are typical of the modern systems.

PART 2
FLUORESCENT SCREENS

Principles of Operation

Fluorescent screens consist of phosphor particles dispersed in a binder and coated on a reflecting, supporting base. As shown in Fig. 2, an X-ray or gamma-ray photon striking the screen perpendicularly along the axis Z excites a grain of phosphor at the point z, which emits a number of light photons. Typical paths followed by the photons are illustrated.

A light photon generated in the screen phosphor has a probability of leaving the emitting surface, depending on:

1. the number of scattering events;
2. the probability of absorption at each collision with phosphor particles;
3. the depth at which the photon originates (the X-ray stopping power of the phosphor); and,
4. the spatial orientation of the free path lengths between collisions.

Construction and Materials

Fluorescent screens employ a variety of supports, depending on the application. Typical screens have a white plastic or cardboard base, approximately 0.4 mm thick, as a support for the luminescent chemical layers. The screen base must be chemically inert so as not to react unfavorably with the luminescent material. It also must be uniformly radiolucent and cannot contain inclusions of radiopaque substances which might cause shadows on the fluorescent image. Finally, the support must be durable enough for use in the radiation fields it encounters.

The luminescent chemical (phosphor) is in the form of small grains, and grain size is a construction parameter affecting performance. The phosphor material is combined with a suitable binder, usually a cellulose derivative, and coated on the support in a uniform layer. The final packing density of the phosphor particles is usually on the order of 50 percent. A protective surface is often added to the screen to help it resist markings and abrasive wear and to permit cleaning.

In X-ray image intensifiers, the fluorescent layer is deposited on the inside of a vacuum tube envelope. In this case, phosphors whose physical properties are unacceptable for use in air [such as hygroscopic CsI(Na)] can be employed. The depositing technique is often proprietary.

Characteristics of Screens

Fluorescent screens are characterized by their efficiency, spectral emission, persistence, unsharpness and gamma. Table 1 contains parameters of typical screen phosphors.

FIGURE 2. Structure of Typical X-ray Intensifying Screen and the Typical Paths Followed by Light Photons

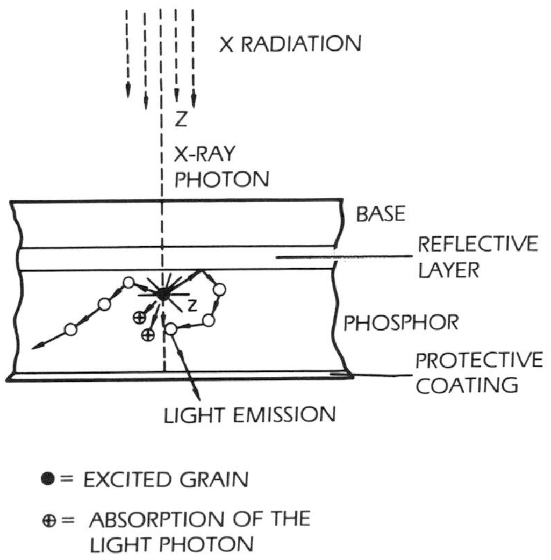

Efficiency

The overall efficiency of the fluorescent screen, in converting X-rays to light, is composed of three terms:

1. the incident X-ray absorption efficiency, n_a;
2. The intrinsic absorbed X ray to light conversion efficiency, n_c; and
3. the light transmission efficiency, n_T, determined by the pathlength of light in the screen coatings.

Hence the overall efficiency can be expressed in eq. 1 as:

$$n = n_a n_c n_T. \qquad \text{(Eq. 1)}$$

The absorption efficiency, n_a, which is the fraction of the incident flux absorbed by the screen, can be calculated when the incident X-ray spectrum and the composition of the screen are known. At low energy, where X-ray absorption predominates over scatter, the absorption efficiency is approximately equal to the attenuation of the radiation beam. Table 2 presents some data on the attenuation of an 80 kVp X-ray beam by several $CaWO_4$ and rare earth commercial screens.[43] A high absorption efficiency is important for maximizing the signal to noise ratio in the detection process.

The efficiency n_c is approximately equal to the efficiency of the phosphor under cathode ray excitation, which can be measured separately. The transmission efficiency n_T can be estimated if the scattering and absorption parameters of the screen are measured and if the surface conditions are known. Experimental measurements of the light photon output for X-ray photon input to typical fluorescent screen materials results in an energy efficiency in the range of 1 to 7 percent at low (20-100) kVp.[27]

Spectral Emission

The spectral emission profiles of phosphors can be an important parameter, depending on how they are employed in a real-time system. Figure 3 is a plot of the spectral emission of four types of screens: $CaWO_4$, LaOBr, Gd_2O_2S and ZnCdS.

The effect of spectral emission is demonstrated in Table 3 where the relative light yield for phototopic response (the human eye) and an S-20 photocathode are compared at 140 kVp X-ray energy. When measured with an S-20 photocathode, the $CaWO_4$ and LaOBr screens (which emit in the blue) show an increase in response over the photopic response, which is maximum in the green.

TABLE 1. Phosphor Parameters

Phosphor	Density (g/cm^3)	Wavelength of Peak Emission (nm)	Decay Constant (ns)
ZnS (Ag:Ni)	4.1	450	60
ZnCdS (Ag:Ni)	4.5	550	85
CsI (Na)	4.5	420	650
$CaWO_4$	6.1	430	6000
Gd_2O_2S(Tb)	7.3	544	480000

TABLE 2. Fluorescent Screen Data (at 80 kVp)

	Screens	Attenuation (%)
Fast $CaWO_4$	Par	21
	TF-2	33
	Hi plus	39
Fast rare-earth	Rarex-BG mid speed	43
	Rarex-BG high speed	52
	Lanex regular	69

From Photographic Science and Engineering. Reprinted with permission.

FIGURE 3. Spectral Emission of Several Fluroescent Screen Phosphors

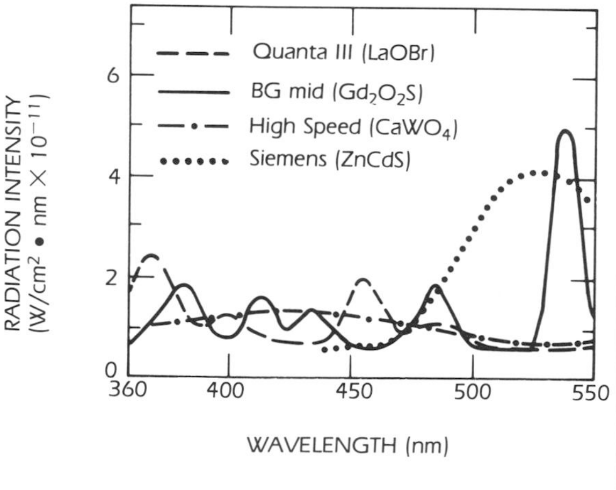

TABLE 3. Relative Light Yield as a Function of Detector

Screen	Photopic	S-20 Photocathode
ZnCdS	100	100
Gd_2O_2S	50	50
$CaWO_4$	7	32
LaOBr	4	25

Persistence

The *persistence* of a fluorescent screen is the time over which it continues to emit light following excitation. Persistence curves are a characteristic of phosphors. Some curves have an exponential decay, whereas others have long decay tails. An estimate of persistence can be made by assuming an exponential decay and then assigning a decay constant (time for the phosphor to decay by a factor of $1/e$, where e is the natural logarithmic base). A few typical decay constants are listed in Table 1. The persistence, particularly with rapid-decay phosphors, can vary significantly depending on the purity and the manufacturing process.

Unsharpness

Unsharpness in images formed by fluorescent screens is primarily a function of the grain size of the phosphor and the screen thickness, increasing as the parameters increase. Light transmission characteristics of the screen can also affect the unsharpness. Figure 4a demonstrates how unsharpness can affect the detection of a sharp-edged defect by spreading the edge shape. Here, C represents the contrast in percentage of brightness change, d represents width of defect and U represents the screen unsharpness. For a fixed value of U, a change in contrast C produces a change in the slope of the unsharp edge. It can be seen from Fig. 4b that when d is smaller than $2U$, the defect will vanish unless C_1 is above the minimum-observable contrast level. The following relationship may be obtained from Fig. 4b:

$$d = \frac{2C_1U}{C} \text{ for } C_1 \leq C. \qquad \text{(Eq. 2)}$$

Typical values of screen unsharpness for commercially available screens vary from $U = 0.50$ mm to $U = 1.0$ mm.

Screen Gamma

The fluorescent screen gamma (γ) is a measure of the contrast ratios between the output screen image brightness, B, and the input X-radiation intensity, I:

$$\frac{\Delta B}{B} = \gamma \frac{\Delta I}{I} \qquad \text{(Eq. 3)}$$

As in film radiography, the output image must have a minimum brightness ratio between adjacent image areas for detection by an observer or an image intensifying component. For most fluorescent screens at industrial radiography energies, the screen gamma is very close to 1.0, so the fluorescent screen itself is very seldom the limiting factor as far as the total imaging system gamma is concerned.

Radiation Energy Effects

The efficiency of fluorescent screens is a function of the energy of the radiation. This is shown in Fig. 5 where the mass absorption coefficient is plotted against the curves of phosphor materials. These

FIGURE 4. Effect of Unsharpness on Defect Detection

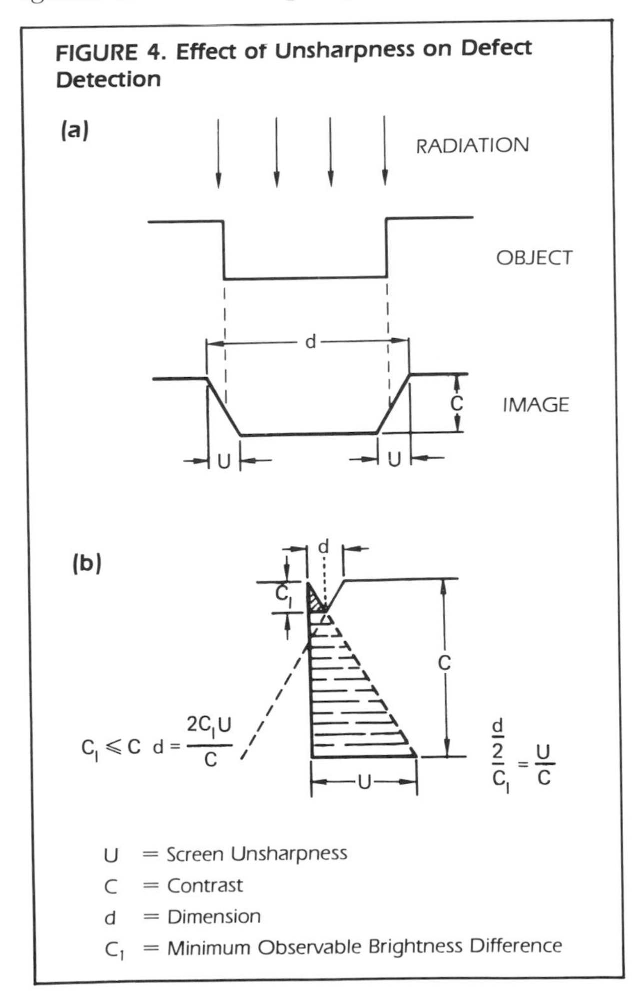

data were calculated from X-ray cross-section tabulations.[40] The K absorption edge is indicated by the step in the coefficient for those phosphors containing sufficiently heavy elements. Other than at the K absorption edge, the absorption coefficient (and hence efficiency) decreases with increasing photon energy.

The absorption efficiency is given by the equation

$$n_a = 1 - e^{-\frac{\mu}{\rho}XP} \qquad \text{(Eq. 4)}$$

where μ/ρ is the mass absorption coefficient, X is the phosphor thickness and P is the packed phosphor density.

When dealing with machine sources of X-rays, the energy spectrum of the radiation must be considered. The total absorption efficiency n_a is found by integrating the absorption efficiency and photon intensity over energy.

FIGURE 5. Mass Absorption Coefficient of Phosphors as a Function of Energy

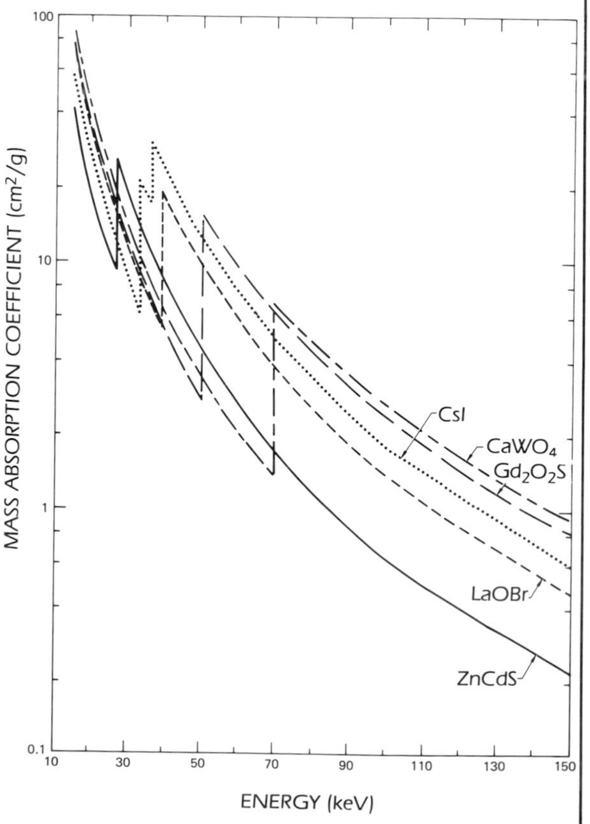

The X-ray photon energy spectrum will be changed by the presence of an object; the direct transmitted fluorescent beam is hardened. At the same time, lower energy scattered radiation will be generated, for which screens generally have greater absorption efficiency.

The photopic brightness output of several screen types, as a function of peak X-ray kilovoltage, is shown in Fig. 6. Similar data have been measured in other experiments.[35]

Special Screens for Use in Real-Time

Fiber Bundle Scintillator

A new type of fluorescent screen now being developed is called the *bunched fiber scintillator*. Many small fibers are loaded with a scintillation material and bundled together to form a light pipe. The length of the fibers determines the screen efficiency, which may easily approach 100 percent in a few millimeters. The resolution is determined by the fiber diameters. The difficulties in the development of such screens are the loading of the scintillation material and crosstalk between fibers.

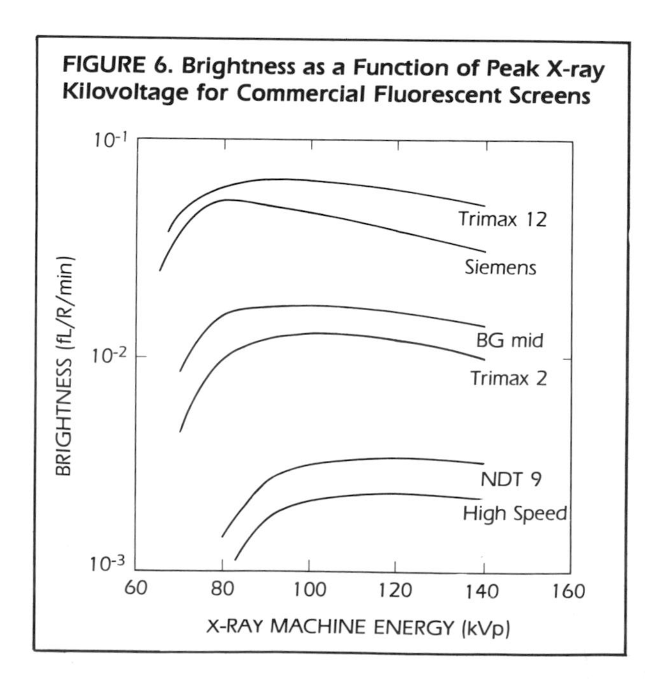

FIGURE 6. Brightness as a Function of Peak X-ray Kilovoltage for Commercial Fluorescent Screens

Scintillating glass fibers offer the best method of perfecting such a screen. Although the glass scintillators have a low light yield compared to many scintillating materials, this disadvantage is not limiting. Fibers of 0.5 mm diameter and smaller may be made.

Neutron Sensitive Screens

Real-time radiography may be performed using neutron beams when the fluorescent screen is sensitive to neutrons. High thermal neutron cross-section elements such as lithium-6, boron-10 and gadolinium are used in neutron-sensitive screens. **Plastic scintillation materials can be used in fast neutron radiography.**

The characteristics of screen composition and construction are more important in neutron imaging than in X-ray imaging because the intensity levels of available neutron sources is generally lower than conventional X-ray sources. It is, therefore, of primary importance for the screen to stop a sufficient quantity of neutrons to obtain an acceptable contrast level. Some data are available on screen brightness versus neutron flux for a variety of fluorescent screens that can be used in neutron radiography. In Fig. 7, curves for these data are shown.[13]

FIGURE 7. Relative Screen Speed Versus Neutron Flux

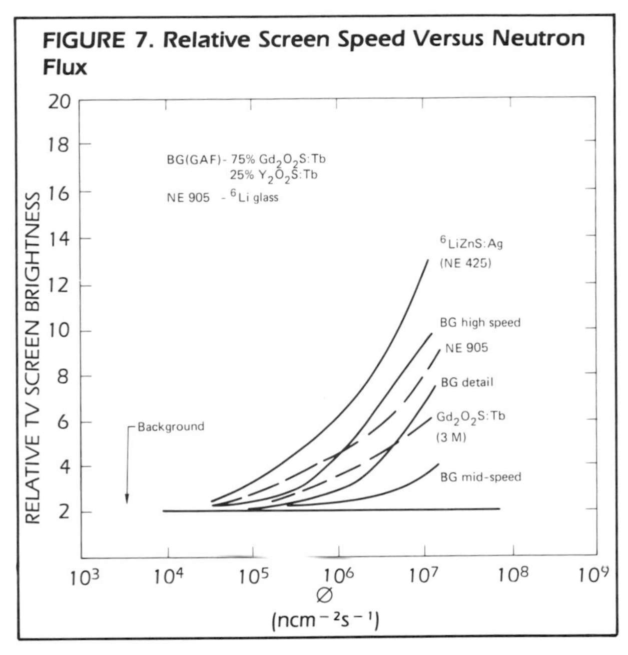

High Energy Screens

Some materials, when they absorb X rays, emit electrons in copious quantities. The phosphor materials used in fluorescent screens are generally more sensitive to electrons than to primary X rays. At high X-ray energies, the electrons from suitable secondary electron emitters can be used to enhance the imaging process. Heavy metals such as lead, tungsten, or tantalum are often useful in MeV radiography, serving as filters in front of the fluorescent screen. Along with the production of secondary electrons to aid high energy detection, the heavy metal will shield the phosphor from lower-energy scattered radiation. This improves the contrast sensitivity of the image.

Normally, thicker layers of fluorescent material are used in high energy applications.

PART 3

IMAGE QUALITY AND RADIOGRAPHIC PARAMETERS

For real-time imaging systems employing fluorescent screens, the factors that limit resolution are similar to those in normal radiography. Table 4 shows some of the parameters which affect image quality.

For film radiography, the X-ray photons absorbed in the recording medium during the long exposure times (minutes) are integrated for image formation. In real-time imaging, only the photons absorbed during the scan time of the image pickup system [or in the case of fluoroscopy, the summation time of the human eye (0.2 s)], will add to each single image. With the increasing sophistication of present image systems, the limits of object thickness penetrated and the contrast of details detected are only restricted by the quantum structure of the radiation and the noise introduced by each stage of the imaging process.

Each detail, as defined by size and contrast, can be described by a number of radiation quanta that is proportional to the intensity of the radiation. The relative statistical fluctuation of radiation intensity (in a two-dimensional element of area in the detector plane) is proportional to the reciprocal value of the square root of the intensity. In general, radiation contrast in the image element must exceed the value of this fluctuation in order to yield detail that may be distinguished from the background fluctuation.

Contrast

Subject Contrast

Subject contrast for fluorescent screens is defined as the fractional change in brightness resulting from a change in absorber thickness, Δx. The approximately linear relationship between screen brightness, B, in millilamberts (candelas per square meter, in SI) and X-ray intensity on the screen, I, in roentgens (grays, in SI) per minute may be written as:

$$B = mI \quad \text{(Eq. 5)}$$

or

$$\Delta B = m\Delta I \quad \text{(Eq. 6)}$$

where m is the proportionality constant. The absorption law for monochromatic radiation is

$$I = I_O e^{-\mu x} \quad \text{(Eq. 7)}$$

or

$$\Delta I = -\mu I \Delta x \quad \text{(Eq. 8)}$$

where μ is the linear absorption coefficient and x is the absorber thickness. By combining the above

TABLE 4. Image Quality

Contrast		Definition	
Subject	System	Geometry	Mottle
Thickness	Screen gamma	Source focal spot size	Quantum fluctuations
Scatter	Intensifier and/or camera	Source-to-object distance	Screen grain
Radiation quality	Television monitor	Object-to-screen distance	Raster scan of the television monitor
		Screen thickness	
		Motion	

equations we have

$$\frac{\Delta B}{B} = -\mu \Delta x = C \qquad \text{(Eq. 9)}$$

where C is the contrast.

The effect of kilovoltage on subject contrast, C, manifests itself through the absorption coefficient μ which varies with kilovoltage. The efficiency of the fluorescent screen affects subject contrast by its ability to convert the incoming X-ray photons to light and manifests itself through the screen brightness response, B.

Observed Contrast

The observed contrast in real-time imaging is affected by several factors beyond the screen response. One must include the effect of all system components. This is done by defining gamma (γ) which is a proportionality factor for the contrast ratios of the output to the input intensity. This can be written as

$$\gamma = \frac{\frac{\Delta B}{B}}{\frac{\Delta I}{I}} \qquad \text{(Eq. 10)}$$

or

$$\frac{\Delta B}{B} = -\gamma \mu \Delta x \qquad \text{(Eq. 11)}$$

where μ is the linear absorption coefficient and Δx is the change in the thickness of the absorber material. The combined system gamma is the product of the individual components in a real-time imaging chain. In a vidicon television chain, for example, the system gamma would be given by:

$$\gamma = \gamma_s \gamma_a \gamma_v \gamma_k, \qquad \text{(Eq. 12)}$$

where

γ = overall X-ray television system gamma;
γ_s = fluorescent input screen gamma;
γ_a = electronic amplifier chain gamma;
γ_v = vidicon tube gamma; and
γ_k = television picture tube gamma.

For a typical case, the fluorescent screen gamma is taken as $\gamma_s = 1.0$, a conventional closed-circuit television amplifier chain provides a $\gamma_a = 1.0$ (maximum), the vidicon tube provides a $\gamma_v = 0.9$, and the television picture tube gamma is typically $\gamma_k = 3.0$ (see ref. 29). Although it appears that considerable contrast gain is possible in a TV chain, the final imaging element, the human eye, must also be considered. The human-eye gamma is nonlinear and less than 1. Typical value of human-eye gamma is 0.3. With increased brightness or glare, this value drops rapidly.

Effects of Scatter

Scattered radiation affects contrast in fluorescent screens by effectively raising the background brightness level. The scattered radiation affects only the primary imaging component of a real-time system which is the fluorescent screen. Here the contrast is defined as

$$C = \frac{\Delta B}{B} = \frac{\Delta I}{I} \qquad \text{(Eq. 13)}$$

If, however, scattering is not eliminated, the equation for contrast becomes

$$C = \frac{\Delta I}{I + I_s} \quad , \qquad \text{(Eq. 14)}$$

where I_s is the scattered radiation intensity. Now if $I_s = KI$, where K is the scattering factor, the equation for screen contrast becomes

$$C = \frac{\Delta I / I}{K + 1} \quad . \qquad \text{(Eq. 15)}$$

As this equation illustrates, it is important to keep the scattered radiation impinging on the fluorescent screen to a minimum in order to keep the contrast at an acceptable level.

Control of Scatter

Scattered radiation comes from many sources in a real-time imaging setup. There is room scatter, object scatter, fixture scatter, air scatter in the primary radiation beam, and scatter from any and all objects placed in the path of the radiation beam. The control of scatter in a real-time imaging system is the same as in normal film radiography. Specific techniques to reduce scatter are listed below.

1. Collimate the primary beam to the minimum viewing area necessary.
2. Shield the setup to reduce room scatter from walls, ceilings and floors.

3. Filter the primary beam to remove the low energy portion of the spectrum, which is more susceptible to scatter.
4. Filter the radiation beam between the object and the fluorescent screen, because many filters preferentially remove the lower-energy scattered radiation.
5. Use antiscatter grids, both fixed and moving, between the object and the fluorescent screen when the resolution required necessitates such use.
6. Use projection magnification to increase the distance of the fluorescent screen from the object scatter.

In general, it is very important in real-time imaging techniques to carefully consider all areas where scattered radiation can be introduced and to then attempt eliminating them or at least reducing their effect.

Definition

Unsharpness

The penumbral image width, due to a finite focal-spot size, q, is defined as geometric unsharpness, U_g. This is illustrated in Fig. 8a, and further defined by eq. 16.

$$U_g = \frac{b}{a} q = q(M - 1) \quad \text{(Eq. 16)}$$

Here M is the magnification factor and is shown schematically in Fig. 8b.

It is generally agreed that penumbral images are not dependable. The limiting case for an umbral image is illustrated in Fig. 8c, where

$$\frac{d}{q} = \frac{b}{a+b} \quad \text{(Eq. 17)}$$

The defect, d, is assumed equal to the defect depth, Δx, where absorber thickness is x. From eqs. 16 and 17 or from Fig. 8c, it follows that for optimum definition Md is much greater than U_g, while

$$Md = U_g \quad \text{(Eq. 18)}$$

is the "minimal" definition on a perfect screen. In a similar fashion, it can be shown that the minimum observable defect size, d, multiplied by the magnification factor M must be equal to or larger than the unsharpness, U_f, due to the fluorescent screen, or:

$$Md \geq U_f \quad \text{(Eq. 19)}$$

FIGURE 8. X-ray Image Projection Geometry: (a) Geometric Unsharpness, U_g; (b) X-ray Projection Enlargement; (c) Limiting Condition for Umbral Image

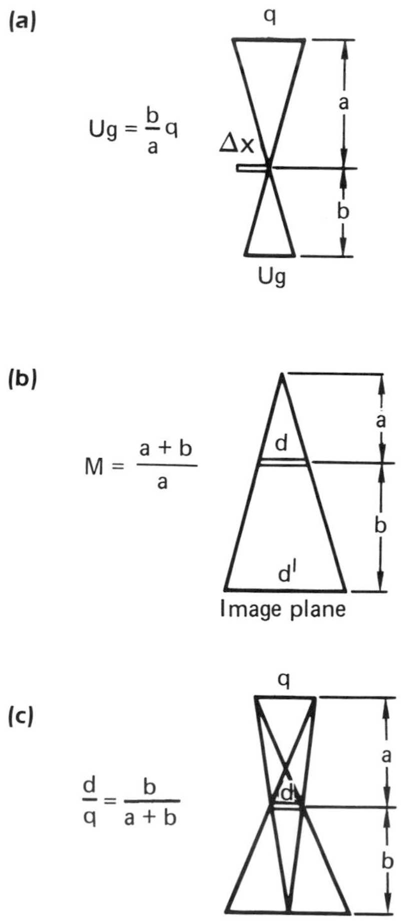

(a) $Ug = \frac{b}{a} q$

(b) $M = \frac{a + b}{a}$ — Image plane

(c) $\frac{d}{q} = \frac{b}{a + b}$

q = Focal spot width
Ug = Penumbral image width
Δx = d
M = Magnification

The limit on definition is actually controlled by both types of unsharpness, so Md is greater than, or equal to, U_O. The total unsharpness U_O is equal to the cube root of the sum of the cubes of the unsharpness due to geometry U_g and the fluoroscopic screen U_f (see ref. 24), or:

$$U_O = (U_g^3 + U_f^3)^{1/3} \quad . \qquad \text{(Eq. 20)}$$

Optimum Magnification

X-ray projection enlargement is shown schematically in Fig. 8b. the magnification factor M is defined as

$$M = 1 + \frac{b}{a} = \frac{a+b}{a} \quad . \qquad \text{(Eq. 21)}$$

where a is the focal-spot-to-object distance and b is the object-to-screen distance. In general, due to the large inherent unsharpness of fluorescent screens, magnification of the X-ray image can be used to improve the radiographic definition. The optimum magnification. M_{opt}, for any screen unsharpness U_f and focal spot width q is given by (see refs. 16 and 28):

$$M_{opt} = 1 + \left(\frac{U_f}{q}\right)^{\frac{3}{2}} \qquad \text{(Eq. 22)}$$

The smallest observable defect d is then given by

$$d = \frac{U_f}{M_{opt}^{\frac{2}{3}}} \qquad \text{(Eq. 23)}$$

As is evident from eq. 22, the focal spot size is an important parameter in defining the optimum magnification as well as the geometric unsharpness. Decreasing the focal spot size will improve the radiographic definition as well as allow larger magnifications. Unfortunately, in most commercially available X-ray machines and radiation sources, a decrease in focal spot size means a decrease in radiation intensity, which causes a corresponding decrease in brightness and screen contrast.

Effects of Motion

Normally in real-time imaging applications, unsharpness due to object movement can limit radiographic definition. Determining factors for this are: the X-ray excitation rate; the decay time of the fluorescent screen phosphor; and, the delay time or scan time of the imaging system components. Typically, real-time systems are used with continuous or rapid excitations (120 pulses per second), rapid decay phosphor (on the order of milliseconds or faster) and frame rates of 30 frames per second. The human eye requires about 0.2 seconds of integration time, so it should be the limiting factor.

Quantum Mottle

The statistical fluctuation of brightness on fluorescent screens, due to the randomness of X ray production and absorption, should be considered whenever radiographic definition is a concern. The following list shows the numerous sources of this fluctuation which must be included in any consideration of the real-time imaging process.

1. X-ray photon production
2. X-ray photon absorption in the specimen
3. X-ray photon absorption in the screen
4. Conversion of X-ray photons to light photons
5. Fraction of light photons reaching the eye after traversing the imaging system
6. Light-photon absorption in the retina

These fluctuations obey Poisson distribution such that the standard deviation s is equal to the square root of the intensity, n:

$$s = n^{1/2} \quad . \qquad \text{(Eq. 24)}$$

Because real-time radiography involves a series of processes, the standard deviation for the total sequence is

$$s^2 = s_1^2 + s_2^2 + \ldots s_n^2 \quad . \qquad \text{(Eq. 25)}$$

Certain stages in the process will cause either an increase or decrease in the intensity. It can be shown that to a first approximation

$$s^2 = gn \quad , \qquad \text{(Eq. 26)}$$

where g is the amplification of the process (from the state of lowest intensity to the final observation by the retina) and n is the final intensity.[41]

To observe detail in the light image reaching the retina, there must be a detail intensity difference greater than the standard deviation. Thus

$$\Delta n = ks \quad , \qquad \text{(Eq. 27)}$$

where Δn is the smallest difference in the number of detectable photons in the retinal image, and k is a constant termed the threshold contrast to standard deviation ratio. Contrast is correctly defined as

$$C = \frac{\Delta n}{n} \quad . \qquad \text{(Eq. 28)}$$

Then

$$C = \frac{kg^{1/2}}{n^{1/2}} \quad .$$

Since n is the total intensity of photons in the retinal camera image, it is equal to the number of photons arriving per second, times the storage time of the detector, t. Therefore,

$$n_r t = n \quad , \qquad \text{(Eq. 29)}$$

where

$$n_r = \frac{\pi d^2 n_0}{4} \quad . \qquad \text{(Eq. 30)}$$

Here d is the diameter of the object under observation and n_0 is the number of photons reaching the detector, per unit area of the fluorescent screen, per second.[41]

Now

$$C = \frac{2kg^{1/2}}{d(\pi t n_0)^{1/2}} \quad , \qquad \text{(Eq. 31)}$$

or

$$d = \frac{2kg^{1/2}}{C\,(\pi t n_0)^{1/2}} \quad . \qquad \text{(Eq. 32)}$$

The smallest discernible object size that can be detected is seen to improve with increasing contrast and number of stimulating photons. Detail sensitivity in an image will be limited by statistical fluctuations as long as d is greater than the unsharpness of the system.

The statistical fluctuations of fluorescent screen brightness due to the randomness of the process is important at low brightness levels, which usually occur with radioactive isotope sources or low intensity neutron sources for real-time imaging. Most industrial X-ray machines produce sufficient intensity to render the statistical fluctuations unimportant for most applications. With the advent of on-line digital processing of the real-time image, much of the effect of the quantum fluctuation from isotope sources can be removed in near real-time by video frame-averaging or summing (see Part 6 of this Section).

Radiation Sources

The radiation source often plays the most important role in real-time imaging. Several points must be considered when choosing a radiation source for a particular real-time application. These include:

1. What thickness and type of material must the source radiation penetrate?
2. What specifically is to be imaged (e.g., hydrogenous material, defects in materials, specific components, etc.)?
3. Is there sufficient radiation intensity to produce an acceptable fluorescent screen contrast?
4. Is the radiation focal spot of an acceptable size for the image definition required?

Conventional hot cathode X-ray machines have been employed in real-time imaging systems due to availability, ease of operation, high radiation intensity and the large range of available focal spots. Half-wave rectified, full-wave rectified and constant potential machines vary in the X-ray output waveform. This waveform variation (X-ray intensity as a function of time) may be a consideration when processing images formed in time intervals less than the frequency of variation.

Linear accelerators, for high energy (2-15 MeV) applications, use resonant waveguides to accelerate electrons to a target. The radiation is produced in pulses typically of 3 to 4 μs widths and at frequencies of 60 to 420 Hz. In real-time imaging, care must be taken to protect components in the detector system from radiofrequency noise and direct high intensity radiation. Television-based imaging systems often show a broadband noise signal in the image at the frequency of accelerator operation. Setup configuration, synchronizing the accelerator to the imaging system, and image frame averaging can be used to reduce or eliminate these effects.

Gamma ray sources such as ^{192}Ir and ^{60}Co are also being used in real-time imaging applications. Advantages for these sources include compactness, portability and the ability to penetrate substantial thicknesses of material. Disadvantages include relatively low output intensity and larger focal spots. The relatively low radiation intensity can be partially compensated for with on-line digital processing techniques applied to the output video signal of the imaging system. Some isotope sources are available with focal spots comparable to those found in conventional X-ray tubes, but again a reduction in focal spot size reduces the effective source strength and therefore radiation intensity.

Real-time neutron imaging is finding increased application in industry with the advent of portable neutron sources, and nuclear reactors with beam ports suited to real-time imaging applications. Neutron sources are still expensive compared to other conventional radiation sources and those sources other than nuclear reactors suffer from relatively low intensity neutron outputs. Also, the effective focal spots of neutron sources are large. In most neutron sources, whether reactors, accelerators or isotopes, the focal spot is defined by the collimator opening at the neutron source. In order to transport a reasonable number of neutrons down the beam tube, these collimator openings are necessarily larger than the focal spots possible in X-ray machines and high specific-activity gamma ray sources.

Radiation Energy Selection

The size and material of the part to be imaged will determine the radiation type and energy to be used. As a general rule, the radiation's type and energy should be selected so that the specimen thickness to be penetrated is five half-value layers. One half-value layer reduces the transmitted radiation intensity by 50 percent. Satisfactory results can also be obtained if deviation from the rule is no greater than a factor of 2 (i.e., in the range of 2.5 to 10 half-value layers).

The configuration of the part is of some importance because of the scattering characteristics of most materials. Configuration should also be considered when choosing a radiation source because it will affect the images obtained.

The half-value layer (HVL) can be equated with values of μx from the absorption law stated in eqs. 7 and 8:

$$I = I_O e^{-\mu x}$$

or

$$\frac{\Delta I}{I} = -\mu \Delta x$$

Table 5 presents this data in tabular form. The detection percentage is based on a subject contrast of $\Delta x/x = 2\%$. The image contrast of $\Delta I/I$ is considerably increased for large thicknesses. This is a great advantage, provided that the transmitted energy can be recorded.

TABLE 5. Half-value Layers Associated with Different Values of μx

HVL	Values of μx	Detection $\frac{\Delta I}{I} \times 100$	Transmission $\frac{I}{I_o} \times 100$
1.45	1	2	36
2.17	1.5	3	22
2.89	2	4	13
3.62	2.5	5	8
4.33	3	6	5
5.05	3.5	7	3
5.78	4	8	1.8
6.50	4.5	9	1.1
7.22	5	10	0.67
7.94	5.5	11	0.41
8.66	6	12	0.25
9.40	6.5	13	0.15
10.01	7	14	0.09

PART 4
FLUOROSCOPY

Basic Technique

The basic technique of direct viewing fluoroscopy systems is shown in Fig. 9. The radiation source, object and screen are oriented in the same fashion as in conventional film radiography. Often an optimum magnification is used, as discussed in Part 3 of this Section. Following the fluorescent screen, a radiation barrier is provided to protect personnel. A cabinet arrangement containing the entire system is typical of most designs.

The advantages of fluoroscopy systems are low cost, simplicity and speed of inspection. Compared to film radiography, the fluoroscopic unit has a higher capital investment; however, the operational costs for fluoroscopy are quite small. The fluoroscope also provides the answer to the inspection question immediately. If the quality is in doubt, a technique of higher sensitivity may be applied to the area of interest in the object.

Fluoroscopic units do not produce permanent records and suffer from limitations in detail sensitivity and resolution relative to film. The operator must be dark-adapted to see the dimly illuminated screen and special care must be taken for operator safety. Finally, fluoroscopy systems are normally restricted to applications below 200 kV (63 mm of aluminum or 3 mm of steel).

Radiation Barrier Windows

The radiation exposure of the personnel operating a fluoroscopic unit is a primary concern. The whole-body dose limit to persons working occupationally with radiation equipment is 1 mSv (0.1 rem) per week with a maximum of 13 mSv (1.3 rem) per 13-week period.[33] The reduction of radiation exposure is accomplished with shielding windows which transmit the light of the fluorescent screen, but stop the highly ionizing radiation.

In considering a radiation barrier window, the important characteristics are: X-ray absorption, optical transmittance, resistance to discoloration, and scattering properties. Table 6 lists the properties of some common barrier windows.

Lead glass is the most common window material. High density lead silicate glass has been found to be the most efficient window, while lead oxide (PbO) glasses have a protective coefficient directly proportional to the percentage of PbO. The *protection coefficient,* defined as the ratio of the thickness of lead to the thickness of material providing equivalent X-ray absorption, is used to evaluate the radiation attenuation. The *lead equivalence* is the commercial method of specifying lead glass. For example, a piece of lead glass 4 mm thick that has the same absorptive properties as 2 mm of metallic lead would be designated as 2 mm lead equivalent glass.

FIGURE 9. Basic Fluoroscopic Cabinet Unit

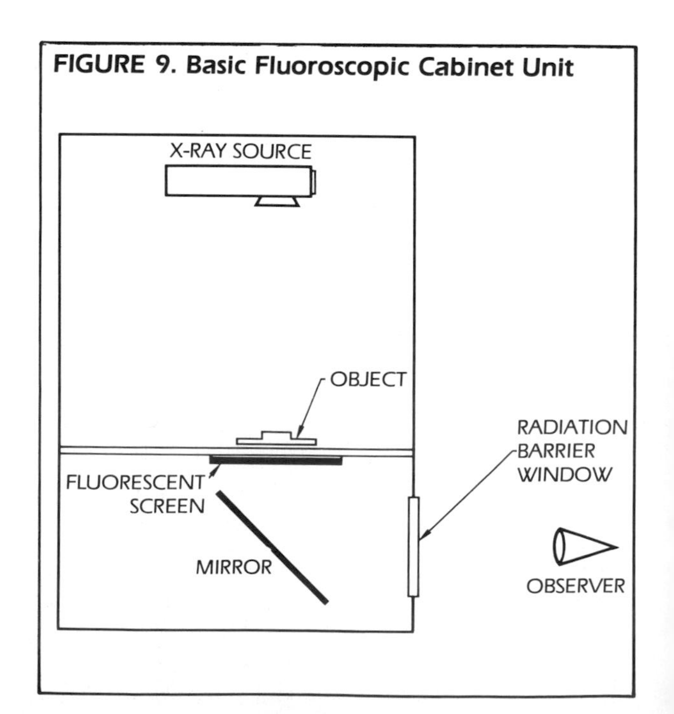

TABLE 6. Radiation Barrier Window Parameters

Material	Density (g/cm^3)	Protective Coefficient	Scattering Ratio $\frac{I_s + I_p}{I_p}$	Color
Lead silicate glass	6.2	0.55	1.01	yellow
Lead oxide glass	5.1	0.26	1.06	slightly yellow
Lead perchlorate (liquid window)	2.6	0.11	1.08	clear
Lead plexiglass	1.6	0.05	—	similar to acrylic resin (clear)

The optical transmittance of the window must be high and remain high in service. High radiation intensity will discolor PbO glasses as well as common glasses. This normally occurs in the range of 10^3 to 10^4 grays (10^5 to 10^6 rad), depending on the specific material. Fluoroscopy generally does not require dose levels high enough for this to be a serious problem. Discoloration can be removed in many cases by exposure to heat or infrared and visible radiation. The high density lead silicate glass will discolor under high radiation (excessive by fluoroscopic standards) but will recover in a matter of hours. Cerium stabilization can be used to prevent browning of lead glass under irradiation. Photosensitivity (permanent retention of an image) is a problem with some glasses.

X-ray scattering by the window is a further consideration. A few values for this were provided in Table 6, where I is the scattered radiation and I_p is the direct radiation.[34] The X rays scattered from the barrier back to the fluorescent screen will result in a general increase in brightness with a subsequent loss in detail visibility due to small brightness changes. Some evidence suggests higher lead content barriers may reduce the backscatter effect.[16] A separation between the fluorescent screen and the window is one possible method for reducing this noise.

Liquid barriers are another possibility. A saturated solution of lead perchlorate is optically clear and resistant to discoloration, although it is hazardous.[5] Potassium iodide has been used successfully as a substitute for lead perchlorate.[11]

Mirrors are another method of avoiding personal exposure from fluoroscopy units. A loss of light and visual acuity does result. In cabinet devices, the mirror and a lead glass barrier are both employed. Front-surface mirrors must be used to avoid ghost images which will develop due to multiple reflections in back-surface mirrors. A front-surfaced mirror commonly consists of a vacuum-deposited aluminum coating with a protective layer of silicon monoxide or magnesium fluoride. A visual reflectance of 88 percent is typical.[39] Periodic cleaning of the mirror, lead glass shield, and fluorescent screen is advised because a significant light loss can result from buildup on the surfaces.

Operational Characteristics

Factors Affecting Imaging

In fluoroscopy systems, the information obtained may be affected by two kinds of factors; radiographic and physiological. Table 7 lists these factors.

Visual Acuity

The operation of fluoroscopy systems typically occurs with light levels in the range of 3 to 0.0003 cd/m^2 (1.0 to 0.0001 millilambert). The human eye uses two types of vision mechanisms: *photopic* (cone) vision and *scotopic* (rod) vision. The crossover from photopic to scotopic vision occurs around 0.06 cd/m^2 (0.02 millilambert). Industrial fluoroscopy is often performed with scotopic vision at less than 0.06 cd/m^2 (0.02 millilambert) brightness level. The importance of the use of photopic or scotopic vision is shown in Fig. 10, which is a plot of visual acuity versus brightness.[6] Visual acuity (defined as the reciprocal of the angle between two contours, in minutes) is shown to improve with brightness and achieves a marked improvement with the transition

TABLE 7. Factors Affecting Fluoroscopic Image Formation

Radiographic	Physiological
Source intensity	Sensitivity of the eye to low light levels
Focal spot size	Dark adaptation
X-ray energy spectrum	Response to color spectrum
Scattered radiation	Experience of operator
Geometric unsharpness	
Screen unsharpness	
Conversion efficiency	
Light emission frequency	
Light interference with protective barriers	

FIGURE 10. Graph of Visual Acuity Versus Brightness

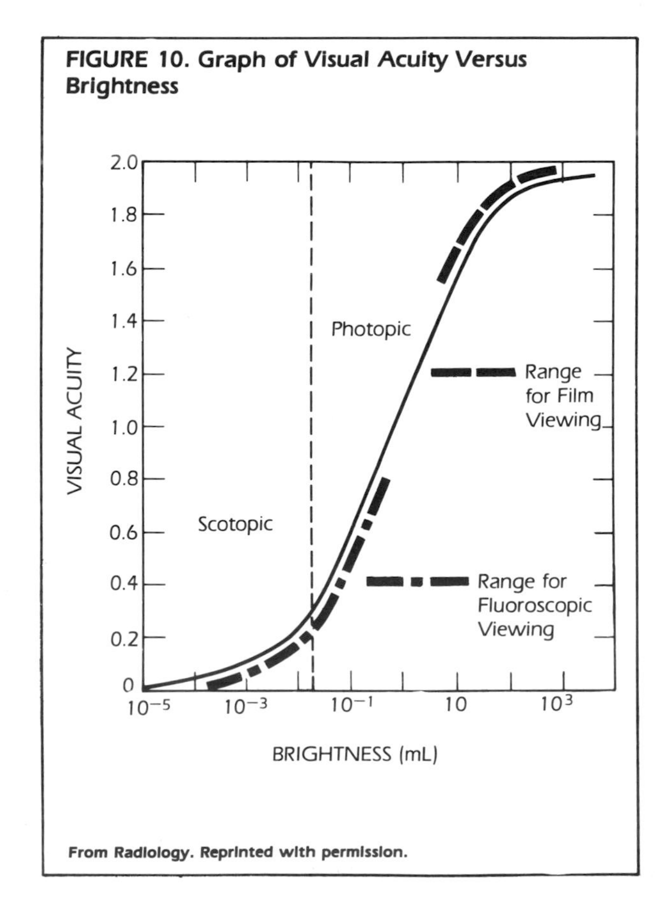

From Radiology. Reprinted with permission.

from scotopic to photopic vision. At a brightness level of 0.3 cd/m² (0.1 millilambert), a visual angle of 0.03 degrees must be subtended by two separate points to be distinguished. At a viewing distance of 0.25 m (10 in.) this would correspond to resolving 0.13 mm.

Visual acuity is a function of the light wavelength. The eye has a maximum response to 550 nm radiation, yellow light. Visual acuity is normally given for white light. In yellow or yellow green light, the visual acuity of the eye is slightly higher. It is slightly lower for red light and 10 to 20 percent lower for blue or for red light. Figure 11 shows the sensitivity of the eye as a function of wavelength for photopic and scotopic vision.[38]

Dark Adaptation

Dark adaptation is required when the screen brightness is less than 0.3 cd/m² (0.1 millilamberts). One minute of dark adaptation is required to detect a 1 percent contrast at 0.1 millilambert viewing level.[20] Dark adaptation is not just dilation of the pupil, but an increase in retinal sensitivity. Scotopic vision is adapted in 10 minutes, but the eye will continue to improve for at least 30 minutes. Twenty

FIGURE 11. Relative Sensitivity of the Human Eye as a Function of Wavelength

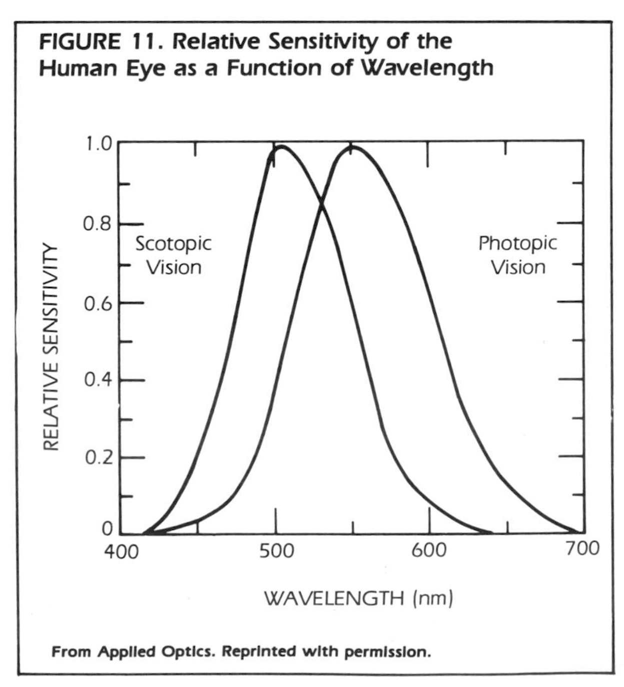

From Applied Optics. Reprinted with permission.

minutes of dark adaptation is considered good practice. Red light, up to thirty times brighter than white, will not affect dark adapted eyes. The use of red goggles outside the viewing area and the use of red light in the viewing rooms are useful techniques to maintain eye sensitivity.

Contrast Sensitivity

The contrast sensitivity in fluoroscopy is a function of the X-ray absorption, screen response, and visual response. The contrast is given by $\Delta B/B = -\mu\Delta x$ as derived in eqs. 5 through 9. The observed perceptible value of $\Delta B/B$ for brightness in the range of 0.003 to 3 cd/m² (0.001 to 1 millilambert) is [30]

$$100\frac{\Delta B}{B} = 1.6B^{-0.38} \qquad \text{(Eq. 33)}$$

Substituting for $\Delta B/B$ and solving for the perceptible object contrast $\Delta x/x$, the percent sensitivity S in absolute value is

$$S = 100\frac{\Delta x}{x} = \frac{1.6}{\mu x B^{0.38}} \qquad \text{(Eq. 34)}$$

Including the effect of scatter where $I_S = KI$, the sensitivity becomes

$$S = 100\ \frac{1.6(K+1)}{\mu x B^{0.38}} \qquad \text{(Eq. 35)}$$

This result indicates that the contrast sensitivity is improved as the brightness is increased. High power X-ray tubes can increase fluoroscopic sensitivity by increasing the brightness of the screen.[10]

The sensitivity in eq. 25 is contrast sensitivity only and does not include detail. An equation for sensitivity[16] assuming cavities of depth = diameter = Δx is

$$S = \frac{3.65}{x}\left(\frac{1+K}{\mu M B^{1/2}}\right)^{1/2} \qquad \text{(Eq. 36)}$$

where M is magnification.

Scattered Radiation

With the object adjacent to the screen, the loss in image quality due to scatter effects can be severe. Grids could be used, but the brightness is reduced by factors of three to seven, depending on the grid type and radiation energy. Of course, good collimation of the beam should always be employed. Movable masks around the X-ray source (to reduce the radiation field to the object size) will remove bright glare from unshielded areas of the fluorescent screen.

Table 8 lists values of the scattering factor K for various aluminum thicknesses and two types of fluoroscopic systems.

When magnification is employed, scattered radiation effects are less significant because the object is separated from the screen.

TABLE 8. Scattering Factor as a Function of Thickness

	Object Close to Screen		With Magnification	
Aluminum Thickness (mm)	K	Brightness (millilambert)	K	Brightness (millilambert)
13	0.64	0.084	0.16	0.018
19	0.83	0.057	0.19	0.015
25	1	0.039	0.23	0.013
38	1.5	0.024	0.37	0.010
51	2	0.013	0.49	0.0051
64	2.7	0.0075	0.65	0.0025
76	3.4	0.0050	0.81	0.0012
89	4.1	0.003	1.0	0.0005

From R. Halmshaw. Reprinted with permission.

Magnification

Magnification can play an important role in improving real-time image sensitivity because of the large inherent screen unsharpness. The relation between magnification M, geometric unsharpness U_g, and focal spot size q were presented in an earlier Part of this Section. The optimum magnification was given as:

$$M_{opt} = 1 + \left(\frac{U_f}{q}\right)^{3/2},$$

where U_f is the fluorescent screen unsharpness.

An interesting experiment was performed[16] to evaluate the optimum magnification. The fluoroscopic image of a series of wires on specimens of varying thickness were viewed at various magnifications. The optimum magnification, where the image detail was determined to be best, is shown in Fig. 12. The results, which are an average of subjective evaluations, are only applicable to the technical conditions of the equipment employed. Suggested reasons for these results are as follows:

1. If the magnification causes the unsharpness to be greater than the width of the image detail, the contrast is reduced.
2. The width of unsharpness for the image of an edge is different for scotopic or photopic vision; photopic vision has better definition.
3. The increase in image size that is useful may depend more on screen brightness than on unsharpness because of the brightness versus visual response of the eye.

In practice, large magnifications are frequently not practical unless microfocused X-ray sources are used. A reasonable magnification of 2X is often employed in equipment that will be used on a variety of objects. The use of magnification also provides flexibility in the selection of the X-ray source-to-screen distance. The source-to-screen distance may be selected to provide the brightest screen possible with the optimum projected image size.

Experimental Sensitivities

Typically the sensitivity of fluoroscopy is in the range of two to five percent. Using a high power fluoroscopic unit (4000 kVp-mA), sensitivities as low as 1.5 percent were obtained on aluminum samples from 19 to 76 mm thick.[10] A 0.6 × 0.5 mm focal spot, 2X magnification, and 460 mm source-to-screen distance were employed.

Movement

Movement of the object may take place not only to speed inspection, but to assist inspection. Often movement serves as a stimulus for observation of

FIGURE 12. Experimentally Determined Curves of Optimum Values of Magnification Versus Material Thickness for 0.3 mm Focal Spot X-ray Source: (a) Steel Specimens; (b) Aluminum Specimens

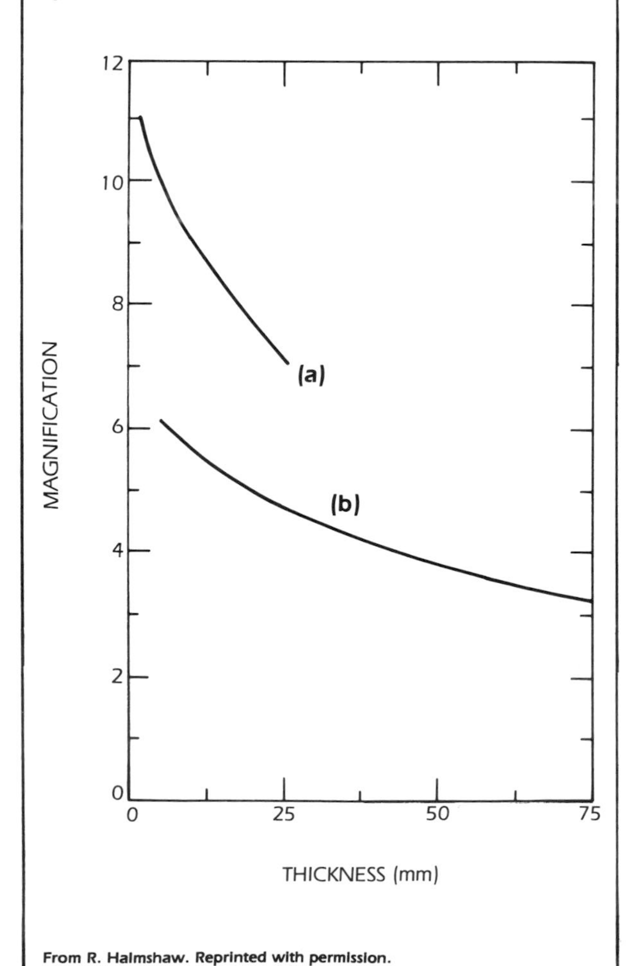

From R. Halmshaw. Reprinted with permission.

contrast changes that would not be observed in a still image. Comparative motion attracts attention; this fact is easily demonstrated with a fluoroscopy system. For scanning, the maximum practical speed for high quality fluoroscopy is around 50 millimeters per second.[1] Most fluorescent screens have decay times that are rapid (10^{-2} seconds or less) with respect to the image retention of the eye. The critical flicker frequency at the light intensity levels used in fluoroscopy is around 60 Hz.

Statistical Considerations

Statistical fluctuations exist in the processes of X-ray production, X-ray transmission, absorption in the fluorescent screen, light emission from the screen, and light absorption in the retina of the eye, all of which enter into the consideration of sensitivity.[5] Table 9 lists a statistical evaluation of an industrial fluoroscopy system. The final statistic is 1.1×10^3 photons/mm^2. In this case the lowest intensity is at the last stage so that the statistical fluctuation is 3 percent. To achieve higher contrast sensitivity over a square millimeter area, for this system, would require changes in the statistical operation of one or more of the components.

TABLE 9. Statistics of Industrial Fluoroscopy System

X-ray production (2R/s at screen)	1×10^9 photons/mm^2/s
X-ray transmission of object (6%)	6×10^7
X-ray screen absorption (50%)	3×10^7
Light photons emitted by screen	1.5×10^{11}
Light photons reaching eye (250 mm distance)	1.9×10^5
Photons absorbed by retina (3%)	5.7×10^3
Photons absorbed per 0.2 s	1.1×10^3 photons/mm^2

Recommended Practice

Good technique in fluoroscopy may proceed along two lines. One method uses a large focal spot tube because of the large inherent unsharpness of the screen. The other method uses a small focal spot and magnification. The geometry of both these methods should be optimized for the detail required in the object of interest. Typical X-ray tube-to-screen distance in a fluoroscopic unit is 380 to 500 mm. Collimation of the X-ray beam for reduction of scatter is very important. Variable diaphragms at the X-ray source and the viewing screen are useful. Of course a quality fluorescent screen is essential. High brightness with low unsharpness and emission in the yellow-green region are desirable.

For the inspector, constant-brightness viewing is desirable. Changes in the background light intensity of the screen can affect the observer's interpretation. Constant-brightness viewing of an object which may change in dimensions or material is achieved by adjustment of the X-ray kilovoltage and/or tube current. The control for the X-ray machine should therefore be located at the viewing location. Automatic systems have been developed to adjust the X-ray tube controls to a predetermined screen brightness for whatever object is placed in the fluoroscope.

The operator must of course be protected from the radiation field. A 250 mm distance from the screen to the operator is normal with sufficient shielding to reduce the radiation to an acceptable level.

Applications

Fluoroscopy systems are used in industry today for rapid inspection of a variety of items. Light alloy castings are a prime application. Fluoroscopy is more often applied to determine product uniformity rather than flaws. The electrical, food, and explosives industries have many applications for the method. Mail sorting and baggage inspection are other areas where fluoroscopy is quite useful.

PART 5
REMOTE VIEWING SYSTEMS

Remote viewing systems using a television pickup offer significant advantages over fluoroscopy systems. The television display allows screen brightness and contrast to be adjusted by the operator. This improves visual acuity and permits use in a lighted room where dark adaptation is not necessary. Figure 13 is a block diagram of a remote viewing system. The fluorescent screen converts the radiation to light and an image intensifier boosts the light intensity to a level suitable for pickup by a television camera. The signal from the camera is sent to a television monitor located in a safe control location. A videorecorder is often used to provide a permanent record of the inspection — another advantage of this type of real-time imaging system over a fluoroscopy unit.

Shielding and mirrors are used to protect the electronic components from radiation. At X-ray energies below 1 MeV, this is not a serious consideration for reasons other than good radiographic practice. Below 300 kV the intensifier and camera can be placed in the direct line of the radiation. Typically, radiation damage occurs in electronics such as transistors at 1 kilogray (10^5 rad), and will discolor optical components at 1 to 10 kilograys (10^5 to 10^6 rad). Table 10 lists radiation damage thresholds for a few components. Given time outside the radiation field, most materials will recover from the damage. Specialized equipment such as camera tubes can be made to tolerate up to 1 megagray (10^8 rad).

FIGURE 13. Remote Real-Time Radiographic Viewing System

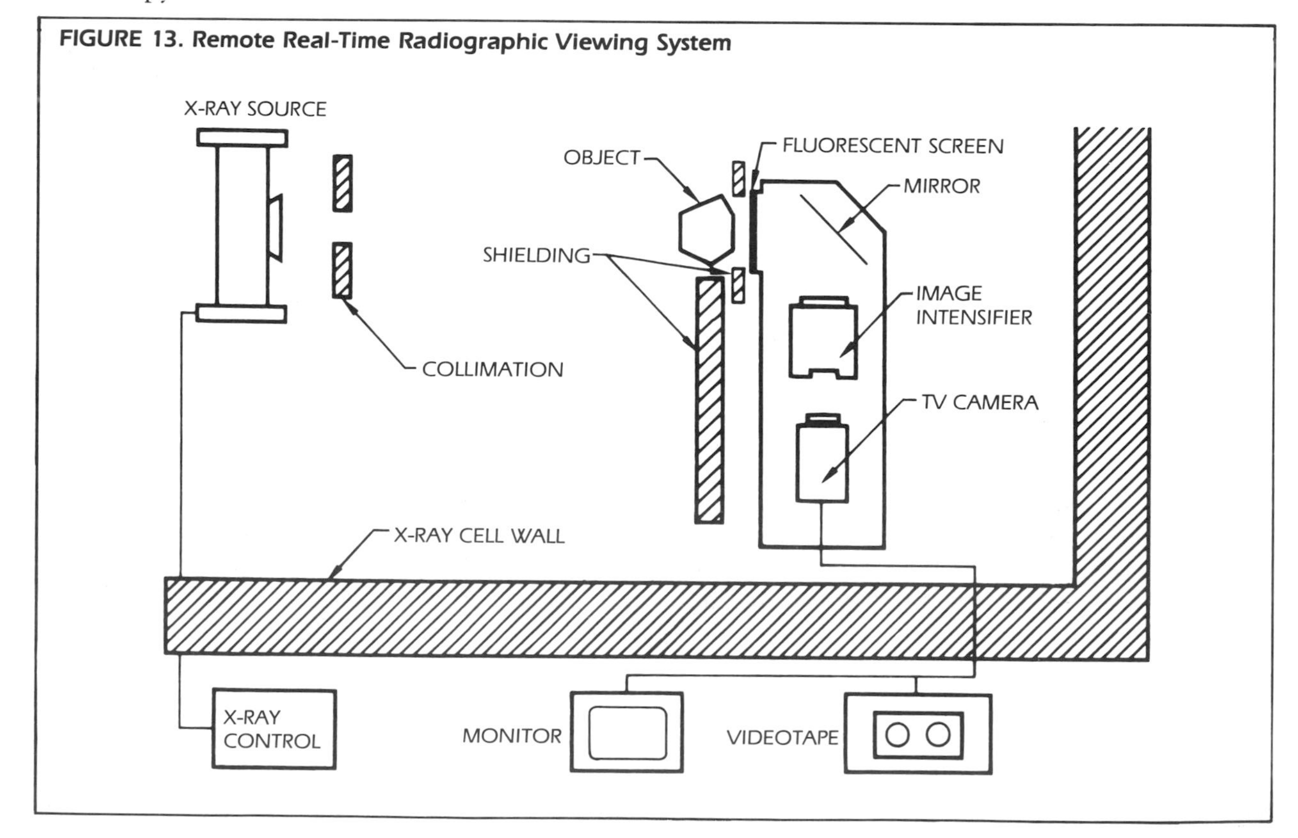

TABLE 10. Radiation Damage Thresholds

Component	Threshold grays (rads)
Silicon semiconductor	7×10^4 (7×10^6)
Germanium semiconductor	5×10^2 (5×10^4)
Capacitors	3-7 $\times 10^6$ (3-7 $\times 10^8$)
Resistors	2-5 $\times 10^6$ (2-5 $\times 10^8$)
Ceramic, glass, optical	2×10^2 (2×10^4)
Plastics	2×10^2 (2×10^4)

Glass may fluoresce under strong irradiation which can result in undesired light signals. Noise may also be generated in the electronics, increasing with the radiation intensity. Good shielding practice is desirable for mid-energy systems and very important in high energy operations. Of course, shielding for personal protection is a necessary feature. The remote viewing system allows real-time radiography at very high energies where personnel shielding for direct observation of a fluoroscopic screen would be impractical.

Image Intensifiers

X-ray Image Intensifiers

The *image intensifier tube* converts photons to electrons, accelerates the electrons, and then reconverts them to light. Figure 14 shows a generalized diagram of an intensifier tube. Intensifiers typically operate in the range of 30 to 10,000 light amplification factors. The intensification is not necessarily solely electronic, but may also include a reduction in image area (electrons from a large area input screen are focused on a small area output screen).

The earliest type of image intensifiers for X-ray applications used a zinc-cadmium sulphide (ZnCdS) layer inside the glass envelope to convert the X rays to light. The photocathode adjacent to the fluorescent layer converted the light to electrons. The original Philips tube used a 25 kV potential between the photocathode and output phosphor. Even though only 10 percent of the light photons from the fluorescent screen would generate electrons at the photocathode and only 10 percent of

FIGURE 14. X-ray Image Intensifier Tube Design

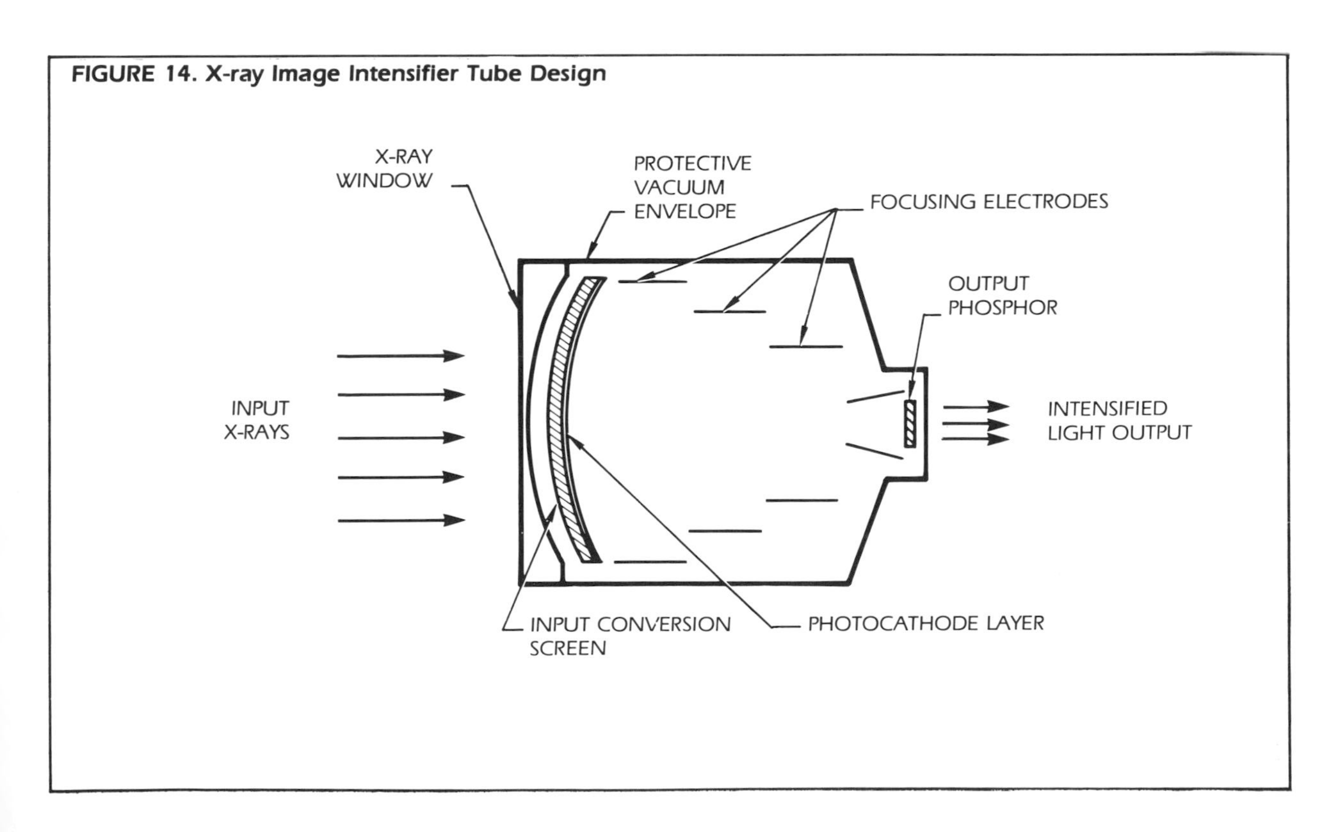

the accelerated electrons would produce light at the output phosphor, a 10 to 15 times increase in luminous flux was generated by the acceleration. The tube had a curved input screen with a 13 cm diameter and an output screen with a 1.5 cm diameter. The nine-times reduction in diameter from the fluorescent screen to the viewing screen provided an additional factor of 80 in brightness gain. The total gain was between 800 and 1200 (see ref. 42).

Technical improvements in electronic gain, fluorescent and photocathode layer efficiencies, and electron optics have made modern X-ray image intensifiers very useful in both medical and industrial applications (see refs. 2, 14, 36, 44, 49). CsI(Na) is now commonly used as the fluorescent layer because it has twice the X-ray absorption of ZnCdS and the crystalline structure minimizes lateral light diffusion. Rare earth phosphors such as Gd_2O_2S are also found to be superior to ZnCdS. It appears that at X-ray energies below 100 kV, CsI(Na) is very good; at higher kilovoltages, the rare earths may be useful.[49]

Modern tubes are available with 10 to 40 cm input diameters, multiple modes (which electronically select variable field size of the input), and fiberoptic output for direct camera coupling. A typical 21 cm tube performs with resolution on the order of 4 line pairs per millimeter and gains on the order of 10,000. Resolution is at a maximum at the center of these intensifiers and decreases somewhat at the edges.

The advantages of these tubes are the relatively low cost, generally compact size and high resolution/contrast. A disadvantage is that minification will increase image unsharpness. Also, because a length-to-diameter ratio of 1.0 to 1.5 is required for the electron optics, large diameter inputs require large tubes. This not only increases bulk but creates a potential implosion hazard. The curved input screens in these tubes cause distortion. The tubes are sensitive to voltage drifts, stray magnetic fields, and space-charge defocusing at high dose levels.[44] Electron scattering, thermonic emission or light reflection on interior surfaces are causes for loss of contrast from intensifiers. Fabrication techniques in the latest generation tubes minimize these problems.

Tubes with 36 cm input have been manufactured.[26] The advanced vacuum tube technology requires a metal tube body. A titanium membrane is used for the entrance window to withstand atmospheric pressure and maintain transparency to X rays. The titanium produces less scatter than a glass window which improves contrast. An acceleration voltage of 35 kV is used. Limiting resolution in the large format is specified at 3.6 line pairs per millimeter. Tubes as large as 40 cm diameter have also been marketed.

Image Converters and Light Amplifiers

Another type of image intensifier, often called the *image converter,* uses a light input and light output system. The electrons from the photocathode are accelerated by high voltage before striking a phosphor screen. Luminous gains of 50 to 100 times are typical. Coupling of the tubes in stages can result in gains of 10^5 to 10^6. Figure 15 shows three categories of image intensifiers. Typical input diameters for commercial tubes are 18, 25 and 40 mm. In Fig. 15a and 15b, an electrostatic field directs the electrons. A fiberoptic faceplate minimizes the decrease in resolution near the edge of the tube. The fiberoptics are also good for coupling to other components of an optical system. Shown in Fig. 15c is a magnetically focused image tube. In this tube, an axial magnetic field is used to provide good resolution and low distortion over the entire screen. The magnetically focused tubes are more expensive than the electrostatic tubes.

For radiography, the light from a fluorescent screen is input to the photocathode of the intensifier using a lens system. In some cases direct contact between the photocathode and fluorescent screen is used. With a proper lens system, the image converter can be used with any size fluorescent screen. For a large input screen, the loss of intensity experienced in focusing to the smaller diameter intensifier may be limiting, depending on the radiographic parameters. The term image converter is used because the light input and light output may not necessarily be at the same wavelength. Gain characteristics are approximate values in this case because the input and output spectrums are different.

Proximity Focus Intensifier

A *proximity focus intensifier* tube is one in which the separation between the photocathode and output phosphor screen is small. With a potential difference of several kilovolts, the electrons are focused directly across the gap. Resolutions up to 40 line pairs per millimeter are possible. Proximity focusing X-ray intensifier tubes have been developed, providing a one-to-one, input-to-output size representation over large area (25 cm) images.[44] These tubes use a CsI fluorescent input screen ahead of the photocathode.

Light intensity gains of 500 to 4000 are possible with resolutions between 1 and 2 line pairs per millimeter. An X-ray image intensifier of this type is much smaller and lighter than the tubes described earlier. In principle, the unsharpness with this design could be limited to the input fluorescent screen unsharpness.

FIGURE 15. Image Converters. (a) Single-Stage Electrostatically Focused Image Intensifier Tube; (b) Three-Stage Electrostatically Focused Image Intensifier Tube; (c) Three-Stage Magnetically Focused Image Intensifier Tube.

Channel Electron Multiplier

The *channel electron multiplier* or *microchannel array* is an assembly of small tubes for amplifying an electron signal using secondary emission. The channels are glass- or ceramic-coated with a high resistance material on the inside. A potential difference of 500 to 1000 V is applied across the channel. An electron entering the channel will strike a wall causing one or more secondary electrons to be released. These will continue to strike the channel

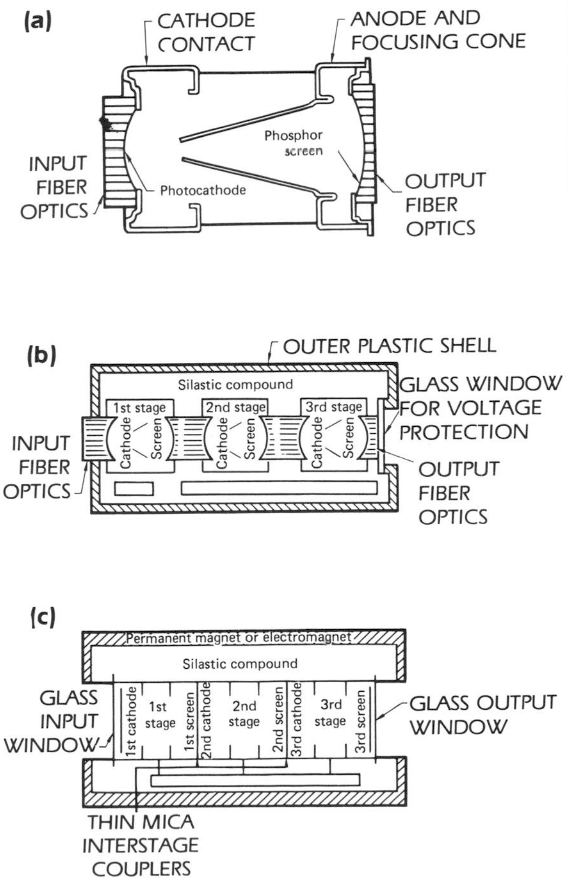

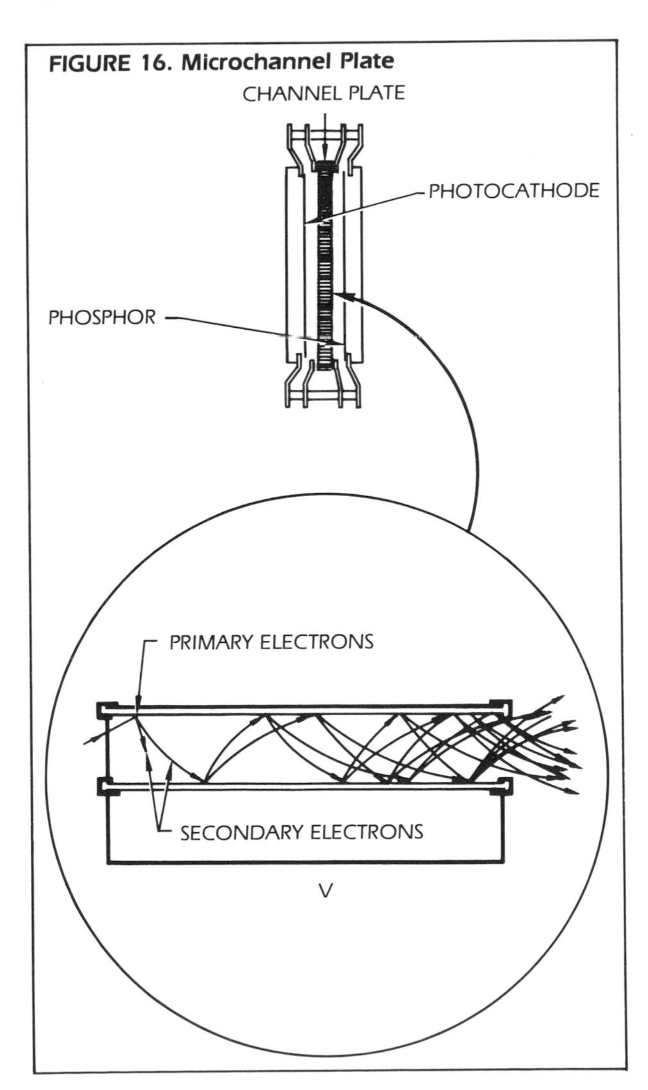

FIGURE 16. Microchannel Plate

walls yielding more electrons as they are accelerated by the electric field along the channel. The gain of the channel multiplier depends on the applied voltage and the ratio of length to diameter.

These multipliers are limited only by the technology for fabricating small diameter channels which will not break down under the electron field. With 1000 V and a length-to-diameter ratio of 50, a typical gain is about 10^4. The resolution of the device is limited by the size of the channel spacing.[15,47]

Figure 16 shows a diagram for a microchannel plate. The microchannel plate is often used in conjunction with the electrostatic image intensifier tubes to form what is called a *second generation image converter*. The microchannel array follows a first stage electrostatic lens.

Direct X-ray sensitive microchannel arrays can be made. A metallic converter that emits secondary electrons following a high energy photon excitation (200 kV and above) serves as the input. These electrons are amplified in the channels.[7]

Spectral Matching

Image intensifiers rely on a photocathode to convert input light radiation to electrons. The X-ray image intensifiers have a fluorescent screen ahead of the photocathode to convert X rays to light. The spectral response and sensitivity varies among photocathode materials. (See Fig. 17 for the spectral response of several typical photocathodes.) Desirable characteristics include high efficiency at the wavelength of light being observed, and a low dark current (the signal level when no light is falling on the photocathode).

The light emitted from the intensifier is generated by the action of electrons on a phosphor. The spectral emission characteristics of some common phosphors are shown in Fig. 18.

FIGURE 17. Photocathode Response Spectrum

SENSITIVITY

S-11 S-20 S-25 S-1

0 200 400 600 800 1000 1200

Violet Blue Green Yellow Red

WAVELENGTH (nm)

PHOTOCATHODE	MATERIAL	DARK CURRENT (A/cm^2)
S-1	$Ag+Cs_2O$	10^{-10} to 10^{-13}
S-11	Cs_3Sb	10^{-14} to 10^{-15}
S-20	$Na_2KSb+Cs$	10^{-15}
S-25	$Na_2KSb+Cs_3Sb$	10^{-15}

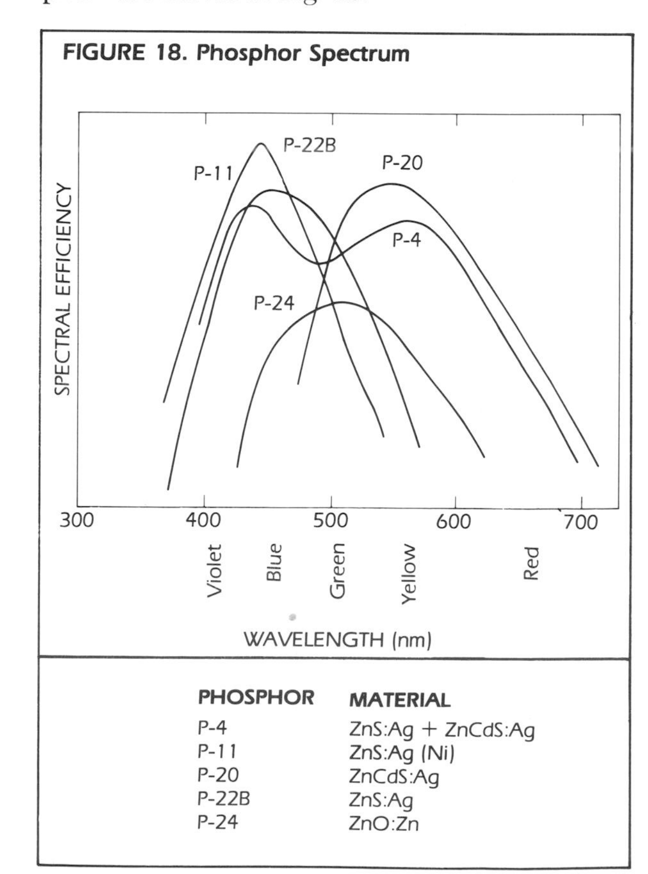

FIGURE 18. Phosphor Spectrum

PHOSPHOR	MATERIAL
P-4	ZnS:Ag + ZnCdS:Ag
P-11	ZnS:Ag (Ni)
P-20	ZnCdS:Ag
P-22B	ZnS:Ag
P-24	ZnO:Zn

The spectral performance of phosphors and photocathodes is important for real-time radiography systems. The photocathode of the intensifier must be capable of functioning over the range of light emitted by the fluorescent screen. Good matching of these spectra can make a significant improvement in system performance. Selection of the X-ray fluorescent screen must be based on knowledge of the photocathode response of the light pickup system. Alternatively, the input phosphor of the light pickup should respond well to the light produced by the selected fluorescent screen. A further consideration may be made to suppress certain wavelengths, since some phosphors emit promptly at one wavelength and decay with time at another. An advantage might be gained by matching well to the prompt light and poorly to the decaying signal. Generally, good spectral matching exists between commercial intensifiers and television systems.

Statistics

The image intensifier system can improve imaging by boosting the light output so that the statistical limitation in the image process is not at the eye, as in direct-viewing fluoroscopy, but at the input fluorescent screen. The intensifier itself operates on a statistical process for the generation of electrons and the regeneration of light. The sources of fluctuation are essentially independent, so that eqs. 25 and 26 apply.

Amplification (g in eq. 26: $s^2 = gn$) may be used for improving detail sensitivity in an intensifier system. This is done by choosing that amplification which makes the number of light quanta (utilized by the observer's eye) equal to the number of radiation quanta (utilized by the input fluorescent screen).

Television Cameras, Image Tubes and Peripherals

A wide variety of television cameras and image tubes are available for use on real-time imaging systems. Many different camera configurations can be used to accommodate inspection requirements. On one extreme is the small compact camera with no user adjustments. At the other extreme is the larger, two-piece camera, with many controls for optimizing image quality. Solid state cameras are less common

FIGURE 19. Vidicon Television Camera

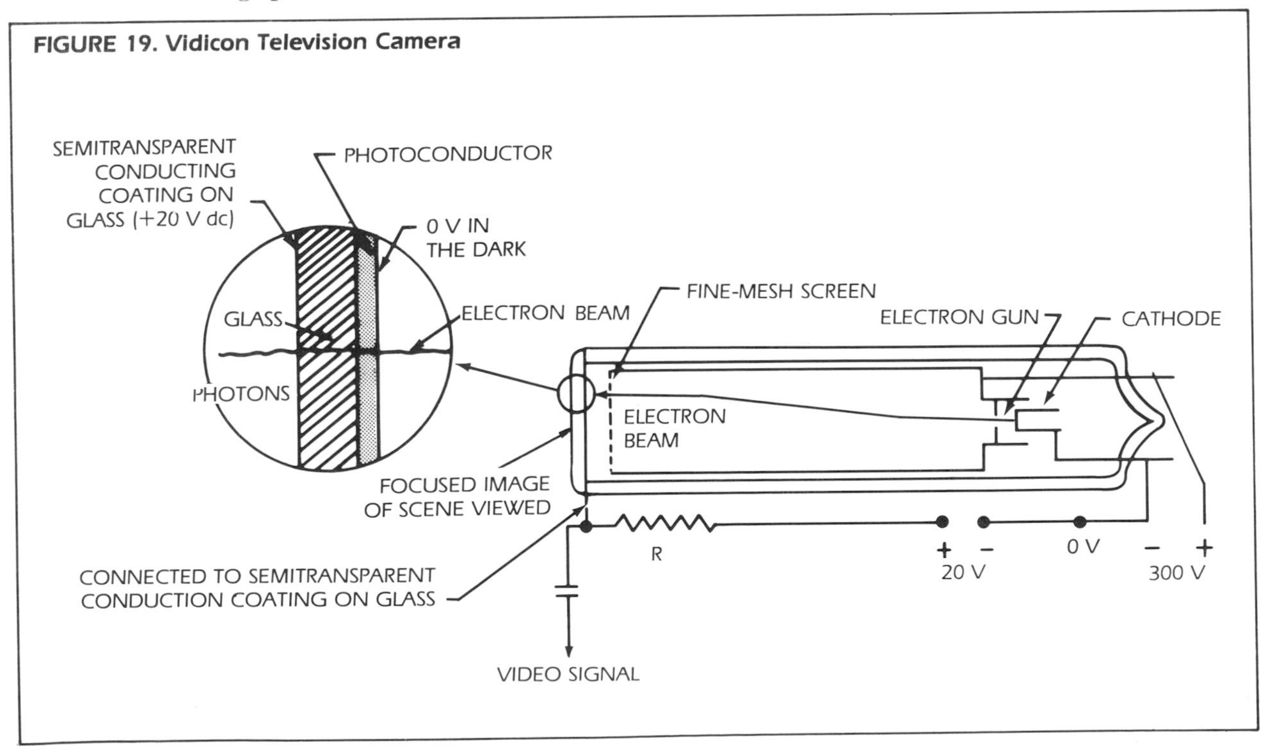

now but will see greater application in the future. A variety of image tubes are also available. The most common types for radiographic applications are vidicon, silicon-intensifier target (SIT), image isocon and X-ray sensitive tubes.

Image Tubes

The *vidicon* is a small, rugged and simple tube. An electron beam scans a light-sensitive photoconductive target. A signal electrode of transparent material is coated onto the front of the photoconductor. The scanning electron beam charges the target to the cathode potential. When light is focused on the photoconductor, the target conductivity increases, changing the charge to more positive values. The signal is read by the electron beam which deposits electrons on the positively charged areas, causing a capacitively coupled signal at the signal electrode (see Fig. 19).

The vidicon has a number of variations, depending on the selection of the photoconductive material. The standard vidicon uses an antimony trisulphide layer. The plumbicon uses a lead oxide junction layer, the newvicon uses cadmium and zinc telluride, and the silicon diode tube uses a silicon diode array target structure.

Another type of tube called the *silicon intensifier target* (SIT) uses a photocathode as an image sensor and focuses the photoelectrons onto a silicon mosaic diode target. Readout is similar to the vidicon. The design allows for very high light gains in the pickup by accelerating the photoelectrons to high energies (perhaps 10 keV) before they strike the target. The SIT tubes and *intensified SIT tubes* (ISIT) are used extensively for low-light-level application.

Similar to the SIT tube is the *secondary electron conduction* (SEC) tube. This tube is no longer manufactured but still may be found in some systems. An advantage of these tubes is their ability to integrate the light signal at the target. A photocathode and electrostatic focusing section are used

FIGURE 20. Image Isocon Television Camera

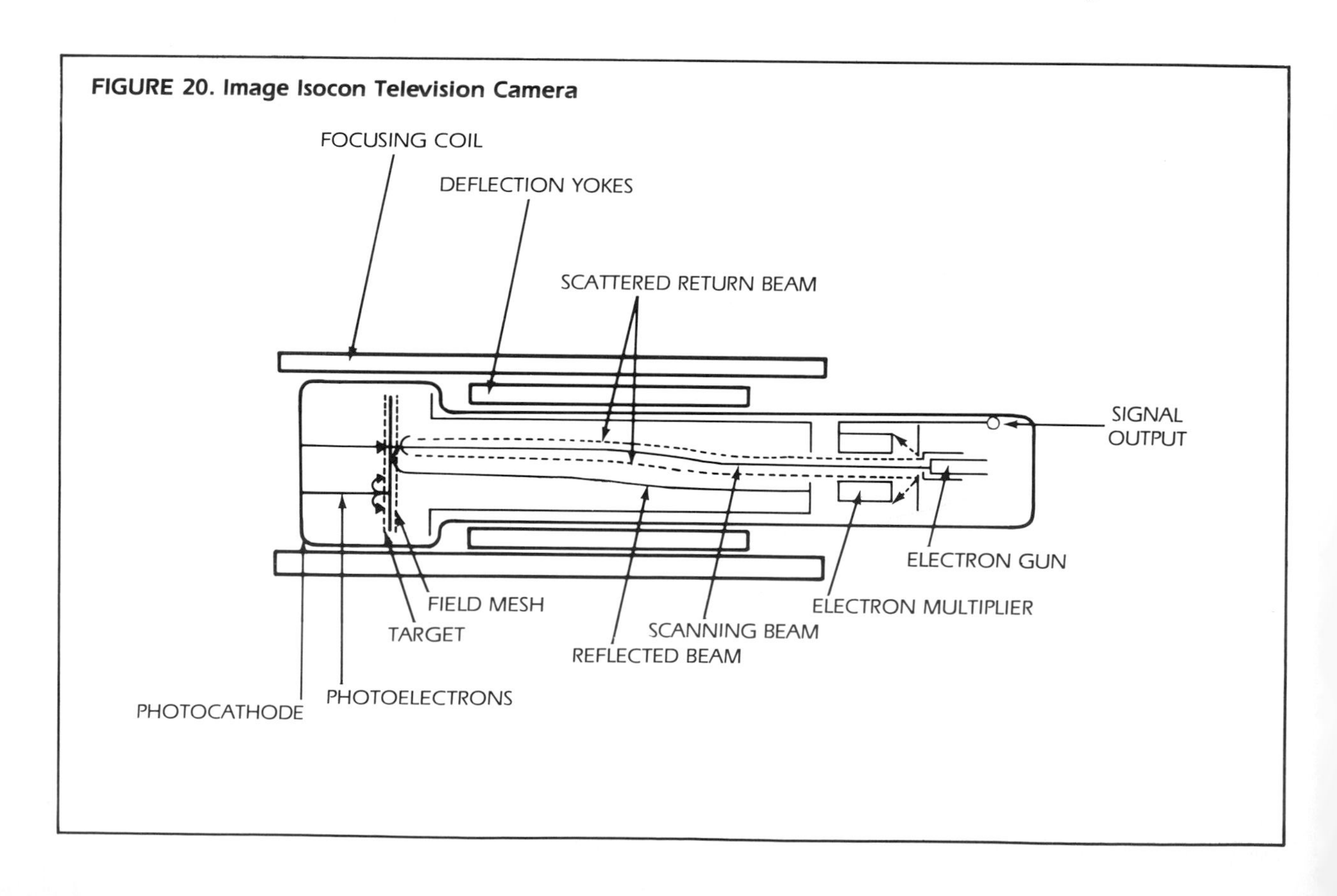

ahead of the target, which is a porous layer of KCl deposited on an aluminum and aluminum oxide base. Photoelectrons arrive at the target with about 8 keV of energy, which is dissipated by the generation of secondary electrons in the KCl layer. The secondary electrons migrate to a signal plate, leaving positive charges in the KCl. A scanning beam returns the KCl to cathode potential. This current pulse is capacitively coupled to the signal plate whose current is used to develop a voltage across a load resistor, forming the video signal.

The *image isocon tube*, shown in Fig. 20, is widely used in radiographic applications. The image on its photocathode forms a photoelectron pattern that is focused by an axial magnetic field onto a thin, moderately insulating target. The photoelectrons striking the target cause secondary emission electrons which are collected in a nearby mesh, leaving a net positive charge on the target. The beam from an electron gun scans the target, depositing electrons on the positively charged areas. The scattered and reflected components in the return beam are separated. Only the scattered component enters the electron multiplier surrounding the electron gun. This signal is amplified to become the video output.

The *image orthicon* is the predecessor of the image isocon and may be found in older imaging systems. However, the image orthicon could not differentiate scattered and reflected electrons in the return beam, and this resulted in noise in the dark areas of the image.

Camera System Characteristics

The performance criteria for camera tubes are based on the sensitivity, dynamic range, resolution, dark current and lag. A plot of signal output versus faceplate illuminance for some typical camera tubes is shown in Fig. 21 and the slope of these curves is called the *tube gamma*. The light source for illuminance is important in the response characteristics of the tubes.

Table 11 lists characteristics for television tubes used in real-time radiographic applications. Lag is given as the percentage of the original signal present after 50 milliseconds.

In X-ray imaging applications, image isocon television tubes are commonly used because of their low light-level sensitivity and high dynamic range; unfortunately, isocons are very expensive. Vidicons often are used in combination with X-ray sensitive image-intensifier tubes; vidicons are simple as well as inexpensive. The newvicon is more sensitive than the plumbicon or the Sb_2S_3 vidicon and is finding more usage with image intensifiers. SIT tubes are being used with low light-level systems when high dynamic range is not required.

FIGURE 21. Television Camera Output Versus Light Input

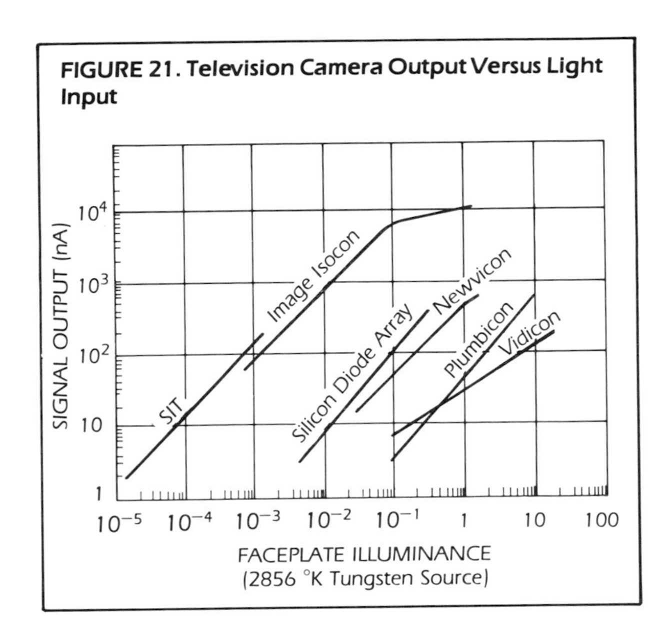

Camera Matching

The standard scanning rate for camera tubes is 30 frames per second. Each frame image is created in a 33 millisecond exposure time. The frame is composed of two fields in which the electronically scanned 525 vertical lines are interlaced. The first field in 1/60th of a second contains the odd numbered lines and the second field contains the even numbered lines.

In some applications, where very low radiation intensities are experienced, it is necessary to use a slower scanning rate. The target of the television camera can be made to integrate the incoming signal for several minutes and then scan it to provide one frame of information. SEC cameras are often used in this way. Faster scanning rates may be used to image rapid dynamic systems, provided a sufficient light intensity is present and the lag features are acceptable.

TABLE 11. Typical Characteristics of Television Camera Tubes

Tube Type	Dynamic Range	Typical Resolution TV Lines	Dark Current (nA)	Lag (%)	Gamma
Image Isocon	2000	1000	0	7	1.0
Sb_2S_3 Vidicon	300	900	20	20	.65
Newvicon	100	800	8	20	1.0
PbO Vidicon	300	700	3	4	.95
SIT	100	700	8	12	1.0

Television cameras require electrical adjustments to set up the operating parameters of the image tube and signal processing electronics. Cameras may be one- or two-piece systems. The one-piece camera is usually self-contained with few (or no) user adjustable controls.

The two-piece camera is much more versatile. The camera head can be made much smaller, decreasing bulk and weight for restricted mounting requirements. The camera control unit will have all controls readily accessible to the user for optimizing image quality. Common adjustable controls or switches found in most cameras are described in Table 12.

The beam, target and focus controls optimize the image tube's performance. Other control features (such as pedestal, gamma correction, and polarity reversal) enhance image quality and ease of operation. The polarity reverse feature, for example, is a comparatively simple and advantageous option: small detail against a bright background is difficult to detect; however, inverting the polarity and having details appear against a dark background makes them more visible to the observer.

TABLE 12. Common Adjustments for Television Cameras

BEAM CURRENT
Controls the electron beam in the image tube. Usually set to just discharge the picture highlights.

TARGET VOLTAGE
Sets the positive potential on the image tube target. This will be a fixed voltage for most tubes. On sulfide vidicons this voltage is variable and controls sensitivity.

FOCUS
Electrostatic focus adjustment for the electron beam.

PEDESTAL LEVEL
Voltage adjust for the black level of the picture.

GAMMA CORRECTION
Electronic change of the slope characteristic (gamma) of the tube.

POLARITY REVERSE
Inverts the black and white areas of the image.

X-ray Sensitive Vidicon

Although the usual input to a television camera is a light signal, for radiographic purposes it is possible to make the camera sensitive directly to X radiation. The vidicon camera in particular may be modified for X-ray sensitivity and has been found useful for obtaining direct real-time radiographic images. Two alterations of the vidicon are needed for good results: an X-ray window and an efficient target. Thin glass or beryllium X-ray windows located close to the target replace the heavy optical glass windows in conventional tubes. Although the normal vidicon photoconductive targets will respond to X ray, they are so thin that absorption of radiation is minimal. Suitable thick targets must be used. Selenium has been found to be very effective, having adequate response and low lag.[29] Lead oxide targets are more common, having a high density for good X-ray absorption and resulting sensitivity.[23]

The X-ray sensitive vidicon is an imaging system for small objects and low kilovoltages, 150 kV or less. The X-ray intensity must be high, in the range of 50 to 500 R/min. The vidicon tube typically has a sensing area of only 9.5 × 12.5 mm. Presentation of the image on a 48 cm (19 in.) television screen results in better than 30 times magnification. With a 525-line scan rate, the resolution in the object is better than 0.02 mm. The X-ray-sensitive vidicon camera has a gamma on the order of 0.7 to 1.0. Penetrameter sensitivities of 2 percent have been obtained. The cameras have experienced problems with deterioration, possibly due to: local overheating in the target layer; poor bonding to the heat sink layer; substrate irregularities; or incompatibility between beryllium and target materials.

Solid State Cameras

Solid state cameras use an array of photodiodes or charged coupled devices as the sensitive layer. These arrays may be linear or area arrays of individually addressable elements. The solid state cameras are much smaller and have wider spectral response (see Fig. 22), reduced lag, higher quantum yields (50 percent), and (depending on the application) may have equivalent resolution capabilities when compared to commercially available vidicon cameras. Solid state cameras are rugged, are not damaged by intense light images, and do not require the scanning electron beam found in vidicons.

The photodiode arrays in solid state cameras are simple photon detectors, typically reverse-biased silicon photodiodes, which absorb incident photons and liberate current carriers. This gives rise to a current referred to as the *photocurrent signal*, which is proportional to the arrival rate of the incident photon. The efficiency of the photodiode strongly depends on its material and construction as well as the wavelength of the incident photons. The diodes consist of p-type islands in an n-type substrate. Standard arrays are available with 128 × 1024 diodes with a center spacing as small as 25 μm. A dynamic range of 100:1 is typical.[21]

Charge coupled devices (CCD) work in a manner very similar to photodiodes.[3,22] A photon, incident on the depletion region of a CCD, will create an electron-hole pair if absorbed. This creates a current flow which, in a CCD, is stored in the potential well of the device. The amount of charge collected at the potential well is in direct proportion to the amount of local light intensity. The CCD is fabricated with a combination of thin film technology and silicon technology. Arrays are available with 576 rows of 384 points with a photoelement center spacing of 23 μm. Dynamic range is typically 1500:1.

The image on the solid state array is coupled to video circuitry by the horizontal and vertical scan generators, which read the charge level at the detector elements. Figure 23 shows the schematic of a CCD array. The output of the video can be specified to fit a particular video format. The clock sequence, which sequentially reads the charge level on each device, is started after a suitable image integration time. This integration time can be adjusted, making solid state cameras useful for low light level applications, provided the detection element can retain the charge over the integration period.

The interesting feature of solid state cameras is that individual pixel elements may be addressed and the signals processed digitally. With individually

FIGURE 22. Sensitivity of a Photodiode Array Camera Versus a Vidicon Camera

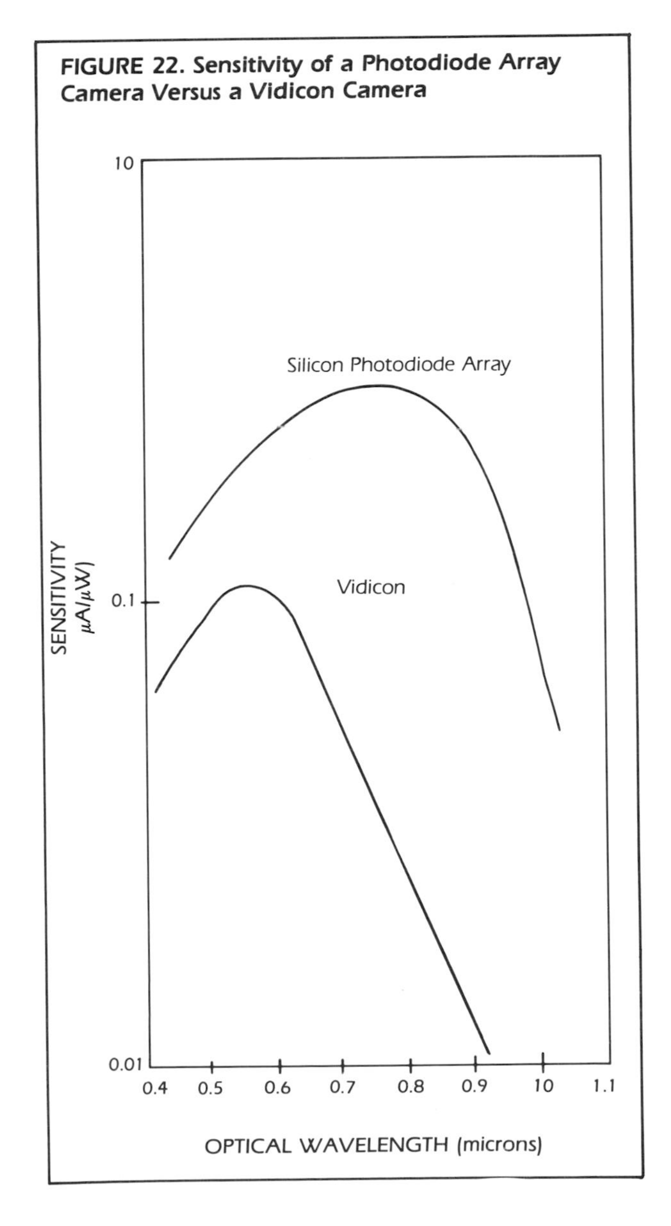

FIGURE 23. Schematic of a Charged Coupled Device Array

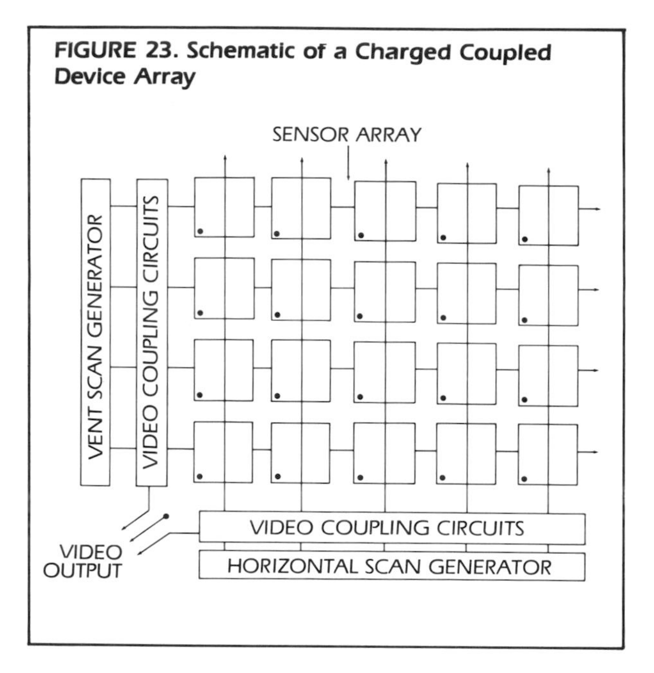

addressable elements, the integration time and the video output format may be simply specified. The cost of solid state cameras increases significantly with increasing array size and the electronic circuit complexity required to scan large arrays. Each element must be individually calibrated for uniform response throughout the field. This feature can be used to correct nonuniform fields in radiography. Since each element is independent, blooming can be controlled; a bright element does not spill over into a neighboring dark element.

The imaging in real-time radiographic applications may be accomplished with optical focusing of the light from a scintillation screen onto the solid state detector. In this case, the image format and resolution are similar to conventional television cameras. Also, scintillation materials may be deposited directly on the array. Each element becomes an independent radiation detector and the resolution is dependent on the element spacing.

Optical Coupling

Real-time imaging systems using fluorescent screens for X ray to light conversion, or systems with intensifier to TV camera chains, require coupling of the optical signals between components. The use of mirrors or lenses is the most common technique.

Front-faced silver mirrors must be used to avoid ghost images caused by multiple reflections in back-faced mirrors.

Optical lenses provide good coupling, depending on the f number and transmission characteristics. The illuminance, E, on a pickup surface, coupled by a lens in a simple optical system, is given by

$$E = \frac{\pi B T}{4f^2 (1 + M)^2} , \quad \text{(Eq. 37)}$$

where B is the luminance of the output phosphor, T is the lens transmission, M is the magnification, and f is the relative aperature of the lens (or f number). The lower the f number, the more light that is collected for imaging.

Simple lenses are not often used in coupling because of the low optical efficiency. For example, if a reduction in image size by a factor of 2 is required from the coupling, the phosphor-to-lens distance must be twice the lens-to-pickup-surface distance. From the lens formula, the phosphor-to-lens distance will be three times the focal length of the lens, giving poor light collection efficiency. With a one-to-one coupling, this distance is still twice the focal length.

Collimated optics are superior because the objective lens, focused at infinity, is located at its focal length from the phosphor. Collection efficiency is nine and four times greater, respectively, for the two magnifications discussed above (two-to-one, one-to-one). The second lens will determine the image size at the pickup surface by the ratio of its focal length to the focal length of the objective lens. Vignetting, which is a reduction in light intensity at the edges, does occur in collimated optics. This is minimized when the lenses are close to each other. Some real-time imaging systems use specialized optics such as Bouwers/Schmidt lenses, where concentric mirrors are employed to provide a low (0.65) f number.[37]

Fiberoptics may also be used for coupling. Intensifier tubes often have fiberoptic input and output surfaces. These may be coupled directly to other components with similar surfaces by using an optical gel. Fiberoptic light-guides can be considered for moving light from one location to another (i.e., from the fluorescent or phosphor screen to the camera). The potential advantages of this are greater retention of the light, one-to-one size transfer, improved contrast by suppression of undesirable reflections and shortening of the system dimensions. The use of fiberoptics over considerable distance or around unusual obstructions is also a possibility.

Fibers for fiberoptic systems are glass or plastic with diameters of 1 or 2 μm to 50 μm. The fiberoptic array operates on total internal reflection. To reduce leakage each fiber is coated with a material having a lower index of refraction. Losses in the fiberoptic system are due mainly to the opacity of the fibers in long systems or the acceptance angle of the light in short lengths.

Viewing and Recording Systems

Television Monitors

Television monitors for observing video signals are cathode ray tubes using a modulated electron beam to write on the output phosphor. In North America the standard format is 525 lines, interlaced. *Interlacing* means that the total picture frame is composed of two fields: the first field uses every other line on the screen; the second writes between the lines of the first. The television camera provides the monitor with the appropriate operating format in the video signal.

Other systems are used routinely in Europe and have various numbers of lines which can be as high as 1200, triple interlaced. The tubes vary in size, deflection system and component design. Contrast controls can be adjusted to increase or decrease the gamma over a range of values, typically 0 to 10.

To see an object, there must be a sufficient number of scan lines in the television image; following is one description of scan-line requirements.[37] To visualize n objects there must be $2n$ lines. Allowing for random orientation, this should be increased by a factor of $2^{1/2}$. To see a mesh with 50 holes/cm would require 140 lines/cm scan rate. Horizontally, the resolution is determined by the bandwidth of the signal. One cycle of bandwidth is required to see the mesh (one-half cycle for the holes and one-half cycle for the spacing between holes). In a conventional system, the viewing matrix uses a 3×4 aspect ratio, the horizontal being larger than the vertical. If 525 scan lines are used, then the horizontal will require $4(525 \div 2^{1/2})$ half-cycles or about 250 cycles to maintain the same resolution. Using 30 frames per second scanning and a factor of 1.2 for retrace time, the required bandwidth is on the order of $525 \times 250 \times 30 \times 1.2 = 4.7$ MHz.

Although it may appear that resolution could be improved by increasing the number of scan lines, two problems result: the charge capacity on the camera's target elements will be reduced in proportion to the area change, which may reduce the sensitivity; and, an increase in bandwidth will be required, which increases the noise in the electronics in proportion to the square root of the bandwidth. The result is that the standard number of scan lines is often as good as or superior to higher scan line systems.[37,46]

Many cameras and television monitors are designed for higher bandwidth operation (10 MHz and greater). This provides greater resolution horizontally (800 lines or more) than vertically. Since the 525-line vertical is standard, the resolution of video systems is often quoted by the horizontal resolution value, which is a function of bandwidth.

Recording Equipment

Recording the video signal provides a permanent record of the radiographic process. Videotape records are the most common method. These recorders may be either reel-to-reel or cassette records. The reel-to-reel recorders offer long-time recording; reels with lengths of many hours may be used. The tapes can be cut and spliced for presentation. Cassette videotape recorders are more easily operated than reel-to-reel machines when loading, unloading, and storing tapes, but cassettes are usually limited to 1 hour of playback time. The signal-to-noise ratio of the record will improve with the width of the tape, which varies from one-half to two inches.

Both reel-to-reel recorders and cassette videotape recorders may be equipped with pause and slow motion modes. The pause mode, however, only shows one field rather than the full two-field frame. This results in a reduction of information by one-half.

In place of the second field, some expensive recording equipment repeats the first field of information in the pause mode. This improves the visual display, but still represents a reduction in information. The reason for showing only one field is to eliminate interfield jitter caused by movement between field scans of the camera.

Slow motion modes are quite useful for replaying rapid events, unfortunately in slow motion replay a broadband synchronizing signal is reproduced on the video screen, sweeping by at each frame. this is not easily removed and can be very distracting when evaluating images; removal of the synchronizing signal noise is possible with some of the more expensive video replay equipment.

A videodisk is another possibility for storage of real-time radiographic information. Videodisks are costly and are normally used only for recording short periods of data, such as 600 frames. The quality of the videodisk can be excellent. It will play back either forward or backward, slow or fast, and step forward or backward one frame or one field at a time. The full frame of information is available in the stop mode of some data handling videodisks; however most instruments play back only one field, producing it twice in order to avoid interfield jitter. For short duration events requiring a high quality record (for slow motion analysis), the videodisk is an excellent method.

All video recording devices alter the video signal in some way. When linking the output of the recording camera to monitors and recorders, the strength of the signal is reduced in proportion to the number of devices. This results in a weaker signal at the recorder and a poorer record. Repeated copying results in a degradation of the image as the signal becomes successively reduced. Playback from recorded images will be inferior to the original image. It is also important to consider the bandwidth capability of recording equipment. It should be matched to the bandwidth required in the original image, to maintain as much information as possible. The standard television broadcast is 4.2 MHz, to which many video recording devices are matched. However, videotape units up to 10 MHz and videodisks up to 6 MHz are available and may later be extended in bandwidth for specialized applications.

Hardcopy

Hardcopy of a single-frame image is often desirable. If the real-time radiography is of a static object, photography of the television monitor is suitable. If the photographic exposure time integrates over many frames, the result will be superior in image quality to the video monitor display. Also available are gray scale printers which reproduce video pictures on paper. Twelve to sixteen gray levels are typical.

These printers use a cathode ray tube with a fiberoptic faceplate to transfer an image line to a dry silver photographic paper. As the paper passes the CRT, it is exposed to successive lines of image information. The dry silver paper is then developed by heating in a processor.

For events involving motion, the frame of interest must be stopped for hardcopy. One way of accomplishing this is through use of a *kinescope,* which records video information onto 16 mm movie film. The self-contained unit synchronizes the film frames to the video frames displayed on its internal monitor. The movie record is the hardcopy.

Similarly, movie records may be made using *cineradiography.* A motion picture camera is focused on the output phosphor of the image intensifier for this technique. Optics which split the output light signal can be used so that both television viewing and cineradiography may be performed simultaneously. Cineradiography generally provides better image quality than that obtained with a kinescope.[45] A kinescope, because it does view a bright video screen, can generate output onto film at lower X-ray exposure levels than can be obtained with cineradiography.

Also available are a number of electronic devices which will grab frames. This can be performed in an analog mode or may involve digitization of the frame. Hardcopy can be obtained when the output of these devices is displayed on appropriate storage scopes and then processed through printer units.

System Evaluation

Penetrameters

The most common method for evaluation of radiographic sensitivity is the use of a penetrameter or image quality indicator. The penetrameter is a thin plaque of the same material as the object. The plaque thickness is a certain percentage of the object thickness. The penetrameter has holes with diameters one, two and four times its thickness. The radiographic quality of the imaging system is then listed as the smallest percentage penetrameter detectable and the smallest hole detectable. Other types of penetrameters are in use, such as the DIN penetrameter in Europe. This penetrameter uses wires of graded diameters, made of the same material as the object. The smallest wire detected is the quality level.

Many experimenters use their own measurement systems for evaluating the quality of an imaging system. These are generally called image quality indicators (IQI). IQIs vary in design, but commonly they contain a step wedge of material from which the smallest percentage change (in material or contrast level) can be determined. Resolution may be measured by the smallest hole size detectable. Since the ability to observe a certain diameter hole is a function of the depth of the hole, it is common to specify resolution using a high contrast object.

The use of wire meshes to evaluate real-time imaging systems is also a common technique. The wire mesh is an object whose fine structure is repetitive. Wire meshes are often used to indicate the quality level of inspection at certain speeds of movement.

Modulation Transfer Function

Another method of system evaluation, which is more rigorous in its approach, is the *modulation transfer function* (MTF). The MTF is the ratio of the image amplitude to the object amplitude, as a function of sinusoidal frequency variation in the object:

$$R(\omega) = \frac{I(\omega)}{O(\omega)} \qquad \text{(Eq. 38)}$$

where $R(\omega)$ is the sine wave response, $I(\omega)$ is the image amplitude and $O(\omega)$ the object amplitude.

This can be visualized by considering a bar pattern. As the pattern becomes finer and finer, the image response begins to lose contrast. A plot of this response is called the *square wave response* (when a bar pattern is used), and is very similar to the MTF. Square wave response factors can be used to evaluate imaging systems, and under certain conditions may be corrected to the sine wave response or MTF equivalence.[4,9]

The true MTF, generated by an object having sinusoidal variations in intensity, is measured routinely for optical components such as lenses, intensifiers and cameras. Normally, in radiography, determining the modulation transfer function for such an object is prohibitively difficult. Instead, the MTF is derived by generating the edge spread function, differentiating to obtain the line spread function, and Fourier transforming to yield the MTF. Derivations of this and examples may be found in the literature (see refs. 25, 30, 31, 32, 38).

The importance of the MTF in evaluating systems is that the total system MTF is the product of the individual MTFs of the components:

$$MTF_{system} = MTF_1 \times MTF_2 \times MTF_3 \quad . \qquad \text{(Eq. 39)}$$

This is shown in Fig. 24, where the modulation transfer functions for the fluorescent screen, optical coupling, image orthicon camera, and total system are shown.[17,18] The components of a system are analyzed quantitatively and the poorest is easily determined. This method allows for the prediction of a proposed system's performance from data on the individual components.

FIGURE 24. Modulation Transfer Function of a System

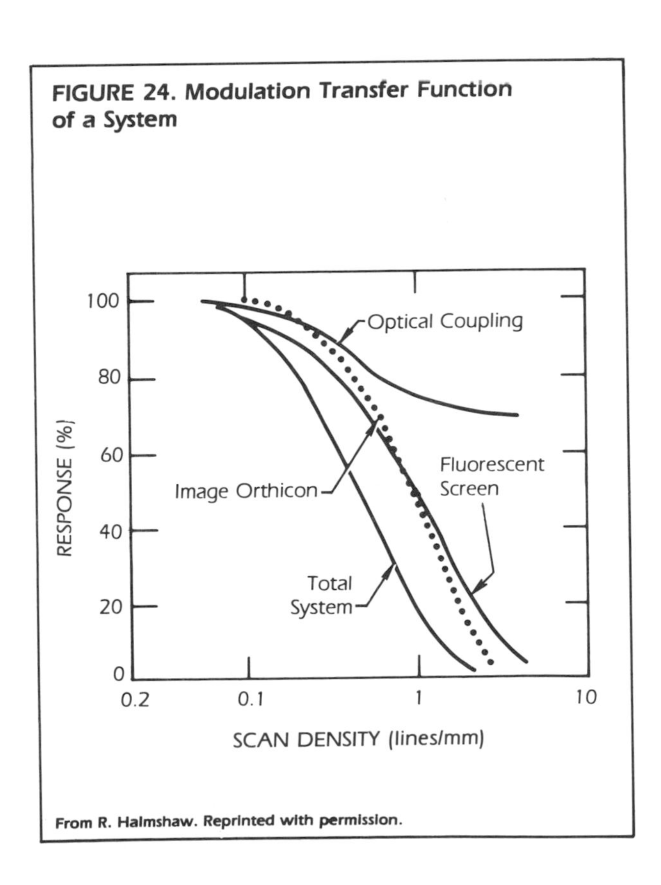

From R. Halmshaw. Reprinted with permission.

System Design

Modern general purpose remote-reviewing systems are usually of two types: X-ray image-intensifier and vidicon camera; or a fluorescent screen and low light-level isocon camera combination. Additional equipment should include a high quality video monitor, a videotape recorder, a videodisk for dynamic recording, a frame digitizer, an image-enhancement device (512 × 512 pixels with 8 bits, available for real-time summing, running average, or contrast adjustments plus frame grabbing for edge enhancement and subtraction techniques), and a hardcopy system.

The X-ray image-intensifier systems are usually less expensive. They operate best at the low and intermediate X-ray energies for which they are designed, though there are systems suitable for mega-electron volt radiography.

The isocon camera systems are more versatile. Since the fluorescent screen is changeable, the system can be adjusted for optimum performance at any energy. Isocon camera systems may also be used for X ray or neutron detection.

Both systems can perform in the 1.5 to 2 percent sensitivity range. The resolution limit will generally depend on the input field size, the video bandwidth employed and the input phosphor. In the not too distant future, solid state imaging will compete favorably with, or possibly exceed, the performance of the intensifier and isocon based systems.

For more information on the design and implementation of RTR handling systems, see Section 19, Part 12, "Handling Systems for Real-time Radiography."

PART 6

IMAGE ENHANCEMENT IN REAL-TIME RADIOGRAPHY

Image enhancement is a general term referring to techniques or processes which modify an original image so that important information may be more easily detected and displayed.

Most often the purpose of image enhancement in nondestructive testing is to display low contrast detail relating to cracks, boundaries, part orientation, voids and inclusions. In some cases the objective is to locate edges more precisely in order to improve measurement accuracy.

Conventional real-time radiographic systems compare unfavorably with film-based radiography in two major respects: low contrast sensitivity and limited resolution. Modern real-time image enhancement equipment can compensate for both these limitations.

Initially, real-time radiographic images are analog or digital video signals, generated by television cameras or solid state devices (see Part 5 of this Section). Equipment used to perform image enhancement must be compatible with the scanning formats of the imaging system. While the emphasis of the following discussion is on real-time radiography, it should be pointed out that many of these techniques are also useful in nonreal-time applications. For example, by reviewing an X-ray film on a light box with a television camera, it is possible to extract and display information that might otherwise be missed. It is also possible to apply most image enhancement operations to videotape recorded signals (see Volume 3 of the second edition *Nondestructive Testing Handbook*, Sec. 15, *Image Data Analysis*).

Digital Processing

Most modern image enhancement systems use digital processing techniques; though there are exceptions, discussed later, where analog signal processing offers some advantages. With digital processors the first step is to convert the analog video input to digital form using a high speed analog to digital converter (ADC), and the last step is to reconvert the digital signals back to analog form with a digital to analog converter (DAC).

Standard practice is to sample only the active portion of each scan line. For the most common TV scanning format (525 lines per frame and 30 frames per second), the line period is 62.5 microseconds, approximately 50 microseconds of which contains video. A frequently used sampling rate is slightly under 10 MHz which provides 512 samples per scan line. Each of these video samples is converted by the ADC into a digital code.

The number of bits in the code determines the number of amplitude (gray) levels that can be initially discriminated according to the following relation:

$$N = 2^B \qquad \text{(Eq. 40)}$$

where N is the number of gray levels and B is the number of bits, i.e. the word length. $B=8$ ($N=256$) is the most commonly used.

While the typical observer is unable to distinguish the difference between a 6-bit encoding image and one encoded with a higher number of bits, higher resolution encoding is often needed to minimize artifacts introduced by the processing operations. Real-time image enhancement requires the arithmetic processor to output a processed picture element approximately every 100 nanoseconds. This is too fast for the typical general purpose computer. Consequently, real-time image processors most commonly use hardwired special purpose arithmetic units. While this imposes some limitations on the selection of processing algorithms, the most useful types of processing are now incorporated in commonly available real-time equipment. High speed array processors have also been applied to these problems.

Real-time image enhancement processes can be conveniently grouped into two major categories: (1) point processing operations and (2) neighborhood processing operations.

TABLE 13. Image Enhancement Procedures

Point Processing	Neighborhood Processing
Summing	Differentiation
Averaging (temporal)	Gradient operations
Subtracting (differencing)	Edge detection
Gray level mapping	Spatial filtering
Histogram equalizing	

Point processors perform the same operation on each pixel in the image, independent of the values of neighboring pixels. As the name implies, neighborhood operations are performed on each pixel, based on the values of neighboring pixels. Table 13 lists typical operations for each category. Each of these will be discussed and some examples of their use will be presented.

To improve the capability for displaying low contrast information, it is often necessary to improve the signal-to-noise ratio of the video. The video output of a real-time radiography system is a mixture of signals representing the object being examined, systematic noise and random noise. Examples of the second component, systematic noise, include: fluorescent conversion screen nonuniformities, camera tube target burns, light leaks, power line interferences, etc. The random noise component is usually a mixture of signal-induced noise caused by the statistical nature of the quantum processes involved, plus an additive component due to thermal and bias current noise originating in the video amplifiers. Random noise may be greatly reduced by summing or averaging successive TV frames while some types of fixed pattern noise may be reduced using subtraction techniques.

Summation

When a series of TV frames is added together, the repetitive component representing the object will sum linearly while the random noise component sums quadratically. After N frames, the signal will be larger by a factor N while the random noise will increase by a factor of $N^{1/2}$. As a result, signal-to-noise ratio also improves by $N^{1/2}$.

Figure 25 is a block diagram of a representative real-time digital video processor. Composite video is first separated into video and sync components. The video is then sliced and scaled to match the ADC range. The ADC samples and encodes the applied video, outputting an N-bit digital word to the arithmetic processor. In a real-time processor with standard input of 525 lines, 30 frames per second, and with 512 picture elements (pixels) per line resolution, the arithmetic processor will receive a new digital word every 100 nanoseconds.

An essential element of the system is a full frame image memory in which the intermediate or end result of the processing operation may be stored. In the summing mode, the contents of memory are read out and sent to the arithmetic processor in synchronism with the next incoming digitized frame; in this way, matching pixels arrive at the adder simultaneously. The result of the addition is returned to memory while other picture elements are being processed. The memory output signal is also sent to the DAC and then to the output for display.

The number of frames which can be summed depends on (1) the capacity of the memory, (2) the alignment of the ADC output with the memory and (3) the maximum encoded level of the input video. Memory capacity is defined in terms of bits per pixel. An example will help clarify these concepts.

Suppose the memory capacity or depth is 12 bits and ADC resolution is 8 bits. If the ADC is aligned with the memory such that the most significant bit (MSB) is aligned with the eighth bit of the memory and if the brightest pixel is encoded at the maximum value, then after 16 summations the memory will be full. This occurs because

$$\frac{2^{12}}{2^8} = 2^{12\text{-}8} = 2^4 = 16$$

or

$$\frac{2^{12}}{2^8} = \frac{4096}{256} = 16$$

Because it is quite possible that the maximum video level can be encoded at less than the maximum encoder output, it is often possible to sum a larger number of frames.

When the video is very noisy, using a high resolution encoder is wasteful. By shifting the ADC output downward relative to the memory, the less significant bits are discarded, leading to coarser resolution but allowing more frames to be summed before overflow occurs. As long as the magnitude of the random noise component is larger than the effective encoding interval, the gray scale resolution of the summed or averaged image will be governed by the final signal-to-noise ratio rather than the effective ADC resolution.

Continuous Image Averaging

An alternative to the summing process just described is to perform a moving average on the input video. In this mode the input to the arithmetic unit from the ADC is divided by N and added to $(N-1)/N$ of the current contents of memory. For ease of implementation, N is restricted to powers of 2. The result of this procedure is to accumulate in memory an exponentially weighted sum of previously input frames. Figure 26 shows this weighting for the case of $N=4$ and $N=8$. The improvement of the signal-to-noise ratio (S/N) is $2N-1$. Figure 27 shows how low contrast elements of an image can be pulled out of the noise and made detectable.

The advantage of this process is the availability of a continuous dynamic view of an image with improved S/N, rather than an intermittent static image obtained at the termination of a summation. The amount of image motion which can be tolerated is a trade-off with S/N improvement. Large values of N introduce lag and loss of spatial resolution when movement occurs, but they also bring the largest improvement in the S/N. With suitable design modifications it is possible in some cases to average moving images without loss of resolution.

Subtraction Operations

An additive fixed pattern noise component can be removed or greatly reduced with digital image subtraction. This can be combined with summation so that the dual benefits of random noise reduction and fixed pattern noise reduction can be achieved.

FIGURE 25. Block Diagram of a Real-Time Digital Video Processor

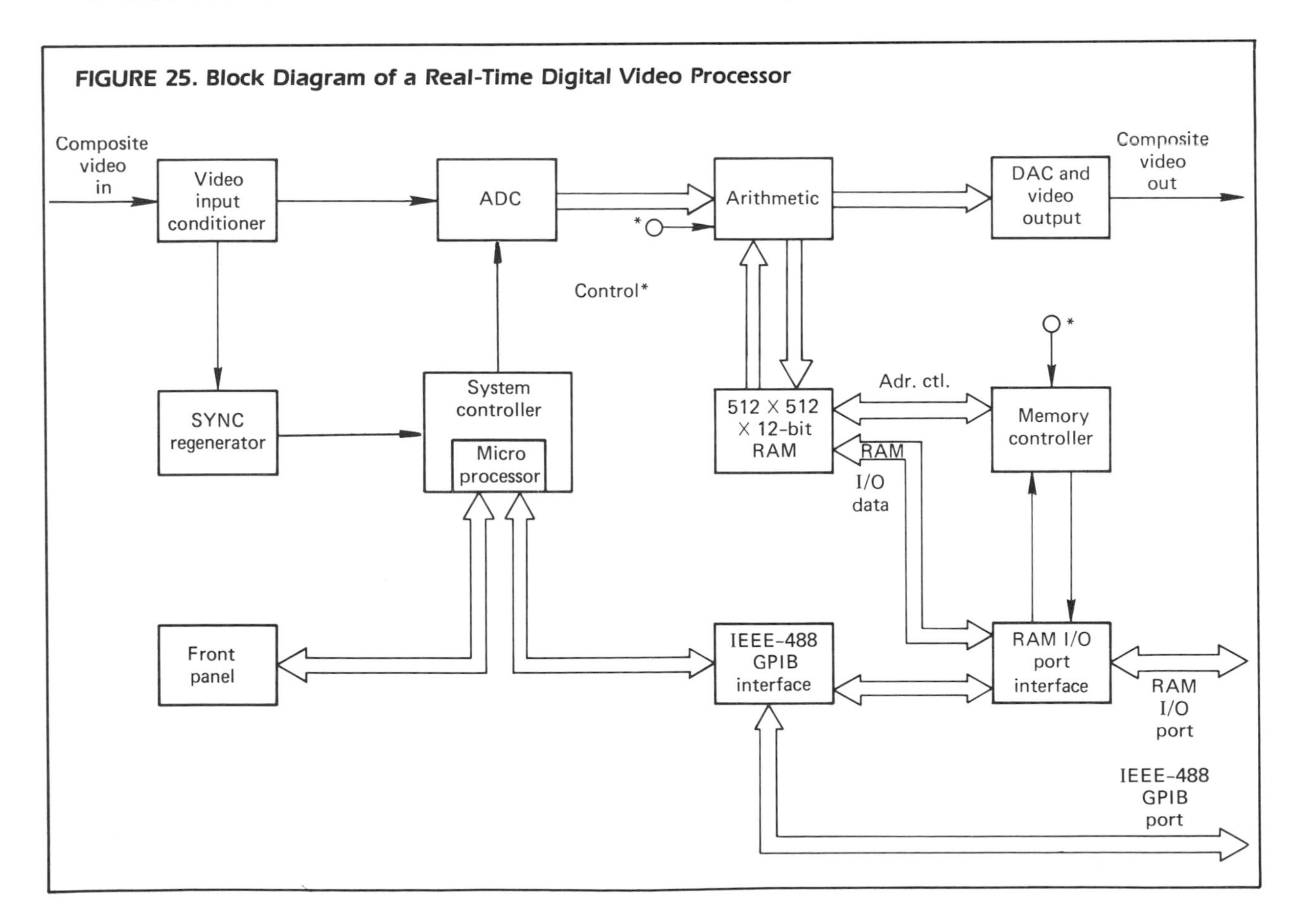

A procedure for doing this is to sum a series of frames that contain signal, random noise and fixed pattern noise.

After summation has ended, the data in memory is subtracted from one frame and then the optical input is removed so that the output from the sensor is random noise and fixed pattern noise only. An equal number of frames are then added to the complemented data in memory. Finally, the memory contents are inverted (complemented) once more. If the fixed pattern noise is independent of input level, then it will be subtracted from the previously summed fixed pattern noise in equal amounts and thereby cancelled. This process does add some additional random noise to the resulting image.

Edge Enhancement

Many radiographic images consist of areas having high brightness combined with areas having very low brightness, with both areas containing detail structure of significance. Neither the display nor the human eye has the dynamic range needed to cope with these images. This situation and the method for solving it are illustrated in Figure 28. One operational strategy is to examine these regions separately. While this is commonly done when viewing radiographic film, the method is less practical with real-time radiographic systems.

Examination of these images shows that transitions from very bright to very dark do not occur rapidly, a result to be expected from a consideration of typical system modulation transfer function (MTF). Expressed in other terms, one can say that

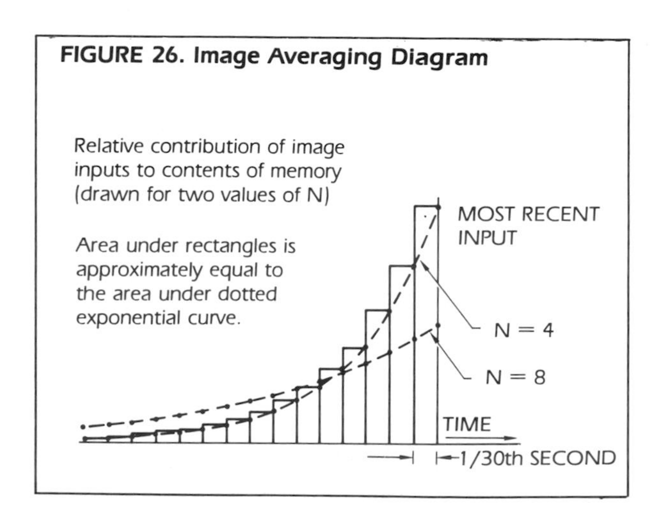

FIGURE 26. Image Averaging Diagram

FIGURE 27. Four 25 mm Thick Steel Pieces with 1% Penetrameters: (a) Unprocessed Video Display; (b) Image Processing Using a 32 Frame (about 1 second) Continuous Averaging; (c) Averaging Plus Real-Time Digital Image Filtering.

(a)

(b)

(c)

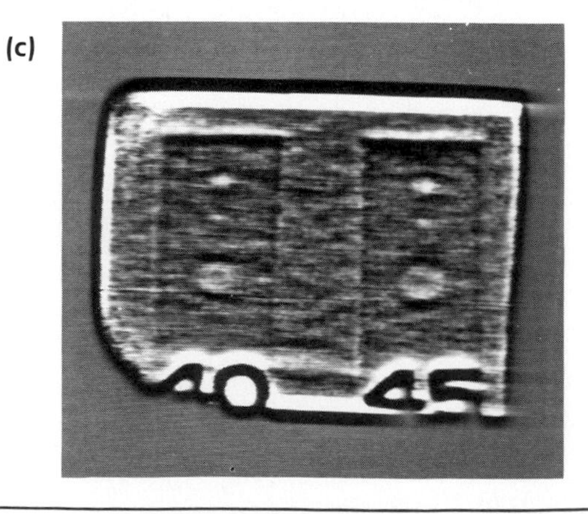

the lower spatial frequency components of the image demand the most dynamic range. On the other hand, the most significant image information is usually contained in the higher spatial frequency components. It is these components which characterize small cracks, voids, inclusions, material interfaces, edges, etc. From this point of view, edge enhancement operations are essentially image filtering operations, where the low spatial frequencies are attenuated while high spatial frequency components are emphasized.

Subtraction operations provide a method for accomplishing one-dimensional edge enhancement. One commonly used scheme is to subtract a slightly delayed video signal (image shift) from the original, undelayed signal. In regions of the picture where there is no change taking place (where there is no detail), subtraction is complete. However, regions of the picture containing detail do not exactly match and subtraction leaves a residue which serves to emphasize these regions. This technique can be employed in real-time when a system has a memory or a partial memory that will store part of the picture momentarily and then subtract it from itself. Mathematically, this process is called *differentiation*. Thus the signal produced is proportional to the rate of change for brightness with distance. Figure 29 is an example.

FIGURE 28. Image Detail Enhancement Procedure

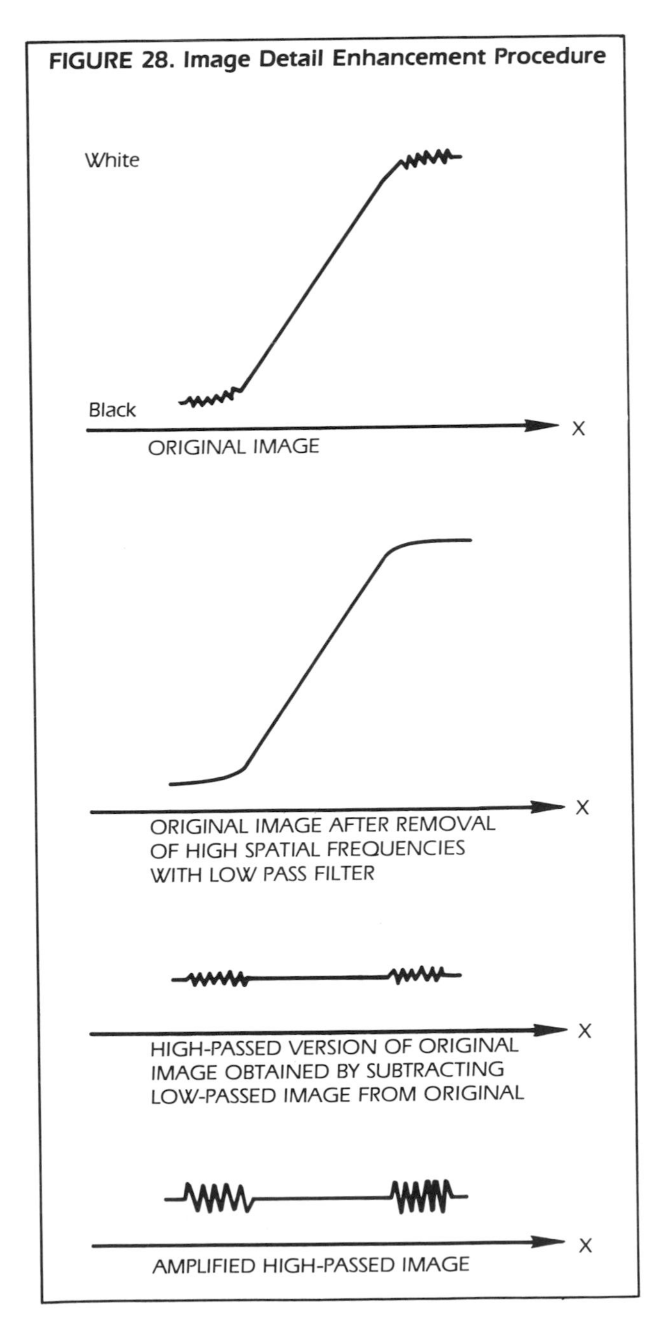

FIGURE 29. Edge Enhancement Obtained Through Image Shift and Subtraction (Differentiation)

Another procedure makes use of subtraction capability: performing an electronic analog of photographic unsharp masking. In this procedure a blurred image is subtracted from a sharp image. As with the differentiated image, the broad variations in brightness and darkness are removed and only edges are displayed. The diagrams of Fig. 30 illustrate the concept in both the space and frequency domains.

Figure 31 is the diagram of a unit specifically designed to do edge enhancement or, more generally, image filtering. In this instrument, the video is convolved in real-time with an aperture using digital delay techniques. The aperture size may be selected by the operator to be N pixels horizontally by 1 pixel vertically, or $1 \times N$ or $N \times N$. All pixels are equally weighted.

The effect of this initial operation is to low-pass the image (transmit only those values below an established cutoff) in the horizontal or vertical direction (or both simultaneously), depending on aperture shape. Image detail is thus removed, resulting in a soft image of reduced resolution. In some situations this operation, which in effect averages the values of neighboring pixels, can be useful for improving the signal-to-noise ratio.

FIGURE 30. Edge Enhancement Obtained by Subtracting Blurred Image from Original: (a) Space Domain, (b) Partial Frequency Domain

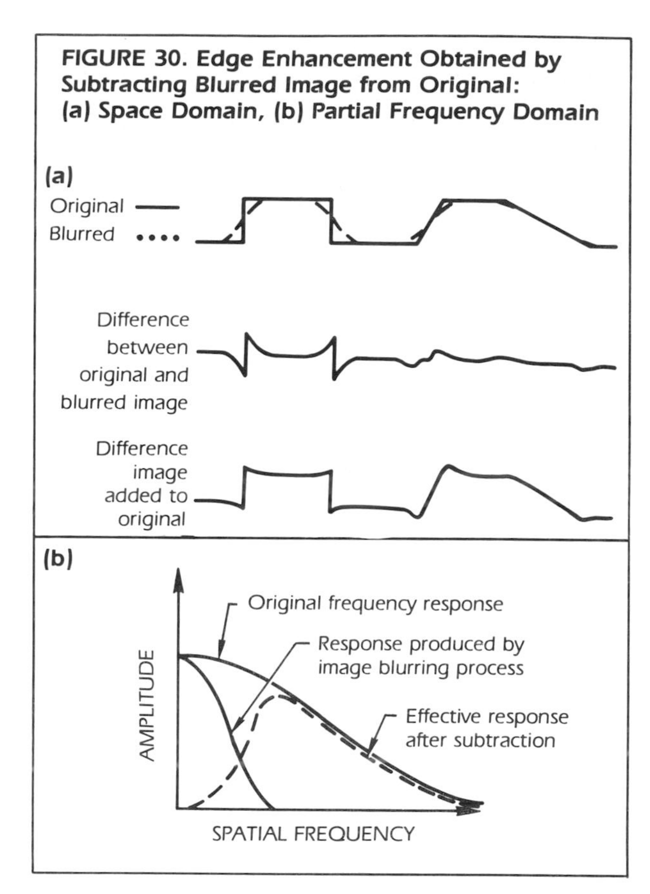

FIGURE 31. Block Diagram of a Real-Time Digital Video Filter

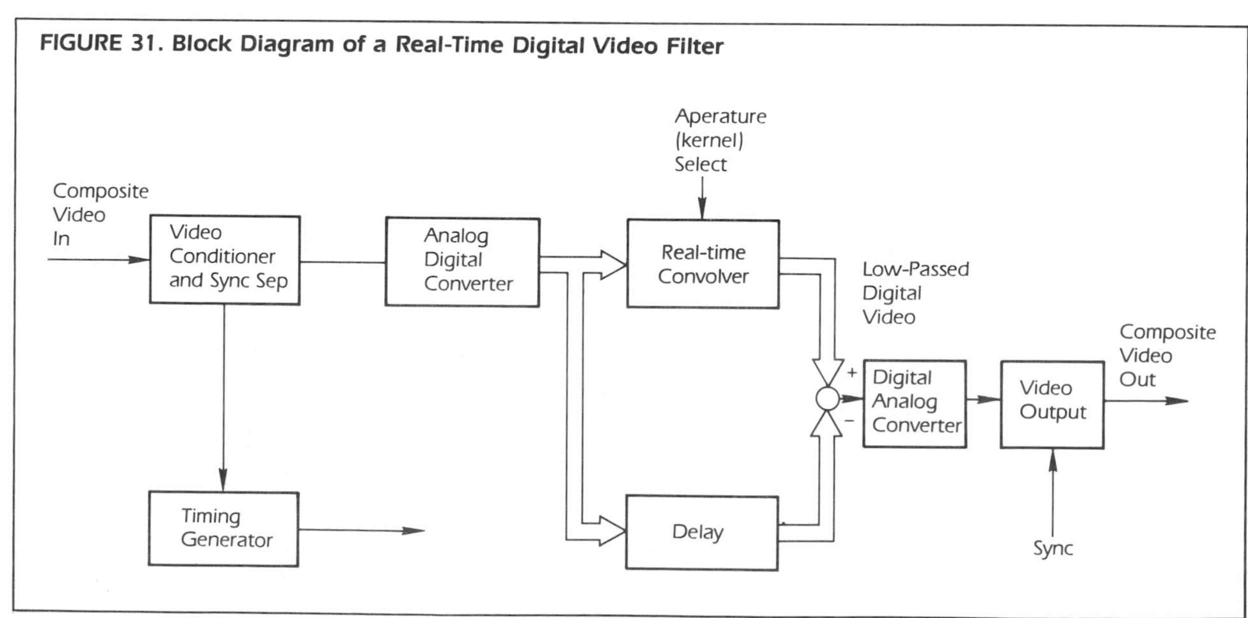

FIGURE 32. Linear High-Pass Filtering of an X-ray Image: (a) Original Image, (b) Filtered Image

(a)

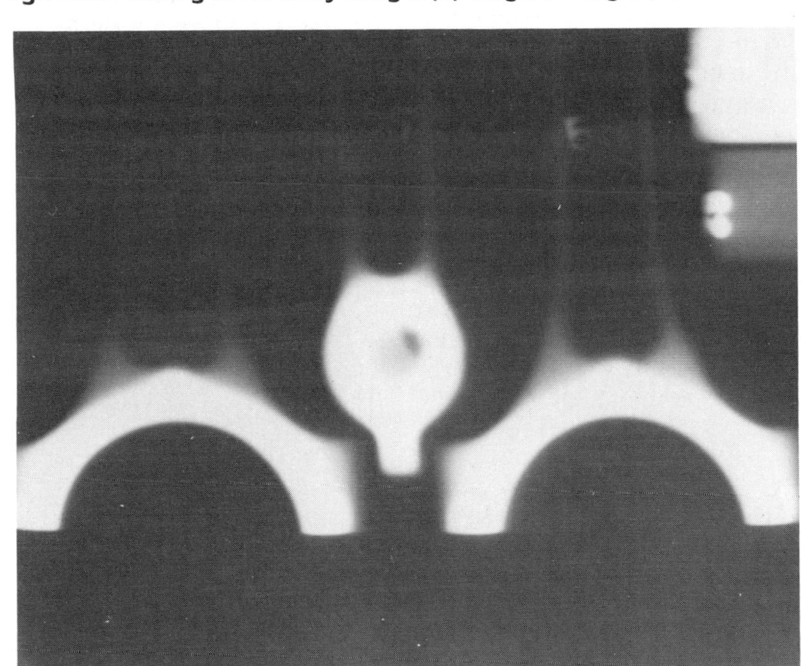

(b)

More frequently, the low-passed image is subtracted from the original image to obtain a linear high-pass filtered image. Figure 32 shows the results obtained with this technique. In the original unfiltered image, the dynamic range of the display is inadequate for showing detail in either the bright or dark parts of the image. After filtering, significant image detail is visible in both areas.

Figure 33 is an interesting example of filtering out image structure so that underlying details can be more readily detected. In Fig. 33b, two-dimensional real-time linear filtering has been applied; in Fig. 33c all horizontally aligned components of the image have been removed along with the low-frequency vertically aligned components.

Nonlinear high pass filtering is performed when the original image is divided by the low-passed image. This type of filtering is also called *homomorphic filtering* and is especially effective in enhancing low contrast detail in dark areas of an image. In addition to enhancing low contrast detail, high pass filtering can be effective in compensating for nonuniformities in the X-ray beam, vignetting effects of the relay optics, and nonuniform response of the television sensors.

Edge enhancement is an important tool for improving the measurement of part dimensions from X-ray images.[48] Linear filtering in which the resultant image is proportional to the second derivative rather than the first derivative is much more appropriate for this purpose.

A system has been described that utilizes two digital image memories for performing shading correction in real-time.[12] One memory stores the additive dark signal which the system outputs with zero optical input. The second memory stores the signal produced when a presumably uniform input is presented to the system. This signal is a measure of system gain as a function of location in the field of view. X-ray beam nonuniformities, lens vignetting, sensor shading, etc., will combine to produce a nonuniform output. After these two memories are loaded, the system then views an actual scene. Prior to display, the additive component is subtracted and the so-called white reference signal is divided into the remainder, with the result that a true first order correction is made.

FIGURE 33. Comparison of Linear High-Pass Filtering in One and Two Dimensions: (a) Original Image, (b) Image After Application of Two-Dimensional Real-Time Linear Filtering, (c) Image After Removal of Horizontally Aligned Components and Low-Frequency, Vertically Aligned Components

(a)

(b)

(c)

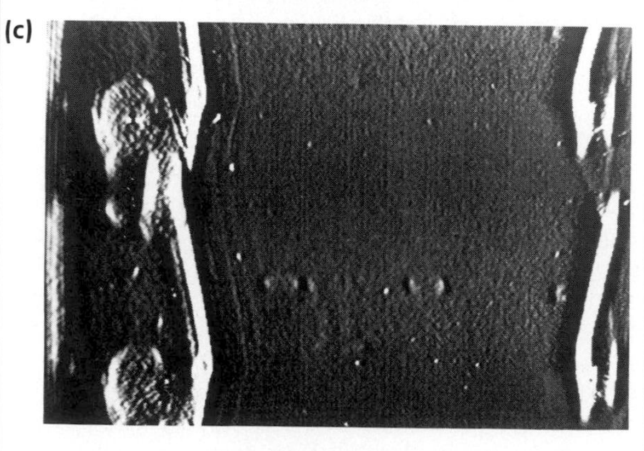

Contrast Stretching

There are frequent occasions where the important information in an image is concentrated in a small area having a narrow range of brightness levels. The ability to stretch the contrast in such a case can make the difference between detection and nondetection of a critical detail.

One method of stretching contrast is to subtract a selected value from each pixel, clip all negative values, increase the gain of the positive values and limit the gray values to a fixed maximum (white clip). This procedure variously referred to as *offset and gain* or *windowing*, can be very effective for bringing out low contrast information. A shortcoming of the technique is that those parts of the image outside the "window" are completely lost because they are displayed as either black or white. In some situations an improvement would be to suppress but not eliminate contrast outside the window while stretching contrast within the window.

With digital technology, such contrast stretching is easy to accomplish. One technique, sometimes referred to as *gray scale mapping,* uses an 8-bit (256 gray level) look-up table which can be modified by the operator. This table is inserted in the signal path preceding the DAC. Each digital word representing the brightness level of a pixel is routed to the look-up table which then outputs a new value prescribed by the table. This new value is sent to the DAC and from there, in analog form, to the display. The table can be quickly modified so that the operator can experiment with changes that might optimize the display.

Histogram Equalization

A more automated approach to contrast modification is known as histogram equalization. The basic idea of this approach is to optimize the dynamic range of the display and the observer. The procedure requires the construction of a histogram of the image or a selected region of the image. This graph shows the number of pixels lying within each brightness interval. A look-up table is then computed which redistributes the brightness levels to produce an image having a histogram with more desirable properties. The two most useful histogram functions are the *equalized histogram* in which the pixels are uniformly distributed from black to white and the *hyperbolized histogram* in which a hyperbolic distribution favoring more black pixels is produced. Many systems are available which compute the required look-up table in real-time or near real-time.

Logarithmic Processing

A logarithmic amplifier provides higher gain for low level signals and thus stretches contrast in the black. The signal output of an X-ray sensor to the first order is

$$I = I_O e^{-\mu x} = I_O e^{-s} \qquad \text{(Eq. 41)}$$

where I_O is the signal obtained with zero absorption of the beam, μ is the absorption coefficient and x the path length. If μ, x or both are slightly changed, as might be the case for a part with a foreign material inclusion or a part with a void, then the change in the signal is given by

$$dI = -Ids \qquad \text{(Eq. 42)}$$

That is, the signal modulation is proportional to the signal level. Thus a void or inclusion would produce a larger signal modulation in a thin section than in a thick section.

By producing a new signal obtained with a logarithmic amplifier, the relation becomes

$$I_L = lnI = -(lnI_O)\mu x = -K\mu x = -Ks \qquad \text{(Eq. 43)}$$

In this case a small change in s produces a change in I_L that is independent of signal level since $dI_L = -Kds$. Disregarding second order noise and scattering effects due to thicker sections, the result is that an inclusion or void will be displayed with the same contrast whether it is in a thick or thin part of an object.

Analog Enhancement

There are several types of analog image enhancement equipment that have found application in real-time radiography. One of these types is the *analog scan converter.* These units employ a special storage tube whose target can store a charge pattern representing the image. In addition to the target, the tube has an electron gun and appropriate beam control elements. The target consists of a finely spaced array of nonconducting islands on a doped silicon conductive substrate. By suitable control of the beam energy, these islands can be charged in a positive or negative direction so that either an old charge pattern can be erased or a new one written. After writing a charge pattern, the potentials can be changed so that these islands act as a barrier or control grid. This grid limits the amount of reading beam current which can land on the interstitial areas of the substrate and produce an output signal. In this reading mode, the beam is unable to land on the island, so that the stored image is not destroyed.

These units have been most commonly used with flash X-ray and transient applications such as baggage inspection. They can also be used as signal integrators. By repetitively writing a low level, noisy signal on the target, the signal-to-noise ratio can be increased. The principal attraction of these units has been low acquisition cost which is typically around 20 percent of a digital unit. Another advantage is the relative ease with which differences between a write and read scan format can be accommodated. Among the main limitations are:

1. the tendency for performance to degrade because of instability in various critical potentials;
2. limited dynamic range set by the ratio of maximum storage capacity to target fixed pattern noise and readout amplifier noise;
3. nonlinear transfer function; and
4. nonuniform response and nonuniform resolution throughout the image area.

An additional drawback of these units is the difficulty of performing a read-while-write operation.

Another analog processing technique which is often effective is *pseudo 3-D display.* This type of display is accomplished through the use of special, wide bandwidth CRT deflection yokes driven by circuitry which mixes the X and Y axis deflections and also modulates the Z axis (the display brightness). The result of this mixing is to display an isometric view in which the vertical displacement from the plane represents brightness.

Pseudocolor

Pseudocolor is a technique used with video systems wherein a brightness level is displayed as a color. For example, dark areas of an image might be displayed as blue, moderate brightness as green and high brightness as red. Depending on the capabilities of the unit and the application, the number of brightness levels translated into distinctive colors may vary from as few as four to more than 100.

With good signal-to-noise ratio, pseudocolor can aid in the discrimination of small brightness differences since the eye can distinguish a much larger number of different colors than it can brightness differences. A second value of this technique is that parts of an image having the same brightness can be more easily recognized even when located some distance apart. The eye is unable to judge absolute brightness levels and is easily fooled by the brightness of adjacent areas.

Conclusion

Real-time radiography has been found useful for the inspection of static objects as well as for imaging dynamic events. Fluorescent screens play an important role in real-time radiography, converting radiation to light which is then observed directly or intensified and detected. The radiographic parameters for real-time radiography are similar to conventional film radiography with special emphasis on statistics and magnification.

Direct-viewing fluoroscopy uses the human eye as a detector of fluorescent light or the light from an intensifier. Remote-viewing systems replace the human observer with a television camera, image intensifying or solid state imaging components. The remote-viewing systems have many advantages over direct viewing conditions, such as: safety; image enhancement; and the ability to produce permanent records.

Image enhancement techniques are used to highlight the information available in real-time images and to thereby expand the overall range of applications.

Real-time radiography will continue to be a quickly expanding field. New developments in detection and display promise to increase utilization of this method throughout the industry.

REFERENCES

1. Bates, C.W., Jr. "New Trends in X-ray Image Intensification." Society of Photo-Optical Instrumentation Engineers (SPIE) Proceedings, Vol. 47 (1974): pp 152-158.
2. Bates, C.W., Jr. "Concepts and Implementation in X-ray Image Intensification." *Real-Time Radiologic Imaging; Medical and Industrial Applications,* ASTM STP 716, (D.A. Garrett and D.A. Bracher, editors). American Society for Testing and Materials (1980): pp 45-65.
3. Biberman, L.M., and S. Nudelman. *Photoelectric Imaging Device.* New York: Plenum Press. Vol. 2 (1971).
4. Bossi, R.H., J.L. Cason, and C.N. Jackson, Jr. "The Modulation Transfer Function and Effective Focal Spot as Related to Neutron Radiography." *Materials Evaluation,* Vol. 30 (May 1972): p 103.
5. Cassen, B., and R.C. McMaster. "Fluoroscopic Methods of Inspection of Metallic Materials, Part VII." WPB Contract W-138, California Institute of Technology, Pasadena, CA (1945).
6. Chamberlain, W.E. "Fluoroscopes and Fluoroscopy." *Radiology,* Vol. 38 (1942): pp 383-412.
7. Chalmetron, V. "Microchannel X-Ray Image Intensifiers." *Real-Time Radiologic Imaging: Medical and Industrial Applications,* ASTM STP 716, (D.A. Garrett and D.A. Bracher, editors) (1980): pp 66-89.
8. Coltman, J.W. "Fluoroscopic Image Brightening by Electronic Means." *Radiology,* Vol. 51 (1948): pp 359-367.
9. Coltman, J.W. "The Specification of Imaging Properties by Response to a Sine Wave Input." *Journal of the Optical Society of America,* Vol. 44 (June 1954): p 468.
10. Criscuolo, E.L., and D. Polansky, "Improvements in High-Sensitivity Fluoroscopic Technique." *Nondestructive Testing,* Vol. 14 (January 1956): pp 30-31.
11. Curtis, L.R., "Fluoroscopic Inspection of Aluminum and Magnesium Castings." *Nondestructive Testing,* Vol. 7 (March 1948-49): p 25.
12. Dalberg, R.C. "Real-Time Digital Image Filtering and Shading Correction." SPIE Proceedings, Vol. 207 (1979).
13. Dahlke, L.W., and C.T. Oien. Personal communication. Sandia National Laboratories, Livermore, CA (1980).
14. Diakides, N.A. "Phosphors." Proceedings SPIE, Vol. 42 (1973): pp 83-92.
15. Fink, D.G., editor. *Electronic Engineers' Handbook* (1975).
16. Halmshaw, R. *Physics of Industrial Radiography.* New York: American Elsevier Publishing Co. Inc. (1966).
17. Halmshaw, R. "Direct-View Radiological Systems." *Research Techniques in Nondestructive Testing,* R.S. Sharpe, editor. Academic Press (1970).
18. Halmshaw, R. "Fundamentals of Radiographic Imaging." *Real-Time Radiologic Imaging: Medical and Industrial Applications,* ASTM STP 716, (D.A. Garrett and D.A. Bracher, editors) (1980): pp 5-21.
19. Hampe, W.R. "Modern Fluoroscopy Practices." *Nondestructive Testing,* Vol. 14 (January 1956): pp 36-40.
20. Hecht, S., and Yun Hsia. "Dark Adaptation Following Light Adaptation to Red and White Light." *Journal of the Optical Society of America,* Vol. 35 (1945): p 261.
21. Hobson, G.S. *Charged Transfer Devices.* New York: Holsted Press (1978).
22. Howe, M.J., and D.V. Morgan. *Charge Couple Devices and Systems.* New York: Interscience (1979).
23. Jacobs, J.E. "X-Ray-Sensitive Television Camera Tubes." *Real-Time Radiologic Imaging: Medical and Industrial Applications,* ASTM STP 716, (D.A. Garrett and D.A. Bracher, editors) (1980): pp 90-97.

24. Klasens, H.A. "Measurement and Calculation of Unsharpness Combinations in X-ray Photography." *Philips Research Reports* (1946): p 241.

25. Klingman, E. "Theory and Application of an Edge Gradient System for Generating Optical Transfer Functions." NASA TN D-6424 (1971).

26. Kuhl, W., and J.E. Schrijvers. "A New 14-Inch X-ray Image Intensifier Tube." *Medicamundi,* Vol. 22 (March 1977).

27. Ludwig, G.W., and J.S. Prener. "Evaluation of Gd_2O_2S:Tb as a Phosphor for the Input of X-ray Image Intensifier." Schenectady, NY: General Electric Corporate Research and Development (1972).

28. McMaster, R.C. "Fluoroscopy and X-ray Imaging Devices." *Nondestructive Testing Handbook,* Vol. 1 (1959).

29. McMaster, R.C., L.R. Merle, and J.P. Mitchell. "The X-Ray Vidicon Television Image System." *Materials Evaluation,* Vol. 25 (March 1967): p. 46.

30. Mees, C.K. *Theory of the Photographic Process.* New York: MacMillan Co. (1942).

31. Morgan, R.H. "The Frequency Response Function." *The American Journal of Roentgenology and Radium Therapy,* Vol. 88 (January 1962): p. 175.

32. Morgan, R.H., L.M. Bates, U.V. Gopalarao, and A. Marinaro. "The Frequency Response Characteristics of X-ray Films and Screens." *The American Journal of Roentgenology, Radium Therapy and Nuclear Medicine,* Vol. 92 (February 1964): p 426.

33. "Permissible Dose from External Sources of Ionizing Radiation." National Bureau of Standards, Handbook 59 (1954).

34. O'Connor, D.T. "Industrial Fluoroscopy." *Nondestructive Testing,* Vol. 11, No. 2 (1952).

35. Rossi, R.D., and W.R. Hendree. "Some Physical Characteristics of Rare-Earth Imaging Systems." SPIE Proceedings, Vol. 70 (1975): p 224.

36. Schagen, P. "X-ray Imaging Tubes." *NDT International,* Vol. 14, No. 1 (1981): pp 9-14.

37. Siedband, M. "Electronic Imaging Devices II." *Physics of Diagnostic Radiology,* USDHEW Publication No. (FDA) 74-8006 (1971).

38. Smith, F.D. "Optical Image Evaluation and the Transfer Function." *Applied Optics,* Vol. 2, No. 4 (1963): p 335.

39. Smith, W.J. *Modern Optical Engineering; The Design of Optical Systems.* McGraw Hill (1976).

40. Storm, E., and H.E. Israel. "Photon Cross Sections from 0.001 to 100 MeV for Elements 1 through 100." Report LA-3753, Los Alamos National Laboratory, Los Alamos, NM (1967).

41. Sturm, R.E., and R.H. Morgan. "Screen Intensification Systems and Their Limitations." *The American Journal of Roentgenology and Radium Therapy,* Vol. 62, No. 5 (1949).

42. Teves, M.C., and T. Tol. "Electronic Intensification of Fluoroscopic Images." *Phillips Technical Review,* Vol. 14, No. 2 (1952): pp 33-43.

43. Wagner, R.E. "Noise Equivalent Parameters in General Medical Radiography: The Present Picture and Future Pictures." *Photographic Science and Engineering,* Vol. 21, No. 5 (1977).

44. Wang, S.P., D.C. Robbins, and C.W. Bates, Jr. "A Novel Proximity X-ray Image Intensifier Tube." SPIE Proceedings, Vol. 127, *Optical Instruments in Medicine VI* (1977): pp 188-194.

45. Webster, E.W., R. Wipfelder, and H.P. Prendergrass. "High Definition versus Standard Television in Televised Fluoroscopy." *Radiology,* Vol. 88, No. 2 (1967): pp 355-357.

46. Webster, E.W., "Electronic Imaging Devices I." *Physics of Diagnostic Radiology,* USDHEW Publications No. (FDA) 748006 (1971).

47. Woodhead, A.W., and G. Eschard. "Microchannel Plates and Their Applications." *Acta Electronica,* Vol. 14, No. 2 (1971): pp 181-200.

48. Vary, A., and K.J. Bowles. "Application of an Electronic Image Analyzer to Dimensional Measurements from Neutron Radiographs." Los Angeles, CA: ASNT Spring Conference (1973).

49. Vosburg, K.G., R.K. Swank, and J.M. Houston. "X-ray Image Intensifiers." *Advances in Electronics and Electron Physics* (L. Marton, editor), Vol. 43, Academic Press (1977).

SECTION 15

IMAGE DATA ANALYSIS

M.H. Jacoby, Lockheed Missiles & Space Company, Inc., Sunnyvale, CA

PREFACE

Digital image processing methods provide a convenient, mathematically based approach to improve the extraction of data from X-ray film, as well as from fluoroscopic and acoustic images. This Section discusses some of the basic enhancement techniques that have been developed over the past several years. It also describes how these methods have been incorporated into a computerized system, developed at the Lockheed Palo Alto Research Laboratory (LPARL), for the automatic evaluation of inspection data in image form.

The field of image data processing got its start a little over forty years ago, when the old and distinguished discipline known as optics was joined to the then-emerging discipline of information science. The "bonding agent" was communication theory as developed by Shannon[1] and Whittaker.[2]

The advances in data processing that have occurred since then — spurred in part by continuing improvements in hardware and software — have made possible the computer-based analysis of quite complicated imagery. In fact, digital methods for image analysis have become common in a variety of different fields: radar, astronomy, medicine, and geophysics to name a few. Perhaps the most familiar applications of these analysis techniques are the widely distributed pictures of the moon, Mars, Jupiter, and Saturn prepared by NASA.[3,4]

It is expected that, as the costs associated with digital data processing continue to decline, applications that are now just talked about will become commonplace. Therefore, a familiarity with the basic methods of image processing will be helpful to many engineers and scientists.

When scanning a large, complex field through a broad window, as this report surely does, many of the high-technology features are inevitably smoothed over. And there are many omissions. Computerized tomography, to take a specific example, is mentioned only in passing.

Nevertheless, this Section will serve as a useful introduction to those who are just beginning to work in the field and it is hoped that sophisticated practitioners will find many topics of interest as well.

ACKNOWLEDGEMENTS

The application of digital processing techniques to nondestructive test imagery began at LPARL about seven years ago. Initial funding for this work came from the Special Projects Office of the Navy under the direction of M. Baron, and his consultant, D. Polansky, now with the National Bureau of Standards. The bulk of the funding for the development of automatic X-ray image analysis came from the Army Armament Research and Development Command (AARADCOM), Dover, New Jersey, under contract DAAK10-78-C-0434. J. Argento, E. Barnes, J. Moskowitz and R. Streubing were the technical monitors.

The two-dimensional "signals" that make up an image are not in any way the same as two channels of a one-dimensional, time domain signal. Two-dimensional images have properties completely different from signals in one dimension. It follows that the theory of image analysis is not just an extension of the processes used on time domain signals. New ground needs to be broken.

At LPARL, the algorithms for automatic analysis were first formulated by M.J. Fischler. His basic work was expanded by O. Firschein and W.E. Eppler.

Because the work on automatic analysis of X-ray imagery proved to be so stimulating and so readily applicable to a wide variety of image processing problems, other members of the Signal Processing Laboratory were given the opportunity to work on various aspects of the project. These contributors included: P.A. Dondes; A. Blumenthal; M. Heideman; S.M. Jaffe; J. Keng; R.S. Loe; H.N. Massey; D. Milgram; G. McCulley; R.L. Saperstein; J.L. Schmidt; P. Wong; and C.D. Kuglin.

As project leader on the AARADCOM program, I gratefully acknowledge the contributions of these scientists and engineers who came up with imaginative solutions to knotty problems and produced an operating system for the automatic evaluation of X-ray films of artillery shells.

Special thanks also go to J.J. Pearson, manager of the Signal Processing Laboratory, and D.C. Pecka, Director of Communication Sciences, for their executive support.

INTRODUCTION

Visual inspection has always been and continues to be the most important technique for evaluating material. All systems that rely on people to observe and evaluate images — labels on containers, cathode-ray tube displays, X-ray film — are in effect visual inspection systems. Today, government regulations, extended warranties, and the need for high reliability in complex operations require that an inspector make rapid and accurate assessments of large quantities of inspection data. It is often difficult and always costly to have an inspector look at everything. Sampling techniques can be used in place of 100% inspection to show that a population of components is acceptable. Automatic machines (like those that measure the fluid level in bottles) have been used for simple, repetitive tasks where 100% inspection is required. These methods are effective, but it turns out that the weakest link in the entire inspection process is often the subjective evaluation of data by the inspector. To overcome this troublesome but well documented fact, automatic computer-based inspection systems can be used to provide accurate and, above all, consistent evaluations. These systems must be easily reprogrammable so that they can perform a wide variety of tasks.

In nondestructive testing there is a large class of inspections that can benefit from these multifunctional systems. X-ray, both film and fluoroscopic, and acoustic images lend themselves to automated analysis, as do the time-domain waveforms associated with ultrasonic, eddy current, and acoustic emission testing.

These inspection systems can improve the various types of evaluation in two important ways. First, the image or waveform can be enhanced to emphasize subtle details that are not readily apparent to the inspector. Second, the enhanced image or waveform can be evaluated automatically. Flaws can be located and measured and acceptance decisions made by a computer in a consistent and objective manner.

This Handbook Section reviews some of the more familiar enhancement processes currently in use or now being developed, as well as the hardware necessary to these processes. The enhancement methods form a basis for the automatic, computer-based evaluation of images which is discussed in some detail.

Review articles by B.R. Hunt,[5] E.L. Hall, et al.,[6] Trussell,[7] and Janney[8, 9] provide a comprehensive discussion of the field of digital image processing, and list many references and citations. Harrington and Doctor[10] have documented enhancement methods which are useful in ultrasonic, eddy current and acoustic emission technology. Other industrial techniques continue to be explored.[11-13]

The mathematics underlying the processes are not described in this report. Many excellent texts[14-17] are available for those readers seeking further information on the specific digital processes used in two-dimensional image analysis.

Current research in the X-ray field is aimed at the development of computerized X-ray scanners. The most innovative of these techniques is Computer Assisted Tomography (CAT) for radiographic and radioisotope imaging. The impact of CAT scanning on medical X-ray diagnosis has been documented.[18] These techniques have been applied to industrial inspection,[19] and there are also plans to use CAT scanners to inspect solid-propellant rocket motors up to 2.4 m (8 ft.) in diameter.[20] Scanned Projection Radiography (SPR) systems which develop data in point-scan, line-scan, or area-scan mode have been described.[21] All of these new techniques have the digitizing step built into the detector. At present, images formed on X-ray film (conventional radiography), X-ray luminescent screens (fluoroscopy), or cathode-ray tubes (acoustic imaging) need to be digitized and put into a two-dimensional array of picture elements (pixels) before computer-based enhancement or automatic analysis can take place. The digitizing step is crucial in image processing and will be the first topic considered.

PART 1

DIGITIZING THE IMAGE

For the purposes of illustration a particular example will be used here — that of digitizing the fluoroscopic image of a 7 × 12 cm (3 × 5 in.) cylinder. First, the video signal from a TV camera focused on the X-ray conversion screen must be digitized and stored in appropriate form in computer memory. This procedure will convert the analog TV image into a digitized image that is quantized in space and intensity. In the first step, the timing clock in a 10MHz analog-to-digital converter selects the image elements that will be digitized. Here the image of the figure (the cylinder) together with the ground is quantized into 512 rows and 512 columns giving a square array of 262,144 picture elements (pixels) with a spatial resolution of 0.25 mm (0.01 in.) per pixel — about 500 pixels in the 12 cm (5 in.) direction, and 300 pixels in the 7 cm (3 in.) direction. If more spatial resolution is needed, the TV camera optics could be set to place the 512 × 512 array over a smaller region, say 6 × 6 cm (2.5 × 2.5 in.), which would double the spatial resolution. The fluoroscopic image of the cylinder would then be digitized in two sections.

The digitized array might have been set up as a 300 × 500 pixel rectangle to save on storage space in the computer memory. It turns out that many computations in image processing are greatly simplified if the matrix is square with dimensions that are powers of two. In this example each side of the array is chosen to be $2^9 = 512$ pixels.

The image also must be quantized in intensity. Each analog intensity level selected by the clock is converted to the closest of $2^9 = 512$ quantization levels in a nine bit digitization. For example, a scaled analog intensity level of 257.31 is stored as 257. This basic quantization method gives 512 contrast steps when going from black (lowest intensity) to white (highest intensity). Just as spatial resolution can be changed by modifying the TV camera optics, the contrast resolution can be improved by rescaling the digitized image. If the data of interest lie in the intensity range of 200 to 350, then the image can be rescaled so that 200 is set at zero and 350 is set at 511. This scheme allows for 512 "steps" in the area of interest and improves contrast resolution. (See Fig. 1 for a pictorial representation of the digitization process.)

In any case, contrast should never be the limiting factor in detectability. The signal-to-noise ratio should always set this limit.

Signal-To-Noise Ratio (S/N)

Contrast and resolution are the two major parameters by which detail can be enhanced (literally, made visible) on radiographs. The resultant contrast C caused by a change in transmission Δ in an environment free of noise with original X-ray intensity I is expressed as:

$$C = \frac{\Delta \times I}{I} = \Delta$$

When the objects of interest differ only slightly in X-ray attenuation from the surrounding medium, the result is a low-contrast radiograph. Similarly, when the object of interest comprises a very small percentage of the volume under investigation, the result again is a low-contrast radiograph. Objects imaged on low-contrast radiographs are not easily perceived.

When noise is present (all information transmitting systems contain some noise), the ability to detect features of interest depends on the ratio of the calculated signal intensity to the random intensity variations caused by the noise in the system. This ratio is defined as:

$$\frac{\text{Signal}}{\text{Noise}} = \frac{\Delta \times \bar{I}}{\sigma_I} \qquad \text{(Eq. 1)}$$

where $\bar{I}$ is the mean value of the measured intensity and σ_I is the standard deviation of $\bar{I}$, the root-mean-square (rms) fluctuation in the intensity.

FIGURE 1. Sampling and Digitizing an Image

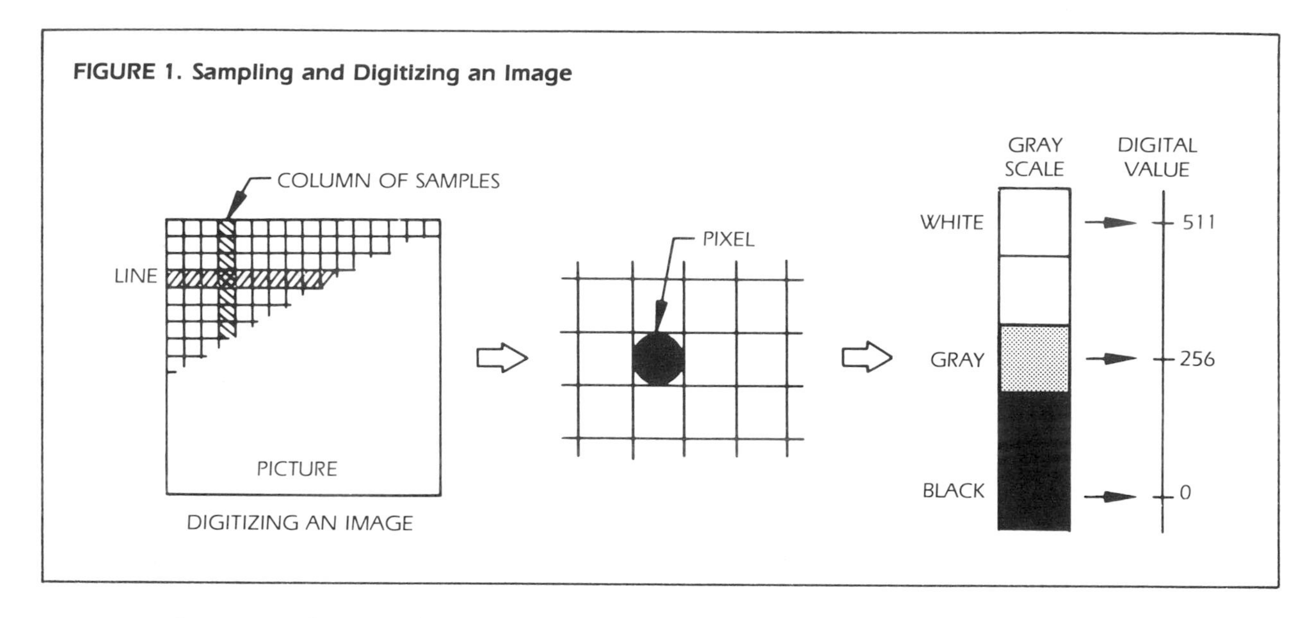

Noise is the primary limiting factor in any image enhancement procedure. Most enhancement techniques also enhance the noise, sometimes more than the signal. Care must be taken to use imaging techniques that minimize noise, and give the maximum signal-to-noise (S/N) ratio.

In X-ray fluoroscopic systems or X-ray systems that use film as a detector, a major source of noise is the statistical uncertainty in the parameter N_S, defined as the number of photons emitted from the X-ray source. It can be shown[22] that N_S is a random variable that follows the Poisson distribution. The transmission of X-ray photons through an object is distributed binomially. The combination of a Poisson and binomial distribution gives a Poisson distribution, which in this case describes the X-ray photons leaving the object. These photons have an average value N given by the familiar equation

$$N = N_O e^{-\mu x} \quad \text{(Eq. 2)}$$

where N_O is the number of photons impinging on the object, μ the linear attenuation coefficient, x the distance traversed, and e the natural logarithmic base.

In a Poisson process with mean value N, the variance is also equal to N and thus the standard deviation is equal to $N^{1/2}$. The signal-to-noise ratio of an anomaly — a void or a low density region in the object — having a contrast Δ is given by

$$\frac{\text{Signal}}{\text{Noise}} = \frac{\Delta N}{N^{1/2}} = \Delta N^{1/2} \quad \text{(Eq. 3)}$$

where the signal is ΔN, defined as the difference between the number of detected photons that traverse the anomoly and the average number of photons, N, that emerge from nearby regions of the object. The noise is $N^{1/2}$, the standard deviation of the photons emerging from the object.

This shows that the S/N ratio for a feature of interest, such as a void which gives a contrast Δ, is directly proportional to the number of photons detected. This implies that detector efficiency must be taken into account, modifying the equation by η; the capture efficiency of the detector, so that

$$\frac{S}{N} = \Delta (N\eta)^{1/2} \quad \text{(Eq. 4)}$$

Thus, S/N can be manipulated in accordance with one's requirements, excepting for limits imposed by radiation dose requirements (film gets black, patients get sicker, plastics become brittle) and the stability of the electronics in fluoroscopic systems.

There are other sources of signal-dependent noise, but analysis[22] shows that the S/N of the X-ray system, film or fluoroscopic, is determined by eq. 4.

One of the other sources of signal-dependent noise is associated with the high frame rate (30 frames/second and the wide band width (10^7 Hz) of the TV camera used to acquire the image. This noise source is uncorrelated from frame to frame and thus can be reduced to insignificant levels by averaging a number of frames together. Off-the-shelf equipment averages 256 frames in less than nine seconds.

Quantization error is another source of noise. It arises when, for example, the analog value 257.31 is stored as 257. Compared with the statistical errors this source is insignificant.

Scatter and Noise

Scattered radiation is a deterministic component and enters into the S/N value in an additive manner. Scatter produces two detrimental effects in X-ray imaging systems: a reduction in contrast, and an increase in noise. With regard to contrast, the scattered radiation produces an intensity, I_{st}, which adds to the transmitted intensity recorded at the detector. Calling this component I_d, the resultant contrast due to a change Δ in transmission is

$$C = \frac{\Delta I_d}{I_d + I_{st}} = \frac{\Delta}{1 + \frac{I_{st}}{I_d}} \qquad \text{(Eq. 5)}$$

The denominator in eq. 5 is called the *contrast reduction factor.* This reduction in contrast is readily apparent in regions of low transmission where I_d is of the same magnitude as I_{st}.

Scattered X-ray photons also degrade the S/N value. This can be shown as follows:

N = number of photons due to the signal

N_{st} = number of scattered photons.

Since these two terms are independent, their variances combine and eq. 4 becomes

$$\frac{S}{N} = \frac{\Delta \eta N}{(\eta N + \eta N_{st})^{1/2}} = \frac{\Delta (\eta N)^{1/2}}{1 + \left(\frac{N_{st}}{N}\right)^{1/2}} . \qquad \text{(Eq. 6)}$$

The denominator in eq. 6 is called the *S/N reduction factor caused by scatter.*

This information on sources of noise in imaging systems is included here to emphasize the need for controlled procedures. The acquired image must contain certain data; enhancement is no substitute for good X-ray technique.

PART 2

IMAGE ENHANCEMENT METHODS

Brightness Transfer Functions

Our theoretical digitized image is stored in the computer memory with a S/N value that depends on the X-ray photon flux and the detector efficiency. Enhancement can now begin.

Figure 2 shows how low-contrast radiographs can be "stretched" to allow for the improved perception of defects. In this particular case, the optical density variations on the radiograph have been split into 512 separate gray shades, that is, a nine-bit digitization ($2^9=512$). The data of interest lie in the range 200 to 350. These input values have been mapped into the output domain to cover the full dynamic range of 0 to 511. Pixels whose values lie below 200 are mapped to 0 (black) and pixels whose values lie above 350 are mapped to 511 (white). This is an example of pixel-based enhancement where the pixel value in the output image depends only on the value of the corresponding pixel in the input image.

Other examples of pixel-based enhancement are shown in Fig. 3. The gain (ratio of output increase to input) can be changed to give a negative slope with truncation. The film's gamma (slope of linear portion of H&D curve) can also be changed to alter and

FIGURE 2. Contrast Stretching

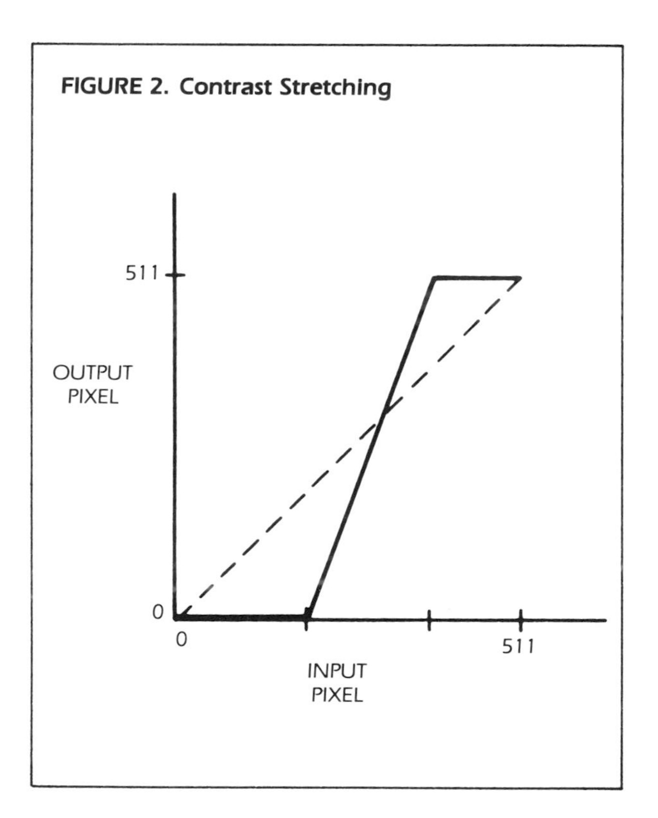

FIGURE 3. Three Brightness Transfer Characteristics (BTC) Functions

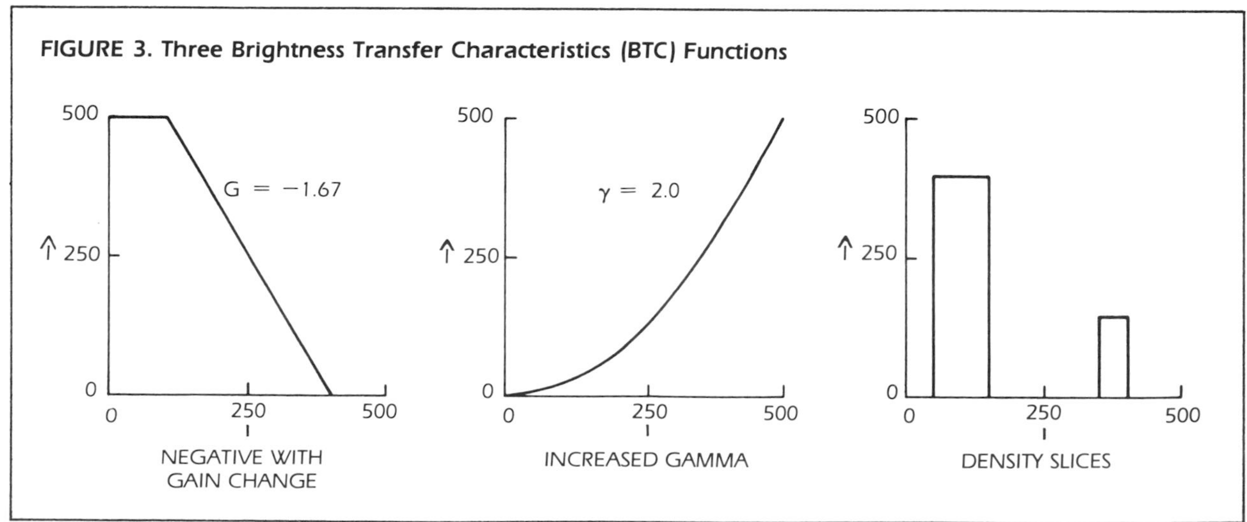

possibly improve output results. In the second example, the value of gamma is set at 2. The third example is a type of thresholding. Input pixel values in the range 50 to 150 are output as 400, and input pixel values in the range 350 to 400 are output as 150. All other input pixel values are given an output value of zero (black). These curves fall into a category of functions called the Brightness Transfer Characteristic (BTC) functions. These BTC functions can be hardwired into the enhancement system or called from the computer as a subroutine.

Gradient Removal

Along with low contrast, another troublesome feature on some radiographs is the presence of an optical density gradient caused by the shape of the part being imaged and the intensity distribution of the X-ray beam. Gradients tend to obscure detail, but they can be removed by curve-fitting and rescaling.

Figure 4 shows how these gradients are removed by field-flattening, a technique which uses polynomial fitting to remove intensity trends. The first graph gives a plot of distance versus intensity across a cylindrical object. The variations caused by the gradient are large when compared with intensity changes caused by flaw indications. So the gradient must be removed before the flaw can be reliably detected either visually by an inspector or automatically by a computer.

The technique fits a polynomial to the gradient, subtracts the original and fitted curves, and displays the remainder on a line-by-line basis. There are problems that surface with this approach:

1. Large flaws influence the fit of the polynomial, and when the curves are subtracted, the flaw will disappear from view.
2. Steep gradients in the intensity curve can cause large residuals.
3. The fitting procedure must be performed at high speed to be practical.

FIGURE 4. Field-Flattening

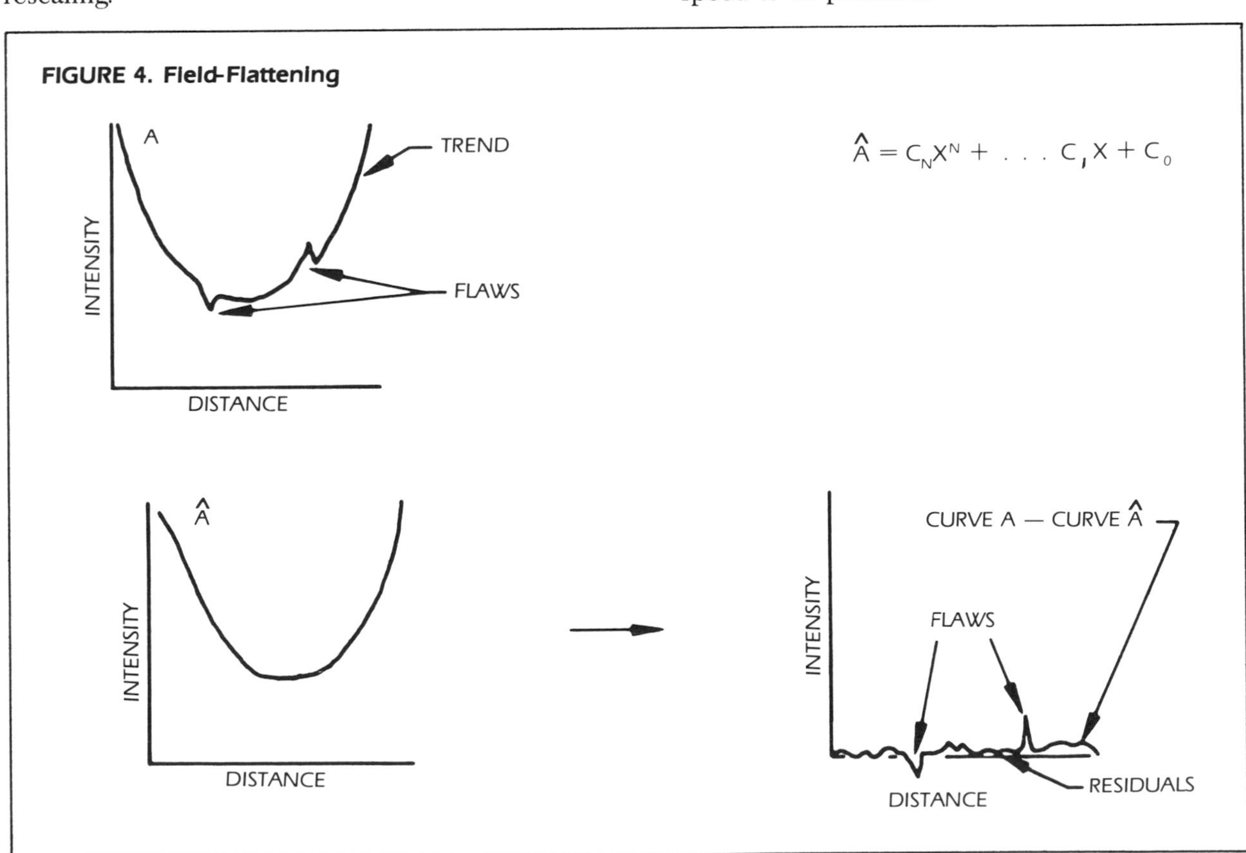

Field-flattening can also be performed by a suppression of low spatial frequencies. This topic is covered in more detail in the section on automatic analysis, where it is shown that field-flattening is an absolutely crucial phase of computer-based analysis.

Digital Filters

The term *filter* in this discussion means a two-dimensional system that has an input and an output that are related by convolution; this implies that the filter is linear and time and space invariant.

Convolution in two dimensions works as follows: take one of the two given functions, the filter function, and rotate it 180° about the origin; translate it to a new location by an amount x, y; multiply the two functions together point by point; and integrate the product over the whole plane to obtain the value of the convolution for that particular x, y. Repeat for other values of x, y until the region of interest is covered.

This process is shown in Fig. 5a, where the value of the convolution for pixel I_8 is calculated. Typical low-pass (smoothing) and high-pass (edge detector) digital filters are shown in Figs. 5b and 5c respectively.

Square filters with two-fold symmetry, such as those whose elements sum to unity, are examples of enhancement methods where the pixel value in the

FIGURE 5. Digital Filters: (a) Convolution Calculation; (b) Smoothing; (c) Edge Detection; and (d) Example of Distribution Invisible Under Convolution

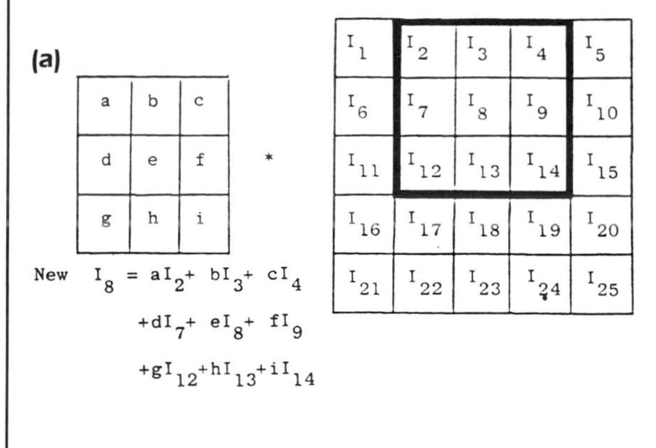

(b)

1/9	1/9	1/9
1/9	1/9	1/9
1/9	1/9	1/9

*

1	1	1	1	1
1	1.3	.8	1.1	1
1	.9	1.2	.7	1
1	1.1	1.1	.8	1
1	.6	1.4	1.1	1
1	1	1	1	1

=

1.03	1.01	1.02	.99	1.01
1.02	1.02	1.00	.98	.98
1.03	1.04	1.00	.97	.96
.96	1.03	.99	1.03	.96
.96	1.00	1.01	1.06	1.01

(c)

-1	-1	-1
-1	8	-1
-1	-1	-1

*

1	1	1	0	0	0
1	1	1	0	0	0
1	1	1	0	0	0
1	1	1	0	0	0

=

0	0	3	-3	0	0
0	0	3	-3	0	0
0	0	3	-3	0	0
0	0	3	-3	0	0
0	0	3	-3	0	0

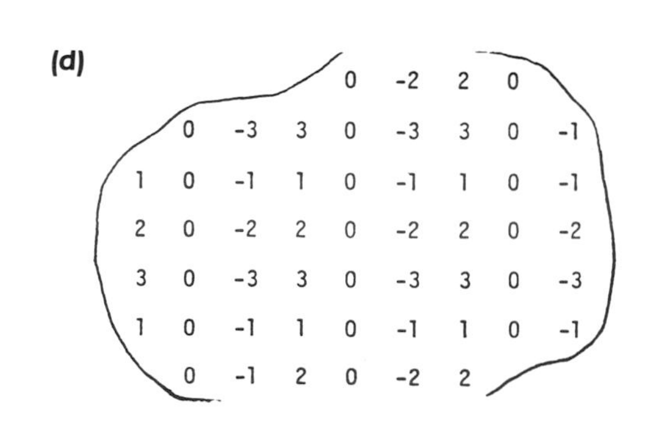

output image depends on the values of a number of pixels in the input image.

Bracewell [17] points out an interesting feature of the 3×3 convolving array used in smoothing. This array can be represented by

$$\frac{1}{9}\ rect\left(\frac{x}{3}\right)\ rect\left(\frac{y}{3}\right) \qquad \text{(Eq. 7)}$$

where *rect* is the rectangle function

$$rect\ (x) = \begin{cases} 1\ when\ x \leqq ½ \\ 0\ Otherwise \end{cases}$$

The Fourier Transform of eq. 7 is

$$9\ sinc\ 3u\ sinc\ 3v \qquad \text{(Eq. 8)}$$

where *sinc* is the sinc function[16]:

$$sinc\ u = \frac{\sin \pi u}{\pi u}$$

The first nulls of eq. 8 are at $u, v = \pm ⅓$ so that in the horizontal and vertical directions spatial periods of 3 units are suppressed. For example, the distribution given in Fig. 5d is invisible when convolved with a 3 × 3 digital filter consisting of nine constants. This exercise opens up the topic of transforms in enhancement.

Transforms

Some computational operations on images are made more efficient if the image is transformed to the spatial frequency domain. When the discrete Fourier Transform (DFT) is used, the smoothing operation would be performed as follows: take the DFT of the two-dimensional array of pixels; attenuate the higher spatial frequencies by multiplication of the DFT by a suitable low-pass transfer function (that given in eq. 8 for example); finally, invert the attenuated DFT. In general, convolution is more efficiently performed in the transform domain, especially if the Fast Fourier Transform[23] is used. An example of this and other processes is given in the next part of this section.

PART 3

EXAMPLES OF IMAGE ENHANCEMENT APPLICATIONS

The results of applying methods such as contrast enhancement, gradient removal, smoothing, and thresholding are shown in the next series of figures. Figure 6a shows the radiograph of a void in a large cylindrical mass of material. The first processing step, after digitizing the image into 512 gray shades, expands the contrast and reveals the gradient shown in Fig. 6b. The gradient is removed and the result smoothed as shown in Fig. 6c. The final step, Fig. 6d, is thresholding, which changes the enhanced radiograph into a binary image where the void is shown as all white and the remainder as all black. The binary image is important in operator-interactive enhancement; measurements can be reproduced more reliably from this type of image.

Another example is given in the next series of images. The original radiograph shown in Fig. 7a is digitized to nine bits. Contrast enhancement, gradient removal, smoothing and thresholding are performed. The result is shown in Fig. 7e. The image of the washer-like indication is clearly seen. It should be noted that the radiograph used in this process was given one-tenth the normal exposure.

FIGURE 6. Image Enhancement: (a) Original Radiograph of Void; (b) Contrast Enhanced; (c) Field Flattened Image; and (d) Binary Image

(a)

(b)

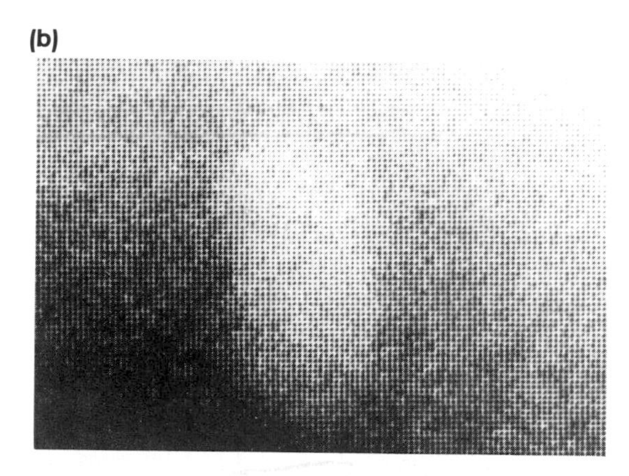

(c)

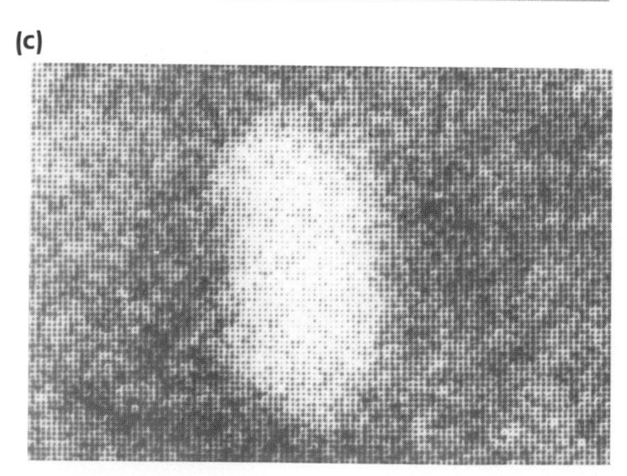

(d)

Edge Enhancement

One of the most prevalent problems in radiographic image enhancement is the lack of edge sharpness and the consequent difficulty in interpretation and measurement. As has been shown, to enhance a given image a variety of techniques have evolved, each of which depends on treating the image as a two-dimensional matrix array of film brightness values $P(x,y)$.

The image $P'(x,y)$ produced by simply taking the first derivative is one of the most common and effective enhancement techniques. The resulting image P' has the psychological effect of a three-dimensional object.

A second technique is to use a linear combination of the Laplacian operator and the original image,

$$P'(x,y) = \left[1 + k \left(\frac{\partial^2}{\partial x^2} + \frac{\partial^2}{\partial y^2} \right) \right] P(x,y)$$

where k is a numerical constant. The result produces a slight overshoot at the edges of an object, which serves to sharpen the definition on the

FIGURE 7. Applications of Image Enhancement Methods to an Underexposed (one-tenth normal) Radiograph.

(a) Original Low-Contrast Radiograph

(b) Contrast Enhanced

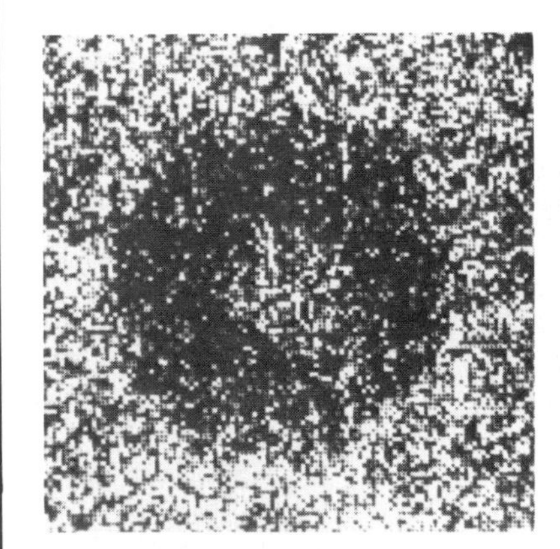

(c) Field-Flattened Image

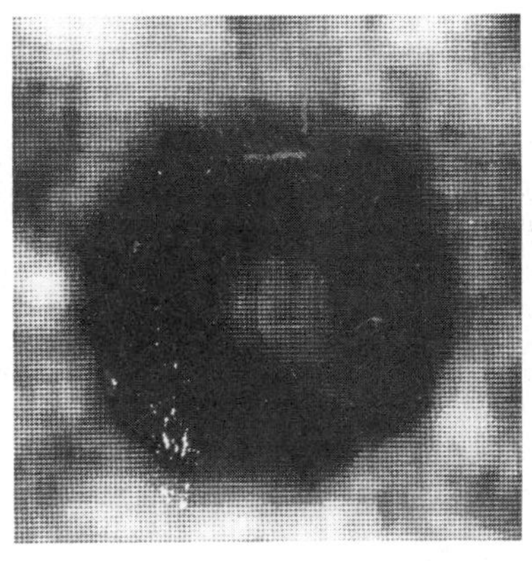

(d) Smoothing Applied

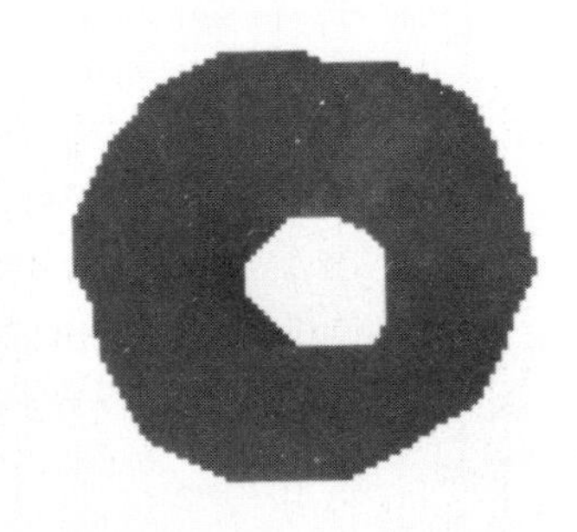

(e) Binary Image

radiograph. Higher order derivatives may be applied, although considerations of noise level and computation complexity limit practical matrix operations to first- and second-order derivatives. Additional techniques such as line, edge, or shape detection operations may also be performed if the nature of the radiographic image's features can be anticipated. In general, these techniques may be treated as special cases of the first derivative operation. These all depend on various types of differencing methods. Fig. 8b shows the results of high-pass filtering on the image in Fig. 8a. These images will be discussed further in the text on automatic analysis.

FIGURE 8. High-Pass Filtering: (a) Original Image of Artillery Shell; (b) Result of High-Pass Filtering

(a)

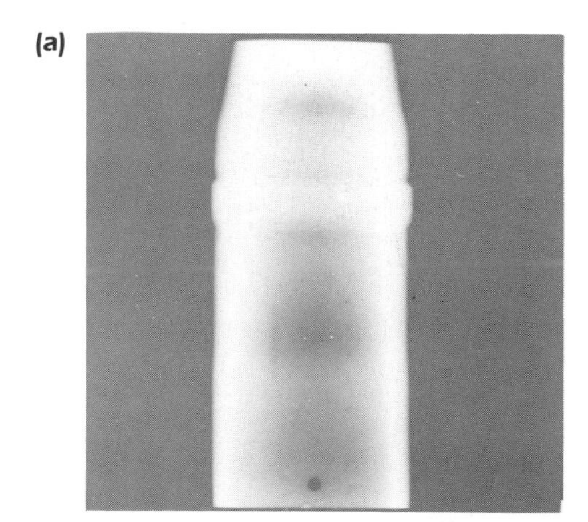

(b)

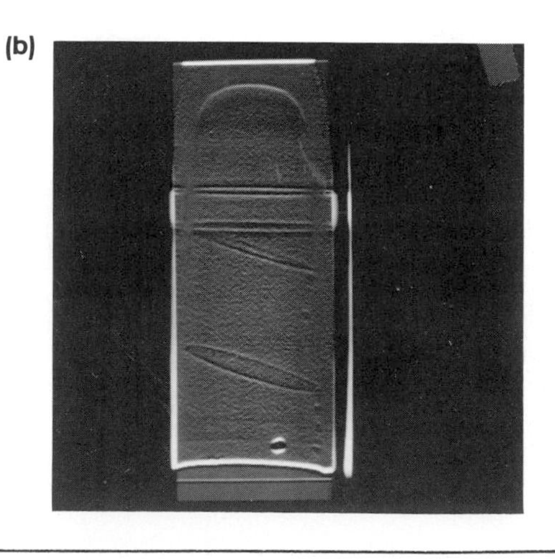

Spatial Filtering

Figure 9 reproduces a radiograph of a glass-wrapped, rubber insulated, cylindrical pressure vessel. The requirement was to enhance a straight cut on the surface of the insulation. The indication was barely discernible on the original radiograph, even with the aid of magnification. Transform domain (spatial) filtering was used to enhance the image.

FIGURE 9. Radiograph of Glass-Wrapped Pressure Vessel

FIGURE 10. Result After Field-Flattening and Contrast Enhancement

The original 10 × 13 cm (4 × 5 in.) section of the radiograph was digitized with a flying-spot scanner over a 512×512 array of brightness values, each 45 micrometers (0.0018 inch) in diameter. Fig. 10 shows the results of the gradient-removal process and the contrast-enhancement process. The cut in the insulation shows up clearly in this view.

Because of the way this chamber is processed, cuts in the insulation can occur only in the vertical and horizontal direction. With this knowledge it is possible to develop a spatial filtering technique to remove the indications of the glass case (i.e. the cross-hatched lines) and just display any horizontal and vertical indications.

Figure 11 is a photograph of the magnitude of the spatial frequency components of the original radiograph. Since $F(u,v)$ is a complex number, there is also a spatial phase present. This phase is not shown in the diagram, but it is contained in the values stored in the computer memory.

The two-dimensional Fourier Transform of one particular spacing in the image plane is given by the magnitude of the impulse pair in the spatial frequency domain. To remove that particular spacing from the image plane, simply filter out the impulse pair and transform the filtered spectrum back into the image plane. It should be noted that no matter what section of a uniform scene is being considered, the magnitude of the spectrum is always the same and thus the filter function is always the same.

The filtered spectrum of the image is shown in Fig. 12. All the impulse pairs that make up the cross-hatch indications have been removed except one, the pair at $u=0$, $v=0$. This can be considered as the steady-state value of the image, and sets the background brightness level of the image.

Figures 13 and 14 show what happens when the filtered spectrum is transformed back into the image domain. The cross-hatch is removed and the cuts in the insulation are quite clear. Figure 14, in which indications of noise are somewhat reduced, is a smoothed version of Fig. 13. It should be noted that the small vertical cut evident on the left-hand side of Fig. 13 was not perceived at all on the original radiograph. A high-density inclusion is also evident in the enhanced image.

Histograms

Histogram equalization is an enhancement technique that reallocates the pixel values in an image so that equal ranges of intensity values have equal numbers of pixels. Plotting intensity values against frequency of occurrence, a histogram can be used to set threshold levels that are useful in flaw detection.

Recent work[24] has described another use for histograms, that of categorizing structures without

FIGURE 11. Magnitude of Spatial Frequency Components

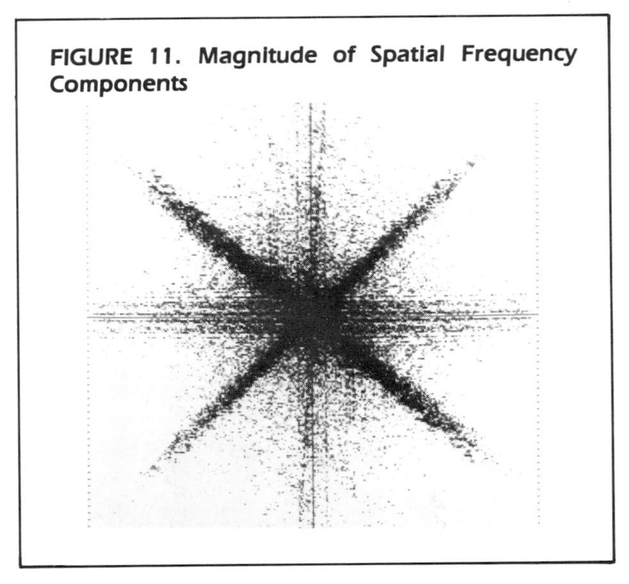

FIGURE 12. Filtered Spectrum

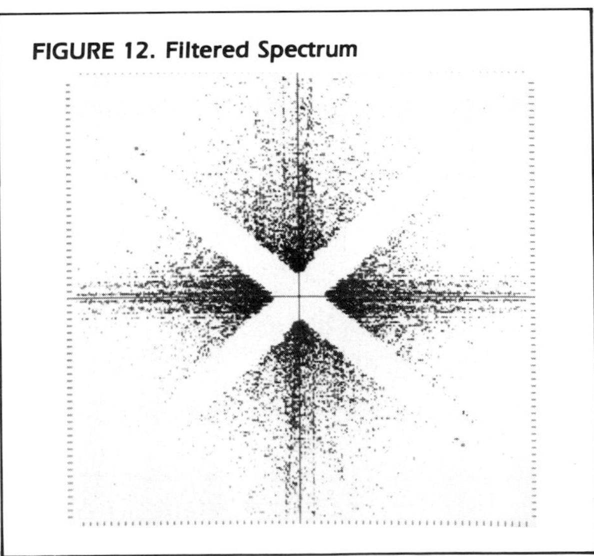

enumerating flaw-like indications. Unlike homogeneous metallic structures, adhesive bonds of graphite-epoxy composites appear to be a conglomeration of flaw-like indications — often as many as 40 to 50 per unit area as seen on an acoustic image. While it makes sense to count flaws in homogeneous metallic structures — the number of voids in an aluminum casting, for example — counting flaw-like indications may not be the best method for evaluating composite structure bonds; a global parameter such as the spread of intensity values, as measured by the coefficient of variation, might give more information about the quality of a bond than counting flaw-like indications. In Fig. 15 the acoustic image of a bonded graphite-epoxy panel is shown. The area of interest has been outlined in white. The histogram of the intensity values is superimposed on the image. The abscissa plots intensity values; the ordinate plots frequency of occurrence.

FIGURE 13. Filtered Spectrum Transformed Back to Image Domain

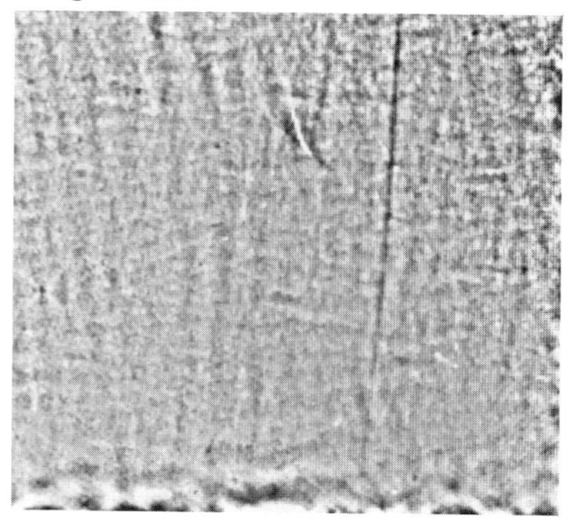

FIGURE 14. Smoothed Version of Figure 13

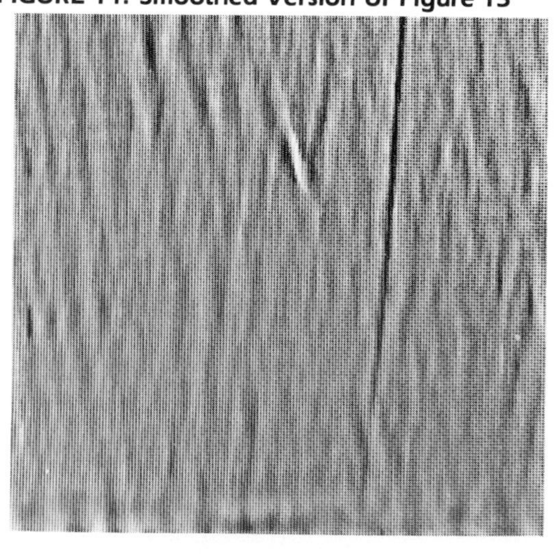

Pseudocolor Enhancement

The coding of several scale intervals with a different color results in an enhancement technique that is quite striking to the viewer but according to some workers has a drawback: sudden changes in color caused by the coding scheme often do not correspond to sudden or abrupt changes in optical density on the image. For example, a small change in the optical density along a slowly varying gradient can trigger a color change, while, on the other hand, a steep change in optical density that occurs within a gray scale interval will not cause a color change.

Pseudocolor does provide visually striking and useful imagery. The technique can be used for the analysis of film density variations when it is realized that color boundaries represent density contour lines.

FIGURE 15. Acoustic Image with Histogram

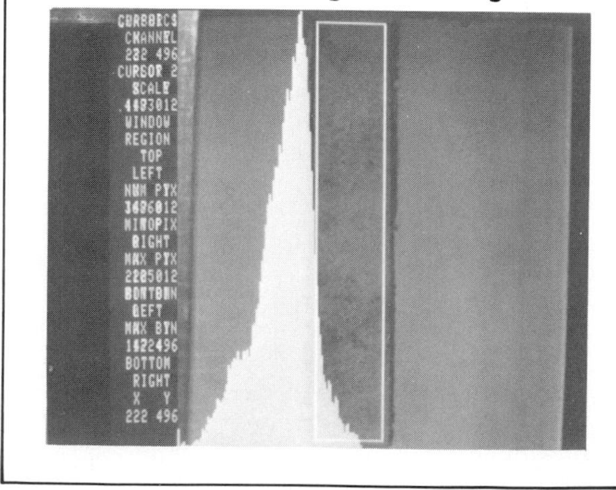

PART 4
ALGORITHMS FOR THE AUTOMATIC EVALUATION OF RADIOGRAPHS

A system has been developed under the sponsorship of the U.S. Army Armament Research and Development Command, Dover, New Jersey, for the automatic evaluation of defects found on radiographs of the high-explosive charge region of 105 mm artillery shells. Performance requirements stated that no more than ten seconds would be allowed for the evaluation of each shell image. There are normally three images on each 35 × 43 cm (14 × 17 in.) film. The analysis methods that have been developed on this program are general enough to apply to a wide variety of image analysis problems.[25-29]

The objective of the Automatic X-ray Inspection System (AXIS) program is to develop a system capable of automatically examining radiographs of high explosive shells and making accept/reject decisions based on the presence or absence of certain classes of defects in the high-explosive region of the shells. Among the defect classes are small and large cavities, cracks, pipes, annular rings, foreign inclusions, regions of high porosity, and base separation. The defect indications have low contrast against a high-contrast background, and some have dimensions comparable to the grain size of the film.

Techniques to identify and measure these defects have been developed and demonstrated on radiographs of 105 mm M-1 artillery shells using a general purpose digital image-processing system. More recently, a speedup in analysis by some 350 times has been achieved through the use of an array processor.

The equipment configuration for implementing the techniques at the required speeds is shown in Fig. 16a. The limiting blocks in the diagram are the scanner-integrator for image quality and the array processor for computational speed. Experiments indicate that use of a high-resolution 4 cm (1.5 in.) vidicon camera, followed by logarithmic digitization and nine frames of digital integration, can reduce the entire TV contribution to less than 0.006 H&D units (Hurter-Driffield units express the ratio of photo density to logarithm of exposure). Thus the system can provide sufficient image quality for successful operation of the detection algorithms. A system for analyzing acoustic images is shown in Fig. 16b. The actual hardware is pictured in Fig. 16c.

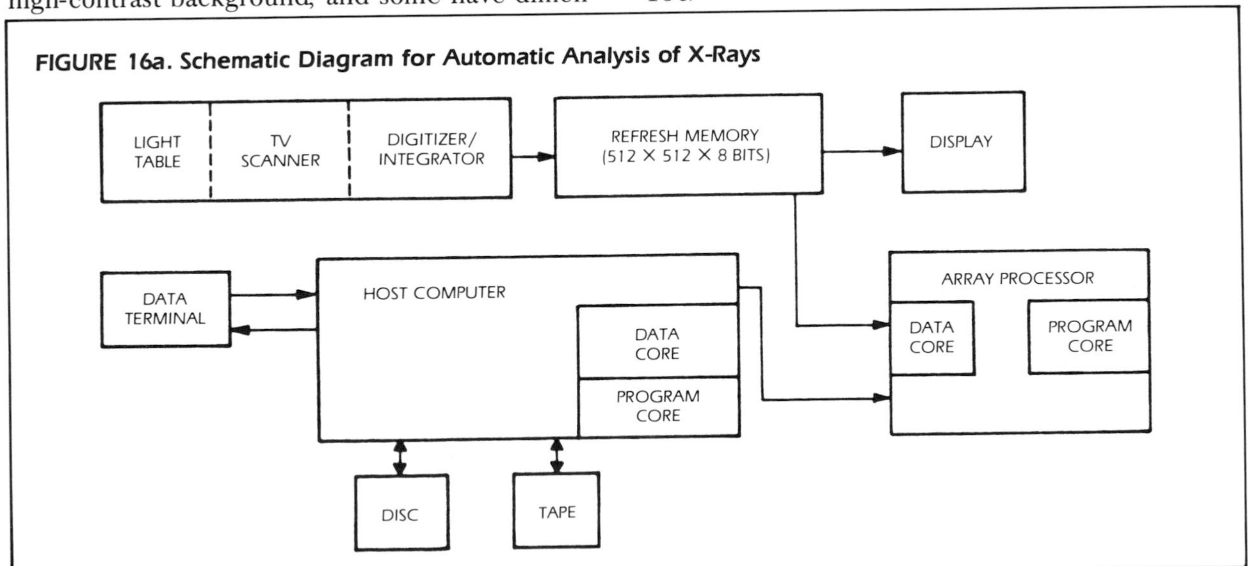

FIGURE 16a. Schematic Diagram for Automatic Analysis of X-Rays

FIGURE 16b. Schematic Diagram for Automatic Analysis of Acoustic Images

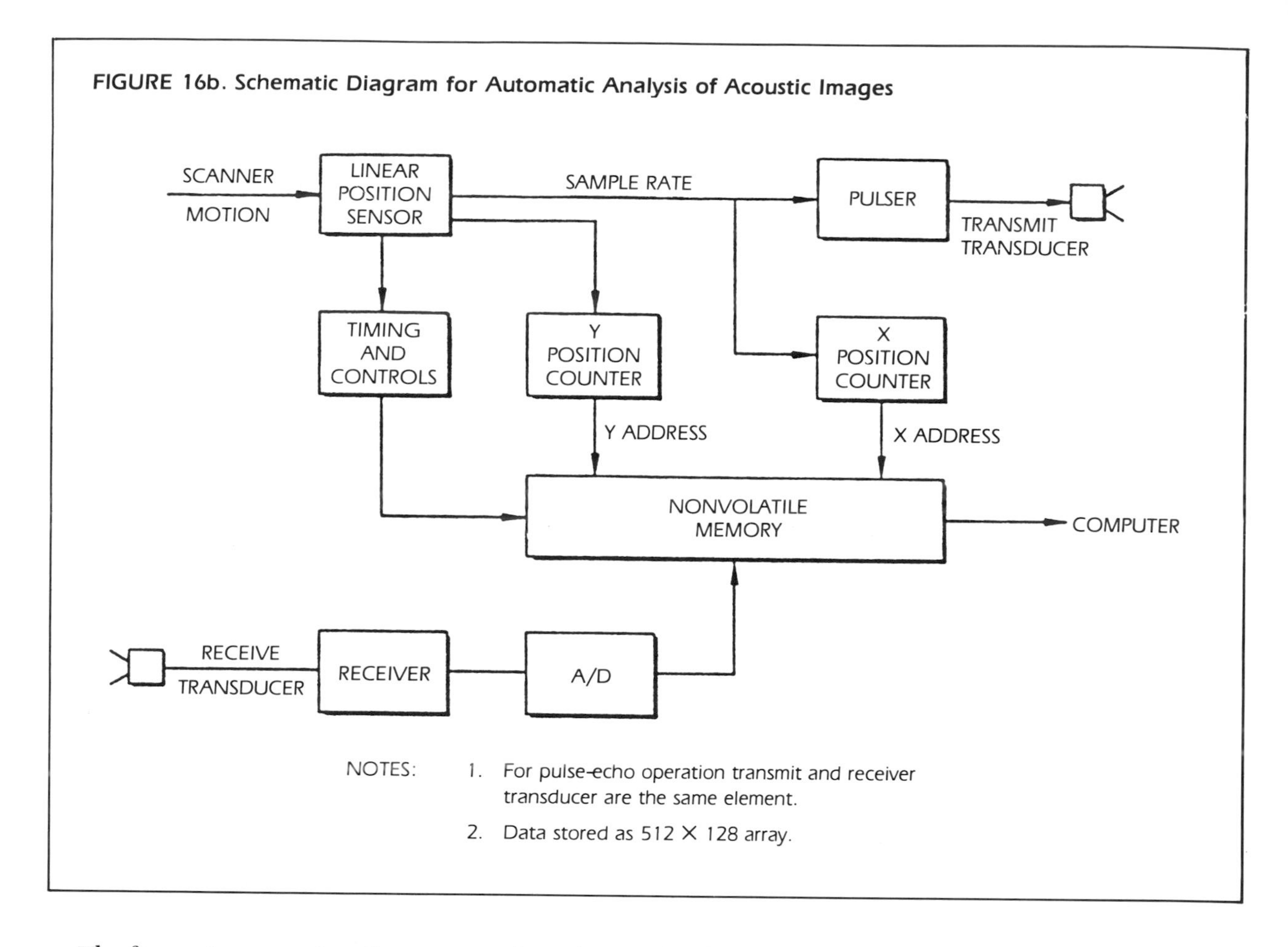

The five main categories of processing algorithms are:

1. field-flattening, to remove the intensity trend using polynomial fitting;
2. high-pass and low-pass filtering, for noise removal and for enhancement of certain defects;
3. variable threshold setting (based on histogram analysis), used to separate flaws from noise.
4. measurement of the maximum width, maximum length, and orientation of defects;
5. detection of base separation.

Field-Flattening

The image density is sampled at 500 μm (0.020 inch) intervals and is digitized to 512 levels to form a 200 by 300 pixel array; the result for a typical

FIGURE 16c. Automatic Analysis System

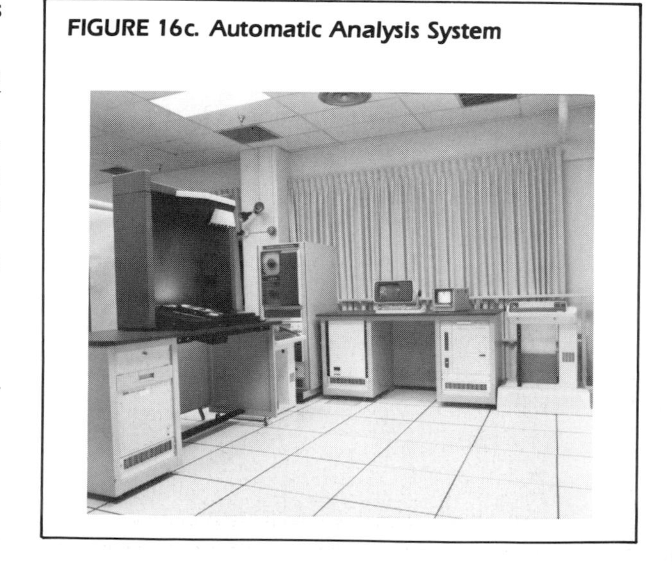

radiograph is shown in Fig. 17. Large density variations occur within the charge region and these are caused by the casing diameter and thickness of the shell. Figure 18 shows the intensity variation along the row designated by the horizontal cursor in Fig. 17. These systematic intensity varaitions are extremely large compared with intensity changes caused by flaws, and therefore must be removed before the flaws can be detected. The intensity variation with horizontal distance is much better behaved (in the sense of being approximated by a low-order polynomial) than in the vertical direction. For this reason, the field-flattening algorithm takes as its input one horizontal row at a time; it fits the intensity variation in the charge region with a polynominal, subtracts the estimated intensity from the actual intensity, and displays the difference (i.e. the field-flattened image).

The result of the field-flattening operation is shown in Figs. 19 and 20. In Fig. 19, the estimated systematic trend has been subtracted and the difference multiplied by a factor of eight; the flaws are quite apparent in this figure, while they could not be observed at all in the input image (Fig. 17). Fig. 20 gives the column average for 16 successive rows; it shows that the residual trend is less than +2

FIGURE 17. Digitized Radiograph of 105 mm shell

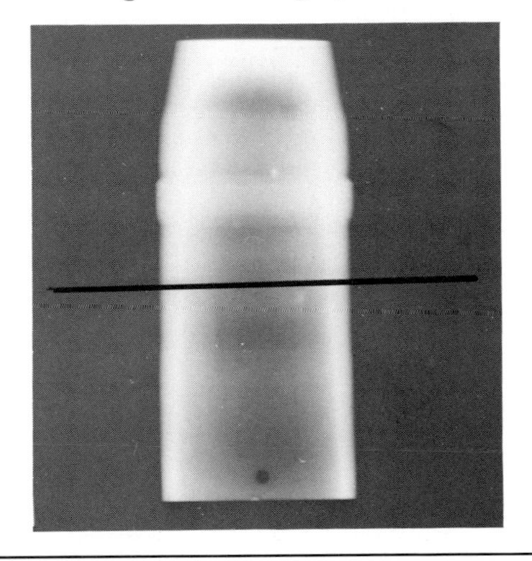

FIGURE 19. Result of Field-Flattening

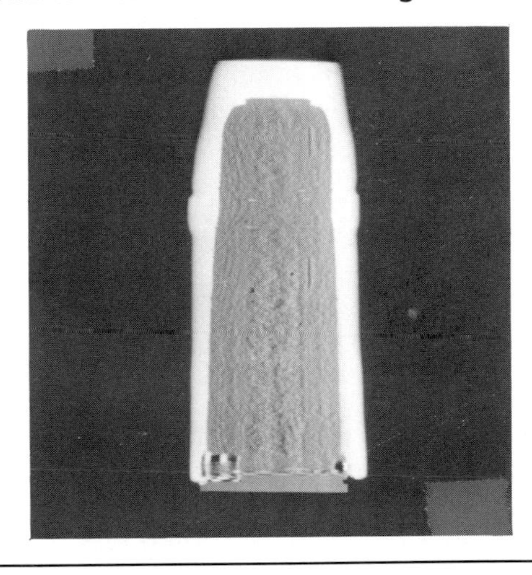

FIGURE 18. Intensity Values Along Horizontal Cursor in Figure 17

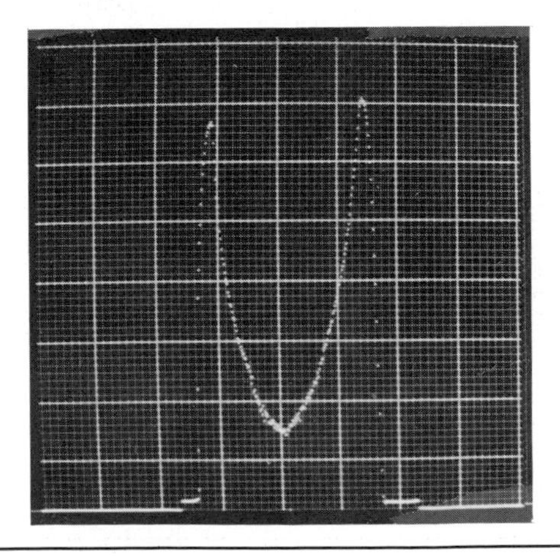

FIGURE 20. Residual Trend Remaining After Field-Flattening

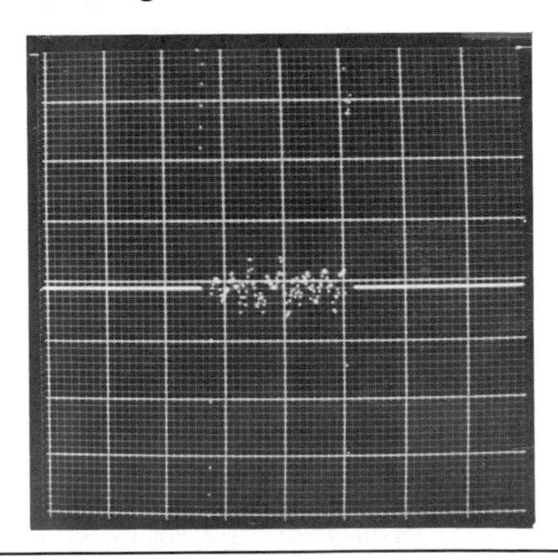

counts referred to the input image, which represents approximately ±0.006D film density, or 0.75 mm (0.03 in.) of flaw thickness.

Because of the iterations of curve-fitting required for each line, the field-flattening process is time consuming. Time savings were achieved in this example by precomputing values, for the least squares operation as much as possible, and by restructuring the fitting algorithm so that the iterations did not require entirely new computations.[28]

Low-Pass Filtering

The noise caused by film granularity and television electronics is large compared with that resulting from field-flattening residuals. The noise is characterized by high spatial frequency (i.e. only a small amount of correlation between adjacent pixels), and therefore can be reduced by spatial low-pass filtering. However, care must be taken to choose filter parameters which reduce the noise as much as possible without significantly reducing the amplitude of variations caused by flaws. The filter chosen was a 3 × 3 pixel moving-window average. The result of low-pass filtering is given in Figs. 21 and 22, each of which shows that the flaws are still clearly visible and the noise is markedly reduced. The grid superimposed on Fig. 21 will be explained in the section on threshold computation.

The reasons for choosing a 3 × 3 pixel filter are:

1. The moving-window average can be computed faster than other more general filter functions.
2. The 3 × 3 pixel size is compatible with the resolution of the television camera (the diameter of the television image spread function is 2.5 pixels).
3. The 3 × 3 pixel low-pass filter reduces the random noise by a factor of (almost) 3, resulting in less noise than that caused by field-flattening residuals.

Thresholding Operation

Pixels revealing cracks, cavities, and abnormal porosities have intensity values which are lower than "normal"; pixels revealing foreign inclusions have intensity values which are higher than "normal" (see Figs. 21 and 22). This is a problem which is complicated by the fact that "normal" values (i.e. threshold values outside of which a pixel would be considered to reveal a defect) vary throughout the charge region. The principal causes of the threshold variations are the spatial variation of the noise and residual trends left after field-flattening.

To compute these spatially dependent thresholds, the charge region is divided into 25 × 25 pixel areas, as shown by the grid in Fig. 21. The window was sized large enough to allow statistics that are

FIGURE 21. Result of Low-Pass Filtering on Field-Flattened Image Using a 3 by 3 Pixel Moving-Window Average

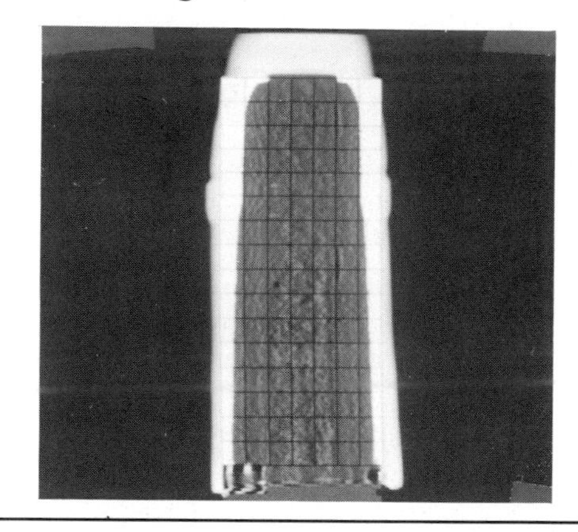

FIGURE 22. Intensity Values Along Horizontal Cursor After Low-Pass Filtering and Field-Flattening

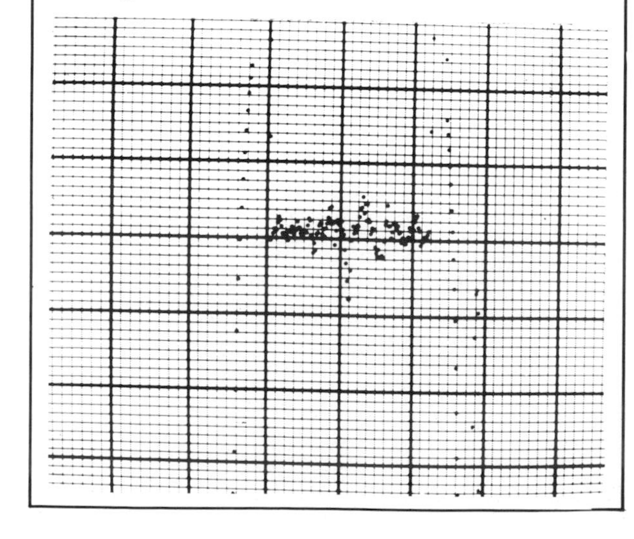

not unduly influenced by flaws, and small enough to allow statistics that do not vary widely within the window. For each window of the field-flattened image, the histogram of the 625 intensity values is computed and then used to determine which intensity values are greater than 10% of the viewed pixels.

The thresholding operation takes as its inputs the (filtered) field-flattened image and the position-dependent threshold values, and produces as output an image in which all pixels with intensity less than the lower threshold are marked with value 501; those above the upper threshold are given the value 502. The Defect Processing System uses these marks to distinguish between foreign inclusions, cracks, cavities, etc. Pixels having values between these two thresholds (those in the "normal" range) are given the value zero.

FIGURE 23. Histogram and Cumulative Histogram Used to Obtain Threshold Setting

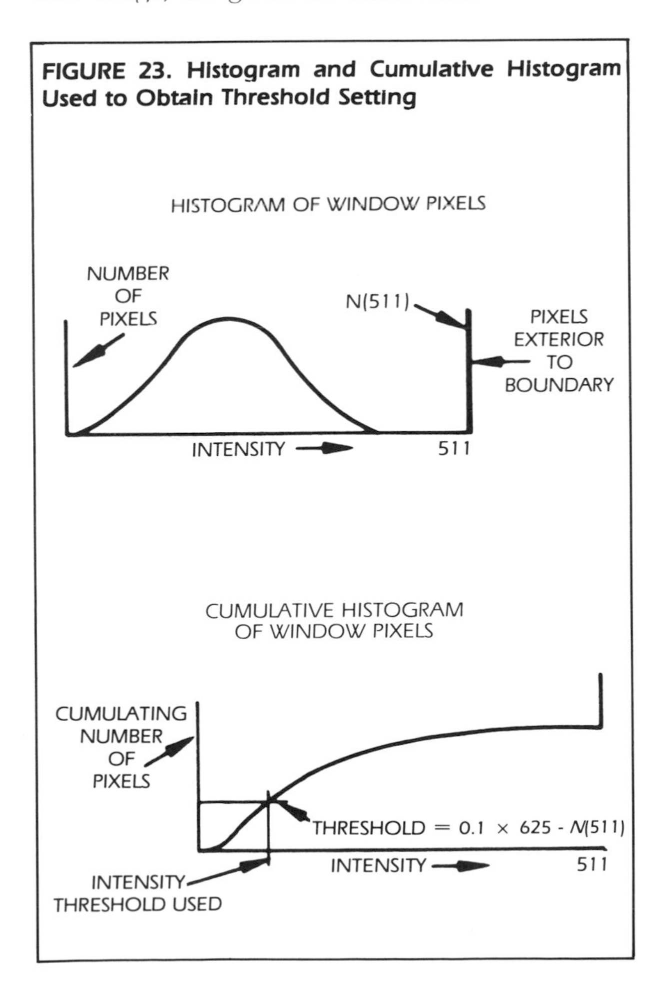

Figure 23 shows a typical histogram and cumulative histogram of intensities in a 25 × 25 pixel window. The threshold is set at that intensity value at which the cumulative histogram reaches 10% of the pixels in the window (where ten percent of the pixels are considered as candidate defect pixels).

To handle the case where the boundary of the shell cuts through portions of the window, the region exterior to the boundary is set at a value 511, which results in a spike in the histogram at 511 (as shown in Fig. 23). The number of pixels marked 511 in the window is then taken into account in the threshold computation by setting the (cumulative number of pixels) threshold at $0.1 \times 625 - N(511)$.

Defect Measurement

Defect pixels are grouped together based on a *line adjacency criterion* in which contiguous defect pixels which overlap from row to row are considered to belong to the same group. The perimeter pixels of connected groups are obtained in the host computer, starting with the binary representation shown in Fig. 24. Each group is then analyzed for maximum width, maximum length, and orientation, using the array processor and the *dot product* approach.

The dot product approach is based on the following observation: if a line connecting two boundary pixels is projected onto a ray of unit vectors

FIGURE 24. Binary Image Showing Pixels Exceeding Local Thresholds

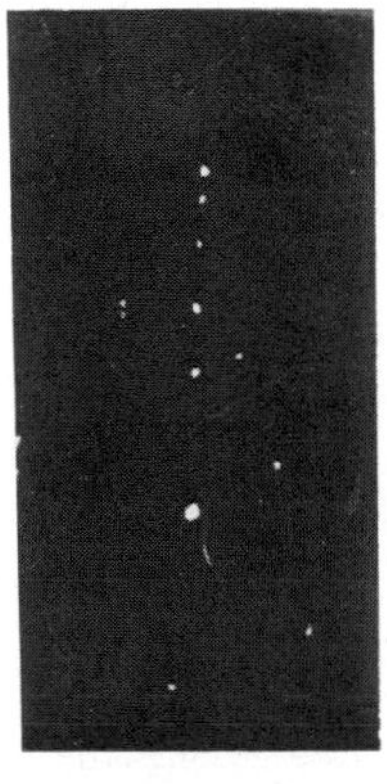

FIGURE 25. Association Algorithm

Adjacent pixels and pixels that overlap from row to row are considered to be in same defect group.

ORIGINAL MARKED PIXELS			ASSOCIATED DEFECT GROUPS	
XX	XX		1 1	2 2
XX			1 1	
		→		
X			3	
X			3	
X			3	

radiating from the origin, then the largest projection of the line is obtained for the unit vector most parallel to the line. Thus, a set of unit vectors is used, 10 degrees apart and radiating from an origin, taking the dot product of every boundary pixel in a defect group and each unit vector. Then the maximum absolute difference between the smallest and the largest dot product for each unit vector is computed. The largest difference gives the maximum length of the defect; the unit vector normal to the orientation gives the maximum width. A report on defects whose measurements exceed specifications is printed out (see Figs. 25 and 26).

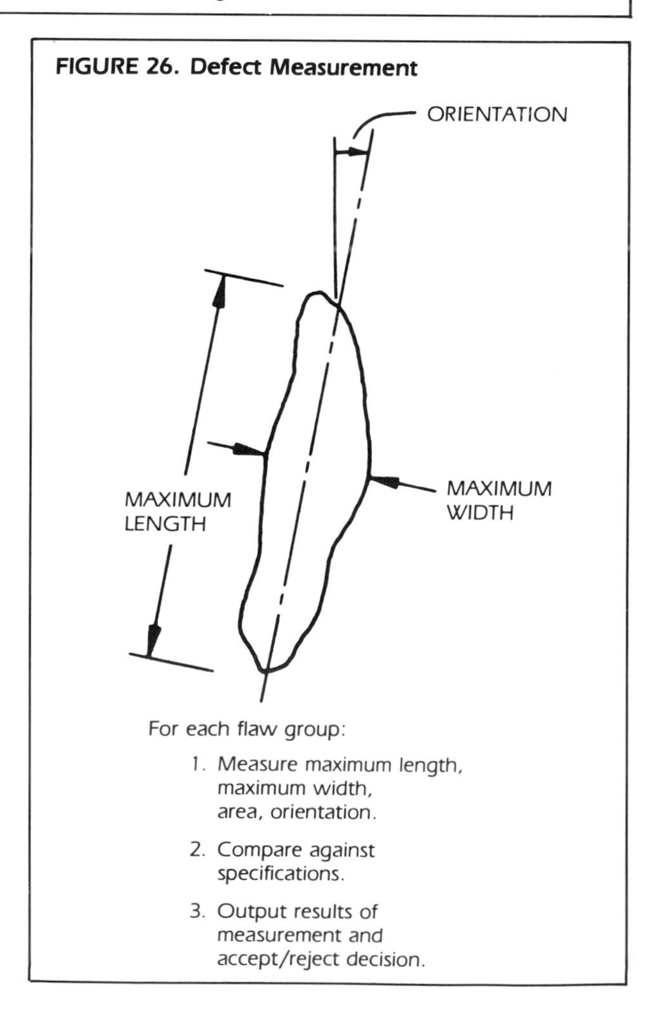

FIGURE 26. Defect Measurement

Detection of Base Separation

Base separation is a type of defect which occurs in artillery shells when the explosive separates from the base of the metal casing. This subtle flaw must be detected in the presence of a dominant intensity trend caused by the explosive/metal transition and random noise caused by film granularity and television electronics. The detection is complicated by the fact that the flaw is quite narrow, approximately 0.25 mm (0.01 in.) wide. For this reason, a magnified television image is used in which the pixel spacing is 125 μm (200 pixels/inch). The base-separation algorithm is described in Fig. 27.

One problem which can occur in the case of very wide flaws is that the field-flattener mistakes the flaw for the intensity trend and removes it. For this reason another set of operations, referred to as Major Defect Detection, is necessary to find horizontal flaws such as transverse cracks and annular rings.

Unsharp Masking for Major Defect Detection

The primary difference between unsharp masking and the operations previously described is the substitution of a spatial high-pass filter for the field-flattener. The high-pass filter develops an output image in which the output intensity is the difference between the input intensity (at the corresponding pixel) and the average of input intensities in a window *W* wide and *H* high. For the results which follow, the values used were $W = 1$, $H = 7$. This imaging process has the effect of differentiating in the vertical direction so that the process is sensitive to horizontal flaws. Figure 28 shows the result for large transverse cracks. Note that it is only the outlines (rather than the included areas) which are detected; this is because the differentiation process is blind to steady-state values.

Figures 29 and 30 show the trend and the noise, respectively, for the same 16 rows that were used for Fig. 20. Like the field-flattener, the high-pass filter gain was set to eight. Comparison of Figs. 20 and 29 shows that the high-pass filter is more effective in eliminating the trend than the polynomial field flattner. Figure 30 shows that the random noise outputs are almost identical.

As explained previously, the random noise component can also be reduced by a low-pass filter. Moving-window averages with window sizes of 3×3 pixels and 9×1 pixel were investigated. The wide/narrow window was expected to enhance horizontal flaws, but its performance was no better than the 3×3 pixel window; therefore, the square window was used.

FIGURE 27. Base Separation Detection

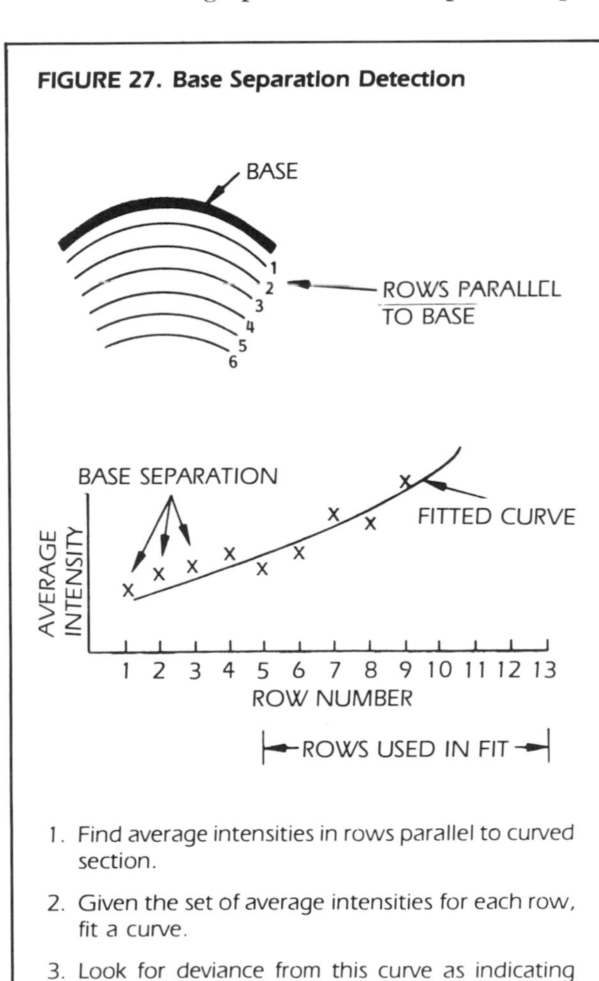

1. Find average intensities in rows parallel to curved section.
2. Given the set of average intensities for each row, fit a curve.
3. Look for deviance from this curve as indicating base separation.

FIGURE 28. Result of High-Pass Filter

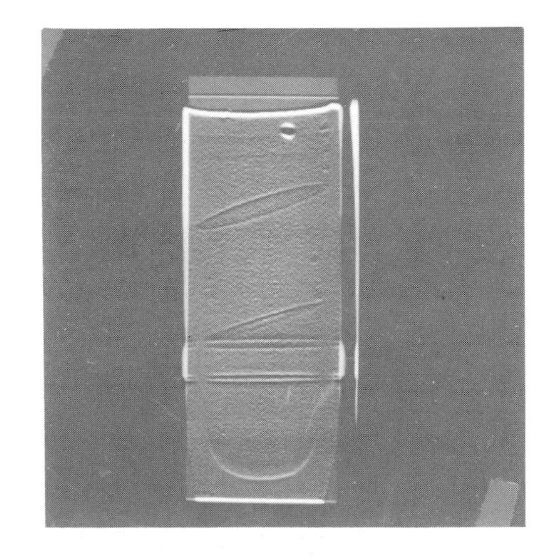

FIGURE 29. Residual Trend Remaining After High-Pass Filtering

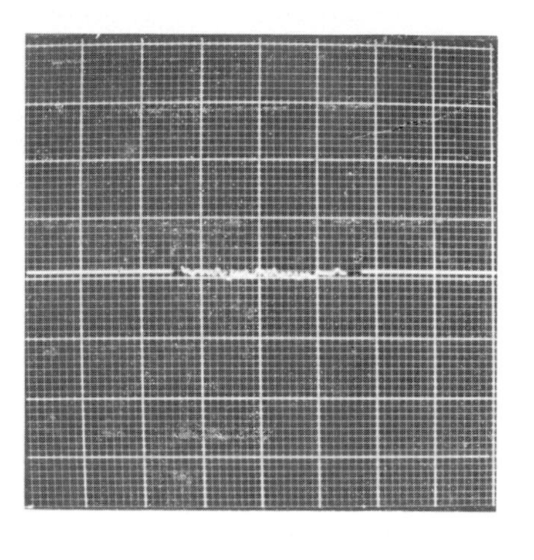

FIGURE 30. Standard Deviation of Noise Remaining After High-Pass Filtering

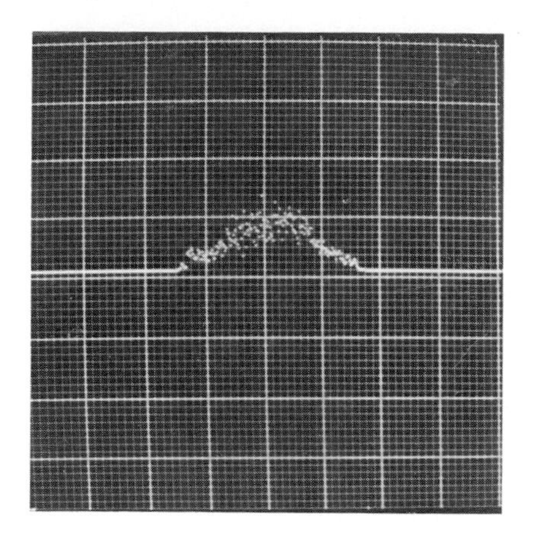

PART 5

HARDWARE FOR IMAGE DATA PROCESSING

The image processing techniques and algorithms for automatic analysis described here can be carried out entirely by software routines programmed into any one of a number of computers, from relatively slow minicomputers to extremely fast array processors. It is also possible to hardwire the processing routines into modules that become a permanent part of the image processing system. What is lost with this approach is versatility; what is gained is a good balance between hardware cost and processing speed. Changing, expanding, and debugging is often easier with a software program than a hardwired module. Hardwired modules cost much less than general purpose array processors; the goal is a proper mix of hardware and software to minimize cost and maximize computational speed and efficiency.

Input Device

By far the most important piece of hardware in a digital image processing system is the input device (scanner). The role of the scanner is to convert the image into a digital signal that can be stored in memory for further processing. These hardware units come in a wide variety of shapes and sizes, including: television cameras; flying spot scanners; solid state array scanners; optomechanical scanners; and scanning microdensitometers.

The characteristics of some of these units are compared in Table 1. As shown in the table, for a more leisurely acquisition of images the scanning microdensitometer can be used to make precise measurements of data contained in transparencies. A spot size as small as 12 μm (0.0005 in.) is imaged by a microcope onto a detector. The image to be analyzed is mounted on a flatbed, and it is this bed that is moved back and forth through the light path. To maintain dimensional stability the optical train remains stationary. The analog signal that represents optical density on the film is digitized and stored in a memory where it will remain available for further processing. The main purpose of a scanning microdensitometer is to optimize positional accuracy and photometric accuracy. This requirement results in equipment that is expensive and slow.

The TV camera, or its solid state array counterpart, scans images quite rapidly— but a poor signal-to-noise ratio is the result. As a compromise between speed and accuracy, the drum scanner has

TABLE 1. Scanner Characteristics (Typical Units)

Type	Scanner Resolution	Dynamic Range	Acquisition Speed	Signal Noise	Field of View
Scanning microdensitometer	40-400 lines/mm (10^3-10^4 lines/in.)	0-4 H&D	30 lines/min	EXCELLENT	10 × 2.5 cm (4 × 1 in.)
Optomechanical scanner	10-40 lines/mm (250-1000 lines/in.)	0-3 H&D	150 lines/min	EXCELLENT	25 × 25 cm (10 × 10 in.)
CRT flying-spot scanner	5-50 lines/mm (125-375 lines/in.)	0-3 H&D	1200 lines/min	GOOD	10 × 10 cm (4 × 4 in.)
Television camera	2-125 lines/mm (50-3200 lines/in.)	0-3 H&D	15,000 lines/sec	POOR	25 × 25 cm (10 × 10 in.)

some advantages over the flying-spot scanner. The drum scanner consists of a spinning drum on which the film to be scanned is mounted. A collimated light source and detector can be moved parallel to the drum axis to scan consecutive lines on the film. (If it is necessary to scan many films in the shortest possible time, then the TV scanner, despite its drawbacks, must be used.)

A typical radiograph has a density range of 0.8 to 3.5 H&D units. The noise contributed by the preamp of the TV camera increases exponentially with density. Integrating many images reduces the TV noise. A faint defect is about 0.0006 H&D units; thus, for a radiograph having a range of 0.8 to 3.5 H&D units, 256 integrations are required to detect the faintest defects.

TV Scanner Requirements

A TV scanner consists of a TV camera, TV tube, analog-to-digital converter, and a light source to illuminate the film.

The scanner must have sufficient dynamic range to simultaneously measure light intensities transmitted through film optical densities in the range of 0.8 to 3.5 H&D units. Also, the camera output voltage should be roughly proportional to optical film density. This ensures that each portion of the radiograph receives adequate gray shade counts after analog-to-digital conversion. Finally, TV scanner noise contributed by the TV scanner should be less than 0.006 H&D units when scaled to optical density. This performance parameter may be obtained by integrating several TV frames, thereby increasing the signal-to-noise ratio.

Display Memory and Monitor

Once the image is acquired and digitized, it has to be stored and then made available for further processing. This can be done with a display memory, a stack of random-access, non-volatile core memories, each of which stores an image in digital form and can be used to refresh a TV display of the stored image.

A useful format for the memory is one that stores 512 lines at 512 pixels per line with 8 to 9 bits per pixel, and can accommodate from one to five images.

Table 2. Timing for Major Processing Steps (200 × 300 pixel image)

Processing Step	Minicomputer Time	Array Processor Times
Field-Flattening	44 min	3 - 6 sec
Filtering	2 min	1 sec
Thresholding	9 min	1 sec
Defect Measurement	3 min	4 sec
Base Separation	1 min	1 sec

A TV monitor or other CRT device is needed to display the image stored in memory. Resolution of 500 to 1000 lines is usually adequate. The display may be either black and white or color.

Digital Computer and Peripherals

A computer with core memory of 32-64K words and a cycle time of less than 1 μs is required to control the processing and to perform mathematical transformations on the image data.

A *terminal/keyboard* is used by the operator to input commands and parameters to the computer. *Disc memory* is used to store digital data representing processing routines along with original and processed images. A *tape drive* and *magnetic tape* can also be used to store digital images.

Array Processor

The dramatic speedup obtained through the use of an array processor is shown in Table 2. A comparison of the time needed to execute the algorithms for analysis of 105mm shells is presented, first for a minicomputer programmed in FORTRAN and then for an array processor.

The array processor achieves its speed in three ways:

1. the use of extremely fast hardware floating point arithmetic operations;
2. the use of "pipelining" or parallel operations on serial data;
3. a systematic data arrangement which eliminates the need for software indexing operations.

FIGURE 31. Typical Hardwired Image-Enhancement Module

Many of the more useful image processing routines are available as hardwired modules.[30] A typical unit that contains nine functions is pictured in Fig. 31.

Requirements for Analog Video and Digital Interfaces

An additional capability is often required in an image processing facility — that of inputting and outputting image data on videotape.

The videotape input capability is especially useful when image data is to be collected at a remote site for later processing in the laboratory. Upon playback, with a videotape recorder, an analog video signal is produced which can be sampled, quantized, and stored in the digital image memory.

There are two required characteristics of the videotape recorder. First, the reproduced signal must be in a 525-line interlaced format. Composite picture and sync signal should be in EIA RS-170 format (U.S. commercial TV). The videotape recorder also must have a capstan servo to control tape speed. This servo must be capable of synchronization of the tape motion to an external sync signal if the videotape data is to be transferred to computer memory.

PART 6

MEASUREMENTS BETWEEN FUZZY INTERFACES

Automatic techniques to measure distances between well-defined interfaces have been in use for some time. The purpose of the work described here was to develop a computer-based technique to measure the distance between two fuzzy interfaces imaged on a radiograph. The term "fuzzy" has received some attention in the literature that deals with ill-defined sets[31, 32]; also in this category is the problem of precisely locating an edge on a fuzzy radiograph.

The term "unsharpness," has been used in radiography to quantify fuzzy images. Recent work,[33] however, has emphasized the more exacting method of measuring the optical transfer function (OTF) or the modulation transfer function (MTF) of the radiographic system as a means of quantifying the spatial frequency. Every radiographic system has a given frequency response. At the response limit, variations in X-ray exposure, caused by variations in subject contrast at very closely spaced increments along a spatial coordinate, produce no corresponding changes in photographic density on the X-ray film. The MTF is a measure of the spatial frequency response and is usually presented as a plot of the range of spatial frequencies, in lines per millimeter, versus the change in photographic density produced. MTF's can vary widely from system to system depending on source and film parameters. Low MTF values imply fuzzy edges. Note: MTF = OTF when the point spread function of a system is symmetrical.

The problem of defining the edge of an object imaged on a fuzzy radiograph is basically one of, first, deciding what value of photographic density (or photographic density ratio) should be used as an edge designator, and, second, devising a scheme for locating around the fuzzy edge those points that have the designated values.

Various experimentally derived photographic density values have been used as edge designators. Two recent reports[34, 35] approach the problem from fundamentals, in effect calculating the photographic density that designates an edge based on radiographic considerations. No matter how the edge designator value is determined, there is still the problem of finding these density values on the radiograph and then using them to define the edge. When the radiograph has good spatial frequency response (a high MTF value), the process of finding the edge designator points is relatively simple. However, when the opposite is true (i.e., when the MTF as well as the overall contrast is low), the problem of locating the edge designator points becomes somewhat more complex.

Measurement Technique

The requirement is to measure the liquid level in a sealed container. The measurement, to be repeated at regular time intervals, is needed to detect fluid loss and must be accurate to ± 0.4 mm (0.02 in.). To overcome any subjectivity in locating the liquid boundaries on the X-rays, which would make comparison from measurement to measurement somewhat misleading, a consistent and rapid computer-based method is needed.

The container is positioned so that the radiograph gives a clear, uncluttered view of the liquid-gas interface. In Fig. 32, the area of interest is the triangular region of the container's upper left. The solid line in the figure, which looks like the perpendicular bisector of the triangle, runs from the *apex* to the *liquid-gas boundary.*

The radiograph of the container is scanned with a TV camera, digitized, and stored in memory. Measurement of the liquid level consists first in finding the liquid-gas interface and the apex of the container, and then in measuring the minimum length between them (Fig. 32). Accurate location of the edge designator points comprised the major diffi-

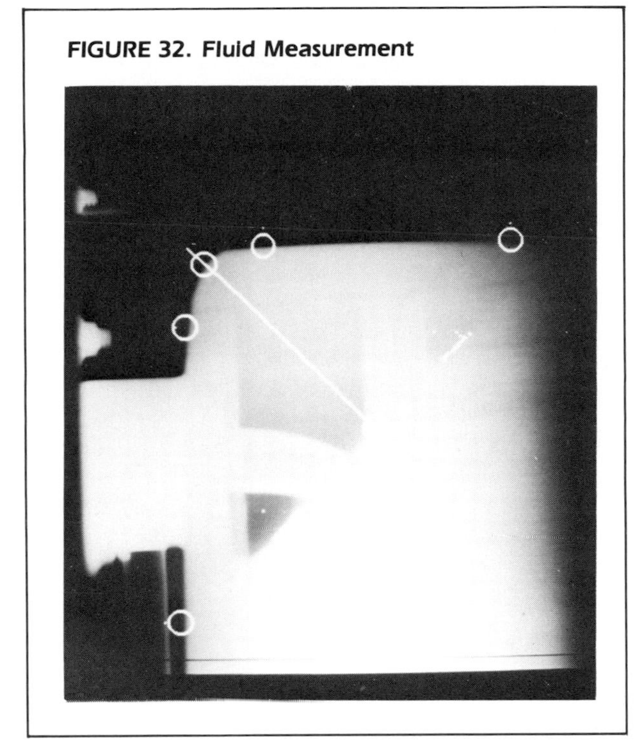
FIGURE 32. Fluid Measurement

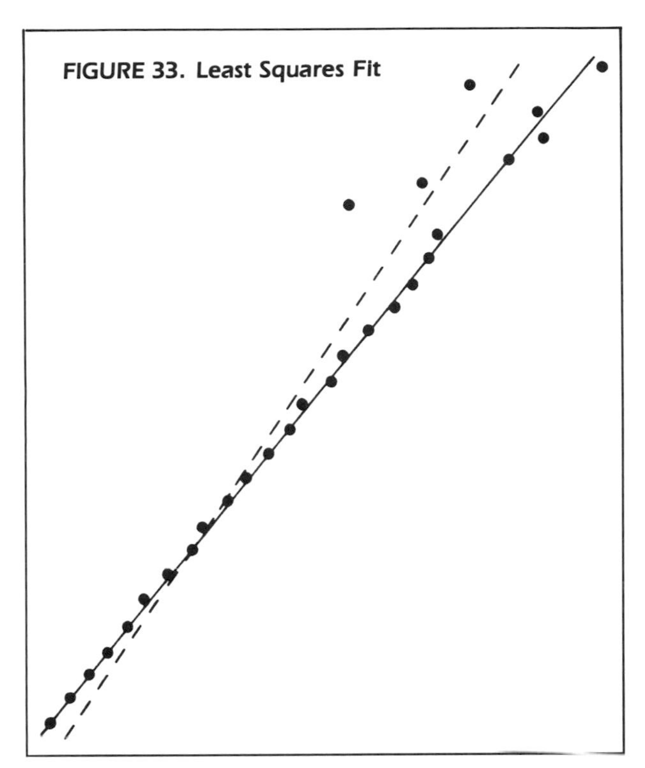
FIGURE 33. Least Squares Fit

culty in obtaining precise measurements. Several methods were devised for each of the two edges until algorithms were obtained that gave consistent measurements.

A maximum-derivative detection scheme was used first to locate the liquid-gas boundary. The algorithm compensated for noise by using more points to calculate the derivative as the signal-to-noise ratio (S/N) decreased. This worked well on high-contrast images with good S/N. With low S/N images, the algorithm broke down. When a large number of points were used to estimate the derivative, the liquid boundary was rotated into the high contrast region. When fewer points were used to overcome this bias, the method gave rise to noisy point estimation.

This algorithm calculates derivatives in two swaths where the liquid boundary has the highest contrast. A least-squares fit on this data is used to define the boundary as a linear function. Wild points induced by image noise were always located near one swath on the high contrast side of the boundary. This bias caused the calculated boundary line to rotate at a small angle with respect to the true boundary (see Fig. 33). An attempt was made to eliminate the biased wild points by establishing a threshold for the maximum distance of a wild point from the least-squares line. Any point beyond this threshold would be rejected and a new fit would be made using the remaining points. This threshold had to be narrow enough to eliminate wild points that lay close to the boundary, but it also had to be wide enough so that none of the actual points that made up the boundary would be eliminated. No single threshold could be found, since one wild point far from the actual boundary could influence the line-fitting to such an extent that any threshold based upon the line would be meaningless.

The next attempt sets up multiple thresholds. In the first step, only the remotest wild points are rejected and a new line fitted. The second and third steps refined the fit. Experiments showed that a three-step threshold provided satisfactory liquid-gas boundary definition in low S/N images (see Figs. 34-36). The next problem was to locate the apex point of the container.

The first step was to establish a corner point where the edges of the container intersected; this corner was then used as a starting point to find the actual apex.

FIGURE 34. Multiple Thresholding Step 1

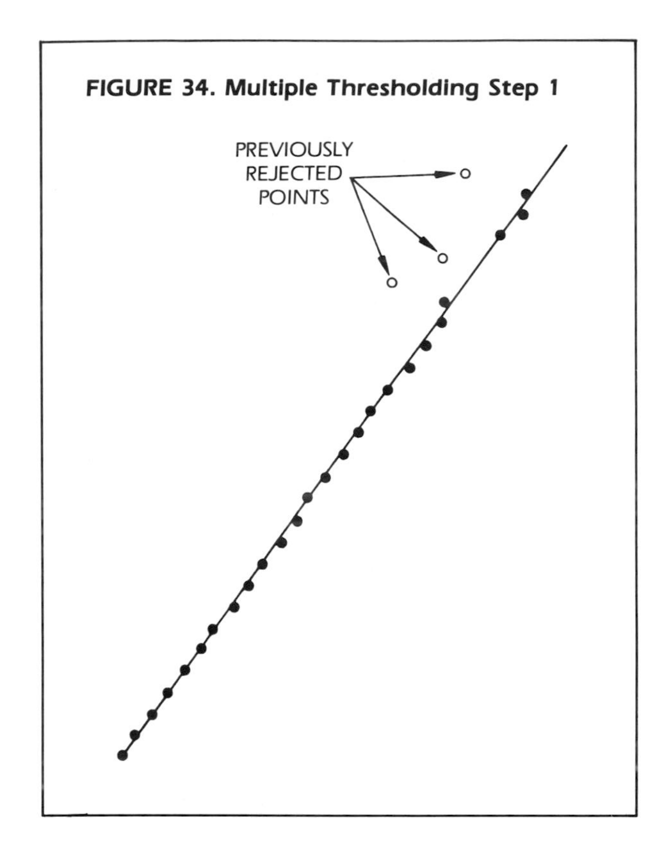

FIGURE 35. Multiple Thresholding Step 2

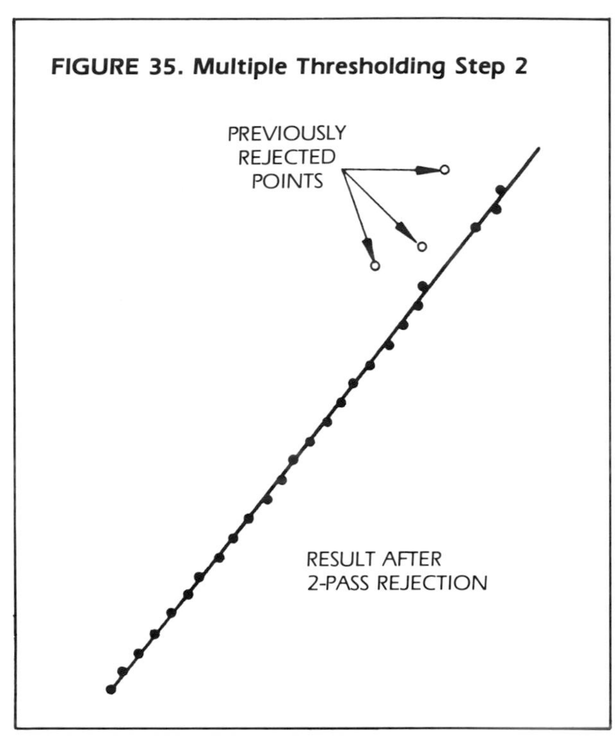

FIGURE 36. Multiple Thresholding Step 3

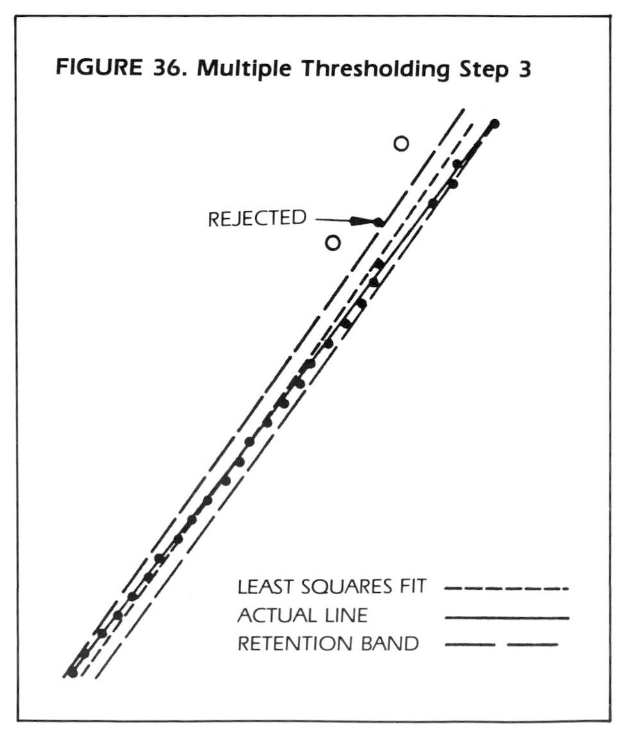

A perpendicular was constructed from the corner point to the liquid boundary and the maximum derivative along this line was calculated. The location of the maximum derivative can be used as the apex point on images with a relatively high MTF. When the MTF is low, the derivative across the boundary is almost constant. As before, several scans through the region of interest were used to establish a series of maximum derivative points. These were used to form a least squares quadratic approximation, and the point of minimum slope on this quadratic was used as the apex point designator. While this algorithm was more stable than the single line scan, it still contained errors.

The next algorithm that was tried was based on the observation that the boundary lines (used to establish the corner) separate two regions of constant density. A histogram of the pixel values near the boundary edge showed two distinct peaks corresponding to these two regions. The trough between these two peaks contained the boundary edge; the average value of the points in the trough

Figure 37. Histogram of Intensity Levels

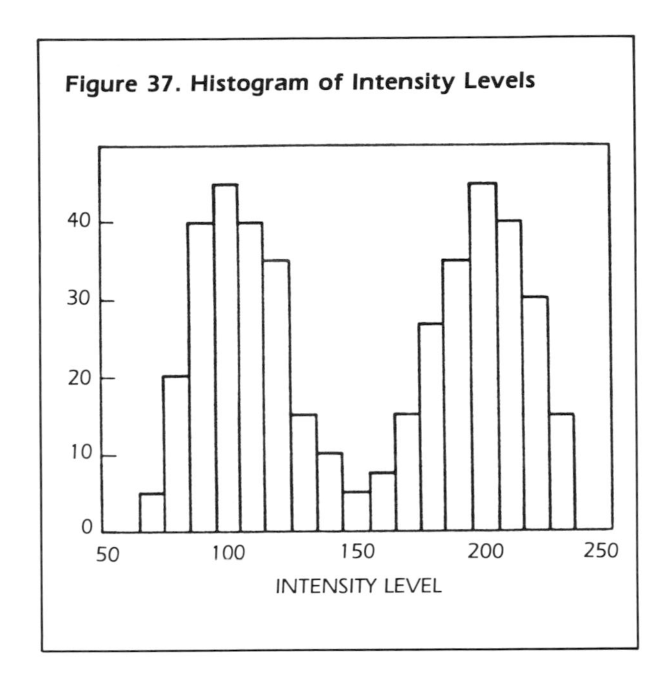

was used to locate the boundary edge (see Fig. 37). Scans were then made through the corner to map out these points to sub-pixel accuracy. A quadratic least-squares fit was made to these points and, as before, the point of minimum slope on the quadratic was used as the apex point designator. A measurement was then made along a perpendicular from apex point to liquid-gas boundary.

This method for apex point location was successful because it was not sensitive to the character of the second derivative across the edge. Simple maximum derivative techniques require that this second derivative be a smoothly varying, non-zero function.

Another area of concern when attempting to make precise measurements by using a television camera in the data-gathering system is that of fixed distortions induced by the camera system. These distortions include simple bilinear distortions which are generally produced by the camera electronics and pincushion distortion which is generated in the optical portion of the camera. Pincushion distortion is particularly noticeable near the edge of images where the paraxial lens approximations fall apart, and this constitutes a limit to the maximum area which is usable for making precise measurements.

Bilinear distortion is eliminated through the use of a nonlinear coordinate transformation. The points required for calculating coefficients in the transform are generated by viewing a grid with the camera system and then finding vertices of the grid to sub-pixel accuracy by using quadratic interpolation. The x, y pixel positions are then related to actual object positions through knowledge of the spacing on the grid.

The transform does not compensate for pincushion distortion; therefore, it was necessary to add two cubic terms to ensure an accurate transformation.

When a radiograph is viewed, each x-y pixel location is converted to the transform coordinate system to ensure linearity in least-squares fits and to ensure accurate and calibrated measurements.

Conclusion

There is no question that image data processing, microelectronics, robotics and artificial intelligence are growth industries of the future. All of these disciplines fit into the wider sphere of computer-related technologies which sooner or later will influence how we all do our jobs.

More and more industries, service companies, medical facilities, space and military operations are making use of images obtained in large quantities from many sources. These images must be analyzed, and the computers used for this analysis could eventually become as commonplace as telephones in the work place.

The field of nondestructive testing, in some of its applications, will gradually change from a specialist centered occupation—one that relies on experienced operators to interpret visual data, much as the medical profession relies on specialists for diagnosis—to a computer-based occupation, centered on interactive expert systems for image data analysis.

REFERENCES

1. Shannon, C.E. "Communication in the Presence of Noise." Proc. Institute of Radio Engineers, 37:10 (1949).
2. Whittaker, E.T. "On the Functions Which Are Represented by the Expansions of the Interpolation Theory." Proc. Royal Society, Edinburgh, Sect. A, 35:181 (1915).
3. Hartman, W.K. "Moons of the Outer Solar System." *Smithsonian* (January 1980): pp 36-47.
4. Gore, R. "What Voyager Saw." *National Geographic.* Vol. 157, No. 1 (January 1980): pp 2-29.
5. Hunt, B.R. "Digital Image Processing." Proc. Institute of Electrical and Electronics Engineers (IEEE). Vol. 63 (April 1975): pp 693-708.
6. Hall, E.L., et al. "A Survey of Preprocessing and Feature Extraction Techniques for Radiographic Images." IEEE Trans. Computers, C-20 No. 9 (September 1971): pp 1032-1044.
7. Trussell, H.J. "Processing of X-Ray Images." Proc. IEEE. Vol. 69 (February 1981). Contains an extensive bibliography.
8. Janney, D.H. "Image Processing in Nondestructive Testing." Nondestructive Evaluation of Materials (1976 Sagamore Army Materials Research Conference Proceedings, Burke and Weiss, Eds.). Vol. 23 (1979): pp 409-420.
9. Janney, D.H., and R.P. Kruger, "Digital Image Analysis Applied to Industrial Nondestructive Evaluation and Automated Parts Assembly," *International Advances in Nondestructive Testing.* Vol. 6 (1979): pp 39-93.
10. Harrington, T.P., and P.G. Doctor. "Data Analysis Methods for Nondestructive Evaluation." Battelle Pacific Northwest Laboratories Report No. BN-SA-1056 (October 1979).
11. Nevatia, R., ed. "Image Understanding Systems and Industrial Applications." Conference held August 30-31, 1978, San Diego, California. Proc. Society of Photo Interpretive Engineers (SPIE). Vol. 155 (1978): pp 177 ff.
12. Demuth, H.B. "Analysis of Fuel-Bundle Radiographs Using Modeling." Los Alamos National Laboratory Report LA-6011-MS (July 1975).
13. Hunt, B.R. "Digital Image Processing: the End of the Home-Builts." *Optical Engineering.* Vol. 19, No. 1 (Jan.-Feb 1980): pp SR-009 and SR-010.
14. Pratt, W.K. *Digital Image Processing.* New York: John Wiley & Sons Inc. (1978).
15. Goodman, J.W. *Introduction to Fourier Optics.* New York: McGraw Hill (1968).
16. Bracewell, R. *The Fourier Transform and Its Applications,* 2nd ed. New York: McGraw Hill (1978).
17. Bracewell, R. *Two Dimensional Imaging.* Class notes, EE 261, Stanford Univ. (Autumn 1981).
18. Brooks, R.A., and G. Dichiro. "Principles of Computer Assisted Tomography (CAT) in Radiographic and Radioisotopic Imaging," *Physical Medical Biology,* Vol. 21, No. 5 (1976): pp 689-732.
19. Kruger, R.P., and T.M. Cannon. "The Application of Computed Tomography, Boundary Detection, and Shaded Graphics Reconstruction to Industrial Inspection." *Materials Evaluation.* Vol. 36, No. 5 (April 1978).
20. Peck, H. "Computed Tomography, a Major Advance for Solid Propulsion Motor Evaluation." Proc. Joint Army, Navy, Air Force Symposium (November 1982).

21. Arnold, B.A., H. Eisenberg, D. Borger, and A. Metherell. "Digital Radiography. An Overview." Proc. SPIE. Vol. 273, Application of Optical Instrumentation in Medicine, IX (1981).

22. Macovski, A. *Medical Imaging Systems.* New York: Prentice-Hall (1982).

23. Cooley, J.W., and J.W. Tukey. *Mathematical Computations.* Vol. 19 (April 1965).

24. Jacoby, M.H. "Automatic Evaluation of X-ray and Acoustic Images." Proc. Spring Conf., ASNT (March 1982).

25. Pearson, J.J., M.A. Fischler, O. Firschein, and M.H. Jacoby. "Application of Image Processing Techniques to Automatic Radiographic Inspection." Proc. Electronic Institute of America, Seventh Annual Symposium (1977).

26. Pearson, J.J., W.G. Eppler, O. Firschein, M.H. Jacoby, J. Keng, and S.M. Jaffey. "Automatic Inspection of Artillery Shell Radiographs." Twenty-Second Annual International Technical Symposium, SPIE, San Diego (1978).

27. Firschein, O., and M.A. Fischler. "Associative Algorithms for Digital Imagery." Twelfth Annual Asilomar Conference on Circuits, Systems, and Computers (November 1978).

28. Firschein, O., W.G. Eppler, and M.A. Fischler. "A Fast Defect Measurement Algorithm Suitable for Array Processor Mechanization." IEEE Computer Society Conference on Pattern Recognition and Image Processing, Chicago, Illinois (August 1979).

29. Eppler, W.G., O. Firschein, G. McCulley, J. Keng, and J.J. Pearson. "Speedup of Radiographic Inspection Algorithms Using an Array Processor." Proc. SPIE, Vol. 207 (1979): pp 96-103.

30. Dahlberg, R.C. "Real Time Image Filtering and Shading Correction." Paper summaries, Fall Conference of the American Society for Nondestructive Testing (October 1979): pp 305-310.

31. Zadeh, L.A. "Fuzzy Sets and Their Application to Pattern Classification and Cluster Analysis." Memo University of California, Berkeley/Engineering Research Lab, Berkeley, California (1976).

32. Kandel, A., and S.C. Lee. *Fuzzy Switching and Automata: Theory and Applications.* New York: Crane Russak. London: Edward Arnold (1979).

33. Halmshaw, R. "Fundamentals of Real-Time Imaging," *Real Time Radiographic Imaging: Medical and Industrial Applications.* ASTM STP 716. D.A. Garrett and D.A. Bracher, eds. American Society for Testing and Materials (1980).

34. Harms, A.A., M. Heindler, and D.L. Lowe. "The Physical Basis for Accurate Dimensional Measurements in Neutron Radiography." *Materials Evaluation,* Vol. 36 (April 1978).

35. Fishman, A., S. Wajnberg, A. Notea, and Y. Segal. "Extraction of Dimensions From Radiographs." *Materials Evaluation*, Vol. 39 (July 1981).

SECTION 16

RADIATION GAGING

Sam G. Snow, Martin Marietta Energy Systems, Inc., Oak Ridge, TN
Roger A. Morris, Los Alamos National Laboratory, Los Alamos, NM

Part 9 of this section included with permission from Fischer Technology, Inc.

INTRODUCTION

Radiation gaging does not use shadow-image formation, but it is nevertheless a radiometric nondestructive testing method by which density, thickness and composition can be determined using the interaction of radiation with a test material. Applications of radiation gaging range from high accuracy measurements of coating thickness to detection of termite damage. Radiation gaging may be used online to enable real-time control of processing equipment or may involve extensive scanning and analysis to construct three-dimensional images of internal density variations in materials.

Radiation gaging includes a wide variety of measurement types. Gamma rays, X rays, beta particles, neutrons, and positive ions can all be used for radiation gaging. These radiations interact with the test material in a number of useful ways.

Despite the wide diversity of methods which can be used, one method, gamma or X-ray attenuation gaging, has found the widest application because of its general applicability to all materials and many component configurations.

Gamma and X-ray attenuation methods are particularly well suited for in-process monitoring applications such as control of thickness in a rolling mill or monitoring density of solution in process piping. As an inspection tool for fabricated components, attenuation gaging can be used to ensure that density, composition and thickness of a wide variety of materials have been kept under control. Attenuation gaging can achieve extremely high accuracies for some applications; or conversely, operational parameters can be adjusted for rapid inspection on an assembly-line basis.

Because there is such a large number of radiation gaging methods with limited applicability, this Handbook Section will cover many of the methods in overview. The greater part of this Section will be devoted to gamma and X-ray attenuation gages.

PART 1

OVERVIEW OF RADIATION GAGING METHODS

Those interactions of radiation with matter that are useful for gaging are listed in Table 1, along with an indication of measurement applications. For convenience, these may be classed in two categories: those involving the use of gamma and X rays, and those based on interactions of nuclear particles (neutrons, positive ions and beta particles).

Gamma and X-ray Gaging

Gamma and X-ray Attenuation

Gaging by measuring the attenuation of gamma and X-ray photon beams is used to determine the density-thickness product ($D \times T$). In many applications, the density may be assumed to be non-varying and this gaging method is used for thickness measurements, often in continuous automatic systems for control of production equipment. Conversely, if thickness is held constant, density can be measured. Use of more than one radiation energy makes possible the measurement of the density-thickness product of individual elemental components in a multi-component material such as a layered structure or a composite.

Tomography

Tomography is a specialized form of attenuation gaging by which multi-direction attenuation measurements are computer-analyzed to reconstruct images of the internal distribution of mass in the test object. This method is extremely valuable when internal defects need to be located as well as detected, or when the shape of the test object is irregular enough to obscure internal density variations in conventional radiography or radiation attenuation gaging.

X-ray Fluorescence

X-ray fluorescence gaging (XRF) is based on excitation and detection of characteristic X-ray emission (*K* or *L* X rays). Because the energy of the radiation emitted is unique to the emitting element, XRF is commonly used for elemental analysis. XRF as an analytical tool is described in Section 17 of this Volume. The use of XRF for thickness gaging or composition monitoring will be briefly described in Part 8 of this section.

Compton Scattering

Compton scattering is used to measure material density, including detection of high and low density defects, and can also be used for one-sided thickness gaging. The equipment used for Compton scatter gaging is very similar to that used for X-ray fluorescence gaging, and in fact X-ray fluorescence and Compton scattering measurements can be combined in a single gage.

Gaging with Nuclear Particles

Beta Backscatter

Beta backscatter is a well-established method for measuring coating thickness. Its applicability is based on the increased backscattering of beta particles as a function of increasing atomic number. As coating thickness increases from zero, the backscatter response varies smoothly from the response characteristic of the substrate atomic number to the response characteristic of the coating atomic number. Intermediate values of backscatter response can be calibrated directly in terms of coating thickness. The sensitivity of this method improves as the difference between the atomic numbers of the coating and the substrate increases.

TABLE 1. Radiation Gaging Techniques

Gaging Method	Principle Measurement Applications
Gamma/X-ray attenuation	Density-thickness product (D × T)
Two-energy attenuation	Composition of two element system
Tomography	Three-dimensional imaging
X-ray fluorescence	Composition, coating thickness
Compton scattering	Density uniformity, one-sided D × T
Beta backscatter	Coating thickness
Neutron attenuation, scattering and moderation	Hydrogen content
Rutherford scattering	Surface composition
Nuclear interactions	Isotope specific analysis

Neutron Gaging

Neutron gaging is a very specialized technique most often used for measuring hydrogen content (usually in the form of water), or less often, the content of other isotopes with high neutron cross sections. Neutron gages can be simple, inexpensive systems for measuring moisture in food, soil or other bulk materials. The basis of the measurement technique can be attenuation of thermal or fast neutrons, moderation of fast neutrons, or scattering of thermal neutrons.

For some applications, such as moisture measurements, neutron gaging is the best or only available method. However, neutron gages are of limited and specialized use. A published review provides an excellent summary of neutron gaging.[1]

Rutherford Scattering

Rutherford scattering, elastic scattering of positively charged particles from atomic nuclei, is a useful surface analysis technique. The energy loss upon scattering at a specific angle is an easily calculable function of atomic weight of the scattering material. This method can be used to identify surface contaminants and determine composition of materials at the surface. However, applicability of this technique is generally limited to samples because positive ions are readily absorbed in air and measurements must therefore be made in vacuum.

Nuclear Reactions

Nuclear reactions are used for gaging when a specific element or specific isotope of an element must be monitored. There exists a convenient nuclear reaction which can be used for this purpose, and other simpler techniques are not applicable. Neutrons, high-energy gamma rays, and charged particles can all be used to excite these reactions. Applications range from determining oxygen content (by measuring the gamma intensity induced by neutron activation) to monitoring for beryllium (by measuring the neutron intensity produced by photonuclear reaction). Applications are limited to very special cases.

PART 2

PRINCIPLES OF GAMMA AND X-RAY ATTENUATION GAGING

The basic gamma or X-ray attenuation gage, in its simplest form, consists of an X-ray or gamma source, source shielding and collimation, an air gap where samples can be introduced, and a collimated detector as shown in Fig. 1. The principle of operation is simple and is described by a single basic equation

If the intensity of the radiation measured by the detector with no sample in place is I_o, then when a sample of thickness T is introduced into the air gap so as to intercept the radiation beam, the intensity I measured by the detector is given by the exponential attenuation law:

$$I = I_o e^{-\mu\rho T} \qquad \text{(Eq. 1)}$$

where ρ is the density of the sample and μ is the mass attenuation coefficient of the sample material for gamma or X rays of the energy used.

Measurements of radiation intensity I can be used to determine characteristics of the sample which depend on the product $\mu\rho T$. Most commonly, radiation attenuation gaging is used to measure thickness when μ and ρ are controlled or known.

Other applications include monitoring density when thickness is held constant and monitoring composition variation through its effect on μ and/or ρ. While the basic principles of these measurements are simple, they must be thoroughly understood in order to select the proper approach to a given measurement problem. The general principles of radiation physics on which attenuation gaging is based are presented in Section 1 of this volume, *Radiation and Particle Physics*.

By rewriting eq. 1 in differential form, the basic sensitivity equation is obtained:

$$\frac{\Delta I}{I} = -\mu\Delta(\rho T) \qquad \text{(Eq. 2)}$$

That is, eq. 2 shows that the relative change in radiation intensity, $\Delta I/I$, due to a small change in the density thickness product ρT, is directly proportional to $\Delta(\rho T)$ and the proportionality constant is the mass attenuation coefficient, μ. This leads to the guideline that μ be as large as possible. Of course, as μ increases, the portion of the radiation transmitted through the sample decreases exponentially, setting a practical limit on the magnitude of μ.

Attenuation Coefficient

The value of the mass attenuation coefficient, μ, in eqs. 1 and 2 is a function of absorber material and radiation energy. A complete table of attenuation coefficients is found in Section 20 of this volume.

Figure 2, which shows μ as a function of energy for a few selected materials, illustrates the dependence of μ on material and energy, characteristics that are important to attenuation gaging. First, at low energies, the attenuation coefficients decrease with increasing energy approximately proportional to the inverse third power of energy, except for the discrete step changes in attenuation coefficient at the atomic absorption edges. Attenuation increases with increasing atomic number of the absorbing material. These characteristics are determined by the predominant attenuation mechanism, photoelectric absorption.

At higher energies, as photoelectric absorption continues to decrease, Compton scattering becomes the predominant mechanism for attenuation. Attenuation at these energies is characterized by continued decrease in attenuation coefficient with increasing energy, but at a much lower rate. Attenuation is relatively independent of atomic number. Above energies of 1.02 MeV, photons can interact with atomic nuclei to produce electron-positron pairs. At higher energies, this becomes the predominant mechanism for attenuation of the radiation beam. Attenuation in this region is characterized by increase in attenuation coefficient with increasing energy. There is very little reason to attempt radiation gaging at these energies because the same attenuation coefficients can be achieved at lower energies, which are generally easier to produce.

Statistical Accuracy Limits

The accuracy of an attenuation gage is limited by the statistical variation in the detected radiation. The measurement of the intensity I is accomplished by detecting a number of photons N during a time τ so that if I is in photons per second, $N = I\tau$. For large values of N, the distribution of the number of detected photons approaches a normal distribution with mean N and standard deviation $N^{1/2}$. This means that, with no sources of error other than the random variation in beam intensity, the error (one standard deviation) in a measurement of N is equal to $N^{1/2}$. Hence, the relative error in measuring the intensity I in eq. 1 is given by:

$$\frac{\Delta I}{I} = \frac{\Delta N}{N} = \frac{\pm N^{1/2}}{N} = \frac{\pm 1}{(\tau I)^{1/2}} \qquad \text{(Eq. 3)}$$

The error in density-thickness product can be expressed as a function of energy by combining eqs. 1, 2 and 3:

$$\Delta(\rho T) = \frac{\pm 1}{\mu(E)} \frac{e^{\mu(E)\rho T/2}}{[\tau I_o(E)]^{1/2}} \qquad \text{(Eq. 4)}$$

showing the dependence of μ and I_o on energy (E).

FIGURE 1. Basic Gamma Attenuation Gage

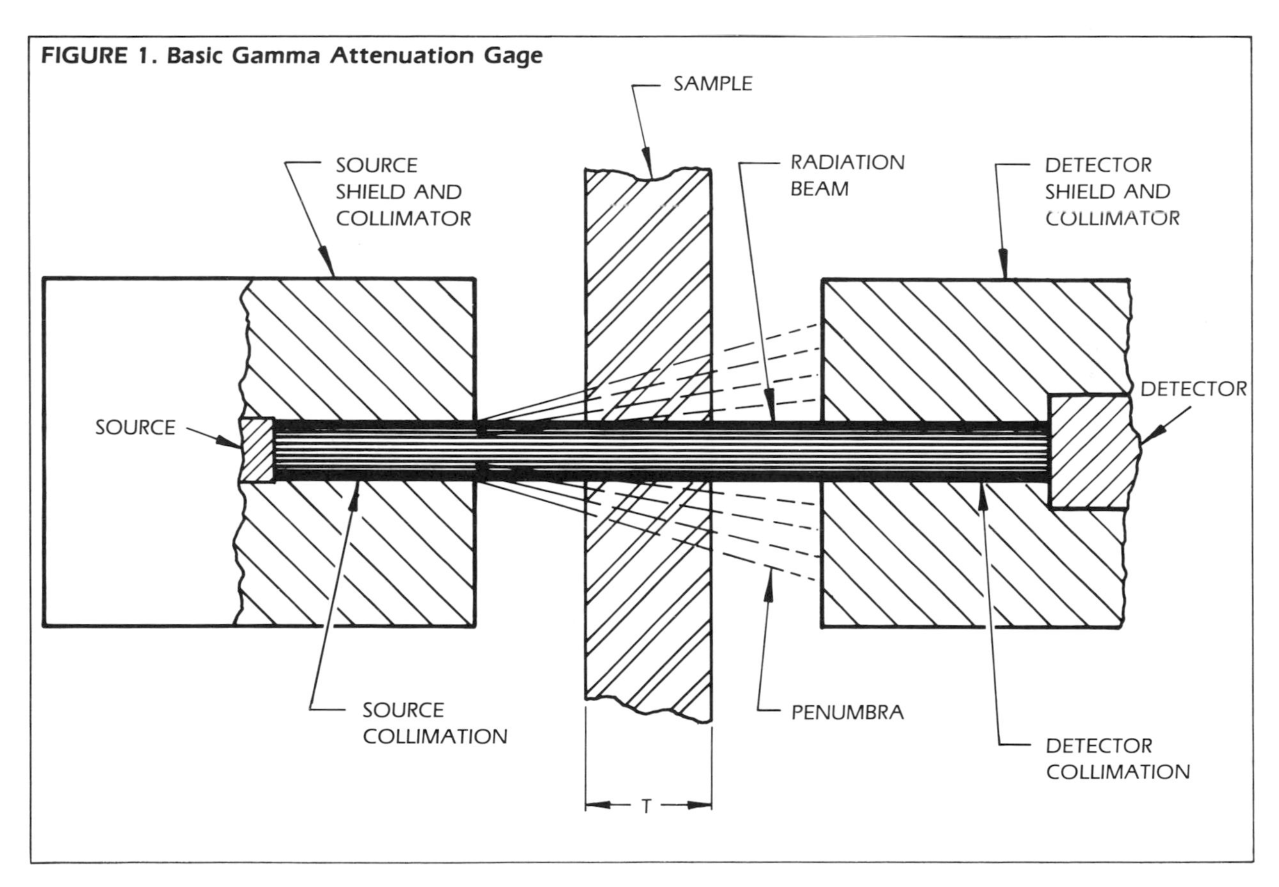

FIGURE 2. Attenuation Coefficients of Selected Materials

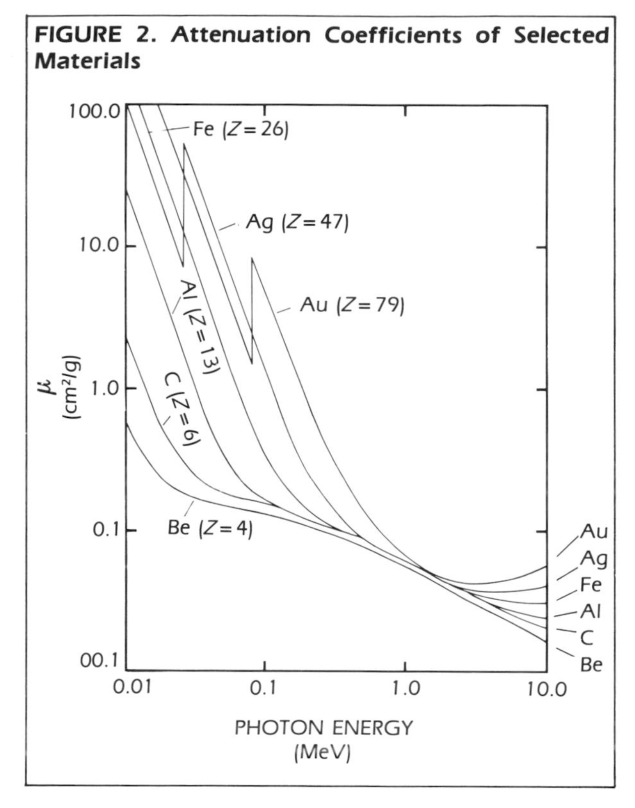

FIGURE 3. Optimum Energy Curve

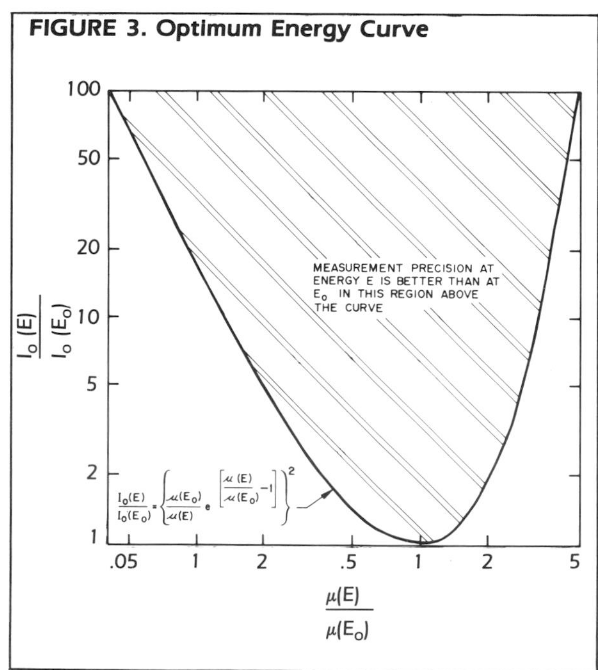

Optimum Gaging Energy

The optimum gaging energy is that energy for which $\Delta(\rho T)$ is a minimum, which occurs when

$$\left(\frac{\rho T}{2} - \frac{1}{\mu}\right) \frac{d\mu}{dE} - \left(\frac{1}{2I_o}\right) \frac{dI_o}{dE} = 0$$

If I_o were independent of energy, then $dI_o/dE = 0$. The optimum gaging energy would be that energy for which $\mu = 2/\rho T$.

In actuality, the available intensity (I_o) increases with energy for both X-ray generators and for isotopic gamma sources. X-ray energy is increased by raising the X-ray tube accelerating voltage, which also increases intensity. Gamma sources are selected from a limited number of isotopes, with fixed gamma energies. Because of the limitation on intensity at low energies due to self-absorption, greater intensity generally can be obtained with higher energy sources.

Therefore, dI_o/dE is, for practical gaging applications, a positive number, and a better estimate for optimum energy is needed.

In order to obtain an improved estimate of the optimum gaging energy, consider the behavior of eq. 4 around the energy, E_o, for which $\mu = 2/\rho T$. The ratio of error $\Delta(\rho T)$, at energy E, to the error, $\Delta(\rho T)_o$, at energy E_o is:

$$\frac{\Delta(\rho T)}{\Delta(\rho T)_o} = \frac{\mu(E_o)}{\mu(E)} \frac{e^{\left[\frac{\mu(E)}{\mu(E_o)} - 1\right]}}{[I_o(E)/I_o(E_o)]^{1/2}}$$

For E to be a better gaging energy than E_o, $\Delta(\rho T)/\Delta(\rho T)_o$ must be less than unity, or

$$\frac{I_o(E)}{I_o(E_o)} > \left(\frac{\mu(E_o)}{\mu(E)} e^{\left[\frac{\mu(E)}{\mu(E_o)} - 1\right]}\right)^2 \quad \text{(Eq. 5)}$$

This is presented graphically in Fig. 3 and is the criterion for deciding whether error in (ρT) at energy E will be less than the error at energy E_o.

A practical method for selecting the gaging energy for measuring a density-thickness product which is nominally $(\rho T)_o$, is as follows:

1. From the tables in Section 20, find the energy, E_o, for which:

$$\mu(E_o) = \frac{2}{(\rho T)_o} \quad \text{(Eq. 6)}$$

2. Estimate the available intensity, I_o, as a function of energy as described in the following discussion of sources.
3. To determine whether another energy E is more nearly optimum than E_o, calculate the ratios $\mu(E)/\mu(E_o)$ and $I_o(E)/I_o(E_o)$ and plot this point on Fig. 3.
4. If $I_o(E)/I_oE_o$ is above the curve in Fig. 3, then E is a better gaging energy than E_o.

For example, based on the relationship $\mu = 2/\rho T$, the optimum energy for gaging 2.5 cm (1 in.) thick aluminum is approximately 60 keV. The intent here is to determine whether 100 keV is a better gaging energy if twice the radiation intensity is available at 100 keV or when:

$$\frac{I_o(E)}{I_o(E_o)} = 2$$

For aluminum, $\mu = 0.27\ cm^2/g$ at 60 keV and $0.17\ cm^2/g$ at 100 keV. Therefore:

$$\frac{\mu(E)}{\mu(E_o)} = 0.63$$

This point $\mu(E)/\mu(E_o) = 0.63$ and $I_o(E)/I_o(E_o) = 2$ is above the curve in Fig. 3, and 100 keV is therefore the better (less statistical error) gaging energy.

PART 3

GAMMA AND X-RAY SOURCES

The radiation source used for radiation attenuation gaging will either be an X-ray generator or a radioisotopic source. The X-ray generator produces a continuous energy spectrum of bremsstrahlung as well as characteristic X rays from the target material. A radioisotope emits one or more discrete energies of gamma and/or characteristic X rays.

Gamma sources are more convenient for many gaging applications because the source and source encapsulation can be relatively small; shielding, collimation, and fixturing can be smaller and simpler. The discrete energies from a radioisotope source are often more suitable than the continuous output from a generator. Radioisotopes have absolute stable output, the only variability being the statistical variability described in the previous discussion and the gradual decrease in activity as the source decays.

In general, radioisotope gages are less expensive than X-ray generators.

X-ray generators have advantages also, and are used for applications that require versatility and high intensity. Gages used for a variety of inspections requiring different gaging energies utilize X-ray generators which can be readily adjusted to the desired energy. X-ray generators are also capable of producing extremely high intensities compared to many of the radioisotope sources, particularly at low energies where self-absorption limits radioisotope output. When measurement precision must be achieved in a time shorter than can be achieved with the optimum radioisotope source, then an X-ray generator should be considered.

Radioisotope Sources

Radioisotope sources commonly used for radiation attenuation gaging are listed in Table 2.[2,3] Other sources can be used for special applications, but the listed sources cover most practical gaging applications.

Estimation of Available Intensity

Determination of the intensity I_o which can be expected from a given radioisotope source is straightforward. Some vendors of encapsulated sources specify the effective activity in photons per second per steradian, which is equal to the total source activity divided by 4π (solid angle) and corrected for attenuation in the source and source encapsulation. From the geometrical relationship of source, detector and collimation, the solid angle subtended by the detector is easily determined. For the most common geometry shown in Fig. 1, the solid angle is simply equal to the area of the detector collimator divided by the square of the source-detector separation. The actual intensity measured by the detector is then:

$$I_o = ASE \qquad \text{(Eq. 7)}$$

where A represents the effective source activity in photons per second per steradian; S represents the solid angle from source to detector in steradians; and E represents detector efficiency.

If not specified by the detector manufacturer, the detector efficiency can be determined by using the exponential attenuation law (eq. 1) to estimate the loss of intensity in the detector front window or dead layer and then the fraction of the remaining intensity deposited in the detector. Selection of the best isotope source for a given application may then be determined by calculating the optimum energy as described in the preceding discussion on principles.

Source Selection

The first choice for the best source is that which has photon energy closest to $\mu = 2/\rho T$.

The maximum practical I_o should be determined for this source and for sources that have higher photon energies. The maximum practical effective source activity for each radioisotope may be determined by cost consideration or by self-absorption limitations. The corresponding values of I_o are determined from eq. 7 and compared to Fig. 3 to determine the best source for the application.

TABLE 2. Radioisotopes Commonly Used for Radiation Attenuation Gaging

Isotope	Approximate Gaging Energy (keV)	Gaging Energy Makeup*	Half-life**
^{109}Cd	22	Ag *K* X rays	453 days
^{153}Gd	42	Eu *K* X rays	242 days
^{241}Am	60	Single gamma (59.6 keV)	433 days
^{153}Gd	100	(98 and 102 keV)	242 days
^{57}Co	122	(+136 keV)	271 days
^{192}Ir	~300	(295, 308 and 317 keV)	74.2 days
^{137}Cs	662	Single gamma from Ba-137m daughter	30.2 years
^{60}Co	1,170	(+1,330 keV)	5.27 years

* See reference 2. ** See reference 3.

Electronic X-ray Sources

Electronic X-ray generators are covered in Section 2 of this Handbook volume. This section addresses those characteristics which are of special importance for radiation attenuation gaging.

Voltage and Current Stability

Voltage and current stability are critical parameters of an X-ray generator used for radiation gaging. Variation in tube current translates directly to variation in radiation intensity, resulting in an error in the measured density-thickness product, as given by eq. 2.

Voltage variation affects beam intensity even more than tube current does. In addition, voltage variation causes a variation in attenuation coefficient, μ, which affects the transmitted intensity. For the case when X-ray intensity varies as the third power of voltage and attenuation coefficient varies as the inverse third power of voltage (approximately true for a range of practical gaging energies), it can be shown[4] that the error in ρT due to variation in voltage V is:

$$\frac{\Delta(\rho T)}{\rho T} = -3\left(1 + \frac{1}{\mu\rho T}\right)\frac{\Delta V}{V}$$

Therefore the error due to variation in tube current and accelerating voltage is:

$$\frac{\Delta(\rho T)}{\rho T} = -3\frac{\Delta V}{V}\left(1 + \frac{1}{\mu\rho T}\right) - \frac{1}{\mu\rho T}\frac{\Delta i}{i} \quad \text{(Eq. 8)}$$

where i is the X-ray tube current.

From eq. 8, it can be seen that the relative error in ρT is least for large $\mu\rho T$, where it approaches three times the relative error in voltage. The best voltage and current stability available for an X-ray generator is on the order of 0.1%. Because electronic X-ray generators can produce high enough intensities to reduce statistical variations to much less than this in very short measurement times, voltage and current stability, rather than statistical variations, usually determine the measurement accuracy for electronic X-ray sources.

Energy Selection

The best gaging energy is not as easily derived for an electronic source as it is for a radioisotopic source. Taken alone, eq. 8 would suggest that error continues to decrease as μ increases without limit (and energy decreases to zero), which is not reasonable. This is because eq. 8 ignores the statistical variation which becomes dominant as energy, and hence, transmitted intensity, approaches zero.

The analysis based on statistical accuracy limits suggests that the best energy is somewhat greater than that for which $\mu = 2/\rho T$, while eq. 8 suggests that the best energy is somewhat less than this. An X-ray generator tube used for gaging material with thickness $(\rho T)_o$ should have the capability of operating at the energy for which $\mu_o = (2/\mu T)_o$.

Selection of the actual gaging energy is best done experimentally by determining the precision for repeated measurements at a number of energies. This can be accomplished quickly and easily for electronic X-ray generators since the high voltage is usually continuously variable from zero to the maximum.

PART 4

DETECTORS

The operation of all gamma and X-ray detectors is based on the effect of energy deposited in the detector. Photons deposit their energy primarily by three types of interaction with the detector material: photoelectric absorption, scattering from atomic electrons, and pair production.

In all three mechanisms, some or all of the photon energy is transferred to kinetic energy of electrons and some of the energy may escape from the detector. It is in the means by which different types of detectors transfer the kinetic energy of these electrons into measurable voltage or current signals that detectors differ. The different types of detectors include: gas-filled detectors, scintillation-phototube systems, and semiconductor detectors.[5]

FIGURE 4. Gas-filled Chamber Voltage Regions

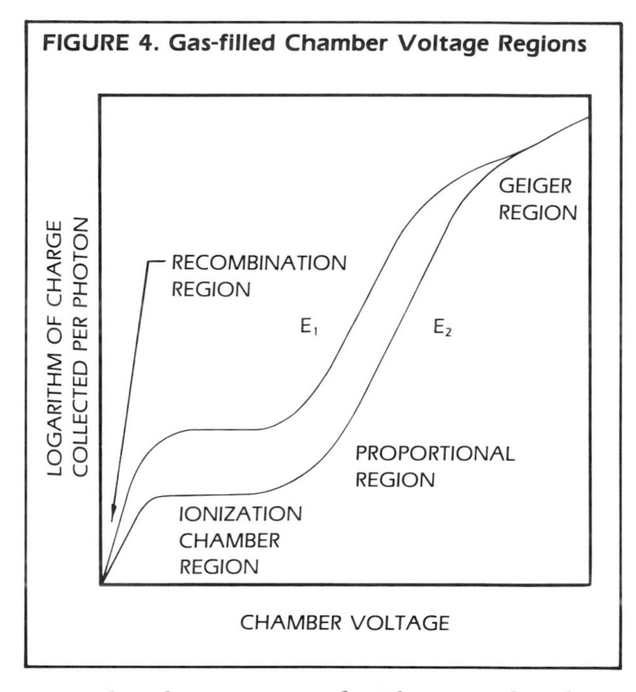

Gas-filled Detectors

A gas-filled detector consists of a chamber containing gas in which an electric field is maintained. Geometry may be planar or coaxial. The primary electrons from the interaction of photons with the detector lose their energy by ionizing gas molecules along their paths, leaving tracks of electron-positive ion pairs. The electrons are quickly swept to the anode by the electric field. The relationship between the negative charge collected at the anode and the energy deposited in the detector by the radiation is dependent on the electric field strength in the chamber, as shown in Fig. 4. At very low voltages the field strength is insufficient to separate all of the electrons from the positive ions before recombination occurs. In this *recombination region*, the relationship between energy deposited and charge collected is complicated and difficult to predict.

Ionization Chambers

For voltages sufficiently high to prevent recombination of electrons and positive ions, the charge collected is independent of the voltage and proportional to the energy deposited. This is the *ionization chamber* region of operation. Ionization chambers are used in the *current mode*. That is, rather than trying to detect the extremely small charge collected from a single photon interaction, the integrated current of many interactions is measured. This current is proportional to the energy deposited in the detector per unit time and hence, for constant photon energy, to the radiation flux.

Proportional Counters

As the electric field in a gas-filled chamber is increased further, secondary electrons are accelerated to energies that ionize other molecules in the gas, producing additional electron-ion pairs. In this *proportional region*, the charge collected is proportional to the energy deposited and increases with increasing voltage. By measuring the charge collected due to a single interaction, proportional counters are used for energy spectrometry. That is, individual photons are detected, and sorted according to charge. Because charge is proportional to energy, this sorting results in an energy spectrum.

Geiger Counters

As voltage is further increased, the gas multiplication continues to increase until each photon interaction produces a complete electrical breakdown of the gas, providing a large pulse of charge which is independent of the deposited energy. This is the *Geiger region* of a gas-filled tube. In this region, individual photon interactions are detected as with proportional counters, but energy information is lost.

Applicability of Gas-filled Detectors

Gas-filled chambers are relatively inefficient because of the low density of gas. For radiation attenuation gaging, only the ionization chamber has an offsetting advantage for specific applications. An ionization chamber can be an extremely stable detector for measuring very intense radiation fields which would damage other types of detectors. For highest efficiency, a high atomic number gas should be used, such as xenon.

Scintillator-Phototubes

As the primary electrons from a photon interaction pass through a material, they ionize atoms and excite atomic or molecular energy levels. Some materials give up a portion of their excitation energy as light, a process called *scintillation*. Scintillation materials which are transparent to their own light can be coupled to a photodetector and used as a radiation detector.

The phototube has a glass front window through which light passes to a photoemissive surface on the inside of the tube. Because the photoemissive surface is at negative potential relative to other parts of the tube, it is called the *photocathode*. In the presence of the impressed electric field, light striking the photocathode causes electrons to be ejected and accelerated to the anode. The charge transferred from cathode to anode is proportional to the light intensity on the photocathode and hence to the energy deposited in the scintillator.

Current Mode Operation

When the scintillator-phototube detector is used in the current mode, the anode current is measured to determine the average radiation flux. The currents generated are on the order of 10^{-10} to 10^{-7} amperes. In order to measure these small currents accurately, the dark current must be small (dark current is the background current measured in the absence of radiation). Because tubes not designed for low-level measurements may have large dark currents, care should be exercised in selecting a phototube. Phototubes with dark currents less than 10^{-11} amperes should be used.

Spectrometry Mode Operation

A scintillator can also operate in a spectrometry mode by using a photomultiplier tube. In a photomultiplier tube, the charge from the photocathode is accelerated to the first dynode of a dynode chain, ejecting more electrons from the surface of the dynode. These electrons are accelerated to the second dynode, ejecting additional electrons. The process continues down the dynode chain, multiplying the charge at each dynode, so that the charge reaching the anode is many times that ejected from the cathode. The charge collected due to a single pulse of light in the scintillator can be measured and related to the energy deposited by the initial photon interaction. Thus, a scintillator-photomultiplier can be used for energy spectrometry.

Semiconductor Detectors

Unlike gas-filled detectors and scintillation detectors which can either be operated in a current mode or spectrometry mode, semiconductor detectors are always used in the spectrometry mode. The advantage of semiconductor detectors is the superior resolution for energy spectrometry. A comparison of the energy spectra of a cobalt-57 source obtained using a scintillation detector and semiconductor detectors is shown in Fig. 5. Semiconductor detectors have the additional advantage of being more stable than scintillation-photomultiplier detectors.

A semiconductor radiation detector can be described as the solid state analogy to a gas-filled detector. Charge carriers in a semiconductor are electron-hole pairs created by ionizing radiation, much as electron-positive ion pairs are created in the gas detector. In semiconductors, electrons exist in energy bands that are separated by band gaps as shown in Fig. 6. In the absence of any excitation, the valence band is completely filled with electrons and the conduction band is empty. The band gap is an energy range in which electrons cannot exist. Electrons can be excited, either by thermal energy or by an interaction with radiation, leaving a positive hole in the valence band and an electron in the conduction band. In the presence of an applied field, the mobile electrons and holes migrate through the semiconductor, creating a current.

Silicon and germanium detectors can be made to incorporate large volumes in which electron-hole pairs are formed and can readily migrate to electrical contacts. By collecting the charge generated by a detected photon, the photon's energy can be measured with precision. In order to minimize thermally generated noise, detectors are cooled to liquid nitrogen temperature. In addition to cooling the detector, the liquid nitrogen operates a pump which maintains an insulating vacuum around the detector. An integral preamplifier collects the small charge generated by individual photons and outputs a corresponding voltage pulse to the external electronics.

Selection of Detector Type

Radiation detection can be grouped in two categories: current mode types and spectrometers. Depending on the specific detector and detector electronics, the upper limit of count rates which can be handled by a spectrometer system is between 10^4 and 10^5 counts per second. As a general rule, a radiation gaging application that requires intensities greater than 10^4 counts per second (to achieve the required accuracy in acceptable measurement time) should not be attempted with a spectrometer system. For most of these applications, a scintillator-phototube system, operated in the current mode, is best.

FIGURE 5. Comparison of Scintillator and Semiconductor Detector Spectra

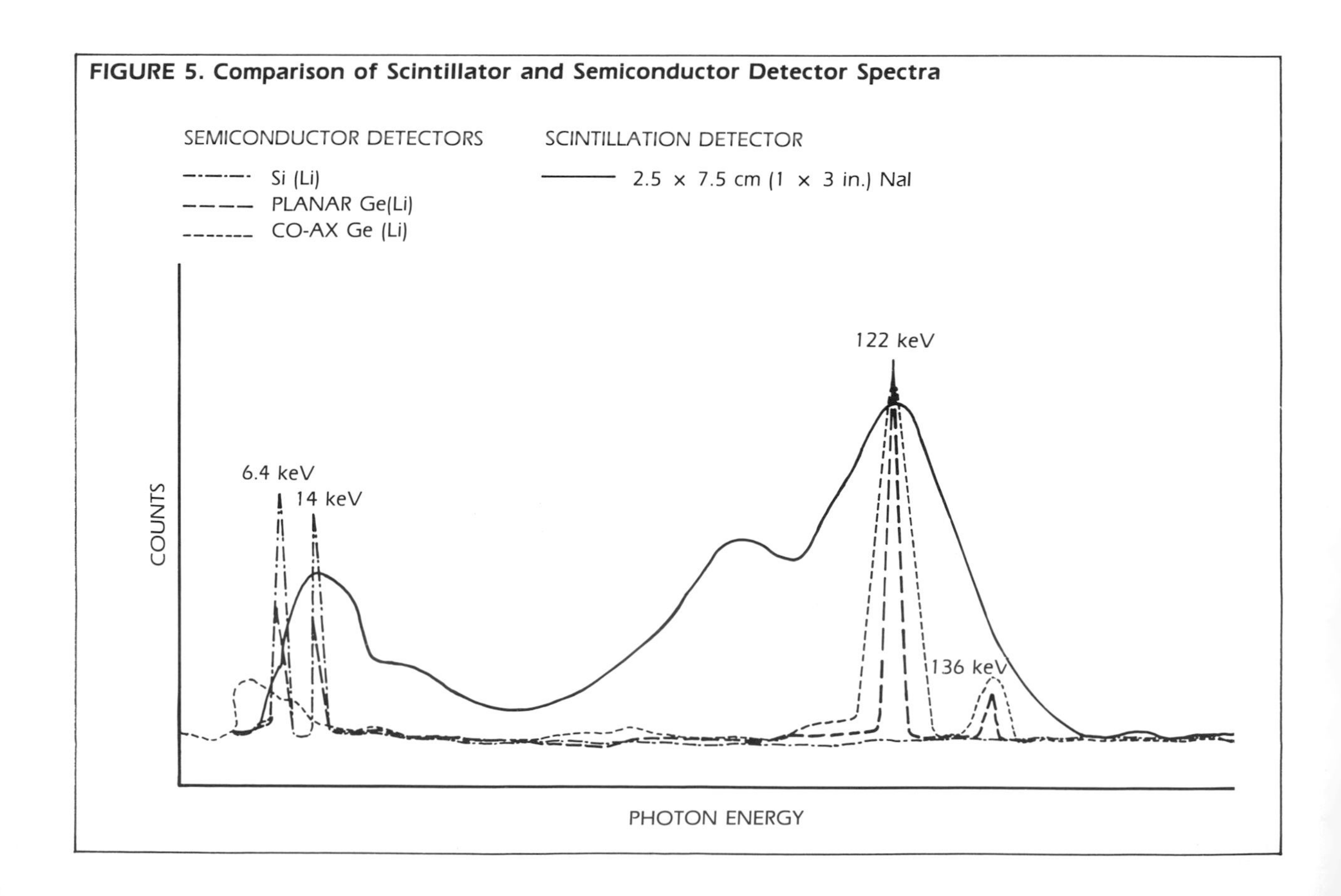

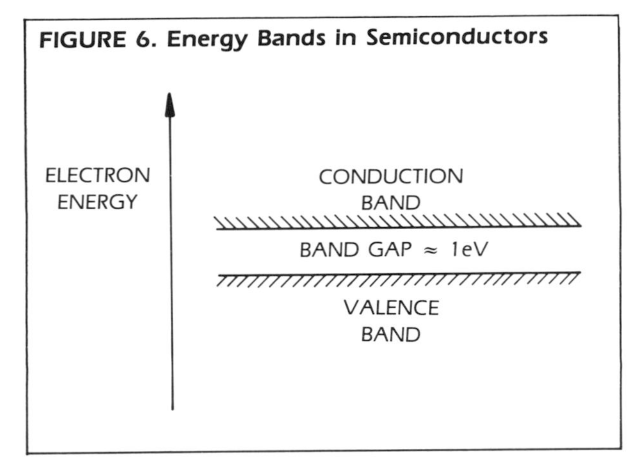

FIGURE 6. Energy Bands in Semiconductors

Current Mode Detectors

The most commonly used scintillator for gamma detection is thallium-activated sodium iodide [NaI(Tl)]. NaI(Tl) has two characteristics that make it unsuitable for attenuation gaging applications requiring current mode operation.

First, the light emission from NaI(Tl) due to deposition of a pulse of gamma or X-ray energy has two components: (1) a relatively intense light pulse with a decay constant of less than a microsecond; and (2) a lower intensity phosphorescence with a 0.15 second decay constant.[6]

At high radiation intensities, the long decay constant portion of the NaI(Tl) builds up a background level of light output. This is not a problem for the

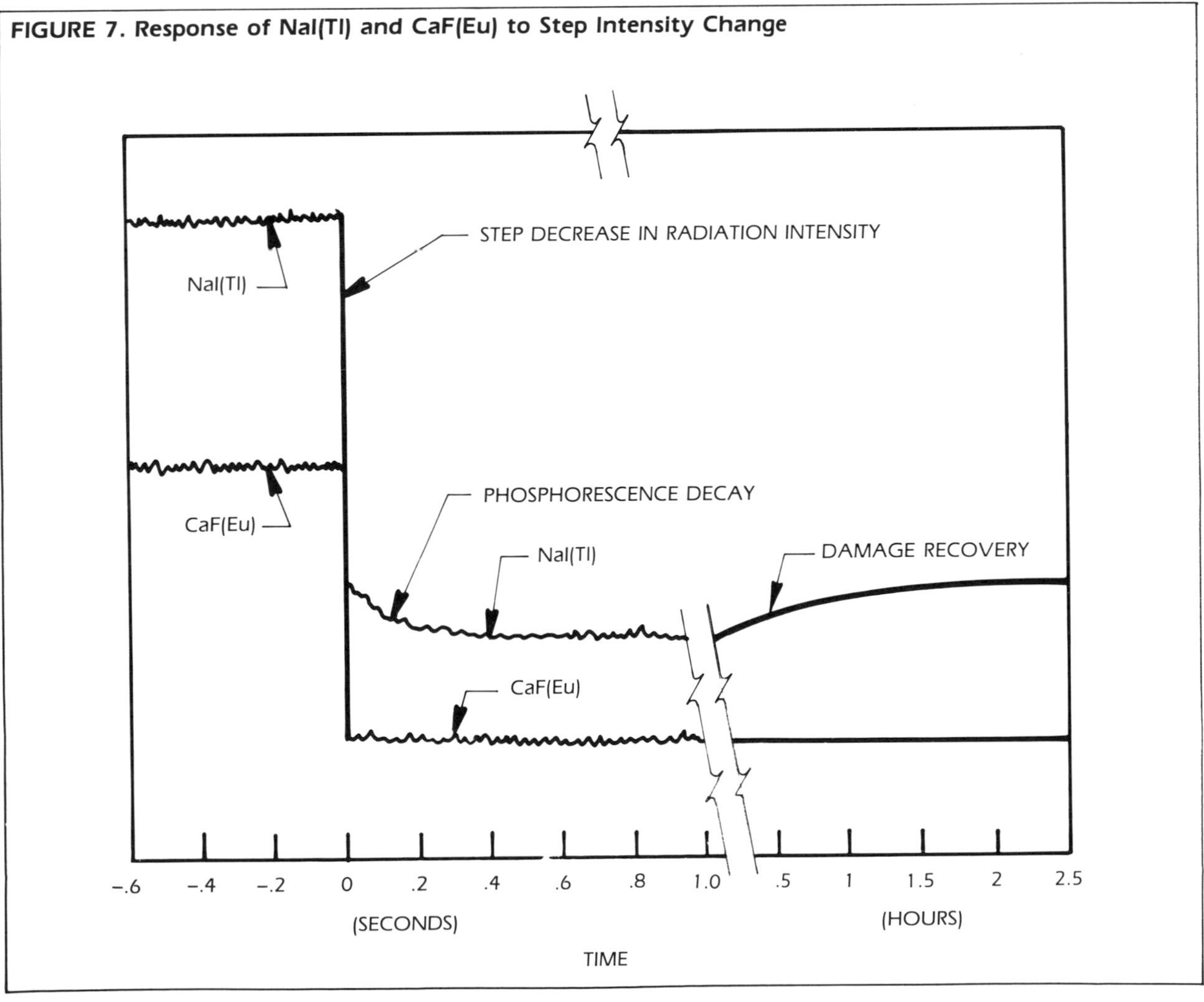

FIGURE 7. Response of NaI(Tl) and CaF(Eu) to Step Intensity Change

NaI(Tl) spectrometer because this direct current level can be discriminated against when measuring voltage pulse heights. In the current mode, however, the DC level adds to the measured signal and contributes to an extremely slow detector response to changes in signal level.

The second unsuitable characteristic is that NaI(Tl) darkens under intense gamma exposure, reducing the amount of light reaching the photocathode. NaI(Tl) recovers from this darkening very slowly, adding an even longer time-constant component to the NaI(Tl) response.

Europium-activated calcium fluoride CaF_2(Eu) is one scintillator that has been found to exhibit neither of these effects. Figure 7 compares the response of NaI(Tl) and CaF_2(Eu) to step changes in radiation intensity. The only drawback of CaF_2(Eu) compared to NaI(Tl) is a lower efficiency because of its lower atomic number, lower density and somewhat less light output for the same energy deposition. Current mode measurements should use either a CaF_2(Eu) scintillator or an alternative that has also been demonstrated to be free of slow response effects.

Spectrometer Mode Detectors

Spectrometer detectors can be used when high intensity is not required. Their energy selection capability makes it possible to choose the desired gaging energy. For example, use of a spectrometer type detector makes it possible to gage with 43 keV X rays from gadolinium-153 while discriminating against the 100 keV gammas. Energy spectrometry also allows discrimination against lower-energy scattered radiation that can be a source of measurement error.

Germanium and silicon detectors offer the best energy resolution. For energies up to about 100 keV, Si(Li) detections are satisfactory. Germanium detectors are preferred for higher energies because of their higher efficiency. Germanium detectors may be either planar or coaxial. Coaxial detectors are preferred for higher energies because, with their greater detector volume, they are more efficient at high energies. For applications where 15 to 20% energy resolution is adequate, a scintillator photomultiplier detector provides a less expensive alternative that does not require liquid nitrogen cooling. For spectrometry, NaI(Tl) has been shown to be an acceptable scintillator.

PART 5

DETECTOR ELECTRONICS

Current Measurements

Currents in the range from 10^{-10} to 10^{-7} amperes can be measured quite accurately with commercial picoammeters. However, careful attention to detail is required. Low-noise, shielded signal cable should be used and any motion of the signal cable during measurement must be avoided to prevent generation of transient signals. In order to minimize leakage currents, insulating surfaces need to be clean and bias voltages should be no greater than necessary for complete charge collection. Phototube bias less than 100 volts is usually adequate. The voltage required for an ionization chamber depends on the chamber geometry and the radiation flux. Higher fluxes require higher voltages to prevent recombination.

The most accurate measurements will be obtained by digitizing the output voltage from the picoammeter. Figure 8 is a diagram of a current measuring system with digital output. This type of system is capable of measuring radiation intensities with a precision of about 0.01%. The output voltage from the picoammeter is digitized by a digitizer for which the output frequency is proportional to voltage. This frequency is then measured with a counter-timer. Count times for a single measurement can be set to obtain the required statistical accuracy within the limitation of equipment stability.

Spectrometry Instrumentation

Figure 9 shows an energy spectrometry system for radiation attenuation gaging. Components of a radiation gaging spectrometry system are summarized below.

Power Supplies

The power supply for the bias voltage of a semiconductor detector needs a voltage capacity no greater than about 1,000 V and current capacity of only a few microamperes. For a scintillator-photomultiplier detector, the power supply provides a current through a resistor string that maintains a voltage gradient down the dynode chain. Voltages up to 3 kV are used. Because the current through the dynode chain must be large compared to the signal current for stable operation, the power supply must be able to provide currents of several milliamperes.

Amplifiers

The preamplifier and amplifier are designed to shape and amplify voltage pulses from the detector. The preamplifier is the first stage of the pulse processing and is kept in close proximity to the detector; it provides pulses suitable for transmission through shielded signal cable.

The only aspect of gamma spectrometry that requires special attention for radiation attenuation gaging is the effect of high count rates. Radiation attenuation gaging measurements require high count rates, often near the limits of instrument capability. Amplifiers should be designed for operation at high count rates without pulse distortion.

Pulse Height Analyzers

Pulse height analyzers sort amplified voltage pulses by pulse height. A single channel analyzer (SCA) outputs a logic pulse if an input pulse has peak height (1) above a lower level discriminator setting and (2) below an upper level discriminator setting. These logic pulses are accumulated over a fixed period of time by the counter and timer. The resulting count is the gage measurement of the selected energy's transmitted radiation intensity.

Multichannel analyzers (MCA) sort pulses with many narrow pulse height intervals to obtain a complete energy spectrum. Counting and timing functions are included in the MCA. Most modern MCAs are computer-based and can provide some degree of analysis. Integration of regions of interest is a common function that is used to obtain the count rate in selected energy peaks, much as the SCA-counter-timer combination does. The advantage of the SCA is lower cost and the ability to handle somewhat higher count rates. Even when the SCA is used, an MCA is extremely useful in set up and for ensuring that the instrument is operating as expected.

FIGURE 8. Current Measurement Electronics

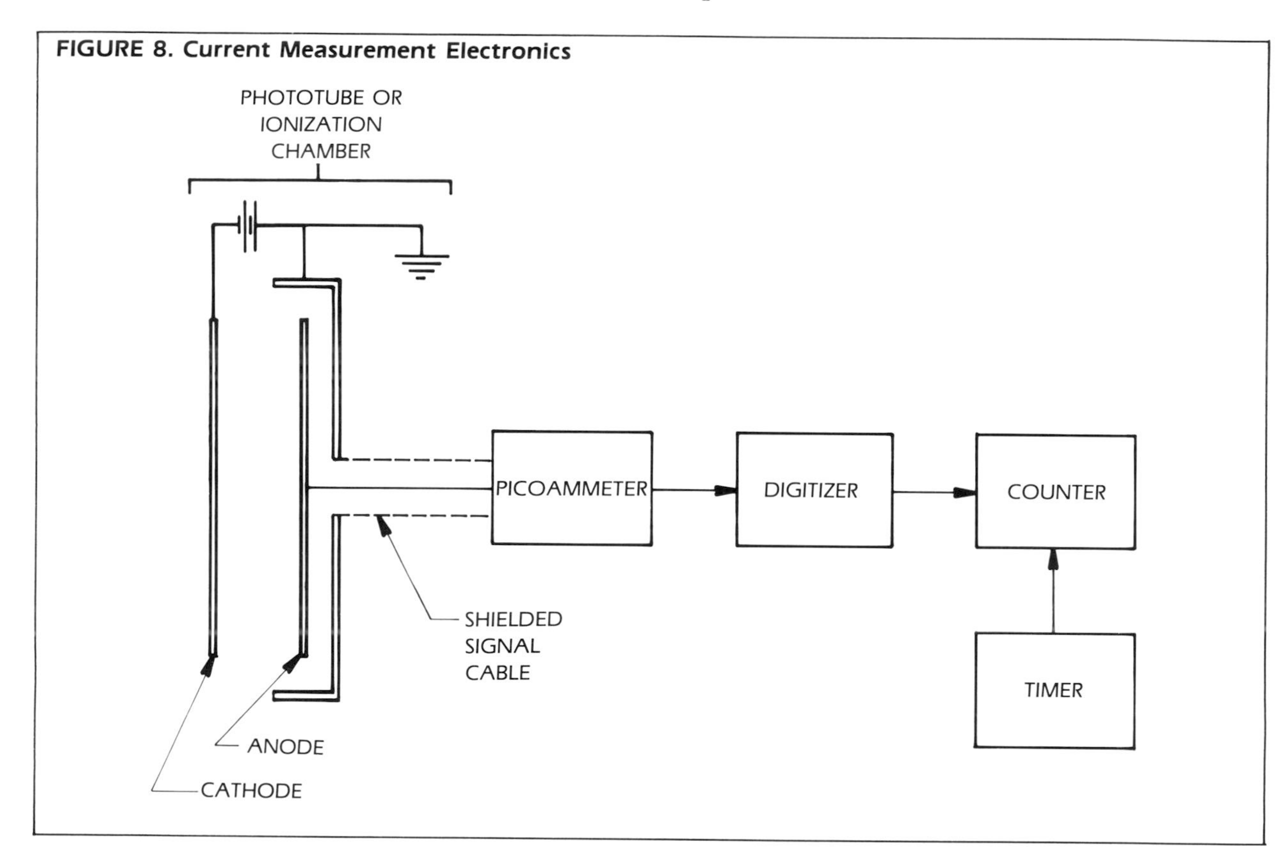

FIGURE 9. Energy Spectrometer Electronics

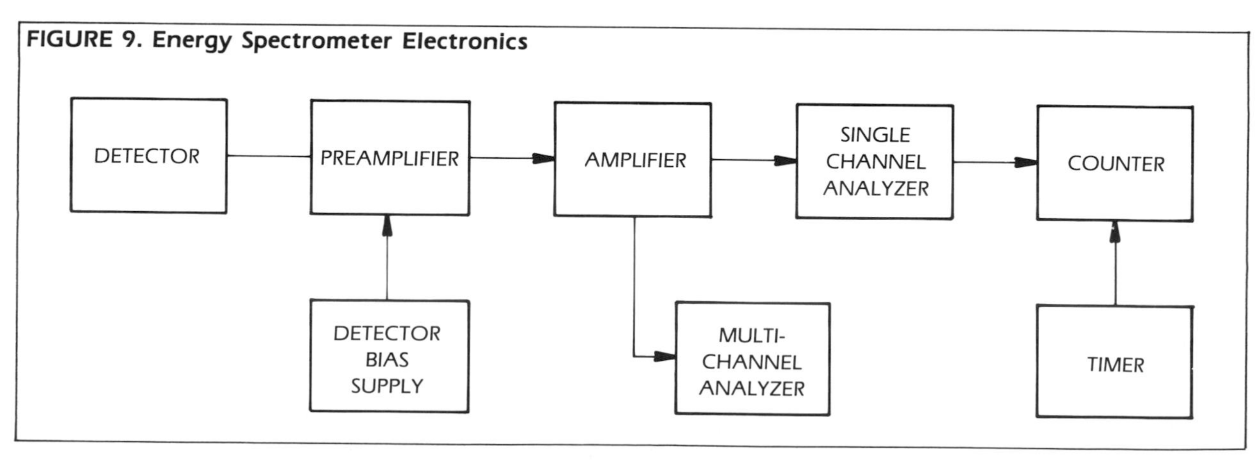

PART 6

FIXTURING AND CONTROLS

Collimation

A collimator is an aperture in the shielding. It defines the area of the test object being sampled and reduces the unwanted (scattered) radiation reaching the detector.

Figure 1 illustrates some elements of collimator design. The source collimator usually has a larger diameter than the source itself because radiation emitted outside the collimator diameter is wasted. Under these conditions, the area of the test object being sampled is a function of the detector collimator diameter, the source diameter, and the position of the test object between the source and detector.

The intensity of the radiation is not uniform across the beam cross section for any but a point source. The radiation intensity is greatest along the collimator axis, falling off near the edges. This means that a radiation attenuation measurement is not equally sensitive across its cross section, but is most sensitive to material on the collimator axis. This is usually not important; however, for applications where a uniform sensitivity across the sampled area is needed, the source diameter should be as small as possible when compared to source-to-detector separation.

Minimizing Detection of Scattered Radiation

One of the most important functions of collimation is to minimize the contribution of scattered radiation to the measured intensity. The radiation reaching the detector is comprised of (1) the primary radiation beam reaching the detector unattenuated, and (2) a much smaller component of scattered radiation. This scattered radiation needs to be minimized because it is not useful for gaging and can introduce error. As the position of the sample between the source and detector changes, the scatter contribution to the signal changes. It is imperative that the change in signal intensity due to the change in sample position (allowed by the positioning equipment) be small compared to the accuracy needed.

The scatter contribution to the total signal is a complicated function of the collimator size and separation, the sample material thickness and position, and the radiation energy used. Scatter contribution is minimized by maximizing collimator length-to-diameter ratio and separation between the source and detector. The detected radiation intensity should be measured as a function of sample position (1) to determine the sample location for which signal change with change in sample position is minimized and (2) to verify that signal variation due to the expected positioning variability is acceptable. Scatter is much less a concern with spectrometer type detectors that can discriminate against much of the lower energy scatter radiation. This is particularly true for the high resolution germanium detector.

Collimator Materials

Uranium and lead are commonly used materials for collimators and shields at higher energies. Uranium, depleted in the uranium-235 isotope, has the highest density and attenuation coefficient of any relatively inexpensive material that can be machined to make collimators. The only drawback of depleted uranium is its slight radioactivity. Precautions must be taken in fabricating components from depleted uranium to avoid personnel exposure. Finished uranium shields and collimation can be used quite safely.

The radioactivity of uranium makes it unsuitable for use as a detector collimator when measuring low radiation intensities. Uranium is the best material for shielding and collimating iridium-192, cesium-137 and cobalt-60 sources and for collimating X-ray sources above 400 kV.

Lead is not as efficient an attenuator of radiation as uranium and is so soft that it is easily damaged. Lead is often used with steel casing for protection. Tungsten is a good attenuator, but needs to be alloyed to be machinable. Machinable alloys of tungsten, such as WNiFe, are still difficult to machine so geometry should be kept simple. The main drawback of tantalum is its cost. It can be quite useful for small collimators. Below 100 kV, it is possible to use steel for some applications; however, the heavier metals make possible more compact, and often simpler, designs. The thickness of the collimator should be great enough that the fraction of radiation which can be transmitted through the shielding, at either the source or the detector collimator, is not large compared to the fractional error in radiation intensity dictated by eq. 2.

Fixturing and Positioning Equipment

Maintaining stable alignment between source and detector is critical. Very slight relative motion between source and detector can cause large signal changes. For this reason, it is always preferred to fix the source and detector on heavy, stable supports, and accomplish scanning by moving the test object. For on-line gages, the test object motion is often provided by the equipment being monitored, as with a rolling mill. In other cases, the radiation gage includes a positioner. In either case, motion parallel to the collimator axes should be minimized to minimize error due to changes in scatter contribution. The apparent increase in thickness due to tilting a sample (cosine error) needs to be kept in mind as should the fact that, for a variable thickness test object, an error in position is equivalent to an error in thickness. Except for these basic considerations, test object sizes, shapes, conditions, etc., vary so greatly that there are no generally applicable designs.

When it is not possible to accomplish scanning by moving the test object, and the source and detector must be moved, they must be specially designed. This often means a heavy yoke holding the source and detector at the designed separation. These will stay in alignment more easily if they are only translated and not tilted. A system for which one of the collimators has a larger diameter than the other is much less affected by misalignment than a system with equal sized collimators. Usually, in order to minimize scatter reaching the detector, this means that the detector collimator is smaller than the source collimator.

Data Analysis and Controls

The calibration equation for radiation attenuation gaging is shown in eq. 1. Calibration is accomplished by determining the constants (I_o and μ) in eq. 1 from measurements on standards. Because the scatter contribution to the measured intensity is not included, this equation usually cannot be fitted to experimental data over a wide range. Calibration over a wide range of thickness is accomplished by dividing the range into separately calibrated subranges, or more often, by modifying eq. 1 to include second order effects. Following this approach, experimental data are fit to an equation of the form:

$$\rho T = A + B[\ln(I)] + C[\ln(I)]^2 \qquad \text{(Eq. 9)}$$

where A, B and C are arbitrary constants related to I_o and μ.

Output data from a radiation attenuation gage may be needed in a variety of different forms for different applications. These include (1) a simple alarm when results are outside a specified tolerance band; (2) signals to be used for control of operating equipment; and (3) hard copy recording of the results. Most often, any or all of these data handling needs can best be satisfied with a computer-based system.

Use of computers for data analysis can range from an inexpensive microcomputer for simple applications, to the large computer systems used in analysis of tomographic data. Advances in capabilities and diversity of computers and reduction in prices for any given capability indicate that use of a computer for control, data acquisition, data analysis and data outputting should be considered for any radiation gaging system.

PART 7

TOMOGRAPHY

Tomography, in the most general sense, is any technique that produces an image of a parameter in a plane of an object without interference from adjacent planes. The physical parameter of the object is represented by the gray levels of the image and the appearance of these gray levels is determined by the type of radiation field that produces the image. In the case of X or gamma radiation, the image represents the linear attenuation coefficient of the object to the radiation.

One type of tomographic imaging (classical tomography) has been used since the 1940s. These images are characterized by the fact that the uninteresting planes are blurred in the final image. The plane of interest is a sharply imaged area in a sea of blurred images. This technique, called *laminography*, is achieved by moving the X-ray source and the detector during the exposure in such a way that only the shadowgraph of the plane of interest is sharply imaged. The image plane is also normal or perpendicular to the radiation beam axis.

The other type of tomography, *computed tomography*, is the present subject. While again relative motion between the detector, object and X-ray source is used to produce the image, the radiation is confined to the plane of interest to produce an image without any interference from adjacent planes. In computed tomography, the image plane is parallel to the beam axis, perpendicular to the image produced by classical tomography. There is a penalty to be paid for this uncluttered image; the image must be produced by computer techniques.

Basic Principles

Figure 10 depicts the scanning and reconstruction process for tomography. Imagine a narrow beam of radiation traversing the object, with angle ϕ as a parameter, to produce a single scan of the object. Each such scan can be represented by the function f_ϕ (x,y) where x and y are the Cartesian coordinates within the object. The value of the function f_ϕ represents the attenuated beam intensity through one ray of the object and can be expressed as:

$$f_\phi = f_{\phi o}\exp\left[-\int_\ell \mu(r)d\ell\right] \qquad \text{(Eq. 10)}$$

Where $f_{\phi o}$ is the unattenuated beam intensity; $\mu(r)$ is the linear attenuation coefficient at the point r within the object; and $d\ell$ is the differential path length along the path ℓ, the ray through the object. Equation 10 is a representation of Beer's Law.

If we divide eq. 10 by $f_{\phi o}$ and take the logarithum of both sides we get a new and more useful function λ_ϕ:

$$\lambda_\phi = \ln\left(\frac{f_\phi}{f_{\phi o}}\right) = \int_\ell \mu(r)\,d\ell \qquad \text{(Eq. 11)}$$

The major mathematical problem in tomography is inverting eq. 11 and solving for $\mu(r)$ as a function of the projection data λ_ϕ.

The quantity λ_ϕ is called the *Radon transformation* of $\mu(r)$ and the inversion of the equation was solved in principle by 1917. However, as a practical imaging technique, the transformation remained a curiosity until modern computer technology permitted fast solutions to be made for large images.

There are a number of methods for solving eq. 11 that have different advantages and disadvantages depending upon operational differences. Among these are the filtered back-projection technique and the iterative techniques. There are a large number of variations of each one but all the techniques within each general category have a basic similarity.[7]

The filtered back-projection technique takes a function derived from the projection data and back-projects (i.e., smears) the data back over the image plane. To show how this works, assume that you have a uniformly absorbing circular object with a very small, highly absorbing point in the center (a delta function). If you back-project the raw data, the resulting image is not the same as the original object; the image consists of a highly absorbing center point but it drops by $1/r$ as you move away from the center, instead of a sharp drop and then a uniform value. Because the input function (the object) was a delta function, the output function (the image) is the point-spread function of the reconstruction process.

The effect of the point-spread function can be corrected by filtering the projection data before back-projecting. This is normally done by taking the Fourier transform of the projection data, multiplying by a ramp function combined with some type of apodizing function (filtering), taking the inverse Fourier transform and then back-projecting.

Another technique that is equivalent to the above is based on the central slice theorem. This theorem states that the one-dimensional Fourier transform of each set of projection data is equal to the two-dimensional transform of the object, evaluated along a slice through the center of the two-dimensional Fourier plane at a particular angle. The image can be reconstructed by assembling all of the Fourier transforms of the projection data into a two-dimensional Fourier transform, multiplying by a filter to correct for the unequal spacing of data and taking the two-dimensional inverse Fourier transform to retrieve the original object.

The other basic reconstruction technique, the iterative method, is based on solving the following set of equations:

$$ {}_{I}\lambda_j = \sum_k W_{jk} f_k \qquad \text{(Eq. 12)} $$

where the ${}_{I}\lambda_j$ is the jth value of the projection data taken at a particlar angle; W_{jk} is a weighting function of the image pixel at coordinates j and k; and f_k is proportional to the linear attenuation coefficient in the pixel.

The problem is to solve eq. 12 for a set of f'_k that best approximates the true values f_k. We can define ${}_{o}\lambda_j$ as the calculated projection data from our trial object derived from f'_k such that:

$$ {}_{o}\lambda_j = \sum_k W_{jk} f'_k \qquad \text{(Eq. 13)} $$

We will say we have the best approximation to the true object function f_k when $({}_{o}\lambda_j - {}_{I}\lambda_j)$ is a minimum. The search for the best f'_k is done

FIGURE 10. Typical Tomographic Configuration

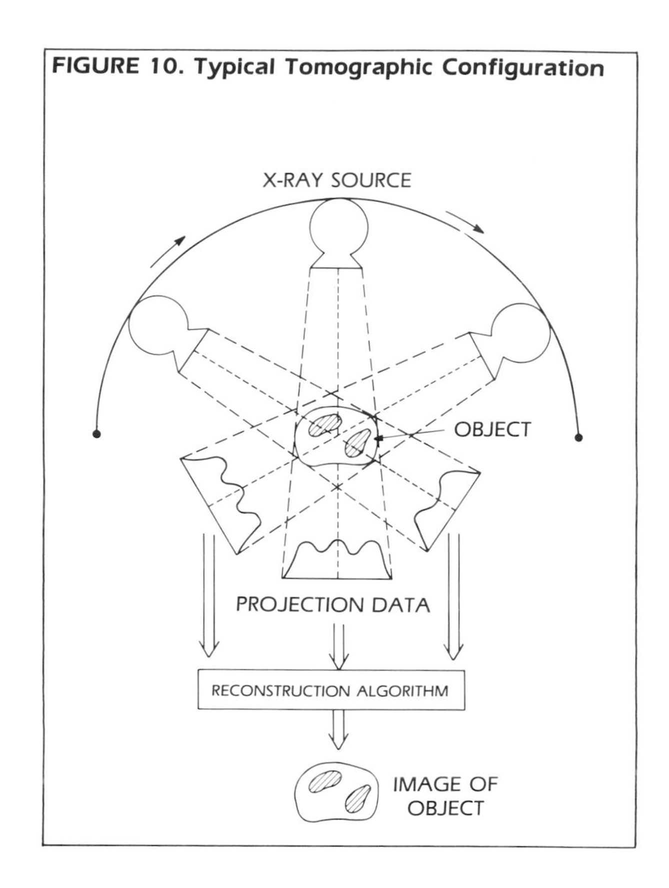

iteratively by choosing a trial set of f'_k, computing ${}_{o}\lambda_j$, comparing ${}_{o}\lambda_j$ to ${}_{I}\lambda_j$, correcting the trial f'_k, and starting over. The process stops when ${}_{o}\lambda_j$ and ${}_{I}\lambda_j$ are close enough. The different iterative techniques differ in (1) how they calculate a correction term to f'_k and (2) the criteria used to stop the iterations.

The *algebraic reconstruction technique* (ART) simply uses a weighted value of the difference between ${}_{o}\lambda_j$ and ${}_{I}\lambda_j$ as the correction term to be added to f_k. Each projection data point is addressed in turn and each pixel is updated each time. If, instead, each image pixel is addressed once and correction values are calculated for all projection values that intersect that pixel, the process is called the *simultaneous iterative reconstruction technique* (SIRT).

Another iterative technique minimizes the sum of the squares of the differences between ${}_{o}\lambda_j$ and ${}_{I}\lambda_j$. This is called the *iterative least squares technique.* Convergence criteria based on entropy (MENT) are also used.

The details of each of these reconstruction techniques vary depending upon the detector geometry (fan-beam or parallel-beam; curved or straight detectors) and the data-sampling density. In general, the filtered back-projection techniques are faster than the iterative techniques because each set of projection data is handled once and in any sequence permitting on-line reconstructions. The iterative techniques offer better noise properties and artifact rejection than the back-projection techniques, particularly when there are a small number of projection data sets.

There are numerous ways to generate artifacts in the reconstructed image. Among these are inaccuracies in the geometry, beam hardening, aliasing, and partial penetration.[8-10]

Tomographic Equipment

The earliest and simplest tomographic scanner consisted of a single source and detector that rotated about the object (a human head). This arrangement is slow but produces the best elimination of scattered radiation. Because the early development of tomography was driven by medical applications, equipment design was constrained by problems unique to the medical field. One factor was absorbed dose in the object (patient). Another factor was exposure time. In medical tomography, the object is living and hence moving. A third constraint in medical tomography that does not apply as much to industrial applications is the need to visualize small changes in density. In general, industrial applications do not need to visualize small, subtle density changes. High spatial resolution is more important.

The first advance after the single source, single detector scanner was the single source, multiple detector in a fan beam. Again, both source and detector rotated about the object. This arrangement speeded up the scanning but had poorer scatter rejection because the multiple detectors all operated at the same time.

Another arrangement consisted of a rotating source and a fixed array of detectors arranged in a circle about the object. These arrangements were developed with the goal of collecting the same type of data in the shortest time with minimum dose to the object.

The data collected from the detectors eventually ends up as digital data stored in mass memory in a computer. The computer controls the movement of the scanner, collects the data, applies whatever reconstruction algorithm is desired and finally displays the results.

Because the design constraints on an industrial tomographic scanner are quite different from those on a medical scanner, it can be very inappropriate to use a medical scanner to perform a feasibility study for an industrial problem. A negative result may not mean anything and a positive result may mean only that a much better result might be obtained if the equipment were properly designed for the application. Because of the diversity of industrial problems, it is unlikely that a general purpose industrial scanner will ever be built.

Problems that have been successfully solved with special purpose scanners range from large concrete structures[11] (inspected with cobalt-60 and requiring 2 mm resolution) down to aluminum capacitors (inspected at 50 kVp with 0.1 mm resolution).

Applications

Almost anything can be tomographically inspected.[12-16] The literature abounds with examples: lumber, rocket motors, small capacitors, etc. As an example, Fig. 11 shows a series of adjacent scans through an aluminum and plastic capacitor. The X-ray energy was 50 kVp while the aperture size and sampling interval were both 0.1 mm. Going from Fig. 11a to 11f, the high density mass near the axis can be seen to grow larger and then recede as the scans proceed along the axis. The resolution of these scans was not sufficient to resolve the individual aluminum foils. The scanning time was about six hours per image.

In general, tomography can be a practical NDT technique, if:

1. the consequences of equipment or material failure are very high (economically or in loss of life);
2. there are a very large number of identical inspections to be made; and
3. the data cannot be obtained in any other way.

As might be expected, the nuclear reactor and aerospace industries are leaders in the application of tomography.

FIGURE 11. Adjacent Tomographic Scans Through a 2.5 cm (1 in.) Diameter Capacitor

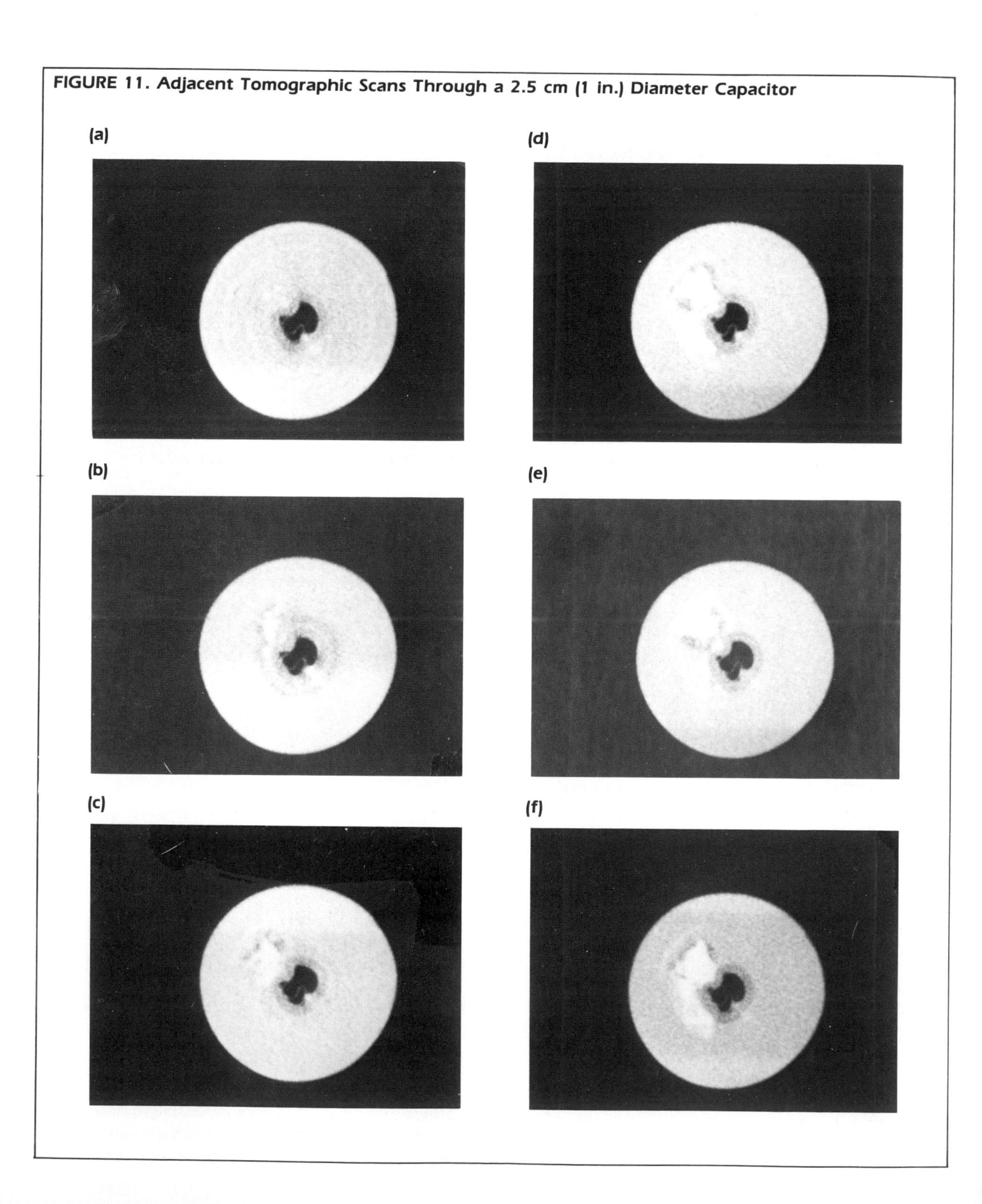

PART 8

OTHER X-RAY AND GAMMA GAGES

Absorption Edge Densitometry

The very good energy resolution of semiconductor detectors makes possible the use of absorption edge spectrometry in a practical gage.[17] This can be used for coating thickness measurements or composition of a two-element composite material. The basis for this measurement is the abrupt change in attenuation that occurs at the photoelectric absorption edge of one of the two elements. By gaging with two energies, one above and one below the absorption edge, it is possible to determine the mass per area of both elements. One of the low energy sources listed in Table 2 can be used together with a transmission target as shown in Fig. 12 to supply the two energies needed.

X-ray Fluorescence

When used for gaging, X-ray fluorescence generally employs an isotopic gamma source and a semiconductor (energy dispersive) detector as illustrated in Fig. 13. Cadmium-109, americium-241, gadolinium-153, and cobalt-57 are suitable excitation sources, the choice depending on the elements to be excited. The semiconductor detector usually used is lithium-drifted silicon [Si(Li)].

An X-ray fluorescence gage can be used to measure the thickness of coatings for a variety of coating-substrate material combinations. The increase in X-ray intensity from (1) the coating material with increasing coating thickness, (2) the attenuation of substrate X-rays by the coating, or (3) a combination of both[18] can be used to measure coating thickness.

Figure 14 shows the energy spectrum obtained from nickel-plated uranium when excited by a cadmium-109 source. When the gage is set up to measure the ratio of the nickel K_α intensity to the uranium L_α intensity, the response is sensitive to nickel thickness and very insensitive to sample positioning variation. Similar spectra can be obtained for gaging composition of alloys or mixtures.

Compton Scattering

Compton scattering can be used to measure density or thickness of low atomic number materials. Source-detector-sample configurations are similar to those used for X-ray fluorescence (Fig. 13). Source energy must be high enough to ensure that (1) Compton scattering is the predominant interaction with the material, and (2) the radiation penetrates adequately to the desired gaging depth. Best sensitivity is obtained for the lowest energy that satisfies these requirements.

The energy of the scattered radiation, E', is related to the energy of the incident radiation, E, and the scattering angle Θ according to the relationship:[19]

$$\frac{1}{E'} - \frac{1}{E} = \frac{1}{m_o C^2} (1 - \cos \Theta) \qquad \text{(Eq. 14)}$$

where m_o is the electron rest mass and C is the speed of light ($m_o C^2 = 0.51$ MeV).

Thus, the energy of the scattered radiation is selected by choice of scattering angle and source energy. This can be important for separating Compton-scattered radiation from interfering X-ray fluorescence radiation.

The intensity of the scattered radiation is usually low enough that a spectrometer type detector can be used. A scintillation detector has adequate energy resolution for many applications.

Examples of the application of Compton scattering gages include measuring the thickness of composite structures with a low energy system[20] (40 and 100 keV from gadolinium-153) and detection of termite damage to railway sleepers with a system using 662 keV gamma rays from cesium-137.[21]

FIGURE 12. Transmission Target Source Assembly

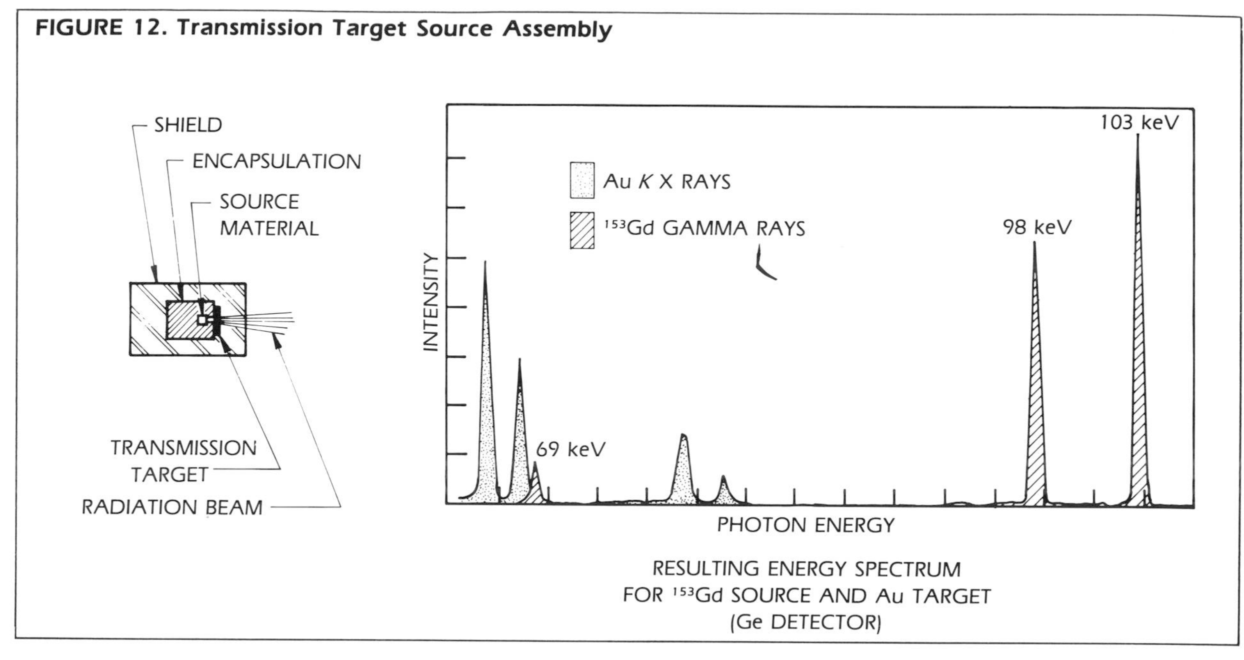

FIGURE 13. X-ray Fluorescence Gage Configuration

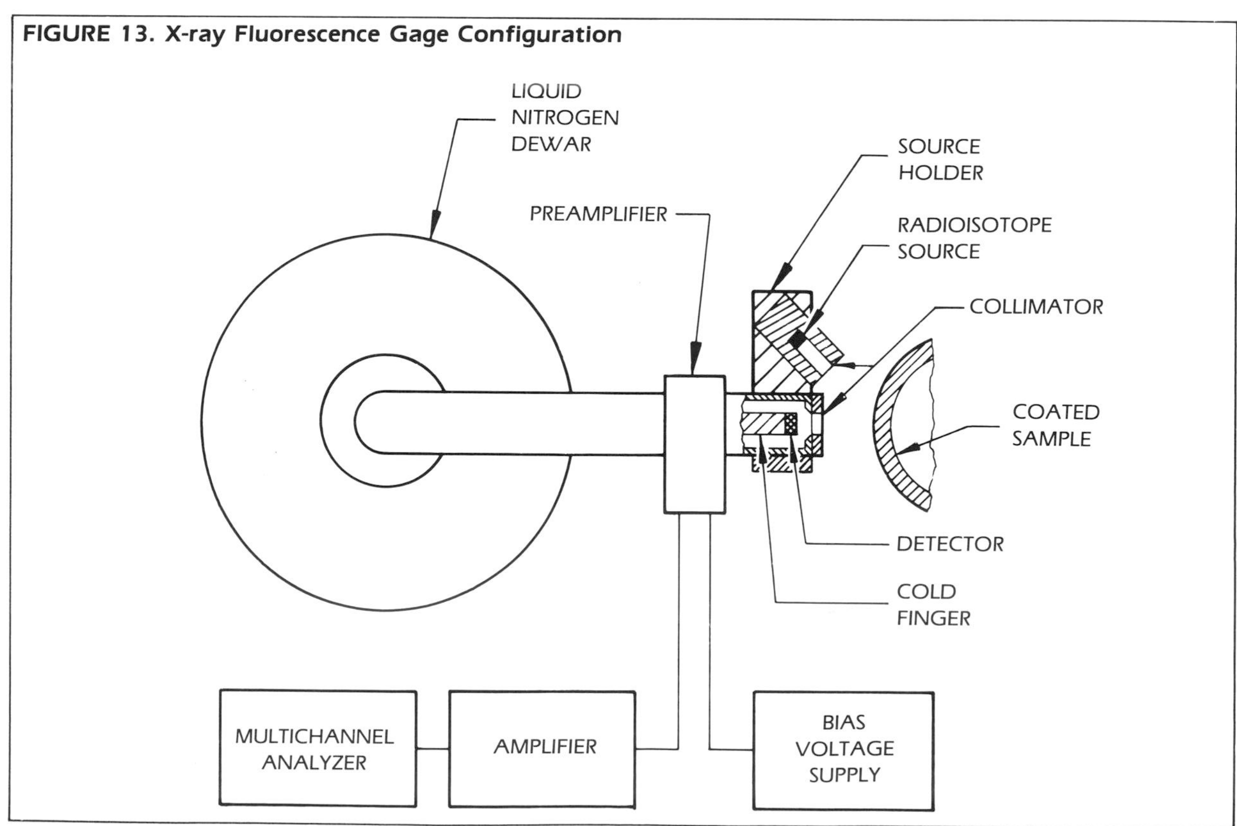

FIGURE 14. X-ray Fluorescence Spectrum

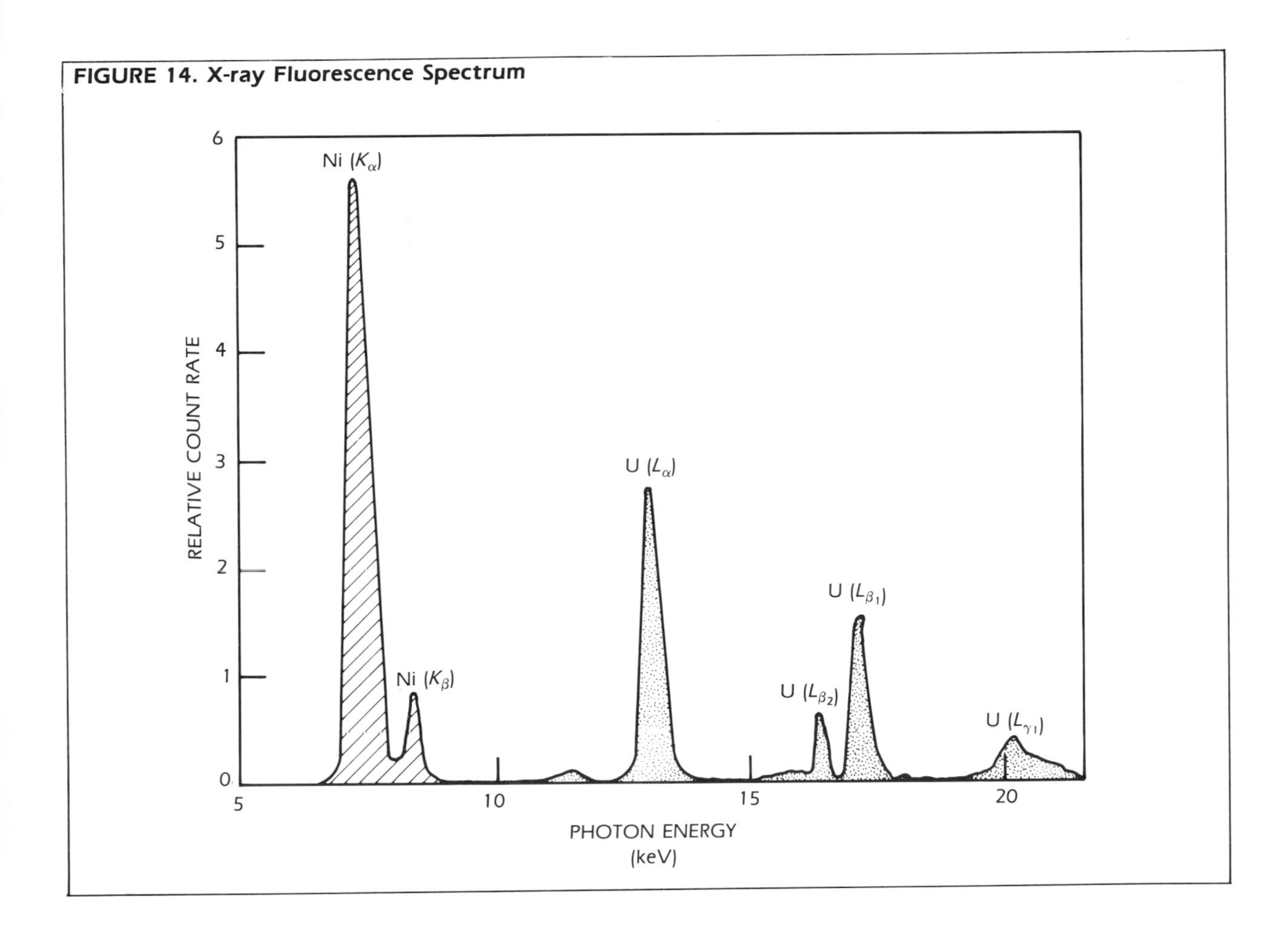

PART 9

BETA BACKSCATTER

Beta Backscatter Principles

Consider the collimated electron beam impinging on a material. At any random point below the surface of the material, an electron may interact in one of three ways. It can be transmitted and moved further into the material, absorbed and re-emitted as a photon, or it can be scattered. Some of the scattered electrons will travel back along the beam axis and re-emerge at the surface, where they can be counted. This is the backscatter effect.

The electrons counted at the surface have arrived from different points within the material. For a given material, the probability that an electron will be backscattered is independent of depth. Because the beam is attenuated as it penetrates the material, the total number of electrons backscattered from a particular plane decreases with the depth of that plane. Therefore, there is a characteristic point within the material where electrons no longer return to the surface; this is known as the *saturation depth.* Seen in terms of thickness, an increase in thickness beyond this point will produce no further increase in the backscatter count rate. To the beam, the material appears infinitely thick; the count rate is known as the *saturation value.*

Backscatter is found to be dependent on atomic number and electron energy. If the ratio of backscattered electrons to impinging electrons (the *backscatter coefficient*) is plotted against atomic number, a curve like that in Fig. 15 is found. As atomic number increases the backscatter coefficient, and thus the number of electrons counted at the surface, increases. If the energy of the electrons is increased, they are less likely to be absorbed, and the saturation thickness increases; the beam can "see" further into the material. Increasing the number of electrons has no effect on the backscatter coefficient or the saturation thickness.

Probe Configuration

In order to utilize beta backscatter for thickness measurements, three basic modules are required: a source of electrons, a detector, and an analyzer. A typical probe configuration is shown in Fig. 16.

The source is a beta-emitting radioactive isotope mounted in a platen. Table 3 lists radioisotope sources commonly used for beta backscatter. The beta particles pass through an aperture, impinge on the test object and are backscattered through the aperture to be detected by a Geiger-Muller tube positioned beneath the source. The electronic pulses generated by the GM-tube are processed in the analyzer.

If the case of a coating-substrate combination is now considered, the count rate can adopt two extremes. With no coating, the count rate X_o corresponds to that of the substrate; with a coating of saturation thickness or more, the count rate X_s will correspond to that of the coating material. For a coating less than saturation thickness, the count rate X_m will lie somewhere between X_o and X_s, the actual value being dependent upon the thickness; this is the basis of the measurement. In practice, it is found that all substrate-coating combinations have dependencies similar to that shown in Fig. 17. In this case, instead of the absolute count rate X_m, the normalized count rate has been plotted.

FIGURE 15. Backscatter Coefficient as a Function of Atomic Number

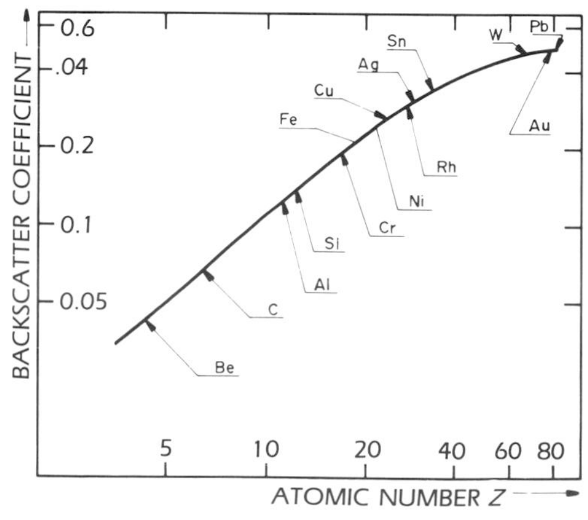

Normalization

With a stable geometry, the only probe parameters that can influence the count rate are:

1. the gradual decrease in isotope activity;
2. the temperature coefficient of the GM-tube; and
3. the aging of the GM-tube.

These can be accounted for by normalizing the count rates using the formula:

$$X_n = \frac{X_m - X_o}{X_s - X_o}$$

All normalized count rates are between 0 and 1, and because the measured quantity is a ratio in which any external influence appears both in the numerator and denominator, for a given isotope, the measurement becomes independent of all probe parameters.

The normalized count rate curve, as shown in Fig. 17, can be characterized by two parameters L and D for a particular material and isotope. Changing L is equivalent to changing the slope of the logarithmic range and changing D causes a parallel shift in the entire curve.

Measurement Error

Statistical Error

Because the electron source is a radioactive isotope, values of X_o, X_s and X_m are all subject to statistical scatter. This means that in practice the normalized curve is not a line but rather a band (see Fig. 18) in which the measurement is likely to fall. This may at first seem to be an inaccurate method, but the statistics of radioactive decay are well

FIGURE 16. Cross-sectional View of a Beta Backscatter Probe

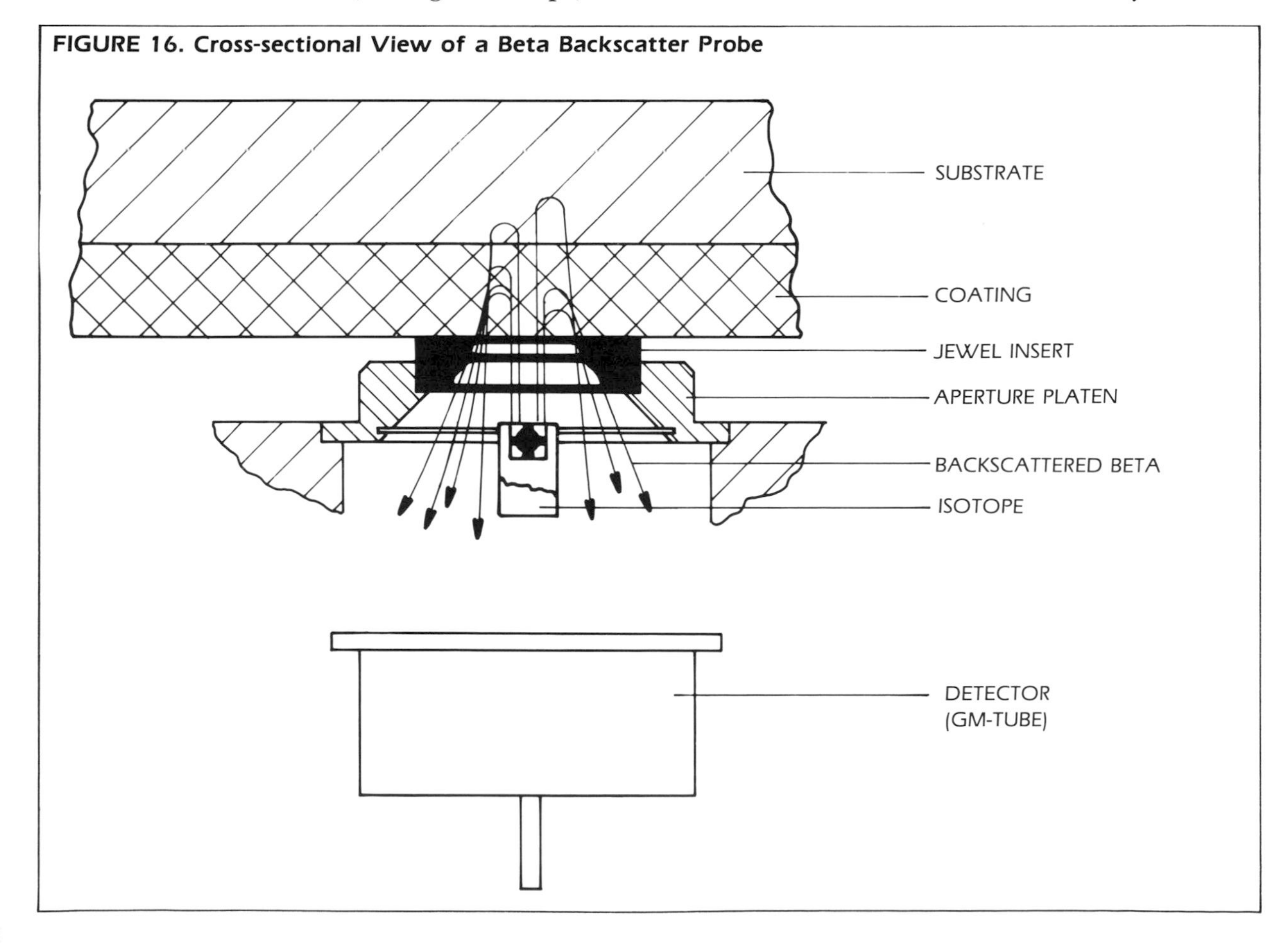

TABLE 3. Radioisotopes Used for Beta Backscatter

Source	Energy (MeV)	Half-life (years)
^{109}Cd	0.20	1.2
^{147}Pm	0.22	2.65
^{204}Tl	0.76	3.65
^{210}Bi	1.17	19.4
^{90}Sr	2.27	28.0
^{106}Ru	3.5	1.0

FIGURE 17. Coating Thickness as a Function of Normalized Count Rate and Parameters *L* and *D*

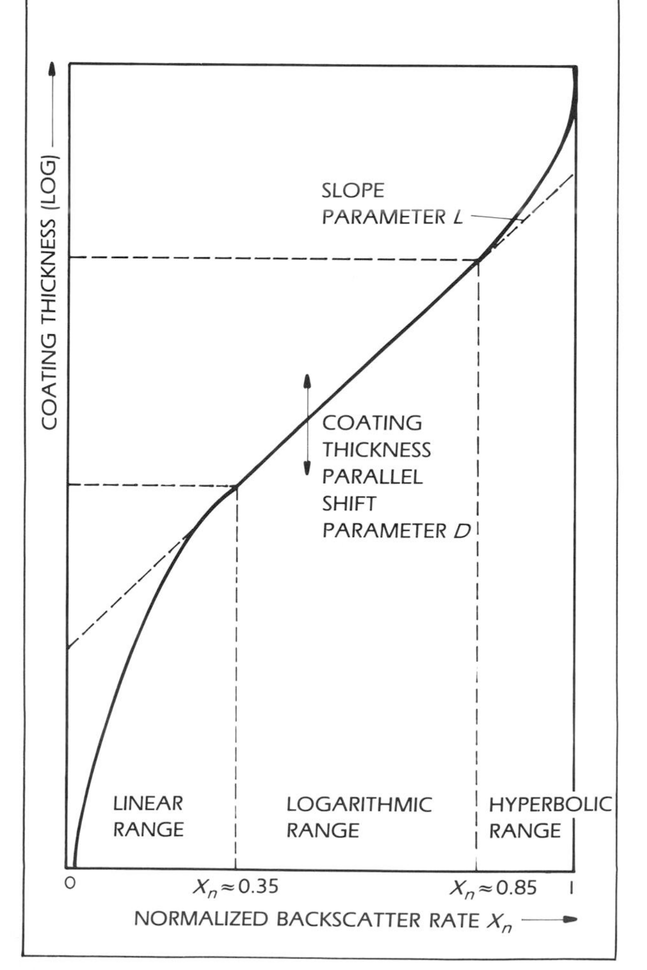

defined and, provided that a sufficient number of measurements are taken, the accuracy of the results is very high.

Density Correction

Density is a parameter which ultimately concerns the number of atoms packed into a particular volume. Although a change in density will produce no difference in saturation count rate, it will make a difference to the saturation depth. Thus, the thickness values obtained for a coating of less than theoretical density must be corrected by the factor *ρ theoretical/ρ practical.* This correction may be made separately on all thickness calculations or may be built into the calibration curve by multiplying the *D* parameter by the same factor. The results displayed are automatically corrected for density.

Other Parameters

The beta backscatter method relies on the difference between two count rates. In order to achieve a sufficient change in count rate for a small change in coating thickness, the count rates X_o and X_s must differ considerably. In practice, a difference in atomic number of at least 20% is required if the

FIGURE 18. Scatter Band of the Calibration Curve

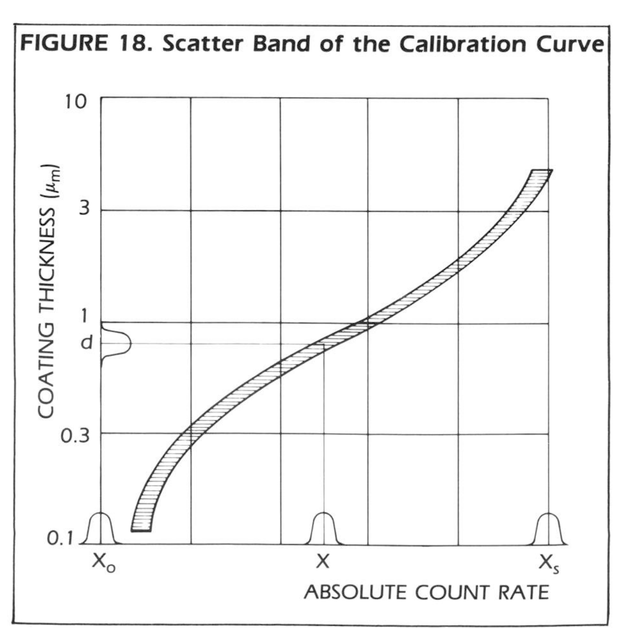

method is to be accurate.

X_o and X_s are also influenced by the size of the probe aperture. As the aperture decreases in diameter, the number of electrons that can return to the counter is decreased and the absolute difference in X_o and X_s becomes smaller, thus increasing the relative error of measurement. For this reason, it is advisable to use the largest possible aperture for a particular set of conditions.

The method is also affected by composition changes in the test material. These are assumed to be constant for the technique, and any inhomogeneities lead to experimental error. Conversely, the backscatter method also can be used to determine alloy compositions.

Conclusion

Radiation gaging encompasses a wide variety of techniques for nondestructive evaluation of materials, processes and finished products. These techniques are used to measure and monitor elemental composition, density and thickness, and are unaffected by other properties or conditions of the material being examined. The processes involved in radiation gaging are well understood and theoretical models for gages can be quite accurate. The high reliability of radiation measuring instrumentation, and the insensitivity of properties other than those of interest, make radiation gages extremely reliable nondestructive testing tools.

REFERENCES

1. Reynolds, G.M. "Neutron Gaging Systems." *Practical Applications of Neutron Radiography and Gauging.* ASTM STP 586. American Society for Testing and Materials (1976): pp 58.
2. Adams, F. and R. Dams. *Applied Gamma-ray Spectrometry,* 2nd ed. Oxford: Pergamon Press (1970).
3. Lederer, C.M., V.S. Shirley, et al. *Table of Isotopes,* 7th ed. New York: John Wiley & Sons (1978).
4. Gignoux, D. "Theory of Radiation Gaging" (private communication).
5. Knoll, G.F. *Radiation Detection and Measurement.* New York: John Wiley & Sons (1979).
6. Koicki, S., A. Koicki and V. Aydacic. *Nuclear Instrumentation and Methods* (1973): pp 297.
7. Barrett, H. and W. Swindell. *Radiological Imaging; the Theory of Image Formation, Detection and Processing.* Vol. 2. New York: Academic Press (1981).
8. Brooks, R.H., et al. "Aliasing: A Source of Streaks in Computed Tomograms." *Journal of Computer Assisted Tomography.* Vol. 3, No. 4 (1979): pp 511-518.
9. Chesler, D.A., et al. "Noise Due to Photon Counting Statistics in Computed X-ray Tomography." *Journal of Computer Assisted Tomography.* Vol. 1, No. 1 (1977): pp 64-74.
10. Hanson, K.M. "Detectability in Computed Tomographic Images." *Medical Physics.* Vol. 6, No. 5 (1979).
11. Morgan, I.L., et al. "Examination of Concrete by Computerized Tomography." *ACI Journal* (January-February 1980).
12. Hopkins, F.F., et al. "Industrial Tomography Applications." *IEEE Transactions on Nuclear Science.* Vol. NS-28, No. 2 (April 1981).
13. Boyd, D.P., A.R. Murgulir and M. Korobkin. "Comparison of Translate-Rotate and Pure Rotary Computed Tomography (CT) Body Scanners." *Optical Instrumentation in Medicine VI.* SPIE Vol. 127 (1979).
14. Kak, A.C., et al. "Computerized Tomography Using Video Recorded Fluoroscopic Images." *IEEE Transactions on Biomedical Engineering.* Vol. BME-24, No. 2 (March 1977).
15. Kruger, R.P., G.W. Wecksung and R.A. Morris. "Industrial Applications of Computed Tomography at Los Alamos Scientific Laboratory." *Optical Engineering.* Vol. 19, No. 3 (1980).
16. Koeppe, R.A., et al. "Neutron Computed Tomography." *Journal of Computer Assisted Tomography.* Vol. 5, No. 1 (February 1981): pp 79-88.
17. Whittaker, J.W. and S.G. Snow. "A Radiation Attenuation Technique for Simultaneous Determination of Layer Thickness in a Bi-Layered Structure." *Materials Evaluation.* Vol. 34, No. 10 (October 1976): pp 224.
18. Coulter, J.E. *X-ray Fluorescence Technique for Measuring Coating Thickness.* Y-1927. Oak Ridge Y-12 Plant (1974).
19. Evans, R.D. *The Atomic Nucleus.* New York: McGraw-Hill Book Company (1955): pp 676.
20. White, J.D. *Compton Scattering Technique for Measuring the Area Density of Glass-Reinforced Structures.* Y-1714. Oak Ridge Y-12 Plant (1970).
21. Foukes, R.A., J.S. Watt, B.W. Seatonberry, A. Davison, R.A. Greig, H.W.G. Lowe and A.C. Abbott. *International Journal of Applied Radiation and Isotopes.* Vol. 29, No. 12 (1978): pp 721.

SECTION 17

X-RAY DIFFRACTION AND FLUORESCENCE

Ron Jenkins, Philips Electronic Instruments, Mahwah, NJ

INTRODUCTION

Lawrence E. Bryant
University of California
Los Alamos National Laboratory
Los Alamos, New Mexico, USA 87545

Everyone expects a nondestructive testing organization to have the capability of performing radiographic inspection, ultrasonic examination and probably penetrant, magnetic particle and eddy current testing. How many people, even NDT persons, expect X-ray diffraction and X-ray fluorescence to be one of the areas of expertise in the NDT organization? Perhaps not many and in some cases this is because in large organizations these techniques may be performed in the analytical chemistry department or elsewhere.

We would like to take the opportunity, as this section of the Handbook is introduced, to encourage the idea that these techniques *can* have a role in NDT by emphasizing their complementary nature to the more conventional nondestructive testing techniques. For example, radiography can find a foreign inclusion, but cannot identify it. If this identification is pertinent, X-ray fluorescence may be able to identify the inclusion and even estimate its concentration.

Whereas some of the techniques and equipment described in this section may indeed be thought of as the tools of an analytical chemist, there are also methods described that are appropriate for the type of information, quick turn-around time and portability that is often expected of a nondestructive testing organization.

It is hoped that this presentation of principles, description of equipment, and the examples of applications will be of value to those already working with X-ray diffraction and fluorescence techniques. Equally important, we hope those involved in "conventional" techniques will see opportunities to broaden their capabilities as well as to enhance and complement existing expertise.

PART 1

X-RAY ANALYTICAL METHODS

Basic Principles

X-rays were discovered by Wilhelm Roentgen in 1895[1]. They are a form of electromagnetic radiation produced in an atom when inner orbital electrons are ejected and the outer electrons move to fill the positions near the nucleus. This transition takes the outer electrons from states of high to low energy, energy that is released in the form of X-rays.

When a beam of this X-ray energy falls onto a specimen, three basic phenomena may result: absorption; scatter; or fluorescence. These phenomena form the basis of several important X-ray analytical methods.

Absorption Analysis

The absorption of X-rays increases with the atomic number of the absorbing matter. This property of X-rays was quickly established and applied to medical diagnosis. At one time, it was also used for the analysis of materials, but these methods have now been superceded by X-ray fluorescence. Today, absorption methods are only found in more specialized fields, such as X-ray absorption-edge fine structure analysis (EXAFS). The absorption effect is still important in establishing a relationship between X-ray intensity and element composition or phase. In X-ray diffraction and fluorescence, *phase* is any chemically homogenous, physically distinct constituent of a substance.

X-ray Fluorescence Spectrometry

Fluorescence occurs when an intense X-ray beam irradiates a specimen and characteristic X-ray spectra are emitted. *Spectrometry* may be defined as the recording of this emission spectrum and the separating of it into its component parts, each part being characteristic of an element.

X-ray fluorescence spectrometry consists of two methods. The first of these is *wavelength dispersive spectrometry*, which uses the diffracting power of a crystal to isolate narrow wavelength bands from the polychromatic characteristic radiation excited in the sample. The second, *energy dispersive spectrometry*, uses a proportional detector to isolate the energy bands (see Part 5).

Because of the known relationship between emission wavelength and atomic number, identification of an element can be made by isolating individual characteristic lines, and elemental concentrations can be estimated from characteristic line intensities. Thus, these two techniques are means of materials characterization in terms of chemical composition.

X-ray Diffraction

The discovery of X-ray diffraction was made by Max von Laue in 1913[2] and may be defined as changes in the scattering characteristics of X-rays due to collision with some object in their path. Diffraction is a special case of X-ray scattering that can be used for the identification of elemental phases.

Scattering occurs when an X-ray photon interacts with the loosely bound outer electrons of an element. When this collision is elastic (no energy is lost in the collision process), the scatter is said to be coherent (or Rayleigh) scatter. Coherently scattered photons may undergo subsequent interactions with other scattered photons, causing reinforcement or interference.

Under certain geometric conditions, scattered wavelengths which are exactly in phase, or exactly out of phase*, may add to one another (reinforcement), or cancel one another out. The coherently scattered photons that constructively interfere with each other give diffraction maxima (peaks in the X-ray diffraction diagram).

*From this point in the text to the next heading, *phase* is used in its physical sense: a uniform motion varying according to simple harmonic laws.

A crystal lattice, for example, consists of a regular arrangement of atoms and, when a monochromatic beam of radiation falls onto these atomic layers, scattering will occur. In order to satisfy the requirements for interference, it is necessary that the scattered waves originating from the individual atoms (the scattering points) be in phase with one another.

The geometric conditions necessary for the waves to be in phase are illustrated in Fig. 1. Two parallel rays strike a set of crystal planes at an angle θ (theta) and are scattered. (Diffraction angles are labeled θ for the specimen-source angle, and 2θ for the source-specimen-detector angle.)

Reinforcement will occur when the difference in the path lengths of the two waves is equal to a whole number of wavelengths. This path length difference is equal to CB + BD (Fig. 1). For reinforcement to occur, CB must equal BD; and if CB = BD = X, then 2X must equal $n\lambda$, where n is an integer, and λ is the wavelength.

It will also be seen that $X = d \times \sin\theta$, where d is the interplanar atomic spacing. Hence, the overall condition for reinforcement is:

$$n\lambda = 2d \times \sin\theta \qquad \text{(Eq. 1)}$$

which is a statement of Bragg's Law.

Role of Crystal Structure in X-ray Scattering and Diffraction

All substances are built up of individual atoms and nearly all substances have some degree of order, or periodicity, in the arrangement of these atoms.

A crystal is a highly ordered substance which can be defined as a homogeneous, anisotropic body (exhibiting properties with values that vary when measured on different axes), having the natural shape of polyhedron.

In practical terms, determining the homogeneity of a substance depends on the means available for measuring the crystallinity. In general, the shorter the diffracted wavelength, the smaller the recognizable crystalline region.

Even noncrystalline materials have a degree of order and each will give some sort of a diffraction pattern. For example, glassy materials and liquids will generally give diffraction patterns in the form of one or more broad diffuse peaks or halos. In X-ray powder diffractometry (see Part 2), one is usually dealing exclusively with crystalline materials. Since every ordered material is made up of a unique arrangement and number of atoms, every ordered

FIGURE 1. Geometric Conditions for Diffraction of X-rays

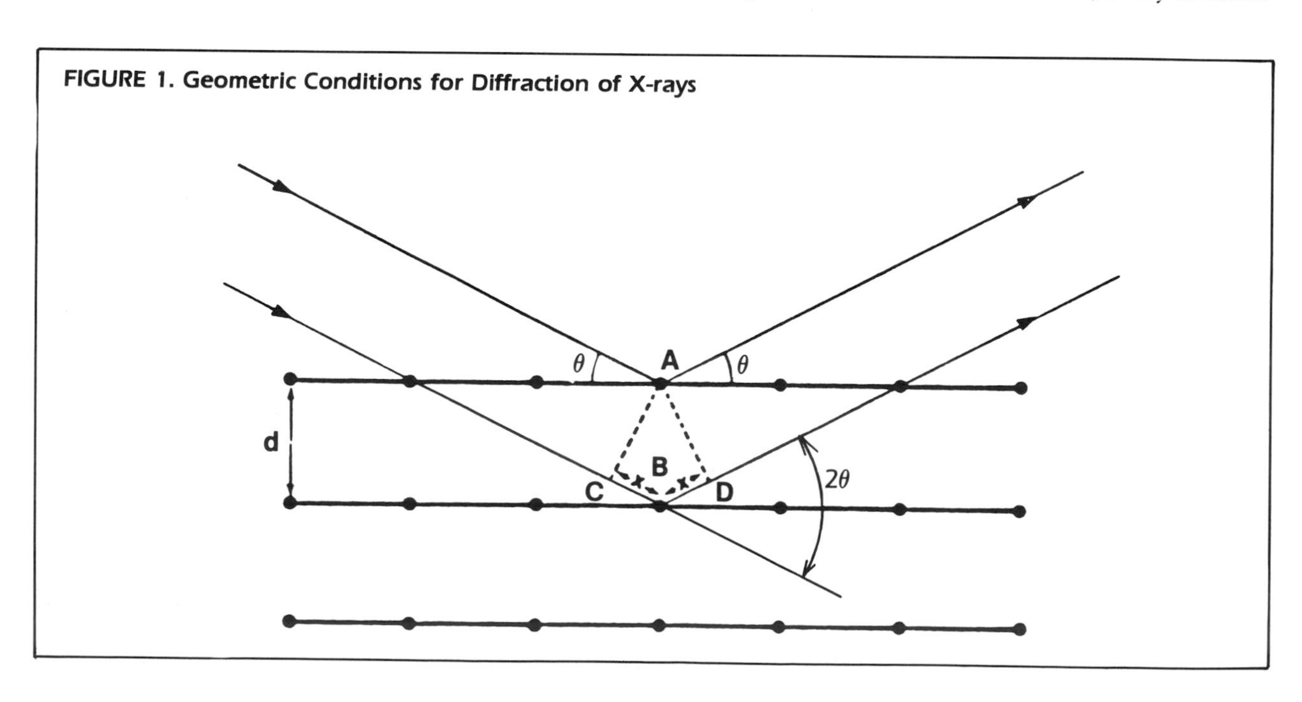

material will give a diffraction pattern which is, to all intents and purposes, also unique. The diffraction pattern can, moreover, be used to determine the degree of crystallinity; that is, the dimensions of the crystalline regions in otherwise amorphous substances.

X-ray Spectra

X-rays are manifested in the form of continuous or characteristic spectra. Continuous spectra are generated when high voltage electrons impinge on a target element, as in an X-ray tube. The electrons are decelerated as they approach the target atom and X-ray energy is emitted as a continuous band of radiation. The limits of this continuum are defined by the maximum energy of the electrons (the voltage of the X-ray tube) and a long wavelength cut-off, defined by the transmission characteristics of the X-ray tube window.

Superimposed on top of this continuous or "white" radiation are sharp lines whose wavelengths are characteristic of the atomic number of the target element.

A typical diagram of the spectral output from a diffraction tube, operated at 50 kV and 20 mA, is found in Fig. 2. For diffraction experiments, this radiation is often filtered to reduce the $K\beta$ emission which would otherwise complicate the diffraction diagram. ($K\beta$ emission is part of the K series, an element's characteristic wavelengths, and usually occurs in pairs with $K\alpha$; the K series is caused by disruption of electrons in the element's K shell and the resulting spectral lines are unique to each element.)

When an atom is bombarded with high energy particles, for example X-ray photons, an inner orbital electron may be displaced, leaving the atom in an excited state. The atom can regain stability by rearrangement of the atomic electrons and the energy excess is emitted as a characteristic X-ray photon. Not all vacancies result in the production of characteristic X-ray photons since there is a competing internal rearrangement process known as the Auger effect[3].

The ratio of the number of vacancies resulting in the production of characteristic X-ray photons to the total number of vacancies created in the excitation process is called the *fluorescent yield*. The

FIGURE 2. Diffraction Tube Spectral Output from (a) Cobalt-Target Tube at 50 kV, and (b) Filtered X-ray Tube

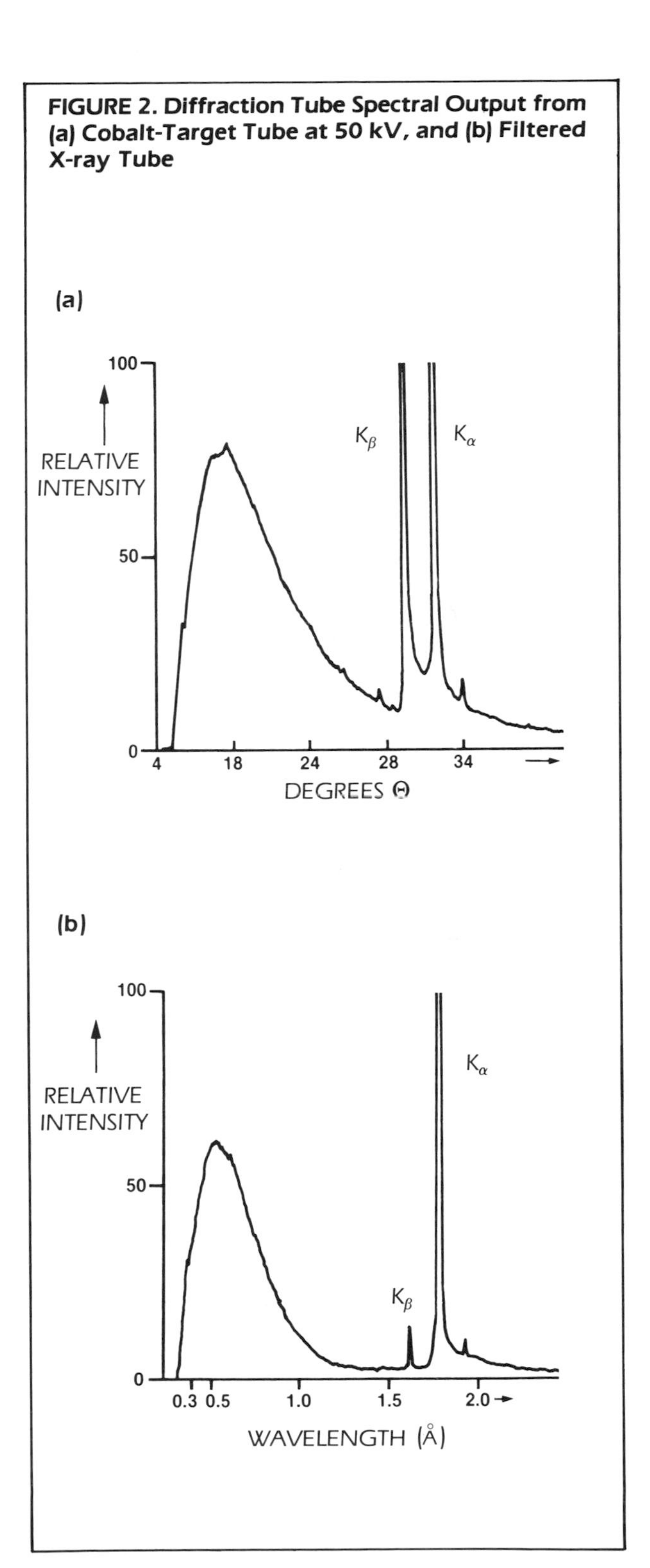

fluorescent yield takes a value around unity for the high atomic number elements, and drops to as low as 0.01 for the elements with low atomic numbers (such as sodium, magnesium and aluminum).

For those vacancies giving rise to characteristic X-ray photons, a set of selection rules can be used to define which electrons can be transferred: their *principal quantum number* (the number which determines the electron's energy level and average distance from the nucleus), n, must change by 1; their *angular quantum number*, l, must change by 1; and the total change in momentum given by the vector sum of $l + s$ must be a positive number changing by 1 or 0.

In effect this means that for the K series, only p → s transitions (s vacancy filled by p level electrons) are allowed, yielding two lines for each principal level change.

Vacancies in the L level follow similar rules and give rise to *L series* lines. There are more of the L lines since p → s, s → p and d → p transitions are all allowed within the selection rules. Found in Fig. 3 is an example of a typical high atomic number K spectrum for the element tin, atomic number 50.

Seven lines are shown representing two pairs of normal transitions, one unresolved doublet and two "forbidden" transitions.

The two pairs of lines for the normal transitions are the $\alpha 1/\alpha 2$ (2p → 1s), and the $\beta 1/\beta 3$ (3p → 1s). The unresolved pair is shown as the $\beta 5$ (4p → 1s), and the two forbidden transitions as $\beta 5$ (3d → 1s), and $\beta 4$ (4d → 1s).

The K series for the lower atomic number elements is somewhat simpler. Very low atomic numbers (less than 18) will, however, show some satellite lines due to dual ionization of the excited atom.

FIGURE 3. The K Emission Spectrum of Tin

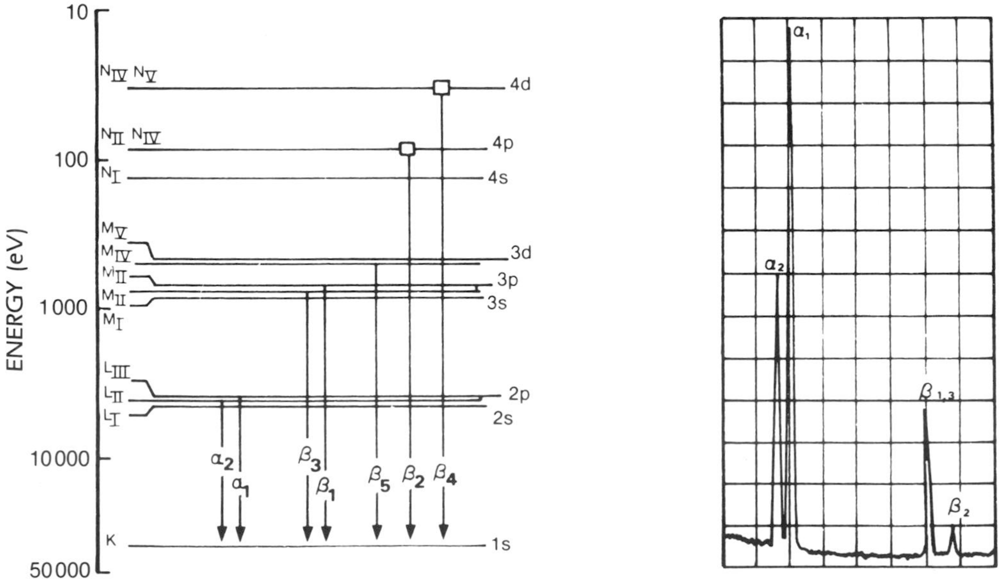

Positioning of Analytical Instrumentation

There are many similarities between the instrumentation used for diffraction and spectrographic measurements, particularly in the case of the wavelength dispersive spectrometer.

As examples, the high voltage generator used to power the X-ray tube is generally of the same type; the digital and analog counting electronics and goniometer (angle measurement or control) circuitry is almost identical. It is often found that in a given laboratory, the high voltage generator and counting electronics are actually shared between the diffractometer and wavelength dispersive spectrometer.

Shown in Fig. 4 arc layout diagrams of the instrumentation used in the three different analytical methods.

FIGURE 4. X-ray Analytical Instrumentation

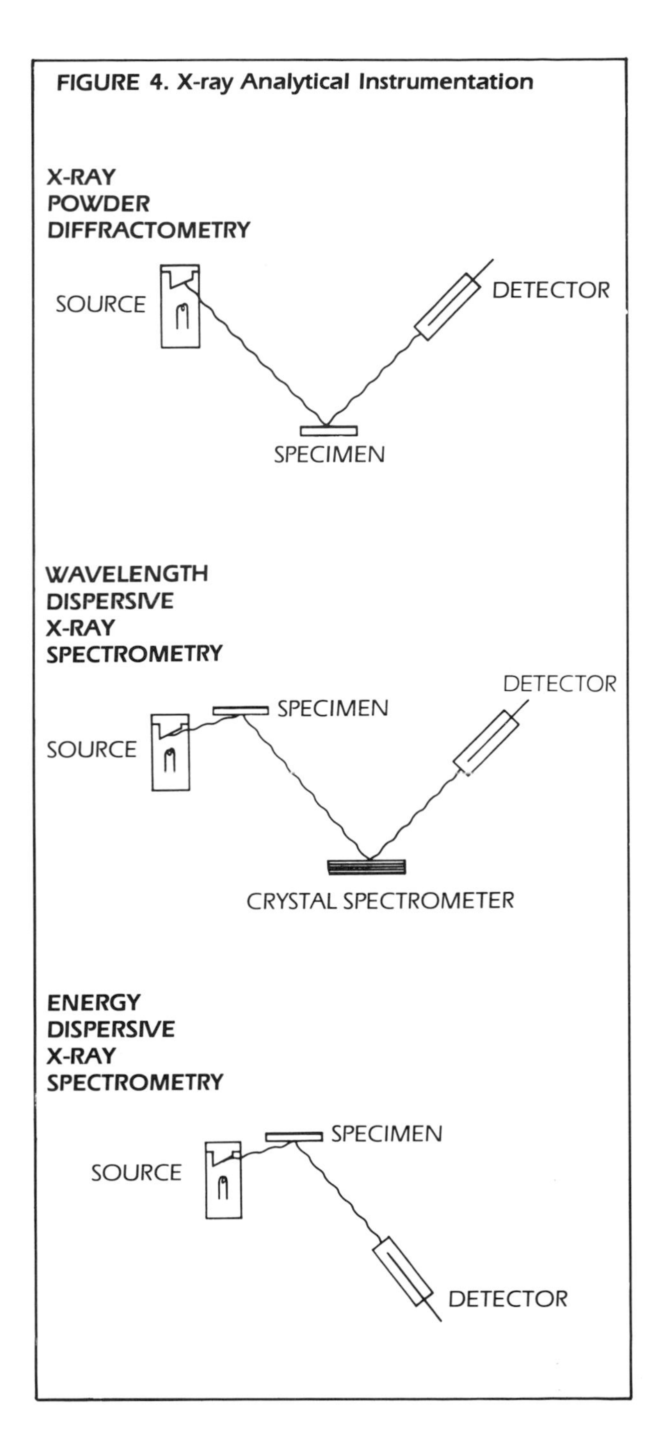

PART 2

INSTRUMENTATION FOR POWDER DIFFRACTION

Basis of the Powder Method

The powder method derives its name from the fact that the specimen is typically in the form of a microcrystalline powder, although as has been indicated, any material which is made up of an ordered array of atoms will give a diffraction pattern. Of all the methods available to the analyst, only X-ray diffraction is capable of providing general purpose qualitative and quantitative information on the presence of phases in an unknown mixture. A diffraction pattern is characteristic of the atomic arrangement within a given phase, and to this extent it acts as a fingerprint of that particular phase. Thus, by using a library of powder data patterns, a series of potential matches can be obtained.

FIGURE 5. The Debye-Scherrer Powder Camera

The Powder Camera

The Debye-Scherrer Camera

The simplest device for the measurement of a powder diffraction diagram is the Debye-Scherrer powder camera, pictured in Fig. 5. A cylindrical camera body carrying an entrance pinhole collimator and an exit beam collimator lies along the diameter of the device. The specimen is mounted in a thin cylinder (often a capillary tube) at the camera's central axis. A piece of film is placed inside the cylindrical wall of the camera and small holes are cut in the film for the entrance and exit collimators (see Fig. 6a).

X-rays are directed onto the specimen via the entrance collimator and the diffracted radiation falls onto the film. The film is then developed and laid flat for interpretation (Fig. 6b).

In order to simply estimate the 2θ values from the film, the diameter of the camera is fashioned so that one or two mm of film correspond to exactly one degree of the 2θ angle.

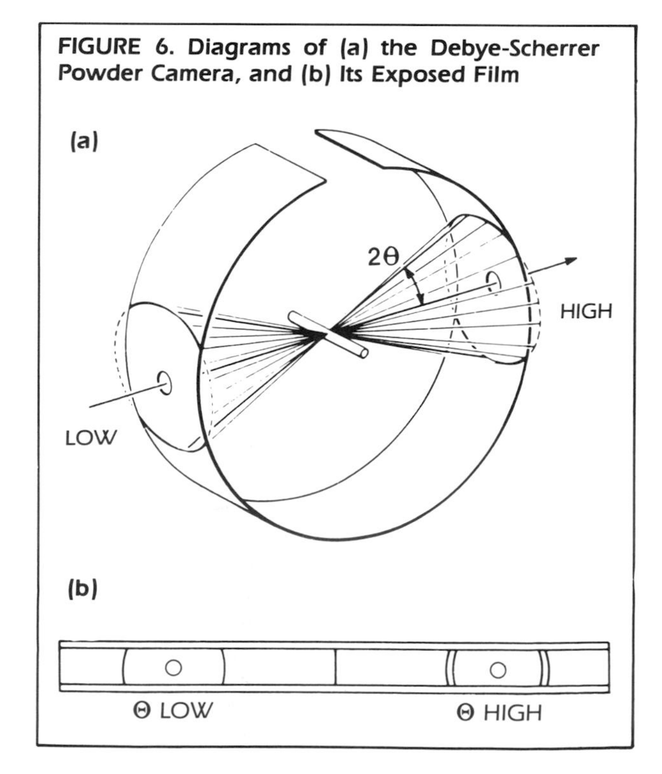

FIGURE 6. Diagrams of (a) the Debye-Scherrer Powder Camera, and (b) Its Exposed Film

The powder camera finds great application in the analysis of very small specimens and good patterns can be obtained from only a few mg of specimen. This camera technique has the disadvantage of being rather slow, with exposure times typically in the range of one to six hours.

Specialized Powder Cameras

In addition to the Debye-Scherrer powder camera, there are also cameras available for more specialized applications. As an example, the Guinier camera is a powder camera which incorporates a focusing monochromator (which can transmit a single wavelength) and works in a transmission mode (providing measurements of radiation transmitted through the specimen). Although the intensities are rather unreliable and it is more difficult to align than the Debye-Scherrer camera, the Guinier camera can give very high quality patterns in a relatively short period of time.

Another useful camera is the Gandolphi camera, which is essentially a Debye-Scherrer camera with an acentric specimen rotation movement added, allowing the generation of a simulated powder pattern from a small single crystal.

The Powder Diffractometer

The powder diffractometer is a device which allows a range of 2θ values to be scanned by rotating the photon detector at twice the angular speed of the specimen and thus maintaining the required geometrical conditions.

The specimen consists of a random distribution of crystallites so that the appropriate planes will be in the orientation needed to diffract the wavelength each time the Bragg condition is satisfied.

In X-ray diffraction, each peak angle value (value of the 2θ angle for which maximum peaks occur in the diagram) corresponds to a specific d spacing. (See Fig. 7 for the geometry of a typical vertical powder diffractometer system.) This geometric arrangement is known as the Bragg-Brentano parafocusing system, and is typified by a diverging beam from a line source, F, falling onto the specimen, S, being diffracted and passing through the receiving **slit, G, which is also on the focusing circle, of radius r. A detector is placed behind the receiving slit to collect the diffracted photons. Distances FS and SG must be the same and equal to the goniometer circle radius. R.**

Diffractometer Divergence

The amount of divergence is determined by the effective focal width of the source and the aperture of a divergence slit which is typically placed between the source and the specimen. Axial divergence is controlled by two sets of parallel plate collimators (Soller slits) placed between (1) focus and specimen, and (2) specimen and scatter slit. A photograph of a typical powder diffractometer system is shown in Fig. 8.

Use of the narrower divergence slit will give a smaller specimen coverage at a given diffraction angle, thus allowing the attainment of lower diffraction angles and larger values of d when the specimen has a large apparent surface. This is achieved, however, only at the expense of intensity loss.

The Diffractogram

Choice of the divergence slit, plus its matched scatter slit, is governed by the angular range to be covered. The decision on whether or not to increase the slit size (at a given angle) will be determined by the available intensity. A detector, typically a scintillation counter, converts the diffracted X-ray photons into voltage pulses.

These pulses may be integrated in a rate meter to give an analog signal on a strip-chart recorder. By synchronizing the scanning speed of the goniometer with the recorder, a plot is obtained for degrees 2θ vs. intensity. This is called the *diffractogram*.

A timer/scaler is also provided for quantitative work and this is used to obtain a measure of the integrated peak intensity of the selected line(s) from each analyte phase in the specimen. A diffracted beam monochromator may also be used in order to improve signal-to-noise characteristics.

Shown in Fig. 9 is a typical diffractogram for zinc oxide recorded over the range of 25 to 75 degrees 2θ. The diffraction pattern is essentially a plot of degrees 2θ vs. intensity and each phase is characterized by a series of peaks on the diagram, each peak representing, in the simplest case, a unique d spacing (a unique interplanar atomic spacing). Each phase will give a characteristic set of

FIGURE 7. Geometry of the Parafocusing Diffractometer

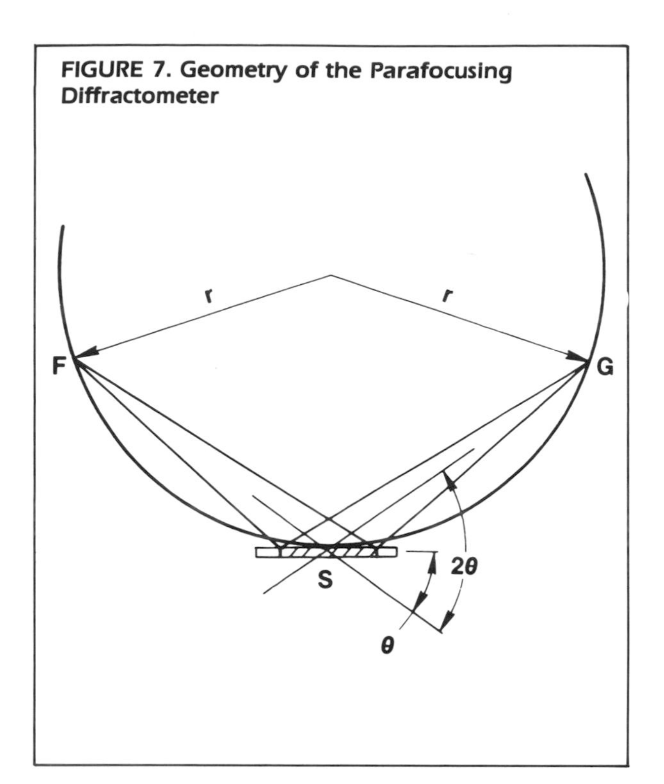

FIGURE 8. Typical Powder Diffractometer System

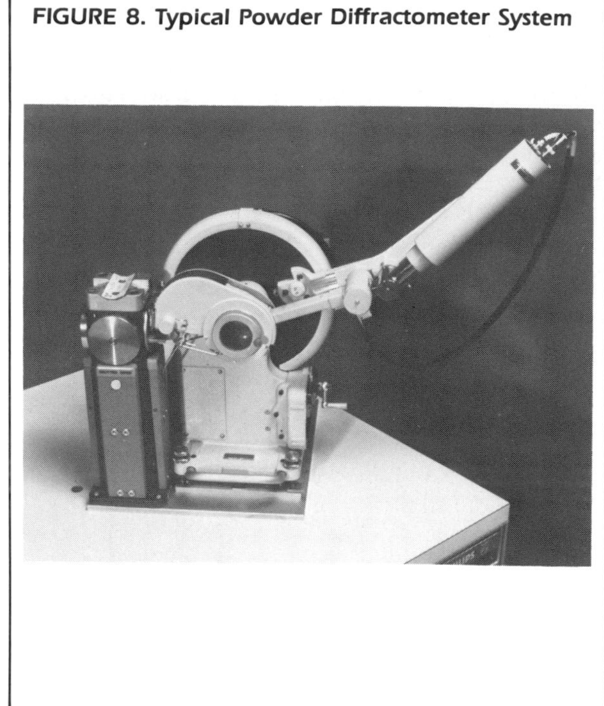

FIGURE 9. Diffractogram of Zinc Oxide

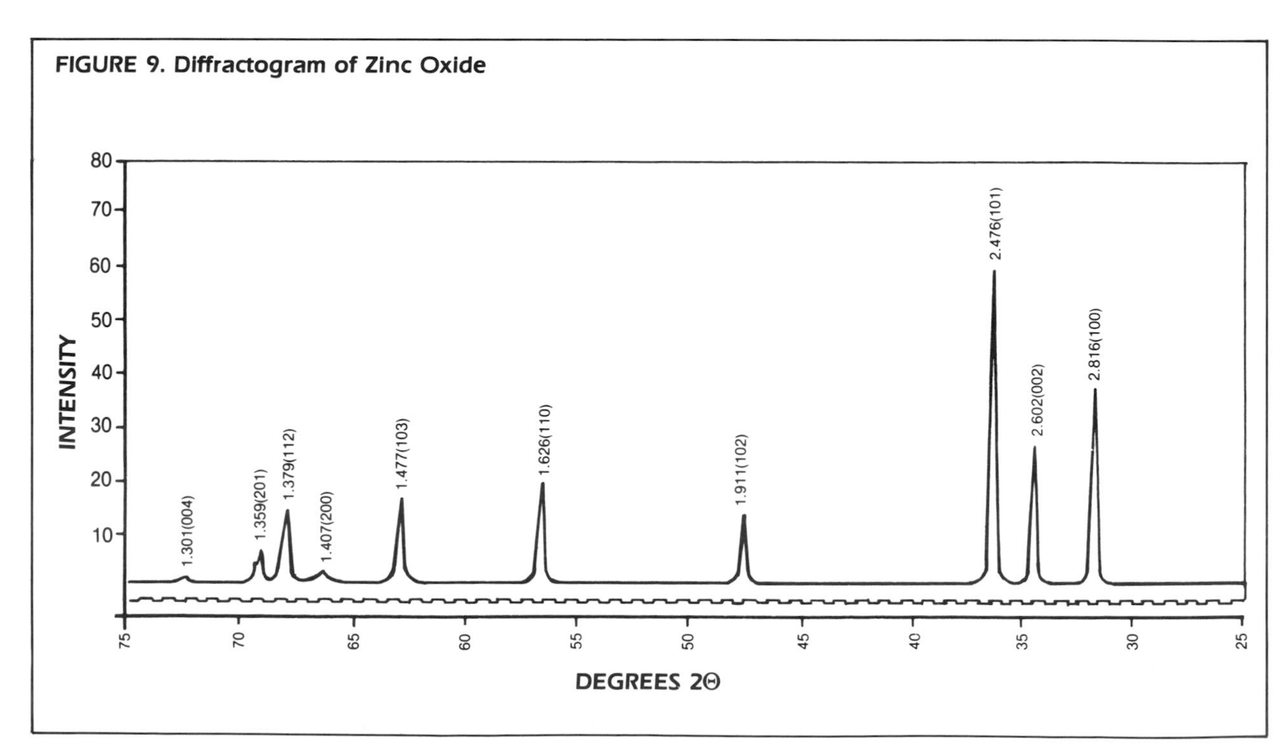

lines and where the specimen is a mixture of phases, the patterns will be superimposed one on top of another. At low angles, single sharp reflections are observed, but at higher angles, these peaks are manifested as doublets because the increased angular dispersion of the diffractometer resolves the $\alpha 1/\alpha 2$ doublet of the diffracted radiation. In order for qualitative identification to be applied to the pattern, the diagram is reduced to a set of d spacing intensity tables.

Special Attachments for the Diffractometer

In addition to the use of the powder diffractometer under normal ambient conditions, it is also possible to obtain special attachments to record diffractograms under nonambient conditions. As an example, attachments are available to record patterns at temperatures ranging from 2000 °C down to the temperature of liquid nitrogen. Such attachments are invaluable for the study of phase changes under experimental conditions such as those which might be found in a catalyst bed.

Exactly the same thermal cycle can be reproduced in the high temperature camera as in a catalyst bed, including rate of temperature rise, holding the catalyst sample at a given temperature for a given time, and so on. Similarly, the use of a low temperature attachment allows the study of low temperature phases in waxes and similar hydrocarbons. Special chambers can be used to hold the specimen in special inert or corrosive atmospheres, or under vacuum. Diffraction patterns can also be recorded under high pressure; in fact, the possibilities are almost unlimited and depend only on the ingenuity of the equipment designer and diffractionist.

PART 3

PHASE CHARACTERIZATION BY X-RAY DIFFRACTION

Qualitative Analysis of Polycrystalline Materials

The possibility of using a diffraction pattern as a means of phase identification was recognized in the 1930s. Later in the same decade, a systematic means was proposed for unscrambling the superimposed diffraction patterns that occur in the analysis of multi-phase materials[4].

The technique was based on (1) the use of a file of single phase patterns (characterized in the first stage by their three strongest reflections) and (2) a search technique for matching strong lines in the unknown pattern with these standard pattern lines.

A potential match was confirmed using the full pattern in question. The identified pattern was then subtracted from the experimental pattern and the procedure repeated on the residual pattern, until all lines were identified. Techniques for this search/matching process have changed little over the years. In the hands of experts, manual search/matching is an extremely powerful tool; for the less experienced user, though, it can be rather time consuming. Round-robin tests have indicated that two to four hours are typically required for the complete identification of a four-phase mixture[5].

A growing complication for this method is that the file of standard patterns increases by about 1,500 each year and currently stands at about 38,000 entries. The responsibility for the maintenance of the Powder Data File lies with the International Centre for Diffraction Data (JCPDS). The Powder Data File is a unique assembly of good quality single-phase patterns and is used by thousands of chemists, geologists and materials scientists all over the world.

Essentially two different approaches are used to identify phases in an unknown mixture and these are illustrated in Figs. 10a and 10b.

The first method (Fig. 10a) is based on a series of guesses based, in turn, on preconceived ideas of what phases might be present. The two basic parameters being used in this search/match process are the d values (which have been calculated from the measured 2θ values in the diffractogram) and the relative intensities of the lines in the pattern. The d values can be accurately measured — perhaps with an accuracy of better than 0.5 percent in routine analysis — but the intensities are, by comparison, unreliable and subject to errors, sometimes running into tens of percent.

The second method (Fig. 10b) is an analytical approach in which no basic assumptions are made about the sample being analyzed. The three strongest lines in the pattern are used to locate potential matches in the JCPDS index. Each time a potential candidate is found, a match is made with the complete pattern. If all lines agree, a phase confirmation is assumed and the lines for the match are subtracted from the original pattern. This process is repeated until all significant lines in the pattern are identified.

FIGURE 10. Two Basic Techniques for Matching Diffraction Patterns

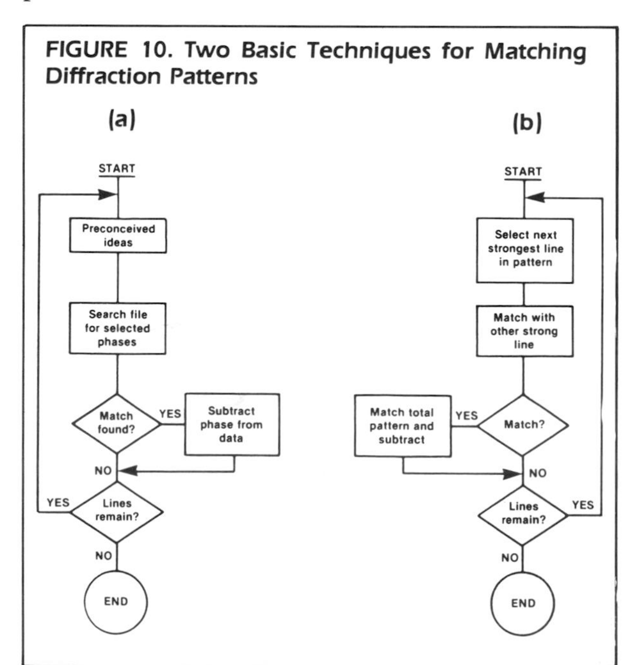

Quantitative Methods

Once the presence of a phase has been established in a given specimen, the analyst can in principle determine how much of that phase is present by using the intensities of one or more of the phase's diffraction lines. However, it may be difficult to obtain an accurate value for these intensities. The intensities of the diffraction peaks are dependent on a number of factors which fall roughly into three categories.

1. *Structure dependent:* a function of atomic size and atomic arrangement, plus some dependence on the scattering angle and temperature.
2. *Instrument dependent:* a function of diffractometer conditions, source power, slit widths, detector efficiency, and so on.
3. *Specimen dependent:* a function of phase composition, specimen absorption, particle size, distribution and orientation.

For a given phase, or selection of phases, all structure-dependent terms are fixed and have no influence on the quantitative procedure. Provided that the diffractometer terms are constant, its effect can also be ignored. Thus, if the analyst calibrates the diffractometer with a sample of the pure phase of interest and then uses the same conditions for the analysis of that phase in an unknown mixture, only the sample orientation and random errors associated with a given observation of intensity have to be considered.

As an example, Fig. 11 shows a series of diffractograms of a quenched steel specimen in which two phases have been identified, the alpha phase and the gamma phase. The integrated area under the peak of the gamma (retained austenite) phase can be used as a measure of the retained austenite in the specimen. For powder specimens, the biggest problem in the quantitative analysis of multi-phase mixtures remains the specimen-dependent terms: particle size; distribution; plus

FIGURE 11. Diffractograms for Measurement of Retained Austenite in Steel

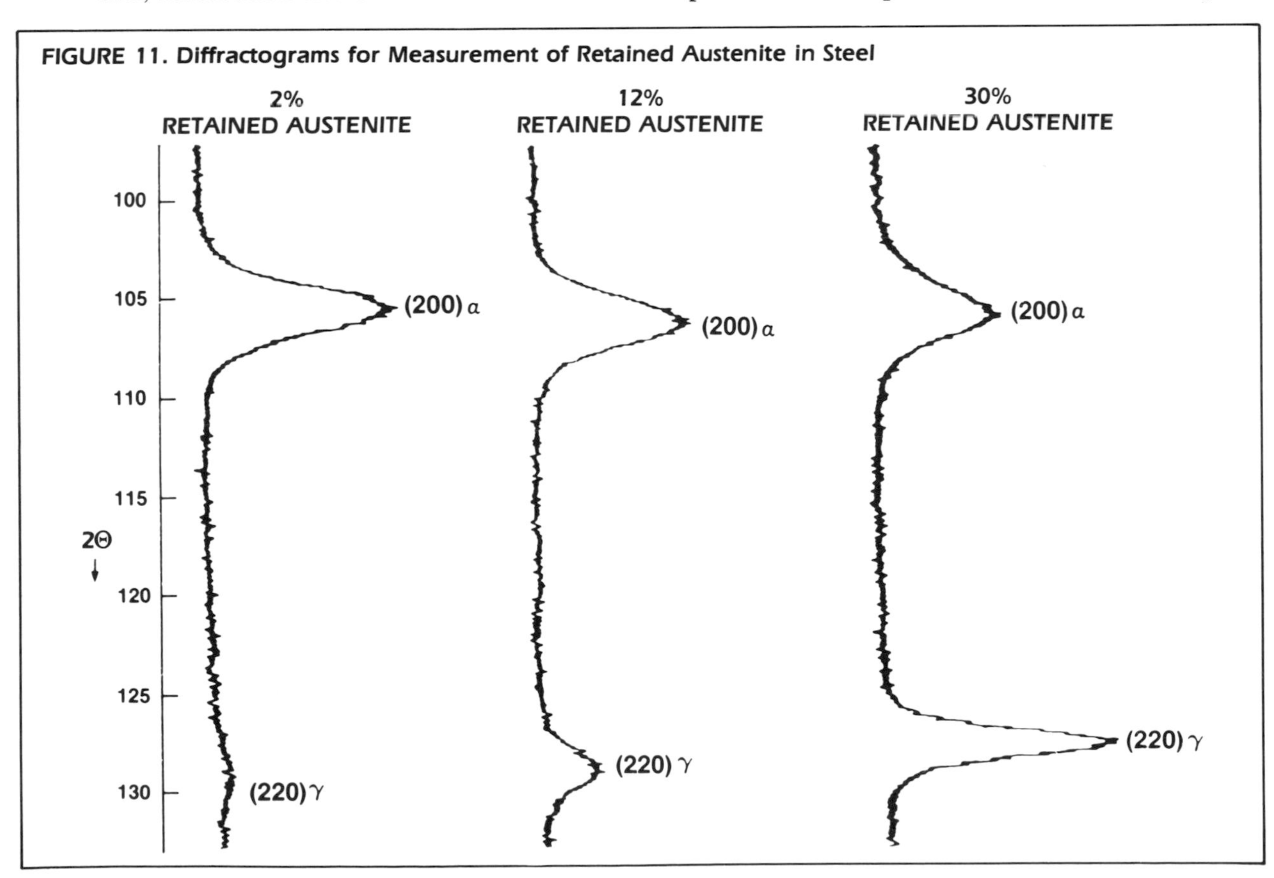

TABLE 1. Typical Detection Limits for Diffraction Samples

Phase Diluent	Kaolinite Talc	α-SiO_2 Calcite	KBr Calcite	Calcium Oxalate Talc
Counts per percent	5,300	7,100	7,150	2,500
Background	98,000	45,000	30,000	122,000
Lower limit of detection	0.12%	0.060%	0.048%	0.28%
Time	8 min	8 min	8 min	8 min
Blank counts	8,000	−570	−1570	2,764
Phase peak	24.40-25.34°	26.20-27.14°	26.64-27.58°	14.50-15.46°
Diluent peak	28.10-29.00°	28.98-29.90°	28.96-29.90°	28.10-29.00°
Divergence slit	1°	1°	1°	1/2°
Receiving slit	1/2°	1/2°	1/2°	1/2°

The data above were gathered using the following technique factors:
Cukα 45 kV 40 mA — Long fine focus tube
Pyrolytic graphite monochromator — 6° take-off angle

effects of absorption. In practice, this problem can usually be reduced to acceptable proportions by extremely careful specimen preparation.

Limitations of the Diffraction Method

There is little doubt that the diffraction method is without parallel as a general purpose tool for phase identification, but as with all analytical techniques, it is not without some shortcomings. Two of these are of particular importance.

The first is related to the chance of misinterpretation during the course of qualitative analysis procedure. Although a diffraction is in principle unique, in practice there are sufficient similarities between patterns to cause confusion. This is particularly so in the case of multi-phase specimens. In addition to this there are probably more than two million possible unique phases, of which 40,000 or less are on the JCPDS file as single phase patterns. There is, as a result, a certain chance that a given phase is not in the file.

The second problem relates to the sensitivity of the powder diffraction method. The diffraction method is not comparable in sensitivity (rate of change in measured signal, per change in analyte phase concentration) to the other X-ray based techniques. By comparison, detection limits in the low parts-per-million can be obtained in X-ray spectrometry, while the powder method has difficulty in identifying several tenths of one percent. The diffraction method is less sensitive than the fluorescence method by about three orders of magnitude.

Table 1 lists some data which were obtained in a study[6] to establish detection limits of typical multiphase mixtures representing a range of crystalline and partially crystalline materials. It will be seen that detection limits typically are in the 0.05 to 0.25 percent region, for an analysis time of eight minutes.

PART 4

OTHER APPLICATIONS OF THE POWDER METHOD

The areas of application for quantitative X-ray powder diffraction are many and varied. Hundreds of analysts are using this technique on a daily basis. Some of the more common applications would include ore and mineral analysis, quality control of rutile/anatase mixtures, retained austenite in steels, determination of phases in airborne particulates, various thin film applications, study of catalysts and analysis of cements. The current state of the quantitative analysis of multi-phase materials is that accuracies on the order of a percent or so can be obtained in those cases where the particle orientation effect is nonexistent or has been compensated for adequately.

Phase Changes

In addition to the identification and quantitation of multi-phase mixtures already discussed, there are many types of more specialized analyses which can be performed with the powder diffraction technique. Small shifts in line position or changes in the shape of a diffracted profile can reveal equally small changes in structural parameters. The modern powder diffractometer is capable of measuring *d* spacings to an accuracy of better than one part in a thousand. Where many lines are used to establish unit cell dimensions, such parameters can be determined to an accuracy of better than one part in ten thousand. (The unit cell is the smallest division of a crystal that maintains the original's characteristics; or, the smallest crystallographic repeat unit). At this level of accuracy, lattice parameter changes due to inclusion of substitutional atoms are observable.

It is also possible to follow the course of a phase change, in a dynamic way, by performing the reaction in the specimen chamber of the diffractometer. Such a procedure is particularly useful where the need is to follow a phase change at high or low temperature. Suitable stages are now available for recording diffraction patterns at temperatures ranging from −180 to 2000 °C.

As an example, the room temperature form of iron is the body-centered cubic (alpha) form. At a temperature of 910 °C, this is transformed into the face-centered cubic (gamma) form, and above 1400 °C iron returns to the body-centered cubic (delta) form. A series of diffractograms in which a sample of alpha iron has been heated in a high temperature furnace (mounted on a diffractometer), with the temperature cycled between 850 and 1100 °C, is shown in Fig. 12.[7]

At the same time, the goniometer has been set at the diffraction angle for the (110) line of alpha iron. It will be seen that as the specimen temperature reaches about 910 degrees, the count rate drops to a noise level indicating the absence of the alpha phase. As the temperature is brought back down below 1000 °C, the alpha phase reappears. Even though there is some hysteresis, the phase transformation is nonetheless clearly apparent.

Study of Chemical Reactions

When atoms of two or more elements occupy similar sites of the same crystal lattice, a *solid solution* is formed. An example of this would be the replacement of nickel atoms by copper atoms giving a so-called substitutional solid solution. There are many examples of such solid solutions which, in practice, may be formed by atoms with atomic sizes within 15 percent of each other, or between compounds of similar structure. Many chemical reactions which take place in the solid state do so by interchange of atoms at the boundary layers of the reacting species, and the powder diffraction method is particularly useful for following this type of chemical reaction. Such reactions are generally classified as solid state *addition and replacement* reactions.

As an example, for many years, tetraethyl lead (TEL) was commonly employed in commercial gasolines as an octane improver. Essentially, the TEL acted as a chain-branching inhibitor, giving a much more regular burning process. In order to remove the lead from the engine after it had done its job, scavengers such as ethylene dibromide and ethylene dichloride were added to the fuel. These converted the lead to volatile halides, which were then swept from the engine with the hot exhaust gases. Within the engine, a whole series of very complex addition and replacement reactions take place and study of the diffraction patterns of engine deposits can yield valuable information about combustion conditions within the combustion zone.

Use of Line Profile Data

In the study of engine deposit materials, one special problem arises during interpretation of the diffraction patterns: the diffraction lines are very broad. This is because the reaction products are formed at temperatures well below their melting points. As a result, the individual crystallites of the various compounds are very small and highly stressed. Both of these factors (size and stress) can be used to gain additional information about the sample under examination and the diffractometer can be used for the measurement of residual stress as well as the estimation of particle size.

The effect of stress/strain on the specimen is illustrated in Fig. 13. The upper diagram shows an unstrained sample, consisting of a series of individual unit cells, and the diffraction line corresponding to the interplanar spacing, d_0. Where a uniform stress is applied to the sample, in the direction of the arrows, a uniform strain results (illustrated in Fig. 13b). A corresponding shift toward a lower angle will occur in the diffraction line because the appropriate planes in the cells are forced closer together. Where a nonuniform stress is applied (Fig. 13c), the cells become distorted. The effect on the diffraction pattern is to produce a broadening of the line profile. By studying line position and shape, and by comparing these data with similar measurements on unstrained specimens, information about residual uniform and nonuniform stress can be obtained[8].

Line broadening can also occur in unstrained specimens when the particle size is small, less than one micron or so. There is a known relationship[9] between the mean crystallite dimension, D, line broadening, b, and the diffraction angle, θ, which can be stated as

$$D = \frac{K \times \lambda}{b \times \mathrm{Cos}\theta} \qquad \text{(Eq. 2)}$$

where K is the shape-factor constant which takes a value between 1 (for spherical particles) and 0.1 (for plate-like particles). The value of b has to be corrected for instrumental broadening since all

FIGURE 12. Series of Diffractograms for Phase Transformation in Iron

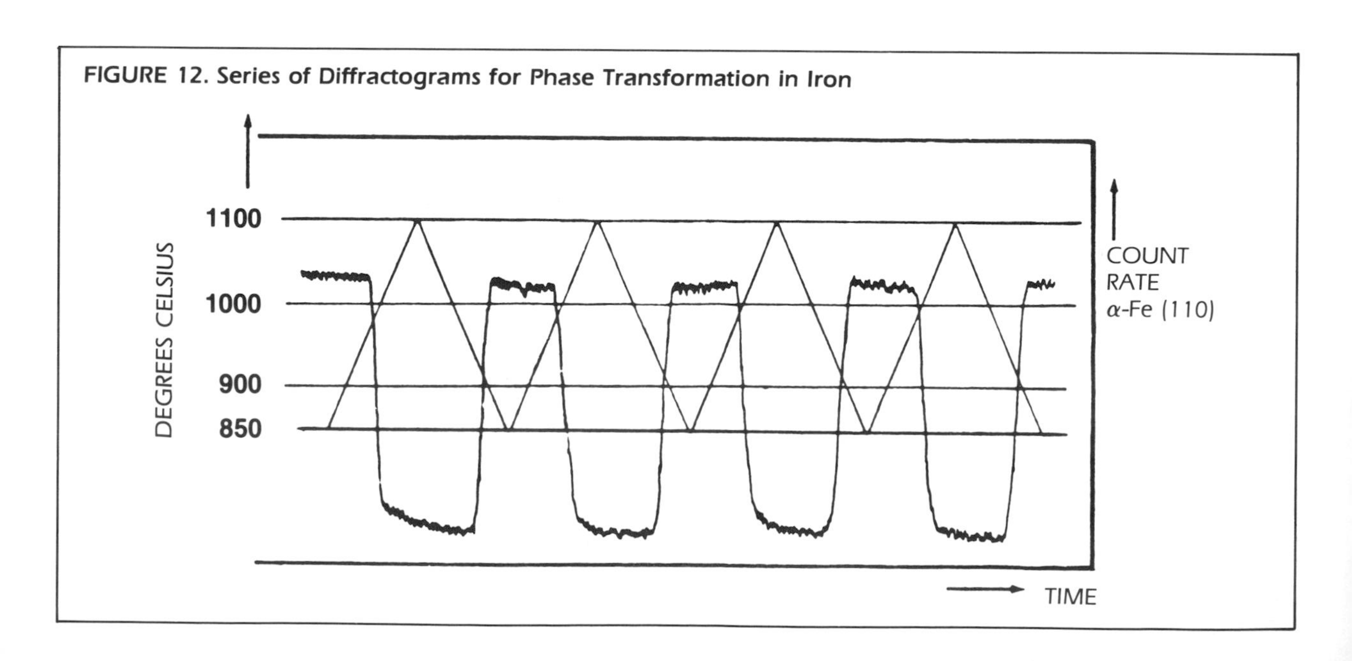

FIGURE 13. Effect of Lattice Strain on Diffraction-Line Width and Position

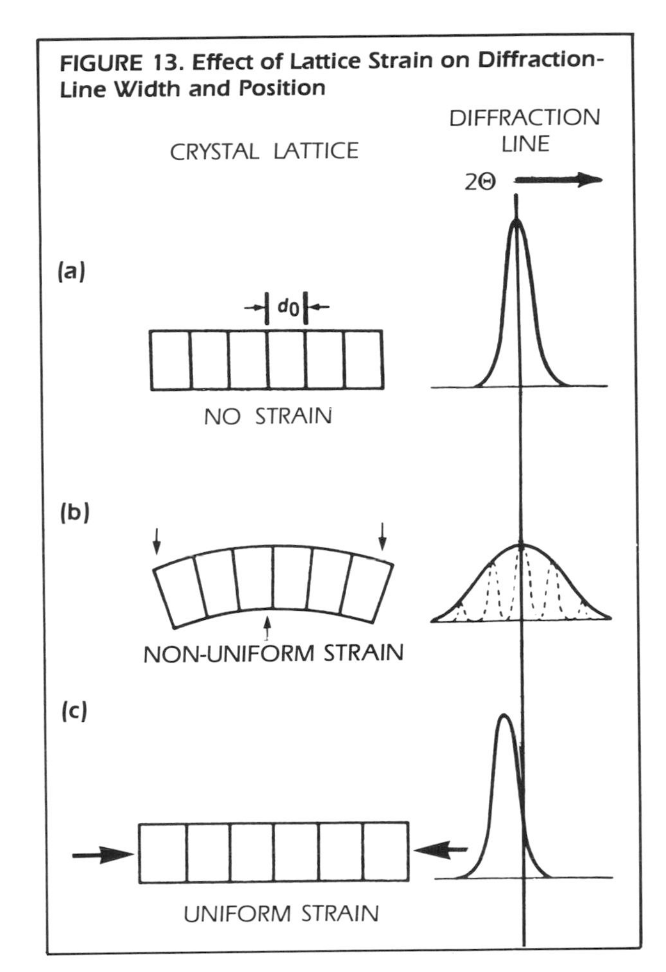

diffraction lines have a finite width of a few tenths of a degree (due mainly to the finite width of the receiving slit). Estimation of dimensions using this technique has been successfully applied to particles in the range of 10 to 100 nanometers.

Measurement of Solid Solutions and Eutectics

X-ray diffraction generally is used for the study of systems involving a combination of discrete phases, but it can also be employed in the study of combined phases. Combined phases may be the result of solid-state addition and replacement reactions where the product has been formed by movement of atoms from one crystallite boundary to another. As noted earlier, formation of the solid state reaction product occurs at a temperature significantly below the phase's melting point and the product is not able to form well-defined crystallites. Usually, the diffraction patterns from such a reaction product are poorly defined and the lines tend to be broad and diffuse.

In practice, a simpler application for X-ray diffraction is the study of alloys, where two metals have combined to give a solid solution or a eutectic (an alloy with the lowest possible constant melting point). The presence of a newly formed phase within such a material becomes immediately apparent from the appearance of new lines in the phase diagram. Also, formation of a continuous range of solid solutions will be indicated by a gradual and predictable shift in the position of certain lines in the diffraction pattern. As an example, in the copper/nickel system, there is an almost linear correlation between the size of the unit cell and the atomic percentages. The usefulness of the X-ray diffraction method is enhanced by its ability to gather data at elevated temperatures, and under special or inert atmospheres.

Measurement of Stress

One of the important metalurgical applications of X-ray diffraction is in the determination of residual stress. When a polycrystalline piece of metal is deformed, the lattice plane spacings will change their values, the new values corresponding to the applied stress. If the stress is uniform, this value change will manifest itself as a shift in the observed 2θ angle of the appropriate diffraction line. If the stress is non-uniform, the effect will appear as a broadening of the appropriate diffracted line profile.

As far as uniform strain is concerned, it is a rather simple procedure to correlate the line shift with the stress in the system, by making use of measured elastic constants or by calibration with materials of known stress. It should be realized, however, that in either instance the diffraction method is actually measuring strain, not stress; it is thus an indirect method of measurement. When an applied stress is removed and the deformation persists, the material is said to have *residual stress*. Determination of residual stress in metals is one of the most common applications of X-ray stress measurement. Automated X-ray machines are available for the rapid measurement of stress, both on laboratory specimens as well as on large samples, such as weldments in pipe lines, aircraft wings, and so on.

PART 5

ELEMENTAL ANALYSIS BY X-RAY FLUORESCENCE

Principles of X-ray Fluorescence

The basis of the X-ray fluorescence technique lies in the relationship, established by Moseley[10], between the atomic number, Z, and the wavelength, λ, (or energy *E*) of the X-ray photons emitted by the sample element. This is expressed in eq. 3 as:

$$\frac{E}{12.4} = \frac{1}{\lambda} = K(Z\text{-}s)^2 \qquad \text{(Eq. 3)}$$

where *K* and *s* are constants dependent on the spectral series of the emission line in question.

Most commercial X-ray spectrometers have a range of about 0.2 to 20 Å (60-0.6 keV), which will allow measurement of the K series from fluorine ($Z = 9$) to lutetium ($Z = 71$), and for the L series from manganese ($Z = 25$) to uranium ($Z = 92$).

Other line series can occur from the M and N levels, but these have little use in analytical X-ray spectrometry. Although almost any high energy particle can be used to excite characteristic radiation from a specimen, an X-ray source offers a reasonable compromise between efficiency, stability and cost. Almost all commercial X-ray spectrometers use an X-ray source. Since primary (source) X-ray photons are used to excite secondary (specimen) radiation, the technique is referred to as *X-ray fluorescence spectrometry*.

Commercially available X-ray spectrometers fall roughly into two categories: wavelength dispersive instruments and energy dispersive instruments. The wavelength dispersive system was introduced in the early 1950s and has developed into a widely accepted analytical tool. Around 10,000 such instruments have been supplied commercially, roughly half of these in the U.S. Energy dispersive spectrometers became commercially available in the early 1970s and today there about 1,000 units in use.

Qualitative Analysis with the X-ray Spectrometer

The output from a wavelength dispersive spectrometer may be either analog or digital. For qualitative work, an analog output is traditionally used and in this instance a ratemeter integrates the pulses over short time intervals, typically on the order of a second or so. The output from the rate meter is fed to a strip-chart recorder which scans at a speed conveniently coupled with the goniometer scan speed. The recorder displays an intensity/time diagram, which becomes an intensity/2θ diagram. Tables are used to interpret the resulting wavelengths.

For quantitative work, it is more convenient to employ digital counting. A timer/scaler combination is provided, which will allow pulses to be integrated over a period of several tens of seconds and then displayed as count or count-rate.

Most modern wavelength dispersive spectrometers are controlled in some way by a minicomputer or microprocessor. By use of specimen changers, they are capable of very high specimen throughput. Once they are set up, the spectrometers will run virtually unattended for several hours.

Quantitative Methods

The great flexibility and range of the various X-ray fluorescence spectrometers, coupled with their high sensitivity and good inherent precision, makes them ideal for quantitative analysis. As with all instrumental analysis methods, high precision can be translated into high accuracy only by compensating for the various systematic errors in the analysis process.

The precision of a well-designed X-ray spectrometer is typically on the order of one-tenth of a percent, the major contributor to the random error being the X-ray source (the high-voltage generator plus the X-ray tube). In addition, there is an error arising from the statistics of the actual counting process.

Systematic errors in quantitative X-ray spectrometry arise mainly from absorption and specimen-related phenomena (matrix effects). This is also the case in X-ray powder diffraction, except that in spectrometry, the systematic errors are much more complicated. In diffraction, one is dealing with a single wavelength: for example, the diffracted, monochromatic line scattered from the primary source (tube line). In spectrometry, many wavelengths are involved. Although these so-called matrix effects are somewhat complicated, many excellent methods have been developed for handling them. The advent of the minicomputer-controlled spectrometer has done much to enhance the application of correction procedures. Today, in most cases, one is able to quantify elements in the periodic table, of atomic number 9 and upwards, to an accuracy of a few tenths of a percent. The areas of application for the X-ray fluorescence technique now cover almost all areas of inorganic analysis.

Trace Analysis

The wavelength dispersive X-ray fluorescence method is a reasonably sensitive technique with detection limits (for most elements) in the low parts per million range. Fig. 14 shows a curve for the lower limit of detection as a function of atomic number, for a typical system. The curve is a smooth "U" shape, which is repeated above atomic number 50 (tin) and displaced upward by about one order of magnitude. Below atomic number 13 (aluminum), the sensitivity drops quite sharply until at the conventional low atomic number limit of the method (fluorine, Z = 9) the achievable detection limit is only about 0.05 percent. With energy dispersive spectrometers, detection limits are typically five to ten times worse than the wavelength dispersive instruments. The lower limit of detection is defined as that concentration equivalent to two standard deviations of the background count rate[11]. In practical terms, the lower limit of detection (LLD) is given by eq. 4:

FIGURE 14. Lower Limits of Detection Obtainable with the Wavelength Dispersive Spectrometer

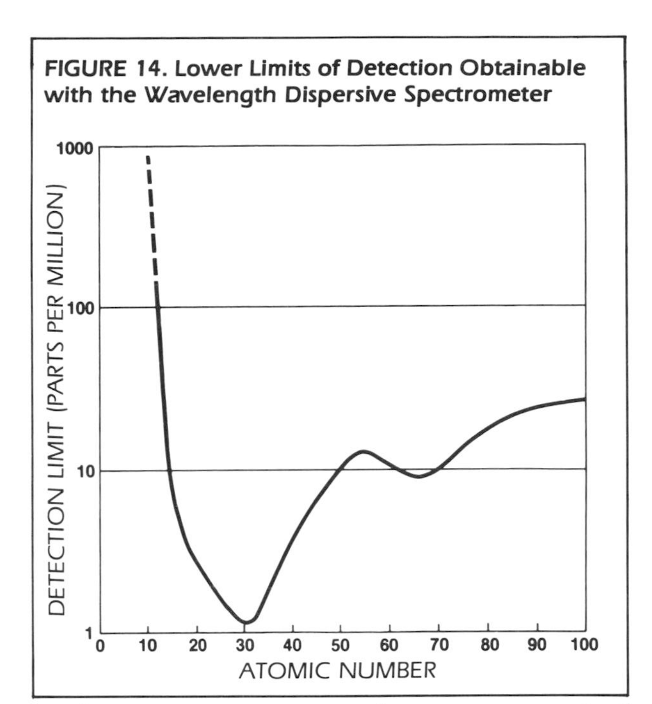

$$LLD = \frac{3}{m}\left(\frac{R_b}{t_b}\right)^{1/2} \qquad \text{(Eq. 4)}$$

where m is the counting rate per unit concentration of the analyte element; R_b the background counting rate; and t_b the analysis time spent counting the background (effectively one half the total analysis time).

For example, in steel, the element phosphorus would give a sensitivity of about 2000 counts/second for each percent, over a background of 35 counts/second. For a total analysis time of 100 seconds, t_b would equal 50 seconds, giving a detection limit of about 12 parts per million.

Generally, the ultimate detection limits can only be obtained where a large sample is available, typically several tenths of a gram. Very small samples can be handled, though with much poorer element detection limits. Samples as small as a few mg will give measureable signals for concentration levels of a percent of more. Special instrumentation has been employed for ultra trace analysis and under favorable circumstances, detection limits down to 10^{-15} grams can be obtained.

PART 6

INSTRUMENTATION FOR X-RAY FLUORESCENCE ANALYSIS

The Wavelength Dispersive Spectrometer

All conventional X-ray spectrometers comprise three basic parts: the primary source unit; the spectrometer itself; and the measuring electronics.

The primary source unit consists of a sealed X-ray tube, plus a very stable high-voltage generator, capable of providing up to 3 kW of power at a typical potential of 60-80 kV. The sealed X-ray tube has an anode of chromium, rhodium, tungsten, silver, gold or molybdenum, and delivers an intense source of continuous radiation which impinges on the specimen, producing characteristic radiation.

In the wavelength dispersive spectrometer, a single crystal of known *d* spacing is used to disperse the polychromatic beam of characteristic wavelengths coming from the sample, such that each wavelength will diffract at a discrete angle. A portion of the characteristic "fluorescence" radiation is then collected by the actual spectrometer, where the beam is passed, via a collimator or slit, onto the surface of an analyzing crystal. Individual wavelengths are then diffracted in accordance with the Bragg law.

A photon detector, typically a gas flow or a scintillation counter, is used to convert the diffracted characteristic photons into voltage pulses which are integrated and displayed as a measure of the characteristic line intensity. In order to maintain the required geometric conditions, a goniometer is used to ensure that the source-to-crystal and the crystal-to-detector angles are kept the same.

The Energy Dispersive Spectrometer

Like the wavelength dispersive spectrometer, the energy dispersive spectrometer also consists of the three basic units: excitation source; spectrometer; and detection system. In this case, however, the detector itself acts as the dispersion agent. The detector is typically a lithium-drifted silicon, *Si(Li)*, detector which is a proportional detector of high intrinsic resolution. The Si(Li) detector diode serves as a solid state version of the gas flow detector in the wavelength dispersive system.

When an X-ray photon is stopped by the detector, a cloud of ionization is generated in the form of electron/hole pairs. The number of electron/hole pairs created (or the total electric charge released) is proportional to the energy of the incident X-ray photon. The charge is swept from the diode by a high voltage applied across it. A preamplifier is responsible for collecting this charge on a feedback capacitor to produce a voltage pulse proportional to the original X-ray photon energy. Thus when a range of photon energies is incident upon the detector, an equivalent range of voltage pulses is produced as detector output. A multichannel analyzer is used to sort the arriving pulses to produce a histogram representation of the X-ray energy spectrum.

The output from an energy dispersive spectrometer is generally given on a visual display unit. The operator is able to display the contents of the various channels as an energy spectrum. Provisions are often made to allow zooming, to overlay spectra, to subtract background, and so on, in a rather interactive manner. As is the case with modern wavelength dispersive systems, nearly all energy dispersive spectrometers will incorporate some form of minicomputer for spectral stripping, peak identification, quantitative analysis, and a host of other useful functions.

Materials Certification

The energy dispersive spectrometer plays an important role in the area of materials certification, covering the range from scrap metal sorting to alloy identification and certification.

In the identification of specific product types, it may be necessary only to identify a few key elements and use these to 'fingerprint' the product.

Sometimes this is done by deliberately adding tracer elements at low concentration. The technique has been used successfully in the sorting of finished polymer pieces by tagging them with the elements chlorine and chromium, at concentration levels of about 0.1 percent. A very short counting time, perhaps a few seconds, is all that the energy dispersive spectrometer requires to establish the source of the polymer product.

Another example of the spectrometer's use is in the analysis of plating on sheet steel, where it is necessary to control the thickness of a coated layer. A calibration curve of X-ray line intensity, as a function of coating thickness, can be easily established from a few standard samples. This curve is then used for quality control purposes by comparison with on-line count data from the energy dispersive system.

PART 7

COMPARISON OF X-RAY FLUORESCENCE WITH OTHER ELEMENTAL METHODS

There are many analytical methods available today and the versatility of these different techniques makes choosing one of them difficult. What is certainly true is that there is no single technique which will offer all of the features of every technique and to this extent, one's choice will be predicated on individual requirements. The factors generally taken into consideration include accuracy, speed, cost and sensitivity.

Table 2 lists some of the more important features of the X-ray fluorescence technique and gives an overall view of the capabilities of the method.

The cost of the modern X-ray spectrometer system varies significantly with the type of system required. There is a wide variety of instruments available, including both energy and wavelength dispersive based systems. Further subdivisions include single channel or multi-channel, sequential or simultaneous units, microprocessor controlled, with or without data processing computer, and so on. Prices vary widely, with six to eight major manufacturers offering and supporting these products.

Both the simultaneous wavelength dispersive and the energy dispersive spectrometers lend themselves admirably to the qualitative analysis of materials. Because the characteristic X-ray spectra are so simple, the actual process of allocating atomic numbers to the emission lines is a relatively simple process, and the chance of making a gross error is small.

As a comparison, qualitative analysis of multiphase materials with the X-ray powder diffractometer is a much more complex business. There are, after all, only 100 or so elements, and within the range of the conventional spectrometer, each element gives, on the average, only a half dozen lines. In diffraction, on the other hand, there are perhaps several million possible compounds, each of which can give an average of 50 or more lines.

Using X-ray emission spectra for qualitative analysis has several benefits. Compared with the ultraviolet (UV) spectrum for example, the X-ray spectrum is much easier to interpret. This is because the X-ray spectrum arises from inner orbital transitions and these are by nature limited in number. The UV spectrum occurs with transitions of electrons to outer, unfilled levels which can be numerous enough to severely complicate interpretation. Also, because X-rays are produced by inner orbital transitions, the effect of chemical combination (valence) is greatly reduced.

TABLE 2. Features of the Wavelength Dispersive X-ray Spectrometer

Accuracy:	Tenths of a percent
Sensitivity:	Low ppm to 100%
Range:	F ($Z=9$) and upwards
Speed:	Scanning 30 secs/element/sample
	Sequential 30 elements $<$ 1 minute
Reliability:	Downtime $<$ 2%
Certainty:	Difficult to make a gross error
Applicability:	Both qualitative and quantitative

PART 8

COMBINATION OF X-RAY DIFFRACTION AND FLUORESCENCE

X-ray fluorescence analysis allows the quantitation of all elements in the periodic table, from fluorine (atomic number 9) and upward. Accuracies of a few tenths of one percent are possible and elements are detectable in most cases to the low parts-per-million level. X-ray powder diffractometry is applicable to any ordered (crystalline) material and, although much less accurate or sensitive than the fluorescence method, it is virtually unique in its ability to differentiate phases.

The fluorescence and diffraction techniques are to a large extent complementary. Fluorescence provides accurate quantitation of elements; and diffraction provides qualitative and semi-quantitative estimations of the ways in which the matrix elements are combined to make up the phases in the specimen.

In an earlier example, we referred to the problem of the analysis of engine deposit materials. Because of the limited amount of material available and the generally poor quality of the diffraction patterns, it is difficult to get more than semi-quantitative analysis from diffraction data alone. As shown in Table 3, however, combination of diffraction and fluorescence yields a good material analysis.

TABLE 3. Material Analysis Data from Engine Deposit Sample

Phase Analysis (X-ray Diffraction)	Elemental Analysis		
	Element	From X-ray Diffraction (percent)	From X-ray Spectrometry (percent)
40% $PbO \bullet PbSO_4$	Pb	76.3	75.0
25% $PbBr_2$	S	2.4	2.6
20% $2PbO \bullet PbCl_2$	Br	10.7	10.3
15% PbClBr	Cl	3.6	3.8
	O	7.0	-
	P	-	0.5
	Fe	-	0.12

PART 9

SAFETY CONSIDERATIONS

TABLE 4. Radiation Doses

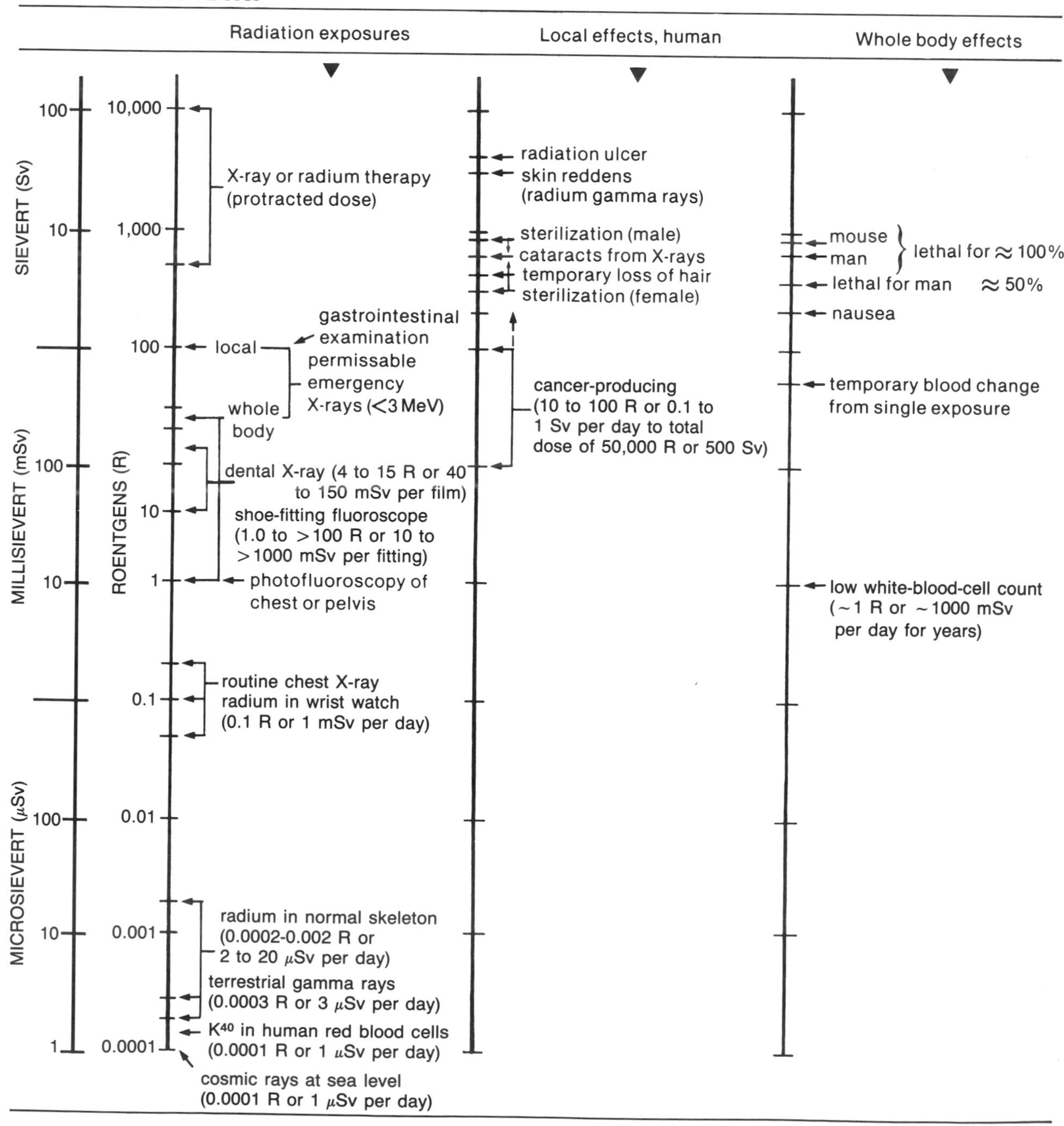

X-ray photons are energetic electromagnetic radiation that ionizes matter by ejecting electrons from its atoms. The extent of the ionization, absorption or molecular change of a given material depends on the quantity (radiation flux or intensity) and the quality (the spectral distribution of the photon energy) of the radiation[12].

The units for measuring the quantity of ionizing radiation are the sievert (Sv) and the roentgen (R). The roentgen is defined as the radiation flux which will produce 2.08×10^9 ion pairs per cubic centimeter in air, this being equivalent to an energy of 88 ergs. Radiation effects are expressed in the SI system by grays (Gy) and sieverts, where 1 Sv ~ 1 Gy ~ 0.01 R. The sievert, like the rem, is variable according to the type of radiation. The Roentgen Absorbed Dose (RAD) is defined as the amount of ionizing radiation which produces 100 ergs of energy per gram in the absorber. This quantity is expressed in the SI system by the gray, which is equivalent to 0.01 RAD or 1 joule per kilogram, and which results from ~ 0.01 R.

The Roentgen Equivalent Man (REM) is the absorbed dose in RADs corrected for the equivalent absorption in living tissue. The SI unit for this is the sievert, with 1 SV ~ 0.01 REM. One sievert results from 0.01 R unless corrected for type of radiation.

Living organisms can be injured by exposures to various doses of ionizing radiation. Death may result from severe overexposure. It is therefore imperative that all operators of X-ray instruments be aware of the potential health hazard[13].

Fortunately, most of the accidents involving X-ray analytical instrumentation result in local rather than whole body effects. The extremities of the body, such as fingers and hands, are able to survive dose rates at levels which, if delivered to the whole body or to sensitive parts of the body, might be fatal.

Biological effects of radiation generally follow a definite pattern, commencing with a latent period, the time lag between the exposure and the appearance of biological symptoms. This time may be hours or days, depending on the magnitude of the exposure. There follows a period of demonstrable effect on cells and tissues which generally arise from cell death or cessation of cell division. The last stage is the recovery period. The extent of the recovery depends mainly on the dose received, but for doses in excess of several hundred REM, recovery is rarely complete. The residual damage may then give rise to long-term effects. Table 4 lists typical radiation doses along with associated local and whole body effects.

PART 10

FUTURE TRENDS FOR X-RAY DIFFRACTION AND FLUORESCENCE

Since the 1950s, X-ray analytical methods have played an increasing role in materials research, quality control and certification. The wide use of the three analytical techniques seems likely to continue for many years. Although there are alternatives to the X-ray fluorescence method for elemental analysis, the ability of the technique to give fast, accurate analyses over wide concentration ranges, coupled with its excellent qualitative capability, makes the technique very attractive. As far as phase analysis is concerned, the X-ray diffraction method's capacity for identification of multi-phase mixtures is unique and this is enhanced by its ability to give data on stress, strain, texture, topography and so on.

More recently, the minicomputer has played an important role in the development of instrumentation, both for X-ray diffraction and X-ray fluorescence. Modern diffractometer and spectrometer systems are highly automated and most of the tedium and complexity of data interpretation are being reduced by the computer. Fast and efficient programs are available for peak hunting, profile deconvolution, data plotting, etc. The rapid growth in computer technology will undoubtedly continue to give impetus to the X-ray analytical field.

In the realm of instrumentation, one of the most fruitful areas for future progress appears to be the use of high power sources. Rotating anode X-ray tubes, operating at 15-20 kW, are now available for diffraction measurements. These give much higher X-ray intensities (by an order of magnitude), when compared to the classic sealed tubes. Two-dimensional position-sensitive detectors are beginning to find applications in stress and low angle scatter measurements. The X-ray analytical field has traditionally been quick to implement newer technologies and this trend certainly continues today, making the role of X-ray methods in materials characterization secure for years to come.

REFERENCES

1. Roentgen, W.C. *Annals of Physics and Chemistry* (Ger.), Vol. 64 (1898): pp 1.

2. von Laue, M. "International Tables for Crystallography" (Historical Introduction), *Symmetry Groups*, Vol. 1. IUCR: Kynoch Press (1952): pp 1-5.

3. Auger, P. Thesis. Paris (1926).

4. Hanawalt, J.D., and H.W. Rinn. *Industrial and Engineering Chemistry*, Vol. 8 (1936): pp 244.

5. Jenkins, R. and C.R. Hubbard. *Advances in X-ray Analysis*, Vol. 22 (1978): pp 133.

6. Nieman, R.L. and R. Jenkins. *X-ray Fluorescence Analysis of Environmental Samples*. T.C. Dzubay, ed. Ann Arbor Publications (1977): pp 77-92.

7. *HTK High Temperature Camera Manual*. Graz, Austria: Anton Paar KG.

8. Cullity, B.D. "Measurement of Residual Stress" (Chapter 16), *Elements of X-ray Diffraction*, 2nd ed. Addison-Wesley (1978).

9. Scherrer, P. *Gottinger Nachrichten*, Vol. 2 (1918): pp 98.

10. Moseley, H.G.J. *Philosophical Magazine*, Vol. 26 (1912): pp 1024.

11. Jenkins, R., R.W. Gould, and D.A. Gedcke. "Trace Analysis" (Chapter 11), *Quantitative X-ray Spectrometry*. Dekker: New York (1981).

12. Jenkins, R. and D.J. Haas. *X-ray Spectrometry*, Vol. 2 (1973): pp 135.

13. Jenkins, R. and D.J. Haas, *X-ray Spectrometry*, Vol. 4 (1975): pp 33.

SECTION 18

RADIATION PROTECTION

William D. Burnett—Sandia Laboratories, Albuquerque, NM

INTRODUCTION

This Handbook chapter summarizes the considerations involved in setting up a safe radiographic facility.

Commercial consulting firms specializing in personnel dosimetry and radiation protection may help with this goal. But regardless of who establishes or monitors the program, it is vitally important that radiation exposures to personnel be reduced to as low a level as is practical.

To this end, each radiographic facility should appoint a *radiation safety officer*, who is responsible for systematically assuring management that a safe operation exists.

Many publications are specifically written to describe in detail the requirements and techniques involved (see references to this Section) and it is not the function of this chapter to duplicate that information.

This text is an *overview* of radiation safety and personnel protection. Every effort was made to provide accurate, current information at the time of publication, but because of potential changes in safety requirements, radiation safety officers and all personnel active in the field of radiography should consult the most up-to-date sources before making a determination on the safety of a radiographic facility.

For example, unsealed radioactive sources and the associated health protection requirements, internal dosimetry, instrumentation, and so on, are not covered in this discussion. Note also that there may be variations in safety regulations, depending on locality.

Special thanks to the National Council on Radiation Protection and the National Bureau of Standards for supplying much of the information used in this chapter.

PART 1

DOSE DEFINITIONS AND EXPOSURE LEVELS

Radiation Quantities and Units

Absorbed dose is the mean energy imparted to matter by ionizing radiation per unit mass of irradiated material. The unit of absorbed dose in the cgs system is the *rad.* One rad equals 100 ergs per gram. In the SI system, the unit of absorbed dose is the *gray*. The gray equals one joule per kilogram. One gray equals 100 rads.

Dose equivalent is a quantity used for radiation protection purposes. It expresses, on a scale common for all radiations, the irradiation incurred by exposed persons. It is defined as the product of the absorbed dose and certain modifying factors. A common unit of dose equivalent is the *rem*. In the SI system, the unit of dose equivalent is the *sievert*, which equals one joule per kilogram. One sievert equals 100 rem.

Exposure[11] is a measure of X or gamma radiation based on the ionization produced in air by the X or gamma rays. In the cgs system, the unit of exposure is the *roentgen*. Sieverts are used for measuring exposure in the SI system.

Quality factor (QF)[1] is a modifying factor used in determining the dose equivalent. The QF corrects for the dependence of biological factor on the energy and type of the radiation. A formerly commonly used term, relative biological effect (RBE), is now restricted in use to radiobiology. For practical purposes the following quality factors are conservative.

Radiation Type	QF
X rays, gamma rays, electrons, and beta rays	1
Neutrons, energy <10 keV	3
Neutrons, energy >10 keV	10
Protons	10
Alpha particles	20
Fission fragments, recoil nuclei	20

Example: consider an absorbed dose in the lens of the eye of 1 milligray (0.1 rad) from 2 MeV neutrons. The dose equivalent is as follows:

$$\begin{aligned} H &= \text{(dose in milligrays) (QF)} \\ &\ (1)\ (10) \\ &\ \text{10 millisieverts (1 rem)} \end{aligned}$$

Disintegration rate is the rate at which a radionuclide decays. The original unit used to describe this decay was the *curie*. The curie equals 37×10^9 disintegrations per second. The SI unit for disintegration rate is the *becquerel* and is defined as one disintegration per second.

Permissible Doses

Concept of ALARA (As Low As Reasonably Achievable).[10]

All persons should make every reasonable effort to maintain radiation exposures as low as is reasonably achievable, taking into account the state of technology and the economics of improvements in relation to benefits to the public health and safety. In this sense, the term *permissible dose* is an administrative term mainly for planning purposes. At the present time no radiation dose, other than zero, is accepted as having no biological effect.

Prospective Annual Limit for Occupationally Exposed Personnel

The maximum permissible prospective dose equivalent for *whole body irradiation* is 50 mSv (5 rem) in any one year.[1] The Nuclear Regulatory Commission[10] has further restricted for its licensees the rate at which this planned annual dose may be received by averaging over calendar quarters rather than calendar years. This and limits for other parts of the body are summarized in Table 1.

TABLE 1.

Radiation Workers[10]	mSv (rem) per Calendar Quarter*
Whole body; head and trunk; Active blood-forming organs; lens of eyes; or gonads	12 (1¼)
Hands and forearms; feet and ankles**	188 (18¾)
Skin of whole body	75 (7½)

*These numbers are obtained by dividing the annual doses by 4.
**All reasonable efforts should be made to keep exposure of hands and forearms within the general limit for skin.[1]

Retrospective Annual Occupational Dose Equivalent[1,10]

There will be occasions when the measured or estimated actual dose equivalent exceeds the prospective limit of 50 mSv (5 rem) in a year. No deviation from sound protection is implied, provided (1) the whole body calendar quarter dose does not exceed 30 mSv (3 rem), and (2) the accumulated whole body dose does not exceed 50 mSv or 10 $(N - 18)$ mSv (5 rem or $[N - 18]$ rem), whichever is smaller, where N equals the individual's age. Thus, substantial knowledge of an individual's work history and dose record from all previous employment is implied before the 50 mSv (5 rem) per year individual exposure can be exceeded. In practice the simpler route is to endeavor to stay within the whole body calendar quarter limit of 12 mSv (1¼ rem).

Permissible Levels of Radiation in Unrestricted Areas[10]

Non-occupationally exposed personnel or all personnel in unrestricted areas (see below) shall not receive more than 5 mSv (0.5 rem) to the whole body in any period of one calendar year.

Restricted Areas

A *restricted area* needs to be established where either (1) a dose in excess of 20 microsieverts (2 millirem) can be received in any one hour or (2) a dose in excess of 1000 microsieverts (100 millirem) can be received in seven consecutive days.

Exposure of Minors[10]

An individual under 18 years of age cannot be exposed to greater than 10 percent of the limits for occupationally exposed workers, i.e., 10 percent of 12 mSv (1¼ rem) per quarter to the whole body, and similarly for the hands, forearms, feet and ankles, and skin of the whole body.

Exposure of Females

During the entire nine-month gestation period, the maximum permissible dose equivalent to the fetus from occupational exposure of the expectant mother should not exceed 5 mSv (0.5 rem).[1] For many X and gamma radiation energies encountered, this is equivalent to about 15 mSv (1.5 rem) to the pregnant woman, because absorption of radiation by the abdominal wall usually reduces the fetal dose to 5 mSv (0.5 rem) or less.[3] The woman who is able to plan her pregnancy should request a monthly radiation dosimeter badge reading to be sure that her radiation dose has remained within safe limits from the period immediately before the beginning of pregnancy and thereafter.

PART 2

RADIATION PROTECTION MEASUREMENTS

Choice and Use of Personnel Dosimetry[4]

Requirements

Personnel monitoring must be performed on all occupationally exposed persons who may receive in a calendar quarter more than one-fourth of the applicable doses in Table 1. Occasional visitors to restricted areas, including messengers, servicemen, and deliverymen, can be regarded as non-occupationally exposed persons who do not need to be provided personnel monitors when it is improbable that they would receive in one year a dose equivalent exceeding the non-occupational limit of 5 mSv (0.5 rem). Long-term visitors in an installation should be regarded as occupationally exposed if they are likely to receive a dose equivalent greater than 5 mSv (0.5 rem) per year.

X Rays, Gamma Rays, and Electrons

At present, the choice lies between ionization chambers, film badges, photoluminescent glasses, and thermoluminescent dosimeters.

1. Ionization Chambers. The principal advantages of ionization chambers, particularly those of the self-reading type, are the simplicity and speed with which readings are made. They are useful, therefore, particularly for monitoring exposures during non-routine operations or during transient conditions, or for monitoring short-term visitors to an installation. Chambers should be tested for leakage periodically and those that leak more than a few percent of full-scale over the period of use should be removed from service. Most of these ionization chambers are small, about the size of a pencil, and are charged on a separate device. They read from a few tens to a few hundred milliroentgens of exposure.

2. Film Badges. Small badges containing special X-ray films are popular personnel dosimeters. The sensitivity of available emulsions is sufficient to detect about 10 mR of cobalt-60 gamma radiation and a few mR of 100 keV X rays. A useful range from a few mR to 2000 R can be covered by two commonly available films or two emulsions of different sensitivity on one film base. For energies below 200 keV, film overresponds where, for example, the photographic density per roentgen at 40 keV is about 20 times higher than for 1 MeV photons. Metallic filters covering portions of the film provide additional readings that help determine the incident radiation energy and afford a means of computing a dose from appropriate calibration curves. Film has several undesirable characteristics. Fogging may result from mechanical pressure, elevated temperatures, or exposure to light. Fading of the latent image may result in decreased sensitivity but may be minimized by special packaging to exclude moisture and by storage in a refrigerator or freezer prior to distribution. Film dosimeters also exhibit directional dependence, particularly for the densities recorded behind metal filters.

3. Photoluminescent Glasses. Silver-activated metaphosphate glasses, when exposed to ionizing radiation, accumulate fluorescent centers that emit visible light when the glass is irradiated with ultraviolet light. The intensity of the light is proportional to radiation exposure up to 1000 R or more. Glass dosimeters exhibit energy dependence below 200 keV and are also subject to fading. They are useful down to only 1R.

4. Thermoluminescence. Thermoluminescent dosimetry (TLD) is rapidly replacing film dosimetry as the most common method. The desirable characteristics include its wide linear range; short readout time; relative insensitivity to field conditions of heat, light, and humidity; reusability; and for some phosphors, energy independence. Response is rate-independent up to 10^{11} R per second, which can

be useful in flash X-ray radiographic installations. Very small TLDs can be used to measure exposure to specific parts of the body. They probably represent the method of choice for measurement of finger, hand, or eye dose. They have a useful range down to several mR for LiF and even lower for more exotic TLD materials.

Neutrons

For neutron fields the practical devices are nuclear-track film, thermoluminescent dosimeters containing ^{6}LiF, and fission-track counting systems. The nuclear-track films do not respond to neutrons below 0.5 MeV in energy; in practice, a substantial fraction of the neutrons may be below this energy. Track counting is a relatively insensitive method of neutron dosimetry. For low doses, counting of a statistically significant number of tracks is too time consuming to be warranted. On the other hand, at high doses it is difficult to distinguish tracks from one another so that they can be counted. Fading occurs, and, as a result, short tracks may disappear. For these reasons, nuclear-track film is more useful in demonstrating that large neutron doses have not been received than in measuring actual low doses.

The ^{6}LiF and fission track-counting systems do not suffer from these disadvantages and will provide measurements at permissible dose levels. These methods are sensitive down to doses of a few millirad and down to thermal neutron energies.

Radiation Detection and Measurement[4]

In an area survey, measurements are made of radiation fields to provide a basis for estimating the dose equivalents that persons may receive. Changes in operating conditions (such as beam orientations and source outputs) can cause changes both in field intensity and pattern. The number of measurements depends on how much the radiation field varies in space and time, and how much people move about in the field. Measurements made at points of likely personnel occupancy under the different operating conditions are usually sufficient to estimate dose equivalent adequately for protection purposes.

Detection

Detection instruments are used to warn of the existence of radiation or radiation hazard and, as distinct from measuring instruments, usually indicate count rate rather than dose rate or exposure rate. Therefore, they should be used only to indicate the existence or, at most, establish relative rates.

Measurement

At points of particular interest, individual determinations of dose or exposure rate should be made with calibrated measuring instruments. Dose integrating devices (dosimeters) may be mounted at points of interest and left for an extended period of time to improve the accuracy of the measurement.

Information concerning the dimensions, dose rate, and location of primary beams of radiation in relation to the source is important in determining direct external exposure from the beam and the adequacy of protective measures. The dose or exposure rates within the beam at specific distances from the source should be measured and compared with expected values.

Measurements close to radiation sources of small dimensions or of radiation transmitted through holes or cracks in shielding require special attention. The general location of defects in shielding should be determined by scanning with sensitive detection instruments. More precise delineation of the size and configuration of the defects can be obtained by using photographic film or fluorescent screens for X-ray, gamma-ray, or electron leakage. Measurements may then be made in any of three ways:

1. An instrument may be used that has a detector volume small enough to ensure that the radiation field throughout the sensitive volume is substantially uniform.

2. An instrument with a large sensitive volume may be used, if an appropriate correction factor is applied. Multiply the reading by the ratio of the instrument chamber cross-sectional area to the beam cross-sectional area:

$$\text{Reading} \times \frac{A_{\text{chamber}}}{A_{\text{beam}}}$$

This correction is only used when A_{chamber} is larger than A_{beam}.

3. Film may be used at the point of interest, provided it has been properly calibrated for the types and energies of the radiations present.

Choice and Use of Instruments[4]

General Properties To Be Considered

1. Energy Response. If the energy spectrum of the radiation field differs significantly from that of the calibration field, a correction may be necessary.
2. Directional Response. If the directions from which the radiations arrive at the instrument differ significantly from those in the calibration field, correction may be necessary. If the dose equivalents being determined are small in comparison to permissible doses, large errors are acceptable and correction may not be necessary.
3. Rate Response. Instruments that measure dose or exposure are called *integrating instruments*; those that measure dose rate or exposure rate are called *rate instruments* or *meters*. If the dose rate or exposure rate differs significantly from that in the calibration field, correction may be necessary. Ordinarily, an integrating instrument should be used only within the rate ranges for which the reading is independent of the rate. Rate instruments, similarly, should be used only within the rate ranges in which the reading is proportional to the rate. A few instruments will "jam" at very high rates; that is, they will cease to function and the reading will drop to zero or close to zero. It is particularly necessary to know the rate response of instruments to be used near machines that produce radiation in short pulses. Rate instruments used near repetitively pulsed machines need only to indicate the average rate for radiation protection purposes.
4. Mixed Field Response. Since some radiations (such as neutrons) have higher quality factors than others, mixed field monitoring is necessary. This can be done either by using two instruments that are each sensitive to only one radiation or by using two instruments that are sensitive to both but to a different extent.
5. Unwanted Response. Interference by energy forms that an instrument is not supposed to measure can be a problem. Response to heat, light, radiofrequency radiations, and mechanical shock are examples.
6. Fail-Safe Provision. To avoid unknowingly exposing personnel to radiation, malfunctions of an instrument should be readily recognizable or should always result in readings that are too high.
7. Precision and Accuracy. Typically, *precision* of a few percent should be obtained on successive readings with the same survey instrument. At the level of a maximum permissible dose a measurement *accuracy* of ±30 percent should be achieved. At levels less than ¼ the maximum permissible dose a lower level of *accuracy* (say, a factor of 2) is acceptable.
8. Calibration. Instruments used for radiation protection are not absolute instruments; that is, they require calibration in a known radiation field or comparison with instruments whose response is known. Many users of radiation protection instruments must rely on the manufacturer to calibrate their instruments properly. Users should arrange a reproducible field in which the instruments are placed and read frequently—at least quarterly. The possibility of reading error due to lack of precision is minimized by computing the mean of several readings. If changes in the mean reading are detected, the instruments should be recalibrated promptly.
9. Time Constant. An important characteristic of a rate instrument is the time constant, which is an indication of the time necessary for the instrument to attain a constant reading when suddenly placed in a constant radiation field. Time constants are generally given as the time required to arrive at $1 - e^{-1}$ (i.e., 0.63) of the final reading. Typical time constants of good rate meters are one second or less. The *response time* of a rate instrument is defined as the time necessary for it to reach 90 percent of full response. It is equal to 2.3 time-constants.

Instruments for Radiation Surveying and Area Monitoring

1. Sealed gamma ray sources and sources of X rays (including flash X-ray machines).

   a. Ionization Chambers (Fig. 1,2). Most gamma-ray and X-ray exposure rate measurements are made with small, portable ionization chambers that weigh less than 2 kg (5 lb). Ionization chambers with separate readers are useful for measuring either very high or very low exposure rates. Ion chambers made of plastic or other low atomic number materials usually give exposure readings that are independent of photon energy down to 50 keV. Chambers designed for high photon energies usually have walls so thick that photon attenuation requires large corrections for low energies. Ionization chambers are available for exposure rates up to over 20 Sv/h (a few thousand R/h).

   b. C M Counters (Fig. 3). The dead time in G-M counters sets a limit to their count rate that, in turn, limits their use to exposure rates up to about 20 μSv/h (a few mR/h). The counters respond to the number of ionizing events within them independent of the energy and thus do not yield equal count rates for equal exposure rates of different energies. G-M counters can be used to *estimate* exposure rates where the photon energy is the same as that used for calibration. Otherwise, their use generally should be for radiation detection, *not* measurement.

   c. Scintillation Instruments. Scintillation devices also have count-rate limitations due to the duration of the light flashes, but they can count much faster than G-M counters. In the same exposure field, scintillation count rates are higher than G-M count rates, so scintillation counters are useful for locating weak X-ray and gamma-ray fields.

2. Beta-Particle and Electron Fields. Most survey measurements of beta-particles and low-energy (<2 MeV) electron fields are made with small, portable ionization chambers with an entrance window covered with a thin, electrically conducting plastic sheet (about 7 mg/cm^2). Because of their short ranges in solids, electrons that arrive nearly tangentially to the entrance window and have to penetrate a long, slanted path through the window are largely or completely absorbed in it. Thus these instruments demonstrate a high directional dependence.

   G-M counters and scintillation instruments similarly can be equipped with thin walls or entrance windows for electron surveys. The G-M counter can be fitted with a removable cover sufficient to stop electrons completely so that by making a reading with the cover on and another with the cover off, one can discriminate between electrons and X or gamma rays. Scintillators must be fitted with a window thick enough to be opaque to visible light. This in turn limits their use for low-energy electrons.

3. Reactors, Sealed Neutron Sources, and Neutron Generators. Frequently the chief problem here is to measure neutrons and gamma rays separately in the same radiation field, as neutrons are almost always accompanied by gamma rays. Neutron doses should be measured separately from gamma-ray doses because of the large difference in quality factors.

   a. Ionization Chambers. One method of determining the fast neutron dose in a mixed field is to use two instruments with different neutron sensitivities and equal gamma-ray sensitivities. The difference in readings is due to neutrons. Ionization chambers whose inner surfaces are lined with a boron compound are very sensitive to thermal neutrons. Readings from a similar, unlined chamber are used to correct for the effects of gamma rays.

   b. Proportional Counters. Fast-neutron survey instruments have been developed that are based on slowing down (moderating) the fast neutrons and measuring the resulting low-energy neutrons.

FIGURE 1. Victoreen 440

A. Description: Sensitive, relatively accurate ion chamber survey meter. Nonsealed ion chamber with very thin end window permits measurement of photon, beta, and alpha radiations. Very stable circuit with reasonably long useful battery life.

B. Applications: Used for reliable measurements of photon radiation levels over wide range of energies and to show the presence of energetic beta and alpha radiations.

C. Use Instructions: Warm-up time for most applications is approximately ½ minute; for greatest accuracy allow about 3 minutes. Directional response and effect of end cap are shown in response curves in item F (chart below). Requires four "D" size cells; has approximately 100 hours battery life at four hours operation per day. For temperatures below 0 °C use alkaline batteries.

D. Specifications:

Detector: Unsealed air ionization chamber with 3 mg/cm^2 end window. End window cap for higher photon energies and for alpha and beta discrimination.

Detects: γ, X-ray—7 keV to several MeV. β-ray—>100 keV. α-ray—>4 MeV.

Ranges: 0–3, 0–10, 0–30, 0–100, 0–300 mR/hr.

Accuracy: Within ±10% of full scale, exclusive of energy dependence.

Response Time: (0–90% of final reading): 20 sec on 0–3 mR/hr. 12.5 sec on all other ranges.

Controls: Single switch: Power-battery check-range. No external zero or calibration.

Weight: 2.26 kg

Dimensions: 25 cm L × 10 cm W × 19 cm H.

E. Calibration: On calibration range to ^{60}Co radiation at four points on scale for each range. All points must indicate within ±20% of actual levels. Routine recalibration scheduled for every six months.

F. Energy Response

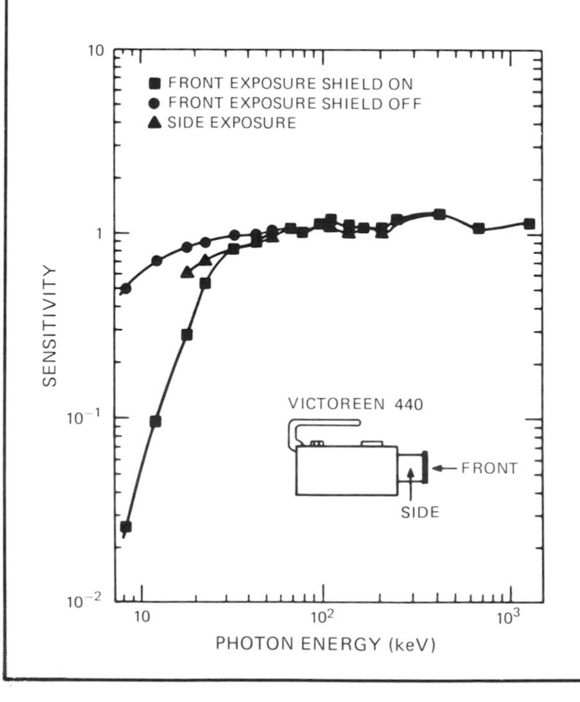

FIGURE 2. Baird Atomics 904-416 CP; Victoreen 740C

A. Description: "Cutie pie" gun-type survey meter using a nonsealed ion chamber with very thin end window to permit measurement of photon, beta, and alpha radiations.

B. Applications: Used for general monitoring around reactors, accelerators, and X-ray machines. Large-area, thin end window allows meaningful measurements of activities of many isotopes. Removable end cap provides beta discrimination and electron equilibrium for high energy gamma radiation.

C. Use Instructions: Approximately 1 minute warm-up time; zero control allows subsequent drift to be essentially eliminated. Gamma-ray energy response and effect of end cap are shown in item F below. Requires four 22.5 V batteries and one 1.34 V mercury cell; has approximately 200 hours battery life. Unsealed ion chamber uses standard air density correction factors.

D. Specifications:

Detector: Unsealed ion chamber with 0.7 mg/cm^2 end window. End window cap for higher photon energies and for alpha and beta discrimination.

Detects: γ, X-ray—20 keV to several MeV. β-ray—>40 keV. α-ray—>3.5 MeV.

Ranges: 0–100, 0–1000, 0–10k, 0–100k mR/hr.

Accuracy: Within $\pm$10% of full scale, exclusive of energy dependence.

Response Time: (0-90% of final reading): 3.5-7.5 sec, depending on range.

Controls: Rotary switch: Off-Batt-Zero-Range. Rotary zero adjust.

Weight: 3.6 kg

Dimensions: 29 cm L × 8.9 cm W × 22.7 cm H.

E. Calibration: On calibration range to ^{60}Co radiation at four points on scale for each range. All points must indicate within $\pm$20% of actual levels. Routine recalibration scheduled for every three months.

F. Energy Response

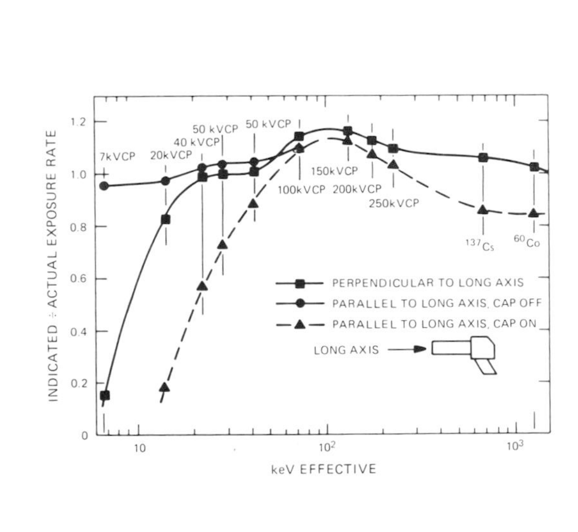

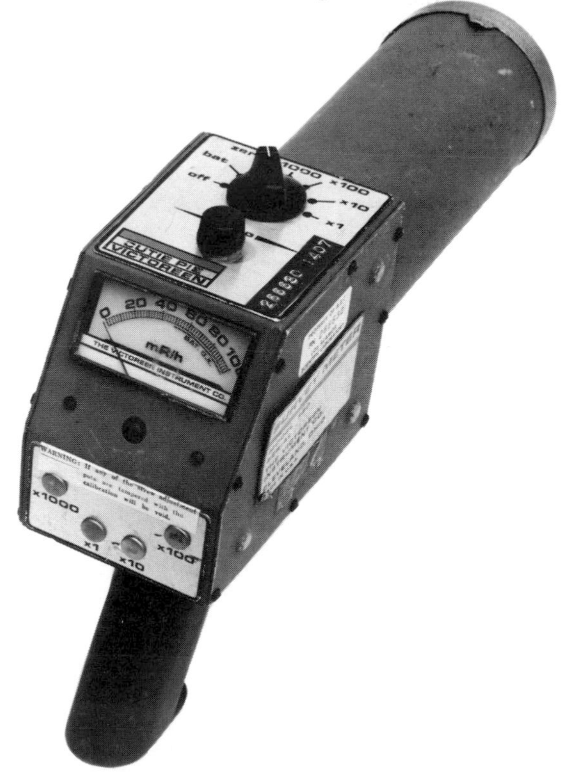

FIGURE 3. Eberline E-112B

A. Description: Sensitive, rugged survey meter using thin-walled, halogen GM tube in a hand probe with a beta discriminating shield. Pulse count-rate circuit does not paralyze in radiation levels up to ten times maximum measurable level. Phone jack provided.

B. Applications: Used for low-level radiation measurements from background up. External probe provides versatility for area and surface monitoring. Movable shield is used to discriminate between beta and gamma radiations. Earphones or speaker device can be attached for fast detection of low-level contamination.

C. Use Instructions: No warm-up time. Gamma-ray energy response is given in item F below. Interpretation of beta-ray response is dependent on source material, distribution, and geometry. Response time of meter can be varied to suit the application. Requires one 67.5 V battery and five 1.34 V mercury cells; approximate battery life is 200 hours. Operation is not reliable in environments below 0°C.

D. Specifications:

Detector: Halogen-filled GM tube with 30 mg/cm² stainless steel wall. Dimensions of probe approximately 3 cm diameter × 16 cm long.

Detects: γ, X-ray—20 keV to several MeV. β-ray—>200 keV.

Ranges: 0–0.2, 0–2, 0–20 mR/hr.

Accuracy: Within ±15% of full scale, exclusive of energy dependence.

Response Time: Controllable between approximately 2 and 10 sec.

Controls: Switch: power-range. Pot: variable meter response.

Weight: 1.8 kg.

Dimensions: 15 cm L × 7.6 cm W × 16.5 cm H.

E. Calibration: On calibration range to ^{60}Co radiation at four points on scale for each range. All points must indicate within ±20% of actual levels. Routine recalibration scheduled for every six months.

F. Energy Response

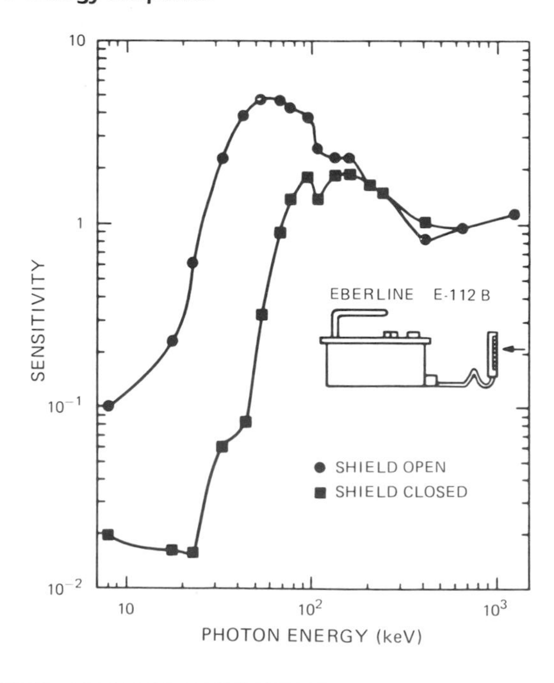

FIGURE 4. Eberline PNR-4

A. Description: Portable, neutron detecting instrument using a cadmium loaded, polyethylene sphere with BF_3 proportional counter in the center. Energy response roughly parallels tissue rem response. Relatively insensitive to gamma rays.

B. Applications: Used for detection and measurement of neutrons with energies ranging from thermal to approximately 10 MeV. Useful exposure rates are from approximately 1 mrem/hr to 5 rem/hr. Insensitive to gamma rays up to 500 R/hr, depending on certain instrument adjustments. Earphones or speaker device can be attached for detection of low neutron levels.

C. Use Instructions: Scale switching and use of multiplying factors are eliminated by LIN-LOG presentation with two pointers, each covering two "linear" decades; reading is given by pointer that is on scale. Relative rem response vs energy and radiation protection guide (RPG) dose curves are given below in item F. Detector may be detached from rest of instrument for remote monitoring. Requires five standard "D" cells; battery life is about 200 hours. For temperatures below 0°C use alkaline batteries.

D. Specifications:

Detector: Nine-in. diameter, cadmium loaded, polyethylene sphere with a BF_3 proportional counter tube in the center.

Detects: Neutrons from thermal to about 10 MeV in rem/hr.

Ranges: Four "linear" decades with full-scale readings of 5, 50, 500, 5k millirem.

Accuracy: Within ±8% of full scale of decade in which it is reading, exclusive of energy dependence and ±10% directional response.

Response Time: From 12 sec to 0.3 sec, depending on decade.

Controls: Switch: Off-On-Batt.

Weight: 8.9 kg.

Dimensions: 24 cm L × 23 cm W × 43 cm H.

E. Calibration: On calibration range to Pu-Be neutrons up to 800 mrem/hr; electronic pulser verifies scale linearity at higher readings. All points must indicate within ±20% of actual levels. Checked for lack of gamma-ray response at 10 R/hr. Routine recalibration scheduled for every three months.

F. Energy Response

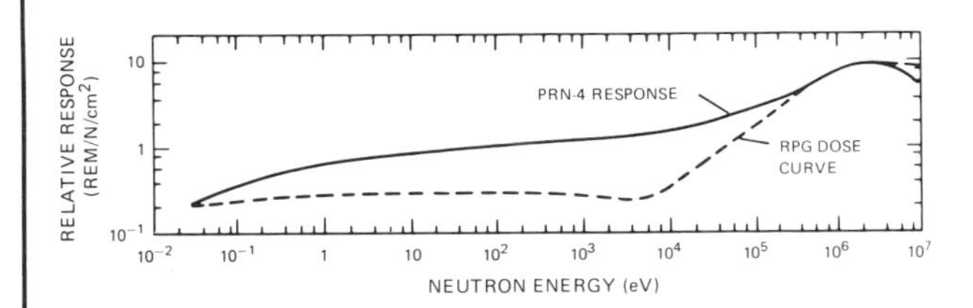

A BF_3 proportional counter surrounded by the proper thickness of paraffin or polyethylene gives a counting rate proportional to the neutron dose rate. The interference due to gamma rays is very small; gamma-ray exposure rates of several Sv/h (several hundred R/h) can be tolerated.

c. Scintillation Instruments (Fig. 4). Fast neutrons can be measured with a sphere of polyethylene having a scintillation crystal enriched in ^{6}Li mounted at the center of the sphere. A light pipe extends to a photomultiplier mounted outside the sphere. If the diameter of the sphere is about 25 cm (10 in.), the counting rate is proportional to the neutron dose equivalent rate and is reasonably independent of neutron energy (±15 percent) in the range from 0.2 to 7 MeV. Response is high by a factor of 2 at 10 keV. Unfortunately these instruments are heavy, typically weighing over 8.5 kg (20 lb).

4. Accelerators. Accelerators are usually equipped with monitors that permit the determination of output and energy of the useful beam with sufficient accuracy for hazard evaluation. Ionization chambers are most commonly employed for monitoring the useful beam. If the beam is defocused or displaced, it may produce intense beams of radiation at unexpected locations. Stray radiation may include leakage of the primary radiation beam, scattered radiation, and X rays, gamma rays and neutrons generated by the primary beams. Survey of the radiation due to electrons, X rays, or gamma rays from particle accelerator interactions should be carried out with ionization chambers. Counter-type instruments have a limited usefulness for measurements of radiation from pulsed accelerators except for the measurement of induced activity.

Instrument Calibration

The National Bureau of Standards (NBS) calibrates laboratory-standard instruments for X rays from 10 to 250 kVP by comparison with a standard free air ionization chamber. Laboratory-standard instruments for measuring exposure from photons of higher energies from 1 to 1,000 mSv (0.1 R to 100 R) or exposure rate from 0.1 to 150 mSv/min (0.01 to 15 R/min) can be calibrated by NBS by comparison with either cesium-137 or cobalt-60 calibrated sources. These laboratory-standard instruments or secondary standards may then be used to calibrate radiation protection survey instruments by intercomparison in radiation fields of similar quality. Due consideration must be given to beam width, uniformity of radiation over the calibration area, and changes in radiation quality caused by scattered radiation.

Neutron-instrument calibration can be afforded by exposure to fields from NBS calibrated neutron sources. One type of such a source is made by mixing a radionuclide such as plutonium, polonium, or radium with a material such as beryllium or boron. The neutrons are produced in (α, n) reactions in the latter materials. Radium sources are difficult to use because they also emit intense gamma radiation.

Leak Testing of Sealed Sources[7]

All sealed sources must be tested for leakage of radioactive material before initial use, at intervals not exceeding six months, whenever damage or deterioration of the capsule or seal is suspected, or when contamination of handling or storage equipment is detected.

The leak test should be capable of detecting the presence of 20 Bq (0.0005 μCi) of removable activity from the source. Sources leaking greater than 200 Bq (0.005 μCi) of removable activity, based on the test methods described below, should be removed from service immediately. Records of leak test results should be specific in units of disintegrations per minute or microcuries. Leak test records should be kept until final disposition of the source is accomplished.

A small sealed capsule may be tested by washing for a few minutes in a detergent solution. An aliquot of this solution should then be counted. An absorbent liner in the storage container normally in contact with the source will also reveal leakage if it is contaminated.

The leak test is usually performed by wiping the entire surface of the sealed source with either dry or wet filter paper or cotton swab. Removable contamination that is transferred to the paper or cotton swab from the source should be determined by measurement with appropriate radiation counting equipment. In the case of wet wipe, the paper or swab should be dried prior to counting.

Leak tests of devices from which the encapsulated source cannot be removed should be made by wiping the accessible surface of the device nearest to the storage position of the source.

Leakage of radon-222, the gaseous daughter product of radium-226, may occur from a radium capsule through small cracks not permitting leakage of the radium element. Radium sources should be leak-tested by detecting radon leakage directly or through the alpha or beta-gamma activity of its daughters. The radium leak test may be performed by placing the radium source in a glass jar with either a cotton ball or filter paper. The jar is sealed for approximately 24 hours to permit radon to build up in the jar. After the 24-hour waiting period, the jar lid is removed and checked immediately for alpha contamination. Tongs or forceps should be used to remove the cotton ball or filter paper from the jar. The cotton ball or filter paper should be counted for beta-emitting radon-daughters. The presence of abnormal radon concentrations in a radium safe may be detected by periodic or continuous air sampling.

Detection of contaminants on the housing or surface of a neutron source may not indicate source leakage, but may be due to induced activity. Confirmation of leakage may require identification of the contaminant.

In leak testing radioactive sources, special equipment may be necessary for radiation exposure control. Depending on the activity of the source, shielding may be required to keep the leak tester's exposure as low as possible. The actual leak test wipe should be done by using tongs or forceps and not the fingers. Rubber gloves should be used to minimize hand contamination. The wipes should be taken quickly and the source returned to its designated container.

PART 3

BASIC EXPOSURE CONTROL—PRINCIPLES AND TECHNIQUES

Physical Safeguards and Procedural Controls[4]

As long as the radiation source remains external, exposure of the individual may be terminated by removing the individual from the radiation field, by removing the source, or by switching off a radiation-producing machine. If the external radiation field is localized, exposure to individuals may be limited readily by shielding or by denying access to the field of radiation.

Physical Safeguards

Physical safeguards include all physical equipment used to restrict access of persons to radiation sources or to reduce the level of exposure in occupied areas. These include *shields, barriers, locks, alarm signals,* and *source shutdown mechanisms.*

Planning and evaluation of physical safeguards should begin in the early phases of design and construction of an installation. Detailed inspection and evaluation of the radiation safety of equipment are mandatory at the time of the initial use of the installation. Additional investigations are necessary periodically to assure that the effectiveness of the safeguards has not decreased with time or as a result of equipment changes.

Procedural Controls

Procedural controls include all instructions to personnel regarding the performance of their work in a specific manner for the purpose of limiting radiation exposure. *Training programs* for personnel often are necessary to promote observance of such instructions. Typical instructions concern *mode of use* of radiation sources, limitations on *proximity* to sources, *exposure time,* and *occupancy* of designated areas, and the sequence or kinds of *actions permitted* during work with radiation sources.

Periodic *area surveys* and *personnel monitoring* are necessary to assure the adequacy of and compliance with established procedural controls.

Classes of Installations (X and Gamma Rays)

NBS Handbook 114[2] classifies several types of nonmedical X-ray and gamma-ray installations: *protective, enclosed, unattended,* and *open.*

Protective Installation

This class provides the highest degree of inherent safety because the protection does not depend on compliance with any operating limitations. The requirements include:

1. Source and exposed objects are within a permanent enclosure within which no person is permitted during irradiation.
2. Safety interlocks are provided to prevent access to the enclosure during irradiation.
3. If the enclosure is of such a size or is so arranged that occupancy cannot be readily determined by the operator, the following requirements should also be provided: (a) fail-safe audible or visible warning signals to indicate the source is about to be used; (b) emergency exits; (c) effective means within the enclosure of terminating the exposure (sometimes called "Scram").
4. The radiation exposure 5 cm (2 in.) outside the surface of the enclosure cannot exceed 5 μSv (0.5 mR) in any one hour.
5. Warning signs of prescribed wording at prescribed locations.
6. No person can be exposed to more than the

permissible doses. The low allowable exposure level necessitates a higher degree of inherent shielding. At high energies in the megavolt region with high workloads, the required additional shielding may be extremely expensive. For example, in the case of cobalt-60, the required concrete thickness will have to be about 0.3 m (1 ft) greater than for the enclosed type.

Enclosed Installation

This class usually offers the greatest advantages for fixed installations with low use and occupancy. With *proper supervision* this class offers a degree of protection similar to the protective installation. The requirements for an enclosed installation include: 1, 2, 3, 5, 6, above, and

4. The exposure at any *accessible* and *occupied* area 0.3 m (1 ft) from the outside surface of the enclosure does not exceed 100 μSv (10 mR) in any one hour. The exposure at any accessible and *normally unoccupied* area 0.3 m (1 ft) from the outside surface of the enclosure does not exceed 1 mSv (100 mR) in any one hour. This class of installation requires administrative procedures to avoid exceeding the permissible doses. The trade-off between (1) the built-in, but potentially initially expensive, safety of a protective installation and (2) the required continuing supervision of consequences of an overexposure in an enclosed installation should be carefully considered in the planning stages of a new facility.

Unattended Installation

This class consists of automatic equipment designed and manufactured by a supplier for a specific purpose that does not require personnel in attendance for its operation. The requirements for this class include:

1. Source is installed in a single-purpose device.
2. Source is enclosed in a shield, where the "closed" and "open" positions are identified and a visual warning signal indicates when the source is "on."
3. The exposure at any accessible location 0.3 m (1 ft) from the outside surface of the device cannot exceed 20 μSv (2 mR) in any one hour.
4. The occupancy in the vicinity of the device is limited so that the exposure to any individual cannot exceed 5 mSv (0.5 R) in a year.
5. Warning signs.
6. Service doors to areas where potential exposure can exceed the measurements in items 3 and 4 above must be locked or secured with fasteners requiring special tools available only to qualified service personnel.

Open Installation

This class can only be used when operational requirements prevent the use of the other classes, such as in mobile and portable equipment where fixed shielding cannot be used. Mobile or portable equipment used routinely in one location should be made to meet the requirements of one of the fixed installation classes. Adherence to safe operating procedures is the main safeguard to overexposure. The requirements include:

1. The perimeter of any area in which the exposure can exceed 1 mSv (100 mR) in any one hour must be posted as a *high radiation area*.
2. No person may be permitted in the high radiation area during irradiation. In cases of unattended operation, positive means, such as a locked enclosure, shall be used to prevent access.
3. The perimeter of any area in which the radiation level exceeds 50 μSv (5 mR) in any one hour must be posted as a *radiation area*.
4. The equipment essential to the use of the source must be inaccessible to unauthorized use, tampering, or removal. This shall be accomplished by the attendance of a knowledgeable person or other means such as a locked enclosure.
5. No person can be exposed to more than the permissible doses.

TABLE 2. Gamma Ray Sources

Radionuclide	Atomic Number (Z)	Half-life	Gamma-ray Energy (MeV)	Gamma-ray Constant (R/curie-h at 1 m)
Cesium 137	55	30yr	0.662	0.32
Chromium 51	24	28d	0.323	0.018
Cobalt 60	27	5.2yr	1.17, 1.33	1.3
Gold 198	79	2.7d	0.412	0.23
Iridium 192	77	74d	0.136, 1.065	0.5
Radium 226	88	1622yr	0.047 to 2.4	0.825
Tantalum 182	73	155d	0.066 to 1.2	0.6

Source: 2.

TABLE 3. Forward X-ray Intensity from an Optimum Target

Peak Voltage	Intensity (R/min-mA at 1 m)
50 kV	0.05
70	0.1
100	0.4
250	2.0
1 MV	20.0
2	280.0
5	5000.0
10	30,000.0
15	100,000.0
20	200,000.0

Source: 2, 5.

Output of Radiation Sources

Table 2 lists some data on gamma-ray sources of interest for industrial purposes. Table 3 lists some typical radiation machine outputs for varying voltages.

Working Time

This is the allowable working time in hours per week for a given exposure rate. For example, for an exposure rate of 100 μSv/h (10 mR/h) to the whole body

$$\text{Working time} = \frac{\text{permissible occupational dose per week}}{\text{exposure dose rate}}$$

$$= \frac{1000\ \mu\text{Sv/week}}{100\ \mu\text{Sv/hour}} \left(\frac{100\ \text{mR/week}}{10\ \text{mR/hour}} \right)$$

$$= 10\ \text{h/week}$$

Working Distance

The inverse square law applied to radiation states that the dose rate from a point source is inversely proportional to the square of the distance from the origin of the radiation source provided that (1) the dimensions of the radiation source are small compared with the distance, and (2) no appreciable scattering or absorption of the radiation occurs in the media through which the radiation travels. In practice, the first requirement is satisfied whenever the distance involved is at least 10 times greater than the largest source dimension. In situations where there is insignificant scattering or absorption, the primary beam is the total radiation field.

Example: Consider a 100 millicurie iridium-192 source in air in the shape of a pencil, 0.0063 m (¼ in.) diameter and 0.127 m (5 in.) long. What would the working time be at 3 m?

Solution: 3 m is obviously more than 10 times 0.127 m (5 in.), so the inverse square law applies. Also, scattering is not a problem. From Table 2, the gamma-ray constant for iridium-192 is 0.5 R/h per curie at 1 m. Therefore:

$$\text{Exposure rate at 1 m} = (0.5\ \text{R/h-curie})\ (0.1\ \text{curie}) = 0.05\ \text{R/h}$$

$$\text{Exposure rate at 3 m} = (0.05\ \text{R/h})\ (1\text{m}/3\text{m})^2 = 0.0055\ \text{R/h} = 5.5\ \text{mR/h}$$

$$\text{Working time} = \frac{100\ \text{mR/week}}{5.5\ \text{mR/h}} = 18\ \text{h/week}$$

FIGURE 5. The maximum range of beta particles as a function of energy in the various materials indicated. (From SRI Report No. 361, "The Industrial Uses of Radioactive Fission Products." With permission of the Stanford Research Institute and the U.S. Atomic Energy Commission.)

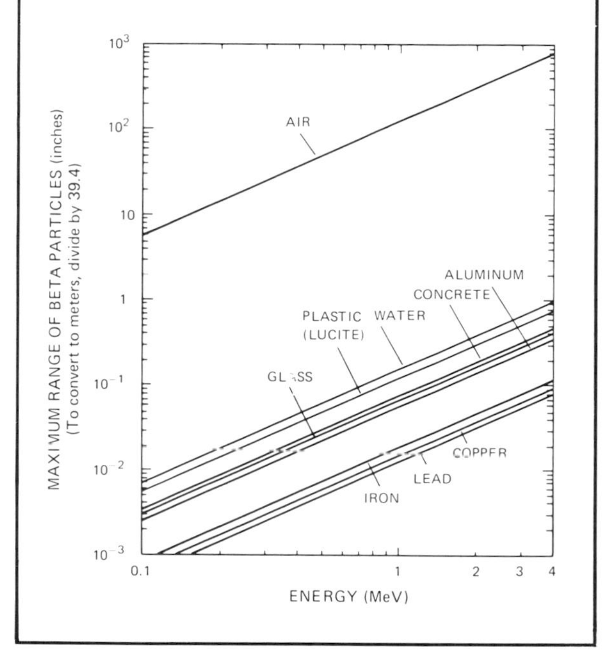

TABLE 4. Shielding Equivalents: Approximate Tenth (TVL) and Half-Value (HVL) Layer Thicknesses in Lead and Concrete for Several Gamma-Ray Sources

Source	Data in inches*			
	Lead		Concrete	
	TVL	HVL	TVL	HVL
Ra 226	2.2	0.65	9.2	2.7
Co 60	1.6	0.49	8.6	2.6
Cs 137	0.84	0.25	6.2	1.9
Ir 192	0.79	0.24	5.5	1.6
Au 198	0.43	0.13	5.5	1.6

*To convert to meters, divide by 39.4

FROM NATIONAL BUREAU OF STANDARDS (SEE REFERENCE 2).

TABLE 5. Shielding Equivalents: Approximate Half-Value (HVL) and Tenth-Value (TVL) Layers for Lead and Concrete for Various X-Ray Tube Potentials

Peak Voltage (kV)	Lead (mm)		Concrete (mm)*	
	HVL	TVL	HVL	TVL
50	0.05	0.16	4.32	15.10
70	0.15	0.5	8.38	27.95
100	0.24	0.8	15.10	50.80
125	0.27	0.9	20.30	66.00
150	0.29	0.95	22.35	73.60
200	0.48	1.6	25.40	83.80
250	0.9	3.0	27.95	94.00
300	1.4	4.6	31.21	104.00
400	2.2	7.3	33.00	109.10
500	3.6	11.9	35.55	116.80
1000	7.9	26.0	44.45	147.10
2000	12.7	42.0	63.50	210.40
3000	14.7	48.5	73.60	241.20
4000	16.5	54.8	91.40	304.48
6000	17.0	56.6	104.00	348.00
10,000	16.5	55.0	116.80	388.50

*To convert to inches, divide by 25.4

FROM NATIONAL BUREAU OF STANDARDS (SEE REFERENCE 2).

Shielding

Common materials such as concrete and lead can be used as absorbers or shields to reduce personnel exposures. *Beta* or *electron radiation* is completely stopped by the thicknesses of material shown in Figure 5. The thickness of any material that will reduce the amount of radiation passing through the material to one-half is referred to as the *half-value layer* (HVL). Similarly, the thickness that will reduce the radiation to one-tenth is referred to as the *tenth-value layer* (TVL). (See Tables 4, 5, Figs. 6, 7.)

The use of these terms implies an exponential function for transmitted radiation in terms of shield thickness. An inspection of Figures 6 and 7 shows that this is not strictly correct because the transmission curves are not completely linear on a semilogarithmic plot. Hence, the listed HVLs and TVLs in Tables 4 and 5 are approximate values obtained with large attenuation.

Table 6 lists densities of commercial building materials. For X and gamma radiation, the absorption process depends to a large degree on Compton

TABLE 6. Densities of Commercial Building Materials

Material	Average density*	
	g/cm³	lb/ft³
Aluminum	2.7	169
Bricks:		
Fire clay	2.05	128
Kaolin	2.1	131
Silica	1.78	111
Clay	2.2	137
Cements:		
Colemanite borated	1.95	122
Plain (1 Portland cement: 3 sand mixture)	2.07	129
Concretes:		
Barytes	3.5	218
Barytes-boron frits	3.25	203
Barytes-limonite	3.25	203
Barytes-lumnite-colemanite	3.1	194
Iron-Portland	6.0	375
MO (ORNL mixture)	5.8	362
Portland (1 cement: 2 sand: 4 gravel)	2.2	137
Glass:		
Borosilicate	2.23	139
Lead (hi-D)	6.4	399
Plate (avg.)	2.4	150
Iron	7.86	491
Lead	11.34	708
Lucite (polymethyl methacrylate)	1.19	74
Rocks:		
Granite	2.45	153
Limestone	2.91	182
Sandstone	2.40	150
Sand	2.2	137
Sand plaster	1.54	96
Type 347 stainless steel	7.8	487
Steel (1% carbon)	7.83	489
Uranium	18.7	1,167
Uranium hydride	11.5	718
Water	1.0	62

FROM NATIONAL BUREAU OF STANDARDS (SEE REFERENCE 2).

absorption and scattering which, in turn, increases with the atomic electron density. As a first approximation, electron density varies directly with the mass density of a material. Hence, the denser building materials are usually better shielding materials for a given thickness of material. On a mass basis, shielding materials are much the same above about 500 keV. Where space is a problem, lead is often used to achieve the desired shield attenuation. Lead, however, requires extra structural support since it is not self supporting. Concrete is by far the most commonly used shielding material for economic, structural, and local availability reasons—in addition to desirable shielding characteristics. Where space considerations are of prime importance, depleted uranium shields, while expensive, offer excellent solutions to difficult problems.

FIGURE 6. Transmission through lead of gamma rays from selected radionuclides

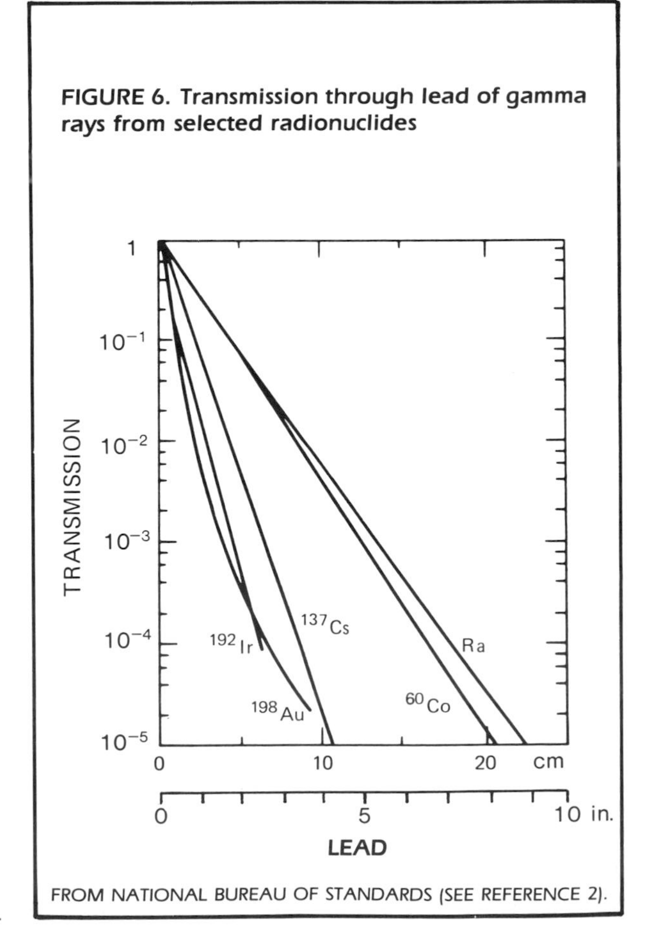

FROM NATIONAL BUREAU OF STANDARDS (SEE REFERENCE 2).

Table 4 lists HVLs and TVLs for several commonly used gamma-ray emitting radionuclides. Table 5 lists similar information for X-ray peak voltages. Figures 6 and 7 show actual transmission through lead and concrete for the gamma-ray emitting radionuclides. Figures 8-12 show transmission through concrete, steel, and lead for X-ray beams of various peak energies. These graphs present *broad-beam shielding* information, which includes all scattered radiation resulting from deflection of the primary gamma or X rays within the shield as well as absorption of the primary radiation. Most engineering applications need to consider broad beam geometry. *Narrow beam geometry*, where only the primary beam needs consideration, is seldom encountered in practice.

FIGURE 7. Transmission through concrete (density: 147 lb/ft^3 or 2.35 g/cm^3) of gamma rays from radium; cobalt 60, cesium 137, gold 198, irridium 192

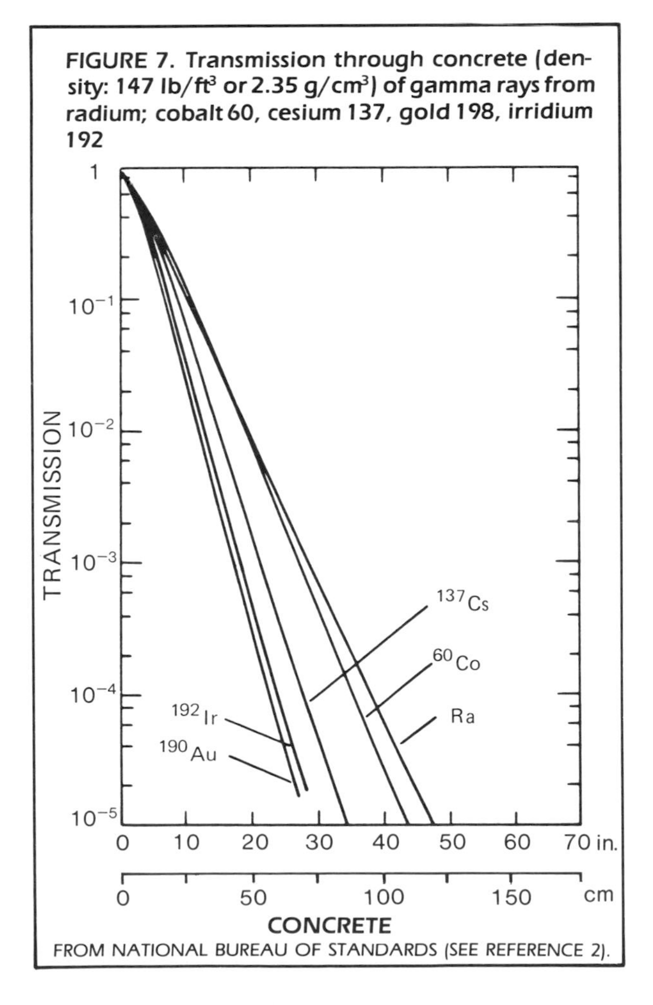

FROM NATIONAL BUREAU OF STANDARDS (SEE REFERENCE 2).

FIGURE 8. Transmission through concrete (density 2.35 g cm^{-3}) of x rays produced by 0.1 to 0.4 MeV electrons, under broad-beam conditions. Electron energies designated by an asterisk (*) were accelerated by voltages with pulsed wave form; unmarked electron energies were accelerated by a constant potential generator. Curves represent transmission in dose-equivalent index ratio.

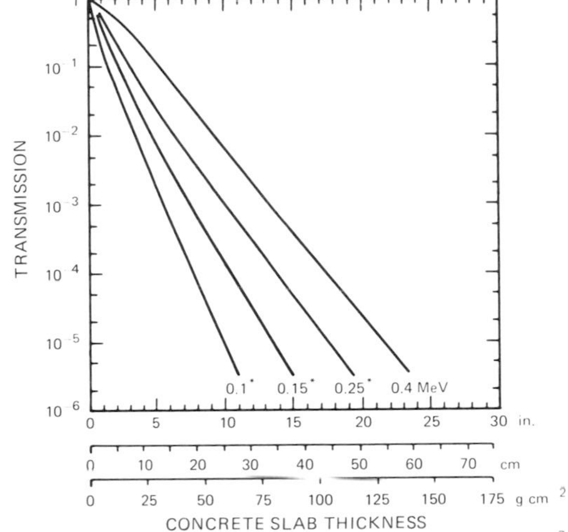

The bottom scale indicates the required 'mass thickness' (g cm^{-2}) Different density concrete may be used provided the required mass thickness is achieved. Additionally, where weight is a consideration, this scale is useful in selecting the optimum shielding material.

FROM NATIONAL COUNCIL ON RADIATION PROTECTION (SEE REFERENCE 8).

Protective Exposure Enclosures

Because of scattered radiation, protection for the operator and other personnel working in the neighborhood often requires shielding of the part being radiographed and any other material exposed to the direct beam, in addition to the shield for the source itself. Preferably the source and materials being examined should be enclosed in a room or hood with the necessary protection incorporated into the walls. These concepts are discussed under Classes of Installations.

Shields can be classified as either primary or secondary. Primary shields are designed to shield against the primary radiation beam; secondary shields are only thick enough to protect against tube housing leakage and scatter radiation. Therefore, the X-ray tube or source should *not* be pointed toward secondary shields. For this reason, mechanical stops should be used to restrict tubehousing orientations toward primary barriers. Operating restrictions, such as not pointing the beam at certain walls or the ceiling, should be spelled out in the operating procedures.

When changes in operating conditions are contemplated, the radiation safety officer (RSO) should be contacted for consultation and any appropriate surveys to determine additional shielding requirements.

For design purposes, the primary beam should not be pointed at a high personnel occupancy space and the distance from the radiation source to any occupied space should be as great as is practical. *Scattered radiation* usually has a lower effective energy than the primary beam and may, therefore, be easier to shield.

Thickness of Shielding Walls

The shielding in the walls of the enclosures should be of sufficient thickness to reduce the exposure in all occupied areas to *as low* a value *as* is *reasonably achievable* (ALARA). In the design the desired thickness can be determined with reasonable accuracy by reference to tables or by calculations. (See ref. 2.)

In many cases an additional TVL can be included at little extra cost and will increase the margin of safety considerably. A series of measurements of transmitted radiation in occupied areas, called a *radiation survey*, is necessary to document the adequacy of the facility's design. Such a radiation survey can be derived from a combination of portable instrument readings and personnel dosimeters placed at appropriate locations in the facility (called *badge plants*).

FIGURE 9. Transmission of thick-target X rays through ordinary concrete (density 2.35 g cm^{-3}), under broad-beam conditions. Energy designations on each curve (0.5 to 176 MeV) refer to the monoenergetic electron energy incident on the thick X-ray producing target. Curves represent transmission in dose-equivalent index ratio.

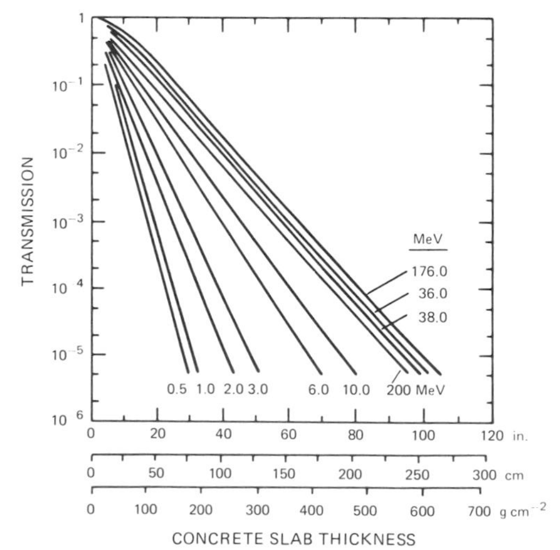

The bottom scale indicates the required 'mass thickness' (g cm^{-2}) Different density concrete may be used provided the required mass thickness is achieved. Additionally, where weight is a consideration, this scale is useful in selecting the optimum shielding material.

FROM NATIONAL COUNCIL ON RADIATION PROTECTION (SEE REFERENCE 8).

FIGURE 10. Broad-beam transmission through steel (density 7.8 g cm^{-3}) of X rays produced by 1- to 31-Mev electrons. Energy designations on each curve refer to the monoenergetic electron energy incident on the thick X-ray producing target. Curves represent transmission in dose-equivalent index ratio.

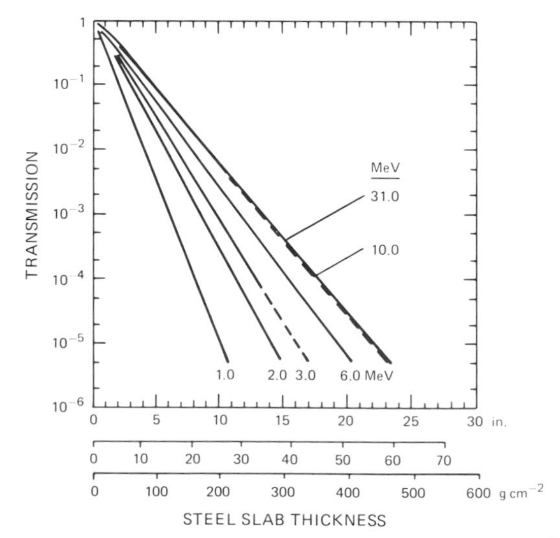

The bottom scale indicates the required 'mass thickness' (g cm^{-2}) which is useful in selecting the optimum shielding material where weight is a consideration.

FROM NATIONAL COUNCIL ON RADIATION PROTECTION (SEE REFERENCE 8).

FIGURE 11. Broad-beam transmission through lead of thick target X rays produced by 0.1- to 0.4-MeV electrons. Electron energies designated by an asterisk (*) were accelerated by voltages with pulsed wave form; unmarked electron energies were accelerated by a constant potential generator. Curves represent transmission in dose-equivalent index ratio.

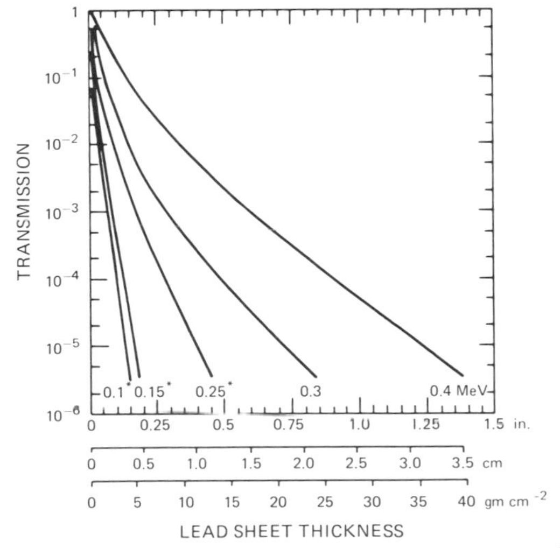

The bottom scale indicates the required 'mass thickness' (g cm^{-2}) which is useful in selecting the optimum shielding material where weight is a consideration.

FROM NATIONAL COUNCIL ON RADIATION PROTECTION (SEE REFERENCE 8).

FIGURE 12. Broad-beam transmission through lead of X rays produced by 0.5- to 86-MeV electrons. Energy designations on each curve (0.5 to 86 MeV) refer to the monoenergetic electron energy incident on the thick X ray producing target. Curves represent transmission in dose-equivalent index ratio.

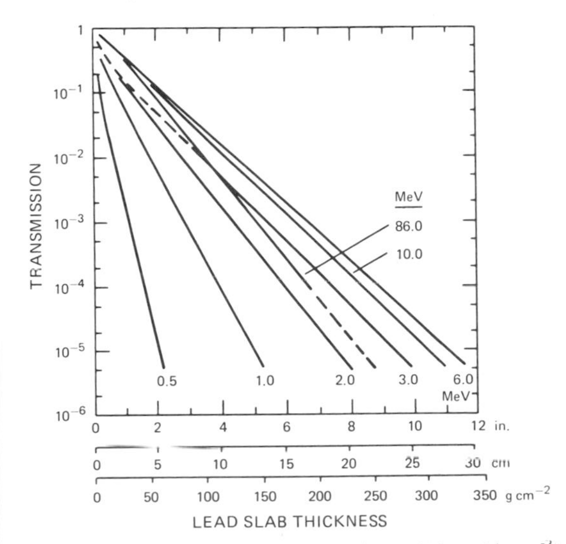

The bottom scale indicates the required 'mass thickness' (g cm^{-2}) which is useful in selecting the optimum shielding material where weight is a consideration.

FROM NATIONAL COUNCIL ON RADIATION PROTECTION (SEE REFERENCE 8).

PART 4

NEUTRON RADIOGRAPHY

Introduction

Neutrons are of interest in radiography because their interaction with matter is significantly different from X or gamma rays. Neutrons are absorbed and scattered more in low atomic-numbered (low Z) materials than high Z materials. Thus, plastics, explosives, and some organic materials can be examined for defects with little interference from encapsulating metals and electronic parts and wiring.

Neutron Sources[6]

Radioactive Neutron Sources

These are of two types, (α, n) and (γ, n). In the (α,n) source, the alpha emitter is mixed with the target material or may even be alloyed with it. In (γ,n) sources the gamma-ray emitter is usually surrounded by the material [e.g., Be or heavy water (deuterium oxide)] in which the neutron-producing reaction takes place. Polonium-210, plutonium-239, and americium-241 in combination with beryllium or other target materials are suitable radioactive alpha-neutron sources from the standpoint of large neutron yield and low gamma-ray yield. Radium-226 and all photoneutron (γ, n) sources produce high gamma-ray intensities that constitute serious hazards and may create undesirable interference with the neutron radiographic process itself.

Spontaneous Fission Neutron Sources

These sources are attractive because of their fission-like spectrum, relatively low gamma-ray yield and their small mass. A nuclide of especial interest is californium-252, which is becoming available in appreciable quantities.

Accelerator Sources

Constant-voltage accelerators such as Van de Graaff and Cockcroft-Walton accelerators can produce energies up to about 20 MeV for protons and deuterons, and still higher energies for alpha particles and heavy ions. Small accelerators using deuterons of 100 to 200 keV energy can produce large numbers of 14 MeV neutrons when using a tritiated target. High-frequency positive-ion accelerators include the cyclotron, synchrocyclotron, proton synchrotron, and heavy-ion linear accelerator. These are capable of producing a wide range of neutron energies. Protons above 10 MeV will produce neutrons when striking almost any material. High-frequency electron accelerators such as the betatron produce X rays through the interaction of the accelerated electrons with the target. The X rays in turn produce photoneutrons, most with energies of a few MeV but with some neutrons having energies up to near the maximum energy of the accelerator.

Nuclear Reactor Sources

Neutron production in reactors occurs as a result of the fission process. In the usual operating mode the number of fissions (and neutrons) is essentially constant in time. The neutron energies range from thermal to 15 MeV with the number over 10 MeV being small.

Shielding

Fast Neutrons

Adequate shielding against neutrons will often attenuate gamma radiation to acceptable levels at both reactors and accelerators. Water and other hydrogenous shields may constitute an important exception to this rule. Ordinary or heavy aggregate concrete or earth are the recommended materials in most installations. Any economy achieved by the use of water-filled tanks is likely to be offset by maintenance difficulties. Both paraffin and oil, although they are good neutron absorbers, are fire hazards, and neither should be used in large stationary shields. Methods of shielding calculations are discussed in detail in NCRP 38.[6]

The importance of concrete as a structural and shielding material merits special mention. Its use for gamma- and X-ray shielding has been previously discussed. Because of its relatively high hydrogen and oxygen content, it is also a good neutron shield. The subject of shielding calculations for neutrons is complex and should be performed by specialists. For some benchmarks, however, one can note approximate TVLs of 0.25 m (10 in.) of concrete for 14 MeV neutrons and 0.15 m (6 in.) for 0.7 MeV neutrons.

Thermal Neutrons

Generally the energies associated with thermal neutrons are less than 1 eV. For radiation protection the most important interaction of thermal neutrons with matter is radioactive capture. In this process, the neutron is captured by the nucleus with the emission of gamma radiation. A shield adequate for fast neutrons usually will be satisfactory for thermal neutrons. The low quality factor (QF = 2) for thermal neutrons (0.025 eV) makes their biological consequence considerably less than for fast neutrons.

PART 5

SKYSHINE[8]

In the design of facilities, there is often a question concerning the magnitude of shielding required for the roof over the building. As an ordinary weather roof provides little if any attenuation for upward-directed radiation, there is a significant probability that radiation reflected back from the atmosphere will be unacceptable in the immediate area of the facility. Looking at Figure 13, for X and gamma rays this radiation (1) increases roughly as $\Omega^{1.3}$, where Ω is the solid angle subtended by the source and shielding walls, (2) decreases with $(d_s)^2$, where d_s is the horizontal distance from the source to the observation point, and (3) decreases with $(d_i)^2$, where d_i is the vertical distance from the source to about 2 m above the roof. The shield thickness necessary to reduce the radiation to an acceptable level may be calculated according to the methods in NCRP Report No. 51[8] and may alternatively be designed into the roof structure, or mounted over the source with a lateral area sufficient to cover the solid angle, Ω. Similar statements apply to neutron skyshine, except that the functional dependences of the radiation at d_s are slightly different for Ω and d_s.

FIGURE 13. Radiation Shielding—Skyshine Method

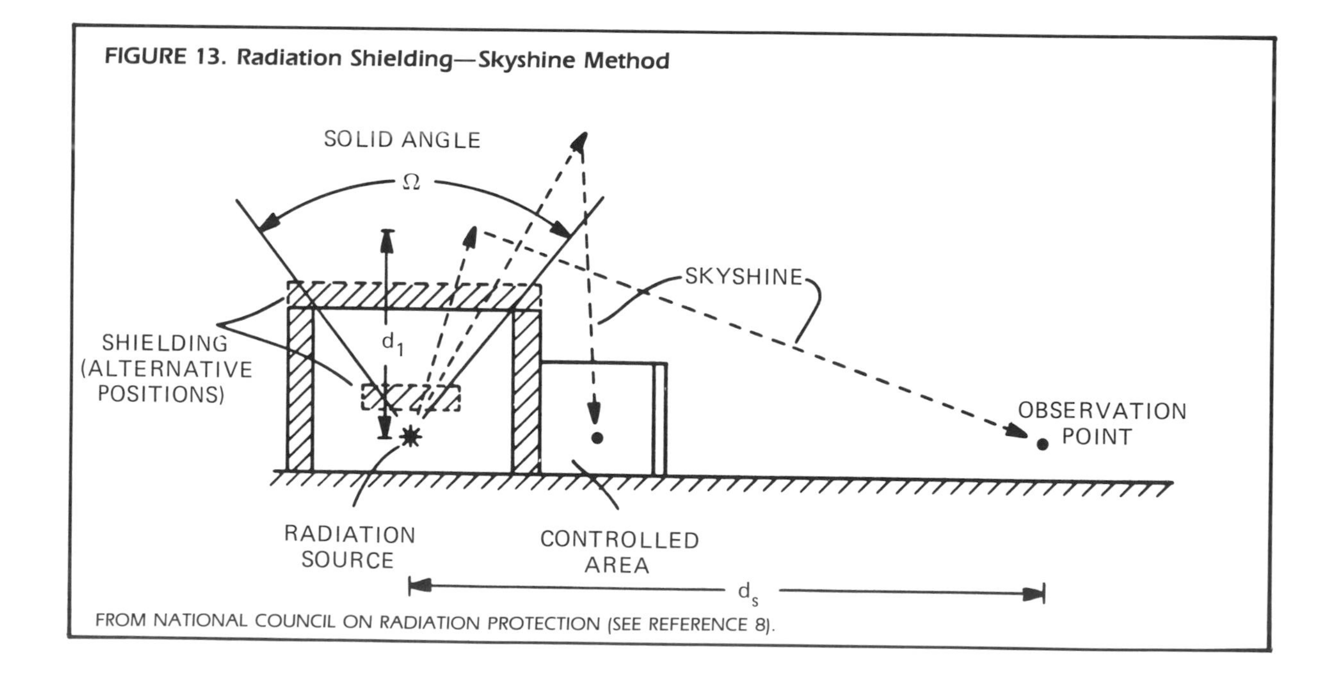

FROM NATIONAL COUNCIL ON RADIATION PROTECTION (SEE REFERENCE 8).

PART 6

MANAGEMENT CONSIDERATIONS

Government Licensing[9]

Most manufacturers specify that radiation-producing devices should be operated only by qualified personnel. Most states require the *registration* of radiation machines and provide survey services during compliance audits. *Licenses to possess by-product materials* (radioisotopes other than radium) are issued by the Nuclear Regulatory Commission (NRC) or states operating under the rules of the NRC (so-called agreement states).

Radiation Safety Officer (RSO)

Personnel responsible for work with radiation are also responsible for radiation safety. A *radiation safety officer* (RSO) needs to be appointed if fields may be experienced in excess of 1 millisievert (100 millirem) per work week in accessible regions inside or outside externally applied shielding. The RSO is responsible for:

1. Technical assistance in planning and execution of work insofar as radiation safety is concerned.
2. Appraisal of safe operation of the radiation source through surveys and personel monitoring.
3. Notification of personnel working around the source of any special hazards.
4. Reporting of radiation hazards or unsafe practices to the proper authorities.
5. Seeking advice from qualified experts when necessary.
6. Keeping records of personnel exposures and area dose levels.
7. Keeping informed of any changes in the mode of operation of the source.
8. Periodically providing radiation safety training.

Transportation of Radioactive Materials

Radioactive material is considered a hazardous material. As a result its shipment within the United States is controlled by the Department of Transportation under Title 49 Code of Federal Regulations Parts 171-177. These regulations prescribe the rules and procedures for packaging, marking, labeling, placarding, and shipping.

Additional requirements for the international shipment of such materials by air are set forth by the International Air Transport Association (I.A.T.A.).

Except for very minor quantities, use of the Postal Service for transport of radioactive materials is prohibited.

Finally the Inter-Governmental Maritime Consultative Organization (I.M.C.O.) represents the collection of nations around the world that regulates the international transport of dangerous goods by sea.

Disposal

The disposal of leaking sources, contaminated equipment, or sources decayed below useful levels must be according to CFR Title 10. Generally, an NRC licensed commercial radioactive waste-disposal service is utilized for this purpose, either directly by the owner of the source or indirectly by returning the source to the manufacturer.

References

1. "Basic Radiation Protection Criteria." National Council on Radiation Protection and Measurements (NCRP) Report No. 39 (1971). Available from NCRP Publications, 7910 Woodmont Ave., Suite 1016, Bethesda, MD 20814.
2. "General Safety Standard for Installations Using Non-Medical X-Ray and Sealed Gamma-Ray Sources, Energies up to 10 MeV." NBS Handbook 114, U.S. Department of Commerce/ National Bureau of Standards (1975). Available from U.S. Government Printing Office, Washington, DC 20402 (SD Catalog No. C13.11:114).
3. "Guidelines on Pregnancy and Work." The American College of Obstetricians and Gynecologists, DHEW (NIOSH) Publication No. 78-118 (1972), U.S. Government Printing Office, Washington, DC 20402.
4. "Instrumentation and Monitoring Methods for Radiation Protection." NCRP Report No. 57 (1978).
5. Patterson, H.W., and R.H. Thomas. *Accelerator Health Physics*. New York: Academic Press (1973).
6. "Protection against Neutron Radiation." NCRP Report No. 38 (1971).
7. "Protection Against Radiation from Brachytherapy Sources." NCRP Report No. 40 (1972).
8. "Radiation Protection Design Guidelines for 0.1-100 MeV Particle Accelerator Facilities." NCRP Report No. 51 (1977).
9. "Rules of General Applicability to Licensing of Byproduct Material." Code of Federal Regulations, Title 10 (Energy) Part 30. U.S. Government Printing Office, Washington, DC 20402.
10. "Standards for Protection against Radiation." Code of Federal Regulations, Title 10 (Energy) Part 20. U.S. Government Printing Office, Washington, DC 20402.
11. "Structural Shielding, Design, and Evaluation for Medical Use of X Rays and Gamma Rays of Energies up to 10 MeV." NCRP Report No. 49 (1976).

SECTION 19

SPECIALIZED RADIOGRAPHIC METHODS

Robert Buchanan, Lockheed Missles and Space Company, Sunnyvale, CA
Dana E. Elliott, Los Alamos National Laboratory, Los Alamos, NM
James D. Geis, Ridge, Inc., Tucker, GA
James W. Guthrie, Sandia National Laboratories, Albuquerque, NM
Roger Hadland, Hadland Photonics, Hemel Hempstead, U.K.
Donald J. Hagemaier, McDonnell-Douglas Corporation, Long Beach, CA
John F. Landolt, Rockwell International, Golden, CO
Claude Laperle, General Electric, St. Petersburg, FL
Gregory A. McDaniel, Presbyterian Hospital of Dallas, Dallas, TX
Edward H. Ruescher, Southwest Research Institute, San Antonio, TX
Lawrence E. Bryant, Los Alamos National Laboratory, Los Alamos, NM

PART 1

HIGH-SPEED VIDEOGRAPHY FOR OPTICAL AND X-RAY IMAGING

High-speed Video System

The SP-2000 Motion Analysis System, or high-speed video (HSV), can record up to 2,000 full frames per second or up to 12,000 partial (hex) frames per second with a playback speed of 60 frames per second. This allows a slowing down of the recorded action by a factor of up to 200 times. The system is portable, yet capable of supporting two cameras simultaneously. The two resulting images could be: (1) views of the same event from different angles; and/or (2) different fields of view or different depths of field. These can be viewed on the same TV monitor by use of picture inset techniques with variable size and position of the inset. Other useful features of the HSV system include a data frame (as seen in Fig. 1) and *X* and *Y* reticles that can be activated on replayed images to give accurate position data for any frame.

These reticle lines are illustrated in Fig. 2a, where they are positioned on the center of a small sphere, giving a horizontal *X* and vertical *Y* position with the *X* and *Y* position values indicated on the right side of the monitor (data frame).

Also visible in the data frame are indications of: (1) the particular frame being viewed; (2) identification number; (3) *X* and *Y*; (4) recording rate in frames per second; (5) frame count; (6) tape count; (7) status messages such as *stopped* and *still image*; (8) time of day; (9) date; and (10) elapsed time.

The appeal of the HSV system is twofold. First, the live camera set-up conditions increase the probability of success on the initial recording; secondly, there is the immediate playback feature common to all video systems.

The fact that the system's data is in digital form means that, with the optional computer interface, information such as *X* and *Y* data can be directly input to a computer.

These convenient features are the result of a number of technological advances, including: (1) a solid state video sensor; (2) specialized microgap recording heads; (3) high-density magnetic recording tape; (4) microprocessors for a wide range of sophisticated yet simple controls that are essentially immune to operator error.

FIGURE 1. Sequence of High-speed Video Images of a Water-filled Balloon Falling on Pavement

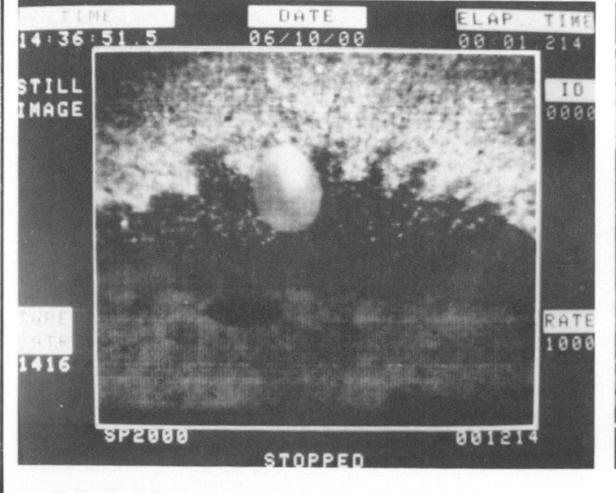

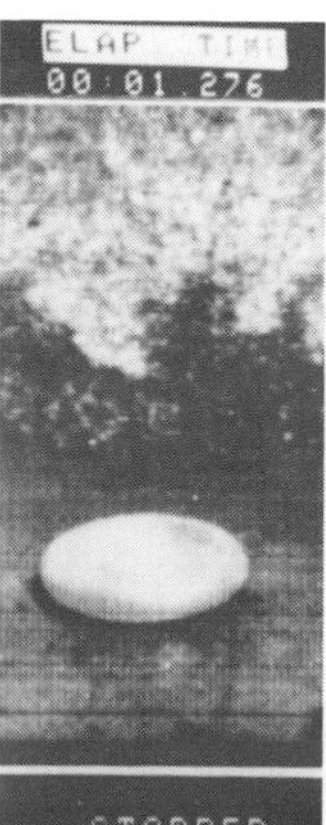

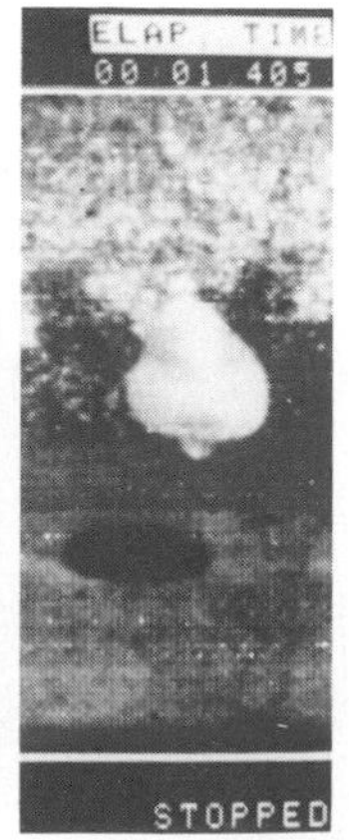

Optical Applications

High-speed video can be used to document a wide range of dynamic events, ranging from small-scale laboratory experiments to large field tests.

Laboratory applications have included observation of pressure tests, such as: destruction of small diameter spheres to learn the failure mechanism; ignition and combustion of heat powder samples to determine burn-rate data; and circulation of beads in various liquid media. The latter example is illustrated in Fig. 2, where two frames (0.037 second apart and recorded at 1,000 frames per second) show the relative movement of various beads and allow determination of object velocity, acceleration and spin rate. Other laboratory tests include diagnosis of laser welding, arcing of large electric motors and many types of high-velocity mechanical actions.

Among larger scale field tests using high-speed video is the observation of rocket-assisted drop-tower experiments, where the two cameras record a metal shearing process and projectile progress in order to determine acceleration data.

FIGURE 2. High-speed Video Images: (a) Spheres Circulating in Liquid Medium with X-Y Reticle Activated; (b) Image of Spheres 0.037 Second after Fig. 2a

(a)

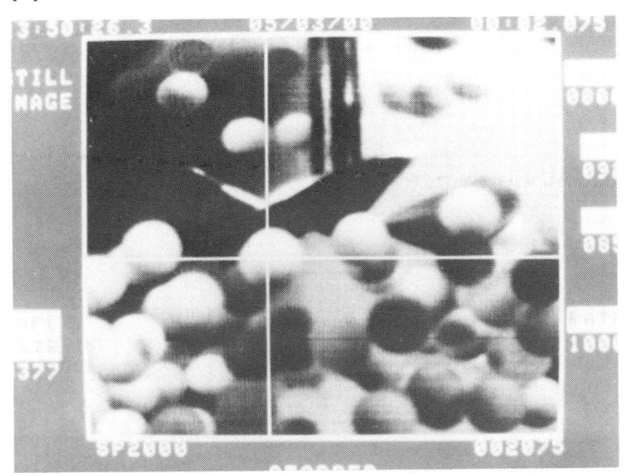

(b)

High-speed Video Combined with Dynamic Radiography

The system has been described thus far in terms of direct optical observation of high-speed events. However, the same camera can be used to monitor the output fluor of an X-ray sensitive image intensifier. When combined with an X-ray generator, the high-speed video/X-ray image intensifier allows recording of dynamic radiographic images up to the maximum 12,000 partial frames per second.

The X-ray image intensifier consists of: (1) an input fluor that converts X-rays to light; (2) a photoelectron layer that converts the light to electrons;

FIGURE 3. Set-up for High-speed Video Imaging: (a) X-ray Tubehead; (b) Object to be Inspected; (c) X-ray Image Intensifier; and (d) High-speed Video Camera

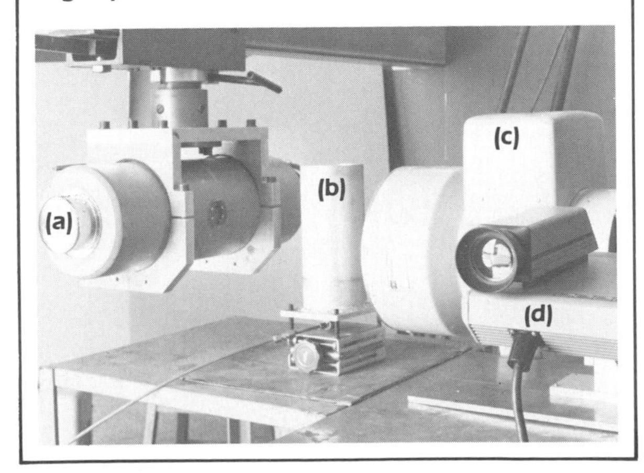

and (3) an electron acceleration (or intensification) stage. The electrons are converted back to light at the output fluor. Both the input and output fluors have decay constants of 650 nanoseconds or less, eliminating motion blur when recording up to the system's maximum frame rate.

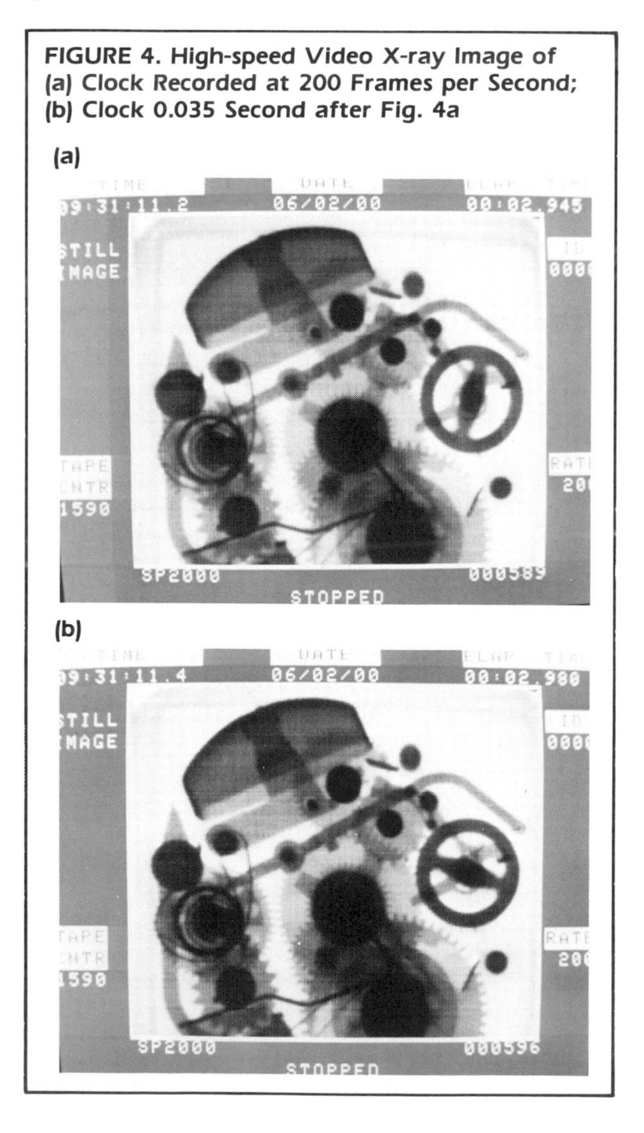

FIGURE 4. High-speed Video X-ray Image of (a) Clock Recorded at 200 Frames per Second; (b) Clock 0.035 Second after Fig. 4a

(a)

(b)

Application of High-speed Video/Dynamic Radiography

A typical set-up is seen in Fig. 3 (left to right): X-ray tubehead; object to be inspected, in this case, a cylinder; X-ray image intensifier; and high-speed video camera. Dynamic radiographic imaging of the mechanical actions of a clock and a compressor have been demonstrated using a similar set-up. In Fig. 4, two frames of a clock, 0.035 second apart, show an obvious change in the balance wheel rotation at the right center region of the frame.

Figure 5 illustrates combined optical and X-ray imaging of a fan operating at 1,200 revolutions per minute (seen optically in the lower left corner). The remainder of the image is the X-ray visualization of a lead marker mounted on one blade of the fan. The frame rate is 4,000 pictures per second; the upper and lower images are separated by 1/4000 second. This illustration shows both the second image inset capability (optical image inset in X-ray image) and split frame feature (two images separated by a small time interval and shown on the same video frame).

FIGURE 5. Combined Optical and X-ray Image of Fan at 4,000 Frames per Second

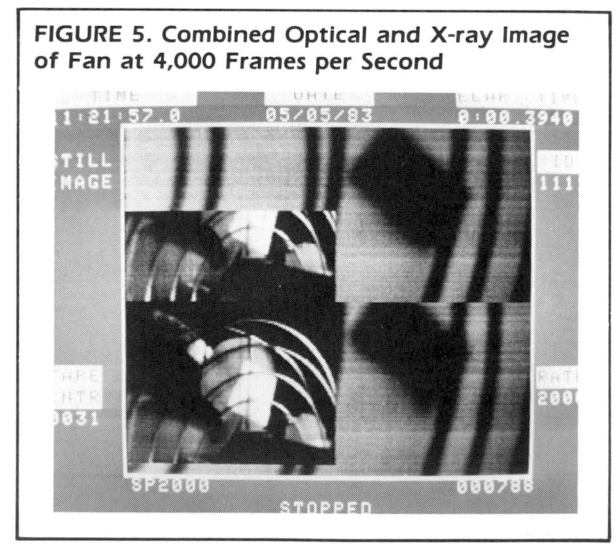

PART 2

HIGH-SPEED CINE X-RAY DEVELOPMENTS

Flash X-ray techniques are the primary method of recording fast mechanical transients when surrounding material, debris or flame prevents the use of conventional high-speed photography. Flash X-ray provides a high quality, large format image of an event and, if several images are required, several sources can be used, each projecting onto a film cassette. This is ideal for observing events such as a projectile separating from its sabot because the subject moves a considerable distance between each picture, avoiding image overlap. Flash X-ray is not so suitable for studying events that remain in place, such as metal flow in casting molds, projectiles penetrating targets and transient events in rotating machinery. These all require a sequence of images projected from either a single X-ray source or multiple sources located sufficiently close that variations in image aspect are minimized.

In 1975, Rolls-Royce had the problem of observing the passage of metallic foreign bodies in a gas turbine engine. Initial work with a repetitive pulse X-ray source and an intermittent camera operating up to 500 frames per second indicated a speed up to 10,000 frames per second was required. As a result, a system was developed that operated up to 10,000 frames per second with exposures down to 10 microseconds using a continuous X-ray source.[1]

The arrangement is shown in Fig. 6 and comprises a 380 kV, 10 mA constant potential source, an X-ray image intensifier with gating control and a high-speed rotating prism camera.

The most sensitive prism cameras have a limiting aperture of f2.8, which collects less than 0.5% of the light emitted from an isotropic source. The most critical aspect of the system was to convert X rays to visible light with the highest possible efficiency. Initial tests with scintillators, optically imaged into visible light intensifier tubes, proved to be between 1 and 2 orders less sensitive than medical X ray intensifiers. The largest available intensifier (with 320 mm input window) was chosen to maximize recording format.

The X-ray image intensifier had a cesium iodide scintillator and S20 cathode (optimized for 80 kV) but was lower in sensitivity by a factor of two times at 380 kV (in terms of electron to X-ray photon

FIGURE 6. Cine X-ray System with Repetitive Pulse X-ray Source and Intermittent Camera

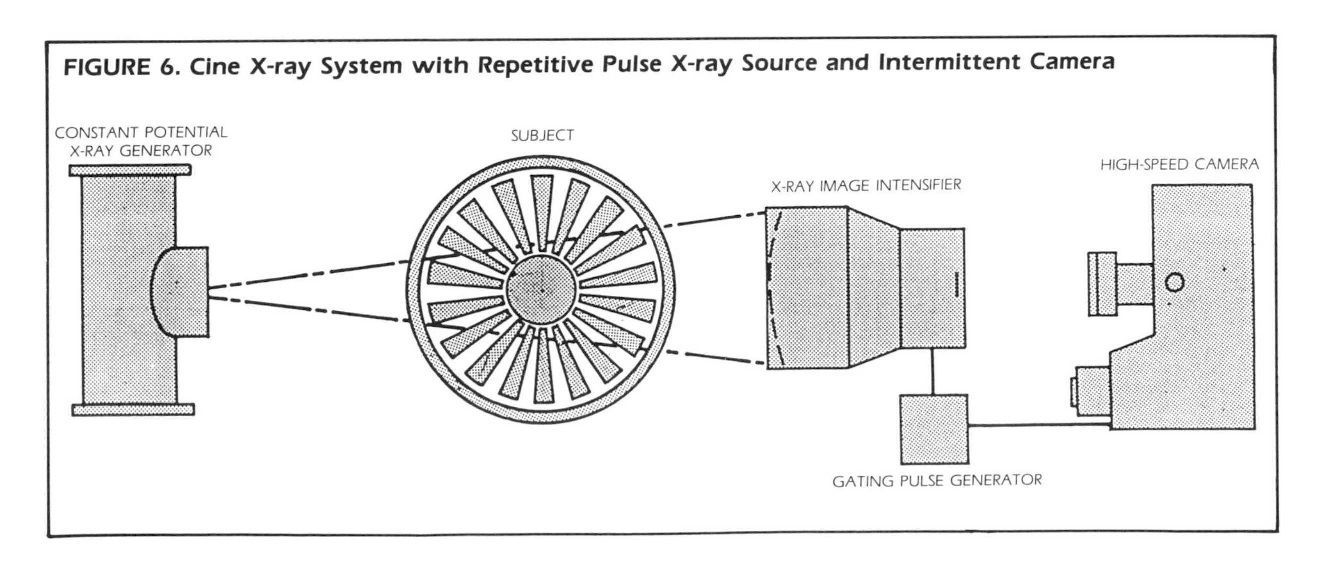

ratio). The phosphor was P20, which exhibited a characteristic decay of about 150 microseconds, as determined by image blur in a photographic system. This blur was unacceptable for recording small, high velocity particles and a short duration gating unit was develped to gate on the X-ray tube from one of its focus electrodes in synchronism with each frame of the prism camera (see Fig. 7). This produced a static image on the intensifier screen, although still with a 150 microsecond decay time. This gating unit could be driven up to 10 kHz and had pulse widths variable from 1 to 100 microseconds. A 16 mm prism camera with transistor-transistor logic (TTL) was used, providing synchronization pulses at the start of each frame.

In practice, the system was limited by the X-ray source. At 380 kV and 10 mA, penetrating 11 mm (0.4 in.) of aluminum, a maximum framing speed of 5,000 frames per second was achieved with 1 m (3.2 ft) source-to-intensifier separation. At slower speeds, effective penetration increased dramatically (see Fig. 8). Recent improvements in cameras and intensifiers increase the system sensitivity by three stops, which would allow the separation to increase 2.5 to 3 m (8 to 10 ft). The limiting resolution of these systems is around 15 line pairs per millimeter on 16 mm film (processed to 2000 ASA) but in a practical situation this degrades to 10 or 12 line pairs per millimeter, mainly because of contrast loss in the phosphor image.

Subjects obscured only by flame or smoke, such as rocket vane movement, are best recorded with lower voltage X rays. The C-X 130 system is designed for the same purpose and uses a 130 kV, 1,000 mA source, which can operate with an 8 m (26 ft) separation at 10,000 frames per second.

X-ray source size on the above system varies from 1.5 to 4 mm (0.06 to 0.16 in.), which limits practical resolution in the subject plane to around 1 mm (0.04 in.). Recently, a microfocus source has been used to improve resolution for recording fuse rupture and crack propagation in opaque solids. The demountable X-ray head is continuously pumped and electrostatically focuses a beam of electrons onto a tungsten target, generating a 30 micron source. The tube is fed from a bipolar supply up to 160 kV, with a maximum sustainable current of 10 mA for 100 milliseconds. The main problem is cratering of the anode, which results in a larger spot size and vignetting of the beam. This effect is reduced to a minimum by driving the tube as a constant potential unit but gating the beam on and off with a bias in the focusing cup.

FIGURE 7. Synchronization of Gating Pulse with Camera Shutter

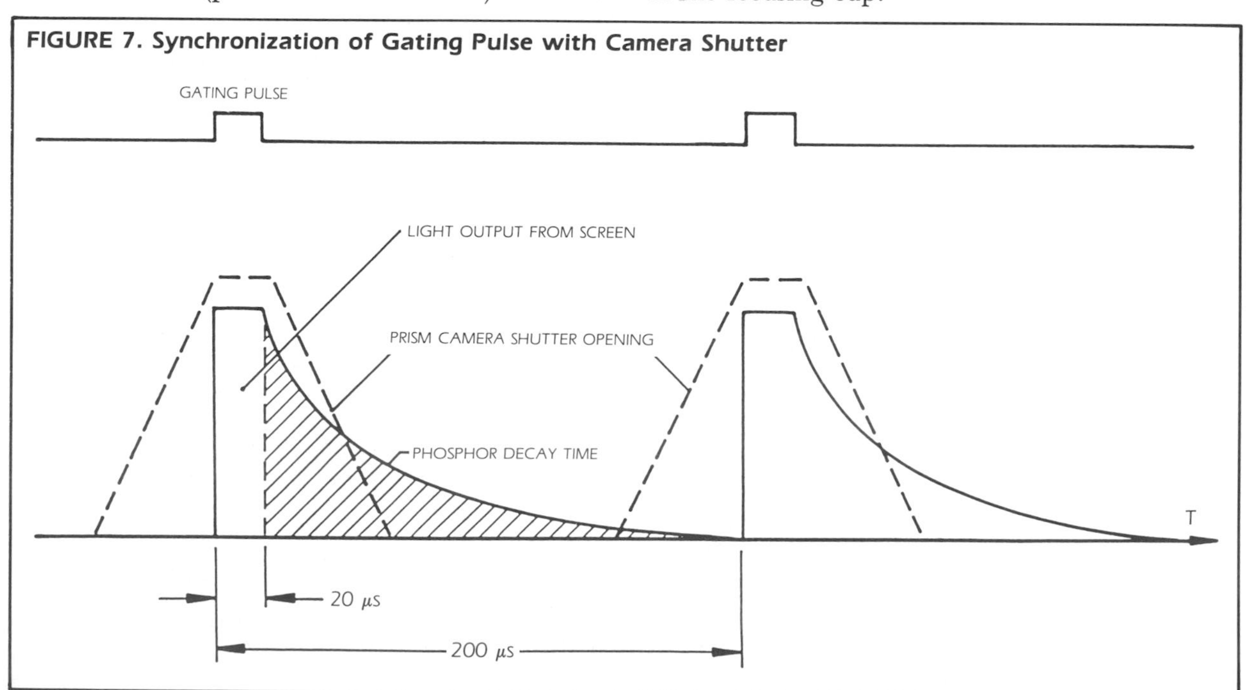

Various anode geometries can be used and it is possible to position the subject within 10 mm (0.4 in.) of the source. The maximum standoff at 10,000 frames per second is 1.5 m (5 ft); very high geometric magnification can be achieved, allowing subjects down to 2 mm (0.08 in.) across to be recorded on a 16 mm frame with 30 micrometer resolution.

Many events require formats larger than 320 mm

FIGURE 8. Casting of Engine Blades; Excerpt from 16 mm Cine Film at 300 Frames per Second

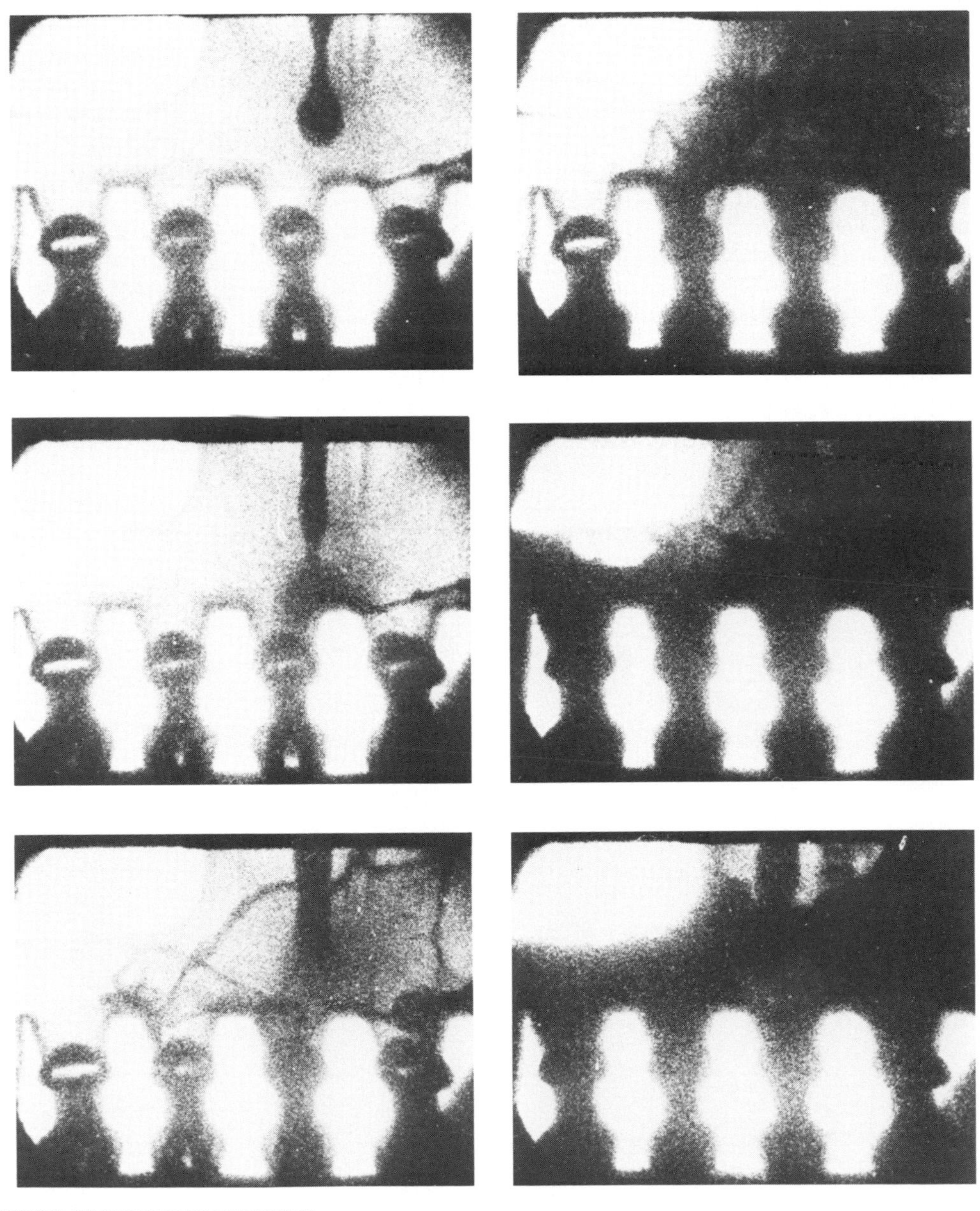

FROM ROLLS-ROYCE. REPRINTED WITH PERMISSION.

or generate conditions too violent for large vacuum tubes. A system developed for recording the launch and blast effects of a large solid-fuel rocket enables areas of several square meters to be radiographed, depending on required framing speed. The system comprises a high power constant potential source, scintillator, intensifier and a high-speed intermittent or prism camera (see Fig. 9).

The X-ray source is a modified medical unit capable of generating up to 130 kV at 1,000 mA for up to 10 microseconds or 700 mA for 100 microseconds. The scintillator comprises a matrix of standard intensifying screens, each sandwiched between a protective layer of carbon fiber or glass-fiber bonded sheet on the event side and clear polycarbonate sheet on the camera side. The assembly is screwed together around the edges and the matrix is held on a single polycarbonate sheet.

The scintillator type depends on required framing speed because the most efficient scintillators have longer decay times. Gadolinium oxysulphide with praeseodymium doping is an efficient green emitter (well matched for S20 photocathodes) and, depending on doping levels, can decay rapidly enough to operate at 10,000 frames per second.

The optical intensifier increases the photographic speed of a 16 mm cine camera 100 to 500 times, depending on the spectral content of the light source. The intensifier has a 25 mm input window with S20 cathode and a 25 mm rapid-decay phosphor which peaks in the blue to optimize spectral matching to film. Modification to the electron-optical design reduces geometric distortion to approximately 2%. A special, high voltage generator is required to drive the tube; the high phosphor brilliance necessary to record on film requires a peak current in excess of 10 mA.

The intensifier is lens-coupled to a high-speed 16 mm cine camera and the assembly records the scintillator image from a protected housing via a mirror.

Ambient light reflected off the scintillator can be brighter than the luminance of the X-ray image, so a black plastic tunnel must be erected to keep the

FIGURE 9. Large Format Cine Radiography up to 10,000 Frames per Second

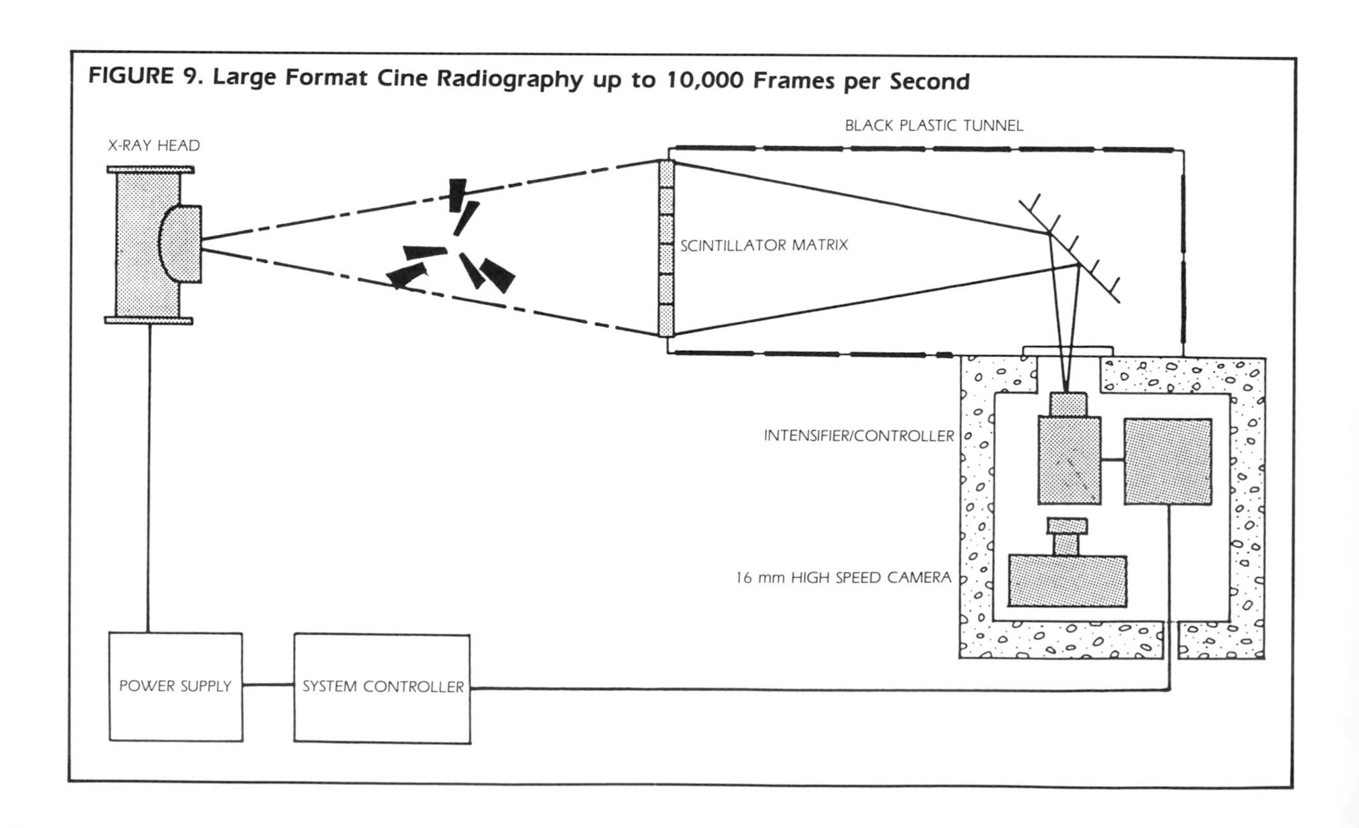

scintillator in the dark. This also extends over the front of the scintillator to prevent direct light breakthrough.

The source-to-scintillator distance can be as much as 2 m (6 ft) at a speed of 10,000 frames per second, using an f2.8 prism camera. Using an intermittent high-speed camera with f1.4 optics at 1,000 frames per second, source-to-scintillator separation can be increased to 12 m (38 ft). The included beam angle exceeds 30°, giving maximum scintillator formats of 1 m (3 ft) square and 6 m (19 ft) square, respectively, for the two distances mentioned.

This system has not yet been used on a live event so the effectiveness of the black plastic tunnel and scintillator matrix is not known. Blast from a rocket launch will probably not be sufficient to disrupt the matrix but explosive events certainly will. When the shock wave hits the matrix, it will be driven toward the camera but may retain integrity sufficient for giving qualitative data for a few milliseconds. Small perforations of the scintillator and plastic tunnel will only produce bright spots but, as the scintillator matrix finally separates, the light level dramatically increases. In order to prevent damage to the intensifier phosphor, a photodetector senses excess luminance and crowbars the high voltage supply.

The framing speed of the systems described is limited by the high-speed camera to 10,000 frames per second. Shorter exposures can be obtained by shorter gate pulses on the X-ray intensifier or by selecting a narrow angle shutter on the prism camera. Both techniques require a proportional increase in X-ray flux but the sampling rate remains the same.

Development of the IMAX System

Recordings have been made[2] of target penetration by long rod projectiles; this requires at least 100,000 frames per second to accurately determine the penetration dynamics. An image converter camera was used to record the output of an X-ray intensifier and to generate a 10 millisecond pulse of X rays from a modified 200 kV AC unit. The system, known as CADEX (Fig. 10), was synchronized using foil shorting screens. The first foil was positioned a distance equivalent to 2 ms uprange of the target to give the X-ray source time to reach peak intensity. Current into the transformer primary was modulated by a bank of transistors and a profiling circuit to generate a nearly constant X-ray output over a 10 millisecond period. The second foil was placed on the

FIGURE 10. CADEX/IMAX Cine X-ray System

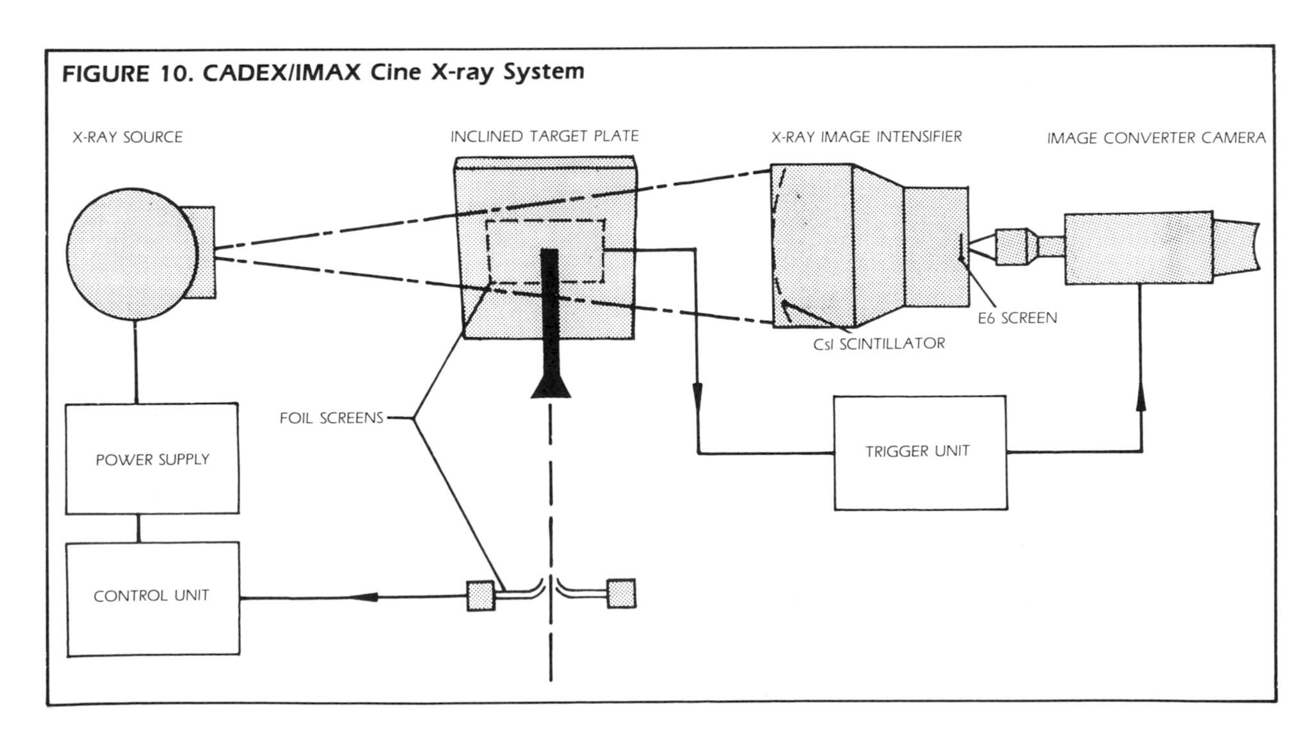

surface of the inclined target plate. After receiving a trigger pulse, the image converter camera starts to record the first frame in less than one interframe period and takes a sequence of 8 to 16 pictures at a rate determined by the plug-in module of the Imacon which determines frame speed.

The P20 phosphor of the X-ray intensifier tube has too long a decay time to be used above 10,000 frames per second, regardless of the gate-on duration. An E6 phosphor was used instead, which has a sub-microsecond main decay component and peak spectral emission of 420 nm. The E6 phosphor has effective light output similar to P20 for film recording but is four times less sensitive as detected by an S20 cathode of the Imacon, due to spectral mismatch. High aperture relay optics between intensifier and camera were therefore essential to maintain a realistic source-to-intensifier separation.

The system (see Fig. 11) produced reasonable pictures of target impact events (Fig. 12) and demonstrated its ability to withstand hostile environments when suitably housed. For Fig. 13, the intensifier

FIGURE 11. CADEX Set-up for Recording 105 mm Artillery Shell at Muzzle

FROM P&EE ESKMEALS. REPRINTED WITH PERMISSION.

FIGURE 12. A 105 mm Artillery Shell in Muzzle Flash Region; Imaged at 10^5 Frames per Second

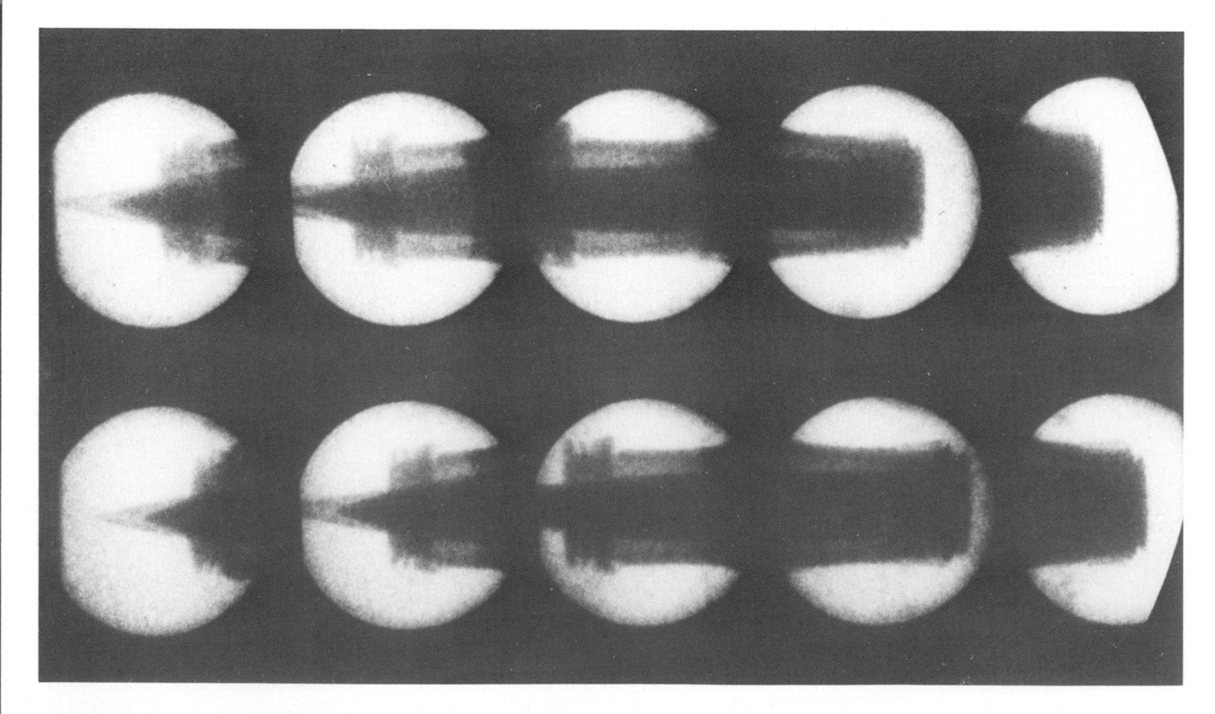

FROM P&EE ESKMEALS. REPRINTED WITH PERMISSION.

tube was located 0.5 m downrange and 0.5 m to the side of a 105 mm tank gun.

The system, now designated IMAX, was further developed by simplifying the X-ray power supply and modifying the cesium iodide layer to improve its output for higher energy X rays. The X-ray source output is now 275 kV at 60 mA for 1 millisecond. These improvements allow the X-ray source and intensifier to be separated by up to 4 m (13 ft) at 100,000 frames per second. Typical results are shown in Figs. 14 and 15. This corresponds to an X-ray dose of approximately 0.015 μGy (1.5 μR) per exposure received by the intensifier scintillator. Other X-ray sources can be used, the most promising being the 160 kV microfocus source previously described with the 16 mm cine camera.

The maximum current of 10 mA limits source-to-intensifier separation to 1.5 m (4.8 ft) at 100,000 frames per second or 0.5 m (1.6 ft) at 1 million frames per second, unless an intensifier is fitted to the image converter camera. However, this should be avoided if possible because of image degradation due to quantum noise. The small source size allows large geometric magnifications without noticeable penumbral unsharpness and, because the subject can

FIGURE 13. Rod Penetrating 20 mm (0.75 in.) Inclined Armor Plate; Imaged at 10^5 Frames per Second

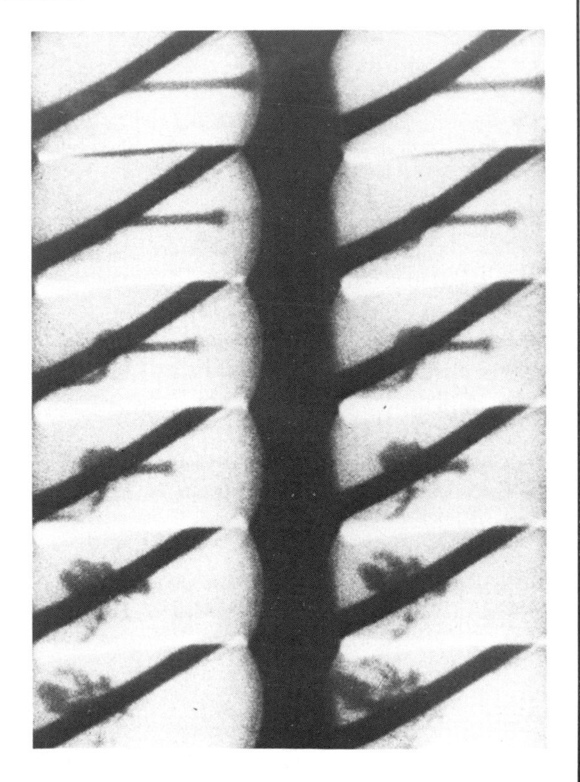

FROM RARDE, FORT HALSTEAD. REPRINTED WITH PERMISSION.

FIGURE 14. A 40 mm Artillery Shell Penetrating Metal Plate; Imaged at 25,000 Frames per Second

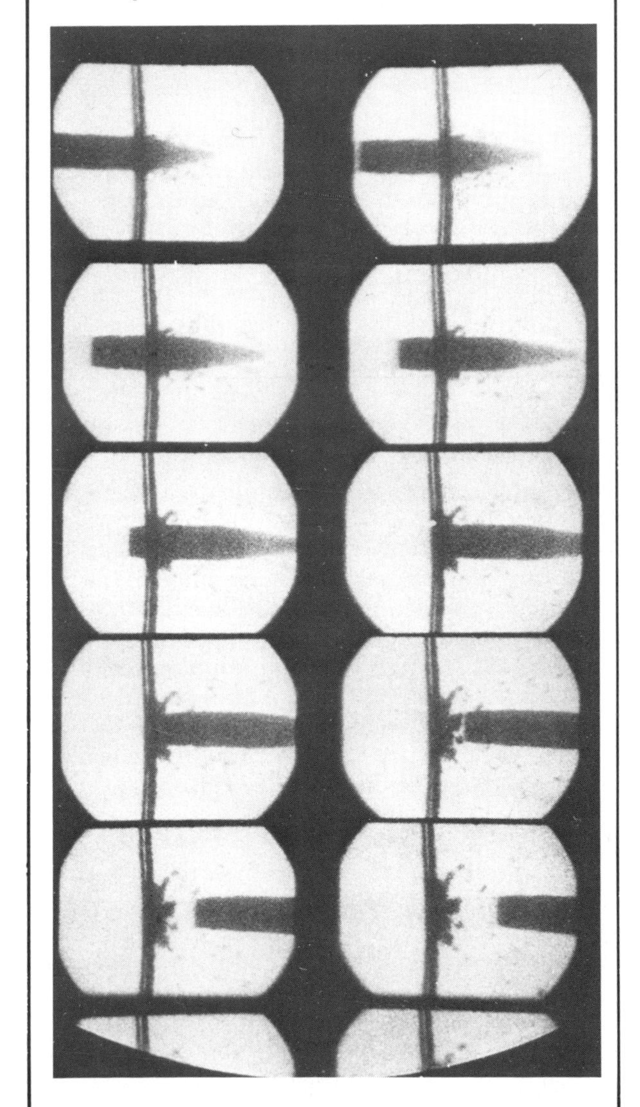

FROM RARDE, FORT HALSTEAD. REPRINTED WITH PERMISSION.

be placed within 6 mm (0.25 in.) of the source, subject areas down to 2 mm (0.08 in.) diameter can be recorded with 30 micron resolution.

Applications include imaging of fuse rupture, crack propagation and electrically exploding wires. When using the system on small events, changes occur very rapidly within the limited field of view and corresponding high framing rates are required to stop motion blur. Two million frames per second is the maximum speed an IMAX system can be operated before phosphor decay causes unacceptable blur.

Use of Multi-cathode Heads

The largest X-ray intensifiers have an input window of 400 mm (16 in.) diameter and larger formats must use a separate scintillator, lens-coupled to a high-speed camera. This requires a far greater X-ray flux at the scintillator, for which a multi-cathode X-ray head with a corresponding number of Marx pulsers (see Fig. 16) has been proposed. A typical image converter camera requires about 2 milliergs/ cm^2 of blue or green light to make an exposure on 3000 ASA film or 4.1^{-5} ergs/cm^2 with a 50× gain intensifier. This can be achieved with a flash X-ray pulse equivalent to 80 μGy (8 mR) at 1 m (3 ft), irradiating a quenched gadolinium oxysulphide scintillator at 6 m (19 ft) distance and imaged with an f1.4 lens. Similar precautions as observed for the large format prism camera system must be taken to avoid ambient light falling on the scintillator, superimposing light on the dynamic image.

Depending on the type of scintillator used, framing rates up to 1 million frames per second may be achieved. At these rates synchronization is difficult because, in addition to varying jitter in the separate pulsers, the camera should ideally give a sync pulse, in advance of each frame, equal to the X-ray pulser delay.

Another system has been developed[3] in which the image converter camera is replaced by a number of gated channel plate intensifiers, each intensifier being gated on in synchronism with a corresponding Marx pulser. Because both intensifier and pulser can be triggered on command, with a known delay, a pulse sequence generator can be used to control the system and set the framing rate. The intensifiers are 40 mm proximity-focus type, which gives a 3× improvement in the number of line pairs per frame compared with an image converter camera.

Both systems require the X-ray heads to be as close together as possible. Six sealed tubes are immersed in an oil-filled housing, spaced 44 mm (1.7 in.) apart in each direction, and operated at 150 kV. An alternative approach uses a single demountable head

FIGURE 15. A 7.62 mm Artillery Shell Generating Void in Gelatin Block; Imaged at 50,000 Frames per Second

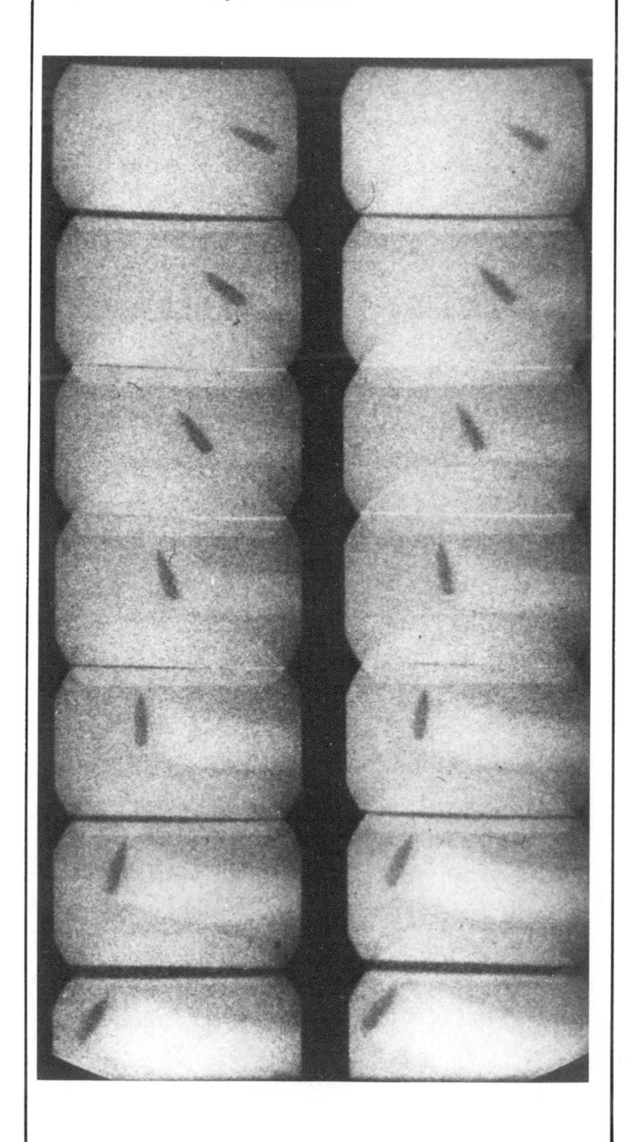

with six cathodes beamed onto a single central anode of 50 mm (2 in.) diameter. Six pulses are generated, spaced equally around the circumference of a 50 mm (2 in.) diameter circle, each generating a 300 kV, 25 nanosecond pulse with a dose of 200 μGy (20 mR) at 0.5 m (1.6 ft).

Use of Channel Plate Intensifiers

Channel plate intensifiers combine fast shuttering ability with high optical sensitivity. With maximum voltage across the channel plate element, the light amplification is so high the output image loses quality due to the limited number of photons required at the input window (cathode). Resolution drops by up to 50% under these conditions when 3000 ASA film is placed in contact with the fiber optic output screen. If this loss can be tolerated, low power flash X-ray units can be used.

Figure 17 shows an array of units irradiating a scintillator up to 5 m (16 ft) away (shadowgraph separation) and the images being recorded by gated intensifiers. These units are self contained in 10 cm (4 in.) diameter armored housing and produce a 15 μGy (1.5 mR) dose of 50 ns duration at 50 cm (20 in.)

The 5 m (16 ft) separation can be achieved using an f1.4 lenses on the intensifier cameras and doped gadolinium oxysulphide screens, limiting the maximum framing rate to 100,000 frames per second (see Fig. 18). Faster decaying scintillators, such as E6, are available, but usually have a lower quantum efficiency (ratio of visible photons out per X-ray photon in) and emit blue light, to which an intensifier cathode is less sensitive. This reduces X-ray source-to-scintillator separation to around 2.5 m (8 ft), but the framing rate can exceed one million frames per second without image retention problems.

System for Study of Shaped Charge Development

A modification of the above system can be used to study shaped charge development. Here, an array of small flash X-ray units form an array of images on separate parts of a single scintillator (Fig.

FIGURE 16. Large Format Recording up to 1 Million Frames per Second

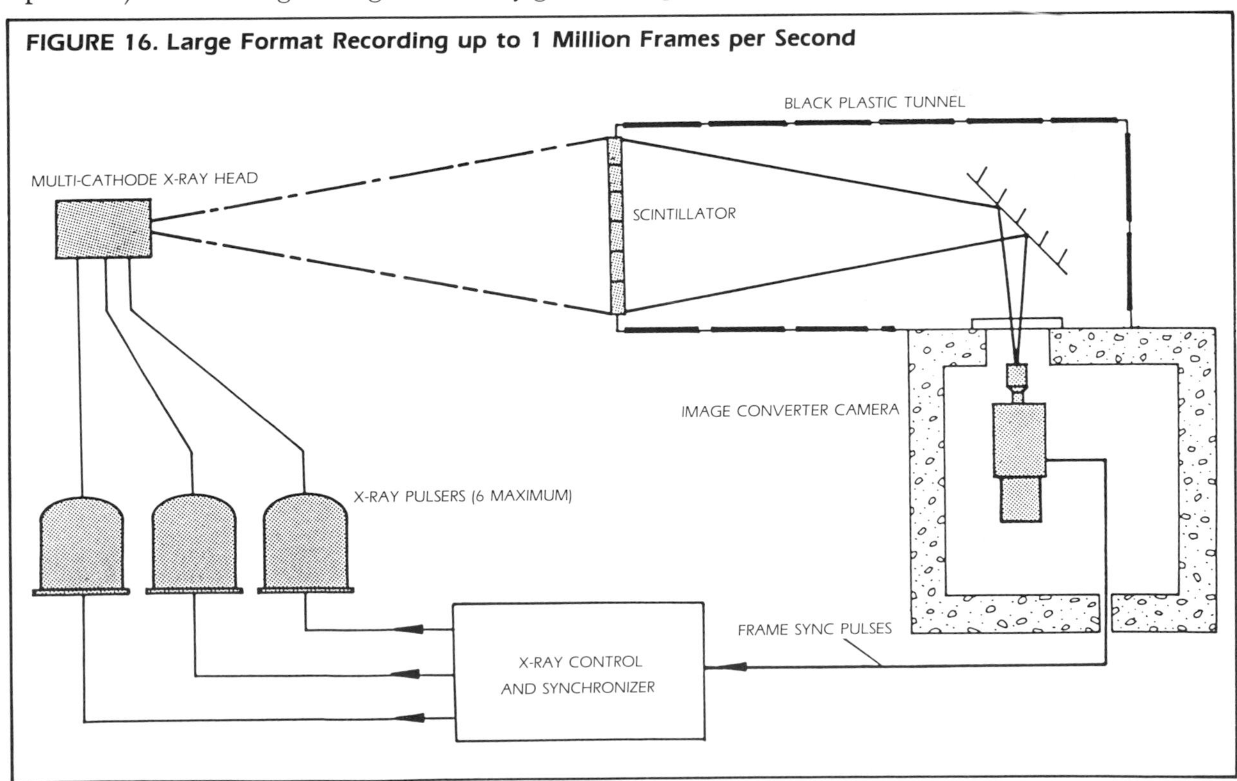

19). The scintillator is photographed by a single channel plate intensifier or camera (see Fig. 20). Lead masks are used to prevent image overlap. As each part of the scintillator receives only one X-ray pulse, decay time is not important. A high sensitivity, long decay screen can be used, such as zinc cadmium sulphide or standard gadolinium oxysulphide, and the intensifier is gated on for the total decay period (see Fig. 21). With 6 m (20 ft) separation between X-ray sources and scintillator, the intensifier can be operated at a lower gain level, retaining high resolution. Angular separation between successive images depends on subject size. To image an 8 cm (3 in.) diameter shaped charge with 2:1 source-to-subject/subject-to-scintillator ratio, there will be a 3.5° separation. Smaller sections reduce proportionally.

In general, the systems described provide data complementary to flash radiographs. They all generate a sequence of images without parallax, which is essential for understanding the dynamics of unsymmetrical events (see Table 1 for a summary of these systems). However, such systems cannot compete in terms of image quality and total picture resolution. These parameters are largely determined by the available X-ray dose and the spatial resolution of the high speed camera. Future work will concentrate on these two areas.

FIGURE 18. Synchronization Between X-ray Sources and Intensifier Shutter Periods: (a) X-ray Pulses; (b) Scintillator Light Output; (c) Intensifier Shutter Period

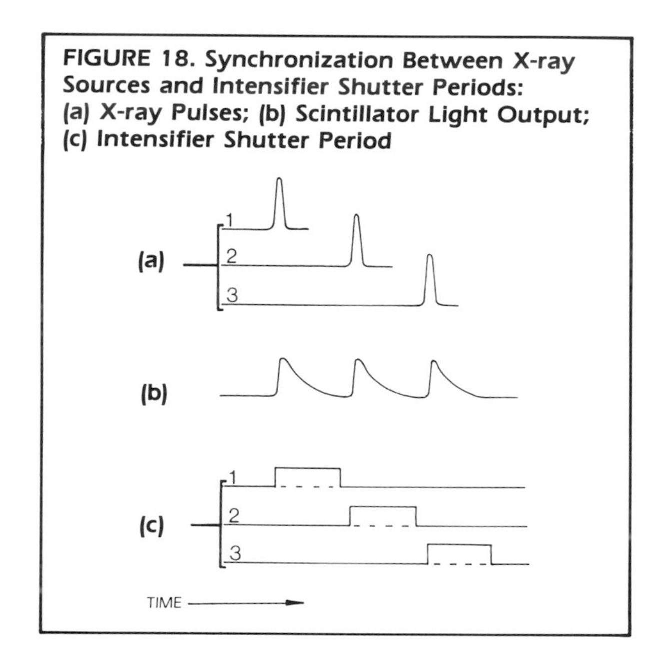

FIGURE 17. Large Format Cine X-ray with Improved Resolution

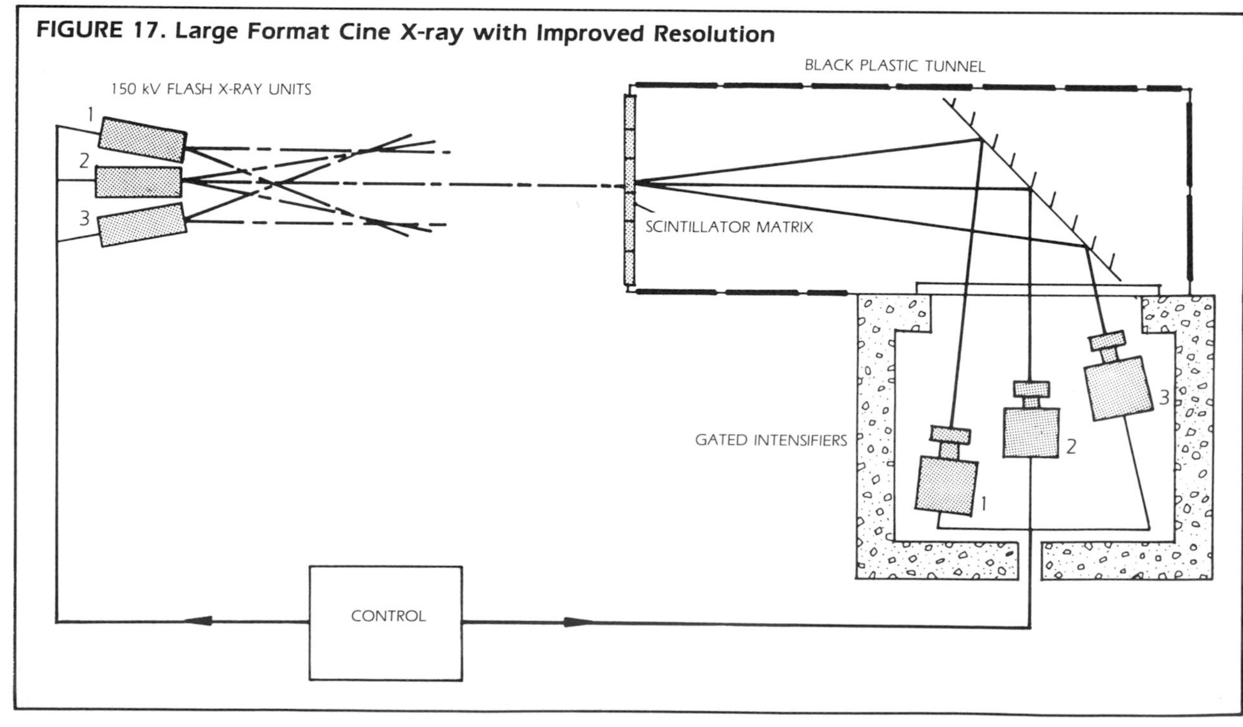

FIGURE 19. Shaped Charge Jet Recording with Small Flash X-ray Units; Exposure Made at 150 kV for 50 ns

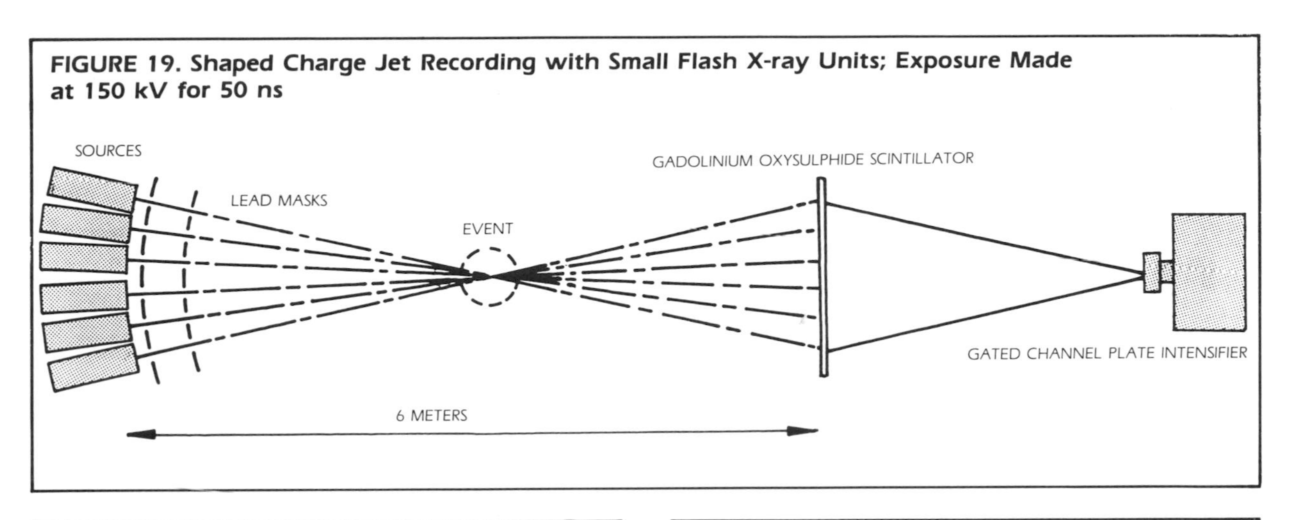

FIGURE 20. Image Sequence Recorded by Intensifier

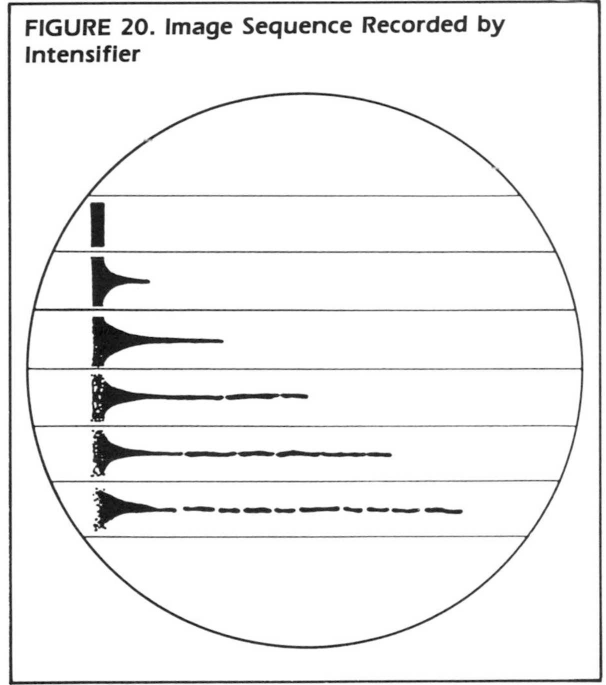

FIGURE 21. Relationship Between X-ray Pulses and Scintillator Output as Recorded by Intensifier Camera: (a) Flash X-ray Pulses; (b) Light Output of Corresponding Scintillator Segments; (c) Intensifier Gate-on Period

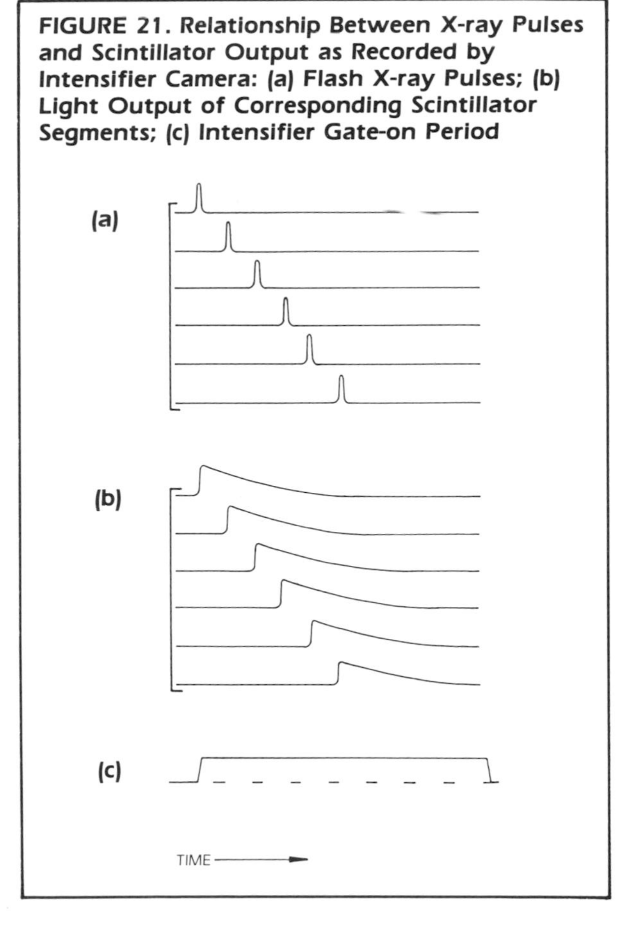

TABLE 1. Summary of Cine X-ray Systems

Designation Source Specifications	X-ray to Visible Converter	Shadowgraph Distance and Performance Limits	Camera Specifications
C-X 320 10 to 320 kV Up to 10 mA Constant potential 3.5 mm source	X-ray image intensifier (optional gating)	2 m at 10,000 fps 6 m at 1,000 fps	16 mm high-speed
C-X MICRO 30 to 160 kV Up to 10 mA 1 to 100 ms pulse 30 μm source	X-ray image intensifier (optional gating)	1.5 m at 10,000 fps 4.5 m at 1,000 fps	16 mm high-speed
C-X 130 130 kV 700 to 1000 mA 1 to 100 ms 1.3 mm source	X-ray image intensifier (optional gating)	8 m at 10,000 fps (2.5 m at 10,000 fps with 5 μs gate)	16 mm high-speed
Large format cine X-ray 130 kV 700 to 1000 mA 1 to 100 ms 1.3 mm source	Fast decay scintillator (type depends on framing speed)	2 m at 10,000 fps 9 m at 1,000 fps	16 mm high-speed optical intensifier
IMAX 275 kV 60 mA 1 ms 1.8 mm source	X-ray image intensifier	1.5 m at 10^6 fps 4 m at 10^5 fps (10 m at 10^6 fps)	IMACON 790 50X gain intensifier
MICRO IMAX 30 to 160 kV Up to 10 mA 1 to 100 ms 30 μm source	X-ray image intensifier	0.6 m at 10^6 fps 1.4 m at 10^5 fps (4 m at 10^6 fps)	IMACON 790 50X gain intensifier
HSX-300 300 kV 50 ns Up to 6 sources	Fast decay scintillators	3 m at 10^6 fps 5 m at 10^5 fps	Gated intensifier synchronization box
HX-150 150 kV 50 ns Up to 6 sources	Long decay scintillator	6 m at any speed to 10^7 fps	High-resolution gated intensifier

PART 3

GEOMETRIC ENLARGEMENT RADIOGRAPHY

Definition and Principles

Geometric enlargement radiography involves creating an enlarged radiographic image of a subject by utilizing the divergent nature of the radiation from a source.

Enlargement radiography occurs any time the recording medium is moved away from the subject and source in a direction parallel to that of radiation passage. The process is straightforward shadow projection as illustrated in Fig. 22. A successful, direct radiographic enlargement requires that the radiation emitting area be as small as possible, a true point source being the most desirable. Departure from a point source introduces penumbral unsharpness as seen in Fig. 23. With the conventional small focal spot of dual focus X-ray tubes (about 0.5 mm [0.02 in.] square), only a very modest direct enlargement can be tolerated without severe loss of resolution due to penumbral unsharpness. For the above target size, the geometric unsharpness on a 2:1 enlargement equals the largest (diagonal) target

FIGURE 22. Illustration of (a) Contact Object Scatter on Recorder; (b) Enlargement with Much Less Scatter on Recorder

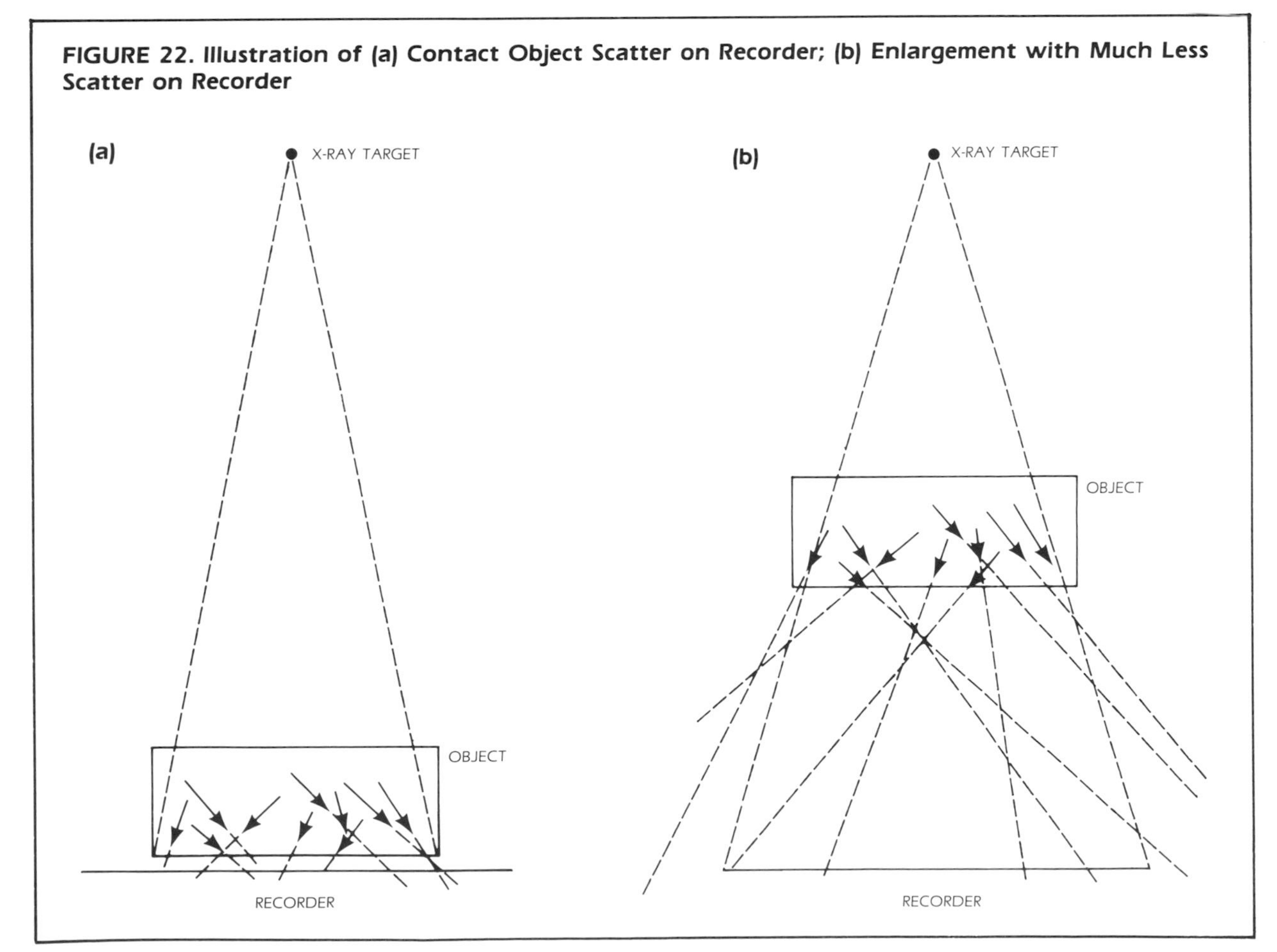

dimension, 0.7 mm (0.027 in.). Even with the very small focal spot of a betatron, 0.2 mm maximum dimension, enlargements beyond 3:1 are generally quite unsharp and of dubious utility.

Specialized low energy X-ray generators with extremely small emitting surfaces have been developed; for example, a unit has been reported which can put one watt on a one micrometer square target. This generator is the energy source for an X-ray microscope. Another version of the same source permits 15 W on a somewhat enlarged emitting area, utilizing a rotating target.

Reasons for Direct Enlargement Radiography

The primary reason for doing enlargement radiography is to enhance the resolution of small details in the radiograph. There are at least two ways in which this enhancement is achieved. The natural microscopic effect is the first and perhaps most obvious means of enhancement. On this basis, the greater the enlargement, the greater the enhancement, but the geometry of the radiographic set-up must be good enough to sustain this premise. The recorder also must be adequate for the task.

A second means of enhancement by direct enlargement is through the reduction of object scatter from the radiograph. When the recorder is in contact with or very near the object, the object scatter, even at relatively wide angles, can reach the recorder, degrading the image quality. Separation of the object and recorder will permit portions of the scatter to miss the recorder; greater separation eliminates more scatter.

FIGURE 23. Diagram of How Departure from Point Source Introduces Unsharpness

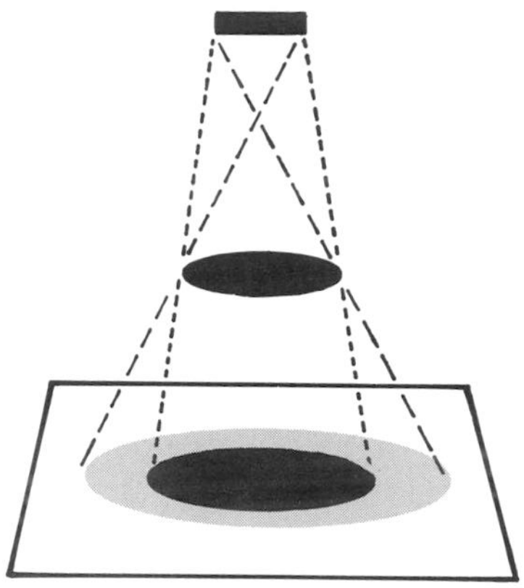

Applications

Direct radiographic enlargement has been used successfully in assembly radiography. Enhancement of fine detail is provided throughout the image.

Sometimes enlargement radiography is used to enhance resolution in the inspection of material or solid parts. As previously stated, the geometry and recorder must be adequate to permit this.

Direct enlargement is of great importance in nearly all kinds of real-time radiography, including fluoroscopy and the newest real-time imaging systems. The technique takes better advantage of the recording medium, which otherwise has resolution appreciably lower than fine grained radiographic film. It also provides the separation of specimen and recorder to permit specimen manipulation, one of the prime advantages of this technique over conventional radiography.

PART 4

RADIOGRAPHY OF RADIOACTIVE SUBJECTS

Definitions

Radioactive material has been defined as any material undergoing 74 or more disintegrations per second. The peculiarity in radiographing such material is that the radiation emanating from the subject will contribute to the total radiation reaching the detector, thereby reducing the signal-to-noise ratio and degrading the radiographic image. The quantity and type of radiation from the specimen influences the radiographic technique and determines how it must be altered from that for nonradioactive subjects.

Subject Information

Radioactive materials are radiographed to determine their internal conditions. Many radioactive materials are encountered in nuclear energy programs and many of these produce particle radiation, alpha or beta, that can be stopped with very little filtration. Slight filtration can also reduce the effect of low energy gamma or characteristic X rays coming from a radioactive subject.

The nature of such radiography is unusual only in the care necessary for handling the material and preventing personnel exposure and area contamination. Many materials that emit low-energy gammas have very high density, and high atomic numbers, so that the gammas emitted come from the surface or very slightly beneath it. The screens normally used in high-energy X-ray radiography will eliminate these low-energy emanations, so that, except for handling precautions, radiography is little changed from that for nonradioactive materials.

Some materials, usually coming from nuclear reactor environments, produce high-energy gamma rays in copious quantities. These are the materials and assemblies that provide a radiographic challenge.

Requisites for X-Ray Radiography

To do X-ray radiography on a highly radioactive subject, it is necessary to find an X-ray generator with an output greater than the gamma ray output of the subject. This is true because the X-ray target will have to be farther from the recorder than the subject.

The output of the generator must be of sufficient intensity to penetrate the subject and still produce a signal sufficiently stronger than the gamma noise from the subject.

It is also advantageous to have the X-ray emitting area of the target small enough to permit some direct radiographic enlargement, moving the subject away from the recorder and toward the target. This reduces the gamma intensity reaching the recorder from the subject, just as direct radiographic enlargement reduces the effect of subject scatter (see Fig. 22, Part 3). This method has been used to evaluate fuel elements after they have been used in nuclear reactors.[4,5]

With highly radioactive subjects, it is necessary to develop remotely controlled means of positioning the subject for the radiography. This positioning must be accomplished rapidly to control autoradiographic noise. Masking and heavy, fast-moving shutters also help improve the ratio of X-ray signal to gamma-ray noise by eliminating gammas from parts of the subject not in the picture, and by blocking the gammas until X rays are produced.

Filters are of very little value in radiography of subjects emitting hard gamma rays because the X-ray beam will be attenuated about as much as the gamma autoradiation and the signal-to-noise ratio will not be improved.

Neutron Radiography

If the neutron absorbing characteristics of the radioactive subject are amenable to neutron imaging, transfer-foil neutron radiography provides an excellent means of inspection. The neutron image is formed by neutron activation of the transfer foil, which is insensitive to the gamma autoradiation from the subject. The activation image is then transferred to a film or other suitable recorder as described in the next part of this Handbook Section.

PART 5

AUTORADIOGRAPHY

Definition and Basic Principles

Autoradiography is the technique of producing an image of a radioactive specimen on a recording medium using the radiation emanating from that specimen. The basic principles of the test are essentially those which apply to ordinary radiography, as far as the energy interaction with the recording medium is concerned. Autoradiography can be used to determine which parts of a subject are actually radioactive in instances where radioactive material is distributed in an inert matrix. It can also be used to differentiate materials of different radioactivity levels in a subject. The thickness and uniformity of protective coatings on radioactive material have been determined autoradiographically. Autoradiography is used to confirm and record the presence and location of fuel in nuclear reactor fuel elements.

Specimen Information

The specimen must be radioactive but these materials have wide ranges of radioactivity. The emanations may be electromagnetic or gamma rays from a wide spectrum of energies. They may also be particulate radiations, beta or alpha particles. It is also possible that there may be combinations of more than one type of radiation. The autoradiographic procedure will depend on the nature of the radiation and the information sought.

The level or intensity of the radiation from the specimen may also vary over a wide range and this will certainly affect the autoradiographic procedure.

The information desired from the test will determine how the autoradiography is done. If only the mere presence of radioactive material in sizable quantity needs to be recorded, the procedure will be much less demanding than delineating small regions of radioactive material in a matrix.

Another factor to be considered is the radioactive cleanliness of the specimen. In many instances, specimens of radioactive material can slough off debris which is radioactive. The handling of such specimens must be done so that area contamination and eventually radioactive contamination of personnel will not occur; the use of autoradiography on such specimens may be limited or totally impossible.

The size and shape of the specimen will also influence the autoradiographic procedure. A precise technique with intimate contact between recording medium and specimen surface can be difficult on specimens without flat, smooth surfaces.

Recording Media

Standard industrial radiographic film serves very well for many autoradiographic applications. The wide variety of speeds and resolution characteristics ensure a wide applicability. Although there are limitations on the contours to which an ordinary industrial radiographic film will conform, extraordinary actions have extended these limits. Film has been cut in shapes that permit it to conform closely to 25 mm (1 in.) diameter spherical surfaces, for example.

If the autoradiograph is to reveal aspects of the specimen in the micrographic scale, it may be necessary to use the finer grained emulsion of spectrographic plates. This is possible with relatively flat, smooth specimens. Special emulsions are available which can be mixed in liquid form and cast onto the surface of the specimen. They are certainly expeditious where intimate contact between the specimen and recording medium is needed and the specimen surface is neither flat nor smooth. The handling of such emulsions requires considerable skill and care and must be done under suitable dark room conditions.

Specialized Area Requirements

When close contact between the specimen and the recording medium is required, they will have to be held together in darkroom conditions. When ample space is available, and the room can be left in its darkened condition long enough, the task can be performed without any other light-tight enclosure. However, if ordinary light is needed in the room, specialized light-tight containers can be made to hold the specimen and recording medium.

If the requirements are less stringent, so that the material can be placed between the specimen and recording medium, a light-tight film holder can be employed and the autoradiography can be performed almost anywhere that the radioactive material can be handled.

If the level of the radioactivity of the specimen is great enough, handling may also require a special location, such as a hot cell with remote handling capability.

Maintaining Specimen-Recorder Contact

When precision autoradiography is to be accomplished, it is necessary to hold the specimen and recorder in intimate contact throughout the exposure. Forms designed to hold radiographic film in position can be spring-loaded to maintain suitable pressure. This feature can be built into exposure boxes when the number of specimens justifies the expense. Such forms or fixtures can also be made of resilient plastic material more easily and economically than with spring loading arrangements.

Vacuum holders are good for maintaining contact when specimens have a flat, smooth surface and a relatively regular shape. Such holders must be made of material sufficiently flexible to achieve the desired conformation to specimen contour. This method is particularly suited to the autoradiography step in the transfer plate method of neutron radiography.

The use of liquid emulsions, cast onto the specimen, is another method for maintaining specimen-recorder contact. Such material sets up on the specimen surface and is then floated off the surface in a bath after the exposure has been accomplished. The specimen must be amenable to immersion in a liquid for this method to be used.

Handling Radioactive Materials

Before any attempt is made to handle radioactive material, the nature of the material and radioactivity must be thoroughly known. The personnel safety of such an undertaking is always assessed first. Many materials must not be handled outside protective enclosures, due to sloughing off of their radioactive material. Not only must such materials be handled in protective enclosures, such as dry boxes with ample protective filters, but any recorder that has been in contact with such a material may also be too contaminated to handle outside a dry box. Unless some material can be introduced between the subject and the recorder to protect the latter from contamination, autoradiography may be completely impractical.

It may also be impossible to handle highly active specimens outside a hot cell. This is particularly true of hard gamma emitters because large material masses are required to attenuate their gamma ray energies. More modest attenuators will suffice for betas and only very thin attenuators are needed for alphas; high activity specimens that emit alphas or betas exclusively can be handled safely with modest equipment, if there is no sloughing problem.

With any highly active subject, it may be necessary to provide some means to permit quick positioning of the recorder and specimen to prevent motion blur as the two are brought together.

Subjects with low activity are relatively simple and safe to handle if there is no sloughing hazard. If the activity is low enough, such objects can be safely handled with gloved hands. Those of a little greater activity can be handled with tongs. Generally, there is no need for speed in bringing the low activity specimen and recorder together, but for precise work it is necessary to keep them together and undisturbed for a long exposure time.[6,7]

Exposure Time

The time required to obtain the autoradiograph will be an inverse function of the intensity of the radiation coming from the subject. Because autoradiography is more frequently used on low activity subjects, exposure times tend to be long compared to conventional radiography. Overnight exposures and exposure times of two or three days are common.

Combining Radiography and Autoradiography

Sometimes it is desirable to produce gross material-location autoradiographs, in conjunction with a radiographic image, to identify a specimen's more radioactive components. This is done by making a radiograph of lower than normal film density, and leaving the specimen in contact with the loaded film holder long enough to produce the desired autoradiographic exposure. This method has been used to locate uranium-235/plutonium-239 fuel pellets assembled with uranium-238 insulator pellets in fast reactor fuel elements.

Autoradiograph Marking

Because the image produced in autoradiography is generally confined to the contact area between the recording medium and the specimen, added steps must be taken to preserve their relative positions.

This is particularly true (1) if precise matching of the autoradiograph to the surface of the specimen is needed for complete evaluation, and (2) if the specimen has a regular shape, such as a disk. Such schemes as notching the film and marking the specimen can be used. It is difficult to intervene identifying markers because good film-specimen contact is usually required.

PART 6

ELECTRON RADIOGRAPHY

Principles

The electron radiography technique uses high-energy secondary photoelectrons instead of X rays for registering a specimen's image on radiographic film (see Bibliography entries 1-7).

True transmission electron radiography can be used for paper-thin, low atomic number specimens when hard X-ray photons (about 250 kV) produce secondary photoelectrons. These electrons are typically from lead foil and are used (after differential absorption in the specimen) to register a latent image on film. The normal result is an image of the specimen volume similar to that obtained from X-ray photons. In electron radiography, however, the specimen material and thickness are limited by the range of the photoelectrons.

A related electron radiography procedure called *specimen electron emission* uses hard X-ray photons to produce secondary photoelectrons at the surface of a suitable specimen, enabling a material-related surface image to be registered on film.

The experimental arrangements for the two techniques are shown in Figs. 24a and 24b. In the transmission technique, the photoelectrons are produced from the lead foil above and adjacent to the specimen, whereas in the emission technique, the specimen itself is the source of the photoelectrons. Note also that the hard X rays used in both techniques have direct access to the radiographic film; however, these hard X rays ideally result only in an increase in background film density.

Electron Transmission Radiography

The number of emitted photoelectrons from the foil depends on the wavelength and intensity of the incident X rays and increases with the atomic number of the foil. As the penetration of the X-ray beam increases, the photographic effect of the electrons from the lead foil increases relative to the photographic effect of the direct X rays that penetrate the lead foil and the specimen. The selected radiographic film or plate should therefore have a low absorption coefficient for the hard, penetrating X rays and a high absorption coefficient for the photoelectrons.

In low voltage X-ray microradiography, the transmission of photons is sensitive to atomic number variations in the specimen; in electron transmission radiography, photoelectron transmission is influenced mainly by material density variations.

Film enlargements from electron radiography (transmission and emission techniques) are lower than those for X-ray microradiography with the same film type, due mainly to the diffuse nature of the photoelectrons.

Electron Emission Radiography

Electron emission might be used if a scanning electron microscope (SEM) or other analytical systems are not available, or when a bulk specimen may not be compatible with the SEM. Factors to be considered for an SEM specimen include vacuum compatibility and size (about 2.5 cm [1 in.] maximum); surface roughness is a factor in both techniques. Also, if the specimen must be examined in the field, then a portable X-ray system would be the choice.

With the specimen emission technique, it is primary X-ray photons on the specimen surface and secondary photoelectrons (emission-sensitive to atomic number) off the surface that are important. In the SEM, it is primary electrons on the specimen surface and backscattered electrons off the surface (emission-sensitive to atomic number) or secondary electrons off the surface (not very emission-sensitive to atomic number) that are critical. These differences in emission sensitivity can produce differences in information about the specimen.

In the sophisticated SEM procedure, electronic means are used to position and control the primary electron beam on the sample, to detect backscattered or secondary electrons or to measure sample current (net current from electrons on and off the sample).

FIGURE 24. Electron Radiography: (a) Transmission Technique; (b) Specimen Emission Technique

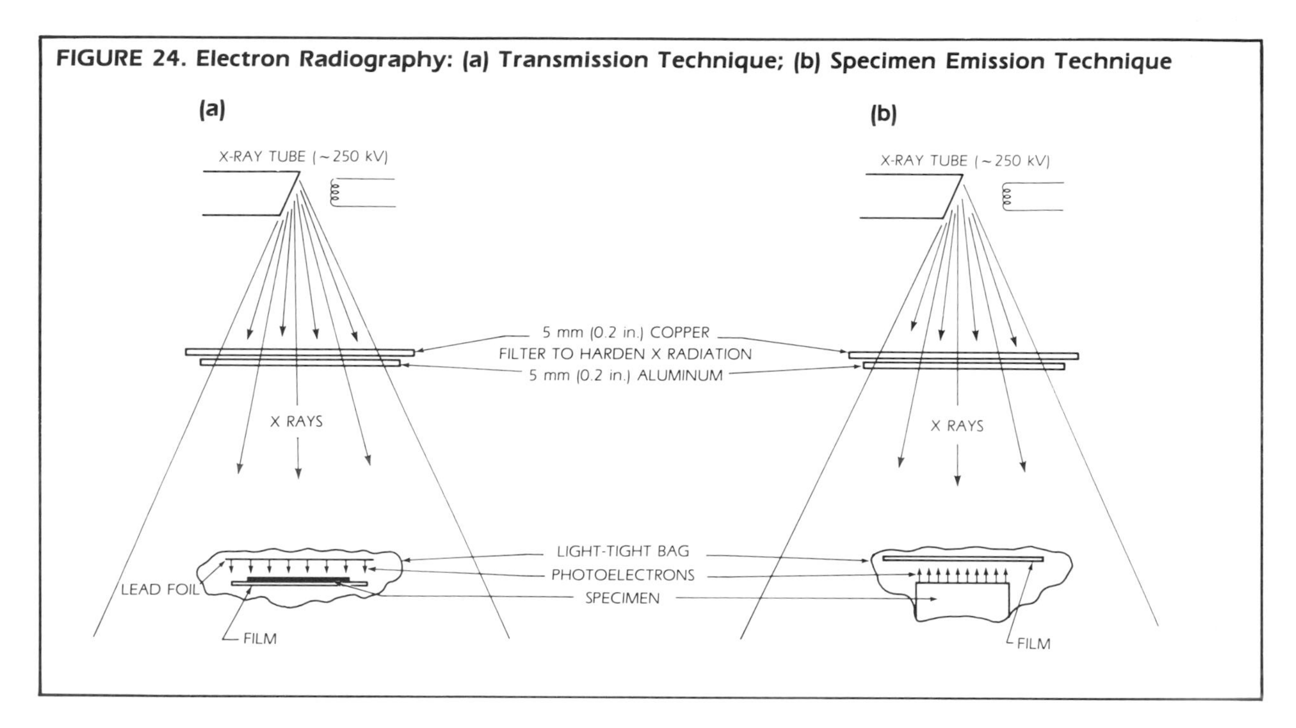

The result is a sharp, magnification-controllable output not available by enlargement of the X-ray film or other adjustments of the electron emission technique. Both the specimen emission and SEM techniques are able to differentiate surface elements that photographic or optical examination cannot.

Techniques

Both transmission and emission techniques use a conventional X-ray apparatus capable of operating around 250 kV. A filter consisting of 5 mm (0.2 in.) copper plus 5 mm (0.2 in.) aluminum is satisfactory for hardening the X-ray beam (see Fig. 24). The aluminum is on the sample side of the copper to absorb characteristic radiation and scatter from the copper.

Transmission Technique

The lead foil must be thin (25 μm [0.001 in.]) in order to contact all areas of interest on one side. The photographic film or plate must contact the other side of the sample. The lead foil contact requirement may not be a nondestructive procedure for some fragile samples. The lead foil, specimen and photographic material (single sided) are enclosed in a light-tight bag which may be air-evacuated for increased sample-film contact. Exposure parameters are determined by testing. Suitable specimens include plastics and organic matter. Electron transmission testing of paper has also been done; various compositions of printing inks which may not be radiographically imaged can be imaged with the electron technique when impurities in the ink or paper become electron emitters and produce a positive image instead of a negative.

Emission Technique

For best results, the specimen should be polished in the area that will be in intimate contact with the single emulsion radiographic film. Again, a light-tight bag is used to cover the film and the sample. Exposure parameters are determined by experiment.

PART 7

IN-MOTION RADIOGRAPHY

Introduction

In stationary, conventional radiography, a series of separate exposures is made and either the object or the X-ray unit is moved to a new position for each exposure. Each radiograph is separate and discrete, although they may be made to overlap. With in-motion radiography (IMR), the X-ray unit and the part are in constant relative motion to each other and a narrow irradiating beam traverses the area being inspected. The result is a continuous radiograph that can be almost any desired length, limited mainly by the size of the film. Furthermore, repeated passes may be made over the same area to get the desired density in a finished radiograph.

The techniques used for IMR are the same as conventional radiography except for exposure time. The exposure time is converted to speed of travel and is recorded as millimeters or inches per minute. To decrease the density of the film, the speed is increased; to increase the film density, the speed is decreased. Other controlling factors of film density versus speed are: (1) type of material being radiographed; (2) thickness of material; (3) film speed; (4) use of intensifying screens; and (5) source-to-film distance.

IMR extends the usefulness of radiography as a tool for nondestructive testing and has several advantages when compared with conventional radiography.

FIGURE 25. Illustration of Geometric Unsharpness for Static Condition

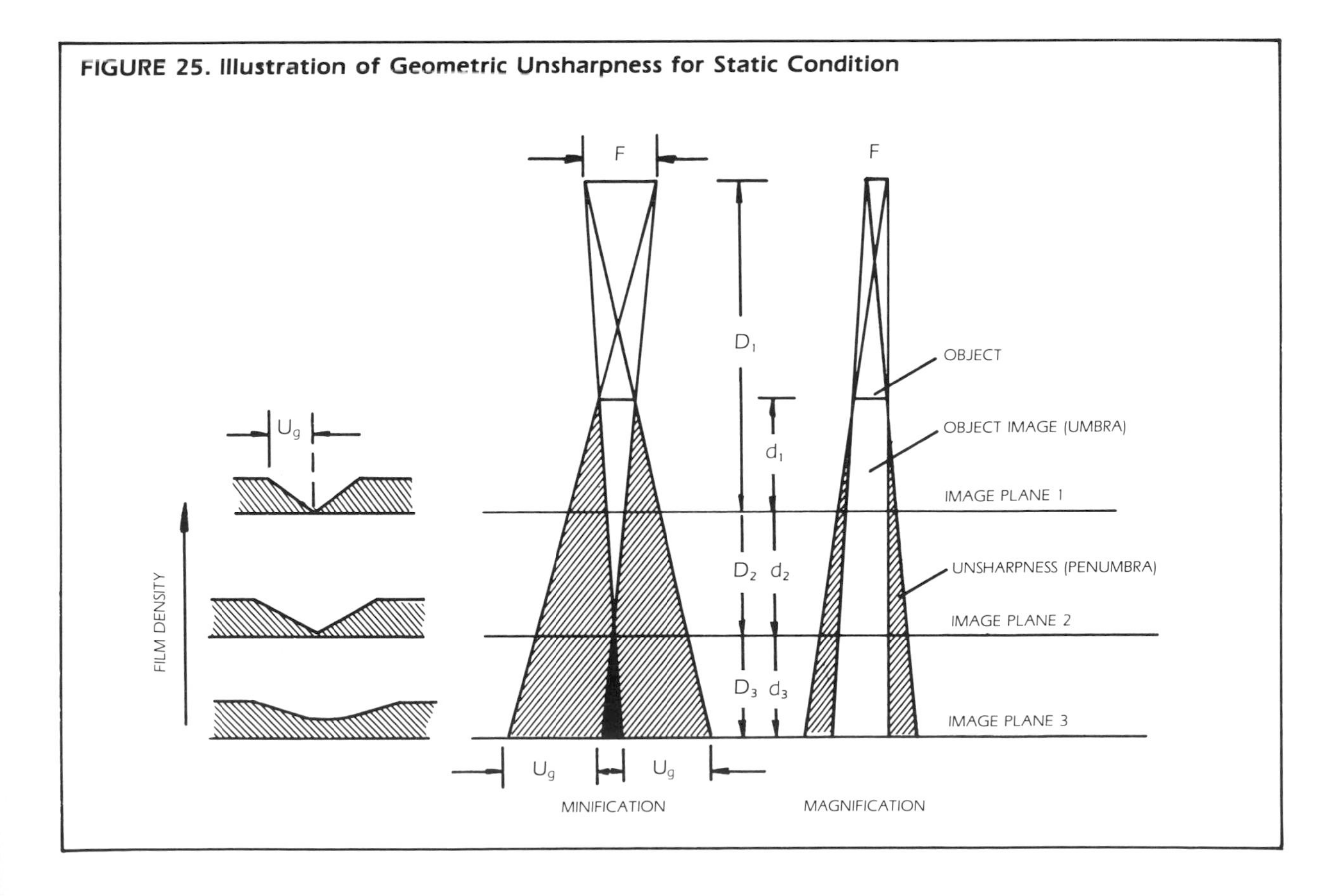

1. IMR permits X-ray inspection of almost unlimited part sizes with modest radiographic equipment.
2. It permits on-the-spot radiography in production shops without disturbing other operations in the area. This is possible because radiation hazard is less than with stationary radiography.
3. The image of the object is recorded on one continuous sheet of film.

In this discussion, the mathematical formulas used to perform IMR are given, as well as descriptions of various applications using this technique. In all cases, radiographic film is the recording medium.

Definitions

IMR utilizes existing X-ray and handling equipment with limited modification to accommodate the in-motion features. The X-ray exposure is made (1) as the part is moved in relation to a collimated beam of radiation emitted by a stationary X-ray source or (2) as the collimated X-ray source is moved in relation to a stationary part.

Preliminary investigation of in-motion radiography discloses that image blurring is always greater in the direction of motion. Therefore, it is necessary to establish the in-motion variables by first determining the influence of the blurring factors when the object is moved through the X-ray beam.

In-motion Radiography Theory

Static Unsharpness

X-ray tubes always produce a certain amount of geometric unsharpness (U_g) because of the finite dimensions of the focus or source and the source-to-film distance. The value of this unsharpness (U_g) is

FIGURE 26. Conditions for Unsharpness Due to Motion

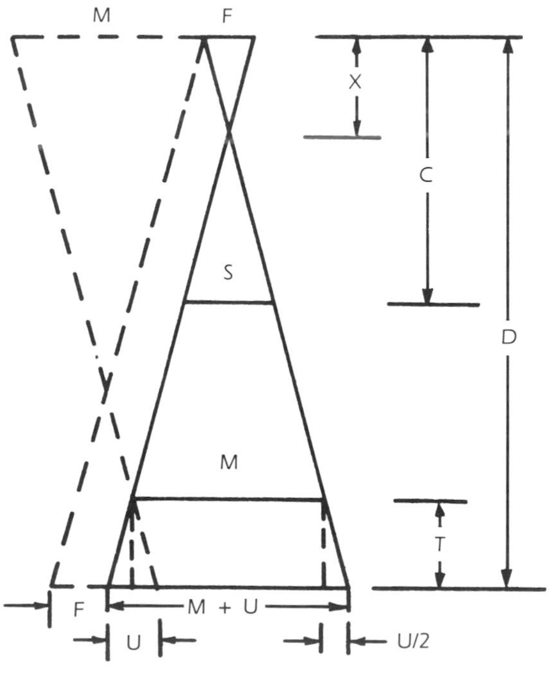

F = EFFECTIVE FOCAL SPOT SIZE
S = SLIT WIDTH
X = RELATED TO F AND S
C = SOURCE-TO-SLIT DISTANCE
D = SOURCE-TO-FILM DISTANCE
T = SOURCE SIDE OF PART-TO-FILM DISTANCE
M = MOTION AT SOURCE SIDE OF PART
U = UNSHARPNESS
M + U = MOTION PLUS UNSHARPNESS AT FILM PLANE

FIGURE 27. Diagram of Parameters Affecting Minification

F + M = EFFECTIVE FOCAL SPOT SIZE DUE TO MOTION
D = SOURCE-TO-FILM DISTANCE
O = OBJECT SIZE
T = OBJECT-TO-FILM DISTANCE
t = DISTANCE OF APEX ABOVE OR BELOW FILM PLANE
C = SOURCE-TO-SLIT DISTANCE
S = SLIT WIDTH

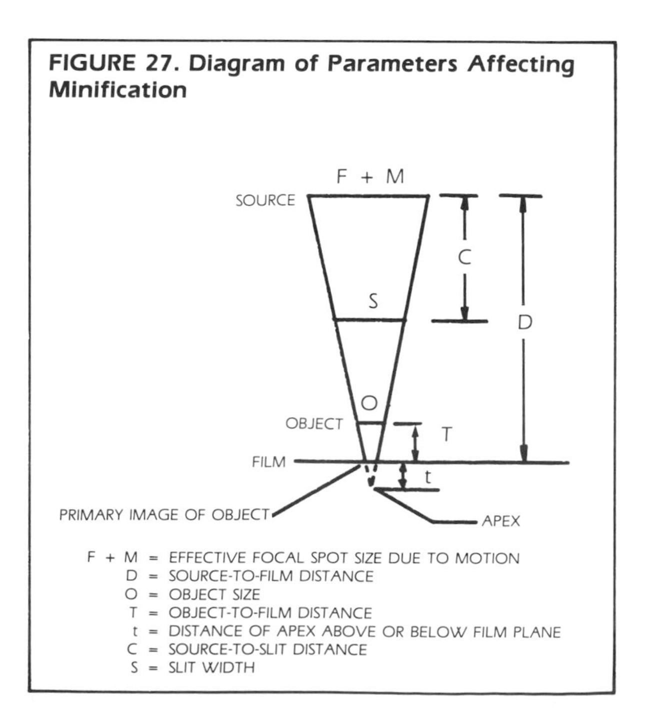

given in eq. 1:

$$U_g = \frac{F \times d}{D - d} \qquad \text{(Eq. 1)}$$

where F is the diameter of the effective focal spot; D is the source-to-film distance; and d is the defect-to-film distance.

This relationship is illustrated in Fig. 25. The maximum value of U_g is related to a defect situated at a maximum distance from the film when $d = T$, where T is the part thickness.

Static Magnification and Minification

Graininess is always present in a photographic image and can sometimes make it difficult to distinguish very small defects. It is possible, by increasing the object-to-film distance and decreasing the overall focus-to-film distance, to enlarge the radiographic image of discontinuities (Fig. 25) so that they are easier to detect. This is, however, only possible with a very small focal spot. When the focal spot is large, the image of the discontinuity is minified and the unsharpness increases drastically. In the latter case, an image may not appear on the radiograph.

Unsharpness Due to Motion

A geometric representation of the in-motion variables which establish the conditions for unsharpness was first developed in 1961.[8] The unsharpness equation (eq. 1) was derived[1] using the relationships in

FIGURE 28. Apex Distance Above or Below Film Plane as a Function of Effective Focal Spot Size (In-Motion)

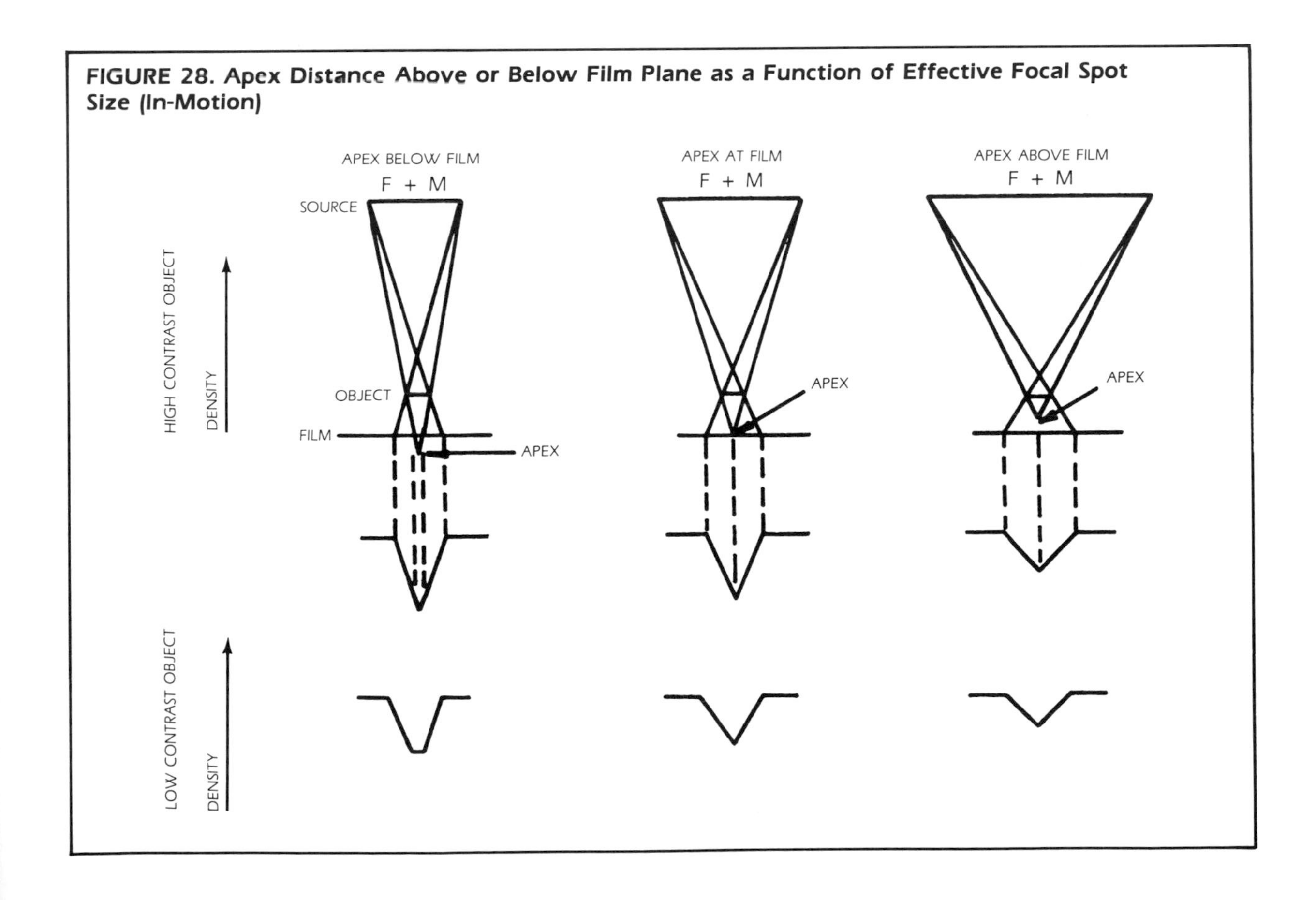

eq. 2 (see Fig. 26), and may also be derived using the relationships defined in eq. 3 and eq. 4:

$$\frac{F}{X} = \frac{S}{(C - X)} = \frac{M}{D - (X + T)} = \frac{M + U}{D - X} \quad \text{(Eq. 2)}$$

$$U = \frac{T(F + S)}{C} \quad \text{(Eq. 3)}$$

$$\frac{(F + M)}{(F + S)} = \frac{(D - T)}{C} \quad \text{(Eq. 4)}$$

FIGURE 29. Geometry Technique Curves; Minification Apex at Film

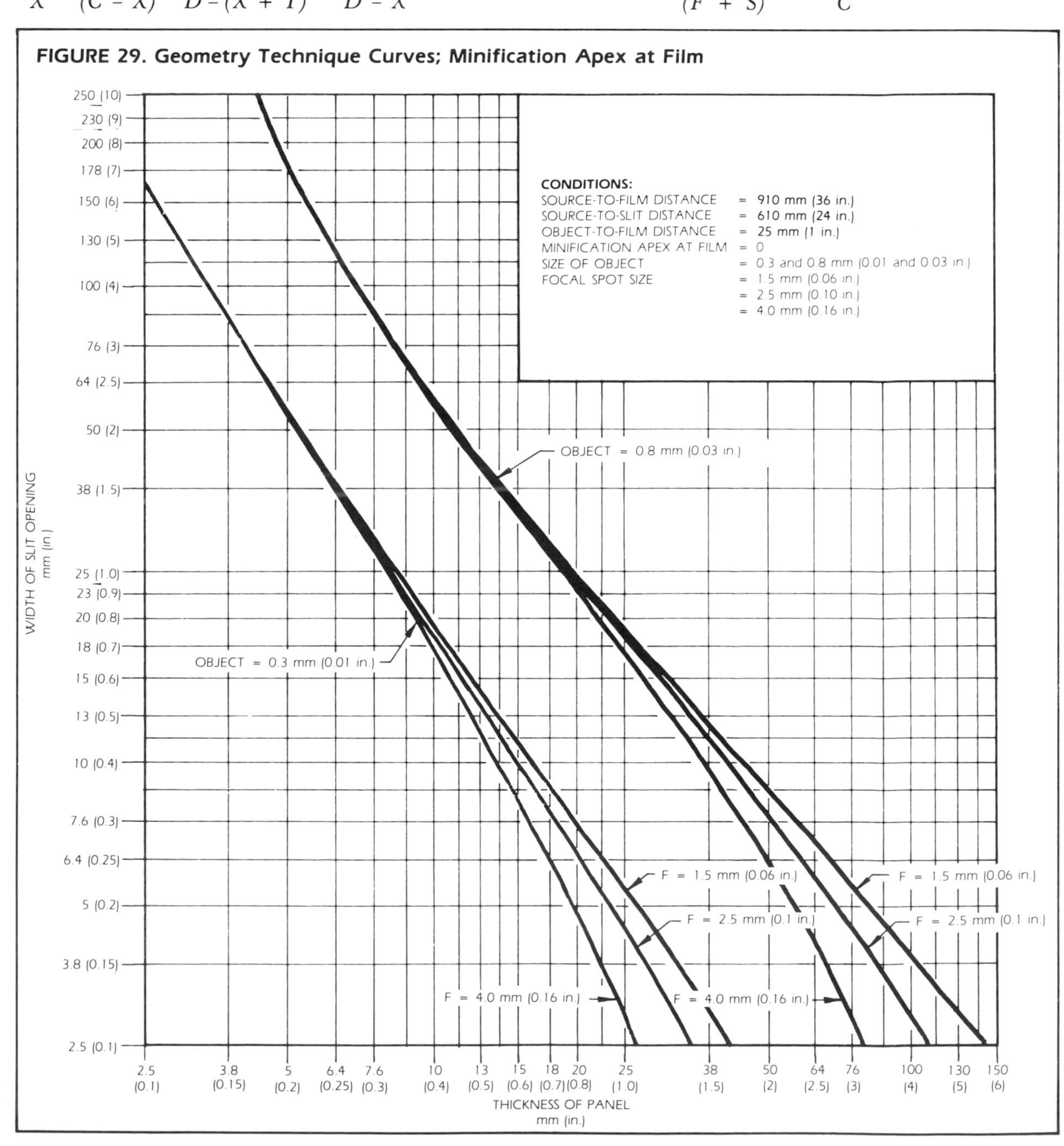

where F is the effective focal spot size; S is the slit width; X is the image displacement (related to F and S); C is the source-to-slit distance; D is the source-to-film distance; T is the source side of part-to-film distance; M is the motion at the source side of the part; U is the unsharpness; and $M + U$ is the motion plus unsharpness at the film plane.

To calculate the unsharpness expected with in-motion radiography in the direction of motion, refer to Fig. 26. Solving eq. 3 for U yields:

$$U = \frac{(F + S)T}{C} \qquad \text{(Eq. 5)}$$

This shows that unsharpness in the direction of motion is not related to source-to-film or source-to-object distance, but rather to the slit-width, object-to-film distance, and source-to-slit distance. Unsharpness perpendicular to the direction of motion (parallel with the slit) is dependent on the same geometry factors as stationary radiography shown in Fig. 25.

Because there is an apparent enlargement of the focal spot for in-motion radiography, there may be occasions when it is necessary to measure the extent of this effective size. This can be determined by solving for $(F + M)$ in eq. 4.

$$(F + M) = \frac{(F + S)(D - T)}{C} \qquad \text{(Eq. 6)}$$

Solving eq. 6 for $(F + S)$ yields:

$$(F + S) = \frac{(F + M)C}{(D - T)} \qquad \text{(Eq. 7)}$$

In-motion unsharpness can be expressed in terms of the effective in-motion focal spot size $(F + M)$ by substituting the expression $(F + S)$ in eq. 5:

$$U = \frac{(F + S)T}{C} = \frac{(F + M)T}{(D - T)} \qquad \text{(Eq. 8)}$$

In-motion Minification[8]

Image minification is the basic mechanism by which certain details in the object image can be prevented from appearing on the radiograph. This is done by reducing the size of the primary image of objects smaller than the effective focal spot. In Fig. 25, the large focal spot causes image minification, whereas a small focal spot causes image magnification. At Image Plane-1, the large focal spot produces

FIGURE 30. Image Displacement as a Function of Thickness (T) and Angle (Θ)

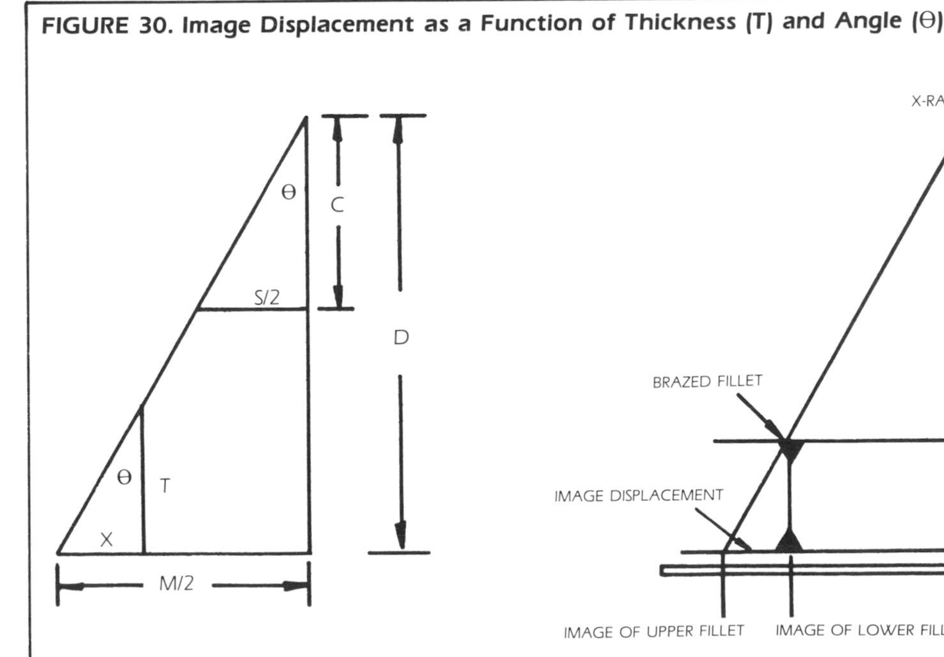

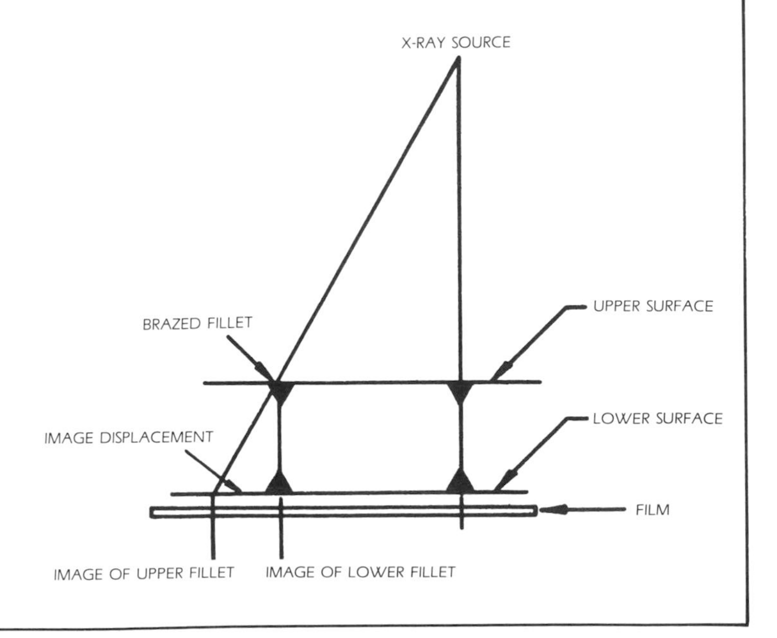

FIGURE 31. Techniques for In-Motion Radiography: (a) Linear In-Motion Radiography; (b) Rotation In-Motion Radiography

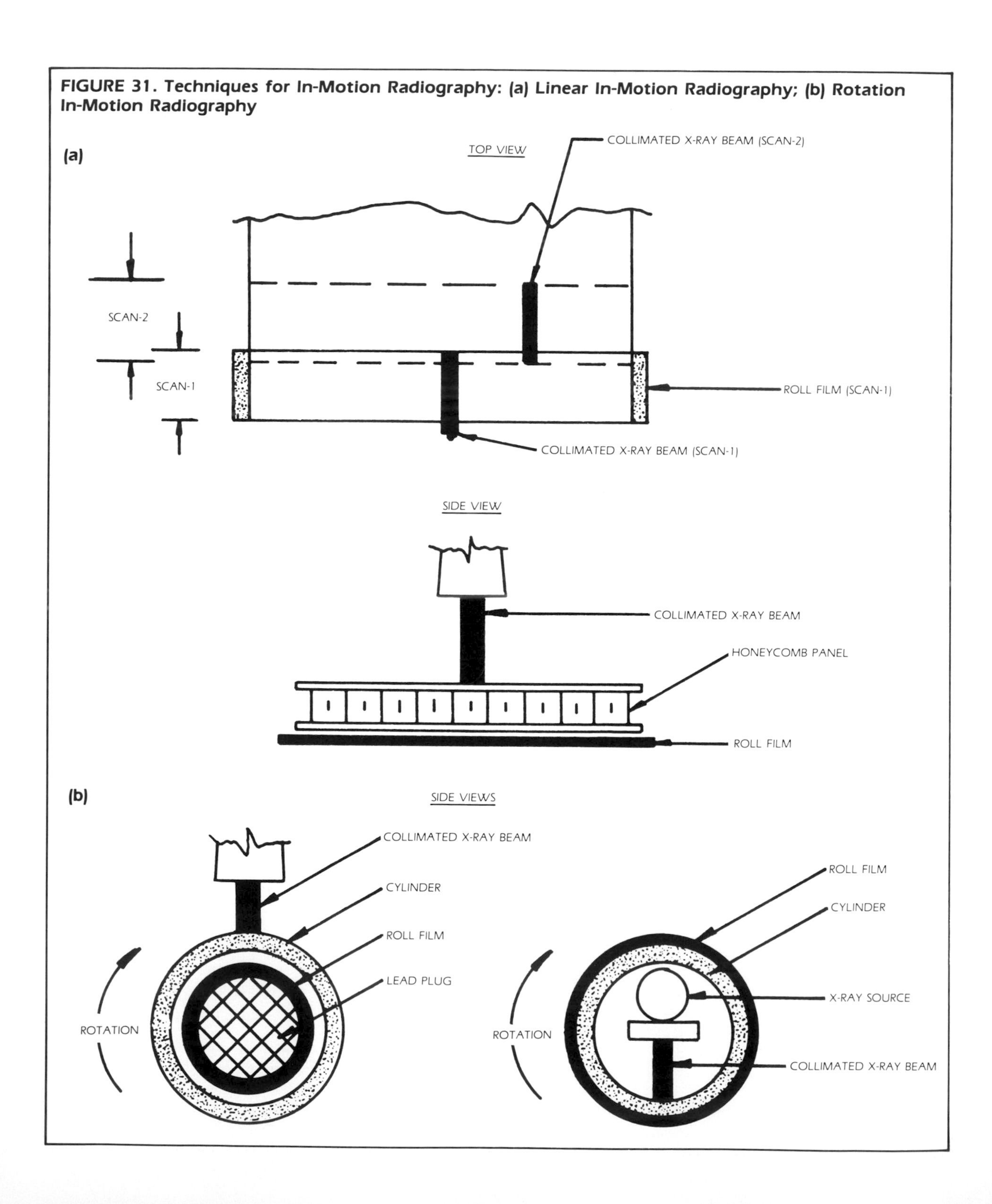

FIGURE 32. Typical Collimator for Honeycomb Structure Inspection

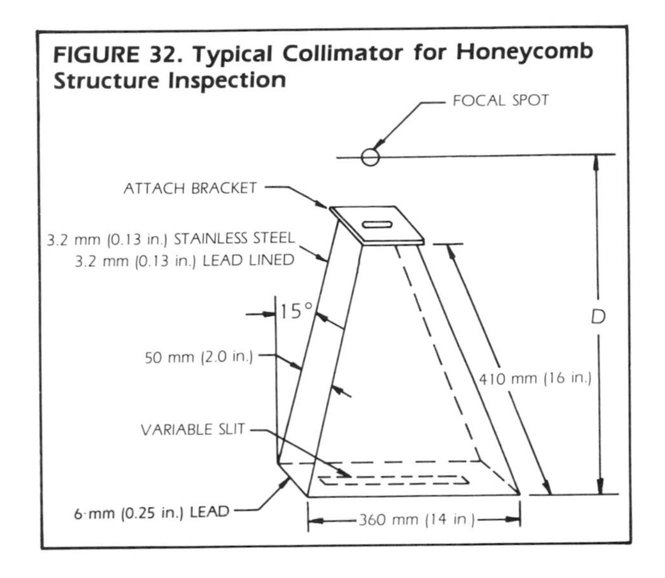

considerable unsharpness (penumbra) and reduction in the image of the object (umbra). At Image Plane-2, the image of the object ceases to exist, and there is complete unsharpness. At Image Plane-3, the situation gets worse in that unsharpness overlaps in the center, producing a further reduction in image contrast and an increase in film density. Figure 25 also shows that objects larger than the effective focal spot will never disappear.

These results for the large focal spot are directly related to in-motion radiography. The conditions governing minification are given in a general equation derived from the geometric representation shown in Fig. 27. By using similar triangles, the specific equation for the minification apex below the film is:

FIGURE 33. Schematic View of In-Motion X-ray Technique with Longitudinal Welds in Constant Motion

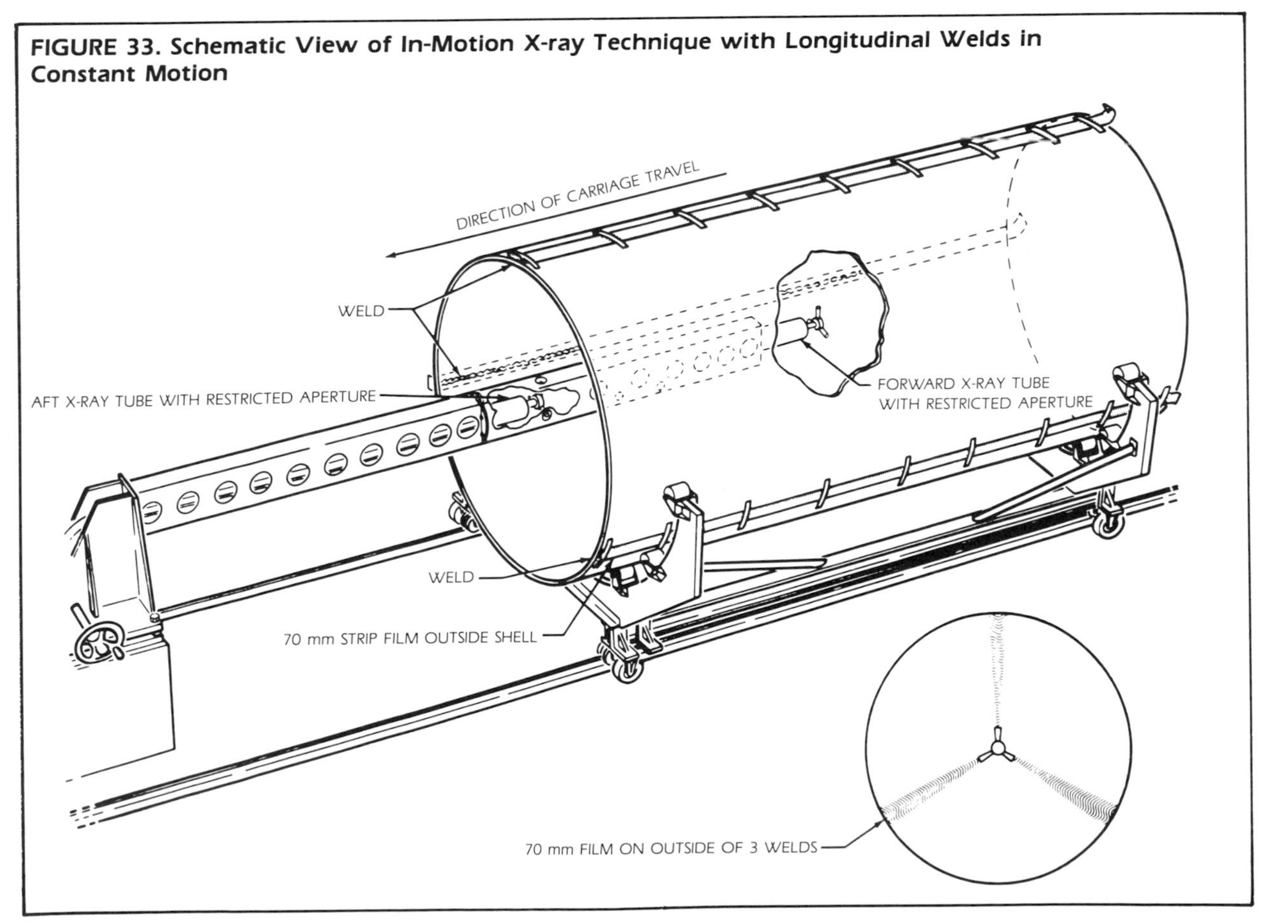

$$O = \frac{T + t}{D + t}(F + M) \qquad \text{(Eq. 9)}$$

Greater minification could raise the apex above the film, in which case t would have a minus value. Therefore, the equation for any degree of minification is:

$$O = \frac{T \pm t}{D \pm t}(F + M) \qquad \text{(Eq. 10)}$$

And from eqs. 6 and 10, we derive the general equation:

$$O = \frac{T \pm t}{D \pm t}\left[\frac{(D - T)(S + F)}{C}\right] \qquad \text{(Eq. 11)}$$

where $(F + M)$ is the effective focal spot size due to motion; D is the source-to-film distance; O is the object size; T is the source side of object-to-film distance; t is the distance of the apex above or below the film plane; C is the source-to-slit distance; and S is the slit width.

This can be used to determine or control the precise location of the apex (point of complete minification). However, it does *not* show where the apex should be located to produce the desired degree of indistinctness. In addition, the degree of inherent contrast between the object and its surroundings affects distinctness for any given location of the apex (see Fig. 28).

Therefore, the first step in an experimental IMR investigation should be to determine the desired location for the apex at the expected contrast level. For example, high contrast (subject or object) should require a higher apex (negative value) than low contrast subject or object.

FIGURE 34. Schematic Drawing of IMR System: (a) Tank Carriage Drive; and (b) Synchronized X-ray Actuator

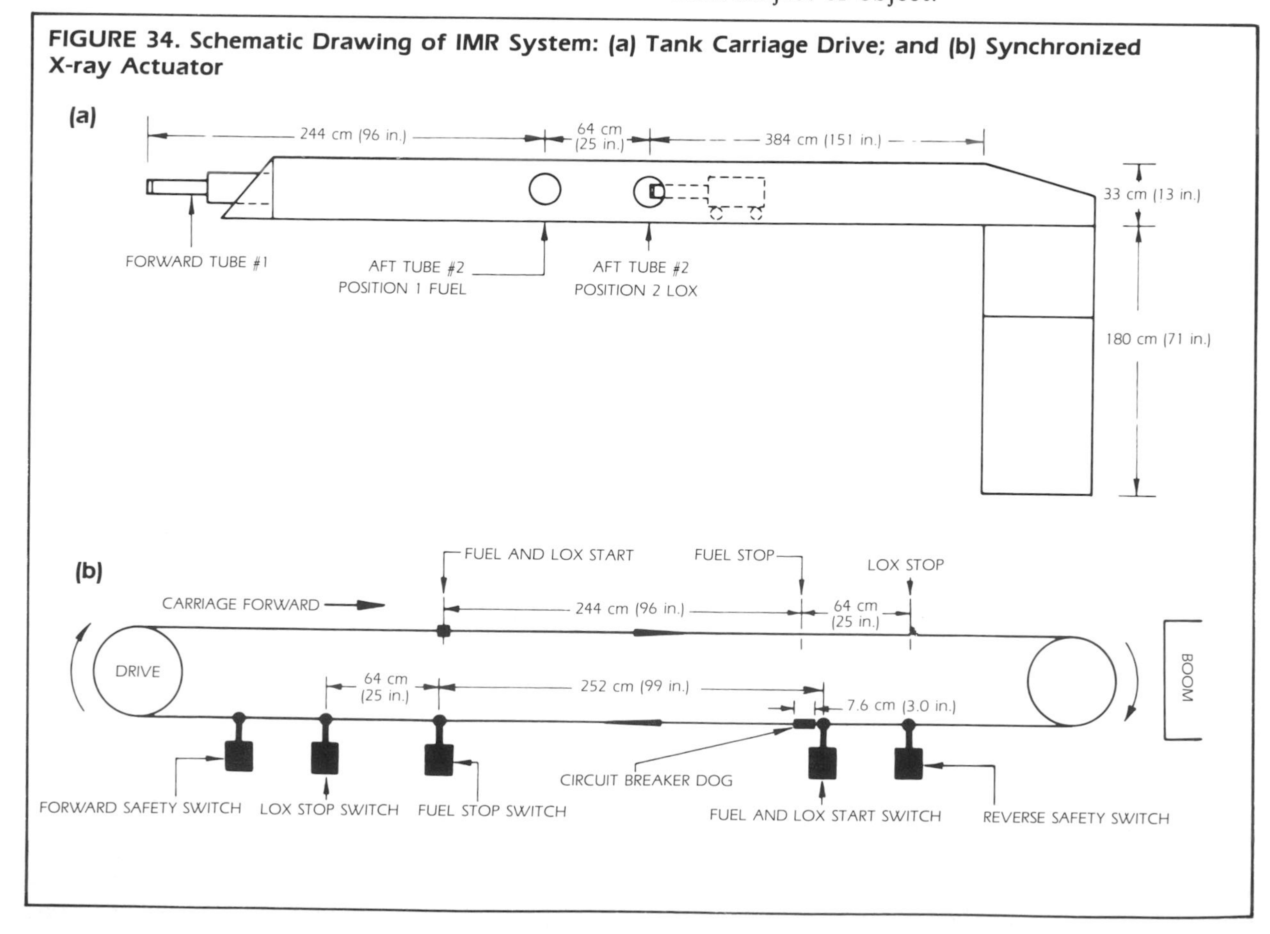

Minification Graph

Typical geometry technique curves (Fig. 29) were derived for a given apex at the film plane where $t = 0$. Similar graphs were made for $t = +0.01$ and -0.01 mm ($+0.3$ and -0.3 in.) for typical brazed honeycomb panels. Additional graphs can then be made showing only the apex location line for as many different source-to-film distances as are necessary. The direct relationship between object size and $(F + M)$ requires no additional graphs. This sum is simply multiplied by the same factor as the object size.

Image Displacement

When the object is in line with the center of the X-ray beam, the upper object surface is superimposed over the lower object surface image. However, when the X-ray beam is at an angle to the object, the upper and lower object images are separated or displaced as illustrated in Fig. 30. The amount of displacement, X, is a function of object height, T, and beam angle, Θ:

$$X = \frac{T(M/2)}{D} \quad \text{(Eq. 12)}$$

$$\tan \Theta = \frac{(M/2)}{D} = \frac{X}{T} = \frac{(S/2)}{C} \quad \text{(Eq. 13)}$$

and from eqs. 12 and 13, we derive the general equation:

$$X = T \tan \Theta \quad \text{(Eq. 14)}$$

During in-motion radiography, the upper object image will be cast to the right as it enters the X-ray beam, and to the left as it leaves the X-ray beam (see Fig. 30). This image displacement is somewhat akin to the parallax displacement technique used to determine the depth of defects (see Part 9 of this Section).

Total Image Blurring Due to Motion

Total image blurring occurs parallel to the direction of motion and includes: (1) in-motion image unsharpness (eqs. 5 and 8); (2) in-motion image minification (eq. 11); and (3) in-motion image displacement (eq. 14). The unsharpness of U_g is directly proportional to focal spot size F; slit width S; and thickness T. The image minification is directly proportional to T, S, F, and D (source-to-film distance). The image displacement X is directly proportional to T and beam angle Θ. Note that the thickness value T appears in all the equations. This is the reason why thick specimens do not lend themselves to inspection by in-motion radiography. However,

FIGURE 35. In-Motion Radiographic Facility for Welded Solid-Propellant Motor Cases

FIGURE 36. In-Motion Radiographic Facility for Brazed Honeycomb Structures

FIGURE 37. Conventional (Still) Double Surface Radiographic Technique

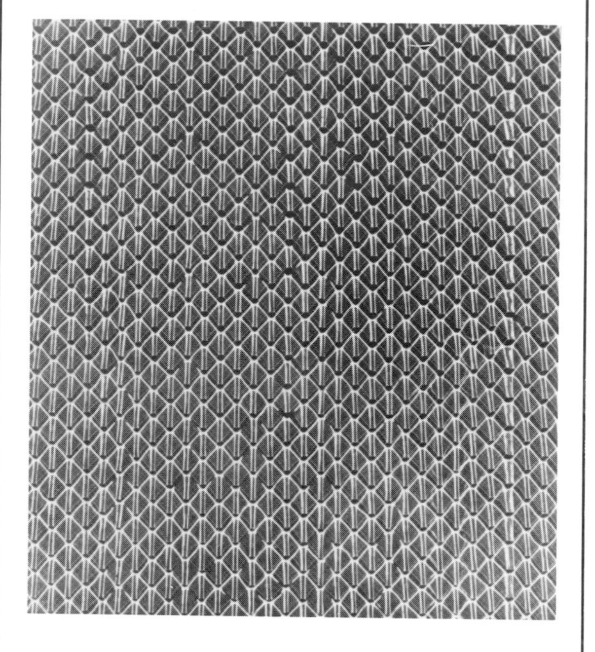

EXPOSURE TECHNIQUE: 150 kV, 12 mA, 90 seconds, 244 cm (96 in.) FFD
PANEL DEPTH: 2.5 cm (1.0 in.)
FACE SHEET THICKNESS: 0.03 cm (0.01 in.)

FIGURE 38. In-Motion (Single Surface) Radiographic Technique

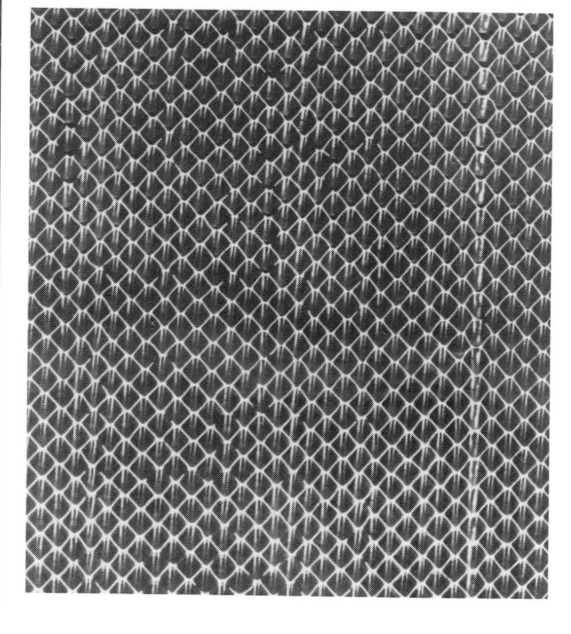

EXPOSURE TECHNIQUE: 150 kV, 12 mA, 60 IPM, 91 cm (36 in.) FFD
PANEL DEPTH: 2.5 cm (1.0 in.)
FACE SHEET THICKNESS: 0.03 cm (0.01 in.)

IMR does work well for honeycomb panels: the filleting images on the source side are blurred out, whereas the filleting images on the film side are sharply defined.

Examples of In-motion Radiography

Linear Motion

In this technique, either the part under inspection is moved past the collimated X-ray beam or the collimated X-ray beam is scanned over the surface of the part (see Fig. 31a). Roll films of various lengths and widths are used. The size of the film is related to the length of the part and the width of the collimated X-ray beam at the film plane. For inspection of longitudinal fusion welds, 35 mm film can be used. For inspection of adhesively bonded or brazed bonded honeycomb structures, the film width may be 25 to 35 cm (10 to 14 in.). For wide structures, multiple scans are required as illustrated in Fig. 31a.

Rotary Motion

In the rotary motion technique, the X-ray source and slit are usually stationary and the cylindrical part rotates one or more revolutions through the collimated X-ray beam. The technique is illustrated in Fig. 31b. Usually, the X-ray source is external, with the film wrapped around a lead plug at the inner diameter of the cylinder. If the cylinder has a fairly large diameter, the X-ray source may be placed inside the cylinder and the film wrapped around the outer surface of the part. In both cases, the film should be in intimate contact and firmly fixed to rotate with the part. A typical collimator for linear or rotary motion inspections is illustrated in Fig. 32.

Synchronous Radiography[9]

This method of in-motion radiography can be most easily compared to a stroboscope. It is uniquely applicable to cyclical motion, as opposed to the linear motion in most slit-scanning applications.

To perform synchronous radiography, a short-pulse X-ray generator is needed. The length of the X-ray pulse depends on the speed of the cyclical motion. Just as one can view a rotating shaft with stroboscopic, synchronized light, a radiograph of a cycling object can be made by applying X rays only at a given position in the object's cycle. This method has been used with vibration testing to observe the condition of enclosed components, such as relay contacts at various points in the vibration cycle. Synchronous radiography has also been done with a betatron to observe conditions in reciprocating engines at various points in the engine cycle.

This IMR technique requires (1) a recording medium of some kind; (2) an X-ray source; (3) an exposure area; and (4) a means of synchronizing the production or transmission of X-ray bursts with the cyclical motion. Generators that produce X rays in short bursts are generally easier to synchronize with a cyclic motion. For example, the expander of a betatron can be energized only when the cyclic motion is at the proper point. To accomplish this with a low-energy generator, it is better to arrange a shutter so that X rays are transmitted only at the proper times, to synchronize with the position selected for viewing the subject's cycle.

Because that actual time of each burst is a very small fraction of the total elapsed time, synchronous radiographs require great amounts of elapsed time.

FIGURE 39. In-Motion Radiography Exposure Room

With real-time radiography systems, the results obtainable with synchronous radiography may be available in real-time.

Circumstances for Utilizing In-Motion Radiography

In-motion radiography provides a way to automate the radiographic inspection process and thereby save time and money. Linear IMR is attractive when the subjects to be radiographed are long and of uniform shape, longitudinal fusion welds in cylinders, nuclear fuel elements, and bonded honeycomb structures, for example. The rotary IMR technique

FIGURE 40. Gooseneck Tube Support and IMR Cone Assembly

is useful for obtaining a single-wall exposure of thin cylinders. If a rod-anode (360°) X-ray tube is not available, the rotary IMR technique may be used to conveniently radiograph a circumferential weld in a tank or cylinder. It is not practical to set up a system to scan short or thick objects. If the thickness becomes too large, the unsharpness due to motion becomes intolerable, and the source side of the object becomes blurred. The obvious way to control unsharpness in these systems is by varying the slit width. For static (stationary) radiography, the generally accepted unsharpness, using eq. 1, is 0.08 to 0.1 mm (0.003 to 0.004 in.). For IMR, the generally accepted unsharpness, using eq. 3, is 0.1 to 0.25 mm (0.004 to 0.01 in.). The maximum acceptable value for unsharpness must be determined by trial exposures.

FIGURE 41. Control Panel for IMR of Honeycomb Structures

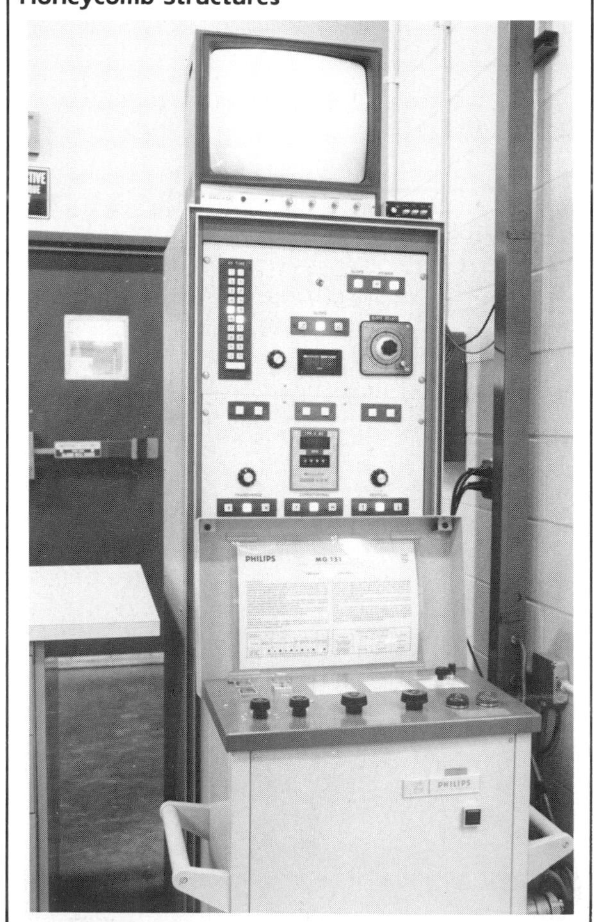

In most cases, the unsharpness at the source side of the part must not produce unacceptable blurring for items such as weldments or nuclear fuel rods. However, for brazed honeycomb structures, it is beneficial to blur out the fillets on the source side of the panel to obtain a clear view of the fillets on the film side. This technique requires radiographing the panel twice to see the fillets on both sides of the panel.

Practical items for IMR are thin plates, rods or cylinders. The subject matter might include weldments, nuclear fuel, ceramic or plastic material, and bonded joints where the bonding material is X-ray opaque.

The main advantage of IMR is cost savings and reduction of the radiation field. One long piece of roll film, of the required width, can be used instead of many pieces of cut film. The long piece of roll film reduces the number of view markers. Roll film can be processed in automatic film processors, reviewed more conveniently on viewers with special adapters, and stored in canisters. Because the collimator drastically reduces the size of the radiation beam, the radiation is easier to shield, allowing IMR to be performed in the shop rather than in a lead-lined exposure room.

FIGURE 42. In-Motion Radiographic Cabinet for Inspecting Nuclear Fuel Elements

Applications of IMR

Weldments

The radiography of weldments in-motion was conceived in 1956.[10] Before the project could begin, an X-ray laboratory had to be built and special equipment designed or purchased; because packaged strip film was not available, a machine was designed and built to package 70 mm roll film.

The radiographic technique consisted of placing two rod-anode X-ray tubes on a boom and porting the X rays through an aperture that projected a narrow beam to the longitudinal welds of a tank. The welded tank was placed on a carriage that rode a track anchored to the floor of the exposure room (Fig. 33). An electrical variable-speed drive propelled the carriage and tank along the track at a speed of 15 cm (6 in.) per minute. Simultaneously, the three longitudinal weldments were exposed to the X-ray beams.

Two 150 kV X-ray units were used to reduce the exposure time to one-half. One tube was located at the end of the boom and a second was located half the tank-length behind it. With this arrangment, it was necessary to propel the tank only half its length to obtain complete exposure of the three longitudinal welds. The aft X-ray tube was positioned inside the boom to compensate for the two different tank lengths; the two locations are shown in Fig. 34. The tube was locked in position 1 to radiograph the 4.5 m (15.5 ft) fuel tank and in position 2 to radiograph the 6 m (20 ft) liquid oxygen tank.

The carriage drive mechanism and X-ray machines were synchronized to operate simultaneously when the start button was pushed in the control room. A circuit breaker *dog* was attached to the same continuous drive cable as the carriage and moved in the opposite direction, making contact with the stop switch after moving exactly half the tank length. When contact with the stop switch was made, the X-ray machines shut off and the carriage motion stopped. Three lead ports were used to restrict the X-ray beams to the three longitudinal welds. The slits in the port ends were 0.3 cm (0.13 in.) long and 2.5 cm (1.0 in.) wide, producing a beam 2.5 (1.0 in.) long and 20 cm (8.0 in.) wide, at a 1.2 m (4 ft) source-to-film distance.

A second laboratory was built for IMR inspection of longitudinal weldments in steel tanks (Fig. 35). For this purpose, a rod-anode X-ray tube was mounted on the end of an adjustable boom. The boom could be raised or lowered to compensate for differences in tank diameter.

Table 2 shows a comparison of the techniques used. Both laboratories were in operation for a number of years and were dismantled upon completion of contract work.

Brazed Honeycomb Structures

IMR inspection of brazed honeycomb structures for aircraft was developed in 1961.[8] A radiation collimator (similar to Fig. 32) was designed and fabricated with an adjustable aperture for controlling the radiation pattern. This provided the unsharpness gradient necessary for blurring preselected details nearest to the radiation source without causing unsharpness of the image surface adjacent to the film. The focus-to-film distance was established at 0.9 m (3 ft). The optimum in definition, area coverage, radiation

FIGURE 43. X-ray Tube Head, Manipulator and Cone With Slit

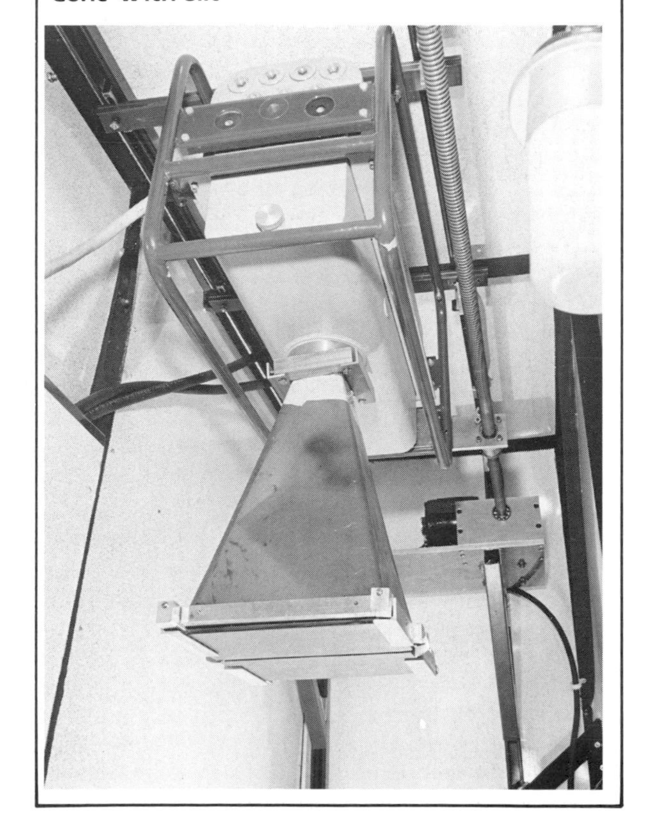

TABLE 2. Comparison of IMR Techniques for Welds

Material	Thickness cm (in.)	Length m (ft)	Voltage (kV)	Amperage (mA)	Focus-Film Distance cm (in.)	Travel per Minute cm (in.)
2014 aluminum	0.95 (0.37)	6.0 (20)	90	15	122 (48)	15 (6.0)
2014 aluminum	0.95 (0.37)	4.5 (15)	90	15	122 (48)	15 (6.0)
4340 steel	0.23 (0.09)	3.6 (12)	150	3.5	46 (18)	10 (4.0)
4340 steel	0.17 (0.06)	2.0 (6.5)	140	3.5	46 (18)	11 (4.5)
4340 steel	0.38 (0.15)	4.2 (14)	160	4.0	51 (20)	9 (3.5)

potential and speed of travel was considered in determining the focus-to-film distance for brazed honeycomb. The X-ray exposure was accomplished by controlled linear movement of the brazed assembly across the area of radiation emitted by the stationary, collimated X-ray source (see Fig. 36).

Fine-grained, high-contrast film was cut in widths of 30 cm (12 in.) and lengths of 60 m (200 ft). Cassettes to accommodate the various lengths of radiographic film were designed and fabricated from a special plastic material suitable for radiographic work. Securing the plastic cassettes in intimate contact with the honeycomb panel was accomplished using precut lengths of magnetic rubber strips placed on the cassette and magnetically attached to the honeycomb panel.

In all phases of the operation, in-motion radiography was less time-consuming than conventional radiography. The ease of operation was particularly noted in the interpretation of the radiographs. Figure 37 is a conventional (still) radiograph showing brazed fillets on the source and film side (double surface) of the panel. Figure 38 is an in-motion radiograph showing only the brazed fillets on the film side (single surface) of the panel. It is much easier to detect lack of braze and to measure the fillet width on the Fig. 38 radiograph.

Adhesive Bonded Honeycomb Structures

In-motion fluoroscopic inspection of metallic, bonded honeycomb structures has been done for many years in the Netherlands and in Sweden[11] and more recently in Texas.[12] IMR techniques, employing roll film, are presently being used to inspect a variety of bonded honeycomb structures. These structures are composed of aluminum core bonded to either aluminum, boron-epoxy or carbon-epoxy skins. The inspections are performed to detect crushed core, core node separation, foreign objects, core splice flaws and core tie-in at closures.

A collimator is used to guide the X-ray beam in the direction of motion. Roll films of required lengths and 70 mm, 12, 25 and 35 cm widths are being used. The films are processed in automatic equipment, and special film reading boxes have been fabricated, some of which provide for viewing up to 1.5 m (5 ft) of film at one time. The assemblies are oriented with the honeycomb cells parallel to the X-ray beam; the direction of motion is parallel to the ribbon direction.

Figure 39 shows IMR being performed on a composite upper-wing skin approximately 8 m (29 ft) tip-to-tip and 2 m (6 ft) forward-to-aft at the centerline. The stainless steel tool is designed to manipulate the assembly in five axes to permit orientation of the contoured surface perpendicular to the X-ray beam. The X-ray tube head has a lead-shielded cone attached to the port to limit radiation onto a narrow line about 12 mm (0.5 in.) wide in the direction of tube motion. The tube support is mounted on the ceiling and has an extension up to 12 m (40 ft).

A more detailed view of the gooseneck tube support and in-motion cone is shown in Fig. 40. This figure shows a bonded honeycomb assembly with boron-epoxy skins on 356 mm (14 in.) wide roll film mounted on 3 mm (0.13 in.) vinyl-lead backup material. The pendant control permits movement of the tube support in three directions.

The control console for the in-motion system has the standard X-ray control panel modified to provide kilovoltage slope control during the in-motion exposure (see Fig. 41). Because many aircraft structures taper in thickness from inboard to outboard, it is impossible to maintain constant film density without adjusting X-ray parameters during the in-motion exposure. Kilovoltage was chosen as the variable because it can be changed to match energy level with the thickness of the part. Another improvement in the in-motion control is constant speed control, which automatically compensates for varying loads and maintains a constant speed which is displayed on a digital tachometer. The television monitor shows the area of the assembly that is being subjected to radiation. It does not provide an X-ray image, but is used to aid in alignment of the assembly.

Nuclear Fuel Elements[14]

Nuclear fuel elements range up to three meters long with a fueled region about 1.2 m (4 ft) long. The fuel is uranium-plutonium carbide in pellets about 7 mm (0.3 in.) in diameter. The pellets are loaded into tubing with a 9.4 mm (0.4 in.) outside

FIGURE 44. Radiographic Jogging Technique: (a) Schematic Drawing of Object (Tube); (b) Cross Section of Tube; (c) Drawing of Processed Radiograph

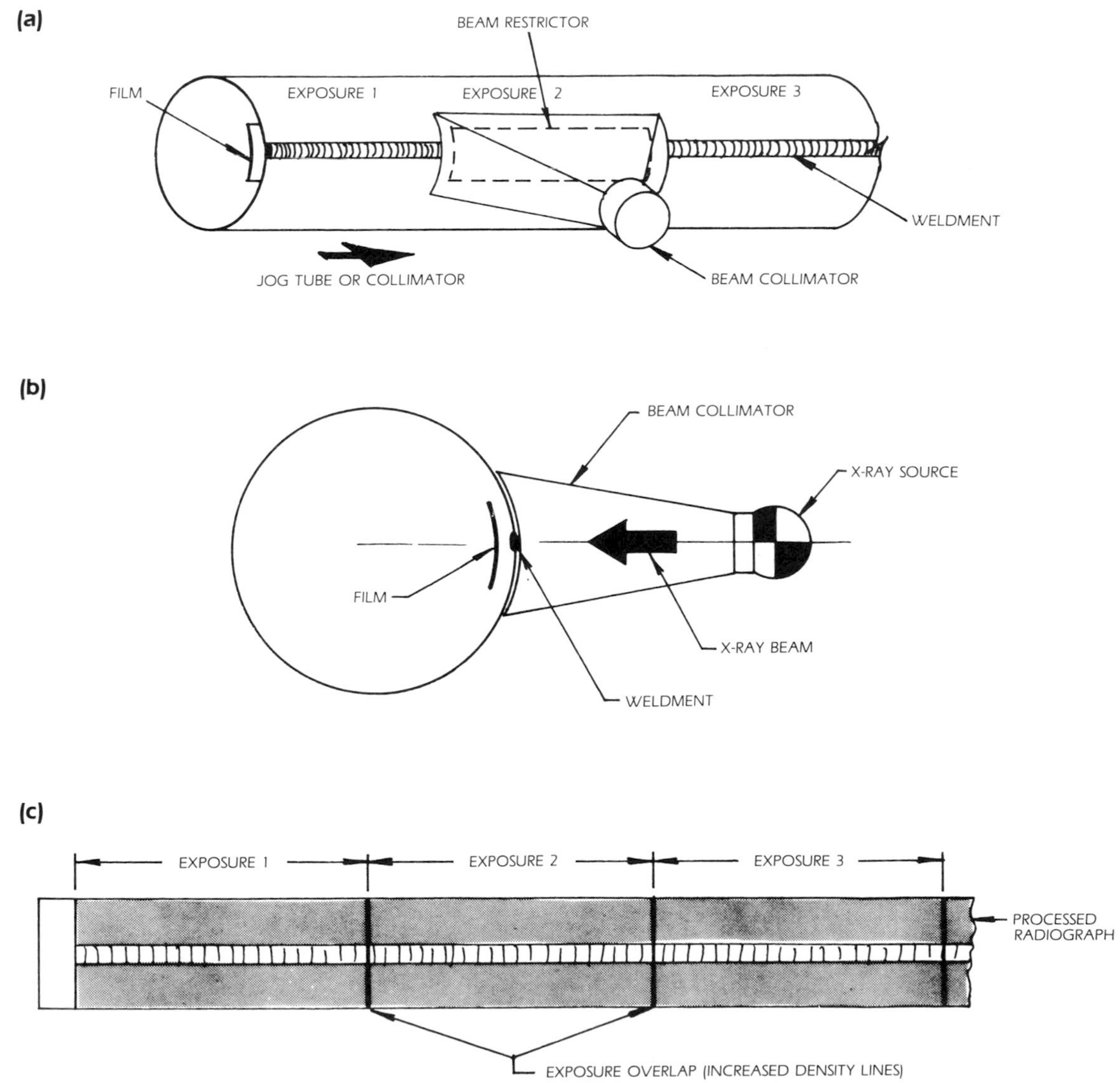

FIGURE 45. Radiographs Showing Folding Characteristics of an Internal Fuel Expulsion Bladder

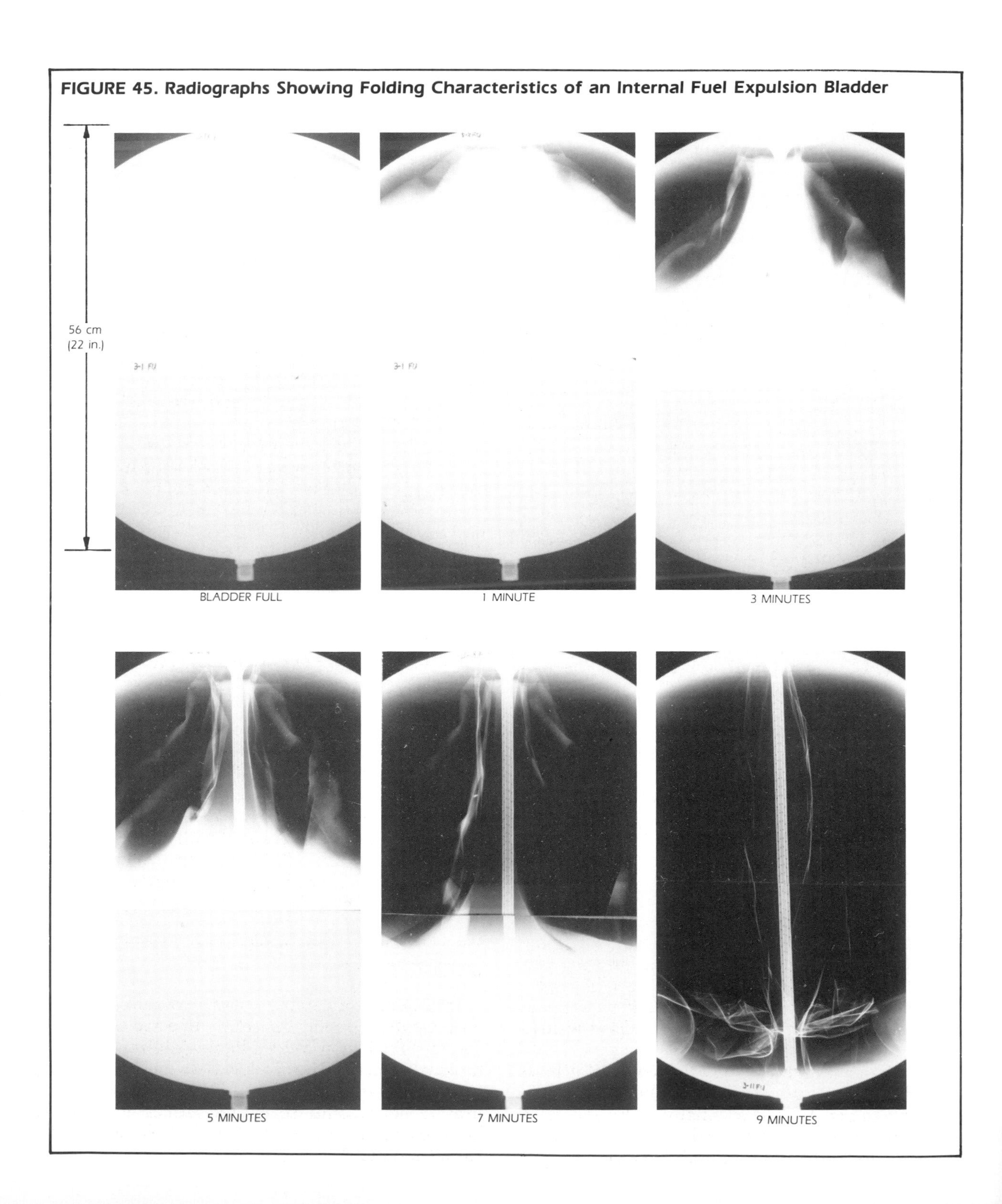

diameter to make up the 1.2 meters of fuel.

The elements are radiographed to confirm proper positioning of all components within the cladding, to detect spacing between pellets, if any, and to detect fuel-pellet chips between the fuel and cladding. This radiography does not evaluate the fuel pellets, because radiation from a 200 kVp X-ray generator does not have sufficient energy. The cabinet is shown in Fig. 42, and the tube head with attached cone and slit is shown in Fig. 43. A ball screw, driven by a variable-speed, gear-reducer motor is used to move the tube head past the stationary fuel-element film array.

The system is set up so that the X-ray target-to-slit distance is about 0.6 m (2 ft) and the target-to-film distance is about 1.2 m (4 ft). The slit is adjustable, but has been maintained at about 6.4 mm (0.25 in.) producing an X-ray beam about 12 mm (0.5 in.) wide at the film.

Fine grain X-ray film of 356 mm (14 in.) width permits radiography of up to 18 elements at a time. The exposed films are processed in automatic processors and interpreted on illuminated viewers with appropriate film handling attachments.

Slit scanning provides several advantages to radiography of these fuel elements. The interpretation of the radiographs on a single, continuous piece of film provides exact locations of all components. The fact that the beam is continuously orthogonal to the axes of the elements makes precise length measurements possible. This feature is also advantageous for showing separations between individual pellets. The scanning radiography procedure is faster than a series of overlapping views to cover the elements, and there is also a saving on radiographic film.

Radiographic Jogging Technique

At times, it may not be possible to obtain the desired sensitivity or exposure time using IMR techniques. When this condition exists, the radiographic jogging technique may be used to advantage. This technique has been used to radiographically inspect thin fusion-welded pneumatic hydraulic tubing, and has replaced the IMR technique for fusion welded motor cases.

To use this technique, the beam collimator must nest close to the surface of the part being inspected as shown in Fig. 44. The collimator width and beam restrictor width are governed by the amount of coverage for each exposure and the source-to-part distance. Repeated exposures are made by jogging either the X-ray source or part a fixed distance for each exposure. The exposure overlap distance should be kept to a small but practical value. These overlap areas receive two exposures resulting in increased film density lines between exposures, as shown in Fig. 44c.

Pseudo In-motion Radiographic Technique

This technique uses standard X-ray equipment and film. It is used in medical radiography in conjunction with synchronization between the rapid film changer and the repeated exposures. It is useful for industrial radiography when slow motion changes in the test object must be recorded on large pieces of radiographic film.

The example shown in Fig. 45 consists of a 55 cm (22 in.) diameter titanium fuel tank with an internal fuel expulsion bladder. The radiographic study was performed to determine how the bladders folded inside the tank as the fuel was expelled. Some bladders had ruptured resulting in poor fuel expulsion. The size of the tanks precluded using cineradiographic techniques employing a 22 cm (9 in.) diameter image intensifier.

Production titanium tanks were filled with water and then the water was expelled using nitrogen at 90 pounds per square inch gage. The source-to-film distance was chosen so that the X-ray beam would cover the full diameter of the tank. Two 35 × 43 cm (14 × 17 in.) films were used for each exposure. The test consisted of filling the tank and radiographing while full and then expelling at one minute intervals with a radiograph being made between each expulsion. Tests were repeated with the tank in the flange-up and flange-down positions. A number of tanks were tested until the design engineers determined how the bladders were being ruptured.

Summary

In-motion radiography was first developed to decrease radiographic inspection time and associated costs for the fusion welded liquid-oxidizer and fuel tanks used on the Thor rocket. Subsequent applications have been used for other fusion welded tanks, brazed honeycomb structures, adhesive bonded honeycomb structures, nuclear fuel elements and rocket motors.[13]

To associate the term *motion* with radiography is contrary to all of the normal radiographic concepts. Usually, a great deal of effort is expended to ensure a minimum amount of relative motion between the subject, recording medium and X-ray source. However, in some specialized cases, the advantages of controlled motion of the X-ray source relative to the subject and recording medium outweigh the disadvantages. Some of the advantages of IMR include:

1. large or extended specimens may be radiographed, reducing set-up time;
2. fewer individual films make interpretation easier;
3. the amount of distortion over large areas of the specimen is minimal; and
4. on relatively thick specimens where the area of interest is near the recording medium, the area of interest will appear sharper than it would with a standard, stationary X-ray examination.

Some disadvantages of IMR include the following.

1. The relative motion between the X-ray source, specimen and recording medium limits the method to relatively thin specimens; the area of interest of the specimen must be close to the recording medium.
2. The exaggerated size of the focal spot in IMR causes the need for tight collimation of the X-ray beam which, in turn, causes the loss of a high percentage of the available X-ray energy.

PART 8

PROJECTION MICROFOCUS RADIOGRAPHY

Projection radiography can be accomplished with a true microfocus X-ray source; that is, an X-ray tube with an electron focal spot smaller than 0.1 mm. In practice, focal spots from 0.002 to 0.025 mm (0.0001 to 0.001 in.) have proven to be the most useful for real-time systems,[15] while spots from 0.025 to 0.075 mm (0.001 to 0.003 in.) have proven satisfactory for film techniques using moderate magnification levels. The film techniques have been documented previously in the literature.[16] Successful

FIGURE 46. Drawing of Real-Time Microfocus X-ray System

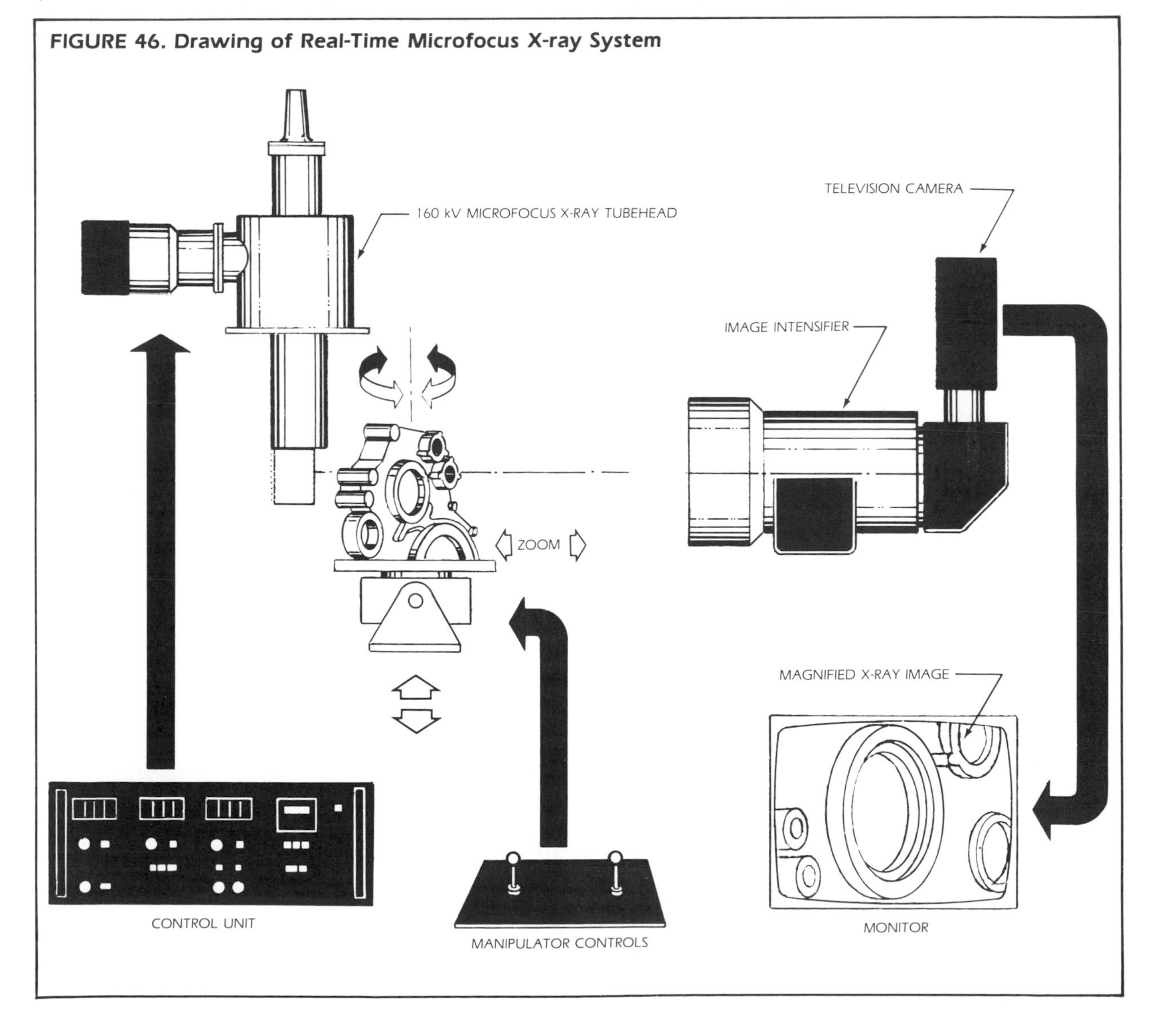

real-time projection radiography using microfocal equipment has, until recently, been limited by a combination of low X-ray output, marginal X-ray system reliability and a total lack of real-time performance specifications other than the quality indicators generally required for film radiography.

Real-time Projection Microfocus Radiography

A typical system for inspection of smaller objects would contain any of the components covered in Section 14, Part 5 (*Real-Time Radiography*, "Remote Viewing Systems") of this Handbook. One other type of system which has shown versatility in roughly 450 applications has been the remote video viewing system shown in Fig. 46. The X-ray tube is a 160 kVCP microfocus unit capable of 300 watt operation at a focal spot size of 0.25 mm (0.01 in.). It can also operate continuously at 80 watts with a focal spot of 0.012 mm (0.0005 in.). This means that the X-ray unit can operate continuously at 160 kV and 0.5 mA with a 12 micron focal spot size and, in this configuration, can resolve details as small as 25 microns or one thousandth of an inch at a 1:1 geometric relationship (without magnification).

Low light level imaging cameras combined with high resolution fluors or X-ray image intensifiers and camera combinations are capable of resolving ten to two line pairs per millimeter, respectively, as measured by resolution test pattern, with good contrast (better than 50% modulation of the composite video signal). Accordingly, these video systems are not capable of resolving the fine details (0.1 mm [0.004 in.] or less) available in the X-ray image at 1:1 magnification. However, if projection magnification techniques of 10 × or greater are employed, even the two line pair per millimeter system is capable of resolving a 20 line pair per millimeter test pattern as shown in Fig. 47 (these images are radiographic positives).

The geometry used for the test data was source-to-detector distance of 1,500 mm (60 in.) and a source-to-object distance of 150 mm (6 in.), producing the 10 × projection magnification. The real-time imaging system employed a 23 cm (9 in.) X-ray image intensifier optically coupled with a 15 MHz CCTV fitted with a 2.5 cm (1 in.) vidicon image tube. With low absorbing materials, projection magnifications of 50 × or more can be obtained; 100 × projections have been achieved. The arrangement shown, or others, can also utilize deposited rare-earth screens or crystal fluors if the camera is equipped with a sufficiently sensitive image tube such as a silicon-intensified target or a doubly intensified newvicon tube. These camera/screen/crystal combinations can be less expensive than a cesium iodide (CsI) image intensifier, but the images produced are usually much noisier.

FIGURE 47. Type 39 Lead Resolution Tester, Showing 20 Line Pairs per Millimeter

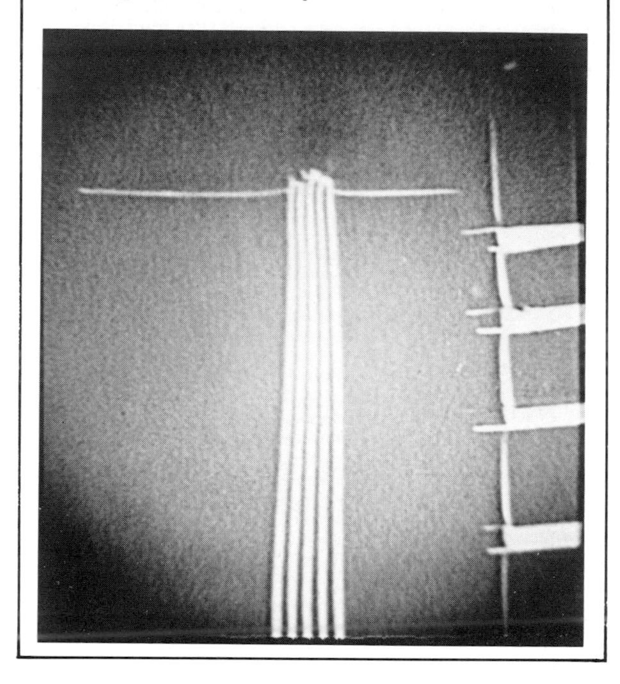

Zoom Technique

A useful technique which can be achieved with real-time projection microfocus radiography is that of *zooming*, or dynamically positioning the object with a manipulator between the X-ray tube and image receptor. In this techinque, the object is moved from roughly half way between the X-ray tube and the receptor toward the X-ray tube. The effect of this motion is shown in Fig. 48. The object illustrated is a single integrated circuit which was initially situated for low projection magnification (about 5 ×). The resultant image is shown as it appeared on the TV monitor in Fig. 48a. The integrated circuit was then zoomed toward the X-ray tube through 10 × as shown in Fig. 48b, 20 × in Fig. 48c and finally to approximately 50 × in Fig. 48d. It is evident that the higher the projection magnification, the more detail one can see in the integrated circuit, even down to the solder joint voids in the silicon chip-to-substrate bond, the individual soldered leads and the etching of the metal substrate. The total length of the metal components of the integrated circuit is 18 mm (0.7 in.).

A similar test, done on a metal jet engine turbine blade, is shown in Fig. 49. The entire blade as shown in Fig. 49a displays no obvious defects. However, on close inspection at approximately 12 × magnification (Fig. 49b), a small crack at the trailing edge of the blade is visible. Note that drilling undercuts

FIGURE 48. Sequence of Radiographic Real-time Magnifications of Integrated Circuit: (a) Low Magnification; (b) Medium Magnification; (c) High Magnification; (d) Ultra-High Magnification

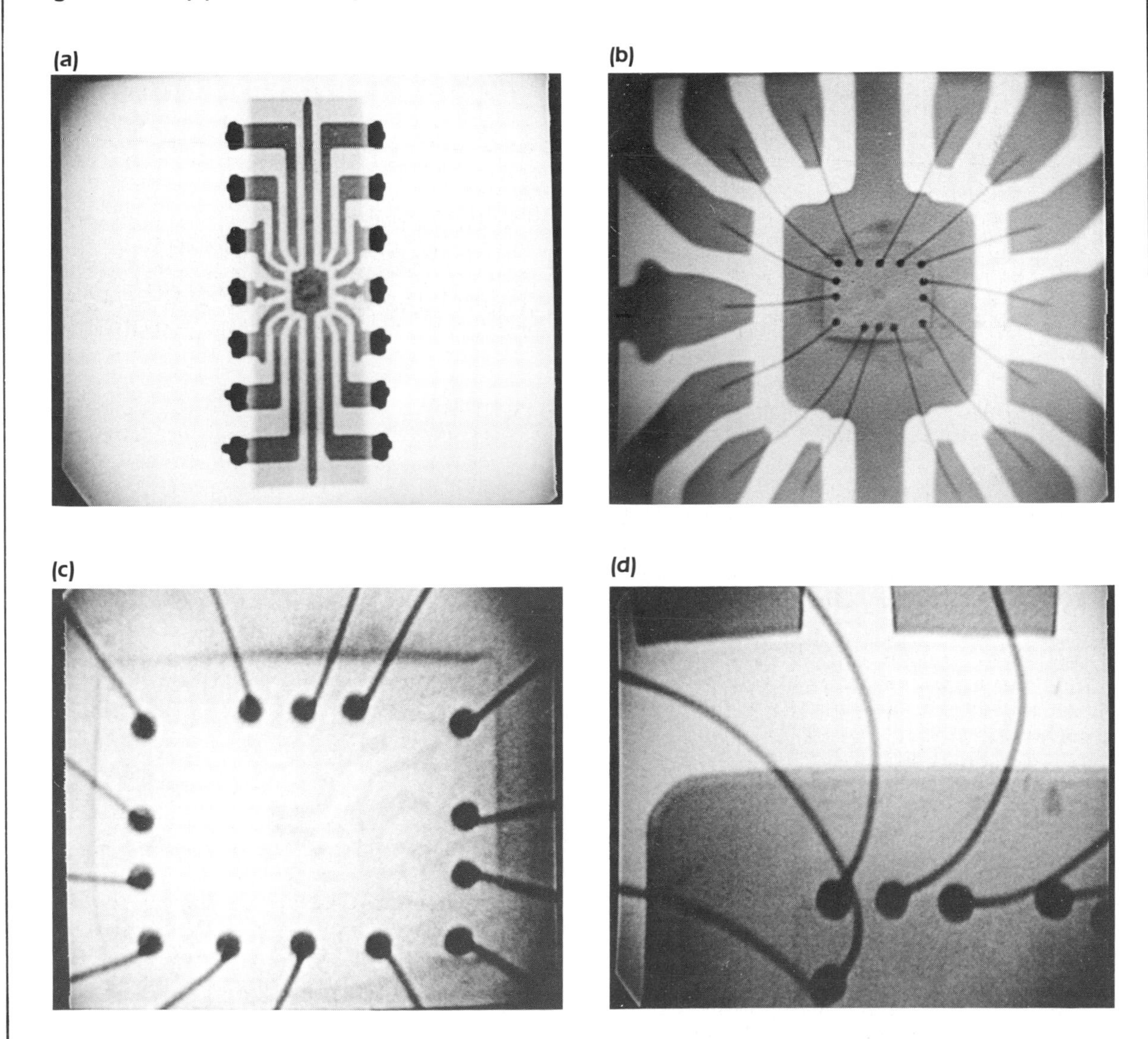

are present at the blade tip in Fig. 49c and that the body of the blade (Fig. 49d) also contains small steel shot sized approximately 0.25 mm (0.01 in.).

Another important benefit of projection techniques is the greatly improved image contrast level which results by eliminating all but very low-angle scatter at the image plane.

Automatic Defect Recognition Applications

Automatic defect recognition (ADR) is applied to parts which must be inspected for the presence or absence of certain components or for the presence or absence of bonding agents such as solder and brazing. ADR may also be used at very high speed for objects that can be scanned and interrogated by intensity statistics, pixel statistics or similar window techniques for voids, inclusions or other anomalies with good contrast against the surrounding material. The picture in Fig. 50 shows the ease of achieving an inspection for the presence, absence or correct location of components in a small armaments arming device. The diameter of the device is approximately 25 mm (1.0 in.), and when imaged at 10 × magnification, it can be inspected by a series of window scans which give the computer a signature for the correct location and presence of components.

FIGURE 49. Sequence of Radiographic Magnifications of Turbine Blade: (a) Low Magnification Real-Time View; (b) High Magnification Real-Time View of Cracked Metal in Turbine Blade; (c) High Magnification Real-Time View of Drilling Faults in Blade; (d) High Magnification Real-Time View of 0.03 cm (0.01 in.) Steel Shot

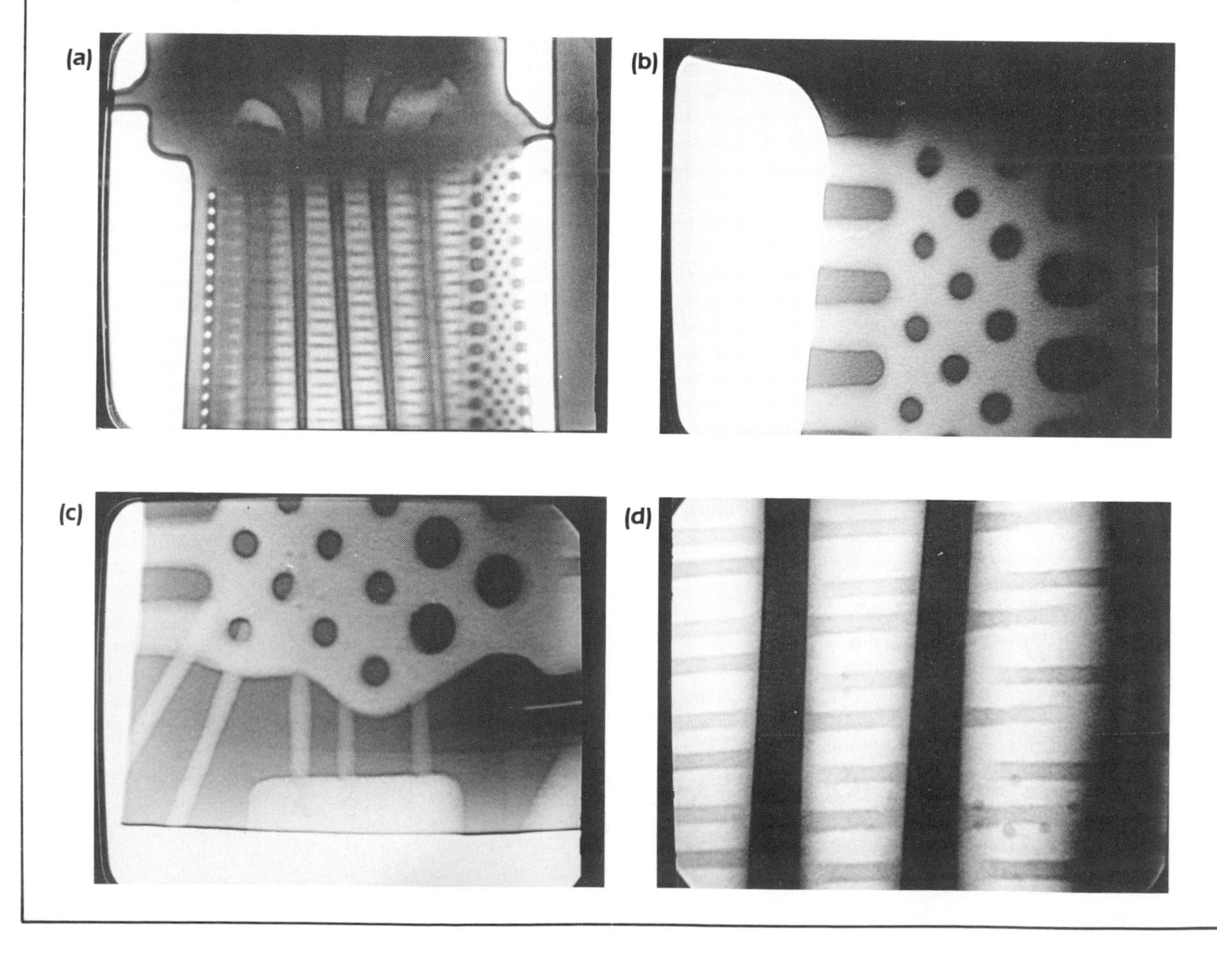

The speed of a standard 525 line TV system equipped with a flash analog-to-digital converter allows the window scans to be done in 1/60 second, giving an automatic inspection capability up to 60 parts per second.

A second example of a specimen that lends itself to ADR is the tantalum capacitor shown in Fig. 51. This capacitor, which is manufactured in batches of ten thousand or more, can be inspected at approximately 20× magnification to reveal centering of the electrode, solder filling, and voids in the hermetic seals around the top of the can and around the lead wire. Here, a video window is positioned in the appropriate area and an automatic intensity comparison (accept/reject) is made.

The third example, which is also manufactured in very large quantities, is the resistor spark plug shown in Fig. 52. The spark plug is typical of parts that can be inspected automatically for homogeneity of core material. In this case, area measurement and intensity measurement can be used to detect voids in the resistive sealing compound inside the ceramic insulating shell. The magnification required for adequate resolution of voids as small as 0.05 mm (0.002 in.) is approximately 25×.

FIGURE 50. Munitions Safety and Arming Device Details Visible with Microfocus

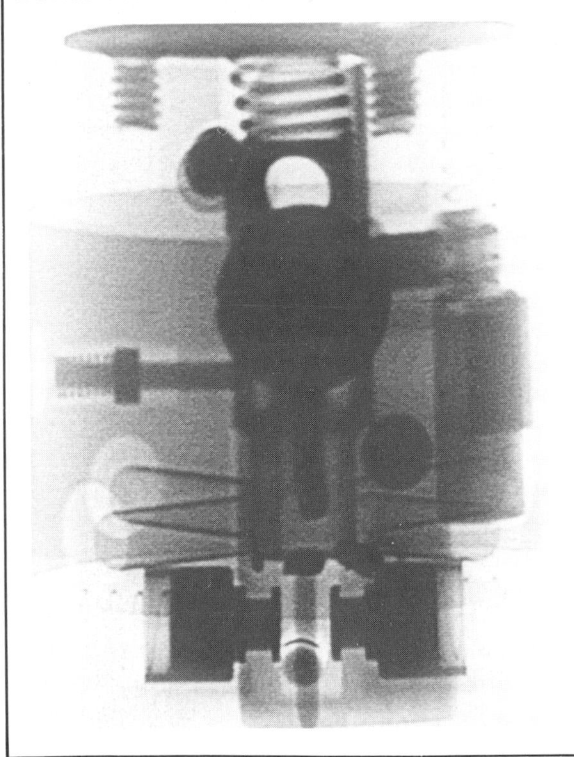

FIGURE 51. Medium Magnification of Solder Faults in Miniature Tantalum Capacitor

FIGURE 52. Medium Magnification View of Electrode Seal in Resistor Spark Plug

Microfocus benefits ADR by greatly increasing the image quality and reducing the band width, stability and repeatability tolerances required of the ADR positioning and imaging equipment.

High Power Applications

The utility of a microfocus X-ray system is greatly enhanced if the system can be used to penetrate dense, thick objects as well as small, highly detailed objects. This requirement for the system to penetrate fairly thick specimens with enough X-ray flux to produce a useful image in the video system forced the development of the high output microfocus equipment used in these experiments.

Limited Field of View

The requirement to simultaneously display an appropriate image quality indicator taxes the system when combined with the requirement to produce an acceptable real-time and hard-copy image or tape recording for archival purposes. The results available for video recording are shown in Fig. 53, a butt welded 12 mm (0.5 in.) steel plate. Figure 53a shows the real-time video display of the weld at 160 kV and 0.5 mA. The weld bead is seen to contain porosity and lack of fusion in the centerline of the weld. This unprocessed real-time image is sufficient for inspection of the weld. However, if one must also simultaneously show a 2%-1T hole in a #10 ASME penetrameter, the magnification of the projected image must be large enough to make the 1T hole appear to occupy about 4 TV raster lines. At this magnification, as shown in Fig. 53b a field of view less than the length of an ASME penetrameter is produced. The same situation exists with the European penetrameters such as the DIN 62. As shown in Fig. 53c, a DIN 62 Fe 6ISO12 penetrameter is very difficult to include in the image at the magnification required for a 1.4% sensitivity. It is still not completely in the field of view at the magnification required for 2% sensitivity. It is, therefore, much easier to inspect a weld at 1.4% sensitivity with microfocus projection techniques than it is to simultaneously display the appropriately sized penetrameter in the field of view of the imaging system. The latter is required by most radiographic codes and other presently standardized techniques.

FIGURE 53. Comparison of Penetrameters in Microfocus Radiography: (a) Radiograph of Steel Butt Weld with Porosity and Lack of Fusion; (b) 2%-1T Penetrameter Sensitivity; (c) DIN 6IS012 and ASME #10 Penetrameters on Steel Weld

(a)

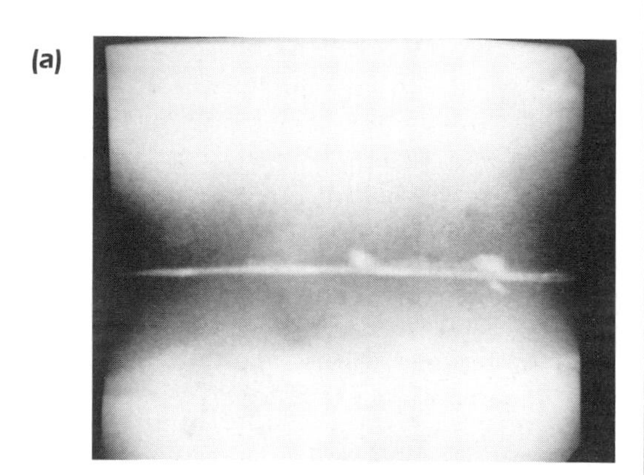

(b)

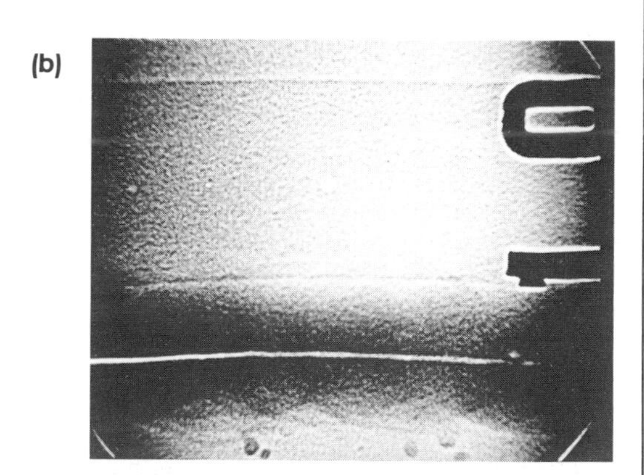

(c)

Because radiographic sensitivity is the requirement, conventional penetrameters will have to be redesigned or code language adjusted if these new technologies are to be used.

Special Applications

Some objects which do not fit in a volume convenient to a system's fixed enclosure can also be inspected using high magnification microfocus and remote real-time video systems.

Systems have been constructed for objects as varied as honeycomb core aircraft parts as shown in Fig. 54 and for dental X-rays as shown in Fig. 55. The aircraft sections naturally require a very large manipulator to properly align the microfocus source/imager and parts to show subtle defects such as crushed core cells. This application places severe demands on the microfocus unit's mechanical and electrical design. Likewise, the dental application requires a special rod anode to permit inter-oral location for panoramic radiography of the teeth. The results of these two applications show the capability of microfocus to detect minute detail: cracked enamel in the teeth and the crushed honeycomb cell ends in the aircraft structure. Also, it shows the flexibility of the equipment to adapt to specialized requirements.

FIGURE 54. Crushed Core Cells in Aluminum Honeycomb

FIGURE 55. View of Dental Work with Microfocus Rod Anode

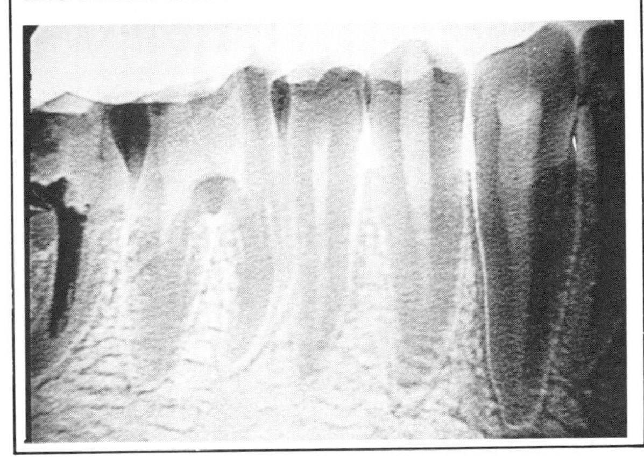

Projection Microfocus Radiography — Film Imaging

The system shown in Fig. 46 can be utilized for film projection microfocus radiography by simply placing a film cassette at the desired point in the image plane. The results attainable on films made in this manner are usually superior to the real-time images seen on the TV monitor. Combining film radiography and real-time techniques has the advantage of sighting the correct view of the part, via the real-time monitor, before exposing a film. A very good description of previous developments in film projection radiography exists in the literature, and contains many high quality photographs showing results achieved with microfocal radiography.[17]

Conclusion

Projection microfocal radiography using both real-time imaging and radiographic film is a very practical method for detecting minute details in objects ranging from tiny integrated circuits to large steel structures having 12 mm (0.5 in.) thick walls. The major impedance to wide acceptance of real-time projection microfocus radiography is the unwillingness of industry to accept a slightly modified penetrameter control. One suggestion is the adoption of smaller penetrameters that can be fitted into the small field of view of a highly magnified projection X-ray image.

PART 9

RADIOGRAPHIC FLAW DEPTH DETERMINATION

Introduction and Background

The importance of radiographic flaw depth determination is directly linked to the growing use of fracture mechanics for determining product serviceability. The usefulness of fracture mechanics is, in turn, directly related to the accuracy of information on the character, size, shape, and location of discontinuities.

Two basic methods are available for determining the depth of a flaw beneath an inspection surface. One of the methods involves radiography, the second employs ultrasonics. This chapter deals with the *radiographic parallax methods* for determining flaw depth within a part or weldment.

In addition to the radiographic parallax methods, the technique of depth determination through radiography also includes such specialized applications as computer assisted tomography (currently common in the medical field and in its infancy for specialized applications in the industrial field) as well as stereo radiography.

Computer assisted tomography (CAT) is a technique in which the test object is rotated about its axis or the radiographic source is rotated about the central axis of the test object. Data are then collected with regard to position and radiation intensity. This information is compiled in a computer and displayed as a cross-sectional slice of the test object.

Stereo radiography is a radiographic method using two separate radiographs made with a source shift exactly parallel to the film plane; the movement of the source between exposures is approximately the same as the pupilary distance of human eyes. The processed radiographs are viewed on a special table; the right eye sees one radiograph and the left eye sees the other. The brain combines the images, giving the impression of a three-dimensional radiograph. Stereo radiography can and is used today, but as with computer-assisted tomography, industrial applications are limited and very specialized.

The Parallax Principle

The radiographic examination methods most commonly used today are based on parallax methods. These methods are based on the principle that from two exposures made with different positions of the X-ray tube, the depth of the flaw is computed from the shift of the shadow of the flaw.[18]

Figure 56 illustrates the parallax principle, using the shadows of objects in a room illuminated by lights at two different positions. Note that the shadows of objects closest to the light bulbs have the largest shadow projection; the objects closest to the lights also have their shadows projected the longest distance.

An object close to the background does not appear to change position, while an object farther from the background appears to shift a moderate amount.

The amount of left or right movement of the projected shadows is directly proportional to the closeness of the object to the light source. A comparison of visual and radiographic parallax principles is shown in Fig. 57.

Similar-Triangle Relationship

A similar-triangle relationship is the basis for most of the calculations used in the radiographic parallax methods (see Fig. 58); the height D divided by the height $T - D$ is equal to base B divided by base A. All of the radiographic parallax methods discussed here maintain this fundamental relationship.

Radiographic parallax methods employ three variations of the similar triangle relationship. These three methods are the *rigid formula*, the *single marker approximate formula* and the *double marker approximate formula*. The data for the similar triangle relationship are derived from the displacement of the image from the film plane.

The film plane is used, rather than the depth below the surface, because it is not always possible

FIGURE 56. Drawing of Parallax from Visible Light

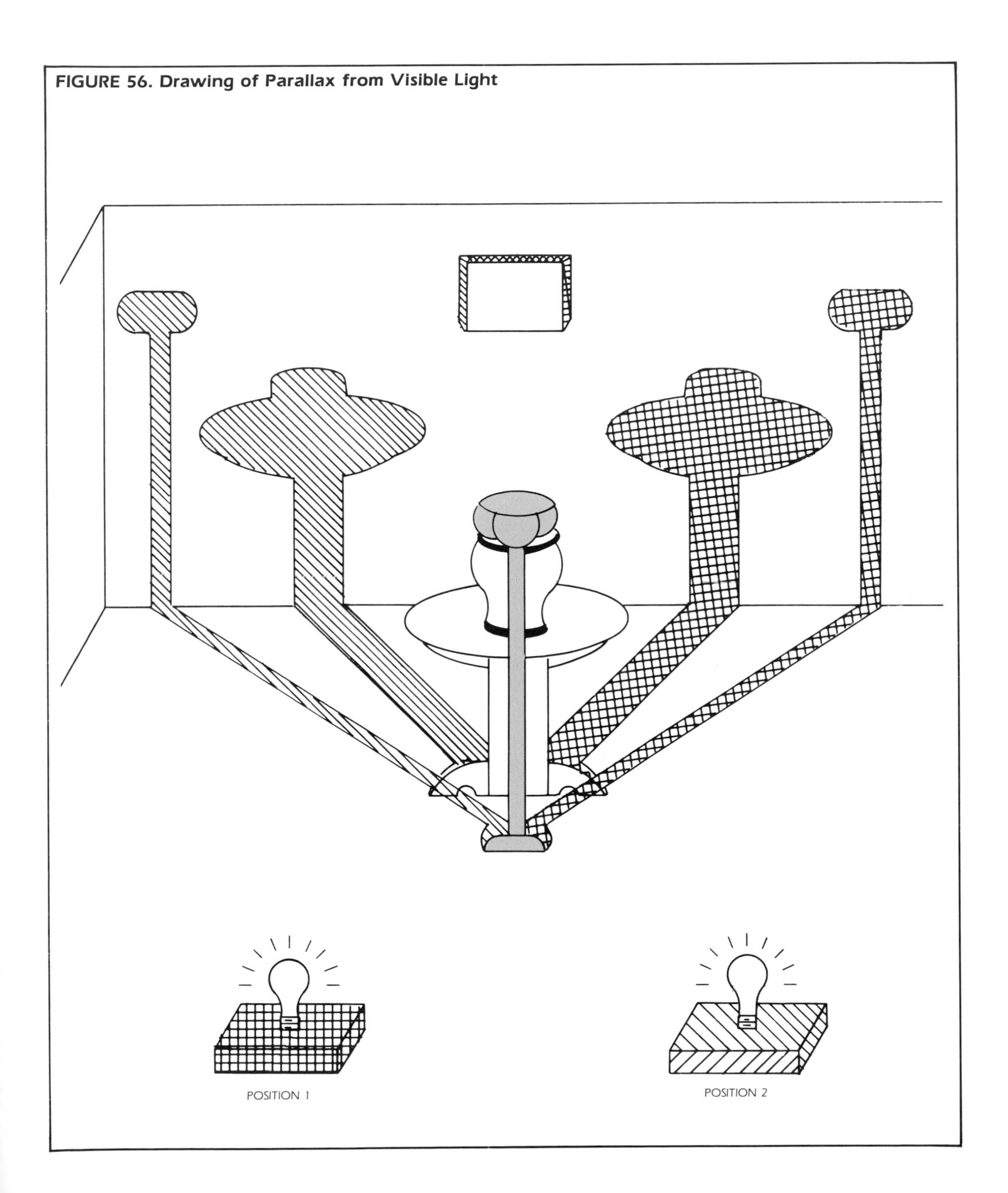

FIGURE 57. Comparison of Visual and Radiographic Parallax

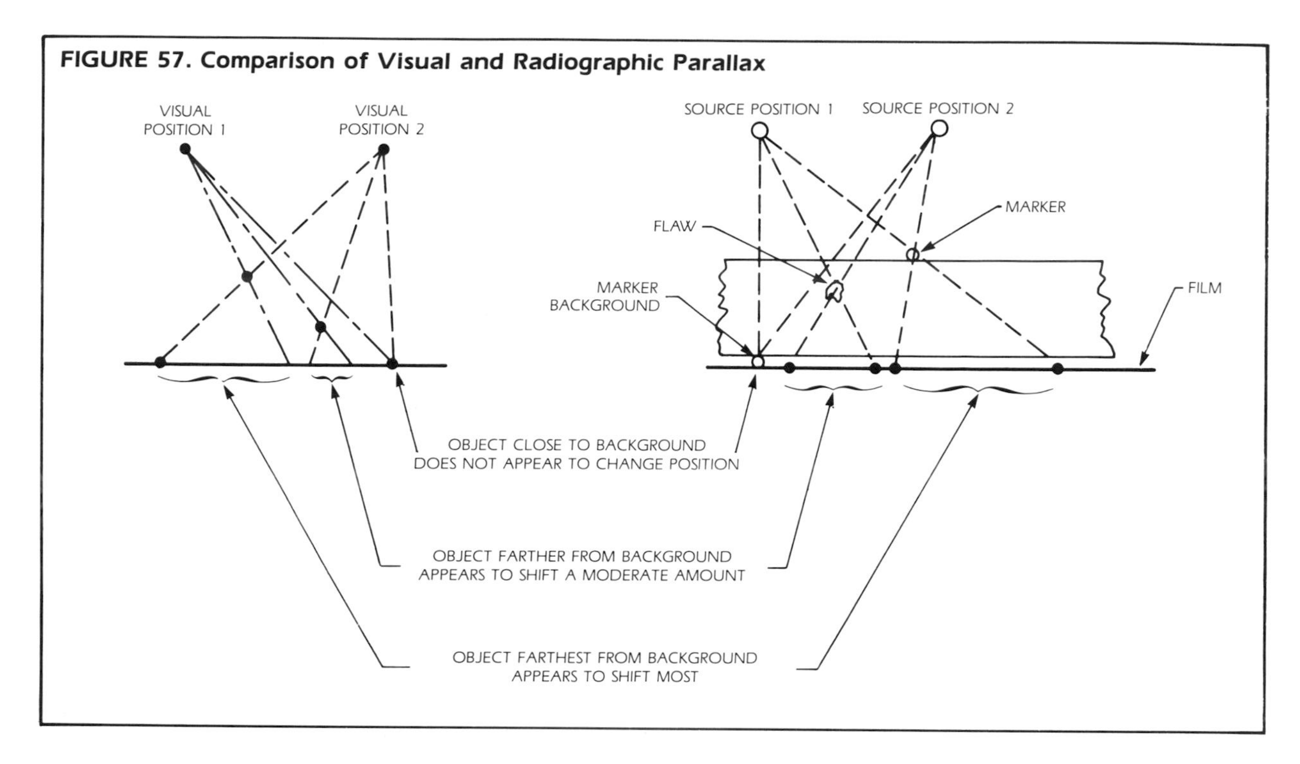

to have the film in intimate contact with the surface of the part.

In addition to problems encountered in calculating the object's height above the film, certain orientation or flaw geometries can cause measurement errors; these are not due to failure of the method, but failure of the radiographer to recognize and compensate for variations in object displacement.

Rigid Formula

Figure 59 is a schematic diagram showing the rigid formula parallax method, which is also defined in eq. 15.

$$\frac{D}{T - D} = \frac{B}{A} \qquad \text{(Eq. 15)}$$

or

$$D = \frac{BT}{A + B}$$

and

$$H = D - K = \frac{BT}{A + B - K}$$

FIGURE 58. Similar Triangle Relationship

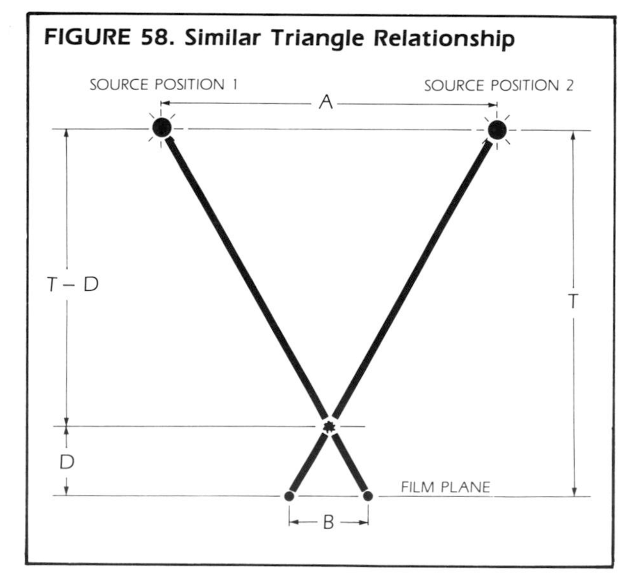

where B is the parallax or image shift of the flaw; A is equal to the source-shift between exposures; T is the source-to-film distance; and D is the distance of the flaw above the image plane.[18,19]

By measuring or knowing the first three parameters, the fourth parameter can be calculated based upon the similar triangle relationship. With the rigid parallax method, no markers are necessary. However, the part thickness, the source-to-film distance and the source-shift must be accurately known. In addition to knowing these measurements, the image of the flaw must be present on a double-exposed radiograph.

FIGURE 59. Diagram of Rigid Formula Parallax Method

FIGURE 60. Single Marker Approximate Method

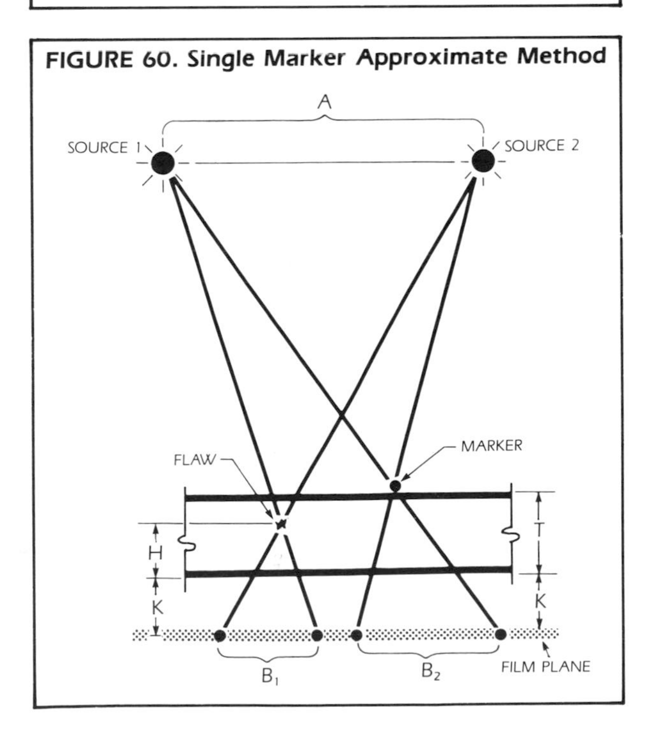

FIGURE 61. Double Marker Approximate Method

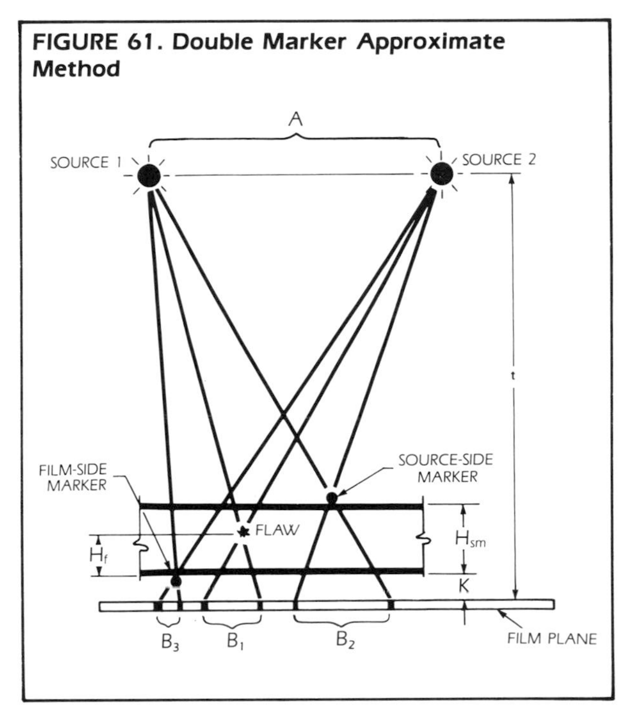

FIGURE 62. Parallax Calculations for Position of Flaw Center

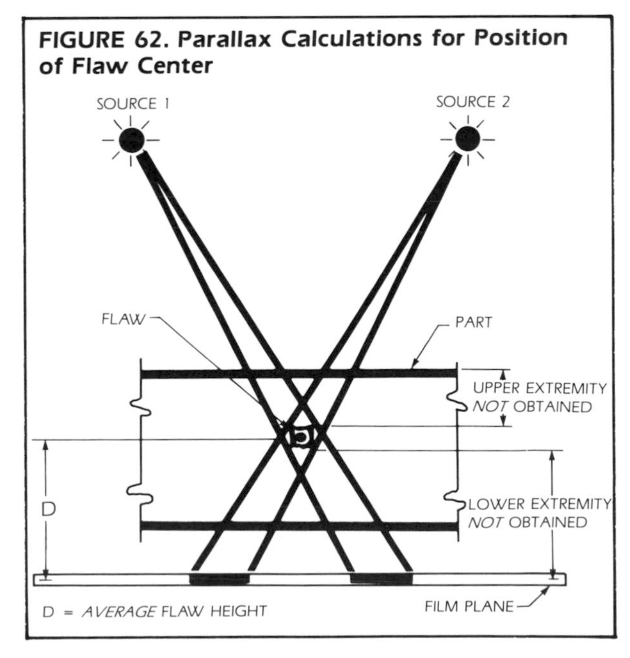

Normally, this radiograph is made by (1) calculating the necessary exposure time; (2) making one part of the radiograph with one-half of this exposure time; (3) moving the source parallel to (and a specified distance along) the film plane; and (4) then making the second half of the exposure. The rigid parallax method can be used when the film is placed in intimate contact with the bottom of the part and when there are no limitations on the height of the source above the film plane. It is important to have significantly large source-to-film versus top of object-to-film ratios when utilizing the rigid parallax method.

Three other important points should be remembered when using the rigid formula parallax radiography.

1. The fundamental relationship between flaw height and image shift is nonlinear.
2. As the flaw height approaches the source-to-film distance, the image shift increases without limit.
3. When the flaw height is small compared to the source-to-film distance, the curve of accuracy approaches linearity.

FIGURE 63. Illustration of Flaw Geometries that Permit Calculation of Average Flaw Displacement

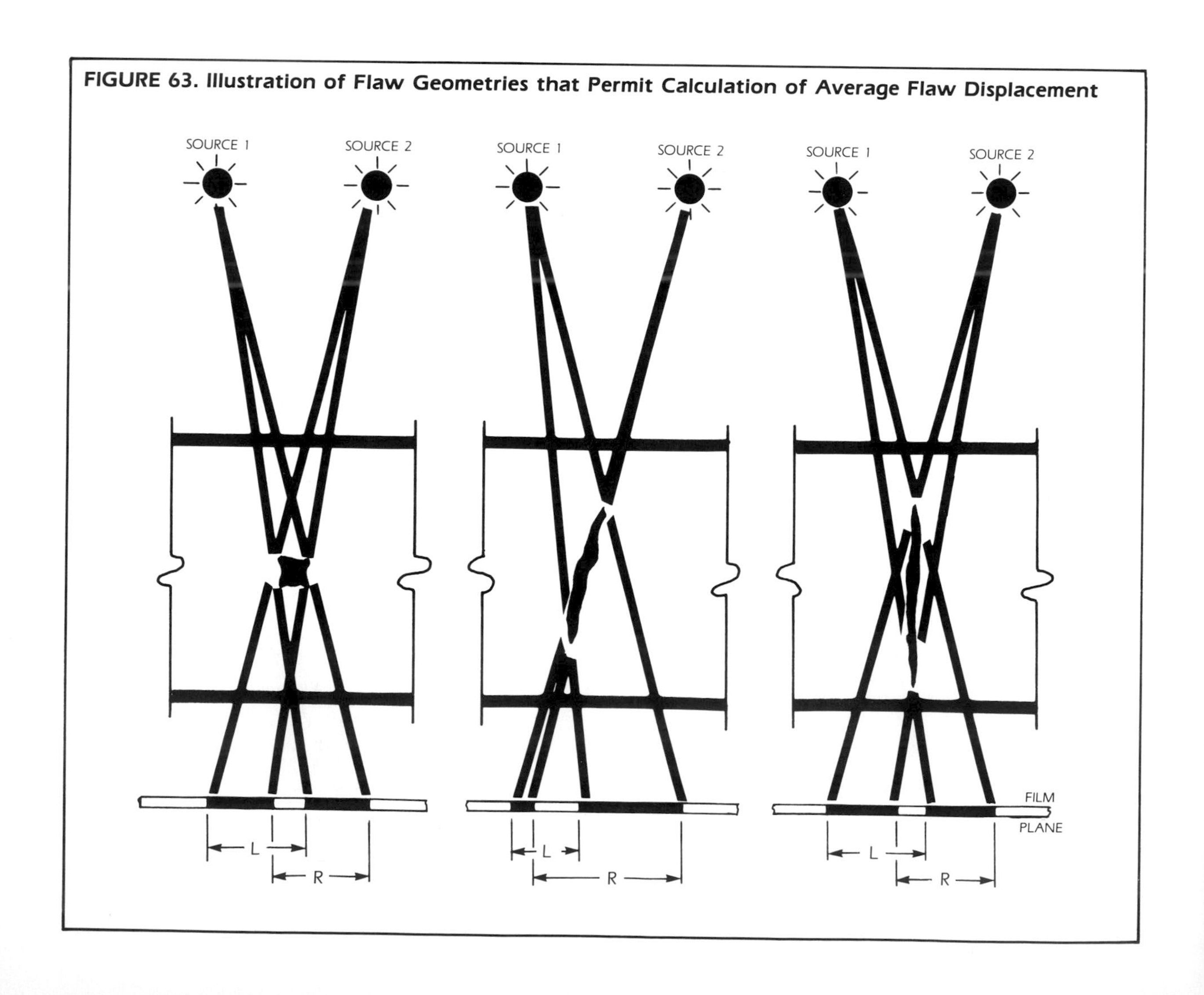

Single Marker Approximate Formula

When the part thickness and flaw height are small relative to the source-to-film distance, the relationship between *D* and *B* approaches linearity and the height of the flaw above the film plane becomes approximately proportional to its parallax. A proportional relationship offers certain advantages in that an artificial flaw or marker can be placed on the source side of an object as shown in Fig. 60.

The height of the defect can be estimated or calculated by comparing the shift of its radiographic image with that of the marker. For example, if the single marker shift is twice the shift of the flaw, this indicates that the flaw is approximately mid-wall. This parallax method eliminates the need for detailed measurement of the part thickness, source-to-film distance and the source-shift as required by the rigid method.

With source-to-film distances at least ten times greater than the part thickness, maximum errors on the order of three percent (of the part thicknesses) can be expected. This is based on the premise that the film is in intimate contact with the part being radiographed. If the film is not in intimate contact with the part, the error will be increased because the proportional ratio is based upon the flaw height above the film plane.

Double Marker Approximate Formula

When the film cannot be placed in intimate contact with the object or the image of the flaw is not

FIGURE 64. Examples of Flaw Geometries Where Centerline of Flaw Is Not Determined by Shift-Averaging

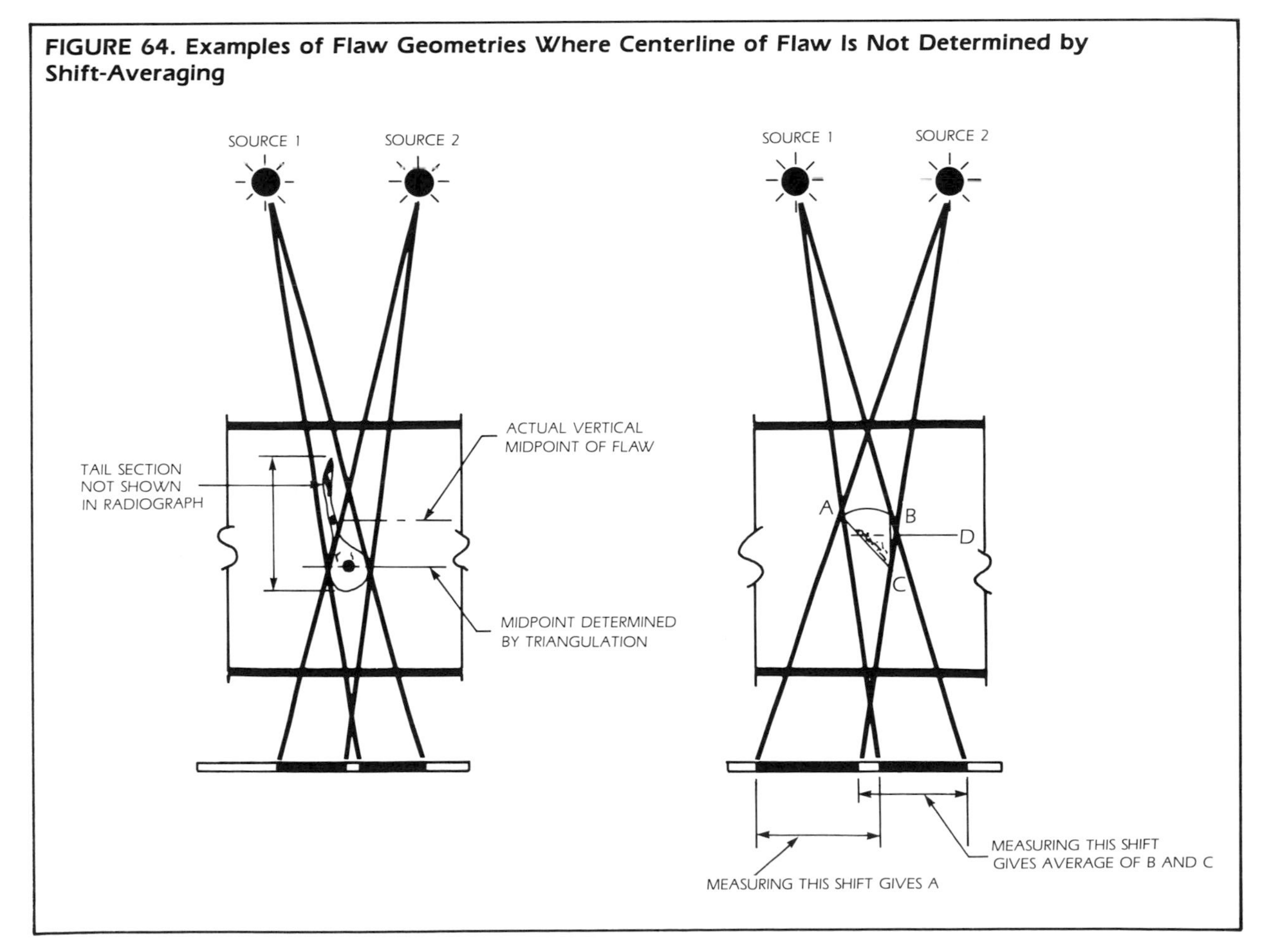

present on a double-exposed radiograph, the *double marker approximate method* should be utilized (see Fig. 61).

If both markers are thin, neglect their thickness and assume that they represent the top and bottom of the test piece. By measuring the parallax or image shift of each marker, as well as that of the flaw, the relative position of the flaw between the two surfaces of the test object can be obtained by linear interpolation, using eqs. 16 and 17.

$$B_1 - B_3 \cong \Delta B_f \qquad \text{(Eq. 16)}$$

and

$$B_2 - B_3 \cong \Delta B_{sm}$$

and

$$\frac{H_f}{H_{sm}} \cong \frac{B_1 - B_3}{B_2 - B_3}$$

FIGURE 65. Diagram of Inclined Plane Parallax Test Situation

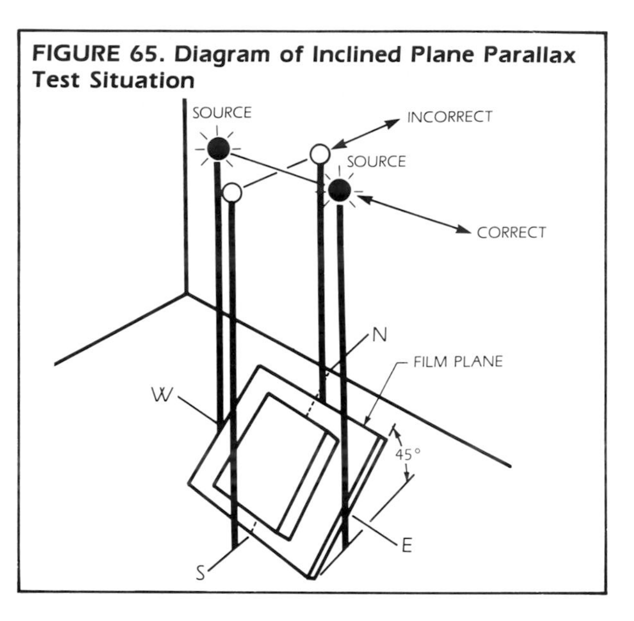

FIGURE 66. Incorrect Source Shifts

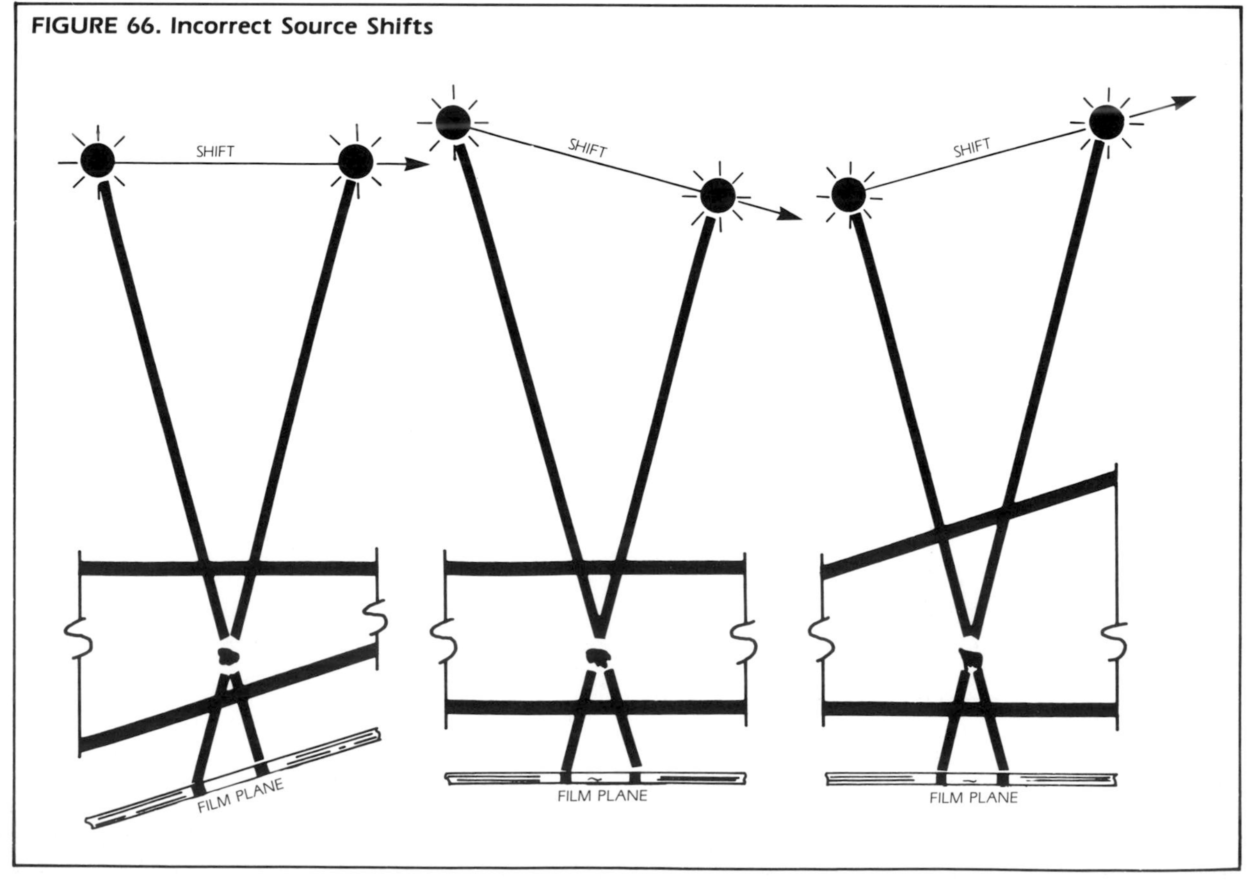

TABLE 3. Triangulation Measurement Requirements

Formula	Flaw and Marker Shifts (B)	Source to Film Distance (T)	Source Shift (A)	Film Separation (K)	General Application Rules
Rigid formula	Yes	Yes	Yes	Yes	1. For relatively short source-to-film distances or where marker placement is difficult. 2. Where part thickness is unknown or difficult to measure.
Approximate formula: source-side marker	Yes	No	No	Yes	1. Also requires the D_2 (part thickness) plus (K) be known. 2. For relatively long distances. 3. For situations where film-side marker placement is difficult.
Approximate formula: source-side and film-side markers	Yes	No	No	No	1. Also requires that H_{sm} (part thickness) be known. 2. Most accurate approximate formula. 3. Best for long source-to-film distances. 4. Simplifies data retrieval.

$$\frac{H_f}{H_{sm}} \cong \frac{\Delta B_f}{\Delta B_{sm}} \qquad \text{(Eq.17)}$$

or

$$H_f \cong H_{sm} \times \frac{\Delta B_f}{\Delta B_{sm}}$$

where H_f is the height of the flaw above the film-side marker; and H_{sm} is the distance between the source-side marker and the film-side marker.

Listed in Table 3 are the various parallax formulas, the triangulation measurement requirements, and the general areas of application for the double-marker, single-marker and rigid formula parallax methods.

Effects of Flaw Geometry on Parallax Accuracy

The effect of flaw geometry on the accuracy of parallax calculations is common to all three methods. Figure 62 illustrates the classic case of flaw geometry found in radiographic parallax methods. Calculations typically indicate the center line dimension of the flaw above the film plane. However, in those cases where the geometry of the flaw in not cylindrical or rectilinear, its shape can influence the accuracy and/or detectability of flaws. If the general shape of the flaw can be determined by viewing a standard radiograph, proper allowances can be made.

Figure 63 shows three cases where the approximate, average displacement of the flaw on the film can be calculated, using eq. 18.

$$\text{parallax shift} = \frac{L + R}{2} \qquad \text{(Eq. 18)}$$

If the flaw geometry is similar to one of those shown in Fig. 64, averaging the flaw shift does not show the true flaw dimension location.

Source Movement

Figure 65 illustrates correct and incorrect source movement for situations when the parallax method must be used on an inclined plane; note that the correct movement is parallel to the film plane. This principle is further illustrated in Fig. 66, which shows incorrect source movements and source-to-film distance changes for a flat object or a tapered surface.

Figure 67 illustrates that, when the source position is not perpendicular to the film plane, the approximate angle of the film plane (to the source) must be known and compensated for in the calculations, even with correct source movement.

Care must also be exercised when using the parallax method of flaw depth determination on cylindrical parts. It is important to maintain the source-to-film-plane angle as closely as possible to 90°, particularly in those cases where flexible film cassettes are used. This is also true when radiography of cylindrical parts is made using a rigid cassette. The radiographer must be aware of, and maintain to the maximum extent possible, a normal relationship of the film plane to source. Additionally, when rigid cassettes are used, the separation distance between the rigid cassettes and the inside or outside of the cylindrical object must be known and compensated for in the calculations.

FIGURE 67. Comparison of Part Thickness in Horizontal and Inclined Plane Tests

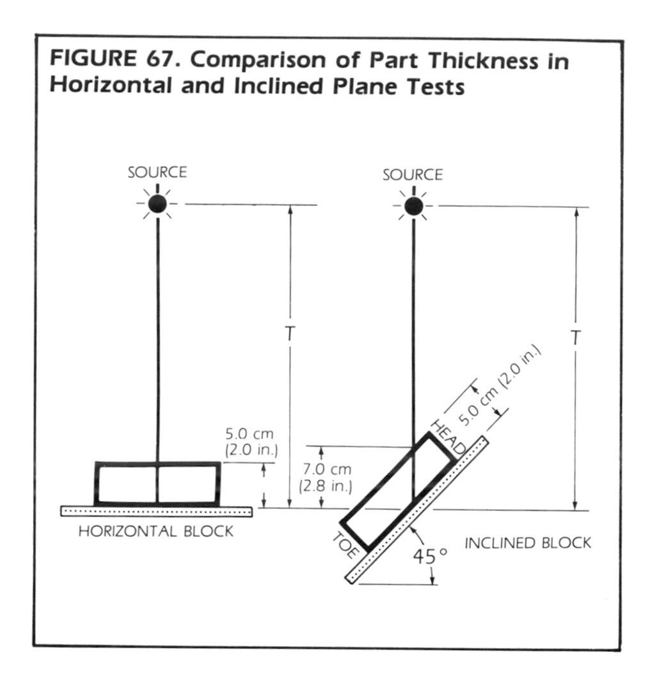

PART 10

THICKNESS MEASUREMENT RADIOGRAPHY

Measurement by radiography is an excellent tool for the determination of thickness when the configuration of an object permits its use. The radiographic technique is not applicable in all cases, but there are many instances when radiographic methods are suitable, such as: (1) when dimensional irregularities or thicknesses in specimens need to be measured, but are inaccessible to the usual measuring devices (gages or calipers); (2) for measuring defects within specimens; (3) when a more efficient NDT method is not available without large, unwarranted expense; and (4) when precise dimensions are not required.

It should be understood that a radiographic method might not be the best technique for making a particular measurement; its use has to be evaluated by weighing the required accuracy against the possible accuracy, and considering all alternatives.

Measurement by radiography can be categorized under two general methods: direct and indirect, the indirect category consisting of many techniques. Direct and indirect measurement will each be covered in the following discussions.

Direct Thickness Measurement

Making a direct thickness measurement from a radiographic image is a useful application of radiography. The procedure involves radiographing an object with the area of interest imaged so that the measurement can be made directly on the film. The process of measuring may be accomplished using a scale, calipers, divider or a measuring magnifier (pocket comparator). Factors affecting the image must be closely controlled to obtain optimum results. These include: geometric factors; scattered radiation; and burnout, due to thinness of the object at boundaries that are involved in the measurement. Objects that lend themselves to the greatest accuracy in direct measurments are those having a curved wall or similar configuration. Cylindrical items such as tubing can be radiographed so that the cross section of the wall is imaged on the film and a measurement may be made by using a pocket comparator.

Those objects best suited for direct measurement are those in which the areas of interest can be oriented very close to the film, so that there is no significant distortion of the image. Fine grained films are also an aid to this method.

The evaluation of defects can be done by the direct method if the geometry of the specimen permits it; radiographic views must be made perpendicular to each other in order to derive three dimensional measurements of the defects. These views not only allow the thickness and shape of defects to be determined, but also allow location of the defect within the specimen.

Indirect Thickness Measurement

If the direct thickness measurement technique cannot be applied, the indirect methods should be considered. The indirect methods are based on the relationship between the thickness of a material being radiographed and the photographic density in the resultant radiograph.

A radiographic image is made up of variations in photographic density that correspond to thickness and/or material density differences in the object being examined. If material density variations are not present, then the photographic density/thickness phenomenon can be utilized to make thickness measurements. There are two general approaches to indirect thickness measurements: (1) densitometric comparison; and (2) visual comparison.

Densitometric Comparison

The densitometric method is best applied to circumstances where relatively large areas of uniform thickness are imaged on film, so that an area of uniform photographic density, at least slightly larger than the aperture of the densitometer, is available for density measurement.

The densitometric comparison procedure begins by producing a radiograph of the specimen and a suitable stepped wedge. A calibration curve is constructed, based on the photographic densities obtained by densitometer from the radiographic image of the wedge's various steps (Fig. 68). The curve is drawn by plotting the photographic density of each step versus the thickness of the step, for the range of thickness desired.

The curve can then be applied to an object with unknown thickness (or thicknesses) by determining the photographic density for the radiograph's area of interest. The density is then located on the calibration curve to obtain the corresponding thickness.

There are several factors that affect the accuracy of this method and it is important to understand what the factors are, and what controls are required in order to obtain the best results from this procedure. To avoid errors that can be introduced by scattered radiation, the stepped wedge should be as wide as possible so that photographic density measurements can be made away from the edge, avoiding significant scatter influence on readings. If it is necessary to use a narrow stepped wedge or if the specimen is narrow, lead masking should be used to avoid errors caused by scatter.

FIGURE 68. Densitometer Reading of the Step Wedge

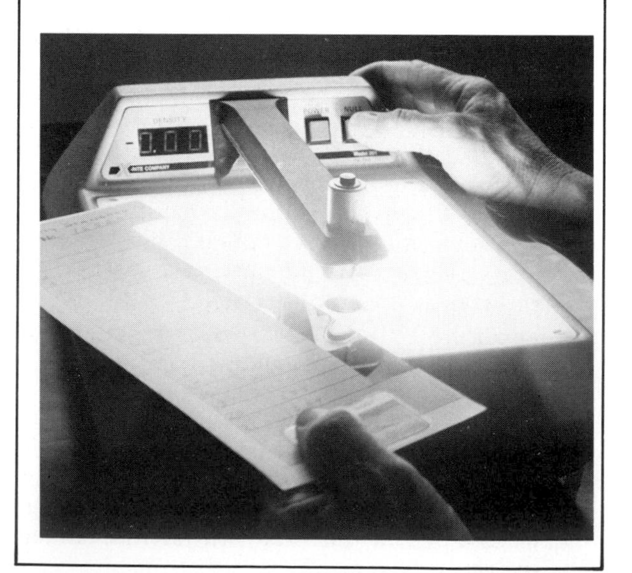

In addition, the stepped wedge must be made of the same material as the measurement specimen. The wedge should also cover the range of thicknesses to be encountered in the specimen.

The data for the calibration curve is best obtained by radiographing the stepped wedge and specimen at the same time and on the same film. If separate film is necessary, then the technique should be identical

Visual Comparison

The densitometric technique works well for relatively large areas of uniform thickness, but when necessary to establish the defect's thickness (dimension in the direction of the beam), a visual comparative technique may be used.

Visual comparison techniques are best suited for the determination of (1) flaw thicknesses, and (2) thickness variations occuring in a manner that makes the use of a densitometer impractical (narrow bands of thinning or thickening, adjacent to or within welds, for example).

Visual comparison techniques include: (1) using a penetrameter of appropriate thickness; (2) using a specially fabricated standard (a plate with slots or flat-bottomed holes) or a stepped wedge; and (3) using a calibrated visual reference standard (CVRS).

To evaluate the thickness of a flaw using a penetrameter, the procedure is to radiograph the specimen with the penetrameter so that the flaw image falls within the boundary of the penetrameter image, but does not overlap the image of the penetrameter hole. The images of the flaw and hole are juxtaposed and the photographic densities of the two can be visually compared. If the highest density noted in the flaw is equivalent (visually) to the density of the hole image, then the maximum flaw thickness is the same as the penetrameter thickness.

To determine the thickness of a flaw using a standard, the procedure is to first fabricate a standard using material of the same type and properties as the specimen, with a thickness determined by the maximum flaw thickness of interest. A series of holes is drilled in the standard; one hole is through the plate and the remainder are flat-bottomed holes of varying depths as required for the circumstances.

Radiograph the standard using the same technique as used for the specimen. The radiograph of the standard is then placed over the radiograph of the specimen. The images of the varying flat-bottomed holes are then moved adjacent to the defect image until the thickness is the same as the flat-bottomed hole depth.

FIGURE 69. Radiographic Film Standard

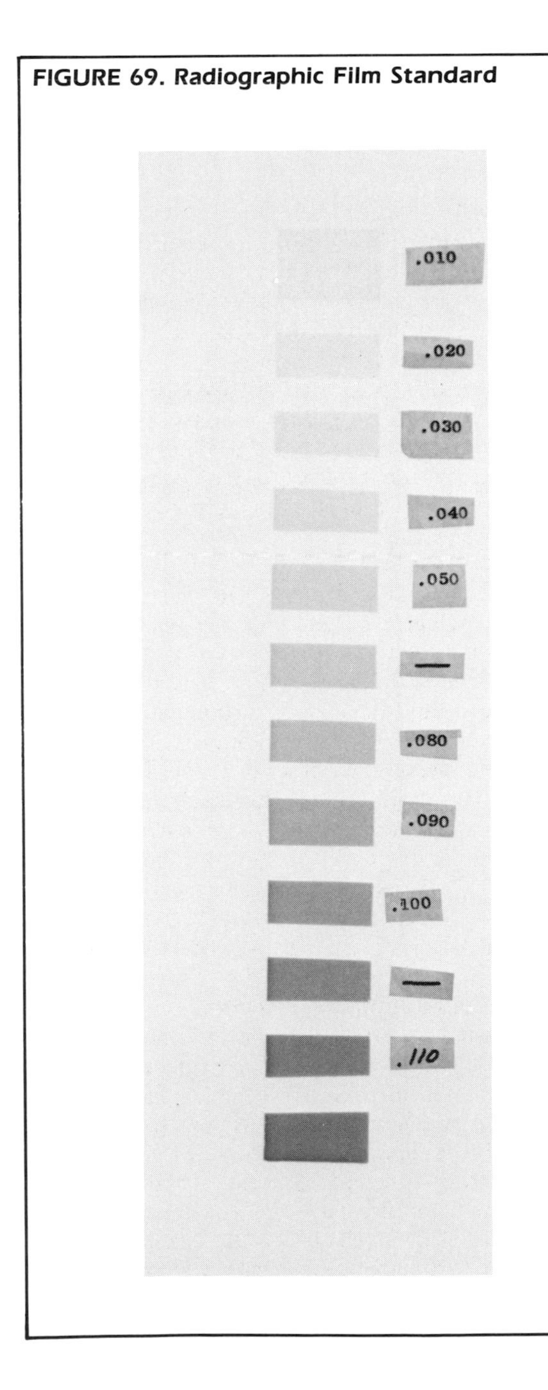

Calibrated Visual Reference Standard Method

The calibrated visual reference standard (CVRS) method provides a way of evaluating the third dimension of defects with reasonable field results even when procedures are not closely controlled. The method uses film with strip densities of known thickness differences (see Fig. 69) superimposed next to the defect. Various strips are used to compensate for film density, kilovoltage and, to a limited extent, processing variations. If the field radiographic conditions are accurately duplicated, accurate measurements are possible.

The visual comparative method is not significantly influenced by location of the defect region within the material or the thickness range of the film standard.

The development of the technique for defect thickness determination begins with the preparation of comparative radiographic film standards. The preparation of standards includes the following steps.

1. The same material type, composition and thickness as the material to be evaluated is fabricated into a thickness standard.
2. Flat-bottomed holes, slots or steps (see Fig. 70) are machined at various depths, starting at the least defect thickness of concern, in known increments such as 0.25 mm (0.01 in.) to the maximum thickness anticipated.
3. This standard is then radiographed, applying the same parameters used when radiographing the material containing the defects. These parameters include kilovoltage or type of source, film, screens, processing, and the film density. Radiographic film density difference is then determined for the parent material and the machined defect thickness of the standard. Any variations greater than ±0.25 density units between the material to be evaluated and the parent material of the standard should not be allowed. Variations in density of ±0.25 do not significantly bias results and reduce the number of reference standards required to a reasonable level. The contrast sensitivity changes with position on the characteristic curve and this change will cause an error in depth evaluation which can be either less than or greater than the actual measurement of the defect.

4. Radiographic film is then exposed to light to create a density step wedge on a clear background. The density variations on the light-exposed film must be equal to the variations noted between parent material of the standard and the known thicknesses of the stepped wedge. Light is used to maintain minimum film density background, so that film can be viewed with normal viewing equipment. It is also important to have $\Delta D/D$ as large as possible for any given thickness variation; D equals the total background density when the standard is superimposed on the film to be evaluated and ΔD equals the increase in density due to a flaw.

Upon completion of the radiographic film standard, a certain amount of practice is required for proficiency in this method.

The film reader tries matching the radiographic film standard against defect images of known thickness to ensure that the film standard is correct and to establish familiarity with this process.

The reader then evaluates the defect thickness of slots and holes which are unknown to him. This type of training continues until a satisfactory confidence level is attained. The training for a specific program can be expedited by limiting the practice to materials, thickness, geometry, and defect types of interest.

Sample specimens containing defects are evaluated for defect thickness. The specimens are then sectioned and the results are reviewed to calculate bias and accuracy limits. The bias and limits of accuracy are then established by qualified personnel.

The determination of defect thickness is performed in the following manner:

1. View the film of the defect using an illuminator with a range of at least 4.0 radiographic density (see Fig. 71).
2. Place the film standard over the film with the density steps adjacent to the defect image.
3. Manipulate the film standard until the density of the defect and the density of the film standard appear to be of equal magnitude. A reduction of the density step area to approximate the defect size can sometimes be helpful in reducing the optical illusion present when visually comparing the photographic density of two different size areas (see Fig. 72).
4. The depth is determined by comparing various segments of the film standard with the defect and then selecting the closest density match.
5. This measurement is recorded and the correction factor and/or bias is applied to provide the corrected depth thickness.

FIGURE 70. Depth Step Wedge

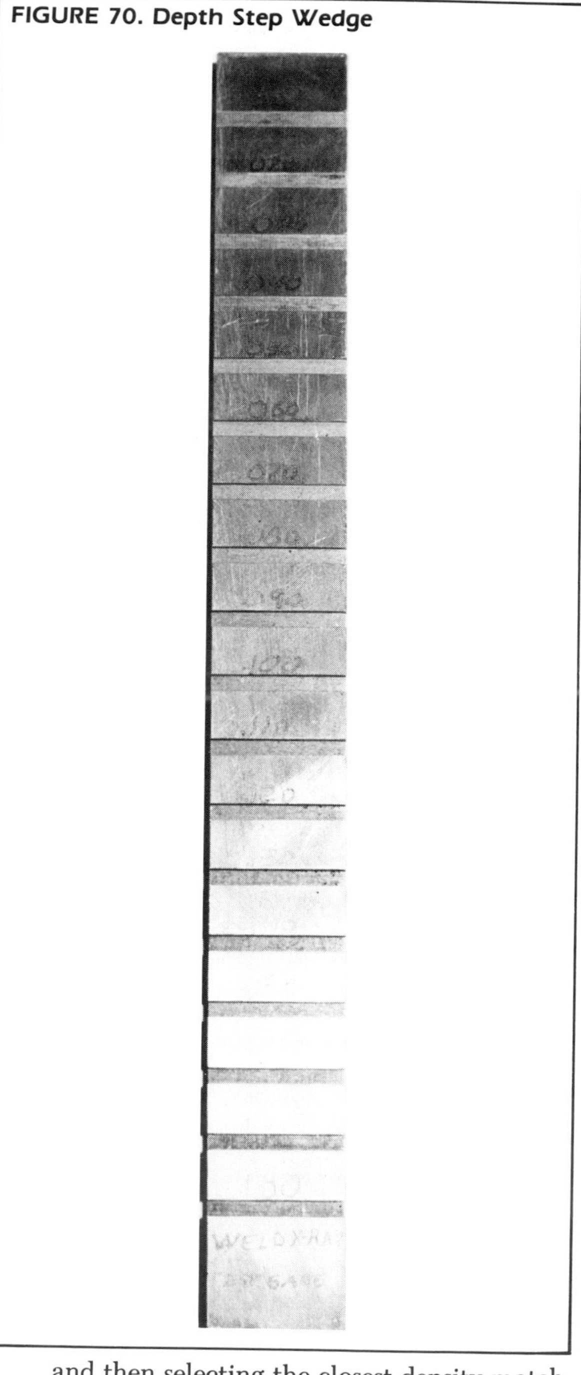

Care must be taken to ensure the proper film density standard is used. When a specimen has wide variations in thickness, use care to determine film density where the maximum thickness difference occurs between the defect and the parent material.

FIGURE 71. Weld with Number 12 Penetrameter

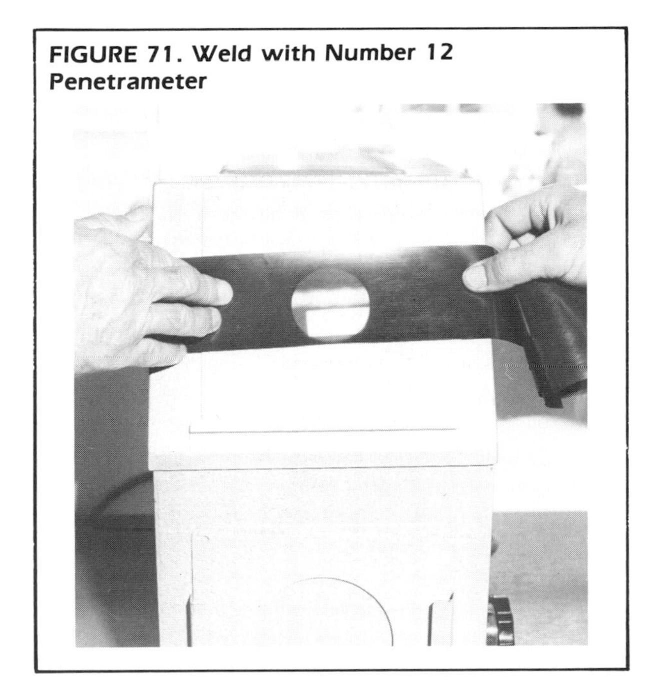

FIGURE 72. Weld with Radiographic Film Standard

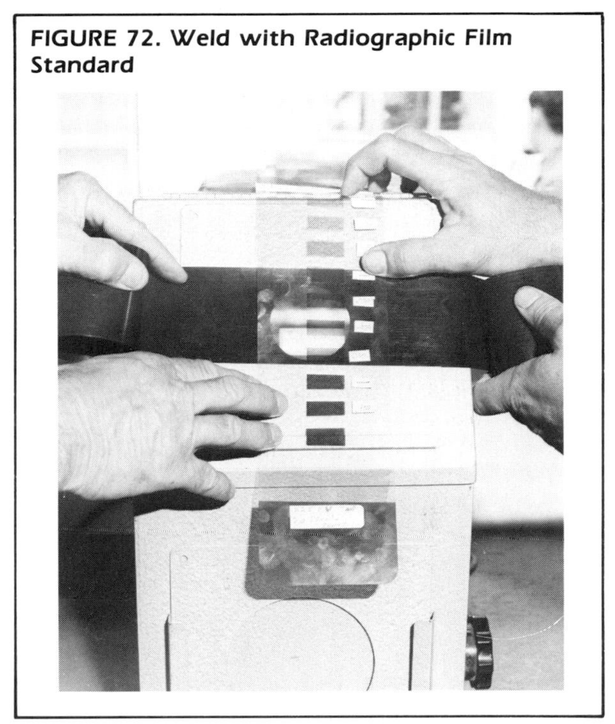

PART 11

ROBOTIZED X-RAY INSPECTION STATION

Development of a 6-axes, jointed robot for real-time X-ray inspection applications comprises a two-phase program.

Phase 1 of the program has two objectives: to automate the existing X-ray inspection process; and to study the effects of long-term radiation exposure on the robot optical encoders.

Inspection Station

The robotized X-ray inspection station (Fig. 73) is laid out so that the X-ray controller, the robot computer/controller, and the visual monitoring system are located outside the shielded X-ray room (see Figs. 74 and 75).

FIGURE 73. Robotized X-ray Inspection Station: (a) Product Palletizing Fixture; (b) Serial Identification Fixture; (c) Mechanized Film Storage Fixture; and (d) Mechanized Flexible Exposure Fixture

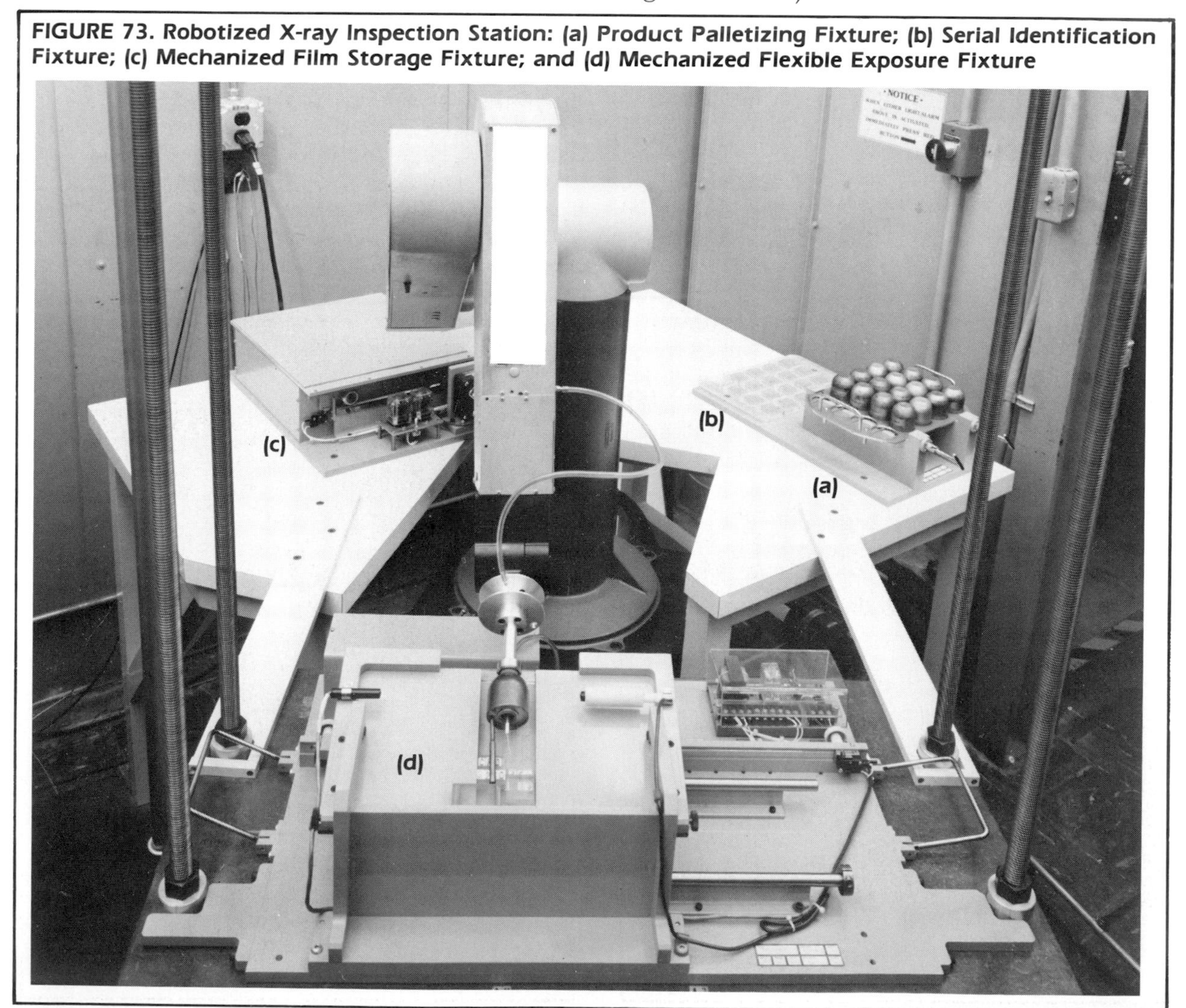

The radiographic source, the robot and its tooling are located inside the shielded X-ray room. The robotized station consists of a product palletizing fixture, a serial identification fixture, a mechanized film storage fixture, and a mechanized flexible exposure fixture. All tools and fixtures were custom designed and fabricated.

The jointed-arm robot was chosen for its close tolerance repeatability, application software and flexibility of motions. It has six revolute axes and is electrically driven by DC servos.

The robot computer/controller has 16K memory and is interfaced with (1) a floppy disk drive for external program storage and (2) an input/output (I/O) module consisting of eight input and eight output signals used for process control. The robot controller is also interfaced with the X-ray controller, which has been retrofitted for automatic operation.

The robot uses two different custom designs for end-effector tooling. The first, shown in the center foreground of Fig. 73, consists of a single-function vacuum tool, used for picking up film packets, serial identification carriers and light-weight products. The second is a two-function tool comprising a vacuum tip and a gripper for heavier products. Both are quick-disconnect assemblies and use a pressure/vacuum sensor to monitor the presence of the product at the tool.

The film storage fixture is a mechanically driven, lead-shielded container with two storage areas. One area has a self-indexing mechanism for the unexposed film packets. The other area is used for storage of exposed film. The cover is driven by a DC servo motor and is interlocked with the X-ray controller to prevent unwanted exposure of the film.

The flexible exposure fixture is a mechanized assembly having adjustable angles of exposure and adjustable shields for exposure width. The fixture has a sliding tray that holds the film packet for exposure. The exposure window holds the serial identification and date carriers. In addition to limit switches used to control the travel of the sliding tray, the fixture has a photoelectric sensor for detecting and monitoring the presence of the product in the window. Should the product not be present, the X-ray controller cannot cycle.

In order to cover a working horizontal envelope greater than 330 degrees during the robotized process, the robot converts from a right-handed configuration to a left-handed configuration and vice

FIGURE 74. X-ray Controller, Robot Computer/Controller and Visual Monitoring System Located Outside Shielded X-ray Room

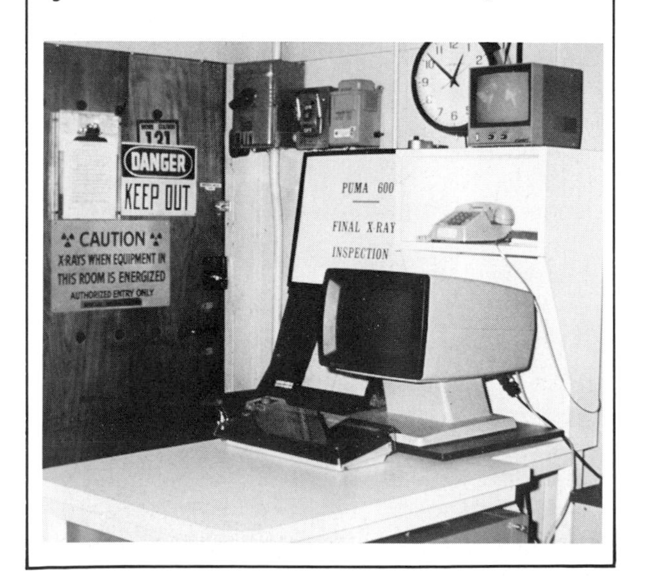

FIGURE 75. Positioning of Diskette for Loading Robot Application Program

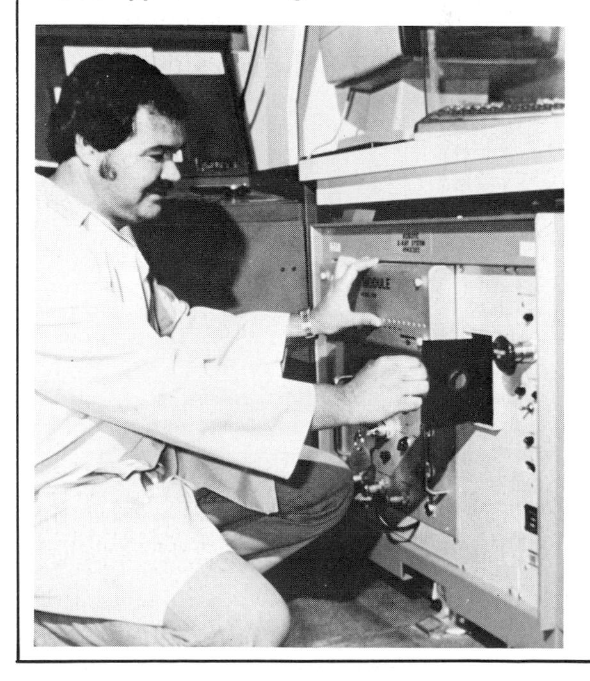

FIGURE 76. Unexposed Film Packet Being Positioned on the Film Exposure Fixture

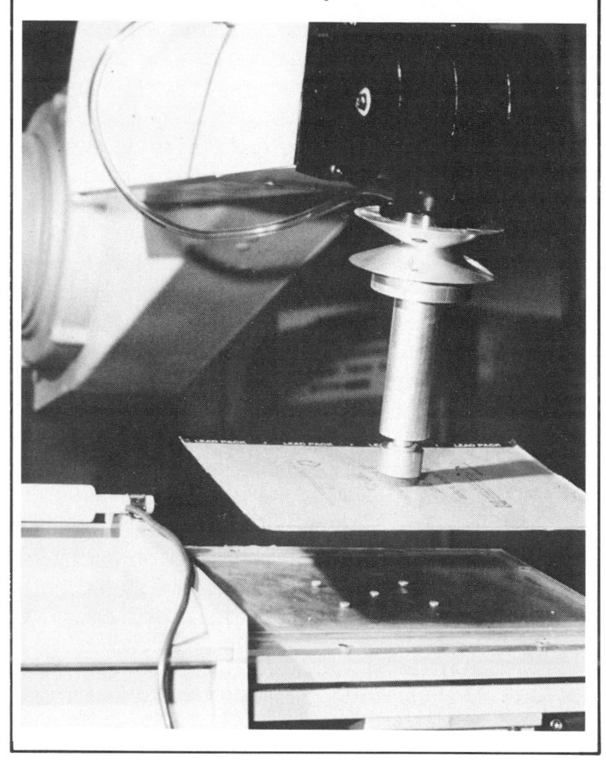

versa through programming manipulation. The differences are shown in Figs. 73 and 76; Fig. 73 depicts a robot right shoulder and Fig. 76 shows a robot left shoulder.

A new application program is developed for each product and is identified with control numbers referenced on the operator CRT.

Robotized Process

To initiate the automated process, the operator first responds to controller inquiries on the application and product, then inserts the appropriate diskette and depresses an auto-start button. The calibration of the robot, loading of the program from the diskette, and execution of the process is automatically performed (see Fig. 75).

Initially, the film storage fixture is opened. The robot picks up an unexposed film packet and places it on the film exposure fixture (Fig. 76). Vacuum on the sliding tray prevents the film packet from falling off. The film packet is then moved under the lead shield, where one section of the film is shown in the window. The robot picks up a serial number for the product to be radiographed and places it in the exposure window of the film exposure fixture (see Fig. 77). The robot then lifts the product, orients it and places it in the exposure window over a section of the film. The robot controller, after satisfying the interlock signals from the film storage fixture, the end-effector and the photoelectric sensor, initiates the exposure cycle.

FIGURE 77. Positioning of Serial Number on Exposure Window

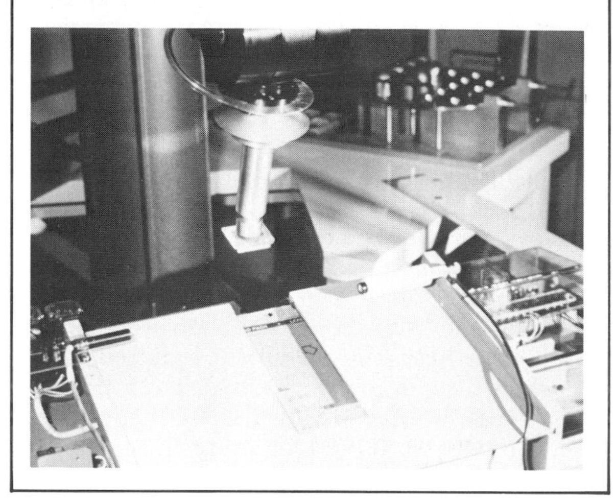

For each additional exposure, the product is reoriented and the film is indexed to an unexposed area. When the required exposures have been completed, the robot returns the product, the serial identification and the exposed film to their appropriate locations. The process is repeated according to the lot size.

Benefits

The robotized station provides a safer environment for X-ray inspection. A repetitive job is performed by the robot, leaving the operator to perform more meaningful functions such as reading and evaluating films for product defects. In addition, as a productivity improvement, the robotized inspection station is a better controlled process.

PART 12

HANDLING SYSTEMS FOR REAL-TIME RADIOGRAPHY

Introduction

A real-time radiography (RTR) system contains two main components: the X-ray imaging system and the handling system. A part to be inspected is positioned in the imaging system's X-ray beam; the handling system is the means used to position and support the part. The handling system may be as simple as a tray, or as complex as an automated, six-axis, computerized manipulator. The degree of complexity depends upon the application.

A handling system can also contain two main components: a manipulator and an automated loader. The automated loader moves the parts, in a prescribed way, to and from the manipulator. The manipulator controls the motion of the part within the X-ray beam.

Economic Considerations

The handling system is an important economic factor in any RTR system. First, the cost of the handling system may be a large percentage of the system's overall cost. Secondly, the throughput (number of parts inspected per unit time) is dependent upon the design of the handling system and

FIGURE 78. Schematic Illustration of a Class 1 Handling System

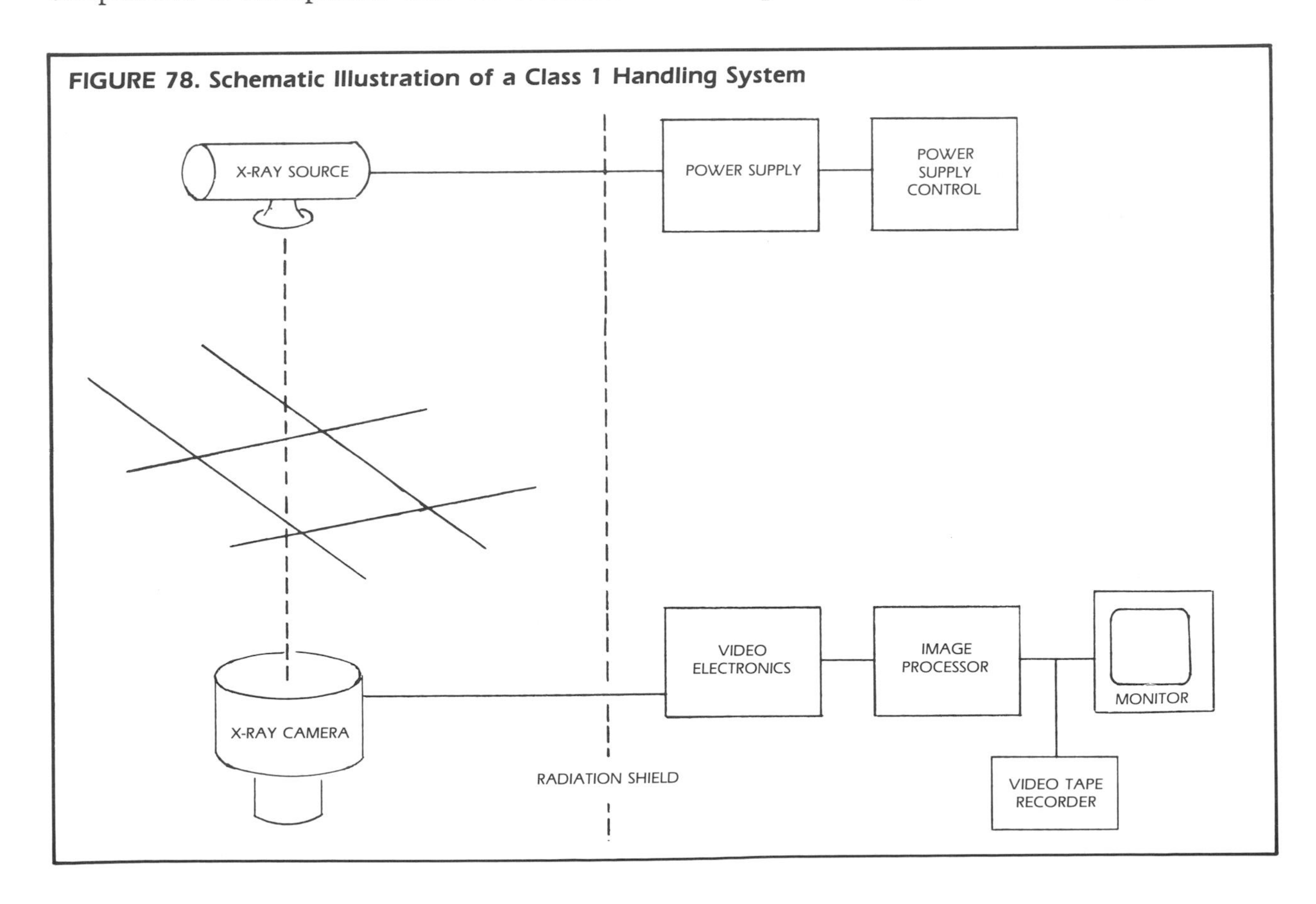

throughput determines the cost effectiveness of a real-time radiography investment. Also, the daily cost of operation is partly determined by the number of people required to operate the system and staffing is in turn affected by the design of the handling system.

Another aspect of the handling system's economic impact is safety and product damage. Any handling system that damages the products, or is unsafe, ultimately results in added cost.

Handling System Classification

There are basically three categories or classifications of handling systems as determined by complexity.

FIGURE 79. Class 1 Handling System Used to Inspect Space Shuttle Tiles; System's Computer Is Capable of Making Accept/Reject Decisions

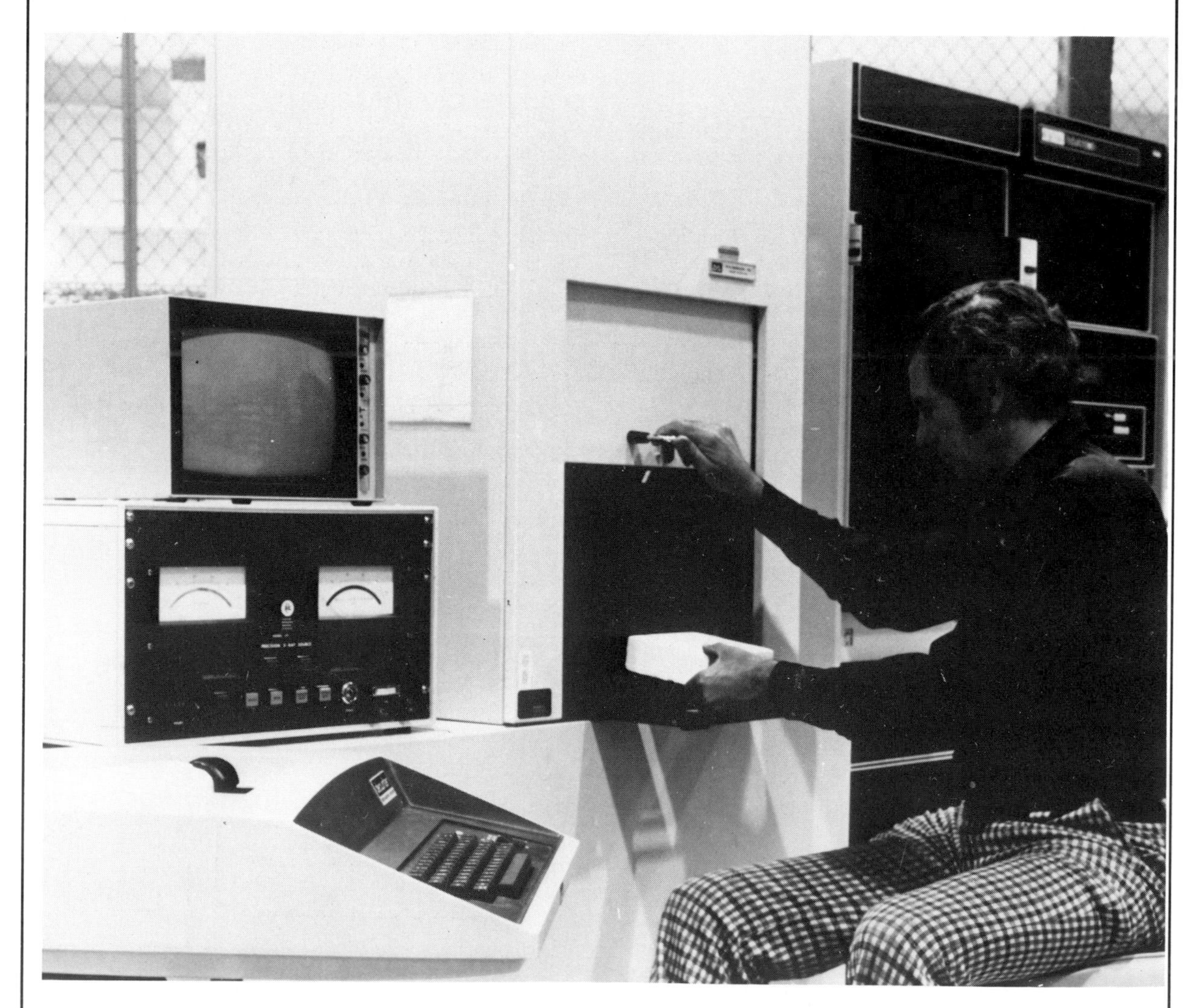

FROM THE LOCKHEED MISSLES AND SPACE COMPANY. REPRINTED WITH PERMISSION.

A *class 1* handling system is static — the part does not move during inspection. In a *class 2* handling system, the radiographic object moves but is controlled manually by the operator. A *class 3* handling system offers completely automated control of the object.

Figure 78 illustrates a *class 1* handling system in schematic form. On the left side of the figure is the radiation enclosure containing an X-ray source and X-ray imaging camera. The X-ray source is connected to its power supply, and the X-ray camera is connected to the real-time radiography imaging system, in the usual way. The cross-hatched lines between the X-ray source and the X-ray camera represent the handling system.

With a *class 1* system, the operator simply opens a door in the radiation shield, places the part in the X-ray path, closes the door, and then turns on the X-rays, thus producing the real-time radiographic image. This system is often used as an alternative to conventional radiographic film. It is most appropriate for applications where low throughput is acceptable and no part manipulation is required.

Figure 79 shows a *class 1* system in use. Another application of this system might include a conventional exempt enclosure with a part tray and image amplifier tube (an exempt enclosure has a radiation level in its vicinity so low that workers are not required to wear film badges).

A *class 2* handling system is illustrated in Fig. 80. The *class 2* system contains a manipulator that moves the part during inspection. The manipulator may have up to three axes of translation and three axes of rotation. There may be an automated loader associated with this type system, but the operator manually controls the movement of each part during the inspection process. A *class 2* handling system is appropriate for applications where high throughput is not required, but where part manipulation during inspection is a necessity. Figure 81 shows a *class 2* handling system in use.

Figure 82 illustrates a *class 3* handling system in schematic form. The heart of the *class 3* system is the microprocessor, which links the imaging system controls, the X-ray source power supply controls and the handling system controls. The microprocessor

FIGURE 80. Schematic Illustration of a Class 2 Handling System

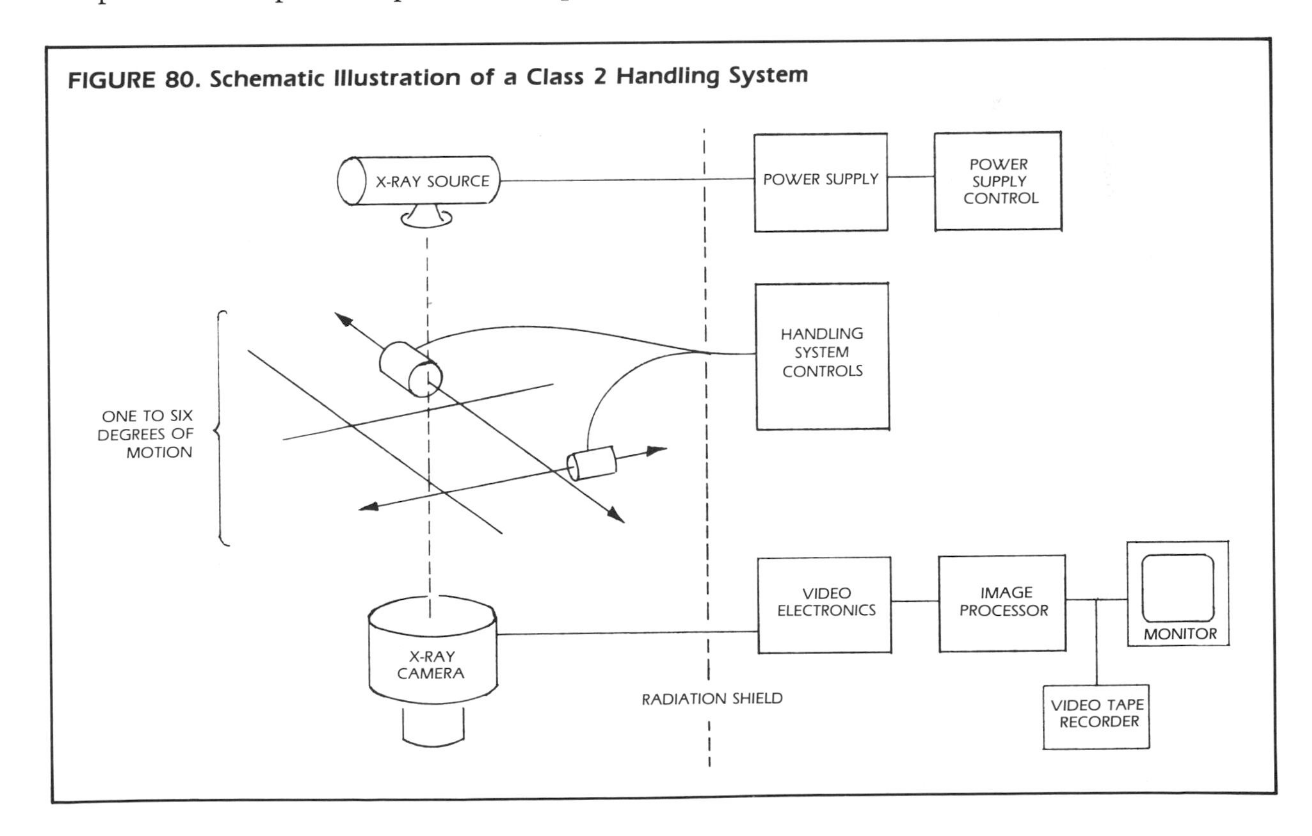

issues orders electronically to each of these units and causes them to act. Operators can interact with the microprocessor through a terminal and data can be stored on a disk drive; hardcopy output can be obtained on a printer.

The microprocessor controls the X-ray kilovoltage, the milliamperage, the image processor and the video tape recorder. The *class 3* handling system is distinguished by its automation. It is appropriate for applications where both high throughput and part manipulation are required. Figure 83 shows some components of a *class 3* handling system.

FIGURE 81. Class 2 Handling System Used to Inspect Rocket Motors; the High-energy X-ray Source (at right) and the Real-time X-ray Camera (at left) are Positioned Manually in the Horizontal and Vertical Directions; Movement of the Motor is Manually Controlled during Inspection

FROM HERCULES, INC.

Handling System Options

Certain options are available with the different classes of handling systems. In a *class 1* system, the options are limited. An exempt or a walk-in radiation enclosure may be used. A load/unload indicator (pressure sensor indicating the presence or absence of a part) may also be employed.

In the *class 2* system, position coordinates are displayed on the main monitor, or a closed circuit TV may be used for observation of the part's motion. The automated loader is also a *class 2* option.

Class 3 systems offer the most options, including automated motion drives. An automated motion drive can, for example, provide a helical scan of an object at the push of a button. Computer-generated inspection reports are also possible with *class 3* handling hardware.

The *class 3* system often includes automated image processing (providing integration or subtraction, edge enhancement or any other features available in the image processor). The operator can retrieve from the system's memory a complete list of instructions or inspection requirements for each type of test object. Automated sequence data (scan plans) may be stored on the system and recalled for use with specific parts.

Scan Plans

Figure 84 is an example of a scan plan. Scan plans are used when (1) the field of view is smaller than the object being inspected, or (2) when many samples are inspected at one time.

In this example, the unit to be inspected is a rocket motor 1.5 m (5 ft) long and 150 mm (6 in.) in diameter. A two-plane inspection is required. The field of view was chosen to meet the inspection requirements of the motor and is shown in the figure by dotted lines. This field of view allows scanning of about 150 mm (6 in.) of the motor at a time; the length of the motor has therefore been divided into 10 inspections stations.

FIGURE 82. Schematic Illustration of a Class 3 Handling System

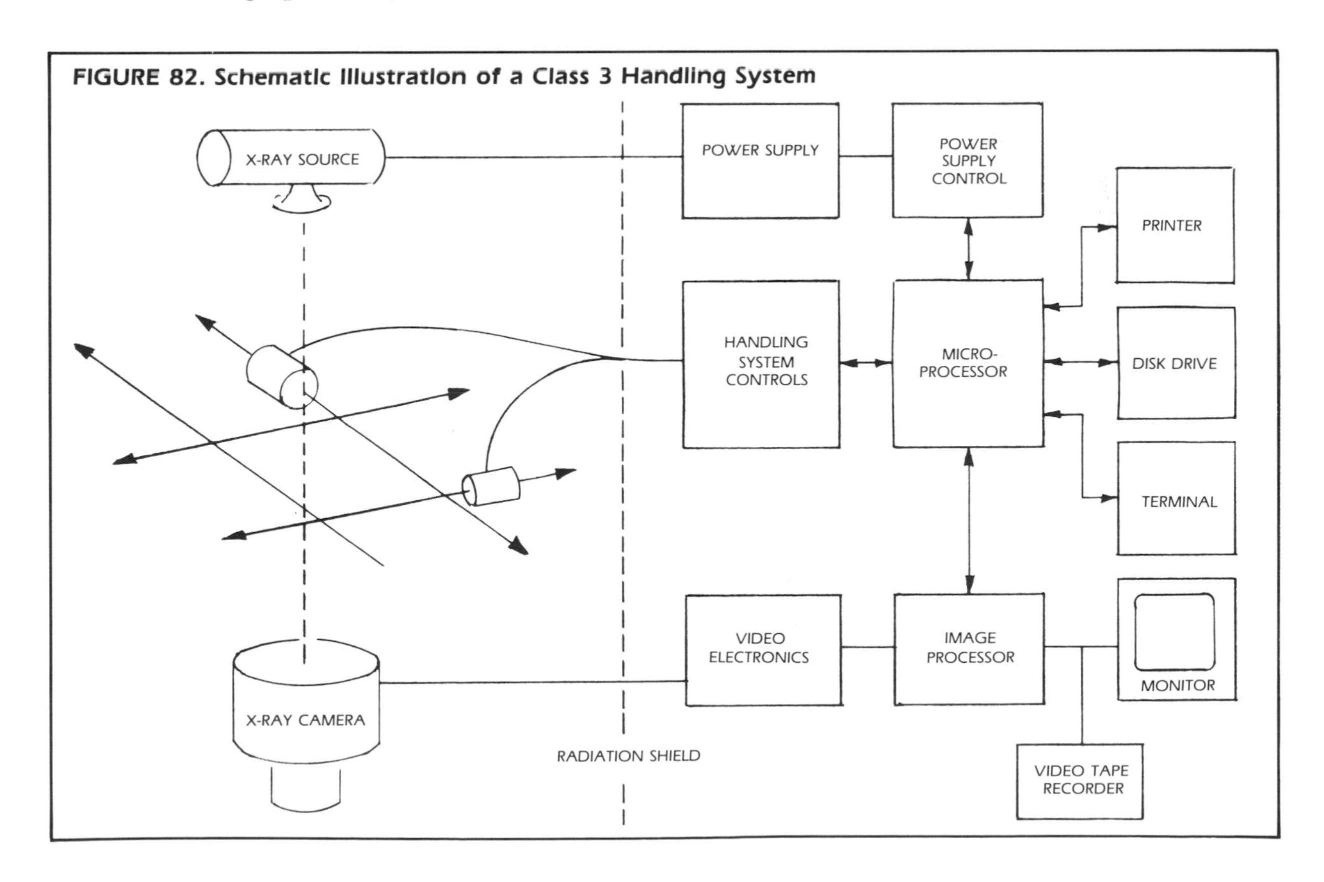

The scan begins when the system gives the signal to load the motor onto the manipulator. When the motor is loaded, as indicated by appropriate electronic signals, the system will turn on the X-ray generator and the video tape recorder. It then automatically moves the rocket motor to zero degrees and to inspection station 1. At this point, the motion stops and the microprocessor signals the image processor to integrate the first frame and transfer it to the image processor's refresh memory for viewing by the operator.

While the operator is viewing this image, the handling system automatically moves to station 2 and integrates the next image. When the operator is finished viewing image 1, image 2 will immediately be displayed. The imaging system, in the

FIGURE 83. Class 3 Handling System Used to Inspect Turbine Blades; Tray Is Positioned Automatically; X-ray Source Kilovoltage Changes Controlled by Computer; Computer Collects and Records the Accept/Reject Data on All Test Objects

FROM ROLLS ROYCE, INC.

meantime, will go on to produce a third image. In this way, the handling system is always one step ahead of the operator, preparing the next image for viewing.

If, during the sequencing of these images, the operator locates a discontinuity, the system can be switched from automatic mode to manual mode. At this time, object manipulations are performed manually while the image processor records all the images on video tape. When the system is returned to the automated mode, it returns to the point at which it left the scan plan and continues from that point.

After completing the ten stations in the zero degree plane, the handling system automatically rotates the rocket motor to the 90 degree plane and then traces back from station 10 to station 1, producing integrated images at each station.

The scan plan then calls for a helical real-time scan of the motor. This occurs automatically and allows the operator to inspect the motor for cracks which might not have been visible (due to orientation) in the 90 degree or the zero degree scan. The helical scan also allows complete inspection of the case-to-liner and the liner-to-propellant bonds. After completing the helical scan, the system automatically returns to the home position, turns off the X-ray generator and the video tape recorder, unloads the motor and lets the operator know that the next inspection may begin.

The scan plan is an important and flexible part of a *class* 3 handling system. If, for example, there are several different types of rocket motors to be inspected, a scan plan can be developed for each type. Each plan is stored on disk and called up by the operator as needed.

Handling System Design

There are four design considerations for RTR handling systems: (1) the inspection requirements; (2) the throughput requirements; (3) the labor requirements; and (4) the safety requirements.

Inspection Requirements

Before a handling system can be designed, it is necessary to identify the specific objects to be inspected. The inspection requirements for these objects must also be defined, including the types of defects to be detected and the required penetrameter sensitivity. These requirements are used to determine the inspection's field of view.

There is a fundamental relationship between field of view and the minimum defect size to be detected. A useful rule is that three or four television scan lines must cover the minimum defect dimension to ensure detection. The field of view is also important because it strongly affects the throughput that can be obtained with a real-time radiography system.

Next, it is necessary to specify how many degrees of motion are required for the inspection. It is important to realize that every degree of freedom (or every axis of movement) designed into a handling system represents one more degree of complexity and added cost to the system.

Additional considerations include: (1) the maximum scan speed which can be used successfully with a particular handling system; and (2) how much flexibility will be allowed the operator in determining scan speeds.

Determining Scan Speed

Following is an example of how to determine the maximum scan speed when a specific 2T hole must be detected. It has already been determined that the images viewed by the operator in a single frame-time must overlap if a given penetrameter hole is to be observed.

Assume that: (1) the 2T hole is to be observed and that it is 0.5 mm (0.02 in.) in diameter; (2) a frame-time of one-half second is used with 15 frame averages to improve sensitivity.

The maximum velocity that the object can travel is the distance of half the diameter of the penetrameter hole (0.25 mm [0.01 in.]) divided by the frame-time in seconds (0.5 s). This gives a maximum velocity of 0.5 mm (0.02 in.) per second.

Scanning parts too quickly can lessen sensitivity. The designer must also consider potential operator errors in scanning speed.

Throughput Requirements

It is essential to establish the number of inspections per unit time (per shift, for example). This is a critical value because it directly influences the payback period of the real-time radiography investment.

Often the throughput requirements are closely related to production rates. Keeping up with production is usually necessary to make the system a cost-effective investment. After establishing the throughput, it is necessary to consider the impact it will have on the handling system's complexity. This in turn determines the type of system that is needed.

It is then necessary to determine if the required throughput is compatible with the conceptual design of the entire handling system. It is recommended that a conservative estimate be used in establishing the inspection time. Include the time necessary to load the parts onto the automated handler, onto the manipulator or into the real-time system itself. Also include some time for maintenance and repair.

The designer must work back and forth, between the required throughput and the conceptual design, to arrive at the optimum system for the job, at the most reasonable cost.

Labor Requirements

Ideally, one person can load a real-time radiography handling system, do the inspection, and maintain the throughput requirements. Sometimes, however, it is necessary to have one person loading and another inspecting.

The operator-fatigue factor is also a consideration here. Experience shows that an operator can maintain complete concentration on a CRT screen for no more than 45 minutes. However, if the operator is not concentrating 100 percent of his time on the screen, but during the operation can take his eyes off the screen (to load a part or do some other task), this 45 minute period can be extended.

Another important consideration is involving operators in the conceptual design of the RTR handling system. They can contribute, sometimes more effectively than anyone, ideas that will improve the efficiency and cost-effectiveness of the design. Also, operators might fear that they are being replaced by the new system. This results from an incomplete understanding of what RTR is intended to do and

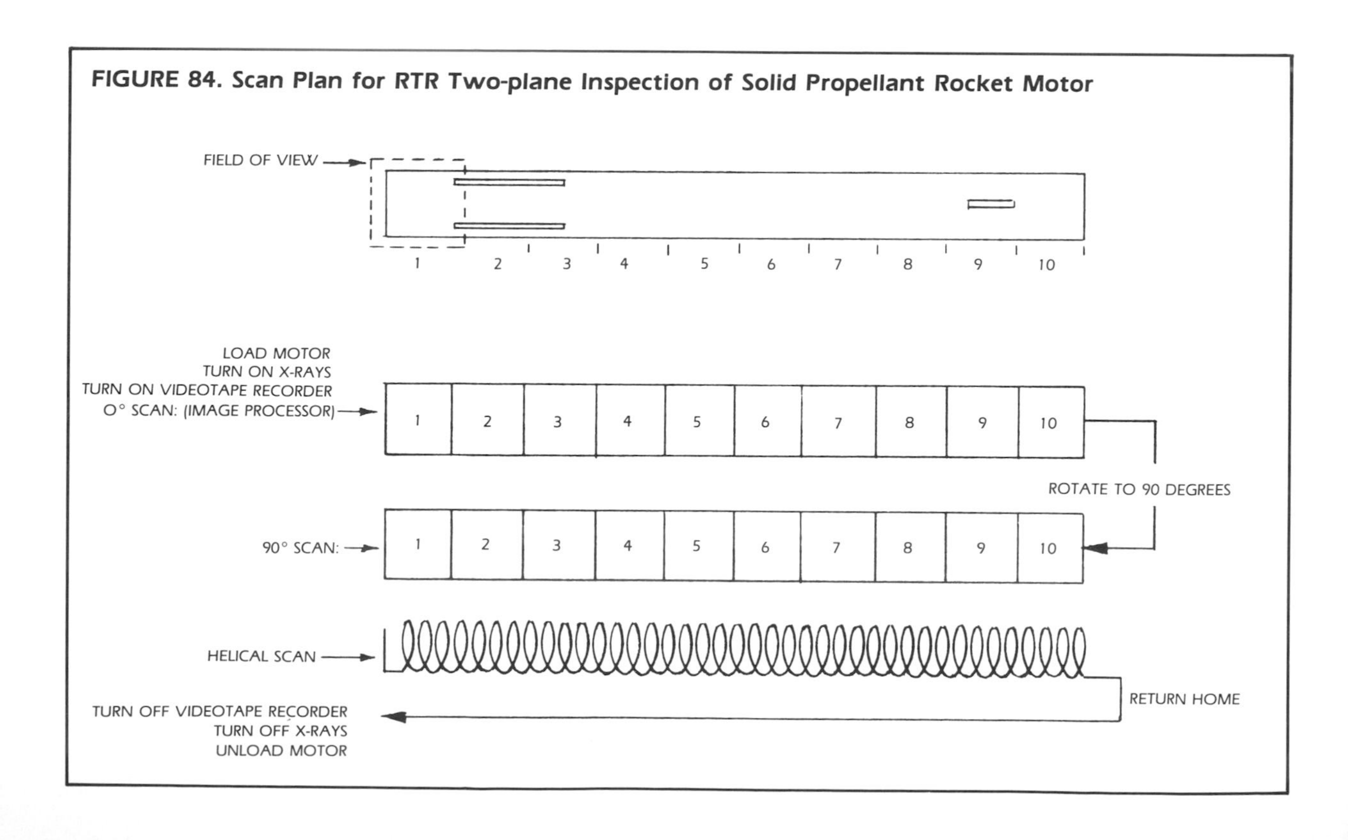

FIGURE 84. Scan Plan for RTR Two-plane Inspection of Solid Propellant Rocket Motor

will be alleviated once the operator fully understands the capabilities of real-time radiography.

Safety Requirements

Because safety is a primary concern in any radiographic facility, it is essential that the radiation safety officer be included very early in the conceptual design phase, to monitor the safety requirements for personnel and for the system itself.

Class 3 handling systems present a new safety concern: the software. This is an area in which many safety officers have no training or experience, yet it is vital to any *class 3* system. It is important to remember that software design is a new and critical area for safety review.

Conclusions

The economics of the RTR system are greatly affected by the design of the handling system. A real-time handling system can be very simple, or very complex, and this is determined by the system's application. The design used to implement this application must include (1) the specific inspection requirements; (2) the required throughput; (3) the labor requirements; and (4) safety considerations. The design of the RTR handling system should be given the same consideration as image quality or any other aspect of a real-time radiography system.

REFERENCES

1. Houston, A.E. and P.A.E. Stewart. *Cineradiography with Continuous X-ray Sources.* Proceedings of the Twelfth Congress on High Speed Photography: pp 140-145.

2. Bracher, R.J. and A.E. Houston. *High Speed Cineradiography of Projectiles.* Proceedings of the Twelfth Congress on High Speed Photography: pp 532-537.

3. Trimble, J.J. and C.L. Aseltine. *Flash X-ray Cineradiography at 100,000 Frames per Second.* Proceedings of the Fifteenth Congress on High Speed Photography.

4. Elliott, D.E. and J.F. Torbert. *Radiographic Inspection of Irradiated Fuel Elements Using the 22 MeV Betatron.* Proceedings of the Sixteenth Conference on Remote Systems Technology (1868).

5. Tenney, G.H. "Radiography of Radioactive Heavy Metals." *Industrial Radiography and Non-Destructive Testing.* Vol. V, No. 4 (Spring 1947).

6. Elliott, D.E. and G.H. Tenney. "Radiography and Autoradiography of Plutonium." *The Metal Plutonium.* Chicago, IL: University of Chicago Press (1961).

7. Wagner, G.F., K.L. Bahl and G.H. Tenney "Autoradiography of Fuel Elements of the KIWI Reactor—An Experimental Nuclear Rocket Engine." *Materials Evaluation.* Vol. 23, No. 10 (October 1965): pp 502.

8. Phillips, C., et al. *In-Motion Parameters for Application in the Nondestructive Testing of Brazed Honeycomb Structures.* North American Aviation Report No. CD-3739 (June 1961).

9. Pulk, R.A. and C. Mitchell "Applications and Problems in Stroboradiography." *Materials Evaluation.* Volume 14, No. 5 (September 1956): pp 24.

10. Hitt, W.C. and D.J. Hagemaier, *Radiography of Weldments In-Motion.* ASTM STP278 (1959).

11. Hoppins, W.K. "In-Motion Radiography." *Precision Metal* (August 1968).

12. Bulban, E.J. "Fluoroscopic Inspection Assures Quality." *Aviation Week* & *Space Technology* (August 1979).

13. Elson, B.J. "X-ray System Inspects Rocket Motors." *Aviation Week* & *Space Technology* (August 1980).

14. Elliott, D.E. *Scanning Radiography of Fuel Element.* Proceedings of the ASNT National Conference (March 1980)..

15. McDaniel, G.A. *Recent Developments in High Output Microfocus X-ray Systems.* Proceedings of the ASNT Automated Nondestructive Testing Seminar (June 1983).

16. Fontin, L.A. and R.S. Peugeot. *An Operational 150 kV Microfocus Rod Anode X-ray System for Nondestructive Testing.* Proceedings of the 37th National ASNT Conference (October 1978): pp 229-232.

17. Ely, R.V. *Microfocal Radiography.* New York, NY: Academic Press (1980).

18. *Radiography in Modern Industry,* 3rd edition. Rochester, NY: Eastman Kodak Company (1969): pp 106-107.

19. *Nondestructive Testing Handbook,* 1st edition. Robert C. McMaster, ed. Columbus, Ohio: The American Society for Nondestructive Testing. Vol. 1, Sec. 20 (1959): pp 49.

BIBLIOGRAPHY

1. Trillat, J.J. "On the Photographic Action of Secondary Electrons Resulting from the Action of X-rays on Metals." Academy of Sciences, Paris. Vol. 216, No. 4 (1943): pp 179-181.
2. Tasker, H.S. and S.W. Towers. "Electron Radiography Using Secondary Beta Radiation from Lead Intensifying Screens." *Nature*. Vol. 156, No. 50 (July 1945).
3. Trillat, J.J. "Electron Radiography and Microradiography." *Journal of Applied Physics*. Vol. 19 (September 1948): pp 844-52.
4. Seemann, H.E. Comments on paper listed as item 3 above. *Journal of Applied Physics*. Vol. 20 (February 1949): pp 231.
5. Berman, A.I. *Electron Radiography*. Atomic Energy Commission Report AECU-1853. Los Alamos, NM: Los Alamos Scientific Laboratory (December 1950).
6. McGonnagle, W.J. *Nondestructive Testing*. McGraw-Hill Book Company, Inc. (1961) pp 168-170.
7. *Radiography in Modern Industry*, fourth edition. R.A. Quinn and C.C. Sigl, eds. Rochester, NY: Eastman Kodak Company (1980).
8. *Consideration of Fracture Mechanics Analysis and Defect Dimension Measurement Assessment for the Trans-Alaska Oil Pipeline Girth Welds*. Berger, H. and J.H. Smith, eds. Washington, DC: National Bureau of Standards. NBSIR 76-1154, Vol. 1 (1976): pp 80-88.
9. Landolt, J.F., W.D. Stump and J.L. Summers. "A Visual Comparative Method for Radiographic Determination of Defect Thickness." *Materials Evaluation*. Vol. 36, No. 11 (October 1978): pp 33.

SECTION **20**

ATTENUATION COEFFICIENT TABLES

INTRODUCTION

The absorption of an X ray or gamma ray beam passing through matter (and the beam's resulting attenuation) is the consequence of a series of single events; during each event a photon is removed from the beam after interaction with a nucleus or an orbital electron in the absorbing element. The total probability (per atom) for scattering or absorption of a photon of the original energy is given by a proportionality constant, σ. This is often referred to as the *cross-section* because it has the dimensions of an area. Such cross-sections are usually measured in *barns*; one barn is equivalent to 10^{-28} m^2 or 10^{-24} cm^2.

The total attenuation coefficient is the sum of the attenuation coefficients due to scattering, the photoelectric effect, and pair production. The *photoelectric effect* is that process in which a photon transfers its total energy to an electron in some shell of an atom. It is most significant at lower photon energies. As photon energy increases, *scattering* becomes the main process contributing to attenuation. Very high energy photons are absorbed by *pair production,* in which a photon is converted into an electron and a positron. This process occurs in the electrical field of a nucleus and requires a minimum photon energy of 1.02 MeV.

The total attenuation coefficient can be expressed in three different forms:

The *atomic attenuation coefficient* measures the probability of absorption, per atom of absorbing material, in barns.

The *mass attenuation coefficient* measures the probability of absorption per gram of absorbing material in a square centimeter of the beam.

The *linear attenuation coefficient* measures the probability of absorption per centimeter of the absorbing material's thickness.

Linear Attenuation Coefficient

The linear attenuation coefficient, μ, can be expressed as:

$$\mu = \mu/\rho \times \rho \qquad \text{(Eq. 1)}$$

where μ/ρ is the mass attenuation coefficient, and ρ is the density of the absorbing material. The linear attenuation coefficient has a dimension of cm^{-1}.

The linear attenuation coefficient of water, for example, is:

$$0.0705 \times 1 = 0.0705 \text{ cm}^{-1}$$

where 1 is the density (ρ), in grams per cubic centimeter at standard temperature and pressure; 0.0705 is the mass attenuation coefficient (μ/ρ) for water.

The linear attenuation coefficient for air is:

$$0.761 \times 0.0012 = 0.91 \times 10^{-3} \text{ cm}^{-1}$$

where 0.0012 is the density of air in grams per cubic centimeter and 0.761 is its mass attenuation coefficient (calculation of the mass attenuation coefficients for air and water is shown below).

Mass Attenuation Coefficient

The mass attenuation coefficient (μ/ρ) of a compound or mixture is the sum of the mass attenuation coefficients of the constituent *elements*, weighted in proportion to their relative abundance (R).

$$\mu/\rho_{total} = \mu/\rho_a R_a + \mu/\rho_b R_b + \ldots \qquad \text{(Eq. 2)}$$

The total mass attenuation coefficient for a compound, water, at 1 MeV, is:

$$\mu/\rho_{water} = 0.126\ (2/18) + 0.0636\ (16/18) = 0.0705\ \text{g/cm}^2$$

where 0.126 and 0.0636 are the mass attenuation coefficients of hydrogen and oxygen at 1 MeV. The relative abundance is figured using the atomic weights of 1 for hydrogen and 16 for oxygen.

The same method can be used to calculate the mass attenuation coefficient at 0.02 MeV for air (a mixture), which consists in percentages (by weight) primarily of N_2 (75.6%), O_2 (23.1%) and A (1.3%). The mass attenuation coefficients are: nitrogen, 0.598 g/cm^2; oxygen, 0.840 g/cm^2; and argon, 8.87 g/cm^2. Therefore the total mass attenuation coefficient for air at 0.02 MeV is

$$\mu/\rho_{air} = (0.598 \times 0.756) + (0.840 \times 0.231) + (8.87 \times 0.013) = 0.761\ \text{g/cm}^2$$

K Absorption Edge

Some of these tables include information on the element's *K* absorption edge. When the transmitted photon energy reaches the binding energy of a particular shell of electrons, there is an abrupt increase in the absorption. The energy at which this sharp change occurs for *K* electrons is called the *K absorption edge* and is used to identify the situation where kinetic energy of the ejected *K* electron is zero. Further increase of the photon energy causes the absorption to decrease almost inversely with the cube of the energy.

Preparation of the Tables

Tables 1-40 are based upon a narrow-beam absorption. The calculated atomic, mass and linear attenuation coefficients for various elements are given, in the energy range of 0.01 to 30 MeV.

The tables were prepared by the radiation physics committee of the American Society for Nondestructive Testing, under the direction of C. Robert Emigh, then with Los Alamos National Laboratory and presently with the University of New Mexico. The tabulations provide data for the photoelectric component; for the pair-production component which includes both nuclear and orbital electron contributions; for the scattering component, and a correction for electronic binding energies. Many of the values were obtained from G.R. White's calculated values in the *Handbook of Radiology*. Corrections to these values and values for other elements were obtained by graphical interpolation. For convenience, the values were presented to no more than three significant figures, although the estimated probable error is no larger than one-half unit in the last place, or three percent, whichever is greater. The linear attenuation coefficients were calculated with the density most commonly used for the given element.

TABLE 1. Hydrogen (Z = 1)

Energy (MeV)	Cross Sections in Barns (10^{-28} m^2)			Attenuation Coefficients		
	Scattering	Photo-electric	Pair	Atomic (barns)	Mass* (g/cm^2)	Linear † (1/cm)
0.01	0.637	0.005	—	0.642	0.384	32.1×10^{-6}
0.015	0.627	0.001	—	0.628	0.375	31.4×10^{-6}
0.02	0.616	—	—	0.616	0.368	30.8×10^{-6}
0.03	0.596	—	—	0.596	0.356	29.8×10^{-6}
0.04	0.578	—	—	0.578	0.345	28.9×10^{-6}
0.05	0.561	—	—	0.561	0.335	28.1×10^{-6}
0.06	0.546	—	—	0.546	0.326	27.3×10^{-6}
0.08	0.517	—	—	0.517	0.309	25.9×10^{-6}
0.10	0.493	—	—	0.493	0.295	24.7×10^{-6}
0.15	0.444	—	—	0.444	0.265	22.2×10^{-6}
0.20	0.407	—	—	0.407	0.243	20.4×10^{-6}
0.30	0.354	—	—	0.354	0.212	17.8×10^{-6}
0.40	0.317	—	—	0.317	0.189	15.8×10^{-6}
0.50	0.289	—		0.289	0.173	14.5×10^{-6}
0.60	0.268	—	—	0.268	0.160	13.4×10^{-6}
0.80	0.235	—	—	0.235	0.140	11.7×10^{-6}
1.0	0.211	—	—	0.211	0.126	10.6×10^{-6}
1.5	0.172	—	—	0.172	0.103	8.63×10^{-6}
2.0	0.146	—	—	0.146	0.0873	7.31×10^{-6}
3.0	0.115	—	0.001	0.116	0.0693	5.80×10^{-6}
4.0	0.0960	—	0.0010	0.0970	0.0580	4.86×10^{-6}
5.0	0.0828	—	0.0014	0.0842	0.0503	4.21×10^{-6}
6.0	0.0732	—	0.0019	0.0751	0.0449	3.76×10^{-6}
8.0	0.0599	—	0.0027	0.0626	0.0374	3.13×10^{-6}
10	0.0510	—	0.0033	0.0543	0.0325	2.72×10^{-6}
15	0.0377	—	0.0046	0.0423	0.0253	2.12×10^{-6}
20	0.0302	—	0.0056	0.0358	0.0214	1.79×10^{-6}
30	0.0220	—	0.0071	0.0291	0.0174	1.46×10^{-6}

* Calculated using atomic weight 1.008
† Calculated using density 0.0838×10^{-3} gram/cm^3

TABLE 2. Beryllium (Z = 4)

Energy (MeV)	Cross Sections in Barns (10^{-28} m²)			Attenuation Coefficients		
	Scattering	Photo-electric	Pair	Atomic (barns)	Mass* (g/cm²)	Linear † (1/cm)
0.01	3.54	5.42	—	8.96	0.599	1.09
0.015	3.01	1.39	—	4.40	0.294	0.535
0.02	2.77	0.52	—	3.29	0.220	0.400
0.03	2.53	0.13	—	2.66	0.178	0.324
0.04	2.38	0.05	—	2.43	0.162	0.295
0.05	2.28	0.02	—	2.30	0.154	0.280
0.06	2.21	0.01	—	2.22	0.148	0.269
0.08	2.09	—	—	2.09	0.140	0.255
0.10	1.99	—	—	1.99	0.133	0.242
0.15	1.78	—	—	1.78	0.119	0.217
0.20	1.63	—	—	1.63	0.109	0.198
0.30	1.41	—	—	1.41	0.0943	0.172
0.40	1.27	—	—	1.27	0.0849	0.155
0.50	1.16	—	—	1.16	0.0775	0.141
0.60	1.07	—	—	1.07	0.0715	0.130
0.80	0.940	—	—	0.940	0.0628	0.114
1.0	0.845	—	—	0.845	0.0565	0.103
1.5	0.686	—	0.001	0.687	0.459	0.0835
2.0	0.586	—	0.003	0.589	0.0394	0.0717
3.0	0.460	—	0.008	0.468	0.0313	0.0570
4.0	0.384	—	0.014	0.398	0.0266	0.0484
5.0	0.331	—	0.019	0.350	0.0234	0.0426
6.0	0.293	—	0.024	0.317	0.0212	0.0386
8.0	0.240	—	0.031	0.271	0.0181	0.0329
10	0.204	—	0.039	0.243	0.0162	0.0295
15	0.151	—	0.051	0.202	0.0135	0.0246
20	0.121	—	0.061	0.182	0.0122	0.0222
30	0.0880	—	0.075	0.163	0.0109	0.0198

* Calculated using atomic weight 9.013
† Calculated using density 1.82 grams/cm³

TABLE 3. Carbon (Z = 6)

Energy (MeV)	Cross Sections in Barns (10^{-28} m²)			Attenuation Coefficients		
	Scattering	Photo-electric	Pair	Atomic (barns)	Mass* (g/cm²)	Linear † (1/cm)
0.01	6.90	38.6	—	45.5	2.28	5.06
0.015	5.30	10.2	—	15.5	0.778	1.73
0.02	4.64	3.91	—	8.55	0.429	0.952
0.03	4.04	0.99	—	5.03	0.252	0.559
0.04	3.71	0.38	—	4.09	0.205	0.455
0.05	3.53	0.18	—	3.71	0.186	0.413
0.06	3.38	0.10	—	3.48	0.175	0.389
0.08	3.18	0.04	—	3.22	0.162	0.360
0.10	3.02	0.02	—	3.04	0.153	0.340
0.15	2.69	—	—	2.69	0.135	0.300
0.20	2.46	—	—	2.46	0.123	0.273
0.30	2.13	—	—	2.13	0.107	0.238
0.40	1.90	—	—	1.90	0.0953	0.212
0.50	1.74	—	—	1.74	0.0873	0.194
0.60	1.61	—	—	1.61	0.0808	0.179
0.80	1.41	—	—	1.41	0.0707	0.157
1.0	1.27	—	—	1.27	0.0637	0.141
1.5	1.03	—	—	1.03	0.0517	0.115
2.0	0.878	—	0.006	0.884	0.0444	0.0986
3.0	0.691	—	0.018	0.709	0.0356	0.0790
4.0	0.576	—	0.031	0.607	0.0305	0.0677
5.0	0.497	—	0.042	0.539	0.0270	0.0599
6.0	0.439	—	0.052	0.491	0.0246	0.0546
8.0	0.359	—	0.068	0.427	0.0214	0.0475
10	0.306	—	0.083	0.389	0.0195	0.0433
15	0.226	—	0.110	0.336	0.0169	0.0375
20	0.181	—	0.130	0.311	0.0156	0.0346
30	0.132	—	0.160	0.292	0.0146	0.0324

* Calculated using atomic weight 12.010
† Calculated using density 2.22 grams/cm³

TABLE 4. Nitrogen (Z = 7)

Energy (MeV)	Cross Sections in Barns (10^{-28} m^2)			Attenuation Coefficients		
	Scattering	Photo-electric	Pair	Atomic (barns)	Mass* (g/cm^2)	Linear † (1/cm)
0.01	9.0	79.4	—	88.4	3.80	44.3×10^{-4}
0.015	6.7	21.2	—	27.9	1.20	14.0×10^{-4}
0.02	5.73	8.21	—	13.9	0.598	6.97×10^{-4}
0.03	4.84	2.15	—	6.99	0.301	3.51×10^{-4}
0.04	4.45	0.81	—	5.26	0.226	2.63×10^{-4}
0.05	4.14	0.38	—	4.52	0.194	2.26×10^{-4}
0.06	3.98	0.21	—	4.19	0.180	2.10×10^{-4}
0.08	3.73	0.08	—	3.81	0.164	1.91×10^{-4}
0.10	3.54	0.04	—	3.58	0.154	1.79×10^{-4}
0.15	3.15	0.01	—	3.16	0.136	1.58×10^{-4}
0.20	2.87	—	—	2.87	0.123	1.43×10^{-4}
0.30	2.48	—	—	2.48	0.107	1.25×10^{-4}
0.40	2.22	—	—	2.22	0.0955	1.11×10^{-4}
0.50	2.02	—	—	2.02	0.0869	1.01×10^{-4}
0.60	1.87	—	—	1.87	0.0804	0.937×10^{-4}
0.80	1.65	—	—	1.65	0.0710	0.827×10^{-4}
1.0	1.48	—	—	1.48	0.0637	0.742×10^{-4}
1.5	1.20	—	—	1.20	0.0516	0.601×10^{-4}
2.0	1.03	—	0.01	1.04	0.0447	0.521×10^{-4}
3.0	0.806	—	0.025	0.831	0.0357	0.416×10^{-4}
4.0	0.672	—	0.042	0.714	0.0307	0.358×10^{-4}
5.0	0.580	—	0.057	0.637	0.0274	0.319×10^{-4}
6.0	0.512	—	0.071	0.583	0.0251	0.292×10^{-4}
8.0	0.419	—	0.092	0.511	0.0220	0.256×10^{-4}
10	0.357	—	0.111	0.468	0.0201	0.234×10^{-4}
15	0.264	—	0.148	0.412	0.0177	0.206×10^{-4}
20	0.212	—	0.174	0.386	0.0166	0.193×10^{-4}
30	0.154	—	0.213	0.367	0.0158	0.184×10^{-4}

* Calculated using atomic weight 14.088
† Calculated using density 1.165×10^{-3} gram/cm^3

TABLE 5. Oxygen (Z = 8)

Energy (MeV)	Cross Sections in Barns (10^{-28} m^2)			Attenuation Coefficients		
	Scattering	Photo-electric	Pair	Atomic (barns)	Mass* (g/cm^2)	Linear † (1/cm)
0.01	11.3	146	—	157	5.91	78.7×10^{-4}
0.015	8.3	39.6	—	47.9	1.80	24.0×10^{-4}
0.02	6.9	15.4	—	22.3	0.840	11.2×10^{-4}
0.03	5.77	4.09	—	9.86	0.371	4.94×10^{-4}
0.04	5.18	1.55	—	6.73	0.253	3.37×10^{-4}
0.05	4.86	0.73	—	5.59	0.211	2.81×10^{-4}
0.06	4.62	0.40	—	5.02	0.189	2.52×10^{-4}
0.08	4.31	0.15	—	4.46	0.168	2.24×10^{-4}
0.10	4.06	0.07	—	4.13	0.156	2.08×10^{-4}
0.15	3.61	0.02	—	3.63	0.137	1.82×10^{-4}
0.20	3.29	0.01	—	3.30	0.124	1.65×10^{-4}
0.30	2.84	—	—	2.84	0.107	1.43×10^{-4}
0.40	2.54	—	—	2.54	0.0957	1.27×10^{-4}
0.50	2.31	—	—	2.31	0.0870	1.16×10^{-4}
0.60	2.14	—	—	2.14	0.0806	1.07×10^{-4}
0.80	1.88	—	—	1.88	0.0708	0.943×10^{-4}
1.0	1.69	—	—	1.69	0.0636	0.847×10^{-4}
1.5	1.37	—	—	1.37	0.0516	0.687×10^{-4}
2.0	1.17	—	0.01	1.18	0.0444	0.591×10^{-4}
3.0	0.921	—	0.033	0.954	0.0359	0.478×10^{-4}
4.0	0.768	—	0.054	0.822	0.0310	0.413×10^{-4}
5.0	0.663	—	0.074	0.737	0.0278	0.370×10^{-4}
6.0	0.586	—	0.091	0.677	0.0255	0.340×10^{-4}
8.0	0.479	—	0.119	0.598	0.0225	0.300×10^{-4}
10	0.408	—	0.143	0.551	0.0208	0.277×10^{-4}
15	0.302	—	0.190	0.492	0.0185	0.246×10^{-4}
20	0.242	—	0.224	0.466	0.0175	0.233×10^{-4}
30	0.176	—	0.273	0.449	0.0169	0.225×10^{-4}

* Calculated using atomic weight 16.000
† Calculated using density 1.332×10^{-3} gram/cm^3

TABLE 6. Sodium (Z = 11)

Energy (MeV)	Cross Sections in Barns (10^{-28} m^2)			Attenuation Coefficients		
	Scattering	Photo-electric	Pair	Atomic (barns)	Mass* (g/cm^2)	Linear † (1/cm)
0.01	20.6	588	—	609	16.0	15.5
0.015	14.0	169	—	183	4.79	4.65
0.02	11.3	67.5	—	78.8	2.06	2.00
0.03	8.91	18.1	—	27.0	0.707	0.686
0.04	7.71	7.0	—	14.7	0.385	0.374
0.05	7.07	3.3	—	10.4	0.272	0.264
0.06	6.67	1.90	—	8.57	0.225	0.218
0.08	6.08	0.74	—	6.82	0.179	0.174
0.10	5.66	0.32	—	5.98	0.157	0.152
0.15	5.01	0.09	—	5.10	0.134	0.130
0.20	4.54	0.04	—	4.58	0.120	0.117
0.30	3.92	0.01	—	3.93	0.103	0.100
0.40	3.50	—	—	3.50	0.0917	0.0890
0.50	3.19	—	—	3.19	0.0836	0.0812
0.60	2.94	—	—	2.94	0.0770	0.0748
0.80	2.59	—	—	2.59	0.0679	0.0659
1.0	2.32	—	—	2.32	0.0608	0.0590
1.5	1.89	—	—	1.89	0.0495	0.0481
2.0	1.61	—	0.02	1.63	0.0427	0.0415
3.0	1.27	—	0.06	1.33	0.0348	0.0338
4.0	1.06	—	0.10	1.16	0.0304	0.0295
5.0	0.911	—	0.139	1.05	0.0275	0.0267
6.0	0.805	—	0.170	0.975	0.0255	0.0248
8.0	0.659	—	0.221	0.880	0.0231	0.0224
10	0.561	—	0.266	0.827	0.0217	0.0211
15	0.415	—	0.351	0.766	0.0201	0.0195
20	0.333	—	0.413	0.746	0.0195	0.0189
30	0.242	—	0.500	0.742	0.0194	0.0188

* Calculated using atomic weight 22.997
† Calculated using density 0.971 gram/cm^3

TABLE 7. Magnesium (Z = 12)

Energy (MeV)	Cross Sections in Barns (10^{-28} m^2)			Attenuation Coefficients		
	Scattering	Photo-electric	Pair	Atomic (barns)	Mass* (g/cm^2)	Linear † (1/cm)
0.01	24.3	851	—	875	21.7	37.8
0.015	16.4	244	—	260	6.44	11.2
0.02	13.0	99.0	—	112	2.77	4.82
0.03	10.1	27.4	—	37.5	0.929	1.62
0.04	8.71	10.5	—	19.2	0.476	0.829
0.05	7.88	5.13	—	13.0	0.322	0.561
0.06	7.37	2.84	—	10.2	0.253	0.440
0.08	6.70	1.10	—	7.80	0.193	0.336
0.10	6.25	0.53	—	6.78	0.168	0.292
0.15	5.48	0.14	—	5.62	0.139	0.242
0.20	4.96	0.06	—	5.02	0.124	0.216
0.30	4.28	0.02	—	4.30	0.107	0.186
0.40	3.82	0.01	—	3.83	0.0949	0.165
0.50	3.48	—	—	3.48	0.0862	0.150
0.60	3.22	—	—	3.22	0.0798	0.139
0.80	2.82	—	—	2.82	0.0699	0.122
1.0	2.53	—	—	2.53	0.0627	0.109
1.5	2.06	—	0.01	2.07	0.0513	0.0893
2.0	1.76	—	0.02	1.78	0.0441	0.0768
3.0	1.38	—	0.08	1.46	0.0362	0.0630
4.0	1.15	—	0.12	1.27	0.0315	0.0548
5.0	0.994	—	0.165	1.16	0.0287	0.0500
6.0	0.878	—	0.201	1.08	0.0268	0.0467
8.0	0.719	—	0.261	0.980	0.0243	0.0423
10	0.612	—	0.314	0.926	0.0229	0.0399
15	0.452	—	0.415	0.867	0.0215	0.0374
20	0.362	—	0.490	0.852	0.0211	0.0367
30	0.264	—	0.593	0.857	0.0212	0.0369

* Calculated using atomic weight 24.32
† Calculated using density 1.741 grams/cm^3

TABLE 8. Aluminum (Z = 13)

Energy (MeV)	Cross Sections in Barns (10^{-28} m^2)			Attenuation Coefficients		
	Scattering	Photo-electric	Pair	Atomic (barns)	Mass* (g/cm^2)	Linear † (1/cm)
0.01	29	1170	—	1200	26.8	72.4
0.015	19	343	—	362	8.08	21.8
0.02	15	141	—	156	3.48	9.40
0.03	11.7	39.0	—	50.7	1.13	3.05
0.04	9.7	15.2	—	24.9	0.556	1.50
0.05	8.7	7.3	—	16.0	0.357	0.964
0.06	8.1	4.0	—	12.1	0.270	0.729
0.08	7.34	1.60	—	8.94	0.200	0.540
0.10	6.82	0.78	—	7.60	0.170	0.459
0.15	5.96	0.21	—	6.17	0.138	0.373
0.20	5.39	0.08	—	5.47	0.122	0.329
0.30	4.64	0.02	—	4.66	0.104	0.281
0.40	4.14	0.01	—	4.15	0.0927	0.250
0.50	3.78	—	—	3.78	0.0844	0.228
0.60	3.49	—	—	3.49	0.0779	0.210
0.80	3.06	—	—	3.06	0.0683	0.184
1.0	2.75	—	—	2.75	0.0614	0.166
1.5	2.23	—	0.01	2.24	0.0500	0.135
2.0	1.90	—	0.03	1.93	0.0431	0.116
3.0	1.50	—	0.09	1.59	0.0355	0.0959
4.0	1.25	—	0.14	1.39	0.0310	0.0837
5.0	1.08	—	0.19	1.27	0.0284	0.0767
6.0	0.952	—	0.237	1.19	0.0266	0.0718
8.0	0.778	—	0.311	1.09	0.0243	0.0656
10	0.663	—	0.365	1.03	0.0230	0.0621
15	0.490	—	0.484	0.974	0.0217	0.0586
20	0.393	—	0.570	0.963	0.0215	0.0581
30	0.286	—	0.690	0.976	0.0218	0.0589

* Calculated using atomic weight 26.98
† Calculated using density 2.70 grams/cm^3

TABLE 9. Silicon (Z = 14)

Energy (MeV)	Cross Sections in Barns (10^{-28} m^2)			Attenuation Coefficients		
	Scattering	Photo-electric	Pair	Atomic (barns)	Mass* (g/cm^2)	Linear † (1/cm)
0.01	33	1580	—	1610	34.5	81.1
0.015	22	470	—	492	10.6	24.9
0.02	17	194	—	211	4.53	10.6
0.03	12.8	54.4	—	67.2	1.44	3.38
0.04	10.8	21.4	—	32.2	0.691	1.62
0.05	9.6	10.3	—	19.9	0.427	1.00
0.06	8.9	5.8	—	14.7	0.315	0.740
0.08	8.0	2.3	—	10.3	0.221	0.519
0.10	7.38	1.10	—	8.48	0.182	0.428
0.15	6.44	0.29	—	6.73	0.144	0.338
0.20	5.82	0.12	—	5.94	0.127	0.298
0.30	5.01	0.04	—	5.05	0.108	0.254
0.40	4.46	0.02	—	4.48	0.0961	0.226
0.50	4.07	—	—	4.07	0.0873	0.205
0.60	3.75	—	—	3.75	0.0804	0.189
0.80	3.30	—	—	3.30	0.0708	0.166
1.0	2.96	—	—	2.96	0.0635	0.149
1.5	2.40	—	0.01	2.41	0.0517	0.121
2.0	2.05	—	0.04	2.09	0.0448	0.105
3.0	1.61	—	0.10	1.71	0.0367	0.0862
4.0	1.34	—	0.16	1.50	0.0322	0.0757
5.0	1.16	—	0.23	1.39	0.0298	0.0700
6.0	1.03	—	0.28	1.31	0.0281	0.0660
8.0	0.84	—	0.35	1.19	0.0255	0.0599
10	0.714	—	0.426	1.14	0.0245	0.0576
15	0.528	—	0.565	1.09	0.0234	0.0550
20	0.423	—	0.663	1.09	0.0234	0.0550
30	0.308	—	0.793	1.10	0.0236	0.0555

* Calculated using atomic weight 28.09
† Calculated using density 2.35 grams/cm^3

TABLE 10. Argon (Z = 18)

Energy (MeV)	Cross Sections in Barns (10^{-28} m²)			Attenuation Coefficients		
	Scattering	Photo-electric	Pair	Atomic (barns)	Mass* (g/cm²)	Linear † (1/cm)
0.01	56	4280	—	4340	65.4	10.9×10^{-2}
0.015	36	1320	—	1360	20.5	3.41×10^{-2}
0.02	27	561	—	588	8.87	1.48×10^{-2}
0.03	19	164	—	183	2.76	0.459×10^{-2}
0.04	15.6	64.5	—	80.1	1.21	0.201×10^{-2}
0.05	13.6	31.6	—	45.2	0.682	0.113×10^{-2}
0.06	12.4	18.0	—	30.4	0.458	0.0762×10^{-2}
0.08	10.8	7.2	—	18.0	0.271	0.0451×10^{-2}
0.10	9.85	3.60	—	13.5	0.204	0.0339×10^{-2}
0.15	8.43	0.98	—	9.41	0.142	0.0236×10^{-2}
0.20	7.57	0.41	—	7.98	0.120	0.0200×10^{-2}
0.30	6.48	0.12	—	6.60	0.0995	0.0165×10^{-2}
0.40	5.76	0.05	—	5.81	0.0876	0.0146×10^{-2}
0.50	5.24	0.03	—	5.27	0.0795	0.0132×10^{-2}
0.60	4.84	0.02	—	4.86	0.0733	0.0122×10^{-2}
0.80	4.24	—	—	4.24	0.0639	0.0106×10^{-2}
1.0	3.81	—	—	3.81	0.0575	0.00956×10^{-2}
1.5	3.09	—	0.02	3.11	0.0469	0.00780×10^{-2}
2.0	2.64	—	0.06	2.70	0.0407	0.00677×10^{-2}
3.0	2.07	—	0.17	2.24	0.0338	0.00562×10^{-2}
4.0	1.73	—	0.27	2.00	0.0302	0.00502×10^{-2}
5.0	1.49	—	0.37	1.86	0.0280	0.00466×10^{-2}
6.0	1.32	—	0.45	1.77	0.0267	0.00444×10^{-2}
8.0	1.08	—	0.59	1.67	0.0252	0.00419×10^{-2}
10	0.918	—	0.691	1.61	0.0243	0.00404×10^{-2}
15	0.679	—	0.913	1.59	0.0240	0.00399×10^{-2}
20	0.544	—	1.06	1.60	0.0241	0.00401×10^{-2}
30	0.396	—	1.29	1.69	0.0255	0.00424×10^{-2}

* Calculated using atomic weight 39.944
† Calculated using density 1.663×10^{-3} gram/cm³

TABLE 11. Calcium (Z = 20)

Energy (MeV)	Cross Sections in Barns (10^{-28} m^2)			Attenuation Coefficients		
	Scattering	Photo-electric	Pair	Atomic (barns)	Mass* (g/cm^2)	Linear † (1/cm)
0.01	69	6380	—	6450	96.9	149
0.015	44	2010	—	2050	30.8	47.4
0.02	33	859	—	892	13.4	20.6
0.03	24	254	—	278	4.18	6.44
0.04	19	102	—	121	1.82	2.80
0.05	15.8	50.6	—	66.4	0.098	1.54
0.06	14.3	28.8	—	43.1	0.648	0.998
0.08	12.3	11.6	—	23.9	0.359	0.553
0.10	11.2	6.0	—	17.2	0.259	0.399
0.15	9.48	1.60	—	11.1	0.167	0.257
0.20	8.47	0.67	—	9.14	0.137	0.211
0.30	7.23	0.20	—	7.43	0.112	0.172
0.40	6.42	0.09	—	6.51	0.0978	0.151
0.50	5.84	0.05	—	5.89	0.0885	0.136
0.60	5.38	0.03	—	5.41	0.0813	0.125
0.80	4.72	0.01	—	4.73	0.0711	0.109
1.0	4.24	—	—	4.24	0.0637	0.0981
1.5	3.43	—	0.02	3.45	0.0518	0.0798
2.0	2.93	—	0.07	3.00	0.0451	0.0695
3.0	2.30	—	0.21	2.51	0.0377	0.0581
4.0	1.92	—	0.33	2.25	0.0338	0.0521
5.0	1.66	—	0.45	2.11	0.0317	0.0488
6.0	1.46	—	0.55	2.01	0.0302	0.0465
8.0	1.20	—	0.72	1.92	0.0289	0.0445
10	1.02	—	0.84	1.86	0.0280	0.0431
15	0.755	—	1.12	1.88	0.0283	0.0436
20	0.605	—	1.31	1.92	0.0289	0.0445
30	0.440	—	1.57	2.01	0.0302	0.0465

* Calculated using atomic weight 40.08
† Calculated using density 1.54 grams/cm^3

TABLE 12. Titanium (Z = 22)

Energy (MeV)	Cross Sections in Barns (10^{-28} m^2)			Attenuation Coefficients		
	Scattering	Photo-electric	Pair	Atomic (barns)	Mass* (g/cm^2)	Linear † (1/cm)
0.01	84	9150	—	9230	116	527
0.015	53	2900	—	2950	37.1	168
0.02	39	1250	—	1290	16.2	73.5
0.03	27	374	—	401	5.04	22.9
0.04	22	154	—	176	2.21	10.0
0.05	18.3	76.3	—	94.6	1.19	5.40
0.06	16.3	43.9	—	60.2	0.757	3.44
0.08	14.0	17.9	—	31.9	0.401	1.82
0.10	12.5	9.2	—	21.7	0.273	1.24
0.15	10.6	2.5	—	13.1	0.165	0.749
0.20	9.40	1.04	—	10.4	0.131	0.595
0.30	7.99	0.31	—	8.30	0.104	0.472
0.40	7.09	0.13	—	7.22	0.0908	0.412
0.50	6.43	0.07	—	6.50	0.0818	0.371
0.60	5.94	0.05	—	5.99	0.0754	0.342
0.80	5.19	0.02	—	5.21	0.0655	0.297
1.0	4.66	0.01	—	4.67	0.0587	0.266
1.5	3.78	0.01	0.02	3.81	0.0479	0.217
2.0	3.22	—	0.09	3.31	0.0416	0.189
3.0	2.53	—	0.25	2.78	0.0350	0.159
4.0	2.11	—	0.41	2.52	0.0317	0.144
5.0	1.82	—	0.54	2.36	0.0297	0.135
6.0	1.61	—	0.67	2.28	0.0287	0.130
8.0	1.32	—	0.86	2.18	0.0274	0.124
10	1.12	—	1.02	2.14	0.0269	0.122
15	0.829	—	1.34	2.17	0.0273	0.124
20	0.664	—	1.58	2.24	0.0282	0.128
30	0.484	—	1.90	2.38	0.0299	0.136

* Calculated using atomic weight 47.9
† Calculated using density 4.54 grams/cm^3

TABLE 13. Vanadium (Z = 23)

Energy (MeV)	Cross Sections in Barns (10^{-28} m^2)			Attenuation Coefficients		
	Scattering	Photo-electric	Pair	Atomic (barns)	Mass* (g/cm^2)	Linear † (1/cm)
0.01	92	10,700	—	10,800	128	763
0.015	58	3,430	—	3,490	41.3	246
0.02	43	1,490	—	1,530	18.1	108
0.03	29	449	—	478	5.65	33.7
0.04	23	185	—	208	2.46	14.7
0.05	19.6	92.7	—	112	1.32	7.87
0.06	17.4	53.3	—	70.7	0.836	4.98
0.08	14.8	21.8	—	36.6	0.443	2.58
0.10	13.3	11.1	—	24.4	0.289	1.72
0.15	11.1	3.1	—	14.2	0.168	1.00
0.20	9.85	1.27	—	11.1	0.131	0.781
0.30	8.36	0.38	—	8.74	0.103	0.614
0.40	7.41	0.16	—	7.57	0.0896	0.534
0.50	6.73	0.09	—	6.82	0.0807	0.481
0.60	6.20	0.06	—	6.26	0.0741	0.442
0.80	5.44	0.03	—	5.47	0.0647	0.386
1.0	4.88	0.02	—	4.90	0.0580	0.346
1.5	3.96	0.01	0.03	4.00	0.0473	0.282
2.0	3.37	—	0.09	3.46	0.0409	0.244
3.0	2.65	—	0.28	2.93	0.0347	0.207
4.0	2.21	—	0.44	2.65	0.0313	0.187
5.0	1.90	—	0.60	2.50	0.0296	0.176
6.0	1.68	—	0.73	2.41	0.0285	0.170
8.0	1.38	—	0.94	2.32	0.0274	0.163
10	1.17	—	1.12	2.29	0.0271	0.162
15	0.867	—	1.46	2.33	0.0276	0.164
20	0.695	—	1.74	2.44	0.0289	0.172
30	0.506	—	2.06	2.57	0.0304	0.181

* Calculated using atomic weight 50.95
† Calculated using density 5.96 grams/cm^3

TABLE 14. Chromium (Z = 24)

Energy (MeV)	Cross Sections in Barns (10^{-28} m^2)			Attenuation Coefficients		
	Scattering	Photo-electric	Pair	Atomic (barns)	Mass* (g/cm^2)	Linear † (1/cm)
0.01	101	12,500	—	12,600	146	1050
0.015	64	4,040	—	4,100	47.5	342
0.02	47	1,760	—	1,810	21.0	151
0.03	32	533	—	565	6.54	47.0
0.04	25	221	—	246	2.85	20.5
0.05	21	111	—	132	1.53	11.0
0.06	18.5	63.9	—	82.4	0.954	6.86
0.08	15.7	26.3	—	42.0	0.486	3.49
0.10	14.0	13.5	—	27.5	0.318	2.29
0.15	11.7	3.75	—	15.5	0.179	1.29
0.20	10.3	1.55	—	11.9	0.138	0.992
0.30	8.74	0.46	—	9.20	0.107	0.769
0.40	7.75	0.20	—	7.95	0.0921	0.662
0.50	7.03	0.11	—	7.14	0.0827	0.595
0.60	6.48	0.07	—	6.55	0.0758	0.545
0.80	5.67	0.03	—	5.70	0.0660	0.475
1.0	5.09	0.02	—	5.11	0.0592	0.426
1.5	4.13	0.01	0.03	4.17	0.0483	0.347
2.0	3.51	0.01	0.11	3.63	0.0420	0.302
3.0	2.76	—	0.30	3.06	0.0354	0.255
4.0	2.30	—	0.48	2.78	0.0322	0.232
5.0	1.99	—	0.65	2.64	0.0306	0.220
6.0	1.76	—	0.79	2.55	0.0295	0.212
8.0	1.44	—	1.02	2.46	0.0285	0.205
10	1.22	—	1.21	2.43	0.0281	0.202
15	0.905	—	1.59	2.50	0.0290	0.209
20	0.725	—	1.87	2.60	0.0301	0.216
30	0.528	—	2.24	2.77	0.0321	0.231

* Calculated using atomic weight 52.01
† Calculated using density 7.19 grams/cm^3

TABLE 15. Manganese (Z = 25)

Energy (MeV)	Cross Sections in Barns (10^{-28} m^2)			Attenuation Coefficients		
	Scattering	Photo-electric	Pair	Atomic (barns)	Mass* (g/cm^2)	Linear † (1/cm)
0.01	110	14,400	—	14,500	159	1180
0.015	70	4,690	—	4,760	52.2	388
0.02	51	2,051	—	2,100	23.0	171
0.03	34	626	—	660	7.24	53.8
0.04	27	263	—	290	3.18	23.6
0.05	22	132	—	154	1.69	12.6
0.06	19.7	76.2	—	95.9	1.05	7.80
0.08	16.6	31.4	—	48.0	0.527	3.92
0.10	14.7	16.2	—	30.9	0.339	2.52
0.15	12.2	4.51	—	16.7	0.183	1.36
0.20	10.8	1.88	—	12.7	0.139	1.03
0.30	9.13	0.56	—	9.69	0.106	0.788
0.40	8.09	0.24	—	8.33	0.0914	0.679
0.50	7.33	0.13	—	7.46	0.0818	0.608
0.60	6.76	0.08	—	6.84	0.0750	0.557
0.80	5.91	0.04	—	5.95	0.0653	0.485
1.0	5.30	0.03	—	5.33	0.0585	0.435
1.5	4.30	0.01	0.03	4.34	0.0476	0.354
2.0	3.66	0.01	0.12	3.79	0.0416	0.309
3.0	2.88	—	0.33	3.21	0.0352	0.262
4.0	2.40	—	0.52	2.92	0.0320	0.238
5.0	2.07	—	0.70	2.77	0.0304	0.226
6.0	1.83	—	0.86	2.69	0.0295	0.219
8.0	1.50	—	1.11	2.61	0.0286	0.212
10	1.28	—	1.31	2.59	0.0284	0.211
15	0.943	—	1.72	2.66	0.0292	0.217
20	0.755	—	2.02	2.78	0.0305	0.227
30	0.55	—	2.43	2.98	0.0327	0.243

* Calculated using atomic weight 54.93
† Calculated using density 7.43 grams/cm^3

TABLE 16. Iron (Z = 26)

Energy (MeV)	Cross Sections in Barns (10^{-28} m^2)			Attenuation Coefficients		
	Scattering	Photo-electric	Pair	Atomic (barns)	Mass* (g/cm^2)	Linear † (1/cm)
0.01	120	16,500	—	16,600	179	1410
0.015	75	5,380	—	5,460	58.9	464
0.02	55	2,380	—	2,440	26.3	207
0.03	37	729	—	766	8.27	65.1
0.04	29	308	—	337	3.64	28.6
0.05	24	155	—	179	1.93	15.2
0.06	20.9	90.7	—	112	1.21	9.52
0.08	17.5	38.0	—	55.5	0.599	4.71
0.10	15.4	19.1	—	34.5	0.372	2.93
0.15	12.8	5.4	—	18.2	0.196	1.54
0.20	11.3	2.2	—	13.5	0.146	1.15
0.30	9.50	0.66	—	10.2	0.110	0.866
0.40	8.42	0.29	—	8.71	0.0940	0.740
0.50	7.63	0.16	—	7.79	0.0841	0.662
0.60	7.03	0.10	—	7.13	0.0769	0.605
0.80	6.15	0.05	—	6.20	0.0669	0.527
1.0	5.52	0.03	—	5.55	0.0599	0.471
1.5	4.46	0.02	0.03	4.51	0.0487	0.383
2.0	3.81	0.01	0.12	3.94	0.0425	0.334
3.0	2.99	—	0.35	3.34	0.0360	0.283
4.0	2.50	—	0.57	3.07	0.0331	0.260
5.0	2.15	—	0.76	2.91	0.0314	0.247
6.0	1.90	—	0.92	2.82	0.0304	0.239
8.0	1.56	—	1.20	2.76	0.0298	0.235
10	1.33	—	1.41	2.74	0.0296	0.233
15	0.981	—	1.86	2.84	0.0306	0.241
20	0.786	—	2.17	2.96	0.0319	0.251
30	0.572	—	2.61	3.18	0.0343	0.270

* Calculated using atomic weight 55.85
† Calculated using density 7.87 grams/cm^3

TABLE 17. Cobalt (Z = 27)

Energy (MeV)	Cross Sections in Barns (10^{-28} m^2)			Attenuation Coefficients		
	Scattering	Photo-electric	Pair	Atomic (barns)	Mass* (g/cm^2)	Linear † (1/cm)
0.01	130	18,800	—	18,900	193	1720
0.015	82	6,170	—	6,250	63.9	569
0.02	60	2,760	—	2,820	28.8	256
0.03	40	848	—	888	9.08	80.8
0.04	31	360	—	391	4.00	35.6
0.05	25	181	—	206	2.11	18.8
0.06	22	106	—	128	1.31	11.7
0.08	18.5	43.8	—	62.3	0.637	5.67
0.10	16.3	22.5	—	38.8	0.397	3.53
0.15	13.4	6.40	—	19.8	0.202	1.80
0.20	11.8	2.65	—	14.5	0.148	1.32
0.30	9.91	0.80	—	10.7	0.109	0.970
0.40	8.75	0.34	—	9.09	0.0929	0.827
0.50	7.94	0.19	—	8.13	0.0831	0.740
0.60	7.30	0.12	—	7.42	0.0758	0.675
0.80	6.39	0.06	—	6.45	0.0659	0.587
1.0	5.73	0.04	—	5.77	0.0590	0.525
1.5	4.64	0.02	0.03	4.69	0.0479	0.426
2.0	3.96	0.01	0.14	4.11	0.0420	0.374
3.0	3.11	0.01	0.38	3.50	0.0358	0.319
4.0	2.59	—	0.61	3.20	0.0327	0.291
5.0	2.24	—	0.82	3.06	0.0313	0.279
6.0	1.98	—	1.00	2.98	0.0305	0.271
8.0	1.62	—	1.29	2.91	0.0297	0.264
10	1.38	—	1.53	2.91	0.0297	0.264
15	1.02	—	2.00	3.02	0.0309	0.275
20	0.815	—	2.35	3.17	0.0324	0.288
30	0.594	—	2.82	3.41	0.0349	0.311

* Calculated using atomic weight 58.94
† Calculated using density 8.90 grams/cm^3

TABLE 18. Nickel (Z = 28)

Energy (MeV)	Cross Sections in Barns (10^{-28} m^2)			Attenuation Coefficients		
	Scattering	Photo-electric	Pair	Atomic (barns)	Mass* (g/cm^2)	Linear † (1/cm)
0.01	141	21,300	—	21,400	220	1950
0.015	89	7,020	—	7,110	73.0	646
0.02	65	3,160	—	3,230	33.2	294
0.03	43	984	—	1,030	10.6	93.8
0.04	33	418	—	451	4.63	41.0
0.05	27	210	—	237	2.43	21.5
0.06	23	123	—	146	1.50	13.3
0.08	19.4	51.1	—	70.5	0.724	6.41
0.10	17.1	26.4	—	43.5	0.447	3.96
0.15	14.0	7.52	—	21.5	0.221	1.96
0.20	12.3	3.12	—	15.4	0.158	1.40
0.30	10.3	0.95	—	11.3	0.116	1.03
0.40	9.10	0.41	—	9.51	0.0977	0.865
0.50	8.24	0.22	—	8.46	0.0869	0.769
0.60	7.58	0.14	—	7.72	0.0793	0.702
0.80	6.63	0.07	—	6.70	0.0688	0.609
1.0	5.94	0.04	—	5.98	0.0614	0.543
1.5	4.81	0.02	0.04	4.87	0.0500	0.443
2.0	4.11	0.01	0.15	4.27	0.0439	0.389
3.0	3.22	0.01	0.41	3.64	0.0374	0.331
4.0	2.69	0.01	0.65	3.35	0.0344	0.304
5.0	2.32	—	0.88	3.20	0.0329	0.291
6.0	2.05	—	1.07	3.12	0.0320	0.283
8.0	1.68	—	1.39	3.07	0.0315	0.279
10	1.43	—	1.64	3.07	0.0315	0.279
15	1.06	—	2.14	3.20	0.0329	0.291
20	0.846	—	2.52	3.37	0.0346	0.306
30	0.616	—	3.02	3.64	0.0374	0.331

* Calculated using atomic weight 58.69
† Calculated using density 8.85 grams/cm^3

TABLE 19. Copper (Z = 29)

Energy (MeV)	Cross Sections in Barns (10^{-28} m^2)			Attenuation Coefficients		
	Scattering	Photo-electric	Pair	Atomic (barns)	Mass* (g/cm²)	Linear † (1/cm)
0.01	150	23,600	—	23,800	226	2010
0.015	96	8,000	—	8,100	76.8	684
0.02	70	3,580	—	3,650	34.6	308
0.03	46	1,120	—	1,170	11.1	98.8
0.04	35	474	—	509	4.83	43.0
0.05	29	242	—	271	2.57	22.9
0.06	24	143	—	167	1.58	14.1
0.08	20.5	60.2	—	80.7	0.765	6.81
0.10	17.9	30.7	—	48.6	0.461	4.10
0.15	14.5	8.9	—	23.4	0.222	1.98
0.20	12.8	3.7	—	16.5	0.156	1.39
0.30	10.7	1.1	—	11.8	0.112	0.997
0.40	9.43	0.48	—	9.91	0.0940	0.837
0.50	8.54	0.26	—	8.80	0.0834	0.742
0.60	7.86	0.16	—	8.02	0.0760	0.676
0.80	6.87	0.08	—	6.95	0.0659	0.587
1.0	6.16	0.05	—	6.21	0.0589	0.524
1.5	4.98	0.02	0.04	5.04	0.0478	0.425
2.0	4.25	0.02	0.16	4.43	0.0420	0.374
3.0	3.34	0.01	0.44	3.79	0.0359	0.320
4.0	2.78	0.01	0.71	3.50	0.0332	0.295
5.0	2.40	0.01	0.95	3.36	0.0319	0.284
6.0	2.12	—	1.16	3.28	0.0311	0.277
8.0	1.74	—	1.48	3.22	0.0305	0.271
10	1.48	—	1.75	3.23	0.0306	0.272
15	1.09	—	2.29	3.38	0.0320	0.285
20	0.877	—	2.69	3.57	0.0339	0.302
30	0.638	—	3.23	3.87	0.0367	0.327

* Calculated using atomic weight 63.54
† Calculated using density 8.90 grams/cm³

TABLE 20. Zinc (Z = 30)

Energy (MeV)	Cross Sections in Barns (10^{-28} m^2)			Attenuation Coefficients		
	Scattering	Photo-electric	Pair	Atomic (barns)	Mass* (g/cm^2)	Linear † (1/cm)
0.01	164	26,400	—	26,600	245	1750
0.015	103	8,920	—	9,020	83.1	593
0.02	75	4,060	—	4,140	38.2	272
0.03	49	1,280	—	1,330	12.3	87.7
0.04	37	549	—	586	5.40	38.5
0.05	29	276	—	305	2.81	20.0
0.06	26	163	—	189	1.74	12.4
0.08	21.5	68.5	—	90.0	0.829	5.91
0.10	18.7	35.5	—	54.2	0.499	3.56
0.15	15.2	10.2	—	25.4	0.234	1.67
0.20	13.2	4.28	—	17.5	0.161	1.15
0.30	11.1	1.29	—	12.4	0.114	0.83
0.40	9.77	0.56	—	10.3	0.0949	0.677
0.50	8.85	0.30	—	9.15	0.0843	0.601
0.60	8.14	0.19	—	8.33	0.0768	0.548
0.80	7.11	0.09	—	7.20	0.0663	0.473
1.0	6.38	0.06	—	6.44	0.0593	0.423
1.5	5.16	0.03	0.04	5.23	0.0482	0.344
2.0	4.40	0.02	0.17	4.59	0.0423	0.302
3.0	3.45	0.01	0.47	3.92	0.0361	0.258
4.0	2.88	0.01	0.74	3.63	0.0335	0.239
5.0	2.48	0.01	1.01	3.50	0.0323	0.230
6.0	2.20	0.01	1.22	3.43	0.0316	0.225
8.0	1.80	—	1.59	3.39	0.0312	0.223
10	1.53	—	1.87	3.40	0.0313	0.223
15	1.13	—	2.45	3.58	0.0330	0.235
20	0.906	—	2.87	3.78	0.0348	0.248
30	0.660	—	3.45	4.11	0.0379	0.270

* Calculated using atomic weight 65.38
† Calculated using density 7.133 grams/cm^3

TABLE 21. Germanium (Z = 32)

Energy (MeV)	Cross Sections in Barns (10^{-28} m^2)			Attenuation Coefficients		
	Scattering	Photo-electric	Pair	Atomic (barns)	Mass* (g/cm^2)	Linear † (1/cm)
0.01	189	3,690	—	3,880	32.2	173
0.01112	170	3,000	—	3,170	26.3	141
K0.01112 ‡	170	27,500	—	27,700	230	1230
0.015	119	11,100	—	11,200	93.0	498
0.02	86	5,130	—	5,220	43.3	232
0.03	56	1,640	—	1,700	14.1	75.6
0.04	42	708	—	750	6.23	33.4
0.05	34	356	—	390	3.24	17.4
0.06	29	212	—	241	2.00	10.7
0.08	23.7	89.9	—	114	0.946	5.07
0.10	20.5	46.6	—	67.1	0.557	2.99
0.15	16.4	13.5	—	29.9	0.248	1.33
0.20	14.3	5.70	—	20.0	0.166	0.890
0.30	11.9	1.74	—	13.6	0.113	0.606
0.40	10.5	0.76	—	11.3	0.0938	0.503
0.50	9.45	0.41	—	9.86	0.0818	0.438
0.60	8.70	0.26	—	8.96	0.0744	0.399
0.80	7.59	0.13	—	7.72	0.0641	0.344
1.0	6.80	0.08	—	6.88	0.0571	0.306
1.5	5.51	0.04	0.05	5.60	0.0465	0.249
2.0	4.69	0.03	0.19	4.91	0.0408	0.219
3.0	3.69	0.01	0.54	4.24	0.0352	0.189
4.0	3.07	0.01	0.86	3.94	0.0327	0.175
5.0	2.65	0.01	1.15	3.81	0.0316	0.169
6.0	2.34	0.01	1.39	3.74	0.0310	0.166
8.0	1.92	0.01	1.81	3.74	0.0310	0.166
10	1.63	—	2.13	3.76	0.0312	0.167
15	1.21	—	2.78	3.99	0.0331	0.177
20	0.966	—	3.25	4.22	0.0350	0.188
30	0.704	—	3.91	4.61	0.0383	0.205

* Calculated using atomic weight 72.60
† Calculated using density 5.36 grams/cm^3
‡ K = K absorption edge

TABLE 22. Selenium (Z = 34)

Energy (MeV)	Cross Sections in Barns (10^{-28} m^2)			Attenuation Coefficients		
	Scattering	Photo-electric	Pair	Atomic (barns)	Mass* (g/cm^2)	Linear † (1/cm)
0.01	215	4,410	—	4,630	35.3	170
0.01268	172	2,530	—	2,700	20.6	99.1
K0.01268 ‡	172	22,300	—	22,500	172	827
0.015	137	13,600	—	13,700	105	505
0.02	98	6,360	—	6,460	49.3	237
0.03	63	2,050	—	2,110	16.1	77.4
0.04	47	895	—	942	7.19	34.6
0.05	38	454	—	492	3.75	18.0
0.06	32	270	—	302	2.30	11.1
0.08	26	116	—	142	1.08	5.19
0.10	22.3	60.2	—	82.5	0.629	3.03
0.15	17.7	17.7	—	35.4	0.270	1.30
0.20	15.1	7.45	—	22.6	0.172	0.827
0.30	12.7	2.28	—	15.0	0.114	0.548
0.40	11.2	1.00	—	12.2	0.0931	0.448
0.50	10.1	0.54	—	10.6	0.0809	0.389
0.60	9.26	0.34	—	9.60	0.0732	0.352
0.80	8.07	0.17	—	8.24	0.0629	0.303
1.0	7.23	0.11	—	7.34	0.0560	0.269
1.5	5.85	0.05	0.06	5.96	0.0455	0.219
2.0	4.99	0.03	0.22	5.24	0.0400	0.192
3.0	3.92	0.02	0.61	4.55	0.0347	0.167
4.0	3.26	0.01	0.96	4.23	0.0323	0.155
5.0	2.82	0.01	1.30	4.13	0.0315	0.152
6.0	2.49	0.01	1.57	4.07	0.0311	0.150
8.0	2.04	0.01	2.04	4.09	0.0312	0.150
10	1.73	0.01	2.39	4.13	0.0315	0.152
15	1.28	—	3.12	4.40	0.0336	0.162
20	1.03	—	3.66	4.69	0.0358	0.172
30	0.748	—	4.39	5.14	0.0392	0.189

* Calculated using atomic weight 78.96
† Calculated using density 4.81 grams/cm^3
‡ K = K absorption edge

TABLE 23. Zirconium (Z = 40)

Energy (MeV)	Cross Sections in Barns (10^{-28} m^2)			Attenuation Coefficients		
	Scattering	Photo-electric	Pair	Atomic (barns)	Mass* (g/cm^2)	Linear † (1/cm)
0.01	309	9,220	—	9,530	62.9	411
0.015	196	2,820	—	3,020	19.9	130
0.01760	171	1,790	—	1,960	12.9	84.2
K0.01760 ‡	171	15,800	—	16,000	106	692
0.02	140	11,000	—	11,100	73.3	479
0.03	88	3,710	—	3,800	25.1	164
0.04	65	1,650	—	1,720	11.4	74.4
0.05	51	850	—	901	5.95	38.9
0.06	43	512	—	555	3.67	24.0
0.08	33	226	—	259	1.71	11.2
0.10	28	117	—	145	0.958	6.26
0.15	21.8	35.3	—	57.1	0.377	2.46
0.20	18.5	15.2	—	33.7	0.223	1.46
0.30	15.2	4.68	—	19.9	0.131	0.855
0.40	13.3	2.08	—	15.4	0.102	0.666
0.50	11.9	1.13	—	13.0	0.0859	0.561
0.60	11.0	0.71	—	11.7	0.0773	0.505
0.80	9.53	0.36	—	9.89	0.0653	0.426
1.0	8.53	0.23	—	8.76	0.0579	0.378
1.5	6.89	0.11	0.08	7.08	0.0468	0.306
2.0	5.88	0.07	0.32	6.27	0.0414	0.270
3.0	4.61	0.04	0.85	5.50	0.0363	0.237
4.0	3.84	0.03	1.35	5.22	0.0345	0.225
5.0	3.31	0.02	1.80	5.13	0.0339	0.221
6.0	2.93	0.02	2.17	5.12	0.0338	0.221
8.0	2.40	0.01	2.79	5.20	0.0343	0.224
10	2.04	0.01	3.28	5.33	0.0352	0.230
15	1.51	0.01	4.26	5.78	0.0382	0.249
20	1.21	—	5.00	6.21	0.0410	0.268
30	0.88	—	6.00	6.88	0.0454	0.296

* Calculated using atomic weight 91.22
† Calculated using density 6.53 grams/cm^3
‡ K = K absorption edge

TABLE 24. Niobium (Z = 41)

Energy (MeV)	Cross Sections in Barns (10^{-28} m^2)			Attenuation Coefficients		
	Scattering	Photo-electric	Pair	Atomic (barns)	Mass* (g/cm^2)	Linear † (1/cm)
0.01	326	10,300	—	10,600	68.7	589
0.015	208	3,140	—	3,350	21.7	186
0.01902	152	1,490	—	1,640	10.6	90.8
K0.01902 ‡	152	13,000	—	13,200	85.6	734
0.02	148	11,900	—	12,000	77.8	667
0.03	93	4,070	—	4,160	27.0	231
0.04	69	1,810	—	1,880	12.2	105
0.05	54	936	—	990	6.42	55.0
0.06	45	563	—	608	3.94	33.8
0.08	35	251	—	286	1.85	15.9
0.10	29	130	—	159	1.03	8.83
0.15	22.5	39.3	—	61.8	0.401	3.44
0.20	19.1	16.9	—	36.0	0.233	2.00
0.30	15.6	5.23	—	20.8	0.135	1.16
0.40	13.6	2.32	—	15.9	0.103	0.883
0.50	12.3	1.25	—	13.6	0.0882	0.756
0.60	11.2	0.79	—	12.0	0.0778	0.667
0.80	9.78	0.40	—	10.2	0.0661	0.566
1.0	8.75	0.26	—	9.01	0.0584	0.500
1.5	7.08	0.13	0.09	7.30	0.0473	0.405
2.0	6.02	0.08	0.33	6.43	0.0417	0.357
3.0	4.73	0.05	0.90	5.68	0.0368	0.315
4.0	3.95	0.03	1.41	5.39	0.0350	0.300
5.0	3.40	0.03	1.89	5.32	0.0345	0.296
6.0	3.00	0.02	2.27	5.29	0.0343	0.294
8.0	2.46	0.02	2.93	5.41	0.0351	0.301
10	2.09	0.01	3.43	5.53	0.0359	0.308
15	1.55	0.01	4.48	6.04	0.0392	0.336
20	1.24	0.01	5.24	6.49	0.0421	0.361
30	0.90	—	6.27	7.17	0.0465	0.398

* Calculated using atomic weight 92.91
† Calculated using density 8.57 grams/cm^3
‡ K = K absorption edge

TABLE 25. Molybdenum (Z = 42)

Energy (MeV)	Cross Sections in Barns (10^{-28} m^2)			Attenuation Coefficients		
	Scattering	Photo-electric	Pair	Atomic (barns)	Mass* (g/cm^2)	Linear † (1/cm)
0.01	340	11,400	—	11,700	73.5	750
0.015	220	3,480	—	3,700	23.2	240
0.02004	160	1,510	—	1,670	10.5	107
K0.02004 ‡	160	13,000	—	13,200	82.9	846
0.03	98	4,390	—	4,490	28.2	288
0.04	71	1,960	—	2,030	12.7	130
0.05	56	1,030	—	1,090	6.85	69.9
0.06	47	620	—	667	4.19	42.7
0.08	36	274	—	310	1.95	19.9
0.10	30	144	—	174	1.09	11.1
0.15	23.2	43.4	—	66.6	0.418	4.26
0.20	19.8	18.7	—	38.5	0.242	2.47
0.30	16.1	5.8	—	21.9	0.138	1.41
0.40	14.0	2.6	—	16.6	0.104	1.06
0.50	12.6	1.4	—	14.0	0.0879	0.897
0.60	11.5	0.88	—	12.4	0.0779	0.795
0.80	10.0	0.45	—	10.5	0.0659	0.672
1.0	8.96	0.29	—	9.25	0.0581	0.593
1.5	7.25	0.14	0.09	7.48	0.0470	0.479
2.0	6.15	0.09	0.35	6.59	0.0414	0.422
3.0	4.83	0.05	0.94	5.82	0.0365	0.372
4.0	4.03	0.04	1.50	5.57	0.0350	0.357
5.0	3.48	0.03	1.98	5.49	0.0345	0.352
6.0	3.08	0.01	2.38	5.47	0.0344	0.351
8.0	2.52	0.01	3.06	5.59	0.0351	0.358
10	2.14	0.01	3.59	5.74	0.0360	0.367
15	1.59	0.01	4.68	6.28	0.0394	0.402
20	1.27	0.01	5.47	6.75	0.0424	0.432
30	0.92	—	6.57	7.49	0.0470	0.479

* Calculated using atomic weight 95.95
† Calculated using density 10.2 grams/cm^3
‡ K = K absorption edge

TABLE 26. Silver (Z = 47)

Energy (MeV)	Cross Sections in Barns (10^{-28} m^2)			Attenuation Coefficients		
	Scattering	Photo-electric	Pair	Atomic (barns)	Mass* (g/cm^2)	Linear † (1/cm)
0.01	443	18,500	—	18,900	106	1110
0.015	285	5,670	—	5,960	33.3	349
0.02	202	2,460	—	2,660	14.9	156
0.02559	156	1,180	—	1,340	7.48	78.5
K0.02559 ‡	156	10,000	—	10,200	57.0	598
0.03	125	6,590	—	6,720	37.5	393
0.04	91	2,960	—	3,050	17.0	178
0.05	70	1,580	—	1,650	9.22	96.7
0.06	57	952	—	1,010	5.64	59.2
0.08	44	427	—	471	2.63	27.6
0.10	36	225	—	261	1.46	15.3
0.15	27.1	68.6	—	95.7	0.534	5.60
0.20	22.6	30.4	—	53.0	0.296	3.11
0.30	18.3	9.49	—	27.8	0.155	1.63
0.40	15.8	4.27	—	20.1	0.112	1.17
0.50	14.2	2.32	—	16.5	0.0922	0.967
0.60	13.0	1.46	—	14.5	0.0810	0.850
0.80	11.3	0.75	—	12.1	0.0676	0.709
1.0	10.1	0.48	—	10.6	0.0592	0.621
1.5	8.13	0.24	0.12	8.49	0.0474	0.497
2.0	6.91	0.15	0.45	7.51	0.0419	0.440
3.0	5.43	0.09	1.20	6.72	0.0375	0.393
4.0	4.52	0.06	1.87	6.45	0.0360	0.378
5.0	3.90	0.05	2.50	6.45	0.0360	0.378
6.0	3.44	0.04	2.99	6.47	0.0361	0.379
8.0	2.82	0.03	3.81	6.66	0.0372	0.390
10	2.40	0.02	4.47	6.89	0.0385	0.404
15	1.77	0.01	5.81	7.59	0.0424	0.445
20	1.42	0.01	6.79	8.22	0.0459	0.481
30	1.03	0.01	8.14	9.18	0.0513	0.538

* Calculated using atomic weight 107.88
† Calculated using density 10.49 grams/cm^3
‡ K = K absorption edge

TABLE 27. Cadmium (Z = 48)

Energy (MeV)	Cross Sections in Barns (10^{-28} m^2)			Attenuation Coefficients		
	Scattering	Photo-electric	Pair	Atomic (barns)	Mass* (g/cm^2)	Linear † (1/cm)
0.01	466	20,200	—	20,700	111	960
0.015	322	6,220	—	6,540	35.1	304
0.02	212	2,700	—	2,910	15.6	135
0.02676	159	1,170	—	1,330	7.13	61.7
K0.02676 ‡	159	9,700	—	9,860	52.8	457
0.03	131	7,080	—	7,210	38.6	334
0.04	95	3,210	—	3,310	17.7	153
0.05	73	1,710	—	1,780	9.54	82.5
0.06	60	1,030	—	1,090	5.84	50.5
0.08	46	461	—	507	2.72	23.5
0.10	38	245	—	283	1.52	13.1
0.15	27.9	74.9	—	103	0.552	4.77
0.20	23.3	33.4	—	56.7	0.304	2.63
0.30	18.7	10.4	—	29.1	0.156	1.35
0.40	16.2	4.66	—	20.9	0.112	0.969
0.50	14.5	2.55	—	17.1	0.0917	0.793
0.60	13.3	1.60	—	14.9	0.0799	0.691
0.80	11.5	0.83	—	12.3	0.0659	0.570
1.0	10.3	0.53	—	10.8	0.0579	0.501
1.5	8.30	0.26	0.13	8.69	0.0466	0.403
2.0	7.06	0.17	0.47	7.70	0.0413	0.357
3.0	5.54	0.10	1.25	6.89	0.0369	0.319
4.0	4.62	0.07	1.95	6.64	0.0356	0.308
5.0	3.98	0.05	2.61	6.64	0.0356	0.308
6.0	3.51	0.05	3.12	6.68	0.0358	0.310
8.0	2.88	0.03	3.98	6.89	0.0369	0.319
10	2.45	0.03	4.66	7.14	0.0383	0.331
15	1.81	0.02	6.05	7.88	0.0422	0.365
20	1.45	0.01	7.08	8.54	0.0458	0.396
30	1.06	0.01	8.47	9.54	0.0511	0.442

* Calculated using atomic weight 112.41
† Calculated using density 8.65 grams/cm^3
‡ K = K absorption edge

TABLE 28. Tin (Z = 50)

Energy (MeV)	Cross Sections in Barns (10^{-28} m^2)			Attenuation Coefficients		
	Scattering	Photo-electric	Pair	Atomic (barns)	Mass* (g/cm^2)	Linear † (1/cm)
0.01	510	24,000	—	24,500	124	905
0.015	330	7,410	—	7,740	39.3	287
0.02	240	3,220	—	3,460	17.6	128
0.02925	150	1,050	—	1,200	6.09	44.5
K0.02925 ‡	150	8,580	—	8,730	44.3	323
0.03	143	8,150	—	8,290	42.1	307
0.04	103	3,700	—	3,800	19.3	141
0.05	79	1,990	—	2,070	10.5	76.7
0.06	65	1,210	—	1,280	6.50	47.5
0.08	49	539	—	588	2.98	21.8
0.10	40	286	—	326	1.65	12.0
0.15	29.6	88.8	—	118	0.599	4.37
0.20	24.6	39.3	—	63.9	0.324	2.37
0.30	19.7	12.4	—	32.1	0.163	1.19
0.40	17.0	5.6	—	22.6	0.115	0.840
0.50	15.2	3.0	—	18.2	0.0924	0.675
0.60	13.8	1.9	—	15.7	0.0797	0.582
0.80	12.0	1.0	—	13.0	0.0660	0.482
1.0	10.7	0.64	—	11.3	0.0574	0.419
1.5	8.65	0.32	0.14	9.11	0.0462	0.337
2.0	7.36	0.20	0.51	8.07	0.0410	0.299
3.0	5.76	0.12	1.35	7.23	0.0367	0.268
4.0	4.80	0.08	2.14	7.02	0.0356	0.260
5.0	4.14	0.06	2.82	7.02	0.0356	0.260
6.0	3.66	0.05	3.37	7.08	0.0359	0.262
8.0	2.99	0.04	4.29	7.32	0.0372	0.272
10	2.55	0.03	5.04	7.62	0.0387	0.283
15	1.89	0.02	6.54	8.45	0.0429	0.313
20	1.51	0.01	7.63	9.15	0.0464	0.339
30	1.10	0.01	9.15	10.3	0.0523	0.382

* Calculated using atomic weight 118.70
† Calculated using density 7.30 grams/cm^3
‡ K = K absorption edge

TABLE 29. Antimony (Z = 51)

Energy (MeV)	Cross Sections in Barns (10^{-28} m^2)			Attenuation Coefficients		
	Scattering	Photo-electric	Pair	Atomic (barns)	Mass* (g/cm^2)	Linear † (1/cm)
0.01	535	25,900	—	26,400	131	867
0.015	344	8,010	—	8,350	41.3	273
0.02	244	3,490	—	3,730	18.5	122
0.03050	148	1,000	—	1,150	5.69	37.7
K0.03050 ‡	148	8,140	—	8,290	41.0	271
0.04	108	3,970	—	4,080	20.2	134
0.05	82	2,140	—	2,220	11.0	72.8
0.06	67	1,300	—	1,370	6.78	44.9
0.08	51	580	—	631	3.12	20.7
0.10	42	311	—	353	1.75	11.6
0.15	30.4	93.5	—	124	0.614	4.06
0.20	25.1	42.8	—	67.9	0.336	2.22
0.30	20.1	13.5	—	33.6	0.166	1.10
0.40	17.3	6.11	—	23.4	0.116	0.768
0.50	15.5	3.34	—	18.8	0.0930	0.616
0.60	14.1	2.09	—	16.2	0.0802	0.531
0.80	12.3	1.09	—	13.4	0.0663	0.439
1.0	10.9	0.71	—	11.6	0.0574	0.380
1.5	8.82	0.34	0.15	9.31	0.0461	0.305
2.0	7.51	0.22	0.54	8.27	0.0409	0.271
3.0	5.89	0.13	1.42	7.44	0.0368	0.244
4.0	4.91	0.09	2.21	7.21	0.0357	0.236
5.0	4.23	0.07	2.95	7.25	0.0359	0.238
6.0	3.73	0.06	3.51	7.30	0.0361	0.239
8.0	3.06	0.04	4.47	7.57	0.0375	0.248
10	2.60	0.03	5.22	7.85	0.0388	0.257
15	1.92	0.02	6.79	8.73	0.0432	0.286
20	1.54	0.01	7.90	9.45	0.0468	0.310
30	1.12	0.01	9.49	10.6	0.0524	0.347

* Calculated using atomic weight 121.76
† Calculated using density 6.62 $grams/cm^3$
‡ K = K absorption edge

TABLE 30. Iodine (Z = 53)

Energy (MeV)	Cross Sections in Barns (10^{-28} m^2)			Attenuation Coefficients		
	Scattering	Photo-electric	Pair	Atomic (barns)	Mass* (g/cm^2)	Linear † (1/cm)
0.01	590	29,800	—	30,400	144	710
0.015	380	9,360	—	9,740	46.2	228
0.02	270	4,130	—	4,400	20.9	103
0.03	164	1,260	—	1,420	6.74	33.2
0.03323	150	933	—	1,080	5.13	25.3
K0.03323 ‡	150	7,510	—	7,660	36.4	179
0.04	117	4,490	—	4,610	21.9	108
0.05	89	2,470	—	2,560	12.2	60.1
0.06	72	1,500	—	1,570	7.45	36.7
0.08	54	677	—	731	3.47	17.1
0.10	44	360	—	404	1.92	9.47
0.15	32	113	—	145	0.688	3.39
0.20	26.5	50.0	—	76.5	0.363	1.79
0.30	21.0	16.0	—	37.0	0.176	0.868
0.40	18.1	7.2	—	25.3	0.120	0.592
0.50	16.2	3.9	—	20.1	0.0954	0.470
0.60	14.8	2.5	—	17.3	0.0821	0.405
0.80	12.8	1.3	—	14.1	0.0669	0.330
1.0	11.4	0.84	—	12.2	0.0579	0.285
1.5	9.18	0.41	0.17	9.76	0.0463	0.228
2.0	7.81	0.26	0.59	8.66	0.0411	0.203
3.0	6.10	0.16	1.53	7.79	0.0370	0.182
4.0	5.09	0.11	2.41	7.61	0.0361	0.178
5.0	4.39	0.08	3.17	7.64	0.0363	0.179
6.0	3.88	0.07	3.78	7.73	0.0367	0.181
8.0	3.17	0.05	4.81	8.03	0.0381	0.188
10	2.70	0.04	5.63	8.37	0.0397	0.196
15	2.00	0.02	7.30	9.32	0.0442	0.218
20	1.60	0.01	8.51	10.1	0.0479	0.236
30	1.17	0.01	10.2	11.4	0.0541	0.267

* Calculated using atomic weight 126.92
† Calculated using density 4.93 grams/cm^3
‡ K = K absorption edge

TABLE 31. Cesium (Z = 55)

Energy (MeV)	Cross Sections in Barns (10^{-28} m^2)			Attenuation Coefficients		
	Scattering	Photo-electric	Pair	Atomic (barns)	Mass* (g/cm²)	Linear † (1/cm)
0.01	633	34,700	—	35,300	160	299
0.015	411	11,000	—	11,400	51.7	96.7
0.02	290	4,830	—	5,120	23.2	43.4
0.03	177	1,480	—	1,660	7.52	14.1
0.03603	146	881	—	1,030	4.67	8.73
K0.03603 ‡	146	7,100	—	7,250	32.9	61.5
0.04	127	5,160	—	5,290	24.0	44.9
0.05	96	2,840	—	2,940	13.3	24.9
0.06	78	1,720	—	1,800	8.16	15.3
0.08	58	765	—	823	3.73	6.98
0.10	47	414	—	461	2.09	3.91
0.15	34	129	—	163	0.739	1.38
0.20	27.8	58.4	—	86.2	0.391	0.731
0.30	21.9	18.6	—	40.5	0.184	0.344
0.40	18.8	8.46	—	27.3	0.124	0.232
0.50	16.8	4.64	—	21.4	0.0970	0.181
0.60	15.3	2.93	—	18.2	0.0825	0.154
0.80	13.3	1.54	—	14.8	0.0671	0.125
1.0	11.8	1.00	—	12.8	0.0580	0.108
1.5	9.53	0.49	0.18	10.2	0.0462	0.0864
2.0	8.10	0.31	0.64	9.05	0.0410	0.0767
3.0	6.36	0.18	1.66	8.20	0.0372	0.0696
4.0	5.29	0.13	2.59	8.01	0.0363	0.0679
5.0	4.56	0.10	3.42	8.08	0.0366	0.0684
6.0	4.04	0.08	4.07	8.19	0.0371	0.0694
8.0	3.30	0.06	5.17	8.53	0.0387	0.0724
10	2.81	0.05	6.04	8.90	0.0403	0.0754
15	2.07	0.03	7.82	9.92	0.0450	0.0842
20	1.66	0.02	9.12	10.8	0.0490	0.0916
30	1.21	0.01	10.9	12.1	0.0548	0.102

* Calculated using atomic weight 132.91
† Calculated using density 1.87 grams/cm³
‡ K = K absorption edge

TABLE 32. Barium (Z = 56)

Energy (MeV)	Cross Sections in Barns (10^{-28} m^2)			Attenuation Coefficients		
	Scattering	Photo-electric	Pair	Atomic (barns)	Mass* (g/cm^2)	Linear † (1/cm)
0.01	662	37,000	—	37,700	165	624
0.015	371	11,900	—	12,300	53.9	204
0.02	303	5,190	—	5,490	24.1	91.1
0.03	185	1,590	—	1,780	7.81	29.5
0.03748	145	837	—	982	4.31	16.3
K0.03748 ‡	145	6,720	—	6,870	30.1	114
0.04	132	5,510	—	5,640	24.7	93.4
0.05	100	3,030	—	3,130	13.7	51.8
0.06	81	1,834	—	1,920	8.42	31.8
0.08	60	815	—	875	3.84	14.5
0.10	49	444	—	493	2.16	8.16
0.15	35	138	—	173	0.759	2.87
0.20	28.4	62.8	—	91.2	0.400	1.51
0.30	22.4	20.0	—	42.4	0.186	0.703
0.40	19.2	9.14	—	28.3	0.124	0.469
0.50	17.1	5.02	—	22.1	0.0969	0.366
0.60	15.6	3.18	—	18.8	0.0825	0.312
0.80	13.5	1.67	—	15.2	0.0667	0.252
1.0	12.1	1.09	—	13.2	0.0579	0.219
1.5	9.70	0.53	0.19	10.4	0.0456	0.172
2.0	8.25	0.34	0.67	9.26	0.0406	0.153
3.0	6.47	0.20	1.73	8.40	0.0368	0.139
4.0	5.39	0.14	2.68	8.21	0.0360	0.136
5.0	4.65	0.11	3.59	8.35	0.0366	0.138
6.0	4.11	0.09	4.21	8.41	0.0369	0.139
8.0	3.35	0.07	5.35	8.77	0.0385	0.146
10	2.86	0.05	6.26	9.17	0.0402	0.152
15	2.11	0.03	8.09	10.2	0.0447	0.169
20	1.69	0.02	9.45	11.2	0.0491	0.186
30	1.23	0.02	11.3	12.6	0.0553	0.209

* Calculated using atomic weight 137.36
† Calculated using density 3.78 grams/cm^3
‡ K = K absorption edge

TABLE 33. Thulium (Z = 69)

Energy (MeV)	Cross Sections in Barns (10^{-28} m^2)			Attenuation Coefficients		
	Scattering	Photo-electric	Pair	Atomic (barns)	Mass* (g/cm^2)	Linear † (1/cm)
0.015	705	27,400	—	28,100	100	935
0.02	496	12,200	—	12,700	45.2	423
0.03	300	3,790	—	4,090	14.5	136
0.04	210	1,690	—	1,900	6.76	63.2
0.05	156	884	—	1,040	3.70	34.6
0.05945	126	497	—	623	2.22	20.8
K0.05945 ‡	126	3,880	—	4,010	14.3	134
0.06	124	3,860	—	3,980	14.2	133
0.08	89	1,750	—	1,840	6.54	61.1
0.10	74	970	—	1,040	3.70	34.6
0.15	48	313	—	361	1.28	12.0
0.20	38	144	—	182	0.647	6.05
0.30	29.0	47.2	—	76.2	0.271	2.53
0.40	24.4	22.1	—	46.8	0.166	1.55
0.50	21.6	12.5	—	34.1	0.121	1.13
0.60	19.6	8.05	—	27.7	0.0985	0.921
0.80	16.9	4.35	—	21.3	0.0758	0.709
1.0	15.0	2.85	—	17.9	0.0637	0.596
1.5	12.0	1.38	0.33	13.7	0.0487	0.455
2.0	10.2	0.89	1.11	12.2	0.0434	0.406
3.0	7.99	0.51	2.73	11.2	0.0398	0.372
4.0	6.65	0.36	4.15	11.2	0.0398	0.372
5.0	5.73	0.28	5.41	11.4	0.0405	0.379
6.0	5.05	0.23	6.36	11.6	0.0413	0.386
8.0	4.14	0.17	7.97	12.3	0.0438	0.410
10	3.52	0.13	9.27	12.9	0.0459	0.429
15	2.60	0.08	11.9	14.6	0.0519	0.485
20	2.08	0.06	13.9	16.0	0.0569	0.532
30	1.52	0.04	16.5	18.1	0.0644	0.602

* Calculated using atomic weight 169.4
† Calculated using density 9.35 grams/cm^3
‡ K = K absorption edge

TABLE 34. Tantalum (Z = 73)

Energy (MeV)	Cross Sections in Barns (10^{-28} m^2)			Attenuation Coefficients		
	Scattering	Photo-electric	Pair	Atomic (barns)	Mass* (g/cm^2)	Linear † (1/cm)
0.015	808	34,000	—	34,800	116	1930
0.02	569	15,200	—	15,800	52.6	873
0.03	342	4,750	—	5,090	17.0	282
0.04	238	2,120	—	2,360	7.86	130
0.05	176	1,110	—	1,290	4.30	71.4
0.06	140	638	—	778	2.59	43.0
0.06751	124	440	—	564	1.88	31.2
K0.06751 ‡	124	3,360	—	3,480	11.6	193
0.08	101	2,140	—	2,240	7.46	124
0.10	79	1,180	—	1,260	4.20	69.7
0.15	53	387	—	440	1.47	24.4
0.20	41	179	—	220	0.733	12.2
0.30	31.1	59.3	—	90.6	0.302	5.01
0.40	26.1	28.4	—	54.5	0.182	3.02
0.50	23.0	15.9	—	38.9	0.130	2.16
0.60	20.9	10.3	—	31.2	0.104	1.73
0.80	17.9	5.60	—	23.5	0.0783	1.30
1.0	15.9	3.69	—	19.6	0.0653	1.08
1.5	12.7	1.78	0.39	14.9	0.0496	0.823
2.0	10.8	1.15	1.28	13.2	0.0440	0.730
3.0	8.45	0.66	3.08	12.2	0.0406	0.674
4.0	7.04	0.47	4.67	12.2	0.0406	0.674
5.0	6.06	0.36	6.06	12.5	0.0416	0.691
6.0	5.35	0.29	7.08	12.7	0.0423	0.702
8.0	4.37	0.21	8.86	13.4	0.0446	0.740
10	3.72	0.17	10.3	14.2	0.0473	0.785
15	2.75	0.11	13.2	16.1	0.0536	0.890
20	2.20	0.08	15.4	17.7	0.0590	0.979
30	1.61	0.05	18.4	20.1	0.0670	1.11

* Calculated using atomic weight 180.88
† Calculated using density 16.6 grams/cm^3
‡ K = K absorption edge

TABLE 35. Tungsten (Z = 74)

Energy (MeV)	Cross Sections in Barns (10^{-28} m^2)			Attenuation Coefficients		
	Scattering	Photo-electric	Pair	Atomic (barns)	Mass* (g/cm^2)	Linear † (1/cm)
0.015	840	36,000	—	36,800	121	2260
0.02	590	16,000	—	16,600	54.4	1020
0.03	350	5,040	—	5,390	17.7	331
0.04	245	2,220	—	2,470	8.09	151
0.05	180	1,160	—	1,340	4.39	82.1
0.06	145	674	—	819	2.68	50.1
0.06964	122	437	—	559	1.83	34.2
K0.06964 ‡	122	3,230	—	3,350	11.0	206
0.08	104	2,250	—	2,350	7.70	144
0.10	80	1,250	—	1,330	4.36	81.5
0.15	54	408	—	462	1.51	28.2
0.20	42	186	—	228	0.747	14.0
0.30	31.5	63.1	—	94.6	0.310	5.80
0.40	26.5	29.8	—	56.3	0.184	3.44
0.50	23.4	16.7	—	40.1	0.131	2.45
0.60	21.2	11.0	—	32.2	0.105	1.96
0.80	18.2	5.9	—	24.1	0.0790	1.48
1.0	16.1	3.9	—	20.0	0.0655	1.22
1.5	12.9	1.9	0.40	15.2	0.0498	0.931
2.0	10.9	1.2	1.32	13.4	0.0439	0.821
3.0	8.57	0.71	3.20	12.5	0.0410	0.767
4.0	7.10	0.50	4.81	12.4	0.0406	0.759
5.0	6.13	0.38	6.23	12.7	0.0416	0.778
6.0	5.42	0.31	7.34	13.0	0.0426	0.797
8.0	4.43	0.23	9.06	13.7	0.0449	0.840
10	3.77	0.18	10.6	14.6	0.0478	0.894
15	2.79	0.11	13.5	16.4	0.0537	1.00
20	2.24	0.08	15.7	18.0	0.0590	1.10
30	1.63	0.06	18.8	20.5	0.0672	1.26

* Calculated using atomic weight 183.92
† Calculated using density 18.7 grams/cm^3
‡ K = K absorption edge

TABLE 36. Platinum (Z = 78)

Energy (MeV)	Cross Sections in Barns (10^{-28} m^2)			Attenuation Coefficients		
	Scattering	Photo-electric	Pair	Atomic (barns)	Mass* (g/cm^2)	Linear † (1/cm)
0.015	940	43,800	—	44,700	138	2950
0.02	670	19,700	—	20,400	63.0	1350
0.03	400	6,240	—	6,640	20.5	439
0.04	280	2,720	—	3,000	9.26	198
0.05	188	1,440	—	1,630	5.03	108
0.06	163	836	—	999	3.08	65.9
0.07858	117	381	—	498	1.54	33.0
K0.07858 ‡	117	2,830	—	2,950	9.10	195
0.08	115	2,680	—	2,800	8.64	185
0.10	88	1,500	—	1,590	4.91	105
0.15	59	498	—	557	1.72	36.8
0.20	45	226	—	271	0.836	17.9
0.30	34	77.3	—	111	0.343	7.34
0.40	28.3	37.1	—	65.4	0.202	4.32
0.50	24.8	21.1	—	46.0	0.142	3.04
0.60	22.5	13.9	—	36.4	0.112	2.40
0.80	19.2	7.6	—	26.8	0.0827	1.77
1.0	17.0	4.9	—	21.9	0.0676	1.45
1.5	13.6	2.4	0.47	16.5	0.0509	1.09
2.0	11.6	1.5	1.51	14.6	0.0451	0.965
3.0	9.04	0.90	3.55	13.5	0.0417	0.892
4.0	7.52	0.63	5.37	13.5	0.0417	0.892
5.0	6.46	0.48	6.92	13.9	0.0429	0.918
6.0	5.71	0.39	8.07	14.2	0.0438	0.937
8.0	4.67	0.29	9.98	14.9	0.0460	0.984
10	3.98	0.22	11.7	15.9	0.0491	1.05
15	2.94	0.14	14.9	18.0	0.0555	1.19
20	2.36	0.10	17.3	19.8	0.0611	1.31
30	1.72	0.07	20.6	22.4	0.0691	1.48

*Calculated using atomic weight 195.23
† Calculated using density 21.4 grams/cm^3
‡ K = K absorption edge

TABLE 37. Gold (Z = 79)

Energy (MeV)	Cross Sections in Barns (10^{-28} m^2)			Attenuation Coefficients		
	Scattering	Photo-electric	Pair	Atomic (barns)	Mass* (g/cm^2)	Linear † (1/cm)
0.015	977	45,500	—	46,500	142	2740
0.02	681	20,700	—	21,400	65.4	1260
0.03	412	6,550	—	6,960	21.3	412
0.04	285	2,890	—	3,180	9.71	188
0.05	210	1,510	—	1,720	5.25	101
0.06	165	880	—	1,050	3.21	62.0
0.08	118	382	—	500	1.53	29.6
0.08091	117	368	—	485	1.48	28.6
K0.08091 ‡	117	2,740	—	2,860	8.74	169
0.10	92	1,570	—	1,600	5.07	98.0
0.15	60	520	—	580	1.77	34.2
0.20	46	241	—	287	0.877	16.9
0.30	34.5	80.9	—	115	0.351	6.78
0.40	28.7	39.4	—	68.1	0.208	4.02
0.50	25.2	22.5	—	47.7	0.146	2.82
0.60	22.8	14.7	—	37.5	0.115	2.22
0.80	19.5	8.03	—	27.5	0.0840	1.62
1.0	17.3	5.29	—	22.6	0.0690	1.33
1.5	13.8	2.56	0.48	16.8	0.0513	0.991
2.0	11.7	1.65	1.56	14.9	0.0455	0.879
3.0	9.17	0.95	3.65	13.8	0.0422	0.815
4.0	7.62	0.67	5.50	13.8	0.0422	0.815
5.0	6.56	0.50	7.10	14.2	0.0434	0.838
6.0	5.80	0.41	8.22	14.4	0.0440	0.850
8.0	4.74	0.30	10.2	15.2	0.0464	0.896
10	4.04	0.24	11.9	16.2	0.0495	0.956
15	2.98	0.15	15.2	18.3	0.0559	1.08
20	2.39	0.11	17.7	20.2	0.0617	1.19
30	1.74	0.08	21.2	23.0	0.0703	1.36

* Calculated using atomic weight 197.2
† Calculated using density 19.32 grams/cm^3
‡ K = K absorption edge

TABLE 38. Lead (Z = 82)

Energy (MeV)	Cross Sections in Barns (10^{-28} m^2)			Attenuation Coefficients		
	Scattering	Photo-electric	Pair	Atomic (barns)	Mass* (g/cm^2)	Linear † (1/cm)
0.02	750	24,000	—	24,800	72.1	818
0.03	450	7,620	—	8,070	23.5	266
0.04	310	3,310	—	3,620	10.5	119
0.05	230	1,740	—	1,970	5.73	65.0
0.06	180	1,040	—	1,220	3.55	40.3
0.08	127	444	—	571	1.66	18.8
0.08823	113	334	—	447	1.30	14.7
K0.08823 ‡	113	2,510	—	2,620	7.62	86.4
0.10	100	1,780	—	1,880	5.47	62.0
0.15	64	596	—	660	1.92	21.8
0.20	49	275	—	324	0.942	10.7
0.30	36.2	93.4	—	130	0.378	4.29
0.40	30.1	45.7	—	75.8	0.220	2.49
0.50	26.3	26.1	—	52.4	0.152	1.72
0.60	23.8	17.3	—	41.1	0.120	1.36
0.80	20.3	9.5	—	29.8	0.0867	0.983
1.0	18.0	6.2	—	24.2	0.0704	0.798
1.5	14.4	3.0	0.5	17.9	0.0521	0.591
2.0	12.2	2.0	1.7	15.9	0.0462	0.524
3.0	9.51	1.1	4.0	14.6	0.0425	0.482
4.0	7.91	0.80	6.02	14.7	0.0427	0.484
5.0	6.79	0.60	7.63	15.0	0.0436	0.494
6.0	6.00	0.49	8.84	15.3	0.0445	0.505
8.0	4.91	0.35	11.0	16.3	0.0474	0.538
10	4.18	0.28	12.8	17.3	0.0503	0.570
15	3.09	0.18	16.3	19.5	0.0567	0.643
20	2.48	0.13	18.9	21.5	0.0625	0.709
30	1.80	0.09	22.6	24.5	0.0712	0.807

* Calculated using atomic weight 207.21
† Calculated using density 11.34 grams/cm^3
‡ K = K absorption edge

TABLE 39. Uranium (Z = 92)

Energy (MeV)	Cross Sections in Barns (10^{-28} m^2)			Attenuation Coefficients		
	Scattering	Photo-electric	Pair	Atomic (barns)	Mass* (g/cm^2)	Linear † (1/cm)
0.03	590	12,000	—	12,600	31.9	597
0.04	400	5,250	—	5,650	14.3	267
0.05	300	2,780	—	3,080	7.79	146
0.06	230	1,640	—	1,870	4.73	88.5
0.08	163	716	—	879	2.22	41.5
0.10	123	374	—	497	1.26	23.6
0.1163	103	239	—	342	0.866	16.2
K0.1163 ‡	103	1,790	—	1,890	4.78	89.4
0.15	78	905	—	983	2.49	46.6
0.20	59	417	—	476	1.20	22.4
0.30	42	146	—	188	0.0476	8.90
0.40	34.7	73.2	—	108	0.273	5.11
0.50	30.2	43.1	—	73.3	0.186	3.48
0.60	27.1	29.2	—	56.3	0.142	2.66
0.80	23.0	16.0	—	39.0	0.0987	1.85
1.0	20.3	10.5	—	30.8	0.0779	1.46
1.5	16.2	5.1	0.8	22.1	0.0559	1.05
2.0	13.7	3.3	2.3	19.3	0.0488	0.913
3.0	10.7	1.9	5.21	17.8	0.0450	0.842
4.0	8.88	1.3	7.62	17.8	0.0450	0.842
5.0	7.62	1.0	9.73	18.4	0.0466	0.871
6.0	6.74	0.81	11.1	18.6	0.0471	0.881
8.0	5.51	0.59	13.5	19.6	0.0496	0.928
10	4.69	0.46	15.7	20.9	0.0529	0.989
15	3.47	0.30	20.0	23.8	0.0602	1.13
20	2.78	0.22	23.1	26.1	0.0661	1.24
30	2.02	0.15	27.6	29.8	0.0754	1.41

* Calculated using atomic weight 238.07
† Calculated using density 18.7 grams/cm^3
‡ K = K absorption edge

TABLE 40. Plutonium (Z = 94)

Energy (MeV)	Cross Sections in Barns (10^{-28} m²)			Attenuation Coefficients		
	Scattering	Photo-electric	Pair	Atomic (barns)	Mass* (g/cm²)	Linear † (1/cm)
0.03	627	13,200	—	13,800	34.8	679
0.04	426	5,700	—	6,130	15.5	302
0.05	312	3,020	—	3,330	8.39	164
0.06	243	1,780	—	2,020	5.09	99.3
0.08	171	778	—	949	2.39	46.6
0.10	130	409	—	539	1.36	26.5
0.12256	101	222	—	323	0.814	15.9
K0.12256 ‡	101	1,660	—	1,760	4.44	86.6
0.15	81	976	—	1,060	2.67	52.1
0.20	61	455	—	516	1.30	25.4
0.30	44	159	—	203	0.512	9.98
0.40	35.7	80.0	—	116	0.292	5.69
0.50	31.0	47.0	—	78.0	0.197	3.84
0.60	27.9	31.9	—	59.8	0.151	2.94
0.80	23.6	17.5	—	41.1	0.104	2.03
1.0	20.8	11.6	—	32.4	0.0817	1.59
1.5	20.6	5.60	0.81	27.0	0.0681	1.33
2.0	14.0	3.63	2.46	20.1	0.0507	0.989
3.0	10.9	2.08	5.45	18.4	0.0464	0.905
4.0	9.09	1.46	8.04	18.6	0.0469	0.915
5.0	7.82	1.09	10.1	19.0	0.0479	0.934
6.0	6.91	0.90	11.5	19.3	0.0487	0.950
8.0	5.65	0.65	14.0	20.3	0.0512	0.998
10	4.80	0.50	16.3	21.6	0.0545	1.06
15	3.54	0.33	20.7	24.6	0.0620	1.21
20	2.84	0.24	23.9	27.0	0.0681	1.33
30	2.07	0.17	27.8	30.0	0.0756	1.47

* Calculated using atomic weight 239
† Calculated using density 19.5 grams/cm³
‡ K = K absorption edge

Index

A

B

C

D

E

F

H

I

G

H

I

J

K

L

M

Q

R

S

X

Z